REVIEWS in MINERALOGY and GEOCHEMISTRY

Volume 45 2001

NATURAL ZEOLITES:

OCCURRENCE, PROPERTIES, APPLICATIONS

Editors:

DAVID L. BISH *LOS ALAMOS NATIONAL LABORATORY LOS ALAMOS, NEW MEXICO*

DOUGLAS W. MING *NASA JOHNSON SPACE CENTER HOUSTON, TEXAS*

COVER: Representation of the crystal structure of heulandite viewed down the *c*-axis, with *b* vertical. Ten-membered and eight-membered tetrahedral rings form channels parallel to *c*. In the channels,

small green spheres represent Na atoms,
larger light-blue spheres represent K atoms,
small yellow spheres represent Ca atoms,
large blue spheres represent H_2O molecules.

[Figure courtesy of M.E. Gunter; cf. Fig. 16, p. 41, this volume.]

Series Editor for MSA: ***Paul H. Ribbe***
Virginia Polytechnic Institute & State University
Blacksburg, Virginia

MINERALOGICAL SOCIETY of AMERICA

REVIEWS IN MINERALOGY AND GEOCHEMISTRY

(Formerly: REVIEWS IN MINERALOGY)

ISSN 1529-6466

Volume 45

Natural Zeolites:
Occurrence, Properties, Applications

ISBN 0-939950-57-X

** This volume is the seventh of a series of review volumes published jointly under the banner of the Mineralogical Society of America and the Geochemical Society. The newly titled *Reviews in Mineralogy and Geochemistry* has been numbered contiguously with the previous series, *Reviews in Mineralogy*.

Additional copies of this volume as well as others in this series may be obtained at moderate cost from:

THE MINERALOGICAL SOCIETY OF AMERICA
1015 EIGHTEENTH STREET, NW, SUITE 601
WASHINGTON, DC 20036 U.S.A.

NATURAL ZEOLITES:

OCCURRENCE, PROPERTIES, APPLICATIONS

FOREWORD

In the year 2000, the Geochemical Society (GS) and the Mineralogical Society of America (MSA) joined efforts in publishing review volumes. As a result, *Reviews in Mineralogy* became *Reviews in Mineralogy and Geochemistry*. This is the seventh in the newly named series. It represents part of of MSA's attempt to revisit subjects of its earliest volumes, i.e., the rock-forming mineral groups. Zeolites were the subject of Volume 4: 1500 copies were published in 1977 as *Short Course Notes*; later printings were circulated as *Reviews in Mineralogy*. More than 4600 copies were sold, but the book has been out of print for a decade or more. MSA is pleased to present this new work, diligently prepared over the past four years by the editors, Dave Bish and Doug Ming.

P.H. Ribbe, Series editor
Blacksburg, Virginia
October 4, 2001

PREFACE

"Mineralogy and Geology of Natural Zeolites" was published in 1977. Dr. Fred Mumpton, a leader of the natural zeolite community for more than three decades, edited the original volume. Since the time of the original MSA zeolite short course in November 1977, there have been major developments concerning almost all aspects of natural zeolites. There has been an explosion in our knowledge of the crystal chemistry and structures of natural zeolites (Chapters 1 and 2), due in part to the now-common Rietveld method that allows treatment of powder diffraction data. Studies on the geochemistry of natural zeolites have also greatly increased, partly as a result of the interests related to the disposal of radioactive wastes, and Chapters 3, 4, 5, 13, and 14 detail the latest results in this important area. Until the latter part of the 20th century, zeolites were often looked upon as a geological curiosity, but they are now known to be widespread throughout the world in sedimentary and igneous deposits and in soils (Chapters 6-12). Likewise, borrowing from new knowledge gained from studies of synthetic zeolites and properties of natural zeolites, the application of natural zeolites has greatly expanded since the first zeolite volume. Chapter 15 details the use of natural zeolites for removal of ammonium ions, heavy metals, radioactive cations, and organic molecules from natural waters, wastewaters, and soils. Similarly, Chapter 16 describes the use of natural zeolites as building blocks and cements in the building industry, Chapter 17 outlines their use in solar energy storage, heating, and cooling applications, and Chapter 18 describes their use in a variety of agricultural applications, including as soil conditioners, slow-release fertilizers, soil-less substrates, carriers for insecticides and pesticides, and remediation agents in contaminated soils. Most of the material in this volume is entirely new, and "Natural Zeolites: Occurrence, Properties, Applications" presents a fresh and expanded look at many of the subjects contained in Volume 4.

We appreciate the patience and assistance of series editor, Paul Ribbe, throughout the preparation of this volume. We are also grateful to our employers, the Los Alamos National Laboratory and the NASA Johnson Space Center, for their support throughout the writing and editing of this volume. We are particularly grateful to all of the authors for their perseverance in preparing this volume. And finally we thank Dr. Fred Mumpton for providing the impetus for beginning this volume. It is our hope that this new, expanded volume will rekindle interest in this fascinating and technologically important group of minerals, in part through the 'Suggestions for Further Research' section in each chapter.

David L. Bish
Los Alamos, New Mexico

Doug Ming
Houston, Texas

September 1, 2001

Reviews in Mineralogy and Geochemistry Volume 45

NATURAL ZEOLITES:
OCCURRENCE, PROPERTIES, APPLICATIONS

Table of Contents

1 Crystal Structures of Natural Zeolites

T. Armbruster, M. E. Gunter

2 The Crystal Chemistry of Zeolites

E. Passaglia, R. A. Sheppard

3 Geochemical Stability of Natural Zeolites

S. J. Chipera, J. A. Apps

4 Isotope Geochemistry of Zeolites

H. R. Karlsson

5 Clinoptilolite-Heulandite Nomenclature

D. L. Bish, J. M. Boak

6 Occurrence of Zeolites in Sedimentary Rocks: An Overview

R. L. Hay, R. A. Sheppard

7 Zeolites in Closed Hydrologic Systems

A. Langella, P. Cappelletti, M. de' Gennaro

8 Formation of Zeolites in Open Hydrologic Systems

R. A. Sheppard, R. L. Hay

9 Zeolites in Burial Diagenesis and Low-grade Metamorphic Rocks

M. Utada

10 Zeolites in Hydrothermally Altered Rocks

M. Utada

11 Zeolites in Soil Environments

D. W. Ming, J. L. Boettinger

12 Zeolites in Petroleum and Natural Gas Reservoirs

Azuma Iijima

13 Thermal Behavior of Natural Zeolites

David L. Bish and **J. William Carey**

14 Cation-Exchange Properties of Natural Zeolites

R. T. Pabalan, F. P. Bertetti

15 Applications of Natural Zeolites in Water and Wastewater Treatment

Dénes Kalló

16 Use of Zeolitic Tuff in the Building Industry

C. Colella, M. de' Gennaro, R. Aiello

17 Natural Zeolites in Solar Energy Heating, Cooling, and Energy Storage

Dimiter I. Tchernev

18 Use of Natural Zeolites in Agronomy, Horticulture, and Environmental Soil Remediation

Douglas W. Ming, Earl R. Allen

1 Crystal Structures of Natural Zeolites

Thomas Armbruster

Laboratorium für chemische und mineralogische Kristallographie
University of Bern
CH-3012 Bern, Switzerland

Mickey E. Gunter

Department of Geological Sciences
University of Idaho
Moscow, Idaho 83844

INTRODUCTION

General aspects

According to the recommended nomenclature for zeolite minerals (Coombs et al. 1998), there are more than eighty distinct zeolite species. Table 1 (see Appendix below) is an alphabetical, page-indexed listing of the zeolite minerals discussed in this chapter. Although zeolites are not as geologically abundant or widespread as many other silicate mineral groups, there is perhaps more interest in the crystal structures of zeolites than in any other mineral group, as evidenced by the number of reported crystal structure refinements (Table 2).

Table 2. Number of crystal structures reported in ICSD (1995) for selected mineral groups.

Zeolites	273
Cation-exchanged zeolites	409
Feldspars	136
Amphiboles	99
SiO_2 polymorphs	76
Micas	75
Pyroxenes	51

Oxygen and silicon are the two most abundant elements in the Earth's crust, followed by Al, Fe, Ca, Na, Mg, K. Along with H, Ba, and Sr, these are also the major elements found in most zeolite minerals. In fact, many similarities can be drawn between the feldspars, the most abundant mineral group in the Earth's crust, and zeolites. The two most important principles in understanding zeolites (or any mineral group) are that charge balance must be maintained (i.e. the sum of the formal valence charges of the ions in the chemical formula must equal zero) and that the atoms must fit together to form a stable structure under the conditions of formation. Zeolites form at low pressures and temperatures in the presence of H_2O and possess channels and voids in their structures. Their frameworks are thus more open (i.e. they have lower densities) than other silicates. Much of the information about the crystal structure of a zeolite, or any mineral, can be gained from a correct interpretation of its chemical formula. With this in mind, it seems worthwhile to develop systematically the chemical formulas of zeolites based upon a progression from quartz to feldspars and finally to zeolites.

Quartz, feldspars, and zeolites are all classified as tetrahedral framework structures; they have a 1:2 ratio of tetrahedra cations to oxygen. This means that every TO_4 tetrahedron (where T is most commonly Si or Al) shares every O with an adjacent tetrahedron. Quartz (or any SiO_2 polymorph) exists because of the charge balance between one Si^{4+} and two O^{2-} and the fact that Si^{4+} and O^{2-} can be assembled to form a framework. Feldspars, in turn, can be viewed as four SiO_2 or Si_4O_8 groups, with Al substituting for Si. This substitution causes a charge imbalance in the framework. For

1529-6466/00/0045-0001$10.00

instance, when one Al replaces one Si in Si_4O_8, $[AlSi_3O_8]^{1-}$ results. Thus, a monovalent cation, such as Na^+ or K^+ is needed for charge balance. This would result in $Na[AlSi_3O_8]$ or $K[AlSi_3O_8]$, or in general $(Na,K)[AlSi_3O_8]$. Enclosing Na and K in parentheses, separated by a comma, indicates that they can substitute for each another in the same structural site in feldspar. Although Si and Al are tetrahedrally coordinated by O, the relatively large Na and K ions occupy structural "voids" surrounded by six to eight oxygen atoms. With continued Al substitution, $[Al_2Si_2O_8]^{2-}$ arises. This composition is often considered the maximum Al substitution which is possible without direct contact of two AlO_4 tetrahedra, named Loewenstein's rule or the aluminum avoidance rule (Loewenstein 1954). This rule also requires that Si and Al must be well ordered in a framework with an Si to Al ratio of 1:1.

There are various methods to determine (Si,Al) ordering in framework silicates. The simplest approach uses the T-O distance calculated from atomic coordinates of an X-ray crystal-structure refinement. In feldspars an SiO_4 tetrahedron has a mean T-O distance of ~1.61 Å whereas an AlO_4 tetrahedron has a mean T-O distance of ~1.75 Å (e.g. Kunz and Armbruster 1990). Intermediate values are characteristic of a mixed (Si,Al) occupancy. In reality, however, Si-O and Al-O mean distances are not constant values but vary with the distortion of an individual tetrahedron or of the whole framework and depend on additional bonds from extraframework cations (e.g. Alberti et al. 1990, Alberti 1991). For neutron diffraction experiments, Si and Al exhibit different scattering powers and their respective populations can directly be refined. Magic-Angle-Spinning Nuclear Magnetic Resonance (MAS NMR) spectroscopy is a useful powder method to study the (Si,Al) distribution in a mineral, for example for the nuclei ^{29}Si or ^{27}Al (e.g. Kirkpatrick 1988). The MAS NMR spectrum provides information on the local environment of a tetrahedron (i.e. the type and number of neighboring T sites). A typical result could be that an SiO_4 tetrahedron is connected to one Al and three Si tetrahedra.

The (Si,Al) distribution within a tetrahedral framework has a strong bearing on the position of the extraframework cations. According to Pauling's (1939) electrostatic valence rule not only must charge neutrality of a chemical formula be maintained, but each ion must approach an electrostatically balanced structural environment. The strength of an electro-static bond may be defined as a cation's valence charge divided by its coordination number. In an (Si,Al) framework, an SiO_4 tetrahedron may be bonded to three Si and to one Al tetrahedra. Each of the Si-O bonds contributes an electrostatic bond strength of $4/4 = 1$ and each of the Al-O bonds contributes a bond strength of $3/4 = 0.75$. An oxygen atom sharing two Si tetrahedra has a bond strength sum of two which agrees with its formal charge (-2) though with opposite sign. In contrast, an oxygen atom sharing an Si and an Al tetrahedron has a bond strength sum of $1 + 0.75 = 1.75$. Such an oxygen is termed underbonded and is capable of accepting a bond from an extraframework cation. If this oxygen additionally bonds to eight-coordinated extraframework Ca^{2+}, the bond strength sum is increased by $2/8 = 0.25$ and oxygen becomes electrostatically neutral. There are other ways that an ion can achieve an electrostatically balanced environment (e.g. underbonding may lead to shortening and overbonding to lengthening of a bond). An electrostatically balanced structural environment, including the bond length variation, can be evaluated using the bond valence concept of Brown (1992). However, even Pauling's (1939) simple bond strength approach indicates that framework structures with a disordered (Si,Al) distribution prefer a rather disordered extraframework distribution, whereas (Si,Al) ordered structures commonly have an ordered extraframework arrangement.

The chemical formulas for the zeolites are similar to the feldspars with the addition of H_2O. The 1:2 ratio of T to O must be maintained (exceptions are interrupted frameworks). The greater the substitution of Al for Si, the larger the charge deficiency which must be compensated by cations entering the structure. An aluminosilicate framework has the general formula $[Al_{nx}Si_{n(4-x)}O_{n8}]^{nx-}$, where n is some multiple of this basic building unit needed to fill the unit cell. These frameworks contain voids or cages (Fig. 1) and channels into which cations, commonly referred to as channel or extraframework cations, enter for charge balance. As an example, starting with $[AlSi_3O_8]^{1-}$ and letting n = 9 and x = 1, the framework composition $[Al_9Si_{27}O_{72}]^{9-}$ would be created. Thus, this particular framework would need nine positive charges.

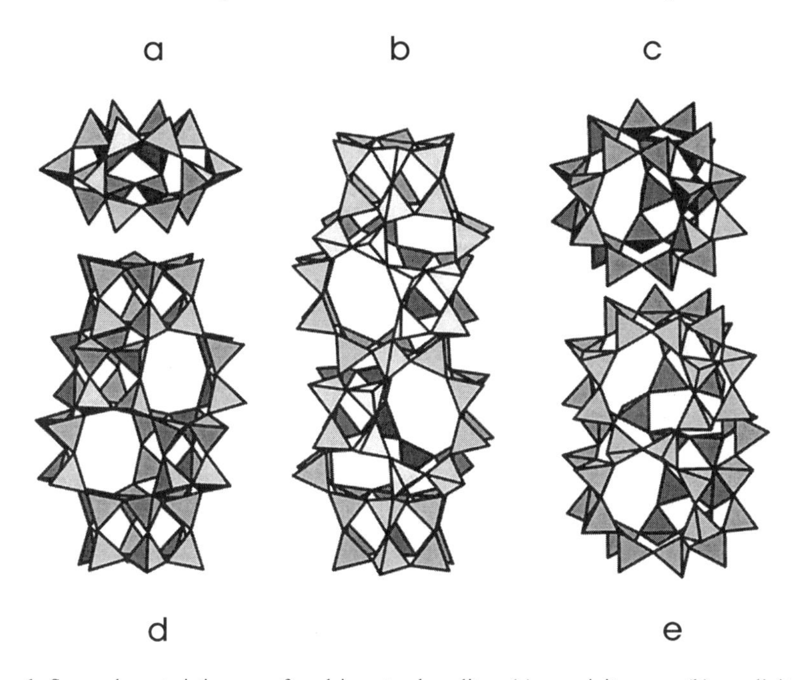

Figure 1. Some characteristic cages found in natural zeolites: (a) cancrinite cage, (b) gmelinite cage, (c) chabazite cage, (d) levyne cage, and (e) erionite cage. The sodalite cage (β-cage) and the α-cage are shown in Figure 13 (below).

One possible way to obtain this would be $(Na,K)Ca_4[Al_9Si_{27}O_{72}]$. The channel cations bond into the frameworks but commonly have one or more sides "exposed" in the channel. These exposed sides lack local charge satisfaction. H_2O can enter the structure to form partial hydration spheres around the channel cations, with the negative dipole of the H_2O molecule pointing toward the exposed positive portion of the channel cation. To complete the above zeolite formula, one could write $(Na,K)Ca_4[Al_9Si_{27}O_{72}]\cdot 24H_2O$. This is the general formula for the zeolite heulandite. Clinoptilolite has a similar framework $[Al_6Si_{30}O_{72}]^{6-}$ but with less Al, thereby requiring a lower channel cation charge (-6 vs -9). Clinoptilolite's general formula can be written as $(Na,K)_6[Al_6Si_{30}O_{72}]\cdot 20H_2O$. Note that clinoptilolite has fewer H_2O molecules than heulandite. Divalent cations, like Ca^{2+} and Mg^{2+}, will in general have larger hydration spheres than monovalent cations, like Na^+ and K^+. Thus, zeolites with divalent channel cations will have more channel H_2O than monovalent cations. There is also more room in the channels for H_2O because one Ca^{2+} occupies less space than two Na^+.

The tetrahedral framework of any zeolite is structurally and chemically much more rigid and stable than the channel cations and H_2O molecules. The extraframework content in zeolites can be exchanged and dehydrated, which is why zeolites have found so many important industrial applications. Heulandite, $(Na,K)Ca_4[Al_9Si_{27}O_{72}]·24H_2O$, and clinoptilolite, $(Na,K)_6[Al_6Si_{30}O_{72}]·20H_2O$, can be used as examples. Their framework structures are identical and a solid solution exists between these two idealized members. The main difference in the above formulas is the ratio of Si to Al. Gunter et al. (1994) and Yang and Armbruster (1996a) performed single-crystal structure refinements on completely exchanged Na, Pb, K, Rb, and Cs heulandites. For these samples the framework remained essentially unchanged. In a similar manner, Armbruster and Gunter (1991) performed crystal structure refinements on a fully hydrated and four partially hydrated phases of a clinoptilolite, yielding a different cation arrangement, but the cation content remained unchanged. Thus the channel occupants can be altered both artificially and in nature.

However, other zeolites exist which have identical framework structures but do not exhibit a continuous solid solution series with respect to the channel cations. An example of this is natrolite $Na_{16}[Al_{16}Si_{24}O_{80}]·16H_2O$, mesolite $Na_{16}Ca_{16}[Al_{48}Si_{72}O_{240}]·64H_2O$, and scolecite $Ca_8[Al_{16}Si_{24}O_{80}]·24H_2O$. From a first glance at the chemical formulas, it might appear there is a solid solution, with two Na cations substituting for one Ca. However, very little variation exists from these ideal formulas found in nature or synthesized in the lab (Ross et al. 1992).

Extended definition of a zeolite mineral

Hey (1930) recognized that zeolites in general have aluminosilicate frameworks with loosely bonded alkali and/or alkali-earth cations and H_2O molecules occupying extraframework positions. More recently, synthetic zeolites have been produced which have tetrahedral sites occupied by elements other than Si and Al. Such examples also exist in the mineral kingdom (e.g. weinebeneite, $Ca[Be_3P_2O_8(OH)_2]·4H_2O$). Furthermore, certain minerals exist where the tetrahedral framework is interrupted by an OH group (e.g. parthéite). On the other hand, some structures exist with tetrahedral frameworks, large cavities and/or channels with a charge-balanced framework, and extraframework ions are not necessary (e.g. clathrasils, ALPOs, SAPOs). A natural representative of clathrasils is melanophlogite with a pure SiO_2 framework (Gies 1985). Because these "zeolite-like" materials occur, the subcommittee on zeolites of IMA CNMMN (Coombs et al. 1998) proposed a revised definition of a zeolite mineral:

A zeolite mineral is a crystalline substance with a structure characterized by a framework of linked tetrahedra, each consisting of four O atoms surrounding a cation. This framework contains open cavities in the form of channels and cages. These are usually occupied by H_2O molecules and extra-framework cations that are commonly exchangeable. The channels are large enough to allow passage of guest species. In the hydrated phases, dehydration occurs at temperatures mostly below 400°C and is largely reversible. The framework may be interrupted by (OH,F) groups; these occupy a tetrahedron apex that is not shared with adjacent tetrahedra.

This more general view of zeolites, without any constraint in chemical composition, has also been adopted in the following review of zeolite structures. Excluded are frameworks in which any channels are too restricted to allow typical zeolitic behavior, such as reversible dehydration, molecular sieving, or cation exchange (e.g. feldspars, feldspathoids, scapolites, and melanophlogite). In addition, cancrinites are not considered because they represent an independent mineral group.

Zeolitic behavior is not limited to tetrahedral framework structures. The mineral cavansite $Ca(VO)(Si_4O_{10})\cdot 4H_2O$ (Evans 1973, Rinaldi et al. 1975a) has a tetrahedral framework similar to the zeolite gismondine but with intercalated square pyramidal VO_5 groups. A review of additional minerals not regarded as zeolites but with zeolitic properties is given by Zemann (1991). Two of his prominent examples are: pharmakosiderite with a porous framework $[(Fe_4(OH)_4(AsO_4)_3]^-$ formed by $FeO_3(OH)_3$ octahedra and AsO_4 tetrahedra with H_2O and cations like Na, K, Ba, and Ag occupying the structural voids; and zemannite, $Na_2[Zn_2(TeO_3)_3]\cdot H_2O$, which has wide channels confined by ZnO_6 octahedra and TeO_3 pyramids. Even carbonates like defernite, $Ca_6(CO_3)_{1.58}(Si_2O_7)_{0.21}(OH)_7[Cl_{0.5}(OH)_{0.08}(H_2O)_{0.42}]$, (Armbruster et al. 1996) and holdawayite $(Mn_6(CO_3)_2(OH)_7[Cl,OH]$ (Peacor and Rouse 1988) have wide channels confined by eight $(Ca,Mn)O_{6-7}$ polyhedra. The channels are lined by OH^- groups with Cl^-, OH^-, or H_2O sitting at the center.

Classification

Currently, three classification schemes are used widely for zeolites. Two of these are based upon specifically defined aspects of the crystal structure, whereas the third has a more historical basis, placing zeolites with similar properties (e.g. morphology) into the same group. Because the physical properties of a mineral (e.g. morphology) are related to its crystal structure, this third method is also indirectly based upon the zeolite's crystal structure.

The first structural classification of zeolites is based upon the framework topology, with distinct frameworks receiving a three-letter code (Table 3; Meier et al. 1996). For instance, the framework for heulandite and clinoptilolite are identical. A framework code of HEU has been assigned to these two zeolites. Heulandite was named before clinoptilolite and was thus given priority in the naming. The frameworks of natrolite, mesolite, scolecite, and gonnardite are all identical, and the code NAT is used to describe these four zeolites. Again, natrolite was the first discovered of the group. Because the channel occupants can be exchanged, a classification based upon framework topology is logical, and this classification scheme works well for those zeolite researchers whose major interests are in cation exchange and synthetic zeolites. Because these codes "should not be confused or related to actual minerals" (Meier et al. 1996), this classification scheme does not work for geologists attempting to name zeolite minerals.

Another structural aspect associated with structure codes is the framework density (FD), which is the number of T-atoms per 1,000 $Å^3$. At a glance, this value reflects the porous nature of a zeolite (i.e. the lower the FD, the larger proportion of the structure is occupied by voids and channels). One criterion Meier et al. (1996) used for inclusion in their Atlas of Zeolite Structure Types is an FD smaller than 21. Other silicates such as scapolite and quartz have FDs of about 22 and 27, respectively. For a complete discussion of framework codes, see Meier et al. (1996).

A second structural method for the classification of zeolites is based upon a concept termed "secondary building units" (SBU). The primary building unit for zeolites is the tetrahedron. SBUs are geometric arrangements of tetrahedra. Here an analogy with general silicate mineralogy will be helpful. In silicate minerals, the tetrahedra can be arranged into groups such as rings, chains, sheets, or frameworks. Thus, in silicate mineralogy we could consider each of these an SBU. (This simplification is not exactly correct, because silicate minerals are mostly composed of additional polyhedral units which are neglected in this SBU model.) In zeolites, groupings of tetrahedra also exist within the framework structure. Quite often, these SBUs tend to control the morphology of the zeolite.

Table 3. Framework codes (Meier et al. 1996) and associated mineral names for zeolites.

ANA (ammonioleucite, analcime, hsianghualite, leucite, pollucite, wairakite)
BEA (tschernichite)
BIK (bikitaite)
BOG (boggsite)
BRE (brewsterite)
CHA (chabazite, willhendersonite)
-CHI (chiavennite)
DAC (dachiardite)
EAB (bellbergite)
EDI (edingtonite, kalborsite)
EPI (epistilbite)
ERI (erionite)
FAU (faujasite)
FER (ferrierite)
GIS (amicite, garronite, gismondine, gobbinsite)
GME (gmelinite)
GOO (goosecreekite)
HEU (clinoptilolite, heulandite)
LAU (laumontite)
LEV (levyne)
LOV (lovdarite)
LTL (perlialite)
MAZ (mazzite)
MER (merlinoite)
MFI (mutianite)
MON (montesommaite)
MOR (mordenite)
-MOR (maricopaite)
NAT (gonnardite, mesolite, natrolite, paranatrolite, scolecite)
NES (gottardiite)
OFF (offretite)
-PAR (parthéite)
PAU (paulingite)
PHI (harmotome, phillipsite)
RHO (pahasapaite)
-RON (roggianite)
STI (barrerite, stellerite, stilbite)
TSC (tschörtnerite)
TER (terranovaite)
THO (thomsonite)
VSV (gaultite)
WEI (weinebeneite)
YUG (yugawaralite)
no framework code (cowlesite, tvedalite)

Breck (1974) lists seven major groups of zeolites based upon the geometry of the SBU (Table 4). For instance, he classifies NAT structures into his Group 5 with a T_5O_{10} SBU. The predominant crystallographic component of this group is linked chains of tetrahedra parallel to the **c**-axis, which in turn causes minerals of this group to be morphologically elongated parallel to the **c**-axis (Fig. 2), sometimes to the point of appearing fibrous. Group 7 zeolites ($T_{10}O_{20}$), which include heulandite (HEU), clinoptilolite (HEU), stilbite (STI), and brewsterite (BRE), tend to be platy in nature. Their form can be explained by the orientation of their SBUs and, in turn, their channels.

Table 4. Zeolite classification scheme developed by Breck (1974) based on SBUs. This is Breck's table; it has not been updated.

Group 1 (S4R - single 4-ring)
- analcime
- harmotome
- phillipsite
- gismondine
- paulingite
- laumontite
- yugawaralite
- (P)

Group 2 (S6R - single 6-ring)
- erionite
- offretite
- levynite
- sodalite hydrate
- (T, omega, losod)

Group 3 (D4R - double 4-ring)
- (A, N-A, ZK-4)

Group 4 (D6R - double 6-ring)
- faujasite
- chabazite
- gmelinite
- (X, Y, ZK-5, L)

Group 5 (T_5O_{10})
- natrolite
- scolecite
- mesolite
- thomsonite
- gonnardite
- edingtonite

Group 6 (T_8O_{16})
- mordenite
- dachiardite
- ferrierite
- epistilbite
- bikitaite

Group 7 ($T_{10}O_{20}$)
- heulandite
- clinoptilolite
- stilbite
- brewsterite

For instance, the channels in heulandite parallel to the **c**-axis (Fig. 16, below) allow the mineral to break easily along (010), thus yielding perfect (010) platy cleavage. Thus, SBUs aid in understanding of the external morphology of the mineral as well as the crystal structure. See Gottardi and Galli (1985) and Meier et al. (1996) for a more complete discussion of SBUs.

The third broad classification scheme is similar to the SBU classification of Breck (1974), except that it includes some historical context of how the zeolites were discovered and named. This scheme uses a combination of zeolite group names which have specific SBUs and is the most widely used by geologists. This is the classification scheme used by Gottardi and Galli (1985), shown in Table 5.

Newcomers to zeolite research may find these multiple classification schemes very confusing, yet they are necessary. Each scheme is used by certain scientists; the structure codes and SBUs are used more by zeolite researchers (both crystallographers and mineralogists), whereas the group names of Gottardi and Galli (1985) are more useful to geologists and descriptive mineralogists. We had originally thought to propose yet a fourth classification scheme based upon the number of channels, their size and orientation, but we chose not to add more confusion to zeolite classification schemes. Instead, we have adopted a slightly modified classification of Gottardi and Galli (1985).

ZEOLITES WITH T_5O_{10} UNITS — THE FIBROUS ZEOLITES

The basic building block for the zeolites in this subgroup is T_5O_{10} chains of $(Al,Si)O_4$ tetrahedra running parallel to the **c**-axis (Figs. 2a-c) with a periodicity along the

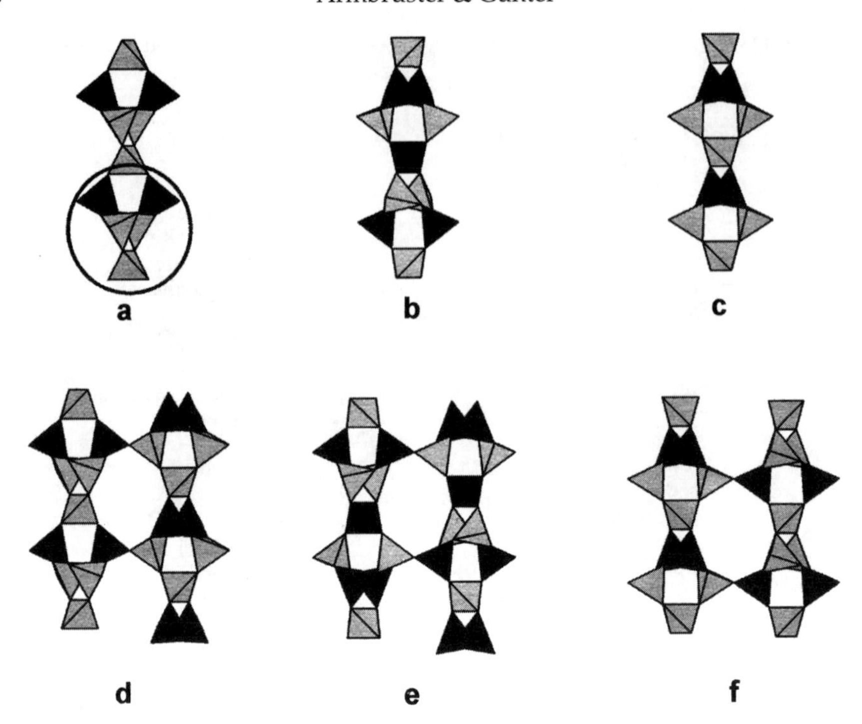

Figure 2. Tetrahedra linking and different types of T_5O_{10} tetrahedra chains for the fibrous zeolites and differing types of cross-linking of tetrahedra chains creating the different framework topologies. The shading of an individual tetrahedron represents its Si (lighter shading) or Al (darker shading) content. The **c**-axis is vertical for all projections. (a) The chain from natrolite viewed down the **a**-axis with an Si/Al ratio of 1.5. The circle outlines a single T_5O_{10} SBU. (b) The chain from thomsonite viewed down the **a**-axis with an ordered Si/Al ratio of 1.0 (i.e. largest amount of Al for any zeolite in this group). (c) A (110) projection of a T_5O_{10} tetrahedra chain from edingtonite viewed down [110] with an Si/Al ratio of 1.5. (d) A $(1\bar{1}0)$ projection showing cross-linking of the T_5O_{10} chains in natrolite. One chain is translated about 1.65 Å in relation to the next. This translation also occurs in (110) (i.e. the plane perpendicular to $(1\bar{1}0)$). (e) A (010) projection showing cross-linking of the T_5O_{10} chains in thomsonite. This translation does not occur in the (100) plane (i.e. the plane perpendicular to (010)). (f) A (110) projection showing cross-linking of the T_5O_{10} chains in edingtonite. No translation occurs in this plane, or in any other plane, in the edingtonite framework.

c-axis of approximately 6.6 Å, or some multiple of 6.6 Å. This 6.6-Å repeat, circled in Figure 2a, contains five tetrahedral sites and ten oxygens, thus the distinction of the T_5O_{10} unit for this group (Breck 1974). Morphologically, this subgroup often appears as needle-like and fibrous crystals extended parallel to the tetrahedral chains. Because of this dominant morphology, this subgroup of zeolites is often referred to as the fibrous zeolites (Gottardi and Galli 1985).

Three different framework topologies exist based upon the cross-linking of the T_5O_{10} tetrahedral chains (Figs. 2d-f), leading to three subgroups: the natrolite group (NAT) (natrolite, mesolite, scolecite, paranatrolite, and gonnardite), thomsonite (THO), edingtonite (EDI) and kalborsite (EDI). Structure projections on the **ab**-plane look very similar for the NAT, THO, and EDI frameworks, with the main difference being the type and location of the channel occupants (Figs. 3 and 4, below). These differences in channel contents are also related to cross-linking of the T_5O_{10} chains, Si/Al content, and (Si,Al) ordering in the chains.

Table 5. Zeolite classification scheme of Gottardi and Galli (1985). [This scheme is used in this chapter.] Listed next to each heading is Breck's associate SBU. This index has not been updated with the new data presented in this chapter; it includes some discredited names.

fibrous zeolites (T_5O_{10})
- natrolite, tetranatrolite, paranatrolite, mesolite, scolecite
- thomsonite
- edingtonite
- gonnardite

single connected 4-ring chains (single 4-ring)
- analcime, wairakite, viséite, hsianghualite
- laumontite, leonhardite
- yugawaralite
- roggianite

doubly connected 4-ring chains
- gismondine, garronite, amicite, gobbinsite
- phillipsite, harmotome
- merlinoite
- mazzite
- paulingite

6-rings (single & double 6 rings)
- gmelinite
- chabazite, willhendersonite
- levyne
- erionite
- offretite
- faujasite

mordenite group (T_8O_{16})
- mordenite
- dachiardite
- epistilbite
- ferrierite
- bikitaite

heulandite group ($T_{10}O_{22}$)
- heulandite, clinoptilolite
- stilbite, stellerite, barrerite
- brewsterite

Unknown structures
- cowlesite
- goosecreekite
- parthéite

The T_5O_{10} chains may be cross-linked in various ways (Alberti and Gottardi 1975, Smith 1983, Malinovskii et al. 1998), but only three connection types have been verified among natural zeolites to form natrolite, thomsonite, and edingtonite group zeolites (Fig. 2). The difference in the three arrangements shown in Figure 2 is the amount of translation of the individual chains with respect to each another as they cross-link to form the three-dimensional framework. Edingtonite is the simplest of the three cases. There are no relative translations between the cross-linking chains (Fig. 2f). Cross-linking of the tetrahedral chains in thomsonite appears similar to that in edingtonite when viewed down the [100] direction, with the difference being that thomsonite has a higher tetrahedral Al content than edingtonite. However, when viewed down the [010] direction, the chains are translated 1/8 *c*, approximately 1.65 Å, (Fig. 2e). In the natrolite group, the chains are also translated about 1.65 Å with respect to each other (Fig. 2d); however, in this structure every adjacent chain is translated with respect to its nearest neighbor, unlike in thomsonite where the translation only occurs in the (010) plane. In the crystal structures analyzed to date, (Si,Al) appears ordered in mesolite and scolecite, ordered to slightly disordered in natrolite, and completely disordered in paranatrolite, and gonnardite (Ross et al. 1992). With this in mind, Alberti et al. (1995) recently proposed some refinements on the classification of this group based on (Si,Al) order-disorder.

All of the framework structures for this group of zeolites contain channels parallel to the **c**-axis. Naturally, these channels must contain sufficient cations to charge balance the framework. The dominant cations occurring in this group are Na, Ca, or Ba. Differing amounts of H_2O also occur within these channels. The channel cations bond both to framework oxygens and to oxygens associated with channel H_2O molecules. The channels are formed by elliptical rings of eight tetrahedra and vary in size for the three different frameworks. As Ross et al. (1992) noted, the size and shape of the [001] channels vary as a function of the cross-linking of the tetrahedral chains, with edingtonite having the largest channels, thomsonite intermediate, and the natrolite group the smallest channels. Thus, edingtonite would be expected to house larger cations than natrolite, for example, which indeed is the case.

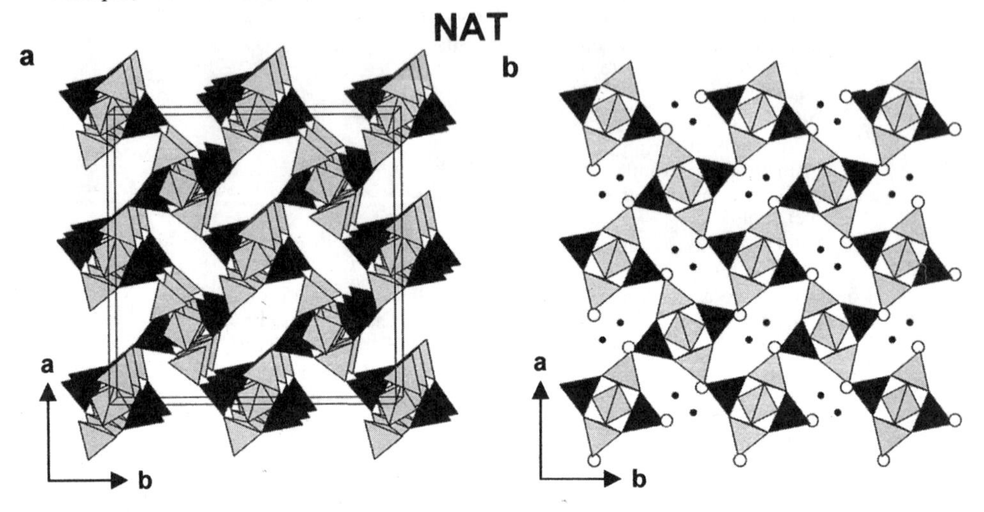

Figure 3. Projections of the natrolite-group structures. (a) The natrolite framework projected on (001) and slightly tilted out of the (001) plane with the **a**-axis vertical and the **b**-axis horizontal. The projection shows the unit cell, the elliptical [001] channels, and a view down the cross-linked T_5O_{10} chains. (b) The natrolite structure projected on (001) with the **a**-axis vertical, showing the framework and channel contents. The Na atoms (small black circles) near the channel center are related to each other by a 2_1 screw in the [001] channel center. H_2O molecules (white circles) are located nearer the channel edges.

Natrolite group (NAT): natrolite, scolecite, mesolite, gonnardite, paranatrolite

The NAT framework of these five zeolites (Fig. 3a) has maximum space group symmetry $I4_1/amd$ assuming complete disorder of (Si,Al). Ordering of Si and Al in the framework lowers this symmetry to $I4_1md$ because previously equivalent tetrahedra become distinct. Distortion of the framework (e.g. tetrahedral rotation caused by addition of channel cations) further reduces this symmetry to $I\bar{4}2d$, the space group for gonnardite. Symmetric and nonsymmetric addition of channel cations further reduces the symmetry to *Fdd*2 (natrolite and mesolite) and *F*1*d*1 (scolecite). The special arrangement of extraframework cations, responsible for the lowered symmetry of these minerals, may also cause characteristic twinning (Akizuki and Harada 1988).

Natrolite (NAT), $Na_{16}[Al_{16}Si_{24}O_{80}]\cdot16H_2O$, is orthorhombic, space group *Fdd*2, *a* = 18.29, *b* = 18.64, *c* = 6.59 Å, Z = 1. The structure of natrolite was first proposed by Pauling (1930) and then determined by Taylor et al. (1933). Since then, many other refinements have been made, ranging from low-temperature neutron studies (Artioli et al. 1984) to dehydration studies (Peacor 1973, Bakakin et al. 1994, Joswig and Baur 1995, Baur and Joswig 1996). (See Gottardi and Galli 1985, Ross et al. 1992 and Alberti et al. 1995, for other reviews). Low-temperature neutron refinements have also allowed for location of the H atoms within the channel and for determination of (Si,Al) ordering (e.g. Artioli et al. 1984). Natrolite's framework consists of ordered cross-linked chains of Si and Al tetrahedra as shown in Figure 3a when viewed down the **c**-axis or when viewed with the **c**-axis vertical (Fig. 2d). Of these three closely related zeolite species, natrolite, mesolite, and scolecite, natrolite shows the highest degree of (Si,Al) disorder. Several refinements have been performed on strongly (Si,Al) disordered natrolites (Alberti and Vezzalini 1981a, Krogh Andersen et al. 1990).

When viewed down the **c**-axis, each channel contains two Na and two H_2O molecules (Fig. 3b). A 2_1 screw axis at the center of each channel leads to a spiral of Na atoms up or down the channel. Each Na is six-coordinated to four framework oxygens and two H_2O molecules. These coordination polyhedra share edges and run parallel to the **c**-axis. A natural K-rich natrolite has K on a new extraframework site close to an H_2O position in regular natrolite (Meneghinello et al. 1999).

Natrolite (unit-cell volume = 2250 Å^3) is completely dehydrated at 548 K leading to anhydrous natrolite with a volume = 1785 Å^3 and space group *F*112 (Joswig and Baur 1995). At 823 K the unit-cell volume increases to 1960 Å^3 (Baur and Joswig 1996) and the space group becomes again *Fdd*2, the same symmetry as found for hydrated natrolite. High-pressure studies are reported by Belitsky et al. (1992). Structures of ion-exchanged natrolite varieties were presented by Baur et al. (1990) and Stuckenschmidt et al. (1992, 1996).

Scolecite (NAT), $Ca_8[Al_{16}Si_{24}O_{80}]\cdot 24H_2O$, is monoclinic, space group *F*1*d*1, a = 18.51, b = 18.98, c = 6.53 Å, β = 90.64°, Z = 1 (Adiwidjaja 1972, Joswig et al. 1984, Stuckenschmidt et al. 1997, Kuntzinger et al. 1998). Neutron diffraction studies at room temperature and at 20 K (Joswig et al. 1984, Kvick et al. 1985) enabled the location of H atoms and showed the framework structure to be ordered. The *F*1*d*1 space group setting (Joswig et al. 1984) permits a direct comparison with the structure of other T_5O_{10} zeolite species because it retains the **c**-axis parallel to the T_5O_{10} chains, whereas the standard *Cc* setting (Kvick et al. 1985) places the **a**-axis parallel to the chains.

The reduction in symmetry from orthorhombic (*Fdd*2) to monoclinic (*F*1*d*1) is related to the symmetry of the channel cation distribution. The 2_1 axes in natrolite do not occur in scolecite because there is only one Ca per channel. The other Na site is replaced by an H_2O molecule, so each channel of scolecite contains one Ca site and three H_2O molecules (Fig. 3c). All Na in natrolite has been replaced by Ca and additional H_2O by the following substitution: Ca + H_2O $\rightarrow$ 2Na, thereby maintaining charge balance. Each Ca is seven-coordinated to four framework oxygens and three channel H_2O molecules. These coordination polyhedra do not share edges with each other, as is the case for the Na polyhedra in natrolite. Phase transformations at high hydrostatic pressures were investigated by Bazhan et al. (1999).

Mesolite (NAT), $Na_{16}Ca_{16}[Al_{48}Si_{72}O_{240}]\cdot 64H_2O$, is orthorhombic, space group *Fdd*2, a = 18.405, b = 56.65, c = 6.544 Å, Z = 1 (Adiwidjaja 1972, Artioli et al. 1986a) with a structure similar to natrolite. Mesolite possesses the same space group and similar unit-cell parameters as natrolite, except the b cell edge is tripled. There is complete (Si,Al) ordering in the framework.

Mesolite has two distinct channel types: one contains two Na's and two H_2O molecules and is similar to the channels in natrolite, and the other channel contains one Ca and three H_2O molecules and is similar to the channels in scolecite (Fig. 3d). One plane of Na channels alternates with two planes of Ca channels. These planes are parallel to (010), and this symmetric arrangement of channels results in the tripling of the b cell

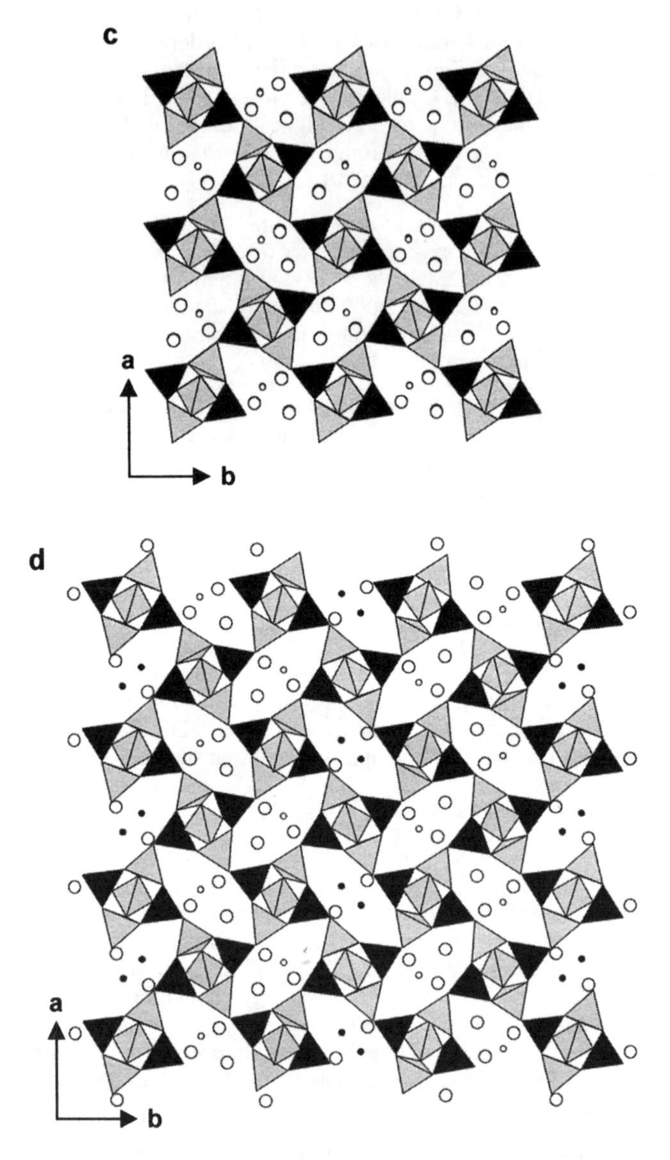

Figure 3, contined. Projections of the natrolite-group structures. (c) The scolecite structure projected on (001) with the **a**-axis vertical, showing the framework and channel contents. A single Ca atom (small white circles) occurs in the [001] channel and has moved closer to the channel center than the Na site in natrolite. Three H_2O molecules (larger white circles) occur in the channel. Two of these are similar to those found in natrolite but have moved away from the channel edge. A new H_2O site occurs in the channel center near the position of one of the Na atoms in natrolite. (d) The mesolite structure projected on (001) with the **a**-axis vertical, showing the Na sites (small black circles), which are similar to those found in natrolite, and the Ca sites (small white circles), which are similar to those found in scolecite. The H_2O positions (larger white circles) are also similar to those found in natrolite and scolecite. There is an ordered arrangement of one "natrolite channel strip" and two "scolecite channel strips" shown vertically, resulting in a tripling of the **b**-axis.

edge, as mentioned above. Structural distortions and extraframework cation arrangement upon dehydration are discussed by Ståhl and Thomasson (1994). Above 598 K mesolite becomes X-ray amorphous.

Gonnardite (NAT) has a highly variable composition, approximated by $(Na,Ca)_{6-8}[(Al,Si)_{20}O_{40}]\cdot 12H_2O$ (Foster 1965). Alberti et al. (1995) proposed a distinction between gonnardite and tetranatrolite (Chen and Chao 1980) based upon their Si/Al content. However, recently tetranatrolite was discredited (Artioli and Galli 1999) because no valid chemical or crystallographic parameter (Coombs et al. 1998) can be used to distinguish gonnardite from tetranatrolite. Gonnardite is tetragonal, space group $I\bar{4}2d$, a = 13.21, c = 6.62 Å, Z = 1 (Mazzi et al. 1986, Mikheeva et al. 1986, Artioli and Torres Salvador 1991). Only one type of channel exists in gonnardite. However, this one channel type is an "average" of the Na-channel in natrolite and the Ca-channel in scolecite. This average channel contains disordered (Ca,Na) atoms on the two metal sites shown in Figure 4a. The atomic environment surrounding any Na in gonnardite is similar to Na in natrolite. The atomic environment surrounding any Ca in gonnardite is similar to Ca in scolecite. However, for a Na vertically adjacent to a Ca parallel to the **c**-axis, the coordination of Na would increase because of extra channel H_2O located nearer the middle of the [001] channels (Mazzi et al. 1986).

Paranatrolite (NAT), $Na_{16}[Al_{16}Si_{24}O_{80}]\cdot 24H_2O$, has a simplified formula identical to natrolite with the exception of additional H_2O (Chao 1980). The symmetry is pseudo-orthorhombic, a = 19.07, b = 19.13, c = 6.580 Å, Z = 1, but may be monoclinic or triclinic (Chao 1980). Separate species status of paranatrolite is debatable and the mineral may be regarded as over-hydrated gonnardite or natrolite (Coombs et al. 1998). The extra H_2O may fit into channel sites which are vacant in natrolite and are located between the Na sites (Mazzi et al. 1986, Gabuda and Kozlova 1997). These Na sites are related by a 2_1 screw; thus, vacancies exist between them parallel to the **c**-axis. These are the same sites occupied by H_2O in scolecite but they are located nearer the middle of the channel (Fig. 3c).

Thomsonite (THO) and edingtonite (EDI) frameworks

Thomsonite (THO), $Na_4Ca_8[Al_{20}Si_{20}O_{80}]\cdot 24H_2O$, is orthorhombic, space group *Pncn*, a = 13.09, b = 13.05, c = 13.22 Å, Z = 1 (Alberti et al. 1981, Ståhl et al. 1990). The structure of thomsonite was first determined by Taylor et al. (1933) and solved in space group *Pbmn* with a = 13.0, b = 13.0, c = 6.6 Å. The doubling of c (Alberti et al. 1981, Ståhl et al. 1990) is a result of (Si,Al) ordering in the T_5O_{10} chains which are parallel to the **c**-axis (Fig. 2b). Gottardi and Galli (1985) and Tschernich (1992) gave *Pcnn* as the space group setting for thomsonite. *Pcnn* and *Pncn* differ by the choice of the **a**- and **b**-axes. For *Pcnn*, $a < b$, whereas for *Pncn*, $a > b$. The framework consists of cross-linked chains of T_5O_{10} tetrahedra as shown in Figure 2e and explained in detail above. The major channels in thomsonite are parallel to [001], and Ca, Na, and H_2O reside in them. Two types of these channels with different channel occupants can be distinguished. In one channel, there is a fully occupied site with either Na or Ca in equal amounts. In the other channel, Ca occupies a split position about 0.5 Å apart and thus can only be 50% occupied (Fig. 4b). Ross et al. (1992) emphasized that the (Na,Ca) polyhedra share edges parallel to the **c**-axis and that the Ca polyhedra alternate with vacancies. Na and Ca are disordered over these edge-sharing polyhedra (Ståhl et al. 1990). The cation in the edge-sharing (Na,Ca) polyhedron is eight-coordinated with four framework oxygens and four channel H_2O molecules, whereas the isolated Ca polyhedron is six-coordinated with four framework oxygens and two channel H_2O molecules. There are four different H_2O sites, two in the channel containing the (Na,Ca) polyhedron and the other two nearer the channel edge (Fig. 4b).

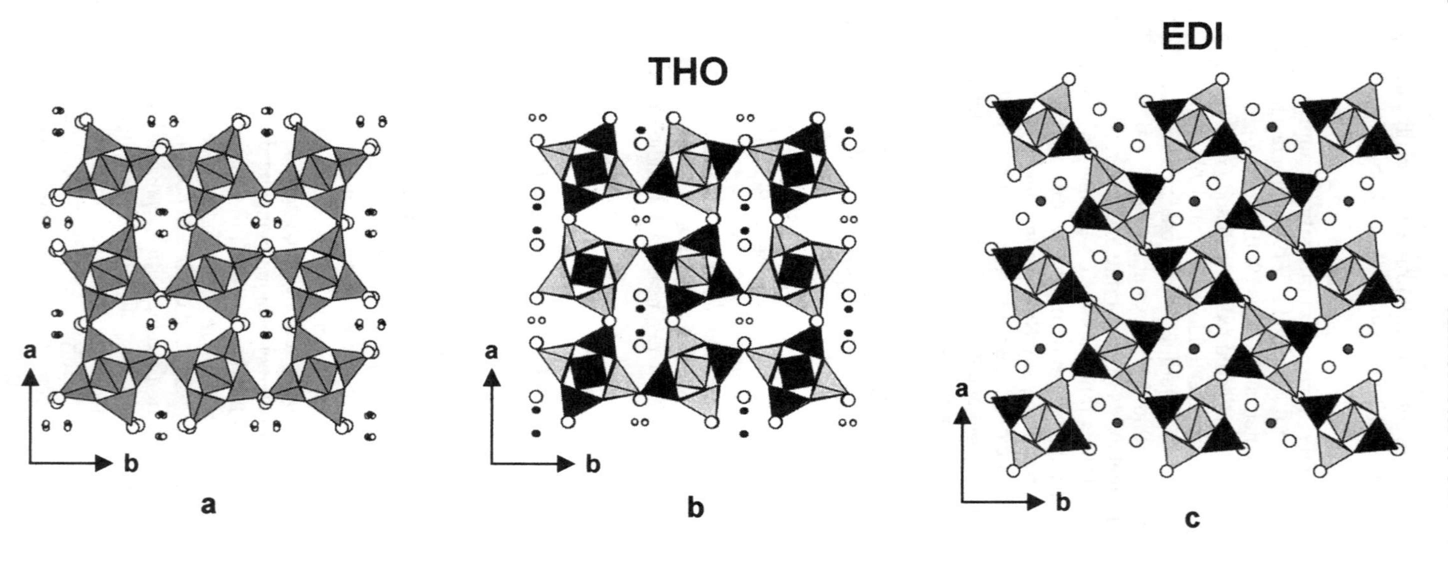

Figure 4. (a) The gonnardite structure projected on (001) with the **a**-axis vertical, showing the disordered framework (i.e. no distinction between the tetrahedra) and channel contents. Each [001] channel can contain both Na atoms (small black circles) and Ca atoms (small white circles) similar to the cation locations in natrolite and scolecite. The H_2O sites (larger white circles) are located nearer the channel edges as in natrolite. (b) The thomsonite structure projected on (001) with the **a**-axis vertical showing the ordered Si/Al framework, which is more Al-rich than the previous fibrous zeolites. There is an even distribution of channel types between Na (small black circles) and Ca (small white circles). H_2O molecules (larger white circles) are nearer the channel center in the Na-containing channels and nearer the edge in the Ca-containing channels. (c) The edingtonite structure projected on (001) with the **a**-axis vertical. Note that the [001] channels are wider in this fibrous zeolite framework than in any of the previous frameworks (i.e. NAT or THO frameworks). This larger-diameter [001] channel houses a Ba (or Sr or Ca) atom (gray-shaded circle) at its center. H_2O sites (white circles) exist both in the [001] channel and near its edge.

Edingtonite (EDI), $Ba_2[Al_4Si_6O_{20}]\cdot8H_2O$, is either orthorhombic, space group $P2_12_12_1$, $a = 9.55$, $b = 9.67$, $c = 6.523$ Å, Z = 1 Å (Taylor and Jackson 1933, Galli 1976, Kvick and Smith 1983, Belitsky et al. 1986) or tetragonal (Mazzi et al. 1984) space group $P\bar{4}2_1m$, $a = 9.581$, $c = 6.526$ Å. Based on optical observations, Akizuki (1986) also discussed triclinic growth sectors in edingtonite. Orthorhombic edingtonites are (Si,Al) ordered, whereas (Si,Al) disorder increases the symmetry to tetragonal. Kvick and Smith (1983) used neutron diffraction to locate the H positions. The framework of edingtonite consists of cross-linked chains of T_5O_{10} tetrahedra as shown in Figure 2f and explained in detail above. The Ba atoms are located in the center of [001] channels on a two-fold axis (Fig. 4c) and are ten-coordinated to six framework oxygens and four H_2O molecules. The coordination polyhedra alternate with vacancies parallel to the **c**-axis similar to the Ca polyhedra in mesolite, scolecite, and thomsonite. There are two H_2O sites in each [001] channel, one located nearer the center and the other nearer the channel's edge (Fig. 4c). Ståhl and Hanson (1998) studied the *in situ* dehydration process using X-ray synchrotron powder-diffraction data and monitored the breakdown of the structure between 660 and 680 K.

Kalborsite (EDI), $K_6B(OH)_4Cl[Al_4Si_6O_{20}]$, is tetragonal, space group $P\bar{4}2_1c$, $a = 9.851$, $c = 13.060$ Å, Z = 2 (Malinovskii and Belov 1980). The framework exhibits a well-ordered (Si,Al) distribution. Doubling of the **c**-axis relative to edingtonite occurs due to ordering of $[B(OH)_4]^-$ tetrahedra and Cl^-, both bonded to extraframework K, along the [001] channels. $[B(OH)_4]^-$ and Cl^- sites correspond to the Ba sites in edingtonite. The different size of $[B(OH)_4]^-$ and Cl^- anions causes different distortions of two adjacent cages.

ZEOLITES WITH CHAINS OF CORNER-SHARING FOUR-MEMBERED RINGS INCLUDING THOSE WITH FINITE UNITS OF EDGE-SHARING FOUR-MEMBERED RINGS AND RELATED STRUCTURES

This group is an extension of that defined by Gottardi and Galli (1985) as "zeolites with singly connected 4-ring chains." It is an extension because of several newly described frameworks for natural zeolites.

Analcime, wairakite, pollucite, leucite, ammonioleucite, hsianghualite (ANA)

All minerals listed in this subgroup have the ANA (analcime) framework topology with the maximum space group symmetry *Ia*3*d* which may be lowered due to tetrahedral cation ordering and the extraframework cation arrangement. Four-membered tetrahedral rings, the SBU for this group, form chains in three dimensions, yielding a complex tetrahedral framework. The channel openings are formed by strongly distorted eight-membered rings (aperture 1.6 × 4.2 Å). The channels run parallel to <110> producing six channel directions.

Analcime (ANA), $Na_{16}[Al_{16}Si_{32}O_{96}]\cdot16H_2O$, crystallizes in various space groups. A stoichiometric analcime according to the above formula was refined in the cubic space group *Ia*3*d* with $a = 13.73$ Å, Z = 1 (e.g. Ferraris et al. 1972). Such an analcime has a statistical (Si,Al) distribution with only one symmetry-independent tetrahedron. All Na sites within the structural voids are occupied. Na is six-coordinated by four framework oxygens and two H_2O molecules (Fig. 5a). Yokomori and Idaka (1998) and Takaishi (1998) provided structural evidence that the ideal topological symmetry of analcime is not cubic but trigonal $R\bar{3}$ with $a = 11.909$ Å, $\alpha = 109.51°$. In general, analcime has split X-ray reflections and non-cubic optical properties (Akizuki 1981a). These non-cubic crystals are always polysynthetically twinned (e.g. Mazzi and Galli 1978). The symmetry

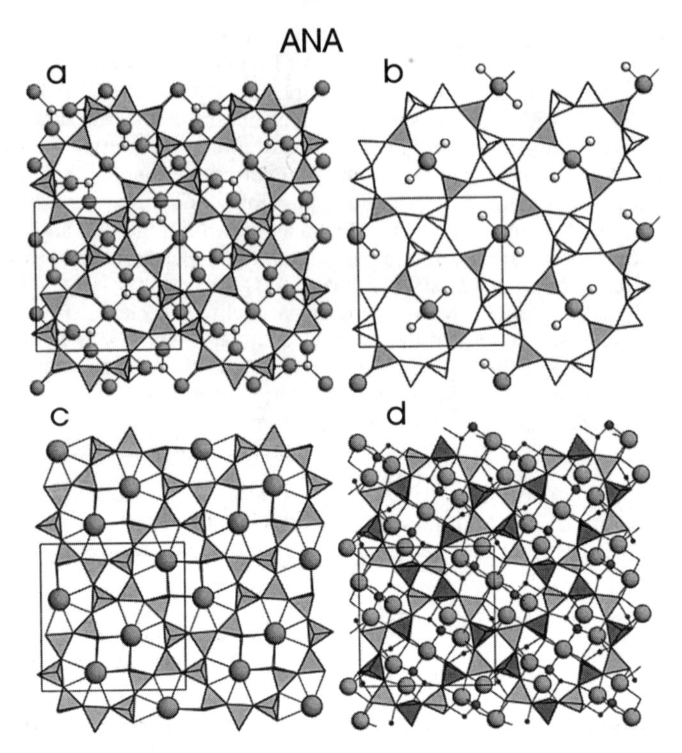

Figure 5. The ANA framework projected parallel to [001] from z ≈ 0 to z ≈ 1/4 with unit-cell outlines. (a) In the cubic analcime structure (*Ia3d*) all tetrahedra show the same shading and are randomly occupied by 2/3 Si and 1/3 Al. Large spheres represent extraframework Na; smaller spheres are H_2O molecules. (b) In the monoclinic (*I2/a*) structure of wairakite two different shadings of tetrahedra are distinguished: dark tetrahedra are occupied by Al, and light tetrahedra bear Si. Between the apices of two AlO_4 tetrahedra, Ca (larger spheres) occupies a cavity position. Ca is additionally bonded to two H_2O molecules (small circles). (c) In the cubic (*Ia3d*) structure of pollucite all tetrahedra have the same shading, characterizing a random (Si,Al) distribution as in cubic analcime. Large spheres representing Cs or K occupy the same sites as H_2O in analcime. (d) In the cubic ($I2_13$) structure of hsianghualite light tetrahedra are occupied by Si, and dark tetrahedra by Be illustrating the perfect alternation between Si and Be. Larger spheres in the structural cavities are Ca, intermediate spheres are F, and small circles are Li. Li is tetrahedrally coordinated by three framework oxygens and one fluorine. However, in this truncated portion of the structure not all ligands coordinating Li are shown.

lowering is caused by partial (Si,Al) ordering (Teertstra et al. 1994a, Kato and Hattori 1998) accompanied by extraframework cation order. Four main types of symmetries are observed: cubic *Ia3d*, tetragonal $I4_1/acd$ with $a > c$, tetragonal $I4_1/acd$ with $c > a$, and orthorhombic (*Ibca*) with all the intermediate varieties (Mazzi and Galli 1978); Hazen and Finger (1979) even reported monoclinic or triclinic symmetry. Due to tetrahedral tilting, analcime undergoes several phase transitions with increasing pressure and becomes triclinic at 12 kbar (Hazen and Finger 1979). The dehydration dynamics and the accompanied structural distortions up to 921 K were studied by Cruciani and Gualtieri (1999).

Wairakite (ANA), $Ca_8[Al_{16}Si_{32}O_{96}]\cdot16H_2O$, with near end-member composition (Aoki and Minato 1980) is monoclinic, *I2/a*, a = 13.699, b = 13.640, c = 13.546 Å, β = 90.51°, Z = 1. As found for most analcimes, wairakite exhibits fine lamellar twinning which probably formed as a consequence of a cubic to monoclinic phase transition (Coombs 1955). Liou (1970) showed the existence of a tetragonal disordered phase between 300 and 460°C. The transformation from this phase to ordered (monoclinic) wairakite is very sluggish. Compared with analcime, wairakite has only half the channel cations due to substitution of Ca^{2+} for $2Na^+$. These eight Ca ions exhibit an ordered distribution on the sixteen available positions (Fig. 5b), which also correlates with increased (Si,Al) order (Takéuchi et al. 1979). Monoclinic wairakite has six symmetry independent tetrahedral sites. Four are occupied by Si and two by Al. Ca is six-coordinated to four oxygen atoms, associated with two AlO_4 tetrahedra, and two H_2O molecules.

Pollucite (ANA), $Cs_{16}[Al_{16}Si_{32}O_{96}]$, is not often regarded as a zeolite mineral (Gottardi and Galli 1985) because the end-member formula is anhydrous (Teertstra and Cerny 1995). However, pollucite forms a complete solid solution series with analcime. Thus, most members of this series are H_2O bearing (e.g. Teertstra et al. 1992). Pollucite possesses a structural framework (space group $Ia3d$, a = 13.69 Å, Z = 1, for the composition $Cs_{12}Na_4[Al_{16}Si_{32}O_{96}]\cdot 4H_2O$) analogous to cubic analcime with (Si,Al) disorder (Beger 1969). Teertstra et al. (1994a,b) reported some degree of short-range (Si,Al) ordering as observed by ^{27}Al and ^{29}Si MAS NMR spectroscopy. Cs in pollucite (Fig. 5c) occupies the same sites as H_2O in cubic analcime. If pollucite has an analcime component, H_2O and Cs are statistically distributed over the same site.

Leucite (ANA), $K_{16}[Al_{16}Si_{32}O_{96}]$, was previously not regarded a zeolite mineral (Gottardi and Galli 1985, Tschernich 1992) because it is anhydrous and in nature does not seem to form extended solid solution series with either one of the hydrous minerals of this group. Leucite undergoes a cubic-tetragonal phase transition at approximately 600°C. The structure of high-temperature cubic leucite (space group $Ia3d$, a = 13.0 Å, Z = 1) has a statistical (Si,Al) distribution (Peacor 1968), as in cubic pollucite and analcime (Fig. 5c). Extraframework K in cubic leucite occupies the same site as Cs in pollucite, leading to a twelve-fold K coordination with six K-O distances of 3.35 Å and additional six K-O distances of 3.54 Å. When a twinned crystal is heated above 600°C, it transforms to cubic symmetry, but when cooled again, the crystal develops twins with the same proportions and twin boundaries as before (e.g. Korekawa 1969, Mazzi et al. 1976, Palmer et al. 1988). Grögel et al. (1984) and Palmer et al. (1989) suggested the formation of an intermediate tetragonal phase (space group $I4_1/acd$). The room-temperature structure of leucite was refined by Mazzi et al. (1976) in the tetragonal space group $I4_1a$ with a = 13.09, c = 13.75 Å, yielding essentially the same (Si,Al) disorder as in the cubic phase. The major difference between the two phases is the coordination of K. At high temperature (cubic leucite), K is coordinated to twelve oxygens. In tetragonal leucite, the K-bearing cavity is compressed, and K becomes six-coordinated with K-O = 3Å. The strong thermal vibration of K at high temperature requires a larger cavity volume. However, with decreasing temperature the thermal motion becomes less dominant, and the structure changes to an energetically more favorable environment for K. Several NMR studies (e.g. Phillips and Kirkpatrick 1994) have shown that the framework of leucite exhibits some degree of short-range (Si,Al) ordering (in analogy to pollucite). The hypothesis that such ordering could trigger the cubic to tetragonal phase transition was rejected by Dove et al. (1993) using static lattice-energy calculations.

Ammonioleucite (ANA), $(NH_4)_{16}[Al_{16}Si_{32}O_{96}]$, was described as a natural transformation product (ion exchange) of analcime (Hori et al. 1986). Thus, the framework characteristics of analcime are still preserved (Moroz et al. 1998). Ammonioleucite is reported to be tetragonal, space group $I4_1ad$, a = 13.214, c = 13.713 Å, Z = 1.

Hsianghualite (ANA), $Li_{16}Ca_{24}$ $F_{16}[Be_{24}Si_{24}O_{96}]$, crystallizes in the cubic space group $I2_13$ with a = 12.864 Å, Z = 1 (Rastsvetaeva et al. 1991), which is due to (Be,Si) order (Fig. 5d) different from the symmetry of other cubic structures ($Ia3d$) of the analcime type. Ca, Li, and F occur as extraframework ions. Ca is eight-fold coordinated by six framework oxygens and two fluorines. Li occupies a site close to the cavity wall and is tetrahedrally coordinated by three framework oxygens and one fluorine. Fluorine is four-coordinated with three bonds to Ca and one to Li. The synthetic compound $Cs_{16}[Be_8Si_{40}O_{96}]$ (ANA) has a pollucite structure and crystallizes in space group $Ia3d$, a = 13.406 Å with (Si,Be) disorder (Torres-Martinez et al. 1984).

Kehoeite, doranite, and viséite, originally ascribed as ANA frameworks, have been discredited (White and Erd 1992, Teertstra and Dyer 1994, Di Renzo and Gabelica 1995).

Laumontite (LAU)

Fully hydrated laumontite (LAU) has the simplified formula $Ca_4[Al_8Si_{16}O_{48}]\cdot 18H_2O$ (Armbruster and Kohler 1992, Artioli and Ståhl 1993, Ståhl and Artioli 1993). Before these studies, it was assumed that fully hydrated laumontite contained $16H_2O$ pfu (e.g. Coombs 1952, Pipping 1966, Gottardi and Galli 1985). When laumontite is exposed to low humidity, it partially dehydrates at room temperature to a variety named "leonhardite" (Blum 1843, Delffs 1843) with the simplified formula $Ca_4[Al_8Si_{16}O_{48}]\cdot 14H_2O$ (e.g. Coombs 1952, Pipping 1966, Armbruster and Kohler 1992, Artioli et al. 1989). This dehydration can be reversed by soaking the sample in H_2O at room temperature (e.g. Coombs 1952, Armbruster and Kohler 1992) and observed with the polarizing microscope: extinction angle varies directly with degree of hydration (Coombs 1952).

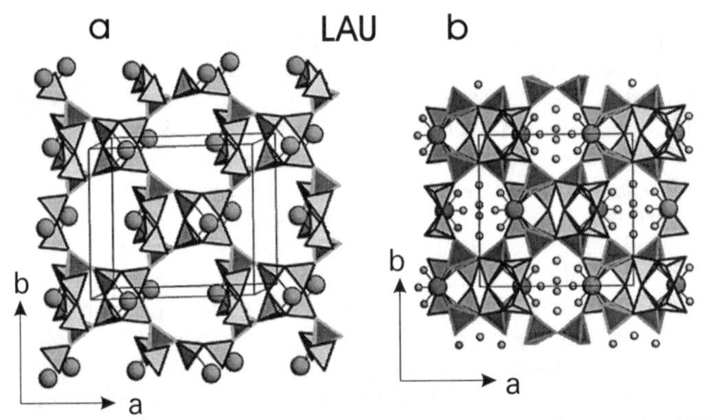

Figure 6. Polyhedral drawings of the laumontite (LAU) framework; AlO_4 tetrahedra have light edges, SiO_4 tetrahedra have dark edges, and spheres represent Ca. (a) Portion of the framework projected approximately on (001) to emphasize the indented shape of the channels parallel to the **c**-axis. (b) Projection along the **c**-axis (channel direction); there are two types of H_2O molecules (small spheres): those coordinating Ca and those in the center of the channels connected to other H_2O molecules and to the channel wall by hydrogen bonds. Note that the AlO_4 tetrahedra are not connected to each other as could be interpreted from this projection.

Various structural studies of laumontite with different degrees of hydration (between 10.8 and 18 H_2O pfu) show that the framework topology remains unchanged and can be described in $C2/m$ symmetry (Artioli et al. 1989, Armbruster and Kohler 1992, Artioli and Ståhl 1993, Ståhl and Artioli 1993). The framework exhibits two different types of four-membered rings, those where Si and Al alternate and those formed by only SiO_4

tetrahedra. Large channels run parallel to the **c**-axis (Fig. 6a) confined by ten-membered rings (aperture 4.0 × 5.3 Å). Ca occupies a four-fold site on a mirror plane within the **c**-extended channels and is coordinated by three H_2O molecules and four oxygens, belonging to AlO_4 tetrahedra (Fig. 6b).

Room-temperature unit-cell parameters for fully hydrated laumontite with $18H_2O$ are a = 14.863, b = 13.169, c = 7.537 Å, β = 110.18°, Z = 1 (Ståhl and Artioli 1993), whereas the $14H_2O$ variant has a = 14.75, b =13.07, c = 7.60 Å, β = 112.7°, Z = 1 (Pipping 1966). Coombs (1952) found no piezoelectric effect for laumontite and leonhardite but a strong pyroelectric effect. Thus, it may be concluded that the structure lacks a center of symmetry and the space group is either *C*2 or *Cm*. Ståhl and Artioli (1993) discussed the possibility of a locally ordered H_2O superstructure leading to lowering of symmetry. The simplest model would lead to doubling of the a-axis with space group *P*2. Armbruster and Kohler (1992) and Ståhl et al. (1996) investigated the dehydration of laumontite. The complicated clustering of H_2O and accompanying phase transitions in laumontite with varying degrees of hydration were studied by Gabuda and Kozlova (1995) using NMR ^1H and ^{27}Al spectroscopy between 200 and 390 K.

Fersman (1909) introduced the term "primary leonhardite" for the composition $Na_{1.24}K_{1.59}Ca_{2.55}[Al_{8.18}Fe^{3+}_{0.03}Si_{15.86}O_{48}]\cdot 14H_2O$, later confirmed by Pipping (1966). This "primary leonhardite," with more than five channel cations pfu, neither dehydrates nor rehydrates at room temperature. The excess of channel cations compared with ordinary laumontite indicates that "primary leonhardite" has additional extraframework cation sites (Baur et al. 1997, Stolz and Armbruster 1997) occupied by H_2O molecules in fully hydrated laumontite. This also explains why "primary leonhardite" cannot be further hydrated and why it shows no indication of weathering as usual for exposed laumontite. The species name "leonhardite" was recently discredited (Wuest and Armbruster 1997) because "leonhardite" is just a partially dehydrated variety of laumontite.

Yugawaralite (YUG)

The framework topology of yugawaralite (YUG), $Ca_2[Al_4Si_{12}O_{32}]\cdot 8H_2O$, space group *C*2/*m* symmetry is reduced to the acentric space group *Pc*, a = 6.72, b = 13.93, c = 10.04 Å, β = 111.1°, Z = 1, due to complete (Si,Al) ordering (Kerr and Williams 1967, 1969; Leimer and Slaughter 1969, Eberlein et al. 1971, Kvick et al. 1986). The mineral is thus piezo- and pyroelectric (Eberlein et al. 1971). Symmetry reduction to *P*1 has been reported on the basis of optical measurements (Akizuki 1987a).

Four-membered rings of tetrahedra are occupied by three Si and one Al in an ordered fashion (Fig. 7a). The structure possesses two types of channel systems confined by eight-membered rings. One channel system extends parallel to the **a**-axis (aperture 2.8 × 3.6 Å), and the other is parallel to the **c**-axis (aperture 3.1 × 5.0 Å). Ca resides at the intersection of the two types of channels and is bonded to four framework oxygens and four H_2O molecules. A neutron single-crystal structure refinement at 13 K (Kvick et al. 1986) located the proton positions. The dehydration of yugawaralite, including an accompanying phase transition at ~200°C, was studied by Gottardi and Galli (1985) and Alberti et al. (1994).

Goosecreekite (GOO)

Goosecreekite, $Ca_2[Al_4Si_{12}O_{32}]\cdot 10H_2O$, is monoclinic, space group $P2_1$, a = 7.401, b = 17.439, c = 7.293 Å, β = 105.44°, Z = 1. The framework consists of almost completely ordered (Si,Al) tetrahedra (Rouse and Peacor 1986). The assignment of goosecreekite to this group is rather arbitrary, as it has a 6-2 SBU (Meier et al. 1996). However, the structure can also be constructed from corner- and edge-sharing four-membered rings. Strongly deformed eight-membered rings (aperture 4.0 × 2.8 Å) confine channels

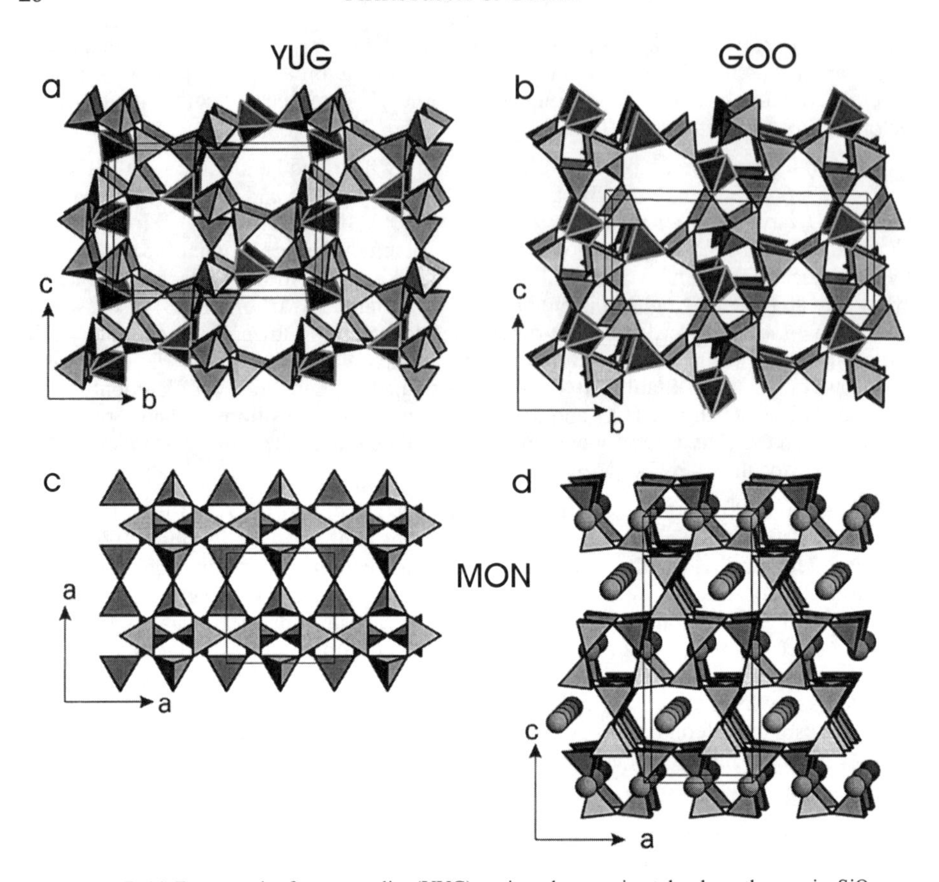

Figure 7. (a) Framework of yugawaralite (YUG) projected approximately along the **a**-axis. SiO_4 tetrahedra have black edges, and AlO_4 tetrahedra have light edges. (b) Framework of goosecreekite (GOO) projected approximately along the **a**-axis. SiO_4 tetrahedra have black edges, and AlO_4 tetrahedra have light edges. (c) A portion of the tetragonal substructure of montesommaite (MON) projected on (001). The framework is built by linked (001) layers consisting of four- and eight-membered rings. (d) The tetragonal substructure of montesommaite (MON) projected approximately along the **a**-axis. Spheres represent (K,Na) sites.

parallel to the **a**-axis (Fig. 7b) which are connected by additional eight-ring channels (aperture 2.7 × 4.1 Å) running zigzag-wise parallel to the **b**-axis. Eight-membered ring channels also run parallel to the **c***-axis. Ca is located roughly at the intersection of these channels and bonds to two framework oxygens and five H_2O molecules. All H_2O molecules are bonded to Ca.

Montesommaite (MON)

Montesommaite (MON), $(K,Na)_9[Al_9Si_{23}O_{64}]\cdot 10H_2O$, was described as being closely allied to merlinoite and gismondine (Rouse et al. 1990, Alberti 1995). However, in the more rigorous classification used in this chapter, montesommaite must be assigned to the group with chains of corner-sharing four-membered rings and not the "edge-sharing four-membered rings." Montesommaite is orthorhombic, space group, *Fdd*2, *a* = 10.099, *b* = 10.099, *c* = 17.307 Å, Z = 1, but it is nearly tetragonal, pseudo $I4_1/amd$ (*a* = 7.141, *c* = 17.307 Å). The tetragonal substructure was solved (Rouse et al. 1990) by analogy to

Smith's (1978) hypothetical model 38 and refined from semi-quantitative powder diffraction data. The framework can be constructed by linking (001) sheets, formed by four- and eight-membered rings (Fig. 7c). The structure (Fig. 7d) has straight eight-membered ring channels running parallel to the **a**- and **b**-axes (aperture 3.6 Å). K is in eight-fold coordination by six framework oxygens and two H_2O molecules.

Roggianite (–RON)

Roggianite (–RON) with the simplified formula $Ca_2[Be(OH)_2Al_2Si_4O_{13}]\cdot 2.5H_2O$ crystallizes in space group *I4/mcm*, *a* = 18.370, *c* = 9.187 Å, Z = 8 (Passaglia and Vezzalini 1988, Giuseppetti et al. 1991). The crystal structure (Giuseppetti et al. 1991) shows a well-ordered (Si,Al,Be) distribution and is characterized by large twelve-membered tetrahedral rings (aperture 4.2 Å) forming channels parallel to the **c**-axis (Fig. 8a). These channels bear three types of partially occupied, disordered H_2O sites but no cations. A special structural and chemical feature of roggianite is the existence of framework OH groups terminating the two corners of a BeO_4 tetrahedron. All interrupted framework structures, including roggianite, are marked by a dash in front of the framework code. Ca occupies a narrow cage and bonds within 2.4 Å to six framework oxygens. Two additional framework oxygens sites are 2.9 Å from Ca; thus, the coordination may be described as 6+2. Voloshin et al. (1986) described an alleged new mineral with the unapproved name "ginzburgite" which is identical to roggianite.

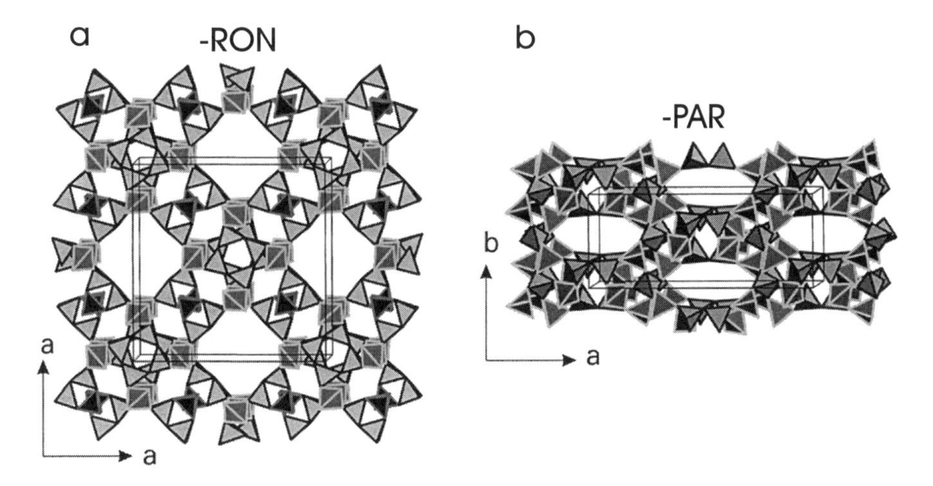

Figure 8. (a) Framework of roggianite (-RON) projected approximately along the **c**-axis (channel direction). SiO_4 tetrahedra are light with dark rims, AlO_4 tetrahedra are gray with light rims, and $BeO_2(OH)_2$ tetrahedra are black with light rims. H_2O molecules (not shown) fill the wide structural channels, whereas Ca (not shown) occupies a narrow cage close to the $BeO_2(OH)_2$ tetrahedra. (b) Framework of parthéite (-PAR) projected approximately along the **c**-axis. SiO_4 tetrahedra have black edges, and AlO_4 tetrahedra have light edges.

Parthéite (–PAR)

Parthéite (–PAR), $Ca_2[Al_4Si_4O_{15}(OH)_2]\cdot 4H_2O$, is monoclinic, space group *C2/c*, *a* = 21.555, *b* = 8.761, *c* = 9.304 Å, β = 91.55°, Z = 4. The structure (Engel and Yvon 1984) exhibits a completely ordered (Si,Al) distribution. Due to the presence of OH groups, the framework is interrupted at every second AlO_4 tetrahedron. The framework can be constructed from corner-sharing four-rings with intercalated finite units of three edge-sharing four-rings. Thus, strictly speaking, the framework of parthéite is a hybrid between two different structural units of Gottardi and Galli (1985). Strongly compressed channels (Fig. 8b) exist parallel to the **c**-axis delimited by ten-membered rings (aperture 6.9 × 3.5 Å). Ca in these channels has a distorted cube-like coordination formed by six framework oxygens and two H_2O molecules.

ZEOLITES WITH CHAINS OF EDGE-SHARING FOUR-MEMBERED RINGS

Zeolite minerals with the gismondine (GIS), phillipsite (PHI), and merlinoite (MER) framework have the same double crankshaft as their main building block (Fig. 9a). A double crankshaft is formed by alternating four-membered rings with four tetrahedral apices pointing up (U) and four tetrahedral apices pointing down (D). Thus, in the chain direction the arrangement can be described as UUDDUUDD. These four-membered rings are connected by edge-sharing to chains leading to a corrugated ribbon resembling a crankshaft with a periodicity of ~10 Å. If this tetrahedral double crankshaft is projected along the direction of extension, it appears as a four-membered ring. Smith (1978) and Sato (1979) showed that there are seventeen possibilities to connect the double crankshaft to a framework. Only three of these have been observed in zeolite minerals. A review of these hypothetical and observed structures is provided by Sato and Gottardi (1982). Different types of double crankshafts are found for mazzite, perlialite, and boggsite (Fig. 9b,c).

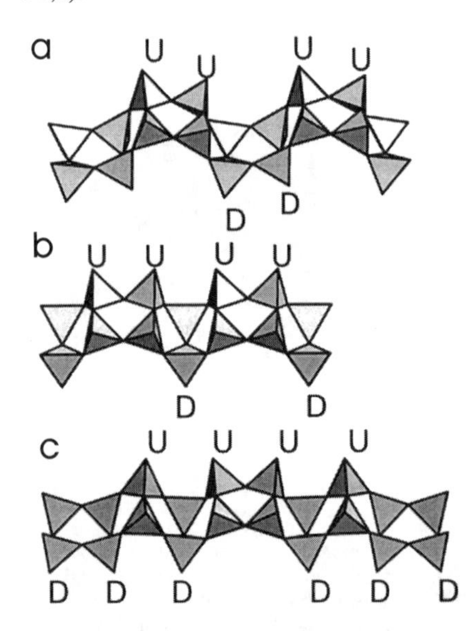

Figure 9. (a) Double crankshaft (GIS, PHI, and MER) formed by alternating four-membered rings with four tetrahedral apices pointing upward and downward. These rings share edges to form infinite chains with a periodicity of approximately 10 Å. Along the chain direction, the orientation of tetrahedral apices is defined as UUDDUUDD (U = up, D = down). A double crankshaft projected along the extension direction appears as a four-membered ring. (b) Chains of edge-sharing four-membered rings in mazzite and perlialite have the sequence of tetrahedral apices UUDUUD (U = up, D = down) leading to a periodicity of approximately 7.5 Å. (c) Chains of edge-sharing four-membered rings in boggsite have the sequence of tetrahedral apices DDUDUUDUDD (D = down, U = up) or vice versa leading to a periodicity of approximately 20 Å.

Gismondine, garronite, amicite, gobbinsite (GIS)

All minerals in this subgroup have the GIS (gismondine) framework topology, which has the maximum space group symmetry $I4_1/amd$. The channels running parallel to the **a**- and **b**-axes are confined by eight-membered rings (aperture 3.0 × 4.7 Å). Only the structure of the high-silica variety of the synthetic zeolite Na-P (GIS) preserves the topological symmetry $I4_1/amd$ (Hakanson et al. 1990). There are two systems of double crankshafts (along the **a**- and **b**-axes) related by a 4_1 axis forming the GIS framework (Fig. 10a). If the tetrahedra are alternately occupied by Si and Al (gismondine and amicite), the symmetry is lowered at least to orthorhombic *Fddd*. Additional symmetry lowering occurs due to the distribution of channel cations and H_2O molecules.

Gismondine (GIS), $Ca_4[Al_8Si_8O_{32}]\cdot16H_2O$, has a completely (Si,Al) ordered framework with space group $P2_1/c$, a = 10.02, b = 10.62, c = 9.84 Å, β = 92.5°, Z = 1, with the pseudo-4_1 axis parallel to the **b**-axis. Ca is displaced from the center of the cavity at the intersection of the eight-membered ring channels and is attached to one side of the eight-membered ring (Fig. 10a). Ca is coordinated to two framework oxygens and four H_2O molecules (Fischer and Schramm 1971, Rinaldi and Vezzalini 1985). Artioli et al. (1986b) located all proton positions by single-crystal neutron diffraction at 15 K and found two statistically distributed configurations for the Ca coordination. In the more frequent configuration (70%), Ca is six-coordinated. In the other variant (30%), one H_2O molecule splits to two new sites; thus, the coordination becomes seven-fold. Upon partial dehydration gismondine undergoes several phase transitions accompanied by symmetry reduction (van Reeuwijk 1971, Vezzalini et al. 1993, Milazzo et al. 1998). Structure refinements on cation-exchanged varieties were performed by Bauer and Baur (1998).

Garronite (GIS), $NaCa_{2.5}[Al_6Si_{10}O_{32}]\cdot13H_2O$, may be considered the (Si,Al) disordered equivalent of gismondine. Artioli (1992) applied the Rietveld method to X-ray powder diffraction data to study the structures of two garronite samples and found the space group to be lowered to $I\bar{4}m2$ with a = 9.9266, c = 10.3031 Å, Z = 1. The symmetry lowering from $I4_1/amd$ to $I\bar{4}m2$ is explained in terms of cation and H_2O arrangements in the structural cavities. There seems to be an indication of partial (Si,Al) ordering. Four partially occupied H_2O sites, two partially occupied nearby Ca sites, and one Na site were located. The short distances preclude simultaneous occupation of Ca and Na sites. Orthorhombic symmetry has been proposed by Nawaz (1983) and Howard (1994). Garronite can only partly be dehydrated and the framework collapses above 254°C. Partially dehydrated phases have decreased symmetry with space groups $I2/a$ and $P4_12_12$ (Schröpfer and Joswig 1997, Marchi et al. 1998).

Amicite (GIS), $Na_4K_4[Al_8Si_8O_{32}]\cdot10H_2O$, has perfect (Si,Al) order like gismondine but with space group $I2$, a = 10.226, b = 10.422, c = 9.884 Å, β = 88.32°, Z = 1 (Alberti and Vezzalini 1979). The pseudo-4_1 axis generates two sets of double crankshafts parallel to the **b**-axis. Na and K are well ordered in two completely occupied sites. Na is six-coordinated by three framework oxygens and three H_2O molecules. K is seven-coordinated by four framework oxygens and three H_2O molecules. In amicite, two of the gismondine Ca sites are occupied by Na with two additional Na sites. K sites in amicite are occupied in gismondine by H_2O. Three filled H_2O sites are bonded to channel cations. One additional H_2O site is only half occupied and shows a long distance (2.9 Å) to Na. Amicite can be completely dehydrated without destruction of the framework (Vezzalini et al. 1999).

Gobbinsite (GIS), $Na_5[Al_5Si_{11}O_{32}]\cdot11H_2O$, could only be studied on the basis of X-ray powder data (Nawaz and Malone 1982, McCusker et al. 1985, Artioli and Foy 1994). McCusker et al. (1985) refined the structure of $Ca_{0.6}Na_{2.6}K_{2.2}[Al_6Si_{10}O_{32}]\cdot12H_2O$ in space group $Pmn2_1$, a = 10.108, b = 9.766, c = 10.171 Å, Z =1, using the Rietveld method and geometric constraints for the framework. The pseudo-4_1 axis in gobbinsite is parallel to the **c**-axis. Ordering of Si and Al is not apparent but cannot be ruled out. The Na sites (more than half occupied) are five-coordinated by three framework oxygens and two H_2O molecules and occupy an eight-membered ring perpendicular to the **a**-axis. K (less than half occupied) is located in an eight-membered ring perpendicular to the **b**-axis and is coordinated by two framework oxygens and one H_2O molecule. Orthorhombic symmetry occurs because of the cation distribution.

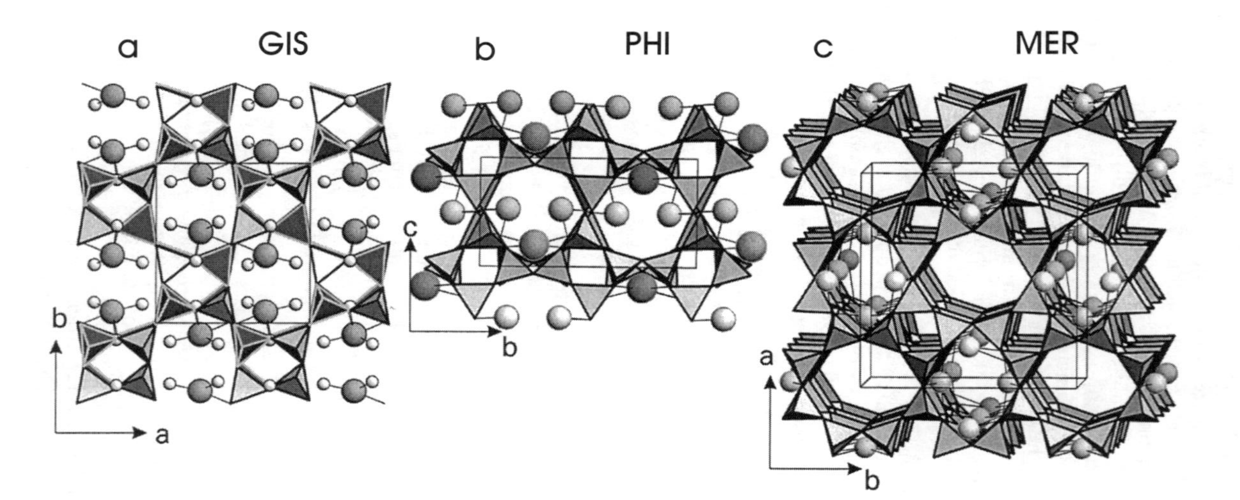

Figure 10. (a) The gismondine (GIS) framework projected along the **c**-axis. The pseudo-4_1 axis is parallel to the **b**-axis. Two double crankshaft systems can be seen: (1) along the direction of projection (four-membered rings) and (2) parallel to the **a**-axis. Tetrahedra with light edges are filled with Al; those with black edges by Si. Large spheres represent Ca; small spheres are fully occupied H_2O. (b) The phillipsite (PHI) framework projected along the **a**-axis, parallel to the extension of double crankshafts. Large spheres represent K (phillipsite) or Ba (harmotome) which block eight-membered ring channels parallel to the **c**-axis. Small spheres are (Na,Ca) obstructing eight-membered ring channels parallel to the **b**-axis. (c) The merlinoite (MER) framework projected approximately along the **c**-axis parallel to the extension of double crankshafts. Spheres represent K sites blocking passages through eight-membered ring channels parallel to the **a**- and **b**-axes.

Phillipsite (PHI) and merlinoite (MER) frameworks

Phillipsite (PHI), $K_2(Ca_{0.5},Na)_4[Al_6Si_{10}O_{32}]\cdot 12H_2O$, and **harmotome** (PHI), $Ba_2(Ca_{0.5},Na)[Al_5Si_{11}O_{32}]\cdot 12H_2O$, form a solid-solution series. The true space group of phillipsite and harmotome is still under debate. Recent X-ray and neutron single-crystal structure refinements between 15 and 293 K confirm the centric space group $P2_1/m$ for harmotome (Stuckenschmidt et al. 1990) proposed by Rinaldi et al. (1974). There are, however, hints of acentricity (space group $P2_1$ or even $P1$), indicated by piezoelectricity (Sadanaga et al. 1961) and optical domains (Akizuki 1985). Room-temperature unit-cell parameters are a = 9.869, b = 14.139, c = 8.693 Å, β = 124.81°, Z = 1, for harmotome, $Ba_{1.92}Ca_{0.46}K_{0.07}[Al_{4.65}Si_{11.26}O_{32}]\cdot 12H_2O$, and a = 9.865, b = 14.300, c = 8.668 Å, β = 124.80°, Z = 1, for phillipsite, $K_{2.0}Ca_{1.7}Na_{0.4}[Al_{5.3}Si_{10.6}O_{32}]\cdot 13.5H_2O$ (Rinaldi et al. 1974). Both structures have similar cation and H_2O distributions within the structural channels (Fig. 10b), and Ba and K occupy more or less the same site. Three types of channels confined by eight-membered rings of tetrahedra exist, one parallel to the **a**-axis (aperture 3.6 Å), one parallel to the **b**-axis (aperture 4.3 × 3.0 Å), and another parallel to the **c**-axis (aperture 3.3 × 3.2 Å). The double crankshafts run parallel to the **a**-axis (Fig 10b). Ordering of (Ca,Na) vacancies combined with (Si,Al) order might be responsible for symmetry lowering as discussed above. Stuckenschmidt et al. (1990) proposed a partially ordered (Si,Al) distribution, whereas Rinaldi et al. (1974) stated that the near uniformity of T-O distances gives little or no suggestion of (Si,Al) order. Ba in harmotome is coordinated to four H_2O molecules and seven framework oxygens. Only one of these four H_2O molecules is weakly fixed by hydrogen bonds; the others are strongly disordered. Two additional strongly disordered H_2O molecules reside in the structural channels and are not bonded to cations (Stuckenschmidt et al. 1990). Similar H_2O disorder is also found in phillipsite (Rinaldi et al. 1974). Stacking faults on (100) and (010) by a slip of $a/2$ or $b/2$ are possible in all frameworks with the double crankshaft. Thus, as a result of $a/2$ faults, sedimentary phillipsite may be composed of phillipsite and merlinoite domains (Gottardi and Galli 1985). Rietveld refinements of Sr- and Cs-exchanged phillipsites were reported by Gualtieri et al. (1999a,b).

Merlinoite (MER), simplified $K_5Ca_2[Al_9Si_{23}O_{64}]\cdot 24H_2O$, is pseudo-tetragonal, space group $I4/mmm$ but the true structure is orthorhombic $Immm$ with a = 14.116, b = 14.229, c = 9.946 Å, Z = 1 (Galli et al. 1979). The framework has a statistical (Si,Al) distribution and contains channels, confined by eight-membered rings, parallel to the **a**-axis (aperture 3.5 × 3.1 Å) and the **b**-axis (aperture 3.6 × 2.7 Å). Eight-membered double and single rings delimit channels (aperture 5.1 × 3.4 Å and 3.3 × 3.3 Å) parallel to the **c**-axis. The double crankshafts also run parallel to the **c**-axis. The framework topology of merlinoite (Fig. 10c) is identical to the synthetic barium chloroaluminosilicate (Solov'eva et al. 1972) which is tetragonal. K and Ba occupy the center of two deformed eight-membered rings connecting a small and a large cage. Additional partially occupied and disordered cation positions were located within the larger cage. Two fully and six partially occupied H_2O sites were located in merlinoite (Galli et al. 1979). Structure refinements of natural merlinoite varieties, $NaK_5Ba_3[Al_{12}Si_{20}O_{64}]\cdot 20H_2O$, $Immm$ with a = 14.099, b = 14.241, c = 10.08 Å, Z =1 (Baturin et al. 1985), and $Na_5K_7[Al_{12}Si_{20}O_{64}]\cdot 24H_2O$, $Immm$ with a = 14.084, b = 14.264, c = 10.112 Å, Z = 1 (Yakubovich et al. 1999) also yielded strongly disordered extraframework cation and H_2O arrangements.

Mazzite (MAZ) and perlialite (LTL)

In contrast to the double crankshafts in which alternately two tetrahedral apices point up and down along the chain direction, the edge-sharing four-ring chains in mazzite and perlialite have the sequence of two tetrahedral apices up and one down or vice versa

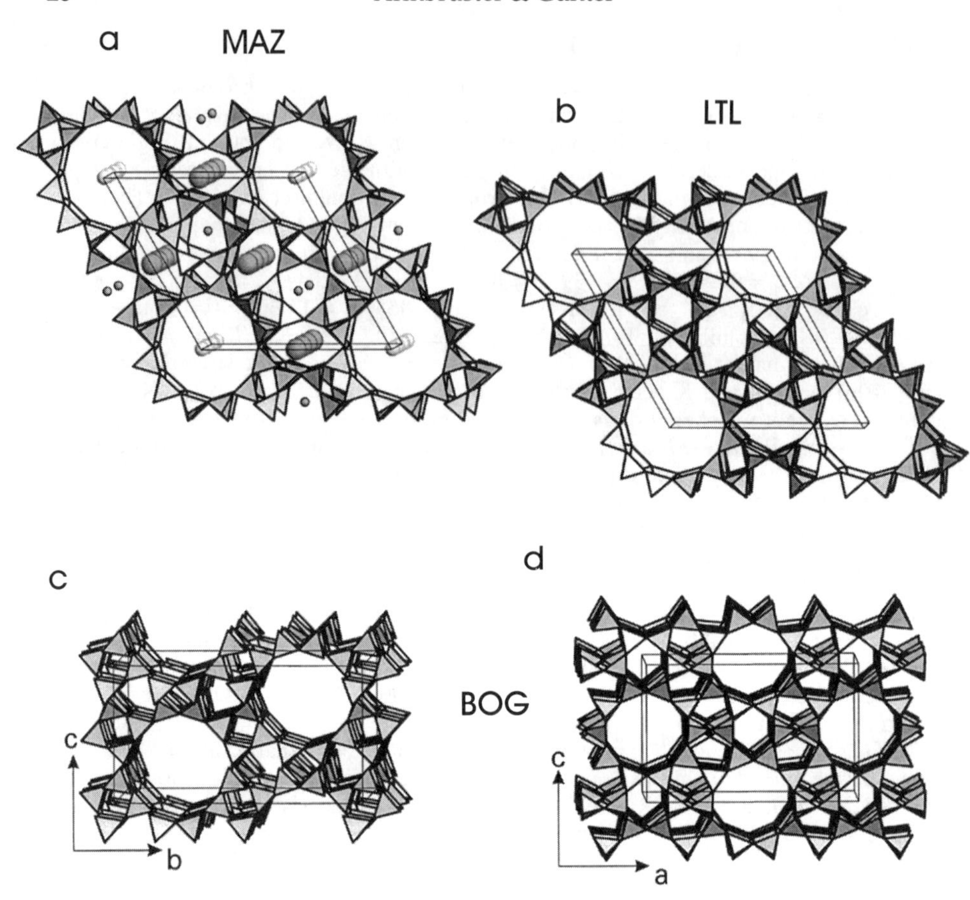

Figure 11. (a) The mazzite (MAZ) framework projected approximately along the **c**-axis, parallel to the edge-sharing four-membered ring chains. Notice the six-membered single ring and the twelve-membered double ring in contrast to perlialite (LTL). Small spheres are Mg, intermediate spheres are Ca, and large spheres are K. (b) The perlialite (LTL) framework projected approximately along the **c**-axis. Notice the six-membered double rings and twelve-membered single rings in contrast to mazzite. (c) The boggsite (BOG) framework projected approximately along the **a**-axis, parallel to the crankshafts (d) The boggsite (BOG) framework projected approximately along the **b**-axis.

(Fig. 9b), leading to a periodicity of about 7.5 Å.

Mazzite (MAZ), $K_{2.5}Mg_{2.1}Ca_{1.4}Na_{0.3}[Al_{10}Si_{26}O_{72}]\cdot 28H_2O$, is hexagonal, space group $P6_3/mmc$, a = 18.392, c = 7.646 Å, Z = 1 (Galli 1975). Two types of channels run parallel to the **c**-axis (Fig. 11a). One is confined by twelve-membered rings of tetrahedra (aperture 7.4 Å), the other by eight-membered rings (aperture 5.6 × 3.4 Å). T-O distances are consistent with (Si,Al) disorder. Mg occurs in a "gmelinite-type" cage (Fig. 1b) and is six-coordinated by H_2O molecules. Galli (1975) speculated that the Mg-H_2O complex acts as a template for the formation "gmelinite-type cages" (Fig. 1b). An additional cation site (occupancy approximately 50%) is mainly occupied by K and is situated in the distorted eight-membered rings. K is octahedrally coordinated by four framework oxygens and two H_2O molecules. Ca is located in the center of the large channels running parallel to the **c**-axis and is surrounded by disordered H_2O molecules. The walls of the large channels are lined with H_2O molecules, and the cations sit at regular intervals in the middle of these H_2O pipes (Galli 1975). Dehydration of mazzite and accompanying cation diffusion was investigated by Rinaldi et al. (1975b) and Alberti and Vezzalini (1981b).

Perlialite (LTL), $K_9Na(Ca,Sr)[Al_{12}Si_{24}O_{72}]\cdot15H_2O$, is hexagonal, space group *P6/mmm*, a = 18.54, c = 7.53 Å, Z = 1. The framework topology (Artioli and Kvick 1990) is the same as for synthetic Linde Type L (LTL) as determined by Barrer and Villiger (1969). Two types of channels run parallel to the **c**-axis. One is bounded by twelve-membered rings (aperture 7.1 Å), and the other is bounded by strongly compressed eight-membered rings (aperture 3.4 × 5.6 Å). T-O distances give no indication of (Si,Al) order. When comparing the structural drawings for mazzite and perlialite (Fig. 11a,b) projected parallel to the **c**-axis, strong similarities are evident. However, in perlialite six-membered double rings connect the sheets, whereas six-membered single rings occur in mazzite. Thus, in perlialite six-membered double rings connect two cancrinite-type cages (Fig. 1a), and in mazzite six-membered single rings link two gmelinite cages (Fig. 1b). The powder sample of perlialite used for structure refinement (Artioli and Kvick 1990) was separated with Tl-malonate (heavy liquid). Thus, perlialite became partly ion-exchanged and approached the composition $K_8Tl_4[Al_{12}Si_{24}O_{72}]\cdot20H_2O$. K was located in the cancrinite cage (fully occupied) and in the center of the six-membered double rings (occupancy 20%). Cations in the large channels bonded to six framework oxygens of the eight-membered rings and additional H_2O molecules which fill the interior of the twelve-membered ring channels. Notice the different arrangement of cations and H_2O in the wide channels compared with mazzite. The second most preferred extraframework site (fully occupied) is in the compressed eight-membered rings, forming channels parallel to the **c**-axis.

Boggsite (BOG)

Boggsite (BOG), $Ca_8Na_3[Al_{19}Si_{77}O_{192}]\cdot70H_2O$, is orthorhombic, space group *Imma*, a = 20.236, b = 23.798, c = 12.798 Å, Z = 1. A third type of chain of edge-sharing four-membered rings is found in boggsite. The up (U) and down (D) orientation of the tetrahedral apices follows the sequence DDUDUUDUDD (Fig. 9c), thus leading to the 20 Å periodicity along the **a**-axis. The structural framework (Pluth and Smith 1990) is characterized by wide channels parallel to the **a**-axis (Fig. 11c) confined by twelve-membered rings (aperture 7.0 Å). Each twelve-membered ring channel along the **a**-axis has offset ten-membered ring windows (Fig. 11d) into left and right channels parallel to the **b**-axis (aperture 5.8 × 5.2 Å). The (Si,Al) distribution is random. Extraframework cations and H_2O molecules form a highly disordered ionic solution that lacks systematic bonding to the framework.

Paulingite (PAU)

Paulingite (PAU) with the simplified structural formula $Na_{14}K_{36}Ca_{59}Ba_2$-$[Al_{173}Si_{499}O_{1344}]\cdot550H_2O$ is cubic, space group $Im3m$, a = 35.09 Å, Z = 1. The complicated and large framework (Gordon et al. 1966, Andersson and Fälth 1983, Bieniok et al. 1996, Lengauer et al. 1997) is characterized by channels confined by eight-membered rings of tetrahedra (aperture 3.8 Å) running parallel to the **a**-axis. No (Si,Al) order was detected. Seven different polyhedral cages occur in paulingite (Bieniok et al. 1996). Bieniok (1997) studied the dehydration and accompanying structural distortion.

ZEOLITES WITH SIX-MEMBERED RINGS

Smith and Bennet (1981) provided a review of the 98 most simple nets built by parallel six-membered rings linked by four-membered rings into the infinite set of ABC-6 nets. In addition to the zeolites gmelinite, chabazite, willhendersonite, levyne, erionite, offretite, and bellbergite, the zeolite-related minerals of the cancrinite group (afghanite, bystrite, cancrinite, cancrisilite, davyne, franzinite, giuseppettite, hydroxycancrinite, liottite, microsommite, pitiglianoite, quadridavyne, sacrofanite, tiptopite, tounkite, vishnevite, and wenkite) and the minerals of the sodalite group (bicchulite, hauyne, kamaishilite, lazurite, nosean, sodalite, and tugtupite) belong to this family. One of the characteristics of minerals in this structural family is stacking faults in the sequence of the six-membered double or single rings. In X-ray single-crystal photographs, these faults lead to some diffuse reflections parallel to the stacking direction. The importance of these stacking faults in this group is evident when one considers that gmelinite and offretite have infinite channels confined by twelve-membered rings; any stacking faults will thus block the channels. Following the classification of Gottardi and Galli (1985), faujasite and pahasapaite are also included in this group. Faujasite has six-membered double-rings not in one but in four planes, and pahasapaite has six-membered single rings.

Because cancrinites and sodalites are not reviewed in this chapter, at least the fundamentals of these frameworks will be briefly explained. Cancrinite can be constructed from six-membered rings stacked in an ab sequence (small letters are used to denote six-membered single rings), whereas sodalite has an abc stacking sequence of the ring units. The term "cancrinite-like" has been applied to feldspathoids that do not have the abc sequence of sodalite (e.g. liottite: ababac, and franzinite: cabcbacb). For further discussion, and references consult Smith and Bennet (1981).

Gmelinite (GME)

Gmelinite (GME), $(Na,K,Ca_{0.5},Sr_{0.5})_8[Al_8Si_{16}O_{48}]\cdot 22H_2O$, is hexagonal, space group $P6_3/mmc$, a = 13.621 to 13.805, c = 9.964 to 10.254 Å, Z = 1, depending on chemistry. The tetrahedral framework (Fig. 12c) consists of parallel stacks of six-membered double rings in the sequence ABAB (Fischer 1966, Galli et al. 1982, Malinovskii 1984, Vigdorchik and Malinovskii 1986, Vezzalini et al. 1990, Sacerdoti et al. 1995). The (Si,Al)-disordered structure exhibits channels parallel to the **c**-axis confined by twelve-membered rings (aperture 7.0 Å) which are connected perpendicular to the **c**-axis by eight-membered rings (aperture 3.9 × 3.6 Å). Two extraframework cation positions occur, with most cations located in the gmelinite cage (C1) between two six-membered double rings stacked parallel to the **c**-axis. These cations are coordinated by three framework oxygens and three H_2O molecules. Another cation site (C2) is in the large channel close to the eight-membered ring and is coordinated to three framework oxygens and four H_2O molecules. H_2O is strongly disordered with up to eight H_2O sites. Depending on the occupancy and type of cation in C2, the gmelinite framework undergoes a deformation involving lengthening parallel to the **c**-axis and shortening parallel to the **a**-axis (Vezzalini et al. 1990). Crystal structure refinements of exchanged varieties and accompanying structural distortions were discussed by Vigdorchik and Malinovskii (1986) and Sacerdoti et al. (1995). If the stacking of the six-membered double rings exhibits faults (e.g. ABCABC), a disordered intergrowth with chabazite results, leading to diffuse streaks in X-ray photographs along the **c**-axis (Fischer 1966).

Chabazite and willhendersonite (CHA)

Chabazite (CHA), $(Ca_{0.5},Na,K)_4[Al_4Si_8O_{24}]\cdot 12H_2O$, has a framework structure consisting of parallel stacks of six-membered double rings in the sequence ABC (Fig. 12b). Thus, a new type of cage (the chabazite cage) is typical of this structure (Fig. 1c). Large channels confined by twelve-membered rings characteristic of gmelinite are not formed in chabazite. The largest channels perpendicular to [001] (hexagonal setting) are confined by eight-membered rings (aperture 3.8 × 3.8 Å). The framework topology of

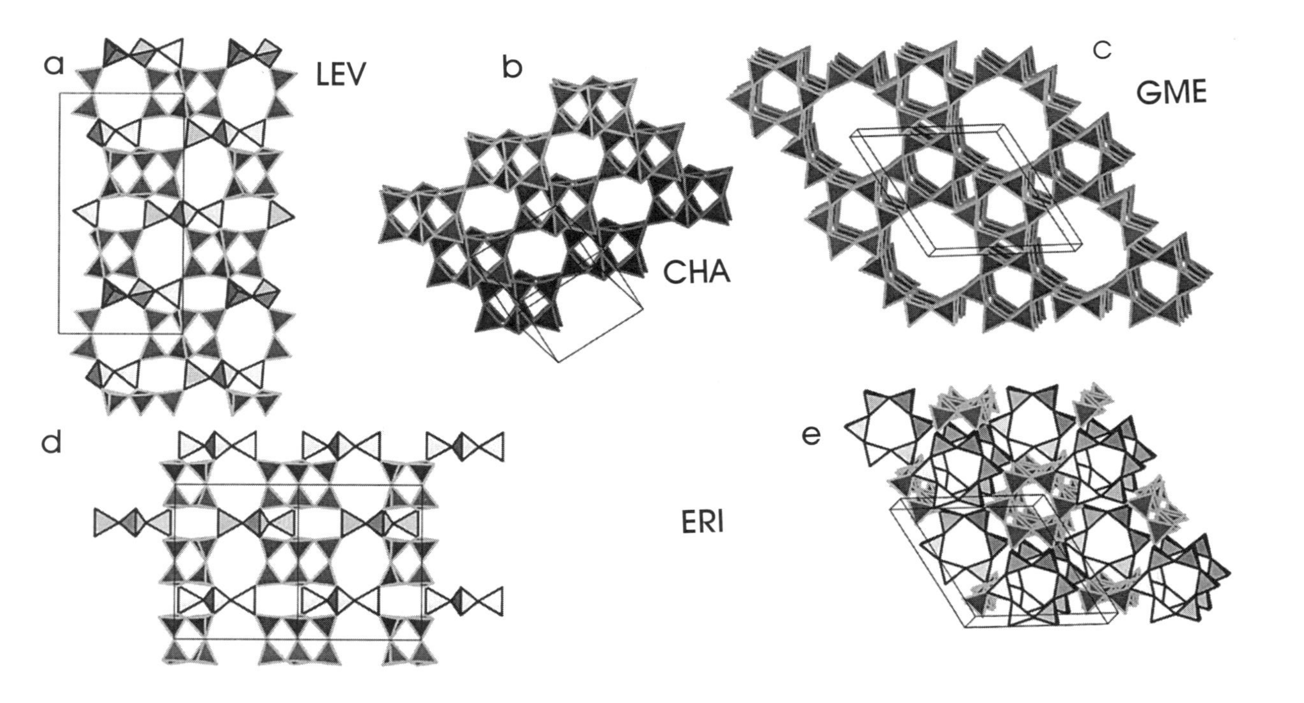

Figure 12. Polyhedral drawings of zeolite structures composed of single (dark rims) and double six-membered rings (light rims). Single rings are denoted with small letters, and double rings with capital letters. (a) The levyne (LEV) framework projected approximately along the **a**-axis. The structure consists of rings stacked in an AbCaBcA sequence parallel to the **c**-axis. (b) The chabazite (CHA) framework with rhombohedral unit-cell outlines. The structure consists of six-membered double rings stacked in an ABC sequence. (c) The gmelinite (GME) framework projected approximately along the **c**-axis. The structure consists of six-membered double rings stacked in an ABAB sequence parallel to the **c**-axis. (d) The erionite (ERI) framework projected parallel to [110]. The stacking sequence of rings is AbAc parallel to the **c**-axis. (e) The erionite framework projected approximately along the **c**-axis.

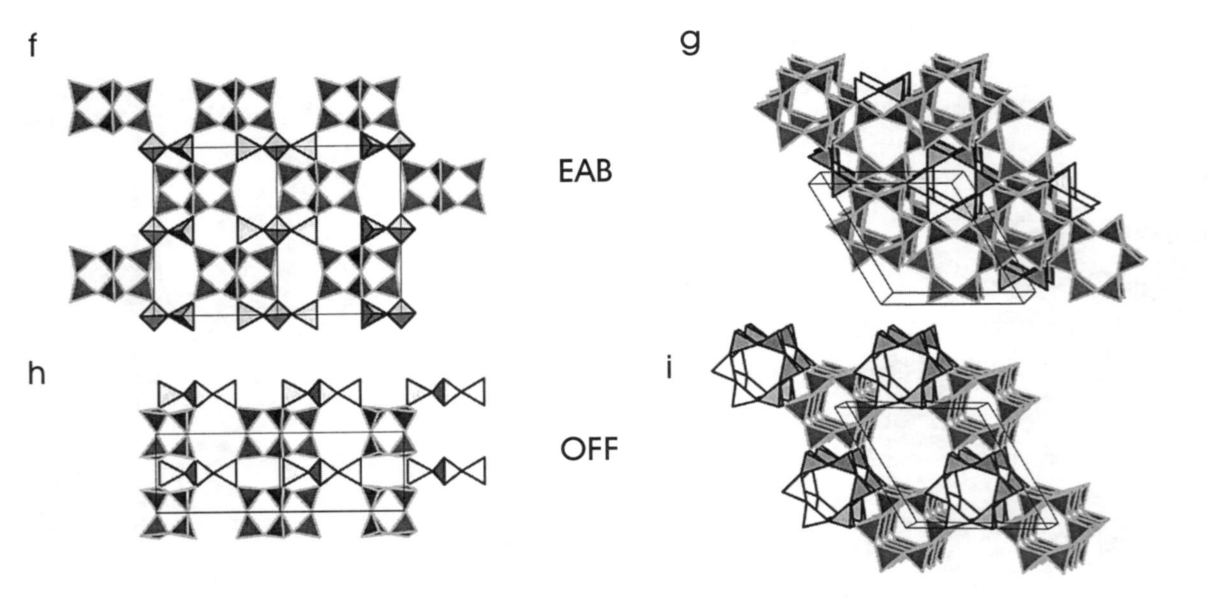

Figure 12, continued. Polyhedral drawings of zeolite structures composed of single (dark rims) and double six-membered rings (light rims). Single rings are denoted with small letters, and double rings with capital letters. (f) The bellbergite (EAB) framework projected parallel to [110]. The structure consists of six-membered rings stacked in an aBaC sequence parallel to the **c**-axis. (g) The bellbergite (EAB) framework projected approximately along the **c**-axis. (h) The offretite (OFF) framework projected parallel to [110]. The structure consists of six-membered rings stacked in an AbAb sequence parallel to the **c**-axis. (i) The offretite (OFF) framework projected approximately along the **c**-axis.

CHA is rhombohedral $R\bar{3}m$, which has also been attributed to chabazite (space group $R\bar{3}m$, a = 13.8 Å, c = 15.02 Å, Z = 1). Based primarily on optical observations, the symmetry of chabazite is known to be lower than rhombohedral, leading to complicated twinning (Akizuki 1981b). Mazzi and Galli (1983) separated optically homogeneous domains of four chabazite crystals from different localities and performed structure refinements in triclinic ($P\bar{1}$) and rhombohedral space groups. In the six independent tetrahedra sites (in $P\bar{1}$) , some (Si,Al) ordering was observed. However, this ordering pattern was different in the four cases investigated. Mazzi and Galli (1983) suggested that chabazite may have randomly arranged domains with perfect (Si,Al) ordering. By use of ^{29}Si NMR spectroscopy, Engelhardt and Michel (1987) confirmed the variable (Si,Al) distribution in natural chabazites.

A series of cation-exchange experiments (Na, K, Ag, Cs, Ca, Sr, Ba, Cd, Mn, Co, and Cu), accompanied by structure refinements of the hydrated and dehydrated chabazites, were reviewed by Smith (1988). Four channel cation sites were distinguished (e.g. Alberti et al. 1982). C1 is located on the three-fold axis in the center of the double six-ring. C1 is not occupied in the Na- and K- forms but is partly occupied in natural, Ca-, and Sr-forms. C2 is the major cation site and is located on the three-fold axis, above or below the double six-membered rings, bonding to three framework oxygens of the ring and additional H_2O sites. C3 with usually low occupancy is located on the three-fold axis at the center of the chabazite cage. C4 is nearly in the center of the eight-membered rings, forming windows between the large cages. H_2O molecules mainly complete the C2 coordination and are additionally disordered over various sites with medium to low occupancies. Butikova et al. (1993) described structure refinements of a natural Ca-rich chabazite and its dehydrated form at 250°C. They observed strong structural distortions at high temperature and diffusion of cations to new sites, thus confirming results of Belokoneva et al. (1985).

Willhendersonite (CHA), $K_2Ca_2[Al_6Si_6O_{24}]\cdot 10H_2O$, also has the CHA tetrahedral framework topology (Peacor et al. 1984), but the structure has an Si/Al ratio of 1.0 and almost perfect (Si,Al) ordering, reducing the symmetry from $R\bar{3}m$ to $R\bar{3}$. A further symmetry reduction to triclinic $P\bar{1}$, a = 9.206, b = 9.216, c = 9.500 Å, α = 92.34, β = 92.70, γ = 90.12°, Z = 1, occurs (Tillmanns et al. 1984), which is caused by the ordered distribution of cations and H_2O molecules within the structural voids, and distortion of the framework due to cations which are too small to fill specific sites. In willhendersonite the six-membered rings are strongly compressed which leads to a different coordination of the C2 site (C2') discussed for chabazite. This C2' site occupied by Ca is seven-coordinated to four framework oxygens and three H_2O molecules. In spite of the Si/Al and K/Ca ratio of 1.0, Na and Ca are strongly disordered. The structure of Ca-endmember willhendersonite (space group $P\bar{1}$) with slightly reduced (Si,Al) ordering was reported by (Vezzalini et al. 1997a).

Levyne (LEV)

Levyne (LEV), $(Ca_{0.5},Na,K)_6\ [Al_6Si_{12}O_{36}]\cdot 18H_2O$, is rhombohedral, space group $R\bar{3}m$, a = 13.338, c = 23.014 Å, Z = 1. The structure (Merlino et al. 1975, Sacerdoti 1996) is built by alternating six-membered double-rings with six-membered single rings (Fig. 12a). The stacking sequence is AbCaBcA, where capital letters represent the double rings and small letters the single rings. The structure exhibits three equivalent channel systems perpendicular to the three-fold axis confined by eight-membered rings (aperture 3.6 × 4.8 Å). There is no indication of (Si,Al) ordering in the data of Merlino et al. (1975). However, Sacerdoti (1996) reported 35% Al on T1 and 25% Al on T2 based on three refinements. Five cation positions are known. The most important, C1, is above or below the six-membered double rings where the cation is six-coordinated to three framework oxygens and three H_2O molecules. C2 is in the center of the cage coordinated only to H_2O molecules. C5 is in the center of the six-membered ring, and C3 and C4 are between C2 and C5.

Erionite (ERI)

Erionite (ERI), $K_2(Na,Ca_{0.5})_8[Al_{10}Si_{26}O_{72}]\cdot 28H_2O$, is hexagonal, space group *P*6_3*/mmc*, *a* = 13.26, *c* = 15.12 Å, Z = 1 (Kawahara and Curien 1969, Schlenker et al. 1977a, Alberti et al. 1997, Gualtieri et al. 1998). The structure is built by alternating six-membered single (small letters) and six-membered double rings (capital letters) stacked in the AbAc sequence (Fig.12d). The single rings are preferred by Al (Gualtieri et al. 1998). The crystals are characterized by offretite stacking faults leading locally to the AbAb sequence (Fig. 12h). Three equivalent channel systems run perpendicular to the **c**-axis and are bounded by eight-membered rings (aperture 3.6 × 5.1 Å; Fig. 12d). Three types of cages characterize this zeolite: a six-membered double ring (empty), a cancrinite cage (preferred by K), and an erionite cage (Fig. 1e) with dispersed Ca, Na, and Mg. Dehydration (Schlenker et al. 1977a) leads to internal ion-exchange, where Ca migrates into the cancrinite cage and K moves to the center of the eight-membered ring, blocking the windows. As in most zeolites constructed by six-membered rings, stacking faults reduce the crystal quality of erionite (Millward et al. 1986). Samples from Sasbach (Kaiserstuhl, Germany) represent continuous transitions from offretite to erionite (Rinaldi 1976).

Bellbergite (EAB)

Bellbergite (EAB), $(K,Ba,Sr)_2Sr_2Ca_2(Ca,Na)_4[Al_{18}Si_{18}O_{72}]\cdot 30H_2O$, is hexagonal, with possible space groups *P*6_3*/mmc*, $P\bar{6}2c$, or *P*6_3*mc*, *a* = 13.244, *c* = 15.988 Å, Z = 1 (Rüdinger et al. 1993). The unit-cell parameters are quite similar to those of erionite. The structure has the EAB framework (Meier and Groner 1981) with the aBaC stacking sequence of six-membered single (small letters) and double rings (capital letters). Thus, the structure is related to erionite (AbAc) in such a way that single and double rings are interchanged (Fig. 12d,f). This leads to the same cages as found in erionite (i.e. six-membered double rings, cancrinite cage, and erionite cage). Three identical channel systems run perpendicular to the **c**-axis (Fig. 12f), bounded by eight-membered rings (aperture 3.6 × 5.1 Å). The structure was refined in space group *P*6_3*/mcc,* and no (Si,Al) ordering was found (Rüdinger et al. 1993). Alberti (1995) speculated that stacking faults or symmetry lower than *P*6_3*/mmc* may explain the (Si,Al) disorder which is unusual for framework silicates with an Si/Al ratio of 1.

Offretite (OFF)

Offretite (OFF), $KCaMg[Al_5Si_{13}O_{36}]\cdot 15H_2O$, is hexagonal, space group $P\bar{6}m2$, *a* = 13.29, *c* = 7.58 Å, Z = 1 (Gard and Tait 1972, Mortier et al. 1976a,b; Alberti et al. 1996a, Gualtieri et al. 1998). The structure is built by alternating six-membered single (small letters) and six-membered double rings (capital letters) stacked in the AbAb sequence (Fig. 12h). The crystals are characterized by erionite stacking faults leading locally to the AbAc sequence. With regard to zeolitic properties, the major difference between offretite and erionite are channels confined by twelve-membered rings of tetrahedra (aperture 6.7 × 6.8 Å) running parallel to the **c**-axis in offretite which are blocked by single six-membered rings in erionite (Fig. 12i). Three equivalent channel systems bounded by eight-membered rings (aperture 3.6 × 4.9 Å) run perpendicular to the **c**-axis and interconnect the main channels. Four types of cages characterize this zeolite: six-membered double rings (empty or very low occupancy), cancrinite cage (preferred by K), a gmelinite cage (Fig. 1b) filled with Mg surrounded by disordered H_2O, and the wide channels whose centers are occupied by Ca-H_2O complexes. Si and Al are disordered over the two tetrahedral sites (Alberti et al. 1996a). Dehydration (Mortier et al. 1976a,b) leads to internal ion-diffusion.

Offretite is commonly intergrown with erionite either in macrodomains (Rinaldi 1976) or as cryptodomains detected by single-crystal study (Mortier et al. 1976a). Application of transmission electron-microscopy combined with structural data clearly showed that the Mg concentration is the major factor controlling whether erionite or offretite is formed (Gualtieri et al. 1998). A chabazite-offretite epitaxial overgrowth was found by Passaglia and Tagliavini (1994).

Faujasite (FAU)

Faujasite (FAU), $Na_{20}Ca_{12}Mg_8[Al_{60}Si_{132}O_{384}]\cdot 235H_2O$, is cubic, space group $Fd3m$, a = 24.60 Å, Z = 1 (Bergerhoff et al. 1958, Baur 1964). Faujasite is actually a rare zeolite, but it is well known, because it has the same framework topology (FAU) as Linde X and Linde Y, synthetic counterparts used as sorbants and catalysts. A review of structural work on synthetic counterparts (Linde X and Linde Y), on ion-exchanged faujasite, and on ion-exchanged and dehydrated faujasite was provided by Smith (1988). Properties of natural and synthetic faujasite were discussed by Stamires (1973). Faujasite corresponds to the most open framework of all natural zeolites. About half of the unit-cell

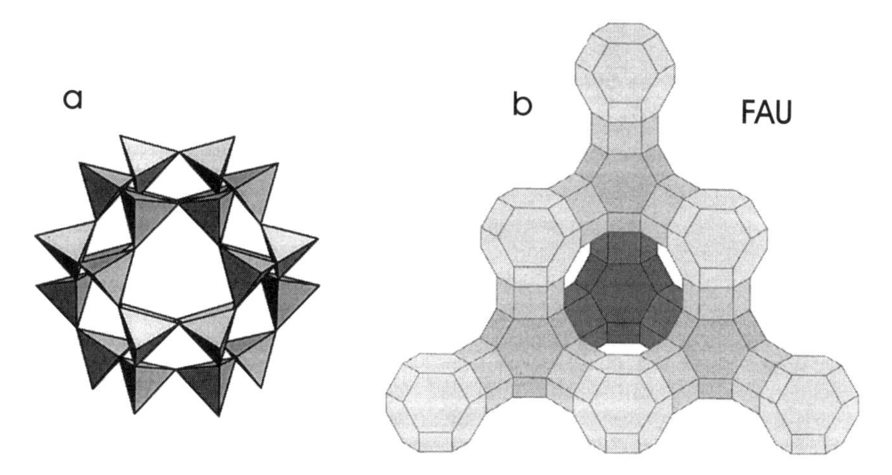

Figure 13. (a) The sodalite cage (β-cage) characteristic of faujasite (FAU) and Linde Type A (LTA). (b) Portion of the faujasite structure viewed parallel to [111] with β-cages represented by polyhedra. The cages are connected by six-membered double rings.

space is void in the dehydrated form. The structure consists of sodalite cages (Fig. 13a; truncated octahedra, β-cages) connected in a cubic manner over six-membered double rings. Thus, wide intersecting channels are formed parallel to <111> with an aperture of 7.4 Å (Fig. 13b). Approximately 50% of the cations reside in the sodalite cage bonded to three framework oxygens of the six-membered rings and additional H_2O molecules. The remaining cations and H_2O molecules are disordered in the large cavities. The same β-cages also exist in the important synthetic product Linde Type A (LTA). In the LTA framework, the β-cages are connected by four-membered double rings in a cubic fashion (Fig. 13c).

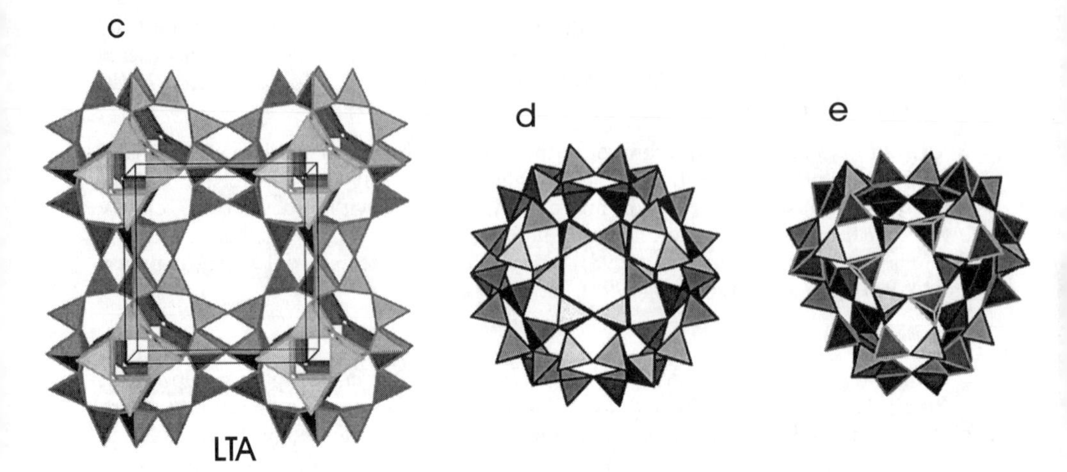

Figure 13, continued. (c) The Linde Type A (LTA) framework viewed approximately along the **a**-axis with β-cages connected by four-membered double rings. (d) The α-cage (symmetry *m*3*m*) as found in hydrated zeolite rho (RHO). (e) The α-cage (symmetry 23) as found in pahasapaite. BeO_4 tetrahedra have light rims and PO_4 tetrahedra have dark rims.

Pahasapaite (RHO)

Pahasapaite (RHO), $(Ca_{5.5}Li_{3.6}K_{1.2}Na_{0.2})Li_8[Be_{24}P_{24}O_{96}]\cdot 38H_2O$, is cubic, space group *I*23, *a* = 13.781 Å, Z = 1 (Rouse et al. 1987). The structure (Rouse et al. 1989) contains ordered BeO_4 and PO_4 tetrahedra forming a three-dimensional array of distorted truncated cubo-octahedra or α-cages (Fig. 13d,e) connected via double eight-membered rings (aperture 2.2 × 2.2 Å). There are two identical, interpenetrating systems of cages related by the *I*-centering of the lattice. Similar α-cages also exist in paulingite. Pahasapaite has a distorted zeolite rho framework. The wide cages have a diameter of about 8 Å. Hydrated zeolite rho has the maximum symmetry *Im*3*m*, which is reduced to $I\bar{4}3m$ in dehydrated zeolite rho and to *I*23 in pahasapaite due to (Be,P) ordering. Eight Li and 32 H_2O molecules reside within the cages. The remaining six H_2O molecules and 10.5 cations block the passages of double eight-rings between two cages. When pahasapaite is dehydrated (Corbin et al. 1991), the unit-cell volume decreases by 14% due to H_2O loss and increases for the dehydrated form by ~1% between 25°C and 400°C (Parise et al. 1994).

ZEOLITES OF THE MORDENITE GROUP

Structures assigned to this subgroup can be built from single five-membered rings with an attached tetrahedron (i.e. SBU: 5-1). These structures contain sheets built by six-membered rings (6^3 nets) which are often highly puckered (Fig. 14a,b,c). In the classical tetrahedral sheet of six-membered rings found in mica, all tetrahedral apices point in the same direction; thus, a framework cannot be constructed. However, frameworks can be built from sheets where half of the tetrahedral apices point upward and half point downward. These are the common structural principles of mordenite (MOR), dachiardite (DAC), epistilbite (EPI), ferrierite (FER), and bikitaite (BIK) (Meier 1978). This sheet concept has the advantage that cation and H_2O diffusion can be more easily explained for these structures because the sheets are not permeable in the temperature range of zeolite applications. Furthermore, the sheets help define the perfect cleavage and morphology of these zeolites. For a more thorough derivation of this group, refer to Meier (1978),

Gottardi and Galli (1985), and van Koningsveld (1992).

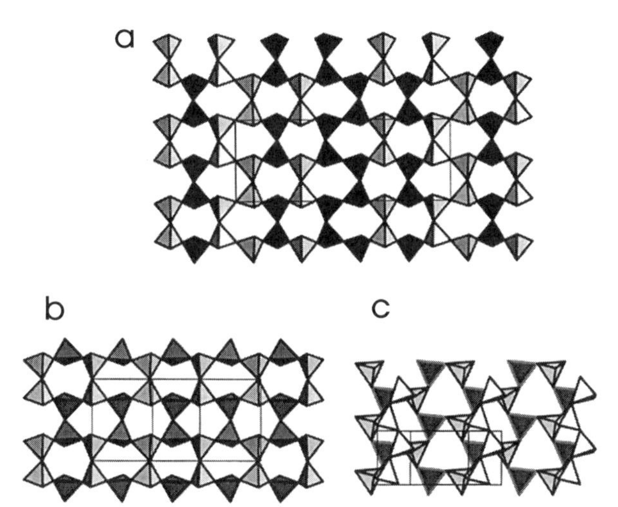

Figure 14. Sheets built by six-membered rings in the mordenite group. (a) (100) sheet in mordenite; half of the tetrahedral apices are pointing up and half are pointing down. (b) (100) sheet in dachiardite. Sheets with the same arrangement of tetrahedral apices also exist parallel to (010) in epistilbite and parallel to (100) in ferrierite. (c) The tridymite-type sheet parallel to (001) in bikitaite. The sheet is decorated with pyroxene-type tetrahedral chains extending parallel to the **b**-axis. Tetrahedra in the chains are filled with Si; those in the sheet of triclinic bikitaite are filled alternately with Si (light tetrahedra with black rims) and Al (dark tetrahedra with light rims).

Mordenite (MOR), $(Na_2,K_2,Ca)_4[Al_8Si_{40}O_{96}]{\cdot}28H_2O$, is orthorhombic, space group $Cmc2_1$ (notice the topological symmetry, space group *Cmcm* which is used in several studies), a = 18.11, b = 20.46, c = 7.52 Å, Z = 1 (Meier 1961, Alberti et al. 1986, Shiokawa et al. 1989, Passaglia et al. 1995). The symmetry lowering from *Cmcm* to $Cmc2_1$ is necessary to properly model a T-O-T angle which is 180° in *Cmcm*. Apparent T-O-T angles of 180° are energetically unfavorable (e.g. Meier and Ha 1980, Gibbs 1982) and are often a result of either an incorrect space group assignment or oxygen disorder (Alberti 1986). The lowering to the acentric space group $Cmc2_1$ in mordenite is also reflected in the distribution of extraframework cations (Alberti et al. 1986). Grammlich-Meier (1981) modeled an (Si,Al)-ordered mordenite framework in the monoclinic space group *Cc* and obtained two distinct conformeric solutions. There is considerable evidence that these structural variants exist side by side as domains, and since twinning is also likely, a mordenite crystal could contain up to eight different domains (Meier et al. 1978).

The mordenite structure can be envisioned as puckered sheets formed by six-membered rings parallel to (100), which define the perfect (100) cleavage (Fig. 14a). The sheets are linked by four-membered rings in a way that twelve-membered rings (aperture 6.5 × 7.0 Å) and strongly compressed eight-membered rings remain at the seam, forming channels parallel to the **c**-axis (Fig. 15a). Another set of compressed eight-membered rings (aperture 2.6 × 5.7 Å) interconnects the wide channels parallel to the **b**-axis. Judging from T-O distances, the four-membered rings are slightly enriched in Al (e.g. Alberti 1997). Cations in mordenite mainly occupy three sites. Two of these sites are close to the four-membered Al-enriched rings and are located in the connecting channels parallel to the **b**-axis; the A site centers the strongly compressed eight-membered ring channels, whereas the D site is near the center of the eight-membered ring, giving access to the wide channel. The E site is in the large channels.

Gottardi and Galli (1985) and Smith (1988) reviewed ion-exchanged mordenites. In dehydrated K- and Ba-exchanged mordenites (Mortier et al. 1978, Schlenker et al. 1978), the symmetry lowers to *Pbcn*. Temperature-dependent (between 20 and 450°C) structural studies on Ca-exchanged natural mordenite were published by Elsen et al. (1987). Song

(1999) studied crystal defects in mordenites which may result in channel blockage.

Maricopaite (-MOR), $Pb_7Ca_2[Al_{12}Si_{36}(O,OH)_{100}]\cdot n(H_2O, OH)$ (Peacor et al. 1988), is orthorhombic, space group *Cm2m*, *a* = 19.434, *b* =19.702, *c* = 7.538 Å, Z = 1. The structure exhibits a rather random (Si,Al) distribution and is closely related to mordenite but has an interrupted framework (emphasized by the structure code –MOR) in which 17% of the TO_4 groups are three-fold connected (Rouse and Peacor 1994). This difference has several consequences; in maricopaite there are no four- and eight-membered rings parallel to (001). A strongly compressed channel apparently bounded by eight-membered rings still appears in projections parallel to the **c**-axis. Inspection of Figure 15b discloses that the apparent rings are composed of staggered half rings, one half at z = 0 and the other half at z = 1/2. Thus, the voids are considerably larger than in mordenite. Notice that the elongation of the compressed channels is parallel to the **a**-axis in mordenite but is parallel to the **b**-axis in maricopaite. The compressed channels confined by the staggered half rings (Fig. 15b) are obstructed by two types of $Pb_4(O,OH)_4$ clusters which appear to act as a template for this unusual type of void. Two Pb atoms bond to the same framework oxygen; thus, the bond valence sum for this oxygen is satisfied without connection to an additional TO_4 tetrahedron, causing an interrupted framework. However, Pb sites are only partially occupied. Thus, if Pb is locally absent, a proton (H) balances the framework oxygen valence sum.

Dachiardite (DAC), $(Na,K,Ca_{0.5})_4[Al_4Si_{20}O_{48}]\cdot 18H_2O$, is monoclinic, space group *Cm*, *a* = 18.676, *b* = 7.518, *c* = 10.246 Å, β =107.87°, Z = 1. However, all available structure refinements (Gottardi and Meier 1963, Vezzalini 1984, Quartieri et al. 1990) have been performed in the higher space group *C2/m*. Vezzalini (1984) and Quartieri et al. (1990) detected two types of domains (called A and B) in an equal ratio, resulting in the average symmetry *C2/m*. These domains form to avoid energetically unfavorable 180° T-O-T angles (e.g. Meier and Ha 1980, Gibbs 1982), similar to the symmetry

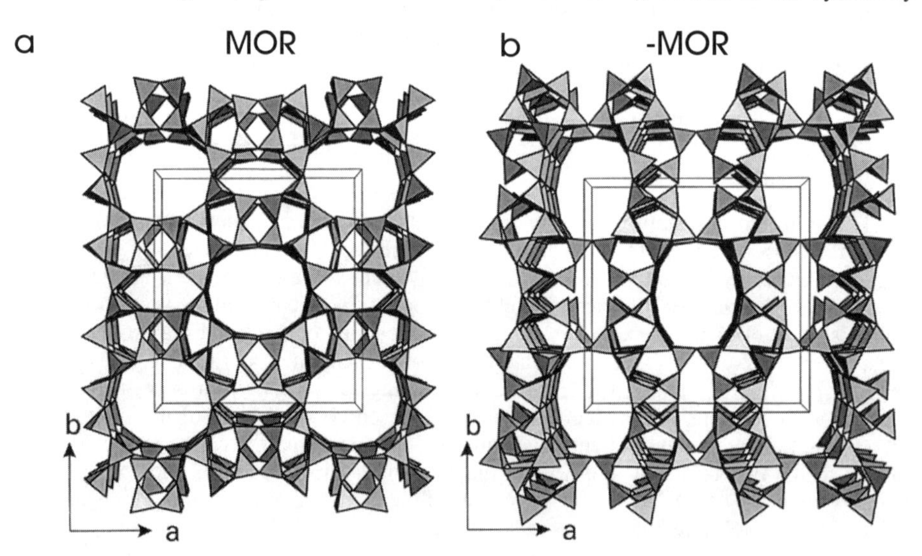

Figure 15. (a) The mordenite (MOR) framework projected along the **c**-axis. The sheets of six-membered rings parallel to (100) are linked parallel to the **a**-axis by four-membered rings. (b) The maricopaite (-MOR) framework projected along the **c**-axis. The unit-cell origin is shifted by 1/2 along the **a**-axis to facilitate better comparison with mordenite. In contrast to mordenite there are no four-membered rings due to the interrupted framework in maricopaite.

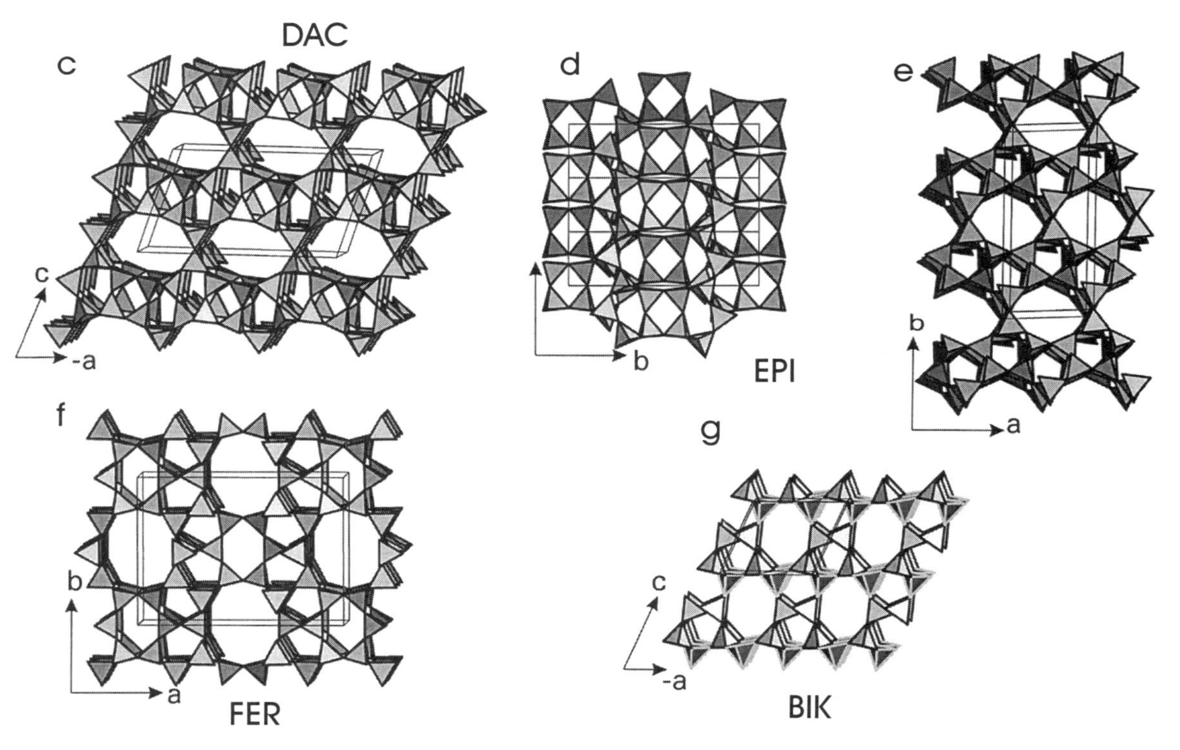

Figure 15, continued. (c) The dachiardite (DAC) framework projected approximately along the **b**-axis. The sheets of six-membered rings parallel to (100) are linked parallel to the **a**-axis by four-membered rings. (d) The epistilbite (EPI) framework projected on (100). The sheets of six-membered rings parallel to (010) are linked parallel to the **b**-axis by four-membered rings obstructing the ten-membered ring channels. (e) The epistilbite (EPI) framework projected approximately along the **c**-axis. (f) The ferrierite (FER) framework projected approximately along the **c**-axis. The sheets of six-membered rings parallel to (100) are linked parallel to the **a**-axis by six-membered rings. (g) The bikitaite (BIK) framework (triclinic variety) projected parallel to the **b**-axis. The sheets of six-membered rings parallel (001) are linked by tetrahedral chains. Light tetrahedra with dark rims are filled with Si; dark tetrahedra with light rims are filled with Al.

lowering in mordenite (described above) and epistilbite (discussed below). Out of the few occurrences of dachiardite (e.g. Tschernich 1992), only the samples from Elba (Gottardi and Meier 1963, Vezzalini 1984) and from Hokiya-dake, Japan (Quartieri et al. 1990) give sharp single-crystal X-ray reflections. All other investigated samples yield diffuse peaks and streaking (Alberti 1975a, Gellens et al. 1982) caused by severe disorder and twinning. Actually, the mineral "svetlozarite" proposed by Maleev (1976) was shown by Gellens et al. (1982) to be a twinned dachiardite with stacking faults and twin domains only a few unit-cells in size. Quartieri et al. (1990) identified two types of dachiardite frameworks (normal dachiardite and modified dachiardite) within the same crystal. They assumed that alternating small domains of different size or possibly a high density of stacking faults caused domain formation.

The dachiardite framework can be constructed from cross-linking slightly puckered sheets parallel to (100) formed by six-membered rings (Figs. 14b and 15c). Two of these sheets are linked parallel to the **a**-axis by four-membered rings. Thus, roughly elliptical channels confined by ten-membered rings (aperture 5.3 × 3.4 Å) are formed parallel to the **b**-axis. These channels are additionally connected by channels through eight-membered rings (aperture 3.7 × 4.8 Å) running parallel to the **c**-axis. The structural difference between mordenite and dachiardite can best be envisioned by the arrangement of tetrahedra pointing up and down within the six-membered ring sheets (compare Fig. 14a and b with Fig. 15a and c). In the Elba dachiardite, two extraframework cation positions are distinguished; C1 is at the intersection of the *b* (ten-membered ring) and *c* (eight-membered ring) channels and is coordinated by three framework oxygens and five H_2O molecules. C2 is in the channel parallel to the **c**-axis, with a distance >3.3 Å to the framework oxygen atoms.

Epistilbite (EPI), $Ca_3[Al_6Si_{18}O_{48}]\cdot 16H_2O$, is monoclinic, space group *C*2, or triclinic, space group *C*1, $a = 9.08$, $b = 17.74$, $c = 10.15$ Å, $\alpha = 90$, $\beta = 124.58$, $\gamma = 90°$, Z = 1. Older refinements (Kerr 1964, Merlino 1965, Perrotta 1967) were done in space group *C*2/*m*, which led to a correct description of the tetrahedral connectivity. Slaughter and Kane (1969) and Alberti et al. (1985) analyzed low-symmetry (*C*2) domains in the structure. Alberti et al. (1985) recognized that these domains (named A and B) form to avoid energetically unfavorable T-O-T angles of 180°. In contrast to dachiardite (Vezzalini 1984), these domains do not occur in a 1:1 ratio. Yang and Armbruster (1996b) showed that these domains can be explained by a twin-like (010) mirror plane. Furthermore, they found that epistilbite from Gibelsbach, Fiesch (Valais, Switzerland) is triclinic (*C*1) as a result of (Si,Al) ordering and extraframework cation distribution. Previously Akizuki and Nishido (1988) suggested triclinic symmetry based on optical studies.

The structure of epistilbite possesses the same up and down orientation of tetrahedra within the sheets of six-membered rings as in dachiardite (Fig. 14b). The sheets parallel to (010), which define the perfect (010) cleavage, are also connected by four-membered rings (Fig. 15d). However, in the **c** direction the four-membered rings alternate at different x levels (shifted by 1/2) and block the ten-membered rings. Open channels are confined by eight-membered rings (aperture 3.7 × 5.2 Å) extending parallel to [001] (Fig. 15e). In triclinic epistilbite, four Ca positions are located in a cage confined by the ten-membered rings of tetrahedra. Two of these sites are related to the other two sites by a pseudo two-fold axis; thus, only two sites can be occupied simultaneously because of short Ca-Ca distances. Ca has a square antiprismatic coordination with five H_2O molecules and three framework oxygens. A strong correlation exists between the Al distribution in the neighboring tetrahedra and the occupancy of the four possible Ca sites.

Ferrierite (FER), $(Na,K)Mg_2Ca_{0.5}[Al_6Si_{30}O_{72}]\cdot 20H_2O$, may be orthorhombic, space group *Immm*, *a* = 19.18, *b* =14.14, *c* = 7.5 Å, Z = 1, which agrees with the maximum symmetry of the framework topology (Vaughan 1966, Gramlich-Meier et al. 1984). Gramlich-Meier et al. (1985) refined the structure of a monoclinic variety of ferrierite (Mg-poor) in space group $P2_1/n$ (standard setting $P2_1/c$), *a* =18.89, *b* = 14.18, *c* = 7.47 Å, β = 90°. In contrast to the orthorhombic structure, monoclinic ferrierite has no T-O-T angles of 180°. In light of the discussion on the mordenite, dachiardite, and epistilbite structures, it may be speculated that lower symmetry is a general feature of all ferrierites (Alberti 1986, Alberti and Sabelli 1987). Alberti and Sabelli (1987) refined the structure of Mg-rich ferrierite from Monastir (Sardinia) in space group *Immm* but provided strong evidence, based on disorder of the $Mg(H_2O)_6^{2+}$ extraframework complex, that the true space group is *Pnnm*, which leads to a relaxation of the 180° T-O-T angle. Thus, the straight T-O-T angle must only be apparent because of fractional statistical occupation. The monoclinic symmetry (Gramlich-Meier et al. 1985) seems to be specific for the Mg-poor variety. Both monoclinic and orthorhombic ferrierite have an essentially random (Si,Al) distribution. Electron diffraction patterns of orthorhombic ferrierite display pronounced streaking parallel to [010]* and [110]* caused by contraction and expansion faults (Gramlich-Meier et al. 1984, Smith 1986).

The structure of ferrierite (Fig. 15f) can be envisioned as corrugated six-membered ring sheets (parallel to (100)) with the same arrangement of up and down tetrahedra as in dachiardite and epistilbite (Fig. 14b). However, the sheets in ferrierite, which also define the perfect cleavage, are connected parallel to the **a**-axis by six-membered rings and not by four-membered rings as in the previous two structures. This arrangement leads to channels parallel to the **c**-axis formed by ten-membered rings (aperture 5.4 × 4.2 Å) interconnected by channels, parallel to the **b**-axis, confined by eight-membered rings (aperture 4.8 × 3.5 Å). Mg forms a disordered $Mg(H_2O)_6^{2+}$ complex wedged in between six-membered rings in the channels parallel to the **b**-axis. Alkali ions are disordered in the wide channels parallel to the **c**-axis.

Bikitaite (BIK), $Li_2[Al_2Si_4O_{12}]\cdot 2H_2O$, is either triclinic, space group *P*1, *a* = 8.607, *b* = 4.954, *c* = 7.597 Å, α = 89.90, β = 114.43, γ = 89.99°, Z = 1 (Bissert and Liebau 1986, Ståhl et al. 1989, Quartieri et al. 1999), or monoclinic, space group $P2_1$, *a* = 8.61, *b* = 4.96, *c* = 7.60 Å, β = 114.5°, Z = 1 (Kocman et al. 1974, Bissert and Liebau 1986). The framework of bikitaite can be constructed from puckered six-membered ring sheets of the tridymite type, where up and down tetrahedra alternate (Fig. 14c). These sheets parallel to (001) have pyroxene tetrahedral chains above and below extending parallel to the **b**-axis which connect two neighboring sheets (Fig. 15g). The orientation of the sheets agrees with the observed morphology and perfect cleavage. The structure is characterized by channels parallel to the **b**-axis delimited by deformed eight-membered rings (aperture 2.8 × 3.7 Å). Half of the tetrahedra in the six-membered ring sheets are occupied by Al (well ordered in triclinic and disordered in monoclinic bikitaite), whereas tetrahedra in the pyroxene chains are only occupied by Si. Its is not understood as yet whether short range (Si,Al) ordering is preserved within the sheets of monoclinic bikitaite (Bissert and Liebau 1986). Li is close to the walls of the *b*-extended channels and bonds to three framework oxygens of Al tetrahedra and to one H_2O molecule. The arrangement of H_2O molecules forming hydrogen-bonded H_2O chains parallel to the **b**-axis was investigated by Ståhl et al. (1989) and Quartieri et al. (1999).

$T_{10}O_{20}$ ZEOLITES: THE TABULAR ZEOLITES

The building blocks for this group of zeolites are chains of $T_{10}O_{20}$ $(Al,Si)O_4$ tetrahedra running parallel to the **a**-axis (Breck 1974). These chains are in turn cross-linked in the (010) plane to form three different framework topologies (Figs. 16-18). This cross-linking results in several types of channels that are all interconnected and are in the (010) plane. For all the $T_{10}O_{20}$ zeolites, the periodicity perpendicular to the (010) plane (i.e. the length of *b*) is similar and approximately 18 Å. The $T_{10}O_{20}$ units are more difficult to visualize than the T_5O_{10} units for the fibrous zeolites, and in a similar manner, these $T_{10}O_{20}$ chains control the morphology of these zeolites. The tabular nature results from a very rigid structure parallel to the **a**-axis, based on the $T_{10}O_{20}$ chains, and their cross-linking parallel to the **c**-axis is much stronger than their cross-linking parallel to the **b**-axis. Thus, all the zeolites in this subgroup exhibit a tabular morphology and in general exhibit perfect (010) cleavage because of the weakness imparted to the structure in the **b** direction. For an in-depth discussion of the SBU's for this group, see Alberti (1979).

Heulandite and clinoptilolite (HEU)

Both heulandite, $(Na,K)Ca_4[Al_9Si_{27}O_{72}]\cdot 24H_2O$, and clinoptilolite, $(Na,K)_6[Al_6Si_{30}O_{72}]\cdot 20H_2O$, possess the same tetrahedral framework (labeled HEU) and form a solid-solution series sometimes referred to as the heulandite group zeolites. Heulandite is defined as a series having Si/Al < 4.0, and clinoptilolite as a series having Si/Al 4.0 (Coombs et al. 1998). A detailed discussion on differences and similarities between heulandite and clinoptilolite is provided by Bish (this volume).

The crystal structures of heulandite and clinoptilolite are mostly described to be monoclinic, space group *C*2/*m*, a = 17.7, b = 17.8, c = 7.4 Å, β = 116.4°, Z = 1 (e.g. Alberti 1975b, Koyama and Takéuchi 1977, Bresciani-Pahor et al. 1980, Alberti and Vezzalini 1983, Hambley and Taylor 1984, Smyth et al. 1990, Armbruster and Gunter 1991, Armbruster 1993, Gunter et al. 1994, Cappelletti et al. 1999). However, lower symmetries such as *Cm*, and $C\bar{1}$ have also been reported (Alberti 1972, Merkle and Slaughter 1968, Gunter et al. 1994, Yang and Armbruster 1996a, Sani et al. 1999, Stolz et al. 2000a). The HEU framework (Figs. 16a,b) contains three sets of intersecting channels all located in the (010) plane. Two of the channels are parallel to the **c**-axis—the A channels are formed by strongly compressed ten-membered rings (aperture 3.0 × 7.6 Å), and the B channels are confined by eight-membered rings (aperture 3.3 × 4.6 Å). The C channels are parallel to the **a**-axis, or [102] and are also formed by eight-membered rings (aperture 2.6 × 4.7 Å).

Alberti (1972) concluded that the true probable lower symmetry of heulandite cannot reliably be extracted from X-ray single-crystal data because of strong correlations of *C*2/*m* pseudo-symmetry related sites during the least-squares procedure. Thus, *C*1, $C\bar{1}$, *Cm*, *C*2, and *C*2/*m* are possible space groups for heulandite and clinoptilolite. Akizuki et al. (1999) determined by optical methods and X-ray diffraction that a macroscopic heulandite crystal is composed of growth sectors displaying triclinic and monoclinic symmetry where the triclinic sectors are explained by (Si,Al) ordering on the crystal faces. Yang and Armbruster (1996a) and Stolz et al. (2000a,b) stated that, due to correlation problems, symmetry lowering in heulandite can only be resolved from X-ray data when investigated in cation-exchanged samples where the distribution of extraframework cations also reflects the lower symmetry.

Differing degrees of (Si,Al) ordering over the five distinct tetrahedral sites (assuming *C*2/*m* space group) have been reported for both heulandite and clinoptilolite. In all refinements, the tetrahedron with the highest Al content, T2, joins the "sheets" of $T_{10}O_{20}$ groups by sharing their apical oxygens (Figs. 16a-c). A neutron diffraction study by Hambley and Taylor (1984) located the majority of the H atoms and found (Si,Al) ordering values similar to other *C*2/*m* refinements. Additional (Si,Al) ordering, due to lower symmetry ($C\bar{1}$ or *Cm*), was resolved by Yang and Armbruster (1996a), Sani et al.

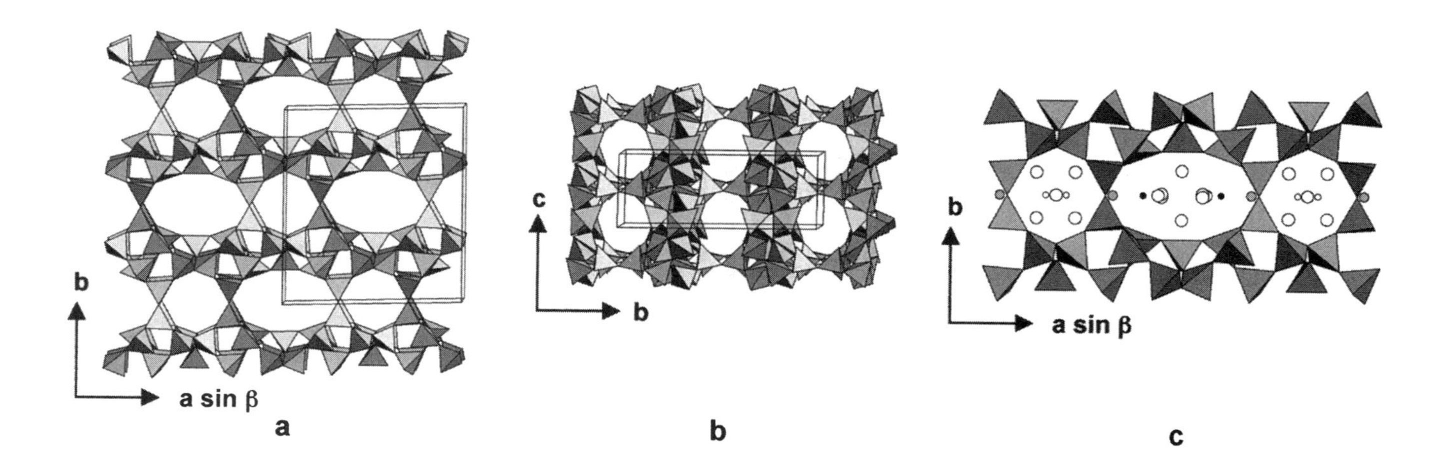

Figure 16. The heulandite structure. (a) A (001) projection looking down the **c**-axis with the **b**-axis vertical, showing the elliptical ten-membered A rings forming one set of [001] channels and the near circular eight-membered rings forming the smaller B channels along [001]. (b) A (100) projection looking down the **a**-axis with the **c**-axis vertical, showing the C channels which are parallel to [100] and are formed by elliptical eight-membered rings. (c) A (001) projection of heulandite looking down the **c**-axis with the **b**-axis vertical, showing three different cation sites and H_2O molecules from a sample refinement. The A channel (elliptical rings) has two different metal sites. The small black dot represents a channel cation site usually occupied by Na, and the larger gray-shaded atom represents a lower-populated site which normally contains K. In the smaller circular B channel, the small white circles represent a site usually occupied by Ca. The larger white circles represent H_2O molecules. In the A channel, these H_2O sites are usually partially occupied and their locations are variable for heulandite and clinoptilolite. (Data for these projections taken from Gunter et al. 1994.)

(1999), and Stolz et al. (2000a,b).

Two main channel cation sites have been reported by all researchers and at least two more sites of lower occupancy have been reported by others (e.g. Sugiyama and Takéuchi 1986, Armbruster and Gunter 1991, Armbruster 1993). These sites usually contain Na, Ca, K, and Mg, with Na and K predominantly close to the intersection of the A and C channels and Ca located in the B channel (Fig. 16b). The Na site in the A channel often also contains Ca, whereas the Ca site in the B channel is usually Na free. K and Na occur at nearby sites but K is more centered in the C channel (Fig. 16b). Both can be distinguished by their different distances from the framework (Fig. 16c). Na, K, and Ca ions are on the (010) mirror plane, present in *C2/m* or *Cm* symmetry, and they coordinate to framework oxygens and channel H_2O molecules. In one refinement, Na was nine-coordinated to four framework oxygens and five strongly disordered and partially occupied channel H_2O molecules, whereas both Ca and K were eight-coordinated to four framework oxygens and four channel H_2O molecules (Gunter et al. 1994). Mg commonly resides in the center of the A channel, coordinated only to six disordered H_2O molecules (Koyama and Takéuchi 1977, Sugiyama and Takéuchi 1986, Armbruster 1993).

Heulandite and clinoptilolite contain differing amounts of H_2O as a function of their extraframework-cation chemistry (Bish 1988, Yang and Armbruster 1996a) and hydration state. The H_2O molecules occurring in the B channel (coordinated to Ca; Fig. 16c) are usually fully occupied, whereas those occurring in the A channel are usually only partially occupied (Koyama and Takéuchi 1977, Armbruster and Gunter 1991). The structural mechanism of dehydration and accompanying framework distortions were studied by Alberti (1973), Alberti and Vezzalini (1983), Armbruster and Gunter (1991) and Armbruster (1993).

Stilbite, stellerite, barrerite (STI)

The three zeolites belonging to this group all possess the same framework, labeled STI in reference to stilbite, which was the first described and most common zeolite in this group. The maximum symmetry possible for this framework is *Fmmm*, which is the space group for stellerite. Barrerite is also orthorhombic but with lower space group symmetry *Amma*, and stilbite is monoclinic, space group *C2/m*. In fact, the difference between minerals in this group is based on symmetry, which is controlled by the channel cations. For instance, the mineral is considered stilbite if it is monoclinic, it is considered stellerite if it is Ca-dominant and orthorhombic, and it is considered barrerite if it is Na-dominant and orthorhombic. Interestingly, Ca-exchanged barrerite increases symmetry to *Fmmm* (Sacerdoti and Gomedi 1984), whereas Na-exchanged stellerite maintains *Fmmm* symmetry (Passaglia and Sacerdoti 1982).

Two sets of connected channels occur in the STI framework. One channel extends parallel to the **a**-axis (Fig. 17) and is confined by a ten-membered ring (aperture 4.9 × 6.1 Å). The other channel (aperture 2.7 × 5.6 Å) is located along [101] for monoclinic STI frameworks or [001] for orthorhombic structures and is confined by an eight-membered ring (Fig. 17). As stated above, both of these channels are in the (010) plane, creating a structural weakness across this plane leading to perfect (010) cleavage and a tabular habit.

Stilbite (STI) has the simplified formula $NaCa_4[Al_9Si_{27}O_{72}]\cdot 30H_2O$ and is usually monoclinic, space group *C2/m*, a = 13.64, b = 18.24, c = 11.27 Å, β = 128.0°, Z = 1 (Galli and Gottardi 1966, Slaughter 1970, Galli 1971). To better compare monoclinic stilbite with orthorhombic stellerite and barrerite, Quartieri and Vezzalini (1987) chose a different space group setting for stilbite, *F2/m*, to obtain a pseudo-orthorhombic unit cell with a = 13.617, b = 18.249, c = 17.779 Å, β = 90.7°. Optical studies of stilbite have shown that within one macroscopic single-crystal, monoclinic (*F*2/*m*) and orthorhombic (*Fmmm*) domains coexist depending on the growth direction (Akizuki and Konno 1985). Akizuki et al. (1993) also studied the symmetry of different growth sectors in stilbite by single-crystal diffraction and refined the structure of an orthorhombic {001} growth sector yielding space group *Fmmm*, a = 13.616, b = 18.238, c = 17.835 Å within a chemically homogeneous crystal. This structure represents a disordered variant of the monoclinic *F*2/*m* structure of stilbite where the higher symmetry occurs on a submicroscopic scale rather than microscopically as in {101} growth sectors. Galli

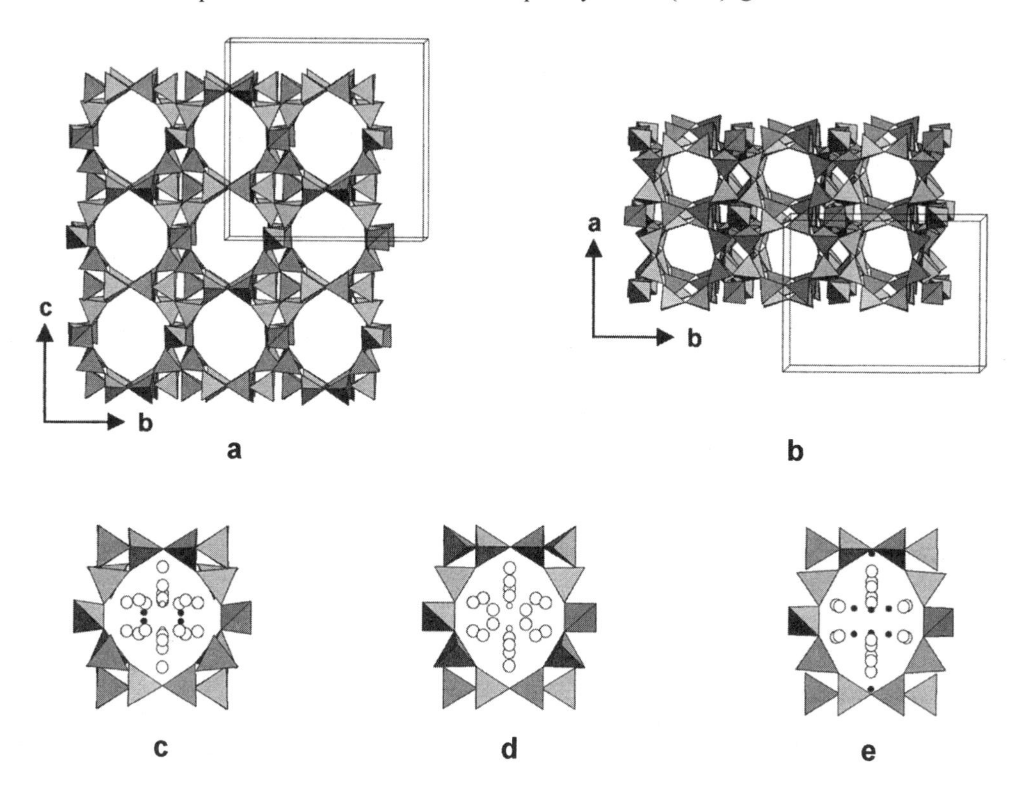

Figure 17. Stilbite-group zeolites framework and channel contents. Projections c-e are on (100) looking down the **a**-axis with the **c**-axis vertical. The small black circles represent Na atoms, the small white circles are Ca atoms, and the larger white circles are H_2O molecules. (a) A (100) projection looking down the **a**-axis with the **c**-axis vertical, showing the elliptical ten-membered rings forming [100] channels. (b) A (001) projection looking down the **c**-axis with the **a**-axis vertical, showing the eight-membered rings forming the smaller set of [001] channels. These drawings are based upon a pseudo-orthorhombic unit cell choice, space group *F2/m* used by Quartieri and Vezzalini (1987) instead of the more standard choice of *C2/m* for stilbite. Using the pseudo-orthorhombic unit cell allows easier comparison among the three zeolites with this framework. Thus, the projected unit cell better fits orthorhombic stellerite and barrerite, than does stilbite. If *C2/m* were chosen, then the [001] channels in the right drawing would be relabeled to [101] channels. (c) Stilbite (Quartieri and Vezzalini 1987) contains both Ca and Na; the Ca sites are nearer the center of the [100] channel and are located on a (010) mirror plane, and the Na sites are located on both sides of the mirror plane near the channel center. (d) Stellerite (Miller and Taylor 1985) contains only Ca; the Ca sites are the same as in stilbite, but H_2O has occupied the Na sites in stilbite. (e) Barrerite (Galli and Alberti 1975b) contains only Na. Na has occupied the Ca sites in stilbite and stellerite. Na sites of stilbite have moved a little, and a new Na site occurs near the channel edge on the (010) mirror plane.

(1971) and Akizuki et al. (1993) implied complete (Si,Al) disorder based upon average T-O distances. Slaughter (1970) and Quartieri and Vezzalini (1987) assumed pronounced (Si,Al) ordering, with Al contents varying between 11 and 40% in the different tetrahedral sites.

All refinements done in monoclinic symmetry yielded one fully occupied Ca site located near the middle of the ten-membered ring parallel to the **a**-axis (Fig. 17c). Ca is bonded only to channel H_2O molecules and not to framework oxygens. Galli (1971) and Quartieri and Vezzalini (1987) found one partially occupied Na site that is seven-coordinated to two framework oxygens and five channel H_2O molecules. Slaughter (1970) interpreted the Ca site to contain a small amount of Na and found three partially occupied Na sites and another undifferentiated partially occupied metal site. The dehydration dynamics of stilbite was studied by Cruciani et al. (1997) using synchrotron X-ray powder diffraction.

Stellerite (STI) is the Ca-dominant end member of this group with the simplified formula $Ca_4[Al_8Si_{28}O_{72}]\cdot28H_2O$ and orthorhombic symmetry, space group *Fmmm*, a = 13.55, b = 18.26, c = 17.80 Å, Z = 1 (Galli and Alberti 1975a, Miller and Taylor 1985). The neutron diffraction data of Miller and Taylor (1985) disclosed slight (Si,Al) ordering with individual tetrahedral Al contents ranging from 10 to 30%.

Stellerite has only one fully occupied, or nearly fully occupied, Ca channel cation site (Fig. 17d) (Galli and Alberti 1975a, Miller and Taylor 1985). Ca is coordinated only to channel H_2O molecules and not to framework oxygens. Similar to stilbite, the $Ca(H_2O)$ complex is hydrogen bonded to the framework oxygens (Miller and Taylor 1985). There are seven partially occupied H_2O sites in the structure with occupancies between 0.2 and 0.8 (Galli and Alberti 1975a, Miller and Taylor 1985).

Barrerite (STI), $Na_8[Al_8Si_{28}O_{72}]\cdot26H_2O$, is orthorhombic, space group *Amma*, a = 13.64, b = 18.20, c = 17.84 Å, Z = 1, and has (Si,Al) disordered over five distinct tetrahedral sites (Galli and Alberti 1975b, Sacerdoti et al. 1999). Five partially occupied channel cations sites can be resolved. Two of these sites are similar to the Ca site in stilbite and stellerite and have the highest Na occupancies, 0.72 and 0.62 (Galli and Alberti 1975b). The other two sites are somewhat similar to the Na sites in stellerite, with lower occupancies, 0.14 and 0.25. The fifth site is specific to barrerite, with an occupancy of 0.25 (Galli and Alberti 1975b). Figure 17e shows the channel cation distribution. Galli and Alberti (1975b) also found 14 channel H_2O sites ranging in occupancy from 0.2 to 0.91. The cation and H_2O distribution in barrerite is thus quite complex and may vary from sample to sample (Galli and Alberti 1975b, Sacerdoti et al. 1999). Upon dehydration barrerite transforms to heat-collapsed phases characterized by major changes in the framework (Alberti and Vezzalini 1978, Sani et al. 1998).

Brewsterite (BRE)

Brewsterite (BRE) with the simplified formula $(Sr,Ba,Ca)_2[Al_4Si_{12}O_{32}]\cdot10H_2O$ is monoclinic, space group $P2_1/m$, a = 6.793, b = 17.57, c = 7.76 Å, β = 94.54°, Z = 1 (Perrotta and Smith 1964, Schlenker et al. 1977b, Artioli et al. 1985, Cabella et al. 1993). As in all tabular zeolites, there are two sets of interconnecting channels in the framework. Eight-membered rings (aperture 2.3 × 5.0 Å) form channels parallel to the **a**-axis (Fig. 18a), and a second set of eight-membered rings (aperture 2.8 × 4.1 Å) forms channels parallel to the **c**-axis (Fig. 18b). Schlenker et al. (1977b) found partial (Si,Al) ordering in the four unique tetrahedral sites; three of the sites contain 30 to 40% Al and the fourth site contains no Al (i.e. it is fully occupied by Si). Neutron diffraction data by Artioli et al. (1985) confirmed this partial (Si,Al) ordering. Based upon optical studies, brewsterite

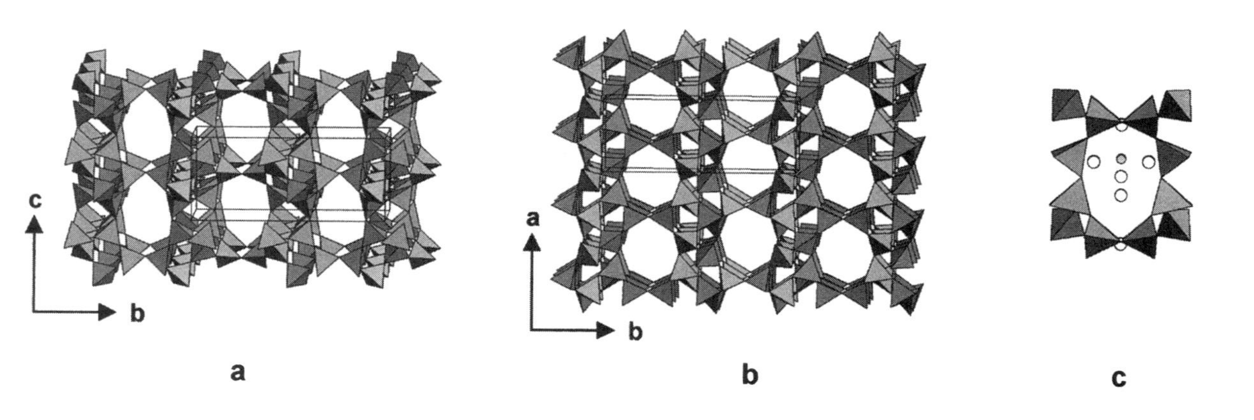

Figure 18. Brewsterite framework and channel contents. (a) A (100) projection looking down the **a**-axis with the **c**-axis vertical, showing the elliptical eight-membered rings forming [100] channels. (b) A (001) projection looking down the **c**-axis with the **a**-axis vertical, showing eight-membered rings which form [001] channels. (c) A (100) projection looking down the **a**-axis with the **c**-axis vertical, showing the channel contents; the single metal site is represented by the smaller gray-shaded circle, and the larger white circles are the H_2O molecules. (Data for these projections obtained from Artioli et al. 1985.)

is triclinic (Akizuki 1987b). This reduction in symmetry may be explained by an ordering of (Si,Al) depending on the growth direction. Recently Akizuki et al. (1996) refined the crystal structures of various growth sectors and found triclinic symmetry (space group $P\bar{1}$). Akizuki et al. (1996) determined slightly different (Si,Al) distributions for tetrahedral sites related to each other by a pseudomirror plane.

All structure refinements located only one fully occupied channel-cation position. The site is located in the middle of the [100] channels and is nine-coordinated to four framework oxygens and five channel H_2O molecules (Fig. 18c). Artioli et al. (1985) also located the H positions in the channels based on neutron diffraction data. Upon dehydration up to 684 K, brewsterite loses eight of its ten H_2O molecules accompanied by diffusion of the channel cations and framework distortion (Ståhl and Hanson 1999). In a dehydration experiment with 24-h equilibration in vacuum at 550 K, Alberti et al. (1999) observed statistical breaking of T-O-T bonds and formation of an altered tetrahedral topology.

OTHER RARE OR STRUCTURALLY POORLY DEFINED ZEOLITES

Lovdarite (LOV), $K_4Na_{12}[Be_8Si_{28}O_{72}]\cdot 18H_2O$, (Men'schikov et al. 1973, Khomyakov et al. 1975) is orthorhombic, space group *Pma*2, a = 39.576, b = 6.9308, c = 7.1526 Å, Z = 1 (Merlino 1990). The structure consists of a framework of ordered Si and Be tetrahedra and is characterized by channels running parallel to the **a**-axis through eight-membered rings (aperture 3.6 × 3.7 Å) and running parallel to the **b**-axis through strongly deformed nine-membered rings (aperture 3.2 × 4.4 Å). A special feature are three-membered rings formed by one BeO_4 and two SiO_4 tetrahedra. Na is five-coordinated by four framework oxygens and one H_2O molecule; K is nine-coordinated by six framework oxygens and three H_2O molecules. Diffraction patterns suggest that lovdarite may consist of distinct disordered domains (Merlino 1990).

Weinebeneite (WEI), $Ca_4[Be_{12}P_8O_{32}(OH)_8]\cdot 16H_2O$ (Walter 1992), space group *Cc*, a = 11.987, b = 9.707, c = 9.633 Å, β = 95.76°, Z = 1, has a tetrahedral framework formed by PO_4 and BeO_4 tetrahedra. Four-membered rings of alternating PO_4 and BeO_4 tetrahedra build a two-dimensional network. These layers are composed of crankshafts forming four- and eight-membered rings (Fig. 19) and resemble layers in the gismondine framework (GIS). In weinebeneite, two such layers are oriented parallel to (100) and are shifted relative to each other by b/2 and connected by additional BeO_4 tetrahedra (Fig. 19). This arrangement leads to three-membered rings (P-Be-Be) which are unknown in framework silicates but were described in the beryllosilicate, lovdarite (Merlino 1990). The framework oxygens connecting two Be tetrahedra are actually OH groups; thus, the framework remains uninterrupted. Ca is situated in structural channels parallel to the **c**-axis and is confined by ten-membered rings and is coordinated by three framework oxygens and four H_2O molecules. One H_2O molecule is positionally disordered.

Gaultite (VSV), $Na_4[Zn_2Si_7O_{18}]\cdot 5H_2O$, (Ercit and van Velthuizen 1994) is orthorhombic, space group *F*2*dd*, a = 10.211, b = 39.88, c = 10.304 Å, Z = 8, and has a framework structure composed of ordered ZnO_4 and SiO_4 tetrahedra forming strongly deformed eight- and nine-membered ring channels along [101]. The eight-membered ring channel is filled with chains of edge-sharing NaO_6 octahedra built by framework oxygens and H_2O molecules. The nine-membered ring channels host disordered Na and H_2O. The framework can be constructed from stacks along **b** of two-dimensional nets composed of four- and eight-membered rings (similar as in weinebeneite, Fig. 19) with periodic insertions of tetrahedra between the sheets. The framework also has characteristic three-membered rings formed by two SiO_4 and one ZnO_4 tetrahedra.

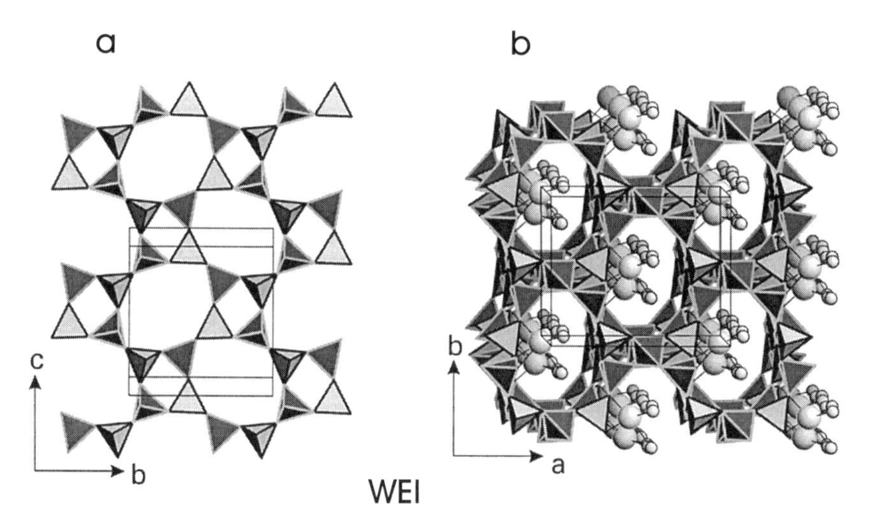

Figure 19. The tetrahedral framework of weinebeneite. BeO_4 tetrahedra have light edges; PO_4 tetrahedra have black edges. (a) The tetrahedral four-ring layer in weinebeneite projected on (100). (b) The weinebeneite framework projected approximately along the **c**-axis. The (100) tetrahedral sheets are linked by intercalated BeO_4 tetrahedra. Large spheres are Ca, and small spheres are fully occupied H_2O sites.

Chiavennite (-CHI), $CaMn[Be_2Si_5O_{13}(OH)_2]·2H_2O$, and **tvedalite**, $(Ca,Mn)_4$-$[Be_3Si_6O_{17}(OH)]·4.3H_2O$ are beryllosilicates (Bondi et al. 1983, Raade et al. 1983, Larsen et al. 1992). Chiavennite (-CHI) is orthorhombic, space group *Pnab*, a = 8.729, b = 31.33, c = 4.903 Å, Z = 4. The crystal structure (Tazzoli et al. 1995) is characterized by an interrupted framework where BeO_4 tetrahedra share only three vertices. Channels parallel to the **c**-axis are confined by nine-membered rings (aperture 3.3 × 4.3 Å). Mn occurs as an extraframework cation and is six-coordinated by four oxygens and two OH groups of the framework. Ca is eight-coordinated by four oxygens, two OH groups, and two H_2O molecules. A possible structure model of tvedalite with a *C*-centered orthorhombic lattice a = 8.7, b = 23.1, c = 4.9 Å derived from chiavennite was given by Alberti (1995).

Tschernichite (BEA), $Ca[Al_2Si_6O_{16}]·8H_2O$ (Boggs et al. 1993, Galli et al. 1995). The X-ray powder pattern could be indexed on a tetragonal unit cell with a = 12.88 and c = 25.01 Å, Z = 8. From the similarity of the X-ray powder patterns of zeolite beta and that of tschernichite, it is assumed that tschernichite is its natural analog (Smith et al. 1991). Zeolite beta occurs in two polymorphs which appear to be stacked in a random sequence in tschernichite (Alberti 1995). Zeolite beta is characterized by two types of channels parallel to the **a**-axis and to the **c**-axis confined by twelve-membered rings.

Gottardiite (NES), $Na_{2.5}K_{0.2}Mg_{3.1}Ca_{4.9}[Al_{18.8}Si_{117.2}O_{272}]·93H_2O$ (Galli et al. 1996) is orthorhombic, space group *Cmca*, a = 13.698, b = 25.213, c = 22.660 Å, Z = 1 (Alberti et al. 1996b), and represents a natural analog of the framework topology (NES) found for synthetic NU-87 (Shannon et al. 1991). The topological symmetry is *Fmmm*, which is reduced to *Cmca* in gottardiite to avoid energetically unfavorable T-O-T angles of 180°. NU-87 is monoclinic, space group $P2_1/c$ (a = 14.32, b = 22.38, c = 25.09 Å, β = 151.5°). The structure (Fig. 20a) consists of sheets parallel to (001) formed by 5^46^2 and 5^4 polyhedral units (e.g. 5^46^2 designates a polyhedron, built by tetrahedra, which is confined by four pentagons and two hexagons). Each (001) sheet is bonded to an analogous sheet through four-membered rings, parallel to the sheets, leading to a two-dimensional channel system parallel to (001). Ten-membered ring channels run parallel to the **a**-axis and twelve-membered ring channels extend zigzag-wise parallel to the **b**-axis. Both channel types are connected by ten-membered ring windows. Extraframework ions and molecules are strongly disordered and positioned close to the center of the channels.

Terranovaite (TER), $Na_{4.2}K_{0.2}Mg_{0.2}Ca_{3.7}[Al_{12.3}Si_{67.7}O_{160}]\cdot{>}29H_2O$ (Galli et al. 1997a) is orthorhombic with average space group *Cmcm*, *a* = 9.747, *b* = 23.880, *c* = 20.068 Å, Z = 1. The structure (Fig. 20b,c) is characterized by chains built of five-membered rings (pentasil chains) and by a two-dimensional ten-membered ring channel system parallel to (010). The structure represents a new topology not found in any other

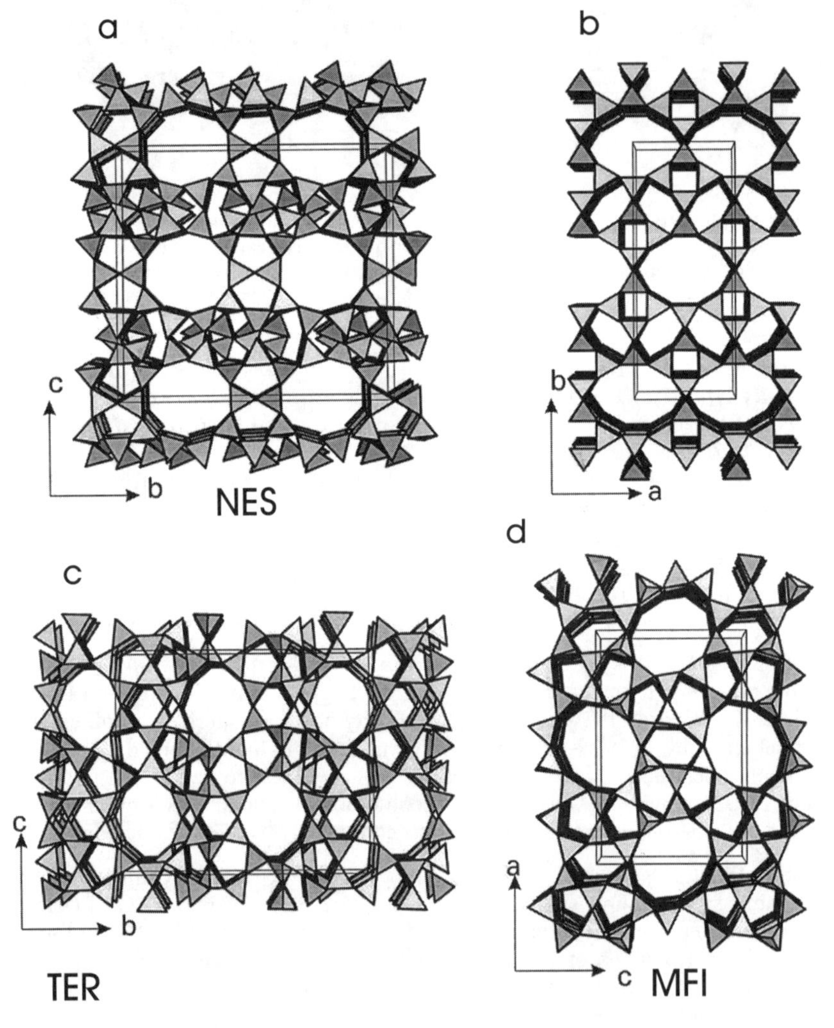

Figure 20. (a) The gottardiite (NES) framework projected approximately along the **a**-axis. Dense sheets extend parallel to (001). (b) The terranovaite (TER) framework projected along the **c**-axis. (c) The terranovaite framework projected along the **a**-axis. (d) The mutinaite (MFI) framework projected along the **b**-axis.

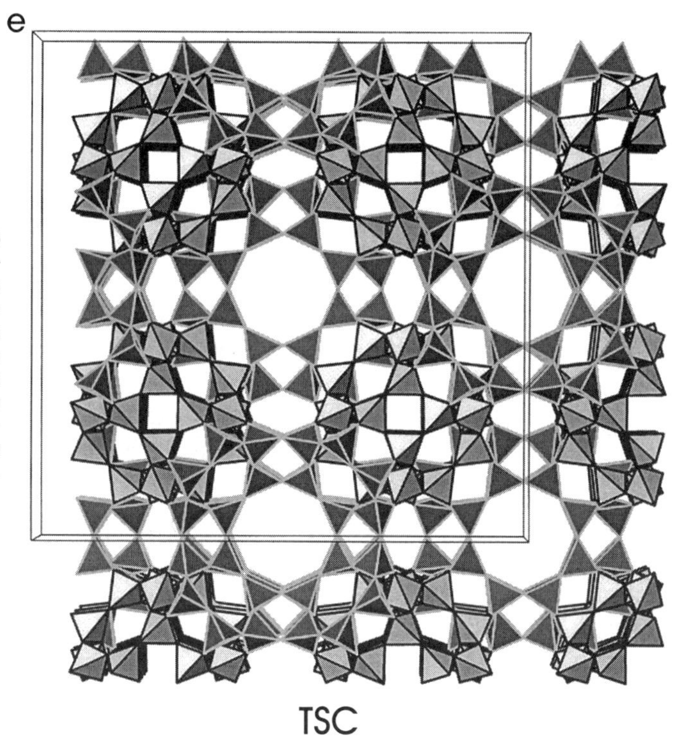

Figure 20, continued. (e) The tschörtnerite framework projected parallel to the **a**-axis. T1-type tetrahedra forming the sodalite cage (β-cage) have dark rims, T2-type tetrahedra forming the α-cage have light rims. The new 96-membered tschörtnerite cage is built by both T1 and T2 tetrahedra.

synthetic or natural zeolite.

Mutianite (MFI), $Na_{2.76}K_{0.11}Mg_{0.21}Ca_{3.78}[Al_{11.20}Si_{84.91}O_{192}]\cdot60H_2O$ (Galli et al. 1997b) is the natural counterpart of ZSM-5 (Kokotailo et al. 1978), space group *Pnm*a, a = 20.201, b = 19.991, c = 13.469 Å, Z = 1 (Vezzalini et al. 1997b). The structure (Fig. 20d) is characterized by chains built of five-membered rings (pentasil chains) and of ten-membered ring channels parallel to [100] and [010]. T-O bond distances are compatible with a disordered (Si,Al) distribution. Channel cations and H_2O molecules are strongly disordered.

Tschörtnerite (TSC), $Ca_4(K,Ca,Sr,Ba)_3Cu_3(OH)_8[Si_{12}Al_{12}O_{48}]\cdot\geq20H_2O$, (Effenberger et al. 1998) is a Cu-bearing new zeolite structure type of cubic symmetry, space group $Fm3m$, a = 31.62 Å, Z = 16. The structure (Fig. 20e) contains interconnections of double six-rings, double eight-rings, sodalite cages, truncated cubo-octahedra, and previously unknown 96-membered cages (tschörtnerite cage) composed of 24 four-rings, 8 six-rings, and 18 eight-rings. Cu has a square-like oxygen coordination forming $Cu_{12}(OH)_{24}$ clusters centered in the truncated cubo-octahedron. In spite of the Si/Al ratio of 1/1 and two symmetry-independent T sites, the T-O bond lengths do not indicate (Si,Al) ordering. Furthermore, the tetrahedral connection implies that Loewenstein's (1954) rule is violated.

Cowlesite (framework unknown), $Ca_6[Al_{12}Si_{18}O_{60}]\cdot33H_2O$, is a Ca-dominate zeolite with minor K and Mg. The crystal structure is unknown, but the X-ray powder pattern of eight samples from different localities could be indexed with an orthorhombic unit cell a = 23.3, b = 30.6, c = 25 Å (Vezzalini et al. 1992).

ACKNOWLEDGMENTS

We thank Dave Bish and Paul Ribbe for their excellent editorial assistance with our chapter; their comments and suggestions greatly improved our work. MEG thanks Kathy Zanetti and Brittany Brown for assisting in very tedious proof reading of the references and tables, Paul Ribbe for two decades of mentoring in crystal chemistry, and finanical support from British Nuclear Fuel Laboratories and National Science Foundation for grant CCLI-9952377.

APPENDIX / INDEX — TABLE 1

(on the following six pages)

NOTE: Each mineral is indexed with the page number on which it is discussed in the text.

Table 1. Alphabetical list of zeolites and zeolite-like minerals covered in this chapter with their, chemical formulas, space group, cell parameters, channel description, and FD (T/1000 Å^3). (Channel descriptions from Meier and Olson 1996; cell parameters and chemical formulas from Gottardi and Galli, 1985; and new data taken from this chapter.)

Meier at al. (1996) use the following symbols for channel descriptions: bold numbers represent the number of tetrahedra defining the channel, the channel free diameter is given in Å, the number of stars represent the number of channels in a given direction, the connectivity of channels is given by "⇔" to represent connected channels or "|" to represent non-connected channels.

Amicite p 23
$Na_4K_4[Al_8Si_8O_{32}]\cdot 10H_2O$
I2
a = 10.226, b = 10.422, c = 9.884, β = 88.32
GIS: {[100] **8** 3.1x4.5 ⇔ [010] **8** 2.8x4.8}***
FD: 15.4 (14.7 expanded)

Ammonioleucite p 17
$(NH_4)_{16}[Al_{16}Si_{32}O_{96}]$
$P4_1ad$
a = 13.214, c = 13.713
ANA: [110] **8** distorted*
FD: 18.6

Analcime p 15
$Na_{16}[Al_{16}Si_{32}O_{96}]\cdot 16H_2O$
Ia3d
a = 13.73
ANA: [110] **8** distorted*
FD: 18.6

Barrerite p 44
$Na_8[Al_8Si_{28}O_{72}]\cdot 26H_2O$
Amma
a = 13.64, b = 18.20, c = 17.84
STI: [100] **10** 4.9x6.1* ⇔ [101] **8** 2.7x5.6*
FD: 16.9

Bellbergite p 32
$(K,Ba,Sr)_2Sr_2Ca_2(Ca,Na)_4[Al_{18}Si_{18}O_{72}]\cdot 30H_2O$
$P6_3/mmc$ or $P\bar{6}2c$ or $P6_3mc$
a = 13.244, c = 15.988
EAB: ⊥ [001] **8** 3.7x5.1**
FD: 15.4

Bikitaite p 39
$Li_2[Al_2Si_4O_{12}]\cdot 2H_2O$
$P2_1$
a = 8.61, b = 4.96, c = 7.60, β = 114.5
BIK: [001] **8** 2.8x3.7*
FD: 20.2

Boggsite p 27
$Ca_8Na_3[Al_{19}Si_{77}O_{192}]\cdot 70H_2O$
Imma
a = 20.236, b = 23.798, c = 12.798
BOG: [100] **12** 7.0x7.0* ⇔ [010] **10** 5.2x5.8*
FD: 15.6

Brewsterite p 44
$(Sr,Ba,Ca)_2[Al_4Si_{12}O_{32}]\cdot 10H_2O$
$P2_1/m$
a = 6.793, b = 17.57, c = 7.76, β = 94.54
BRE: [100] **8** 2.3x5.0* ⇔ [001] **8** 2.8x4.1*
FD: 17.5

Chabazite p 28
$(Ca_{0.5},Na,K)_4[Al_4Si_8O_{24}]\cdot 12H_2O$
R-3*m* or *P*-1
a = 13.2, c = 15.1
CHA: ⊥ [001] **8** 3.8x3.8***
FD: 14.6

Chiavennite p 47
$CaMn[Be_2Si_5O_{13}(OH)_2]\cdot 2H_2O$
Pnab
a = 8.729, b = 31.33, c = 4.903
– CHI: [001] **9** 3.9x4.3*
FD: 20.9

Clinoptilolite p 40
$(Na,K)_6[Al_6Si_{30}O_{72}]\cdot 20H_2O$
C2/m
a = 17.7, b = 17.8, c = 7.4, β = 116.4
HEU: [100] **8** 2.6x4.7* ⇔ {[001] **10** 3.0x7.6* + **8** 3.3x4.6*}
FD: 17.0

Cowlesite (structure unknown) **p 49**
$Ca_6[Al_{12}Si_{18}O_{60}]\cdot 33H_2O$
orthorhombic
a = 23.3, b = 30.6, c = 25.0

Dachiardite p 36
$(Na,K,Ca_{0.5})_4[Al_4Si_{20}O_{48}]\cdot 18H_2O$
Cm
a = 18.676, b = 7.518, c = 10.246, β = 107.87
DAC: [010] **10** 3.4x5.3* ⇔ [001] **8** 3.7x4.8*
FD: 17.3

Edingtonite p 15
$Ba_2[Al_4Si_6O_{20}]\cdot 8H_2O$
$P2_12_12_1$
a = 9.55, b = 9.67, c = 6.523
EDI: [110] **8** 2.8x3.8** ⇔ [001] **8** variable*
FD: 16.6 (14.5 expanded)

Epistilbite p 38
$Ca_3[Al_6Si_{18}O_{48}]\cdot 16H_2O$
*C*1
a = 9.08, b = 17.74, c = 10.15, α = 90, β = 124.6, γ = 90
EPI: [100] **10** 3.4x5.6* ⇔ [001] **8** 3.7 x 5.2*
FD: 18.0

Erionite p 32
$K_2(Na,Ca_{0.5})_8[Al_{10}Si_{26}O_{72}]\cdot 28H_2O$
$P6_3/mmc$
a = 13.26, c = 15.12
ERI: ⊥ [001] **8** 3.6x5.1***
FD: 15.6

Faujasite p 33
$Na_{20}Ca_{12}Mg_8[Al_{60}Si_{132}O_{384}]\cdot 235H_2O$
$Fd\bar{3}m$
a = 24.60
FAU: <111> **12** 7.4***
FD: 12.7

Ferrierite p 39
$(Na,K)Mg_2Ca_{0.5}[Al_6Si_{30}O_{72}]\cdot 20H_2O$
Immm
a = 19.18, b = 14.14, c = 7.5
FER: [001] **10** 4.2x5.4* ⇔ [010] **8** 3.5x4.8*
FD: 17.7

Garronite p 23
$NaCa_{2.5}[Al_6Si_{10}O_{32}]\cdot 13H_2O$
$I\bar{4}m2$
a = 9.9266, c = 10.3031
GIS: {[100] **8** 3.1x4.5 ⇔ [010] **8** 2.8x4.8}***
FD: 15.4 (14.7 expanded)

Gaultite p 46
$Na_4[Zn_2Si_7O_{18}]\cdot 5H_2O$
*F*2*dd*
a = 10.211, b = 39.88, c = 10.304
VSV: [011] **9** 3.3x4.5* ⇔ [101] **9** 3.3x4.5* ⇔ [101] **8** 3.7x3.7*
FD: 17.1

Gismondine p 23
$Ca_4[Al_8Si_8O_{32}]\cdot 16H_2O$
$P2_1/c$
a = 10.02, b = 10.62, c = 9.84, β = 92.5
GIS: {[100] **8** 3.1x4.5 ⇔ [010] **8** 2.8x4.8}***
FD: 15.4 (14.7 expanded)

Gmelinite p 28
$(Na,K,Ca_{0.5},Sr_{0.5})_8[Al_8Si_{16}O_{48}]\cdot 22H_2O$
$P6_3/mmc$
a = 13.75, c = 10.06
GME: [001] **12** 7.0* ⇔ ⊥ [001] **8** 3.6x3.9**
FD: 14.6

Gobbinsite p 23
$Na_5[Al_5Si_{11}O_{32}]\cdot 11H_2O$
$Pmn2_1$
a = 10.108, b = 9.766, c = 10.171
GIS: {[100] **8** 3.1x4.5 ⇔ [010] **8** 2.8x4.8}***
FD: 15.4 (14.7 expanded)

Gonnardite p13
$(Na, Ca)_{6-8}[(Al, Si)_{20}O_{40}]\cdot 12\ H_2O$
I-42*d*
a = 13.21, c = 6.62
NAT: <100> **8** 2.6x3.9** ⇔ [001] **8** variable*
FD: 17.8 (14.5 expanded)

Goosecreekite p 19
$Ca_2[Al_4Si_{12}O_{32}]\cdot 10H_2O$
*P*2$_1$
a = 7.401, *b* = 17.439, *c* = 7.293, β = 105.44
GOO: [100] **8** 2.8x4.0* ⇔ [010] **8** 2.7x4.1* ⇔
[001] **8** 2.9x4.7*
FD: 17.6

Gottardiite p 47
$Na_{2.5}K_{0.2}Mg_{3.1}Ca_{4.9}[Al_{18.8}Si_{117.2}O_{272}]\cdot 93H_2O$
Cmca
a = 13.698, *b* = 25.213, *c* = 22.660
NES: [100] **10** 4.7x6.0**
FD: 17.7

Harmotome p 25
$Ba_2(Ca_{0.5},Na)[Al_5Si_{11}O_{32}]\cdot 12H_2O$
*P*2$_1$/*m*
a = 9.865, *b* = 14.300, *c* = 8.668, β = 124.8
PHI: [100] **8** 3.6* ⇔ [010] **8** 3.0x4.3* ⇔ [001] **8** 3.2x3.3*
FD: 15.8

Heulandite p 40
$(Na,K)Ca_4[Al_9Si_{27}O_{72}]\cdot 24H_2O$
*C*2/*m*
a = 17.7, *b* = 17.8, *c* = 7.4, β = 116.4
HEU: [100] **8** 2.6x4.7* ⇔ {[001] **10** 3.0x7.6* + **8** 3.3x4.6*}
FD: 17.0

Hsianghualite p 18
$Li_{16}Ca_{24}F_{16}[Be_{24}Si_{24}O_{96}]$
*I*2$_1$3
a = 12.864
ANA: [110] **8** distorted*
FD: 18.6

Kalborsite p 15
$K_6B(OH)_4Cl[Al_4Si_6O_{20}]$
P-42$_1$*c*
a = 9.851, *c* = 13.060
EDI: [110] **8** 2.8x3.8** ⇔ [001] **8** variable*
FD: 16.6 (14.5 expanded)

Laumontite p 18
$Ca_4[Al_8Si_{16}O_{48}]\cdot 18H_2O$
*C*2/*m*
a = 14.863, *b* = 13.169, *c* = 7.537, β = 110.18
LAU: [100] **10** 4.0x5.3*
FD: 17.7

Leucite p 17
$K_{16}[Al_{16}Si_{32}O_{96}]$
*Ia*3*d*
a = 13.0
ANA: [110] **8** distorted*
FD: 18.6

Levyne p 31
$(Ca_{0.5},Na,K)_6[Al_6Si_{12}O_{36}]\cdot 18H_2O$
R-3*m*
a = 13.338, *c* = 23.014
LEV: ⊥ [001] **8** 3.6x4.8**
FD: 15.2

Lovdarite p 46
$K_4Na_{12}[Be_8Si_{28}O_{72}]\cdot 18H_2O$
*Pma*2
a = 39.576, *b* = 6.9308, *c* = 7.1526
LOV: [010] **9** 3.2x4.4* ⇔ [001] **9** 3.2x3.7* ⇔
[100] **8** 3.6x3.7*
FD: 18.4

Maricopaite p 36
$Pb_7Ca_2[Al_{12}Si_{36}(O,OH)_{100}]\cdot n(H_2O,OH)$
*Cm*2*m*
a = 19.434, *b* = 19.702, *c* = 7.538
–MOR: [010] **12** 6.5x7.0* ⇔ [010] **8** 2.6x5.7*

Mazzite p 26
$K_{2.5}Mg_{2.1}Ca_{1.4}Na_{0.3}[Al_{10}Si_{26}O_{72}]\cdot 28H_2O$
*P*6$_3$/*mmc*
a = 18.392, *c* = 7.646
MAZ: [001] **12** 7.4* | **8** [001] 3.4x5.6*
FD: 16.1

Merlinoite p 25
$K_5Ca_2[Al_9Si_{23}O_{64}]\cdot 24H_2O$
Immm
a = 14.116, *b* = 14.229, *c* = 9.946
MER: [100] **8** 3.1x3.5* ⇔ [010] **8** 2.7x3.6* ⇔ [001] {**8** 3.4x5.1* + **8** 3.3x3.3*}
FD: 16.0

Mesolite p 11
$Na_{16}Ca_{16}[Al_{48}Si_{72}O_{240}]\cdot 64H_2O$
*Fdd*2
a = 18.41, *b* = 56.65, *c* = 6.55
NAT: <100> **8** 2.6x3.9** ⇔ [001] **8** variable*
FD: 17.8 (14.5 expanded)

Montesommaite p 20
$(K,Na)_9[Al_9Si_{23}O_{64}]\cdot 10H_2O$
*Fdd*2
a = 10.099, *b* = 10.099, *c* = 17.307
MON: [100] **8** 3.2x4.4* ⇔ [001] **8** 3.6x3.6*
FD: 18.1

Mordenite p 35
$(Na_2,K_2,Ca)_4[Al_8Si_{40}O_{96}]\cdot 28H_2O$
*Cmc*2[1]
a = 18.11, *b* = 20.46, *c* = 7.52
MOR: [001] **12** 6.5x7.0* ⇔ [010] **8** 2.6x5.7*
FD: 17.2

Mutianite p 49
$Na_{2.76}K_{0.11}Mg_{0.21}Ca_{3.78}[Al_{11.20}Si_{84.91}O_{192}]\cdot 60H_2O$
Pnma
a = 20.201, *b* = 19.991, *c* = 13.469
MFI: {[010] **10** 5.3x5.6 ⇔ [100] **10** 5.1x5.5}***
FD: 17.9

Natrolite p 10
$Na_{16}[Al_{16}Si_{24}O_{80}]\cdot 16H_2O$
*Fdd*2
a = 18.29, *b* = 18.64, *c* = 6.59
NAT: <100> **8** 2.6x3.9** ⇔ [001] **8** variable*
FD: 17.8 (14.5 expanded)

Offretite p 32
$KCaMg[Al_5Si_{13}O_{36}]\cdot 15H_2O$
P-6*m*2
a = 13.29, *c* = 7.58
OFF: [001] **12** 6.7* ⇔ ⊥ [001] **8** 3.6 x 4.9**
FD: 15.5

Pahasapaite p 34
$(Ca_{5.5}Li_{3.6}K_{1.2}Na_{0.2})Li_8[Be_{24}P_{24}O_{96}]\cdot 38H_2O$
*I*23
a = 13.781
RHO: <100> **8** 3.6*** | <100> **8** 3.6***
FD: 14.3

Paranatrolite p 13
$Na_{16}[Al_{16}Si_{24}O_{80}]\cdot 24H_2O$
pseudo-orthorhombic
a = 19.07, *b* = 19.13, *c* = 6.58
NAT: <100> **8** 2.6x3.9** ⇔ [001] **8** variable*
FD: 17.8 (14.5 expanded)

Parthéite p 21
$Ca_2[Al_4Si_4O_{15}(OH)_2]\cdot 4H_2O$
*C*2/*c*
a = 21.555, *b* = 8.761, *c* = 9.304, β = 91.55
-PAR: [001] **10** 3.5x6.9*
FD: 18.2

Paulingite p 27
formula $Na_{14}K_{36}Ca_{59}Ba_2[Al_{173}Si_{499}O_{1344}]\cdot 550H_2O$
*Im*3*m*
a = 35.09
PAU: <100> **8** 3.8*** | <100> **8** 3.8***
FD: 15.5

Perlialite p 27
$K_9Na(Ca,Sr)[Al_{12}Si_{24}O_{72}]\cdot 15H_2O$
*P*6/*mmm*
a = 18.54, *c* = 7.53
LTL: [001] **12** 7.1*
FD: 16.4

Phillipsite p 25
$K_2(Ca_{0.5},Na)_4[Al_6Si_{10}O_{32}]\cdot 12H_2O$
*P*2₁/*m*
a = 9.869, b = 14.139, c = 8.693, β = 124.81
PHI: [100] **8** 3.6* ⇔ [010] **8** 3.0x4.3* ⇔ [001] **8** 3.2x3.3*
FD: 15.8

Pollucite p 17
$Cs_{16}[Al_{16}Si_{32}O_{96}]$
*Ia*3*d*
a = 13.69
ANA: [110] **8** distorted*
FD: 18.6

Roggianite p 21
$Ca_2[Be(OH)_2Al_2Si_4O_{13}]\cdot 2.5H_2O$
*I*4/*mcm*
a = 18.370, c = 9.187
-RON: [001] **12** 4.2*
FD: 15.6

Scolecite p 11
$Ca_8[Al_{16}Si_{24}O_{80}]\cdot 24H_2O$
*F*1*d*1
a = 18.51, b = 18.97, c = 6.53, β = 90.7
NAT: <100> **8** 2.6x3.9** ⇔ [001] **8** variable*
FD: 17.8 (14.5 expanded)

Stellerite p 44
$Ca_4[Al_8Si_{28}O_{72}]\cdot 28H_20$
F*mmm*
a = 13.55, b = 18.26, c = 17.80
STI: [100] **10** 4.9x6.1* ⇔ [101] **8** 2.7x5.6*
FD: 16.9

Stilbite p 42
$NaCa_4[Al_9Si_{27}O_{72}]\cdot 30H_2O$
*C*2/*m*
a = 13.64, b = 18.24, c = 11.27, β = 127.9
STI: [100] **10** 4.9x6.1* ⇔ [101] **8** 2.7x5.6*
FD: 16.9

Terranovaite p 48
$Na_{4.2}K_{0.2}Mg_{0.2}Ca_{3.7}[Al_{12.3}Si_{67.7}O_{160}]\cdot > 29H_2O$
Cmcm
a = 9.747, b = 23.880, c = 20.068
TER: [100] **10** 5.0x5.5* ⇔ [001] **10** 4.2x7.0*
FD: 17.1

Thomsonite p 13
$Na_4Ca_8[Al_{20}Si_{20}O_{80}]\cdot 24H_2O$
Pncn
a = 13.09, b = 13.05, c = 13.22
THO: [101] **8** 2.3x3.9* ⇔ [010] **8** 2.2x4.0* ⇔
[001] 8 variable*
FD: 17.7 (14.4 expanded)

Tschernichite p 47
$Ca[Al_2Si_6O_{16}]\cdot 8H_2O$
tetragonal
a = 12.88, c = 25.01
BEA: [001] **12** 5.5x5.5* ⇔ <100> **12** 7.6x6.4**
FD: 15.0

Tschörtnerite p 49
$Ca_4(K,Ca,Sr,Ba)_3Cu_3(OH)_8[Si_{12}Al_{12}O_{48}]\cdot \geq 20H_2O$
F*m*3*m*
a = 31.62
TSC: unknown
FD: 12.1

Tvedalite p 47
$(Ca,Mn)_4[Be_3Si_6O_{17}(OH)]\cdot 4.3H_2O$
orthorhombic
a = 8.7, b = 23.1, c = 4.9

Wairakite p 16
$Ca_8[Al_{16}Si_{32}O_{96}]\cdot 16H_2O$
*I*2/*a*
a = 13.699, b = 13.640, c = 13.546, β = 90.51
ANA: [110] **8** distorted*
FD: 18.6

Weinebeneite p 46
$Ca_4[Be_{12}P_8O_{32}(OH)_8]\cdot16H_2O$
Cc
a = 11.987, b = 9.707, c = 9.633, β = 95.76
WEI: [001] **10** 3.1x5.4* ⇔ [100] **8** 3.3x5.1*
FD: 18.1

Willhendersonite p 31
$K_2Ca_2[Al_6Si_6O_{24}]\cdot10H_2O$
P-1
a = 9.206, b = 9.216, c = 9.500, α = 92.34, β = 92.70,
γ = 90.12
CHA: ⊥ [001] **8** 3.8x3.8***
FD: 14.6

Yugawaralite p 19
$Ca_2[Al_4Si_{12}O_{32}]\cdot8H_2O$
C2/m
a = 6.72, b = 13.93, c = 10.04, β = 111.1
YUG: [100] **8** 2.8x3.6* ⇔ [001] **8** 3.1x5.0*
FD: 18.3

REFERENCES

Adiwidjaja G (1972) Strukturbeziehungen in der Natrolithgruppe und das Entwässerungsverhalten des Skolezits. Dissertation, Univ. Hamburg

Akizuki M (1981a) Origin of optical variation in analcime. Am Mineral 66:403-409

Akizuki M (1981b) Origin of optical variation in chabazite. Lithos 14:17-21

Akizuki M (1985) The origin of sector twinning in harmotome. Am Mineral 70:822-828

Akizuki M (1986) Al-Si ordering and twinning in edingtonite. Am Mineral 71:1510-1514

Akizuki M (1987a) An explanation of optical variation in yugawaralite. Mineral Mag 51:615-620

Akizuki M (1987b) Crystal symmetry and order-disorder structure of brewsterite. Am Mineral 72:645-648

Akizuki M, Harada, K (1988) Symmetry, twinning, and parallel growth of scolecite, mesolite, and natrolite. Am Mineral 73:613-618

Akizuki M, Konno H (1985) Order-disorder structure and the internal texture of stilbite. Am Mineral 70:814-821

Akizuki M, Nishido H (1988) Epistilbite: Symmetry and twinning. Am Mineral 73:1434-1439

Akizuki M, Kudoh Y, Satoh Y (1993) Crystal structure of the orthorhombic {001} growth sector of stilbite. Eur J Mineral 5:839-843

Akizuki M, Kudoh Y, Kuribayashi T (1996) Crystal structures of the {011}, {610}, and {010} growth sectors in brewsterite. Am Mineral 81:1501-1506

Akizuki M, Kudoh Y, Nakamura S (1999) Growth texture and symmetry of heulandite-Ca from Poona, India. Can Mineral 37:1307-1312

Alberti A (1972) On the crystal structure of the zeolite heulandite. Tschermaks mineral petrogr Mitt 18:129-146

Alberti A (1973) The structure type of heulandite B (heat-collapsed phase). Tschermaks mineral petrogr Mitt 19:173-184

Alberti A (1975a) Sodium-rich dachiardite from Alpe di Siusi, Italy. Contrib Mineral Petrol 49:63-66

Alberti A (1975b) The crystal structures of two clinoptilolites. Tschermaks mineral petrogr Mitt 22:25-37

Alberti A (1979) Possible 4-connected frameworks with 4-4-1 unit found in heulandite, stilbite, brewsterite, and scapolite. Am Mineral 64:1188-1198

Alberti A (1986) The absence of T-O-T angles of 180° in zeolites. *In* New Developments in Zeolite Science Technology. Proc 7th Int'l Zeolite Conf. Murakami Y, Iijima A, Ward JW (eds) p 437-441

Alberti A (1991) Crystal chemistry of Si-Al distribution in natural zeolites. *In* Chemistry of Microporous Crystals. Proc Int'l Symp on Chemistry of Microporous Crystals, Tokyo; Kodansha Ltd, p 107-122

Alberti A (1995) Crystal structure and chemistry of newly discovered zeolites and dehydration of known zeolites: A review. *In* Natural Zeolites '93 Occurrence, Properties, Use. D.W. Ming FA. Mumpton (eds) Int'l Comm on Natural Zeolites; Brockport, NY, p 159-171

Alberti A (1997) Location of Brønsted sites in mordenite. Zeolites 19:411-415

Alberti A, Gottardi G (1975) Possible structures in fibrous zeolites. N Jahrb Mineral Mh 1975:396-411

Alberti A, Sabelli C (1987) Statistical and true symmetry of ferrierite: Possible absence of straight T-O-T bridging bonds. Z Kristallogr 178:249-256

Alberti A, Vezzalini G (1978) Crystal structures of heat-collapsed phases of barrerite. *In* Natural Zeolites, Occurrence, Properties, Use. Sand LB, Mumpton FA (eds) Pergamon Press, Oxford, p 85-98

Alberti A, Vezzalini G (1979) The crystal structure of amicite, a zeolite. Acta Crystallogr B35:2866-2869

Alberti A, Vezzalini G (1981a) A partially disordered natrolite: Relationships between cell parameters and Si-Al distribution. Acta Crystallogr B37:781-788

Alberti A, Vezzalini G (1981b) Crystal energies and coordination of ions in partly occupied sites: Dehydrated mazzite. Bull Minéral 104:5-9

Alberti A, Vezzalini G (1983) The thermal behaviour of heulandites: A structural study of the dehydration of Nadap heulandite. Tschermaks mineral petrogr Mitt 31:259-270

Alberti, A, Vezzalini G, Tazzoli V (1981) Thomsonite: a detailed refinement with cross checking crystal energy calculations. Zeolites 1:91-97

Alberti A, Galli E, Vezzalini G, Passaglia E, Zanazzi PF (1982) Positions of cations and water molecules in hydrated chabazite. Natural and Na-, Ca-, Sr-, and K-exchanged chabazite. Zeolites 2:303-309

Alberti A, Galli E, Vezzalini G (1985) Epistilbite: An acentric zeolite with domain structure. Z Kristallogr 173:257-265

Alberti A, Davoli P, Vezzalini G (1986) The crystal structure refinement of natural mordenite. Z Kristallogr 175:249-256

Alberti A, Gottardi G, Lai T (1990) The determination of (Si,Al) distribution in zeolites. *In* Guidelines for Mastering the Properties of Molecular Sieves. D. Barthomeuf (ed) Plenum, New York, p 145-155

Alberti A, Quartieri S, Vezzalini G (1994) Structural modifications induced by dehydration in yugawaralite. *In* Zeolites and Related Microporous Materials. State of the Art 1994. Weitkamp J, Karge HG, Pfeifer H, Höldrich W (eds) Studies in Surface Science and Catalysis 84:637-644

Alberti A, Cruciani G, Dauru I (1995) Order-disorder in natrolite-group minerals. Eur J Mineral 7:501-508

Alberti A, Cruciani G, Galli E, Vezzalini G (1996a) A reexamination of the crystal structure of the zeolite offretite. Zeolites 17:457-461

Alberti A, Vezzalini G, Galli E, Quartieri S (1996b) The crystal structure of gottardiite, a natural zeolite. Eur J Mineral 8:69-75

Alberti A, Martucci A, Galli E, Vezzalini G (1997) A reexamination of the crystal structure of erionite. Zeolites 19:349-352

Alberti A, Sacerdoti M, Quartieri, S, Vezzalini G (1999) Heating-induced phase transformation in zeolite brewsterite: new 4- and 5-coordinated (Si,Al) sites. Phys Chem Minerals 26:181-186

Andersson S, Fälth L (1983) An alternative description of the paulingite structure. J Solid State Chem 46:265-268

Aoki M, Minato H (1980) Lattice constants of wairakite as a function of chemical composition. Am Mineral 65:1212-1216

Armbruster T (1993) Dehydration mechanism of clinoptilolite and heulandite: Single crystal X-ray study of Na-poor, Ca-, K-, Mg-rich clinoptilolite at 100 K. Am Mineral 78:260-264

Armbruster T, Gunter ME (1991) Stepwise dehydration of a heulandite-clinoptilolite from Succor Creek, Oregon, U.S.A.: A single crystal X-ray study at 100 K. Am Mineral 76:1872-1883

Armbruster T, Kohler T (1992) Re- and dehydration of laumontite: a single crystal X-ray study at 100 K. N Jahrb Mineral Mh 1992:385-397

Armbruster T, Yang P, Liebich BW (1996) Mechanism of the SiO_4 for CO_3 substitution in defernite, $Ca_6(CO_3)_{1.58}(Si_2O_7)_{0.21}(OH)_7[Cl_{0.5}(OH)_{0.08}(H_2O)_{0.42}]$: A single-crystal X-ray study at 100 K. Am Mineral 81:625-631

Artioli G (1992) The crystal structure of garronite. Am Mineral 77:189-196

Artioli G, Foy H (1994) Gobbinsite from Magheramorne Quarry, Northern Ireland. Mineral Mag 58: 615-620

Artioli G, Galli E (1999) Gonnardite: Re-examination of holotype material and discreditation of tetranatrolite. Am Mineral 84:1445-1450

Artioli G, Kvick, Å (1990) Synchrotron X-ray Rietveld study of perlialite, the natural counterpart of synthetic zeolite-L. Eur J Mineral 2:749-759

Artioli G, Ståhl K (1993) Fully hydrated laumontite: A structure study by flat-plate and capillary powder diffraction techniques. Zeolites 13:249-255

Artioli G, Torres Salvador MR (1991) Characterization of the natural zeolite gonnardite. Structure analysis of natural and cation exchanged species by the Rietveld method. Mater Sci Forum 79-81:845-850

Artioli G, Smith JV, Kvick, Å (1984) Neutron diffraction study of natrolite, $Na_2Al_2Si_3O_{10}·2H_2O$, at 20 K. Acta Crystallogr C40:1658-1662

Artioli G, Smith JV, Kvick, Å (1985) Multiple hydrogen positions in the zeolite brewsterite, $Sr_{0.95}Ba_{0.5}Al_2Si_6O_{16}·5H_2O$. Acta Crystallogr C41:492-497

Artioli G, Smith JV, Pluth JJ (1986a) X-ray structure refinement of mesolite. Acta Crystallogr C42:937-942

Artioli G, Rinaldi R, Kvick, Å, Smith JV (1986b) Neutron diffraction structure refinement of the zeolite gismondine at 15 K. Zeolites 6:361-366

Artioli G, Smith JV, Kvick, Å (1989) Single crystal neutron diffraction study of partially dehydrated laumontite at 15 K. Zeolites 9:377-391

Bakakin VV, Alexeev VI, Seryotkin, YuV, Belitsky IA, Fursenko BA, Balko VP (1994) Crystal chemical surprises of sodium in dehydrated analcime. Abstracts 16th Genl Meeting Int'l Mineral Assoc; Pisa, Italy, p 24-25

Barrer RM, Villiger H (1969) The crystal structure of synthetic zeolite-L. Z Kristallogr 128:352-370

Baturin SV, Malinovskii, YuA, Runova IB (1985) Kristallicheskaya struktura nizkokremnezemistogo merlinoita s Kol'skogo poluostrova (translated title: Crystalline structure of the low-silica merlinoite from the Kola Peninsula). Mineralog Zhur 7;6:67-74

Bauer T, Baur WH (1998) Structural changes in the natural zeolite gismondine (GIS) induced by cation exchange with Ag, Cs, Ba, Li, Na, K and Rb. Eur J Mineral 10:133-147

Baur WH (1964) On the cation and water positions in faujasite. Am Mineral 49:697-704

Baur WH, Joswig W (1996) The phases of natrolite occurring during dehydration and rehydration studied by single crystal X-ray diffraction methods between room temperature and 923 K. N Jahrb Mineral Mh 1996:171-187

Baur WH, Kassner D, Kim CH, Sieber NHW (1990) Flexibility and distortion of the framework of natrolite: Crystal structures of ion-exchanged natrolites. Eur J Mineral 2:761-769

Baur WH, Joswig W, Fursenko BA, Belitsky IA (1997) Symmetry reduction of the aluminosilicate framework of LAU topology by ordering of exchangeable cations: The crystal structure of primary leonhardite with a primitive Bravais lattice. Eur J Mineral 9:1173-1182

Barrer RM, Villiger H (1969) The crystal structure of synthetic zeolite-L. Z Kristallogr 128:352-370

Bazhan IS, Kholdeev OV, Fursenko BA (1999) Phase transformations in scolecite at high hydrostatic pressures. Dokl Akad Nauk 364:97-100

Beger RM (1969) The crystal structure and chemical composition of pollucite. Z Kristallogr 129:280-302

Belitsky IA, Gabuda SP, Joswig W, Fuess H (1986) Study of the structure and dynamics of water in zeolite edingtonite at low temperature by neutron diffraction and NMR-spectroscopy. N Jahrb Mineral Mh 1986:541-551

Belitsky IA, Fursenko BA, Gabuda SP, Kholdeev OV, Seryotkin, YuV (1992) Structural transformations in natrolite and edingtonite. Phys Chem Minerals 18:497-505

Belokoneva EL, Maksimov BA, Verin IA, Sirota MI, Voloshin AV, Pakhomovskii, Ya. A (1985) Crystal structure of iron chabazite at 293 and 570 K and comparison with structures of other natural chabazites. Sov Phys Crystallogr 30:507-510

Bergerhoff G, Baur WH, Nowacki W (1958) Über die Kristallstrukturen des Faujasits. N Jahrb Mineral Mh 1958:193-200

Bieniok A (1997) The dehydration of the zeolite paulingite. N Jahrb Mineral Mh 1997:498-504

Bieniok A, Joswig W, Baur WH (1996) A study of paulingites: Pore filling by cations and water molecules. N Jahrb Mineral Abh 171:119-134

Bish DL (1988) Effects of composition on the dehydration behavior of clinoptilolite and heulandite. *In* Occurrence, Properties and Utilization of Natural Zeolites. D. Kallo HS. Sherry (eds) Akademiai Kiado, Budapest, p 565-576

Bissert G, Liebau F (1986) The crystal structure of a triclinic bikitaite, $Li(AlSi_2O_6)\cdot H_2O$, with ordered Al/Si distribution. N Jahrb Mineral Mh 1986:241-252

Blum JR (1843) Leonhardit, ein N Mineral. Poggendorff Annalen der Physik Chemie 59:336-339

Boggs RC, Howard DG, Smith JV, Klein GL (1993) Tschernichite, a new zeolite from Goble, Columbia County, Oregon. Am Mineral 78:822-826

Bondi M, Griffin WL, Mattioli V, Mottana A (1983) Chiavennite, $CaMnBe_2Si_5O_{13}(OH)_2\cdot 2H_2O$, a new mineral from Chiavenna (Italy). Am Mineral 68:623-627

Breck DW (1974) Zeolite Molecular Sieves. John Wiley and Sons, New York, 771 p

Bresciani-Pahor N, Calligaris M, Nardin G, Randaccio L, Russo E, Comin-Chiaramonti P (1980) Crystal structure of a natural and a partially silver-exchanged heulandite. J Chem Soc Dalton Trans 1980:1511-1514

Brown ID (1992) Chemical and steric constraints in inorganic solids. Acta Crystallogr B48:553-572

Butikova IK, Shepelev, YuF, Smolin, YuI (1993) Structure of hydrated and dehydrated (250°C) forms of Ca-chabazite. Crystallogr Rep 38:461-463

Cabella R, Lucchetti G, Palenzona A, Quartieri S, Vezzalini G, (1993) First occurrence of a Ba-dominant brewsterite: Structural features. Eur J Mineral 15:353-360

Cappelletti P, Langella A, Cruciani G (1999) Crystal-chemistry and synchrotron Rietveld refinement of two different clinoptilolites from volcanoclastics of North-Western Sardiania. Eur J Mineral 11:1051-1060

Chao GY (1980) Paranatrolite, a new zeolite from Mount St-Hilaire, Québec. Can Mineral 18:85-88

Chen TT, Chao GY (1980) Tetranatrolite from Mount St-Hilaire, Québec. Can Mineral 18:77-84

Coombs DS (1952) Cell size, optical properties and chemical composition of laumontite and leonhardite. Am Mineral 37:812-830

Coombs DS (1955) X-ray observations of wairakite and non-cubic analcime. Mineral Mag 30:699-708

Coombs DS, Alberti A, Armbruster T, Artioli G, Colella C, Galli E, Grice JD, Liebau F, Mandarino JA, Minato H, Nickel EH, Passaglia E, Peacor DR, Quartieri S, Rinaldi R, Ross M, Sheppard RA, Tillmanns E, Vezzalini G (1998) Recommended nomenclature for zeolite minerals: Report of the subcommittee on zeolites of the International Mineralogical Association, Commission on New Minerals and Mineral Names. Am Mineral Special Feature, 28 p (a corresponding PDF file may be downloaded from the internet: http://www.minsocam.org/MSA/AmMin/AmMineral.html)

Corbin DR, Abrams L, Jones GA, Harlow RL, Dunn PJ (1991) Flexibility of the zeolite rho framework: Effect of dehydration on the crystal structure of the beryllophosphate mineral, pahasapaite. Zeolites 11:364-367

Cruciani G, Gualtieri A (1999) Dehydration dynamics of analcime by *in situ* synchrotron powder diffraction. Am Mineral 84:112-119

Cruciani G, Artioli G, Gualtieri A, Ståhl K, Hanson JC (1997) Dehydration dynamics of stilbite using synchrotron X-ray powder diffraction. Am Mineral 82:729-739

Delffs W (1843) Analyse des Leonhardits. Poggendorff Annalen Phys Chem 59:339-342

Di Renzo F, Gabelica Z (1995) New data on the structure and composition of the silicoaluminophosphate viséite and discreditation of its status as a zeolite. *In* Natural Zeolites '93 Occurrence, Properties, Use; DW Ming, FA Mumpton (eds) Int'l Comm on Natural Zeolites, Brockport, NY, p 173-185

Dove MT, Cool T, Palmer DC, Putnis A, Salje EK.H, Winkler B (1993) On the role of Al-Si ordering in the cubic-tetragonal phase transition of leucite. Am Mineral 78:486-492

Eberlein GD, Erd RC, Weber F, Beatty LB (1971) New occurrence of yugawaralite from the Chena hot springs area, Alaska. Am Mineral 56:1699-1717

Effenberger H, Giester G, Krause W, Bernhardt H-J (1998) Tschörtnerite, a copper-bearing zeolite from the Bellberg volcano, Eifel, Germany. Am Mineral 83:607-617

Elsen J, King GSD, Mortier WJ (1987) Influence of temperature on the cation distribution in calcium mordenite. J Phys Chem 91:5800-5805

Engel N, Yvon K (1984) The crystal structure of parthéite. Z Kristallogr 169:165-175

Engelhardt G, Michel D (1987) High-Resolution Solid-State NMR of Silicates and Zeolites. John Wiley & Sons, Chichester, UK, 485 p

Ercit TS, Velthuizen J van (1994) Gaultite, a new zeolite-like mineral species from Mont Saint-Hilaire, Québec, and its crystal structure. Can Mineral 32:855-863

Evans HT, Jr (1973) The crystal structures of cavansite and pentagonite. Am Mineral 58:412-424

Ferraris G, Jones DW, Yerkess J (1972) A neutron-diffraction study of the crystal structure of analcime, $NaAlSi_2O_6 \cdot H_2O$. Z Kristallogr 135:240-252

Fersman AE (1909) Etudes sur les zeolites de la Russie. I. Leonhardite et laumontite dans les environs de Simferopolis (Crimee). Trav Musée Géol Pierre le Grand Acad Imp Sci St Pétersbourg 2:103-150 [Extended abstract (in German) N Jahrb Mineral Geol Paläontol (1912) 1:405-408]

Fischer K (1966) Untersuchung der Kristallstruktur von Gmelinit. N Jahrb Mineral Mh 1966:1-13

Fischer KF, Schramm V (1971) Crystal structure of gismondite, a detailed refinement. Adv Chem Ser 101:250-258

Foster MF (1965) Compositional relations among thomsonites, gonnardites, and natrolites. U S Geol Surv Prof Paper 504-D:E1-E10

Gabuda SP, Kozlova SG (1995) Guest-guest interaction and phase transitions in the natural zeolite laumontite. J Inclusion Phenom Mol Recogn Chem 22:1-13

Gabuda SP, Kozlova SG (1997) Anomalous mobility of molecules, structure of the guest sublattice, and transformation of tetra- to paranatrolite. J Struct Chem—Engl tr 38:562-569

Galli E (1971) Refinement of the crystal structure of stilbite. Acta Crystallogr B27:833-841

Galli E (1975) Crystal structure refinement of mazzite. Rendiconti Soc Ital Mineral Petrol 31:599-612

Galli E (1976) Crystal structure refinement of edingtonite. Acta Crystallogr B32:1623-1627

Galli E, Alberti A (1975a) The crystal structure of stellerite. Bull Soc fr Minéral Cristallogr 98:11-18

Galli E, Alberti A (1975b) The crystal structure of barrerite. Bull Soc fr Minéral Cristallogr 98:331-342

Galli E, Gottardi G (1966) The crystal structure of stilbite. Mineralog Petrogr Acta (Bologna) 12:1-10

Galli E, Gottardi G, Pongiluppi D (1979) The crystal structure of the zeolite merlinoite. N Jahrb Mineral Mh 1979:1-9

Galli E, Passaglia E, Zanazzi PF (1982) Gmelinite: Structural refinements of sodium-rich and calcium-rich natural crystals. N Jahrb Mineral Mh 1982:145-155

Galli E, Quartieri S, Vezzalini G, Alberti A (1995) Boggsite and tschernichite-type zeolites from Mt. Adamson, Northern Victoria Land (Antarctica). Eur J Mineral 7:1029-1032

Galli E, Quartieri S, Vezzalini G, Alberti A (1996) Gottardiite, a new high-silica zeolite from Antarctica: the natural counterpart of synthetic NU-87. Eur J Mineral 8:687-693

Galli E, Quartieri S, Vezzalini G, Alberti A, Franzini M (1997a) Terranovaite from Antarctica: A new 'pentasil' zeolite. Am Mineral 82:423-429

Galli E, Vezzalini G, Quartieri S, Alberti A, Franzini M (1997b) Mutianite, a new zeolite from Antartica: The natural counterpart of ZSM-5. Zeolites 19:318-322

Gard JA, Tait JM (1972) The crystal structure of the zeolite offretite, $K_{1.1}Ca_{1.1}Mg_{0.7}[Si_{12.8}Al_{5.2}O_{36}]\cdot 15.2H_2O$. Acta Crystallogr B28:825-834

Gellens LR, Price GD, Smith JV (1982) The structural relation between svetlozarite and dachiardite. Mineral Mag 45:157-161

Gibbs GV (1982) Molecules as models for bonding in silicates. Am Mineral 67:421-450

Gies H (1985) Clathrate mit SiO_2 Wirtstrukturen: Clathrasile. Nachr Chem Techn Lab 33:387-392

Giuseppetti G, Mazzi F, Tadini C, Galli E (1991) The revised crystal structure of roggianite: $Ca_2[Be(OH)_2Al_2Si_4O_{13}]\cdot <2.5H_2O$. N Jahrb Mineral Mh 1991:307-314

Gordon EK, Samson S, Kamb WB (1966) Crystal structure of the zeolite paulingite. Science 154: 1004-1007

Gottardi G, Galli E (1985) Natural Zeolites. Springer-Verlag, Berlin, 409 p

Gottardi G, Meier WM (1963) The crystal structure of dachiardite. Z Kristallogr 119:53-64

Gramlich-Meier R (1981) Strukturparameter in Zeolithen der Mordenitfamilie. Diss Nr 6760, Eidgen Techn Hochschule Zürich

Gramlich-Meier R, Meier WM, Smith BK (1984) On faults in the framework structure of the zeolite ferrierite. Z Kristallogr 169:201-210

Gramlich-Meier R, Gramlich V, Meier WM (1985) The crystal structure of the monoclinic variety of ferrierite. Am Mineral 70:619-623

Grögel T, Boysen H, Frey F (1984) Phase transition and ordering in leucite. Collected Abstracts, 13th Int'l Conf of Crystallography, Acta Crystallogr A40 Suppl:C256-C257

Gualtieri A, Artioli G, Passaglia E, Bigi S, Viani A, Hanson JC (1998) Crystal structure – crystal chemistry relationships in the zeolites erionite and offretite. Am Mineral 83:590-606

Gualtieri AF, Passaglia E, Galli E, Viani A (1999a) Rietveld structure refinement of Sr-exchanged phillipsite. Microporous Mesoporous Mater 31:33-43

Gualtieri AF, Caputo D, Colella C (1999b) Ion exchange selectivity of phillipsite for Cs^+: A structural investigation using the Rietveld method. Microporous Mesoporous Mater 32:319-322

Gunter ME, Armbruster T, Kohler T, Knowles CR (1994) Crystal structure and optical properties of Na- and Pb-exchanged heulandite-group zeolites. Am Mineral 79:675-682

Hakanson U, Fälth L, Hansen S (1990) Structure of a high-silica variety of zeolite Na-P2. Acta Crystallogr C46:1363-1364

Hambley TW, Taylor JC (1984) Neutron diffraction studies on natural heulandite and partially dehydrated heulandite. J Solid State Chem 54:1-9

Hazen RM, Finger LW (1979) Polyhedral tilting:A common type of pure displacive phase transition and its relationship to analcite at high pressure. Phase Transitions 1:1-22

Hey M (1930) Studies on zeolites: Part I. General review. Mineral Mag 22:422-437

Hori H, Nagashima K, Yamada M, Miyawaki R, Marubashi T (1986) Ammonioleucite, a new mineral from Tatarazawa, Fujioka, Japan. Am Mineral 71:1022-1027

Howard DG (1994) Crystal habit and twinning of garronite from Fara Vicentina, Vicenza (Italy). N Jahrb Mineral Mh 1994:91-96

Joswig W, Baur WH (1995) The extreme collapse of a framework of NAT topology: the crystal structure of metanatrolite (dehydrated natrolite) at 548 K. N Jahrb Mineral Mh 1995:26-38

Joswig W, Bartl H, Fuess H (1984) Structure refinement of scolecite by neutron diffraction. Z Kristallogr 166:219-223

Kato M, Hattori T (1998) Ordered distribution of aluminum atoms in analcime. Phys Chem Minerals 25:556-565

Kawahara A, Curien H (1969) La structure cristalline de l'érionite. Bull Soc fr Minéral Cristallogr 92: 250-256

Kerr IS (1964) Structure of epistilbite. Nature 202:589

Kerr IS, Williams DJ (1967) The crystal structure of yugawaralite. Z Kristallogr 125:220-225

Kerr IS, Williams DJ (1969) The crystal structure of yugawaralite. Acta Crystallogr B25:1183-1190

Khomyakov AP, Semenov EI, Bykova AV, Voronkov AA, Smol'yaninova NN (1975) New data on lovdarite. Dokl Akad Nauk SSSR 221:154-157

Kirkpatrick RJ (1988) MAS NMR spectroscopy of minerals and glasses. Rev Mineral 18:341-403

Kocman V, Gait RI, Rucklidge J (1974) The crystal structure of bikitaite, $Li[AlSi_2O_6]\cdot H_2O$. Am Mineral 59:71-78

Kokotailo GT, Lawton SL, Olson DH, Meier WM (1978) Structure of synthetic zeolite ZSM-5. Nature 272:437-438

Koningsveld, van H (1992) Structural relationships and building units in the family of 5-ring zeolites. Zeolites 12:114-120

Korekawa M (1969) Über die Verzwillingung des Leucits. Z Kristallogr 129:343-350

Koyama K, Takéuchi Y (1977) Clinoptilolite: The distribution of potassium atoms and its role in thermal stability. Z Kristallogr 145:216-239

Krogh Andersen E, Krogh Andersen IG, Ploug-Sorensen G (1990) Disorder in natrolites: structure determinations of three disordered natrolites and one lithium-exchanged disordered natrolite. Eur J Mineral 2:799-807

Kuntzinger S, Ghermani NE, Dusausoy Y, Lecomte C (1998) Distribution and topology of the electron density in an aluminosilicate compound from high-resolution X-ray diffraction data: The case of scolecite. Acta Crystallogr B54:819-833

Kunz M, Armbruster T (1990) Difference displacement parameters in alkali feldspars: Effects of (Si,Al) order-disorder. Am Mineral 75:141-149

Kvick, Å, Smith JV (1983) Neutron diffraction study of the zeolite edingtonite. J Chem Phys 79:2356-2362

Kvick, Å, Ståhl K, Smith JV (1985) A neutron diffraction study of the bonding of zeolitic water in scolecite at 20 K. Z Kristallogr 171:141-154

Kvick, Å, Artioli G, Smith JV (1986) Neutron diffraction study of the zeolite yugawaralite at 13 K. Z Kristallogr 174:265-281

Larsen AO, Åsheim A, Raade G, Taftø J (1992) Tvedalite, $(Ca,Mn)_4Be_3Si_6O_{17}(OH)_4{\cdot}3H_2O$, a new mineral from syenite pegmatite in the Oslo Region, Norway. Am Mineral 77:438-443

Leimer HW, Slaughter M (1969) The determination and refinement of the crystal structure of yugawaralite. Z Kristallogr 130:88-111

Lengauer CL, Giester G, Tillmanns E (1997) Mineralogical characterization of paulingite from Vinarická Hora, Czech Republic. Mineral Mag 61:591-606

Liou JG (1970) Synthesis and stability relations of wairakite, $CaAl_2Si_4O_{12}{\cdot}2H_2O$. Contrib Mineral Petrol 27:259-282

Loewenstein W (1954) The distribution of aluminum in the tetrahedra of silicates and aluminates. Am Mineral 39:92-96

Maleev MN (1976) Svetlozarite, a new high-silica zeolite (in Russian). Zap Vse Miner Obshchest 105: 449-453

Malinovskii, YuA (1984) The crystal structure of K-gmelinite. Sov Phys Crystallogr 29:256-258

Malinovskii, YuA, Belov NV (1980) Crystal structure of kalborsite. Dokl Akad Nauk SSSR 252:611-615

Malinovskii, YuA, Burzlaff H, Rothammel W (1998) Method of quantitative crystallochemical comparison of structures. Comparative studies of fibrous zeolites of the natrolite group. Crystallogr Rep 43: 241-255

Marchi M, Artioli G, Gualtieri A, Hanson JC (1998) The dehydration process in garronite: an *in situ* synchrotron XRPD study. Proc IV Convegno Nazionale Scienza e Tecnologia delle Zeoliti, Cernobbio (Como), Italy, p 143-148

Mazzi F, Galli E (1978) Is each analcime different? Am Mineral 63:448-460

Mazzi F, Galli E (1983) The tetrahedral framework of chabazite. N Jahrb Mineral Mh 1983:461-480

Mazzi F, Galli E, Gottardi G (1976) The crystal structure of tetragonal leucite. Am Mineral 61:108-115

Mazzi F, Galli E, Gottardi G (1984) Crystal structure refinement of two tetragonal edingtonites. N Jahrb Mineral Mh 1984:372-382

Mazzi F, Larsen AO, Gottardi G, Galli E (1986) Gonnardite has the tetrahedral framework of natrolite: Experimental proof with a sample from Norway. N Jahrb Mineral Mh 1986:219-228

McCusker LB, Baerlocher C, Nawaz R (1985) Rietveld refinement of the crystal structure of the new zeolite mineral gobbinsite. Z Kristallogr 171:281-289

Meier WM (1961) The crystal structure of mordenite (ptilolite). Z Kristallogr 115:439-450

Meier WM (1978) Constituent sheets in the zeolite frameworks of the mordenite group. *In* Natural Zeolites, Occurrence, Properties, Use. Sand LB, Mumpton FA (eds) Pergamon Press, Oxford, p 99-103

Meier WM, Groner M (1981) Zeolite structure type EAB: Crystal structure and mechanism for the topotactic transformation of the Na, TMA form. J Solid State Chem 37:204-218

Meier R, Ha TK (1980) A theoretical study of the electronic structure of disiloxane $[(SiH_3)_2O]$ and its relation to silicates. Phys Chem Minerals 6:37-46

Meier WM, Meier R, Gramlich V (1978) Mordenite: Interpretation of a superimposed structure. Z Kristallogr 147:329

Meier WM, Olson DH, Baerlocher C (1996) Atlas of Zeolite Structure Types: 4th revised edn. Zeolites 17:1-230 (http://www.iza-structure.org/databases)

Meneghinello E, Martucci A, Alberti A, Di Renzo F (1999) Structural refinement of a K-rich natrolite: evidence of a new extraframework cation site. Microporous Mesoporous Mater 30:89-94

Men'schikov, YuP, Denisov AP, Uspenskaya YI, Lipatova EA (1973) Lovdarite, a new hydrous alkali-beryllium silicate. Dokl Akad Nauk SSSR 213:130-133

Merkle AB, Slaughter M (1968) Determination and refinement of the structure of heulandite. Am Mineral 53:1120-1138

Merlino S (1965) Struttura dell' epistilbite. Atti Soc Toscana Sci Nat A72:480-483

Merlino S (1990) Lovdarite, $K_4Na_{12}(Be_8Si_{28}O_{72})\cdot 18H_2O$, a zeolite-like mineral: Structural features and OD character. Eur J Mineral 2:809-817

Merlino S, Galli E, Alberti A (1975) The crystal structure of levyne. Tschermaks mineral petrogr Mitt 22:117-129

Mikheeva MG, Pushcharovskii DY, Khomyakov AP, Yamnova NA (1986) Crystal structure of tetranatrolite. Sov Phys Crystallogr 31:254-257

Milazzo E, Artioli G, Gualtieri A, Hanson JC (1998) The dehydration process in gismondine: An *in situ* synchrotron XRPD study. Proc IV Convegno Nazionale Scienza e Tecnologia delle Zeoliti, Cernobbio (Como), Italy, p 160-165

Miller SA, Taylor JC (1985) Neutron single-crystal diffraction study of an Australian stellerite. Zeolites 5:7-10

Millward GR, Thomas JM, Terasaki O, Watanabe D (1986) Direct imaging and characterization of intergrowth-defects in erionite. Zeolites 6:91-95

Moroz NK, Seryotkin YV, Afanasiev IS, Belitzky I.A (1998) Arrangement of extraframework cations in NH_4-analcime. J Struct Chem—Engl tr 39:281-283

Mortier WJ, Pluth JJ, Smith JV (1976a) The crystal structure of dehydrated natural offretite with stacking faults of erionite type. Z Kristallogr 143:319-332

Mortier WJ, Pluth JJ, Smith JV (1976b) Crystal structure of natural zeolite offretite after carbon monoxide adsorption. Z Kristallogr 144:32-41

Mortier WJ, Pluth JJ, Smith. J.V (1978) Positions of cations and molecules in zeolites with the mordenite type framework. IV. De-hydrated and re-hydrated K-exchanged "ptilolite." *In* Natural Zeolites, Occurrence, Properties, Use. Sand LB, Mumpton FA (eds) Pergamon Press, Oxford, p 53-63

Nawaz R (1983) New data on gobbinsite and garronite. Mineral Mag 47:567-568

Nawaz R, Malone JF (1982) Gobbinsite, a new zeolite mineral from Co. Antrim N Ireland. Mineral Mag 46:365-369

Palmer DC, Putnis A, Salje EKH (1988) Twinning in tetragonal leucite. Phys Chem Minerals 16:298-303

Palmer DC, Salje EKH, Schmahl WW (1989) Phase transitions in leucite: X-ray diffraction studies. Phys Chem Minerals 16:714-719

Parise JB, Corbin DR, Abrams L, Northrup P, Rakovan J, Nenoff TM, Stucky GD (1994) Structural relationships among some BePO-, BeAsO, and AlSiO-rho frameworks. Zeolites 14:25-34

Passaglia E, Sacerdoti M (1982) Crystal structure of Na-exchanged stellerite. Bull Minéral 105:338-342

Passaglia E, Tagliavini A (1994) Chabazite-offretite epitaxial overgrowths in cornubianite from Passo Forcel Rosso, Adamello, Italy. Eur J Mineral 6:397-405

Passaglia E, Vezzalini G (1988) Roggianite: Revised chemical formula and zeolitic properties. Mineral Mag 52:201-206

Passaglia E, Artioli G, Gualtieri A, Carnevali R (1995). Diagenetic mordenite from Ponza, Italy. Eur J Mineral 7:429-438

Pauling L (1930) The structure of some sodium and calcium aluminosilicates. Proc Nat Acad Sci USA 16:453-459

Pauling L (1939) The Nature of the Chemical Bond and the Structure of Molecules and Crystals. Cornell University Press, Ithaca, New York, 420 p

Peacor DR (1968) A high temperature single crystal diffractometer study of leucite $(K,Na)AlSi_2O_6$. Z Kristallogr 127:213-224

Peacor DR (1973) High-temperature, single-crystal X-ray study of natrolite. Am Mineral 58:676-680

Peacor DR, Rouse RC (1988) Holdawayite, $Mn_6(CO_3)_2(OH)_7(Cl,OH)$, a structure containing anions in zeolite-like channels. Am Mineral 73:637-642

Peacor DR, Dunn PJ, Simmons WB, Tillmanns E, Fischer RX (1984) Willhendersonite, a new zeolite isostructural with chabazite. Am Mineral 69:186-189

Peacor DR, Dunn PJ, Simmons WB, Wicks FJ, Raudsepp M (1988) Maricopaite, a new hydrated Ca-Pb zeolite-like silicate from Arizona. Can Mineral 26:309-313

Perrotta AJ (1967) The crystal structure of epistilbite. Mineral Mag 36:480-490

Perrotta AJ, Smith JV (1964) The crystal of brewsterite, $(Sr,Ba,Ca)(Al_2Si_6O_{16})\cdot5H_2O$. Acta Crystallogr 17:857-862

Phillips BL, Kirkpatrick RJ (1994) Short-range Si-Al order in leucite and analcime: Determination of the configurational entropy from ^{27}Al and variable-temperature ^{29}Si NMR spectroscopy of leucite, its Cs- and Rb-exchanged derivatives and analcime. Am Mineral 79:1025-1031

Pipping F (1966) The dehydration and chemical composition of laumontite. Mineralogical Soc India, Int'l Mineral Assoc Volume, p 159-166

Pluth JJ, Smith JV (1990) Crystal structure of boggsite, a new high-silica zeolite with the first three-dimensional channel system bounded by both 12- and 10-rings. Am Mineral 75:501-507

Quartieri S, Vezzalini G, (1987) Crystal chemistry of stilbites: Structure refinements of one normal and four chemically anomalous samples. Zeolites 7:163-170

Quartieri S, Vezzalini G, Alberti A (1990) Dachiardite from Hokiya-dake: Evidence of a new topology. Eur J Mineral 2:187-193

Quartieri S, Sani A, Vezzalini G, Galli E, Fois E, Gamba A, Tabacchi G (1999) One-dimensional ice in bikitaite: Single-crystal X-ray diffraction, infra-red spectroscopy and ab-initio molecular dynamics studies. Microporous Mesoporous Mater 30:77-87

Raade G, Åmli R, Mladeck MH, Din VK, Larsen AO, Åsheim A (1983) Chiavennite from syenite pegmatites in the Oslo Region, Norway. Am Mineral 68:628-633

Rastsvetaeva RK, Rekhlova OYu, Andrianov VI, Malinovskii, YuA (1991) Crystal structure of hsyanghualite (in Russian). Dokl Akad Nauk SSSR 316:624-628

Reeuwijk LP. van (1971) The dehydration of gismondite. Am Mineral 56:1655-1659

Rinaldi R (1976) Crystal chemistry and structural epitaxy of offretite-erionite from Sasbach, Kaiserstuhl. N Jahrb Mineral Mh 1976:145-156

Rinaldi R, Vezzalini G (1985) Gismondine: Detailed X-ray structure refinement of two natural samples. *In* Zeolites. Drzaj, Hocevar, Pejovnik (eds) Elsevier, Amsterdam, p 481-491

Rinaldi R, Pluth JJ, Smith JV (1974) Zeolites of the phillipsite family. Refinement of the crystal structures of phillipsite and harmotome. Acta Crystallogr B30:2426-2433

Rinaldi R, Pluth JJ, Smith JV (1975a) Crystal structure of cavansite dehydrated at 220°C. Acta Crystallogr B31:1598-1602

Rinaldi R, Pluth JJ, Smith JV (1975b) Crystal structure of mazzite dehydrated at 600°C. Acta Crystallogr B31:1603-1608

Ross M, Flohr MJ.K, Ross DR (1992) Crystalline solution series and order-disorder within the natrolite mineral group. Am Mineral 77:685-703

Rouse RC, Peacor DR (1986) Crystal structure of the zeolite mineral goosecreekite, $CaAl_2Si_6O_{16}\cdot5H_2O$. Am Mineral 71:1494-1501

Rouse RC, Peacor DR (1994) Maricopaite, an unusual lead calcium zeolite with an interrupted mordenite-like framework and intrachannel Pb_4 tetrahedral clusters. Am Mineral 79:175-184

Rouse RC, Peacor DR, Dunn PJ, Campbell TJ, Roberts WL, Wicks FJ, Newbury D (1987) Pahasapaite, a beryllophosphate zeolite related to synthetic zeolite rho, from the Tip Top Pegmatite of South Dakota. N Jahrb Mineral Mh 1987:433-440

Rouse RC, Peacor DR, Merlino S (1989) Crystal structure of pahasapaite, a beryllophosphate mineral with a distorted zeolite rho framework. Am Mineral 74:1195-1202

Rouse RC, Dunn PJ, Grice JD, Schlenker JL, Higgins JB (1990) Montesommaite, $(K,Na)_9(Al_9Si_{23}O_{64})\cdot10H_2O$, a new zeolite related to merlinoite and the gismondine group. Am Mineral 75:1415-1420

Rüdinger B, Tillmanns E, Hentschel G (1993) Bellbergite—a new mineral with the zeolite structure type EAB. Mineral Petrol 48:147-152

Sacerdoti M (1996) New refinements of the crystal structure of levyne using twinned crystals. N Jahrb Mineral Mh 1996:114-124

Sacerdoti M, Gomedi I (1984) Crystal structural refinement of Ca-exchanged barrerite. Bull Minéral 107:799-804

Sacerdoti M, Passaglia E, Carnevali R (1995) Structural refinements of Na-, K-, and Ca-exchanged gmelinites. Zeolites 15:276-281

Sacerdoti M, Sani A, Vezzalini G (1999) Structural refinement of two barrerites from Alaska. Microporous Mesoporous Mater 30:103-109

Sadanaga R, Marumo F, Takéuchi Y (1961) The crystal structure of harmotome. Acta Crystallogr 14: 1153-1163

Sani A, Vezzalini G, Delmotte L, Gabelica Z, Marichal C (1998) ^{29}Si and ^{27}Al MAS NMR study of the natural zeolite barrerite and its dehydrated phases. Proc IV Convegno Nazionale Scienza e Tecnologia delle Zeoliti, Cernobbio (Como), Italy, p 206-207

Sani A, Vezzalini G, Ciambelli P, Rapacciuolo MT (1999) Crystal structure of hydrated and partially NH_4-exchanged heulandite. Microporous Mesoporous Mater 31:263-270

Sato M (1979) Derivation of possible framework structures formed from parallel four- and eight-membered rings. Acta Crystallogr 35A:547-552

Sato M, Gottardi G (1982) The slipping scheme of the double crankshaft structures in tectosilicates and its mineralogical implication. Z Kristallogr 161:187-193

Schlenker JL, Pluth JJ, Smith JV (1977a) Dehydrated natural erionite with stacking faults of the offretite type. Acta Crystallogr B33:3265-3268

Schlenker JL, Pluth JJ, Smith JV (1977b) Refinement of the crystal of brewsterite, $Ba_{0.5}Sr_{1.5}Al_4Si_{12}O_{32}$ $\cdot 10H_2O$. Acta Crystallogr B33:2907-2910

Schlenker JL, Pluth JJ, Smith JV (1978) Positions of cations and molecules in zeolites with the mordenite type framework. VI. De-hydrated barium mordenite. Mater Res Bull 13:169-174

Schröpfer L, Joswig W (1997) Structure analyses of a partially dehydrated synthetic Ca-garronite single crystal under different T, p(H_2O) conditions. Eur J Mineral 9:53-66

Shannon MD, Casci JL, Cox PA, Andrews SJ (1991) Structure of the two-dimensional medium-pore high silica zeolite NU-87. Nature 353:417-420

Shiokawa K, Ito M, Itabashi K (1989) Crystal structure of synthetic mordenites. Zeolites 9:170-176

Slaughter M (1970) Crystal structure of stilbite. Am Mineral 55:387-397

Slaughter M, Kane WT (1969) The crystal structure of a disordered epistilbite. Z Kristallogr 130:68-87

Smith BK (1986) Variations in the framework structure of the zeolite ferrierite. Am Mineral 71:989-998

Smith JV (1978) Enumeration of 4-connected 3-dimensional nets and classification of framework silicates. II. Perpendicular and near-perpendicular linkages from 4.8^2, 3.12^2 and $4.6.^{12}$ nets. Am Mineral 63: 960-969

Smith JV (1983) Enumeration of 4-connected 3-dimensional nets and classification of framework silicates: Combination of 4-1 chain and 2D nets. Z Kristallogr 165:191-198

Smith JV (1988) Topochemistry of zeolites and related materials. 1. Topology and geometry. Chem Rev 88:149-182

Smith JV, Bennett JM (1981) Enumeration of 4-connected 3-dimensional nets and classification of framework silicates: the infinite set of ABC-6 nets; the Archimedean and σ–related nets. Am Mineral 66:777-788

Smith JV, Pluth JJ, Boggs RC, Howard DG (1991) Tschernichite, the mineral analogue of zeolite beta. J Chem Soc Chem Commun 1991:363-364

Smyth JR, Spaid AT, Bish DL (1990) Crystal structures of a natural and a Cs-exchanged clinoptilolite. Am Mineral 75:522-528

Solov'eva LP, Borisov SV, Bakakin VV (1972) New skeletal structure in the crystal structure of barium chloroaluminosilicate $BaAlSiO_2(Cl,OH) \Rightarrow Ba_2[X]BaCl_2[(Si,Al)_8O_{16}]$. Sov Phys Crystallogr 16: 1035-1038

Song SG (1999) Crystal defects of mordenite structures. J Mater Res 14:2616-2620

Ståhl K, Artioli G (1993) A neutron powder diffraction study of fully deuterated laumontite. Eur J Mineral 5:851-856

Ståhl K, Hanson JC (1998) An *in situ* study of the edingtonite dehydration process from X-ray synchrotron powder diffraction. Eur J Mineral 10:221-228

Ståhl K, Hanson JC (1999) Multiple cation sites in dehydrated brewsterite. An *in situ* X-ray synchrotron powder diffraction study. Microporous Mesoporous Mater 32:147-158

Ståhl K, Thomasson R (1994) The dehydration and rehydration processes in the natural zeolite mesolite studied by conventional and synchrotron X-ray powder diffraction. Zeolites 14:12-17

Ståhl K, Kvick, Å, Ghose S (1989) One-dimensional water chain in the zeolite bikitaite: Neutron diffraction study at 13 and 295 K. Zeolites 9:303-311

Ståhl K, Kvick, Å, Smith JV (1990) Thomsonite, a neutron diffraction study at 13 K. Acta Crystallogr C46:1370-1373

Ståhl K, Artioli G, Hanson JC (1996) The dehydration process in the zeolite laumontite: A real-time synchrotron X-ray powder diffraction study. Phys Chem Minerals 23:328-336

Stamires DN (1973) Properties of the zeolite, faujasite, substitutional series: A review with new data. Clays & Clay Miner 21:379-389

Stolz J, Armbruster T (1997) X-ray single-crystal structure refinement of Na,K-rich laumontite, originally designated 'primary leonhardite'. N Jahrb Mineral Mh 1997:131-144

Stolz J, Yang P, Armbruster T (2000a) Cd-exchanged heulandite: Symmetry lowering and site preference. Microporous Mesoporous Mater 37:233-242

Stolz J, Armbruster T, Hennessey B (2000b) Site preference of exchanged alkylammonium ions in heulandite: Single-crystal X-ray structure refinements. Z Kristallogr 215:278-287

Stuckenschmidt E, Fuess H, Kvick, Å (1990) Investigation of the structure of harmotome by X-ray (293 K, 100 K) and neutron diffraction (15 K). Eur J Mineral 2:861-874

Stuckenschmidt E, Kassner D, Joswig W, Baur WH (1992) Flexibility and distortion of the collapsible framework of NAT topology: The crystal structure of NH_4-exchanged natrolite. Eur J Mineral 4:1229-1240

Stuckenschmidt E, Joswig W, Baur WH (1996) Flexibility and distortion of the collapsible framework of NAT topology: the crystal structure of H_3O-natrolite. Eur J Mineral 8:85-92

Stuckenschmidt E, Joswig W, Baur WH, Hofmeister W (1997) Scolecite, Part 1: Refinement of high-order data, separation of internal and external vibrational amplitudes from displacement parameters. Phys Chem Minerals 24:403-410

Sugiyama K, Takéuchi Y (1986) Distribution of cations and water molecules in the heulandite-type framework. Stud Surf Sci Catal 28:449-456

Takaishi T (1998) Ordered distribution of Al atoms in the framework of analcimes. J Chem Soc Faraday Trans 94:1507-1518

Takéuchi Y, Mazzi F, Haga N, Galli E (1979) The crystal structure of wairakite. Am Mineral 64:993-1001

Taylor WH, Jackson WW (1933) The structure of edingtonite. Z Kristallogr 86:53-54

Taylor WH, Meek CA, Jackson WW (1933) The structure of fibrous zeolites. Z Kristallogr 84:373-398

Tazzoli V, Domeneghetti MC, Mazzi F, Cannillo E (1995) The crystal structure of chiavennite. Eur J Mineral 7:1339-1344

Teertstra DK, Cerny P (1995) First natural occurrence of end-member pollucite: A product of low-temperature re-equilibration. Eur J Mineral 7:1137-1148

Teertstra DK, Cerny P Chapman R (1992) Compositional heterogeneity of pollucite from high grade dyke, Maskwa Lake, Southeastern Manitoba. Can Mineral 30:687-697

Teertstra DK, Dyer A (1994) The informal discreditation of "doranite" as the magnesium analog of analcime. Zeolites 14:411-413

Teertstra DK, Cerny P, Sherriff BL, Hawthorne FC (1994a) The crystal chemistry, structure and evolution of pollucite. Abstr 16th General Meeting Int'l Mineral Assoc, Pisa, Italy, p 406

Teertstra DK, Sherriff BL, Xu Z, Cerny P (1994b) MAS and DOR NMR study of Al-Si order in the analcime-pollucite series. Can Mineral 32:69-80

Tillmanns E, Fischer RX, Baur WH (1984) Chabazite-type framework in the new zeolite willhendersonite, $KCaAl_3Si_3O_{12}{\cdot}5H_2O$. N Jahrb Mineral Mh 1984:547-558

Torres-Martinez LM, Gard JA, Howie RA, West AR (1984) Synthesis of $Cs_2BeSi_5O_{12}$ with a pollucite structure. J Solid State Chem 51:100-103

Tschernich RW (1992) Zeolites of the World. Geoscience Press, Phoenix, AZ, 563 p

Vaughan PA (1966) The crystal structure of the zeolite ferrierite. Acta Crystallogr 21:983-990

Vezzalini G (1984) A refinement of Elba dachiardite: Opposite acentric domains simulating a centric structure. Z Kristallogr 166:63-71

Vezzalini G, Quartieri S, Passaglia E (1990) Crystal structure of K-rich natural gmelinite and comparison with the other refined gmelinite samples. N Jahrb Mineral Mh 1990:504-516

Vezzalini G, Artioli G, Quartieri S, Foy H (1992) The crystal chemistry of cowlesite. Mineral Mag 56: 575-579

Vezzalini G, Quartieri S, Alberti A (1993) Structural modifications induced by dehydration in the zeolite gismondine. Zeolites 13:34-42

Vezzalini G, Quartieri S, Galli E (1997a) Occurrence and crystal structure of a Ca-pure willhender-sonite. Zeolites 19:75-79

Vezzalini G, Quartieri S, Galli E, Alberti A, Cruciani G, Kvick, Å (1997b) Crystal structure of the zeolite mutianite, the natural analog of ZSM-5. Zeolites 19:323-325

Vezzalini G, Alberti A, Sani A, Triscari M (1999) The dehydration process in amicite. Microporous Mesoporous Mater 31:253-262

Vigdorchik AG, Malinovskii YA (1986) Crystal structure of Ba-substituted gmelinite $Ba_4[Al_8Si_{16}O_{48}]{\cdot}nH_2O$. Sov Phys Crystallogr 31:519-521

Voloshin AV, Pakhomovskii, YaA, Rogachev DL, Tyusheva FN, Shishkin NM (1986) Ginzburgite—A new calcium-beryllium silicate from desilicated pegmatites. Mineralog Zhur 8;4:85-90 (in Russian). Abstracted in Am Mineral (1988) 73:439-440

Walter F (1992) Weinebeneite, $CaBe_3(PO_4)_2(OH)_2{\cdot}4H_2O$, a new mineral species: Mineral data and crystal structure. Eur J Mineral 4:1275-1283

White JS, Erd RC (1992) Kehoeite is not a valid species. Mineral Mag 50:256-258

Wuest T, Armbruster T (1997) Type locality leonhardite: A single crystal X-ray study at 100K. *In* Program and Abstracts of Zeolite '97, 5th Int'l Conf on the Occurrence, Properties, and Utilization of Natural Zeolites. Ischia, Naples, Italy, p 327-328

Yakubovich OV, Massa W, Pekov IV, Kucherinenko, Ya.V (1999) Crystal structure of a Na,K-variety of merlinoite. Crystallogr Rep 44:776-782

Yang P, Armbruster T (1996a) Na, K, Rb, and Cs exchange in heulandite single-crystals: X-ray structure refinements at 100 K. J Solid State Chem 123:140-149

Yang P, Armbruster T (1996b) (010) disorder, partial Si,Al ordering, and Ca distribution in triclinic ($C\bar{1}$) epistilbite. Eur J Mineral 8:263-271

Yokomori Y, Idaka S (1998) The crystal structure of analcime. Microporous Mesoporous Mater 21: 365-370

Zemann J (1991) Nichtsilikatische Zeolithe. Mitt Österr Mineral Ges 136:21-34

2 The Crystal Chemistry of Zeolites

E. Passaglia

Dipartimento di Scienze della Terra
Università di Modena e Reggio Emilia
Via S. Eufemia 19
41100 Modena, Italy

Richard A. Sheppard

11647 West 37th Place
Wheat Ridge, Colorado 80033

INTRODUCTION

Definition of zeolite

This chapter discusses the crystal chemistry, emphasizing observed chemical variations, for those zeolites that completely fulfill the requirements of Smith (1963) for a zeolite. These requirements include: (a) a three-dimensional framework of tetrahedra occupied more than 50% by Si and Al; (b) an "open" structure with a framework density (i.e. number of tetrahedral atoms per 1000 Å^3) lower than 20 (Brunner and Meier 1989) and hence enclosing cavities connected by windows larger than regular six-membered rings; and (c) an extraframework content represented by cations and water molecules. Thus, this chapter will not deal with those phases which are commonly classified as feldspathoids (leucite, pollucite) and those that can be classified as beryllo-phosphates (pahasapaite, weinebeneite), beryllo-silicates (chiavennite, hsianghualite), or zinc-silicates (gaultite). The requirements of Smith (1963) account for the characteristic properties of zeolites (molecular sieve, reversible dehydration, cation exchange), and although for some zeolite species cation exchange is incomplete or is not yet reported, the presence of large windows (requisite b) reasonably assures its feasibility. Using the above criteria, all minerals known to date which can be classified as zeolites are listed in Table 1.

In the general formula $M_xD_y[Al_{x+2y}Si_{n-(x+2y)}O_{2n}]\cdot mH_2O$, where M are monovalent and D are divalent cations, it is possible to distinguish two parts, which although very different, are mutually dependent and form a homogeneous complex endowed with exclusive chemico-physical properties. The portion in square brackets represents the tetrahedral framework and is characterized by an overall negative charge which increases as the Si/Al ratio decreases. The part outside the square brackets consists of exchangeable extraframework cations, which neutralize the framework negative charge, and, finally, water molecules which often coordinate the extraframework cations.

Classification

Due to the large variability in chemical composition, a reliable classification in the zeolite family is possible only on the basis of structural considerations. In framework silicates (tectosilicates) such as zeolites, the primary building units (TO_4 tetrahedra) are linked so as to form a three-dimensional network where all oxygens are shared by two tetrahedra, the sharing coefficient (Zoltai 1960) being 2 or somewhat smaller. Although numerous theoretical networks can be obtained in this way, only a small number of complex structural units (secondary building units = SBU) formed by a finite (up to 16) number of tetrahedra have been recognized in silicate frameworks so far (Meier 1968, Meier and Olson 1970). The SBU are assembled in different ways to form frameworks

1529-6466/00/0045-0002$10.00

Table 1. Schematic chemical formulae of natural zeolites. STC = Structure Type Code; R_a = Average Si/(Si+Al+Be); R_r = Range of Si/(Si+Al+Be); DEC = Dominant Extraframework Cations; SEC = Subordinate Extraframework Cations. *Italic:* Be-bearing species. Underlined: species with an "interrupted framework"

STC	Zeolite	Schematic chemical formula	R_a	R_r	DEC	SEC
ANA	Analcime	$Na_{16}[Al_{16}Si_{32}O_{96}]\cdot16H_2O$	0.67	0.60-0.74	Na	Ca,K,Mg,Cs
	Wairakite	$Ca_8[Al_{16}Si_{32}O_{96}]\cdot16H_2O$	0.67	0.65-0.70	Ca	Na,Cs
BEA	Tschernichite	$Ca[Al_2Si_6O_{16}]\cdot8H_2O$	0.77	0.73-0.80	Ca	Na,Mg
BIK	Bikitaite	$Li_2[Al_2Si_4O_{12}]\cdot2H_2O$	0.66	-	Li	-
BOG	Boggsite	$Ca_7(Na,K)_4[Al_{18}Si_{78}O_{192}]\cdot70H_2O$	0.81	-	Ca	Na,K,Mg
BRE	Brewsterite	$(Sr,Ba)_2[Al_4Si_{12}O_{32}]\cdot10H_2O$	0.74	0.73-0.74	Sr,Ba	Ca,Na,K
CHA	Chabazite	$(Ca_{0.5},Na,K)_4[Al_4Si_8O_{24}]\cdot12H_2O$	0.67	0.58-0.81	Ca,Na,K	Sr,Mg,Ba
	Willhendersonite	$Ca_2(Ca_{0.5},K)_2[Al_6Si_6O_{24}]\cdot10H_2O$	0.50	-	Ca	K
DAC	Dachiardite	$(Ca_{0.5},Na,K)_4[Al_4Si_{20}O_{48}]\cdot18H_2O$	0.81	0.77-0.86	Na,Ca,K	Cs
EAB	Bellbergite	$Ca_5(Sr,Ba)_3(Na,K)_2[Al_{18}Si_{18}O_{72}]\cdot30H_2O$	0.51	-	Ca	K,Sr,Na,Ba
EDI	Edingtonite	$Ba_2[Al_4Si_6O_{20}]\cdot8H_2O$	0.60	0.59-0.61	Ba	K
EPI	Epistilbite	$Ca_3[Al_6Si_{18}O_{48}]\cdot16H_2O$	0.75	0.73-0.77	Ca	Na,K
ERI	Erionite	$K_2(Ca_{0.5},Na)_7[Al_9Si_{27}O_{72}]\cdot28H_2O$	0.75	0.68-0.79	Ca,Na,K	Mg
FAU	Faujasite	$(Na,K,Ca_{0.5},Mg_{0.5})_{56}[Al_{56}Si_{136}O_{384}]\cdot235H_2O$	0.71	0.68-0.73	Na,Ca	K,Mg
FER	Ferrierite	$(Mg_{0.5},Na,K)_6[Al_6Si_{30}O_{72}]\cdot20H_2O$	0.83	0.78-0.88	Mg,Na,K	Ca,Sr,Ba
GIS	Gismondine	$Ca_4[Al_8Si_8O_{32}]\cdot16H_2O$	0.53	0.50-0.54	Ca	Na,K
	Garronite	$Ca_{2.5}Na[Al_6Si_{10}O_{32}]\cdot13H_2O$	0.63	0.60-0.65	Ca	Na,K
	Gobbinsite	$Na_4Ca[Al_6Si_{10}O_{32}]\cdot11H_2O$	0.64	-	Na	Ca,K,Mg
	Amicite	$Na_4K_4[Al_8Si_8O_{32}]\cdot10H_2O$	0.50	-	Na,K	Ca
GME	Gmelinite	$(Na,K,Ca_{0.5})_8[Al_8Si_{16}O_{48}]\cdot22H_2O$	0.69	0.65-0.72	Na,Ca,K	Sr
GOO	Goosecreekite	$Ca_2[Al_4Si_{12}O_{32}]\cdot10H_2O$	0.75	-	Ca	-
HEU	Heulandite Clinoptilolite	$(Na,K,Ca_{0.5})_7[Al_7Si_{29}O_{72}]\cdot22H_2O$	0.81	0.73-0.85	Ca,Na,K	Mg,Sr,Ba
LAU	Laumontite	$Ca_4[Al_8Si_{16}O_{48}]\cdot16H_2O$	0.67	0.65-0.69	Ca	Na,K

Table 1, continued

LEV	Levyne	$(Ca_{0.5},Na)_6[Al_6Si_{12}O_{36}]\cdot 18H_2O$	0.66	0.62-0.70	Ca,Na	K
LOV	*Lovdarite*	$Na_{13}K_4[Be_8AlSi_{27}O_{72}]\cdot 20H_2O$	0.75	-	Na	K
LTL	Perlialite	$K_9Na(Ca,Mg,Sr)[Al_{12}Si_{24}O_{72}]\cdot 16H_2O$	0.66	0.65-0.67	K	Na,Ca,Sr,Mg
MAZ	Mazzite	$Mg_{2.5}K_2Ca_{1.5}[Al_{10}Si_{26}O_{72}]\cdot 30H_2O$	0.72	-	Mg	K,Ca,Na,Ba
MER	Merlinoite	$K_6Ca_2Na[Al_{11}Si_{21}O_{64}]\cdot 22H_2O$	0.66	0.62-0.71	K	Na,Ca,Ba
MFI	Mutinaite	$Na_3Ca_4[Al_{11}Si_{85}O_{192}]\cdot 60H_2O$	0.88	-	Ca	Na,Mg,K
MON	Montesommaite	$K_9[Al_9Si_{23}O_{64}]\cdot 10H_2O$	0.70	-	K	Na
MOR	Mordenite Maricopaite	$Na_3Ca_2K[Al_8Si_{40}O_{96}]\cdot 28H_2O$ $Pb_7Ca_2[Al_{12}Si_{36}(O,OH)_{100}]\cdot n\ (H_2O,OH)$	0.83 0.75	0.80-0.86 -	Na,Ca,K Pb	Mg,Sr,Ba Ca
NAT	Natrolite Mesolite Scolecite Gonnardite-Tetranatrolite	$Na_{16}[Al_{16}Si_{24}O_{80}]\cdot 16H_2O$ $Na_{16}Ca_{16}[Al_{48}Si_{72}O_{240}]\cdot 64H_2O$ $Ca_8[Al_{16}Si_{24}O_{80}]\cdot 24H_2O$ $Na_{12}Ca_{2.5}[Al_{17}Si_{23}O_{80}]\cdot 20H_2O$ (Z=1/2)	0.60 0.60 0.60 0.57	0.58-0.61 0.58-0.62 0.60-0.61 0.52-0.63	Na Na,Ca Ca Na	Ca - Na Ca
NES	Gottardiite	$Ca_5Na_3Mg_3[Al_{19}Si_{117}O_{272}]\cdot 93H_2O$	0.86	-	Ca	Na,Mg,K
OFF	Offretite	$CaMgK[Al_5Si_{13}O_{36}]\cdot 16H_2O$	0.70	0.69-0.74	Ca,Mg,K	Na
PAR	Partheite	$Ca_8[Al_{16}Si_{16}O_{60}(OH)_8]\cdot 16H_2O$	0.51	0.50-0.52	Ca	Na
PAU	Paulingite	$K_{58}Ca_{42}Na_{10}Ba_{10}[Al_{172}Si_{500}O_{1344}]\cdot 490H_2O$	0.74	0.73-0.77	K,Ca	Na,Ba,Mg,Sr
PHI	Phillipsite Harmotome	$K_2(Na,Ca_{0.5})_3[Al_5Si_{11}O_{32}]\cdot 12H_2O$ $Ba_2(Na,Ca_{0.5})[Al_5Si_{11}O_{32}]\cdot 12H_2O$	0.69 0.71	0.57-0.77 0.68-0.76	K,Na,Ca Ba	Ba K,Na,Ca
ROG	Roggianite	$Ca_{14}(Na,K)[Be_5Al_{15}Si_{28}O_{90}(OH)_{14}](OH)_2\cdot 34H_2O$	0.58	-	Ca	Na,K
STI	Stilbite Stellerite Barrerite	$NaCa_4[Al_9Si_{27}O_{72}]\cdot 30H_2O$ $Ca_8[Al_{16}Si_{56}O_{144}]\cdot 56H_2O$ $Na_{16}[Al_{16}Si_{56}O_{144}]\cdot 52H_2O$	0.75 0.77 0.77	0.71-0.78 0.75-0.78 0.77-0.78	Ca,Na Ca Na	K,Mg Na,K,Mg K,Ca,Mg
	Terranovaite	$Na_4Ca_4[Al_{12}Si_{68}O_{160}]\cdot >29H_2O$	0.85	-	Na	Ca,K,Mg
THO	Thomsonite	$Ca_7Na_5[Al_{19}Si_{21}O_{80}]\cdot 24H_2O$	0.53	0.50-0.58	Ca,Na	K,Mg,Sr
YUG	Yugawaralite	$Ca_2[Al_4Si_{12}O_{32}]\cdot 8H_2O$	0.74	0.74-0.76	Ca	Na
Unk.	Cowlesite	$Ca_{5.5}(Na,K)[Al_{12}Si_{18}O_{60}]\cdot 36H_2O$	0.61	0.58-0.62	Ca	Na,K,Mg
*	Tschörtnerite	$Ca_6Cu_3SrK(OH)_9[Al_{12}Si_{12}O_{48}]\cdot 14H_2O$	0.50	-	Ca	Cu, Sr, K, Ba

* Not yet assigned by IZA

with different topologies (architectures) which generally are coded (Structure Type) after the name of the type material (Meier and Olson 1992). Each structure type is characterized by a topological (i.e. the highest space group) symmetry that is defined (Smith 1974) as the symmetry of a framework when idealized into its most regular shape by movements which leave intact the topologic relationship between nodes (i.e. tetrahedral centers).

Fifty-two zeolite species distributed in 38 different structure types have been identified to date. The topological symmetry (TS) of the framework may be reduced to the real symmetry (RS) of the mineral (Table 2) by one or more of the following causes (Gottardi 1979): (a) ordered distribution of tetrahedral cations; (b) ordered distribution of extraframework ions (cations or water molecules or holes); (c) squeezing of the framework; and d) repulsion of extraframework cations. When reading Table 2, an interesting question is whether phillipsite and harmotome with the same framework (PHI) and symmetry ($P2_1/m$) and heulandite and clinoptilolite with the same framework (HEU) and symmetry ($C2/m$) should be considered different zeolite species. The sharing coefficient of 2 is violated only in three minerals (maricopaite, partheite, and roggianite) which display a so-called "interrupted" tetrahedral framework in which one apex of some tetrahedra is occupied by an (OH) group instead of an O atom and hence is not shared by the adjacent tetrahedron.

Chemistry

The general formula of a zeolite permits a large chemical variability, the only constraint being Lowenstein's rule (Si ≥ Al). A common feature of the chemical compositions is the presence of O, Si, Al, Ca, Mg, Ba, Na, K, and H as fundamental elements and of Fe, Sr, Li, Be, Cs, Cu, and Pb mainly as subordinate or occasional elements. The assignment of Fe to the tetrahedral framework, to extraframework sites, or as micro-impurities of Fe oxides and/or hydroxides is still the subject of open debate because of the very low percentages commonly detected in natural samples and because of the paucity of related studies. Different spectroscopic methods (optical absorption, Mössbauer, electron paramagnetic resonance) show that the Fe in red heulandite crystals from Val di Fassa (Italy) is primarily in hematite and marcasite inclusions and only subordinately as Fe^{3+} in extraframework sites (Bonnin and Calas 1978). X-ray diffraction, scanning electron microscopy, X-ray photoelectron spectroscopy, X-ray absorption spectroscopy, Mössbauer, and energy dispersive X-ray analyses carried out on a clinoptilolite- and mordenite-bearing sample from Cuba revealed that the Fe is present as an extraframework cation (octahedral coordination), and there was no evidence of Fe in tetrahedral coordination (Marco et al. 1995). On the other hand, Fe in tetrahedral sites was detected in synthetic Fe-silicalite and in synthetic Fe-silicate analogs of mordenite (Chandwadkar et al. 1991, 1992). Most zeolites are white or pale colored and show negligible Fe contents, and in many occurrences, Fe-rich smectite formed before or after the crystallization of zeolites. These data suggest that Fe from zeolite analyses should be assigned to impurities and, hence, not considered in the calculation of the chemical formula.

The chemical parameters reported in Table 1 for each zeolite species are averages of compositions from the literature where: (a) data were obtained on crystals, the purity and mineralogical nature of which were previously tested by X-ray analyses; and (b) data are "reliable," i.e. with the content of tetrahedral cations (Si+Al) close to half of the oxygen atoms and the balance error $E = [(Al\text{-}Al_{theor.})/Al_{theor.}] \times 100$ lower than 10%, where $Al_{theor} = (Na+K+Li+Cs) + 2(Ca+Mg+Sr+Ba+Pb)$ (Passaglia 1970). A general overview of the chemistry of the zeolite family is presented in Table 1 in which the framework content is expressed by the ratio, R = Si/(Si+Al+Be), which represents the percentage of tetrahedra occupied by Si, and the extraframework content expressed in terms of the dominant (DEC) and subordinate (SEC) cations.

Table 2. Symmetry and unit-cell parameters of natural zeolites.

Structure Type Zeolite	Topological Symmetry (TS)	Real Symmetry (RS)	*a* (Å)	*b* (Å)	*c* (Å)	β (°)
ANA	Ia3d					
Analcime		$I4_1/acd$, Ibca, I2/a	13.66-13.73	13.68-13.73	13.66-13.74	90.20-90.38
Wairakite		I2/a, Ia	13.66-13.70	13.64-13.66	13.55-13.59	90.20-90.55
BEA	$P4_122$					
Tschernichite		P4/mmm (?)	12.88		25.02	
BIK	Cmcm					
Bikitaite		$P2_1$, P1*	8.61-8.63	4.95-4.96	7.60-7.64	114.42-114.57
BOG	Imma					
Boggsite		Imma	20.24-20.25	23.80-23.82	12.78-12.80	
BRE	$P2_1/m$					
Brewsterite		$P2_1/m$	6.77-6.82	17.45-17.60	7.73-7.75	94.27-94.54
CHA	$R\bar{3}m$					
Chabazite[1]		$R\bar{3}m$	13.69-13.86		14.80-15.42	
Willhendersonite**		$P\bar{1}$	9.18-9.21	9.20-9.22	9.44-9.50	91.42-92.70
DAC	C2/m					
Dachiardite		C2/m	18.62-18.73	7.49-7.54	10.24-10.31	107.87-108.74
EAB	$P6_3/mmc$					
Bellbergite		$P6_3/mmc$	13.24		15.99	
EDI	$P\bar{4}2_1m$					
Edingtonite		$P\bar{4}2_1m$, $P2_12_12$	9.54-9.58	9.66	6.52	
EPI	C2/m					
Epistilbite		C2, C1***	9.08-9.10	17.74-17.80	10.20-10.24	124.55-124.68
ERI	$P6_3/mmc$					
Erionite		$P6_3/mmc$	13.19-13.34		15.04-15.22	

* $\alpha = 89.89°$ $\gamma = 89.96°$

[1] Cell parameters of the pseudo-hexagonal cell

** $\alpha = 91.72\text{-}92.34°$ $\gamma = 90.05\text{-}90.12°$

*** $\alpha = 89.95°$ $\gamma = 90.00°$

Table 2, continued

FAU Faujasite	Fd3m	Fd3m	24.60-24.78			
FER Ferrierite	Immm	Immm(Pnnm), $P2_1/n$	18.90-19.45	14.12-14.28	7.48-7.54	90.0(1)
GIS Gismondine Garronite Gobbinsite Amicite	$I4_1/amd$	 $P2_1/c$ I2/a $Pmn2_1$ I2	 10.01-10.04 9.88 10.10 10.23	 10.60-10.64 10.28 9.77-9.80 10.42	 9.82-9.84 9.876 10.17 9.88	 92.23-92.56 90.11 88.32
GME Gmelinite	$P6_3/mmc$	$P6_3/mmc$	13.62-13.80		9.97-10.25	
GOO Goosecreekite	$C222_1$	$P2_1$	7.40-7.52	17.44-17.56	7.29-7.35	105.44-105.71
HEU Heulandite Clinoptilolite	C2/m	 C2/m C2/m	17.62-17.74	17.81-18.05	7.39-7.53	116.13-116.90
LAU Laumontite	C2/m	C2/m	14.69-14.89	13.05-13.17	7.53-7.61	110-113
LEV Levyne	$R\bar{3}m$	$R\bar{3}m$	13.32-13.43		22.66-23.01	
LOV Lovdarite	$P4_2/mmc$	Pma2	39.58	6.93	7.15	
LTL Perlialite	P6/mmm	P6/mmm	18.49-18.54		7.51-7.53	
MAZ Mazzite	$P6_3/mmc$	$P6_3/mmc$	18.39		7.65	
MER Merlinoite	I4/mmm	Immm	13.86-14.14	14.13-14.23	9.95-10.05	
MFI Mutinaite	Pnma	Pnma	20.22	20.05	13.49	
MON Montesommaite	$I4_1/amd$	$I\bar{4}2d$ ($I4_1md$)	7.14		17.31	
MOR Mordenite Maricopaite	Cmcm	 $Cmc2_1$ Cm2m	 18.05-18.25 19.43	 20.35-20.53 19.70	 7.49-7.55 7.54	

Table 2, concluded

NAT	$I4_1/amd$					
Natrolite		Fdd2	18.28-18.42	18.56-18.70	6.46-6.61	
Mesolite		Fdd2	18.34-18.44	56.52-56.69	6.52-6.56	
Scolecite		F1d1	18.50-18.55	18.90-18.97	6.52-6.53	90.44-90.65
Gonnardite-Tetranatrolite		$I\bar{4}2d$	13.04-13.27		6.58-6.64	
NES	Fmmm					
Gottardiite		Cmca	13.70	25.21	22.66	
OFF	$P\bar{6}m2$					
Offretite		$P\bar{6}m2$	13.27-13.32		7.56-7.61	
PAR	C2/c					
Partheite		C2/c	21.55-21.59	8.76-8.78	9.30-9.31	91.47-91.55
PAU	Im3m					
Paulingite		Im3m	35.05-35.12			
PHI	Bmmb					
Phillipsite[2]		$P2_1/m$	9.86-10.01	14.12-14.34	14.16-14.42	
Harmotome[2]		$P2_1/m$	9.81-9.92	14.10-14.17	14.26-14.34	
ROG	I4/mcm					
Roggianite		I4/mcm	18.33-18.37		9.16-9.19	
STI	Fmmm					
Stilbite[3]		C2/m	13.59-13.66	18.18-18.33	17.71-17.84	90.20-91.15
Stellerite		Fmmm	13.57-13.63	18.16-18.27	17.82-17.87	
Barrerite		Amma	13.59-13.64	18.18-18.20	17.79-17.84	
TER	Cmcm					
Terranovaite		C2cm	9.75	23.88	20.07	
THO	Pmma					
Thomsonite		Pncn	13.00-13.18	13.04-13.16	13.09-13.24	
YUG	C2/m					
Yugawaralite		Pc	6.70-6.73	13.97-14.01	10.04-10.09	111.07-111.20
UNKN.	?					
Cowlesite		Orthorhombic(?)	23.21-23.30	30.60-30.68	24.98-25.04	
?	$Fm\bar{3}m$					
Tschörtnerite		$Fm\bar{3}m$	31.62			

[2] Unit-cell parameters of the pseudo-orthorhombic cell (Z=2) with $\beta = 90°$

[3] Unit-cell parameters of the pseudo-orthorhombic F2/m cell (Z=2)

Framework content. The tetrahedra in zeolites are exclusively occupied by Si and Al, except in roggianite and lovdarite where remarkable amounts of Be have been detected. The average R (R_a) uniformly ranges from 0.50 in amicite, willhendersonite and tschörtnerite to 0.88 in mutinaite; the minimum Si/Al ratio (1) allowed by Lowenstein's rule is observed in only three zeolite species. Mutinaite is the most Si-rich zeolite found to date. Zeolites with R_a of 0.50-0.60 (amicite, bellbergite, edingtonite, gismondine, gonnardite-tetranatrolite, mesolite, natrolite, partheite, roggianite, scolecite, thomsonite, tschörtnerite, willhendersonite) may be classified as Si-poor. Zeolites with R_a > 0.80 (boggsite, dachiardite, ferrierite, gottardiite, heulandite-clinoptilolite, mordenite, mutinaite, terranovaite) may be classified as Si-rich. All others can be considered as "intermediate." The degree of variability of R (R_r) varies greatly among zeolite species. The maximum range (0.23) is displayed by chabazite, and if the isostructural willhendersonite is considered, the structure type CHA exhibits R ranging from 0.50 to 0.81 (R_r = 0.31). High (≥0.10) R_r values are also shown by phillipsite (0.20), analcime (0.14), heulandite-clinoptilolite (0.12), erionite and gonnardite-tetranatrolite (0.11), and ferrierite (0.10). Intermediate (0.09-0.06) R_r values are shown by dachiardite, merlinoite (0.09), levyne, harmotome, thomsonite (0.08), tschernichite, gmelinite (0.07), stilbite and mordenite (0.06). Low (≤0.05) R_r values are shown by the other zeolite species.

Extraframework content. An average water content for each zeolite species is difficult to assess because a comparison of the data from the literature is frustrated by the use of a wide variety of analytical techniques, instruments, and experimental conditions. Furthermore, most recent chemical analyses were performed by electron microprobe and, when reported, the water content is the result of normalization of data to 100%. A representative number of water molecules given in the formulae (Table 1) is taken from Gottardi and Galli (1985). This number is known to be related to both structural and chemical parameters because it increases with the increasing void volume/total volume, Si/Al, and divalent/monovalent cation ratios.

Table 1, where the dominant cations (DEC) are in order of frequency and the subordinate cations (SEC) are in order of abundance, shows that Ca is the most common cation, being the DEC in all analyzed samples of many zeolitic species (wairakite, tschernichite, boggsite, willhendersonite, bellbergite, epistilbite, gismondine, garronite, goosecreekite, laumontite, scolecite, gottardiite, partheite, roggianite, stellerite, cowlesite, yugawaralite, mutinaite, tschörtnerite). It is also one of the DEC in many other species (chabazite, dachiardite, erionite, faujasite, gmelinite, heulandite-clinoptilolite, levyne, mordenite, mesolite, offretite, paulingite, phillipsite, stilbite, thomsonite). Na is the DEC in analcime, barrerite, gobbinsite, gonnardite-tetranatrolite, lovdarite, natrolite, and terranovaite and among the DEC in amicite, chabazite, dachiardite, erionite, faujasite, ferrierite, gmelinite, heulandite-clinoptilolite, levyne, mordenite, mesolite, phillipsite, stilbite, and thomsonite. K is the DEC in merlinoite, montesommaite, and perlialite and among the DEC in amicite, chabazite, dachiardite, erionite, offretite, ferrierite, gmelinite, heulandite-clinoptilolite, mordenite, paulingite, and phillipsite. Mg is the DEC in mazzite and among the DEC in ferrierite and offretite. Ba is the DEC in edingtonite, harmotome, and brewsterite, which also shows Sr as a DEC. Pb is the DEC in maricopaite. Li is the only extraframework cation in bikitaite. Cs has been found as a subordinate cation in some samples of analcime, wairakite, and dachiardite. Although not the DEC, Cu is extraordinarily abundant in tschörtnerite. The distribution of SEC (in order of frequency, Ca, Na, K, Mg, Sr, Ba, Cs, Cu) and the schematic formulae given for each zeolite species allow the definition of epistilbite, gismondine, goosecreekite, laumontite, partheite, roggianite, scolecite, stellerite, tschernichite, willhendersonite, and yugawaralite as Ca-zeolites; barrerite, natrolite, and gonnardite-tetranatrolite as Na-zeolites; montesommaite as a K-zeolite; bikitaite as a Li-zeolite; edingtonite as a Ba-zeolite; maricopaite as a

Pb-zeolite; amicite as a (Na,K)-zeolite; analcime-wairakite, mesolite, mutinaite, and terranovaite as Na,Ca-zeolites; faujasite, ferrierite, gottardiite, mazzite, and offretite as Mg-rich zeolites; harmotome as a Ba-rich zeolite; and tschörtnerite as a Cu-rich zeolite. Chabazite, dachiardite, erionite, gmelinite, heulandite-clinoptilolite, mordenite, and phillipsite show Ca, Na, and K to be the DEC; Na, K, and Mg are the DEC in ferrierite; and Ca, Mg, K in offretite. Considering both the R_r values and the DEC of each zeolite species, the order of variability degree is: chabazite (R_r = 0.23; DEC = Ca, Na, K), phillipsite (R_r = 0.20; DEC = Ca, Na, K), heulandite-clinoptilolite (R_r = 0.12; DEC = Ca, Na, K), analcime-wairakite (R_r = 0.14; DEC = Na, Ca), erionite (R_r = 0.11; DEC = Ca, Na, K), ferrierite (R_r = 0.10; DEC = Mg, Na, K), gmelinite (R_r = 0.07; DEC = Na, Ca, K), levyne (R_r = 0.08; DEC = Ca, Na), and mordenite (R_r = 0.06; DEC = Na, Ca, K). The degrees of variability for the other species are very low and comparable.

Symmetry and unit-cell parameters

With the notable exception of cowlesite, the crystal structures of all zeolite species are known, and the symmetry (RS) is given in Table 2 along with the symmetry of the pertaining structure type (TS). The RS is the same as the TS for about half of the species. In all other species, the RS is slightly or markedly lower due to one or more of the causes described above. These factors explain the existence of crystals (also coexisting) of the same species with different symmetry: tetragonal, orthorhombic, and monoclinic in analcime; monoclinic and triclinic in bikitaite and epistilbite; tetragonal and orthorhombic in edingtonite; orthorhombic and monoclinic in ferrierite. The possible tetragonal $P4/mmm$ symmetry of tschernichite is higher than the TS ($P4_122$) of the pertaining structure type (BEA) and probably is due to an intergrowth of an enantiomorphic pair in the tetragonal space group $P4_122$ and $P4_322$ and a triclinic polymorph with space group $P\bar{1}$ (Boggs et al. 1993). The rhombohedral $R\bar{3}m$ symmetry given for chabazite represents an average structure actually formed by oriented aggregates of triclinic $P\bar{1}$ domains.

Table 2 emphasizes that: (a) the cubic (faujasite, paulingite, tschörtnerite), rhombohedral (levyne), and triclinic (willhendersonite and sometimes bikitaite and epistilbite) symmetries are scarcely represented; (b) the hexagonal (bellbergite, erionite, gmelinite, perlialite, mazzite, offretite) and tetragonal (tschernichite, montesommaite, gonnardite-tetranatrolite, roggianite, and locally analcime and edingtonite) symmetries are moderately represented; and (c) the orthorhombic and monoclinic symmetries are the most common.

Particularly large variations (in Å) in unit-cell parameters occur in chabazite (a = 0.17, c = 0.62), phillipsite (a = 0.15, b = 0.22, c = 0.26), ferrierite (a = 0.55, b = 0.16, c = 0.06), merlinoite (a = 0.28, b = 0.10, c = 0.10), laumontite (a = 0.20, b = 0.12, c = 0.08, β = 3°), gmelinite (a = 0.18, c = 0.28), erionite (a = 0.15, c = 0.18), mordenite (a = 0.20, b = 0.18, c = 0.06), heulandite-clinoptilolite (a = 0.12, b = 0.24, c = 0.14, β = 0.77°), and levyne (a = 0.11, c = 0.35). Crystallochemical studies show that unit-cell parameters are correlated with the framework contents (generally the unit-cell dimensions increase with R decreasing) in chabazite, ferrierite, erionite and mordenite, with the extraframework cations in gmelinite and levyne, and with both in phillipsite. The largest variations are shown by chabazite, which, because of its large variability in chemical composition, appears to be the most "flexible" zeolite species.

Occurrence

Zeolites occur in a wide variety of environments in two major types of occurrences as: (a) macroscopic and microscopic crystals, often euhedral, in veins of ore deposits or in fractures and vugs (amygdules) of plutonic rocks (pegmatite, granite, etc.), metamorphic rocks (gneiss, amphibolite, etc.), and mafic lavas (basalt); and (b) submicroscopic (<20-

30 μm) crystals more or less uniformly distributed in vitroclastic sediments which have undergone diagenetic or low-grade metamorphic processes. The former occurrences (hereafter named "amygdaloidal") include all the species known to date (52), and the zeolites are commonly confined to voids of the host rock, the walls of which are unaltered or locally lined by a thin layer of smectite and/or Fe hydroxides. Up to five or six zeolite species may be associated in such occurrences, also possibly including carbonates (calcite, strontianite, aragonite, siderite, ankerite, rhodochrosite, dolomite, and barytocalcite), sulfates (barite, celestine), phosphates (apatite, fairfieldite), silicates (apophyllite, gyrolite, pectolite, pumpellyite, prehnite, babingtonite, datolite, corrensite, vesuvianite, and eucryptite), oxides and hydroxides (goethite, gibbsite, cuprite, and manganite), sulfides (galena, pyrite, pyrrhotite, chalcopyrite, sphalerite, and cinnabar), silica minerals (quartz, opal, chalcedony, tridymite, and cristobalite), ettringite, strätlingite, mimetite, colemanite, borax, and native copper. Uniquely for these occurrences, the zeolite chemistry does not seem to correlate with that of the host rock. The latter occurrences (hereafter named "sedimentary") can best be detected by X-ray powder diffraction because of the very fine-grained nature of the rocks. Such materials are characterized by a limited number of species (generally analcime, chabazite, heulandite-clinoptilolite, phillipsite, mordenite, erionite, laumontite, and ferrierite; rarely harmotome and faujasite; also gonnardite, gismondine, and natrolite) and the coexistence of at least two different zeolite species and association with other authigenic phases (feldspar, calcite, quartz, opal, tridymite, cristobalite, thaumasite, gypsum, thenardite, dawsonite, searlesite, smectites, iron oxides, and hydroxides), fresh or palagonitized volcanic glass, and lithic fragments. Unlike the former occurrences, the chemistry of sedimentary zeolites is correlated with that of the parent glass and the pore waters, particularly the pH.

DETAILED CRYSTAL CHEMISTRY

The main crystallochemical features of each zeolite species are described below in detail. As noted above, TS refers to topological symmetry, RS to real symmetry, R to [Si/(Si+Al+Be)], DEC to dominant extraframework cations, and SEC to subordinate extraframework cations. Where not specifically reported, the content of extraframework cations is expressed in atoms per formula unit (p.f.u.). For additional information, the reader is referred to the monographs by Gottardi and Galli (1985) and Tschernich (1992); the latter includes a large number of occurrences and descriptions that have not been previously published.

Structure Type: **ANA**

Analcime: $Na_{16}[Al_{16}Si_{32}O_{96}]\cdot 16H_2O$

Wairakite: $Ca_8[Al_{16}Si_{32}O_{96}]\cdot 16H_2O$

The identification of several samples with intermediate chemical compositions and physical properties (refractive indices, density) reveals the existence of a continuous isomorphous series between these two end-members (Surdam 1966, Seki and Oki 1969, Seki 1971, Harada et al. 1972, Harada and Sudo 1976, Aoki and Minato 1980) and permits a classification of the zeolite with Na/(Na+Ca) > 0.5 as analcime and with Na/(Na+Ca) < 0.5 as wairakite. *Analcime* is much more abundant than wairakite and occurs: (a) in vugs of plutonic (pegmatite) and volcanic (basalt, dolerite) rocks; (b) as an alteration product of volcanic glass and "primary" leucite and nepheline from pyroclastic sediments (tuff, tuffite, ignimbrite) diagenetically altered in both continental and marine environments or metamorphosed under low-grade conditions; and (c) as phenocrysts in alkalic igneous rocks. The chemistry varies considerably from the schematic formula, where R = 0.67, Na/(Na+Ca) = 1, and with no other extraframework cations. R ranges from 0.60 in "primary" phenocrysts in theralite from the Square Top intrusion, Australia

(Wilkinson and Hensel 1994), to 0.74 in sedimentary crystals in silicic ash-flow tuffs at Yucca Mountain, Nevada (Broxton et al. 1987). Sedimentary analcime samples are typically more siliceous than the schematic formula. The Na/(Na+Ca) ratio ranges from 1 in many samples to 0.60 in amygdaloidal crystals from Skye, Scotland (Livingstone 1989). The K content is generally lower than 1, but it is anomalously high (1.5-3) in analcime from phonolitic lavas (from a partial "analcimization" of leucite?) from Azerbajan, Iran (Comin Chiaromonti et al. 1979), and from Eifel, Germany (Adabbo et al. 1994). Mg is generally absent or negligible and reaches a maximum (0.62) in analcime from low-grade metamorphosed basaltic tuffs from the Tanzawa Mts., Japan (Seki and Oki 1969). Sr and Ba are almost absent; but Cs is remarkably high in some amygdaloidal samples. An amygdaloidal sample associated with native copper from lava flows at Jalampura (India) contains 0.81% Cu (Talati 1978). Given these different occurrences, several generalizations can be made. Amygdaloidal analcime generally shows R values close to the average (0.67), although a lower value (0.64) has been reported for crystals from a basalt at Skye, Scotland (Livingstone 1989), and higher values (0.72-0.73) have been found in crystals from pegmatite at Bernic Lake, Canada (Cerny 1972) and from basalts at Husa u Marcinova, Bohemia (Pechar 1988), and Boron, California (Wise and Kleck 1988). The Na/(Na+Ca) ratio is generally very close to 1, but much lower values (0.60-0.70) have been reported for Si-poor crystals from Skye (Scotland) and slightly lower values (0.88) in crystals from Procida, Italy (de'Gennaro et al. 1977), and in Cu-bearing crystals from Jalampura, India (Talati 1978). Mg is generally absent, although in a sample from Sagashima (Japan) it reaches 0.42 (Harada et al. 1972). Samples associated with beryl, apatite, cookeite, and lithiophosphate in the Bernic Lake pegmatite show Cs contents up to 5.06 p.f.u., suggesting a possible existence of an almost continuous isomorphous series between analcime and pollucite (Cerny 1974).

Sedimentary analcime exhibits a mean R value of 0.71 with a large range (0.66-0.74). The lower values (0.66-0.68) have been reported for analcime from ferruginous mudstone (Utada and Vine 1984), palagonitic tuff (Iijima and Harada 1968), altered pyroclastic rocks (Barnes et al. 1984, Noh and Kim 1986), and phonolite (Adabbo et al. 1994). The higher values (0.72-0.74) have been reported for analcime from diagenetically altered silicic (mainly rhyolitic) tuffs (Djourova 1976, Walton 1975, Sheppard and Gude 1969a, 1973; Gude and Sheppard 1988, Broxton et al. 1987). The Na/(Na+Ca) ratio is always very close to 1.

In analcime from low-grade metamorphic rocks, R has a mean value of 0.69 and ranges from 0.66 in crystals from metabasites (Cortesogno et al. 1976) to 0.72 in crystals from rhyolitic sediments (Iijima 1978) and from vitric siltstone and sandstone (Vitali et al. 1995). The Na/(Na+Ca) ratio is about 1, although it is 0.66 in a sample associated with wairakite (Seki and Oki 1969).

Analcime "phenocrysts" from alkalic igneous rocks (phonolite, lamprophyre, blairmorite, tephrite, analcimite, theralite, and tinguaite) have been described from at least a dozen occurrences (Coombs and Whetten 1967, Aurisicchio et al. 1975, Pearce 1970, Woolley and Symes 1976, Wilkinson 1977, Comin-Chiaromonti et al. 1979, Giret 1979, Luhr and Kyser 1989, Avdeev 1992, Viladkar and Avasia 1994, Wilkinson and Hensel 1994). The origin ("primary" or "secondary" as replacement pseudomorphs of leucite or nepheline) is uncertain and is still debated in the literature (Wilkinson 1977, Karlsson and Clayton 1991, 1993; Pearce 1993). For these analcimes, R fluctuates near the mean value (0.67); but in analcime from theralites, R is 0.60, and in analcime from tinguaites, R is 0.73 (Wilkinson and Hensel 1994). The Na/(Na+Ca) ratio is close to 1.

Wairakite is much less common than analcime and is found in geothermal areas and in low- to medium-grade metamorphic rocks (Seki and Oki 1969, Seki 1973, Steiner 1955,

Takéuchi et al. 1979, Aoki and Minato 1980, Utada and Vine 1984, Bargar and Beeson 1981). The mean value of R is 0.67, ranging from 0.65 to 0.70; the Na/(Na+Ca) ratio ranges from 0.03 to 0.47. The lowest values of R (0.65) and Na/(Na+Ca) ratio (0.03) are in wairakite from Toi, Japan (Aoki and Minato 1980) and the highest values (0.70 and 0.47, respectively) have been found for wairakite from the Tanzawa Mountains, Japan (Seki and Oki 1969). K and Mg are negligible or absent, and Sr and Ba are absent. Samples from active geothermal areas selectively concentrate Cs, ranging up to 4,700 ppm in crystals from Yellowstone National Park, Wyoming (Keith et al. 1983), and up to 80 ppm in crystals from Wairakei, New Zealand (Steiner 1955).

Optical observations (Akizuki 1980 1981), structure refinements (Mazzi and Galli 1978, Takéuchi et al. 1979, Pechar 1988), and crystallochemical studies (Aoki and Minato 1980, Papezik and Elias 1980) demonstrated that the cubic TS of both zeolite species is lowered to tetragonal, orthorhombic, or monoclinic RS in analcime, and to monoclinic RS in wairakite depending on different Al ordering schemes in the tetrahedral sites and related occupancy of Na and Ca in the extraframework cation sites. The cubic unit-cell parameter of the analcime-wairakite series is 13.66-13.73 Å. The Si/Al ratio correlates with the *a* parameter in analcime (Saha 1959, Coombs and Whetten 1967); correlations between Si, Na, and H_2O contents and monoclinic unit-cell parameters have also been determined for wairakite (Aoki and Minato 1980).

Structure Type: **BEA**

Tschernichite: $Ca[Al_2Si_6O_{16}]\cdot 8H_2O$

Only two occurrences of this zeolite have been described, one at the Neer Road pit, Goble, Oregon, in vugs of basalt associated with boggsite, analcime, offretite, erionite, mordenite, thomsonite, heulandite, levyne, chabazite, okenite, calcite, opal, and native copper (Boggs et al. 1993); and the other at Mt. Adamson, Antarctica, in vugs of basalt, rarely associated with heulandite (Galli et al. 1995). At both localities, tschernichite occurs as large individual crystals (tetragonal dipyramids) and as drusy radiating hemispherical groups. The large crystals are poorer in Si (R = 0.73-0.75), and the Ca content is close to the schematic formula. The drusy crystals are richer in Si (R = 0.79-0.80) and poorer in Ca because the Na and, in the sample from Antarctica, Mg contents are appreciable. The structure is tetragonal *P4/mmm* with *a* = 12.88 Å and *c* = 25.02 Å (Boggs et al. 1993).

Structure Type: **BIK**

Bikitaite: $Li_2[Al_2Si_4O_{12}]\cdot 2H_2O$

Bikitaite occurs in lithium-rich pegmatite dikes at Bikita (Southern Zimbawe) where small transparent crystals are associated with eucryptite, stilbite, quartz, calcite, and allophane (Hurlbut 1957, 1958) and at King's Mountain, North Carolina, where intergrowths of bladed crystals (like jackstraws) are associated with eucryptite, albite, quartz, apatite, and fairfieldite (Leavens et al. 1968). The empirical chemical formulae of the crystals from both occurrences (Kocman et al. 1974, Leavens et al. 1968) are very similar and are close to the schematic formula: R is 0.66, and Li is the only extraframework cation. Crystals from Bikita are monoclinic $P2_1$ with a partially ordered (Si,Al) distribution (Kocman et al. 1974) or triclinic *P*1 with a fully ordered (Si,Al) distribution (Bissert and Liebau 1986). The monoclinic unit-cell dimensions are: *a* = 8.61-8.63 Å, *b* = 4.95-4.96 Å, *c* = 7.60-7.64 Å, β = 114.42-114.57°. In the triclinic crystals: α = 89.89° and γ = 89.96°.

Structure Type: **BOG**

Boggsite: $Ca_7(Na,K)_4[Al_{18}Si_{78}O_{192}]\cdot 70H_2O$

Boggsite occurs with tschernichite from the Neer Road pit, Goble, Oregon (Howard et al. 1990), and from Mt. Adamson, Antarctica (Galli et al. 1995). It occurs in vesicles of basalt as small, colorless, zoned hemispheres embedded in drusy tschernichite at Goble and on a thin layer of smectite at Mt. Adamson. Crystals from the two occurrences display the same tetrahedral content (R = 0.81) and extraframework cations in the same order of abundance (Ca > Na » K > Mg). Nevertheless, in comparison with the holotype sample, where the Ca/Na ratio is 2.6, boggsite from Antarctica is poorer in Ca, richer in Na (Ca/Na = 1.1), and contains more K and Mg. Boggsite is orthorhombic *Imma* (Pluth and Smith 1990) with a = 20.24-20.25 Å, b = 23.80-23.82 Å, c = 12.78-12.80 Å. Optically, it is biaxial negative (Howard et al. 1990).

Structure Type: **BRE**

Brewsterite: $(Sr,Ba)_2[Al_4Si_{12}O_{32}]\cdot 10H_2O$

Brewsterite has been described from five localities: Strontian, Scotland (Mallet, 1859); Burpala pluton, Siberia (Khomyakov et al. 1970); Yellow Lake, Canada (Wise and Tschernich 1978a); Cerchiara, Italy (Cabella et al. 1993); and Harrisville, New York (Robinson and Grice 1993). It occurs in hydrothermal ore veins cutting metamorphic (gneiss, metachert), plutonic (granite, syenite, aplite), and volcanic (porphyritic trachyte) rocks in association with other zeolites (analcime, harmotome, stilbite, heulandite, thomsonite, and mesolite), calcite, barite, strontianite, galena, and pyrite. R is nearly constant (0.73-0.74), and the extraframework content is chiefly Sr and Ba. Ca, Na, K, and Mg are absent or negligible. The Sr/Ba ratio is highly variable and allows the distinction of two types of samples, those with Sr > Ba (Sr-dominant) from Strontian (Schlenker et al. 1977), Burpala pluton, and Yellow Lake (Artioli et al. 1985), and those with Ba > Sr (Ba-dominant) from Cerchiara and Harrisville. Samples from Yellow Lake and Cerchiara are nearly Ba-free and Sr-free, respectively; the remaining samples are intermediate between these two end-members. Brewsterite is monoclinic $P2_1/m$ (Perrotta and Smith 1964, Schlenker et al. 1977, Artioli et al. 1985, Cabella et al. 1993). Unit-cell parameters are: a = 6.77-6.82 Å, b = 17.45-17.60 Å, c = 7.73-7.75 Å, β = 94.27-94.54°. $c > a$ was assessed by Weissenberg X-ray analysis and elongation a was determined by an optical study (Nawaz 1990).

Structure Type: **CHA**

Chabazite: $(Ca_{0.5},Na,K)_4[Al_4Si_8O_{24}]\cdot 12H_2O$

Willhendersonite: $Ca_2(Ca_{0.5},K)_2[Al_6Si_6O_{24}]\cdot 10H_2O$

The TS $R\bar{3}m$ of the tetrahedral framework of chabazite and willhendersonite is lowered to triclinic $P\bar{1}$ as a result of ordering of the (Si,Al) distribution and of the position of the cations and water molecules. Ordering in the tetrahedral sites can be either in domains or complete, depending on the Si/Al ratio of the crystal. Triclinic domains in chabazite (Si/Al > 1) are oriented in such a way that the average structure of the mineral is pseudo-trigonal $R\bar{3}m$ (Smith et al. 1964, Mazzi and Galli 1983). The real triclinic symmetry $P\bar{1}$ in willhendersonite results from a Si/Al ratio of one, with resultant ordering (Tillmanns et al. 1984).

Chabazite is one of the most widespread natural zeolites and has been reported from a variety of occurrences: in vugs of plutonic rocks (granite, syenite, granodiorite, gabbro, and pegmatite), volcanic rocks (tholeiitic and alkalic basalts, including leucitite, and andesite), and metamorphic rocks (gneiss, serpentinite, metabasite, and cornubianite), where it occurs in association with many other zeolites (natrolite, mesolite, thomsonite, phillipsite, stilbite, gismondine, analcime, heulandite-clinoptilolite, mordenite, ferrierite, gonnardite, levyne, gmelinite, laumontite, offretite, and paulingite), apophyllite, quartz, fluorite, barite, carbonates (calcite, aragonite, and siderite), searlesite, colemanite, and pyrite.

Sedimentary chabazite occurs both in palagonitic basalt associated with phillipsite, gonnardite, gmelinite, analcime, thaumasite, apophyllite, and gyrolite and in pyroclastic rocks (phonolitic and trachytic ignimbrites, and rhyolitic and nephelinitic tuff) diagenetically altered in continental environments ("open," "closed," and "geoautoclave" systems) where it coexists with phillipsite, analcime, clinoptilolite, erionite, mordenite, harmotome, dawsonite, and clay minerals.

Chabazite generally exhibits pseudocubic rhombohedral morphology, but lens-shaped aggregates more ("phacolitic" habit) or less ("herschelitic" habit) rounded due to complex twinning are common (Akizuki and Konno 1987, Akizuki et al. 1989). Epitaxial intergrowths with gmelinite are common (Passaglia et al. 1978a, Wise and Kleck 1988), and samples epitaxially overgrown by offretite have also been observed (Passaglia and Tagliavini 1994, Passaglia et al. 1996).

Reliable chemical compositions, including 78 amygdaloidal samples, 5 samples from palagonitic basalt, and 28 sedimentary samples from "open" (22 samples) and "closed" (6 samples) systems, revealed that the schematic formula given above is only an average. R ranges from 0.58 in amygdaloidal chabazite from Kaiserstuhl, Germany (Livingstone 1986), to 0.81 in sedimentary chabazite from the John Day Formation near Monument, Oregon (Sheppard and Gude 1970). Ca, Na, and K are, in order of frequency, the DEC; Sr is generally present in low amounts, but it is remarkable (0.50-0.80) in several amygdaloidal samples (Passaglia 1970, Livingstone 1986, Robert 1988, Birch 1989, Passaglia and Tagliavini 1994). Where present, Ba contents are very low and reach their highest values (ca. 0.13) in some amygdaloidal samples (Passaglia 1970, Argenti et al. 1986). Mg is consistently negligible, but considerable contents (0.5-0.8) have been recognized in the amygdaloidal samples from Narre Warren, Australia (Birch 1989), from Adamello, Italy (Passaglia and Tagliavini 1994), and in a sedimentary sample from the Big Sandy Formation, Arizona (Sheppard and Gude 1973).

In amygdaloidal crystals, R shows a mean value of 0.67 and ranges from 0.58 in the sample from Kaiserstuhl (Livingstone 1986) to 0.75 in the sample from Narre Warren, Australia (Birch 1989). Ca commonly is the DEC, but Na dominates in samples displaying "herschelitic" and "phacolitic" habits (Passaglia 1970, Birch 1988, 1989; Wise and Kleck 1988, Akizuki et al. 1989, Vezzalini et al. 1994). K is dominant in chabazite from Vallerano, Italy (Passaglia 1970), and from Ercolano, Italy (de'Gennaro and Franco 1976). Sr-rich (0.50-0.80) samples are not rare, Ba is negligible, and Mg reaches the highest value (0.79) in the first occurrence of chabazite ("herschelite" habit) epitaxially overgrown by offretite (Passaglia and Tagliavini 1994).

In sedimentary samples from "open systems" and "geoautoclaves," the mean value of R is 0.71, ranging from 0.63 to 0.81. The lower values (0.63-0.67) have been found in crystals from phonolitic, leucititic, and melilitic ignimbrites; intermediate values (0.68-0.74) were determined for crystals from trachytic ignimbrites (de'Gennaro et al. 1980; Passaglia and Vezzalini 1985, Passaglia et al. 1990, Adabbo et al. 1994); and the highest value (0.81) was found in chabazite from a rhyolitic tuff (Sheppard and Gude 1970). K and Ca are, in order of frequency, the DEC, being commonly in similar amounts; Na, Mg, and Sr are consistently subordinate. In sedimentary samples from "closed systems," the mean value of R is 0.77, and it is close to 0.78 in crystals from rhyolitic tuff (Gude and Sheppard 1966, 1978; Sheppard and Gude 1973, Sheppard et al. 1978). It is definitely lower (0.72) in crystals from a nephelinite tuff (Hay 1964). Na is dominant in most "closed-system" samples.

Chabazite formed from the interaction of mafic glass with sea water in palagonitic basalt (Noack 1983, Passaglia et al. 1990) shows R ranging from 0.65 to 0.69 and Na prevailing over K and Ca. The sample from Palagonia (Italy) is an exception, with R of

0.62, Ca dominating, and a remarkable Sr content (0.47) (Passaglia et al. 1990). A sample occurring in "marine volcanics" from the Waitemata Group (New Zealand) has an R value of 0.71 and Na as the DEC (Sameshima 1978).

The unit-cell parameters (pseudo-hexagonal cell) are: a = 13.69-13.86 Å, c = 14.80-15.42 Å. A negative correlation between R and the c parameter was found in natural samples (Passaglia 1970). The unit-cell parameters of cation-exchanged samples showed significant variations primarily in c, dependent on R and on the occupancy of the extraframework sites (Passaglia 1978a, Alberti et al. 1982a).

Willhendersonite has been described as "trellis-like" twinned aggregates from a mafic potassic lava at S. Venanzo, Terni, Italy, in a limestone xenolith from a basalt at Mayen, Eifel, Germany (Peacor et al. 1984), and as vitreous rectangular laths from the breccia ridge of a melilitite plug at Colle Fabbri, Terni, Italy, (Vezzalini et al. 1997). It is associated with phillipsite, thomsonite, apophyllite, chabazite, gismondine, tobermorite, ettringite, thaumasite, calcite, aragonite, and vaterite. For all samples, R is near 0.50, and Ca is the DEC. The K content is very high in the samples from S. Venanzo and Mayen (1.80 and 1.48, respectively) and very low in the sample from Colle Fabbri (0.41). The structure is triclinic $P\bar{1}$ with a = 9.18-9.21 Å, b = 9.20-9.22 Å, c = 9.44-9.50 Å, α = 91.72-92.34°, β = 91.42-92.70°, γ = 90.05-90.12° (Tillmanns et al. 1984, Vezzalini et al. 1997).

Several other aspects of the crystal chemistry of chabazite and willhendersonite are noteworthy. For example, among the natural structure types, the tetrahedral framework CHA shows the highest variation in Si/Al ratio (R = 0.50-0.80), and there is only a narrow gap in R between willhendersonite (0.50) and the lowest value found in chabazite (0.58). Interestingly, crystals with R = 0.50 are entirely triclinic (Tillmanns et al. 1984), whereas crystals with R ranging from 0.66 to 0.72 showed submicroscopic triclinic domains (Mazzi and Galli 1983). These observations lead one to wonder about the degree of triclinicity of chabazite crystals with R lower than 0.66 and of hypothetical crystals with R lower than 0.58.

Structure Type: **DAC**

Dachiardite: $(Ca_{0.5},Na,K)_4[Al_4Si_{20}O_{48}]\cdot 18H_2O$

Dachiardite has been described from a dozen localities in fractures of an aplite pegmatite, in vugs of massive volcanic rocks (basalt, andesite, and porphyrite), and of a hydrothermally altered rhyolite. It is generally associated with mordenite, commonly with heulandite-clinoptilolite, and rarely with epistilbite, stilbite, ferrierite, erionite, phillipsite, yugawaralite, analcime, silica minerals (quartz, chalcedony), and carbonates (calcite, siderite, and ankerite). Most samples occur as simple fibers along b; but simple twinnings are not rare, and the holotype "octagonal beakers" from Elba Island, Italy, are polysynthetic twins of eight individuals (Berman 1925). Svetlozarite, a proposed new zeolite (Maleev 1976), has been interpreted as a multiply twinned and highly faulted dachiardite and was, hence, rejected (Gellens et al. 1982).

R has a mean value of 0.81 in dachiardite but ranges from 0.77 in crystals from Yellowstone National Park, Wyoming (Bargar et al. 1987), to 0.86 in crystals from Altoona, Washington (Wise and Tschernich 1978b). Ca and Na are the common DEC; the K content is highly variable ranging from 0.01 to 1.02 in crystals from Yellowstone National Park where it slightly prevails over Na and Ca (Bargar et al. 1987); Mg, Sr, and Ba are absent or negligible; Cs occurs (0.12) in crystals from Elba Island, Italy (Bonardi 1979, Vezzalini 1984). The Na/Ca ratio uniformly ranges within two extremes. Na-dominant (Na > Ca) crystals are from Alpe di Siusi, Italy (Alberti 1975a), Ogasawara Island, Japan (Nishido et al. 1979), Altoona, Washington, Agate Beach, Oregon (Wise and Tschernich 1978b), Montreal Island, Canada (Bonardi et al. 1981), and Yellowstone

National Park, Wyoming (Bargar et al. 1987). Ca-dominant (Na < Ca) crystals are from Elba Island, Italy (Gottardi 1960, Bonardi 1979, Vezzalini 1984), Cape Lookout, Oregon (Wise and Tschernich 1978b), Kagoshima and Ogasawara Island, Japan (Nishido and Otsuka 1981), Yellowstone National Park, Wyoming (Bargar et al. 1987), and in the rejected "svetlozarite" from Rhodopes, Bulgaria (Maleev 1976). Different samples from the same occurrence may exhibit very different extraframework-cation contents. For example, crystals from Ogasawara Island (Japan) may be either Na-dominant or Ca- dominant, and crystals from Yellowstone National Park (Wyoming) are generally Ca-dominant but are rarely Na-dominant or K-dominant.

The structure is monoclinic *C2/m* (Vezzalini 1984) with a = 18.62-18.73 Å, b = 7.49-7.54 Å, c = 10.24-10.31 Å, and β = 107.87-108.74°. No correlations between Na/Ca and Si/Al ratios were found, but β increases with Si/Al ratio, and refractive indices, $2V_x$, and $c \wedge Z$; all increase with Ca (Nishido and Otsuka 1981).

Structure Type: **EAB**

Bellbergite: $Ca_5(Sr,Ba)_3(Na,K)_2[Al_{18}Si_{18}O_{72}]\cdot30H_2O$

Bellbergite has been described only from Bellberg, Eifel (Germany), in cavities of Ca-rich xenoliths in a leucite-tephrite lava (Rüdinger et al. 1993). It occurs as dipyramids in association with thomsonite, ettringite, pyrrhotite, sanidine, and clinopyroxene. The empirical formula shows an R value of 0.50, Ca as the DEC, high amounts of K and Sr, and subordinate Na and Ba. The structure is hexagonal ($P6_3/mmc$, $P\bar{6}2c$, or $P6_3mc$) with a = 13.24 Å and c = 15.99 Å. Optically, it is uniaxial negative.

Structure Type: **EDI**

Edingtonite: $Ba_2[Al_4Si_6O_{20}]\cdot8H_2O$

Edingtonite has been described from five occurrences in hydrothermal veins in altered mafic rocks and metasediments, associated with sulfides (pyrite, chalcopyrite, galena, and sphalerite), carbonates (calcite and barytocalcite), hematite, hydroxides (goethite and manganite) and, in one occurrence, with analcime and harmotome (Novak 1970, Mazzi et al. 1984, Grice et al. 1984, Belitsky et al. 1986).

R is nearly constant (0.59-0.61), and Ba is the only DEC [Ba/(Ba+Ca+Mg+Na+K) = 0.91-0.99]. K is negligible (<0.10); Na, Ca, and Mg are absent or present in trace amounts; Sr is absent.

Structure refinements of crystals from different localities systematically show two different symmetries. Crystals from the type locality of Old Kilpatrick, Scotland, and from Ice River, Canada, have a disordered (Si, Al) distribution and are tetragonal $P\bar{4}2_1m$ with a = 9.58 Å, c = 6.52 Å (Mazzi et al. 1984). Crystals from Bohlet, Sweden, and Brunswick, Canada, have a nearly ordered (Si,Al) distribution and are orthorhombic $P2_12_12$ with a = 9.54 Å, b = 9.66 Å, c = 6.52 Å (Galli 1976, Kvick and Smith 1983). The (Si,Al) ordering and symmetry (tetragonal or orthorhombic) seem to vary from crystal to crystal from the same occurrence; different crystals from Ice River, Canada, are orthorhombic (Grice et al. 1984) or tetragonal (Mazzi et al. 1984).

Structure Type: **EPI**

Epistilbite: $Ca_3[Al_6Si_{18}O_{48}]\cdot16H_2O$

Epistilbite has been described in vugs of massive rocks along with many other zeolites (mordenite, heulandite, dachiardite, stilbite, scolecite, levyne, laumontite, and chabazite), gyrolite, pumpellyite, quartz, pyrite, and sphalerite. The characterized occurrences (about 20) are mainly from basalt in Iceland and Japan but are also from an aplite pegmatite at Elba Island (Italy), gneiss at Gibelsbach, Switzerland, and dolerite at Mt. Adamson, Antarctica

(Galli and Rinaldi 1974, Mehegan et al. 1982, Akizuki and Nishido 1988, Nishido 1994, Vezzalini et al. 1994).

The chemical composition shows only a slight variability. R has a mean value of 0.75, ranging from 0.73 to 0.77; Ca is the DEC; Na is present in subordinate amounts, reaching a maximum value (~1.0) in a few samples from Iceland. K is occasional or negligible (≤0.30), and Mg, Sr, and Ba are absent.

The structure is monoclinic *C*2 (Alberti et al. 1985) with a = 9.08-9.10 Å, b = 17.74-17.80 Å, c = 10.20-10.24 Å, and β = 124.55-124.68°. A crystal from Gibelsbach shows triclinic *C*1 symmetry (a = 9.08Å, b = 17.74Å, c = 10.21Å, α = 89.95°, β = 124.58°, and γ = 90.0°) due to partial (Si,Al) ordering accompanied by a preferred distribution of the extraframework cations (Yang and Armbruster 1996). Negative correlations between Ca and β and between R and Ca were reported (Galli and Rinaldi 1974). Data on twinnings and optical properties of crystals from Iceland and Japan are reported in Akizuki and Nishido (1988).

Structure Type: **ERI**

Erionite: $K_2(Ca_{0.5},Na)_7[Al_9Si_{27}O_{72}]\cdot 28H_2O$

Erionite from vugs in volcanic rocks (basalt, andesite, limburgite, dolerite, and latite-trachyte) and diagenetically altered vitroclastic rocks (sedimentary) have been investigated. Amygdaloidal crystals are associated with many other zeolites (chabazite, mordenite, stilbite, heulandite, phillipsite, analcime, paulingite, and harmotome), epitaxially intergrown with offretite (Rinaldi 1976, Wise and Tschernich 1976a), or overgrown on levyne (Passaglia et al. 1974). Sedimentary erionite coexists with analcime, chabazite, clinoptilolite, mordenite, and phillipsite, and it has formed in both continental (Staples and Gard 1959, Sheppard et al. 1965, Sheppard and Gude 1969b, Surdam and Eugster 1976, Boles and Surdam 1979, Gude and Sheppard 1981) and marine (Sameshima 1978) environments. In comparison with offretite, erionite is more siliceous, more alkalic, and optically positive (Sheppard and Gude 1969b, Wise and Tschernich 1976a, Rinaldi 1976). R shows a mean value of 0.75 and ranges from 0.68 to 0.79; K, Na, and Ca are the DEC; Mg is generally subordinate but highly variable (0.10-2.0); and Sr and Ba are negligible.

Amygdaloidal specimens show a mean R value of 0.75 which ranges from 0.70 in a sample from Yeongil area, Korea (Noh and Kim 1986), to 0.77 in a sample from Campbell Glacier, Antarctica (Vezzalini et al. 1994). Ca and K are generally the DEC, but Na may also be remarkably high (1.70); Mg is dominant (2.13) in a sample from Sasbach, Germany (Rinaldi 1976). Sedimentary specimens show a mean R value of 0.78 with small variations (0.76-0.79), but a sample from marine sediments (Sameshima 1978) is anomalously Si-poor (R = 0.74); Na and K are the DEC; and Ca and Mg are highly variable.

A general study on the crystal chemistry of erionite (Passaglia et al. 1998) reached the following conclusions: (1) R ranges from 0.68 to 0.79; (2) Ca is the most common DEC, but crystals with Na as the DEC are not rare; K is always present in considerable amounts (1.6-3.4) but is predominant only in a sample from Ortenberg, Germany; Mg is subordinate, reaching a maximum value (0.82) in a sample from Agate Beach (Oregon); (3) crystals epitaxially overgrown on levyne are quite frequent and are substantially poorer in Si and Mg than those associated with other zeolites; (4) the optic sign depends on the Si/Al ratio of the sample (positive in the Si-rich crystals, negative in the Si-poor crystals); and (5) epitaxial intergrowths with offretite are rare. The structure is hexagonal $P6_3/mmc$ with a = 13.19-13.34 Å and c = 15.04-15.22 Å. The unit-cell volume is negatively correlated with the Si/Al ratio (Sheppard and Gude 1969b, Passaglia et al. 1998).

Structure Type: **FAU**

Faujasite: $(Na,K,Ca_{0.5},Mg_{0.5})_{56}[Al_{56}Si_{136}O_{384}]\cdot 235H_2O$

Faujasite has been described from only a limited number of occurrences, including four in vugs in limburgite from Germany (Rinaldi et al. 1975, Wise 1982), three in vesicles of palagonitic mafic tuffs from San Bernardino County, California (Wise 1982), and an occurrence in basaltic tephra at Jabal Hannoun and Jabal Aritayn, Jordan (Ibrahim and Hall 1995). Faujasite commonly occurs with phillipsite but also occurs with offretite in Germany and chabazite in Jordan.

R has a mean value of 0.71 and generally ranges from 0.70 to 0.73; only a sample from Hasselborn, Germany (Wise 1982) is less siliceous (R = 0.68). Ca (average content: 14.2 atoms) and Na (average content: 13.7 atoms) are the DEC, and Mg (average content: 5 atoms) and K (average content: 3.8 atoms) are generally subordinate. All of these cations are quite variable even in different crystals from the same occurrence. For example, Ca ranges from 3.93 to 21.12 and is dominant in a sample from Germany analyzed by Wise (1982) and in three samples from Jordan (Ibrahim and Hall 1995); Na ranges from 7.12 to 22.00 and is dominant in two samples from Sasbach, Germany (Rinaldi et al. 1975), in the sample from San Bernardino County, California (Wise 1982), and in five samples from Jordan. Mg ranges from 1.53 to 11.12, and K ranges from 0.54 to 10.66. Sr is very low (0.32) and has been detected only in a couple of samples from Germany (Wise 1982), and Ba is absent. For both tetrahedral and extraframework contents, no differences are obvious between sedimentary and amygdaloidal specimens, although in the former, Na dominates more commonly. Two occurrences of amygdaloidal Na-rich crystals from basalts of Aci Castello and Aci Trezza (Italy) are reported in Gottardi and Galli (1985, R. Rinaldi's communication) and in Wise (1982), respectively. Unfortunately, no paper describing these occurrences has been published. The structure is cubic *Fd3m* with a = 24.60-24.78 Å.

Structure Type: **FER**

Ferrierite: $(Mg_{0.5},Na,K)_6[Al_6Si_{30}O_{72}]\cdot 20H_2O$

Ferrierite has been described mainly in vugs of volcanic rocks (basalt, andesite, porphyrite, and latite) coexisting with other zeolites (heulandite, dachiardite, mordenite, analcime, chabazite, and harmotome), carbonates (calcite, aragonite, and siderite), barite, apatite, pyrite, cinnabar, and chalcedony (Alietti et al. 1967, Wise et al. 1969, Yajima and Nakamura 1971, Wise and Tschernich 1976b, Birch and Morvell 1978, England and Ostwald 1978, Passaglia 1978b, Orlandi and Sabelli 1983, Sameshima 1986, Birch 1989). It has also been found in sedimentary rocks coexisting with mordenite, clinoptilolite, and quartz (Regis 1970, Noh and Kim 1986). Amygdaloidal samples are generally associated with Fe-hydroxides and calcite.

R shows a mean value of 0.83 and ranges from 0.78 in crystals from Unanderra, New South Wales, Australia (England and Ostwald 1978), to 0.88 in crystals from Agoura, California (Wise et al. 1969). Mg, Na, and K are, in order of frequency, the DEC. Ca is consistently subordinate, reaching a maximum value (~1.0) in a sample from Albero Bassi, Italy (Alietti et al. 1967); Sr and Ba are commonly absent and reach, respectively, maximum values of 0.23 in a sample from Unanderra, New South Wales, and of 0.45 in a sample from Silver Mountain, California (Wise and Tschernich 1976b). Compared with amygdaloidal samples, sedimentary ferrierite samples show a smaller range of R (0.81-0.84), have K as the DEC, and have low Ca and Mg contents.

Structure refinements of Mg-dominant samples from Silver Mountain, California (Gramlich-Meier et al. 1984), and Monastir, Italy (Alberti and Sabelli 1987), and of a Na-dominant, Mg-poor sample from Altoona, Washington (Gramlich-Meier et al. 1985), reveal that where Mg is dominant, the zeolite is orthorhombic *Immm* or *Pnnm*. Where Na is dominant, ferrierite is monoclinic $P2_1/n$. The orthorhombic unit-cell parameters are a = 18.90-19.45 Å, b = 14.12-14.28 Å, and c = 7.48-7.54 Å. The a parameter is negatively correlated with the Si content and positively correlated with the Mg content, which, in turn, is positively correlated with the Al content (Wise and Tschernich 1976b, Gramlich-Meier et al. 1985, Sameshima 1986). A decrease in R is correlated with an increase in Mg and a lengthened a parameter.

Structure Type: **GIS**

Gismondine: $Ca_4[Al_8Si_8O_{32}]\cdot16H_2O$

Garronite: $Ca_{2.5}Na[Al_6Si_{10}O_{32}]\cdot13H_2O$

Gobbinsite: $Na_4Ca[Al_6Si_{10}O_{32}]\cdot11H_2O$

Amicite: $Na_4K_4[Al_8Si_8O_{32}]\cdot10H_2O$

These four zeolite species share the same tetrahedral framework with TS $I4_1/amd$ which is lowered to the RS of the various species as a consequence of: (a) slight distortion of the framework; (b) tetrahedral Si-Al ordering; and (c) the distribution and coordination of the Ca, Na, and K cations and water molecules in the extraframework sites (Alberti and Vezzalini 1979, McCusker et al. 1985, Rinaldi and Vezzalini 1985, Artioli 1992, Artioli and Marchi 1999). These four zeolites all occur in vugs of volcanic rocks in association with phillipsite, in some cases as intimate intergrowths (Walker 1962a,b; Cortesogno et al. 1975, Vezzalini and Oberti 1984, Pöllmann and Keck 1990, Artioli and Foy 1994, Howard 1994). The DEC in all four species are Ca, Na, and K, with Mg, Sr, and Ba in very subordinate amounts or even absent. On the basis of the tetrahedral framework content, gismondine is related to amicite, both having R ≈ 0.50 and, hence, an ordered (Si,Al) distribution. Garronite is related to gobbinsite, both having R ≈ 0.63 and a disordered (Si,Al) distribution. On the basis of extraframework cation content, gismondine is related to garronite, both having Ca as the DEC; and amicite is related to gobbinsite, both having Na as the DEC.

Gismondine (the most common of the group) occurs in vugs of mafic volcanic rocks (basalt, leucitite, nephelinite, and tephrite), but also in fractures at the contact between basalt and metamorphosed serpentinite (Cortesogno et al. 1975, Argenti et al. 1986) and in a palagonitic nephelinite tuff (Iijima and Harada 1968). It occurs in association with thomsonite, chabazite, calcite, aragonite, apophyllite, gyrolite, but mainly with phillipsite, commonly in intimate intergrowths. Morphology, twinning, and optical orientation are reported by Nawaz (1980). R shows a mean value of 0.53 with a small range (0.50-0.54). Only one anomalously high value of R (0.58) has been found in crystals from Hoffil, Iceland (Walker 1962b). Ca is the DEC; Na and K are present in very subordinate amounts, commonly with K > Na in samples from leucitite and Na > K in samples from basalt and nephelinite. The highest Mg content (0.38) found in a sample from Bavaria, Germany, is possibly due to smectite impurities (Pöllmann and Keck 1993). Sr, as traces in samples from leucitite, reaches its maximum value (0.20) in a sample from Lubam, Poland (Vezzalini and Oberti 1984). The sample from Montalto di Castro, Italy, in association with Ca-silicates (vertumnite and tobermorite), has a composition which is very close the schematic formula (R = 0.51, only Ca as the extraframework cation) (Vezzalini and Oberti 1984), whereas the sample from Hofill, Iceland (Walker 1962b) has an unusually high R value (0.58) and Na content (0.90). The structure of gismondine is monoclinic $P2_1/c$ with unit cell dimensions nearly constant: a = 10.01-10.04 Å, b = 10.60-10.64 Å, c = 9.82-9.84 Å, and β = 92.23-92.56°.

The positive correlation (Vezzalini and Oberti 1984) in gismondine between (Na+K)/(Na+K+Ca) and Si/(Si+Al) (i.e. the monovalent cations, mainly Na, increase as the Si content increases) may yield the average chemical composition of *garronite*, where the increase in Si (R ≈ 0.63) is accompanied by a higher Na (and K) and a lower Ca content (~1 and 2.5, respectively). *Garronite* has an R value consistently higher than 0.60 (0.60-0.65), but its Ca content varies from 2.33 in a sample from Island Magee, County Antrim, Ireland, to 2.88 in a sample from Tardree Forest, County Antrim (Nawaz 1982), and its Na(+K) content varies from 0.10-0.30 in samples from Fara Vicentina, Italy (Passaglia et al. 1992), Goble, Oregon (Artioli 1992), and Table Mountain, Colorado (Kile and Modreski 1988), to 1.10-1.40 in samples from Ireland, Iceland (Walker 1962a, Nawaz 1982), and Halap Hill, Hungary (Alberti et al. 1982b). Therefore, a compositional gap between gismondine and garronite seems to exist in the tetrahedral framework content, but not in the extraframework cations.

Garronite occurs only in vugs of basalt in association with phillipsite, gonnardite, chabazite, thomsonite, analcime, and levyne. The morphology and twinning of the sample from Fara Vicentina, Italy, are reported in Howard (1994). The structure is monoclinic *I2/a* (pseudotetragonal $I\bar{4}m2$) with a = 9.88 Å, b = 10.28 Å, c = 9.876 Å, and β = 90.11° (Artioli and Marchi 1999).

Gobbinsite occurs in vugs of basalt at Gobbins and Island Magee, Ireland, associated with gmelinite, chabazite, gonnardite, garronite, and phillipsite (Nawaz 1982, Nawaz and Malone 1982) and at Iki Island, Japan, associated with cowlesite (Kuwano and Tokumaru 1993). In another occurrence at Magheramorne Quarry, Ireland, it is intimately associated with phillipsite (Artioli and Foy 1994). R is approximately constant (0.64); Na is consistently dominant and ranges from 3.61 in a sample from Island Magee (Nawaz 1982) to 5.63 in a sample from Iki Island, Japan. Consequently, Ca ranges from 0.93 to 0.17; Mg is appreciable (0.31) only in the sample from Gobbins (Nawaz and Malone 1982); and K is generally very low, although it is very high (2.11) in a sample from Island Magee (Mc Cusker et al. 1985) where both R (0.62) and Na (2.50) are anomalously low. The structure is orthorhombic $Pmn2_1$ (Mc Cusker et al. 1985) with a ≈ 10.10 Å, b = 9.77-9.80 Å, and c = 10.17 Å.

Amicite occurs in vugs of Na-rich rocks such as nephelinite in association with merlinoite, calcite, and aragonite at Hegau, Germany (Alberti et al. 1979), and ijolite-urtite pegmatite and apatite-bearing nephelinite associated with natrolite at Kola Peninsula, Russia (Khomyakov et al. 1984). The chemistry of specimens from the two occurrences is almost the same: R is about 0.50, Na and K are comparable (~4), and Ca is very low. The structure is monoclinic *I2* (Alberti and Vezzalini 1979) with a = 10.23 Å, b = 10.42 Å, c = 9.88 Å, and β = 88.32°.

Structure Type: **GME**

Gmelinite: $(Na,K,Ca_{0.5})_8[Al_8Si_{16}O_{48}]\cdot 22H_2O$

Gmelinite typically occurs in vugs of basalt mainly associated with analcime and natrolite but also with chabazite, phillipsite, heulandite, and stilbite. R shows a mean value of 0.69 and ranges from 0.65 in a sample from White Head, County Antrim, Ireland (Passaglia et al. 1978a), to 0.72 in a sample from Boron, California (Wise and Kleck 1988). Na is consistently abundant and is commonly the DEC. Ca is highly variable (0-2.5) and is dominant in the samples from Montecchio Maggiore, Italy, and from Great Notch, New Jersey (Passaglia et al. 1978a). K is generally the SEC, but it is the DEC in the samples from Fara Vicentina, Italy (Vezzalini et al. 1990), and from an alkalic rock at Kola Peninsula, Russia (Malinovskii 1984). Mg and Ba are absent or negligible. Where determined, the Sr content is appreciable (0.2-0.5) and reaches 1.35 in the Ca-rich, Na-poor sample from Montecchio Maggiore, Italy (Passaglia et al. 1978a). Large dipyramidal crystals show compositional zoning and are locally intergrown with subordinate chabazite (Passaglia et al. 1978a). Na-rich chabazite crystals from Hayata, Japan, have a "herschelite" habit and a gmelinite rim whose (0001) surface consists of a spiral growth pattern (Akizuki et al. 1989).

The structure is hexagonal $P6_3/mmc$ with a = 13.62-13.80 Å and c = 9.97-10.25 Å, the smaller a values and the larger c values occurring in the K-dominant samples (Malinovskii 1984, Vezzalini et al. 1990). The unit-cell parameters are correlated with the extraframework contents, with a increasing and c decreasing with increasing divalent cations (Passaglia et al. 1978a). Structure refinements of Na-, Ca-, and K-rich natural crystals (Galli et al. 1982, Vezzalini et al. 1990) show that the tetrahedral framework undergoes a deformation involving a lengthening of c and a shortening of a which follows the bond strengths of the extraframework cations (Ca > Na > K) in the large channels (C2 site). Structure refinements of Ca-, Na-, and K-exchanged crystals (Sacerdoti et al. 1995) show a marked lengthening of c and a minor shortening of a in the K-Na-Ca series ($c_K > c_{Na} > c_{Ca}$), correlated with both occupancy factor (100% in the K and Na forms and ca. 50% in the Ca form) and ionic radius of the cation in the gmelinite cage (C1 site).

Structure Type: **GOO**

Goosecreekite: $Ca_2[Al_4Si_{12}O_{32}]\cdot 10H_2O$

The only fully characterized goosecreekite sample is from the type locality at Goose Creek Quarry, Loudoun County, Virginia, where it occurs as euhedral, but highly curved, colorless crystals in vugs of an altered diabase in association with stilbite, quartz, chlorite, actinolite, albite, prehnite, apophyllite, babingtonite, and epidote (Dunn et al. 1980). The empirical chemical formula is very close to the schematic one and shows an R value of 0.75. Ca is the only extraframework cation. The structure is monoclinic $P2_1$ with a = 7.40 Å, b = 17.44 Å, c = 7.29 Å, and β = 105.44° from single-crystal measurements (Rouse and Peacor 1986) and with a = 7.52 Å, b = 17.56 Å, c = 7.35 Å, and β = 105.71° from powder X-ray diffraction data (Dunn et al. 1980).

Structure Type: **HEU**

Heulandite-Clinoptilolite: $(Na,K,Ca_{0.5})_7[Al_7Si_{29}O_{72}]\cdot 22H_2O$

Based on crystallochemical data (X-ray single-crystal photographs, chemical composition, and optical properties), Hey and Bannister (1934) demonstrated that clinoptilolite and heulandite are members of an isomorphous series and, hence, suggested that the "name clinoptilolite is unsuitable and should not be used." Numerous structural studies have revealed that heulandite and clinoptilolite share the same tetrahedral framework and monoclinic $C2/m$ symmetry (Alberti 1972, 1975b; Koyama and Takeuchi 1977). In the past, many efforts were spent to identify chemical and/or physical parameters to justify heulandite and clinoptilolite as two distinct zeolitic species. Clinoptilolite was proposed as a zeolite of the heulandite group with (Na+K) > Ca by Mason and Sand (1960) and with Si/Al > 4 by Boles (1972). A zeolite of the heulandite group with a structure which survives an overnight heating at 450°C was proposed to be named clinoptilolite by Mumpton (1960). The first proposed definition, the only one consistent with mineralogical nomenclature practice, was invalidated by the description of many samples with Ca > (Na+K) and Si/Al > 4 (Kirov 1965, Boles 1972, Sheppard and Gude 1973, Alberti 1975b, Wise and Tschernich 1976b, Alietti et al. 1977, Minato and Aoki 1978, Barrows 1980, Noh and Kim 1986, Altaner and Grim 1990, Tsolis-Katagas and Katagas 1990, Rice et al. 1992, Münch and Cochemé 1993). Studies of several heulandite-clinoptilolite samples revealed three distinct thermal behaviors that were correlated with the Si/Al and (Na+K)/(Ca+Mg+Sr+Ba) ratios (Alietti 1972, Boles 1972, Alietti et al. 1977). Such results, along with those obtained on cation-exchanged samples (Shepard and Starkey 1964, 1966; Alietti et al. 1974), clearly demonstrated that the thermal stability of a sample increased mainly with an increase in the (Na+K)/(Ca+Mg+Sr+Ba) ratio and, to a lesser extent, an increase of the Si/Al ratio. Although a statistical analysis showed the existence of at least five contiguous subgroups that can be distinguished on the basis of Ca (Hawkins 1974), a wealth of chemical data for heulandite-clinoptilolite samples corroborate the pioneering statement by Hey and Bannister (1934) because a consideration of both tetrahedral framework and extraframework cations shows a continuous solid solution along the join between the schematic formulae given in Gottardi and Galli (1985) for heulandite $(Na,K)Ca_4[Al_9Si_{27}O_{72}]$ and for clinoptilolite $(Na,K)_6[Al_6Si_{30}O_{72}]$. Many samples from the literature have been described as heulandite or clinoptilolite on the basis of chemical parameters and/or thermal behavior. Indeed, several definitions are based only on the occurrence, the name heulandite generally being adopted for samples in vugs of igneous rocks and clinoptilolite being used for samples in diagenetically altered vitroclastic sediments. In this review, heulandite and clinoptilolite will be described as a single species (referred to as heulandite-clinoptilolite), although the recommended nomenclature approved by the International Mineralogical Association (Coombs et al. 1997) distinguishes two species on the basis of the framework Si/Al ratio: heulandite has Si/Al < 4 and clinoptilolite has Si/Al > 4. A thorough discussion of clinoptilolite-heulandite nomenclature and chemistry is presented in the chapter by Bish and Boak (this volume).

Heulandite-clinoptilolite is a very common zeolite that has been described from a variety of environments. The zeolite occurs in vugs of massive plutonic (granite, pegmatite, and gabbro), volcanic (basalt, andesite, porphyrite, tephrite, latite, and dolerite) and metamorphic (gneiss) rocks where it is often associated with other zeolites (mordenite, dachiardite, ferrierite, chabazite, stilbite, epistilbite, laumontite, mesolite, scolecite, phillipsite, erionite-offretite, analcime, levyne, gmelinite, thomsonite, and natrolite), calcite, rhodochrosite, prehnite, apophyllite, datolite, quartz, chalcedony, opal, pyrite, chalcopyrite, searlesite, and borax. It is a common alteration product of vitroclastic sediments that have been diagenetically altered in subaerial ("open" and "closed" systems) and marine (deep-sea) environments in association with analcime, mordenite, ferrierite, erionite, phillipsite, chabazite, opal, cristobalite, clay minerals, and carbonates. Heulandite-clinoptilolite also occurs in the "zeolite facies" of low-grade metamorphic rocks.

R ranges from 0.73 in amygdaloidal crystals from basalt of Verandowana, India (Sukheswala et al. 1974), to 0.85 in a sedimentary sample from rhyolitic tuff of Cañadon Hondo, Argentina (Mason and Sand 1960). The monovalent (M = Na+K) and divalent (D = Ca+Mg+Sr+Ba) cations show all possible ratios: M/(M+D) ranges from 0.07 in sedimentary crystals from silicic volcanic sandstone of the Báucarit Formation, Mexico (Münch and Cochemé 1993) to 1.0 in crystals from a burial diagenetic clay-rich horizon of the chalk soil of Sangstrup Klint, Denmark (Nørnberg 1990). The Na/(Na+K) ratio ranges from 0 (Na-free) in the amygdaloidal crystals from Malpais Hill, Arizona (Wise and Tschernich 1976a), to 0.94 (almost K-free) in crystals from Yellowstone National Park, Wyoming (Bargar and Beeson 1981). Ca is generally the most common divalent cation. The Mg content is negligible or low depending on the genetic environment and reaches a high value of 1.46 in the sample from Maldon, Australia (Birch 1989). Sr is generally 0.40-0.50 and is especially high (2.10 and 1.24, respectively) only in amygdaloidal samples from Campegli, Italy (Lucchetti et al. 1982) and Kozakov Hill, Czechoslovakia (Cerny and Povondra 1969). Ba is commonly absent or negligible, but it reaches a value of 1.03 in a sedimentary sample from non-marine sandstone of the Blairmore Group, Canada (Miller and Ghent 1973).

Reliable chemical compositions of a variety of samples from different genetic environments (49 amygdaloidal, 73 subaerial and 39 deep-sea diagenetically altered sediments, and 18 from "burial diagenesis") show the following average values: R = 0.81, M/(M+D) = 0.60, Na/(Na+K) = 0.50, and Ca/(Ca+Mg+Sr+Ba) = 0.70. Amygdaloidalal samples are generally Si-poorer (R = 0.78) than samples from the other genetic environments (R = 0.81-0.82) but they show a high variability in the framework content (R = 0.73-0.83) as do samples from diagenetically altered subaerial sediments (R = 0.74-0.85). The framework contents of the samples from diagenetically altered deep-sea sediments and "burial diagenesis" environments are less variable (R = 0.77-0.84). Monovalent cations commonly are slightly prevalent over divalent cations but are slightly subordinate in amygdaloidal samples, although many crystals are very rich in alkalis (Wise et al. 1969, Alberti 1975b, Yoshimura and Wakabayashi 1977, Barnes et al. 1984, Wise and Kleck 1988). Moreover, M ≈ D in samples from diagenetically altered subaerial sediments, and M » D in samples from both deep-sea (Boles and Wise 1978, Sameshima 1978, Stonecipher 1978, Ogihara and Iijima 1990, Vannucci et al. 1992, Ogihara 1994) and "burial diagenesis" sediments (Iijima 1978, Minato and Aoki 1978, 1979; Minato et al. 1980, Ogihara and Iijima 1989, Tsolis-Katagas and Katagas 1990). Na and K are comparable, but Na is slightly prevalent in amygdaloidal crystals and slightly lower in the other samples where, however, Na-rich samples occur (Sheppard and Gude 1969a, Honda and Muffler 1970, Utada et al. 1972, Minato and Aoki 1978, Bargar and Beeson 1981, Ogihara and Iijima 1989, Vannucci et al. 1992, Ogihara 1994). Both the frequency and abundance of Mg are higher in sedimentary samples, but Sr and Ba are higher in amygdaloidal crystals. Unit-cell parameters show the following ranges: *a* = 17.62-17.74 Å, *b* = 17.81-18.05 Å, *c* = 7.39-7.53 Å, and β = 116.13-116.90°. Hawkins (1974) reported that an increase in R causes *b* to lengthen and *c* to shorten.

Structure Type: **LAU**

Laumontite: $Ca_4[Al_8Si_{16}O_{48}]\cdot18\text{-}14H_2O$

Laumontite is a widespread zeolite that occurs in many environments, including (a) in vugs of plutonic (granite, syenite, monzonite, and gabbro), volcanic (basalt, andesite, and porphyrite), metamorphic (gneiss, amphibolite, hornfels, marble, and skarn) and sedimentary (mudstone, and sandstone) rocks; (b) as a cement of plagioclase-rich sandstone; (c) as a precipitate from spring waters at atmospheric pressure and temperature (McCulloh et al. 1981); and (d) as an alteration product of vitroclastic sediments (tuff, and tuffite) in "burial diagenesis" or in low-grade metamorphism (zeolite facies). Laumontite coexists with other zeolites (natrolite, stilbite, heulandite, scolecite, and thomsonite), calcite, quartz, fluorite, plagioclase, thenardite, gypsum, corrensite, and chlorite.

The chemical composition is generally very close to the schematic formula. R shows a mean value of 0.67, ranging from 0.65 to 0.69. Ca is mainly the DEC inasmuch as the average content of Na and K is usually 0.3 and 0.2, respectively. A sample, originally designated "primary leonhardite" and occurring in fissures of porphyritic andesite at Kurtsy, Crimea, Ukraine, shows low Ca (2.15) and H_2O (13.5) contents and high Na (1.85) and K (1.85) contents (Stolz and Armbruster 1997). The notable content of Be (0.36) in a sample from an altered diabase at Toggiano, Italy (Gallitelli 1928), requires confirmation. Mg, Sr, and Ba are absent or negligible.

The structure (Artioli et al. 1989, Artioli and Ståhl 1993) is monoclinic *C*2/*m* with *a* = 14.69-14.89 Å, *b* = 13.05-13.17 Å, *c* = 7.53-7.61 Å, and β = 110-113°; unit-cell variations depend mainly on the degree of hydration. The uncertainty in the water content given in the schematic formula is due to partial reversible dehydration of the zeolite at room temperature. The resulting phase (named "leonhardite") shows slightly shorter unit-cell parameters and lower refractive indices (Coombs 1952, Pipping 1966) but retains the original symmetry (Artioli et al. 1989) and, hence, must be considered a variety of laumontite. The term "leonhardite" should be avoided as a species name.

Structure Type: **LEV**

Levyne: $(Ca_{0.5},Na)_6[Al_6Si_{12}O_{36}]\cdot 18H_2O$

Levyne has been described only from vugs of massive volcanic rocks (mainly basalt and andesite but also dolerite and tephritic-leucitite) associated with many other zeolites (analcime, gmelinite, natrolite, thomsonite, chabazite, gismondine, cowlesite, phillipsite, and heulandite). It can be epitaxially overgrown by erionite (Galli et al. 1981, Passaglia et al. 1974, Mizota et al. 1974, Wise and Tschernich 1976a) or by offretite (England and Ostwald 1979, Ridkosil and Danek 1983, Birch 1987, 1989; Kile and Modreski 1988). R shows a mean value of 0.66 and ranges from 0.62 to 0.70. The lowest value is in a sample from Oki Island, Japan (Tiba and Matsubara 1977), and the highest R is in a sample from Milwaukie, Oregon (Wise and Tschernich 1976a). Ca is generally the DEC, although Na-dominant samples are not rare (Birch 1979, 1989; Galli et al. 1981). Na is consistently present and prevails over K which is lower than 0.5. Mg, Sr, and Ba may be present in negligible amounts.

The structure is rhombohedral $R\bar{3}m$ (Merlino et al. 1975) with a = 13.32-13.43 Å and c = 22.66-23.01 Å. The monovalent-cation content (mainly Na) is positively correlated with the a parameter and negatively correlated with both the c parameter (Galli et al. 1981) and refractive indices (Tiba and Matsubara 1977).

Structure Type: **LOV**

Lovdarite: $Na_{13}K_4[Be_8AlSi_{27}O_{72}]\cdot 20H_2O$

Lovdarite occurs in association with natrolite in a pegmatite vein at Mount Karnasurt, Lovozero massif, Russia (Men'shikov et al. 1973). Crystallochemical data (Khomyakov et al. 1975) match those given in the original description showing: (a) in the tetrahedral framework, the presence of Be is remarkable, but (Si+Al) are prevalent, with R [Si/(Si+Al+Be)] equal to 0.75; (b) the extraframework content is almost exclusively Na and K in the ratio shown in the schematic formula; and (c) negligible amounts of Ca, Mg, and Ba were reported only in the original description. The structure is orthorhombic *Pma*2 with a = 39.58 Å, b = 6.93 Å, and c = 7.15 Å (Merlino 1990).

Structure Type: **LTL**

Perlialite: $K_9Na(Ca,Mg,Sr)[Al_{12}Si_{24}O_{72}]\cdot 16H_2O$

Only two occurrences of perlialite, both from Russian alkalic rocks, have been described from the Khibiny massif, Kola Peninsula (Men'shikov 1984) and from the Murun massif (Konev et al. 1986). The crystallochemical data for samples from both occurrences are very similar. R shows a mean value of 0.66 and ranges from 0.65 to 0.67 (Khibiny massif); K dominates (~9); Na is close to 1.0; and divalent cations add up to about 1.0, with Ca prevailing over Mg and Sr. Ba was determined in noticeable amounts (0.37) only in the sample from the Murun massif.

The crystal structure, determined on a Tl-rich (following heavy-liquid separation) sample, revealed hexagonal (*P*6/*mmm*) symmetry with a = 18.54 Å and c = 7.53 Å (Artioli and Kvick 1990). The unit-cell parameters given in the original descriptions are: a = 18.49-18.54 Å and c = 7.51-7.53 Å.

Structure Type: **MAZ**

Mazzite: $Mg_{2.5}K_2Ca_{1.5}[Al_{10}Si_{26}O_{72}]\cdot 30H_2O$

Mazzite occurs at Mt. Semiol (France) as clear, transparent hexagonal prisms that form radiating bundles, associated with offretite, phillipsite, chabazite, calcite, and siderite in vugs of an olivine basalt (Galli et al. 1974). The chemical composition reported in the original description is apparently unreliable because it has a large E value (+13.6%). A new electron microprobe analysis carried out on holotype crystals (G.Vezzalini, pers. comm.) is more reliable (E = +6.5%) and corresponds to: $Mg_{2.25}K_{2.18}Ca_{1.30}Na_{0.18}Ba_{0.02}$-$[Al_{10.13}Si_{26.03}O_{72}]\cdot 30H_2O$. R is 0.72, Mg is the DEC, K is high, and Ca is moderate. The structure is hexagonal $P6_3/mmc$ (Galli 1975) with a = 18.39 Å and c = 7.65 Å.

Structure Type: **MER**

Merlinoite: $K_6Ca_2Na[Al_{11}Si_{21}O_{64}]\cdot 22H_2O$

Merlinoite occurrences have been reported from massive volcanic rocks (Passaglia et al. 1977, Alberti et al. 1979, Khomyakov et al. 1981, Hentschel 1986, Della Ventura et al. 1993, Yakubovich et al. 1999), in silicic tephra that was diagenetically altered in a lacustrine environment (Hay and Guldman 1987), and in marine volcanic sediments (Mohapatra and Sahoo 1987). Nevertheless, reliable crystallochemical data have been obtained only from the samples at the type locality (Cupaello, Italy), where merlinoite is associated with phillipsite, chabazite, apophyllite, and calcite in vugs of a nepheline-melilitite (Passaglia et al. 1977), at Sacrofano, Italy, in cavities of a fassaite-rich ejectum (Della Ventura et al. 1993), and at Kola Peninsula, Russia in a pegmatite lode that slices nepheline syenite associated with cancrinite, sodalite and nepheline (Yakubovich et al. 1999). Samples from the three occurrences show different tetrahedral contents (R = 0.71 in the sample from Cupaello, 0.66 in the sample from Sacrofano, and 0.62 in the sample from Kola Peninsula). The extraframework content shows K as DEC. Ca and Ba are present in subordinate amounts only in the samples from Cupaello and Sacrofano. Na content is remarkable (3.16) in the sample from Kola Peninsula, otherwise quite low (0.55-0.67). Negligible amounts of Mg, Sr, and Mn were found only in the sample from Sacrofano. The structure is orthorhombic $Immm$ (Galli et al. 1979) with a = 13.86-14.14 Å, b = 14.13-14.23 Å, and c = 9.95-10.05 Å.

Structure Type: **MFI**

Mutinaite: $Na_3Ca_4[Al_{11}Si_{85}O_{192}]\cdot 60H_2O$

Mutinaite is a newly found zeolite in a few spheroidal cavities of Ferrar dolerite at Mt. Adamson, Antarctica (Galli et al. 1997a). It occurs as subspherical aggregates of tiny radiating lath-like fibers or as aggregates of transparent, colorless to pale-milky, tiny tabular crystals, in association with heulandite and, very rarely, with terranovaite and tschernichite. The empirical formula $(Na_{2.76}K_{0.11}Mg_{0.21}Ca_{3.78})[Al_{11.20}Si_{84.91}O_{192}]\cdot 60H_2O$ shows the highest R value (0.88) found in a natural zeolite to date, Ca fairly prevalent over Na, and Mg and K very subordinate. The structure is orthorhombic $Pnma$ with a = 20.22 Å, b = 20.05 Å, and c = 13.49 Å.

Structure Type: **MON**

Montesommaite: $K_9[Al_9Si_{23}O_{64}]\cdot 10H_2O$

Montesommaite has been described from Pollena, Monte Somma, Italy, where it occurs as colorless, dipyramidal crystals associated with chabazite, natrolite, dolomite, and calcite in vesicles of volcanic scoria (Rouse et al. 1990). The chemical composition is very close to the schematic formula, showing an R value of 0.70 and a very low Na content (0.16). The structure is orthorhombic $Fdd2$, pseudo-tetragonal ($I\bar{4}2d$ or $I4_1md$) with a = 7.14 Å and c = 17.31 Å. On the basis of symmetry, unit-cell parameters, and chemical composition, the zeolite is presumed to be related to merlinoite and members of the gismondine group (Rouse et al. 1990).

Structure Type: **MOR**

Mordenite : $Na_3Ca_2K[Al_8Si_{40}O_{96}]\cdot 28H_2O$

Maricopaite : $Pb_7Ca_2[Al_{12}Si_{36}(O,OH)_{100}]\cdot n(H_2O,OH)$

Mordenite and maricopaite share the same orthorhombic structure type, but they have different space groups (*Cmc*2_1 in mordenite, *Cm*2*m* in maricopaite). Moreover, maricopaite has a so-called "interrupted" framework (see Introduction).

Mordenite has been described in vugs of volcanic (andesite, basalt, dolerite, porphyrite, and rhyolite) and intrusive (granite) rocks (Sukheswala et al. 1974, Passaglia 1975, de'Gennaro et al. 1977, Wise and Tschernich 1978b, Nativel 1986, Vezzalini et al. 1994), and as a diagenetic product of silicic tuff (Sheppard and Gude 1969a, Noh and Kim 1986, Sheppard et al. 1988, Tsolis Katagas and Katagas 1989, Pe-Piper and Tsolis Katagas 1991, Passaglia et al. 1995), pitchstone (Harris and Brindley 1954), and volcanic marine sediments (Bushinsky 1950, Vitali et al. 1995). Amygdaloidal samples are associated with other zeolites (analcime, chabazite, heulandite, dachiardite, stilbite, natrolite, scolecite, epistilbite, ferrierite, erionite, and thomsonite), silica minerals (quartz, cristobalite, and chalcedony), calcite, apophyllite, okenite, and barite. In diagenetically altered volcanic sediments, mordenite occurs in association with clinoptilolite, chabazite, erionite, phillipsite, ferrierite, analcime, tridymite, cristobalite, opal, K-feldspar, and smectite. It is also commonly associated with silica minerals. Many amygdaloidal samples are described as red compact masses with a radial, fibrous structure, the fibers being surrounded by microcrystalline quartz and hematite (Passaglia 1975).

Amygdaloidal and sedimentary samples show the same mean value of R (0.83) with almost the same ranges (0.81-0.86 in amygdaloidal samples, 0.80-0.85 in the sedimentary ones). The lowest R value (0.80) was found in a sample from volcanic marine sediments at Tonga Trench, Southwest Pacific (Vitali et al. 1995); the highest R value (0.86) is in a sample from basalt at Kirkee, India (Passaglia 1975). The extraframework cations are mainly Na, Ca, and K, generally with Na > Ca > K. Ca dominates in the samples from porphyrite at Mt. Civillina, Italy (Passaglia 1975), from basalt at Deccan, India (Sukheswala et al. 1974), and from a diagenetically altered rhyolite at Samos, Greece (Pe-Piper and Tsolis-Katagas 1991). K dominates in an amygdaloidal sample from a pegmatite at Elba Island, Italy (Alberti et al. 1986), and in sedimentary samples from silicic tuffs at Yucca Mountain, Nevada (Sheppard et al. 1988), and near Yeongil, Korea (Noh and Kim 1986). Ca and Na are the same (2.41) in a sample from basalt at Cilaos, Ile de la Reunion (Nativel 1986). The Mg content is commonly low, showing higher values in the sedimentary samples. Sr is generally absent or negligible, although it is relatively high (0.45) in samples from basalts at Penticton, British Columbia ("ashtonite" of Reay and Coombs 1971), and at Cilaos, Ile de la Reunion (Nativel 1986). Ba is generally absent or in traces, but it reaches a remarkable value (0.43) in a sample from basalt at Leucois Paulista, Brazil (Passaglia 1975). The amygdaloidal samples have a mean R comparable with that of the sedimentary samples but are slightly richer in Na and Ca and poorer in K and Mg.

The structure is orthorhombic *Cmc*2_1 (Alberti et al. 1986) with a = 18.05-18.25 Å, b = 20.35-20.53 Å, and c = 7.49-7.55 Å. The b parameter is negatively correlated with R (Passaglia 1975).

Maricopaite has been found only at the Moon Anchor Mine, Tonopah, Maricopa County, Arizona (Peacor et al. 1988). It occurs as sprays of translucent, white, acicular crystals coating and filling fractures in quartz, with mimetite in a vein of a calcite-fluorite gangue. The tetrahedral framework is only partially occupied by Si and Al (R = 0.75), and the extraframework cations are exclusively Pb and Ca, showing amounts close to those given in the schematic formula. Both density (2.94 g/cm^3) and refractive indices (α = 1.56, β = 1.58, γ = 1.59) are higher than those generally found in zeolites but are consistent with the Pb content of the mineral.

The structure (Rouse and Peacor 1994) is orthorhombic *Cm*2*m* with a = 19.43 Å, b = 19.70 Å, and c = 7.54 Å. The tetrahedral framework is "interrupted" by (OH) groups with 17% of TO_4 three-fold connected. $Pb_4(O,OH)_4$ tetrahedral clusters occupy and obstruct cruciform channels. Maricopaite is similar to roggianite and partheite in that it exhibits a framework interrupted by (OH), but it is the only zeolite having Pb as the extraframework cation.

Structure Type: **NAT**

Natrolite: $Na_{16}[Al_{16}Si_{24}O_{80}]\cdot16H_2O$

Mesolite: $Na_{16}Ca_{16}[Al_{48}Si_{72}O_{240}]\cdot64H_2O$

Scolecite: $Ca_8[Al_{16}Si_{24}O_{80}]\cdot24H_2O$

Gonnardite-Tetranatrolite: $Na_{12}Ca_{2.5}[Al_{17}Si_{23}O_{80}]\cdot20H_2O$ (Z = 1/2)

These zeolites share the same tetrahedral framework with TS $I4_1/amd$ which lowers to $I\bar{4}2d$ by an overall rotation of the structural chains by an angle of about 24°. The RS of the zeolite species is determined by the (Si,Al) distribution and occupancy of the extraframework cations and water molecules. The completely ordered (Si,Al) distribution in mesolite, scolecite, and most of the natrolite samples lowers their RS to orthorhombic *Fdd*2, which, in turn, is lowered to monoclinic *F*1*d*1 in scolecite because the sites occupied by Na in natrolite are alternatively occupied by Ca and H_2O in scolecite (Peacor 1973, Pechar 1981, Pechar et al. 1983, Artioli et al. 1984, 1986; Kirfel et al. 1984, Smith et al. 1984). An orthorhombic structure is retained in a few natrolite samples with a partially disordered (Si,Al) distribution (Alberti and Vezzalini 1981, Hesse 1983, Krogh Andersen et al. 1990, Alberti et al. 1995). A strongly disordered (Si,Al) distribution in gonnardite-tetranatrolite prevents their RS from departing from tetragonal $I\bar{4}2d$ (Chen and Chao 1980, Mazzi et al. 1986, Mikheeva et al. 1986, Pechar 1989, Artioli and Torres Salvador 1991). No sound chemical or crystallographic parameter was found to discriminate gonnardite from tetranatrolite; thus, because of time priority, it was suggested to retain gonnardite as a mineral species and to discredit tetranatrolite (Artioli and Galli 1999). Alberti et al. (1982c) assessed the compositional fields of natrolite, mesolite, and scolecite, and found a strong correlation between the unit-cell parameters and the degree of disorder in the Si-Al distribution. A crystallochemical classification of the minerals of the natrolite group, including the non-isostructural mineral thomsonite, was proposed on the basis of new chemical data and data taken from the literature (Nawaz 1988, Ross et al. 1992). Unusual crystallization conditions (high temperature and H_2O pressure) favor the formation of disordered gonnardite-tetranatrolite instead of ordered natrolite. In this context, para-natrolite (Chao 1980) is interpreted as a highly hydrated phase, which under room conditions dehydrates into disordered gonnardite-tetranatrolite (Alberti et al. 1995).

Natrolite is the most common among the zeolites of this group and has been described mainly from vugs of massive plutonic (nepheline syenite, pegmatite, and diorite), volcanic (basalt, phonolite) and metamorphic (gneiss, marble) rocks. It occurs in association with analcime, phillipsite, chabazite, garronite, gismondine, stilbite, gmelinite, pectolite, apophyllite, prehnite, calcite, and aragonite, and also in intergrowths with other "fibrous" zeolites (mesolite, thomsonite, and gonnardite). The chemical composition is very close to the schematic formula. R generally ranges from 0.60 to 0.61, being slightly lower (0.58-0.59) in a few samples (Cortesogno et al. 1975, Pechar et al. 1983, Kile and Modreski 1988, Pechar and Rykl 1989, Ross et al. 1992, Alberti et al. 1995, Meneghinello et al. 1999), which, according to Lowenstein's rule, are partially (Si, Al) disordered (Alberti et al. 1995). The negative charge of the framework is balanced by Na and very small amounts of Ca. K, Mg, Sr, and Ba are absent or negligible ($<$0.5). Crystals with appreciable $Ca\Leftrightarrow Na_2$ substitutions are not rare. In fact, relatively high Ca contents (1-3) and consequently relatively low Na contents (14-10) were found in samples from Voltri, Italy

(Cortesogno et al. 1975), and from several localities in Czechoslovakia (Pechar et al. 1983, Pechar and Rykl 1989). Crystals closely intergrown with gonnardite from Palagonia, Italy, and from Uzabanya, Hungary, have the highest degree (~50%) of disordered (Si,Al) distribution ever found in orthorhombic natrolites (Alberti et al. 1995). The structure is orthorhombic *Fdd*2 and the unit-cell parameters (a = 18.28-18.42 Å, b = 18.56-18.70 Å, and c = 6.46-6.61 Å) do not show any correlations with the chemistry. The (b-a) difference decreases and c increases with increasing (Si,Al) disorder (Alberti and Vezzalini 1981). Correlations between optical properties and crystal chemistry are reported in Gunter and Ribbe (1993).

Mesolite occurs mainly in vugs of massive volcanic rocks (andesite, basalt, and porphyrite) in association with analcime, natrolite, scolecite, thomsonite, chabazite, levyne, laumontite, erionite, and heulandite. R generally shows only small deviations (0.60-0.62) from the theoretical value (0.60) and is slightly lower (0.58-0.59) in crystals occurring in fissures of contact metamorphosed marls in the proximity of a phonolitic intrusion at Horni Jilové, Czechoslovakia (Rychly and Ulrych 1980), and in samples from Carlton Peak, Minnesota (Nawaz et al. 1985), Dunseverick, Ireland (Nawaz 1988), Table Mountain, Colorado (Kile and Modreski 1988), and Coastal Ranges, Oregon (Keith and Staples 1985). Na and Ca are the DEC, and K, Mg, Sr, and Ba are absent or negligible. Most samples have a Na/(Na+Ca) ratio (0.48-0.52) close to the theoretical value (0.50), although some samples with Ca > Na [Na/(Na+Ca) = 0.45-0.47] also have been described (Sukheswala et al. 1974, Alberti et al. 1982c, Artioli et al. 1986). A sample with Ca » Na [Na/(Na+Ca) = 0.41] occurs at Horni Jilové, Czechoslovakia (Rychly and Ulrych 1980). Crystals with Na > Ca have been found in tholeiitic basalt of the Coastal Ranges, Oregon (Keith and Staples 1985), and in altered volcanogenic sediments of Baja California, Mexico [Na/(Na+Ca) = 0.56 and 0.61, respectively] (Barnes et al. 1984). Inasmuch as the structure of mesolite is built by the alternation of one natrolite plane with two scolecite planes, the b parameter is three times larger than that for natrolite and scolecite. The structure is orthorhombic *Fdd*2 with unit-cell parameters showing narrow ranges (a = 18.34-18.44 Å, b = 56.52-56.69 Å, and c = 6.52-6.56 Å) and uncorrelated with chemistry (Alberti et al. 1982c).

Scolecite occurs in vugs of mainly massive volcanic rocks (basalt, andesite, and dolerite), rarely of metamorphic rocks (gneiss, and skarn), and amphibole gabbro. It has been described in association with other zeolites (natrolite, mesolite, thomsonite, heulandite, stilbite, laumontite, chabazite, levyne, and epistilbite), prehnite, calcite, quartz, and pyrite (Kuwano 1977, Alberti et al. 1982c, Johnson et al. 1983, Yamazaki and Otsuka 1989, Ross et al. 1992, Vezzalini et al. 1994). The chemistry is close to the schematic formula: R ranges from 0.60 to 0.61; Ca is by far the DEC; Na is consistently subordinate; and K as well as Mg, Sr, and Ba are negligible or absent. Exceptionally Na-rich (0.87-1.54), and consequently Ca-poor (7.10-7.35), crystals occur in the skarn area of the Italian Mountains, Colorado, basalt of Antrim, Ireland (Alberti et al. 1982c), dolerite of Mt. Adamson, Antarctica (Vezzalini et al. 1994), and andesite of Makinokawa, Japan (Kuwano 1977). The structure is monoclinic *F*1*d*1 with unit-cell dimensions (a = 18.50-18.55 Å, b = 18.90-18.97 Å, c = 6.52-6.53 Å, and β = 90.44-90.65°) correlated with the extra-framework cation contents (Alberti et al. 1982c).

Gonnardite-Tetranatrolite. On the whole, 26 occurrences from 22 different localities have been reported. Out of these, eleven samples have been described as tetranatrolites (Krogh Andersen et al. 1969, Guseva et al. 1975, Chen and Chao 1980, Alberti et al. 1982b, Mikheeva et al. 1986, Nawaz 1988, Pechar 1989, Ross et al. 1992), the other ones as gonnardites (Meixner et al. 1956, Harada et al. 1967, Iijima and Harada 1968, Alberti et al. 1982b,c; Mazzi et al. 1986, Nawaz 1988, Ciambelli et al. 1989, Passaglia et al. 1990 1992, Ross et al. 1992). This zeolite occurs in fractures of pegmatite dikes and in cavities of both massive volcanic rocks (tholeiitic and nepheline basalt, and ijolite) and palagonitic tuff and basalt usually in association with other fibrous zeolites (thomsonite, natrolite), but also with analcime, garronite, phillipsite, chabazite, and calcite. R displays a mean value of 0.57, ranging from 0.52 in the sample from Magnet Cove, Arkansas (Ross et al. 1992) to 0.63 in the sample from Lovozero massif, Russia (Guseva et al. 1975, Rastsvetaeva 1995). Only four samples which were all described as tetranatrolites (Krogh Andersen et al. 1969, Guseva et al. 1975, Chen and Chao 1980, Pechar 1989) show R > 0.60, i.e. Si/Al > 1.5. Na is the DEC and Ca the SEC. The Na/(Na+Ca) ratio has a mean value of 0.84, ranging from 0.65 in the sample from Aci Castello, Sicily (Meixner et al. 1956) to 1 in the sample from Lovozero massif, Russia (Guseva et al. 1975). K is usually found in negligible amounts (<0.10), although it surprisingly reaches a value of 0.88 in the sample from the Khibinsk alkali massif, Russia (Mikheeva et al. 1986). Mg and Ba are absent or negligible. Sr, usually absent, show the maximum value (0.11) in a sample from Mont Saint-Hilaire, Canada (Ross et al. 1992). The water content is on average about 20 molecules, but is higher (about 23 molecules) in the samples described as gonnardites than (about 17 molecules) in the samples described as tetranatrolites. A positive correlation between Na/Ca and Si/Al ratios has been postulated (Passaglia et al. 1992). The structure is tetragonal $I\bar{4}2d$ with a = 13.04-13.27 Å, and c = 6.58-6.64 Å.

Structure Type : **NES**

Gottardiite: $Ca_5Na_3Mg_3[Al_{19}Si_{117}O_{272}]\cdot 93H_2O$

Gottardiite occurs as thin pseudo-hexagonal lamellae in subparallel or wedge-shaped aggregates in vugs lined by Fe-smectite in dolerite at Mt. Adamson, Antarctica (Galli et al. 1996). The empirical formula $(Na_{2.5}K_{0.2}Mg_{3.1}Ca_{4.8})[Al_{18.8}Si_{117.2}O_{272}]\cdot 93H_2O$ shows a high R value (0.86) and Ca, Mg, and Na contents close to the schematic formula. The structure is orthorhombic *Cmca* with a = 13.70 Å, b = 25.21 Å, and c = 22.66 Å (Alberti et al. 1996).

Structure Type: **OFF**

Offretite: $CaMgK[Al_5Si_{13}O_{36}]\cdot 16H_2O$

Offretite has been characterized from a dozen occurrences in vugs of massive volcanic rocks (basalt, limburgite, and cornubianite) in association with mazzite, chabazite, levyne, erionite, phillipsite, and faujasite. Epitaxial intergrowths with erionite (Pongiluppi 1976, Rinaldi 1976, Wise and Tschernich 1976a), and overgrowths on levyne (Sheppard et al. 1974, Wise and Tschernich 1976a, England and Ostwald 1979, Kile and Modreski 1988, Birch 1989) and chabazite (Passaglia and Tagliavini 1994, Passaglia et al. 1996) have been described. Compared with erionite, offretite is typically less siliceous, richer in alkaline earths, and optically negative (Sheppard and Gude 1969b, Rinaldi 1976, Wise and Tschernich 1976a). R shows a mean value of 0.70 with small variations (0.69-0.72), and Ca is the DEC in most samples apart from two occurrences in Australia (Birch 1988 1989) which are Na-dominant. Crystals from Mont Semiol, France (the type locality), and from Sasbach (Germany) are Mg dominant. Mg and Ca are comparable in samples from Adamello, Italy (Passaglia and Tagliavini 1994), and Fittà, Italy (Passaglia et al. 1996). K is consistently close to 1. Sr and Ba are absent. A study of the crystal chemistry (Passaglia et al. 1998) allowed the following conclusions: (1) R ranges from 0.69 to 0.74; (2) Ca, Mg and K are in comparable amounts (~1), and Na, Sr and Ba are generally negligible or absent; (3) in clear disagreement with previous reports, epitaxial overgrowths are found only on chabazite and not on levyne; (4) epitaxial intergrowths with erionite are rare; and (5) the optic sign depends on the Si/Al ratio of the sample (positive in Si-rich crystals, negative in Si-poor crystals). The structure is hexagonal $P\bar{6}m2$ with a = 13-27-13.32 Å and c = 7.56-7.61 Å.

Structure Type: **PAR**

Partheite: $Ca_8[Al_{16}Si_{16}O_{60}(OH)_8]·16H_2O$

Partheite has been described from the Taurus Mts., Turkey, occurring in rodingitic rocks associated with thomsonite and prehnite (Sarp et al. 1979), and from Denezhkin Kamen (Urals), occurring in a gabbro-pegmatite (Ivanov and Mozzherin 1982). The chemical compositions from the two occurrences are very similar and are close to the schematic formula: R averages 0.51, and Ca is essentially the only extraframework cation inasmuch as Na, K, and Mg are negligible. The structure is characterized by an interrupted tetrahedral framework with unshared (OH) vertices at every second AlO_4 tetrahedra. The symmetry is monoclinic *C*2/*c* with a = 21.55-21.59 Å, b = 8.76-8.78 Å, c = 9.30-9.31 Å, and β = 91.47-91.55° (Sarp et al. 1979, Engel and Yvon 1984).

Structure Type: **PAU**

Paulingite: $K_{58}Ca_{42}Na_{10}Ba_{10}[Al_{172}Si_{500}O_{1344}]·490H_2O$

Paulingite has been described only in five localities: three in the USA, one in British Columbia, Canada (Tschernich and Wise 1982), and one in the Czech Republic (Lengauer et al. 1997). It occurs in vugs of basalt mainly associated with phillipsite, erionite-offretite and smectite, but also with chabazite, heulandite, harmotome, calcite, and pyrite. The schematic formula was derived from the average of the available chemical compositions which have a very large variability. The tetrahedral framework content is almost constant (R = 0.73-0.74 in four samples and 0.77 in the holotype sample from Rock Island Dam, Washington). As far as the extraframework content is concerned, K > Ca except in the samples from Ritter, Oregon, and Vinarická Hora, Czech Republic, where Ca is the dominant cation. Na ranges from 6.08 to 15.2, and Ba ranges from 1.60 to 22.24. Mg and Sr were found in crystals from Riggins, Idaho, and in relatively high amounts (1.74 and 1.44, respectively) in crystals from Chase Creek, British Columbia. The structure is cubic *Im*3*m* (Gordon et al. 1966) with a = 35.05-35.12 Å.

Structure Type: **PHI**

Phillipsite: $K_2(Na,Ca_{0.5})_3[Al_5Si_{11}O_{32}]·12H_2O$

Harmotome: $Ba_2(Na,Ca_{0.5})[Al_5Si_{11}O_{32}]·12H_2O$

On the basis of structural and crystallochemical data, phillipsite and harmotome can be described as a single zeolite species. In both structures, the orthorhombic *Bmmb* TS is lowered to monoclinic $P2_1/m$ as a result of a slight distortion induced on the tetrahedral framework by the presence of larger cations (K and Ba) in the two-fold extraframework sites. The other extraframework site is four-fold and accommodates the smaller Ca and Na cations (Rinaldi et al. 1974). The variation in Si/Al, K/Ba, and Na/Ca suggests the absence of a compositional gap, and the unit-cell parameters vary as a function of the Si/Al and (Na+K)/(Na+K+Ca+Ba) ratios. R ranges from 0.57 to 0.77, K/(K+Ba) ranges from 0.03 to 1, Na/(Na+Ca) ranges from 0.02 to 1, and other extraframework cations (Mg, Sr) are absent or very low. The unit-cell parameter (pseudo-orthorhombic setting) ranges are: a = 9.81-10.01 Å, b = 14.10-14.34 Å, and c = 14.16-14.42 Å. The existence of a continuous isomorphous series between phillipsite and harmotome, suggested by Cerny et al. (1977), has been reinforced by later descriptions of samples with intermediate compositions (Tschernich and Wise 1982, Passaglia and Bertoldi 1983, Robert 1988, Hansen 1990, Armbruster et al. 1991). The names phillipsite and harmotome are retained in present nomenclature, and, in light of the above considerations, samples with K/(K+Ba) > 0.50 (i.e. K > Ba) will be descibed as phillipsite, whereas samples with K/(K+Ba) < 0.50 (i.e. Ba > K) will be described as harmotome. The term "wellsite," commonly utilized for Ba-rich phillipsites, is not used. On the basis of structural data, the extraframework two-fold site can be almost completely occupied by K and Ba, and the four-fold site can be partially occupied by Na and Ca. These trends occur in all harmotomes and in most phillipsites, although some K-rich (3-4 atoms p.f.u) phillipsites must accommodate some K in the four-fold site, and the existence of many K-poor (<1 atoms p.f.u.) phillipsites may indicate the presence of Ca and Na also in the two-fold site.

Phillipsite is much more common than harmotome and has been described from vugs of massive volcanic rocks (basalt, nephelinite, leucitite, and melilite), as an alteration product of volcanic glass in palagonitic basalt and tuff ("hyaloclastites"), and in vitroclastic sediments diagenetically altered in continental ("closed," "open," and "geoautoclave" systems) and deep-sea environments. R has an average value of 0.69 and ranges from 0.57 to 0.77. K/(K+Ba) has an average value of 0.99 but ranges from 0.53 to 1, and Na/(Na+Ca) shows an average of 0.6 with a range from 0.02 to 1. Amygdaloidal samples occur as twinned pseudo-orthorhombic prisms or radiating fibrous aggregates with many other zeolites, carbonates, and silica minerals. Intergrowths with gismondine are common (Vezzalini and Oberti 1984, Pöllmann and Keck 1990) but intergrowths with garronite (Alberti et al. 1982b) and with gobbinsite (Artioli and Foy 1994) are rare. R has a mean value of 0.66 and ranges from 0.57 in crystals associated with gismondine and Si-poor chabazite from the leucitite of Vallerano, Italy (Galli and Loschi Ghittoni 1972), to 0.76 in crystals associated with erionite, clinoptilolite, mordenite, and dachiardite in basalt of Cape Lookout, Oregon (Wise and Tschernich 1978b). The K/(K+Ba) ratio shows a mean value of 0.95 and ranges from 1 in most samples to 0.53 in crystals associated with heulandite, gmelinite, analcime, laumontite, and prehnite from igneous rocks of Kurtzy, Crimea (Cerny et al. 1977). The Na/(Na+Ca) ratio averages 0.40 and ranges from 0.02 (almost Na-free) in crystals associated with chabazite from basalt of Ardeche, France (Robert et al. 1988), to 1 (Ca- and K-free) in crystals associated with Na-rich chabazite, gmelinite, clinoptilolite, searlesite, borax, and calcite in basalt of Boron, California (Wise and Kleck 1988). Phillipsites from "hyaloclastites" display an average R value of 0.65, varying from 0.62 in crystals associated with gismondine and chabazite in palagonitic tuff of Oahu, Hawaii (Iijima and Harada 1968), to 0.73 in crystals associated with chabazite in palagonitic trachybasalt of Vivara, Italy (Passaglia et al. 1990). All samples are almost Ba-free [K/(K+Ba) = 0.98-1], and the Na/(Na+Ca) ratio is high (0.8) in crystals associated with analcime, gmelinite, chabazite, gyrolite, thomsonite, and thaumasite in palagonitic basalt of the Mururoa Atoll, South Pacific (Noack 1983), and in crystals from Vivara (Italy). The Na/(Na+Ca) ratio is intermediate (0.3) in crystals from Hawaii and low (0.1) in crystals associated with chabazite and gonnardite in palagonitic basalt of Palagonia, Sicily (Passaglia et al. 1990). Diagenetic phillipsites from "closed systems" have R values ranging from 0.76 to 0.77 for crystals in altered rhyolitic tuff (Hay 1964, Sheppard and Gude 1968, 1969a; Surdam and Sheppard 1978) and lower values (0.71-0.73) in crystals from nepheline phonolitic and trachytic tuffs at Olduvai Gorge, Tanzania (Hay 1964 1980). Ba is absent, and the K content is close to 2, except for crystals from China Lake, California (Hay 1964), and near Barstow, California (Sheppard and Gude 1969a), where it is about 1. Na greatly prevails over Ca [Na/(Na+Ca) = 0.83-1.0]. Diagenetic phillipsites from "open systems" (mainly "geoautoclaves") exhibit an R (average value = 0.72) ranging from 0.70 to 0.74 in crystals from phonolitic-trachytic pyroclastite. A lower value (0.68) is for the sample from trachytic ash-flow tuff of the Canary Islands (Garcia Hernandez et al. 1993), and a higher value is for the sample from rhyodacitic cinerite of Garbagna, Italy (Passaglia and Vezzalini 1985). Ba is absent or negligible [K/(K+Ba) = 0.95-1]. The Na/Ca ratio is extremely variable: Na»Ca in samples from the Neapolitan Yellow Tuff, Italy (Passaglia et al. 1990), from the Canary Islands, and in most specimens from Eifel, Germany (Adabbo et al. 1994); Ca » Na in samples from "phonolitic tephritic ignimbrite with black pumices," Italy (Passaglia et al. 1990); and Na ≈ Ca in sample from Garbagna, Italy. Diagenetic crystals in pelagic volcanic sediments of the Pacific, Indian, and Atlantic

Oceans ("deep-sea" phillipsites) have compositions that are very similar. R values are 0.71-0.74, K contents are close to 2, Ba is absent, and Na greatly prevails over Ca (Sheppard et al. 1970, Kastner and Stonecipher 1978, Stonecipher 1978).

The unit-cell parameters are: a = 9.86-10.01 Å, b = 14.12-14.34 Å, and c = 14.16-14.42 Å. An increase in Al (i.e. a decrease in R) is responsible for an increase in the b and c parameters but does not influence the a parameter which, in turn, is positively correlated with the monovalent cations (Galli and Loschi Ghittoni 1972).

Harmotome [K/(K+Ba) < 0.50] occurs primarly as an amygdaloidal zeolite in metalliferous veins and vugs of plutonic (pegmatite and gabbro) and volcanic (basalt and andesite) rocks (Kostov 1962, Cerny and Povondra 1965, Pokorny 1966, Black 1969, Hoffman et al. 1973, Rinaldi et al. 1974, Cerny et al. 1977, Tschernich and Wise 1982, Passaglia and Bertoldi 1983, Hansen 1990, Armbruster et al. 1991). It also occurs (two occurrences) as a diagenetic product in a lacustrine tuff and a basaltic volcaniclastic sandstone (Sheppard and Gude 1971, 1983). Amygdaloidal samples are associated with other zeolites (analcime, heulandite, laumontite, paulingite, erionite, and ferrierite), apophyllite, prehnite, pectolite, hyalophane, gibbsite, pyrite, galena, quartz, and calcite. R shows an average value of 0.71 and ranges from 0.68 in crystals from trachyandesite of Iskra, Bulgaria (Kostov 1962), to 0.76 in crystals from basalt of Weitendorf, Austria (Armbruster et al. 1991). The (K+Ba) content is close to or slightly higher than 2, and the K/(K+Ba) ratio averages 0.22 and ranges from 0.03 in crystals from Korsnas, Finland (Cerny et al. 1977), to 0.45 in crystals from Selva di Trissino, Italy (Passaglia and Bertoldi 1983). Na generally prevails over Ca [Na/(Na+Ca) = 0.64-1], although it is subordinate in crystals from Selva di Trissino and Chase Creek, Canada (Tschernich and Wise 1982). The only analyzed sedimentary sample is from Wikieup, Arizona, where it is associated with analcime, chabazite, clinoptilolite, and erionite (Sheppard and Gude 1971). R is 0.74, Ba slightly prevails over K [K/(K+Ba) = 0.45], Na prevails over Ca [Na/(Na+Ca) = 0.90], and the Mg content (0.51) is anomalously high for this zeolite species.

The unit-cell parameters are: a = 9.81-9.92 Å, b = 14.10-14.17 Å, and c = 14.26-14.34 Å. The correlations between unit-cell parameters and chemical composition found for phillipsites (Galli and Loschi Ghittoni 1972) also apply to harmotome (Cerny et al. 1977).

Structure Type: **ROG**

Roggianite: $Ca_{14}(Na,K)[Be_5Al_{15}Si_{28}O_{90}(OH)_{14}](OH)_2 \cdot 34H_2O$

Roggianite has been described in two adjacent occurrences from Val Vigezzo, Italy (Passaglia 1969, Vezzalini and Mattioli 1979), both in albitite dikes only rarely associated with thomsonite. The holotype sample from Alpe Rosso was tested later (Passaglia and Vezzalini 1988) for zeolitic behavior and was re-analyzed by electron microprobe and by atomic absorption spectrometry for Be, which was not determined for both the original occurrence and the second occurrence (Pizzo Marcio). The results indicate that roggianite is a Be-bearing zeolite with R [Si/(Si+Al+Be)] equal to 0.58 and Ca as the DEC. Structure refinements show an imperfect tetrahedral framework with some (OH)-vertices not shared by two tetrahedra ("interrupted framework") plus some extraframework (OH) groups (Galli 1980) and one tetrahedral site occupied by Be (Giuseppetti et al. 1991). The structure is tetragonal *I4/mcm* with a = 18.33-18.37 Å and c = 9.16-9.19 Å.

Structure Type: **STI**

Stilbite: $NaCa_4[Al_9Si_{27}O_{72}] \cdot 30H_2O$

Stellerite: $Ca_4[Al_8Si_{28}O_{72}] \cdot 28H_2O$ (Z = 2)

Barrerite: $Na_8[Al_8Si_{28}O_{72}] \cdot 26H_2O$ (Z = 2)

The tetrahedral framework of this group has an orthorhombic *Fmmm* TS (Galli and Gottardi 1966) and shows a minimum anionic charge corresponding to an Al content of 8 atoms (half of the unit-cell of stellerite and barrerite and the full unit cell of stilbite) that is generally neutralized by Ca(+Mg) atoms. Any Al increase is counterbalanced by Na(+K) atoms (Passaglia et al. 1978b). Where Al is very close to 8 and Na ≈ 0 (stellerite), the RS remains orthorhombic *Fmmm* because the framework geometry is unaltered and Ca, completely surrounded by water molecules, fully occupies the site on the mirror plane normal to *a* (Galli and Alberti 1975a). Where Al > 8 atoms and Na > 0 (stilbite), the RS is monoclinic *C2/m* because the framework is rotated clockwise around binary diads and Ca is pushed from the mirror plane by Na located in a site near the framework oxygens (Galli 1971). In the case of Al ≈ 8 and Na(+K) ≈ 8 (barrerite), the mineral is orthorhombic because the lower charge of Na (in comparison with Ca) and its distribution with a low occupancy in the extraframework sites of stellerite and stilbite allows the mirror plane normal to *a* to be maintained. The resulting RS (*Amma*) is lower than the RS of stellerite (*Fmmm*) because some Na atoms occupy a site that has no counterpart in stellerite and stilbite, and they force the framework to rotate around a screw diad parallel to *a* (Galli and Alberti 1975b). Artificially Na-exchanged stellerite has an average orthorhombic *Fmmm* symmetry due to the statistical distribution of Na on both sides of the (001) mirror plane (Passaglia and Sacerdoti 1982), whereas artificially Ca-exchanged barrerite has the expected *Fmmm* symmetry (Sacerdoti and Gomedi 1984). The strong positive correlations between Al and Na(+K) and between Na(+K) and the β angle allow consideration of stellerite (orthorhombic) and stilbite (monoclinic) as a continuous series in which the Ca(+Mg) content is nearly constant and the degree of monoclinicity (the β-angle) increases with Na(+K) and Al content (Passaglia et al. 1978b). Anomalous Na-rich and Ca-poor stilbites were classified as intermediate members between stilbite and barrerite (Passaglia et al.,1978b), and their monoclinicity, markedly lower than expected, was explained on the basis of the structural features (Quartieri and Vezzalini 1987). As a result of the large dependence of the β-angle on chemistry, many samples are obviously heterogeneous, being composed of a number of crystals with different β-angles (Passaglia et al. 1978b, Akizuki and Konno 1985), and other crystals show an association of stilbite and stellerite (Morad et al. 1989). Thus, a reliable classification should be done through an accurate X-ray diffraction analysis inasmuch as the chemical composition and morphology do not precisely define these species.

Stilbite is the most common among the three zeolites and has been described in vugs of plutonic (granite and quartz monzonite), volcanic (basalt, andesite, dacite, and porphyrite), and metamorphic (gneiss, mica schist, and meta-ophiolite) rocks and as sandstone cement (Ueno and Hanada 1982). It occurs in association with other zeolites (heulandite, laumontite, chabazite, mordenite, mesolite, analcime, scolecite, erionite, levyne, natrolite, thomsonite, and stellerite), apophyllite, datolite, prehnite, pectolite, calcite, pyrite, quartz, and opal. Many reliable chemical compositions (if not otherwise quoted in Passaglia et al. 1978b) show that R is in the range 0.73-0.78, very close to the mean value of 0.75, being significantly lower (0.71) only in crystals from Montresta, Italy. Ca(+Mg) contents are commonly very close to 4 but are lower (1-2) in Na-rich crystals from rocks weathered in a foreshore environment at Capo Pula, Italy, Faroer Islands, Kii Peninsula, Japan (Harada and Tomita 1967), Phillip Island, Australia (Birch 1988), Tsuyazaki, Japan (Ueno and Hanada 1982), and Rocky Pass, Alaska (Di Renzo and Gabelica 1997). In samples with Ca(+Mg) close to 4, the Na(+K) content uniformly ranges from ~0.40 in crystals from Oravita, Romania, to 1.75 in crystals from Flodigarry, Sky. A value as high as 2.72 has been found in crystals from Montresta, Italy. Na greatly prevails over K [Na/(Na+K) = 0.70-0.98], except for samples from granite gneiss of some Swiss localities and from Prascorsano, Italy; Mg is negligible or very subordinate, and the only relatively high Mg content (1.13) has been found in a sample from the granite of the Siljank ring structure, Sweden (Morad et al. 1989). Sr and Ba are typically absent or negligible, and Sr is appreciable (0.31) only in a sample from the Siljank ring structure, Sweden. The unit-cell parameters of the pseudo-orthorhombic *F2/m* cell are a = 13.59-13.66 Å, b = 18.18-18.33 Å, c = 17.71-17.84 Å and β = 90.20-91.15°.

Stellerite is less common than stilbite, and it has been described in vugs of the same rock types (granite, granodiorite, dolerite, basalt, andesite, diabase, gneiss, hornfel, and metadolerite) in association with other zeolites (heulandite, epistilbite, chabazite, laumontite, and stilbite), fluorite, prehnite, babingtonite, sphene, fluorapophyllite, calcite, quartz, tridymite, and native copper (Galli and Passaglia 1973, Passaglia et al. 1978b, Alberti et al. 1978, Birch 1989, Morad et al. 1989). It has also been described in fractures in devitrified, non-zeolitic rhyolitic tuffs from Yucca Mountain, Nevada (Carlos et al. 1995). R varies (0.76-0.78) little from the average value (0.77), being slightly lower (0.75) in some samples. Ca(+Mg) is generally very close to 4 with Mg negligible (max. 0.17). Na(+K) ranges from 0.02 in crystals from the metadolerite of Corop, Australia (Birch 1989), to 0.40, but it is higher (0.40-0.60) in samples with Ca(+Mg) slightly lower than 4, and it is anomalously high (1.74) in crystals with Ca(+Mg) equal to 3.46 from andesite of Capo Santa Vittoria, Italy (Passaglia et al. 1978b). K is very subordinate, and Sr and Ba are absent. The unit-cell parameters are: a = 13.57-13.63 Å, b = 18.16-18.27 Å, c = 17.82-17.87 Å, and β = 90° or with statistically insignificant deviations from orthogonality.

Barrerite has been described in vugs of altered andesite that crops out in the foreshore of Capo Pula, Italy, associated with heulandite (Passaglia and Pongiluppi 1974, 1975), and in fractures of basalt exposed in the tidal area of Rocky Pass, Alaska, associated with Na-rich stilbite and epistilbite (Di Renzo and Gabelica 1997, Sacerdoti et al. 1999). R is 0.77-0.78, and the extraframework contents are different from that given in the schematic formula showing Na = 3.92-5.45, K = 1.06-1.73, Ca = 0.84-1.12, and Mg = 0.11-0.18. The unit-cell parameters are within the ranges given for stellerite, being a = 13.59-13.64 Å, b = 18.18-18.20 Å, and c = 17.79-17.84 Å. The X-ray powder pattern does not show any peaks indexable in *Amma* (barrerite) but forbidden in *Fmmm* (stellerite), and, hence, the distinction between the two zeolite species requires careful X-ray single-crystal analyses.

Structure Type: **TER**

Terranovaite: $Na_4Ca_4[Al_{12}Si_{68}O_{160}]\cdot{>}29H_2O$

Terranovaite occurs as transparent, small spheres closely associated with heulandite or as globular aggregates of small prismatic crystals associated with tschernichite in vugs of dolerite at Mt. Adamson, Antarctica (Galli et al. 1997). The empirical formula, $Na_{4.2}Ca_{3.7}K_{0.2}Mg_{0.2}[Al_{12.3}Si_{67.7}O_{160}]\cdot{>}29H_2O$, has a high R value (0.85). Na and Ca are the DEC (with Na slightly prevailing over Ca), and K and Mg are very low. The structure is orthorhombic *C2cm* with a = 9.75 Å, b = 23.88 Å, and c = 20.07 Å and *Cmcm* TS.

Structure Type: **THO**

Thomsonite: $Ca_7Na_5[Al_{19}Si_{21}O_{80}]\cdot 24H_2O$

Thomsonite has been described from many localities where it occurs in vugs of igneous rocks (basalt, andesite, dolerite, and nepheline syenite). It occurs rarely in metamorphic (meta-ophiolite and gneiss) rocks and as a late-stage hydrothermal alteration of lazurite (Hogarth and Griffin 1980). It is associated with other zeolites (analcime, mesolite, laumontite, gonnardite, edingtonite, harmotome, chabazite, phillipsite, gismondine, and natrolite), calcite, aragonite, celestine, pectolite, vesuvianite, and epidote. The schematic formula calculated from reliable chemical analyses in the literature differs from the one which is normally reported, $Ca_8Na_4[Al_{20}Si_{20}O_{80}]\cdot 24H_2O$, and it is slightly richer in Si and Na and poorer in Ca. The mean R value is 0.53 (instead of 0.50), and the mean contents of Na and Ca are 4.5 (instead of 4) and 6.8 (instead of 8), respectively. The chemical composition is variable: R ranges from 0.50 in a few samples to 0.58 in a sample from Hills Port, Antrim, Ireland (Nawaz 1988); the Na content varies from 3.02 in a sample from Procida, Italy (de'Gennaro et al. 1977) to 7.24 in a sample from Mazé, Japan (Harada et al. 1969), and the Ca content ranges from 8.43 in a sample from Shinshiro, Japan (Matsubara et al. 1979) to 3.55 in a sample from Tyamyr, Russia (Yefimov et al. 1966). Samples with chemical compositions close to the schematic formula reported in the literature do exist but are very rare. Ca generally prevails over Na, but samples with Na > Ca have been described from Mazé and Iragawa, Japan (Harada et al. 1969). Samples with comparable Ca and Na have been described from Howitt Plains, Australia (Birch 1988), Table Mountain, Colorado (Kile and Modreski 1988), and Taymyr, Russia (Yefimov et al. 1966). Where present, K and Mg are very low, but a sample in nepheline syenite at Taymyr, Russia (Yefimov et al. 1966) is K and Mg rich (1.54 and 0.83, respectively), and a sample from Ile de la Réunion (Nativel 1972) is the most Mg rich (0.84) known. Where detected, Sr can be appreciable and reaches values as high as 2.12 in the sample associated with brewsterite at Yellow Lake, Canada (Wise and Tschernich 1978a), 2.18 in a sample from Honshu, Japan (Ross et al. 1992), and 3.15 in the K- and Mg-rich and Ca-poor sample from Taymyr, Russia. Ba is absent or negligible.

The different habits and forms (blocky, complex crystals, bladed crystals, botryoidal growths, and waxy balls) of thomsonite have been correlated with the Si content (Wise and Tschernich 1978a). The geometry of the unit cell ($a < b < c$) and optical orientation ($\alpha = b$, $\beta = c$, and $\gamma = a$) have been proposed by Nawaz and Malone (1981). The structure is orthorhombic *Pncn* with a = 13.00-13.18 Å, b = 13.04-13.16 Å, and c = 13.09-13.24 Å. The (Si,Al) distribution in the tetrahedra is ordered in samples with Si/Al ≈ 1 and partially disordered in samples with Si/Al > 1 (Alberti et al. 1981, Pechar 1982, Pluth et al. 1985).

Structure Type: **YUG**

Yugawaralite: $Ca_2[Al_4Si_{12}O_{32}]\cdot 8H_2O$

Yugawaralite has been described from six occurrences in fractures and vugs of volcanic breccia (Wise 1978), effusive rocks (basalt, andesite, trachyandesite, and rhyolite), and pyroclastites (andesitic tuff), all highly altered in active (Seki and Haramura 1966, Bargar and Beeson 1981, Kvick et al. 1986) and possible (Eberlein et al. 1971, Pongiluppi 1977) geothermal areas. It is typically associated with laumontite, calcite, and quartz, but some associations with stilbite, heulandite, stellerite, wairakite, and, rarely, mordenite have also been observed.

The composition is nearly constant. R ranges from 0.74 to 0.76, Ca is the dominant and almost exclusive extraframework cation, Na, K, and Mg are negligible, and Sr and Ba are absent. A sample from a drill hole at Yellowstone National Park (Bargar and Beeson 1981) is unusually poor in Si (R = 0.67), rich in Na (0.17), and it is the only sample documented to occur in association with analcime and dachiardite.

The structure is monoclinic *Pc* (Kvick et al. 1986) with a = 6.70-6.73 Å, b = 13.97-14.01 Å, c = 10.04-10.09 Å, and β = 111.07-111.20°. Optical properties and orientation were given by Harada et al. (1968) and Akizuki (1987).

Structure Type: **Unknown**

Cowlesite: $Ca_{5.5}(Na,K)[Al_{12}Si_{18}O_{60}]\cdot 36H_2O$

Cowlesite has been described from about 20 localities mainly in County Antrim, Ireland (Nawaz 1984, Vezzalini et al. 1992), but also in Australia (Birch 1989), USA, Canada (Wise and Tschernich 1975, Kile and Modreski 1988), and Japan (Fujimoto et al.

1990, Kuwano and Tokumaru 1993). It occurs invariably in vugs of basalt, alone or associated with other zeolites (levyne, erionite-offretite, phillipsite, thomsonite, garronite, chabazite, scolecite, and gismondine). The crystals are very small ($<0.1 \times 0.002$ mm) so that single-crystal studies have failed to date. Such crystals appear to be good candidates for structure study with synchrotron X-rays.

The composition is very close to the schematic formula. R shows a mean value of 0.61 and generally ranges from 0.60 to 0.62 but it is slightly lower (0.58) in samples from Monte Lake, Canada, and Spray, Oregon (Wise and Tschernich 1975). Ca is the DEC, and Na and K are consistently present in subordinate amounts (<1.0) even though Na reaches noticeable values in samples from Flinders, Australia (1.13), Dunseverick, County Antrim (1.31), and Iki Island, Japan (1.86). The sample from the latter occurrence is also richer in K and Mg (0.24 and 0.60, respectively) and is the only one associated with gobbinsite (Kuwano and Tokumaru 1993). The symmetry, derived from powder data and from very faint precession photographs, is orthorhombic (Nawaz 1984) with a = 23.21-23.30 Å, b = 30.60-30.68 Å, and c = 24.98-25.04 Å.

Structure Type: **Not yet assigned**

Tschörtnerite: $Ca_6Cu_3SrK(OH)_9[Al_{12}Si_{12}O_{48}]\cdot 14H_2O$

Tschörtnerite has been found in cavities of a Ca-rich xenolith in leucite tephrite lava at the Bellberg volcano near Mayen, Eifel, Germany (Effenberger et al. 1998). It occurs as light blue, transparent cubes (maximum size of 0.15 mm) in association with chalcopyrite, cuprite, willhendersonite, phillipsite, gismondine, strätlingite, and bellbergite. The empirical formula $(Ca_{5.60}Cu_{2.90}Sr_{1.04}K_{0.70}Ba_{0.30})(OH)_{8.44}[Fe_{0.09}Al_{11.85}Si_{12.06}O_{48}]\cdot 14.01H_2O$ shows an R value very close to 0.50, with Ca as the DEC, and noticeable amounts of Cu and extraframework (OH) groups. The structure is cubic $Fm\bar{3}m$ with a = 31.62 Å and contains a new extremely large cage with 96 tetrahedra and a Cu(OH)-bearing cluster.

REFERENCES

Adabbo M, Langella A, de'Gennaro M, Guerriero A (1994) Sedimentary zeolites from East Eifel volcanic district (Germany). Mater Eng (Modena, Italy) 5:107-118

Akizuki M. (1980) Origin of optical variation in analcime and chabazite. *In* Proc 5th Int'l Conf Zeolites, Naples 1980, LVC Rees (ed) Heyden, London, p 171-178

Akizuki M (1981) Origin of optical variation in analcime. Am Mineral 66:403-409

Akizuki M (1987) An explanation of optical variation in yugawaralite. Mineral Mag 51:615-620

Akizuki M, Konno H (1985) Order-disorder structure and the internal texture of stilbite. Am Mineral 70:814-821

Akizuki M, Konno H (1987) Growth twinning in phacolite. Mineral Mag 51:427-430

Akizuki M, Nishido H (1988) Epistilbite: symmetry and twinning. Am Mineral 73:1434-1439

Akizuki M, Nishido H, Fujimoto M (1989) Herschelite: morphology and growth sectors. Am Mineral 74:1337-1342

Alberti A (1972) On the crystal structure of the zeolite heulandite. Tschermaks mineral petrogr Mitt 18:129-146

Alberti A (1975a) Sodium-rich dachiardite from Alpe di Siusi, Italy. Contrib Mineral Petrol 49:63-66

Alberti A (1975b) The crystal structure of two clinoptilolites. Tschermaks mineral petrogr Mitt 22:25-37

Alberti A, Sabelli C (1987) Statistical and true symmetry of ferrierite: possible absence of straight T-O-T bridging bonds. Z Kristallogr 178:249-256

Alberti A, Vezzalini G (1979) The crystal structure of amicite, a zeolite. Acta Crystallogr B35:2866-2869

Alberti A, Vezzalini G (1981) A partially disordered natrolite: relationships between cell parameters and Si-Al distribution. Acta Crystallogr B37:781-788

Alberti A, Cruciani G, Dauru I (1995) Order-disorder in natrolite-group minerals. Eur J Mineral 7: 501-508

Alberti A, Davoli P, Vezzalini G (1986) The crystal structure refinement of a natural mordenite. Z Kristallogr 175:249-256

Alberti A, Galli E, Vezzalini G (1985) Epistilbite: an acentric zeolite with domain structure. Z Kristallogr 173:257-265

Alberti A, Galli E, Vezzalini G, Passaglia E, Zanazzi PF (1982a) Position of cations and water molecules in hydrated chabazite. Natural and Na-, Ca-, Sr- and K-exchanged chabazites. Zeolites 2:303-309
Alberti H, Hentschel G, Vezzalini G (1979) Amicite, a new natural zeolite. N Jahrb Mineral Mh 1979:481-488
Alberti A, Pongiluppi D, Vezzalini G (1982c) The crystal chemistry of natrolite, mesolite and scolecite. N Jahrb Mineral Abh 143:231-248
Alberti A, Rinaldi R, Vezzalini G (1978) Dynamics of dehydration in stilbite-type structures: stellerite phase B. Phys Chem Minerals 2:365-375
Alberti A, Vezzalini G, Pécsi-Donáth É (1982b) Some unusual zeolites from Hungary. Acta Geol Acad Sci Hung. 25:237-246
Alberti A, Vezzalini G, Tazzoli V (1981) Thomsonite: a detailed refinement with cross checking by crystal energy calculations. Zeolites 1:91-97
Alberti A, Vezzalini G, Galli E, Quartieri S (1996) The crystal structure of gottardiite, a new natural zeolite. Eur J Mineral 8:69-75
Alietti A (1972) Polymorphism and crystal-chemistry of heulandites and clinoptilolites. Am Mineral 57:1448-1462
Alietti A, Brigatti MF, Poppi L (1977) Natural Ca-rich clinoptilolites (heulandite of group 3): new data and review. N Jahrb Mineral Mh 1977:493-501
Alietti A, Gottardi G, Poppi L (1974) The heat behaviour of the cation exchanged zeolites with heulandite structure. Tschermaks mineral petrogr Mitt 21:291-298
Alietti A, Passaglia E, Scaini G (1967) A new occurrence of ferrierite. Am Mineral 52:1562-1563
Altaner SP, Grim RE (1990) Mineralogy, chemistry, and diagenesis of tuffs in the Sucker Creek Formation (Miocene), eastern Oregon. Clays & Clay Minerals 38:561-572
Aoki M, Minato H (1980) Lattice constants of wairakite as a function of chemical composition. Am Mineral 65:1212-1216
Argenti P, Lucchetti G, Penco AM (1986) Zeolite-bearing assemblages at the contact Voltri Group and Sestri-Voltaggio Zone (Liguria, Italy). N Jahrb Mineral Mh 1986:229-239
Armbruster T, Wenger M, Kohler T (1991) Mischkristalle von Klinoptilolith-Heulandit und Harmotom-Phillipsit aus dem Basalt von Weitendorf, Steiermark. Mitt Abt Mineral Landesmus. Joanneum 59:13-18
Artioli G (1992) The crystal structure of garronite. Am Mineral 77:189-196
Artioli G, Foy H (1994) Gobbinsite from Magheramorne quarry, Northern Ireland. Mineral Mag 58:615-620
Artioli G, Galli E (1999) Gonnardite: Re-examination of holotype material and discreditation of tetranatrolite. Am Mineral 84:1445-1450
Artioli G, Kvick A (1990) Synchrotron X-ray Rietveld study of perlialite, the natural counterpart of synthetic zeolite-L. Eur J Mineral 2:749-759
Artioli G, Marchi M (1999) On the space group of garronite. Powder Diffraction 14:190-194
Artioli G, Ståhl K (1993) Fully hydrated laumontite: a structure study by flat-plate and capillary powder diffraction techniques. Zeolites 13:249-255
Artioli G, Torres Salvador MR (1991) Characterization of the natural zeolite gonnardite. Structure analysis of natural and cation exchanged species by the Rietveld method. Mater Sci Forum 79-82: 845-850
Artioli G, Smith JV, Kvick A (1984) Neutron diffraction study of natrolite, $Na_2Al_2Si_3O_{10}·2H_2O$, at 20 K. Acta Crystallogr C40:1658-1662
Artioli G, Smith JV, Kvick A (1985) Multiple hydrogen positions in the zeolite brewsterite $(Sr_{0.95},Ba_{0.05})Al_2Si_6O_{16}·5H_2O$. Acta Crystallogr C41:492-497
Artioli G, Smith JV, Kvick A (1989) Single-crystal neutron diffraction study of partially dehydrated laumontite at 15K. Zeolites 9:377-391
Artioli G, Smith JV, Pluth JJ (1986) X-ray structure refinement of mesolite. Acta Crystallogr C42:937-942
Aurisicchio C, De Angelis G, Dolfi D, Farinato R, Loreto L, Sgarlata F, Trigila R (1975) Sull'analcime di color rosso nelle vulcaniti del complesso alcalino di Crowsnest (Alberta-Canada). Rend Soc Ital Mineral Petrol 31:653-671
Avdeev AG (1992) On the nature of idiomorphic analcime insets within tephrites of the Martuny volcanic complex (Lesser Caucasus). Zap Vses Mineral O-va 121:106-112
Bargar KE, Beeson MH (1981) Hydrothermal alteration in research drill hole Y-2, Lower Geyser Basin, Yellowstone National Park, Wyoming. Am Mineral 66:473-490
Bargar KE, Erd RC, Keith TEC, Beeson MH (1987) Dachiardite from Yellowstone National Park, Wyoming. Can Mineral 25:475-483
Barnes DA, Boles JR, Hickey J (1984) Zeolite occurrence in Triassic-Jurassic sedimentary rocks, Baja California Sur, Mexico. *In* Proc 6th Int'l Zeolite Conf, Reno 1983. D Olson, A Bisio (eds) Butterworths, Guilford, UK, p 584-594

Barrows KJ (1980) Zeolitization of Miocene volcaniclastic rocks, southern Desatoya Mountains, Nevada. Geol Soc Am Bull 91:199-210

Belitsky IA, Gabuda SP, Joswig W, Fuess H (1986) Study of the structure and dynamics of water in the zeolite edingtonite at low temperature by neutron diffraction and NMR-spectroscopy. N Jahrb Mineral Mh 1986:541-551

Berman H (1925) Notes on dachiardite. Am Mineral 10:421-428

Birch WD (1979) Levyne from Clunes, Victoria: a new occurrence. Aust Mineral 25:119-120

Birch WD (1987) Zeolites from Jindivick, Victoria. Aust Mineral 2:15-19

Birch WD (1988) Zeolites from Phillip Island and Flinders, Victoria. Mineral Record 19:451-460

Birch WD (1989) Chemistry of Victorian zeolites: Zeolites of Victoria. Mineral Soc Victoria Spec Pub 2:91-102

Birch WD, Morvell G (1978) Ferrierite from Phillip Island, Victoria. Aust Mineral 15:75-76

Bissert G, Liebau F (1986) The crystal structure of a triclinic bikitaite $Li[AlSi_2O_6]\cdot H_2O$, with ordered Al/Si distribution. N Jahrb Mineral Mh 1986:241-252

Black PM (1969) Harmotome from Tokatoka district, New Zealand. Mineral Mag 37:453-458

Boggs RC, Howard DG, Smith JV, Klein GL (1993) Tschernichite, a new zeolite from Goble, Columbia County, Oregon. Am Mineral 78:822-826

Boles JR (1972) Composition, optical properties, cell dimensions, and thermal stability of some heulandite group zeolites. Am Mineral 57:1463-1493

Boles JR, Surdam RC (1979) Diagenesis of volcanogenic sediments in a Tertiary saline lake, Wagon Bed Formation, Wyoming. Am J Sci 279:832-853

Boles JR, Wise WS (1978) Nature and origin of deep-sea clinoptilolite. *In* Natural Zeolites. Occurrence, Properties, Use. LB Sand, FA Mumpton (eds) Pergamon Press, Oxford, p 235-243

Bonardi M (1979) Composition of type dachiardite from Elba: a re-examination. Mineral Mag 43:548-549

Bonardi M, Roberts AC, Sabina AP (1981) Sodium-rich dachiardite from the Francon quarry, Montreal Island, Quebec. Can Mineral 19:285-289

Bonnin D, Calas G (1978) Études spectroscopiques du fer dans les heulandites rouges de Val Fassa (Tyrol italien). Bull Minéral 101:395-398

Broxton DE, Bish DL, Warren RG (1987) Distribution and chemistry of diagenetic minerals at Yucca Mountain, Nye County, Nevada. Clays & Clay Minerals 35:89-110

Brunner GO, Meier WM (1989) Framework density distribution of zeolite-type tetrahedral nets. Nature 337, p 146

Bushinsky GI (1950) Mordenite in marine sediments, Cretaceous and Paleogene. Dokl Akad Nauk SSSR 73:1271-1274

Cabella R, Lucchetti G, Palenzona A, Quartieri S, Vezzalini G (1993) First occurrence of Ba-dominant brewsterite: structural features. Eur J Mineral 5:353-360

Carlos B, Chipera S, Bish D, Raymond R (1995) Distribution and chemistry of fracture-lining zeolites at Yucca Mountain, Nevada. *In* Natural Zeolites '93, D.W. Ming, F.A. Mumpton (eds) Int'l Comm on Natural Zeolites, Brockport, New York, p 547-563

Cerny P (1972) The Tanco pegmatite at Bernic Lake, Manitoba. VIII. Secondary minerals from the spodumene-rich zones. Can Mineral 11:714-726

Cerny P (1974) The present status of the analcime-pollucite series. Can Mineral 12:334-341

Cerny P, Povondra P (1965) Harmotome from desilicated pegmatites at Hrubsice, Western Moravia. Acta Univ Carol, Geol 1:31-43

Cerny P, Povondra P (1969) A polycationic strontian heulandite: comments on crystal chemistry and classification of heulandite and clinoptilolite. N Jahrb Mineral Mh 1969:349-361

Cerny P, Rinaldi R, Surdam RC (1977) Wellsite and its status in the phillipsite-harmotome group. N Jahrb Mineral Abh 128:312-330

Chandwadkar AJ, Bhat RN, Ratnasamy P (1991) Synthesis of iron-silicate analogs of zeolite mordenite. Zeolites 11:42-47

Chandwadkar AJ, Date SK, Bill E, Trautwein A (1992) Fe^{57}-Mössbauer spectroscopic studies of iron silicates with the mordenite topology. Zeolites 12:180-182

Chao GY (1980) Paranatrolite, a new zeolite from Mont St-Hilaire, Québec. Can Mineral 18:85-88

Chen TT, Chao GY (1980) Tetranatrolite from Mont St-Hilaire, Québec. Can Mineral 18:77-84

Ciambelli P, Franco E, Notaro M, Vaccaro C (1989) Mineralogical and physico-chemical characterization of gonnardite. *In* Zeolites for the Nineties, 8th Int'l Zeolite Conf, Recent Research Reports, p 147-148

Comin-Chiaromonti P, Meriani S, Mosca R, Sinigoi S (1979) On the occurrence of analcime in the northeastern Azerbaijan volcanics (northwestern Iran). Lithos 12:187-198

Coombs DS (1952) Cell size, optical properties and chemical composition of laumontite and leonhardite. Am Mineral 37:812-830

Coombs DS, Whetten JT (1967) Composition of analcime from sedimentary and burial metamorphic rocks. Geol Soc Am Bull 78:269-282

Coombs DS, Alberti A, Armbruster T, Artioli G, Colella C, Galli E, Grice JD, Liebau F, Mandarino JA, Minato H, Nickel EH, Passaglia E, Peacor DR, Quartieri S, Rinaldi R, Ross M, Sheppard RA, Tillmanns E, Vezzalini G (1997) Recommended nomenclature for zeolite minerals: Report of the subcommittee on zeolites of the Intern Mineral Assoc, Commission on New Minerals and Mineral Names. Can Mineral 35:1571-1606

Cortesogno L, Lucchetti G, Penco AM (1975) Associazioni a zeoliti nel "Gruppo di Voltri": caratteristiche mineralogiche e significato genetico. Rend Soc Ital Mineral Petrol 31:673-710

Cortesogno L, Lucchetti G, Penco AM (1976) Mineralogia e minerogenesi di associazioni a tobermorite, zeoliti e smectiti nel settore nord-orientale del gruppo di Voltri. Period Mineral 45:97-108

de'Gennaro M, Franco E (1976) La K-chabasite di alcuni "tufi del Vesuvio." Rend Accad Nazionale Lincei, Ser 8, 60:490-497

de'Gennaro M, Franco E, Paracuollo G, Passarelli A (1980) I "tufi" del Vesuvio con chabasite potassica. Period Mineral 49:223-240

de'Gennaro M, Franco E, Paracuollo G, Porcelli C (1977) Le zeoliti di alcuni blocchi della "Breccia Museo" di Punta della Lingua (Isola di Procida). Rendiconti Accad Sci Fisiche Matematiche, Napoli 44:427-440

Della Ventura G, Parodi GC, Burragato F (1993) New data on merlinoite and related zeolites. Rend Accad Nazionale Lincei 4:303-312

Di Renzo F, Gabelica Z (1997) Barrerite and other zeolites from Kuiu and Kupreanof islands, Alaska. Can Mineral 35:691-698

Djourova E.G (1976) Analcime zeolites from the North-Eastern Rhodopes. C R Acad Bulg Sci 29: 1023-1025

Dunn PJ, Peacor DR, Newberry N, Ramik RA (1980) Goosecreekite, a new calcium aluminum silicate hydrate possibly related to brewsterite and epistilbite. Can Mineral 18:323-327

Eberlein GD, Erd RC, Weber F, Beatty LB (1971) New occurrence of yugawaralite from the Chena Hot Springs area, Alaska. Am Mineral 56:1699-1717

Effenberger H, Giester G, Krause W, Bernhardt H-J (1998) Tschörtnerite, a copper-bearing zeolite from the Bellberg volcano, Eifel, Germany. Am Mineral 83:607-617

Engel N, Yvon, K (1984) The crystal structure of parthéite. Z Kristallogr 169:165-175

England BM, Ostwald J (1978) Ferrierite: an Australian occurrence. Mineral Mag 42:385-389

England BM, Ostwald J (1979) Levyne-offretite intergrowths from Tertiary basalts in the Merriwa district, Hunter Valley, New South Wales, Australia. Aust Mineral 25:117-119

Fujimoto M, Matsubara S, Nishido H (1990) The occurrence of cowlesite from Ikitsuki Island, Nagasaki Prefecture, Japan. Chigaku Kenkyu 39:219-224

Galli E (1971) Refinement of the crystal structure of stilbite. Acta Crystallogr B27:833-841

Galli E (1975) Crystal structure refinement of mazzite. Rend Soc Ital Mineral Petrol 31:599-612

Galli E (1976) Crystal structure refinement of edingtonite. Acta Crystallogr B32:1623-1627

Galli E (1980) The crystal structure of roggianite, a zeolite-like silicate. *In* Proc 5th Int'l Conf Zeolites, Naples 1980, LVC Rees (ed) Heyden, London, p 205-213

Galli E, Alberti A (1975a) The crystal structure of stellerite. Bull Soc fr Minéral Cristallogr 98:11-18

Galli E, Alberti A (1975b) The crystal structure of barrerite. Bull Soc fr Minéral Cristallogr 98:331-340

Galli E, Gottardi G (1966) The crystal structure of stilbite. Mineral Petrogr Acta 12:1-10

Galli E, Loschi Ghittoni A.G (1972) The crystal chemistry of phillipsites. Am Mineral 57:1125-1145

Galli E, Passaglia E (1973) Stellerite from Villanova Monteleone, Sardinia. Lithos 6:83-90

Galli E, Rinaldi R (1974) The crystal chemistry of epistilbites. Am Mineral 59:1055-1061

Galli E, Gottardi G, Pongiluppi D (1979) The crystal structure of the zeolite merlinoite. N Jahrb Mineral Mh 1979:1-9

Galli E, Passaglia E, Zanazzi PF (1982) Gmelinite: structural refinements of sodium-rich and calcium-rich natural crystals. N Jahrb Mineral Mh1982:145-155

Galli E, Rinaldi R, Modena C (1981) Crystal chemistry of levynes. Zeolites 1:157-160

Galli E, Passaglia E, Pongiluppi D, Rinaldi R (1974) Mazzite, a new mineral, the natural counterpart of the synthetic zeolite Ω. Contrib Mineral Petrol 45:99-105

Galli E, Quartieri S, Vezzalini G, Alberti A (1995) Boggsite and tschernichite-type zeolites from Mt. Adamson, northern Victoria Land (Antarctica). Eur J Mineral 7:1029-1032

Galli E, Quartieri S, Vezzalini G, Alberti A (1996) Gottardiite, a new high-silica zeolite from Antarctica: the natural counterpart of synthetic NU-87. Eur J Mineral 8:687-693

Galli E, Quartieri S, Vezzalini G, Alberti A, Franzini M (1997) Terranovaite from Antarctica: a new "pentasil" zeolite. Am Mineral 82:423-429

Galli E, Vezzalini G, Quartieri S, Alberti A, Franzini M (1997a) Mutinaite, a new zeolite from Antarctica: the natural counterpart of ZSM-5. Zeolites 19:318-322

Gallitelli P (1928) La laumontite di Toggiano. Rendiconti Accademia Nazionale Lincei, Ser 6 8:82-87

Garcia Hernandez JE, Notario Del Pino JS, Gonzales Martin MM, Hernan Reguera F, Rodriguez Losada JA (1993) Zeolites in pyroclastic deposits in southeastern Tenerife (Canary Islands). Clays & Clay Minerals 41:521-526.

Gellens LR, Price GD, Smith JV (1982) The structural relation between svetlozarite and dachiardite. Mineral Mag 45:157-161

Giret A (1979) Genèse de roches feldspathoidiques par la destabilisation des amphiboles: massif de Montagnes Vertes, Kerguelen (T.A.A.F.). C R Acad Sci Paris D289:379-382

Giuseppetti G, Mazzi F, Tadini C, Galli E (1991) The revised crystal structure of roggianite: $Ca_2[Be(OH)_2Al_2Si_4O_{13}]\cdot<2.5H_2O$. N Jahrb Mineral Mh 1991:307-314

Gordon EK, Samson S, Kamb WB (1966) Crystal structure of the zeolite paulingite. Science 154:1004-1007

Gottardi G (1960) Sul dimorfismo mordenite-dachiardite. Period Mineral 29:183-191

Gottardi G (1979) Topological symmetry and real symmetry in framework silicates. Tschermaks mineral petrogr Mitt 26:39-50

Gottardi G, Galli E (1985) Natural Zeolites. Springer-Verlag, Berlin-Heidelberg, 409 p

Gramlich-Meier R, Gramlich V, Meier WM (1985) The crystal structure of the monoclinic variety of ferrierite. Am Mineral 70:619-623

Gramlich-Meier R, Meier WM, Smith BK (1984) On faults in the framework structure of the zeolite ferrierite. Z Kristallogr 169:201-210

Grice J.D, Gault RA, Ansell HG (1984) Edingtonite: the first two Canadian occurrences. Can Mineral 22:253-258

Gude AJ 3d, Sheppard RA (1966) Silica-rich chabazite from the Barstow Formation, San Bernardino County, southern California. Am Mineral 51:909-915

Gude AJ 3d, Sheppard RA (1978) Chabazite in siliceous tuffs of a Pliocene lacustrine deposit near Durkee, Baker County, Oregon. U S Geol Surv J Res 6:467-472

Gude AJ 3d, Sheppard RA (1981) Woolly erionite from the Reese River zeolite deposit, Lander County, Nevada, and its relationship to other erionites. Clays & Clay Minerals 29:378-384

Gude AJ 3d, Sheppard RA (1988) A zeolitic tuff in a lacustrine facies of the Gila Conglomerate near Buckhorn, Grant County, New Mexico. U S Geol Surv Bull 1763:1-22

Gunter ME, Ribbe PH (1993) Natrolite group zeolites: correlations of optical properties and crystal chemistry. Zeolites 13:435-440

Guseva LD, Men'shikov YuP, Romanova TS, Bussen IV (1975) Tetragonal natrolite from the Lovozero alkaline massif. Zap Vses Mineral O-va 104:66-69

Hansen S (1990) Harmotome from Odarslov, Skaane, Sweden. Geol. Foeren. Stockholm Foerh. 112:140

Harada K, Sudo T (1976) A consideration on the wairakite-analcime series.—Is valid a new mineral name for sodium analogue of monoclinic wairakite?. Mineral J 8:247-251

Harada K, Tomita K (1967) A sodian stilbite from Onigajo, Mié Prefecture, Japan, with some experimental studies concerning the conversion of stilbite to wairakite at low water vapor pressures. Am Mineral 52:1438-1450

Harada K, Iwamoto S, Kihara K (1967) Erionite, phillipsite and gonnardite in the amygdales of altered basalt from Mazé, Niigata Prefecture, Japan. Am Mineral 52:1785-1794

Harada K, Nagashima K, Sakurai K-I (1968) Chemical composition and optical properties of yugawaralite from the type locality. Am Mineral 54:306-309

Harada K, Tanaka K, Nagashima K (1972) New data on the analcime-wairakite series. Am Mineral 57:924-931

Harada K, Umeda M, Nakao K, Nagashima K (1969) Mineralogical notes on thomsonites, laumontite and heulandite in altered basaltic and andesitic rocks in Japan. J Geol Soc Japan 75:551-552

Harris PG, Brindley GW (1954) Mordenite as an alteration product of a pitchstone glass. Am Mineral 39:819-824

Hawkins DB (1974) Statistical analyses of the zeolites clinoptilolite and heulandite. Contrib Mineral Petrol 45:27-36

Hay RL (1964) Phillipsite of saline lakes and soils. Am Mineral 49:1366-1387

Hay RL (1980) Zeolite weathering of tuffs in Olduvai Gorge, Tanzania. *In* Proc 5th Int'l Conf Zeolites, Naples 1980, LVC Rees (ed) Heyden, London, p 155-163

Hay RL, Guldman SG (1987) Diagenetic alteration of silicic ash in Searles Lake, California. Clays & Clay Minerals 35:449-457

Hentschel G (1986) Paulingit und andere seltene Zeolithe in einem Gefritteten Sandsteineinschluss im Basalt von Ortenberg (Vogelsberg). Geol Jahrb Hessen 114:249-256

Hesse K-F (1983) Refinement of a partially disordered natrolite, $Na_2Al_2Si_3O_{10}\cdot 2H_2O$. Z Kristallogr 163:69-74

Hey MH, Bannister FA (1934) Studies on the zeolites. Part VII. "Clinoptilolite," a new silica-rich variety of heulandite. Mineral Mag 23:556-559

Hoffman E, Donnay G, Donnay JDH (1973) Symmetry and twinning of phillipsite and harmotome. Am Mineral 58:1105

Hogarth DD, Griffin WL (1980) Contact metamorphic lapis lazuli: the Italian Mountain deposits, Colorado. Can Mineral 18:59-70

Honda S, Muffler LJP (1970) Hydrothermal alteration in core from research drill hole Y-1, Upper Geyser Basin, Yellowstone National Park, Wyoming. Am Mineral 55:1714-1737

Howard DG (1994) Crystal habit and twinning of garronite from Fara Vicentina, Vicenza (Italy). N Jahrb Mineral Mh 1994:91-96

Howard DG, Tschernich RW, Smith JV, Klein GL (1990) Boggsite, a new high-silica zeolite from Goble, Columbia County, Oregon. Am Mineral 75:1200-1204

Hurlbut CS (1957) Bikitaite, $LiAlSi_2O_6{\cdot}H_2O$, a new mineral from Southern Rhodesia. Am Mineral 42:792-797

Hurlbut CS (1958) Additional data on bikitaite. Am Mineral 43:768-770.

Ibrahim K, Hall A (1995) New occurrences of diagenetic faujasite in the Quaternary tuffs of north-east Jordan. Eur J Mineral 7:1129-1135

Iijima A (1978) Geological occurrences of zeolite in marine environments. *In* Natural Zeolites. Occurrence, Properties, Use. LB Sand, FA Mumpton (eds) Pergamon Press, Oxford, p 175-198

Iijima A, Harada K (1968) Authigenic zeolites in zeolitic palagonite tuff on Oahu, Hawaii. Am Mineral 54:182-197

Ivanov OK, Mozzherin YuV (1982) Partheite from gabbro-pegmatites of Denezhkin Kamen, Urals (first occurrence in the USSR). Zap Vses Mineral O-va 111:209-214

Johnson GK, Flotow HE, O'Hare PAG, Wise WS (1983) Thermodynamic studies of zeolites: natrolite, mesolite and scolecite. Am Mineral 68:1134-1145

Karlsson HR, Clayton RN (1991) Analcime phenocrysts in igneous rocks: primary or secondary? Am Mineral 76:189-199

Karlsson HR, Clayton RN (1993) Analcime phenocrysts in igneous rocks: primary or secondary? —Reply. Am Mineral 78:230-232

Kastner M, Stonecipher SA (1978) Zeolites in pelagic sediments of the Atlantic, Pacific and Indian Oceans. *In* Natural Zeolites. Occurrence, Properties, Use. LB Sand, FA Mumpton (eds) Pergamon Press, Oxford, UK, p 199-220

Keith TEC, Staples LW (1985) Zeolites in Eocene basaltic pillow lavas of the Siletz River Volcanics, central Coast Range, Oregon. Clays & Clay Minerals 33:135-144

Keith TEC, Thompson JM, Mays RE (1983) Selective concentration of cesium in analcime during hydrothermal alteration, Yellowstone National Park, Wyoming. Geochim Cosmochim Acta 47: 795-804

Khomyakov AP, Kurova TA, Muravishkaya GI (1981) Merlinoite, first occurrence in the USSR. Dokl Akad Nauk SSSR 256:172-174

Khomyakov AP, Cherepivskaya GYe, Kurova TA, Kaptsov VV (1984) Amicchite, $K_2Na_2Al_4Si_4O_{16}{\cdot}5\ H_2O$, first find in the USSR. Dokl Acad Sci USSR, Earth Sci Sect 263:135-137

Khomyakov AP, Katayeva ZT, Kurova TA, Rudnitskaya Ye.S, Smol'yaninova N.N (1970) First find of brewsterite in the USSR. Dokl Acad Sci USSR, Earth Sci Sect 190:146-149

Khomyakov AP, Semenov EI, Bykova AV, Voronkov AA, Smol'yaninova N.N (1975) New data on lovdarite. Dokl Akad Nauk SSSR 221:699-702

Kile DE, Modreski PJ (1988) Zeolites and related minerals from the Table Mountain lava flows near Golden, Colorado. Mineral Rec 19:153-184

Kirfel A, Orthen M, Will G (1984) Natrolite: refinement of the crystal stucture of two samples from Marienberg (Usti nad Labem, CSSR). Zeolites 4:140-146

Kirov GN (1965) Calcium-rich clinoptilolite from the eastern Rhodopes. Ann Univ Sofia Fac Geol Geogr, Livre 1 Geol 60:193-200

Kocman V, Gait RI, Rucklidge J (1974) The crystal structure of bikitaite, $Li[AlSi_2O_6]{\cdot}H_2O$. Am Mineral 59: 71-78

Konev A, Sapozhnikov AN, Afonina GG, Vorob'ev EI, Arsenyuk MI, Lapides IL (1986) Perlialite from Murun alkaline massif. Zap Vses Mineral O-va 115:200-204

Kostov I (1962) The zeolites in Bulgaria analcime, chabazite, harmotome. Ann Univ Sofia Fac Geol Geogr, Livre 2 Geol 55:159-174

Koyama K, Takéuchi Y (1977) Clinoptilolite: the distribution of potassium atoms and its role in thermal stability. Z Kristallogr 145:216-239

Krogh Andersen E, Danø M, Petersen OV (1969) Contribution to the mineralogy of Ilimaussaq. XIII. A tetragonal natrolite. Meddelelser Grønland 181:1-19

Krogh Andersen E, Krogh Andersen IG, Ploug-Sørensen G (1990) Disorder in natrolites: structure determinations of three disordered natrolites and one lithium-exchanged disordered natrolite. Eur J Mineral 2:799-807

Kuwano N (1977) Scolecite from Makinokawa, Ehime Prefecture. Geosci Mag 28:265-270
Kuwano N, Tokumaru S (1993) Some zeolites from Chojabaru, Iki Island, Nagasaki Prefecture. With special reference to the occurrence of cowlesite and gobbinsite. Chigaku Kenkyu 42:159-167
Kvick A, Smith JV (1983) A neutron diffraction study of the zeolite edingtonite. J Chem Phys 79:2356-2362
Kvick A, Artioli G, Smith JV (1986) Neutron diffraction study of the zeolite yugawaralite at 13K. Z Kristallogr 174:265-281
Leavens PB, Hurlbut CS, Nelen JA (1968) Eucryptite and bikitaite from King's Mountain, North Carolina. Am Mineral 53:1202-1207
Lengauer CL, Giester G, Tillmanns E (1997) Mineralogical characterization of paulingite from Vinarická Hora, Czech Republic. Mineral Mag 61:591-606
Livingstone A (1986) A note on strontian chabazite from Kaiserstuhl, Baden, West Germany. Mineral Mag 50:348-349
Livingstone A (1989) A calcian analcime-bytownite intergrowth in basalt from Skye, Scotland, and calcian analcime relationships. Mineral Mag 53:382-385
Lucchetti G, Massa B, Penco AM (1982) Strontian heulandite from Campegli (Eastern Ligurian ophiolites, Italy). N Jahrb Mineral Mh 1982:541-550
Luhr JF, Kyser TK (1989) Primary igneous analcime: the Colima minettes. Am Mineral 74:216-223
Maleev MN (1976) Svetlozarite, a new high-silica zeolite. Zap Vses Mineral O-va 105:449-453
Malinovskii YuA (1984) The crystal structure of K-gmelinite. Kristallografiya 29:256-258
Mallet JV (1859) On brewsterite. The London, Edinburgh and Dublin Phil Mag and J Sci, 4th Ser 18:218-220
Marco JF, Gracia M, Gancedo JR, Gonzales-Carreno T, Arcoya A, Seoane XL (1995) On the state of iron in a clinoptilolite. Hyperfine Interact 95:53-70
Mason B, Sand LB (1960) Clinoptilolite from Patagonia: the relationship between clinoptilolite and heulandite. Am Mineral 45:341-350
Matsubara S, Kato A, Tiba T, Saito Y, Nomura M (1979) Pectolite, analcime, natrolite and thomsonite in altered gabbro from Yanai, Shinshiro, Aichi Prefecture, Japan. Mem Nat Sci Mus Tokyo 12:13-22
Mazzi F, Galli E (1978) Is each analcime different?. Am Mineral 63:448-460
Mazzi F, Galli E (1983) The tetrahedral framework of chabazite. N Jahrb Mineral Mh 1983:461-480
Mazzi F, Galli E, Gottardi G (1984) Crystal structure refinement of two tetragonal edingtonites. N Jahrb Mineral Mh 1984:373-382
Mazzi F, Larsen A.O, Gottardi G, Galli E (1986) Gonnardite has the tetrahedral framework of natrolite: experimental proof with a sample from Norway. N Jahrb Mineral Mh 1986:219-228
McCulloh TH, Frizzel VA, Stewart RJ, Barnes I (1981) Precipitation of laumontite with quartz, thenardite, and gypsum at Sespe Hot Springs, western Transverse Ranges, California. Clays & Clay Minerals 29:353-364
McCusker LB, Baerlocher C, Nawaz R (1985) Rietveld refinement of the crystal structure of the new zeolite mineral gobbinsite. Z Kristallogr 171:281-289
Mehegan JM, Robinson PT, Delaney JR (1982) Secondary mineralization and hydrothermal alteration in the Reydarfjordur drill core, eastern Iceland. J Geophys Res 87:6511-6524
Meier WM (1968) Zeolites structures: Molecular Sieves. Society of Chemical Industry, London, p 10-27
Meier WM, Olson DH (1970) Zeolite frameworks. Molecular Sieve Zeolites-I. Advances in Chemistry Series 101:155-170
Meier WM, Olson DH (1992) Atlas of zeolite structure types. Zeolites 12:1-200
Meixner H, Hey MH, Moss AA (1956) Some new occurrences of gonnardite. Mineral Mag 31:265-271
Meneghinello E, Martucci A, Alberti A, Di Renzo F (1999) Structural refinement of a K-rich natrolite: evidence of a new extraframework cation site. Microporous Mesoporous Materials 30:89-94
Men'shikov YuP (1984) Perlialite $K_9Na(Ca,Sr)[Al_{12}Si_{24}O_{72}]\cdot 15H_2O$—a new potassic zeolite from the Khibiny massif. Zap Vses Mineral O-va 113:607-612
Men'shikov YuP, Denisov AP, Uspenskaya YeI, Lipatova EA (1973) Lovdarite—a new hydrous beryllosilicate of alkalis. Dokl Akad Nauk SSSR 213:429-432
Merlino S (1990) Lovdarite, $K_4Na_{12}(Be_8Si_{28}O_{72})\cdot 18H_2O$, a zeolite-like mineral: structural features and OD character. Eur J Mineral 2:809-817
Merlino S, Galli E, Alberti A (1975) The crystal structure of levyne. Tschermaks mineral petrogr Mitt 22:117-129
Mikheeva MG, Pusheharovskii DYu, Khomyakov AP, Yamnova NA (1986) Crystal structure of tetranatrolite. Kristallografiya 31:254-257
Miller BE, Ghent ED (1973) Laumontite and barian-strontian heulandite from the Blairmore Group (Cretaceous), Alberta. Can Mineral 12:188-192
Minato H, Aoki M (1978) The mode of formation of clinoptilolite from volcanic glass—In the case of Tamatsukuri, Shimane Prefecture, Japan. Sci Pap Coll Gen Educ, Univ Tokyo 28:205-214

Minato H, Aoki M (1979) The genesis of Na-rich clinoptilolite in bentonite deposits—In the case of Myogi deposit in Hojun mine, Gunma Prefecture, Central Japan. Sci. Pap Coll Gen Educ, Univ Tokyo 29:63-74

Minato H, Aoki M, Inoue A, Utada M (1980) Microscopic heterogeneity in chemical composition of clinoptilolite in acidic tuffaceous rock—In the case of Takinoue, Kotooka, Akita Prefecture, Japan. Sci Pap Coll Gen Educ, Univ Tokyo 30:67-78

Mizota T, Shibuya G, Shimazu M, Takeshita Y (1974) Mineralogical studies on levyne and erionite from Japan. Mem Soc Geol Japan 11:283-290

Mohapatra BK, Sahoo RK (1987) Merlinoite in manganese nodules from the Indian Ocean. Mineral Mag 51:749-750

Morad S, Filippidis A, Aldahan AA, Collini B, Ounchanum P (1989) Stellerite and Sr-containing stilbite in granitic rocks from the Siljan Ring structure central Sweden. Bull Geol Inst Univ Uppsala 12:143-149

Mumpton FA (1960) Clinoptilolite redefined. Am Mineral 45:351-369

Münch P, Cochemé J-J (1993) Heulandite-group zeolites in volcaniclastic deposits of the southern Basin and Range province, Mexico. Eur J Mineral 5:171-180

Nativel P (1972) Zonation des zéolites du Cirque de Salazie (Ile de la Réunion). Bull Soc Geol Fr 14:173-178

Nativel P (1986) Découverte de mordénite à Cilaos, Ile de la Réunion, Océan Indien. Bull Minéral 109:337-347

Nawaz R (1980) Morphology, twinning, and optical orientation of gismondine. Mineral Mag 43:841-844

Nawaz R (1982) A chemical classification scheme for the gismondine group zeolites. Irish Naturalists J 20:480-483

Nawaz R (1984) New data on cowlesite from Northern Ireland. Mineral Mag 48:565-566

Nawaz R (1988) Gonnardite and disordered natrolite-group minerals: their distinction and relations with mesolite, natrolite and thomsonite. Mineral Mag 52:207-219

Nawaz R (1990) Brewsterite: re-investigation of morphology and elongation. Mineral Mag 54:654-656

Nawaz R, Malone JF (1981) Orientation and geometry of the thomsonite unit cell: a re-study. Mineral Mag 44:231-234

Nawaz R, Malone JF (1982) Gobbinsite, a new zeolite mineral from County Antrim, Northern Ireland. Mineral Mag 46:365-369

Nawaz R, Malone JF, Din VK (1985) Pseudo-mesolite is mesolite. Mineral Mag 49:103-105

Nishido H (1994) Gyrolite and epistilbite from Kusazumi, Hirado Island, Nagasaki Prefecture, Japan. Shizen Kagaku Kenkyusho Kenkyu Hokoku, Okayama, Rika Daigaku 20:69-75

Nishido H, Otsuka R (1981) Chemical composition and physical properties of dachiardite group zeolites. Mineral J. 10:371-384

Nishido H, Otsuka R, Nagashima K (1979) Sodium-rich dachiardite from Chichijima, the Ogasawara Islands, Japan. Waseda Daigaku Rikogaku Kenkyusho Hokoku 87:29-37

Noack Y (1983) Occurrence of thaumasite in seawater-basalt interaction, Mururoa atoll (French Polynesia, South Pacific). Mineral Mag 47:47-50

Noh JH, Kim SJ (1986) Zeolites from Tertiary tuffaceous rocks in Yeongil Area, Korea. Stud Surf Sci Catal (New Dev Zeolite Sci Technol) 28:59-66

Nørnberg P (1990) A potassium-rich zeolite in soil development on Danian chalk. Mineral Mag 54:91-94

Novak F (1970) Some new data for edingtonite. Acta Univ Carol, Geol 4:237-251

Ogihara S (1994) Ba-bearing clinoptilolite from ODP Leg 127, Site 795, Japan sea. Clays & Clay Minerals 42:482-484

Ogihara S, Iijima A (1989) Clinoptilolite to heulandite transformation in burial diagenesis. Stud Surf Sci Catal (Zeolites: Facts, Figures, Future) 49:491-500

Ogihara S, Iijima A (1990) Exceptionally K-rich clinoptilolite-heulandite group zeolites from three offshore boreholes off northern Japan. Eur J Mineral 2:819-826

Orlandi P, Sabelli C (1983) Ferrierite from Monastir, Sardinia, Italy. N Jahrb Mineral Mh 1983:498-504

Papezik VS, Elias P (1980) Tetragonal analcime from southeastern Newfoundland. Can Mineral 18: 73-75

Passaglia E (1969) Roggianite, a new silicate mineral. Clay Mineral 8:107-111.

Passaglia E (1970) The crystal chemistry of chabazites. Am Mineral 55:1278-1301

Passaglia E (1975) The crystal chemistry of mordenites. Contrib Mineral Petrol 50:65-77

Passaglia E (1978a) Lattice-constant variations in cation-exchanged chabazites. Natural Zeolites. Occurrence, Properties, Use. LB Sand, FA Mumpton (eds) Pergamon Press, Oxford, p 45-52

Passaglia E (1978b) New data on ferrierite from Weitendorf near Wildon, Styria, Austria. Mitt Abt Mineral Landesmus Joanneum 46:565-566

Passaglia E, Bertoldi G (1983) Harmotome from Selva di Trissino (Vicenza, Italy). Period Mineral 52: 75-82

Passaglia E, Pongiluppi D (1974) Sodian stellerite from Capo Pula, Sardegna. Lithos 7:69-73

Passaglia E, Pongiluppi D (1975) Barrerite, a new natural zeolite. Mineral Mag 40:208

Passaglia E, Sacerdoti M (1982) Crystal structural refinement of Na-exchanged stellerite. Bull Minéral 105:338-342

Passaglia E, Tagliavini A (1994) Chabazite-offretite epitaxial overgrowths in cornubianite from Passo Forcel Rosso, Adamello, Italy. Eur J Mineral 6:397-405

Passaglia E, Vezzalini G (1985) Crystal chemistry of diagenetic zeolites in volcanoclastic deposits of Italy. Contrib Mineral Petrol 90:190-198

Passaglia E, Vezzalini G (1988) Roggianite: revised chemical formula and zeolitic properties. Mineral Mag 52:201-206

Passaglia E, Artioli G, Gualtieri A (1998) The crystal chemistry of the zeolites erionite and offretite. Am Mineral 83:577-589

Passaglia E, Galli E, Rinaldi R (1974) Levynes and erionites from Sardinia, Italy. Contrib Mineral Petrol 43:253-259

Passaglia E, Pongiluppi D, Rinaldi R (1977) Merlinoite, a new mineral of the zeolite group. N Jahrb Mineral Mh 1977:355-364

Passaglia E, Pongiluppi D, Vezzalini G (1978a) The crystal chemistry of gmelinites. N Jahrb Mineral Mh 1978:310-324

Passaglia E, Tagliavini MA, Boscardin M (1992) Garronite, gonnardite and other zeolites from Fara Vicentina, Vicenza (Italy). N Jahrb Mineral Mh 1992:107-111

Passaglia E, Tagliavini A, Gutoni R (1996) Offretite and other zeolites from Fittà (Verona, Italy). N Jahrb Mineral Mh 1996:418-428

Passaglia E, Vezzalini G, Carnevali R (1990) Diagenetic chabazites and phillipsites in Italy: Crystal chemistry and genesis. Eur J Mineral 2:827-839

Passaglia E, Artioli G, Gualtieri A, Carnevali R (1995) Diagenetic mordenite from Ponza, Italy. Eur J Mineral 7:429-438

Passaglia E, Galli E, Leoni L, Rossi G (1978b) The crystal chemistry of stilbites and stellerites. Bull Minéral 101:368-375

Peacor DR (1973) High-temperature, single-crystal X-ray study of natrolite. Am Mineral 58:676-680

Peacor DR, Dunn PJ, Simmons WB, Tillmanns E, Fischer RX (1984) Willhendersonite, a new zeolite isostructural with chabazite. Am Mineral 69:186-189

Peacor DR, Dunn PJ, Simmons WB, Wicks FJ, Raudsepp M (1988) Maricopaite, a new hydrated Ca-Pb, zeolite-like silicate from Arizona. Can Mineral 26:309-313

Pearce TH (1970) The analcite-bearing volcanic rocks of the Crowsnest Formation, Alberta. Can J Earth Sci 7:46-66

Pearce TH (1993) Analcime phenocrysts in igneous rocks: primary or secondary?—Discussion. Am Mineral 78:225-229

Pechar F (1981) An X-ray diffraction refinement of the structure of natural natrolite. Acta Crystallogr B37:1909-1911

Pechar F (1982) An X-ray diffraction refinement of the structure of natural thomsonite. Cryst Res Technol 17:1141-1144

Pechar F (1988) The crystal structure of natural monoclinic analcime ($NaAlSi_2O_6{\cdot}H_2O$). Z Kristallogr 184:63-69

Pechar F (1989) An X-ray determination of the crystal structure of natural tetragonal natrolite. Z Kristallogr 189:191-194

Pechar F, Rykl D (1989) Crystal chemistry of selected natrolites of the volcanic Bohemia rocks (CSSR). Chem Erde 49:297-307

Pechar F, Schäfer W, Will G (1983) A neutron diffraction refinement of the crystal structure of natural natrolite, $Na_2Al_2Si_3O_{10}{\cdot}2H_2O$. Z Kristallogr 164:19-24

Pe-Piper G, Tsolis-Katagas P (1991) K-rich mordenite from late Miocene rhyolitic tuffs, Island of Samos, Greece. Clays & Clay Minerals 39:239-247

Perrotta AJ, Smith JV (1964) The crystal structure of brewsterite, $(Sr,Ba,Ca)(Al_2Si_6O_{16}){\cdot}5H_2O$. Acta Crystallogr 17:857-862

Pipping F (1966) The dehydration and chemical composition of laumontite. Mineral Soc India Int'l Mineral Assoc Volume, p 159-166

Pluth JJ, Smith JV (1990) Crystal structure of boggsite, a new high-silica zeolite with the first three-dimensional channel system bonded by both 12- and 10-rings. Am Mineral 75:501-507

Pluth JJ, Smith JV, Kvick A (1985) Neutron diffraction study of the zeolite thomsonite. Zeolites 5:74-80

Pokorny J (1966) Rock-joint minerals of the Ransko basic intrusion. Acta Univ Carol, Geol 1:45-60

Pöllmann H, Keck E (1990) Epitactic intergrowth of the zeolites phillipsite and gismondite from Gr. Teichelberg near Pechbrunn/Bavaria. N Jahrb Mineral Mh 1990:467-479

Pöllmann H, Keck E (1993) Replacement and incrustation pseudomorphs of zeolites gismondite, chabazite, phillipsite, and natrolite. N Jahrb Mineral Mh 1993:529-541

Pongiluppi D (1976) Offretite, garronite and other zeolites from "Central Massif," France. Bull Soc fr Minéral Cristallogr 99:322-327

Pongiluppi D (1977) A new occurrence of yugawaralite at Osilo, Sardinia. Can Mineral 15:113-114

Quartieri S, Vezzalini G (1987) Crystal chemistry of stilbites: structure refinements of one normal and four chemically anomalous samples. Zeolites 7:163-170

Rastsvetaeva RK (1995) Crystal structure of maximum-ordered tetranatrolite $Na_2[Si(Si_{0.5}Al_{0.5})_4O_{10}] \cdot 2H_2O$. Kristallografiya 40:812-815

Reay A, Coombs DS (1971) Ashtonite, a strontian mordenite. Mineral Mag 38:383-385

Regis AJ (1970) Occurrences of ferrierite in altered pyroclastics in central Nevada. Geol Soc Am Ann Meet Programs 2:661

Rice SB, Papke KG, Vaughan DEW (1992) Chemical controls on ferrierite crystallization during diagenesis of silicic pyroclastic rocks near Lovelock, Nevada. Am Mineral 77:314-328

Ridkosil T, Danek M (1983) New physical and chemical data for levyne-offretite intergrowths from Zezice, near Usti nad Labem, Czechoslovakia. N Jahrb Mineral Abh 147:99-108

Rinaldi R (1976) Crystal chemistry and structural epitaxy of offretite-erionite from Sasbach, Kaiserstuhl. N Jahrb Mineral Mh 1976:145-156

Rinaldi R, Vezzalini G (1985) Gismondine: the detailed X-ray structure refinement of two natural samples. Stud Surf Sci Catal (Zeolites: Synth Struct Technol Appl) 24:481-492

Rinaldi R, Pluth JJ, Smith JV (1974) Zeolites of the phillipsite family. Refinement of the crystal structures of phillipsite and harmotome. Acta Crystallogr B30:2426-2433

Rinaldi R, Smith JV, Jung G (1975) Chemistry and paragenesis of faujasite, phillipsite and offretite from Sasbach, Kaiserstuhl, Germany. N Jahrb Mineral Mh 1975:433-443

Robert C (1988) Barian phillipsite and strontian chabazite from the Plateau des Coirons, Ardèche, France. Bull Minéral 111:671-677

Robert C, Goffé B, Saliot P (1988) Zeolitisation of a basaltic flow in a continental environment: an example of mass transfer under thermal control. Bull Minéral 111:207-223

Robinson GW, Grice JD (1993) The barium analog of brewsterite from Harrisville, New York. Can Mineral 31:687-690

Ross M, Flohr, MJK, Ross DR (1992) Crystalline solution series and order-disorder within the natrolite mineral group. Am Mineral 77:685-703

Rouse RC, Peacor DR (1986) Crystal structure of the zeolite mineral goosecreekite, $CaAl_2Si_6O_{16} \cdot 5H_2O$. Am Mineral 71:1494-1501

Rouse RC, Peacor DR (1994) Maricopaite, an unusual lead calcium zeolite with an interrupted mordenite-like framework and intrachannel Pb_4 tetrahedral clusters. Am Mineral 79:175-184

Rouse RC, Dunn PJ, Grice JD, Schlenker JL, Higgins JB (1990) Montesommaite, $(K,Na)_9Al_9Si_{23}O_{64} \cdot 10H_2O$, a new zeolite related to merlinoite and the gismondine group. Am Mineral 75:1415-1420

Rüdinger B, Tillmanns E, Hentschel G (1993) Bellbergite—a new mineral with the zeolite structure type EAB. Mineral Petrol 48:147-152

Rychly R, Ulrych J (1980) Mesolite from Horní Jílové, Bohemia. Tschermaks mineral petrogr Mitt 27:201-208

Sacerdoti M, Gomedi I (1984) Crystal structure refinement of Ca-exchanged barrerite. Bull Minéral 107:799-804

Sacerdoti M, Passaglia E, Carnevali R (1995) Structural refinements of Na-, K-, and Ca-exchanged gmelinites. Zeolites 15:276-281

Sacerdoti M, Sani A, Vezzalini G (1999) Structural refinement of two barrerites from Alaska. Microporous Mesoporous Materials 30:103-109

Saha P (1959) Geochemical and X-ray investigation of natural and synthetic analcime. Am Mineral 44:300-313

Sameshima T (1978) Zeolites in tuff beds of the Miocene Waitemata Group, Auckland province, New Zealand. *In* Natural Zeolites. Occurrence, Properties, Use. LB Sand, FA Mumpton (eds) Pergamon Press, Oxford, p 309-317

Sameshima T (1986) Ferrierite from Tapu, Coromandel Peninsula, New Zealand, and a crystal chemical study of known occurrences. Mineral Mag 50:63-68

Sarp H, Deferne J, Bizouard H, Liebich B.W (1979) La parthéite, $CaAl_2Si_2O_8 \cdot 2H_2O$, un nouveau silicate naturel d'aluminium et de calcium. Schweiz mineral petrogr Mitt 59:5-13

Schlenker JL, Pluth JJ, Smith JV (1977) Refinement of the crystal structure of brewsterite, $Ba_{0.5}Sr_{1.5}Al_4Si_{12}O_{32} \cdot 10H_2O$. Acta Crystallogr B33:2907-2910

Seki Y (1971) Some physical properties of analcime-wairakite solid solutions. J Geol Soc Japan 77:1-8

Seki Y (1973) Distribution and modes of occurrence of wairakites in the Japanese Island arc. J Geol Soc Japan 79:521-527

Seki Y, Haramura H (1966) On chemical composition of yugawaralite. J Jpn Mineral Petrogr Econ Geol 59:107-111

Seki Y, Oki Y (1969) Wairakite-analcime solid solution from low-grade metamorphic rocks of the Tanzawa Mountains, central Japan. Mineral J 6:36-45

Shepard AO, Starkey HC (1964) Effect of cation exchange on the thermal behavior of heulandite and clinoptilolite. U S Geol Surv Prof Paper 475-D:89-92

Shepard AO, Starkey HC (1966) The effects of exchanged cations on the thermal behavior of heulandite and clinoptilolite. Mineral Soc India Int'l Mineral Assoc Vol, p 155-158

Sheppard RA, Gude AJ 3d (1968) Distribution and genesis of authigenic silicate minerals in tuffs of Pleistocene Lake Tecopa, Inyo County, California. U S Geol Surv Prof Paper 597:1-38

Sheppard RA, Gude AJ 3d (1969a) Diagenesis of tuffs in the Barstow Formation, Mud Hills, San Bernardino County, California. U S Geol Surv Prof Paper 634:1-35

Sheppard RA, Gude AJ 3d (1969b) Chemical composition and physical properties of the related zeolites offretite and erionite. Am Mineral 54:875-886

Sheppard RA,Gude, AJ 3d (1970) Calcic siliceous chabazite from the John Day Formation, Grant County, Oregon. U S Geol Surv Prof Paper 700-D:176-180

Sheppard RA, Gude, AJ 3d (1971) Sodic harmotome in lacustrine Pliocene tuffs near Wikieup, Mohave County, Arizona. U S Geol Surv Prof Paper 750-D:50-55

Sheppard RA, Gude, AJ 3d (1973) Zeolites and associated authigenic silicate minerals in tuffaceous rocks of the Big Sandy Formation, Mohave County, Arizona. U S Geol Surv Prof Paper 830:1-36

Sheppard RA, Gude, AJ 3d (1983) Harmotome in a basaltic, volcaniclastic sandstone from a lacustrine deposit near Kirkland Junction, Yavapai County, Arizona. Clays & Clay Minerals 31:57-59

Sheppard RA, Gude, AJ 3d, Edson, GM (1978) Bowie zeolite deposit, Cochise and Graham Counties, Arizona. *In* Natural Zeolites. Occurrence, Properties, Use. LB Sand, FA Mumpton (eds) Pergamon Press, Oxford, p 319-328

Sheppard RA, Gude, AJ 3d, Fitzpatrick JJ (1988) Distribution, characterization, and genesis of mordenite in Miocene silicic tuffs at Yucca Mountain, Nye County, Nevada. U S Geol Surv Bull 1777:1-22

Sheppard RA, Gude AJ 3d, Griffin JJ (1970) Chemical composition and physical properties of phillipsite from the Pacific and Indian Oceans. Am Mineral 55:2053-2062.

Sheppard RA, Gude AJ 3d, Munson EL (1965) Chemical composition of diagenetic zeolites from tuffaceous rocks of the Mojave Desert and vicinity, California. Am Mineral 50:244-249

Sheppard RA,Gude, AJ 3d, Desborough, G.A, White, J.S (1974) Levyne-offretite intergrowths from basalt near Beech Creek, Grant County, Oregon. Am Mineral 59:837-842

Smith JV (1963) Structural classification of zeolites. Mineral Soc Am Spec Paper 1:281-290

Smith JV (1974) Feldspar Minerals 1. Springer, Berlin-Gottingen-Heidelberg, 627 p

Smith JV, Knowles CR, Rinaldi F (1964) Crystal structures with a chabazite framework. III. Hydrated Ca-chabazite at +20 and -150°C. Acta Crystallogr 17:374-384

Smith JV, Pluth JJ, Artioli G, Ross F.K (1984) Neutron and X-ray refinements of scolecite. *In* Proc 6th Int'l Zeolite Conf, Reno 1983, D. Olson, A. Bisio (eds) Butterworths, Guilford, UK, p 842-850

Staples LW, Gard AJ (1959) The fibrous zeolite erionite: its occurrence, unit cell, and structure. Mineral Mag 32:261-281

Steiner A (1955) Wairakite, the calcium analogue of analcime, a new zeolite mineral. Mineral Mag 30:691-698

Stolz J, Armbruster T (1997) X-ray single-crystal structure refinement of Na, K-rich laumontite, originally designated 'primary leonhardite'. N Jahrb Mineral Mh 1997:131-144

Stonecipher SA (1978) Chemistry and deep-sea phillipsite, clinoptilolite, and host sediments. *In* Natural Zeolites. Occurrence, Properties, Use. LB Sand, FA Mumpton (eds) Pergamon Press, Oxford, p 221-234

Sukheswala RN, Avasia RK, Gangopadhyay M (1974) Zeolites and associated secondary minerals in the Deccan traps of western India. Mineral Mag 39:658-671

Surdam RC (1966) Analcime-wairakite mineral series. Geol Soc Am Spec Paper 87:169-170

Surdam RC, Eugster HP (1976) Mineral reactions in the sedimentary deposits of the Lake Magadi region, Kenya. Geol Soc Am Bull 87:1739-1752

Surdam RC, Sheppard RA (1978) Zeolites in saline, alkaline-lake deposits. *In* Natural Zeolites. Occurrence, Properties, Use. LB Sand, FA Mumpton (eds) Pergamon Press, Oxford, p 145-174

Takéuchi Y, Mazzi, F, Haga, N, Galli, E (1979) The crystal structure of wairakite. Am Mineral 64: 993-1001

Talati DJ (1978) Copper bearing aventurine zeolite in an occurrence in India. J Gemmol 16:186-190

Tiba T, Matsubara S (1977) Levyne from Dózen, Oki Islands, Japan. Can Mineral 15:536-539

Tillmanns E, Fischer RX, Baur, H (1984) Chabazite-type framework in the new zeolite willhen-dersonite $KCaAl_3Si_3O_{12}·5H_2O$. N Jahrb Mineral Mh 1984:547-558
Tschernich RW (1992) Zeolites of the World. Geosciences Press, Phoenix, Arizona, 563 p
Tschernich RW, Wise WS (1982) Paulingite: variations in composition. Am Mineral 67:799-803
Tsolis-Katagas P, Katagas C (1989) Zeolites in pre-caldera pyroclastic rocks of the Santorini volcano, Aegean Sea, Greece. Clays & Clay Minerals 37:497-510
Tsolis-Katagas P, Katagas C (1990) Zeolitic diagenesis of Oligocene pyroclastic rocks of the Metaxades area, Thrace, Greece. Mineral Mag 54:95-103
Ueno T, Hanada K (1982) Chemical compositions and geneses of zeolites from Tsuyazaki, Fukuoka Prefecture, Japan. J Mineral Soc Japan 15:259-272
Utada M, Vine JD (1984) Zonal distribution of zeolites and authigenic plagioclase, Spanish Peaks region, southern Colorado. *In* Proc 6th Int'l Zeolite Conf, Reno 1983, D. Olson, A. Bisio (eds) Butterworths, Guilford, UK, p 604-615
Utada M, Minato H, Amano H, Akiyama K (1972) Migration of chemical components of "Oya-ishi" accompanying zeolitization and other alterations. Sci Pap Coll Gen Educ, Univ Tokyo 22:79-97
Vannucci S, Pancani MG, Vasselli O, Coradossi N (1992) Presence of clinoptilolite in the Maastrichtian pelagic sediments of the Barranco del Gredero Section (Caravaca, S-E Spain). Chem Erde 52:165-177
Vezzalini G (1984) A refinement of Elba dachiardite: opposite acentric domains simulating a centric structure. Z Kristallogr 166:63-71
Vezzalini G, Mattioli V (1979) Secondo ritrovamento della roggianite. Period Mineral 48:15-20
Vezzalini G, Oberti R (1984) The crystal chemistry of gismondines: the non-existence of K-rich gismondines. Bull Minéral 107:805-812
Vezzalini G, Artioli G, Quartieri S (1992) The crystal chemistry of cowlesite. Mineral Mag 56: 575-579
Vezzalini G, Quartieri S, Passaglia E (1990) Crystal structure of a K-rich natural gmelinite and comparison with the other refined gmelinite samples. N Jahrb Mineral Mh 1990:504-516
Vezzalini G, Quartieri S, Rossi A, Alberti A (1994) Occurrence of zeolites from northern Victoria Land (Antarctica). Terra Antarctica 1:96-99
Vezzalini G, Quartieri S, Galli E (1997) Occurrence and crystal structure of a Ca-pure willhendersonite. Zeolites 19:75-79
Viladkar SG, Avasia RK (1994) Analcime-phonolite and associated alkaline rocks of Panwad-Kawant Complex, Gujarat, India. Chem Erde 54:49-66
Vitali F, Blanc G, Larqué P (1995) Zeolite distribution in volcaniclastic deep-sea sediments from the Tonga Trench margin (SW Pacific). Clays & Clay Minerals 43:92-104
Walker GPL (1962a) Garronite, a new zeolite, from Ireland and Iceland. Mineral Mag 33:173-186
Walker GPL (1962b) Low-potash gismondine from Ireland and Iceland. Mineral Mag 33:187-201
Walton AW (1975) Zeolitic diagenesis in Oligocene volcanic sediments, Trans-Pecos Texas. Geol Soc Am Bull 86:615-624
Wilkinson JFG (1977) Analcime phenocrysts in a vitrophyric analcimite—Primary or secondary? Contrib Mineral Petrol 64:1-10
Wilkinson JFG, Hensel HD (1994) Nephelines and analcimes in some alkaline igneous rocks. Contrib Mineral Petrol 118:79-91
Wise WS (1978) Yugawaralite from Bombay, India. Mineral Record 9:296
Wise WS (1982) New occurrence of faujasite in southeastern California. Am Mineral 67:794-798
Wise WS, Kleck WD (1988) Sodic clay-zeolite assemblage in basalt at Boron, California. Clays & Clay Minerals 36:131-136
Wise WS, Tschernich RW (1975) Cowlesite, a new Ca-zeolite. Am Mineral 60:951-956
Wise WS, Tschernich RW (1976a) The chemical compositions and origin of the zeolites offretite, erionite, and levyne. Am Mineral 61:853- 863
Wise WS, Tschernich RW (1976b) Chemical composition of ferrierite. Am Mineral 61:60-66
Wise WS, Tschernich RW (1978a) Habits, crystal forms and composition of thomsonite. Can Mineral 16:487-493
Wise WS, Tschernich RW (1978b) Dachiardite-bearing zeolite assemblages in the Pacific Northwest. *In* Natural Zeolites. Occurrence, Properties, Use. L.B Sand, F.A. Mumpton (eds) Pergamon Press, Oxford, p 105-111
Wise WS, Nokleberg WJ, Kokinos M (1969) Clinoptilolite and ferrierite from Agoura, California. Am Mineral 54:887-895
Woolley AR, Symes RF (1976) The analcime-phyric phonolites (blairmorites) and associated analcime kenytes in the Lupata Gorge, Mocambique. Lithos 9:9-15
Yajima S, Nakamura T (1971) New occurrence of ferrierite. Mineral J 6:343-364
Yakubovich OV, Massa W, Pekov IV, Kucherinenko YaV (1999) Crystal structure of a Na,K-variety of merlinoite. Crystallogr Reports 44:776-782

Yamazaki A, Otsuka R (1989) Effects of exchangeable cations on the thermal behavior of fibrous zeolites. Stud Surf Sci Catal (Zeolites: Facts, Figures, Future) 49:533-542

Yang P, Armbruster T (1996) (010) disorder, partial Si, Al ordering and Ca distribution in triclinic (*C*1) epistilbite. Eur J Mineral 8:263-271

Yefimov AF, Ganzeyev AA, Katayeva ZT (1966) Finds of strontian thomsonite in the USSR. Dokl Akad Nauk SSSR 169:148-150

Yoshimura T, Wakabayashi S (1977) Na-dachiardite and associated high-silica zeolites from Tsugawa, northeast Japan. Sci Rep Niigata Univ, Ser Econ Geol Mineral 4:49-65

Zoltai T (1960) Classification of silicates and other minerals with tetrahedral structures. Am Mineral 45:960-973

3 Geochemical Stability of Natural Zeolites

Steve J. Chipera
Los Alamos National Laboratory
Mail Stop D469
Los Alamos, New Mexico 87545

John A. Apps
Lawrence Berkeley National Laboratory
Mail Stop 90-1116, 1 Cyclotron Road
Berkeley, California 94720

INTRODUCTION

Zeolites are used in numerous agricultural, commercial, and environmental applications (e.g. as soil conditioners and fertilizers, and as adsorbents for ammonia, heavy metals, nuclear and organic wastes). It is important to understand their stability to insure their persistence and effectiveness in these applications. An analysis of the occurrence of natural zeolites provides basic data on conditions favorable for zeolite stability and formation. All such environments have neutral to alkaline waters and, with few exceptions, are associated with low-temperature (<300°C) alteration of highly reactive volcanic rocks containing natural glasses. The principal exceptions are certain deep sedimentary sequences, sometimes associated with petroleum maturation where volcanic ash falls are absent, and deep-sea sediments where biogenic opal may be the highly reactive phase. Since initial recognition of the zeolite facies in diagenetic alteration (Coombs et al. 1959), investigators have struggled to develop a quantitative basis for defining zeolite stability fields by development of both the thermodynamic relations and kinetic factors governing zeolite formation. An additional motivation for study in the United States and elsewhere is the occurrence (and use) of zeolites at potential sites for disposing of high-level radioactive waste in underground repositories. Regulatory requirements for a high-level radioactive waste repository, reflecting public concerns over the integrity of subsurface disposal facilities, require unprecedented quantitative predictions of the physical-chemical response of zeolites to heat generated during radioactive decay and to waste-rock interactions over a time period exceeding tens of thousands of years. In light of such requirements, investigators are taking a new look at the thermodynamic and kinetic factors affecting zeolitization.

The geochemical conditions that result in zeolite formation have been outlined in numerous studies (e.g. Hay 1966, 1978; Iijima 1978, Surdam and Sheppard 1978, Sheppard and Hay, this volume; Hay and Sheppard, this volume). Aqueous silica activity, cation concentrations, and pH have all been found to be important fluid parameters that determine which zeolites will form or whether zeolites will form at all. Temperature is an important variable in zeolite stability, partly because of the high water content of zeolites. Pressure plays a lesser role as can be concluded from the similarities in zeolite assemblages in geothermal areas compared with those in burial metamorphic settings. Fluid composition exerts a significant control on zeolite assemblages, as is dramatically illustrated in a comparison of saline-alkaline lakes and open hydrologic systems; very different zeolite assemblages are observed despite similarities in temperature and protolith composition.

The effects of variable water chemistry on zeolite diagenetic reactions were observed during early studies of saline-alkaline lakes (Hay 1966; Sheppard and Gude 1968, 1969). Alkalic, silicic zeolites such as clinoptilolite, erionite, and phillipsite developed on the

1529-6466/00/0045-0003$10.00

margins of the deposits, which in turn grade into analcime followed by potassium feldspar at the centers of the lakes (e.g. the Big Sandy Formation, Sheppard and Gude 1973). It was inferred that alkalic, silicic zeolites formed initially and subsequently transformed into analcime where the Na^+ concentration increased as water evaporated. Boles (1971) experimentally reproduced this result by reacting various clinoptilolite/heulandites in NaOH and Na_2CO_3 solutions at 100°C to form analcime. He concluded that analcime formation is favored by increased pH and Na^+ concentration and that the Si:Al ratio of the analcime is largely a function of the Si:Al ratio of the precursor zeolite regardless of the presence or absence of quartz. The Na^+ concentrations in the analcime-producing experiments (~2700-4600 ppm) were lower than the concentrations that can occur in modern saline-alkaline lakes (>100,000 ppm Na; Jones et al. 1967, Surdam and Sheppard 1978). Iijima (1975) noted that the clinoptilolite-to-analcime reaction occurred at significantly higher temperatures in marine geosynclinal deposits (~84-91°C) compared with ambient temperatures in the saline-alkaline lakes. He also suggested that the chemical composition of the pore water (in particular the Na^+ concentration) was the primary factor in determining reaction temperatures.

In evaluating other factors that may control zeolite diagenesis, several investigators have examined the silica polymorphs coexisting with clinoptilolite and analcime (e.g. Honda and Muffler 1970, Keith et al. 1978, Kerrisk 1983, Duffy 1993). Several of these studies suggest that aqueous silica activity, rather than aqueous Na^+ concentration, is the predominant factor controlling the reaction of clinoptilolite to analcime. Clinoptilolite is frequently observed to coexist with the more soluble silica polymorphs (opal-CT and cristobalite), whereas analcime generally coexists only with quartz. This correlation is well documented in the mineralogic data at Yucca Mountain (Bish and Chipera 1989) and in the altered volcanic tuffs at Yellowstone National Park (Honda and Muffler 1970, Keith et al. 1978, Bargar and Keith 1995).

In most environments, zeolite formation requires a volcanic glass precursor. Dissolution of glass creates a solution that is greatly oversaturated with respect to many minerals. Although zeolite assemblages are normally not the most thermodynamically stable assemblage, they form metastably because of the slower kinetics involved in forming the more stable assemblage (Dibble and Tiller 1981, Moncure et al. 1981). Given sufficient time, the metastable zeolites eventually react to more thermodynamically stable assemblages. The composition of the precursor volcanic glass exerts some control as to which zeolites will form. Silicic zeolites such as clinoptilolite, mordenite, and erionite form from the alteration of silicic glasses, whereas the more aluminous zeolites, phillipsite, natrolite, analcime, and heulandite usually form from the alteration of basaltic glasses. Thermodynamic modeling by Kerrisk (1983) using EQ3/6 showed that high-silica zeolites (clinoptilolite and mordenite) would form along with smectite during the initial stages of silicic-volcanic glass dissolution. Kerrisk found it necessary to suppress quartz precipitation in his calculations (mimicking the supersaturation of silica) in order to form the high-silica zeolites. However, he obtained unrealistic results if cristobalite were also suppressed, causing the silica activity to reach amorphous-silica saturation. If reactions were allowed to continue, silica activity eventually dropped and the reactions progressed to a quartz-analcime-illite assemblage and finally to a quartz-albite-potassium feldspar assemblage with possible calcite.

Although some variation in zeolitic assemblages can be attributed to compositional differences in the volcanic rock or water composition, the effect of temperature often predominates. Figure 1 shows the approximate zeolite zonation associated with alteration of Icelandic tholeiitic basalts, summarized from several investigations (Apps 1983). As rock compositions are essentially identical, temperature appears to be controlling which zeolite

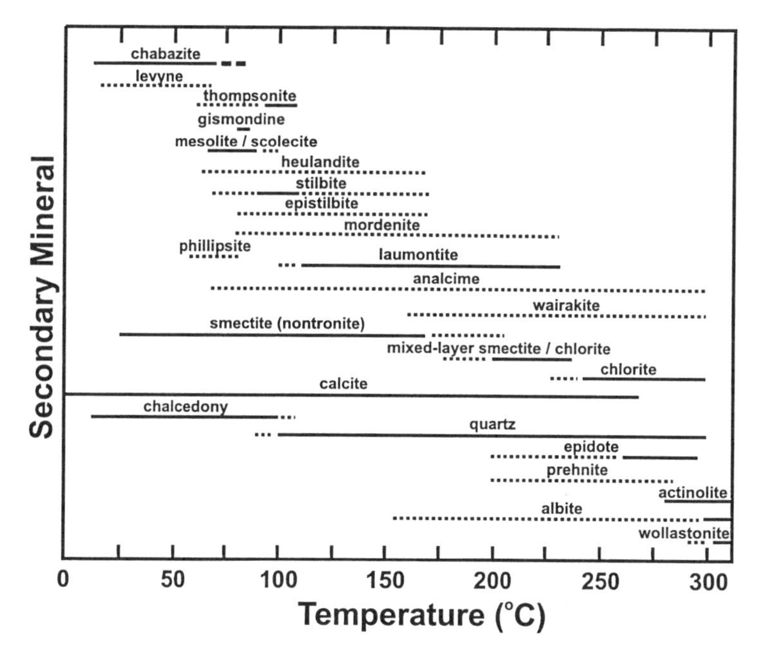

Figure 1. Observed distribution of secondary minerals as a function of temperature in hydrothermally altered Icelandic basalt (modified from Apps 1983).

forms. Similar zonations are observed, often with identical mineral assemblages, in volcanic rocks ranging from basaltic to andesitic in composition. The mineralogical uniformity strongly suggests that a wide range of compositional variability is accommodated by variations in the proportions and compositions of the alteration assemblage present. The composition of groundwater coexisting with altered volcanic rock also shows a similar consistency.

Success in resolving the many complex issues concerning the formation and stability of zeolites has been limited, partly because of the chemical complexity of naturally occurring systems containing zeolites, but also because of the low temperatures at which most zeolites form. At these temperatures, reaction rates are slow and it is difficult to conduct experiments that determine zeolite solubilities or phase equilibria under conditions simulating the natural environment. Despite these difficulties, some progress has been made, and the following pages discuss the approaches that have been taken.

SOLUBILITY OF GLASS PRECURSORS

Most zeolite-bearing assemblages originate from the decomposition of volcanic glass. The mechanisms through which natural glasses decompose in the presence of water are complex and not fully understood. Observations reported in the literature show that the alteration mechanisms are strongly influenced by glass composition, resulting in at least two distinct reaction products (Apps 1987). Rhyolitic glasses form an open, "stuffed tridymite" structure (Taylor and Brown 1979) which can easily hydrate and exchange cations without loss of the aluminosilicate framework. Such glasses can remain in the vitreous state for tens of millions of years (Forsman 1984). In contrast, glasses of basaltic composition do not appear to retain their vitreous aluminosilicate framework upon hydration but decompose to form a proto-smectite structure resulting in the formation of palagonite (e.g. Eggleton and Keller 1982).

Natural silicate glasses contain between 45 and 75 wt % SiO_2. Dissolution of these glasses in groundwater results in supersaturation with respect to the crystalline silica polymorphs. The sequence of polymorphs that crystallize during the hydrolysis of natural glass in a partially closed or closed system is amorphous silica to opal-CT to quartz, with dissolution and precipitation the most likely transformation mechanism (Apps 1987). Although some information is available to estimate the kinetics of dissolution of the silica polymorphs (e.g. Rimstidt and Barnes 1980, Renders et al. 1995), the exact mechanism and rate of conversion among the polymorphs has not yet been established.

The mechanisms of zeolite formation from natural glass are also not fully understood. Field studies, however, provide some insight into the reaction paths (e.g. Hay and Sheppard 1977, Tribble and Wilkens 1994, Münch et al. 1996). The basic observations are the initial formation of smectite coating the glass shards followed by either dissolution of the glass shards with precipitation of zeolites or direct replacement of the glass shard by zeolites. Spherules of opal-CT are often observed accompanying the zeolites as a late-stage product. Although further study of the zeolitization of glassy-rhyolitic ash could modify hypotheses regarding this process, we believe that the alteration sequence in the presence of liquid water proceeds as follows:

- Glass hydration and ion exchange;
- Incipient nucleation of smectite on shard surfaces;
- Glass hydration continues, with increasing alkalinity of the solution and accelerated dissolution of the glass. Clinoptilolite is stabilized in favor of smectite by the increasing alkalinity of the solution.
- Nucleation and rapid growth of clinoptilolite from solution with concurrent dissolution of hydrated glass at the growth interface;
- Clinoptilolite crystallites coarsen as the degree of clinoptilolite supersaturation falls, leading to fewer nucleation sites and slower growth.
- Cations available for clinoptilolite growth are depleted due to exhaustion of these species from the glass, or to leaching and transport by groundwater and formation of opal-CT from the silica residue.

Ostwald (1900) used kinetic arguments to deduce that an unstable system would approach equilibrium through a sequence of progressively lower energy states in which less stable minerals form first followed by more stable phases. In diagenetically or hydrothermally altered vitric volcanic rocks, the so called "Ostwald Step Rule" is well illustrated as a sequence of metastable phases that transform to more stable assemblages over time (Dibble and Tiller 1981). The secondary mineral assemblages observed in altered volcanic rocks are usually metastable assemblages that will eventually recrystallize to progressively lower energy states represented by more stable mineral assemblages (e.g. glass $\Rightarrow$ zeolite + opal-CT $\Rightarrow$ feldspar + quartz). In geothermal systems, the rates of dissolution and precipitation are enhanced because of the elevated temperatures, and consequently the glass and early-formed metastable phases are rapidly destroyed. In diagenetic systems, temperatures typically are much lower and the metastable phases can persist for much greater periods of time.

THERMODYNAMIC APPROACH TO EVALUATING ZEOLITE STABILITY RELATIONS

Derivation of thermodynamic functions and databases

Zeolite facies rocks constitute a multi-component heterogeneous assemblage of minerals. The key to modeling thermodynamic equilibria of any mineral system, including zeolitic assemblages, is an accurate internally consistent thermodynamic database for minerals and the properties of the aqueous phase. Such thermodynamic databases usually consist of tabulations of ΔG°_f, ΔH°_f, S°, C_p, V°, α, and β (Gibbs free energy of formation at standard state, enthalpy of formation at standard state, third law entropy at standard state, heat capacity, volume at standard state, expansivity, and compressibility) for solid phases, partial molar properties for solid phases that possess compositional variability, and a model for the properties of aqueous species as a function of concentration. Standard state conditions are typically reported as 298 K and 1 bar (10^5 Pa) pressure. The data employed in computing the thermodynamic properties of minerals include those obtained through calorimetry, phase equilibria, aqueous solubility measurements, and information obtained using X-ray crystallography. Other sources of information may include electrochemical, spectrometric data, and phase relations and groundwater compositions observed in the field.

Often, the thermodynamic characterization of a phase or species is made using a combination of sources, some of which may be redundant. Reconciling the results has led to various strategies that can have a marked impact on the consistency of the data and their suitability for geochemical modeling. A compilation of thermodynamic properties of minerals normally requires internally consistent thermodynamic data sets that can be linked to reference phases. The traditional method, used by Helgeson et al. (1978), is to compare various data sets and make judgments as to which set(s) are more likely to be correct or of better quality. Statistical regression techniques are then used to refine the data and obtain best values of the thermodynamic parameters, usually starting with reference phases such as corundum, quartz, and lime, and progressing sequentially from minerals with well-established thermodynamic properties to those that are less well characterized. This method is conceptually satisfying in that it permits the investigator to start from a secure base and concentrate on the evaluation of specific subsets of experimental data. However, it possesses a disadvantage in that errors can be propagated rather than distributed. More recent compilations of thermodynamic data have been made with the goal of eliminating or at least mitigating error propagation, and in which internally consistent functions are used to analyze related thermodynamic parameters. All of the data sets are analyzed simultaneously in order to find those that are most consistent. Regardless of the approach used, the most important requirement is that the investigator assumes a critical attitude towards the data and insures that independent checks are available to verify the selected thermodynamic parameters.

Estimation of thermodynamic data for zeolites

Thermodynamic properties of minerals can also be obtained by utilizing an empirical method to estimate the thermodynamic properties. One of the primary reasons for using estimated thermodynamic data for zeolites, even though data have now been measured for many zeolites, is the highly variable chemistry typical of many zeolites. Thermodynamic properties measured for a zeolite in one sample may not be representative of the thermodynamic properties of the same zeolite from other samples. For example, sedimentary analcime is more Si rich (Si:Al ratios of 2.4-2.8) than the analcime used by Johnson et al. (1982) in their calorimetric studies (Si:Al = 2.15). Recent solubility measurements of a stoichiometric analcime and a Si-rich sedimentary analcime yielded standard Gibbs free energies of formation at 25°C of –3089.2 and –3044.4 kJ/mol, respectively (Wilkin and Barnes 1998). Similarly, Si:Al ratios for sedimentary clinoptilolite show considerable variations often ranging from four to five. Many zeolites, such as clinoptilolite, have the added complexity that the exchangeable-cation composition ranges throughout much of the Na-Ca-K ternary.

Various empirical routines have been formulated for estimating thermodynamic data for minerals. Recent methods represent mineral phases as a set of elemental building blocks composed of unique polyhedra (Robinson and Haas 1983, Berman and Brown 1985, Hazen 1985, Chermak and Rimstidt 1989, Holland 1989). By summing the contributions from each polyhedron, thermodynamic properties can be estimated for the complete mineral unit. In the present study, thermodynamic data for the zeolites were estimated using the methods proposed by Berman and Brown (1985) for heat capacity, Chermak and Rimstidt (1989) for Gibbs free energy and enthalpy of formation at 298 K, and Holland (1989) for entropy. Holland's method of estimating entropy from individual polyhedra improves on previous methods by incorporating molar volume data. This improvement is important because entropy depends not only on the individual polyhedra comprising the phase, but also on how tightly the polyhedra are assembled. Because Holland proposed only two choices for entropy contributions of water in the polyhedrally derived structure (20.74 J/(mol•K) for structurally bound and 30.03 J/(mol•K) for "loosely bound" water), a value for "zeolitic" water was determined for the present calculations (in this treatment, "zeolitic" water is more weakly bonded than "loosely bound" water). An average value of 59.1 J/(mol•K) for zeolitic water (41.1 J/(mol•K) if using volume terms) was determined by comparing calculated entropies of anhydrous forms of analcime, heulandite, mordenite, phillipsite, clinoptilolite and stilbite with the measured entropy values for hydrous forms (Johnson et al. 1982, 1985, 1992; Hemingway and Robie 1984, Howell et al. 1990).

The validity of the estimation methods was assessed by comparing estimated and measured thermodynamic data for various zeolites (Tables 1, 2, and Fig. 2). The following

Table 1. Measured zeolite thermodynamic data used for comparison with estimated data.

Mineral	*Data source*	*Chemical formula*
Analcime	Johnson et al. (1982, 1992)	$Na_{0.96}Al_{0.96}Si_{2.04}O_6 \cdot H_2O$
Clinoptilolite	Hemingway and Robie (1984)	$(Na_{0.56}K_{0.98}Ca_{1.50}Mg_{1.23})(Al_{6.7}Fe_{0.3})Si_{29}O_{72} \cdot 22H_2O$
Clinoptilolite[1]	Johnson et al. (1991)	$(Na_{0.954}K_{0.543}Ca_{0.861}Mg_{1.24}Ba_{0.062}Sr_{0.036}Mn_{0.022})(Al_{3.45}Fe_{0.017})Si_{14.533}O_{36} \cdot 10.922H_2O$
Heulandite[2]	Johnson et al. (1985, 1992)	$(Ba_{0.065}Sr_{0.175}Ca_{0.585}K_{0.132}Na_{0.383})Al_{2.165}Si_{6.835}O_{18} \cdot 6.00H_2O$
Laumontite	Kiseleva et al. (1996a)	$Ca_{2.0}Al_{4.0}Si_{8.0}O_{24} \cdot 8H_2O$
Leonhardite	Kiseleva et al. (1996b)	$Ca_{2.0}Al_{4.0}Si_{8.0}O_{24} \cdot 7H_2O$
Merlinoite	Donahoe et al. (1990a,b)	$KAlSi_{1.81}O_{5.62} \cdot 1.69H_2O$
Mesolite	Johnson et al. (1983, 1992)	$(Na_{0.676}Ca_{0.657})Al_{1.99}Si_{3.01}O_{10} \cdot 2.647H_2O$
Mordenite	Johnson et al. (1992)	$(Ca_{0.289}Na_{0.361})Al_{0.940}Si_{5.060}O_{12} \cdot 3.468H_2O$
Natrolite	Johnson et al. (1983, 1992)	$Na_{2.0}Al_{2.0}Si_{3.0}O_{10} \cdot 2.0H_2O$
Phillipsite	Hemingway and Robie (1984)	$(Na_{1.08}K_{0.80})Al_{1.88}Si_{6.12}O_{16} \cdot 6H_2O$
Scolecite	Johnson et al. (1983, 1992)	$Ca_{1.0}Al_{2.0}Si_{3.0}O_{10} \cdot 3H_2O$
Stilbite	Howell et al. (1990)	$(Ca_{1.019}Na_{0.136}K_{0.006})Al_{2.180}Si_{6.820}O_{18} \cdot 7.33H_2O$
Wairakite	Kiseleva et al. (1996a)	$Ca_{2.0}Al_{4.0}Si_{8.0}O_{24} \cdot 4H_2O$
Yugawaralite	Kiseleva et al. (1996a)	$Ca_{2.0}Al_{4.0}Si_{12.0}O_{32} \cdot 8H_2O$

[1] Sr, Ba, and Mn are included as Ca in the estimation methods for this clinoptilolite.

[2] Sr and Ba are included as Ca in the estimation methods for this heulandite.

Table 2. Comparison of estimated thermodynamic data with measured values.

	Gibbs free energy of formation (kJ/mol)			Enthalpy of formation (kJ/mol)		
Mineral	*Measured*	*Estimated*[1]	*% diff*	*Measured*	*Estimated*[1]	*% diff*
Analcime	-3086.1	-3087.8	0.06	-3305.8	-3302.4	-0.10
Clinoptilolite	-19078.4	-19057.8	-0.11	-20645.0	-20641.9	-0.02
Heulandite	-9807.0	-9804.4	-0.03	-10622.5	10625.1	0.02
Laumontite	-13397.8	-13433.6	0.27	-14502.0	-14531.2	0.20
Leonhardite	-13165.3	-13193.8	0.22	-14214.6	-14238.9	0.17
Merlinoite	-3123.3	-3128.2	0.16	-3360.0	-3368.7	0.26
Mesolite	-5527.3	-5522.4	-0.09	-5961.2	-5938.0	-0.39
Mordenite	-6247.6	-6246.3	-0.02	-6756.2	-6766.1	0.15
Natrolite	-5330.7	-5345.5	0.28	-5732.7	-5716.9	-0.28
Scolecite	-5612.0	-5623.0	0.20	-6063.1	-6062.3	-0.01
Stilbite	-10142.0	-10132.1	-0.10	-11033.6	-11025.3	-0.08
Wairakite	-12416.6	-12474.0	0.46	-13293.4	-13361.8	0.51
Yugawaralite	-16805.2	-16849.4	0.26	-18102.6	-18175.2	0.40

[1]Gibbs free energies and enthalpies of formation were estimated using Chermak and Rimstidt (1989).

		Entropy (J/(mol·K))			
Mineral	*Measured*	*Estimated*[3] *w/o volume*	*% diff w/o volume*	*Estimated*[3] *w/ volume*[4]	*% diff w/ volume*
Analcime	226.75	222.6	-1.83	225.5	-0.55
Clinoptilolite[1]	718.08	728.1	1.40	728.0	1.38
Clinoptilolite[2]	1483.06	1480.1	-0.20	1478.2	-0.33
Heulandite	767.18	774.4	0.94	792.8	3.34
Laumontite	971.0	1035.8	6.67	1001.3	3.12
Leonhardite	922.2	976.7	5.91	960.2	4.12
Merlinoite	259.7	269.0	3.58	264.7	1.93
Mesolite	363.00	414.0	14.05	410.5	13.09
Mordenite	486.54	474.3	-2.52	477.4	-1.88
Natrolite	359.73	408.5	13.56	412.1	14.56
Phillipsite	771.90	769.7	-0.29	763.3	-1.11
Scolecite	367.42	418.5	13.90	411.2	11.92
Stilbite	805.54	842.7	4.61	825.2	2.44
Wairakite	801.4	799.4	-0.25	812.9[a]	1.41
Yugawaralite	1219.6	1197.0	-1.85	1196.8[b]	-1.87

[1] Hemingway and Robie (1984). [2] Johnson et al. (1991).

[3] Entropies of formation were estimated using Holland (1989).

[4] Molar volumes were obtained from Smyth and Bish (1988).

[a] Molar volume from Kiseleva et al. (1996a).

[b] Molar volume from the BBG85.MIN database supplied with the Geo-Calc PTA software package.

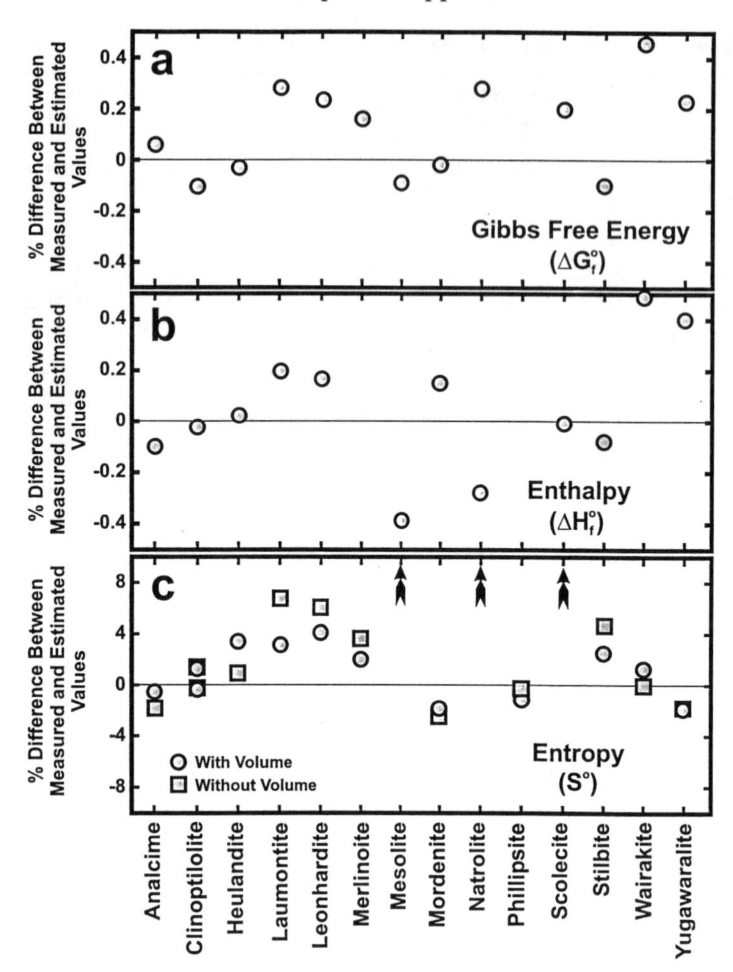

Figure 2. Comparison of measured and estimated thermodynamic data for the zeolites listed in Tables 1 and 2. (a) Gibbs free energy, (b) Enthalpy, (c) Entropy.

zeolites were used for comparison: analcime (Johnson et al. 1982, 1992), clinoptilolite (Hemingway and Robie 1984, Johnson et al. 1991), heulandite (Johnson et al. 1985, 1992), laumontite (Kiseleva et al. 1996a), leonhardite (Kiseleva et al. 1996b), merlinoite (Donahoe et al. 1990a,b), mesolite (Johnson et al. 1983, 1992), mordenite (Johnson et al. 1992), natrolite (Johnson et al. 1983, 1992), phillipsite (Hemingway and Robie 1984), scolecite (Johnson et al. 1983, 1992), stilbite (Howell et al. 1990), and wairakite and yugawaralite (Kiseleva et al. 1996a). The estimation methods of Chermak and Rimstidt (1989) reproduced measured Gibbs free energy and enthalpies of formation at 298 K to within 0.3% (2σ) of the measured values using the published chemical formulas for the zeolites (Fig. 2a,b). Holland's (1989) method of estimating entropy reproduced entropies for all the zeolites except mesolite, natrolite, and scolecite to within 4% (2σ) of the measured values using the published chemical formulas (Fig. 2c). The entropies of mesolite, natrolite, and scolecite (all from the same study, Johnson et al. 1983) did not compare well due to the significantly different value for the entropy of zeolitic water (32

J/(mol•K)) estimated by Johnson et al. (1983).

Thermodynamic data for the zeolites used in this study are provided in Table 3. Measured chemical analyses of zeolites were used wherever possible. If analyses were unavailable for a phase found in a particular environment, an approximate formula was estimated from published analyses (e.g. Gottardi and Galli 1985). No attempt was made to accommodate an additional term related to cation mixing/solid solution effects in the estimated data. It should be noted, however, that the estimated data compare well to the measured data. This may be due in part to the fact that the measured data were obtained in many cases on samples that did not possess endmember compositions; a certain amount of cation mixing is therefore incorporated into the estimations.

In addition to the observed exchangeable- and tetrahedral-cation compositional variability, zeolites exhibit wide variations in water content, readily responding to changes in temperature and humidity. For the present calculations, all zeolites were assumed to be fully hydrated as would be expected for a zeolite below the water table or at 100% relative humidity. The exact amount of water an individual zeolite contains, however, is strongly dependent on the exchangeable cations in the zeolite structure and on the total abundance of cations in the zeolite. For example, under given temperature and humidity conditions, more water will be incorporated in Ca-clinoptilolite/heulandites than in Na- or K-clinoptilolite/heulandites due to the greater hydration energy of Ca^{2+} and the greater number of sites available in the extra-framework cavities (Bish 1988).

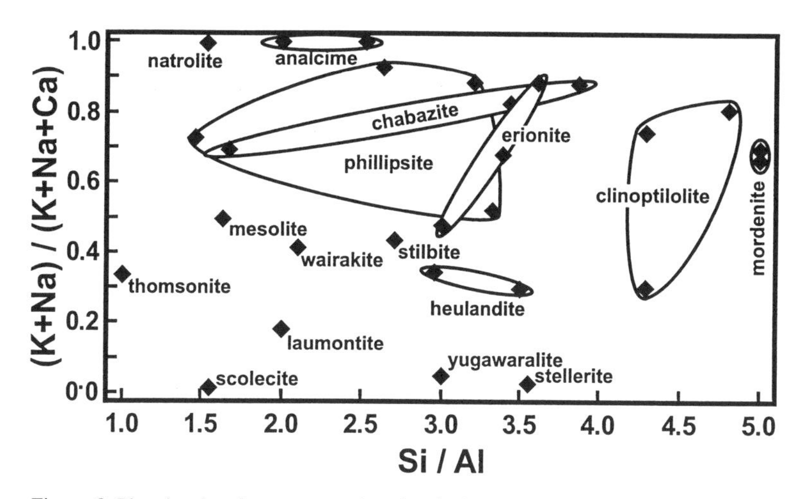

Figure 3. Plot showing the representative chemical compositions of the various zeolites used in this study (Table 3). The same zeolite species formed in different geologic environments often show considerable variation in chemical composition.

Thermodynamic stability calculations

Some aspects of the stability relations of zeolites can be assessed solely from examination of the mineral chemistry. Figure 3 is a plot of the representative chemical compositions (Table 3) for various zeolites from different environments. For example, the silicic zeolites clinoptilolite and mordenite plot at high Si/Al ratios, consistent with environments having high aqueous silica activities. The aluminous zeolites that form in mafic rocks plot on the low Si/Al side of Figure 3. Likewise, the same general trends are apparent in the (K + Na) versus Ca axis on Figure 3. It is interesting to note that that in the

Table 3. Chemical formulae and estimated thermodynamic data for the zeolite phases used in the modeling.

Sample	Chemical formula	[1]ΔG^0_f (kJ/mol)	[1]ΔH^0_f (kJ/mol)	[2]S° (J/(mol·K))	[3]Volume (cm³/mol)	Source for representative chemical analysis
Diagenetic alteration in volcanic tuff						
Analcime	$Na_{10.2}Al_{10.2}Si_{25.8}O_{72}\cdot 12H_2O$	-36660.3	-39247.6	2664.2	1168.1	Broxton et al. (1987)
Chabazite	$(K_{2.0}Na_{3.7}Ca_{1.2})Al_{8.1}Si_{27.9}O_{72}\cdot 36H_2O$	-41888.0	-45774.2	3857.7	1498.0	Carlos et al. (1995b)
Clinoptilolite	$(K_{0.8}Na_{0.4}Ca_{2.8})Al_{6.8}Si_{29.2}O_{72}\cdot 26H_2O$	-39131.4	-42528.8	3087.8	1264.1	Broxton et al. (1986)
Erionite	$(K_{3.0}Na_{1.2}Ca_{2.0})Al_{8.2}Si_{27.8}O_{72}\cdot 30H_2O$	-40533.6	-44117.4	3428.9	1344.0	Gottardi & Galli (1985)
Heulandite	$(K_{0.4}Na_{1.0}Ca_{3.3})Al_{8.0}Si_{28.0}O_{72}\cdot 26H_2O$	-39497.6	-42891.2	3109.9	1266.4	Broxton et al. (1986)
G2-1488 Heul.	$(K_{0.4}Na_{0.3}Ca_{2.5}Mg_{1.0})Al_{7.7}Si_{28.3}O_{72}\cdot 26H_2O$	-39373.0	-42776.5	3081.6	1266.4	unpublished data
G2-1505 Heul.	$(K_{0.5}Na_{0.3}Ca_{2.7}Mg_{0.8})Al_{7.8}Si_{28.2}O_{72}\cdot 26H_2O$	-39413.9	-42815.8	3087.3	1266.4	unpublished data
Laumontite	$(K_{0.6}Na_{0.6}Ca_{5.4})Al_{12.0}Si_{24.0}O_{72}\cdot 24H_2O$	-40293.6	-43577.7	3033.6	1219.2	Gottardi & Galli (1985)
Mordenite	$(K_{0.9}Na_{2.1}Ca_{1.5})Al_{6.0}Si_{30.0}O_{72}\cdot 22H_2O$	-37887.1	-41062.0	2962.5	1273.5	Broxton et al. (1986)
Phillipsite	$(K_{1.2}Na_{1.4}Ca_{2.4})Al_{7.4}Si_{24.6}O_{64}\cdot 24H_2O$	-35409.6	-38462.6	2931.6	1218.2	Carlos et al. (1995b)
Stellerite	$(Ca_{3.9}Na_{0.1})Al_{7.9}Si_{28.1}O_{72}\cdot 28H_2O$	-39960.1	-43468.5	3224.3	1331.0	Carlos et al. (1995b)
Saline-alkaline lakes (closed systems)						
Analcime	$Na_{10.2}Al_{10.2}Si_{25.8}O_{72}\cdot 12H_2O$	-36660.3	-39247.6	2664.2	1168.1	Broxton et al. (1987)
Chabazite	$(K_{0.9}Na_{4.9}Ca_{0.8})Al_{7.4}Si_{28.6}O_{72}\cdot 36H_2O$	-41636.6	-45522.2	3845.4	1498.0	Gottardi & Galli (1985)
Clinoptilolite	$(K_{2.3}Na_{1.7}Ca_{1.4})Al_{6.8}Si_{29.2}O_{72}\cdot 26H_2O$	-39116.6	-42493.7	3157.3	1264.1	Surdam & Eugster (1976)
Erionite	$(K_{2.8}Na_{3.4}Ca_{0.8})Al_{7.8}Si_{28.2}O_{72}\cdot 30H_2O$	-40364.3	-43933.1	3467.5	1344.0	Surdam & Eugster (1976)
Phillipsite	$(K_{2.8}Na_{3.2}Ca_{0.8})Al_{7.6}Si_{24.4}O_{64}\cdot 24H_2O$	-35449.4	-38477.2	3018.0	1218.2	Surdam & Eugster (1976)
Hydrothermal deposits in silicic rocks						
Analcime	$Na_{12.0}Al_{12.0}Si_{24.0}O_{72}\cdot 12H_2O$	-37196.6	-39767.2	2730.4	1168.1	Gottardi & Galli (1985)
Chabazite	$(K_{2.0}Na_{3.7}Ca_{1.2})Al_{8.1}Si_{27.9}O_{72}\cdot 36H_2O$	-41888.0	-45774.2	3857.7	1498.0	Carlos et al. (1995b)
Clinoptilolite	$(K_{0.8}Na_{0.4}Ca_{2.8})Al_{6.8}Si_{29.2}O_{72}\cdot 26H_2O$	-39131.4	-42528.8	3087.8	1264.1	Broxton et al. (1986)
Erionite	$(K_{3.0}Na_{1.2}Ca_{2.0})Al_{8.2}Si_{27.8}O_{72}\cdot 30H_2O$	-40533.6	-44117.4	3428.9	1344.0	Gottardi & Galli (1985)
Heulandite	$(K_{0.4}Na_{1.0}Ca_{3.3})Al_{8.0}Si_{28.0}O_{72}\cdot 26H_2O$	-39497.6	-42891.2	3109.9	1266.4	Broxton et al. (1986)
Laumontite	$(K_{0.6}Na_{0.6}Ca_{5.4})Al_{12.0}Si_{24.0}O_{72}\cdot 24H_2O$	-40293.6	-43577.7	3033.6	1219.2	Gottardi & Galli (1985)

Sample	Chemical formula	[1]ΔG_f^0 (kJ/mol)	[1]ΔH_f^0 (kJ/mol)	[2]S^o (J/(mol·K))	[3]Volume (cm^3/mol)	Source for representative chemical analysis
Mordenite	$(K_{0.9}Na_{2.1}Ca_{1.5})Al_{6.0}Si_{30.0}O_{72}\cdot 22H_2O$	-37887.1	-41062.0	2962.5	1273.5	Broxton et al. (1986)
Phillipsite	$(K_{5.8}Na_{1.6}Ca_{2.8})Al_{13.0}Si_{19.0}O_{64}\cdot 24H_2O$	-37209.2	-40220.4	3138.8	1218.2	Gottardi & Galli (1985)
Stilbite	$(K_{0.1}Na_{2.6}Ca_{3.5})Al_{9.7}Si_{26.3}O_{72}\cdot 28H_2O$	-40483.9	-43969.4	3308.9	1331.0	Gottardi & Galli (1985)
Wairakite	$(K_{0.1}Na_{2.9}Ca_{4.3})Al_{11.6}Si_{24.4}O_{72}\cdot 12H_2O$	-37241.6	-39882.4	2522.1	1147.0 [a]	Gottardi & Galli (1985)
Yugawaralite	$(Na_{0.2}Ca_{3.9})Al_{8.0}Si_{24.0}O_{64}\cdot 17H_2O$	-33935.1	-36637.4	2439.2	1063.4 [a]	Gottardi & Galli (1985)
Vugs / fillings in mafic rocks						
Analcime	$Na_{12.0}Al_{12.0}Si_{24.0}O_{72}\cdot 12H_2O$	-37196.6	-39767.2	2730.4	1168.1	Gottardi & Galli (1985)
Chabazite	$(K_{4.0}Na_{3.1}Ca_{3.2})Al_{13.5}Si_{22.5}O_{72}\cdot 36H_2O$	-43622.6	-47491.2	3974.3	1498.0	Gottardi & Galli (1985)
Erionite	$(K_{1.6}Na_{1.2}Ca_{3.1})Al_{9.0}Si_{27.0}O_{72}\cdot 30H_2O$	-40778.0	-44370.1	3402.8	1344.0	Gottardi & Galli (1985)
Heulandite	$(K_{0.8}Na_{1.1}Ca_{3.6})Al_{9.1}Si_{26.9}O_{72}\cdot 26H_2O$	-39846.8	-43235.1	3138.5	1266.4	Gottardi & Galli (1985)
Laumontite	$(K_{0.6}Na_{0.6}Ca_{5.4})Al_{12}Si_{24}O_{72}\cdot 24H_2O$	-40293.6	-43577.7	3033.6	1219.2	Gottardi & Galli (1985)
Mesolite	$(Na_{5.0}Ca_{5.1})Al_{15.2}Si_{24.8}O_{80}\cdot 23H_2O$	-44396.6	-47821.3	3330.6	1359.0 [b]	Gottardi & Galli (1985)
Mordenite	$(K_{0.3}Na_{2.9}Ca_{1.4})Al_{6.0}Si_{30.0}O_{72}\cdot 22H_2O$	-37868.2	-41041.1	2964.4	1273.5	Gottardi & Galli (1985)
Natrolite	$(K_{0.1}Na_{15.5}Ca_{0.1})Al_{15.8}Si_{24.2}O_{80}\cdot 17H_2O$	-42950.5	-45977.7	3332.1	1359.0	Gottardi & Galli (1985)
Phillipsite	$(K_{5.8}Na_{1.6}Ca_{2.8})Al_{13.0}Si_{19.0}O_{64}\cdot 24H_2O$	-37209.2	-40220.4	3138.8	1218.2	Gottardi & Galli (1985)
Scolecite	$(Na_{0.1}Ca_{7.8})Al_{15.7}Si_{24.3}O_{80}\cdot 24H_2O$	-44886.9	-48401.2	3268.3	1359.0 [b]	Gottardi & Galli (1985)
Stilbite	$(K_{0.1}Na_{2.6}Ca_{3.5})Al_{9.7}Si_{26.3}O_{72}\cdot 28H_2O$	-40483.9	-43969.4	3308.9	1331.0	Gottardi & Galli (1985)
Thomsonite	$(Na_{4.0}Ca_{8.0})Al_{20.0}Si_{20.0}O_{80}\cdot 24H_2O$	-46175.7	-49653.0	3417.8	1359.6	Gottardi & Galli (1985)
Yugawaralite	$(Na_{0.2}Ca_{3.9})Al_{8.0}Si_{24.0}O_{64}\cdot 17H_2O$	-33935.1	-36637.4	2439.2	1063.4 [a]	Gottardi & Galli (1985)
Deep-sea sediments						
Clinoptilolite	$(K_{2.8}Na_{1.4}Ca_{1.0})Al_{6.2}Si_{29.8}O_{72}\cdot 26H_2O$	-38935.4	-42312.3	3155.4	1264.1	Stonecipher (1978)
Phillipsite	$(K_{3.8}Na_{3.8}Ca_{0.6})Al_{8.8}Si_{23.2}O_{64}\cdot 24H_2O$	-35824.6	-38839.2	3075.4	1218.2	Stonecipher (1978)
Analcime	(Used diagenetically altered tuff values)					

[1] Gibbs free energies and enthalpies of formation were estimated using Chermak and Rimstidt (1989).
[2] Entropies of formation were estimated using Holland (1989). [3] Volumes were obtained from Smyth and Bish (1988).
[a] Used the molar volume from Kiseleva et al. (1996a). [b] Used the molar volume of natrolite.

commonly occurring zeolites, there exist dominantly sodic zeolites and dominantly calcic zeolites but there are no zeolites that are dominantly potassic. Potassium occurs mainly in those zeolites that are amenable to significant cation variations (e.g. chabazite, clinoptilolite, erionite, and phillipsite). Although simple examination of mineral chemistries can suggest general stability relationships among the zeolites, thermodynamic analysis provides a more quantitative basis for placing reaction boundaries.

Calculating the position of a reaction boundary involves two steps. The first is to calculate, for every phase involved in the reaction, the apparent Gibbs free energy ($\Delta_a G^{P,T}$) at the temperature and pressure of interest (Helgeson et al. 1978). At a pressure P and temperature T, the apparent free energy of formation of a phase from the elements is defined by the equation

$$\Delta_a G^{P,T} = \Delta_a H^{P,T} - T \cdot S^{P,T} + \Delta G_{\text{phase transitions}} + \Delta G_{\text{disorder}} + \Delta G_{\text{solution}} \tag{1}$$

where

$$\Delta_a H^{P,T} = \Delta_f H^{P_r,T_r} + \int_{T_r}^{T} Cp dT + V^{P_r,T_r} \Delta P \tag{2}$$

and

$$S^{P,T} = S^{P_r,T_r} + \int_{T_r}^{T} \left(\frac{Cp}{T}\right) dT \quad . \tag{3}$$

$\Delta_a H^{P,T}$ is the apparent enthalpy of formation of the pure phases from the elements at the P and T desired and $\Delta_f H^{Pr,Tr}$, $S^{Pr,Tt}$, and $V^{Pr,Tt}$ are the enthalpy of formation, third law entropy, and molar volume at the reference pressure (P_r – usually defined as 1 bar [10^5 Pa]) and reference temperature (T_r – usually defined as 298.15 K). C_p is the heat capacity at constant pressure. $\Delta G_{\text{phase transitions}}$ is the energy contribution due to a phase undergoing a crystal-lographic transition at elevated temperature or pressure, a parameter which has not been characterized for most zeolites. $\Delta G_{\text{disorder}}$ is an additional energy contribution due to the disordering of atoms in the structure which are normally configured in an ordered arrangement; this is a significant component but has also unfortunately, not been characterized well for most zeolites. $\Delta G_{\text{solution}}$ is the energy contribution due to solid solution effects such as mixing of extraframework cations within a zeolite and was assumed in these calculations to be zero as discussed in the previous section.

To calculate the position of the equilibria, the Gibbs free energy of the reaction (ΔG_{rxn}) must equal 0 in accordance with the equation

$$\Delta G_{rxn} = 0 = \Delta G_{rxn}^{P,T} + RT \ln K \tag{4}$$

where $\Delta G_{rxn}^{P,T}$ is the Gibbs free energy of reaction at the desired temperature and pressure, calculated using the apparent Gibbs free energies of the phases discussed above, R is the universal gas constant (8.314 J/(mol·K)), T is the temperature in Kelvin, and *ln*K is the natural log of the equilibrium constant.

Because thermodynamic data were calculated specifically for each zeolite and thereby included any variability in exchangeable cation and Si:Al ratios, the activities for the zeolites were, by definition, 1. The calculations were conducted projecting from water (activity = 1), reactions were balanced on Al, and any reactant or product ionic species were exchanged with the liquid phase. For example, modeling of clinoptilolite-analcime equilibria was conducted assuming the following reaction

$$s \text{ Clinoptilolite} + t \text{ Na}^+ \Leftrightarrow v \text{ Analcime} + w \text{ K}^+ + x \text{ Ca}^{2+} + y \text{ SiO}_2 + z \text{ H}_2\text{O} \, . \tag{5}$$

The equilibrium constant for the clinoptilolite-analcime Reaction (5) is approximated as:

$$K = \frac{(aCa^{2+})^x \, (aK^+)^w \, (aSiO_2)^y}{(aNa^+)^t} \qquad (6)$$

As is obvious from Equation (6), cation concentrations and aqueous silica activity are all important variables in the calculation of zeolite equilibria.

Thermodynamic stability calculations in this study were conducted using the Ge0-Calc PTA-SYSTEM software (Brown et al. 1989). Ge0-Calc runs on a PC and allows rapid calculation of thermodynamic stability fields under conditions of variable temperature, pressure, or activities of phases involved in the reactions. In addition, Ge0-Calc balances complex chemical reactions, a desirable feature when using actual chemical analyses that can deviate significantly from stoichiometric "ideal" formulas. Thermodynamic data for quartz, cristobalite, and tridymite were obtained from the B88.MIN thermodynamic database (Berman 1988) supplied with the Ge0-Calc software package. Likewise, values for aqueous species (aqueous SiO_2, Ca^{2+}, Na^+, K^+, H^+) were taken from the HKF81.AQU database (Helgeson et al. 1981) also supplied with the Ge0-Calc software. The Ge0-Calc database, however, does not contain thermodynamic data for amorphous silica. For this study, amorphous silica data were obtained from the SUPCRT92 database (Johnson et al. 1992). To simplify the calculations, the chemical system was restricted to Na-K-Ca-Al-Si-O-H and pressure was constrained to the liquid side of the water liquid-vapor curve.

Great care must be used in the calculations when selecting appropriate "representative" chemical formulae for the phases of interest and when interpreting the results of the calculations. Figure 4 shows the results of modeling analcime, chabazite, clinoptilolite, erionite, and phillipsite equilibria in $\log[(aK^+)^2/aCa^{2+}]$ vs. $\log[(aNa^+)^2/aCa^{2+}]$ space at 35°C and at an aqueous silica activity in equilibrium with cristobalite. Figure 4a was generated using chemical formulas deemed most representative of those measured for the zeolites that occur at Yucca Mountain, Nevada (an example of a diagenetically altered volcanic tuff). Figure 4b was calculated using chemical formulae believed to be appropriate for zeolites in mafic volcanic assemblages (generally found as amygdule or vug fillings). Figure 4c was calculated using chemical formulae measured for zeolites formed in saline-alkaline lakes, and Figure 4d used a mixture of diagenetic clinoptilolite and chabazite and saline-alkaline lake analcime, erionite, and phillipsite. Depending on the chemical composition of the zeolites used in the calculations, large differences in the calculated stability fields resulted. Consequently, it is important to use well-known representative or measured chemical formulas for the zeolites being modeled in a particular deposit or environment. Given the amount of variability observed using actual zeolite compositions, it is obvious that the use of stoichiometric end-member compositions in thermodynamic calculations, although convenient, can lead to erroneous results.

CALCULATION OF ZEOLITE STABILITY FIELDS FOR VARIOUS GEOCHEMICAL ENVIRONMENTS

Diagenetic alteration of volcanic tuff: Yucca Mountain, Nevada

Yucca Mountain, located in southern Nevada, is being investigated to determine its suitability to host the first high-level radioactive waste repository in the United States. Yucca Mountain is composed of a >1.5 km-thick sequence of tuffs and subordinate lavas that have undergone extensive diagenetic alteration to form thick zeolite-rich horizons. The zeolitic alteration at Yucca Mountain was described by Smyth (1982) and Broxton et al. (1987) as consisting of four characteristic diagenetic zones (as defined by Iijima 1975, 1978). Zone I is the shallowest and is characterized by vitric tuffs containing unaltered volcanic glass and minor smectite, opal-CT, heulandite, and Ca-rich clinoptilolite. Zone II is characterized by the complete replacement of volcanic glass by clinoptilolite ± mordenite

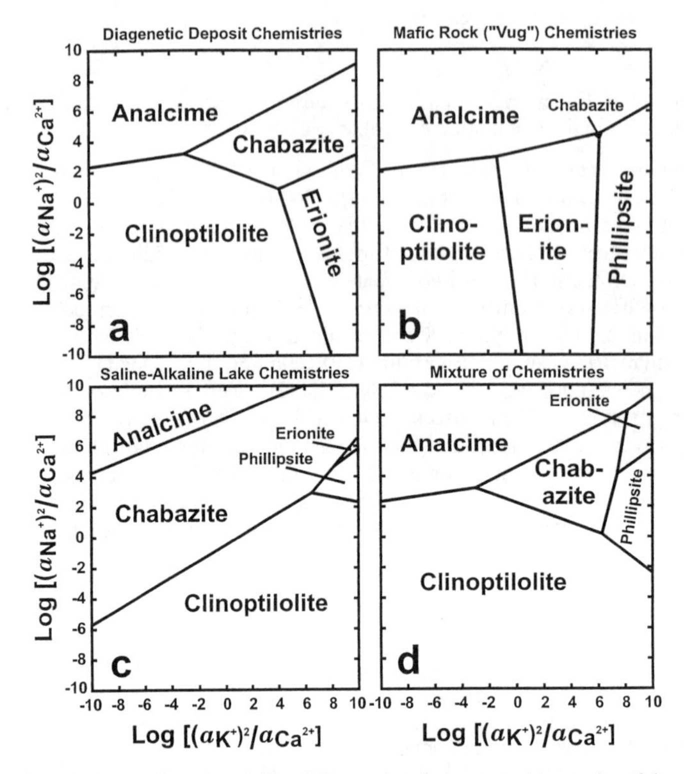

Figure 4. Results of modeling different chemical compositions of analcime, chabazite, clinoptilolite, erionite, and phillipsite in $\log[(a\text{K}^+)^2/a\text{Ca}^{2+}]$ vs. $\log[(a\text{Na}^+)^2/a\text{Ca}^{2+}]$ space, using an aqueous silica activity in equilibrium with cristobalite at 35°C.

(a) Chemical compositions representative of diagenetically altered volcanic tuff.

(b) Chemical compositions representative of zeolites forming in mafic volcanic assemblages (generally found as amygdule or vug fillings).

(c) Chemical compositions representative of zeolites forming in saline-alkaline lakes.

(d) Mixture of diagenetic (clinoptilolite and chabazite) and saline-alkaline lake compositions (analcime, erionite, and phillipsite).

and by lesser amounts of opal-CT, K-feldspar, quartz, and smectite. Zone III consists of analcime, K-feldspar, quartz, and minor calcite and smectite. Albite, K-feldspar, quartz, and minor calcite and smectite characterize zone IV. The potential repository horizon, at ~320 m depth in devitrified tuff, is directly underlain by either Zone-I or Zone-II assemblages.

The bulk-rock zeolite mineralogy at Yucca Mountain consists mainly of clinoptilolite/heulandite, mordenite, and analcime (Bish and Chipera 1989). Significant quantities of stellerite have also been found in the bulk rock in drill core UE-25 UZ#16 (Chipera et al. 1995b). In addition to the above zeolites, chabazite, erionite, and phillipsite have been identified using X-ray powder diffraction (Carlos et al. 1995a, 1995b). These additional zeolites occur sporadically in fractures, generally above the static water level, and were identified only in close spatial association with vitric zones. Erionite is restricted to

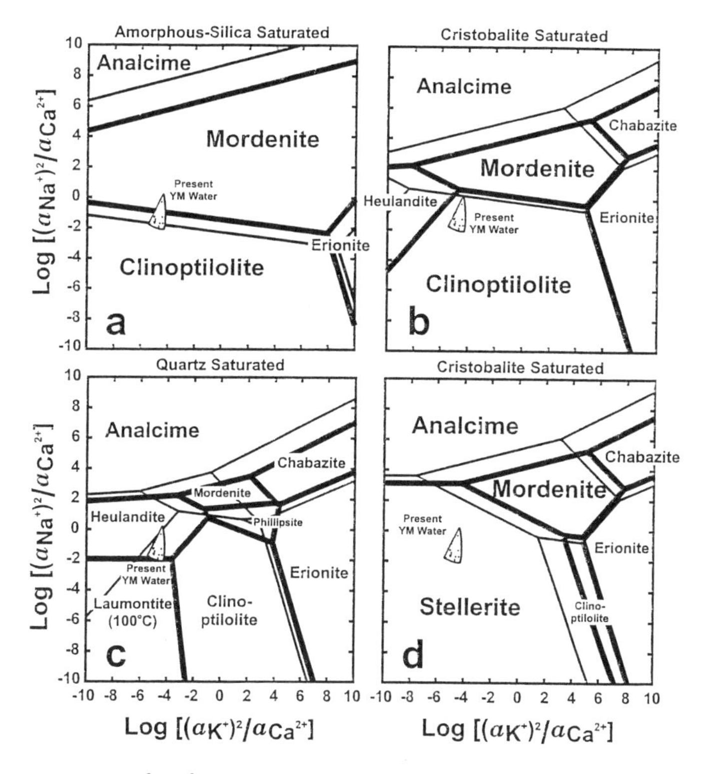

Figure 5. Log[$(a\mathrm{K}^+)^2/a\mathrm{Ca}^{2+}$] vs. log[$(a\mathrm{Na}^+)^2/a\mathrm{Ca}^{2+}$] diagrams for the zeolite phases found at Yucca Mountain (analcime, chabazite, clinoptilolite, erionite, heulandite, laumontite, mordenite, and phillipsite) assuming:
(a) Amorphous-silica saturation at 35°C (thin lines) and 100°C (heavy lines).
(b) Cristobalite saturation at 35°C (thin lines) and 100°C (heavy lines).
(c) Quartz saturation at 35°C (thin lines) and 100°C (heavy lines). Laumontite did not possess a stability field at 35°C.
(d) Same as Figure 5b except that stellerite was included in the calculation.

isolated fractures within the Topopah Spring Tuff lower vitrophyre, although one occurrence with up to 34 wt % erionite has been identified in a 3-m thick sequence of bulk rock in drill core USW UZ#14 (Guthrie et al. 1995). Laumontite, which occurs at higher temperatures than clinoptilolite, was found at the bottom of drill hole UE-25p#1 and is suspected at the bottom of USW G-1 (Bish and Chipera 1989).

Thermodynamic modeling of the zeolites that occur in fractures at Yucca Mountain was conducted by Chipera et al. (1995a) to understand the conditions that resulted in zeolite formation at Yucca Mountain. The modeling was updated for this study using modified chemical compositions that better reflect the chemical compositions observed in the bulk rock at Yucca Mountain. Figures 5a-d show the results of modeling in log[$(a\mathrm{K}^+)^2/a\mathrm{Ca}^{2+}$] vs. log[$(a\mathrm{Na}^+)^2/a\mathrm{Ca}^{2+}$] space using various silica activities and temperatures of 35 and 100°C. Present-day saturated-zone water chemistry (Kerrisk 1987) has been superimposed on the activity/activity diagrams for reference. Ion speciation and activities were calculated using the program PHREEQC (Parkhurst 1995). Although attempts were made to use "representative" chemical analyses, chemical compositions have been shown to vary considerably depending on drill hole location and depth (Broxton et al. 1986). Several calculations were conducted to determine sensitivity of the stability fields to variable input extraframework-cation ratios for the phases. These calculations showed that those zeolites that exhibit a large variation in cation composition (e.g. clinoptilolite-heulandite, mordenite, and erionite) were most sensitive to changes in input cation ratios. Bowers and Burns (1990) also noted that cation composition had an effect on stability and modeled the effects of Ca-Na-K-Mg substitution in clinoptilolite with respect to mordenite and heulandite. They found that increased K or Ca significantly increased the stability field of clinoptilolite.

The effect of silica activity on zeolite equilibria is clearly illustrated by comparing calculations conducted using amorphous silica saturation (Fig. 5a), cristobalite saturation (Fig. 5b), and quartz saturation (Fig. 5c). The stability fields for the more siliceous zeolites (e.g. mordenite and clinoptilolite) are dominant at high silica activity and are greatly reduced at quartz saturation. Less silicic zeolites (phillipsite and laumontite) were calculated to be stable only under conditions of reduced silica activity (e.g. at quartz saturation). Likewise, decreasing silica activity significantly increased the stability of heulandite with respect to clinoptilolite. Although discrete chemical formulae were used in the calculations for clinoptilolite and heulandite, there is a chemical continuum between the two phases and the calculations can be interpreted as two extremes of clinoptilolite/heulandite. Silica activity for the calculations can be constrained by the SiO_2 polymorphs that coexist with the zeolites. For example, opal-CT is generally intergrown with mordenite and clinoptilolite at Yucca Mountain (Bish and Chipera 1989), whereas stellerite and heulandite are intergrown with tridymite and cristobalite (Chipera et al. 1995b) and analcime is intergrown with quartz.

Increasing temperature had several effects on the results of calculations. With increased temperature at a given silica-polymorph saturation, the stability fields for mordenite and clinoptilolite decreased as the surrounding stability fields for heulandite, chabazite, and erionite expanded. For an aqueous silica activity in equilibrium with quartz, increasing temperature also increased the stability field for phillipsite and a stability field for laumontite was produced (Fig. 5c).

Heulandite-stellerite equilibria

In the thermodynamic modeling study of zeolites observed in fractures at Yucca Mountain, Chipera et al. (1995a) found that heulandite (which is isostructural with clinoptilolite and relatively common at Yucca Mountain) had a stability field only when silica activity was reduced to quartz saturation and only when stellerite formation was suppressed. Suppression of stellerite is inconsistent with the common coexistence of heulandite and stellerite in fractures in devitrified tuffs at Yucca Mountain. Although the stability field for heulandite can be enlarged by varying the exchangeable-cation compositions of heulandite, stellerite, or the other bounding phases, a reasonable stability field for heulandite with respect to stellerite could not be computed within the constraints of the observed chemical formulae for the phases at Yucca Mountain. In the current study, the chemical compositions of several of the zeolites (e.g. clinoptilolite) were changed to values that are more representative of the average bulk-rock values at Yucca Mountain. Subsequently, the stability field for stellerite became anomalously large and somewhat unrealistic when compared with clinoptilolite and current Yucca Mountain water compositions (Fig. 5d). Apparently other factors, possibly kinetics, compositional differences, or inaccurate thermodynamic values, contribute to overestimation of stellerite stability.

Yucca Mountain heulandites commonly contain significant Mg and Sr, which were not included in the calculations of Chipera et al. (1995a). Preliminary modeling of heulandite/stellerite stability showed that increasing either the Mg or Sr content in heulandite stabilized it with respect to stellerite (Carlos et al. 1995c). Varying silica activity in the calculations, although exerting a strong control over whether heulandite/stellerite would form relative to other zeolites (e.g. clinoptilolite and mordenite) had little effect on heulandite vs. stellerite stability because they have similar Si:Al ratios. Inclusion of Mg and Sr in the calculations should increase the stability of heulandite with respect to analcime and stellerite, as suggested by related calculations for clinoptilolite by Bowers and Burns (1990). Estimating thermodynamic data, however, is difficult because none of the data-estimation methods accommodate Sr and the Chermak and Rimstidt (1989) method contains data only for 6-fold coordinated Mg.

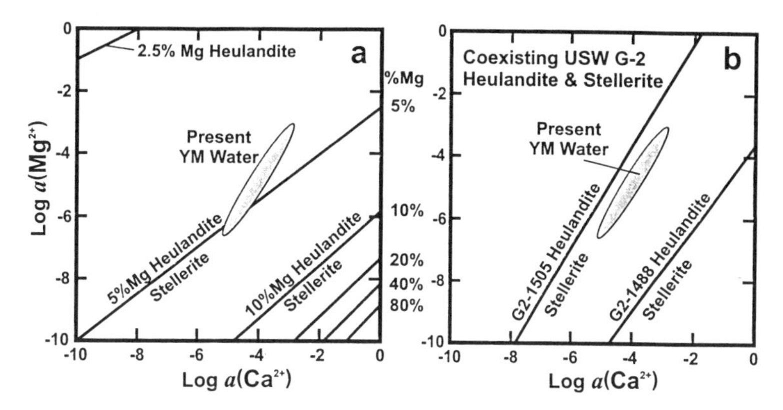

Figure 6. Log $a(Ca^{2+})$ vs. log $a(Mg^{2+})$ diagrams showing heulandite-stellerite equilibria at 100°C. Present-day Yucca Mountain water compositions (Kerrisk 1987) are plotted for reference. Aqueous ion speciation and activities were calculated using the program PHREEQC (Parkhurst 1995).

(a) Effects of variable Mg:Ca ratio in heulandite on heulandite-stellerite equilibria assuming a generalized heulandite formula of $(Ca,Mg)_{3.9}Al_{7.8}Si_{28.2}O_{72}\cdot 26H_2O$ and fixing log(aNa^+) of the groundwater at -2.6.

(b) Calculated equilibria between coexisting heulandite and stellerite with aqueous silica constrained at cristobalite saturation, log(aNa^+) in the groundwater fixed at -2.6, and log(aK^+) fixed at -4.3.

Figure 6a shows the effects of varying the quantity of Mg in heulandite (using available thermodynamic data estimation methods) on its stability with respect to stellerite. Present Yucca Mountain aqueous Ca-Mg ratios (Kerrisk 1987) are plotted for reference. Aqueous ion speciation and activities were calculated using the program PHREEQC (Parkhurst 1995). Heulandites with Mg:Ca ratios as low as 0.05 were calculated to be stable with respect to stellerite. Figure 6b shows the calculated stability for two samples from drill hole USW G-2 where heulandite and stellerite coexist in the same fractures. Although the heulandites have similar chemical compositions (Table 3), the slight variances in their cation abundances result in significant differences in their calculated stabilities. These calculations show that the activities of minor cation species in the groundwater and cation ratios in the solid phases (Mg in particular) appear to exert significant control on the formation of heulandite vs. stellerite. Sr should also stabilize heulandite with respect to stellerite, but the lack of thermodynamic data for Sr in the data-estimation methods precludes any interpretations. Apparently, inclusion of minor elements in the thermodynamic calculations may stabilize zeolites without the need to suppress the formation of other zeolites.

Clinoptilolite-analcime equilibria at Yucca Mountain

The stability of clinoptilolite is an important factor in the assessment of potential changes that could occur in the chemical and bulk-rock properties of Yucca Mountain (e.g. water release, sorption, porosity, permeability, and volume) due to heating by high-level radioactive waste. Consequently, it is necessary to understand the conditions under which clinoptilolite may react to other phases as a result of emplacement of a repository. The mineral distribution observed at Yucca Mountain (Bish and Chipera 1989) and the alteration observed in other volcanic terranes, such as Yellowstone National Park (Honda and Muffler 1970, Keith et al. 1978) suggest that the most probable reaction leading to destruction of clinoptilolite due to heating from a high-level waste repository would be the reaction of clinoptilolite to form analcime. Bowers and Burns (1990) conducted an equilibrium modeling study of the stability of clinoptilolite assuming both current temperature and water chemistry conditions and with respect to potential changes in water chemistry and increased temperature resulting from emplacement of high-level radioactive waste. Bowers and Burns calculated an albite stability field that separated clinoptilolite and analcime. They did not, however, specifically address the clinoptilolite-to-analcime reaction. The reaction of clinoptilolite to analcime is important because analcime has a much smaller cation-exchange capacity for most radionuclides (Vaughan 1978) and because the reaction produces water and gives rise to a net volume decrease of ~17%, assuming the reaction produces quartz as the silica phase. This volume reduction, leading to an increase in porosity, will significantly change the structural and hydrologic properties of clinoptilolite-bearing units.

Chipera and Bish (1997) modeled the clinoptilolite to analcime reaction (Eqn. 5) using 12 observed chemical formulae for Yucca Mountain clinoptilolites and a representative chemical formula for Yucca Mountain analcime (Table 4). The 12 clinoptilolites used in the

Table 4. Chemical formulas of Yucca Mountain clinoptilolites and analcime.

Sample	*Label (Fig. 7)*	*Chemical formula*	*Si/Al ratio*	*Ca/ (Ca+Na+K)*	*Na/ (Ca+Na+K)*	*K/ (Ca+Na+K)*
Clinoptilolite						
USW G1-3706	*a*	$(K_{0.3}Na_{3.2}Ca_{1.0})Al_{5.5}Si_{30.5}O_{72.0}\cdot 26H_2O$	5.55	0.22	0.71	0.07
USW G1-1286	*b*	$(K_{0.5}Na_{0.9}Ca_{2.7})Al_{6.8}Si_{29.2}O_{72.0}\cdot 26H_2O$	4.29	0.66	0.22	0.12
UE-25p#1-3330	*c*	$(K_{0.4}Na_{1.0}Ca_{3.3})Al_{8.0}Si_{28.0}O_{72.0}\cdot 26H_2O$	3.50	0.70	0.21	0.09
USW G1-3598	*d*	$(K_{0.7}Na_{4.3}Ca_{0.9})Al_{6.8}Si_{29.2}O_{72.0}\cdot 26H_2O$	4.29	0.15	0.73	0.12
UE-25p#1-1250	*e*	$(K_{1.7}Na_{1.0}Ca_{1.8})Al_{6.3}Si_{29.7}O_{72.0}\cdot 26H_2O$	4.71	0.40	0.22	0.38
USW H4-1420	*f*	$(K_{1.9}Na_{0.8}Ca_{1.7})Al_{6.1}Si_{29.9}O_{72.0}\cdot 26H_2O$	4.90	0.39	0.18	0.43
USW G2-1691	*g*	$(K_{2.3}Na_{0.9}Ca_{1.6})Al_{6.4}Si_{29.6}O_{72.0}\cdot 26H_2O$	4.63	0.41	0.23	0.59
USW J13-1421	*h*	$(K_{2.5}Na_{1.1}Ca_{1.2})Al_{6.0}Si_{30.0}O_{72.0}\cdot 26H_2O$	5.00	0.25	0.23	0.52
USW GU3-1874	*i*	$(K_{2.8}Na_{1.5}Ca_{0.9})Al_{6.1}Si_{29.9}O_{72.0}\cdot 26H_2O$	4.90	0.17	0.29	0.54
USW G4-1432	*j*	$(K_{2.9}Na_{2.1}Ca_{0.7})Al_{6.4}Si_{29.6}O_{72.0}\cdot 26H_2O$	4.63	0.12	0.37	0.51
USW G1-2166	*k*	$(K_{2.2}Na_{3.4}Ca_{0.4})Al_{6.4}Si_{29.6}O_{72.0}\cdot 26H_2O$	4.63	0.07	0.57	0.37
USW G1-1436	*l*	$(K_{3.3}Na_{2.0}Ca_{0.5})Al_{6.3}Si_{29.7}O_{72.0}\cdot 26H_2O$	4.71	0.09	0.34	0.57
Representative analcime		$Na_{10.2}Al_{10.2}Si_{25.8}O_{72.0}\cdot 12H_2O$	2.53	---	1.00	---

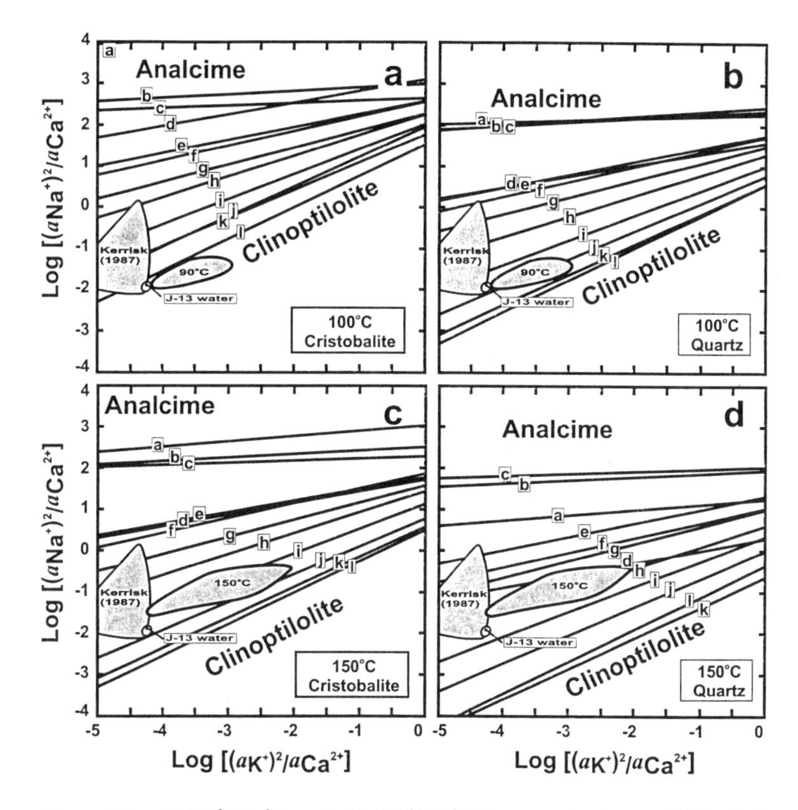

Figure 7. $\text{Log}[(a\text{K}^+)^2/a\text{Ca}^{2+}]$ vs. $\log[(a\text{Na}^+)^2/a\text{Ca}^{2+}]$ diagrams showing equilibria of 12 Yucca Mountain clinoptilolites that exhibit a representative range of exchangeable-cation and Si, Al concentrations (Table 4). Calculations were conducted assuming temperatures of 100 and 150°C and at silica activities in equilibrium with cristobalite or quartz. Chemical compositions of J-13 water, projected 90°C and 150°C J-13 water compositions (see Chipera and Bish 1997 for details and references), and chemical analyses of Yucca Mountain waters from Kerrisk (1987) are plotted on the figure for reference. Aqueous ion speciation and activities were calculated using the program PHREEQC (Parkhurst 1995).

calculations were selected to represent the observed variation in chemical compositions, both in exchangeable-cation ratios and in Si:Al ratios. Using the representative chemical formula of clinoptilolite from USW G1-3598 (Table 4), Reaction (5) is written as:

$$3\,(\text{K}_{0.7}\text{Na}_{4.3}\text{Ca}_{0.9})\text{Al}_{6.8}\text{Si}_{29.2}\text{O}_{72}\cdot 26\text{H}_2\text{O} + 7.5\ \text{Na}^+ \Leftrightarrow$$
$$2\ \text{Na}_{10.2}\text{Al}_{10.2}\text{Si}_{25.8}\text{O}_{72}\cdot 12\text{H}_2\text{0} + 2.1\ \text{K}^+ + 2.7\ \text{Ca}^{2+} + 36\ \text{SiO}_2 + 54\ \text{H}_2\text{O} \quad (7)$$

$\text{Log}[(a\text{K}^+)^2/a\text{Ca}^{2+}]$ vs. $\log[(a\text{Na}^+)^2/a\text{Ca}^{2+}]$ diagrams showing clinoptilolite-analcime equilibria are shown in Figures 7a-d. Figures 7a and 7c show equilibria calculated using cristobalite saturation at 100°C and 150°C respectively. The composition of water from drill hole J-13 (located east of Yucca Mountain) is often used as an example of representative ground water chemistry in Yucca Mountain rocks; water production in this drill hole is from the same geologic unit (Topopah Spring Tuff) as the potential repository host rock at Yucca Mountain. Therefore, projected J-13 water compositions at 90°C and 150°C are also plotted on Figure 7 (see Chipera and Bish 1997 for details and references).

Assuming J-13 water composition, clinoptilolite is the stable phase with respect to analcime for all 12 clinoptilolite compositions at cristobalite saturation and 100°C. However, when temperature was increased to 150°C, several clinoptilolite/analcime equilibria shifted below the Yucca Mountain water compositions on the diagrams, thereby stabilizing analcime. Figures 7b and 7d show results at 100 and 150°C with a silica activity in equilibrium with quartz. The effect of decreasing aqueous silica activity is to further shift the equilibria towards analcime, with several additional equilibria falling below the Yucca Mountain water compositions. At 150°C, with an aqueous silica activity in equilibrium with quartz and assuming J-13 water composition (Fig. 7d), approximately half of the clinoptilolite/analcime equilibria are in the analcime stability field. Because the chemical composition of Yucca Mountain analcime is fairly constant, the effect of varying the analcime Si:Al ratio within the limits observed at Yucca Mountain varied equilibria only slightly for all silica activities and temperatures. It is important to note that although J-13 water is considered representative of water in the Topopah Spring Tuff, other water compositions measured at Yucca Mountain (Kerrisk 1987) result in significantly reduced stability fields for clinoptilolite. Thus, for example, assuming the Na-rich water composition of Kerrisk (1987) at 100°C or 150°C in equilibrium with quartz, only a few clinoptilolite compositions are stable with respect to analcime (Fig. 7b,d).

Figure 7 and Table 4 are identical to those published by Chipera and Bish (1997) apart from a reordering of the clinoptilolite compositions from the most stable to the least stable. Figure 7 and Table 4 show that although trends resulting in increased clinoptilolite stability are observable both in the extraframework-cation and Si/Al ratios, neither factor alone controls clinoptilolite stability. Silica activity, temperature, Si/Al ratio, and the concentrations of K, Na, and Ca in the clinoptilolite and in the ground water are all important in determining the details of clinoptilolite-analcime equilibria.

The observed distribution of clinoptilolite and analcime in rocks at Yucca Mountain suggests that the reaction does not occur at a sharp, well-defined transition as a function of depth or temperature. Clinoptilolite/heulandite and analcime can and do coexist in the same sample, often to depths well beyond the first occurrence of analcime. Although an aqueous silica activity equal to or greater than cristobalite saturation stabilizes clinoptilolite, it appears that reduction in the aqueous silica activity to that of quartz saturation, although generally stabilizing analcime, did not automatically result in conversion of *all* clinoptilolite to analcime at Yucca Mountain. This effect is illustrated well by Figures 7b and d, showing that several clinoptilolite compositions are stable with respect to analcime at 100 and 150°C in Yucca Mountain waters at an aqueous silica activity in equilibrium with quartz, consistent with the mineralogic relations observed at Yucca Mountain.

Zeolite stability in saline-alkaline lake environments

The mineralogy of saline-alkaline lake environments is similar in many ways to that in the diagenetically altered tuff environments. The major difference between the two systems is the alkalinity of the waters from which the zeolites precipitate. Na and K can reach high concentrations (e.g. >100,000 ppm, Jones et al. 1967) in saline-alkaline lakes, whereas Ca is often below detection limits (~ 0.1 ppm) due to its removal in the form of sulfates and carbonates.

Thermodynamic modeling was conducted for a representative suite of zeolite minerals commonly found in saline-alkaline lake environments (clinoptilolite, phillipsite, chabazite, erionite, and analcime). Figures 8a and 8b show modeling results $\log[(a\mathrm{K}^+)^2/a\mathrm{Ca}^{2+}]$ vs. $\log[(a\mathrm{Na}^+)^2/a\mathrm{Ca}^{2+}]$ space assuming silica activities in equilibrium with cristobalite and quartz, respectively. The calculations show that phillipsite is stabilized at lower silica activities whereas erionite requires silica activities at least in equilibrium with cristobalite.

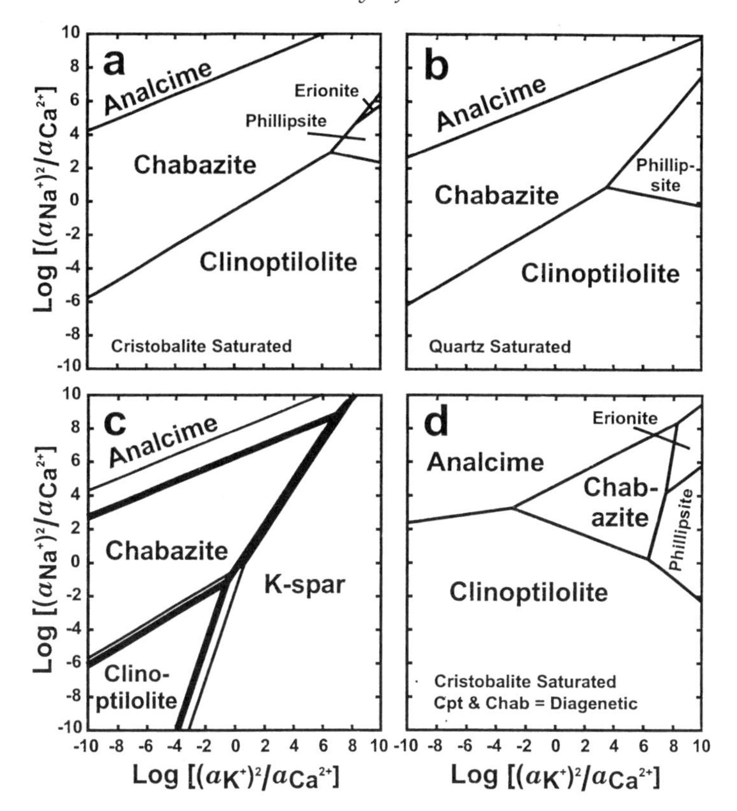

Figure 8. Log[$(a\text{K}^+)^2/a\text{Ca}^{2+}$] vs. log[$(a\text{Na}^+)^2/a\text{Ca}^{2+}$] diagrams for the zeolites occurring in saline-alkaline lake environments (analcime, chabazite, clinoptilolite, erionite, and phillipsite) all calculated assuming 35°C and
(a) An aqueous silica activity in equilibrium with cristobalite.
(b) An aqueous silica activity in equilibrium with quartz.
(c) Same as (a) and (b) except that K-feldspar was included in the calculations. Thin lines: cristobalite saturation. Heavy lines: quartz saturation.
(d) Same as (a) except that diagenetic compositions of clinoptilolite and chabazite were used in the calculations.

When the calculations were conducted using an aqueous silica activity in equilibrium with amorphous silica, a greatly expanded erionite stability field replaced the phillipsite stability field. K-feldspar is found at the center of several saline-alkaline lake deposits, (e.g. the Big Sandy Formation, Sheppard and Gude 1973). Figure 8c shows the results of modeling the same reactions described previously but including K-feldspar. At elevated K^+ concentrations, K-feldspar replaces phillipsite, erionite, and much of the clinoptilolite. As K-feldspar is not ubiquitous in saline-alkaline lake deposits, the formation of K-feldspar must be kinetically limited.

If it is assumed that a saline-alkaline lake has evolved chemically through time, some general stability trends can be inferred. Clinoptilolite is stable in dilute waters that normally contain Ca along with lesser amounts of Na and K. As water evaporates, the lake becomes more alkaline, and chabazite, phillipsite, and erionite increase in stability as clinoptilolite stability decreases with phillipsite and erionite occupying the high-K portions of the diagrams. When the lake becomes significantly enriched in Na, analcime becomes the stable zeolite. The effects of increasing temperature in the calculations was to increase the stability of analcime with respect to phillipsite and chabazite, and the stability of phillipsite with respect to clinoptilolite and chabazite.

As with the earlier modeling results for diagenetically altered tuff, the calculations show that the stability fields for those zeolites that exhibit a large variation in extra-framework-cation composition (e.g. clinoptilolite, chabazite, and erionite) were sensitive to changes in input cation ratios. This may explain the anonymously large stability field for chabazite in these calculations despite its restricted occurrences in zeolite deposits (Figs. 8a and 8b). In light of the fact that metastablity is the general rule in zeolite deposits, additional mechanisms may need to be addressed in the evolution of a saline-alkaline lake. It is possible that zeolites such as clinoptilolite and chabazite formed early in the lake evolution when the lake was less alkaline and the Ca concentration was greater. As the lake evolved by evaporation/concentration, the lake became considerably more alkaline. Cation exchange will naturally take place in the zeolites, which also become more alkalic and no longer possess their initial chemistry. Likewise, the zeolites may no longer be stable with respect to the other phases forming in the evolving saline-alkaline lake, but they may persist due to kinetic limitations. If representative chemical compositions for diagenetic clinoptilolite and chabazite are used in the calculations, the chabazite stability field becomes more restricted and analcime and clinoptilolite again share a common boundary (Fig. 8d).

Hydrothermal deposits in silicic rocks

Zeolite sequences in hydrothermal deposits in silicic volcanic rocks also resemble the sequences seen in diagenetically altered volcanic rocks. The primary difference in the environment is the elevated temperatures under which the zeolites formed in hydrothermal deposits. In addition to clinoptilolite, analcime, mordenite, erionite, phillipsite, and chabazite, which are found in the lower-temperature portions of hydrothermal deposits, stilbite, laumontite, wairakite, and yugawaralite become prominent zeolite species at elevated temperatures.

Bargar and Keith (1995) provided an overview of the research that has been conducted using research drill holes in hydrothermal areas at Yellowstone National Park. The Yellowstone drill cores are unique in that they sample an active system where water composition and temperature can be measured directly and do not have to be inferred. The alteration mineralogy paragenesis shows that zeolites are among the last minerals to form. Radioactive disequilibrium dating studies by Sturchio et al. (1989) indicated that zeolite alteration has taken place for at least the past ~8000 years. Bargar and Keith (1995) listed clinoptilolite and mordenite as the most common zeolites along with moderate amounts of analcime, dachiardite, erionite, heulandite, laumontite, wairakite, and rare yugawaralite and stilbite(?). They suggested that in addition to fluid compositions, temperature exerted a dominant control on zeolite paragenesis in the Yellowstone rocks. Erionite was found only at temperatures below ~110°C, analcime was found at temperatures >80°C, laumontite was found at temperatures from about 160 to 200°C, and wairakite was found at temperatures of 140 to 200°C but only at aqueous silica activities in equilibrium with quartz. Rare yugawaralite occurred at temperatures >200°C.

Thermodynamic modeling was conducted for a representative suite of zeolites commonly found in silicic-rock hydrothermal environments (analcime, chabazite, clinoptilolite, erionite, heulandite, laumontite, mordenite, phillipsite, stilbite, and wairakite). Figures 9a-d show the results of thermodynamic modeling at 25, 100, 150, and 200°C at an aqueous silica activity in equilibrium with quartz. At the lowest temperature (25°C), the stability diagrams resemble those calculated for diagenetically altered volcanic tuffs. However, as temperature was increased, stability fields for clinoptilolite/heulandite, chabazite, erionite, mordenite, and stilbite greatly contracted with expansion of the laumontite, phillipsite, and wairakite fields. Calculations using an aqueous silica activity in equilibrium with cristobalite produced the same general features although the stability-field transitions occurred at significantly higher temperatures. When yugawaralite was included in the calculations, it dominated the calcic region of the stability diagrams, especially at elevated temperatures and at reduced aqueous silica activity (e.g. at quartz saturation). To obtain the stability field for laumontite shown in Figure 9, it was necessary to suppress yugawaralite formation.

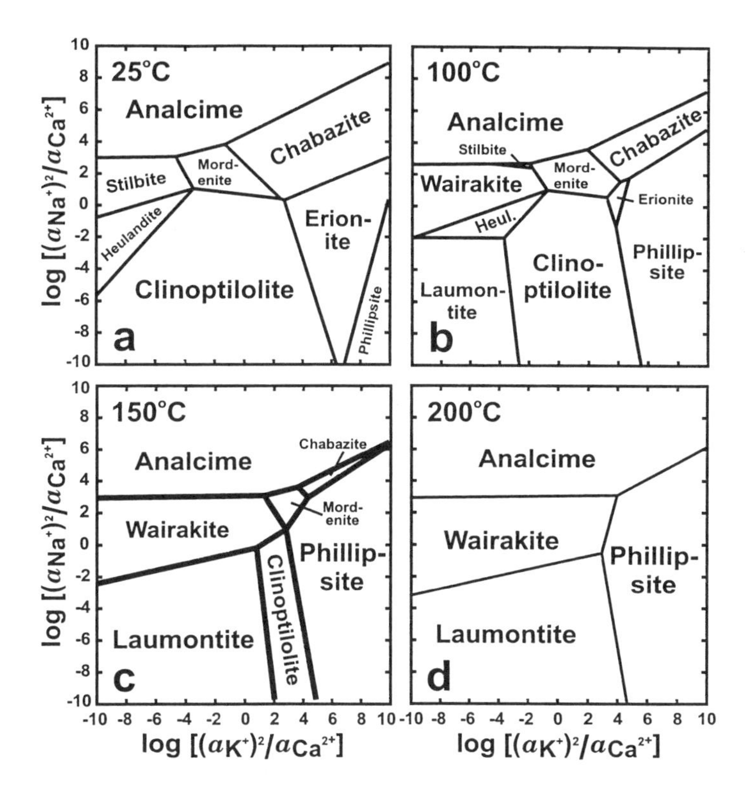

Figure 9. Log$[(a\text{K}^+)^2/a\text{Ca}^{2+}]$ vs. log$[(a\text{Na}^+)^2/a\text{Ca}^{2+}]$ diagrams for zeolites occurring in hydrothermal deposits in silicic-volcanic rocks (analcime, chabazite, clinoptilolite, erionite, heulandite, laumontite, mordenite, phillipsite, stilbite, and wairakite), all calculated assuming an aqueous silica activity in equilibrium with quartz and at (a): 25°C, (b): 100°C, (c): 150°C, (d): 200°C.

Zeolite formation in mafic rocks

In no other environment does such a unique and varied distribution of zeolites occur as in mafic rocks. One of the main reasons for the variety is the reduced aqueous silica activity. Aqueous silica activity in siliceous rocks is controlled by the silica polymorphs present (e.g. quartz, tridymite, cristobalite, and opal). Silica polymorphs are often nonexistent in mafic rocks, resulting in significantly lower aqueous silica activity and formation of stability fields for less silicic zeolites. Aluminous zeolites such as natrolite, mesolite/scolecite, and thomsonite are quite common in altered mafic rocks whereas they are not found in silicic environments.

Zeolites typically form in cavities or vugs in mafic rocks, often as a result of hydrothermal fluids. Kristmannsdóttir and Tómasson (1978) studied zeolite deposition in the basaltic flows and hydroclastic deposits in geothermal areas on Iceland. They subdivided the deposits into two thermal regimes. Rock temperatures <150°C at 1-km depth characterize the low-temperature regime. The low-temperature areas can be further divided into four zeolite zones as temperature increases: (1) chabazite, (2) mesolite/scolecite, (3) stilbite, and (4) laumontite. The high-temperature regions are characterized by rock temperatures >200°C at 1 km depth. With the exception of analcime and wairakite, which continued to form at temperatures >300°C, all zeolites formed at temperatures <250°C. Zeolites in these hydrothermally altered rocks occur mainly as minor constituents in small veins and as amygdule fillings. The zeolites also occur as devitrification products of volcanic glass. The geothermal fluids are generally of meteoric origin with a comparatively low dissolved-solid content and a pH of 9 to 10 (measured at 20°C). In some geothermal areas, however, the water is saline and contains a considerable amount of dissolved solids due either to mixing with seawater or to interaction with chloride trapped in marine formations. Approximate temperature ranges for the zeolites observed in the Iceland geothermal deposits are shown in Figure 1 (from Apps 1983).

Figures 10 and 11 show the results of thermodynamic modeling for the following suite of zeolite minerals reported from cavities or vugs in mafic rocks: analcime, chabazite, erionite, heulandite, laumontite, mesolite, mordenite, natrolite, phillipsite, scolecite, stilbite, thomsonite, and yugawaralite. At aqueous silica activities in equilibrium with quartz, mordenite, analcime, phillipsite and chabazite are calculated as the most stable zeolites. However, when the aqueous silica activity is reduced, stability fields are formed for a new suite of zeolite minerals. Analcime, mordenite, phillipsite, and yugawaralite dominate the stability diagrams at a log $a(SiO_2)$ of -4.0 (Fig. 10a). Reducing the aqueous silica activity (log $a(SiO_2)$ = -4.1, Fig. 10b) created stability fields for natrolite and scolecite. Further reducing the aqueous silica activity (log $a(SiO_2)$ = -4.3, Fig. 10c) created a stability field for thomsonite and greatly expanded the natrolite and scolecite stability fields. Continued reduction of the aqueous silica activity (log $a(SiO_2)$ = -5.0, Fig. 10d) caused natrolite, thomsonite, scolecite, and phillipsite to become the dominant zeolites.

Figures 11a and 11b show the results of modeling the same zeolite phases except in log $a(Na^+)$ vs. log $a(SiO_2)$ space at 35 and 100°C to demonstrate the effects of aqueous silica activity and temperature on the calculations. As expected by their chemistry, the more silicic zeolites (mordenite, analcime, and yugawaralite) occupy regions of elevated aqueous silica activity in the diagrams and the aluminous zeolites (natrolite, scolecite, and thomsonite) occupy regions of lower aqueous silica activity. Temperature, however, exerts a very significant effect on the calculated stability. Phillipsite, chabazite, erionite, and stilbite were found to be the most stable at lower temperatures (Fig. 11a). Increases in temperature to 100°C significantly decreased or totally eliminated the stability fields for these zeolites due to increases in the stability of natrolite, thomsonite, scolecite, and yugawaralite (Fig. 11b).

Zeolite formation in deep-sea sediments

Phillipsite and clinoptilolite are the most dominant zeolites in deep-sea sediments, followed by analcime and by rare erionite, mordenite, and laumontite. Although the factors contributing to phillipsite and clinoptilolite formation can be somewhat complex, Boles and Wise (1978), Kastner and Stonecipher (1978), and Stonecipher (1978) noted some general trends. Phillipsite tends to predominate in clayey sediments and in sediments containing mafic volcanic glass where the accumulation rate is <1 m/10^5 years. It is also found mainly in the Pacific Ocean, generally Upper Miocene to Recent in age. Clinoptilolite predominates in more calcareous sediments and in the presence of rhyolitic glass or excess aqueous SiO_2

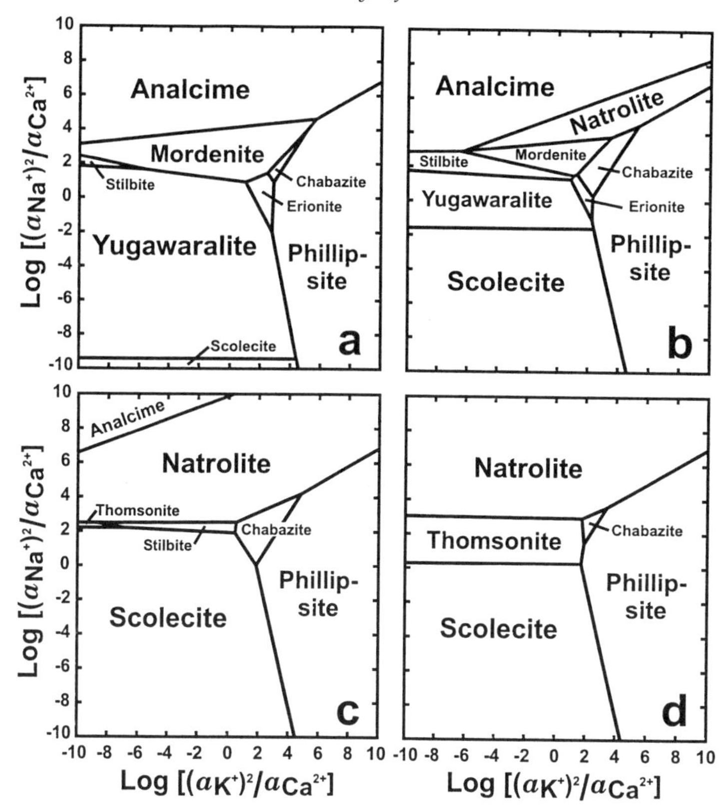

Figure 10. $\text{Log}[(a\text{K}^+)^2/a\text{Ca}^{2+}]$ vs. $\log[(a\text{Na}^+)^2/a\text{Ca}^{2+}]$ diagrams for the zeolites occurring in vugs and cavities in mafic volcanic rocks (analcime, chabazite, erionite, heulandite, laumontite, mesolite, mordenite, natrolite, phillipsite, scolecite, stilbite, thomsonite, and yugawaralite) calculated assuming 35°C and (a) log $a(\text{SiO}_2)$ = -4.0; (b) log $a(\text{SiO}_2)$ = -4.1; (c) log $a(\text{SiO}_2)$ = -4.3; (d) log $a(\text{SiO}_2)$ = -5.0.

Figure 11 (below). Log $a(\text{Na}^+)$ vs. log $a(\text{SiO}_2)$ diagrams for the zeolites occurring in vugs and cavities in mafic volcanic rocks (analcime, chabazite, erionite, heulandite, laumontite, mesolite, mordenite, natrolite, phillipsite, scolecite, stilbite, thomsonite, and yugawaralite) calculated assuming (a) 35°C and (b) 100°C and with log $a(\text{K}^+)$ in the groundwater fixed at -2 and log $a(\text{Ca}^{2+})$ fixed at -3.

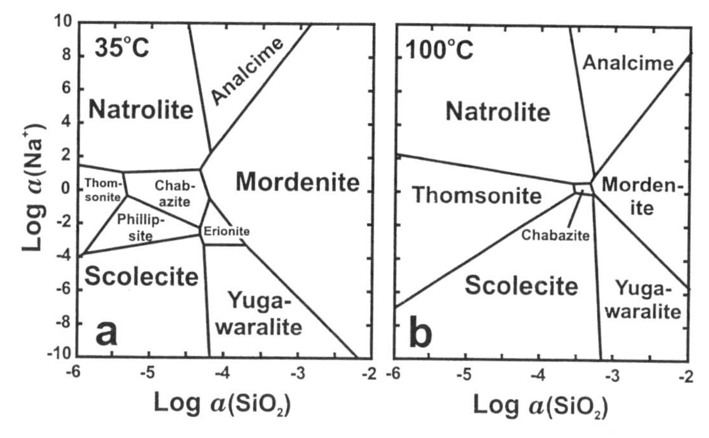

(possibly due to biogenic opal). It is found in both the Pacific and Atlantic Oceans and tends to be older (Lower Miocene to Cretaceous). Phillipsite occurs at the top of the sedimentary column in many places with clinoptilolite occurring deeper. Stonecipher (1978) found that the Si:Al ratio in bulk sediments increased with depth, suggesting that more siliceous material favored the formation of clinoptilolite. Boles and Wise (1978) noted that many clinoptilolite localities occur in microfossil-rich sediments and that volcanic glass is not a prerequisite for clinoptilolite formation. Because phillipsite is the dominant zeolite in many younger sediments, they suggested that it forms first metastably as a silica-deficient phase in the marine pore fluids and is eventually replaced by clinoptilolite.

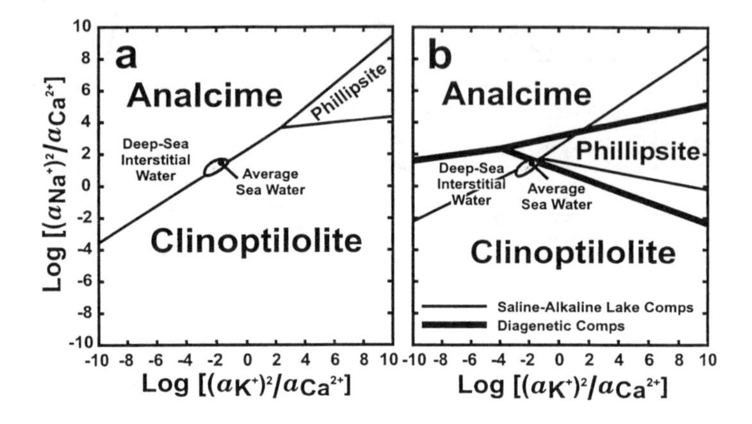

Figure 12. $\log[(a\text{K}^+)^2/a\text{Ca}^{2+}]$ vs. $\log[(a\text{Na}^+)^2/a\text{Ca}^{2+}]$ diagrams for the zeolites common in deep-sea sediments (analcime, clinoptilolite, and phillipsite) calculated using an aqueous silica activity in equilibrium with quartz at 5°C. Average seawater (Drever 1982) and interstitial water compositions from deep-sea sediments (Boles and Wise 1978) are plotted for reference.

(a) Calculations conducted using measured chemical compositions of deep-sea zeolites.

(b) Calculations conducted using chemical compositions representative of zeolites in saline-alkaline lake environment (thin lines) and in diagenetically altered deposits (thick lines).

Figure 12a shows the stability diagram for clinoptilolite-phillipsite-analcime calculated using an aqueous silica activity in equilibrium with quartz at 5°C. Water compositions for average seawater (Drever 1982) and for interstitial water from deep-sea sediments (Boles and Wise 1978) are superimposed on the plot. Aqueous ion speciation and activities were calculated using the program PHREEQC (Parkhurst 1995). The interstitial water in deep-sea sediments can have compositions significantly different from that of seawater if it becomes isolated by the sediments (Drever 1982). Increased silica activity (i.e. at cristobalite saturation) greatly increases the stability field for clinoptilolite with respect to phillipsite, given the water compositions approximated by seawater. Figure 12b shows the stability diagram for clinoptilolite-phillipsite-analcime calculated using aqueous silica activities in equilibrium with quartz and using chemical compositions representative of diagenetic and saline-alkaline-lake environments as a demonstration of the effect of the input chemical compositions on the calculations. The equilibria show a significant difference, with increased stability of phillipsite with respect to clinoptilolite. Although temperature has an effect on stability relationships, it appears that aqueous silica activity and zeolite extraframework-cation ratios exert the most significant control on whether clinoptilolite or phillipsite forms in deep-sea sediments. If the aqueous silica activity increases with time, phillipsite may react to form clinoptilolite, replicating the trends observed in the deep-sea sediments.

The calculations show that some observed water compositions fall within the analcime stability field. It is therefore surprising that analcime is not more common in deep-sea sediments. Perhaps additional kinetic or chemical effects favor the formation of clinoptilolite and phillipsite over analcime or hinder the formation of analcime.

STABILITY CALCULATIONS USING MEASURED THERMODYNAMIC DATA

This section assesses the effects of inconsistencies and errors in thermodynamic data by analyzing the sources of errors and by comparing calculated phase equilibria with experimentally determined stabilities of laumontite and wairakite in the CaO-Al_2O_3-SiO_2-H_2O (CASH) system. The CASH system covers many zeolites that form during diagenesis of mafic extrusives and is chosen not only for its importance, but also because it has been investigated somewhat more thoroughly than other systems

Inconsistencies in thermodynamic data and sources of error

Direct comparison between different thermodynamic compilations often reveals major inconsistencies in properties for various minerals. This arises from differences in the data refinement methods, differences in which data for minerals are chosen, and/or from using different techniques to estimate thermodynamic parameters for minerals where no experimental data are available. The consequences of such differences on model predictions can be small if the data from one compilation alone are used. In such cases, the data are internally consistent and errors introduced in the thermodynamic properties of one species are often inadvertently offset by compensating errors in other species to ensure consistency with experimental observations. However, large and unpredictable errors in model prediction can result from the indiscriminate selection of thermodynamic properties from different databases.

Additional errors in thermodynamic properties can occur due to extrapolation of high-temperature phase equilibria or calorimetric data to predict mineral solubilities under diagenetic conditions or from inadequate experimental characterization of the system under consideration. In many cases, the investigator is unaware of complications that occur. In some situations, the experimental set-up itself can result in wholly erroneous results, such as the phase equilibrium data obtained by a uniaxial press where a vapor phase participates in the reaction. Some of the problems encountered while measuring thermodynamic properties include:

- Inadequate characterization of the starting material;
- Variable and/or poorly characterized hydration state;
- Omission of corrections for configurational entropy;
- Metastable equilibria or equilibrium with unsuspected secondary phases;
- Non-attainment of equilibrium due to slow reaction rates;
- Particle-size effects;
- Unsuspected displacive phase transitions at elevated temperatures and/or pressures;
- Contamination of the system from vessel walls or impure starting materials;
- Equilibration with a surface whose composition differs from that of the bulk composition of a phase;
- Errors in or incomplete analysis of the constituents in the aqueous phase;

- Errors or inaccuracies in the thermodynamic properties of aqueous species, especially at elevated temperatures;
- Problems associated with sampling and analysis of quenched solutions (pH is especially problematic);
- Contamination of aqueous solutions by uptake or loss of CO_2 and other volatiles;

Perhaps the most difficult problem associated with laboratory solubility measurements is verifying attainment of equilibrium. Tests to demonstrate reversibility only prove equilibration with respect to a surface layer, which may have been modified during approach to equilibrium. Furthermore, other metastable phases may have precipitated and controlled the activities of the participating components. Validating laboratory solubility measurements can be attempted by field sampling of groundwaters associated with the mineral of interest, as aqueous equilibrium is much more likely to have been approached in the field, where quasi-equilibrium conditions may have prevailed for hundreds if not thousands of years. Thus, in studying zeolite equilibria with respect to the aqueous phase, there is a strong incentive to sample and analyze groundwaters as a means of obtaining potentially more meaningful data than can be achieved in the laboratory. Unfortunately, validation attempts using groundwater chemical analyses also may be plagued with problems, including:

- Mixing of waters from different horizons;
- Uncertainties regarding the source (i.e. depth and temperature of the original fluid);
- Pore water may be different in composition from water in fractures;
- Chemical evolution of sampled ground water (e.g. exchanging chemical constituents with the wall rocks), resulting in its original source composition being difficult, if not impossible, to reconstruct;
- Boiling of water at depth and loss of volatiles such as CO_2, H_2S, CH_4, etc.;
- Boiling and precipitation of SiO_2, $CaCO_3$, etc., during ascent of fluids in the well bore;
- Condensation of steam in the sample fluid.

It is beyond the scope of this chapter to explore each of these factors. However, their impact is noted in the discussion of the thermodynamic stabilities of laumontite and wairakite below.

Impact of uncertainties in geochemical modeling of zeolitic assemblages

Although the thermodynamic properties of many pure, stoichiometric minerals have been determined, natural systems are composed primarily of minerals that are multi-component solid solutions (e.g. feldspars, pyroxenes, amphiboles, clays, zeolites, carbonates, and sulfates). Many of these minerals do not have accepted solid-solution models and even some of the end-member thermodynamic properties remain in doubt. Zeolites themselves are subject to several factors that confound a simple thermodynamic treatment of their stability relations. Among these are solid-solution behavior, order-disorder, cation exchange, displacive (second order) phase transitions, variable hydration state, and surface free energy.

Typical uncertainties in calorimetric determinations of $\Delta G°_f$ or $\Delta H°_f$ translate into significant uncertainties in mineral solubilities which are generally far larger than the errors in chemical analyses of the aqueous phase components. It is perhaps not appreciated that most phase equilibrium studies have been made at temperatures above 350°C and at pressures between 1 and in excess of 10 kbar. The extrapolation of such data to standard-state conditions at 25°C and 1 atmosphere is therefore substantial. In order to make such extrapolations with the precision reflected in chemical analyses of aqueous solutions (i.e. of the order of 10%), extrapolations must be within ±1.25 kJ/mol for a phase such as anorthite ($CaAl_2Si_2O_8$). Were it not for the many uncertainties associated with solubility measurements in the laboratory or in field interpretations, many of the calorimetric and phase equilibrium data might be resolved by laboratory aqueous solubility measurements. Such phenomena as displacive phase transitions, order-disorder reactions, variable hydration, and variations in stoichiometry might not be detected in the normal course of phase equilibrium investigations. Failure to account for them may make it impossible to reconcile phase equilibrium data with thermodynamic data derived by other methods at different pressures and temperatures. Extrapolating phase equilibrium data to standard-state conditions therefore requires the use of accurate heat capacity functions, incorporation of corrections for compressibility, thermal expansivity, and cation disorder, and recognition of displacive phase transitions and non-stoichiometry. This can be done only through rigorous physical and chemical characterization of the participating phases. Published thermodynamic data for zeolites are presently incomplete and insufficiently accurate to permit geochemical modeling to be conducted with complete assurance. It is commonly necessary, therefore, to retrieve these data from experiments in the literature or elsewhere. If this approach fails to yield the needed information, then various empirical methods must be used to estimate the thermodynamic properties of the zeolites and coexisting minerals, as described above.

Evaluation of the thermodynamic properties of laumontite and wairakite

To illustrate a general approach to the problem, the stabilities of laumontite and wairakite in the CaO-Al_2O_3-SiO_2-H_2O (CASH) system are considered. The most appropriate place to start is through refinement of the thermodynamic properties of wairakite and laumontite. Not only are calorimetric data available for these two minerals, several phase equilibrium studies relate these zeolites to each other and to other phases. In addition, an extensive literature describing field occurrences with chemical analyses of coexisting ground waters also permits validation of the resulting thermodynamic data.

Several attempts have been made to reconcile calorimetric and phase equilibrium data pertaining to laumontite and wairakite (e.g. Helgeson et al. 1978, Hammerstrom 1981, Halbach and Chatterjee 1984, Berman et al. 1985, and Kiseleva et al. 1996a). Table 5 summarizes available thermodynamic data ($\Delta H°_{f,298}$ and $S°_{298}$) for some relevant minerals in the CASH system. With the exception of Helgeson et al. (1978), who used gibbsite as a primary oxide reference instead of corundum, the data are generally referenced to the thermodynamic properties of lime, corundum, quartz, and liquid water. Data of Helgeson et al. (1978) have been adjusted in Table 5 to reflect the CODATA enthalpy value for corundum (-1675.7 kJ/mol) to afford a better basis for comparison. The investigations by Halbach and Chatterjee (1984) and Berman et al. (1985) included refinement of the corundum properties, and as a consequence their values differ from other investigators by approximately +2 kJ/mol.

If allowance is made for differences in the reference oxide properties, a surprising level of agreement is obtained for most rock-forming minerals in the CASH system, despite differing assumptions or corrections of the data to account for heat capacity and volumetric effects, and/or cation-disorder phenomena. However, there are noteworthy exceptions and these unfortunately affect refinement of the thermodynamic properties of the zeolites. Helgeson et al. (1978), Hammerstrom (1981), and Berman et al. (1985) attempted to correlate phase equilibria measurements of laumontite and wairakite with the thermodynamic properties of other participating phases in the CASH system (Table 5). In contrast, other studies by Senderov (1980), Arnorsson et al. (1982), and Kiseleva et al. (1996b) attempted to correlate groundwater composition and field observations with assumed chemical equilibria involving either laumontite and/or wairakite. All researchers described difficulties in attaining satisfactory agreement.

Table 5. Comparison of thermodynamic data for minerals in the CASH system.

Mineral		$\Delta H_f°_{298}$ (kJ/mol) and $S°_{298}$ (J/(mol·K))									
		Helgeson et al. (1978) corrected[1]	*Hammerstrom (1981)*	*Haas et al. (1981)*	*Robinson et al. (1982)*	*Halbach & Chatterjee (1984)*	*Holland & Powell (1985)*	*Berman et al. (1985)*	*Berman (1988)*	*Zhu et al. (1994)*	*Kiseleva et al. (1996 a,b)*
Corundum	$\Delta H_f°_{298}$	-1675.70	-1675.70	-1675.70	-1675.74	-1674.40	-1675.70±1.30	-1673.60	-1675.70	—	—
	$S°_{298}$	50.96	50.92	50.92	50.94	50.92	50.90	51.00	50.82	—	—
α-quartz	$\Delta H_f°_{298}$	-910.65	-910.65	-910.70	-910.67	-910.54	-910.70±1.00	-910.26	-910.70	—	—
	$S°_{298}$	41.34	4.34	41.46	41.05	41.66	41.50	41.21	41.44	—	—
Lime	$\Delta H_f°_{298}$	-635.09	—	-635.09	-635.09	—	—	-635.79	-635.09	—	—
	$S°_{298}$	39.75	—	38.11	38.11	—	—	37.98	37.75	—	—
Wollastonite	$\Delta H_f°_{298}$	-1629.44	-1635.22	-1634.76±0.70	-1634.65	-1633.78	-1634.01±1.76	-1630.60	-1631.50	-1645.84	—
	$S°_{298}$	82.01	82.01	81.03±0.68	81.48±0.69	82.13	81.70	81.76	81.81	—	—
Anorthite	$\Delta H_f°_{298}$	-4230.56	-4232.83	-4227.80±1.10	-4230.69±1.10	-4233.98	-4233.84	-4226.08	-4228.73	-4232.20	—
	$S°_{298}$	207.72	201.02	199.29±0.15	119.27±0.15	199.60	201.00	200.33	200.19	—	—
Laumontite	$\Delta H_f°_{298}$	-7247.70	-7251.99	—	—	—	—	-7273.67	—	—	-7251.00
	$S°_{298}$	485.76	485.76	—	—	—	—	434.55	—	—	—
β-Leonhardite	$\Delta H_f°_{298}$	—	—	—	—	—	—	—	—	—	-7107.32
	$S°_{298}$	—	—	—	—	—	—	—	—	—	—
Wairakite	$\Delta H_f°_{298}$	-6622.90	-6658.89	—	—	—	—	-6663.33	—	—	-6646.70
	$S°_{298}$	439.74	385.17	—	—	—	—	367.40	—	—	—

[1]The Helgeson et al. (1978) data set, which used gibbsite as a primary oxide reference instead of corundum, has been adjusted to reflect the CODATA enthalpy value for corundum (-1675.7 kJ/mol) to allow a more direct comparison with the other data sets.

Most of the available experimental data pertaining to wairakite and laumontite thermodynamic properties involve the definition of univariant reactions in P-T space using phase equilibria. In retrieving thermodynamic data such as $\Delta G^\circ_{f,298}$, adjustments may be made to S°_{298}, ΔH°_f, C_p, V°, α, and β to achieve a fit of the experimental data. However, parameters should be constrained by independent measurements or estimates. Although arbitrary adjustment of both S°_{298} and V° can be made to improve the fit, this commonly leads to questionable values for these parameters. The approach used in this study was to use independent measurements of S°_{298} and V°, while adjusting only $\Delta G^\circ_{f,298}$ to match the phase equilibria as outlined below.

Equilibrium between wairakite and anorthite.

Evaluating the thermodynamic parameters of the univariant reaction,

$$\underset{\text{wairakite}}{CaAl_2Si_4O_{12}{\cdot}2H_2O} \Leftrightarrow \underset{\text{anorthite}}{CaAl_2Si_2O_8} + 2\ \underset{\text{quartz}}{SiO_2} + 2\ H_2O \qquad (8)$$

in P-T space permits the evaluation of the thermodynamic data for wairakite with respect to anorthite (and hence to those for laumontite, heulandite, stilbite, yugawaralite, etc.) in an internally consistent thermodynamic database. Liou (1970) conducted a comprehensive and careful study of this reaction between 500 and 5000 bars pressure and between 300 and 450°C (Fig. 13). Starting materials consisted of natural wairakite and synthetic anorthite in

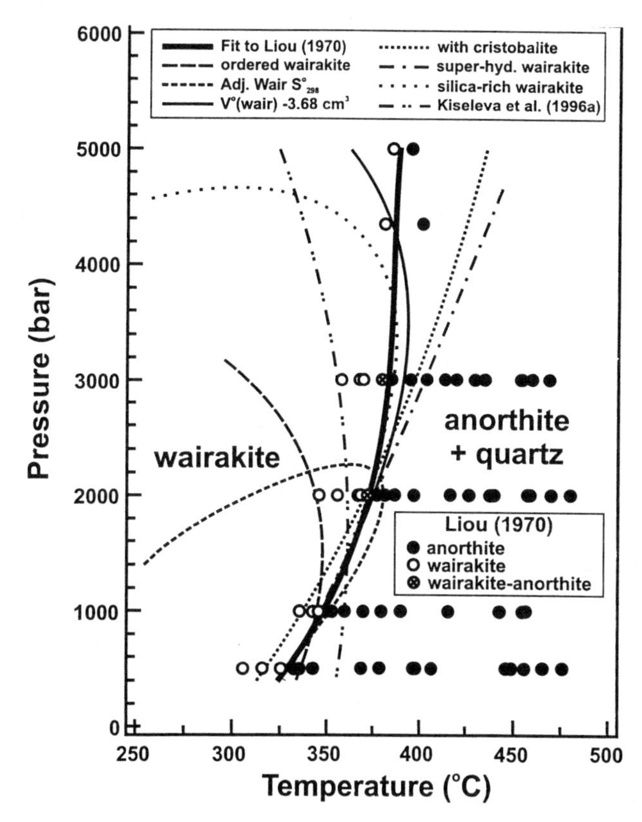

Figure 13. Experimental data from Liou (1970) and calculated positions of the univariant curve in P-T space representing the reaction: wairakite $\Leftrightarrow$ anorthite + 2 quartz + 2 H_2O.

ratios of 9:1 and 1:9. Establishing the position of the univariant curve was difficult due to the tendency for tetragonal wairakite, having a random distribution of Al and Si, to precipitate initially in preference to the more stable ordered monoclinic form normally observed in field samples. Liou estimated that experiments required at least 1500 hours for the more stable monoclinic form to crystallize, but that 5000 hours were required before the degree of order observed in natural wairakite could be achieved. Because his experiments never exceeded 1300 hours, the monoclinic wairakite synthesized in his experiments undoubtedly retained some disorder. Furthermore, the synthesized wairakite was <1 μm in size, suggesting a possible surface free-energy contribution to the $\Delta G°_{f,298}$ of the wairakite. In addition, natural materials prepared for hydrothermal experiments by grinding, separation, and screening invariably contain induced defects in the crystalline solid that can cause initial supersaturation in the aqueous phase. Epitaxial precipitation of disordered wairakite on the natural wairakite surfaces could therefore have occurred in Liou's experiments. The exposed surface of the wairakite in contact with the aqueous phase might also have achieved a lower degree of order than the bulk phase observed through X-ray diffraction. Thus it is expected that the thermodynamic parameters for wairakite derived from Liou's data will reflect a less stable material than that found in nature.

The thermodynamic properties of anorthite are a prerequisite for analyzing the phase equilibrium data presented by Liou (1970). Unfortunately, the literature reveals disagreement in these properties. Differing analyses of phase equilibria involving anorthite lead to differing estimates of $\Delta G°_{f,298}$ and $\Delta H°_{f,298}$ (anorthite), as can be observed by reference to Table 5. At least six calorimetric studies have determined $\Delta H°_f$ for anorthite; two using HF as the solvent, (Kracek and Neuvonen 1952, Barany 1962) and four using lead borate (Charlu et al. 1978, Newton et al. 1980, Carpenter et al. 1985, Zhu et al. 1994). HF calorimetry is prone to potential errors arising from secondary precipitation of calcium fluorides and is therefore considered suspect unless verified by independent methods. $\Delta H°_{f,298}$ for the ordered and chemically pure natural anorthites was determined by lead borate calorimetry by Carpenter et al. (1985; -4232.5 kJ/mol) and by Zhu et al. (1994, -4232.1 kJ/mol), in close agreement with the value of -4233.8 kJ/mol determined by Holland and Powell (1985), but differing substantially from -4228.8 kJ/mol computed by Berman (1988).

Carpenter et al. (1985) showed that anorthite undergoes two changes in standard state between 25°C and ~727°C (from $P\bar{1}$ to $I\bar{1}$ to $C\bar{1}$). The lower-temperature transition from $P\bar{1}$ to $I\bar{1}$ was determined by Brown et al. (1963) by means of high-temperature single-crystal X-ray diffractometry to occur at 243°C. Because the wairakite-anorthite transition is positioned above 300°C, between 500 and 5000 bar in Liou's experiments, the anorthite participating in the reaction should have been in the $I\bar{1}$ rather than the $P\bar{1}$ structural state at the phase boundary, assuming it was ordered and a pure substance. Wairakite is isostructural with analcime which possesses anomalous compressibility and, in some natural samples, undergoes a polyhedral tilt transition at 25°C between 5 and 10 kbar (Yoder and Weir 1960, Hazen and Finger 1979). Recently, Belitskii et al. (1993) confirmed calorimetrically the presence of a second-order phase transition for wairakite in the temperature range 145-150°C. Optical examination further revealed that the transition involved a change from monoclinic to "pseudocubic" symmetry. Although $\Delta H°$ for the transition cannot be quantitatively determined with the limited information provided by Belitskii et al., its magnitude is <1 kJ/mol and is therefore within the confidence level of extant phase equilibrium data.

In the following section, we have attempted to analyze previous studies and existing data for wairakite to determine the important variables in its calculated stability relations with anorthite. A summary of derived thermodynamic properties, based on seven possible interpretations of the position of the subject univariant curve (8) experimentally determined by Liou (1970), is listed in Table 6 and shown on Figure 13. As a starting point, we used the following thermodynamic properties: V°(wairakite) was calculated from the measured unit-cell parameters of wairakite reaction products reported by Liou (1970), along with the values of $S°_{298}$, $\Delta G°_{f,298}$, and V° for quartz from Berman (1988) and the values for anorthite from Zhu et al. (1994). To fit the various interpretations of univariant curve (8), the Helgeson et al. (1978) thermodynamic formulations and the SUPCRT code (Johnson et al. 1992) were used. Because the reaction involves very small changes in both entropy and volume, the position of the curve in P-T space is extremely sensitive to minor adjustments in these parameters. Because both parameters are functions of order-disorder and cation substitution in the wairakite structure, these properties should be rigorously characterized. Unfortunately, insufficient data are available at present to permit such a characterization. The modifications to the thermodynamic parameters used in an attempt to fit the Liou (1970) experimental data are outlined below and displayed on Figure 13.

Ordered wairakite (base case). When using V° for wairakite from Liou (1970) and $S°_{298}$ for wairakite calculated by Helgeson et al. (1978), a poor fit was obtained above 1000 bars by adjustment of wairakite $\Delta G°_{f,298}$ as illustrated on Figure 13 (labeled as "ordered wairakite"). Helgeson et al. also noted that extrapolation of the univariant curve to higher pressures led to a major deviation from experimental data.

Adjustment of wairakite $S°_{298}$. Adjustment of $S°_{298}$ improved the fit between 500 and 2000 bars (illustrated on Fig. 13 as "Adj. Wair $S°_{298}$"). However, the deviation above 2000 bars became even more severe than in the "Base Case" described above.

Adjustment of wairakite V°. An arbitrary adjustment of wairakite V° by -3.68 cm^3/mol to 186.87 cm^3/mol (in conformity with Helgeson et al. 1978) yielded a better fit over a wider range of pressures (shown on Fig. 13 as "V°(wair) -3.68 cm^3"). This adjustment is based on the reasoning that wairakite possesses a high compressibility and may undergo a displacive phase transition at higher pressures. Although this adjustment led to a better fit, it is necessary to understand the displacive phase transition or the compressibility to calculate the $\Delta G°_{f,298}$ for wairakite with the required accuracy.

Metastable equilibrium with cristobalite instead of quartz. An alternative explanation for the difficulties encountered in fitting the phase equilibrium data could be the persistence of a metastable silica polymorph such as cristobalite. Although the temperatures and pressures encountered in Liou's experiments are sufficiently high for quartz to attain equilibrium in <100 hours, it is possible that one or more of the remaining reactants may have produced an aqueous silica activity greater than quartz solubility. An improved fit was obtained by substituting cristobalite for quartz in Reaction (8).

Super-hydration of wairakite. Another possible interpretation of the phase equilibrium data is to assume "super-hydration" of wairakite at high water pressures. Wairakite is isostructural with analcime ($NaAlSi_2O_6 \cdot H_2O$), but only half the extraframework-cation sites are occupied with Ca^{2+} in place of Na^+. The vacant cation site [-] could be occupied by an additional water molecule at higher pressures according to the following reaction

$$Ca[-]Al_2Si_4O_{12} \cdot 2H_2O + H_2O \Leftrightarrow Ca[H_2O]Al_2Si_4O_{12} \cdot 2H_2O \quad (9)$$

Assuming that the additional water is "zeolitic" and possesses the heat capacity and entropy defined by Helgeson et al. (1978) for zeolitic water, we can calculate the properties of this super-hydrated wairakite. Fitting the data by varying $\Delta G°_{f,298}$ yields fair agreement over the pressure range from 500 to 5000 bars.

Table 6. Thermodynamic parameters for wairakite, laumontite, ß-leonhardite, and γ-leonhardite calculated from phase equilibria using various assumptions about the properties of wairakite.

	Maier-Kelley C_p function J/(mol·K)			V°	$S°_{298}$	$\Delta H°_{f,298}$	$\Delta G°_{f,298}$
Assumption	*a*	$b \times 10^3$	$c \times 10^{-5}$	cm^3/mol	*J/(mol·K)*	*kJ/mol*	*kJ/mol*
Wairakite							
Ordered wairakite "Base Case"	420.07	186.06	-68.74	190.55	439.7	-6621.54	-6194.83
Adjustment of wairakite $S°_{298}$	—	—	—	—	453.5	-6613.08	-6190.48
V°(wair) adjusted by -3.68 cm^3/mol	—	—	—	186.87	439.7	-6621.28	-6194.58
Equilibria with respect to cristobalite instead of quartz	—	—	—	190.55	—	-6617.30	-6190.60
Super-hydrated wairakite ($Ca[H_2O]Al_2Si_4O_{12}{\cdot}2H_2O$)	467.81	186.06	-68.74	—	464.8	-6936.72	-6447.96
Silica-rich wairakite ($Ca_{0.75}Al_{1.5}Si_{4.5}O_{12}{\cdot}2.25H_2O$)	423.42	192.51	-72.55	—	456.5	-6555.25	-6120.68
Kiseleva et al. (1996a)	420.07	186.06	-68.74	191.17	400.7	-6646.7±6.3	-6208.30
Laumontite							
$CaAl_2Si_4O_{12}{\cdot}4H_2O$	515.47	186.06	-68.74	208.56	485.8	-7244.95	-6692.88
$CaAl_2Si_4O_{12}{\cdot}4H_2O$ (V°(wair) adj. -3.68 cm^3/mol)	—	—	—	—	—	-7389.92	-6811.94
$CaAl_2Si_4O_{12}{\cdot}4.5H_2O$	539.32	186.06	-68.74	—	515.5	-7389.03	-6811.05
$CaAl_2Si_4O_{12}{\cdot}4.5H_2O$ (V°(wair) adj. -3.68 cm^3/mol)	—	—	—	—	—	-7245.80	-6693.73
Kiseleva et al. (1996a)	515.47	186.08	-68.74	—	485.8	-7251.0±8.5	-6698.9
β-Leonhardite							
$CaAl_2Si_4O_{12}{\cdot}3.6H_2O$	491.62	186.06	-68.74	210.59	461.10	-7197.30	-6572.64
γ-Leonhardite							
$CaAl_2Si_4O_{12}{\cdot}2.9H_2O$	467.77	189.06	-68.74	197.34	431.60	-6945.26	-6446.59

Silica-rich wairakite. As a final interpretation, we considered the effect of equilibration with respect to a non-stoichiometric wairakite by analogy with the analcime binary series ($Na_xAl_xSi_{48-x}O_{96}\cdot(24 - x/2)H_2O$, where $12 < x < 20$). Disordered wairakite also appears to possess a variable composition somewhat similar to that observed in analcime. By analogy with analcime, it is possible that the wairakites described by Liou that equilibrated with anorthite were silica rich. By varying $S°_{298}$ and $\Delta G°_{f,298}$ of silica-rich wairakite, an improved fit to the experimental data was achieved, although the deviation above 4000 bars is significant.

Fit by Kiseleva et al. (1996a). Kiseleva et al. (1996a) used lead borate calorimetry to determine $\Delta H°_{f,298}$ of a well-ordered natural monoclinic wairakite to be -6646.7±6.3 kJ/mol. By adjusting $S°_{298}$(wairakite) to fit the phase equilibrium data of Liou (1970), Kiseleva et al. obtained a value of 400.7 J/(mol·K), compared with a value of 439.7 J/(mol·K) estimated by Helgeson et al. (1978) and the adjusted "Adj. Wair $S°_{298}$" value of 453.5 J/(mol·K) estimated above. The Kiseleva et al. results are plotted on Figure 13 for comparison.

The estimated $S°_{298}$(wairakite) of Kiseleva et al. (1996a) appears to be unreasonably low compared with that of Helgeson et al. (1978), and the error may be attributed to forcing a fit to Liou's experimental data with a wairakite with a less negative $\Delta G°_{f,298}$ and $\Delta H°_{f,298}$. The difference in $\Delta H°_{f,298}$ inferred from our analysis of the wairakite reaction product in Liou's experiment and that of the natural wairakite investigated by Kiseleva et al. is large (>25 kJ/mol). Although partial disorder, small crystallite size, and/or impure water may all contribute to the observed discrepancy with laboratory phase equilibrium data, the magnitude is so great that independent validation is needed.

A number of possible scenarios exist to reconcile the phase equilibrium data of Liou with the wairakite reaction product, although none produce a precise fit to the experimental data. Until the wairakite reaction products, experimental conditions, and compressibility and expansivity of wairakite are more thoroughly characterized, the results will remain indeterminate. Much of the uncertainty may stem from the ill-defined compressibility of wairakite and the possibility that displacive phase transitions affect its thermodynamic properties at elevated pressures. Given the fact that wairakite occurs in nature at depths of <10,000 m, a better approach is to employ thermodynamic properties for wairakite that are optimized to fit Liou's data below 2000 bars.

Laumontite—β-leonhardite—γ-leonhardite—wairakite equilibria

The position of the univariant curve

$$\underset{\text{laumontite}}{CaAl_2Si_4O_{12}\cdot 4H_2O} \Leftrightarrow \underset{\text{wairakite}}{CaAl_2Si_4O_{12}\cdot 2H_2O} + 2\,H_2O \quad (10)$$

has been determined experimentally by Juan and Lo (1971) and by Liou (1971). The former investigation was restricted to 686 and 1382 bars, but both studies are in reasonable agreement. Their results are shown in Figure 14 together with Liou's estimated position of the univariant curve. According to Liou (1971), measurements taken above 3000 bars fall within the lawsonite stability field. Data taken above that pressure therefore define a metastable extension of the univariant curve.

For many years, it was assumed that fully hydrated laumontite possessed the formula $CaAl_2Si_4O_{12}\cdot 4H_2O$ (e.g. see Gottardi and Galli 1985). However, a recent series of crystallographic studies has unequivocally demonstrated that the true formula is $CaAl_2Si_4O_{12}\cdot 4.5H_2O$ (Armbruster and Kohler 1992, Artioli and Stahl 1993, Stahl and Artioli 1993). According to Armbruster and Kohler (1992), three H_2O are bonded to Ca,

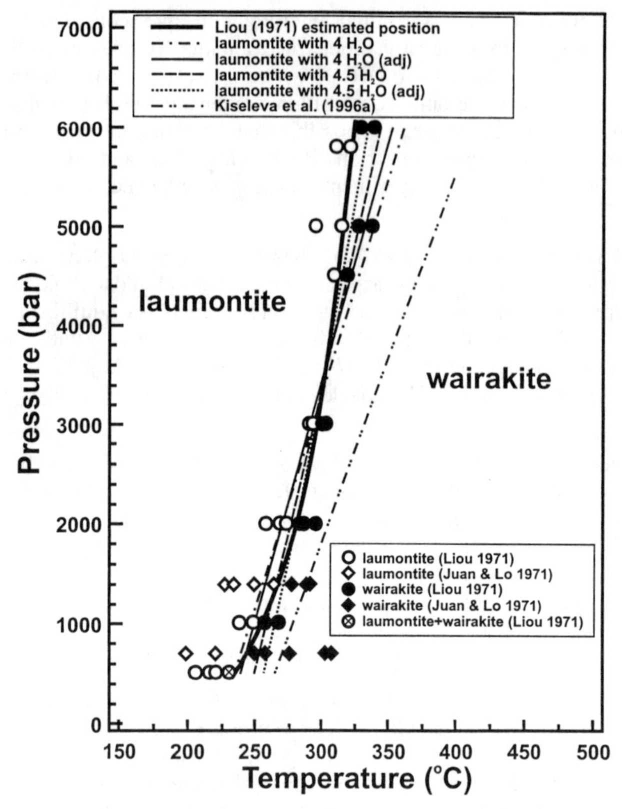

Figure 14. Experimental data from Liou (1971) and Juan and Lo (1971) showing the calculated positions of the univariant curve in P-T space for the reaction: laumontite ⇔ wairakite + 2 H_2O.

and the remaining 1.5 H_2O are linked by hydrogen bonds to Ca-H_2O and to framework O. To calculate $\Delta G°_{f,298}$ for laumontite, the optimized thermodynamic properties derived previously for ordered wairakite ($\Delta G°_{f,298}$ = -6190.48 kJ/mol) were used where the univariant curve was fit to the experimental data of Liou (1971) and Juan and Lo (1971) between 500 and 2000 bars (Fig. 14). $S°_{298}$ values for laumontite, assuming 4.0 and 4.5 H_2O, were calculated using the thermodynamic properties of zeolitic water estimated by Helgeson et al. (1978). V°(laumontite) was calculated from the unit-cell parameter data of Armbruster and Kohler (1992). All fits were quite good between 1000 and 3000 bars, but deviations from the experimental data occur both at higher and lower pressures as can be seen in Figure 14. The deviations at higher pressures are possibly due to inadequate corrections for the effects of the compressibilities of the participating zeolites. Liou (1971) encountered difficulties in attaining equilibrium at 500 bars because of the considerably slower reaction kinetics. However, Liou's data at this pressure are consistent with those of Juan and Lo at 686 bars. No explanation can be provided for the enhanced curvature of the experimental univariant curve below 1000 bar. In addition to the above fits to the experimental data, an additional case was considered using the thermodynamic properties of wairakite calculated on the basis of a corrected molar volume (V°(wair) adjusted by -3.68 cm^3/mol, see Table 6), identified on Figure 14 as "(adj)." For comparison, the univariant curve based on the thermodynamic properties of laumontite and wairakite presented by Kiseleva et al. (1996a) is also plotted on Figure 14.

A characteristic of laumontite is that it dehydrates readily at room temperature when the relative humidity is <100%. The impact of dehydration on the stability of laumontite in water-saturated environments at elevated temperature must be evaluated before further consideration of equilibrium among laumontite and wairakite. Laumontite dehydrates reversibly in three stages between room temperature and 345°C (Gottardi and Galli 1985, Armbruster and Kohler 1992, Kiseleva et al. 1996a). At room temperature and humidity, laumontite is partially dehydrated to β-leonhardite, with the formula $CaAl_2Si_4O_{12}{\cdot}3.6H_2O$. At elevated temperatures, β-leonhardite loses additional water to form γ-leonhardite with 2.9 H_2O at 60°C and 2.7 H_2O at 120°C (Armbruster and Kohler 1992). Precise agreement does not exist on subsequent dehydration steps but it is fairly certain that dehydration to ~2.25 H_2O occurs at about 250°C, and to ~1 H_2O at >345°C. Evidence in the literature is meager as to whether the loss of water is progressive or takes place rapidly over restricted temperature intervals.

For convenience, we assume the formula $CaAl_2Si_4O_{12}{\cdot}xH_2O$, where x = 4.5 for laumontite, 3.5 for β-leonhardite and 3.0 for γ-leonhardite. The first dehydration reaction from laumontite to β-leonhardite can be approximated by

$$\underset{\text{laumontite}}{CaAl_2Si_4O_{12}{\cdot}4.5H_2O} \Leftrightarrow \underset{\beta\text{-leonhardite}}{CaAl_2Si_4O_{12}{\cdot}3.5H_2O} + H_2O \tag{11}$$

Although the partial dehydration of laumontite to β-leonhardite has often been observed at room conditions of humidity and temperature, only two studies have established the temperature at which dehydration occurs in the presence of liquid water and the two results differ significantly. Van Reeuwijk (1974) determined the transition to occur at 57°C, whereas Kirov and Balkanov (1976) reported a temperature of 95±2°C. We have arbitrarily accepted the latter value for the purpose of illustrating the β-leonhardite stability field. Figure 15 shows the stability relationship between laumontite and wairakite when leonhardite is included in the calculations. Assuming V° for β-leonhardite = 201.59 cm^3/mol (from the unit-cell parameter data of Armbruster and Kohler 1992) and $S°_{298}$ = 461.1±5.45 J/(mol·K) (based on the calorimetric study of King and Weller 1961), a Maier-Kelly C_p function for β-leonhardite was calculated from the oxides as described by Helgeson et al. (1978). The thermodynamic properties for β-leonhardite are included in Table 6. For this calculation, the heat capacity of "zeolitic" H_2O was the estimated value from Helgeson et al. (1978) of 47.7 J/(mol·K). The position of Reaction (11) in P-T space was calculated after adjusting $\Delta G°_{f,298}$ for β–leonhardite to ensure that it intersects the water saturation surface at 95°C (Fig. 15). The calculated $\Delta H°_{298}$ for Reaction (11) is +5.87 kJ/mol. However, it should be emphasized that the position of the laumontite dehydration curve is very sensitive to non-stoichiometry of laumontite and the activity of H_2O. Stability of the dehydrated form may be enhanced by the substitution of Na and K for Ca in the structure.

Using the derived thermodynamic properties of β-leonhardite, the univariant reaction of β-leonhardite reacting to wairakite plus water (in both the liquid and vapor state),

$$\underset{\beta\text{-leonhardite}}{CaAl_2Si_4O_{12}{\cdot}3.5H_2O} \Leftrightarrow \underset{\text{wairakite}}{CaAl_2Si_4O_{12}{\cdot}2H_2O} + 1.5\ H_2O_{(v,l)} \tag{12}$$

can be calculated to define the upper temperature and lower pressure limits of β-leonhardite as shown on Figure 15.

The dehydration reaction between β-leonhardite and γ-leonhardite

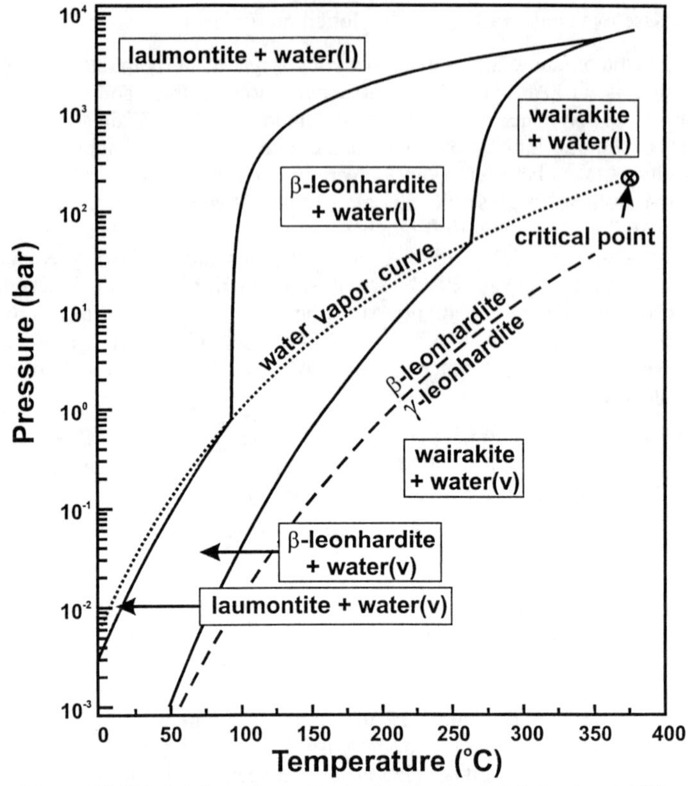

Figure 15. Calculated positions of univariant reactions defining the stability fields of laumontite, β-leonhardite, and wairakite in P-T space.

$$\underset{\beta\text{-leonhardite}}{CaAl_2Si_4O_{12}{\cdot}3.5H_2O} \Leftrightarrow \underset{\gamma\text{-leonhardite}}{CaAl_2Si_4O_{12}{\cdot}3.0H_2O} + 0.5H_2O \quad (13)$$

is also plotted on Figure 15. The entropy (S°_{298} = 431.6 J/(mol·K)) of γ-leonhardite was assumed to equal that of β-leonhardite minus 0.5 $H_2O_{(z)}$, V° of γ-leonhardite is 197.34 cm^3/mol (from the unit-cell parameter data of Armbruster and Kohler 1992), and the heat-capacity function for γ-leonhardite was calculated in the same fashion as for β-leonhardite (Table 6). The reaction is constrained to take place at 120°C at a water-vapor pressure ≤0.032 atm (the saturation partial pressure of water at 25°C), and the calculated position of the univariant Reaction (13) does not intersect the water saturation curve before the critical point of water is reached (Fig. 15).

Despite the substantial quantity of experimental data for wairakite and laumontite, many of the experiments are incompletely constrained or the data are too contradictory to permit a definitive analysis of the zeolites' thermodynamic stabilities. Zeolites commonly display multiple, coupled solid solutions, ion exchange, order-disorder, variable degrees of hydration, second-order displacive phase transitions, anomalous expansivities and compressibilities, and hydration/dehydration reactions. However, these properties are unknown, incompletely studied, or poorly characterized for most zeolitic phases. Unfortunately, even small deviations in the thermodynamic values used in a calculation can lead to significant shifts in calculated equilibria.

CONCLUSIONS AND RECOMMENDATIONS FOR FUTURE STUDIES

Zeolites form metastable diagenetic assemblages in nature, normally through the alteration of volcanic glass in the presence of water at relatively low temperatures (ranging from that of the Earth's surface to ~250°C). Success in resolving the many complex issues concerning the formation and stability of zeolites has been limited, partly because of the chemical complexity of naturally occurring systems containing zeolites, but also because of the low temperatures at which most zeolites form. The long-term stability of natural zeolite assemblages can be qualitatively interpreted by invoking the Ostwald step rule, where successive mineral assemblages have progressively increasing degrees of thermodynamic stability. At diagenetic temperatures, this process can take thousands, if not millions of years to complete.

The traditional approach to understanding the conditions controlling mineral equilibria is to examine their relationships as they occur in nature. Thermodynamic modeling, however, can clarify stability relationships among various zeolites by examining the influences of chemical compositions as a function of temperature, pressure, and chemical potentials of the participating components. For zeolites in particular, the chemical composition of the coexisting aqueous phase is extremely important. Although not directly influencing which zeolite species form, pH is important in determining whether zeolites will form at the expense of other potentially stable assemblages (e.g. Hay 1978, Surdam and Sheppard 1978, Bowers and Burns 1990).

One method of obtaining thermodynamic properties for minerals is by utilizing an empirical method to estimate the thermodynamic properties. Although experimental thermodynamic data now exist for many zeolites, the characteristic highly variable chemistry of zeolites encourages the use of estimated thermodynamic data for zeolites as the thermodynamic properties measured for a zeolite in one sample may not be representative of the thermodynamic properties of the same zeolite species from other samples. Various empirical routines have been formulated for estimating thermodynamic data for minerals. Recent methods represent mineral phases as a set of elemental building blocks composed of unique polyhedra (Robinson and Haas 1983, Berman and Brown 1985, Hazen 1985, Chermak and Rimstidt 1989, Holland 1989). By summing the contributions from each polyhedron, thermodynamic properties can be estimated for the complete mineral unit. The use of estimated thermodynamic data for specifically measured zeolite compositions can clarify the stability relationships among various zeolites within a given deposit or environment.

The majority of zeolites present in saline-alkaline lake or diagenetic deposits formed under conditions of elevated silica activity, consistent with the frequency with which they coexist with opal-CT. Aluminous zeolites such as natrolite, mesolite/scolecite, and thomsonite are common in altered mafic rocks where the aqueous silica activity is not controlled by one of the silica polymorphs and lower silica activity can be attained. Formation of phillipsite is favored in such environments as well as in deep-sea sediments at reduced aqueous silica activities (at or below quartz saturation). The occurrence of erionite, chabazite, and phillipsite is indicative of environments that have elevated potassium concentrations. Elevated temperature, Ca-rich water, and reduced aqueous silica activity all help to produce laumontite and wairakite stability fields in hydrothermally altered silicic rocks. Analcime stability increases as temperature is increased, as the Na^+ concentration in the fluid phase is increased, or when aqueous silica activity is decreased.

Previous studies of Yucca Mountain zeolites showed that heulandite (which is isostructural with clinoptilolite and relatively common at Yucca Mountain) had a stability field only when silica activity was reduced to quartz saturation and only when stellerite formation was suppressed (Chipera et al. 1995a). This is inconsistent with the common occurrence of both heulandite and stellerite in fractures in devitrified tuffs at Yucca Mountain. However, Yucca Mountain heulandites commonly contain significant Mg and Sr, which were not included in the original calculations. Current calculations suggest that the activities of the cation species in the groundwater and the cation ratios in the solid phases (Mg, in particular) exert significant control on the formation of heulandite vs. stellerite. Mg:Ca ratios as low as 0.05 in heulandite appear sufficient to stabilize heulandite with respect to stellerite. These results illustrate the need to expand thermodynamic data estimation methods to include all cations that occur in the zeolites (e.g. Mg, Sr, Ba, etc.).

The significant cation-exchange capabilities of zeolites give rise to an additional problem when modeling zeolite stability. The extraframework-cation composition of many zeolites will change with evolving water composition (e.g. evaporation in a saline-alkaline lake) and may not be representative of the original composition during formation. Empirical cation selectivities such as those formulated by Ames (1964a,b) can provide rough guidelines to the expected evolution in cation composition. However, binary cation-exchange isotherms between a particular zeolite and the solution have been shown to be dramatically affected by such factors as the cationic species, temperature, and the ionic strength of the solution (e.g. Barrer and Klinowski 1974, Murphy and Pabalan 1994, Pabalan 1994, Pabalan and Bertetti 1994). In the future, studies coupling cation-exchange models with the thermodynamic equilibria calculations (e.g. Carey et al. 1996) should provide greatly improved accuracy in the determination of zeolite stability.

The mineralogy and crystal chemistry of naturally occurring zeolites are complex. Zeolites commonly display multiple, coupled solid solutions, ion exchange, order-disorder, variable degrees of hydration, second-order displacive phase transitions, anomalous expansivities and compressibilities, and hydration/dehydration reactions. In any attempt to determine precisely the thermodynamic and kinetic factors controlling zeolite stability, these properties must be considered. However, these properties are unknown or poorly characterized for most zeolitic phases. Experimentally measured thermodynamic properties of zeolites can presently be applied for predictive purposes only to minerals of identical composition. In addition, existing thermodynamic databases do not include data for all zeolites of interest, and the various methods used to create internally consistent databases often do not produce similar results for the phases, even when using the same published data. This problem is compounded by that fact that the different methods used to collect thermodynamic data (e.g. solubility, calorimetry, phase equilibria, field relationships, etc.) often produce conflicting results due to problems inherent in each of the methods. Unfortunately, even small deviations in the thermodynamic values used in an equilibrium calculation can lead to significant shifts in calculated equilibria.

To illustrate the calculation of equilibrium relationships and the limitations inherent in such an approach, the stabilities of two pure end-member calcium zeolites (wairakite and laumontite) were investigated. Despite the substantial quantity of experimental data available for wairakite and laumontite, much of the data are too contradictory to permit a definitive analysis of the relative stabilities of the two minerals. It is clear that significant inconsistencies remain among the various sources of thermodynamic data for wairakite and laumontite. In order to make progress, data acquisition must advance to a higher level of accuracy than has hitherto been common practice. Experimental studies must involve a more comprehensive and rigorous characterization of materials and compositions. In particular, the entropies of wairakite and laumontite should be determined independently. Furthermore, algorithms correcting for disorder, solid solution, cation substitution, displacive phase transitions, changes in molar volume due to hydration/dehydration, and compressibility and expansivity of the participating phases should be used in subsequent thermodynamic analysis.

Although existing thermodynamic data are not of sufficient quality for strict quantitative analyses of zeolite stability, general trends and relative stability relationships can be readily calculated using thermodynamic methods. These trends and relationships can help predict the stability of zeolites in various environmental and industrial applications and can provide a better understanding of the geochemical conditions that produced the observed zeolites even though the original fluids long since vanished.

ACKNOWLEDGMENTS

We thank J.W. Carey and W.M. Murphy for their thorough reviews, comments, and suggestions, which significantly improved this manuscript.

REFERENCES

Ames LL Jr (1964a) Some zeolite equilibria with alkali metal cations. Am Mineral 49:127-145

Ames LL Jr (1964b) Some zeolite equilibria with alkaline earth metal cations. Am Mineral 49:1099-1110

Artioli G, Stahl K (1993) Fully hydrated laumontite: A structure study by flat-plate and capillary powder diffraction techniques. Zeolites 13:249-255

Apps JA (1983) Hydrothermal evolution of repository groundwaters in basalt. *In* NRC Nuclear Waste Geochemistry '83, U S Nuclear Regulatory Commission Report NUREG/CP-0052, p 14-51

Apps JA (1987) Alteration of natural glass in radioactive waste repository host rocks: A conceptual review. Lawrence Berkeley Laboratory Report LBL-22871, 38 p

Armbruster T, Kohler T (1992) Rehydration and dehydration of laumontite: A single-crystal X-ray study at 100 K. Neues Jahrb Mineral Monatsh, 385-397

Arnorsson S, Sigurdsson S, and Svavarsson H (1982) The chemistry of geothermal waters in Iceland. I Calculation of aqueous speciation from 0°C to 370°C. Geochim Cosmochim Acta 46:1513-1532

Barany R (1962) Heats and free energies of formation of some hydrated and anhydrous sodium- and calcium-aluminum silicates. U S Bureau of Mines Report of Investigations 5900, 17 p

Bargar KE, Keith TEC (1995) Calcium zeolites in rhyolitic drill cores from Yellowstone National Park, Wyoming. *In* Natural Zeolites '93: Occurrence, Properties, Use. DW Ming, FA Mumpton (eds) Int'l Comm on Natural Zeolites, Brockport, New York, p 69-86

Barrer RM, Klinowski J (1974) Ion-exchange selectivity and electrolyte concentration. J Chem Soc Faraday Trans 70:2080-2091

Belitskii IA, Fursenko BA, Seryotkin YuV, Goryainov SV, Drebushchak VA (1993) High-pressure, high-temperature behavior of wairakite. 4th Int'l Conf on the Occurrence, Properties, and Utilization of Natural Zeolites, Boise, Idaho, June 20-28, 1993, Program and Abstracts, p 45-48

Berman RG (1988) Internally consistent thermodynamic data for minerals in the system Na_2O-K_2O-CaO-MgO-FeO-Fe_2O_3-Al_2O_3-SiO_2-TiO_2-H_2O-CO_2. J Petrol 29:445-522

Berman RG, Brown TH (1985) Heat capacity of minerals in the system Na_2O-K_2O-CaO-MgO-FeO-Fe_2O_3-Al_2O_3-SiO_2-TiO_2-H_2O-CO_2: Representation, estimation, and high temperature extrapolation. Contrib Mineral Petrol 89:168-183

Berman R, Brown TH, Greenwood HJ (1985) An internally-consistent thermodynamic data base for minerals in the system Na_2O-K_2O-CaO-MgO-FeO-Fe_2O_3-Al_2O_3-SiO_2-TiO_2-H_2O-CO_2. Atomic Energy Canada Ltd., Technical Report TR-377, 62 p

Bish DL (1988) Effects of composition on the dehydration behavior of clinoptilolite and heulandite. *In* Occurrence, Properties and Utilization of Natural Zeolites. D Kallo, HS Sherry (eds) Akademiai Kiado, Budapest, p 565-576

Bish DL, Chipera SJ (1989) Revised mineralogic summary of Yucca Mountain, Nevada. Los Alamos National Laboratory Report LA-11497-MS, 68 p

Boles JR (1971) Synthesis of analcime from natural heulandite and clinoptilolite. Am Mineral 56:1724-1734

Boles JR, Wise WS (1978) Nature and origin of deep-sea clinoptilolite. *In* Natural Zeolites: Occurrence, Properties, Use. LB Sand, FA Mumpton (eds) Pergamon Press, New York, p 235-243

Bowers TS, Burns RG (1990) Activity diagrams for clinoptilolite: Susceptibility of this zeolite to further diagenetic reactions. Am Mineral 75:601-619

Brown WL, Hoffman W, Laves F (1963) Continuous and reversible transformation in anorthite between 25 and 350°. Naturwiss 50:221

Brown TH, Berman RG, Perkins EH (1989) PTA-SYSTEM: A Ge0-Calc software package for the

calculation and display of activity-temperature-pressure phase diagrams. Am Mineral 74:485-487
Broxton DE, Bish DL, Warren RG (1987) Distribution and chemistry of diagenetic minerals at Yucca Mountain, Nye County, Nevada. Clays & Clay Minerals 35:89-110
Broxton DE, Warren RG, Hagan RC, Luedemann G (1986) Chemistry of diagenetically altered tuffs at a potential nuclear waste repository, Yucca Mountain, Nye County, Nevada. Los Alamos National Laboratory Report LA-10802-MS, 160 p
Carey JW, Chipera SJ, Bish DL (1996) The effect of cation exchange and dehydration on clinoptilolite-analcime equilibria: Application to Yucca Mountain, Nevada. Geol Soc Am Program with Abstracts 28:A469
Carlos BA, Chipera SJ, Bish DL, Raymond R (1995a) Distribution and chemistry of fracture-lining zeolites at Yucca Mountain, Nevada. *In* Natural Zeolites '93: Occurrence, Properties, Use. DW Ming, FA Mumpton (eds) Int'l Comm on Natural Zeolites, Brockport, New York, p 547-563
Carlos BA, Chipera SJ, Bish DL (1995b) Distribution and chemistry of fracture-lining zeolites at Yucca Mountain, Nevada. Los Alamos National Laboratory Report LA-12977-MS, 92 p
Carlos BA, Chipera SJ, Snow MG (1995c) Multiple episodes of zeolite deposition in fractured silicic tuff. *In* Proc 8th Int'l Symp on Water-Rock Interaction—WRI-8. YK Kharaka, OV Chudaev (eds) Balkema, Rotterdam, p 67-71
Carpenter MA, McConnell JDC, Navrotsky A (1985) Enthalpies of ordering in the plagioclase feldspar solid solution. Geochim Cosmochim Acta 49:947-966
Charlu TV, Newton RC, Kleppa OJ (1978) Enthalpy of formation of some lime silicates by high-temperature solution calorimetry, with discussion of high-pressure phase equilibria. Geochim Cosmochim Acta 42:367-375
Chermak JA, Rimstidt JD (1989) Estimating the thermodynamic properties (ΔG_f° and ΔH_f°) of silicate minerals at 298 K from the sum of polyhedral contributions. Am Mineral 74:1023-1031
Chipera SJ, Bish DL (1997) Equilibrium modeling of clinoptilolite-analcime equilibria at Yucca Mountain, Nevada. Clays & Clay Minerals 45:226-239
Chipera SJ, Bish DL, Carlos BA (1995a) Equilibrium modeling of the formation of zeolites in fractures at Yucca Mountain, Nevada. *In* Natural Zeolites '93: Occurrence, Properties, Use. DW Ming, FA Mumpton (eds) Int'l Comm on Natural Zeolites, Brockport, New York, p 565-577
Chipera SJ, Vaniman DT, Carlos BA, Bish DL (1995b) Mineralogic variation in drill core UE-25 UZ#16, Yucca Mountain, Nevada. Los Alamos National Laboratory Report LA-12810-MS, 39 p
Coombs DS, Ellis AJ, Fyfe WS, Taylor AM (1959) The zeolite facies, with comments on the interpretation of hydrothermal syntheses. Geochim Cosmochim Acta 17:53-107
Dibble WE Jr, Tiller WA (1981) Kinetic model of zeolite paragenesis in tuffaceous sediments. Clays & Clay Minerals 29:323-330
Donahoe RJ, Hemingway BS, Liou JG (1990a) Thermochemical data for merlinoite: 1 Low-temperature heat capacities, entropies, and enthalpies of formation at 298.15 K of six synthetic samples having various Si/Al and Na/(Na + K) ratios. Am Mineral 75:188-200
Donahoe RJ, Liou JG, Hemingway BS (1990b) Thermochemical data for merlinoite: 2 Free energies of formation at 298.15 K of six synthetic samples having various Si/Al and Na/(Na + K) ratios and application to saline, alkaline lakes. Am Mineral 75:201-208
Drever JI (1982) The Geochemistry of Natural Waters. Prentice-Hall, Englewood Cliffs, NJ, 388 p
Duffy CJ (1993) Preliminary conceptual model for mineral evolution in Yucca Mountain. Los Alamos National Laboratory Report LA-12708-MS, 46 p
Eggleton RA, Keller J (1982) The palagonitization of limburgite glass: A TEM study. Neues Jahrb Mineral Monatsh, 321-336
Forsman NF (1984) Durability and alteration of some Cretaceous and Paleocene pyroclastic glasses in North Dakota. J Non-Crystalline Solids 67:449-461
Gottardi G, Galli E (1985) Natural Zeolites. Springer-Verlag, New York, 409 p
Guthrie GD Jr, Bish DL, Chipera SJ, Raymond R Jr (1995) Distribution of potentially hazardous phases in the subsurface at Yucca Mountain, Nevada. Los Alamos National Laboratory Report LA-12573-MS, 41 p
Haas JL, Robinson GR Jr, Hemingway BS (1981) Thermodynamic tabulations for selected phases in the system CaO-Al_2O_3-SiO_2-H_2O at 101.325 kPa (1 atm) between 273.15 and 1800 K. J Phys & Chem Ref Data 10:575-669
Halbach H, Chatterjee ND (1984) An internally consistent set of thermodynamic data for 21 CaO-Al_2O_3-SiO_2-H_2O phases by linear parametric programming. Contrib Mineral Petrol 88:14-23
Hammerstrom LT (1981) Internally consistent thermodynamic data and phase relations in the CaO-Al_2O_3-SiO_2-H_2O system. MSc Thesis, Univ British Columbia, 62 p
Hay RL (1966) Zeolites and zeolitic reactions in sedimentary rocks. Geol Soc Am Spec Paper 85, 130 p
Hay RL (1978) Geologic occurrence of zeolites. *In* Natural Zeolites: Occurrence, Properties, Use. LB Sand,

FA Mumpton (eds) Pergamon Press, New York, p 135-143
Hay RL, Sheppard RA (1977) Zeolites in open hydrologic systems. *In* Mineralogy and Geology of Natural Zeolites. FA Mumpton (ed) Rev Mineral 4:93-102
Hazen RM (1985) Comparative crystal chemistry and the polyhedral approach. *In* Microscopic to Macroscopic: Atomic Environments to Mineral Thermodynamics. SW Kieffer, A Navrotsky (eds) Rev Mineral 14:317-346
Hazen RM, Finger LW (1979) Polyhedral tilting: A common type of pure displacive phase transition and its relationship to analcite at high pressure. Phase Transitions 1:1-22
Helgeson HC, Delany JM, Nesbitt HW, Bird DK (1978) Summary and critique of the thermodynamic properties of rock-forming minerals. Am J Sci 278-A:1-229
Helgeson HC, Kirkham DH, Flowers GC (1981) Theoretical prediction of the thermodynamic behavior of aqueous electrolytes at high pressures and temperatures: IV Calculation of activity coefficients, osmotic coefficients, and apparent molal and standard and relative partial molal properties to 600°C and 5 kb. Am J Sci 281:1249-1516
Hemingway BS, Robie RA (1984) Thermodynamic properties of zeolites: Low-temperature heat capacities and thermodynamic functions for phillipsite and clinoptilolite. Estimates of the thermochemical properties of zeolitic water at low temperature. Am Mineral 69:692-700
Holland TJB (1989) Dependence of entropy on volume for silicate and oxide minerals: A review and a predictive model. Am Mineral 74:5-13
Holland TJB, Powell R (1985) An internally consistent thermodynamic dataset with uncertainties and correlations: 2 Data and results. J Metamorphic Geol 3:343-370
Honda S, Muffler LJP (1970) Hydrothermal alteration in core from research drill hole Y-1, Upper Geyser Basin, Yellowstone National Park, Wyoming. Am Mineral 55:1714-1737
Howell DA, Johnson GK, Tasker IR, O'Hare PAG, Wise WS (1990) Thermodynamic properties of the zeolite stilbite. Zeolites 10:525-531
Iijima A (1975) Effect of pore water to clinoptilolite-analcime-albite reaction series. J Fac Sci Univ Tokyo, Sec II, 19:133-147
Iijima A (1978) Geologic occurrences of zeolites in marine environments. *In* Natural Zeolites: Occurrence, Properties, Use. LB Sand, FA Mumpton (eds) Pergamon Press, New York, 175-198
Johnson GK, Flotow HE, O'Hare PAG, Wise WS (1982) Thermodynamic studies of zeolites: Analcime and dehydrated analcime. Am Mineral 67:736-748
Johnson GK, Flotow HE, O'Hare PAG, Wise WS (1983) Thermodynamic studies of zeolites: Natrolite, mesolite and scolecite. Am Mineral 68:1134-1145
Johnson GK, Flotow HE, O'Hare PAG, Wise WS (1985) Thermodynamic studies of zeolites: Heulandite. Am Mineral 70:1065-1071
Johnson GK, Tasker IR, Jurgens R, O'Hare PAG (1991) Thermodynamic studies of zeolites: Clinoptilolite. J Chem Thermodynamics 23:475-484
Johnson GK, Tasker IR, Flotow HE, O'Hare PAG, Wise WS (1992) Thermodynamic studies of mordenite, dehydrated mordenite, and gibbsite. Am Mineral 77:85-93
Johnson JW, Oelkers EH, Helgeson HC (1992) SUPCRT92: A software package for calculating the standard molal thermodynamic properties of minerals, gases, aqueous species, and reactions from 1 to 5000 bars and 0° to 1000°C. Computers & Geosciences 18:899-947
Jones BF, Rettig SL, Eugster HP (1967) Silica in alkaline brines. Science 158:1310-1314
Juan VC, Lo HJ (1971) The stability fields of natural laumontite and wairakite and their bearing on the zeolite facies. Proc Geol Soc China 14:34-44
Kastner M, Stonecipher SA (1978) Zeolites in pelagic sediments of the Atlantic, Pacific, and Indian Oceans. *In* Natural Zeolites: Occurrence, Properties, Use. LB Sand, FA Mumpton (eds) Pergamon Press, New York, p 199-220
Keith TEC, White DE, Beeson MH (1978) Hydrothermal alteration and self-sealing in Y-7 and Y-8 drill holes in northern part of Upper Geyser Basin, Yellowstone National Park, Wyoming. U S Geol Surv Prof Paper 1054-A, 26 p
Kerrisk JF (1983) Reaction-path calculations of groundwater chemistry and mineral formation at Rainier Mesa, Nevada. Los Alamos National Laboratory Report LA-9912-MS, 41 p
Kerrisk JF (1987) Groundwater chemistry at Yucca Mountain, Nevada, and vicinity. Los Alamos National Laboratory Report LA-10929-MS, 118 p
King EG, Weller WW (1961) Low temperature heat capacities and entropies at 298.15 K of some sodium- and calcium-aluminum silicates. U S Bureau of Mines Report, Inv 5855, 8 p
Kirov GN, Balkanov I (1976) Rentgenovo izsledvane na suotnoshenieto lomontit-leonkhardit (X-ray study of the laumontite-leonhardite relationship): Geokhim, Mineral Petrol 4:57-61
Kiseleva I, Navrotsky A, Belitsky IA, Fursenko BA (1996a) Thermochemistry and phase equilibria in calcium zeolites. Am Mineral 81:658-667

Kiseleva I, Navrotsky A, Belitsky IA, Fursenko BA (1996b) Thermochemistry of natural potassium sodium calcium leonhardite and its cation-exchanged forms. Am Mineral 81:668-675
Kracek FC, Neuvonen KJ (1952) Thermochemistry of plagioclase and alkali feldspars. Am J Sci Bowen Volume, p 293-318
Kristmannsdóttir H, Tómasson J (1978) Zeolite zones in geothermal areas in Iceland. *In* Natural Zeolites: Occurrence, Properties, Use. LB Sand, FA Mumpton (eds) Pergamon Press, New York, p 277-284
Liou JG (1970) Synthesis and stability relations of wairakite $CaAl_2Si_4O_{12}{\cdot}2H_2O$. Contrib Mineral Petrol 27:259-282
Liou JG (1971) P-T stabilities of laumontite, wairakite, lawsonite, and related minerals in the system $CaAl_2Si_2O_8$-SiO_2-H_2O. J Petrol 12:379-411
Moncure GK, Surdam RC, McKague HL (1981) Zeolite diagenesis below Pahute Mesa, Nevada Test Site. Clays & Clay Minerals 29:385-396
Münch P, Duplay J, Cochemé J (1996) Alteration of silicic vitric tuffs interbedded in volcaniclastic deposits of the southern basin and range province, Mexico: Evidences for hydrothermal reactions. Clays & Clay Minerals 44:49-67
Murphy WM, Pabalan RT (1994) Geochemical investigations related to the Yucca Mountain environment and potential nuclear waste repository. U S Nuclear Regulatory Commission Report NUREG/CR-6288
Newton RC, Charlu TV, Kleppa OJ (1980) Thermochemistry of the high structural state plagioclases. Geochim Cosmochim Acta 44:933-941
Ostwald W (1900) Über die vermeintliche Isomerie des roten und gelben quecksilberoxyds und die Oberflächen-spannung Fester Körper. Zeits Physikal Chem 34:495-503
Pabalan RT (1994) Thermodynamics of ion exchange between clinoptilolite and aqueous solutions of Na^+/K^+ and Na^+/Ca^{2+}. Geochim Cosmochim Acta 58:4573-4590
Pabalan RT, Bertetti FP (1994) Thermodynamics of ion-exchange between Na^+/Sr^{2+} solutions and the zeolite mineral clinoptilolite. Mat Res Soc Symp Proc 333:731-738
Parkhurst DL (1995) User's Guide to PHREEQC—a computer program for speciation, reaction-path, advective-transport, and inverse geochemical calculations. U S Geol Surv Water-Resources Inv Report 95-4227, 143 p
Renders PJN, Gammons CH, Barnes HL (1995) Precipitation and dissolution rate constants for cristobalite from 150 to 300°C. Geochim Cosmochim Acta 59:77-85
Rimstidt JD, Barnes HL (1980) The kinetics of silica-water reactions. Geochim Cosmochim Acta 44:1683-1699
Robinson GR Jr, Haas JL Jr (1983) Heat capacity, relative enthalpy, and calorimetric entropy of silicate minerals: An empirical method of prediction. Am Mineral 68:541-553
Robinson GR Jr, Haas JL Jr, Schafer CM, Haselton HT Jr (1982) Thermodynamic and thermophysical properties of mineral components of basalts. *In* Physical Properties Data for Basalt. LH Gevantman (ed) U S National Bureau of Standards Report NBSIR 82-2587, 425 p
Senderov EE (1980) Estimation of Gibbs energy for laumontite and wairakite from conditions of their formation in geothermal areas. *In* Proc 5th Int'l Conf on Zeolites. LVC Rees (ed) Leyden, London, p 56-63
Sheppard RA, Gude AJ 3rd (1968) Distribution and genesis of authigenic silicate minerals in tuffs of Pleistocene Lake Tecopa, Inyo County, California. U S Geol Surv Prof Paper 597, 38 p
Sheppard RA, Gude AJ 3rd (1969) Diagenesis of tuffs in the Barstow Formation, Mud Hills, San Bernardino County, California. U S Geol Surv Prof Paper 634, 35 p
Sheppard RA, Gude AJ 3rd (1973) Zeolites and associated authigenic silicate minerals in tuffaceous rocks of the Big Sandy Formation, Mohave County, Arizona. U S Geol Surv Prof Paper 830, 36 p
Smyth JR (1982) Zeolite stability constraints on radioactive waste isolation in zeolite-bearing volcanic rocks. J Geol 90:195-202
Smyth JR, Bish DL (1988) Crystal Structures and Cation Sites of the Rock-Forming Minerals. Allen & Unwin, London, 332 p
Stahl K, Artioli G (1993) A neutron powder diffraction study of fully deuterated laumontite. Eur J Mineral 5:851-856
Stonecipher SA (1978) Chemistry of deep-sea phillipsite, clinoptilolite, and host sediments. *In* Natural Zeolites: Occurrence, Properties, Use. LB Sand, FA Mumpton (eds) Pergamon Press, New York, p 221-234
Sturchio NC, Bohlke JK, Binz CM (1989) Radium-thorium disequilibrium and zeolite-water ion exchange in a Yellowstone hydrothermal environment. Geochim Cosmochim Acta 53:1025-1034
Surdam RC, Eugster HP (1976) Mineral reactions in the sedimentary deposits of the Lake Magadi region, Kenya. Geol Soc Am Bull 87:1739-1752
Surdam RC, Sheppard RA (1978) Zeolites in saline, alkaline-lake deposits. *In* Natural Zeolites: Occurrence, Properties, Use. LB Sand , FA Mumpton (eds) Pergamon Press, New York, p 145-174

Taylor M, Brown GE (1979) Structure of mineral glasses: I, The feldspar glasses $NaAlSi_3O_8$, $KAlSi_3O_8$, $CaAl_2Si_2O_8$. Geochim Cosmochim Acta. 43:61-76

Tribble JS, Wilkens RH (1994) Microfabric of altered ash layers, ODP LEG 131, Nankai Trough. Clays & Clay Minerals 42:428-436

Van Reeuwijk LP (1974) The thermal dehydration of natural zeolites. Mededelingen Landbouwhogeschool 74-9, Wageningen, Netherlands, 88 p

Vaughan DEW (1978) Properties of natural zeolites. *In* Natural Zeolites: Occurrence, Properties, Use. LB Sand, FA Mumpton (eds) Pergamon Press, New York, p 353-371

Wilkin RT, Barnes HL (1998) Solubility and stability of zeolites in aqueous solution: I Analcime, Na-, and K-clinoptilolite. Am Mineral 83:746-761

Yoder HS Jr, Weir CE (1960) High-pressure form of analcite and free energy change with pressure of analcite reactions. Am J Sci 258-A:420-433

Zen E-an (1972) Gibbs free energy, enthalpy, and entropy of ten rock-forming minerals: Calculations, discrepancies, implications. Am Mineral 57:524-553

Zhu H, Newton RC, Kleppa OJ (1994) Enthalpy of formation of wollastonite ($CaSiO_3$) and anorthite ($CaAl_2Si_2O_8$) by experimental phase equilibrium measurements and high-temperature solution calorimetry. Am Mineral 79:134-144

4 Isotope Geochemistry of Zeolites

Haraldur R. Karlsson

Departments of Geoscience and *Chemistry and Biochemistry*
Texas Tech University, Box 41053
Lubbock, Texas 79409

INTRODUCTION

Two of the most important variables that can be determined in any geologic study are the timing and conditions of formation. Although the discovery of radioactivity and isotopes at the beginning of this century revolutionized our understanding of the age and formation conditions of rocks and minerals, the isotope geochemistry of certain mineral groups such as the zeolites has remained elusive and poorly understood. To date fewer than 40 papers have been published on the isotope geochemistry of zeolites. This dearth of isotopic data is due to technical difficulties rather than lack of interest. Recently, however, great strides have been made in understanding the isotope geochemistry of zeolites as a result of a better understanding of their crystal structures, coupled with the increased popularity of isotope geochemistry and a rapid growth in the number of isotope laboratories. This chapter reviews the current status of our knowledge of isotopes in natural zeolites, including the methods used to prepare samples and measure isotopic abundances and ratios, some of the problems associated with these measurements, and some of the implications of the results obtained. Because the reader may not be familiar with isotope geochemistry, the fundamentals of the behavior of isotopes will also be reviewed briefly. It should be pointed out that a vast literature exists on the behavior and characteristics of isotopes in synthetic zeolites and on isotopic studies related to industrial applications (see e.g. Chang et al. 1995). This material is, however, outside the scope of this chapter and will only be mentioned when it has a direct bearing on the geochemistry of zeolites.

The earliest published report on isotopes in zeolites may have been that of Taylor and Urey (1938) who determined fractionation of stable Li, K, and N isotopes by an unspecified "zeolite" species. At Columbia University, Taylor and Urey erected columns packed with a synthetic Na-"zeolite" obtained from the Permutit Company and passed chloride solutions (LiCl, KCl, NH_4Cl) through them. The longest column, measuring one hundred feet, was constructed in a stairwell of the Chandler Chemical Laboratory. Isotopic ratios of the elutants were determined using a mass spectrometer specifically built for this purpose.

Background, terminology, and definitions

The isotopes discussed in this chapter along with pertinent information and properties are summarized in Table 1. Isotopes are atoms of the same element that contain a different number of neutrons in their nuclei. Two kinds of isotopes occur naturally - stable isotopes and radioactive isotopes. Stable isotope are nuclei that have not been observed to decay spontaneously. Radioactive isotopes, however, undergo spontaneous decay through emission of subatomic particles (β^- and β^+ decay, α decay) or by fragmentation of the parent nucleus (fission). The decay produces a daughter (or daughters) atom that itself is either stable or radioactive. Nuclei formed by nuclear decay are called radiogenic daughters even if they are stable.

Isotopic abundances in natural samples are not constant but vary for several reasons. First, radioactive and radiogenic isotopes can generally be expected to vary with time as

1529-6466/00/0045-0004$05.00

Table 1. Isotopes of geological significance in zeolites.

Element	*Isotope*	*Abundance*[1]	*Half-life*[2]	*Location*[3]	*Other isotopes*[4]
Hydrogen	^{1}H	99.99		cw	
	^{2}H	0.01		cw	$^{3}H(12.43)^{2}$
Oxygen	^{16}O	99.63		cw, fw	
	^{17}O	0.0375		cw, fw	
	^{18}O	0.1995		cw, fw	
Potassium	^{39}K	93.2581		c	
	^{40}K	0.01167	1.25×10^{9}	c	
	^{41}K	6.3702		c	
Argon	^{36}Ar	0.337		c	
	^{38}Ar	0.063		c	$^{39}Ar\ (269)^{2}$
	^{40}Ar	99.600		c	
Rubidium	^{85}Rb	72.1654		c	
	^{87}Rb	27.8346	48.8×10^{9}	c	
Strontium	^{86}Sr	9.861		c	
	^{87}Sr	6.991		c	$^{88}Sr\ (82.59\%)^{1}$; $^{84}Sr\ (0.557\%)^{1}$
Thorium	^{228}Th		1.9	c	
	^{236}Th		75.4×10^{3}	c	$^{227}Th\ (18.72\ days)^{2}$
	^{232}Th		14.01×10^{9}	c	
Uranium	^{235}U	0.720	7.038×10^{8}	c	
	^{238}U	99.275	4.468×10^{9}	c	$^{234}U\ (0.005\%$ and $244.6 \times 10^{3})^{1,2}$

[1]Natural abundance in atom %. Values for Ar are atom % of atmosphere.

[2]Half life is in years unless otherwise indicated.

[3]Location in zeolite structure. Abbreviations: cw: channel water; c: channel; fw: framework.

[4]Information on abundances and half lives can be found in Faure (1986) and Geyh and Schleicher (1990).

parent abundances decrease and radiogenic daughter abundances increase. Stable isotopes are not affected through decay unless they are themselves radiogenic daughters. Second, chemical, physical, and biological processes can alter isotopic abundances especially in the case of the lighter elements. These processes collectively lead to isotopic fractionation. Because these fractionation effects are linked to the mass differences amongst isotopes of the same elements, the impact will be greatest for the lighter elements where the relative isotopic mass differences are the largest. Beyond Si (mass 28), isotopic fractionation effects are usually quite small.

Zeolites—Special problems

The basic reasons for the difficulties and ambiguities associated with determining isotopic abundances in zeolites lie in their crystal structure. As discussed in the chapter by Armbruster and Gunter (this volume), zeolites have open structures that allow ready access by water, gases, and cations to the framework. This accessibility and the intrinsic labile nature of the channel constituents greatly increase the potential of framework-extraframework interactions compared with other minerals. Moreover, the channel contents can be easily altered or influenced by chemical, isotopic, or physical changes in the surroundings of the zeolite.

STABLE ISOTOPES

Since the advent of the isotope ratio mass spectrometer in the late 1940s, studies of carbon ($^{13}C/^{12}C$), hydrogen (D/H), and oxygen ($^{18}O/^{16}O$) have yielded a bounty of information concerning the origin and conditions of formation of rocks, minerals, and natural waters (Valley et al. 1986). Until recently, however, very few data had been obtained on zeolites due to technical difficulties in their analysis, and only a handful of articles have been published on their stable isotope geochemistry. It is the purpose of this section to review the current status of oxygen and hydrogen isotopic studies of zeolites, to describe some of the methods and problems associated with obtaining these ratios, and to discuss applications of stable isotopes to the study of natural zeolites. This section will summarize experimental and semi-empirical oxygen isotopic geothermometers for zeolites, review exchange rates and mechanism, discuss the potential of zeolites as single-mineral thermometers, evaluate the retention of oxygen isotopes in zeolites, and shed some new light on the oxygen isotopic composition of zeolites in active geothermal systems.

Stable isotope terminology

Isotopic ratios are commonly reported in the δ notation on a per mil (‰) basis rather than as absolute values. This is so because ratios can be measured more precisely than absolute values on a mass spectrometer. The delta value of a sample, δ_{sa}, is defined

$$\delta_{sa}(‰) = \left(\frac{R_{sa} - R_{std}}{R_{std}}\right) \bullet 1000 \tag{1}$$

where R is the ratio of the heavy to light isotope (e.g. D/H, $^{13}C/^{12}C$ or $^{18}O/^{16}O$) and the subscripts sa and std denote the sample and the reference standard, respectively. The reference standard used for oxygen and hydrogen is V-SMOW (Vienna Standard Mean Ocean Water). For carbon isotopes (and often oxygen isotopes in the case of carbonates) the standard is V-PDB (Vienna Pee Dee Belemnite). Absolute ratios for SMOW are 0.002005 for $^{18}O/^{16}O$ (Baretschi 1976) and 0.000156 for D/H (Tse et al. 1980).

Due to slight intrinsic thermodynamic differences between isotopes of an individual element, natural processes such as evaporation, condensation, and diffusion give rise to partitioning of stable isotopes among different phases. The partitioning of stable isotopes between two substances, x and y, is expressed in terms of a fractionation factor $\alpha_{(x-y)}$, which is defined as

$$\alpha_{(x-y)} = \frac{R_x}{R_y} \tag{2}$$

where R is defined as above. Recasting in terms of δ,

$$\alpha_{(x-y)} = \frac{\delta_x + 1000}{\delta_y + 1000} \tag{3}$$

For silicate minerals such as zeolites, the fractionation factor is a function of temperature and, to a lesser extent, of mineral composition (see e.g. O'Neil 1986). Numerous experimental studies have shown that the fractionation factor for most mineral-water systems (m-w) can be approximated over a large temperature range by a linear function of $(1/T)^2$:

$$1000 \ln \alpha_{(m-w)} = A\left(\frac{1}{T}\right)^2 + B \tag{4}$$

where T is temperature (K), and A and B are constants. As a further convenience in notation 1000lnα is replaced by Δ and fractionation curves are generally displayed on a Δ vs. $(1000/T)^2$ plot. It should be noted that some workers prefer to equate Δ(x-y) with

δ_x- δ_y and thus Δ(x-y) $\approx$ 1000lnα. The former notation, however, is preferred by this author because Δ thus defined is a thermodynamically significant quantity and is independent of the choice of standard (R.N. Clayton, pers. comm. 2000).

Analytical methods and difficulties

The goal of isotopic studies of zeolites is to obtain $\delta^{18}O$ and δD values that are representative of the mineral. The measured values may, however, be affected by the analytical procedures employed in ways peculiar to the nature of the zeolite (i.e. chemistry, structure, thermal stability and dehydration behavior). In a hydrated zeolite, oxygen atoms occur in two structurally distinct positions: the aluminosilicate framework (framework oxygen or O_f) and the water occupying the channels (channel water oxygen or O_{cw}). In a fully hydrated zeolite, 14-38% of all oxygens occur as channel water. O_f is either bonded to two silicon atoms or one silicon and one aluminum atom. O_{cw} is bonded to two hydrogen atoms and loosely to one or more alkali or alkaline earth cations (M). However, on average O_{cw} can be bonded to fewer than one channel cation. In an equilibrium system, the isotopic compositions of O_f and O_{cw} differ from one another because these atoms occupy energetically different sites within the zeolite structure. Generally O_f should be more enriched in ^{18}O than O_{cw} because the tetrahedral oxygen is bonded more strongly than the channel water oxygen (Si-O-Si or Si-O-Al vs. M-O<H). This has been confirmed by studies that show $\delta^{18}O$ values of oxygens in the two positions differ by as much as 40‰ (Karlsson and Clayton 1990a). Hence, to determine the oxygen isotope ratios of O_f and O_{cw} in a zeolite, the framework and channel water oxygen must be analyzed separately. There are, however, no methods that currently allow the simultaneous measurement of $\delta^{18}O$ for O_f and O_{cw} with high precision (i.e. better than ±1‰) so these two types of oxygen must be physically separated before they can be analyzed. The separation must be performed in such a manner that insignificant isotopic exchange occurs between framework and channel water. This is important because O_{cw} can undergo rapid isotope exchange with external fluids or vapors even at room temperature, whereas O_f is much more resistant to exchange. Some of the methods that have been applied to achieve this goal are discussed below.

The simplest way to analyze the $\delta^{18}O$ of O_f and O_{cw} in a zeolite is to remove the O_{cw} by dehydration. Zeolites cannot be dehydrated completely by storage at room temperature in a dry atmosphere. Savin (1967) attempted to dry marine phillipsites in a dry box containing P_2O_5 but obtained erratic oxygen isotope results due to incomplete channel water removal. Nor must dehydration be done in the Ni reaction vessel commonly used for oxygen isotope analysis (Clayton and Mayeda 1963) because the channel water adsorbs onto the walls of the reaction vessels (Karlsson and Clayton 1990a; Feng and Savin 1991; Noto 1991). Once zeolites have been loaded into a fluorination line they are reacted overnight at 450-500°C with either BrF_5 or ClF_3 to release O_2. A hot graphite rod is typically used to convert the O_2 to CO_2 for isotopic analysis on a dual-inlet isotope-ratio mass spectrometer (Clayton and Mayeda 1963; Borthwick and Harmon 1982). Care must be taken not to "pretreat" zeolites prior to the main reaction because zeolites will react with the fluorination agents even at room temperature (see e.g. Noto 1991).

Karlsson (1988) designed a system that allows dehydration of zeolites under high vacuum and thus circumvents this problem by isolating the dehydrated zeolite and the channel water. The system is shown schematically in Figure 1. Similar methods for dehydrating zeolites have been described by Feng and Savin (1991).

Some zeolites are stable upon dehydration and exhibit little or no structural change (e.g. analcime, chabazite, clinoptilolite, mordenite), whereas others are unstable and undergo moderate to drastic structural modifications (e.g. heulandite, laumontite, natrolite, stilbite, wairakite). Dehydration may be accompanied by oxygen isotopic exchange between the framework and the channel water resulting in inaccurate $\delta^{18}O$ values, especially if the zeolite structure is modified such that bonds are broken. Extensive studies aimed at determining the extent of exchange and optimum dehydration conditions were carried out by Karlsson (1988), Feng and Savin (1991) and Noto (1991). Accurate $\delta^{18}O$ values ($1\sigma \pm 0.3‰$) can ordinarily be obtained for analcime, clinoptilolite, mordenite, and wairakite, but $\delta^{18}O$ values for chabazite, heulandite, laumontite and stilbite are generally much less accurate ($1\sigma \pm 1$ to 3‰). Noto et al. (1990) took a novel approach in determining $\delta^{18}O_f$ of wairakite by using the ^{17}O method of Matsuhisa et al. (1978). This method consists of allowing the sample to exchange with external waters of known ^{17}O content in a series of experiments and then determining the $\delta^{18}O$ and $\delta^{17}O$ values of the bulk zeolite (framework and channel water). The method is independent of any problems that might be associated with internal isotopic exchange (O_f exchanging with O_{cw}) and one can "extrapolate" back to the original zeolite $\delta^{18}O_f$. Using this method, Noto et al. (1990) were able to estimate the $\delta^{18}O_f$ value of wairakite to within ±2‰. Noto's approach is appropriate for zeolites in which the channel water is retained during room temperature evacuation (e.g. analcime, wairakite). It is not suited to the routine determination of zeolite framework $\delta^{18}O$ values because it is tedious and requires large samples, but may prove useful for optimizing the dehydration conditions of thermally unstable zeolites.

Isotopic partitioning, exchange rates, and thermometry

To interpret and use oxygen and hydrogen isotopic results for zeolites, some knowledge of equilibrium fractionation factors, exchange mechanisms and kinetics is essential. The openness of the zeolite structure allows ready access to framework oxygens and greatly enhances the possibility of exchange with an external medium such as water compared with other silicates such as quartz or feldspar. Early work on synthetic zeolites, most notably by Von Ballmoss (1981) and Von Ballmoss and Meier (1982), supports this notion and was the basis for suggestions by Karlsson and Clayton (1990a) that some zeolites might be excellent low-temperature oxygen isotope geothermometers. The first isotopic fractionation curve determined for a zeolite was that of analcime-water by Karlsson and Clayton (1990b). They measured the equilibrium oxygen isotope partitioning experimentally in the temperature range 300-400°C. Below 300°C, Karlsson and Clayton (1990b) used natural analcimes formed at known temperatures and fluid compositions to constrain the fractionation factors. They obtained two different isotopic fractionation factors; one between the framework and external water ($\Delta^{18}O_{f\text{-}w}$) and another between the channel water and external water ($\Delta^{18}O_{cw\text{-}w}$). According to these results, $\Delta^{18}O_{f\text{-}w}$ for analcime is similar to that of calcite-water (or albite-water) but lower than for quartz-water as shown in Figure 2. Experimentally determined fractionation factors for wairakite-water (Noto and Kusakabe 1997a) and stilbite-water (Feng and Savin 1993b) are also shown. The $\Delta^{18}O_{f\text{-}w}$ value for wairakite is similar to that of analcime-water, suggesting that Ca-Na exchange in an analcime-type structure (Si/Al = 2) has a negligible effect on the oxygen isotopic fractionation factor for the zeolite framework. These data are consistent with results for the structurally related feldspars (O'Neil and Taylor 1967). Variations in Si/Al ratio have a pronounced effect on oxygen isotopic ratios in the plagioclase feldspar solid-solution series. For example, Matsuhisa et al. (1979) found that at 500°C albite is 1.8‰ more enriched in $\delta^{18}O$ than anorthite (i.e. albite (Si/Al = 3), has a greater affinity for ^{18}O than anorthite (Si/Al = 1)). The difference increases to 2.3‰ at 400°C. In contrast, there is no measurable difference in $\delta^{18}O$ between albite and the K-end-member feldspar, orthoclase, for which the Si/Al ratios are identical. Feng and Savin (1993b) determined oxygen isotopic fractionation factors in the stilbite-water system in the temperature range 125 to 300°C. They fit two fractionation curves through the results; one for data below 220°C (not shown in Fig. 2) which is similar to quartz-water and another for data above

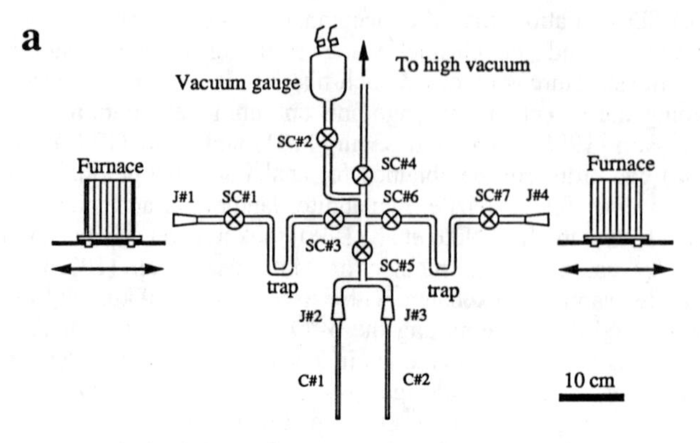

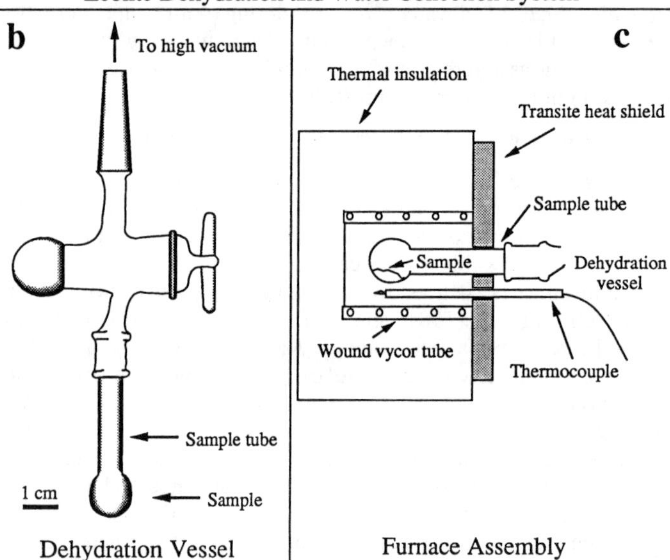

Figure 1. An experimental setup for dehydrating zeolites and collecting water in preparation for isotopic analysis of zeolite frameworks and channel water. (a) Schematic of vacuum system setup. J refers to ground glass joints, SC to high-vacuum stopcock, and C to glass tubing used for isolating water. Furnaces are mounted on rails allowing them to slide laterally. (b) Glass vessel used for dehydrating zeolites and subsequently isolating from the ambient atmosphere. The vessel consists of a detachable sample tube (lower part that is attached to a high-vacuum stopcock (upper part). (c) Resistance furnace for dehydrating zeolites. The furnace cross-section is approximately 7×9 cm. Proportions are to scale.

The procedure for treatment of zeolites is as follows. Powders are weighed into a sample tube which is attached to a high-vacuum stopcock (Fig. 1b). This dehydration vessel is connected by a vertical joint on another part of the vacuum line (not shown) and the sample tube is evacuated. If channel water is to be collected at this point, the bottom of the tube is first frozen in liquid nitrogen for roughly a minute to prevent premature water loss; otherwise the sample tube is simply evacuated. Once evacuation is complete, the vessel is closed and moved to the heating section of the line shown in Figure 1a and attached laterally to joint J#1 (or J#4) in the manner shown in Figure 1c. After cooling the trap in line with the dehydration vessel to liquid-nitrogen temperatures, the zeolite is exposed to the trap by carefully opening the appropriate valves (SC#1 or SC#7) while the valve on the vacuum pump side remains closed (SC#3 or SC#6). This allows collection of channel water from the zeolite either in a single step or batch-wise and with or without heating. The water can either be discarded or transferred to a capillary tube (C#1 or C#2) where it can be isolated for later isotopic analysis. Once dehydration of the zeolite is complete, the sample tube is shut, the dehydration vessel is detached from the line and is weighed to determine weight loss. The dehydration vessel is taken to a glove box containing a dry atmosphere where it is opened and the sample powder is loaded into a Ni reaction tube used for fluorination. From this point on the method follows routine silicate analysis (see e.g. Clayton and Mayeda 1963). From Karlsson (1988).

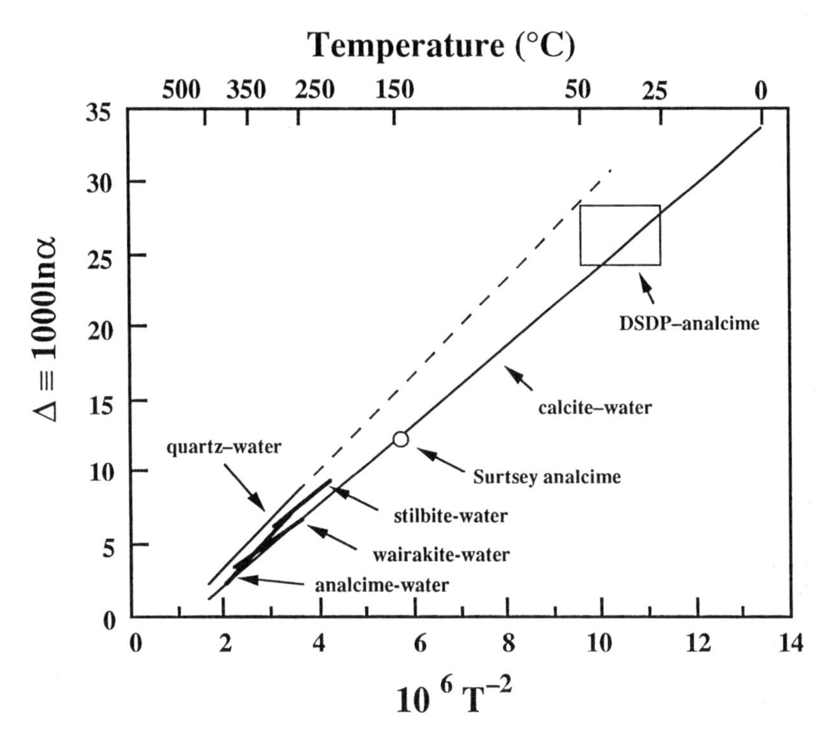

Figure 2. Temperature dependence (T) of oxygen isotope fractionation (α) in the zeolite framework-water system, along with fractionation curves for quartz-water (Matsuhisa et al. 1979) and calcite-water (O'Neil et al. 1969). Experimentally derived zeolite fractionation curves are from Karlsson and Clayton (1990b), Feng and Savin (1993b), and Noto (1991) for analcime, stilbite and wairakite, respectively. Also shown are two empirically derived low-temperature analcime-water fractionations from the Deep Sea Drilling Project (DSDP) and Surtsey, Iceland (circle; Karlsson and Clayton 1990a,b). T is in K.

220°C which is similar to albite-water. They noted that there should only be a single fractionation curve for a given mineral and contended that only the high-temperature fractionation curve was reliable because of the similarity between the Si/Al ratios of stilbite and albite. Feng and Savin (1993b) suggested that the low-temperature curve may have been offset to higher-than-expected $\Delta^{18}O_{f\text{-}w}$ values due to systematic errors caused by the exchange method used. Alternatively, the separate fractionation curves may simply reflect fractionations by different phases. Stilbite undergoes a series of phase transformations with increasing temperature (Rykl and Pechar 1985), transforming into metastilbite (see e.g. Feng 1991) and phase "B" of Alberti et al. (1978) at ~ 200°C.

In the absence of experimental data, Stallard and Boles (1989) developed a semi-empirical expression relating zeolite framework-water fractionation to the Si/Al ratio of the

Table 2. Oxygen isotopic fractionation factors for zeolites[1]

Fractionation	*Mineral*	*A*	*B*	*Range (°C)*	*Source*[3]
$\Delta^{18}O_{f\text{-}w}$ [2]	analcime	2.78	-2.89	25 - 400	1
	stilbite	2.7	-2.4	220 - 300	2
	wairakite	2.46	-1.76	250 - 400	3
$\Delta^{18}O_{f\text{-}cw}$	analcime	2.78	+2.61	25 - 400	1
	wairakite	2.46	-1.03	250 - 400	3

[1]Table contains values for A and B in the general fractionation curve, $\Delta^{18}O = A(10^6/T^2) + B$, where T is temperature (K) and $\Delta^{18}O$ is in ‰. The upper part of the table refers to framework-bulk water fractionation and the lower part refers to framework-channel water fractionation.

[2]Feng and Savin (1993b) used a bond-counting model to estimate $\Delta^{18}O_{f\text{-}w}$ for clinoptilolite. Their expression was obtained by setting A = +4.31 and B = 1.33 – 5.47 x 10^3/T. For other zeolites, estimates of $\Delta^{18}O_{f\text{-}w}$ may be obtained from the semi-empirical expression of Stallard and Boles (1989) by setting A = $(2.64\gamma + 0.93)$ and B = $(2.80\gamma - 5.51)$ where γ = Si/(Si +Al).

[3]Sources: 1: Karlsson and Clayton (1990b); 2: Feng and Savin (1993b); 3: Noto and Kusakabe (1997a).

zeolite (Table 2). Their expression does a fair job of reproducing the experimental data. For example, their predicted analcime (or wairakite) fractionation is too low by 1‰ and 1.8‰ at 250°C and 25°C, respectively. The semi-empirical expression of Stallard and Boles (1989) can therefore be used to estimate isotopic fractionation for zeolites where experimental data are lacking. Another semi-empirical approach is based on bond counting. The basic idea is that each bond type (e.g. Si-O-Si, Si-O-Al, Al-O-M, Si-O-M, etc., where M is a divalent or trivalent cation other than Al) can be represented with its own fractionation curve irrespective of the mineral in which it occurs. Fractionation for a given bond type is obtained from existing experimental data from one or more mineral-water systems (e.g. Si-O-Si is represented by quartz-water, Si-O-Al by a combination of quartz-water and albite-water). A total fractionation curve is then obtained from a weighted sum of the bond types. The approach is limited by the relatively sparse database for mineral-water fractionation curves and is only considered accurate to 1 to 2‰ (e.g. Savin and Lee 1988). Nähr et al. (1998) and Feng et al. (1999) have recently used the bond-counting model to estimate clinoptilolite-water fractionations at low temperature (see footnote of Table 2).

In addition to determining the oxygen isotope fractionation factors between zeolite framework and water, ($\Delta^{18}O_{f\text{-}w}$), channel water-water fractionation factors ($\Delta^{18}O_{cw\text{-}w}$) have also been determined. Results for analcime, stilbite, and wairakite are compiled in Figure 3, along with other pertinent fractionation factors for various hydrous substances. Zeolite channel water, in marked contrast to water ice or water in crystalline hydrates, tends to concentrate ^{16}O relative to bulk water. Stilbite has a $\Delta^{18}O_{cw\text{-}w}$ value of -3‰ at room temperature (Feng and Savin 1993b) and $\Delta^{18}O_{cw\text{-}w}$ for analcime is even more negative and nearly constant (-5 to -6‰) in the temperature range 25-400°C (Karlsson and Clayton 1990b). The $\Delta^{18}O_{cw\text{-}w}$ of wairakite is also negative (-0.7‰) within the measured temperature range 250 to 400°C but is higher than that of analcime (Noto and Kusakabe 1997a). The difference between the isotopic behavior of channel water in zeolites, which tend to concentrate ^{16}O, and other crystalline hydrates, which tend to concentrate ^{18}O, is undoubtedly linked to differences in the bonding environment of water molecules and may be related to the fact that hydrogen bonding is much weaker in zeolite extraframework sites than it is in other crystalline hydrates (Karlsson and Clayton 1990b; Noto and Kusakabe 1997a).

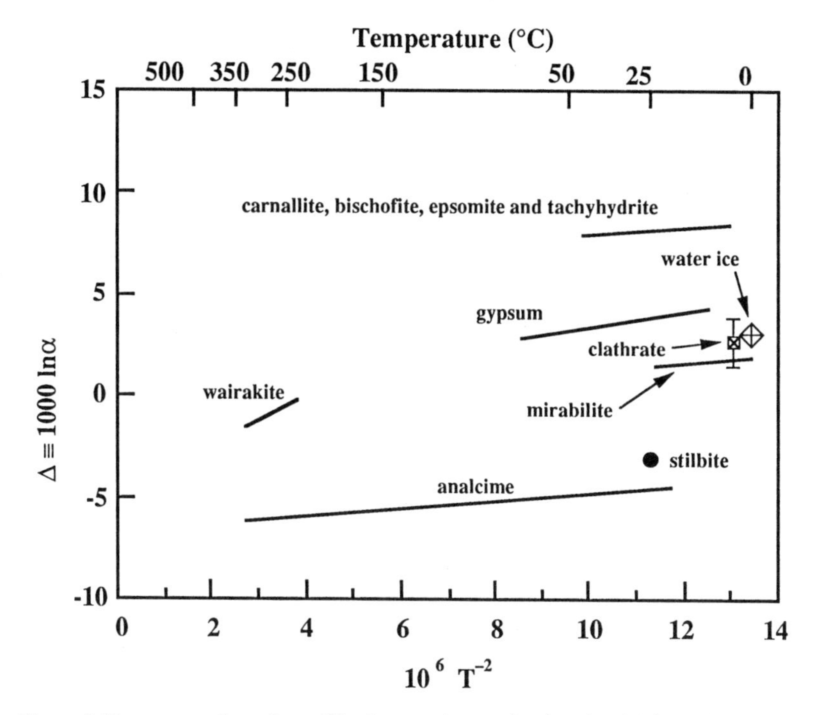

Figure 3. Temperature dependence (T) of oxygen isotope fractionation (Δ) for zeolite-channel water and water in other pertinent crystalline hydrates. Zeolite sources as in Figure 1. For all others, see Karlsson and Clayton (1990a). T is in K.

The fractionation of hydrogen isotopes between channel water and external water has not been determined experimentally for a zeolite. Quantitative inferences about $\Delta D_{cw\text{-}w}$ can, however, be made from measured $\delta D_{cw\text{-}w}$ values if additional constraints are available. For example, when a range of δD_{cw} and δO_{cw} values is available on the same zeolite species, one can estimate $\Delta D_{cw\text{-}w}$ provided $\Delta^{18}O_{cw\text{-}w}$ and the relationship between $\delta^{18}O_{cw}$ and δD_{cw} are known. For example, analcime channel waters display the relationship $\delta D_{cw} = 8\delta^{18}O_{cw} + 10$, so that $\Delta D_{cw\text{-}w} = 8\Delta^{18}O_{cw\text{-}w}$. Because $\Delta^{18}O_{cw\text{-}w}$ is -5‰ we obtain $\Delta D_{cw\text{-}w}$ = -40‰ (Karlsson and Clayton 1990a,b). Similarly by combining δD_{cw} values for geothermal wairakite from Noto (1991) with isotopic compositions of associated fluids, Karlsson (1995) postulated that wairakite $\Delta D_{cw\text{-}w}$ values ranged from -15‰ to -25‰ with smaller $\Delta D_{cw\text{-}w}$ values reflecting higher temperatures. Noto and Kusakabe (1997b) expanded on this idea and compared δD values of coexisting wairakite and thermal waters from several active geothermal systems in Japan and New Zealand. Extrapolating from higher temperatures, Noto and Kusakabe (1997b) estimated that $\Delta D_{cw\text{-}w}$ for wairakite at 25°C could be as low as -45±15‰. There is some additional anecdotal evidence that suggests that other zeolites might also possess negative $\Delta D_{cw\text{-}w}$ values. Árnason and Tómasson (1970), who studied deuterium and chloride in Icelandic geothermal waters, analyzed the D content of water released from two well cores of altered rock when they were heated to 900°C. The expelled water was about 40‰ lower in δD than in groundwater collected from the wells.

In addition to yielding equilibrium fractionation factors, the isotopic exchange experiments discussed above also give important information regarding rates of isotopic exchange and plausible exchange mechanisms for zeolite-water systems. The exchange-rate data for zeolites are summarized in Figures 4 through 9. In the temperature range 300 to 400°C, complete oxygen isotopic exchange takes place between channel water and bulk water and nearly complete exchange (80-96‰) between the framework and bulk water in a matter of a couple weeks for analcime (average grain diameter: 100 µm; Karlsson and Clayton 1990b). Noto and Kusakabe (1997a) obtained similarly rapid exchange rates for wairakite-water. At temperatures between 250 and 400°C, they observed 64-98% exchange in experiments lasting up to 5 months (grain diameter 74-100 µm). Feng and Savin (1993a) performed detailed framework-water vapor experiments for analcime, clinoptilolite, heulandite and stilbite between room temperature and 500°C in order to determine the extent and rate of exchange. In a series of experiments each lasting 4 hr (Fig. 4), they detected up to 4% oxygen isotopic exchange in all of these zeolites except analcime, at temperature as low as 100°C (grain diameter 74-149 µm). Exchange in analcime was only detectable at temperatures >200°C. The analcime results are consistent with those of Karlsson (1988) who observed no exchange after exposing the mineral to water vapor for two weeks at 200°C. As might be expected, the rate of exchange increases with temperature in analcime, clinoptilolite, and wairakite. For heulandite and stilbite, exchange rates actually decrease with increasing temperature above 220°C for stilbite and above 300°C for heulandite (Fig. 4). It is likely that the channel systems in these zeolites become restricted or collapse with dehydration at high temperatures closing off the most rapid paths of exchange and thus effectively reducing the total framework surface area exposed to exchange. In contrast, the channels in analcime and clinoptilolite remain intact and exchange therefore increases continuously with temperature.

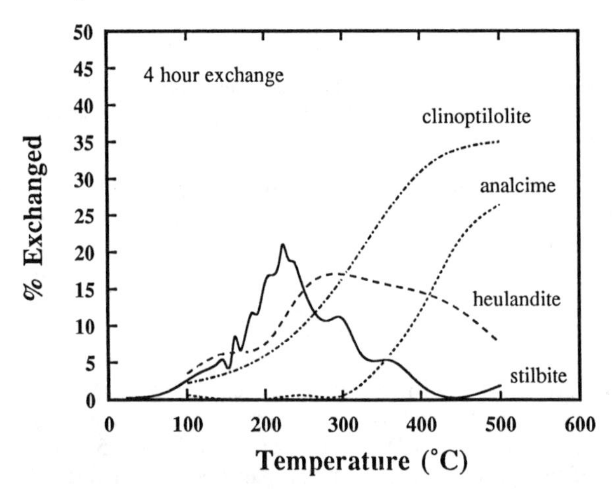

Figure 4. Oxygen isotopic exchange between zeolite framework and external water vapor as a function of temperature. The zeolites were exchanged for four hours with water vapor. Adapted from Feng and Savin (1993a).

The studies discussed in the previous paragraph considered only macroscopic aspects of isotopic exchange. Recently, however, a series of experiments utilizing nuclear magnetic resonance (NMR) have been conducted by Xu and Stebbins (1998a, 1998b) in order better understand oxygen isotope exchange kinetics and mechanisms in zeolites on a microscopic level. Xu and Stebbins (1998a, 1998b) used a solid-state high-resolution ^{17}O NMR technique to distinguish between oxygens in Al-O-Si sites from those in Si-O-Si sites in stilbite from Nasik, India. The stilbite (44-75 µm diameter) was first exchanged with ^{17}O-enriched water at 200°C and 40 bar using conventional hydrothermal techniques and was then "back-reacted" by exchange with water vapor of normal oxygen isotopic composition. The reactions took place at either 157 or 197°C and lasted for 10 to 160 hours (Fig. 5). From the relative peak heights in the NMR spectra, Xu and Stebbins (1998b) were able to deduce that the exchange rate for oxygen in Al-O-Si sites was faster than for oxygen in Si-O-Si sites (Fig. 6). This result is consistent with the notion that Al-O bonds are more easily broken than Si-O bonds in minerals (e.g. O'Neil 1986).

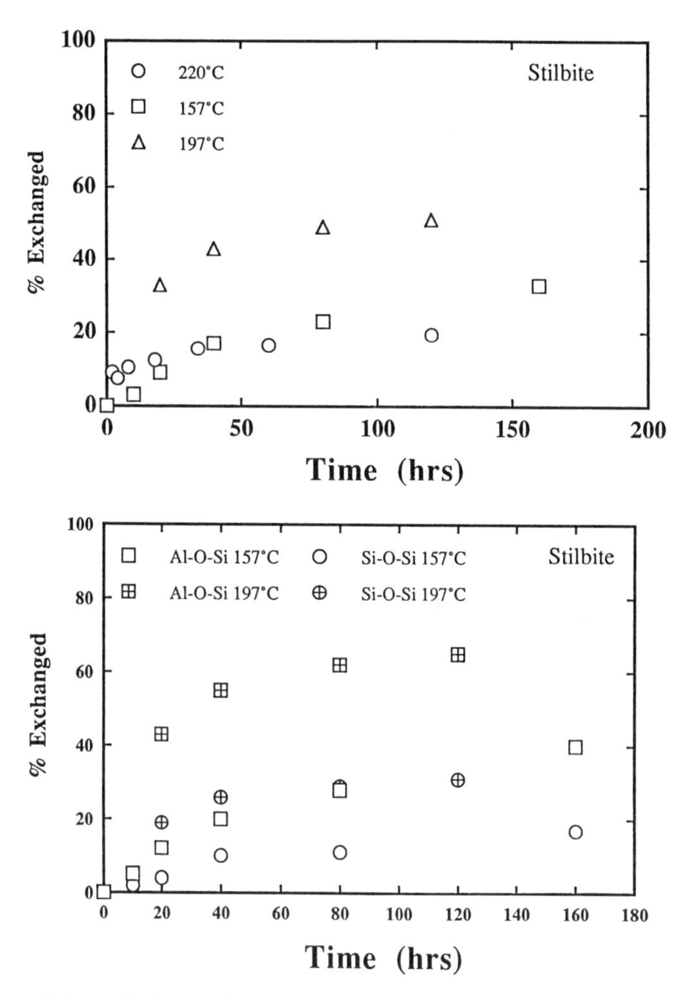

Figure 5 (upper). Oxygen isotopic exchange between stilbite framework and external water vapor shown as the % framework oxygen exchanged as a function of time at fixed temperature. The 220°C data are from Feng and Savin (1993a) and the 157°C and 197°C data are from Xu and Stebbins (1998b). Feng and Savin (1993a) obtained their results by bulk isotopic exchange, whereas Xu and Stebbins (1998b) estimated exchange from high-precision ^{17}O NMR.

Figure 6 (lower). Site-specific oxygen isotope exchange in stilbite. Depicted is % framework oxygen exchanged in Si-O-Si and Al-O-Si sites at 157°C and 197°C with water vapor. The data demonstrate that oxygen in Si-O-Al sites is more readily exchanged than oxygen in Si-O-Si sites. Adapted from Xu and Stebbins (1998b).

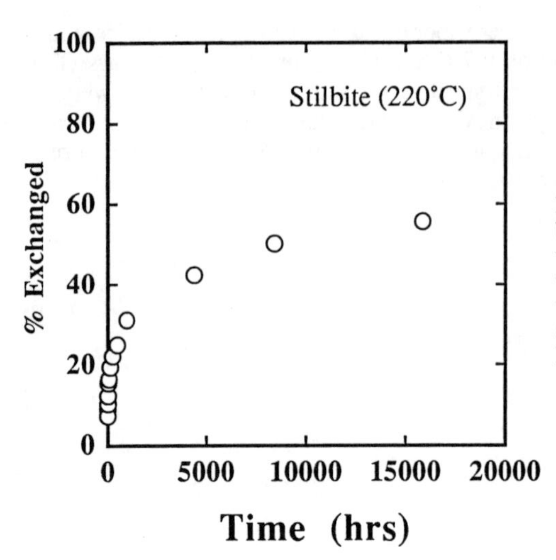

Figure 7. Oxygen isotopic exchange between stilbite framework and external water vapor at 220°C as a function of time. The amount of exchange increases with time but the rate slows after an initial rapid exchange. Adapted from Feng and Savin (1993a).

The rate of oxygen isotope exchange between framework and water at constant temperature has been determined for analcime, stilbite and wairakite. The results are summarized in Figures 7 to 9. As is typically found with silicates, the extent of isotopic exchange increases with time (Fig. 7, see also e.g. Cole, 2000). There are, however, some curious differences between results obtained under different experimental conditions. These differences cannot be accounted for by grain size variations because the grain sizes used in the experiments were similar. For example, exchange in the analcime experiments of Feng and Savin (1993a) at low pressure was less than 100% at 400-450°C even after 3000 hours. This stands in stark contrast to the high-pressure data of Karlsson and Clayton (1990b), who obtained 80-100% exchange and much faster rates for similar or lower temperatures (300-400°C, see Fig. 8). For wairakite (Fig. 9), Noto (1991) achieved the same level of exchange (70-80%) and faster rates in a 100°C experiment at low pressure (1 bar) as for a set of experiments run at higher temperatures (250-270°C) and pressures (0.5-1.0 kbar).

It is well established that increased pressure during hydrothermal experiments can enhance isotopic exchange (e.g. Matthews et al. 1983), and there are examples for analcime where it appears that pressure does indeed accelerate the exchange (Fig. 8). On the other hand, for wairakite, pressure actually appears to retard exchange, as is best seen by comparing the low-pressure 100°C data with the higher-pressure and higher-temperature hydrothermal runs (250, 270 and 300°C data in Fig. 9). It also seems unlikely that pressure can account for the 90% exchange difference between the 100°C (0.03 bar) and 400°C (5 kbar) data for analcime. It should be pointed out that Karlsson and Clayton (1990b) approached isotopic equilibrium from both sides (two isotopically different waters), whereas Feng and Savin (1993a) used the fractionation curve determined by Karlsson and Clayton (1990b) to calculate the equilibrium values for analcime at 400 and 450°C. Noto (1991) produced similar results in 100°C runs by using two isotopically distinct waters. Assuming that there are no intrinsic problems in the framework-water vapor exchange experiments, these data indicate that there are major differences in either exchange rate as a function of pressure and/or that the fractionation factors are pressure dependent. One source for an inverted pressure dependency could be the state of H_2O above the critical point (374°C, 218 atm). In the low-pressure experiments the zeolites exchanged with a low-density fluid whereas in high-pressure experiments the zeolites exchanged with a high-

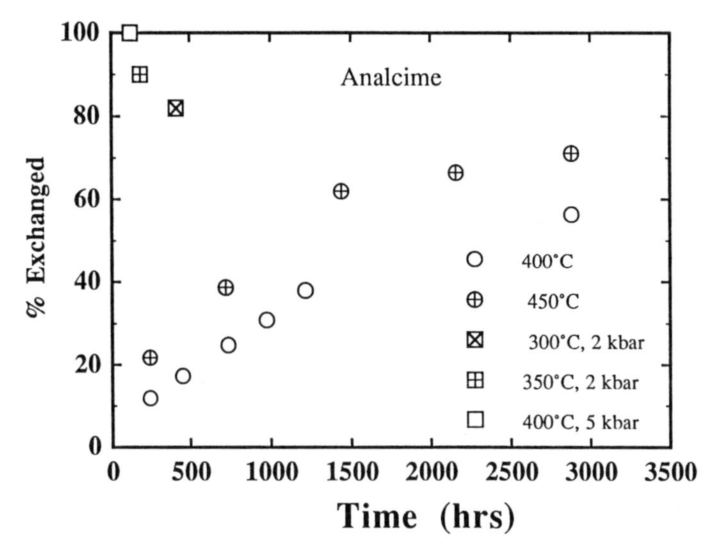

Figure 8. Percent framework oxygen in analcime exchanged with water as a function of time at various temperatures. There are two kinds of exchange data: results of hydrothermal experiments run at high pressures (300°C, 350°C, 400°C from Karlsson and Clayton 1990b) and vapor experiments run at low pressures (400°C, 450°C from Feng and Savin 1993a).

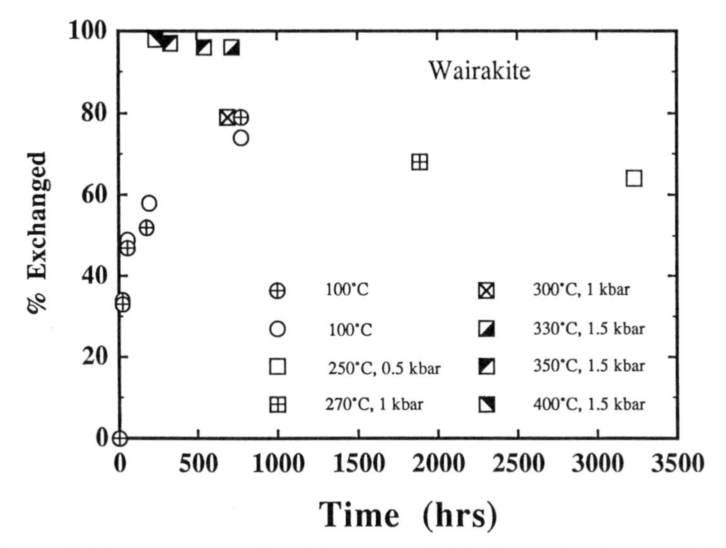

Figure 9. Percent oxygen exchanged between wairakite framework and water during hydrothermal experiments as a function of time at various temperatures. Squares depict high-pressure (0.5-1.5 kbar) experiments (250°C, 270°C, 300°C, 330°C, 350°C, 400°C) and circles represent one-atmosphere experiments. Data from Noto et al. (1990) and Noto and Kusakabe (1997a).

density fluid. This author has no explanation for the apparent discrepancy in the stilbite data obtained by Feng and Savin (1993a) and Xue and Stebbins (1998b) for similar temperatures (197°C vs. 220°C) shown in Figure 5.

Oxygen isotopic exchange between a mineral and a fluid can occur in two different ways. There may be exchange at the mineral surface (site exchange) or exchange involving the whole mineral (bulk exchange). Site exchange may involve sorption-exchange-desorption at the mineral surface followed by solid-state diffusion of fresh supplies from the interior to the surface. Bulk exchange may involve solution-precipitation or a recrystallization front that passes through the crystal. Normally, site exchange is limited by solid-state diffusion whereas bulk exchange rates are governed by the rate of recrystallization (precipitation). The mechanism of oxygen isotopic exchange during zeolite framework-water experiments can be elucidated with scanning electron microscope photomicrographs of mineral grains before and after exchange. Oxygen isotopic exchange has little effect on grain morphologies of analcime, stilbite, and wairakite, suggesting that framework exchange was dominated by the site-exchange process. Researchers disagree as to the rate limiting-step in these reactions. For analcime and wairakite, Karlsson and Clayton (1990b) and Noto and Kusakabe (1997a), respectively, argued that sorption-exchange-desorption is rate-limiting. For stilbite, Feng and Savin (1993a) surmised that the sorption-exchange-desorption process was rate-limiting only during the initial stages of exchange (<20 hrs) but that diffusion became the primary rate-limiting step for longer exchange times. On the other hand, Xue and Stebbins (1998b), using a similar experimental setup and similar conditions to Feng and Savin (1993a), concluded that sorption-exchange-desorption process was the rate-controlling at elevated temperatures (200°C) up to at least 40 hours of exchange. Their conclusion was based on the observation that exchangeability did not vary with grain size and that, if diffusion were rate-limiting, then more exchange would be detected for the small particle-size sample (<20μm) than for the large particle-size sample (>45 μm). However, the connection between grain size and surface area for zeolites is not as obvious as it might seem. One might deduce that because zeolites have such a large internal surface area that the overall surface area is independent of grain size. This not the case. Whereas the internal surface area is indeed unaffected by grain size the outer surface area is affected. The overall surface area thus varies with grain size in the usual manner. Smaller grain sizes have larger overall surface areas than do larger grain sizes. A greater number of atoms are therefore available for exchange on the outer surface of a zeolite that has a smaller particle size than with a larger particle size. The link between overall exchange rate and grain size is therefore not as simple as it might appear and needs further study. For a more detailed discussion of modeling isotopic exchange and kinetics in zeolites the reader is referred to Feng and Savin (1993a, 1993b).

Relationship with associated minerals

In many instances the fluid that originally equilibrated with a zeolite is no longer available for sampling. However, under such circumstances it is still possible to obtain useful information by comparing zeolite $\delta^{18}O$ values with those of coexisting or associated mineral phases (including other zeolites). It can be shown from Equation (4) that in cases where two minerals, x and y, are in isotopic equilibrium with the same fluid, that the fractionation factors between those minerals are in fact in equilibrium with each other such that:

$$\Delta_{x\text{-}y} = A'(1000/T)^2 + B' \qquad (5)$$

where A′ and B′ are constants and T is temperature (K). Often $B' \approx 0$ such that Equation (5) simplifies to:

$$\Delta_{x\text{-}y} = A'(1000/T)^2 \qquad (6)$$

Furthermore we know that $\Delta_{x\text{-}y} \approx \delta_x\text{-}\delta_y$ and thus if one plots δ_x versus δ_y, then all x-y mineral pairs that are in isotopic equilibrium at the same temperature will lie on a straight line with slope A′. Small $\Delta_{x\text{-}y}$ values on a $\delta_x\text{-}\delta_y$ plot imply either a high temperature or similar fractionation curves for the two minerals. Moreover, even when the two minerals are not in isotopic equilibrium, it still possible to obtain useful information (cooling history) if the fractionation curves for either one or both minerals are known. Using the δ_x-δ_y plot approach, Karlsson and Clayton (1990a) compared the oxygen isotopic values of hydrothermal analcime and associated calcite occurring in vugs in concluded that the fractionation curves for analcime-water and calcite-water are probably similar. They also deduced that the order of crystallization, in those cases where analcime and calcite had not coprecipitated, was in every instance consistent with cooling. Noto and Kusakabe (1997b) did a similar analysis for wairakite-calcite and wairakite-quartz pairs for samples from two active hydrothermal areas in Japan in order to study the thermal histories of the geothermal systems.

APPLICATIONS OF STABLE ISOTOPES

Oxygen and H isotope measurements of zeolites have been used to: (1) constrain their formation conditions; (2) assess the extent of water-rock interaction; (3) study the nature and origin of the channel water; (4) estimate reactivity/stability of zeolites in the presence of water; (5) measure channel water mobility; and (6) trace dealumination reactions. Aspects of the first four topics are discussed below. Prospects for single-mineral geothermometry involving zeolites are also discussed. The reader is referred to Barrer (1978) on channel water mobility and to von Ballmoss (1981) concerning dealumination reactions.

Formation conditions and the zeolite framework

It is widely accepted that most zeolites in nature form as a result of fluid-rock interactions at temperatures <~350°C and pressures <~3 kbar (Hay 1978). In active hydrothermal systems it is possible to directly measure pressure, temperature, and fluid compositions, but quantifying these variables for fossil systems has proven difficult. Estimates based on field observations coupled with thermodynamic data or experimentally determined phase relations for zeolites are fraught with difficulties (see e.g. Donahoe et al. 1990, and references therein).

The oxygen isotope ratio of a zeolite framework reflects formation conditions such as temperature and fluid composition and possibly the isotopic composition of precursor materials. To date, $\delta^{18}O_f$ values have been reported for natural occurrences of analcime, chabazite, clinoptilolite, heulandite, laumontite, mordenite, natrolite, phillipsite, scolecite, stilbite and wairakite (Savin 1967; Böhlke et al. 1984; Lee 1987; Kita and Honda 1987; Lambert et al. 1988; Luhr and Kyser 1989; Stallard and Boles 1989; Karlsson and Clayton 1990a; Noto et al. 1990; Noto and Kusakabe 1997b; Feng and Savin 1991; Kitchener et al. 1993; Nähr 1997;Nähr et al. 1998; Feng et al. 1999). These data are summarized in Figure 10 which shows the observed range in $\delta^{18}O_f$ values for each zeolite. As a group, zeolites span an extensive range in $\delta^{18}O_f$, extending from -0.4 ‰ for hydrothermal wairakite (Noto and Kusakabe 1997a,b) to +35 ‰ for deep-sea phillipsite (Savin 1967). Setting aside possible underestimation of $\delta^{18}O_f$ due to incomplete channel water removal in some of the early studies and post-formation isotopic exchange, the overall wide range in isotopic compositions indicates a correspondingly large variation in temperature and/or fluid composition.

Of all the zeolites, the oxygen isotope composition of analcime, wairakite, and clinoptilolite are currently best documented. Karlsson and Clayton (1990a) reported $\delta^{18}O_f$ values for about 50 analcime samples from 34 locations and a variety of geological settings and ages. They observed an apparent correlation between $\delta^{18}O_f$ values and mode of occurrence (Fig. 11). Sedimentary analcimes (S-type), which formed at low temperatures (0-50°C), have high $\delta^{18}O_f$ values (16.6-24.5 ‰), whereas igneous analcimes (P-type),

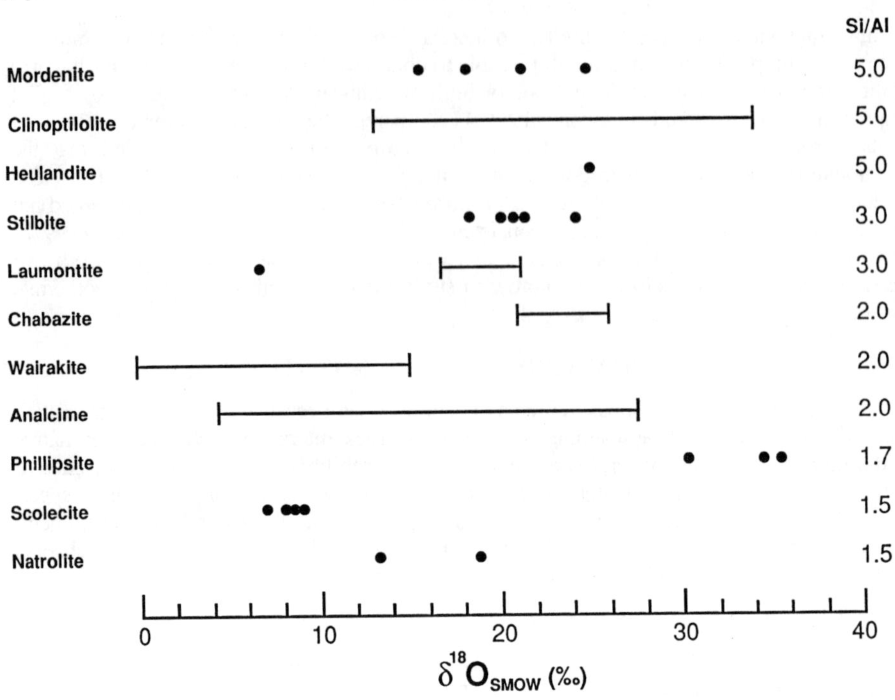

Figure 10. Variations of framework-oxygen isotopic composition ($\delta^{18}O_f$) in natural zeolites. Individual analyses are shown as points and ranges as bars. Bars were used if more than four analyses were available. See text for sources.

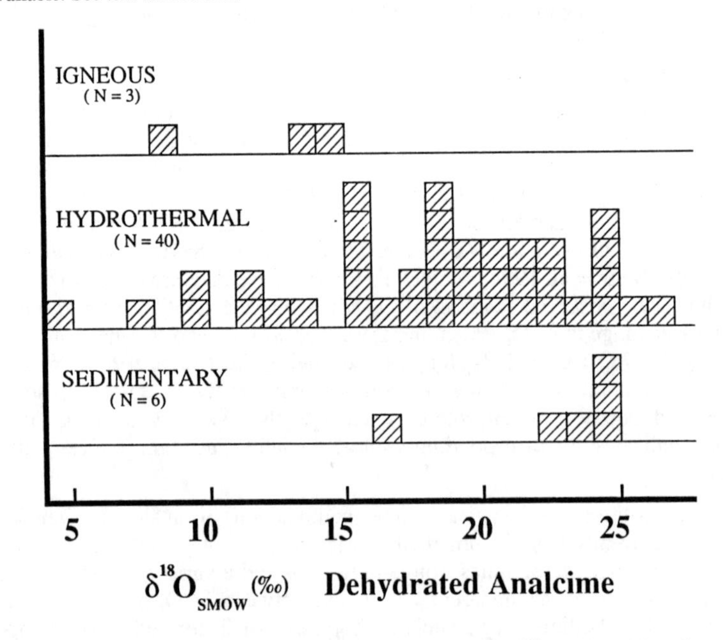

Figure 11. Histogram of framework-oxygen isotopic composition ($\delta^{18}O_f$) of analcime from various sources. Values in brackets give number of samples analyzed. From Karlsson and Clayton (1990a).

which may have formed at high temperatures (600-650°C), have low $\delta^{18}O_f$ values (8.7 to 14.3‰). Hydrothermal analcimes (H-type), which formed at intermediate temperatures (50-350°C), have $\delta^{18}O_f$ values that span the entire range from 4.3 to 26.6‰. Given the temperature dependence of oxygen isotope partitioning between the analcime framework and water oxygen determined by Karlsson and Clayton (1990a, 1990b) from laboratory experiments and field observations, the fluid compositions may be constrained if the temperature is well known. For example, Karlsson and Clayton (1990a) estimated that the fluid compositions in the sedimentary Aquarius Cliffs, Barstow, and Big Sandy Formations (Pliocene-Miocene) were similar (-3 to +5 ‰), whereas $\delta^{18}O$ values of fluids in the Green River Formation (Eocene) were significantly lower (-8 to -3 ‰). It is reasonable to assume that the source fluid was meteoric and hence the initial $\delta^{18}O$ of the lake water must have been similar to that of local precipitation. However, the pore fluid $\delta^{18}O$ values for these formations are higher than estimates for present-day meteoric waters at these sample sites (Karlsson 1988). Recently, Amundson et al. (1996) and Feng et al. (1999) suggested that atmospheric circulation patterns for mid-North America in the upper middle Eocene differed significantly from what they are today. On the basis of soil carbonate $\delta^{18}O$ data Amundson et al. (1996) surmised that in the past a greater component of ^{18}O-enriched precipitation derived from the Gulf of Mexico reached the mid-continent of North America than today.

Noto and Kusakabe (1997a,b) used $\delta^{18}O$ values of wairakite (both O_f and O_{cw}), calcite, and quartz from boreholes to study thermal histories of active geothermal areas in Japan and New Zealand. They also obtained wairakites from several extinct Japanese hydrothermal systems. Their $\delta^{18}O_f$ values for wairakites ranged widely from –0.4 to +14.1‰ but fell into two groups; low $\delta^{18}O_f$ values were observed for the active geothermal areas (-0.4 to +7.8‰) and high $\delta^{18}O_f$ values were observed in the extinct systems (8.0 to 14.1‰). Noto and Kusakabe (1997b) demonstrated that wairakites in the active geothermal areas were in oxygen isotopic equilibrium at current temperatures. By combining oxygen isotope compositions of wairakite and associated minerals (calcite and quartz) they were able to reconstruct the thermal history of two active Japanese geothermal systems. They attributed the higher $\delta^{18}O_f$ values observed in the fossil hydrothermal systems to re-equilibration with local meteoric waters at cooler temperatures.

Clinoptilolites were extensively studied by Nähr et al. (1998) and Feng et al. (1999). Both attempted to reconstruct the environmental conditions during clinoptilolite formation. Nähr et al. (1998) separated clinoptilolite from three sediment cores recovered at ODP sites 672 (Barbados Ridge Complex), 762 (Exmouth Plateau), and 797 (Yamato Basin). They analyzed $\delta^{18}O_f$ for 25 clinoptilolites (4-30 μm fraction) that came from depths ranging from 163.5 to 517 m below the seafloor. Clinoptilolite $\delta^{18}O_f$ values ranged from 18.7 to 32.8‰. The results fall into two distinct groups (Fig. 12a). High $\delta^{18}O_f$ values (26.9 to 32.8‰) were found at sites 672 and 762, whereas low values (18.7 to 24.0‰) were observed at site 797. The high isotopic values are consistent with the low-temperature environment expected in deep sea sediments. However, the general increase observed in the isotope profiles is not consistent with a simple temperature control but implies that the pore fluid composition plays an important role. Increased temperature alone would lead to higher $\delta^{18}O_f$ values. For two of the cores (672 and 797), pore water $\delta^{18}O$ values are available from previous workers (Vrolijk et al. 1990; Brumsack et al. 1992) and range from -3.3 to -6.5‰. $\Delta^{18}O$(clinoptilolite-fluid) values calculated from available sample pairs are shown in Figure 12b together with lines showing how $\Delta^{18}O$ would change with depth if isotopic equilibrium was maintained under the current thermal regime at each site. These lines were generated by obtaining the temperature at any given depth from the current geothermal gradient (66 and 121°C/km) and then solving the theoretical equation for the clinoptilolite-water fractionation (see footnote in Table 2). The clinoptilolites in the upper part of the

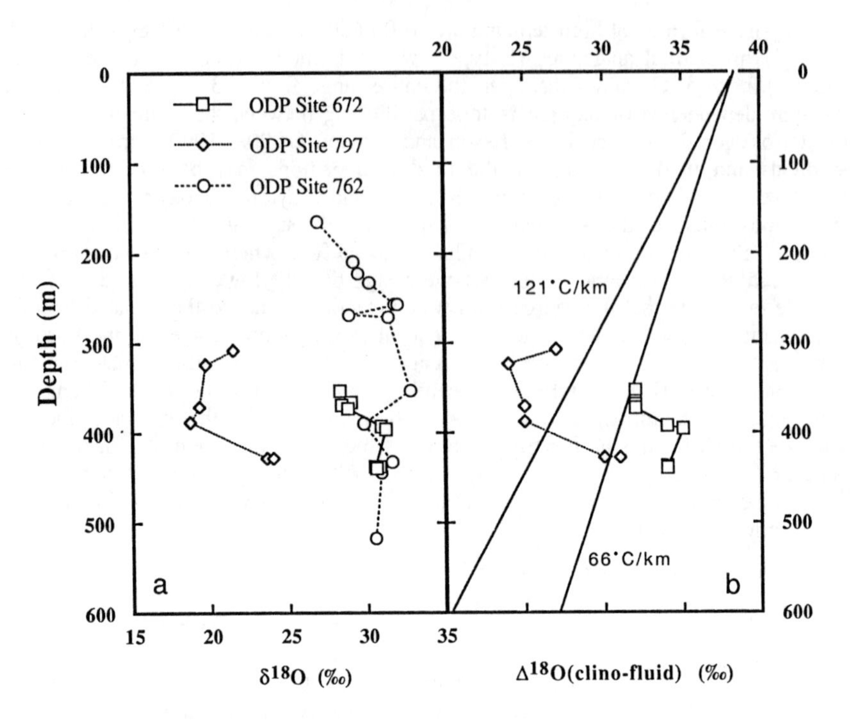

Figure 12. Oxygen isotopic compositions of deep-sea clinoptilolites. (a) Variations in $\delta^{18}O_f$ values of clinoptilolite with depth below the seafloor. The clinoptilolites come from three different ODP drill sites. Observe the difference between Site 797 and the other sites. (b) Calculated fractionation ($\Delta^{18}O$) between clinoptilolite framework and pore fluid for Sites 672 and 797. The two lines superimposed on the plot are the equilibrium fractionations (clinoptilolite-water) as a function of depth assuming geothermal gradients of 66 and 121°C/km and the theoretical clinoptilolite-water oxygen isotope fractionation curve (see Table 2). Data from Nähr et al. (1998).

sections at both sites appear close to isotopic equilibrium but those in the lower portions are not. One possible explanation for this apparent lack of equilibrium is that clinoptilolites originally precipitated in equilibrium in the higher and/or cooler part of sections and were subsequently buried and/or subjected to higher temperatures but did not re-equilibrate (Nähr et al. 1998). Alternatively, the clinoptilolites may have formed under closed-system conditions. Nähr et al. (1998) showed that simple Rayleigh fractionation could account for the low $\delta^{18}O$ values observed in clinoptilolite and pore water at Site 797 (Fig. 12a).

Feng et al. (1999) measured $\delta^{18}O_f$ values for 20 hydrothermal clinoptilolites (mostly 1-3 μm) in volcanic tuffs from Yucca Mountain. Two of the clinoptilolites were from surface outcrops and the remaining were recovered from drill cores (232-1092 m depth). Outcrop clinoptilolite $\delta^{18}O_f$ values ranged from 20.0 to 20.3‰, whereas borehole clinoptilolites varied from 13.4 to 22.1‰. In general, $\delta^{18}O_f$ values decreased with depth in the drill holes reflecting an increasing temperature (Fig. 13). Ambient fluid temperatures during hydrothermal alteration at Yucca Mountain may have been as high as 260°C below 1100 m about 11 Ma ago (Bish and Aronson, 1993). However, the clinoptilolites studied by Feng et al. (1999) came from an apparently cooler environment with paleotemperatures ranging from 14°C at the surface to 75°C at depth (Fig. 6 in Bish and Aronson 1993). Using the theoretical fractionation equation given in Table 2, Feng et al. (1999) estimated

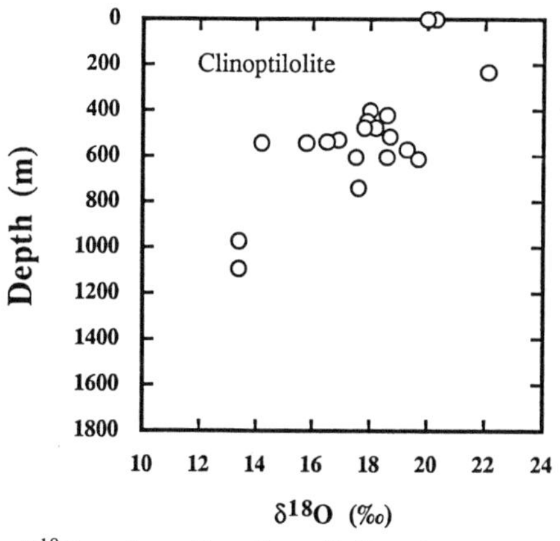

Figure 13. Variation in $\delta^{18}O_f$ of clinoptilolites with depth at Yucca Mountain. Data adapted from Feng et al. (1999).

that $\delta^{18}O_f$ values for clinoptilolites forming under modern conditions were comparable to observed values and that Yucca Mountain clinoptilolite had lost its original isotopic signature over the last 11 Ma.

The two clinoptilolite studies described above show some internal consistencies. Overall, $\delta^{18}O_f$ values for clinoptilolites formed in marine sedimentary environments are higher than those for clinoptilolites formed in terrestrial diagenetic environments (Fig. 14). These differences can be attributed to variations in (a) higher temperatures and (b) more ^{18}O depleted fluids in the terrestrial hydrothermal (meteoric water) than in the marine (sea water) environments.

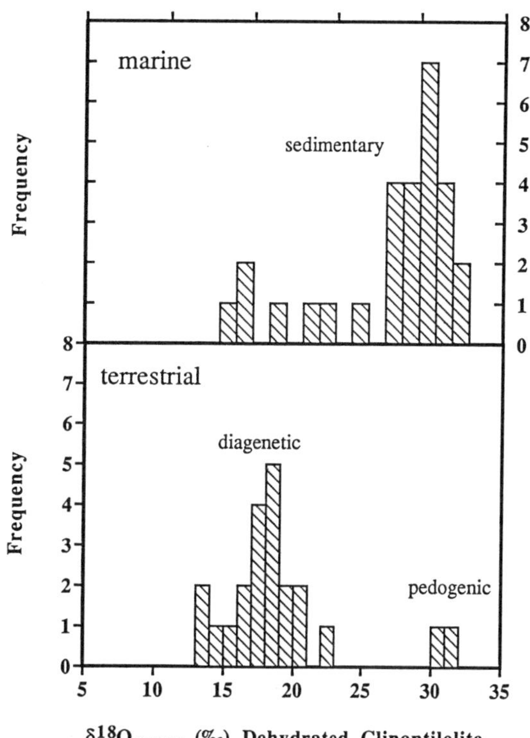

Figure 14. Histogram of the framework-oxygen isotopic composition ($\delta^{18}O_f$) of clinoptilolite from marine and terrestrial environments. Marine clinoptilolites are from ODP Sites 672, 762, and 797 (Nähr et al. 1998). Two genetic types of terrestrial clinoptilolites are shown, diagenetic clinoptilolites from Yucca Mountain, Nevada (Feng et al. 1999), and pedogenic (soil) clinoptilolites from Houla, Texas (Karlsson and Clayton 1990a). The spread in the marine values reflects variations both in temperature and fluid composition. The terrestrial soil clinoptilolites have considerably higher $\delta^{18}O_f$ values than the diagenetic clinoptilolites at Yucca Mountain. It is possible that the difference between the highest clinoptilolite values at Yucca Mountain (>20 ‰) and the Texas soil clinoptilolites simply reflects differences in the average $\delta^{18}O$ of modern meteoric water between the two sites (Texas –4 ‰ vs. Nevada –13.5 ‰).

Channel water

In a study of water diffusion of water in zeolites, Pemsler (1953) obtained the following water mobility sequence: mesolite > chabazite ~ stilbite > scolecite > natrolite > heulandite > analcime. To date, $\delta^{18}O$ and δD values for channel water in zeolites have been obtained for analcime, chabazite, clinoptilolite, heulandite, laumontite, mordenite, natrolite, stilbite, and wairakite (Luhr and Kyser 1989; Karlsson and Clayton 1990a; Feng and Savin 1991; Noto and Kusakabe 1997b). With the exception of analcime, chabazite, and wairakite, these studies have been limited to a few samples and localities.

Self-diffusion coefficients of channel water are known for about a dozen zeolites and range from 10^{-4} cm^2/s to 10^{-17} cm^2/s with corresponding activation energies of 3 to 23 kcal/mol at 30 to 50°C (for a compilation see Barrer 1978). With the exception of analcime, the measured diffusion rates in natural zeolites at room temperature are between those of liquid water (10^{-5} cm^2/s at 45°C) and water ice just below the melting point (10^{-10} cm^2/s at -2°C), with a rough correlation between diffusion rate and channel size. Zeolites with large channels have fast diffusion rates (e.g. faujasite), whereas those with small channels have slower diffusion rates (e.g. analcime). Because diffusion rates are high for all zeolites studied except analcime, the channel water in most zeolites will quickly re-equilibrate isotopically even at low temperatures (<100°C). In detail, retention times vary according to species. Generally, zeolites with wide channels and/or high cation-hydration numbers exchange channel water isotopes more readily than zeolites with narrow channels and/or low cation-hydration numbers.

$\delta^{18}O_{cw}$ values for zeolites are within the $\delta^{18}O$ range observed for global precipitation outside of polar regions, which is -1. to -25 ‰ (Yurtsever and Gat 1981). Given the rapid rates of exchange between channel water and the external water, a relationship between the isotopic compositions of the channel water and meteoric water should be expected. Craig (1961) showed that the oxygen and hydrogen isotope composition in world-wide precipitation and meteoric waters can be related to one another with the following simple equation:

$$\delta D = \delta^{18}O + 10 \qquad (7)$$

The expression of this relationship on a δD-$\delta^{18}O$ plot is known as the meteoric water line (MWL). Locally, there can be minor deviations from the MWL in which case a local form of the expression may be used. For analcime (Fig. 15), chabazite, laumontite, natrolite and wairakite, the channel water isotopic compositions fall on or close to the MWL clearly demonstrating that the channel water is meteoric in origin (Luhr and Kyser 1989; Karlsson and Clayton 1990a; Noto and Kusakabe 1997b). Samples from high latitudes tend to have lower $\delta^{18}O$ and δD analogous to what has been observed for precipitation. A plot of analcime $\delta^{18}O_{cw}$ as a function of latitude also suggests that channel water retains the isotopic signature of precipitation from the sample locality (Fig. 16). As observed in precipitation (Yurtsever and Gat 1981), the $\delta^{18}O_{cw}$ values of islands and coastal samples of a given latitude tend to be lower than for continental samples. Samples from Western Europe fall above the solid curve whereas North American samples fall below it. Most striking is the trend for the North American samples, which lie to the low $\delta^{18}O$ side of the North American Continent Line (NACL). The NACL delineates the change in mean $\delta^{18}O$ of precipitation over North America. If the channel waters are in equilibrium with local precipitation, then their $\delta^{18}O_{cw}$ values should plot along a line parallel to the NACL but shifted by -5‰ (dashed line in Fig. 16) to account for $\Delta^{18}O_{cw-w}$ of analcime. Thus, analcime channel water appears to retain a memory of meteoric waters at the original sample locality under sample formation conditions. The length of this memory will depend on crystal size and postformational thermal history. Theoretical calculations based on self-

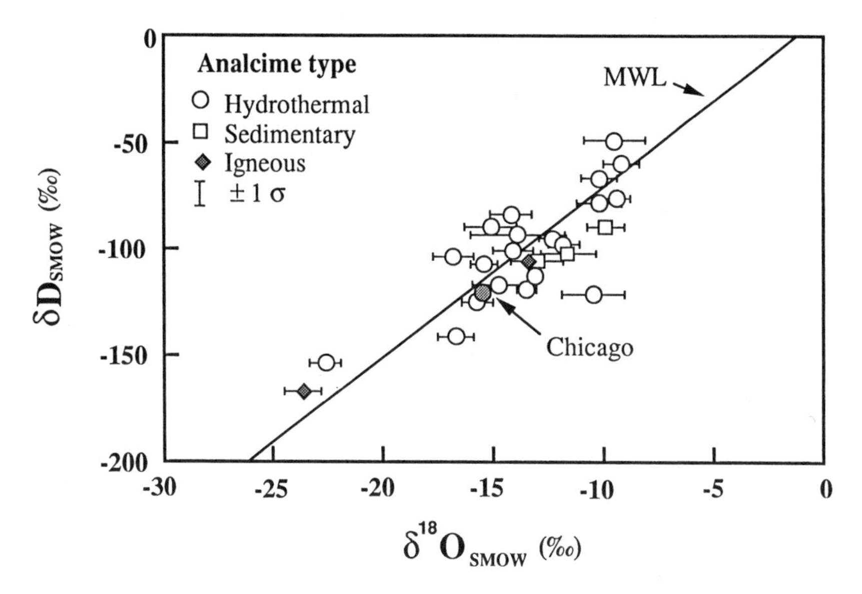

Figure 15. Relationship between $\delta^{18}O$ and δD in analcime channel water. MWL is the meteoric water line. "Chicago" (stippled circle) is the weighted mean composition of meteoric waters in Chicago, where the samples were stored prior to analysis. From Karlsson and Clayton (1990a).

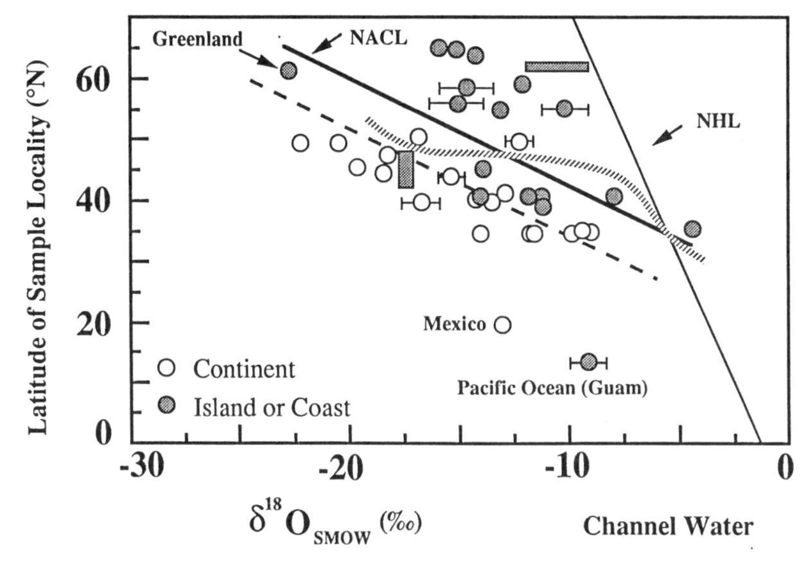

Figure 16. Relationship between channel-water oxygen isotopes ($\delta^{18}O_{cw}$) of analcime and latitude of sample locality. Boxes are shown for those samples where the latitude of the locality was uncertain. NHL: Mean variation in $\delta^{18}O$ of precipitation over the Northern Hemisphere. NACL: Mean variation in $\delta^{18}O$ of precipitation over the North American continent. Dashed line indicates composition of analcime channel waters in equilibrium with NACL waters. A stippled curve separates samples from North America (below curve) from those of Western Europe (above curve). From Karlsson and Clayton (1990a).

diffusion rates indicate that at low temperatures large crystals of analcime could retain the original isotopic composition for hundreds of thousands of years. For example, the homogenization time for a 2.5-cm diameter analcime crystal would be about 10^5 and 1600 yr at 45°C and 100°C, respectively (Karlsson and Clayton 1993). The $\delta^{18}O_{cw}$ of natrolite channel water may also reflect the isotopic composition of meteoric waters at the sample site (Karlsson and Clayton 1990a). Channel waters in chabazite, clinoptilolite and laumontite, however, re-equilibrate so rapidly that the isotopic composition is that of the water vapor in the laboratory just prior to analysis.

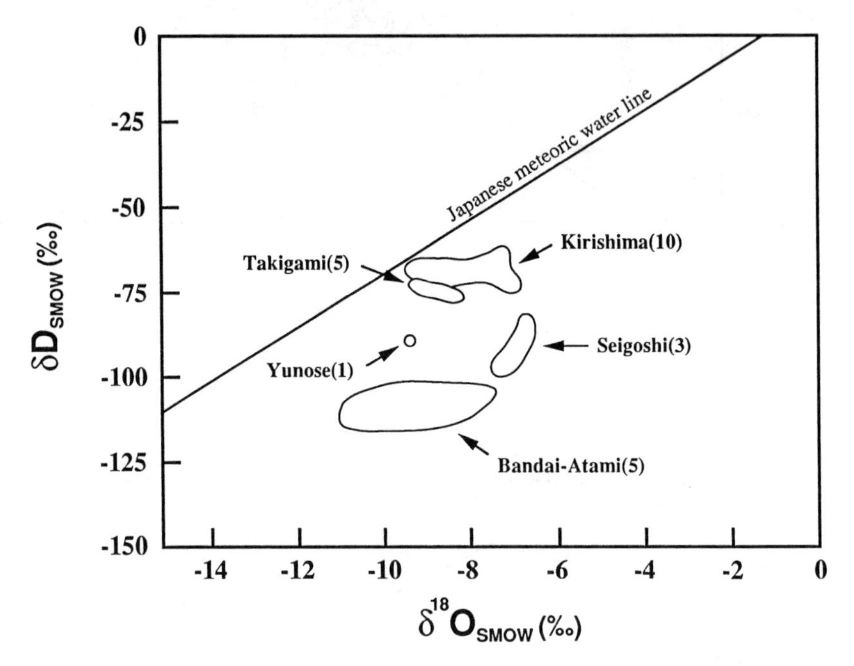

Figure 17. Relationship between $\delta^{18}O$ and δD of wairakite channel water in Japanese hydrothermal systems. Fields have been drawn around the data. Numbers in brackets give number of samples in each field. The Japanese meteoric water line ($\delta D = 8\delta^{18}O + 13.5$) is shown for reference. Adapted from Noto (1991).

A more detailed study of the variability of channel water isotopic compositions over a geographic area smaller than that of the area investigated by Karlsson and Clayton (1990) was carried out by Noto and Kusakabe (1997a,b). They obtained $\delta^{18}O_{cw}$ and δD values of hydrothermal wairakite from five different Japanese localities [the Takigami and Kirishima geothermal fields (Kyushu), Yunose (Yamaguchi Prefecture), the Seigoshi mine (Shizuoka Prefecture), and Bandai-Atami (Fukushima Prefecture)], and two New Zealand localities (the Wairakei and Ngatamariki geothermal fields). Noto and Kusakabe (1997a,b) concluded that the wairakites in the active geothermal areas were in isotopic equilibrium with current hydrothermal fluids by observing that measured down-hole temperatures and calculated temperatures from the fractionation of oxygen isotopes between framework and channel water for wairakite are in good agreement. Noto's channel water results are reproduced in Figure 17. The wairakite waters plot to the right of the MWL and, furthermore, each locality has a distinct isotopic composition (Fig. 17). There is an apparent correlation between δD and latitude of sample locality. The decrease in δD with increasing latitude suggests a genetic relationship to local meteoric waters as in the case of

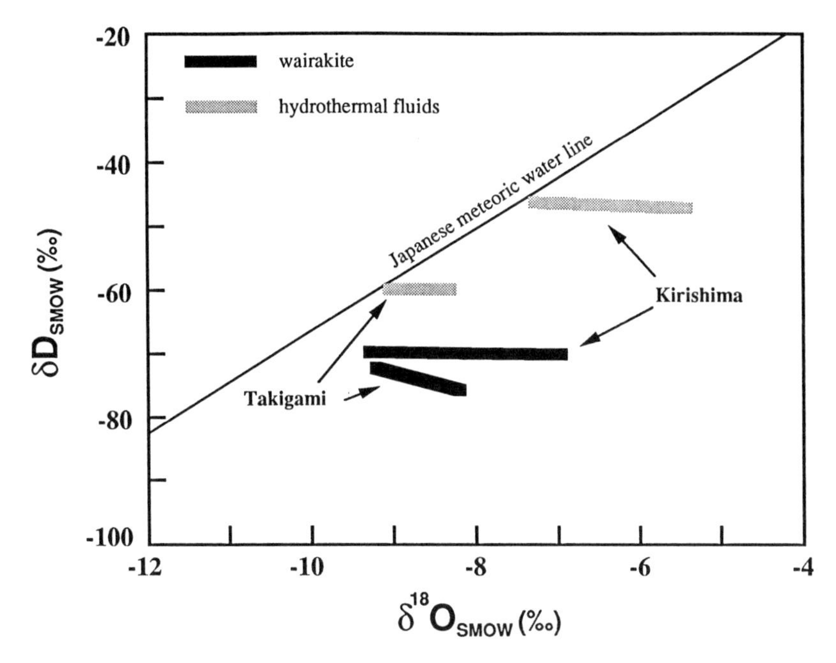

Figure 18. Isotopic composition of wairakite channel water and geothermal fluids at Kirishima and Takigami geothermal systems, Japan. Lines have been drawn through the data to show general trends. The Japanese meteoric water line ($\delta D = 8\delta^{18}O + 13.5$) of Matsubaya et al. (1973) is shown for reference. Local surface waters (not shown) lie at the intersection between the geothermal fluid lines and the meteoric water line. Wairakite data are from Noto (1991) and geothermal data from Kodama and Nakajima (1988) and Takenaka and Furuya (1991).

analcime. At Kirishima, Takigami, and Bandai-Atami, there is a noticeable variation in $\delta^{18}O_{cw}$ whereas δD remains relatively constant. This variation in $\delta^{18}O$ is reminiscent of the so-called $\delta^{18}O$-shifts often observed in geothermal waters. Geothermal waters which originate as local meteoric waters frequently experience a greater change in $\delta^{18}O$ than in δD due to isotopic exchange during water-rock interaction. Most rocks are rich in oxygen but poor in hydrogen and isotopic exchange therefore has little effect on δD of waters percolating through them whereas $\delta^{18}O$ can change dramatically. Rocks are typically enriched in ^{18}O relative to meteoric waters and, as a result, isotopic compositions of geothermal waters are shifted to higher $\delta^{18}O$ values. The magnitude of the $\delta^{18}O$ shift is related to water/rock ratio; the lower the ratio, the greater the shift (see e.g. Sheppard 1986). If these waters were in isotopic equilibrium with local meteoric waters they would lie on a line parallel to the MWL, which they do not. Recent isotopic results obtained for surface and geothermal waters at the Kirishima (Kodama and Nakajima 1988) and Takigami (Takenaka and Furuya 1991) geothermal fields suggest that wairakite channel waters indeed preserve the isotopic compositions of the geothermal fluids. These data are summarized in Figure 18. Wairakite channel waters lie parallel to those of geothermal fluids in terms of $\delta^{18}O$ but are shifted to lower δD values by about 15‰ and 25‰ for Takigami and Kirishima, respectively. The difference in δD shift may be due to temperature differences at the two geothermal systems. Overall, these data suggest that the wairakite channel water is in isotopic equilibrium with geothermal fluids. The Bandi-Atami and Seigoshi wairakites (Fig. 18) show even greater $\delta^{18}O$ shifts than those at Kirishima and Takigami indicating a correspondingly greater shift in the geothermal fluids. These shifts imply that water/rock ratios were lower at Bandai-Atami and Seigoshi than at Kirishima and Takigami. Noto's work on wairakite suggests that it may be possible to estimate water/rock ratios in fossil hydrothermal systems based on the $\delta^{18}O$ shifts of zeolite channel waters.

Single-mineral geothermometry

Hamza and Epstein (1980) suggested the possibility of single-mineral geothermometers based on the partitioning of oxygen isotopes between different sites in a mineral. In principle, zeolites could be used in this way by taking advantage of the equilibrium fractionation of oxygen isotopes between framework and channel water, $\Delta^{18}O_{f\text{-}cw}$. Karlsson (1988) and Karlsson and Clayton (1990b) showed that $\Delta^{18}O_{f\text{-}cw}$ can be related to $\Delta^{18}O_{f\text{-}w}$ and $\Delta^{18}O_{cw\text{-}w}$ by the simple relation:

$$\Delta^{18}O_{f\text{-}cw} = \Delta^{18}O_{f\text{-}w} - \Delta^{18}O_{cw\text{-}w}. \qquad (8)$$

Provided $\Delta^{18}O_{f\text{-}w}$ and $\Delta^{18}O_{cw\text{-}w}$ can be estimated experimentally and/or empirically over the same temperature range, $\Delta^{18}O_{f\text{-}cw}$ can be calculated. Noto and Kusakabe (1997a) used this approach to derive a framework-channel water fractionation curve for wairakite assuming a constant $\Delta^{18}O_{cw\text{-}w}$ value of -0.73‰. Similarly, a constant $\Delta^{18}O_{cw\text{-}w}$ value of -5.5‰ has been used to generate a framework-channel water fractionation relation for analcime (see Table 2). Karlsson and Clayton (1990a) argued that single-mineral thermometry would not work for zeolites except in the case of active geothermal systems such as Surtsey, Iceland, because the channel water is much more readily exchanged than the framework. Their notion was that the closure temperature for oxygen isotopic exchange (T_c: final temperature of equilibration) would be lower in the channel water than in the framework allowing the channel water to equilibrate to lower temperatures than the framework. However, because $\Delta^{18}O_{cw\text{-}w}$ does not change noticeably with temperature for either analcime or wairakite, $\Delta^{18}O_{f\text{-}cw}$ will attain a constant value at and below the framework T_c when the fluid composition is constant. In other words, the last equilibrium value of $\Delta^{18}O_{f\text{-}cw}$ will be preserved during rapid cooling (or heating) at temperatures below the framework T_c even if the channel water is exchanged as long as the bulk fluid composition remains constant. Noto and Kusakabe (1997a,b) found that the wairakite single-mineral thermometer yields temperatures close to measured down-hole temperatures in two active Japanese geothermal systems. Their data demonstrate that negligible oxygen isotopic re-equilibration occurs in the wairakite framework during retrieval of samples from drill holes in active geothermal areas.

Primary magmatic zeolites

The origin of analcime phenocrysts in volcanic rocks continues to be a subject of controversy (see e.g. Karlsson and Clayton 1993; Pearce 1993). Some workers argue, based primarily on field and petrographic evidence, that analcime crystallized directly from a silicate melt and is thus an igneous mineral (P-type), whereas others contend that the analcime formed from a pre-existing igneous mineral such as leucite and is, therefore, a secondary mineral (X-type). Karlsson et al. (1985) and Luhr and Kyser (1989) suggested that it might be possible to separate igneous analcime from other types on the basis of the framework oxygen isotope ratio. Igneous analcime formed at magmatic temperatures should have low $\delta^{18}O_f$ values analogous to those of other igneous minerals such as albite (~7‰; see Karlsson and Clayton 1991), whereas secondary analcime formed at low temperatures should have high $\delta^{18}O_f$ values. Luhr and Kyser (1989) and Karlsson and Clayton (1990a, 1991) analyzed $\delta^{18}O_f$ values of "igneous" analcime from two well-known localities, Crowsnest, Canada, and Colima, Mexico, and as Figure 11 shows, it is possible to distinguish "igneous" analcime from most sedimentary analcime (S-type), but igneous analcime cannot be separated from hydrothermal analcime (H-type) on the basis of oxygen isotopic composition.

Further support for the leucite-analcime transformation idea comes from Sgarbi et al. (1998), who studied a suite of highly potassic mafic rocks (kamafugites) from Brazil. They found close chemical similarities between Cretaceous age leucitites and analcimites, which led them to conclude that the analcimites had originally been leucitites. Oxygen isotopic analyses of diopside isolated from the analcimites yield typical magmatic values ($\delta^{18}O$: 5.4 to 6.0‰) suggesting that the leucite-analcime transformation did not reset the original pyroxene $\delta^{18}O$ values. It thus appears that useful information regarding original magmatic $\delta^{18}O$ values can still be garnered from analcime-bearing igneous rocks.

The genesis of leucocratic globules in volcanic rocks has been attributed variously to amygdales, segregation of a late-stage magmatic fluid into vesicles, and liquid immiscibility (Foley 1984). Calcite-analcime bearing gobules ("ocelli") occur in Mesozoic age alkaline basaltic dikes erupted into a submarine environment in the Mecsek Mountains of southern Hungary. Based on the chemistry, oxygen isotopic compositions, and petrography, Demeny et al. (1997) concluded that the analcime in the "ocelli" is intermediate between P and H type. These authors contend that the ocelli represent carbonate-rich droplets that separated from a mafic melt via liquid immiscibility. Calcite $\delta^{13}C$ and $\delta^{18}O$ values range from -7.9 to -8.8‰ (V-PDB) and 13.0 to 13.3‰ (V-SMOW), respectively. Demeny et al. (1997) noted that the $\delta^{18}O$ values were within the upper limit of what is observed for carbonatite melts and argued on that basis that the ocelli represented primary melts. However, the fact that very similar isotopic compositions have been obtained for hydrothermal calcites found in submarine volcanic plugs off the coast of Norway (Prestvik et al. 1999) weakens this argument. Demeny et al. (1997) rejected low-temperature alteration by meteoric water on the basis of the isotopic composition of the ocellar calcites which are similar to those of carbonatites. These isotopic compositions, however, are also similar to those of submarine hydrothermal calcites with a high mantle-derived C component (Stakes and O'Neil 1982). Furthermore, elevated $\delta^{18}O$ values for feldspars (7.6 to 15.0‰) from closely related volcanic rocks are suggestive of low-temperature alteration. It therefore appears premature to rule out conventional hydrothermal origin for the Hungarian analcimes.

Preservation of stable isotope ratios in natural zeolites

A crucial question in the use of stable isotopes in zeolites is the degree of retention of original isotopic signatures, especially $\delta^{18}O_f$ values. Based on their experiments with analcime, Karlsson and Clayton (1990b) argued that zeolites exposed to waters hotter than 300°C, such as in high-temperature geothermal areas, would not retain their original $\delta^{18}O_f$ values. They suggested that some zeolites might retain their original $\delta^{18}O_f$ values in low-temperature geothermal systems (<150°C) and in sedimentary or soil environments. Based on their experimental data, Feng and Savin (1993a) concluded that clinoptilolite, heulandite, and stilbite exchange much too readily even at low temperatures (100°C) to be resistant to postformational alteration. However, analcime can resist isotopic exchange at temperatures below 200-250°C and thus can yield useful geological information. Feng and Savin (1993a) estimated analcime $\delta^{18}O_f$ could remain relatively unaffected for millions of years even at 100°C. Noto and Kusakabe's (1997a,b) work on wairakite in active geothermal systems indicates that it too will retain its original $\delta^{18}O_f$ for long periods of time.

Two recent studies could have direct bearing on the retention of original $\delta^{18}O_f$ by clinoptilolite. Nähr et al. (1998) analyzed the oxygen isotopic compositions of authigenic clinoptilolites occurring in marine sediments of late Miocene to early Eocene age (5.3 to 57.8 Ma). At ODP Site 672 (Barbados) they concluded that clinoptilolite, in the lower part of zeolite-bearing interval, was not in isotopic equilibrium with the pore fluid under current thermal conditions (<60°C), indicating that clinoptilolite could retain its original isotopic signature for extended periods of time in a low-temperature environment (Fig. 12a,b). Feng et al. (1999) used drill hole data to test the retention of $\delta^{18}O_f$ in clinoptilolites of the 11 Ma old Yucca Mountain tuffs. By boot-strapping from the isotopic compositions of other "well-behaved" authigenic minerals (clay) they inferred the initial formation conditions for the clinoptilolite – namely the paleo-temperature and paleo-fluid $\delta^{18}O$. Using these results as input parameters, they calculated $\delta^{18}O_f$ for the clinoptilolite from the theoretical isotopic fractionation curve for clinoptilolite-water (Table 2). The theoretical paleo-$\delta^{18}O_f$ values were markedly higher than the observed $\delta^{18}O_f$. When they repeated the calculation using modern conditions as input parameters the calculated $\delta^{18}O_f$ values were closer to the observed $\delta^{18}O_f$ values. Feng et al. (1999) concluded that clinoptilolite at Yucca Mountain had not retained its original $\delta^{18}O_f$ but re-equilibrated to modern conditions. The deepest clinoptilolites measured occur at roughly 1 km depth which yields a current temperature of roughly 40°C (25°C/km and 14°C at surface). The implication of the work of Feng et al. (1999) is that clinoptilolite could re-equilibrate within 11 Ma at temperatures under 40°C.

The conclusions reached in these two studies regarding the retention of original $\delta^{18}O_f$ values of clinoptilolite in the low-temperature environment (<60°C) are apparently at odds with each other. These differences could in part be reconciled if the Yucca Mountain clinoptilolites underwent isotopic resetting during the initial cooling stages of the geothermal system. The scatter in $\delta^{18}O_f$ values observed by Feng et al. (1999) in samples from ~600 m depth could reflect local differences in the environment (permeability, fluid composition, temperature) and the extent of resetting (Fig. 13). Conversely, the marine clinoptilolite could have formed under closed-system conditions giving the appearance of a lack of isotopic equilibrium. These issues can only be resolved if useful information about the time of formation is obtained. Clearly, zeolites that exchange their oxygen isotopes more easily than analcime and wairakite may be used to study postformational alteration histories in settings where the conditions and/or timing of alteration can be independently determined (Feng and Savin 1993a).

RADIOACTIVE AND RADIOGENIC ISOTOPES

Zeolites are found in geologic formations of all ages ranging from newly erupted basalts, such as those of Surtsey island off the southern coast of Iceland where zeolites are forming during palagonitization, to those of the 1.3 Ga old basalts in the Keeweenaw Peninsula, Upper Michigan (Jakobsson 1978; Karlsson 1988). Zeolites reported to occur in meteorites may even be older (Rubin 1997a, b). Estimates of the ages of zeolites and timing of zeolitization can be obtained indirectly by dating associated minerals or surrounding rocks that themselves are datable. Unless the zeolite formed at the same time as associated mineral formation, the dates obtained will only give an upper limit on the initial zeolite crystallization age. Zeolites may form shortly after emplacement of the geologic formation such as in the case of Surtsey or they may form considerably later. For example, zeolites in Eastern Iceland are likely to be much younger than their host basalts because they occupy flat-lying zeolite zones formed after the lava pile had been tilted and uplifted (Walker 1960). It is therefore more desirable to date zeolites directly using radioactive isotopes present in the mineral. Thus far, radioactive dating has been limited to the content of the channels such as K-Ar dating, Rb-Sr dating, and fission track dating. The choice of dating method will, as in other mineral systems, depend on the expected age and chemistry of the zeolite. Table 1 gives the pertinent parameters for the systems that have been used. The same assumptions are made when zeolites are dated as with other minerals, namely, that (1) the zeolite contained little or no radiogenic daughter (initial daughter) when it formed, (2) the zeolite remained a closed system until the age was determined, and (3) sufficient amounts of the radioactive parent were present initially to elemental mobility. Zeolites in which radioactive and radiogenic isotopes have been studied include analcime, chabazite, clinoptilolite, heulandite, mordenite, natrolite, phillipsite, and stilbite. Recently, interest in other radioactive isotopes in zeolites such as ^{137}Cs and ^{87}Sr has been on the rise because of the potential importance of zeolites in trapping radionuclides escaping from potential nuclear waste repositories such as Yucca Mountain, Nevada (Vaniman and Bish 1995). These issues, however, are outside the scope of this paper.

Radiometric dating—General principles

When a mineral forms (crystallizes), it incorporates radioactive parent atoms from its surroundings into its crystal structure. Some or all of these parent atoms subsequently decay to form radiogenic daughter atoms. Thus, over time, the number of parent atoms will diminish while the number of daughter atoms increases. Radiometric dating is based on the premise that this change in concentration is systematic and measurable.

The decay of parent nuclei is governed by a simple statistical principle that is the same for all radioactive atoms:

$$P_t = P_0 e^{-\lambda t}, \quad (9)$$

where t is time, P_t is the number of parent atoms present at time t, P_0 is initial number of parent nuclei present (i.e. at $t = 0$), and λ is the decay constant. The decay constant is unique for each radioactive isotope and it represents the probability that decay will occur within a given time period. Another way of expressing the rate of decay is in terms of a half-life ($t_{1/2}$). One half-life is the time it takes a given number of parent atoms to decrease by one-half. The decay constant and half-life are related by:

$$t_{1/2} = \log_e \frac{2}{\lambda} \quad (10)$$

It is clear from Equation (9) that if P_t, P_0 and λ are known, it is possible to determine t or an age for the mineral or rock of interest. Generally, λ is known independently, P_t is measured directly, but P_0 is unknown. It is, however, possible to obtain P_0 by determining the number of radiogenic daughter atoms formed at time t (D_t) because for every parent atom that decays a single daughter atom is formed. The number of daughter atoms is, therefore, related to the number of parent atoms decayed or $D_t = P_0 - P_t$ and we can P_0 replace in Equation (9) with ($D_t + P_t$) to obtain:

$$P_t = (D_t + P_t)e^{-\lambda t} \quad (11)$$

Equation (11) can be solved for t:

$$t = (1/\lambda)\log_e [(D_t/P_t)+1] \quad (12)$$

In the event that the rock or mineral incorporated some daughter atoms at formation, the initial amount of non-radiogenic daughter atoms (D_0) must be subtracted from D_t, and D_t in Equation (12) must be replaced with $D_t - D_0$. Equation (12) yields a single date that is a so-called model age and is most commonly used in cases where D_0 is negligible such as is usually the case for K-Ar dating. Where D_0 is significant and cannot be estimated or determined directly, an isochron method is used. This procedure, which yields an isochron age , circumvents the problem of the initial daughter content by analyzing a number of different rocks and/or minerals that formed at the same time in the same geologic unit, but whose concentrations of the parent and daughter elements differ. Because chemical processes such as crystallization do not discriminate between different isotopes among the heavier elements, the ratio of daughter atoms (D_0) over that of a non-radiogenic stable isotope (S) will be identical in all the samples when they form. The amount of radiogenic daughter atoms produced in each sample will be directly related to the amount of radioactive parent present initially in the sample, and thus the isotope ratio D_t/S will also change proportionally. When such data are plotted on a D_t/S versus P_t/S isotope ratio plot, a straight line or isochron emerges whose slope is a function of age. This method is commonly used in Rb-Sr dating but can be applied to any radiometric system where multiple samples are available (see e.g. Faure 1986).

A basic assumption common to all radiometric dating methods is that the system (rock or mineral) has remained closed since its formation so that no parent or daughter atoms have been added or removed. Testing of this requires more than one age determination from the same formation. Agreement between model ages for different samples may then indicate that the system has remained closed. Best results are obtained when the graphic isochron method is used because a straight line results with a closed system. It is, however, possible to obtain such a line as a result of mixing between two different isotopic reservoirs.

APPLICATIONS OF RADIOACTIVE AND RADIOGENIC ISOTOPES

K-Ar dating

The potassium-argon system is based on the decay of ^{40}K to ^{40}Ar. The half-life of ^{40}K, which also decays to ^{40}Ca (branched decay), is 1.25 Ga, which limits age determinations to dates older than several hundred thousand years. Total potassium content is determined using methods such as atomic absorption or flame photometry, and the amount of ^{40}K in the sample is calculated from the total K concentration assuming that the ^{40}K occurs in natural abundance (Table 1). In most minerals, Ar loss (i.e. an open system) is the major concern and greatest limitation of the method.

The first published dates for zeolites are those of Dymond (1966) who reported K-Ar ages for volcanogenic sediments in seven deep-sea cores of the North Pacific. Dymond was interested in estimating the formation rate of marine phillipsite by comparing its age with the age of the formation containing it. He obtained dates for five mineral separates that ranged from 0.67 to 6.6 Ma, the apparent age increasing with depth below the surface. The age of the formation was obtained by dating volcanic minerals such as feldspars, biotite, and amphibole. In two instances where the dates could be compared, the phillipsite was 21 Ma younger than associated volcanic minerals. Dymond asserted that this difference was too large and he attributed the apparent age discrepancies to differences in either the time of mineral formation or to differences in argon retention. Phillipsite could have lost argon while still in the formation or during the sample treatment process, resulting in younger apparent ages. In an attempt to resolve the questions raised by Dymond's results, Bernat et al. (1970) obtained K-Ar dates for approximately 50 additional Pacific Ocean phillipsites. However, they also included a large number of surface samples in addition to samples from different depths. The ages of surface samples ranged from 1.5 to 10 Ma, with most of the samples clustering around 3 Ma. These high apparent ages suggest that the phillipsites are either relict older minerals or that they have lost ^{40}K or gained ^{40}Ar. Bernat et al. (1970) ruled out the possibility of older minerals because it contradicts the notion of continuous sedimentation of the deep-sea floor. The high K content of these phillipsites negates any appreciable K loss, leaving Ar gain as the only plausible explanation. Bernat et al. (1970) argued that Ar gain was due to incorporation of impurities into the phillipsite during growth and not to initial radiogenic Ar. Their conclusion was based on the observation that the oldest ages were obtained on the more magnetic inclusion-rich fractions. Ages obtained from phillipsite sampled at different depths in a 600-cm core range from 1.9 to 3.6 Ma. The age variation with depth, however, was not a simple increase as observed by Dymond (1966). Instead, the apparent ages also varied with grain size. Ages for the 5-37 μm fraction decreased in the first 100 cm and remained relatively constant below that depth. For the larger grain fraction (>37 μm), ages decreased in the first 100 cm and then remained constant through the remainder of the core (Fig. 19). Bernat et al. (1970) suggested that the simplest explanation was that phillipsite continued to grow within the sediment after deposition. They concluded that K-Ar systematics of phillipsite are too complex to allow simple dating of these minerals.

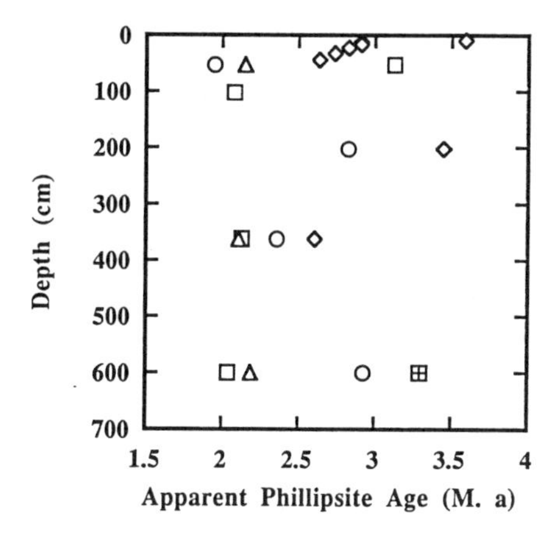

Figure 19. Variation in apparent K-Ar ages for marine phillipsite with depth in Pacific core LSDH 96. Symbols refer to different grain size fractions: Square with cross (<23 μm); diamond (5-37 μm); circle (23-30 μm); triangle (30-37 μm); blank square (>37 μm). Note the marked decrease in K-Ar ages for the 5-37 μm size fraction (diamonds) in the first 50 cm. Also note the considerable variation in apparent ages with grain size for the same depth. At shallow depths (<100 cm) the coarsest grain size has the oldest apparent age, whereas below that depth the finest grain size has the oldest apparent age. Experimental error in ages is less than ±0.2 Ma. Data from Bernat et al. (1970).

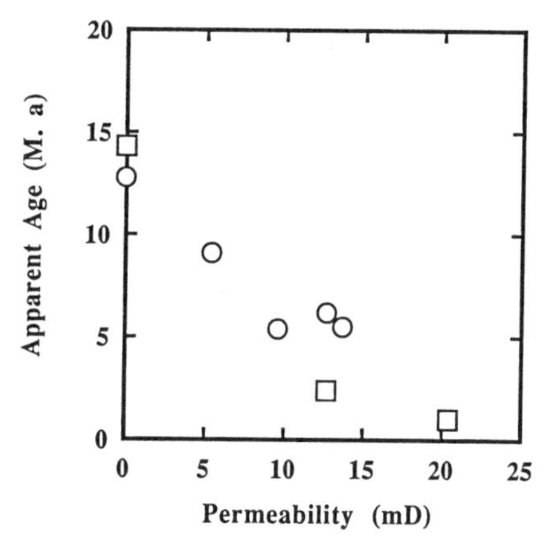

Figure 20. Relationship between apparent K-Ar ages of clinoptilolite and permeability of host rock. Symbols refer to different grain size fractions: Squares (10-75 μm); circles (100-160 μm). Samples are from the Emirier Tuff Member of the Bigadic Formation, northwestern Anatolia, Turkey. Adapted from Gundogdu et al. (1989).

Gundogdu et al. (1989) dated authigenic clinoptilolite from a lacustrine volcanogenic sedimentary sequence of Miocene age from northwestern Anatolia, Turkey. The K-Ar ages scatter between 1.0 to 14.3 Ma with no clear correlation to stratigraphic depth. In one instance, a biotite age of 18.4 Ma was obtained on a rock containing clinoptilolite with a 13.4 Ma age. Volcanic rocks in the stratigraphic unit below the zeolite deposits were dated at 22.3 Ma. Gundogdu et al. (1989) found an inverse correlation between the apparent clinoptilolite age and whole-rock permeability, the oldest samples occurring in the least permeable rocks and youngest samples in the most permeable rocks (Fig. 20). This finding suggests that the clinoptilolite K-Ar dates were reset in the more permeable rocks by interaction with interstitial fluids leading to argon loss. Argon loss is the most likely explanation because the K contents do not vary systematically with permeability. From the results of Gundogdu et al. (1989), it is clear that K-Ar data for clinoptilolite are highly sensitive to fluid-rock interaction and are thus easily reset. K-Ar ages for clinoptilolite may thus be useful only for rocks with low permeability or those that have experienced no significant water-rock interaction since their formation.

WoldeGabriel et al. (1992) presented K-Ar ages for 21 clinoptilolite-rich samples from five drill holes in Miocene tuffs at Yucca Mountain, Nevada. The samples dated were not pure but contained at least 75% clinoptilolite. Resulting K-Ar dates fell into three groups: 2-3 Ma, 4-5 Ma, and 7-11 Ma, with an overall increase in age with depth. The oldest ages were comparable to smectite/illite ages obtained for these formations by Bish and Aronson (1993). WoldeGabriel et al. (1992) suggested that the oldest dates represent true crystallization ages whereas the younger ages were reset due to water-rock interaction in a manner analogous to what Gundogdu et al. (1989) found in the Turkish clinoptilolites. In a follow-up study, WoldeGabriel (1995) obtained dates for 21 zeolites (20 clinoptilolites and one mordenite) from both surface outcrops and drill cores at Yucca Mountain. Dates were also obtained for coexisting illite/smectite (I/S) for comparison. The apparent K-Ar ages for clinoptilolite ranged from 2 to 13 Ma and generally increased with depth (Fig. 21). WoldeGabriel (1995) reasserted his earlier conclusions from 1992 and added that impurities were probably not a factor in affecting the K-Ar ages of the zeolites. He suggested that K-Ar dating of K-rich zeolites might be a useful tool in constraining the timing of diagenetic processes in low-temperature environments.

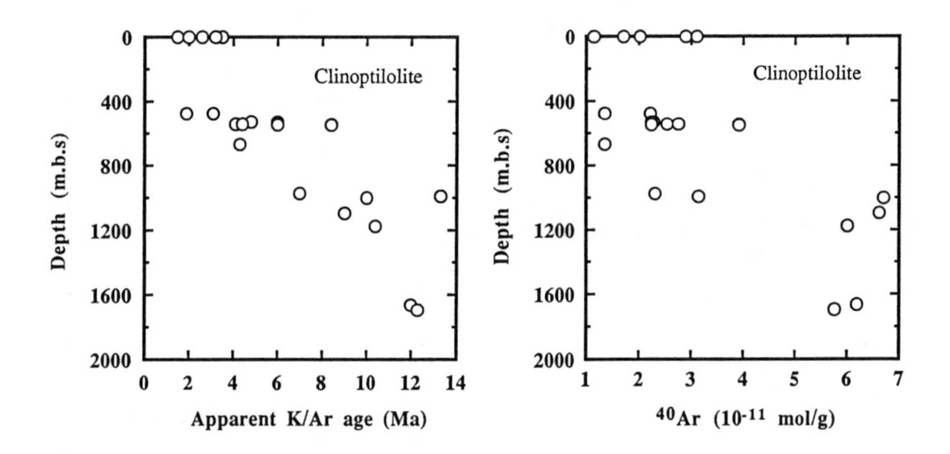

Figure 21 (left). Variation in apparent K-Ar ages for clinoptilolite with depth (m below surface) at Yucca Mountain. The age increases steadily with depth such that the youngest samples occur in surface outcrops (2 Ma) and the oldest (12 Ma) are found at 1.7 km depth. The 13 Ma "outlier" at 1 km may result from low K_2O contents rather than excess ^{40}Ar. Adapted from WoldeGabriel (1995).

Figure 22 (right). Variation in ^{40}Ar content for clinoptilolites with depth (m below surface) at Yucca Mountain. Note the apparent gap between $4{\times}10^{-11}$ and $6{\times}10^{-11}$ mol/g. Data from WoldeGabriel (1995).

Figures 22 and 23 display additional analysis of WoldeGabriel's results for clinoptilolite in order to clarify the relationship between ^{40}Ar and depth or K_2O. Although, ^{40}Ar increases with depth, there appears to be a "jump" in ^{40}Ar at roughly 1000 m (Fig. 22). Above that depth $^{40}Ar < 4 \times 10^{-11}$ mol/g, whereas at greater depths $^{40}Ar > 5.5 \times 10^{-11}$ mol/g. Why there should be such a discontinuity in ^{40}Ar is unclear. This difference between the high and low Ar samples cannot be attributed to differences in K_2O contents because they overlap (Fig. 23). Interestingly, the depth of the ^{40}Ar change coincides with the depth where Bish and Aronson (1993) suggested a boundary existed between an upper (cool) and

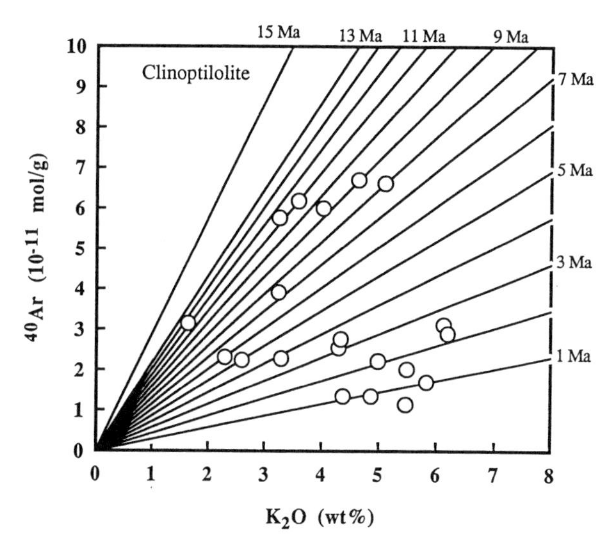

Figure 23. The relationship between ^{40}Ar and K_2O content of clinoptilolites from Yucca Mountain. Superimposed on this plot are K-Ar isochrons (straight lines). Ages in Ma are given for every other isochron. It is apparent from this graph that clinoptilolites from Yucca Mountain fall into two populations that are distinct in terms of ^{40}Ar content but have similar K_2O contents. The age spread within each population could result from a postformational change in K_2O. The graph illustrates that the 13 Ma old sample ("outlier" in Fig. 21) is also low in K_2O and thus its age is very sensitive to small changes in K_2O. Data from WoldeGabriel (1995).

lower (hot) convective systems 11 Ma ago.

In a recent study, WoldeGabriel et al. (1996) determined K-Ar ages for authigenic zeolites occurring in lacustrine deposits of the western US to further test the validity of the dating method. They collected clinoptilolite and phillipsite from surface outcrops in the altered tuffs of the Barstow Formation (Miocene), Spanish Canyon Formation (Miocene), and the Lake Tecopa Basin (Plio-Pleistocene and Miocene). Although, the ages of these formations are well constrained from tephrochronology and stratigraphic relationships, WoldeGabriel et al. (1996) also dated associated primary feldspar phenocrysts and "well-behaved" authigenic minerals such as K-feldspar and interstratified illite/smectite. The associated minerals allowed a direct comparison of the zeolite ages with the timing of eruptive and diagenetic events. Ages of authigenic K-feldspar separated from the Barstow Formation's Skyline Tuff range from 13.8 to 15.0 Ma, which is similar to the age of tuff itself (15 Ma). This similarity suggests that diagenesis began shortly after deposition of the tuff. Illite/smectite, however, yielded a somewhat lower age of 10.7±2.0 Ma. Three authigenic clinoptilolites ranged from 6.7 to 13.5 Ma. Because petrographic features suggest that the K-feldspar postdates the clinoptilolite WoldeGabriel et al. (1996) concluded that the young clinoptilolite ages are due to Ar loss. Moreover, they contended that the Ar loss was more severe in the two clinoptilolites (6.7 and 8.7 Ma) that were taken near the axis of the Barstow syncline due to enhanced permeability.

Plagioclase from the Spanish Canyon Formation had an Ar-Ar age of 18.9±0.8 Ma, whereas two clinoptilolite samples yielded K-Ar ages of 10.9±1 and 16.2±3 Ma. The plagioclase age is similar to known eruption ages (18.5 to 19.3 Ma) obtained from nearby tuffs. The older clinoptilolite age suggests that diagenesis occurred shortly after tuff deposition but the younger clinoptilolite age was thought by WoldeGabriel et al. (1996) to result from Ar loss caused by later diagenesis or dehydration.

Lake Tecopa primary sanidine from the basal unit of altered Miocene-age tuff yielded an Ar-Ar of 11.2±0.2 but authigenic clinoptilolite from the same unit had a K-Ar age of 15.2±5 Ma. The older-than-expected age for the clinoptilolite was thought to stem from Ar excess (WoldeGabriel et al. 1996). Authigenic K-feldspar from Tuff C and phillipsite from Tuff A yielded K-Ar ages of 0.5±1.3 Ma and 0.4±0.03 Ma, respectively. These ages are very similar to the age of their host tuffs.

WoldeGabriel et al. (1996) interpreted these results to indicate that diagenesis had occurred shortly after deposition of the volcanic tuffs. Moreover, they found it encouraging that at least in some instances the zeolite K-Ar dates were within the age range of better-established authigenic minerals. They dismissed "abnormal" ages as due to either Ar losses or gains.

RB-SR DATING AND SR ISOTOPIC ABUNDANCES

The Rb-Sr dating method is based on the β^- decay of ^{87}Rb to ^{87}Sr with a half-life of 48 Ga. A non-radiogenic stable Sr isotope, ^{86}Sr, is used when Rb-Sr isotopic data are normalized to yield isochrons. In addition to dating, the initial ^{87}Sr/^{86}Sr ratio itself is also of great significance because it may be used to trace the previous geochemical history of rocks and minerals. The initial ^{87}Sr/^{86}Sr ratio is set when the system closes (zero age) and its value reflects the Rb/Sr and the ^{87}Sr/^{86}Sr ratios of the source.

Rb is a common trace element in zeolites containing Na and K because it substitutes for those elements. Sr substitutes for Ca and should, therefore, occur in trace amounts in Ca-bearing zeolites. Some zeolites (e.g. brewsterite and gmelinite; see Chapter 1 by Armbruster and Gunter, this volume) even contain Sr as a major element. As a result, Rb-Sr ages for zeolites are probably useful only if the isochron method is applied because significant amounts of initial ^{87}Sr are likely to be present.

Rb-Sr dates for zeolites have been published only by Clauer (1982) who studied phillipsites from south Pacific sediments of middle Eocene age (40 Ma). The eight samples came from two different environments, the core of a poly-metallic nodule lying on the seafloor surface and from an indurated volcaniclastic sediment located between 12 and 20 cm beneath the nodule. In both occurrences the phillipsite was interstitial between smectite-coated volcanic glass grains and acted as a cement for the sediment. Following conventional practice, Clauer (1982) acid leached the phillipsites in order to remove unwanted surface contaminants. His leaching studies indicate that little radiogenic ^{87}Sr was selectively removed from the phillipsite residue. Moreover, ^{87}Sr/^{86}Sr ratio of the leachate varied inversely with its Sr content implying the presence of two Sr reservoirs of different ages. Clauer (1982) suggested modern seawater for the younger source and Tertiary age Sr for the older source because all the leachate ^{87}Sr/^{86}Sr ratios fell near or below that of modern seawater (^{87}Sr/^{86}Sr = 0.70910±0.00035). When plotted on a ^{87}Sr/^{86}Sr versus ^{87}Rb/^{86}Sr diagram, five of the phillipsite residues fell on a 14.7±3.3 Ma isochron with an initial ^{87}Sr/^{86}Sr ratio of 0.70441±0.00036. The remaining three residues lay above this isochron (Fig. 24). As Clauer (1982) pointed out, the initial Sr ratio is very similar to that obtained for oceanic basalts in the region. The older Sr source is therefore most likely basalt, and the phillipsite derived its Sr from the weathering of the basalts. The isochron age, which is approximately 25 Ma younger than the formation age, suggests either that phillipsite grew at a much later date or, more likely, remained open with respect to Rb and

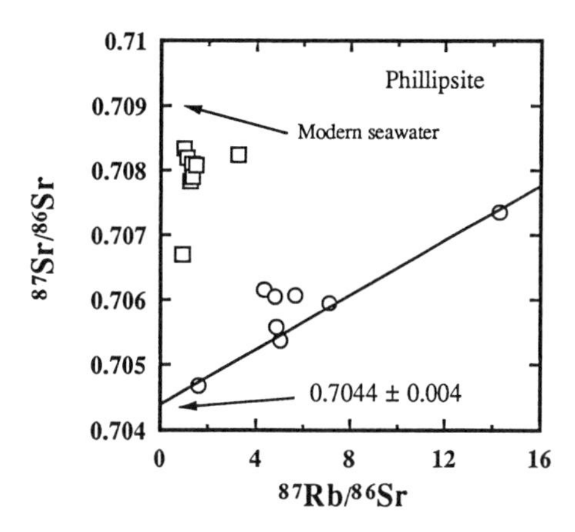

Figure 24. Rb-Sr isochron diagram for marine phillipsite. Data illustrate the effect of acid leaching on phillipsite isotopic values. The component that is removed from each bulk phillipsite (□) has a Sr isotopic composition similar to that of seawater. An isochron fit through the bulk-phillipsite data yields an unrealistic age (older than the formation itself) whereas most of the leached-phillipsite data (○) can be fit to an isochron with a 14.7 Ma age. This plot indicates that it may be possible to restore Rb-Sr ages for some contaminated samples. Modified from Clauer (1982).

Sr isotopes for a long time. Clauer (1982) speculated that the induration event led to isotopic closure by limiting fluid access to the formation.

Lambert et al. (1988) measured the Rb-Sr isotopic abundances of an analcime and a natrolite occurring in a porphyritic alkali basalt sill mid-Tertiary age from west Texas. Previous studies had revealed that igneous rocks of the province show a large range in initial Sr isotope ratios (0.7031-0.7120) and this was interpreted to reflect magmatic processes such as assimilation of more radiogenic crust during magmatic ascent (see e.g. Barker et al. 1977). Lambert et al. (1988) measured Rb and Sr isotopes of coexisting calcite, analcime, and natrolite from the same vug in the alkali basalt and obtained initial Sr isotope ratios of 0.70370±9, 0.70428±48, and 0.70655±41, respectively, for these minerals. The dissimilar initial Sr isotope ratios cannot result from variable Rb/Sr because this ratio is low for all three minerals. Either these minerals did not co-precipitate or the two zeolites remained open to exchange of their Sr subsequent to their formation. According to Lambert et al. (1988), who favored the second interpretation, the initial calcite Sr is consistent with that of silicate minerals - a likely source for the Sr. Moreover, Lambert et al. (1988) pointed out that the zeolites contain Sr that can be represented by a mixture of Sr from the surrounding silicate or magmatic Sr on the one hand and Sr obtained from the Cretaceous marine limestone intruded by the sill on the other. Acid leaching experiments performed by Lambert et al. (1988) on basalt and a phonolite show that leaching removes the secondary minerals along with their high Sr isotope ratios thereby lowering the initial Sr of the rock residue. Lambert et al. (1988) suggested that some of the initial Sr isotope variations observed by Barker et al. (1977) were probably due to fluid-rock interaction.

Strontium isotope ratios have also been used by Gundogdu et al. (1989) and Pearce (1993) to aid in deciphering zeolite genesis. Gundogdu et al. (1989) used $^{87}Sr/^{86}Sr$ ratios of clinoptilolite and associated minerals to determine the origin of Sr in the Turkish clinoptilolite described in the K-Ar section. The $^{87}Sr/^{86}Sr$ ratios of clinoptilolite were assumed to be initial values because the Rb contents of all of the minerals were low. Carbonates and smectites occurring stratigraphically above and below the phillipsite yielded $^{87}Sr/^{86}Sr$ ratios with averages of 0.70837±0.00003 and 0.70856±0.00002, respectively. Sr isotope values of four phillipsites were somewhat higher; ranging from 0.70861 to 0.70893. The limited range in Sr isotope values suggests that these minerals grew in a relatively closed environment, the Sr source being associated volcanic material. Higher $^{87}Sr/^{86}Sr$ values for the phillipsite, which crystallized later than the carbonates and the smectite, may stem from an increased supply of volcanogenic material.

Pearce (1993) argued that initial $^{87}Sr/^{86}Sr$ values for analcime phenocrysts from the Crowsnest Formation, Alberta, Canada, proved a primary igneous origin for these minerals because their Sr isotopic values were clearly mantle like. His assertion was based on the observation that the $^{87}Sr/^{86}Sr$ ratios for the two analcimes studied were 0.7034 and 0.7046 which falls within range of mantle (0.702-0.706) rather than crustal values (0.712-0.726). Karlsson and Clayton (1993), however, pointed out that there were significant $^{87}Sr/^{86}Sr$ differences between the analcime phenocrysts and their respective matrices. They suggested that the analcime was secondary and that carbonatite tuff layers occurring within the formation were the source of the Sr. The question of primary igneous analcime has already been discussed in the stable isotope part of this chapter.

Fission-track dating

Spontaneous fission of uranium atoms inside rocks and minerals leads to the ejection of particles and leaves structural damage or so-called fission tracks that can be viewed and quantified using a petrographic microscope. Samples are etched with a suitable solvent to make the tracks visible and the track density is determined. The number of tracks generated is a function of the overall U concentration and time. From the observed fission track density, total U concentration, and half-life of ^{238}U, the most common isotope of U, it is possible to calculate an age for the sample. The fission-track method is most commonly used for dating young geologic samples because the tracks fade at elevated temperatures. The rate at which the fission tracks fade depends on the material and temperature, but it can be quite rapid. For example in glasses fission tracks disappear in less than 1 million years at 100°C.

Only one study, Koul et al. (1981), has utilized the fission track method to date zeolites. These authors obtained ages for twelve hydrothermal zeolites from as many localities in the Faero Islands. The six chabazites, five stilbites, and one heulandite were collected from amygdules and a fracture filling in a Tertiary age flood basalt sequence that spans over 3 km stratigraphically. Dating of the basalt sequence indicates that the exposed lavas were all emplaced between 50 and 60 Ma ago, but the apparent zeolite ages increase progressively with depth from 40 Ma at the top of the pile to 55 Ma at the bottom. On the face of it, the age data imply formation of the zeolites millions of years after the lava pile was extruded. An alternative espoused by Koul et al. (1981) is that the fission-track dates represent the temperature at which the zeolites stopped annealing. In this scenario, the lavas erupted and zeolitization occurred shortly thereafter throughout the entire pile, but cooling was slowest in the top of the pile allowing greater annealing of the original fission tracks in the upper part of the lava sequence. Until the annealing behavior of fission tracks in zeolites is better understood the interpretation of fission track data will remain unclear.

Decay series isotopes of U and Th

The decay of naturally occurring ^{238}U, ^{235}U, and ^{232}Th isotopes generates radiogenic daughters of new isotopes and new elements that are themselves radioactive and decay to produce yet other radioactive daughters. Each of the three starting parent isotopes thus generates a decay chain containing a series of intermediate radioactive isotopes and elements (10 to 14 intermediate isotopes) and an end-product of the decay, a stable isotope of the element Pb. The chains are completely independent of each other in that no intermediate isotope appears in two chains and as such can be used in two fundamentally different ways to obtain age information on natural samples. This is a consequence of the fact that the half lives of ^{238}U, ^{235}U, and ^{232}Th (10^8-10^{10} a) are very much longer than those of any of the radiogenic daughters (seconds – 10^5 a). Thus in very old rocks (> several tens of Ma), it will appear as if the intermediate daughters do not exist and the U and Th isotopes are in effect decaying directly to stable Pb isotopes. The age can then be determined in the conventional manner. For very young samples (<10^6 a), it is also possible to take advantage of the short half lives of the intermediate radioactive daughters when there is radioactive disequilibrium. The disequilibrium comes about in two ways. First, in a new mineral or rock where no radiogenic daughter exists, 400-500 Ka are required for all of the radioactive daughters to reach equilibrium in the sense that just as many atoms decay as are formed. Second, even after radioactive equilibrium is established, a chemical or biological process can separate parents from daughters or change their relative abundances, thereby initiating a new disequilibrium. In either case, ratios of two isotopes in one of the chains can be used to calculate the timing of the disturbance. This technique has shown great promise for the dating of young sedimentary formations and determination of sedimentation rates (see e.g. Geyh and Schleicher 1990), but there are few studies in which the method was applied to zeolites.

Bernat and Goldberg (1969) reported thorium concentrations and isotopic abundances of Th, Ac, and Pa in phillipsite and detrital mineral samples taken from a 475-cm deep sedimentary core in the North Pacific. The purpose of their study was to investigate the mobility of Th isotopes in marine sediments in the hope that it might explain why ^{228}Th in seawater is higher than can be accounted for by simple radioactive equilibrium. ^{228}Th, a member of the ^{232}Th decay series, is supported by the β^- decay of ^{228}Ra ($t_{1/2}$ = 5.76 a) to ^{228}Ac which in turn quickly ($t_{1/2}$ = 6.15 h) decays to ^{228}Th. ^{228}Th itself ($t_{1/2}$ = 1.913 a) decays to ^{224}Ra through α emission and would thus quickly disappear if unsupported. Bernat and Goldberg (1969) measured Th concentrations and isotopic abundances in three different isolates of the core, namely phillipsite, a clay fraction, and acid leachates of total sediment. The phillipsite represents the authigenic mineral portion, the clay fraction represents the detrital component, and the leachate represents the phillipsite and Th removed from surface of detrital grains. Overall the sediment contains about 3 wt % phillipsite. Bernat and Goldberg (1969) observed a decrease in the ^{232}Th/^{230}Th ratio with depth in the top 20 cm of the core for all isolates with a leveling off below 20 cm. In the top 20 cm, ^{232}Th/^{230}Th of the three isolates were within experimental error of each other but below that depth, the ratio for phillipsite was significantly higher than for the clay or the leachate fractions (1 vs. 0.5). Furthermore, the ^{228}Th/^{232}Th ratio was relatively constant with depth but highest in the phillipsite (6 to 10 versus 1). Th concentrations remained constant with depth in the clay fraction and leachate (9 to 15 ppm) but decreased in the phillipsite in the top 100 cm from 13 ppm reaching a constant level of 3 ppm below that depth. These data suggest that thorium isotopes in the phillipsite are derived entirely from seawater and that the contribution from detrital minerals is insignificant. Thus the phillipsite ^{232}Th/^{230}Th ratio variation in the upper 20 cm could be used to calculate sedimentation rates. Bernat and Goldberg (1969) obtained a sedimentation rate of 0.7 mm per 1000 years for the upper 20 cm which is consistent with measurements using other minerals in the area. However, the decrease in the concentration of Th in phillipsite with depth suggests that the mineral has continued to grow at depth after it formed at the sediment-water interface. This assumes that the phillipsite removed most of the Th from the interstitial solution at the time of initial formation, and subsequent crystal growth therefore was more and more depleted in Th resulting in a overall lowering of the Th content of the phillipsite. Assuming a sedimentation rate of 0.7 mm/1000 years, the phillipsite at 20 cm would have formed 700 Ka ago and thus the growth would have continued for at least that many years. This idea of continued growth of phillipsite was explored further by Bernat et al. (1970) in three additional Pacific sediment cores from which they obtained isotopic and concentration data on Th and U along with K-Ar ages. Bernat et al. (1970) also obtained the abundance of phillipsite (wt %) as a function of depth. The data reveal a decrease in Th and U concentrations in phillipsite with depth whereas the mineral abundance increases. Such trends can, according to Bernat et al. (1970), only be explained by continued growth of phillipsite within the sediment column following formation. Bernat and Church (1978) further analyzed the Th isotope data of phillipsite in deep-sea sediment cores in light of trace element distributions. They concluded that phillipsite must form rapidly near the sediment surface and incorporate Th and rare-earth elements from both detrital and authigenic sources. Constancy concentration of soluble ions with depth and excesses in the $^{228}Th/^{232}Th$ ratios suggested that the phillipsites are in constant chemical communication with interstitial fluids.

APPLICABILITY OF RADIOGENIC ISOTOPES — DISCUSSION

There is little doubt that most analyses described above are of good quality and are thus reliable. The interpretation of these data is, however, often ambiguous and some of the key assumptions may be flawed. For example, an implicit assumption of K-Ar dating is that the K isotopes in zeolites occur in their natural abundances and are thus unfractionated (i.e. the $^{40}K/K$ ratio is constant). However, Taylor and Urey (1938) showed that zeolites retain ^{41}K in preference to ^{39}K, such that the $^{39}K/^{41}K$ ratio for zeolites may be as much as 10% lower than the natural abundance ratio (13.856). In addition the initial ^{40}K content could be higher than assumed because ^{40}K is preferentially taken up by the zeolite relative to ^{39}K. Calculations by the author that assume mass fractionation behavior for K isotopes yield a roughly 4.6% increase in the ^{40}K abundance which translates into a 4.6% overestimate in the age if the natural abundance of ^{40}K is used. Other problems, aside from contamination by other minerals and inclusions, include ion exchange, loss or addition of radiogenic daughters, and continued crystal growth. We consider each of these factors below.

Ion exchange can either increase or decrease the parent and/or daughter concentrations, thereby altering or completely resetting the apparent age. Rb and Sr are clearly mobile elements in analcime, natrolite, and phillipsite as shown by Clauer (1982) and Lambert et al. (1988). In the zeolite-bearing Tertiary lava pile of Eastern Iceland, evidence for mobility of these elements during hydrothermal alteration was so pervasive that Wood et al. (1976) seriously questioned the use of Sr isotopes in the basalts as tracers of igneous processes. WoldeGabriel and Levy (1996) studied experimentally the effect of ion exchange on the "original" Ar content of a natural clinoptilolite sample. The original K_2O content (5.28 wt %) of the clinoptilolite was altered by exchange with various salt solutions (1M $CaCl_2$, CsCl, KCl or NaCl) for three to five days at 50°C. Although, the exchange resulted in a wide range of K_2O contents (1.04-9.22 wt %) it had little effect on the original Ar content of clinoptilolite except in a single case involving a NaCl solution. After five days of exchange with that solution the Ar content of clinoptilolite had fallen by 50%. WoldeGabriel and Levy (1996) suggested that the apparent Ar loss might be related to the size of the substituting ion. They surmised that Na^+ (1.10 Å) was less effective in blocking the escape of Ar from clinoptilolite than large ions such as K^+ (1.46 Å) or Cs^+ (1.78 Å). However, ion size can not be the only factor because Ca^{2+} (1.08 Å) has effectively the same size as Na^+ and K-Ca ion exchange does not result in Ar loss. Moreover, charge balance requires that a single Ca^{2+} ion replaces two K^+ ions and thus the cation population is reduced which should "unblock" more Ar atoms. The effect of the cation on the zeolite framework is also important. The exchange of Ca^{2+} probably leads to distortion of the clinoptilolite framework thereby restricting the movement of Ar. Moreover, because Ca and Na ions occupy sites crystallographically distinct from K (Armbruster and Gunter, this volume), it is also plausible that complex effects on Ar mobility may arise. Although the Ar content of clinoptilolite may be insensitive to cation exchange, the age of the mineral will still be affected because the original ^{40}K content will either be over or underestimated.

Loss of radiogenic daughters such as Ar may or may not be a problem during sample processing in the laboratory but there is no question that it can be significant in a natural setting where the zeolite may exist for long periods of time and be exposed to multiple heating events. In the simplest scenario, the diffusive loss of Ar is related to crystal radius (X), the diffusion coefficient (D), and time (t) namely :

$$X^2 = 2\ Dt \tag{13}$$

The diffusion coefficient in turn is temperature dependent as follows:

$$D = D_0 \exp(-Q/RT) \tag{14}$$

where D_0 is the pre-exponent, Q the activation energy, R the gas constant, and T the temperature (K). Given the parameters D_0 and Q, it is possible to calculate D for any given temperature assuming that the diffusion mechanism does not change. Freer (1981) has compiled such parameters for a number of silicates including Ar in phillipsite and mordenite. Using Equation (14) and tabulated data in Freer (1981), the Ar diffusion coefficients (cm^2 /sec) at 25°C are 1.8×10^{-7} and 7.6×10^{-5} for the K and Rb end-members of phillipsite. According to Equation (13), a phillipsite grain of 100 μm diameter would "equilibrate" isotopically with atmospheric Ar in a matter of minutes at room temperature, a result consistent with Breck's (1974) contention that Ar implanted in phillipsite is easily removed. Thus all of this Ar should be lost during the initial steps of degassing experiments in the laboratory. These results appear at first glance to be at odds with measurements indicating that phillipsite retains Ar quite well and the diffusion coefficient is far lower. For example, Dymond (1966) derived a diffusion coefficient from his marine phillipsites of 10^{-22} cm^2/sec, 15-17 orders of magnitude lower than the values calculated above. Bernat et al. (1970) found experimentally that negligible Ar was lost from phillipsite when heated to 150°C for several hours. WoldeGabriel and Levy (1996) examined the Ar retention of natural clinoptilolite in the laboratory as function of temperature and water saturation conditions. They found that samples heated in air for 16 hours retained their original Ar content up to 150°C. Between 150°C and 200°C a significant amount of Ar was lost. Samples heated hydrothermally at 100°C retained their original Ar for five months. WoldeGabriel and Levy (1996) suggested that differences in water saturation conditions at Yucca Mountain might explain why younger apparent ages are observed for clinoptilolites at shallower depths. Those clinoptilolites are located above the current water table and thus under-saturated with respect to water. They suggested that these clinoptilolites had lost Ar.

One possible explanation for the variability observed in Ar retention is that phillipsite and clinoptilolite contain loosely bound and strongly bound Ar. The loosely bound Ar resides inside channels that are open and free of obstructions. It is removed during the initial vacuum pump down used to clean the systems for K-Ar analysis. The strongly bound Ar is, however, trapped behind obstructions blocking the channels (Vaughan et al. 1995; Newsam and Deem 1995) or derived from a foreign host in the zeolite such as inclusions. This second type of Ar comes off only during more intense degassing in the laboratory and may slowly leak out of phillipsite crystals in their natural settings. Although the long-term effects on Ar migration at low temperature are unknown, as noted by WoldeGabriel (1995), it seems reasonable to assume that Ar migration does occur.

One of the great paradoxes for K-Ar dating of sediment surface samples of marine phillipsite is high apparent ages. One plausible explanation is that the phillipsites contain excess ^{40}Ar inherited from another source rather than produced through decay of ^{40}K within the mineral. In a marine environment, likely sources include seawater, older phillipsites, and other minerals that have been dissolved. Bernat et al. (1970) discounted seawater because it would require more dissolved Ar than can be obtained from deep seawater. Because the labile Ar quickly diffuses out of phillipsite, it is possible for younger phillipsites growing above to trap the Ar before it escapes through the sediment-water interface. Argon released during alteration of magmatic silicate and glass could become incorporated by phillipsite in a similar manner. Bernat et al. (1970) suggested that the excess Ar comes from metallic inclusions that served as nucleation sites for the phillipsite crystals, but it is clear from their data that even phillipsites with few metallic inclusions have non-zero K-Ar ages.

A final problem area in the use of radioactive isotopes is the issue of continued crystal growth. This was mentioned above as the simplest explanation for Th and U isotope systematics and of K-Ar ages of phillipsite in marine cores. Bernat et al. (1970) developed a mathematical model that explained the data reasonably well using a combination of radioactive decay and continuous growth. So far, the problem of continued crystal growth has been addressed only for zeolites in marine sedimentary environments but it is likely that this process is significant in other settings.

It is clear from the discussion above that fluids must play a large role in influencing the isotopic ratios and abundances of radioactive and radiogenic elements in zeolites. Fluid-mineral interaction can affect ion exchange, dissolution, and reprecipitation of zeolites thereby resetting the "original" isotopic compositions. The extent to which these interactions affect a given sample depends on permeability and the availability of fluids over time. For example, Gundogdu et al. (1989) found an inverse linear relationship between clinoptilolite K-Ar age and whole-rock permeability (Fig. 20). The fact that no plateau was reached in K-Ar ages suggests that clinoptilolite was reset even at very low permeability (i.e. <1 millidarcy) and that, therefore, original K-Ar ages for zeolites such as clinoptilolite are preserved only when fluids are largely absent. Fluid-mineral interaction may also cause recrystallization of zeolites and obscure the original isotopic compositions. In order to obtain information on formation conditions it is therefore desirable to use samples that have not experienced recrystallization.

COMBINED APPLICATIONS OF STABLE AND RADIOACTIVE ISOTOPES

There are presently no published examples of studies where both stable and radiogenic isotope systems have been used together on zeolites. Therefore, I considered it useful to present an example illustrating the potential power of a combined approach. The data are from Yucca Mountain where both $\delta^{18}O_f$ and K-Ar results are available for clinoptilolite (WoldeGabriel 1995; Feng et al. 1999). As previously discussed, Feng et al. (1999) argued that clinoptilolites at Yucca Mountain did not retain their original $\delta^{18}O_f$ values but had reached equilibrium with the modern environment over the last 11 Ma. When the $\delta^{18}O_f$ data are viewed in light of the apparent K-Ar ages, however, a different picture emerges (Fig. 25). The "oldest" clinoptilolites have the lowest $\delta^{18}O_f$ whereas the "youngest" clinoptilolites have the highest $\delta^{18}O_f$ values. If the clinoptilolites acted as a closed system with respect to oxygen, potassium, and argon, these data would imply that clinoptilolites formed at different times under different environmental conditions. Because apparent K-Ar ages for clinoptilolites at Yucca Mountain increase with depth, that would imply that the alteration ceased early for the deeper samples but continued for the samples closer to or at the surface. This scenario is, however, inconsistent with mineralogical data suggesting that no significant hydrothermal alteration has occurred over the last 10.7 Ma at Yucca Mountain (Bish and Aronson 1993). An alternative explanation is that the clinoptilolites all formed initially at ~10 Ma but have since been reset to varying degrees both in terms of oxygen and K-Ar (open system). It seems unlikely that oxygen isotopes in the framework could be exchanged with an external fluid without simultaneously perturbing the K-Ar system. If this assumption is correct, then the data imply that the oldest clinoptilolites were relatively unaffected by alteration and thus preserve their original $\delta^{18}O_f$ values, whereas the

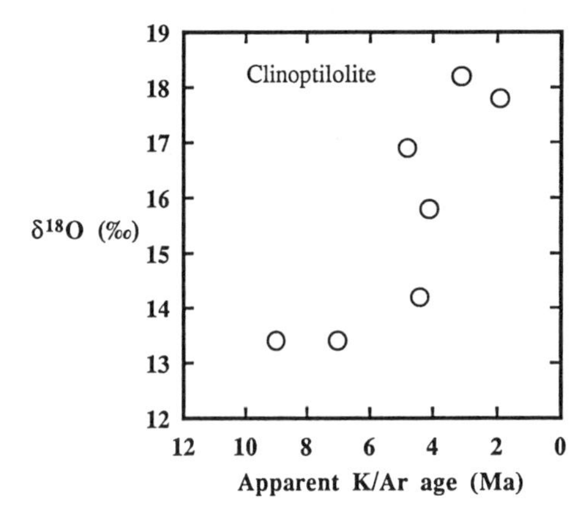

Figure 25. Relationship between framework-oxygen isotope compositions and apparent K-Ar ages for clinoptilolites at Yucca Mountain. If the clinoptilolites have remained a closed system (O, K, Ar), then the data suggest an abrupt rise in clinoptilolite $\delta^{18}O_f$ 4 Ma ago due to a change in environmental conditions. Alternatively, the data could indicate an open system, such that the oldest ages reflect the most pristine $\delta^{18}O_f$ values, whereas the youngest ages are produced for the most affected clinoptilolites. The graph was generated with data from WoldeGabriel (1995) and Feng et al. (1999).

"youngest" clinoptilolites were reset both in terms of $\delta^{18}O_f$ and K-Ar dates. The K-Ar dates would then be related to the age of disturbance at a given depth. From a reaction kinetics perspective one would expect that the deepest and, thus, hottest clinoptilolites would be the most affected. However, this is not borne out by the data—the opposite appears true. This observation is difficult to reconcile with the interpretation of Feng et al., (1999), who suggested that the clinoptilolites originally formed 11 Ma ago, under hotter conditions, in oxygen isotopic equilibrium with a paleofluid whose $\delta^{18}O$ was -9.5‰. Subsequently, the clinoptilolites re-equilibrated, under cooler conditions, with a fluid of lower $\delta^{18}O$ (-13.5‰). Therefore, either the estimates of paleoenvironmental conditions (fluid $\delta^{18}O$, geothermal gradient) upon which Feng et al. (1999) based their conclusions are imprecise and/or the theoretical oxygen isotope clinoptilolite-water fractionation equation is inaccurate.

CONCLUSIONS AND FUTURE DIRECTIONS

The primary question in all isotopic studies of zeolites is whether or not zeolites remain closed systems following their formation. The answer is probably negative in many if not most cases due to the open nature of the zeolite crystal structure, which allows rapid movement of channel constituents and ready access to the zeolite framework. Experimental evidence demonstrates that the more open zeolites exchange their framework oxygen easily, even at low temperatures (100°C), whereas more closed zeolites such as analcime and wairakite resist isotopic exchange to as high as 250°C. As the oxygen isotopic exchange of the framework oxygen is relayed through the channel water oxygen, it follows that the channel water itself and other channel constituents (major and trace elements) are also quite mobile. This mobility poses a major difficulty to dating zeolites accurately because radioactive and radiogenic isotopes reside in the zeolite channels. It is therefore apparent that "true" formation ages of zeolites may only be obtainable under special circumstances, namely where thermal events and postformational interaction with fluids are absent. Nevertheless, isotopic data obtained for zeolites can still yield useful information concerning postformational alteration histories (see e.g. Lambert et al. 1988; Feng and Savin 1993; WoldeGabriel 1995; WoldeGabriel et al. 1996; Feng et al. 1999).

Many outstanding questions remain that limit our understanding of isotopes and their behavior in zeolites. Partitioning studies of both stable and radioactive isotopes between zeolites and their surrounding media (liquid or air) should be expanded for species already studied (analcime, clinoptilolite, heulandite, stilbite, wairakite) and begun for new species.

There is also a need for improved analytical methods for stable isotopes that allow routine analysis of zeolites. Current methods are cumbersome and need to be simplified. The ^{40}Ar-^{39}Ar dating method should be attempted for K-bearing zeolites as it may shed light on ^{40}Ar loss and excess ^{40}Ar. Where possible, two or more dating methods should be used on zeolites and more studies are needed that combine stable and radioactive isotopes, and that tie zeolites in with well quantified minerals such as coexisting silicates and carbonates.

ACKNOWLEDGMENTS

I am grateful to David Bish for inviting me to contribute to this book. Earlier versions of the manuscript benefited from reviews and suggestions by John Beckett, Fred Longstaffe and David Bish. Dave skillfully and patiently edited the final chapter version. I am indebted to Dave for sticking with me through "thick-and-thin" during the writing of this chapter. The writing of this chapter stretched over two centuries. During the writing of this chapter my wife gave birth to twins. One of the twins subsequently developed a life-threatening heart problem at the age of six months, but is now cured at the age of almost four. I dedicate this chapter to my children, Eirik (3), Eva (3), and Karl (11), and my wife Abbie.

REFERENCES

Alberti A, Rinaldi R, Vezzalini G (1978) Dynamics of dehydration in stilbite-type structures; stellerite phase B. Phys Chem Minerals 2:365-375

Amundson R, Chadwick O, Kendall C, Wang Y, DeNiro M (1996) Isotopic evidence for shifts in atmospheric circulation patterns during the late Quaternary in mid-North America. Geology 24:23-26

Árnason B, Tómasson J (1970) Deuterium and chloride in geothermal studies in Iceland. Geothermics, Spec Issue 2. U N Symp Development and Utilization of Geothermal Resources, Pisa. Vol 2:1405-1415

Ballmoss R von (1981) The ^{18}O-exchange Method in Zeolite Chemistry: Salle-Sauerlander, 235 p

Ballmoss R. von, Meier VW (1982) Oxygen-18 exchange between ZSM-5 and water: J Phys Chem 86:2698-2700

Baretschi P (1976) Absolute ^{18}O content of Standard Mean Ocean Water. Earth Planet Sci Lett 31:341-344

Barker DS, Longe LE, Hoops GK, Hodges FN (1977) Petrology and Rb-Sr isotope geochemsitry of intrusions in the Diabolo plateau, northern Trans-Pecos magmatic province, Texas and New Mexico. Geol Soc Am Bull 88:1437-1466

Barrer RM (1978) Zeolites and Clay Minerals as Sorbents and Molecular Sieves. Academic Press, New York, 497 p

Bernat M, Goldberg ED (1969) Thorium isotopes in the marine environment. Earth Planet Sci Lett 5:308-312

Bernat M, Bieri RH, Kodie M, Griffin JJ, Goldberg ED (1970) Uranium, thorium, potassium and argon in marine phillipsites. Geochim Cosmochim Acta 34:1053-1071

Bernat M, Church TM (1978) Deep-sea phillipsites: trace geochemistry and modes of formation. in Natural Zeolites, Occurrence, Properties, Use. Sand LB, Mumpton FA (eds) Pergamon Press, Elmsford, New York, p 259-267

Bish DL, Aronson JL (1993) Paleogeothermal and paleohydrologic conditions in silicic tuff from Yucca Mountain, Nevada. Clays & Clay Minerals 41:148-161

Böhlke JK, Alt, J. C, and Muehlenbachs, K (1984) Oxygen isotope-water relations in altered deep-sea sediments: Low-temperature mineralogical controls: Can J Earth Sci 21:67-77

Boles JR, Wise WS (1978) Nature and origin of deep-sea clinoptilolite. in Natural Zeolites, Occurrence, Properties, Use. Sand LB, Mumpton FA (eds) Pergamon Press, Elmsford, New York, 235-243

Borthwick J, Harmon RS (1982) A note regarding ClF_3 as an alternative to BrF_5 for oxygen isotope analysis. Geochim Cosmochim Acta 46:1665-1668

Breck DW (1974) Zeolite Molecular Sieves: Structure, Chemistry, and Use. Wiley, New York, 771 p

Brumsack H-J, Zuleger E, Gohn E, Murray RW (1992) Stable isotope and radiogenic isotopes in pore waters from Leg 127, Japan Sea. Proc ODP, Sci Results 127/128:635-650

Chang Y-F, Somorjai GA, Heinemann H (1995) An $^{18}O_2$ temperature-programmed isotope exchange study of transition-metal-containing ZSM-5 zeolites used for oxydehydrogenation of ethane. J Catalysis 154:24-32

Clauer N (1982) Strontium isotopes of Tertiary phillipsites from the southern Pacific: Timing of the geochemical evolution. J Sed Pet 52:1003-1009

Clayton RN, Mayeda TK (1963) The use of bromine pentafluoride in the extraction of oxygen from oxides and silicates for isotopic analysis: Geochim Cosmochim Acta 27:43-52

Cole DR (2000) Exchange in mineral-fluid systems. IV. The crystal chemical controls on oxygen isotope exchange rates in carbonate-H_2O and layer silicate-H_2O systems. Geochim Cosmochim Acta 64: 921-931

Craig H (1961) Isotopic variations in meteoric waters: Science 133:1702-1703

Demeny A, Harangi S, Forizs I, Nagy G (1997) Primary and secondary features of analcimes formed in carbonate-zeolite ocelli of alkaline basalts (Mecsek Mts, Hungary): textures, chemical and oxygen isotope compositions. Geochem J 31:37-47

Donahoe RJ, Liou JG, Hemingway BS (1990). Thermochemical data for merlinoite: 2. Free energies of formation at 298.15 K of six synthetic samples having various Si/Al and Na/(Na+K) ratios and application to saline, alkaline lakes. Am Mineral 75:201-208

Dymond JR (1966) Potassium-argon geochronology of deep-sea sedimentary materials. PhD dissertation, Univ California, San Diego, California, 58 p

Faure G (1986) Principles of Isotope Geology. Wiley, New York, 589 p

Feng X, Savin SM (1991) Oxygen isotope studies of zeolites: Stilbite, analcime, heulandite and clinoptilolite—I. Analytical technique. *In* Stable Isotope Geochemistry: A Tribute to Samuel Epstein. HP Taylor, JR O'Neil, IR Kaplan (eds) Geochemical Society, Spec Publ 3:271-283

Feng X, Savin SM (1993a) Oxygen isotope studies of zeolites: Stilbite, analcime, heulandite, and clinoptilolite—II. Kinetics and mechanisms of isotopic exchange between zeolites and water vapor. Geochim Cosmochim Acta 57:4219-4238

Feng X, Savin SM (1993b) Oxygen isotope studies of zeolites: Stilbite, analcime, heulandite, and clinoptilolite—III. Oxygen isotope fractionation between stilbite and water or water vapor. Geochim Cosmochim Acta 57:4239-4247

Foley SF (1984) Liquid immiscibility and melt segregation in alkaline lamprophyres from Labrador. Lithos 17:127-137

Freer R (1981) Diffusion in silicate minerals and glasses: A data digest and guide to the literature. Contrib Mineral Petrol 76:440-454

General Electric Company (1989) Nuclides and Isotopes: General Electric Nuclear Energy Operations, San Jose, California, 57 p

Geyh MA, Schleicher H (1990) Absolute Age Determination: Physical and Chemical Dating Methods and Their Application. Springer-Verlag, New York, 503 p

Gundogdu MN, Bonnot-Courtois C, Clauer R. (1989) Isotopic and chemical signatures of sedimentary smectite and diagenetic clinoptilolite of a lacustrine Neogene basin near Bigadie, western Turkey. Applied Geochem 4:635-644

Hamza MS, Epstein S (1980) Oxygen isotopic fractionation between oxygen of different sites in hydroxyl-bearing silicate minerals. Geochim Cosmochim Acta 44:173-182

Hay RL (1978) Geologic occurrence of zeolites. *In* Natural Zeolites, Occurrence, Properties, Use. Sand LB, Mumpton FA (eds) Pergamon Press, Elmsford, New York, 135-145

Jakobsson SJ (1978) Environmental factors controlling the palagonitization of the Surtsey tephra, Iceland. Bull Geol Soc Denmark 27:91-105

Karlsson HR (1988) Oxygen and hydrogen isotope geochemistry of zeolites. PhD dissertation, Univ Chicago, Chicago, Illinois, 288 p

Karlsson HR (1995) Application of oxygen and hydrogen isotopes to zeolites: in Natural Zeolites '93. Ming DW, Mumpton FA (eds) Int'l Comm Nat Zeolites, Brockport, New York, p 125-140

Karlsson HR, Clayton RN (1990a) Oxygen and hydrogen isotope geochemistry of zeolites: Geochim Cosmochim Acta 54:1369-1386

Karlsson HR, Clayton RN (1990b) Oxygen isotope fractionation between analcime and water: An experimental study: Geochim Cosmochim Acta 12:133-149

Karlsson HR, Clayton RN N (1991) Analcime in igneous rocks: Primary or secondary? Am Mineral 76:189-199

Karlsson HR, Clayton RN (1993) Analcime phenocrysts in igneous rocks: Primary or secondary?—A reply. Am Mineral 78:230-232

Kita I, Honda S (1987) Oxygen isotopic difference between hydrothermally and diagenetically altered rocks from Tsugaru-Yunosawa area, Aomori, Japan. Geochem J 21:35-41

Kitchener GR, Bell BR, Fallick AE (1993) Influence of secondary zeolites on oxygen isotope patterns in a single basalt lava, Isle of Skye, Scotland. Zeolite '93. Fourth Int'l Conf. Occurrence, Properties, and Utilization of Natural Zeolites. Abstr with Prog, p 135-137

Kodama M, Nakajima T (1988) Exploration and exploitation of the Kirishima geothermal field. J Geotherm Res 25:201-230 (in Japanese with English abstract)

Koul SL, Chadderton LT, Brooks CK (1982) Fission track dating of zeolites. Nature 294:347-350

Lambert DD, Malek DJ, Dahl DA (1988) Rb-Sr and oxygen isotopic study of alkalic rocks from the Trans-Pecos magmatic province, Texas: Implications for the petrogenesis and hydrothermal alteration of continental alkalic rocks. Geochim Cosmochim Acta 52:2357-2367

Lee YI (1987) Isotopic aspects of thermal and burial diagenesis of sandstones at DSDP site 445. Daito Ridge, northwest Pacific Ocean: Chem Geology (Isotope Geosci Sect) 60:95-102

Luhr JF, Kyser TK (1989) Primary igneous analcime: The Colima minettes. Am Mineral 74:216-223

Matsubaya O, Sakai H, Kusachi I, Sakate H (1973) Hydrogen and oxygen isotopic ratios and major element chemistry of Japanese thermal water systems. Geochem J 7:123-151

Matsuhisa Y, Goldsmith JR, Clayton RN (1978) Mechanism of hydrothermal recrystallization of quartz at 250°C and 15 kbar. Geochim Cosmochim Acta 42:173-183

Nähr T (1997) Authigener Klinoptilolith in Mariner Sedimenten—Mineralchemie, Genese and Mögliche Anwendung als Geothermometer. Geomar Report 66. PhD dissertation, Christian Albrechts Univ, Kiel, 119 p (in German with English summary)

Nähr T, Botz R, Bohrmann G, Schmidt M (1998) Oxygen isotopic composition of low-temperature authigenic clinoptilolite. Earth Planet Sci Lett 160:369-381

Newsam JM, Deem MW (1995) Effect of faulting on sorption capacities of microporous solids. J Phys Chem 99:8379-8381

Noto M (1991) An experimental study of oxygen isotope fractionation between wairakite and water, and its application to wairakite from some Japanese geothermal systems. PhD dissertation, Okayama Univ, Okayama, Japan, 133 p

Noto M, Kusakabe M, Kometani M (1990) $^{18}O/^{16}O$ ratio determination of framework oxygen of apophyllite and wairakite by the preferential isotopic exchange of their water of crystallization: Chem Geol 80:231-241

Noto M, Kusakabe M (1997a) An experimental study of oxygen isotope fractionation between wairakite and water. Geochim Cosmochim Acta 61:2083-2093

Noto M, Kusakabe M (1997b) Oxygen isotope geochemistry of geothermal wairakite. Geochim Cosmochim Acta 61:2095-2104

O'Neil JR (1986) Theoretical and experimental aspects of isotopic fractionation. Rev Mineral16:1-40

O'Neil JR, Taylor HP Jr (1967) The oxygen isotope and cation exchange chemistry of feldspar. Am Mineral 52:1414-1437

O'Neil JR, Clayton RN, Mayeda TK (1969) Oxygen isotope fractionation in divalent carbonates. J Chem Physics 51:5547-5558

Pearce TH (1993) Primary or secondary analcime—Discussion. Am Mineral 78:225-229

Pemsler P (1953) Diffusion of heavy water into hydrated crystalline zeolites: The mobility of water in zeolites. PhD dissertation, New York Univ, New York, 115 p

Prestvik T, Torske T, Sundvoll B, Karlsson HR (1999) Petrology of early Tertiary nephelinites off mid-Norway. Additional evidence for an enriched endmember of the ancestral Iceland plume. Lithos 46:317-330

Rubin AE (1997a) Mineralogy of meteorite groups. Meteorit Planet Sci 32:231-247

Rubin AE (1997b) Mineralogy of meteorite groups: An update. Meteorit Planet Sci 32:733-734

Rykl D, Pechar F (1985) Thermal decomposition of the natural zeolite stilbite. Zeolites 5:389-392

Savin SM (1967) Oxygen and hydrogen isotope ratios in sedimentary rocks and minerals. PhD dissertation, California Inst Technology, Pasadena, California, 220 p

Savin SM, Lee M (1988) Isotopic studies of phyllosilicates. Rev Mineral 19:189-219

Sheppard SMF (1986) Characterization and isotopic variations in natural waters. Rev Mineral 16:165-183

Stakes DS, O'Neil JR (1982) Mineralogy and stable isotope geochemistry of hydrothermally altered rocks. Earth Planet Sci Lett 57:285-304

Stallard M, Boles JR (1989) Oxygen isotope measurements of albite-quartz-zeolite mineral assemblages, Hokonui Hills, Southland, New Zealand. Clays & Clay Minerals 37:409-418

Takenaka T, Furuya S (1991) Geochemical model of the Takigami geothermal system, northeast Kyushu, Japan. Geochem J 25:267-281

Taylor TI, Urey HC (1938) Fractionation of the lithium and potassium isotopes by chemical exchange with zeolites. J Chem Phys 6:429-438

Tse RS, Wong SC, Yuen CP (1980) Determination of deuterium/hydrogen ratios in natural waters by Fourier transform nuclear magnetic resonance spectrometry. Anal Chem 52:2445

Valley JW, Taylor HP Jr, O'Neil JR (eds) (1986) Stable Isotopes in High Temperature Geological Processes: Mineralogical Society of America Reviews in Mineralogy, Vol 16, Washington, DC, 570 p

Vaniman DT, Bish DL (1995) The importance of zeolites in the potential high-level radioactive waste repository at Yucca Mountain, Nevada. *In* Natural Zeolites '93: Occurrence, Properties, Use. Ming DW, Mumpton FA (eds) Int'l Comm Nat Zeolites, Brockport, New York, p 533-546

Vaughan DEW, Strohmaier KG, Treacy MMJ, Rice S.B, Leonowicz ME (1995) The influence of intergrowths on zeolite properties. *In* Natural Zeolites '93: Occurrence, Properties, Use. Ming DW, Mumpton FA (eds) Int'l Comm Nat Zeolites, Brockport, New York, p 187-198

Vrolijk P, Chambers SR, Gieskes JM, O'Neil JR (1990) Stable isotope ratios of interstitial fluids form the Northern Barbados accretionary prism. Proc ODP, Sci Results 110:155-178

Walker GPL (1960) Zeolite zones and dike distribution in relation to the structure of the basalts of Eastern Iceland. J Geol 68:515-528

WoldeGabriel G (1995) K/Ar dating of clinoptilolite, mordenite and coexisting illite/smectite from Yucca Mountain, Nevada. *In* Natural Zeolites '93: Occurrence, Properties, Use. Ming DW, Mumpton FA (eds) Int'l Comm Nat Zeolites, Brockport, New York, p 141-156

WoldeGabriel G, Broxton DE, Bish DL, Chipera SJ (1992) Preliminary assessment of clinoptilolite K/Ar results from Yucca Mountain, Nevada, USA: A potential high-level radioactive waste repository site. *In* Proc. 7th Int'l Symp Water-Rock Interaction, Park City, 1992, YK Kharaka, AS Maest (eds) Balkema, Rotterdam, p 457-461

WoldeGabriel G, Levy S (1996) Ion exchange and dehydration effects on potassium and argon contents of clinoptilolite. Mat Res Soc Symp Proc 412:791-798

WoldeGabriel G, Broxton DE, Myers FM (1996) Mineralogy and temporal relations of coexisting authigenic minerals in altered silicic tuffs and their utility as potential low-temperature dateable minerals. J Volc Geotherm Res 71:155-165

Wood DA, Gibson IF, Thompson RN (1976) Elemental mobility during zeolite facies metamorphism of the Tertiary basalts of Eastern Iceland. Contrib Mineral Petrol 55:241-254

Xu Z, Stebbins JF (1998a) Oxygen sites in the zeolite stilbite: a comparison of static, MAS, VAS, DAS and triple quantum MAS NMR techniques. Solid State Nucl Mag Res 11:243-251

Xu Z, Stebbins JF (1998b) Oxygen site exchange kinetics observed with solid state NMR in a natural zeolite. Geochim Cosmochim Acta 62:1803-1809

Yurtsever Y, Gat JR (1981) Atmospheric waters. *In* Stable Isotope Hydrology: Deuterium and Oxygen-18 in the Water Cycle. JR Gat, R Gonfiantini (eds) Int'l Atomic Energy Agency. Tech Rept Ser 210:103-142

5 Clinoptilolite-Heulandite Nomenclature

David L. Bish and Jeremy M. Boak

Los Alamos National Laboratory
Mail Stop D469
Los Alamos, New Mexico 87545

INTRODUCTION

Heulandite was first named in 1822 for the English mineral collector, J. H. Heuland (see Dana 1914, p. 574-576). Although it is typically found as macroscopic (commonly 0.2-2 cm in size) crystals in cavities in mafic igneous rocks and is volumetrically minor on the Earth's surface, heulandite also occurs in larger amounts in some sedimentary deposits, often in association with clinoptilolite. Due to the large size of typical heulandite crystals, its chemical composition and crystallographic and optical properties were easily characterized. Mineralogists recognized early that heulandite is monoclinic and that heating changes its optical properties in a predictable manner (Slawson 1925).

Unlike heulandite, the mineral known today as clinoptilolite typically occurs as microscopic crystals, commonly 2-20 μm in size and commonly intimately admixed with other fine-grained minerals. Although clinoptilolite is known today as the most common natural zeolite, occurring in large amounts (millions of tons) in altered volcanic tuffs and saline, alkaline-lake deposits, it is generally mentioned only in passing in beginning mineralogy texts. Clinoptilolite was not described as a distinct mineral species until 1932 (Schaller 1932). Prior to that time, the platy material described by Pirsson (1890) in amygdules in weathered basalt from the Hoodoo Mountains, Wyoming, had been classified as "mordenite," based primarily on its chemical similarity with mordenite from Nova Scotia. In spite of the chemical similarities, Pirsson did recognize the crystallographic similarity between the Wyoming "mordenite" and heulandite.

Dana's sixth edition (1914, p. 572-573) lists both ptilolite and mordenite as members of the mordenite group. Ptilolite (which we now recognize as mordenite) is listed by Dana as commonly occurring in aggregates of needles having parallel extinction, but "mordenite" is reported by Dana to be monoclinic, having a form closely approximating that of heulandite. Schaller (1932) proposed the name *clino*ptilolite based on the chemical similarity between clinoptilolite and mordenite (ptilolite) and on the fact that clinoptilolite had inclined extinction and was monoclinic. Obviously, the chemical similarity of clinoptilolite and mordenite resulted in their classification together into the mordenite group long before the crystal structures of either mineral had been solved.

In some ways, it appears that the distinction between clinoptilolite and heulandite arose historically because of the mistaken linkage of clinoptilolite with mordenite (ptilolite), leading to a divergence between the species clinoptilolite and heulandite before their zeolite frameworks were known to be identical. The very different crystal sizes and modes of occurrence for clinoptilolite and heulandite only served to perpetuate the distinction. Showing remarkable prescience, Hey and Bannister (1934) concluded shortly after Schaller proposed the name clinoptilolite, that clinoptilolite was simply a silica-rich heulandite, based on similarities in their X-ray diffraction patterns. Since then, considerable confusion has existed concerning differentiation between the two species. Although, Merkle and Slaughter (1968) and Alberti (1975) solved the crystal structures of heulandite and

1529-6466/00/0045-0005$05.00

clinoptilolite, respectively, more than 25 years ago, considerable effort continues to go into characterizing and understanding the differences between these two species. Although their framework structures are identical, it is clear that clinoptilolite and heulandite have different modes of occurrences, thermal behaviors, and compositions.

As described in a chapter by Armbruster and Gunter (this volume; see Fig. 16), both species share an identical Al-Si framework structure containing three sets of intersecting channels. Exchangeable cations and water molecules occupy these channels in variable amounts. As with other natural zeolites, the framework constituents are not exchangeable under normal laboratory conditions, so the Si/Al ratio and structural distribution are essentially fixed. In contrast, the exchangeable cations and their structural arrangement in the channels can be modified in nature and in the laboratory by a variety of means (e.g. Barrer 1978). Indeed, the channel cation composition may be largely exchanged (but generally not completely, particularly for Ca in heulandite) in the laboratory through cation-exchange experiments. Thus, nearly end-member alkali and alkaline-earth compositions of clinoptilolite or heulandite of given Si/Al ratio and distribution can often be created. For example, Gunter et al. (1994) exchanged Na and Pb, individually, onto a sample of heulandite from Poona, India, and performed single-crystal X-ray diffraction measurements on both natural and exchanged samples. They showed that the positions of both the extraframework-cations and the water molecules depend on the nature of the exchangeable cation. Note, however, that the exact structural arrangement in the channels also depends on the specific Si/Al framework distribution and the water content (which is dependent on the water vapor pressure surrounding the sample) (e.g. Armbruster and Gunter 1991; Gunter et al. 1994). Thus, the exchangeable cations and channel water molecules are interdependent. Changes in cation composition cause changes in the amount and structural distribution of water molecules; likewise, changes in channel water content affect the positions of the channel cations.

The chemistry of these two species varies almost continuously between about

$(Na,K,Ca_{0.5})_{10}(Al_{10}Si_{26}O_{72})\cdot 24H_2O$ (heulandite; Si:Al = 2.6) and

$(Na,K,Ca_{0.5})_{5.4}(Al_{5.4}Si_{30.6}O_{72})\cdot 20H_2O$ (clinoptilolite; Si:Al = 5.7);

Si/Al ratios outside this range are rare. Mumpton (1960) and others suggested that there is not complete solid solution between heulandite and clinoptilolite, but the abundance of electron microprobe analyses available today suggest compositional continuity between these two species. Indeed, Alietti (1972) and Boles (1972) showed that there is no gap in framework or extraframework cation compositions between clinoptilolite and heulandite. We have written the Na, K, and Ca contents in these formulae as continuously variable within the specified totals to reflect the fact that these cations can be partially to completely exchanged in the laboratory or in nature without affecting the Si/Al ratio (e.g. Araya and Dyer 1981).

Although Gottardi and Galli (1985) defined heulanditic minerals as containing predominant Ca and clinoptilolites containing dominant Na and K with little Ca, several investigators (e.g. Alietti et al. 1977; Broxton et al. 1987) have shown that silica-rich species of this group (i.e. clinoptilolite) can contain Ca > (Na + K). In addition, Mumpton (1960) prepared a Ca-exchanged clinoptilolite and predicted the existence of Ca-rich clinoptilolite. This situation emphasizes the terminology questions that arise when dealing with these two species. Should silica-rich species containing Ca as the predominant exchangeable cation be termed heulandite or clinoptilolite? Likewise, what shall we call Al-rich species containing significant Na and/or K?

The Structure Commission of the International Zeolite Association has answered these questions from a *structural* point of view by using the single framework code HEU for both heulandite and clinoptilolite. This framework topological classification, however, does not solve the terminology problem for geologists and mineralogists who are concerned with the complex compositional variations of these two species. Likewise, it does not satisfactorily address the observations that heulandites are distinct from clinoptilolites in many ways, including properties, mode of occurrence, and thermal and geologic stability. Thus, a variety of means have been proposed over the years to identify whether a particular sample is heulandite or clinoptilolite.

DISTINGUISHING BETWEEN CLINOPTILOLITE AND HEULANDITE

Numerous methods have been proposed for distinguishing empirically between the two species, but most of them are based on the exchangeable-cation composition (Mason and Sand 1960), thermal stability (Mumpton 1960), or framework chemistry (Boles 1972). Other methods or criteria that may be used for differentiation include optical properties (Boles 1972), thermogravimetric analysis or differential scanning calorimetry (Bish 1988), and proton nuclear magnetic resonance (NMR) spectroscopy (Ward and McKague 1994). Ward and McKague showed that the water in heulandite possesses long-range order, whereas that in clinoptilolite has no long-range order.

Commonly, the two species cannot be easily or accurately distinguished solely on the basis of X-ray powder diffraction data. Several authors (e.g. Boles 1972) have shown that the unit-cell parameters for heulandite are distinct from those for clinoptilolite, but the sensitivity of the unit-cell parameters to changes in water content (e.g. Bish 1984) makes accurate determination difficult. Although thermal stability and spectroscopic and optical properties may be used to distinguish between heulandite and clinoptilolite, they are all derivative properties of crystal structure and composition and are therefore not acceptable bases for definition of mineral species.

Both thermodynamic modeling and field observations strongly suggest that clinoptilolite and heulandite have significantly different stability fields and equilibria (i.e. they *do* behave like different minerals). For example, thermodynamic modeling by Chipera and Bish (1997a) showed that clinoptilolite is typically stable (or metastable) in equilibrium with higher-solubility silica polymorphs, such as cristobalite and opal-CT, whereas heulandite has both theoretical and observed stability fields in equilibrium with quartz. This particular difference in stability fields has important ramifications when attempting to interpret apparent "intermediate" compositions that may represent overgrowths or zoned crystals (see below).

Any discussion of differentiation between clinoptilolite and heulandite must be twofold. First, the nomenclatural distinction between heulandite and clinoptilolite must be based on structural and compositional trends between the two minerals (not, for example, on a derivative property such as thermal behavior). Inasmuch as the crystal structures of their Si-Al frameworks are essentially identical, this distinction must logically be based on chemical differences. Although several (fictitious) end-member compositions can be envisioned, such as a purely silicic clinoptilolite or a heulandite having all possible exchange sites filled, such materials have not been found in nature.

Several authors have defined heulandite and clinoptilolite based solely on their extraframework-cation composition, but this criterion has several shortcomings. The extraframework (exchangeable) cations are largely exchangeable, and if this criterion were adopted, one could transform a clinoptilolite to heulandite simply by performing a cation-exchange experiment in the laboratory. For example, Shepard and Starkey (1966) performed cation-exchange experiments on clinoptilolites and heulandites, resulting in materials with greatly modified thermal behaviors. As shown below, heulandite and clinoptilolite may have overlapping extraframework-cation compositions in nature.

The tetrahedral framework chemistry (primarily Al and Si) is an alternative criterion on which to base clinoptilolite-heulandite nomenclature. Numerous authors have, indeed, concluded that this is the preferred criterion for delineating heulandite-clinoptilolite minerals (e.g. Boles 1972). Most recent publications, however, have proposed using some combination of extraframework and framework compositions to delineate between heulandite and clinoptilolite. Mumpton (1960) defined clinoptilolite as a zeolite having a structure similar to heulandite, with an exchangeable-cation composition close to $(Na_{1.4}Ca_{0.1}K_{0.30})$ and an Si/Al ratio between 4.25 and 5.25. Although he concluded that thermal treatment was the best method for *differentiating* between clinoptilolite and heulandite in the laboratory, he did not define either mineral based on its thermal stability. Alietti (1967) distinguished clinoptilolite and heulandite based on their extraframework cation composition, but Alietti (1972) divided clinoptilolite and heulandite based on their Si/Al ratio and their divalent cation content. Like Mumpton, he made the experimental distinction between the two minerals based on their thermal stability. Alietti (1972) and others divided heulandites and clinoptilolites into three groups based on their thermal stabilities. Heulandites (type 1) are defined as those minerals that undergo an irreversible transformation to the so-called B phase during heating and whose structure is completely destroyed after heating to 450°C. Type-2 heulandites may or may not transform to a B phase, and they undergo only partial destruction at 450°C. Type-3 heulandites or clinoptilolites do not transform to a B phase and are stable commonly up to ~650°C.

Similarly, Valueva (1994) used the chemical and thermal stability data of Alietti et al. (1977) to relate the three distinct thermal behaviors of these minerals to their compositions. She described an empirical parameter, $\Gamma = Ca/Al + 0.115*Al$, which accurately distinguished between heulandites and clinoptilolites according to their thermal stabilities. In addition, she also showed that the dehydration enthalpies could be conveniently classified using this parameter. Samples with $\Gamma > 1.15$ belong to type 1, those with $\Gamma < 1.0$ belong to type 3, and type 2 varieties have a Γ range from 1.0-1.15.

All of these studies (Mumpton 1960; Alietti 1972; Valueva 1994) have one important aspect in common, namely they all *discriminated* first between heulandite and clinoptilolite based on their thermal properties and then *defined* compositional breaks based on these thermal properties. In the absence of some other measure such as thermal behavior, heulandite and clinoptilolite typically have not been distinguished *a priori* from each other simply on the basis of chemical composition. It is important to define clinoptilolite-heulandite based on chemical composition, although the mineral properties, such as thermal behavior, are obviously a measure or indication of chemical and/or structural variability.

Hawkins (1974) was one of the first to recognize the existence of several subgroups in heulandites and clinoptilolites based entirely on chemical grounds. Using cluster analysis of published chemical data, Hawkins identified at least five compositional subgroups within the heulandite family. Two larger subgroups corresponding to what were called clinoptilolite and heulandite were discriminated, and he concluded that calcium is the most important compositional variable distinguishing between the subgroups. Hawkins did not have access to the Ca-clinoptilolite analyses published recently.

More recently, Boak et al. (1991) examined chemical variations in clinoptilolite and heulandite and proposed the existence of two distinct compositional trends. They described a "clinoptilolite trend" with the formula $M_6Al_6Si_{30}O_{72}{\cdot}12H_2O$, where M = ($Na,K,Ca_{0.5}Mg_{0.5}$) and Si/(Al+$Fe^{3+}$) varied between 4 and ~5.8. Their distinct "heulandite trend" follows the formula $M_6(MAl,Si)_6Al_6Si_{24}O_{72}{\cdot}12H_2O$, with Si/(Al+$Fe^{3+}$) ranging from 2.7 to 5.0 and (Ca+Mg)/(Na+K) ≥ 1. They concluded that increasing Al substitution in heulandite lies along an exchange vector that connects silica to the other tectosilicates rather than along a plagioclase-type (NaSi ⇔ CaAl) substitution. Here we see the first indications that two distinct groups may be discerned for clinoptilolite and heulandite based solely on composition.

Nakamuta (1993) came to an essentially identical conclusion, suggesting that two types of solid solution exist with these two minerals. He proposed that so-called type-1 and type-2 heulandites form a solid solution with an exchange vector R^{2+} + 2 Al ⇔ 2 Si, whereas clinoptilolites (type-3 heulandites) form a different solid solution with an exchange vector 2 R^+ ⇔ R^{2+} in the region of low Al content.

Sufficient chemical analyses of clinoptilolite-heulandite minerals are available to formulate a useful criterion on which heulandite-clinoptilolite nomenclature can be based. The analyses of Boak et al. (1991) plus other analyses from the literature (Alietti 1972; Boles 1972; Stonecipher 1978; Broxton et al. 1987) are plotted in Figure 1. The thermal behavior for some of the minerals (Boles and Alietti analyses) and several compositional limits are also noted on the figure. This illustration also shows formulae limits based on crystal-chemical grounds and the approximate upper and lower limits of Si:Al in clinoptilolite. The left-hand limit is simply the Si ⇔ $Ca_{0.5}$Al vector, and the right-hand limit is a similar Si ⇔ NaAl (or Si ⇔ KAl) vector. The two vectors, or formula limits, meet at the origin (0,0; no substitution of Al for Si, and no extraframework cations).

Two distinct compositional trends are apparent in this figure: (1) a horizontal divalent-univalent exchange trend ("clinoptilolite trend") for samples having Si:Al ratios between 4.0 and 5.7; and (2) a tetrahedral (Al + Fe^{3+}) substitution ("heulandite trend") along an Si ⇔ $Ca_{0.5}$Al or Si ⇔ NaAl exchange vector (divalent charge:(total charge) ratios of 0.7 to 0.9). As Boak et al. (1991) concluded, the "clinoptilolite trend" conforms to the formula $M_xAl_xSi_{36-x}O_{72}{\cdot}12H_2O$, where M = (Na, K, $Ca_{0.5}Mg_{0.5}$) and Al+Fe^{3+} ranges from 5.37 to 7.2 [Si/(Al+Fe^{3+}) of 4.0 to 5.7]. The "clinoptilolite trend" can be described primarily by a $Ca_{0.5}$ ⇔ Na, K exchange vector, *with no Si or Al component necessary*. The "clinoptilolite trend" is consistent with the occurrence of cation-exchange reactions, i.e. cation exchange in the natural system would simply move a clinoptilolite within this trend (and would not modify a clinoptilolite to heulandite).

This figure supports assignment of clinoptilolite to a region on the figure with an Si:(Al+Fe^{3+}) ratio greater than four. Indeed, Boles (1972) defined the compositional break at an Si:Al ratio of 4.0, and other authors (e.g. Alietti et al. 1977) placed the break at similar Si:Al ratios, generally about 3.7-3.8. These breaks were based primarily on thermal behavior, that is, samples exhibiting a transition to a collapsed "B" phase were defined as heulandite. However, the two trends are obvious on Figure 1 irrespective of thermal behavior.

There are several noteworthy aspects to this figure. First, there is a notable overlap region in the lower left-hand part of the figure. It is not clear, based solely on composition, whether a sample in this region is clinoptilolite or heulandite. Second, the comparison with the limited thermal data shown on the figure is striking. The figure correctly classifies most samples having either clinoptilolite or heulandite thermal behavior. However, the samples having intermediate thermal behavior typically occur either in the lower overlap region or even outside the above-noted compositional limits. Four of the ten intermediate samples have balance errors (Gottardi and Galli 1985) greater than 10%, suggesting that the

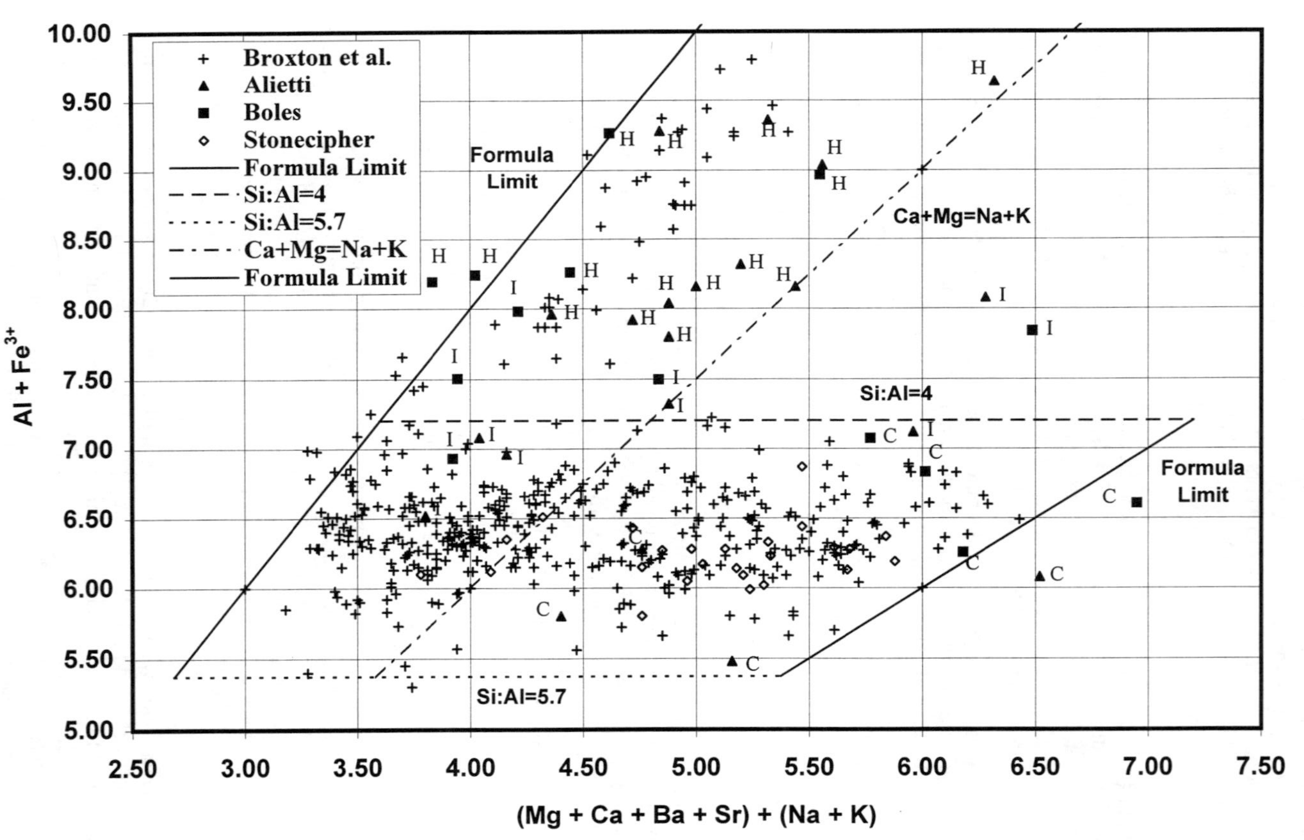

Figure 1. Plot of ($Al+Fe^{3+}$) vs. (sum of extraframework cations) from literature analyses of clinoptilolite and heulandite (Alietti 1972; Boles 1972; Stonecipher 1978; Broxton et al. 1987) [cations based on a 72-oxygen formula]. The thermal behavior of some samples (C = clinoptilolite; H = heulandite; I = intermediate) is plotted on the figure, as are the formula limits required by charge balance, the Si:Al = 4 and Si:Al = 5.7 horizontal lines, and the line depicting compositions where the number of divalent and univalent extraframework cations is equal. Compositions to the right of the latter line must have univalent cations > divalent cations. The left-hand "Formula Limit" is the Si ⇔ $Ca_{0.5}Al$ vector and the right-hand "Formula Limit" is an Si ⇔ NaAl or Si ⇔ KAl vector.

distribution of intermediate thermal-behavior samples on Figure 1 may result partly from unreliable chemical analyses. However, this distribution appears to be significant and suggests that it is worthwhile to examine intermediate samples for evidence of zonation and/or mixtures. Intermediate samples may represent zoned crystals or clinoptilolites that have a heulandite overgrowth due to prograde diagenetic reactions. Alternatively, retrograde reactions, perhaps from the thermal decay of an aureole around rhyolitic ash-flow tuffs, could conceivably produce overgrowths of clinoptilolite on heulandite, although, to the writers' knowledge, this has not been reported. Of course, such intermediate samples could also represent material with a composition intermediate between heulandite and clinoptilolite.

Note that the left-hand formula limit in Figure 1 is based on the presence of only divalent and univalent extraframework cations and assumes that all Fe is present as Fe^{+3} in the framework. The presence of trivalent *extraframework* cations would move the left-hand formula limit to the left. Such a substitution may explain the existence of samples to the left of this limit. Analyses to the right of the right-hand formula limit require a different explanation. Although never identified to the writers' knowledge, an anionic substitution, such as found in sodalite-group minerals (e.g. $Cl^-$and SO_4^{2-}), would potentially explain such analyses.

The Zeolite Subcommittee of the Commission on New Minerals and Mineral Names of the International Mineralogical Association (IMA) has proposed several rules governing zeolite nomenclature (Coombs et al. 1997). In particular, their rule 3 states that "Zeolite mineral species shall not be distinguished solely on the basis of the framework Si:Al ratio. An exception is made in the case of heulandite and clinoptilolite; *heulandite is defined as the zeolite mineral series having the distinctive framework topology of heulandite and the ratio Si : Al <4.0. Clinoptilolite is defined as the series with the same framework topology and Si : Al ≥ 4.0.*" They based their argument for this compositional criterion on the entrenched usage of the two names and convenience of these names for recognizing an important chemical feature. The committee recognized that the usual 50% rule in mineral nomenclature could not be applied because there are no clearly defined Si-Al end-member compositions for heulandite and clinoptilolite. Although the committee stated that the cutting line was arbitrary, it was coincidentally defined in a manner entirely consistent with the data presented in this chapter. The committee also emphasized that thermal stability is a derivative property and, although useful as an aid to identification, it is not appropriate as the basis for definition.

CONCLUSIONS AND RECOMMENDATIONS FOR FUTURE WORK

Both the history and definition of the clinoptilolite-heulandite series are complex. As with other mineral groups, multiple ambiguous and compositionally overlapping mineral names were created over the years, ultimately driving the creation of the IMA conventions governing mineral terminology today. Using the chemical analyses now available, two distinct chemical trends can be defined for heulandite and clinoptilolite. Although the accepted IMA definition of clinoptilolite and heulandite does not (and should not) consider their compositional variations within a petrologic setting, minerals whose composition is silica rich (Si/Al > 4.0) and whose variations within a petrologic setting are mainly in the extraframework cation content are clinoptilolites. Minerals whose composition is aluminous and calcic, and whose variations within a petrologic setting are mainly in the framework cations (Si and Al) are heulandites. It would be interesting to determine why clinoptilolites have not been found with Si:Al ratios greater than ~5.7 or why heulandites with Si:Al ratios <2.6 do not occur. The lack of more aluminous heulandites may be a result of gradual filling of all available extraframework sites. Alternatively, the absence of more aluminous heulandites or more siliceous clinoptilolites may simply reflect the encroachment by other, more stable minerals, outside the composition space outlined in Figure 1.

Compositional variations in the "clinoptilolite trend" can be described primarily by a $Ca_{0.5} \Leftrightarrow$ Na, K exchange vector, with no Si or Al component necessary. The "heulandite trend" follows the formula $M_6(MAl,Si)_6Al_6Si_{24}O_{72}\cdot 12H_2O$, with Si/(Al+$Fe^{3+}$) ranging from 2.7 to 5.0 and (Ca+Mg)/(Na+K) $\geq$ 1. Increasing Al substitution in heulandite lies along an exchange vector that connects silica to the other tectosilicates (Si $\Leftrightarrow$ $Ca_{0.5}$Al or Si $\Leftrightarrow$ NaAl). Figure 1 emphasizes these distinct trends for the two minerals, and as with other minerals (e.g. feldspars), the two trends are not mutually exclusive.

Naturally the IMA definition of clinoptilolite and heulandite does not allow a region of compositional overlap. However, there is an intriguing compositional region characterized by an Si:Al ratio of approximately 4 ± 0.5 (Al + Fe^{3+} of 6.5 to 8 on Fig. 1), with divalent > univalent extraframework cations. This region may represent analyses of mixtures or of zoned or overgrown materials, and future work should be done to resolve the compositional, structural, and physical-property details in this part of composition space. The presence of mixtures of distinct mineral phases (e.g. overgrowths) or zoning might indicate a more complete separation between the two series. The IMA definitions are quite consistent with observed thermal behaviors, although they are in no way based on thermal behavior. It is also noteworthy that specimens with intermediate thermal behavior lie in the region noted above, at the junction between the two mineral series.

The need for additional research is also suggested by the presence of analyses outside rational formula limits. For analyses to the left of the Si $\Leftrightarrow$ $Ca_{0.5}$Al vector, two possible explanations are possible. The left-hand limit in Figure 1 is based on the presence of only divalent and univalent extraframework cations, and the presence of trivalent *extraframework* cations such as Fe^{3+} would move the left-hand limit to the left. Some studies have attributed Fe^{3+} to impurities, whereas others have placed it in either extraframework or framework sites. Clearly it is worthwhile to study Fe-containing clinoptilolites in more detail to resolve this question.

Alternatively, the analyses outside this formula limit could result from the presence of other, unanalyzed extraframework cations. Such cations, if measured, would shift compositions to the right toward the formula limit. One compositional feature of the clinoptilolite/heulandite series not discussed in this paper is the tendency for available analyses to show charge balance errors resulting from a deficiency of extraframework cations in comparison with aluminum. The deficiency amounts to as much as 20% in otherwise apparently reasonable compositions. For the large database of analyses in Broxton et al. (1987), the compositions are clearly biased toward a positive balance error. One explanation for the deficiency is loss of sodium under the electron microprobe beam. Broxton et al. (1987) addressed this concern in their study, however, using a defocused beam and showing progressive increase in Na content with increasing beam diameter, up to an asymptotic limit. For their data, charge balance is not correlated to sodium content, providing further confirmation that charge balance problems are not due to this analytical difficulty.

Another possible explanation for the deficiency is that hydronium ions occupy some of the extraframework cation sites, but demonstration of this relationship may be difficult. The calcic clinoptilolites have the lowest site occupancy in the extraframework-cation sites (e.g. lower left-hand portion of Fig. 1). There is no correlation evident between their site occupancy and balance error. Similarly, the deficiency may be due to the presence of NH_4^+

or Li^+ in extraframework sites. Similar substitutions have been found in feldspars and illite/smectites. Future work should address the issue of charge balance and the possible existence of other extraframework cations not normally included in chemical analyses, such as hydronium, NH_4^+, Cs^+, and Li^+. It is noteworthy that all of the analyses to the left of the left-hand formula limit have positive balance errors, consistent with this explanation.

Similarly, further research is warranted to determine whether an anionic substitution, such as found in sodalite-group minerals (e.g. $Cl^-$and SO_4^{2-}), may be present in the few samples whose analyses fall to the right of the right-hand formula limit. The presence of any anionic species would require additional extraframework cations not balanced by Al + Fe^{3+}, giving rise to a negative balance error (assuming the anionic species were not measured). Interestingly, both analyses in Figure 1 falling outside the right-hand formula limit have large negative balance errors.

An additional area of potential future work is to search zeolitized alkalic mafic rocks for alkali-rich heulandites, which have been prepared in the laboratory but are not yet known in nature (e.g. Gunter et al. 1994). Heulandites commonly reflect lower $a(SiO_2)$ than clinoptilolites. Other phases, including zeolites, may thermodynamically or kinetically preclude the formation of sodic or potassic heulandites in natural environments. For example, based on thermodynamic modeling, Chipera and Bish (1997b) concluded that Mg in heulandite stabilizes it with respect to stellerite, thereby expanding the distribution of heulandite and restricting stellerite in many zeolitic rocks. At present, the absence of aluminous materials with a predominance of univalent over divalent extraframework cations is striking, and Figure 1 shows only two analyses (both with "I" behavior) to the right of the "Ca + Mg = Na + K" line for aluminous compositions.

REFERENCES

Alberti A (1975) The crystal structures of two clinoptilolites. Tschermaks mineral petrogr Mitt 22:25-37

Alietti A (1967) Heulanditi e clinoptiloliti. Mineral Petrogr Acta 13:119-138

Alietti A (1972) Polymorphism and crystal-chemistry of heulandites and clinoptilolites. Am Mineral 57:1448-1462

Alietti A, Brigatti MF, Poppi L (1977) Natural Ca-rich clinoptilolite (heulandites of group 3): New data and review. Neues Jahrb Mineral Monatsh 493-501

Araya A, Dyer A (1981) Studies on natural clinoptilolite-I. J Inorganic Nuclear Chem 43:589-594

Armbruster T, Gunter ME (1991) Stepwise dehydration of heulandite-clinoptilolite from Succor Creek, Oregon, U.S.A.: A single-crystal X-ray study at 100 K. Am Mineral 76:1872-1883

Barrer RM (1978) Cation-exchange equilibria in zeolites and feldspathoids. *In* Natural Zeolites: Occurrence, Properties, Use. LB Sand, FA Mumpton (eds) Pergamon Press, Oxford, p 385-395

Bish DL (1984) Effects of exchangeable cation composition on the thermal expansion/contraction of clinoptilolite. Clays & Clay Minerals 32:444-452

Bish DL (1988) Effects of composition on the dehydration behavior of clinoptilolite and heulandite. *In* Occurrence, Properties and Utilization of Natural Zeolites. D Kallo, HS Sherry (eds) Akademiai Kiado, Budapest, p 565-576

Boak JM, Cloke P, Broxton D (1991) Mineral chemistry of clinoptilolite and heulandite in diagenetically altered tuffs from Yucca Mountain, Nye County, Nevada. Geol Soc Am Program with Abstracts 23:A186

Boles JR (1972) Composition, optical properties, cell dimensions, and thermal stability of some heulandite group zeolites. Am Mineral 57:1463-1493

Broxton DE, Bish DL, Warren RG (1987) Distribution and chemistry of diagenetic minerals at Yucca Mountain, Nye County, Nevada. Clays & Clay Minerals 35:89-110

Chipera SJ, Bish DL (1997a) Equilibrium modeling of clinoptilolite-analcime equilibria at Yucca Mountain, Nevada. Clays & Clay Minerals 45:226-239

Chipera SJ, Bish DL (1999) Thermodynamic modeling of natural zeolite stability. *In* Clays for Our Future. Kodama H, Mermut A, Torrance JK (eds) Proc 11th Int'l Clay Conf, ICC Organizing Comm, Ottawa, p 596-602

Coombs DS, Alberti A, Armbruster T. Artioli G, Colella C, Galli E, Grice JD, Liebau F, Mandarino JA, Minato H, Nickel EH, Passaglia E, Peacor DR, Quartieri S, Rinaldi R, Ross M, Sheppard RA, Tillmanns E, Vezzalini G (1997) Recommended nomenclature for zeolite minerals: Report of the Subcommittee on Zeolites of the International Mineralogical Association, Commission on New Minerals and Mineral Names. Can Mineral 35:1571-1606

Dana ES (1914) The System of Mineralogy of James Dwight Dana. 6th Edition. John Wiley & Sons, London

Gottardi G, Galli E (1985) Natural Zeolites. Springer-Verlag, Berlin, 409 p

Gunter ME, Armbruster T, Kohler T, Knowles CR (1994) Crystal structure and optical properties of Na- and Pb-exchanged heulandite-group zeolites. Am Mineral 79:675-682

Hawkins DB (1974) Statistical analyses of the zeolites clinoptilolite and heulandite. Contrib Mineral Petrol 45:27-36

Hey MH, Bannister FA (1934) Studies on the zeolites. Part VIII. "Clinoptilolite," a silica-rich variety of heulandite. Mineral Mag 23:556-559

Mason B, Sand LB (1960) Clinoptilolite from Patagonia: The relationship between clinoptilolite and heulandite. Am Mineral 45:341-350

Merkle AB, Slaughter M (1968) Determination and refinement of the structure of heulandite. Am Mineral 53:1120-1138

Mumpton FA (1960) Clinoptilolite redefined. Am Mineral 45:351-369

Nakamuta Y (1993) Crystal chemistry of diagenetic heulandite-group zeolites from Tertiary sedimentary rocks. *In* Zeolite '93: 4th Int'l Conf: Occurrence, Properties, and Utilization of Natural Zeolites, Program and Abstracts, Boise, Idaho, June 20-28, 1993, p 146-148

Pirsson LV (1890) On mordenite. Am J Sci 40:232-237

Schaller WT (1932) The mordenite-ptilolite group: Clinoptilolite, a new species. Am Mineral 17:128-134

Shepard AO, Starkey HC (1966) The effects of exchanged cations on the thermal behavior of heulandite and clinoptilolite. Mineral Soc India, IMA Vol, p 155-158

Slawson CB (1925) The thermo-optical properties of heulandite. Am Mineral 10:305-331

Stonecipher SA (1978) Chemistry of deep-sea phillipsite, clinoptilolite, and host sediments. *In* Natural Zeolites: Occurrence, Properties, Use. LB Sand, FA Mumpton (eds) Pergamon Press, Elmsford, New York, p 221-234

Valueva GP (1994) Chemical parameters for classification of heulandites. Russian Geol Geophys 35:1-4

Ward RL, McKague HL (1994 Clinoptilolite and heulandite structural differences as revealed by multinuclear nuclear magnetic resonance spectroscopy. J Phys Chem 98:1232-1237

6

Occurrence of Zeolites in Sedimentary Rocks: An Overview

Richard L. Hay

Department of Geosciences
University of Arizona
Tucson, Arizona 85721

Richard A. Sheppard

11647 West 37th Place
Wheat Ridge, Colorado 80033

INTRODUCTION

Zeolites have been known since the mid-1750s, but prior to the early 1950s, most reported occurrences of zeolites were in fracture fillings and amygdules in igneous rocks, particularly basaltic lava flows. Indeed, most of the large attractive zeolite specimens in museum collections were obtained from lavas. In recent years, zeolites have been recognized as important rock-forming constituents in low-grade metamorphic rocks and in a variety of sedimentary rocks. Most zeolites in sedimentary rocks are finely crystalline, that is they occur as microscopic or submicroscopic crystals, and they are therefore of little appeal to mineral collectors; however, deposits of this type are voluminous and have great geologic significance and economic potential.

Zeolites are among the most common authigenic silicate minerals that occur in sedimentary rocks, and they form in sedimentary rocks of diverse lithology, age, and depositional environment. About twenty species of zeolites have been reported from sedimentary rocks, but only eight zeolites commonly make up the major part of zeolitic rocks. These are analcime, chabazite, clinoptilolite, erionite, heulandite, laumontite, mordenite, and phillipsite.

This chapter will consider chiefly the zeolites in sedimentary rocks, with emphasis on volcaniclastic deposits, which contain the largest concentrations of zeolites. The occurrence of zeolites in lava flows is mentioned only briefly. Journal articles on the occurrence and origin of natural zeolites have multiplied at a rapid rate since the *Mineralogy and Geology of Natural Zeolites* was first published in 1977 as Volume 4 of the Mineralogical Society of America's *Reviews in Mineralogy*. The present review will highlight areas of more recent research on the occurrence and origin of zeolites and some of the coexisting minerals.

ZEOLITE MINERALOGY

A wide variety of zeolites has been identified in sedimentary deposits, with the most common being clinoptilolite, analcime, heulandite, laumontite, and phillipsite. Less abundant zeolites include chabazite, erionite, mordenite, natrolite (and gonnardite), and wairakite. Merlinoite is a K-rich low-silica zeolite rather similar in structure to phillipsite, and the two have almost identical X-ray diffraction patterns. Only recently discovered (Passaglia et al. 1977), its distribution and host rock are now known to vary widely, and its compositional limits probably will be expanded by further research. More complete listings of zeolite mineralogy are given in this volume in a chapter by Armbruster and Gunter and by Iijima (1988a), Gottardi and Galli (1985), and Tschernich (1992). The highest concentrations of zeolites are generally found in glass-rich volcaniclastic deposits.

1529-6466/00/0045-0006$05.00

As might be expected, low-silica zeolites are most common in mafic volcaniclastic deposits, and high-silica zeolites are most common in silicic volcaniclastic deposits (Table 1).

Table 1. Principal zeolites in volcaniclastic deposits. [1]

Zeolite species	$Si/(Al+Fe^{3+})$	H_2O [2]	Cations	Host rock [3]
Thomsonite	1.0-1.08	1.2	Ca	B
Gonnardite	1.08-1.38	1.0-1.2	Na	B
Phillipsite	1.08-3.35	1.7-3.3	Ca, Na, K	A + B
Natrolite	1.5	1.0	Na	B
Scolecite	1.5	1.5	Ca	B
Analcime	1.5-2.9	1.0-1.3	Na	A + B
Chabazite	1.4-4.1	2.7-4.1	Ca, Na	A + B
Laumontite	1.9-2.4	2.0	Ca	A + B
Wairakite	2.0	1.0	Ca	A + B
Merlinoite	2.6	2.7	K, Ba	A + B
Yugawaralite	3.0	2	Ca	A + B
Stilbite	2.6-3.5	2.8-3.5	Ca, Na	B
Heulandite	2.7-4.0	2.5-3.1	Ca, Sr, K, Na	A + B
Erionite	2.6-3.8	3.0-3.5	K, Na, Ca	A + B
Clinoptilolite	4.0-5.6	3.5-4.0	Na, K, Ca	A
Mordenite	4.1-5.7	3.0-3.5	Na, Ca, K	A
Ferrierite	4.9-5.6	3.3	Mg, K, Na	A

[1] Compositional data are modified after Iijima (1988a).
[2] Number of water molecules per aluminum atom.
[3] A refers to silicic and B to mafic volcaniclastic rocks.

ORIGIN OF NATURAL ZEOLITES

Zeolite-forming reactions

Zeolites can originate from a variety of precursor materials including volcanic and impact glasses, aluminosilicate gels, and aluminosilicate minerals including other zeolites, smectite, kaolinite, feldspars, and feldspathoids. Volcanic glass is the major precursor of zeolites. Volcanic glass reacts to form zeolites by a dissolution-precipitation process in which a gel-like material may be an intermediate phase. The reaction of glass may be a complex, multi-stage process, as shown by a study of altered dacitic perlite (Noh and Boles 1989). The early diagenetic stages in the perlite they studied involved a series of incongruent dissolution reactions from glass via smectite to clinoptilolite and a K-rich aluminosilicate gel-like phase. Still later reactions resulted in clinoptilolite + mordenite and crystallization of the gel-like phase to K-feldspar.

Early formed zeolites are commonly altered to other zeolites. These reactions can result from a change in the physico-chemical environment or simply from the availability of sufficient time for a less-stable phase to transform to a more stable phase. Phillipsite, mordenite, chabazite, and clinoptilolite can be replaced by analcime or K-feldspar, and analcime can be replaced by laumontite, K-feldspar, or albite. For example, the concentration of phillipsite at shallower burial depths than clinoptilolite in deep-sea sediments is widely attributed to the coupled dissolution of phillipsite and crystallization of clinoptilolite with depth of burial, although replacement textures have not been demonstrated (e.g. Couture 1977; Boles and Wise 1978).

Role of water chemistry

Zeolites and clay minerals can form from the same aluminosilicate precursor materials; whether a zeolite or clay mineral is formed depends on the activities of dissolved species such as H^+, alkali and alkaline-earth ions, H_4SiO_4, and $Al(OH_4)^-$. The single most important requirement for zeolite formation is a high activity ratio of $(Na^+ + K^+ + Ca^{2+})/H^+$. Thus, zeolites are formed principally in alkaline environments, and large concentrations of relatively pure zeolites are found chiefly in the altered vitric tephra deposits of saline and highly alkaline lakes (pH = 9.5-10). pH also affects zeolite-forming reaction rates, which are much more rapid at a pH above 9 than at lower pH values. This increase in reaction rates is attributable to the increasing solubility of SiO_2 and Al species above a pH of about 9 (Taylor and Surdam 1981). Water activity also plays an important role in determining zeolite stability. Because the chemical activity of water is reduced by the presence of any solute, an increase in salinity can lower the temperature of dehydration reactions and consequently lower the temperatures at which the less hydrous zeolites are stable (Hay 1966).

The nature and proportions of extra-framework cations are obviously important in determining which cationic types of zeolite will form. For example, a mono-ionic zeolite, such as analcime or laumontite, cannot be formed in the absence of Na^+ or Ca^{2+}, respectively. Differences in the cationic composition of thermal water seem to have been a major control on whether Ca-rich or (Na+K)-rich zeolites formed in the rhyolites of thermal areas in Yellowstone National Park (Bargar and Keith 1995).

Role of temperature

Temperature exerts a major control on both the rates of reaction and the species of zeolite formed. Reaction rates are increased by higher temperatures, and stability fields of zeolites can differ greatly as a function of temperature. Less hydrous zeolites, such as laumontite, analcime, and wairakite, are stable at higher temperatures than the more hydrous zeolites, such as clinoptilolite, chabazite, and stilbite. Stability fields of zeolites are discussed more fully in the chapter on thermodynamic modeling of zeolites.

Role of pressure

Zeolites are hydrous phases with open structures of low specific gravity and they are therefore highly sensitive to pressure. Elevated pressures should favor the zeolites of higher specific gravity, which are also the less hydrous phases. Hence, the vertical pressure-temperature gradient in a rock column can be expected to result in a vertical zonation of zeolite types from most hydrous at the surface to least hydrous at depth. The relation of $P(H_2O)$ to P_{load} can be a significant factor in determining the mineralogy of zeolites and associated minerals in the deeper zones of burial diagenesis (Coombs et al. 1959). $P(H_2O)$ is related to permeability, and laumontite may form under hydrostatic $P(H_2O)$ in permeable beds at depths where heulandite is formed (or preserved) at near-lithostatic $P(H_2O)$ in relatively impermeable beds.

Reaction rates

The principal factors affecting zeolite reaction rates are the nature of the reactants (e.g. glass), the presence of H_2O, rock permeability, chemical environment, and temperature. Zeolites form more rapidly from glass than from most crystalline materials, and the reaction rate of glass varies inversely with its silica content. Alkali-rich low-silica glass is probably the most reactive common natural material.

The youngest documented zeolitic alteration from alkalic, low-silica glass may be phillipsite and chabazite in phonolitic tephra that erupted from Monte Nuovo volcano near Naples, Italy, in 1538 A.D (de' Gennaro et al. 1995). A somewhat older example is phillipsite cementing the lower part of a leucitite lahar(?) from Vesuvius that was erupted in 472 A.D., described below. Still another Italian example is chabazite in the leucite phonolite tephra that buried Ercolano in 79 A.D (de' Gennaro et al. 1980). de' Gennaro et al (1995) inferred that the zeolites of the Monte Nuovo tephra and Ercolano were probably formed at elevated temperatures prior to cooling of the eruptive deposits. Phillipsite of the Vesuvius lahar(?) of 472 A.D. probably formed at low temperature. The most rapid low-temperature zeolitic alteration of rhyolitic glass thus far documented is in Teels Marsh, Nevada, a saline highly alkaline playa in which an ash layer deposited about 1,000 years ago has been partly to wholly altered to phillipsite and less commonly to searlesite ($NaBSi_2O_6 \cdot H_2O$), analcime, and clinoptilolite (Taylor and Surdam 1981). In contrast, zeolites alter to alkali feldspar much more slowly than glass is altered to zeolites in contact with saline, highly alkaline fluids. Phillipsite and/or merlinoite (Donahoe et al. 1984) have altered to K-feldspar in highly alkaline brine of Searles Lake, California, over a time span of about 100,000 years at temperatures of about 25°C (Hay and Guldman 1987).

Phillipsite can form from mafic glass in sea-floor sediments at the sediment-water interface, and crystals may grow to their full size (~45 μm) in as little as 150,000 years (Czynscinski 1973). Alteration rates in sea-floor sediments are highly variable, and unaltered mafic vitric shards are preserved in some sediments as old as Cretaceous (Churkin and Packham 1973). In silicic ash layers, complete alteration to smectite and either phillipsite or clinoptilolite at temperatures of about 20°C generally requires at least 3 to 4.5 m.y (Hein and Scholl 1978, Desprairies et al. 1991).

Iijima and Ogihara (1995) made estimates of the time required for different zeolite reactions at elevated temperatures based on time-temperature relationships in drill cores of silicic tuffs in thick marine sequences in Japan. The reaction time for alteration of silicic glass to clinoptilolite at 40°-50°C ranges widely, and the minimum is about 0.4 m.y. The minimum time for reaction of clinoptilolite to analcime at 80°C is 0.9 m.y. In another study, the reaction time of analcime to albite at 120°C was estimated at about 0.4 m.y (Iijima 1988b).

TYPES OF ZEOLITE OCCURRENCE

Most zeolite occurrences in sedimentary rocks can be grouped into one of several types of geologic environments or hydrogeologic systems. These are (1) saline, alkaline lakes; (2) soils and land surfaces; (3) deep-sea sediments; (4) low-temperature open to closed tephra systems; (5) burial diagenesis; and (6) hydrothermal alteration. All of these occurrences generally exhibit characteristic patterns of zeolite zoning in tephra (Fig. 1). Zoning of ash layers in highly saline and highly alkaline lakes is chiefly lateral and reflects chemical gradients in the original lake water. The statistically differing distribution of phillipsite and clinoptilolite as a function of sediment age and burial depth in deep-sea sediments can be considered a vertical zoning. The open-system type of occurrence refers to accumulations of tephra, 8 m to 1 km or more in thickness, that may show a vertical and/or lateral zoning of authigenic silicate minerals. Burial diagenesis refers to the vertical zonation of authigenic minerals caused by the increase of temperature and pressure with zonation in thick sedimentary accumulations. Hydrothermal alteration implies zeolite formation in association with a high geothermal gradient as in a geothermal area having an igneous intrusion as the heat source.

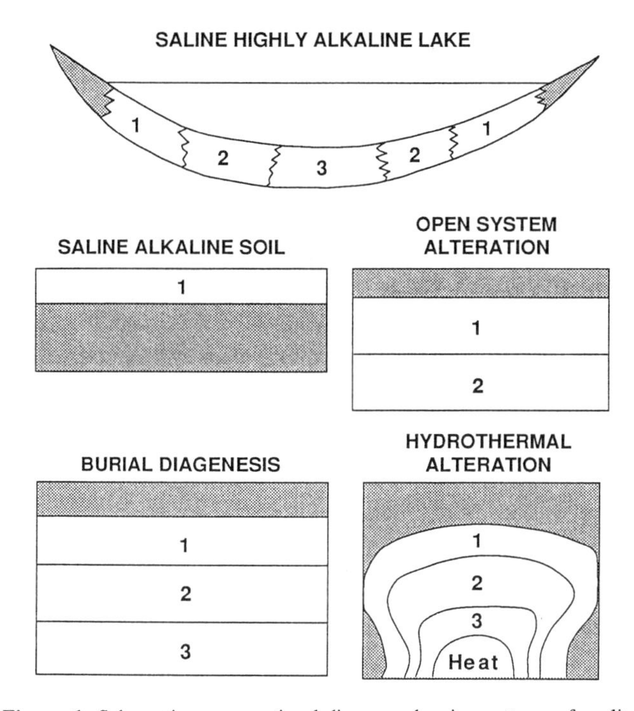

Figure 1. Schematic cross-sectional diagrams showing patterns of zeolite zoning in tephra deposits for different types of alteration. Dashed pattern indicates the zone of fresh glass; Zone 1 is characterized by non-analcimic alkali-rich zeolites, Zone 2 by analcime, and Zone 3 by alkali feldspars.

ZEOLITES IN DEPOSITS OF SALINE, ALKALINE LAKES

Zeolites are both common and widespread in tephra deposits of saline, alkaline lakes (pH = 9.5-10), and the largest relatively pure concentrations of natural zeolites are found here. Highly alkaline saline lakes owe their high pH to dissolved sodium carbonate-bicarbonate and/or sodium borate. The most common zeolites formed from silicic vitric ash are phillipsite, clinoptilolite, and erionite; less common are mordenite and chabazite, and seemingly rare is merlinoite. Nearly monomineralic beds of clinoptilolite, phillipsite, erionite, chabazite, and analcime occur in Miocene to Pleistocene lake deposits of the western United States (Sheppard and Gude 1968, 1969, 1973; Gude and Sheppard 1988). Pore-fluid composition is a significant factor in determining which zeolites form directly from silicic glass. One well-documented example is the formation of phillipsite from rhyolitic glass in relatively low-silica brine in Teels Marsh, Nevada. The boron content of the brine was sufficiently high that the precipitation of searlesite ($NaBSi_2O_6{\cdot}H_2O$) effectively buffered the silica content of the brines at low levels, thus favoring phillipsite over the more siliceous clinoptilolite, mordenite, and erionite (Surdam and Sheppard 1978).

Tuff beds in and adjacent to saline-lake deposits may be mineralogically zoned with (a) an outer zone of glass and/or smectite in which pore fluid was relatively dilute; (b) a zeolite zone of alteration to phillipsite, clinoptilolite, erionite, etc., reflecting elevated salinities and pH; (c) a zone of higher salinity with analcime; and (d) the zone of highest salinity, generally in the lowest part of the basin, where tuffs may be altered to K-feldspar. An example of such zonation is the Big Sandy Formation of Pliocene age in Arizona, in which nonanalcimic zeolites, analcime, and K-feldspar zones form a concentric "bullseye" pattern (Sheppard and Gude 1973). This concentric mineralogic zoning represents a series of progressive reactions: from glass to open-framework zeolite; open-framework zeolite to analcime; and analcime to K-feldspar as a function of increasing salinity. The analcime zone is missing in some lake deposits, for example in Lake Tecopa, California (Sheppard and Gude 1968), where the zone of open-framework zeolites grades directly into the K-feldspar zone, which contains searlesite. On the Greek island of Samos, albite coexists with K-feldspar, generally as a minor constituent, in the central mineralogic zone (Stamatakis 1989).

Tuffaceous beds of the Upper Jurassic Brushy Basin Member of the Morrison Formation are mineralogically zoned in a bullseye pattern over an area of 150,000 km^2 in the Colorado Plateau and San Juan Basin. From outer to inner, the zones are characterized by clinoptilolite, analcime ± K-feldspar, and albite. This zoning has been attributed to early diagenesis of vitric ash in a saline, highly alkaline playa-lake termed Lake T'oo'dichi' (Turner and Fishman 1991).

Authigenic analcime is common in non-tuffaceous claystones of saline and highly alkaline lakes. The Green River Formation of Utah (Remy and Farrell 1989) is an example of the formation of analcime by reaction of detrital clays with saline and highly alkaline lake and pore water. As far as we are aware, authigenic analcime has not been identified in clays of saline but not highly alkaline lakes such as Great Salt Lake (Spencer et al. 1984) and the lowest lacustrine unit of Searles Lake, California (Hay et al. 1991).

Analcime-rich deposits termed analcimolites are the thickest and most extensive monomineralic sedimentary zeolite deposits of the geologic record. Most of them are of Mesozoic age and include Triassic deposits of the western United States, Upper Jurassic and Lower Cretaceous rocks of the Congo Basin, and Lower Cretaceous deposits of the Sahara Desert (Hay 1966). Analcimolites of the Sahara, for example, total about 40 m in thickness and cover an area of more than 13,000 km^2 (Joulia et al. 1959). Analcimolites are typically interbedded with reddish-brown, ferruginous analcimic argillite or claystone. A saline, highly alkaline playa lake or series of playa lakes seem a likely environment for these deposits, and Joulia et al (1959) mentioned the presence of pseudomorphs after a saline mineral, possibly pirssonite or gaylussite. Evidence of volcanic material is generally lacking, and clay minerals seem the most likely precursor for analcime. As a modern example, kaolinite has reacted to form analcime at shallow depths in Willcox Playa, Cochise County, Arizona (Pipkin 1965).

ZEOLITES IN SOILS AND SURFACE DEPOSITS

Soils contain a wide variety of zeolites, and their occurrence was reviewed by Ming and Mumpton (1989), Boettinger and Graham (1995), and Ming and Boettinger (this volume). Most zeolites in soils are inherited from zeolitic parent materials, but the *in situ* (pedogenic) formation of zeolites in soils is clearly established for many examples. Zeolites that formed in soils include analcime, phillipsite, clinoptilolite, chabazite, and natrolite. These can form in highly alkaline soils from both glass and crystalline aluminosilicate materials, including clay minerals. Zeolites are destroyed in neutral and acidic soils.

The highest contents of authigenic zeolites are formed at and near the land surface where the pH is 9.5 as a result of the concentration of sodium carbonate-bicarbonate by evapotranspiration. At Olduvai Gorge, Tanzania, soils have a high pH, and alkaline salt efflorescences occur widely on sheltered outcrops of tuff in the dry season. Minerals formed in volcanic ash soils and on outcrops include phillipsite, chabazite, natrolite, analcime, and dawsonite ($NaAl(OH)_2CO_3$) (Hay 1976, 1980). Volcanic ash apparently altered rapidly to zeolites in this highly alkaline surface environment, and nephelinite ash deposited about 1,250 years ago is widely cemented by phillipsite. The present high soil pH and efflorescences of sodium carbonate are at least partly attributable to natrocarbonate tephra erupted from the Oldoinyo Lengai volcano.

Authigenic analcime is a common zeolite of saline and highly alkaline non-tuffaceous soils. One well-documented example is an area in the San Joaquin Valley of southern California where montmorillonite has reacted to form analcime and probably illite at and near the land surface (Baldar and Whittig 1968). Renaut (1993) described the formation of analcime, illite, and minor natrolite from poorly ordered clay minerals in a lake-margin paleosol bordering Lake Bogoria, a saline and highly alkaline lake in northern Kenya. The zeolites were formed during a dry period of low lake level from 20,000-13,000 yr. B.P. when sodium-carbonate-rich pore water was concentrated close to the land surface by evaporative pumping

ZEOLITES IN DEEP-SEA SEDIMENTS

Zeolites have formed widely at relatively low temperatures in deep-sea sediments. Most of the information about zeolites in present-day ocean basins has been obtained from drill cores of the Deep Sea Drilling Project (DSDP), from 1968-1975, and of the Ocean Drilling Program (ODP) from 1976 to 1996. The present review is based chiefly on the reviews of DSDP results (Kastner and Stonecipher 1978, Iijima 1978, Boles and Wise 1978). Results of the ODP have not substantially modified the conclusions reached from DSDP studies, although they provide much new chemical data on pore-fluid compositions.

Phillipsite and clinoptilolite are the principal zeolites in deep-sea sediments, and their average amounts have been estimated at 1.5% for phillipsite and about 2% for clinoptilolite (Kastner 1979). Analcime is next in abundance, and many other zeolites have been identified including harmotome, heulandite, mordenite, erionite, chabazite, gmelinite, thomsonite, natrolite, laumontite, wairakite, and merlinoite. Wairakite, heulandite, mordenite, and some others of the latter group are probably related to abnormally high geothermal gradients caused by localized heat sources such as intrusions of basaltic magma (e.g. Vitali et al. 1995). Merlinoite has only recently been identified in deep-sea sediments, where it occurs in pelagic clays (Vitali et al. 1995) and manganese nodules (Mohapatra and Sahoo 1987).

The Si/Al ratio of deep-sea phillipsite (2.3-2.8) is intermediate between the nonmarine phillipsites of mafic and silicic volcaniclastic deposits, and K is the principal exchangeable cation (Sheppard et al. 1970). Deep-sea clinoptilolite has Si/Al ratios of 4.2-5.2, but 80% of the clinoptilolite has ratios of 4.5-5.0 (Stonecipher 1978). The cation composition is generally $K > Na > Ca$.

Distribution

Phillipsite can be found at or near the sediment-water interface, is most common at sub-bottom depths of less than 150 m, and rarely occurs at depths of >500 m (Kastner and Stonecipher 1978). It is most abundant in Miocene and younger sediments and decreases in abundance in older sediments. Its distribution in sediment types is clayey > volcaniclastic > calcareous > siliceous. Phillipsite is commonly associated with smectite and/or evidence of palagonite, a form of altered basaltic glass. It is rare in younger sediments of the Atlantic and Pacific margins consisting of siliceous volcaniclastic sediment and detritus from continents and islands.

Clinoptilolite is relatively rare at sub-bottom depths less than 100 m and increases in abundance with depth and age of the host rock. It is overall most common in sediments of Eocene through Cretaceous age. Its distribution as a function of sediment type is calcareous > clayey > volcaniclastic > siliceous. It is commonly associated with one or more of the following: opal-CT, smectite, sepiolite, and palygorskite. It is common in terrigenous muds from continents and islands and relatively rare in pelagic brown clay of the central ocean basins. Clinoptilolite is more common in sediments of the Atlantic and Indian oceans than in those of the Pacific Ocean.

Analcime has been identified in deep-sea sediments of all ages and is generally associated with basaltic materials. Its abundance increases with age and burial depth. Analcime commonly occurs with phillipsite and clinoptilolite but has not been observed to replace either zeolite.

Origin of zeolites

Although vitric material is important, it is apparently not required for the formation of phillipsite and clinoptilolite at low temperatures in deep-sea sediments. Indeed, only 20-25% of deep-sea clinoptilolite occurs in dominantly volcaniclastic sediment (Kastner and Stonecipher 1978). Biogenic silica can be a significant factor in forming both phillipsite and clinoptilolite and has contributed to the formation of clinoptilolite from mafic vitroclastic material (Kastner and Stonecipher 1978) and probably has contributed somewhat to the formation of marine phillipsite from mafic materials (Czynscinski 1973). Siliceous sediment is much more common at high latitudes than at low latitudes, and this difference in the content of biogenic silica may account for the formation of clinoptilolite from silicic ash in the Bering Sea (Hein and Scholl 1978) and for the formation of phillipsite from silicic ash in the southwest Pacific and adjacent areas (e.g. Desprairies et al. 1991). The general occurrence of phillipsite rather than clinoptilolite in deposits of younger age and shallower depths has been attributed to the dissolution of phillipsite and the precipitation of clinoptilolite, a more stable phase, with the addition of biogenic silica (Couture 1977), i.e. phillipsite + biogenic silica $\Rightarrow$ clinoptilolite

Vitric ash layers interbedded in fine-grained deep-sea sediments commonly alter in a closed hydrologic system. This results in zeolitic alteration of silicic glass at sub-bottom depths of as little as 150 m and temperatures of 20°C or less, compared with depths of hundreds of meters and temperatures of 40°-50°C for zeolites in many marine silicic tephra deposits on land (Iijima 1978 1988a). One example of closed-system alteration is provided by drill core of Site 767 in the Celebes Sea (Desprairies et al. 1991). Pore fluids in silicic to intermediate tuffs and tuffaceous sediments of the drill core vary systematically in SiO_2, K^+, $\delta^{18}O$, $^{87}Sr/^{86}Sr$ ratio, and other chemical parameters with regard to the age and progressive alteration of glass (Table 2; von Breymann et al. 1991). The concentration of SiO_2 in pore fluids increases downward into a "glass hydrolysis" zone (Zone II) in which small amounts of opal-CT and smectite have formed from glass. Silica remains high in the underlying zone (Zone IIIa) where phillipsite, smectite, and glass coexist. Silica then decreases downward through tuffs and tuffaceous sediments in which most or all glass is altered to smectite and phillipsite (Zones IIIb, IIIc, and IV). Concentrations of K^+ decrease downward, mostly reflecting its uptake by phillipsite (Zones III and IV). The $\delta^{18}O$ values of pore fluids decrease downward due to the selective incorporation of pore fluid ^{18}O into smectite and phillipsite at low temperature. The downward decrease in $^{87}Sr/^{86}Sr$ values was attributed by von Breymann et al (1991) to dilution of sea water Sr (~0.709) by Sr with lower 87/86 ratios (~0.705?) from volcanic glass during alteration to smectite and zeolite. About 4 m.y. were required for complete alteration of silicic glass (Zone IIIa).

Table 2. Pore fluid data for ODP Hole 767, Celebes Sea.[1]

Diagenetic zone[2]	Depth (mbsf)	Age (m.y.)	µM SiO_2	mM K	$\delta^{18}O$ (‰)	$^{87}Sr/^{86}Sr$
I. Fresh glass (G)	0-71.5	0-0.90	536-689	10-12.5	0.985 to 0.215	0.709141 to 0.708765
II. Glass hydrolysis (G >> S, O)	71.5-158	0.90-2.42	642-809	10-7.9	0.428 to -0.919	0.708695 to 0.708731
III. Phillipsite						
IIIa (G, P, S)	158-233	2.42-3.79	721, 910	5.1	-0.230 to -0.821	0.708671 to 0.708404
IIIb (P, S)	233-242	3.79-3.96	390	n d	-1.638	0.707856
IIIc (P, S, G)	242-329	3.96-5.80	167-187	2.9	-2.232	0.706309
IV. Smectite (S >> P, G)	329-397	5.80-7.91	152, 159	2.2	-2.455	0.706128

[1] Data from von Breymann et al (1991)
[2] Abbreviations are G: glass, S: smectite, P: phillipsite, O: opal-CT, mbsf: m below sea floor; and n d: not determined.

ZEOLITES IN LOW-TEMPERATURE, OPEN TO CLOSED TEPHRA SYSTEMS

Hydrogeology of tephra alteration

It has long been recognized that hydrolysis of glass to smectite raises the a(SiO_2), pH, and activity ratio of (Na^+ + K^+)/H^+ to a field where zeolite crystallization is favored. This process can be viewed in terms of open and closed hydrologic systems. In closed-system alteration of tephra, reactions proceed to completion without substantial ionic diffusion or interchange of pore fluids from outside the reacting system. In open systems, fluids moving through tephra deposits are changed progressively by the same water-rock reactions as in closed systems. The two systems can, however, differ in the distribution of phases formed in glass-water reactions. Closed systems contain both early and late-stage reaction products, whereas in open systems the early formed phases, such as smectite, should be concentrated near the entry of fluid to the system, and the zeolites should be concentrated farther along the flow path. Closed systems in the strict sense require impermeable deposits, but the concept can be applied to hydrologic systems of low permeability with pore fluids that are static or moving at low rates in comparison with reaction rates. For example, bentonitic mudstone in an appropriate hydrogeologic setting can form a relatively closed system. The usage of "closed hydrologic system" adopted here differs from that of Surdam (1977), in which "closed hydrologic system" referred solely to the lakes of closed hydrologic basins in arid or semiarid regions.

The extent of change in the whole-rock chemical composition is one line of evidence reflecting the "openness" of the system. Significant chemical change is evidence of open-system alteration; lack of chemical change except for hydration is compatible with, but not necessarily proof of, a closed system. It should be stressed that truly closed systems are extremely rare in geology, and most tephra systems termed "closed" have experienced some ionic or fluid interchange with adjacent sediments.

Open-system tephra alteration

Open hydrologic systems vary greatly in scale and in their relationship to the water table, depending to a considerable extent on tephra composition. Alkalic, low-silica tephra deposits only 8-20 m thick in the vadose zone may be altered to zeolites and clay minerals or palagonite. In this type of system, hydrolysis is accomplished by downward percolation of meteoric water. Phillipsite and chabazite are the most common zeolites in alkalic, low-silica tephra. On the contrary, nonmarine accumulations of silicic tephra generally must reach thicknesses of several hundred meters or more to develop zeolites by open-system hydrolysis. In thick basinal accumulations of silicic tephra such as the John Day Formation of Oregon, most of the zeolitic deposits generally lie below the water table where most water movements have a lateral component (Hay 1963). Many zeolitic deposits of open-system type are relatively thick, of high grade, and are being mined or are suitable for mining (see chapter on zeolites in open hydrologic systems).

Of interest because of its recent age is a K-rich, low-silica lahar(?) that was erupted from Vesuvius in 472 A.D. The lahar is about 8 m thick where exposed near Pollena, on the northeast slope of Vesuvius. It consists of lava clasts, mostly of lapilli and block size, in an ash matrix rich in vesicular vitroclasts with crystals of leucite and clinopyroxene (i.e. leucitite). The upper part, poorly exposed and about 1 m thick, is unconsolidated and appears unaltered. The lower 6-7 m are cemented by phillipsite which also lines and fills vesicles in the vitroclasts. Phillipsite crystals are as much as 40 μm long. The surface of the lahar is vegetated, and limited exposure prevented sampling adequate to determine whether or not clay minerals had formed in the soil. This lahar overlies permeable deposits, and the lower portion appears to have zeolitized through the action of downward-percolating rainwater in a largely open system. As in many zeolitized rocks, alteration in an open system may have gradually given way to closed-system alteration as the permeability decreased as a result of zeolitization.

Koko Crater, Hawaii, is an example of open-system alteration of alkalic, low-silica mafic (basanite) glass (Hay and Iijima 1968). It is an eroded cone formed about 35,000 years ago and consists mainly of tuff and lapilli tuff that are mineralogically zoned through thicknesses of 10-20 m or more. Beneath a thin soil, the surface zone, 1.5-12 m thick, contains fresh glass, opal, and smectite. This overlies a zeolitic palagonite zone, the upper 1.5 to 10 m of which contains phillipsite and chabazite, and analcime predominates in the lower part of the palagonite zone. The contact between zones parallels the topography and cuts across stratigraphy, indicating that the zonation is a diagenetic, post-eruptive feature (see Fig. 4 in the chapter "Formation of Zeolites in Open Systems").

Faujasite, a relatively rare zeolite, forms as much as 30% of basaltic vitric air-fall tuffs of Quaternary age in northeast Jordan (Ibrahim and Hall 1995). The tephra deposits are of alkali olivine basalt composition and comprise three diagenetic zones. The upper zone contains fresh glass and is 8-40 m thick, the middle zone is palagonitic tuff 10-20 m thick, and the lower zone, as much as 40 m thick, contains faujasite, phillipsite, and chabazite, which form a cement and fill vesicles. Contacts between zones are sharp, roughly follow topography, and cut across stratification. Ibrahim and Hall (1995) attribute the palagonite and zeolites to open-system alteration of basaltic glass by meteoric water. The sharp contacts may reflect differences in permeability, flow rate, and pore-fluid composition.

The John Day Formation of Oregon (Eocene to lower Miocene) is a mineralogically simple example of large-scale open-system alteration of silicic tephra. Where studied by Hay (1963), it is about 600-900 m thick and consists largely of smectitic tuffaceous claystone and silicic tuff. An upper zone as much as 450 m thick contains fresh glass or glass altered to smectite and opal. A lower zone, also as much as 450 m thick, contains clinoptilolite, smectite, and less commonly celadonite and K-feldspar. The transition between upper and lower zones is generally sharp and occurs through a zone of glass dissolution 2-20 mm thick. Where the formation is folded, the transition zone cuts across bedding, showing that much, if not most, of the zeolitic alteration occurred during and/or after folding (see Fig. 1 in the chapter "Formation of Zeolites in Open Systems"). K-Ar dates on authigenic K-feldspar and celadonite indicate that much and possibly all of the zeolitic alteration was completed before the full thickness of the formation was deposited. Chemical changes in rhyolitic ignimbrite (ash-flow tuff) perhaps best document the extent of open-system alteration in the John Day Formation. The ignimbrite consists of multiple flow units near the middle of the formation, where it is interbedded in dacitic tuffs and tuffaceous claystones. In the lower zeolite diagenetic zone, the rhyolitic vitric flow unit(s) are altered to smectite, clinoptilolite, and less commonly to celadonite and K-feldspar, and chemical analyses document losses of SiO_2 and Na, and gains of Fe, Mg, and Ca, resulting for most samples in a water-free composition close to that of dacite.

The Oligocene White River Sequence of the Great Plains and Rocky Mountains exhibits a more complex pattern of clinoptilolite distribution. These deposits consist mainly of fine-grained reworked silicic tephra deposits and are commonly 100-200 m thick. Only 7 of 20 localities studied contain more than minor zeolitic deposits, providing an opportunity to determine the controls on zeolitic alteration. The dominant lithology is bentonitic mudstone that is 80-90 percent vitric ash or its alteration products, smectite, clinoptilolite, and opal-CT (Lander and Hay 1993). Tephra, from which the mudstones formed, was deposited at average rates of 4 mm-4 cm/1000 yr and was partly weathered to clay minerals before burial. Relatively coarse-grained tuff beds comprise about 2% of the formation. Clinoptilolite-rich deposits in the White River Sequence in Wyoming are attributed to long fluid residence times associated with hydrologic discharge zones (see Fig. 3 in the chapter "Formation of Zeolites in Open Systems"), and clinoptilolite in the northern Great Plains is attributed to low hydrologic head and low permeability resulting in long fluid residence time (Lander and Hay 1993). Patterns of alteration seem to indicate open-system alteration for ash layers and dominantly closed-system alteration in the finer-grained mudstones. The two systems are not closely linked, and unaltered ash layers are interbedded in zeolitic mudstones.

Closed-system tephra alteration

Fine-grained sediments of the deep-sea floor can constitute systems that are sufficiently closed to give rise to hydrolysis of silicic ash layers to smectite (± opal-CT) followed by crystallization of zeolites. Permeabilities are low and pore fluids are static except where deformation or local heat sources cause fluid movements. The chemistry of pore fluids can differ significantly in tephra zones representing different stages in the alteration of glass, as discussed below for ODP Site 767, showing that reaction rates of glass are sufficiently rapid compared with diffusion rates; glass reaction progress thus controls the pore-fluid composition.

Drill core from Searles Lake, California, provides a lacustrine example of ash alteration exhibiting some degree of closed-system behavior (Hay and Guldman 1987). Unit I (3.18-2.56 Ma) is the lowermost lacustrine unit, at a depth of 541.6-693.4 m, which was deposited in a moderately saline but not highly alkaline lake. It consists of muds, minor amounts of sandstone and anhydrite, and a few rhyolitic ash layers. Ash beds are altered to widely varying degrees, depending principally on grain size. The finer-grained, most-altered ash beds are smectitic bentonites that contain a trace to 5% each of clinoptilolite and opal. Smectite was the first to form, as expected in a closed system; however, in comparison the contents of clinoptilolite and opal are much too small for truly closed-system alteration of the chemically analyzed rhyolitic shards. Diffusion and/or advection may account for the loss of SiO_2 and alkalis and probable gain of Mg by smectite in the alteration of the ash layer.

The Miocene Chalk Hills Formation of southwestern Idaho is an example of closed-system alteration of tephra that produced a large, mineable deposit of clinoptilolite (Sheppard 1991). Zeolitic alteration is of special interest here because of the documented relationship between lake-water chemistry and diagenetic alteration. The mostly lacustrine formation is about 100 m thick and consists of mudstone, siltstone, sandstone, and numerous thin volcanic ash beds or tuffs. Ash beds deposited in water having a pH of 7-9 and low to moderate salinity (0.3-3.0 per mil), as inferred from fossil ostracode species, were altered to smectite, clinoptilolite, opal-CT, and minor phillipsite, erionite, mordenite, and K-feldspar (Fig. 2). Ash beds are unaltered where they were deposited in fresh water, as indicated by fossil diatom species. The proportions of smectite and zeolites vary laterally in the saline zone, and bentonite is mined only a short distance (1.6 km) from the Castle Creek zeolite (clinoptilolite) deposit, both of which formed from the same tuff (Sheppard 1993). This lateral mineralogic variation is suggestive of fluid movements or ionic diffusion within the saline zone.

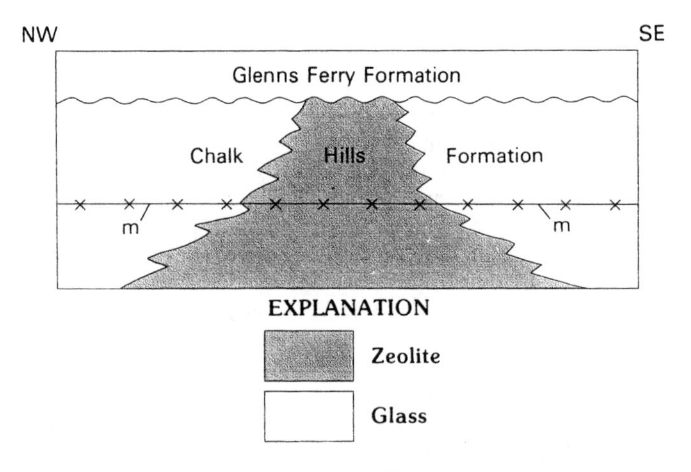

Figure 2. Diagrammatic cross-section showing the relationship of fresh-glass and zeolite zones in tuffaceous rocks of the Chalk Hills Formation, Idaho. A marker tuff is indicated by the letter m. The Chalk Hills Formation is about 100 m thick (Sheppard 1991).

BURIAL DIAGENESIS

Burial diagenesis, also termed burial metamorphism, applies to those zeolites formed on a regional scale in thick accumulations of sedimentary rock. Large amounts of zeolites are formed only in the burial diagenesis of volcaniclastic deposits, which are commonly vertically zoned, reflecting primarily the increase in temperature with depth. The regional occurrence and vertical zoning of zeolites in thick accumulations of sediment was first documented by Coombs (1954) in Southland, New Zealand. This sequence is 9-10 km thick and consists of Triassic and Jurassic marine strata with numerous tuffs. Heulandite and analcime characterize the upper part of the sequence and laumontite the lower part, which also contains pumpellyite and prehnite and, thus, represents a transition into the greenschist facies of metamorphism (Coombs et al. 1959).

Most reported examples of burial diagenesis are in the Circum-Pacific area, and probably the most instructive are in the upper Cenozoic silicic volcaniclastic marine strata of the Green Tuff region of Japan where zeolites are presently being formed. Four zones are now widely recognized in the Japanese tephra sequences: I, fresh glass; II, clinoptilolite-mordenite; III, analcime; and IV, albite. Zonation is controlled primarily by temperature, as shown by study of drill cores from areas in which zeolites are presently being formed (Iijima 1988a, and references therein), although later studies of burial diagenetic deposits show that water chemistry also affects the zonation. Glass is transformed to clinoptilolite and/or mordenite at about 41°-55°C; clinoptilolite and mordenite react to form analcime in Zone III, at temperatures of 84°-91°C; and analcime reacts to form albite in Zone IV at 120°-124°C. Zone III is subdivided into Zone IIIa, with analcime and heulandite, and Zone IIIb, with analcime and laumontite.

Pore-fluid chemistry affects both the rate and the temperature of zonal transformations. Reaction of glass to clinoptilolite and mordenite at 40°-50°C may require as little as 400,000 years or as much as tens of millions of years (Iijima and Ogihara 1995). Unusually saline and alkaline pore fluids can lower the temperature of the I-II (glass-zeolite) transition to about 21°C and the II-III (clinoptilolite/mordenite-analcime) transition to about 37°C (Iijima 1995).

The upper zones of burial diagenesis can be mineralogically similar to those of open hydrologic systems. The two types can generally be distinguished by (1) the thickness of the glass-zeolite transitional zone, which is much greater in burial diagenesis; and (2) the time span of zeolitic alteration, which in open systems is generally short-lived compared with burial diagenesis. The distinction between burial diagenetic and open hydrologic systems is to some extent artificial in thick accumulations of tephra spanning a significant temperature range.

HYDROTHERMAL ALTERATION

Zeolites are common in active geothermal areas with steep geothermal gradients, most of which are associated with volcanism and/or igneous intrusions. Zeolites and associated minerals are zoned in relation to temperature, but the zoning may not be well defined at higher gradients. In Iceland, for example, mineralogic zoning is much better defined in areas in which the gradient is <150°C/km than in areas in which the gradient is >200°C/km (Kristmannsdóttir and Tómasson 1978). Mineral assemblages differ greatly between different thermal areas, reflecting differences in host rock, temperature, and fluid chemistry.

The occurrence of hydrothermal zeolites differs to some extent from that of burial-diagenetic zeolites. Wairakite and yugawaralite, for example, appear to be much more common as hydrothermal minerals than as products of burial diagenesis (Iijima and Utada 1972). Hydrothermal crystallization temperatures can also differ considerably from those of burial diagenesis. Mordenite in the Wairakei geothermal area, for example, occurs at temperatures of 150°-230°C, compared with temperatures of about 40°-90°C for burial diagenesis in the Neogene Green Tuff region of Japan (Iijima 1988a). These temperature differences are undoubtedly due to a combination of kinetic factors and differences in pore fluid and source compositions.

Hydrothermal alteration can yield a much more diverse mineral assemblage than low-temperature alteration. For example, hydrothermal alteration of basaltic tephra by sea-water-dominated fluid on the volcanic island of Surtsey has produced smectite, phillipsite, analcime, tobermorite, and minor amounts of opal, chabazite, and xonotlite (Jakobsson and Moore 1986). Tobermorite and xonotlite are hydrated calcium silicates. This mineralogy compares with the assemblage smectite, phillipsite, and iron and manganese oxides formed by low-temperature alteration of basaltic glass on the sea floor.

Zeolites and clay minerals are associated with Kuroko ("black smoker") polymetallic sulfide deposits in the Green Tuff region of Japan (Iijima 1974, Utada 1988). Zeolites and clays exhibit a complex zonation produced by submarine hydrothermal alteration and low-temperature diagenesis. An argillaceous halo envelops the ore deposits and is immediately surrounded by a distinctive analcime zone and a Na-mordenite zone, both of which are superimposed on the clinoptilolite-mordenite (II) zone of burial diagenesis.

ZEOLITES IN LAVA FLOWS AND IGNIMBRITES

The zeolites of lava flows are of interest both for museum display and for determining zeolite crystal structures. Although these zeolites have been collected by mineralogists since the 1750s, their origin is to some extent still controversial. Zeolites in relatively fresh lavas are commonly assumed to have formed from magmatic solutions (e.g. Wise and Kleck 1988) and/or interaction of lava with meteoric water during the cooling of a flow (e.g. Tschernich 1992). Merino et al (1995) recently proposed an origin involving magmatic solutions at elevated temperatures for those zeolites associated with banded agate in flood basalts. As hypothesized by Merino et al (1995), agates in flood basalts did not crystallize from aqueous solutions but from lumps of polymeric silica containing trace elements and water. In this model, the silica lumps crystallize in an oscillatory fashion, producing the banding of agate through self-organization. Crystallization of silica to quartzose agate leads to oxidation of Fe^{2+} to Fe^{3+} by H_2O, which raises alkalinity and thus promotes the formation of zeolites.

Crystallization of zeolites during cooling of a flow seems not unreasonable as they readily form during post-cooling hydrothermal alteration of lavas. However, zeolites appear to be absent in newly cooled lavas, and Walker (1960a,b) has convincingly shown that zeolites deposited in sequences of basaltic lavas on Ireland and Iceland post-date the cooling of the lavas. Mineralogic zoning of the Icelandic lavas is sub-horizontal and cuts across the inclined lava sequence, reflecting the geothermal gradient. Hence, the Icelandic zoning can be considered an example of burial diagenesis. The absence (or scarcity) of zeolites formed in lavas during cooling may reflect the short time that the upper, more vesicular parts of most lavas are in the temperature range (~100°-250°C?) for rapid growth of zeolites to the sizes found in the vesicles of lava flows. For example, more than a year at temperatures of 210° to >250°C is required for growth of wairakite crystals as much as 1.5 mm in diameter in the Ngatamariki geothermal field in New Zealand (Browne et al. 1989).

Some of the analcime "phenocrysts" in volcanic rocks are demonstrably secondary replacements of leucite phenocrysts (Luhr and Giannetti 1987). A truly igneous origin for some other analcime phenocrysts remains controversial. The late Quaternary minette lavas of Colima, Mexico, seem to represent the strongest case for primary igneous analcime, but Luhr and Kyser (1989) and Karlsson and Clayton (1991) reached opposing conclusions on the basis of chemical and stable-isotope data (see chapter on "Isotope Studies of Zeolites").

Several studies have concluded that the zeolites in some ignimbrites are the result of alteration at elevated temperatures during cooling, and the term "geoautoclave" (see chapter on Formation of Zeolites in Open Hydrologic Systems) has been applied to this process (Aleksiev and Djourova 1975). Most studies infer that alteration was caused by vapor either trapped or evolved from the cooling ignimbrites. Pérez-Torrado et al (1995), for example, proposed that the Roque Nublo ignimbrites of Gran Canary, Canary Islands, were produced by phreatomagmatic eruptions, and the eruptive clouds at about 100°C were too poorly expanded for the water vapor to be dissipated during transport. Meteoric water is an alternative or additional source of fluid for an ignimbrite geoautoclave.

The products of geoautoclave alteration are chiefly or wholly smectite and zeolite and thus mineralogically resemble ignimbrites altered at low temperatures in open hydrothermal systems. The most common line of evidence used for distinguishing geoautoclave alteration in an ignimbrite is the presence of co-genetic air-fall tephra deposits that are either unaltered or contain less zeolite than the zeolitic ignimbrite (e.g. Garcia Hernandez et al. 1993). A lack of vertical mineralogic zoning has also been used as a criterion for identifying geoautoclave zeolites in an ignimbrite (de' Gennaro et al. 1995). Neither of the above criteria, however, is rigorous proof of a geoautoclave origin.

The oxygen-isotope composition of quartz has been used to show that clinoptilolite/heulandite, smectite, and chalcedonic quartz in a welded tuff of Yucca Mountain, Nevada, formed at temperatures of 40°-100°C during cooling of the ignimbrite (Levy and O'Neil 1989). These minerals are concentrated in a narrow transition zone between densely welded vitrophyre and tuff devitrified to alkali feldspar and cristobalite within the same cooling unit. This rather special example of ignimbrite alteration shows that zeolites can be formed in ignimbrites and lava flows during cooling. Oxygen-isotope analysis of authigenic zeolites and smectites offers a potential method of evaluating the geoautoclave hypothesis for zeolite-rich ignimbrites, at least those of late Quaternary age.

FUTURE RESEARCH

Much remains to be learned about conditions under which zeolites are formed in nature. In particular, more work is needed to establish the temperatures at which zeolites were formed in ignimbrites and in other relatively recent tephra deposits such as those that buried Ercolano, Italy, in 79 A.D. and those that formed Monte Nuova in 1538 A.D (de' Gennaro et al. 1995).

The origin of zeolites in lavas clearly needs further work. The apparent absence of zeolites in cooling lavas may seem to suggest that zeolites do not crystallize in lavas prior to or during cooling, yet Robert et al (1988) reported the cutting of phillipsite amygdules by columnar jointing of a basanite flow in France. One approach to this problem is the systematic search for zeolites in cooling lavas and those that have only recently cooled. Silica-poor, alkali-rich lavas would be ideal for this purpose, as they commonly contain high contents of zeolites.

Regrettably, the framework oxygen isotope composition of zeolites appears an unreliable means of estimating their crystallization temperatures (Karlsson 1995). Two limiting factors are (1) the difficulty in measuring the framework oxygen isotopic composition; and (2) the rapid oxygen isotopic exchange between water and framework oxygen. Analcime and wairakite are exceptions and may retain their original isotopic composition even in active geothermal systems. Nevertheless, oxygen isotopic analysis of some other minerals commonly associated with zeolites can be used for thermometry (e.g. opal-CT, quartz, smectite, and calcite).

ACKNOWLEDGMENTS

We are indebted to Dr. Giuseppe Rolandi (Dipt. Geofis. Volcanol., Univ. Napoli) for taking one of us (RLH) to the zeolitic lahar(?) erupted from Vesuvius in 472 A.D.

REFERENCES

Aleksiev B, Djourova EG (1975) On the origin of zeolitic rocks. C R Acad Bulg Sci 28:517-520

Baldar NA, Whittig LD (1968) Occurrence and synthesis of soil zeolites. Proc Soil Sci Am 32:235-238

Bargar KE, Keith TEC (1995) Calcium zeolites in rhyolitic drill cores from Yellowstone National Park. *In* Natural Zeolites '93: Occurrence, Properties, Use. DW Ming, FA Mumpton (eds) Int'l Comm on Natural Zeolites, Brockport, New York, p 87-98

Boles JR, Wise WS (1978) Nature and origin of deep-sea clinoptilolite. *In* Natural Zeolites: Occurrence, Properties, Use. LB Sand, FA Mumpton (eds) Pergamon Press, Elmsford, New York, p 235-244

Boettinger JL, Graham RC (1995) Zeolite occurrences in soil environments: an updated review. *In* Natural Zeolites '93: Occurrence, Properties, Use. DW Ming, FA Mumpton (eds) Int'l Comm on Natural Zeolites, Brockport, New York, p 23-38

Browne PRL, Courtney SF, Wood CP (1989) Formation rates of calc-silicate minerals deposited inside drillhole casing, Ngatamariki geothermal field, New Zealand. Am Mineral 74:759-763

Churkin M, Packham GH (1973) Volcanic rocks and volcanic constituents in sediments, Leg 21, Deep Sea Drilling Project. Initial Reports of the Deep-Sea Drilling Project 21, RE Burns (ed) U S Govt Printing Office, Washington, DC, p 481-493

Coombs DS (1954) The nature and alteration of some Triassic sediments from Southland, New Zealand. Trans Roy Soc New Zealand 82:65-103

Coombs DS, Ellis AS, Fyfe WS, Taylor AM (1959) The zeolite facies, with comments on the interpretation of hydrothermal syntheses. Geochim Cosmochim Acta 17:53-107

Couture IA (1977) Composition and origin of palygorskite-rich and montmorillonite-rich zeolite-containing sediments from the Pacific Ocean. Chem Geology 19:113-130

Czyscinski K (1973) Authigenic phillipsite formation rates in the central Indian Ocean and the equatorial Indian Ocean. Deep-Sea Res 20:555-559

de' Gennaro M, Adabbo M, Langella A (1995) Hypothesis on the genesis of zeolites in some European volcaniclastic deposits. *In* Natural Zeolites '93: Occurrence, Properties, Use. DW Ming, FA Mumpton (eds) Int'l Comm on Natural Zeolites, Brockport, New York, p 51-67

de' Gennaro M, Franco E, Paracuollo G, Passarelli G (1980) I "tufi" del Vesuvio con cabasite potassica. Period Mineral, Roma 49:223-240

Desprairies A, Riviere M, Pubellier M (1991) Diagenetic evolution of Neogene volcanic ashes (Celebes and Sulu Seas). Silver EA, Rangin C, von Breymann MT et al., Proc Ocean Drilling Program, Scientific Results 124:489-503

Donahoe RJ, Liou JG, Guldman S (1984) Synthesis and characterization of zeolites in the system Na_2O-K_2O-Al_2O_3-SiO_2-H_2O. Clays & Clay Minerals 32:433-443

Garcia Hernandez JE, Notario del Pino JS, Gonzalez Martin MM, Hernan FG, Rodriguez Losada JA (1993) Zeolites in pyroclastic deposits in southeastern Tenerife (Canary Islands). Clays & Clay Minerals 41:521-526

Gottardi G, Galli E (1985) Natural Zeolites. Springer-Verlag, Berlin, 409 p

Gude AJ III, Sheppard RA (1988) A zeolitic tuff in a lacustrine facies of the Gila Conglomerate near Buckhorn, Grant County, New Mexico. U S Geol Survey Bull 1763, 22 p

Hay RL (1963) Stratigraphy and zeolitic diagenesis of the John Day Formation of Oregon. Univ Calif Publ in Geol Sciences 42:199-262

Hay RL (1966) Zeolites and zeolitic reactions in sedimentary rocks. Geol Soc Am Spec Paper 85, 130 p

Hay RL (1976) Geology of the Olduvai Gorge. Univ California Press, Berkeley, 203 p

Hay RL (1980) Zeolitic weathering of tuffs in Olduvai Gorge, Tanzania. *In* Proc 5th Int'l Conf on Zeolites, Naples. LVC Rees (ed) Hayden & Son, London, p 155-163

Hay RL, Guldman SG (1987) Diagenetic alteration of silicic glass in Searles Lake, California. Clays & Clay Minerals 35:449-457

Hay RL, Guldman SG, Matthews JC, Lander RH, Duffin ME, Kyser TK (1991) Clay mineral diagenesis in core KM-3 of Searles Lake, California. Clays & Clay Minerals 39:84-96

Hay RL, Iijima A (1968) Nature and origin of the palagonite tuffs of the Honolulu Group on Hawaii. Geol Soc Am Mem 116:331-376

Hein JR, Scholl DW (1978) Diagenesis and distribution of late Cenozoic volcanic sediment in the southern Bering Sea. Geol Soc Am Bull 89:197-210

Ibrahim K, Hall A (1995) New occurrences of diagenetic faujasite in the Quaternary tuffs of northeast Jordan. Eur. J. Mineral. 7:1129-1135

Iijima A (1974) Clay and zeolitic alteration zones surrounding Kuroko deposits in the Hokuroku district, northern Akita, as submarine hydrothermal-diagenetic alteration products. Mining Geol Special Issue 6:267-289

Iijima A (1978) Geological occurrences of zeolite in marine environments. *In* Natural Zeolites: Occurrence,

Properties, Use. LB Sand, FA Mumpton (eds) Pergamon Press, Elmsford, New York, p 175-198
Iijima A (1988a) Diagenetic transformation of minerals as exemplified by zeolites and silica minerals—a Japanese view. *In* Diagenesis II, Developments in Sedimentology 43, GV Chilingarian, KH Wolf (eds) Elsevier Science Publishers, Amsterdam, p 147-211
Iijima A (1988b) Application of zeolites to petroleum exploration. *In* Occurrence, properties, and utilization of natural zeolites, D Kalló, HS Sherry (eds) Akadémiai Kiadó, Budapest, p 29-37
Iijima A (1995) Zeolites in petroleum and natural gas reservoirs in Japan: a review. *In* Natural Zeolites '93: Occurrence, Properties, Use. DW Ming, FA Mumpton (eds) Int'l Comm on Natural Zeolites, Brockport, New York, p 99-114
Iijima A, Ogihara S (1995) Temperature-time relationships of zeolitic reactions during burial diagenesis in marine sequences. *In* Natural Zeolites '93: Occurrence, Properties, Use. DW Ming, FA Mumpton (eds) Int'l Comm on Natural Zeolites, Brockport, New York, p 115-124
Iijima A, Utada M (1972) A critical review on the occurrence of zeolites in sedimentary rocks in Japan. Jap J Geol Geog 42:61-84
Jakobsson SP, Moore JG (1986) Hydrothermal minerals and alteration rates at Surtsey volcano, Iceland. Geol Soc Am Bull 97:648-659
Joulia F, Bonifas M, Camez T, Millot G, Weil R (1959) Découverte d'un important niveau d'analcimolite greseuse dans le Continetal intercalaire de l'ouest de l'Air (Sahara central). Dakar Service de Géologie et de Prospection Miniere, 40 p
Karlsson HR (1995) Application of oxygen and hydrogen isotopes to zeolites. *In* Natural Zeolites '93: Occurrence, Properties, Use. DW Ming, FA Mumpton (eds) Int'l Comm on Natural Zeolites, Brockport, New York, p 125-140
Karlsson HR, Clayton RN (1991) Analcime phenocrysts in igneous rocks: primary or secondary? Am Mineral 76:189-199
Kastner M (1979) Zeolites. *In* Marine Minerals. R.G. Burns (ed) Rev Mineral 6:111-120
Kastner M, Stonecipher SA (1978) Zeolites in pelagic sediments of the Atlantic, Pacific, and Indian Oceans. *In* Natural Zeolites: Occurrence, Properties, Use. LB Sand, FA Mumpton (eds) Pergamon Press, Elmsford, New York, p 199-220
Kristmannsdóttir H, Tómasson J (1978) Zeolite zones in geothermal areas in Iceland. *In* Natural Zeolites: Occurrence, Properties, Use. LB Sand, FA Mumpton (eds) Pergamon Press, Elmsford, New York, p 277-284
Lander RH, Hay RL (1993) Hydrogeological controls on zeolitic diagenesis of the White River Sequence. Geol Soc Am Bull 105:361-376
Levy SS, O'Neil JR (1989) Moderate-temperature zeolitic alteration in a cooling pyroclastic deposit. Chem. Geol. 76:321-326
Luhr JF, Giannetti B (1987) The brown leucitic tuff of Roccamonfina Volcano (Roman Region, Italy). Contrib Mineral Petrol 95:420-436
Luhr JF, Kyser TK (1989) Primary igneous analcime: The Colima minettes. Am Mineral 74:216-223
Merino E, Wang Y, Deloule É (1995) Genesis of agates in flood basalts: twisting of chalcedony fibers and trace-element geochemistry. Am J Sci 295:1156-1176
Ming DW, Mumpton FA (1989) Zeolites in soils. *In* Minerals in Soil Environments, 2nd Edn. JB Dixon, SB Weed (eds) Soil Sci Soc Am, Madison, Wisconsin, p 873-911
Mohapatra BK, Sahoo RK (1987) Merlinoite in manganese nodules from the Indian Ocean. Mineral Mag 51:749-750
Noh JH, Boles JR (1989) Diagenetic alteration of perlite in the Guryongpo area, Republic of Korea. Clays & Clay Minerals 37:47-58
Passaglia E, Pongiluppi D, Rinaldi R (1977) Merlinoite a new mineral of the zeolite group. Neues Jahrb Miner Mh (1977), 355-364
Pérez-Torrado FJ, Martí J, Queralt I, Mangas J (1995) Alteration processes of the Roque Nublo ignimbrites (Gran Canaria, Canary Islands). J Volc Geotherm Res 65:191-204
Pipkin BW (1965) Mineralogy of 140-foot core from Willcox Playa, Cochise, Arizona. Am Assoc Petrol Geol Bull 51:470
Remy RR, Ferrell RE (1989) Distribution and origin of analcime in marginal lacustrine mudstones of the Green River Formation, south-central Uinta Basin, Utah. Clays & Clay Minerals 37:419-432
Renaut RW (1993) Zeolitic diagenesis of Late Quaternary fluviolacustrine sediments and associated calcrete formations in the Lake Bogoria basin, Kenya Rift Valley. Sedimentology 40:271-301
Sheppard RA (1991) Zeolitic diagenesis of tuffs in the Miocene Chalk Hills Formation, western Snake River Plain, Idaho. U S Geol Survey Bull 1963, 27 p
Sheppard RA (1993) Geology and diagenetic mineralogy of the Castle Creek zeolite deposit and the Ben-Jel bentonite deposit, Chalk Hills Formation, Oreana, Idaho. *In* Zeo-Trip '93, An Excursion to Selected Zeolite and Clay Deposits in Southeastern Oregon and Southwestern Idaho, June 26-28, 1993, FA

Mumpton (ed) Int'l Comm on Natural Zeolites, Brockport, New York, p 1-13
Sheppard RA, Gude AJ III (1968) Distribution and genesis of authigenic silicate minerals in tuffs of Pleistocene Lake Tecopa, Inyo County, California. U S Geol Survey Prof Paper 597, 38 p
Sheppard RA, Gude AJ III (1969) Diagenesis of tuffs in the Barstow Formation, Mud Hills, San Bernardino County, California. U S Geol Survey Prof Paper 634, 35 p
Sheppard RA, Gude AJ III (1973) Zeolites and associated authigenic silicate minerals in tuffaceous rocks of the Big Sandy Formation, Mohave County, Arizona. U S Geol Survey Prof Paper 830, 36 p
Sheppard RA, Gude AJ III, Griffin JJ (1970) Chemical composition and physical properties of phillipsite from the Pacific and Indian Oceans. Am Mineral 55:2053-2062
Spencer R, Baedecker MJ, Eugster HP, Forester RM, Goldhaber B, Jones BF, Kelts K, Mckenzie J, Madsen DB, Rettig SL, Rubin M, Bowser CJ (1984) Great Salt Lake, and precursors, Utah: The last 30,000 years. Contrib Mineral Petrol 86:321-334
Stonecipher SA (1978) Chemistry and deep-sea phillipsite, clinoptilolite, and host sediments. *In* Natural Zeolites: Occurrence, Properties, Use. LB Sand, FA Mumpton (eds) Pergamon Press, Elmsford, New York, p 221-234
Stamatakis MG (1989) Authigenic silicates and silica polymorphs in the Miocene saline-alkaline deposits of the Karlovassi basin, Samos, Greece. Econ Geol 84:788-798
Surdam RC (1977) Zeolites in closed hydrologic systems. Rev Mineral 4:65-91
Surdam RC, Sheppard RA (1978) Zeolites in saline, alkaline lake deposits. *In* Natural Zeolites: Occurrence, Properties, Use. LB Sand, FA Mumpton (eds) Pergamon Press, Elmsford, New York, p 145-174
Taylor MW, Surdam RC (1981) Zeolite reactions in the tuffaceous sediments at Teels Marsh, Nevada. Clays & Clay Minerals 29:341-352
Tschernich RW (1992) Zeolites of the World. Geoscience Press, Phoenix, Arizona, 563 p
Turner CE, Fishman NS (1991) Jurassic Lake T'oo'dichi: A large alkaline, saline lake, Morrison Formation, eastern Colorado Plateau. Geol Soc Am Bull 103:538-558
Utada M (1988) Occurrence and genesis of hydrothermal zeolites and related minerals from the Kuroko-type mineralization areas in Japan. *In* Occurrence, Properties, and Utilization of Natural Zeolites. D Kallő, HS Sherry (eds) Akadémiai Kiadó, Budapest, p 39-48
Vitali F, Blanc G, Larqué P (1995) Zeolite distribution in volcaniclastic deep-sea sediments from the Tonga Trench margin (SW Pacific). Clays & Clay Minerals 43:92-104
von Breymann MT, Swart PK, Brass GW, Berner U (1991) Pore water chemistry of the Sulu and Celebes seas: extensive diagenetic reactions at sites 767 and 768. Silver EA, Rangin C, von Breymann MT et al., Proc Ocean Drilling Program, Scientific Results 124:203-215
Walker GPL (1960a) Zeolite zones and dike distribution in relation to the structure of the basalts of eastern Iceland. J Geology 68:515-528
Walker GPL (1960b) The amygdale minerals in the Tertiary lavas of Ireland. III. Regional distribution. Mineral Mag 32:503-527
Wise WS, Kleck WD (1988) Sodic clay-zeolite assemblage in basalt at Boron, California. Clays & Clay Minerals 36:131-136

7 Zeolites in Closed Hydrologic Systems

A. Langella[*], P. Cappelletti[§], M. de' Gennaro[§]

[*] *Facoltà di Scienze*
Università del Sannio
Via Port'Arsa, 11
82100 Benevento, Italy

[§] *Dipartimento di Scienze della Terra*
Università "Federico II" di Napoli
Via Mezzocannone, 8
80134 Napoli, Italy

INTRODUCTION

Zeolites were first described from vugs and fissures in basaltic flows in the mid-eighteenth century (Cronstedt 1756), and they are now known to be widespread in a variety of geological environments. More zeolite species occur in vesicles and fractures of basaltic rocks than in any other geologic setting. However, zeolites from sedimentary rocks, in particular, represent the most important occurrences both in terms of aerial extent of deposits and in terms of the abundance of certain zeolite species. The introduction of new analytical techniques in the latter part of the 20th century provided a significant stimulus to zeolite research; these techniques allowed identification of large quantities of zeolites in widespread deposits of sedimentary rocks and, in particular, in sediments from saline, alkaline lakes of the western United States (Sheppard and Gude 1968, 1969a, 1973). Based on geological and hydrologic environments, zeolite occurrences in sedimentary rocks can be classified in the following framework: (a) saline, alkaline lakes; (b) alkaline soils and land surfaces; (c) deep-sea sediments; (d) open hydrologic systems; (e) as products of hydrothermal alteration; and (f) burial diagenetic or low-metamorphic environments. From a broader perspective, zeolite deposits can be classified into two main groups, namely closed hydrologic systems and open hydrologic systems. The work of Sheppard and Gude (1968, 1969a, 1973) and later Surdam (1977), discussing not only the mineralogy of the deposits but also the geology, hydrology, and chemistry of the depositional basins, represent milestones in the study of closed hydrologic system formation of zeolites. Surdam (1977) considered closed hydrologic basins in two different tectonic settings: (a) block-faulted regions in arid and semiarid regions; and (b) trough valleys associated with rifting. Examples of the first setting are the closed lakes of the Basin and Range province of the western United States whereas the East Rift Valley of Kenya is the outstanding example of rift areas.

The present chapter is a detailed analysis of the both "old" and the more recent literature on zeolite occurrences in closed hydrologic systems. Through analysis of new occurrences and by considering the concept of "closed hydrologic systems" in association with known zeolite deposits which have been previously classified in different genetic categories, this chapter aims to expand our present understanding of closed-system zeolitization.

GEOLOGICAL SETTINGS OF SALINE, ALKALINE LAKES

Hydrologic closed-system mineral-forming environments commonly include two distinct geologic contexts: *playa-lake systems* and *rift-type systems* (Surdam 1977).

1529-6466/00/0045-0007$05.00

Significant hydrogeologic and morphologic differences exist between these two systems which give rise to physico-chemically unique surface and pore waters.

Playa-lake systems can be considered in every respect as sedimentary basins whose geomorphological conditions support the migration of both surface and pore waters towards the most depressed area of the basin. The presence of a substantially semi- or impermeable substratum gives rise to the formation of a lake. Such playa lake waters can change their composition by normal evaporative processes and/or by hydrolysis of crystalline and amorphous phases present in the sediments.

Rift-type system can be distinguished from playa-lake systems because the waters feeding the impounded lake derive from underground springs. The waters therefore have minimal sediment load due to the reduced surface stream influx. Spring waters in these systems are commonly characterized by high salinities as a consequence of subsurface water-rock interactions.

In both systems, evaporation plays an important role in modifying the chemical composition of waters, and an essential condition is that evaporation exceeds precipitation during dry periods. The evolution of the water chemical composition in a closed system brings about the formation of a brine which, according to Eugster (1970) and Hardie and Eugster (1970), may or may not be alkaline. The alkalinity is primarily a consequence of Ca^{++}–HCO_3^- equilibria in solution which can control the precipitation of calcite. However, if one considers in addition the mineral-fluid interactions occurring at the bottom of the basin, it is clear that aqueous equilibria are much more complex, particularly when the sediments include silicate minerals or volcanic glass. It therefore appears that the hydrolysis of these phases, most importantly of volcanic glass, controls the solution chemistry and gives rise to an elevated pH as a consequence of the following exchange reaction ("s" indicates solid or glass and "l" indicates liquid):

$$Na^+_{(s)} + H_2O_{(l)} \Leftrightarrow H^+_{(s)} + Na^+_{(l)} + OH^-_{(l)}$$

The resulting aqueous conditions favor alkaline attack of the glass framework (de' Gennaro et al. 1988):

$$\equiv Si - O - \underset{Na^+}{\overset{\ominus}{Al}} \equiv + OH^- \Rightarrow \equiv Si - O^{\ominus} - \underset{Na^+}{\overset{\ominus}{Al}} \equiv$$

The first product of such reactions could be a gel-like phase which evolves with time to hydrated aluminosilicates such as zeolites.

Zeolite-bearing deposits of saline alkaline lakes

Zeolites are widely dispersed in most saline, alkaline lakes of the world, ranging in age from Early Carboniferous to Holocene (Hay 1966, Surdam and Sheppard 1978). In this sedimentary environment, characterized in some instances by very thick (up to 1 km) and extensive (thousands of km^2) sequences, zeolites can represent a considerable portion of the intercalated volcanic beds, of pelites (clays and mudstones), and of carbonates (mainly dolomite) (Table 1). Locally, zeolites form almost pure beds of authigenic minerals replacing the vitric tuff. Zeolites forming in this particular environment have a low-temperature origin. Volcanic material falling into closed basins (saline, alkaline lakes) slowly reacts with solutions entering the basins. In arid or semiarid areas, the evaporation rate is usually high and is not balanced by sufficient inflowing waters, giving rise to an increase in salinity and pH. The concentric zonation of waters resulting from evaporation and shrinkage of the water-filled portion of the basin, causing an increase in salinity and pH from the periphery to the center of the basin, is reflected in zonation of

Table 1. Important features of selected saline, alkaline lake deposits.

Age	*Location*	*Formation*	*Zeolites*	*Associated minerals*	*References*
Miocene	Samos Island, Greece	Lower Neogene Unit (LNU), rhyolitic volcanic tuffs	CLI, ANA	KF, Opal-CT, SM	Stamatakis (1989)
Burdigalian-Helvetian	Slanci Basin, Serbia	Slanci Series, rhyodacitic tuffs	ANA	I/S	Obradovic & Dimitrijevic (1987)
Middle Miocene	Vranje basins, Serbia	Zlatakop Fm., dacitic tuffs	CLI	Q	Obradovic & Dimitrijevic (1987)
Middle Miocene	Valjevo-Mionica basins, Serbia		ANA	KF, SLS	Obradovic (1977)
Miocene	Beypazari, Turkey	Çayirhan Fm., rhyodacitic volcanics in coal	CLI, ANA	KF, Q, PYR, DOL, I/S	Whateley et al. (1996)
Late Permian	Sub-Urals, Russia	Red Fm., pyroclastics in sandstones	ANA		Kossowskaya (1973)
Late Quaternary	Lake Bogoria Basin, Kenya	Loboi Silts Fm., trachytic pyroclastics in claystones and mudstones	ANA, CHA, NAT	CC, I/S, OX	Renaut (1993)
Pleistocene-Holocene	Lake Magadi region, Kenya	High Magadi and Orolonga Beds, trachytic pyroclastic ashes	ANA, ERI, PHI, CHA, MOR, CLI	Gels, CC, I/S, GYL, Q, MGD	Surdam & Eugster (1976)
Pliocene	Grant County, New Mexico, USA	Gila Conglomerate, siliceous tuff in lacustrine facies	CLI, CHA, ERI, MOR, ANA, PHI	I/S, Opal-CT, KF, Q	Gude & Sheppard (1988)
Upper Jurassic	Plateau region, Colorado, USA	Brushy Basin member, Morrison fm., rhyodacitic pyroclastic ashes	CLI, ANA	I/S, KF, ALB, Q	Turner & Fishman (1991)
Miocene	Western Snake River plain, Idaho, USA	Chalk Hills fm., rhyolitic volcanic tuffs	CLI, ERI, MOR, PHI	Opal-CT, SM, Q, KF, CC, GYP	Sheppard (1989)
Miocene	Sonora State, Mexico	Baucarit Fm., Las Palmas Devisaderos, rhyolitic ash-flow tuff	CHA, ERI, HEU	I/S, GYP, THE	Cochemé et al. (1996)
Carboniferous	W Newfoundland, Canada	Rocky Brook Fm., silicoclastic and calcareous sediments	ANA	SM, I, C/S	Gall & Hyde (1989)
Late Carboniferous	Werris Creek, NSW, Australia	Escott unit of Currabubula Fm., rhyolite to rhyodacitic tuffs	HEU, MOR	KF, Q	Flood & Taylor (1991)

Note: CLI = clinoptilolite; ANA = Analcime; CHA = Chabazite; ERI = Erionite; PHI = Phillipsite; MOR = Mordenite; HEU = Heulandite.

KF = K-feldspar; Q = Quartz; PYR = Pyrite; DOL = Dolomite; CC = Calcite; OX = Iron oxides; GYL = Gaylussite; MGD = Magadiite; SM = Smectite; I = Illite; I/S = mixed-layer Illite/Smectite; C/S = Mixed-layer Chlorite/Smectite; ALB = Albite; THE = Thenardite; SLS = Searlesite

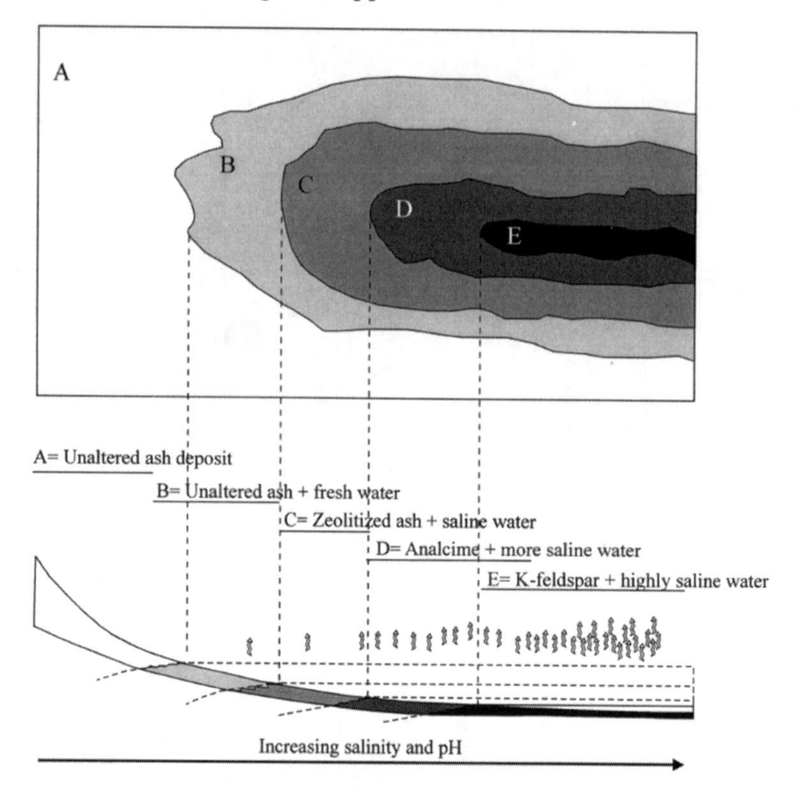

Figure 1. Distribution of authigenic mineral zones in saline, alkaline playa-lake systems.

the resulting mineral deposits. Such deposits display a concentric pattern, particularly in playa-lakes as a result of closed-system alteration. Naturally, as evaporation proceeds in such basins, the volume of tephra interacting with saline, alkaline solutions undergoes a progressive and often rhythmic reduction in volume as the water-filled portions of the basins shrink (Fig. 1). Volcanic ash remains unaltered on the land surface surrounding the basin and above the water table; proceeding inwards, glass is increasingly altered due to the presence of a capillary fringe with a consequent increase in alkalinity. After an intermediate stage characterized by the formation of an aluminosilicate gel (Mariner and Surdam 1970), zeolites begin to crystallize when the pH reaches alkaline values (~9-10). A further increase in salinity and alkalinity brings about the formation of zeolites such as analcime, which later evolves to potassium feldspar in the very central portions of lakes. These reactions can be summarized according to the following scheme:

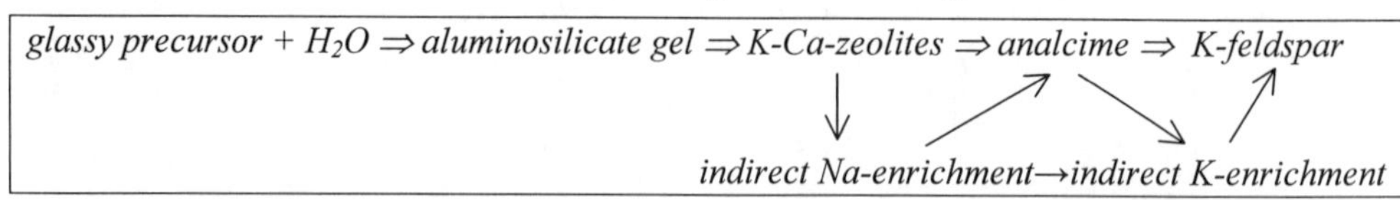

It should be remarked that the boundaries between the authigenic mineral zones are gradational in part because of climate variations. The latter cause changes in hydrologic parameters and, as a consequence of dilution/concentration processes, a variation in the chemistry of the system. However, in some instances, closed hydrologic basins lack any significant mineral zoning. This is the case in valleys in rift areas where the broad and flat areas necessary for this kind of lateral variation are absent. For example, the presence of very steep basin flanks, such as in a narrow fault trough, inhibits the development of a recognizable mineral zonation sequence. A further difference between the two kind of geological settings is the chemical composition of the volcanic products in the basins. In the Basin and Range province, volcanic materials are predominantly rhyolitic in composition, whereas tectonic settings such as rift valleys are mainly characterized by trachytic pyroclastic materials. The following sections describe details of a variety of saline, alkaline lake deposits organized geographically, beginning with Eurasia, through Africa, North America, and Australia (Fig. 2). In these sections, playa-lake and rift-system deposits are discussed jointly.

Samos Island, Greece. Samos Island is characterized by two main Neogene basins (Mytilinii and Karlovassi) situated on a basement consisting of four tectonic units (Papanikolaou 1979, Theodoropoulos 1979, Mezger and Okrusch 1985). The western basin of the island (Karlovassi) consists of a basal conglomerate overlain by the Lower Neogene Unit, a thick sequence (about 400 m) of limestones and marly limestones, tuffaceous rocks and volcanic ash layers, dolomitic clays and porcelaneous limestones. Stamatakis (1986, 1989a, 1989b) reported a clear zoning of authigenic silicates in volcanic layers of this sequence. A peripheral non-authigenic mineral area consisting of carbonate rocks and epiclastic minerals surrounds an inner zone where ash tuffs and tuffites were deposited, characterized by an inward progression from K-zeolite (clinoptilolite) to Na-zeolite (analcime) and finally to K-feldspar (Fig. 3). The evidence supporting a saline, alkaline lake genesis of the observed assemblages is given by the occurrence of evaporite minerals within the entire sequence, the existence of concentric mineral zoning, and the high B content of K-feldspars. The solutions permeating the deposit progressively increased their salinity and alkalinity with time, reaching their highest values during the Messinian salinity crisis. At pHs between 8 and 9, clinoptilolite began to form directly from the unaltered volcanic glass, along with smectite (Surdam and Parker 1972, Sheppard and Gude 1973). A further increase in salinity and alkalinity led to the crystallization of analcime. K-feldspar formed at the expense of analcime and clinoptilolite when K^+/Na^+ or K^+/H^+ ratios and SiO_2 activities increased, and H_2O and partial pressure of CO_2 decreased. At this late stage of diagenesis, remaining B in pore fluids entered the structure of authigenic K-feldspar (Sheppard and Gude 1973). This process explains the anomalous high B content in K-feldspar (up to 2500 ppm) compared with very low values (up to 12 ppm) in pyrogenic sanidine (Stamatakis 1989a). The abundance of K-feldspar, never co-existing with volcanic glass, and its spatial relationship with other phases such as zeolites are further evidence of late-stage authigenic crystallization.

Slanči basin, Serbia. Tectonically active lake basins of Miocene age are located in the marginal part of the Pannonian Sea. The stratigraphic series of the Slanči basin (Helvetian) near Belgrade consists of dolomitic marls, tuffaceous marls and clays, and tuffs and tuffites of rhyodacitic composition. Authigenic minerals have been identified only in the tuffs and tuffites which correspond to the middle part of the series (Obradovic 1988). The alteration sequence of the volcanic products begins with a Ca-montmorillonite zone which laterally and downwards evolves to clinoptilolite and then analcime. Chemical analyses of analcime revealed wide compositional variability, depending primarily on composition of the host-rock. For example, anomalous Mg- and Ca-bearing analcimes have been found in tuffaceous dolomitic marls.

Although lateral mineralogical zonation has not been observed in the above sequences, the typical paragenesis of volcanic glass $\rightarrow$ montmorillonite $\rightarrow$ clinoptilolite

Figure 2. Worldwide distribution of zeolite deposits described in this chapter.

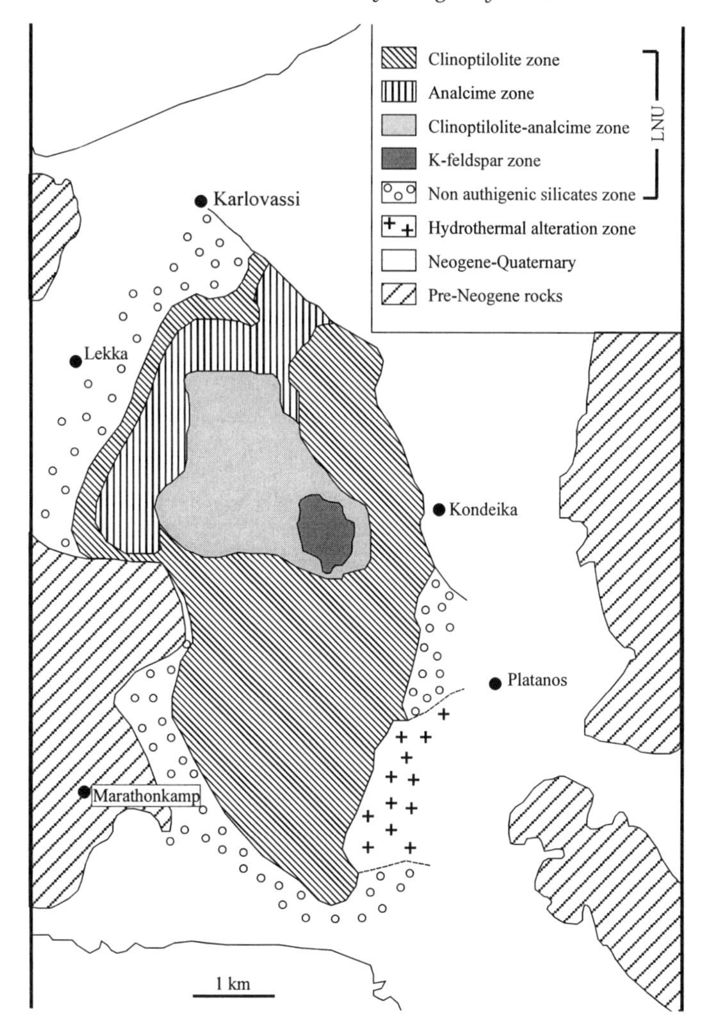

Figure 3. Karlovassi basin (Samos Island, Greece). Modified after Stamatakis (1989b). LNU = Lower Neogene Unit.

→ analcime suggests alteration in a saline, alkaline lake under conditions of low temperature and pressure. This is confirmed by comparison with another lake in Serbia belonging to the same basin. In this lake, a combination of climatic change, giving rise to conditions favorable for evaporation, along with shallowing of the basin itself due to tectonic movements caused a gradual shift to saline, alkaline conditions (Obradovic 1988).

Miocene Basins of southern Serbia. Intramountain valleys of southern Serbia (Vranje, Vladicin Han, Kamenica-Toponica) are characterized by the presence of zeolite-bearing dacite tuffs (Obradovic and Dimitrijevic 1987). Clinoptilolite from Zlatokop (Vranje basin) is the prevailing authigenic phase, followed by analcime and minor amounts of mordenite and erionite. According to Obradovic (1977), the clinoptilolite is particularly Ca-enriched and the Si/Al ratio is 4.3-4.8. The transition of volcanic glass →

clinoptilolite → analcime took place in a saline, alkaline lake at the time of sedimentation and diagenesis. Chalcedony and quartz in the uppermost part of the deposit substantiate the circulation of later silica-rich solutions. There is no evidence that these fluids participated to the zeolitization process.

The Valjevo-Mionica basin contains a series of oil shales about 200 m thick. The middle part of this series consists of tuffs and tuffaceous pelites bearing thin beds of searlesite with minor amounts of analcime and K-feldspar. This assemblage of authigenic minerals suggests that the deposit formed in a saline, alkaline lake. Leaching of the volcanic precursor provided elements such as B and Na which were responsible for the formation of searlesite. According to Obradovic (1977), analcime does not form directly from volcanic glass but from a pre-existing zeolite. Inasmuch as the only assemblage in this area is analcime, searlesite, and K-feldspar (with no other zeolites), Obradovic hypothesized that probable precursor zeolites have totally altered to analcime. At the same time, Obradovic did not exclude the possibility that analcime could have formed directly by precipitation from solution. More recent research on water-rock interaction (de'Gennaro et al. 1993) shows that analcime can crystallize directly from solution in Na-rich environments by interaction with a wide range of glass compositions. Therefore, the presence of a zeolitic precursor need not necessarily be invoked to explain the occurrence of analcime. K-feldspar appears to be the last phase to crystallize from analcime under the saline, alkaline conditions in these rocks.

Çayirhan mine, Beypazari, Turkey. Occurrences of authigenic zeolites in the sedimentary sequences of the Beypazari basin (Turkey) were first described by Ataman (1976). Later, a detailed study (Whateley et al. 1996) was carried out on zeolites of the Çayirhan coalfield, which developed in the upper Miocene as a consequence of extensional tectonic events. The basement of this basin consists mainly of Paleozoic metamorphic rocks (schists) intruded by granite and Cretaceous limestones, ophiolites, and Paleocene sediments. The coal-bearing sequence contains two distinct seams: the first has clinoptilolite as the dominant zeolite and very subordinate analcime, whereas the second seam contains these zeolites in reversed proportion. The predominance of zeolites and feldspars along with low content of quartz and clay minerals within the mineral fraction of the coal led Whateley et al. (1996) to conclude that volcanic glass, likely derived from the Neogene Teke volcanics, was the major detrital component in the precursor matter of the Çayirhan coalbed.

Zeolite genesis in this sedimentary sequence can be summarized as follows:

- Alloctonous organic matter was deposited in a closed fresh-water basin with neutral to slightly acidic pH; the major detrital component of this peat was the volcanic glass derived from the Teke volcanics.
- The peat deposit was altered as a consequence of a rise in alkalinity of the syngenetic saline pore water (pH > 9) contributed from local volcanic sources.
- Biological, biochemical, and geochemical processes necessary to increase the rank of the peat to lignite occurred over a range of temperature from 39° to 68°C.
- Late syngenetic formation of zeolites took place within this temperature range; higher temperatures would have raised the coal to higher rank.
- The occurrence of different zeolites is likely related to different proportions of Na:Ca and/or K:Ca, and Al:Si in the precursor volcanic ash and/or in the pore fluids. Analcime crystallized from high Na/Ca-ratio solutions and relatively high Al/Si ratio of volcanic ash, whereas clinoptilolite is present where lower Na/K+Ca and Al/Si ratios occurred.

Red Analcime Formation of the sub-Urals. Kossowskaya (1973) used the crystal chemical variations of a particular zeolite in different modes of occurrence and the paragenetic relationship with other zeolites and clay minerals as a novel indicator of geological facies. As far as Russian Paleozoic deposits are considered, analcime was observed in the terrigenous Permian red formation of the sub-Urals. Its total thickness is about 600 m, characterized by a mixed terrigenous and chemical precipitate lower complex formed in saline lakes and lagoons, a middle complex of marine origin, and an upper complex again of saline lake and lagoon environments. In spite of the similarity in conditions of formation and the compositions of detrital material in these three complexes, analcime occurs only in the upper 150- to 300-m thick beds, as a cement in sandstones or filling pores of the many chemical precipitate rocks associated with them. The primary difference between the lower and upper complex consists of the presence of heavy minerals (pyroxene and amphibole) in the upper complex only. According to Kossowskaya, the stratigraphic boundary of authigenic analcime as a sandstone cement coincides almost exactly with the heavy mineral boundary. These phases reveal the former presence of pyroclastic material, thereby suggesting that volcanic activity in this area began no earlier than late Permian.

Lake Bogoria Basin, Rift Valley, Kenya. Occurrences of zeolites at Lake Bogoria were first reported by Hay (1970, 1976), but a detailed study was not carried out until later (Renaut 1993). This basin occupies a depression in the central part of the Kenya Rift Valley, currently characterized by sodium carbonate rich waters (pH ~10) and thus representing a recent analogue to ancient lake margin zeolite occurrences. The depositional area is underlain mainly by late Tertiary and Pleistocene basalt, trachyte, and phonolite. Zeolite alteration occurs on the Sandai Plain (northern area) where the prevailing siltstone, mudstone, and claystone (Loboi Silts) contain up to 40% analcime and minor natrolite, both concentrated mostly in the upper 1 m of the Loboi Silts. Most of the detrital mineral fraction of this formation was derived from the weathering of the above-mentioned volcanic rocks. In the marginal areas of the depression, these sediments are altered, and analcime is the dominant authigenic mineral along with minor natrolite and possible chabazite. Four possibilities were considered by Renaut (1993) to explain the origin of analcime: (a) direct precipitation from interstitial pore fluids or lake water; (b) formation from a gel; (c) alteration of precursor zeolites; and (d) alteration of precursor detrital silicates (feldspars or clay minerals) by reaction with pore waters. Because analcime is preferentially concentrated in claystone and mudstone, Renaut (1993) suggested that poorly crystalline detrital clays and amorphous materials reacted with interstitial sodic pore waters concentrated by evaporation to form analcime. Other silicates (plagioclase) and minor volcanic glass may also have undergone alteration. The overall reaction may have been as follows: detrital silicates + Na-rich solutions + mafic minerals + oxygen $\Rightarrow$ analcime + iron oxides. The limited occurrence of zeolites can be ascribed to environmental instability including, in addition to climatic variations, tectonic instability, changes in drainage basin morphology, and hydrothermal recharge as well as insufficient available reaction time. Because the only prolonged period of relatively dry conditions in that region was during the late Pleistocene, between 20,000 and 12,000 years BP, as also reported for other East African lakes, Renaut (1993) concluded that zeolite formation occurred during this period from the alteration of poorly ordered clay minerals by sodium carbonate-bicarbonate pore waters that were concentrated close to the land surface by evaporation and evapotranspiration.

Lake Magadi region, Rift Valley, Kenya. Lake Magadi occupies the lowermost area of the Eastern Rift Valley. It is a trona-precipitating saline lake fed by alkaline hot springs. Evaporite beds are currently accumulating over an area of about 60 km^2. Older lake deposits belong to the High Magadi Beds (9,100 yr.) which overlay the Oloronga Beds (780,000 yr.). The sedimentary rocks of these formations are essentially bedded chert and trachytic tuffaceous materials. A thorough investigation of these tuffaceous sediments of Pleistocene to Holocene age was carried out by Surdam and Eugster (1976), and their studies showed that the sediments are highly zeolitized. Authigenic zeolites are predominantly erionite and analcime along with minor amounts of chabazite, clinoptilolite (Fig. 4), mordenite, and phillipsite. These minerals all represent the common alteration products derived from the interaction of trachytic glass with alkaline lake waters. The study of present-day glass and brine chemistry in the Magadi basin represents a novel and productive approach to the interpretation of this particular suite of authigenic minerals. Erionite derives directly from trachytic glass interacting with water in an environment characterized by high sodium and silica activities and very low activities of alkaline earths. Clinoptilolite can form when some Ca and Mg are available along with a high silica activity. Chabazite crystallization also requires alkaline earths but lower silica activities. Phillipsite formation requires a low silica activity, and this zeolite is concentrated in soil horizons. Analcime is found both in surface and core samples. In the former case it represents an alteration product after erionite and can only form when silica activity is further lowered. In some cases analcime from core samples can derive by direct crystallization from a Na-Al-Si gel. This zeolite precursor gel is widespread in the Magadi basin and is mainly located at or adjacent to hot springs where alkaline brines are in contact with trachytic debris. No authigenic feldspars have been found at Lake Magadi because the kinetic and temperature conditions necessary for the alteration of detrital feldspar fragments did not exist. Apart from isolated occurrences, albite has not been reported in tuffaceous rocks younger than Miocene. Williamson (1987) described authigenic albite in several silicic tuffs of the Kramer Beds, the Miocene lacustrine formation that is host for the large borate deposit in the Mojave Desert. Other authigenic minerals in the tuffs include analcime, clinoptilolite, K-feldspar, and searlesite.

Figure 4. Polarizing micrograph of clinoptilolite crystals on felted erionite from Lake Magadi, Kenya (Surdam and Eugster 1976).

Surdam and Eugster (1976) attempted to explain erionite's prevalence as a product of trachytic glass alteration at Lake Magadi as follows:

- The trachytic glass is rich in alkalis and poor in alkaline earths.
- The existence of high activities of SiO_2 and Na_2O, necessary for erionite crystallization, is shown by the presence of silica and silica phases, such as magadiite and hydrous sodium aluminosilicate gels.
- The current Magadi basin inflowing waters and brines are rich in sodium and silica and devoid of alkaline earths.

They hypothesized that analcime crystallized through a zeolite precursor, typically erionite, according to the following reaction:

$$\underset{\text{Magadi erionite}}{Na_{0.5}K_{0.5}AlSi_{3.5}O_9 \cdot 3H_2O} + 0.5\ Na^+ \rightarrow \underset{\text{analcime}}{NaAlSi_2O_6 \cdot H_2O} + 0.5\ K^+ + 1.5\ SiO_2 + 2\ H_2O$$

The high silica content of analcime from Lake Magadi is further support for its derivation from erionite, because analcime associated with other zeolites (e.g. phillipsite) generally has lower SiO_2 contents (Sheppard and Gude 1969b).

Buckhorn, Grant County, New Mexico, USA. Siliceous tuffs in a lacustrine facies of the Gila Conglomerate of probable Pliocene age occur at Buckhorn, Grant County, New Mexico (Surdam et al. 1972, Sheppard and Gude 1974, 1982; Eyde 1982, Mumpton 1984, Gude and Sheppard 1988). Field studies showed that a small closed basin (25-30 km^2) existed, where a saline, alkaline soda-lake environment developed under low-temperature conditions. In the lower part of these lacustrine sediments, a silicic vitric marker tuff (0.45 – 2.75 m thick) was recognized and characterized by the presence of at least ten authigenic minerals, including six zeolites. Clinoptilolite and chabazite are the most widespread zeolites, whereas erionite, phillipsite, and mordenite are rare. Within the marker tuff, the large lateral and vertical variability in relative clinoptilolite content and in mineral assemblage can be explained by chemical variations in the shallow, closed-system, saline alkaline environment (Gude and Sheppard 1988). Analcime occurs at this location only as an alteration product of a precursor zeolite. Three concentric diagenetic zones were identified in this deposit. Chabazite occurs where the lake water was less saline and alkaline, namely in the outer portion of the basin influenced by the presence of inflowing fresh waters. The zeolite zone is characterized by a glass-smectite-chabazite assemblage associated with erionite, clinoptilolite (Fig. 5) and/or mordenite, but without analcime. The innermost zone begins where analcime is found (Fig. 6).

An initially Na-poor and presumably K-Ca rich system favored the crystallization of chabazite. The consequent K-depletion changed the chemical composition and pH of the solution, and new zeolites formed. Only when the system was sufficiently enriched in Na was analcime formed. These findings are supported by laboratory experiments (de′ Gennaro et al. 1993, 1999) where synthetic mono/polycationic glasses and natural glasses were interacted with monocationic high salinity solutions. In both experiments (natural and synthetic glasses), the resultant mineral assemblages were determined primarily by the glass composition. Analcime occurred only when Na in the system was the dominant cation.

The hypothesized paragenesis of the zeolites in the marker tuff at Buckhorn is as follows:

glass ⇒ smectite ⇒ zeolites (except analcime) ⇒ analcime (from a zeolite precursor)

Although K-feldspar was not identified in the marker tuff, it was identified in tuffs stratigraphically higher in the same formation (Gude and Sheppard 1988).

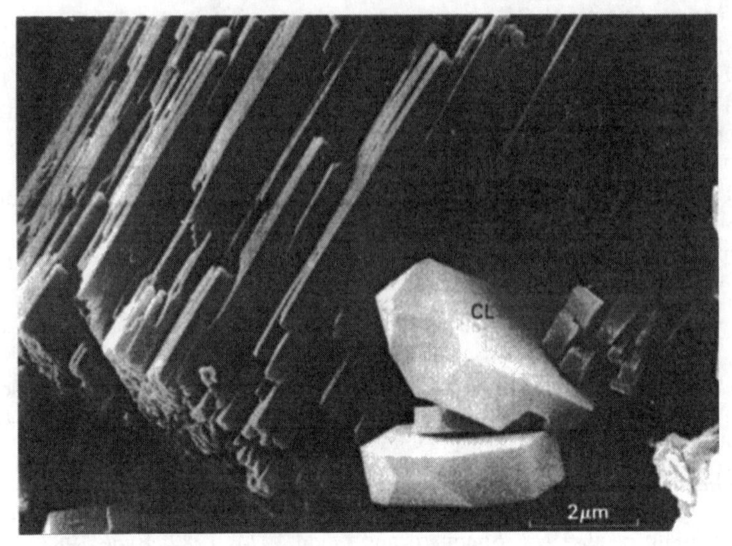

Figure 5. Erionite needle bundles with clinoptilolite in a lacustrine tuff near Buckhorn, New Mexico, USA (from Gude and Sheppard 1988).

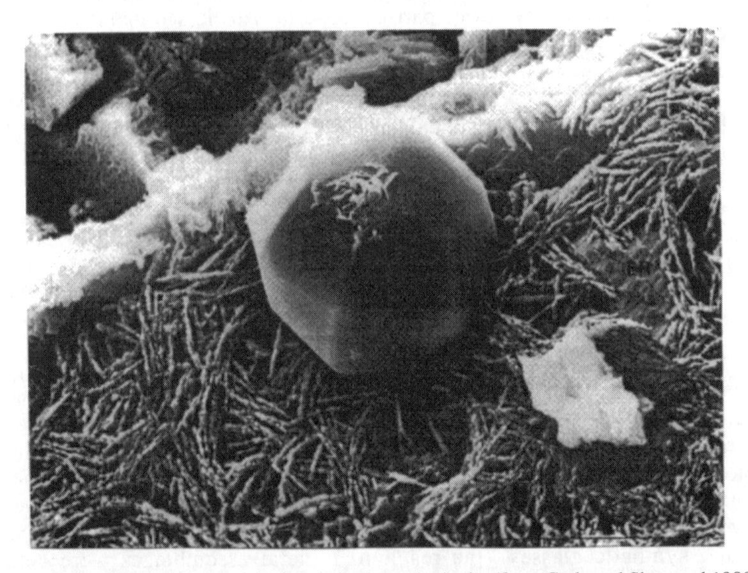

Figure 6. Large analcime crystal with chabazite and erionite (from Gude and Sheppard 1988).

Morrison Formation, eastern Colorado Plateau, Colorado, USA. The upper Jurassic Morrison Formation was deposited in one of the largest known ancient saline, alkaline lakes (Lake T'oo'dichi') which occupied a large part (about 150,000 km^2) of the eastern Colorado Plateau region. Early studies on the depositional environment of the Brushy Basin Member of the Morrison Formation revealed a monotonous sequence of mudstones interpreted as distal fluvial or overbank in origin (Craig et al. 1955). More detailed studies began to recognize these deposits as occurring in a shallow lacustrine environment, and Bell (1981, 1983) first recognized a lateral authigenic mineral zonation typical of a saline, alkaline basin. The basinward progression of diagenetic mineral zones is smectite → clinoptilolite → analcime → potassium feldspar → albite. Zonation was also observed within the clay minerals, with highly smectitic material in the outermost areas grading to highly illitic clays in the central zones. The predominant lithologies of the Brushy Basin Member are mudstone and claystone, with subordinate sandstone and limestone lithofacies. Rhyolitic to dacitic volcanic ashes, products of late Jurassic volcanism, were carried into the basin by prevailing winds and formed the tuff beds (2-50 cm thick) of the formation. These products were the main source of the authigenic minerals identified in this lacustrine sequence (Turner and Fishman 1991). The smectite diagenetic mineral zone is characterized by thin (2-25 cm thick) layers of randomly mixed-layer illite/smectite (70-100% expandable layers) derived from airfall ash or slightly reworked tuffaceous material (Turner and Fishman 1991). This outer zone, representing the margin of the ancient lake, was affected by fresh water recharge which altered the volcanic glass to smectite with a consequent increase in pH and silica activity of solutions. As the $(K^+ + Na^+)/H^+$ ratio increased in pore waters, conditions favorable for clinoptilolite crystallization occurred. In fact, the inner diagenetic zone is characterized by clinoptilolite, along with randomly mixed-layer illite/smectite clays and authigenic quartz or chalcedony. The vitroclastic texture of all the clinoptilolite-bearing samples and the abundance of shard pseudomorphs reflect the large contribution of volcanic ash during deposition. The still-preserved shard texture, not obliterated by early clay formation, supports the formation of clinoptilolite from ash (Turner and Fishman 1991).

Analcime and K-feldspar, occurring either separately or together, along with mixed-layer illite/smectite (0-100% expandable layers) define the inner diagenetic zone. In this part of the basin, a further increase in solution pH, silica activity, and Na^+/H^+ ratio all favor replacement of clinoptilolite by analcime (Surdam and Sheppard 1978). These transformations result in a silica excess which may explain the abundant authigenic quartz and chalcedony cement of the tuff. The continuously increasing salinity characterizing this part of the basin further lowered the water activity, thereby favoring the formation of anhydrous K-feldspar from zeolites and analcime.

The innermost diagenetic mineral zone is defined by albite and highly illitic mixed-layer illite/smectite (0-30% expandable layers). Chlorite is also present. The formation of highly illitic clays likely depleted waters of K^+ with respect to Na^+, thereby favoring the formation of albite rather than K-feldspar. According to Turner and Fishman (1991), the following reaction forms albite and illite from smectite, analcime, and K-feldspar precursors:

$$\underset{\text{smectite}}{2\,Na_{0.5}Al_{2.5}Si_{3.5}O_{10}(OH)_2} + \underset{\text{analcime}}{NaAlSi_2O_6{\cdot}H_2O} + \underset{\text{K-feldspar}}{2\,KAlSi_3O_8} \Rightarrow$$

$$\underset{\text{albite}}{2\,NaAlSi_3O_8} + \underset{\text{illite}}{2\,KAl_3Si_3O_{10}(OH)_2} + \underset{\text{quartz}}{3\,SiO_2} + H_2O$$

The production of silica from this reaction resulted in the precipitation of quartz or chalcedony.

Western Snake River plain, Idaho, USA. The Miocene Chalk Hills Formation in southwestern Idaho reveals a sedimentary succession characterized by mudstones, siltstones, and diatomite interbedded with at least 15 vitric tuffs (Sheppard 1989). These sediments unconformably overlie Miocene volcanic rocks and are themselves unconformably overlain by Pliocene sedimentary rocks. The tuff layers are unaltered or partly altered to smectite throughout most of the formation where the depositional environment was a freshwater lake (fresh-glass zone). These same tuffaceous beds are completely altered in the zeolite zone and, in some areas, the beds gradationally pass into the fresh-glass zone. Clinoptilolite is the most abundant zeolite, generally associated with smectite and, to a lesser extent, with opal-CT. Other zeolites only occasionally occurring are erionite, mordenite, and phillipsite.

The pattern of alteration, the lateral gradation, and the observed mineral assemblage (smectite, clinoptilolite, opal-CT) suggest that water was trapped in the tuffs during lacustrine deposition. Sheppard (1989) reported an alteration scheme in which the rhyolitic glass of the tuff was altered by hydrolysis to form smectite which raised the pH, the activity of SiO_2, and the $(Na^+ + K^+)/H^+$ activity ratio of the pore water, thereby creating conditions favorable for clinoptilolite and smectite crystallization. In this hypothesized closed system, the crystallization of these two phases led to an SiO_2 excess, resulting in the formation of ubiquitous opal-CT. The rare phillipsite seems to have preceded the crystallization of other zeolites. Traces of erionite and mordenite, crystallizing after clinoptilolite, can be explained by minor local changes in pH and cation ratios. According to Sheppard, the lower pH and alkalinity of the lake compared with similar saline, alkaline Cenozoic deposits accounts for the abundance of clinoptilolite over other zeolites.

Baucarit Formation, Sonora State, NW Mexico. This formation of Miocene age consists of two superimposed sequences of continental sediments. Sandstone and conglomerate characterize the lower sequence (about 300-400 m thick), whereas the upper sequence consists primarily of claystone (Demant et al. 1988). Zeolites of different species and origin occur in all the volcanic layers intercalated within the entire succession and in the lower sequence (Cochemé et al. 1996, Munch et al. 1996). In the latter case, the sandstone is pervasively cemented by heulandite. Also in the lower sequence many intercalated basaltic flows occur, and quartz, clay minerals, calcite, and several species of zeolites crystallize in their amygdales and veins. Predominantly rhyolitic ash-fall deposits (10 cm to 1 m thick) are regularly intercalated both in the lower and upper sequence. The predominant feature of these ash-falls is that each single outcrop is characterized by a dominant zeolite species. Heulandite-group zeolites occur in the lower sequence whereas chabazite and erionite, in addition to heulandite-clinoptilolite, have been identified in the upper sequence. Cochemé et al. (1996) proposed the existence of diffuse hydrothermal alteration in the lower sequence resulting from the high heat flow from basaltic-andesitic volcanism, and the genesis of heulandite was linked to this hydrothermal system. As far as zeolites of the upper sequence are concerned, Cochemé et al. (1996) suggested that they crystallized from brines after leaching of lake-deposited volcanic glass. This genesis is in accord with the presence of unreacted glass and the occurrence of saline minerals (gypsum, thenardite). In addition, these zeolites (clinoptilolite, erionite, chabazite) and the associated authigenic minerals are characteristic of a low-temperature environment. The variety of zeolite species was inferred to result from alteration of glasses having fairly uniform composition.

Carboniferous Rocky Brook Formation, western Newfoundland, Canada. Analcime has been found in lacustrine and pedogenic lake-margin sediments of the Rocky Brook Formation (Visean) in the Deer Lake Basin of western Newfoundland (Gall and Hyde 1989). Analcime is the only zeolite in the entire formation and, unlike other similar deposits of the same age, it is a rare example of Paleozoic diagenetic analcime not associated with volcanogenic products. The analcime-bearing formation is characterized by a sequence of interfingered calcareous/dolomitic mudstone and crystalline dolostone and limestone. Thicknesses range from a few tens of centimeters to 1 km. Analcime occurs both as a microscopic cement in open or calcite-occluded vugs and as submicroscopic crystals in finer-grained lake and lake-margin sediments. Among the hypothesized origins of these two different occurrences of authigenic analcime, Gall and Hyde suggested precipitation from sodic lake water or pore fluids within partly lithified sediments for the microscopic analcime. Evidence for this genesis includes: (a) occurrence in unmetamorphosed sediments; (b) good euhedral crystalline form without replacement textures; (c) fairly uniform chemistry. The authors concluded that the submicroscopic analcime formed either by direct precipitation from a fluid phase or by reaction with clay minerals (smectite or illite) in the Rocky Brook Formation.

Werris Creek, New South Wales, Australia. Flood (1991) and Flood and Taylor (1991) provided a new interpretation to the diagenetic alteration in some silicic ash-fall tuffs and associated lacustrine sediments outcropping at Werris Creek, in New South Wales, Australia. Repeated explosive volcanic emissions emplaced rhyolitic to rhyodacitic ignimbrites interbedded with fluvial and fluvio-glacial conglomerate, sandstone, and mudstone, representing the prevailing deposits of a late Carboniferous stratigraphic succession up to 2400 m thick. Sedimentation of these volcanic products in distal areas gave rise to the Currabubula Formation (McPhie 1983), characterized by thin water-lain ash-fall tuffs, each representing a single eruptive event and commonly overlain by a thicker epiclastic unit consisting of reworked ash-fall material. Within this formation, the Escott Unit (~16 m thick) is pervasively altered to zeolite (mainly heulandite and subordinate mordenite). Regional studies have revealed at least 1000 m of overburden cover on this zeolitic sequence; nevertheless, during the 300 Ma history of the deposit, temperatures never exceeded 50°C as suggested by the presence of mordenite and the absence of diagenetic albite, laumontite, and analcime. These low-temperature zeolitization conditions cannot be considered as belonging to the zeolite facies metamorphism because in this case thermodynamic equilibrium, not recorded in the low-temperature zeolites, should be. attained. Flood and Taylor (1991) hypothesized a pervasive *in-situ* alteration of the silicic glass component to zeolite in an alkaline lake which provide a favorable chemical setting for diagenesis. The absence of analcime or albite in such old sedimentary successions is remarkable, as these phases are generally found as late diagenetic alteration phases of silicic volcanic products.

Chemical data

Literature data regarding the chemistry and mineralogy of zeolites in closed hydrologic systems are often difficult to compare because they often are not representative of individual mineral species (e.g. bulk-rock analyses vs. single-mineral analyses). In order to interpret correctly the mineral-forming processes which give rise to the crystallization of a specific zeolite, the following information is required:

1- composition of starting materials;
2- composition of solutions interacting with them;
3- hydrologic conditions;
4- kinetic factors.

It is well known that silica-poor glasses quickly react in most environments, whereas rhyolitic glasses react relatively slowly (Hay 1966). However, if the latter glasses interact with saline, alkaline solutions, the zeolitization rate is considerably faster, probably as a result of the relatively higher solubility of silica-rich glasses at pH values above 9. Rhyolitic glasses can produce different zeolites as a function of the solution composition. In general terms, any glass-to-zeolite transformation is characterized by a silica loss, and this rule is valid either in open or closed systems. In open systems, the leached silica is removed from the system, whereas silica remains in solution in closed systems. The higher pHs, typical of closed systems, increase silica solubility and the resulting zeolites display lower Si/Al ratios. For example, in the John Day Formation, the reaction of a rhyolitic glass to form an equal volume of clinoptilolite was accompanied by a loss of about 20% of the silica and alkali ions and a considerable gain in Ca and Mg. If the silica loss increases to about 30%, the same glass can form phillipsite as in saline, alkaline lakes of California (Hay 1964). A possible explanation for this mineralogic variability can be found in the silica activity; for example, Hay (1966) suggested that more silica-rich zeolites will crystallize when the aqueous silica activity reaches elevated values, such as at Lake Magadi, where erionite is the dominant phase. Chipera and Apps (this volume) have a complete description of the effects of water composition and temperature on zeolite formation.

Mariner and Surdam (1970) proposed a model in which the relation between solution pH and Si/Al ratio of the zeolite is controlled by the reaction between silicic glass and alkaline solution, which takes place in two steps: (1) formation of a gel, with an Si/Al ratio controlled by the Si/Al ratio of the solution; and (2) zeolite nucleation from the gel, with the zeolite Si/Al ratio depending on the composition of the gel itself. This model was successfully tested (Surdam and Mariner 1971) for zeolites from an alkaline lake at Teels Marsh, Nevada. In this environment, a gel phase occurs, together with a rhyolitic glass and phillipsite. Reaction of the gel, carefully separated from the system, with a sodium carbonate solution gave phillipsite. The possible path of crystallization from a rhyolitic glass can be summarized in two reactions (Surdam and Sheppard 1978):

$$\text{glass} + H_2O \Rightarrow \text{gel} \Rightarrow \text{zeolite}; \qquad \text{glass} + H_2O \Rightarrow \text{gel} \Rightarrow \text{solution} \Rightarrow \text{zeolite}$$

The first reaction will result in a zeolite with an Si/Al ratio similar to that of the gel, whereas the second would not be subject to this constraint. Surdam and Mariner also found that no gel-like phases were observed in the pH range 7-9, where siliceous zeolites should form. The lack of zeolites may be explained by the low Al solubility over this pH range.

Molar diagrams of $SiO_2/(Al_2O_3+Fe_2O_3)$ vs. $(CaO+MgO)/(Na_2O+K_2O)$ provide a useful graphical representation of the extra-framework cationic compositional variation of zeolites from closed systems. This type of diagram, used by Surdam and Eugster (1976) and by Surdam and Sheppard (1978), is useful for visualization because many different species of zeolites plot in a small area. Figure 7 illustrates the compositions of clinoptilolites, phillipsites, and erionites from Lake Magadi, Kenya (Surdam and Eugster 1976), including other published and recently obtained chemical data. Phillipsites are distributed in the low-silica and very low alkaline-earth region; erionite analyses reported by Hay (1964) occupy a region of very low alkaline-earth and intermediate silica content; and those from Cochemé et al. (1996) have a higher divalent cation (primarily Ca) content.

Chabazites from closed hydrologic systems (not included on Fig. 7) display silica contents comparable to erionites but are generally characterized by much higher alkaline-earth contents. Clinoptilolites are scattered over a large portion of the diagram, in the region of high silica and high alkaline earths. A large number of analcime analyses from other authors (Gude and Sheppard 1988, Gall and Hyde 1989, Renaut 1993, Whateley at al. 1996) are also included on the figure. As expected for analcime the analyses plot in a very narrow area, in the field of low silica and very low alkaline-earth contents. All analyses are very similar, despite the different provenance and starting materials. Given the wide variability in clinoptilolite chemistry, its cationic composition can be better represented using Figure 8 proposed by Alietti (1972) and later modified by Gottardi and Galli (1985). An alternate representation of clinoptilolite-heulandite chemistry is provided by Bish and Boak (this volume).

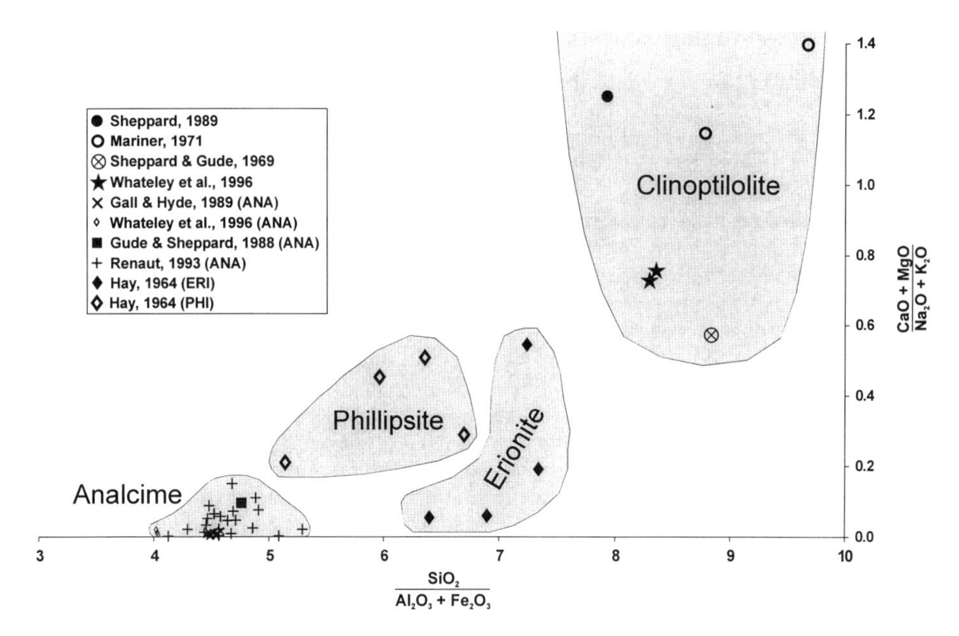

Figure 7. $SiO_2/(Al_2O_3+Fe_2O_3)$ vs. $(CaO+MgO)/(Na_2O+K_2O)$ molar diagram for zeolites from closed hydrologic systems.

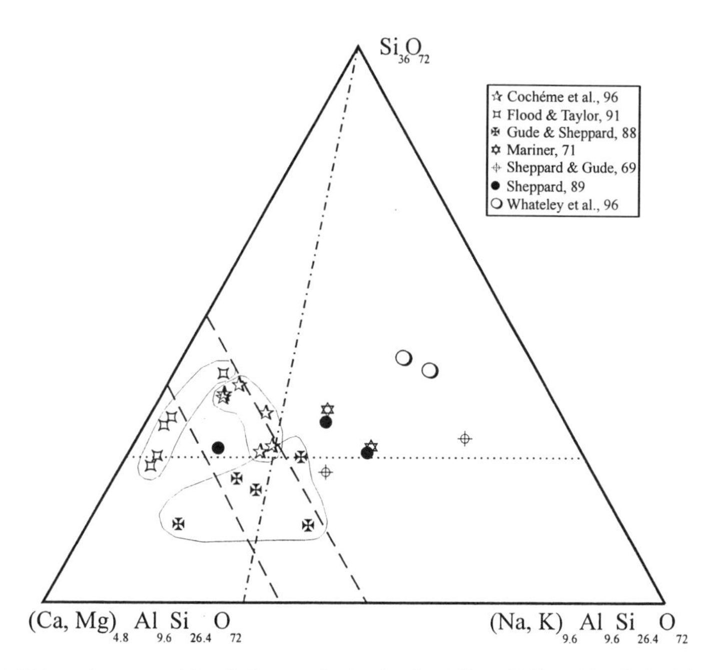

Figure 8. Triangular compositional diagram for heulandites/clinoptilolites. The dashed-dotted line identifies zeolites with (Na+K) = (Ca+Mg); the dotted line refers to a Si:Al = 4.

Questionable zeolitization systems

The zeolite occurrences described above are all associated with well-defined closed hydrologic systems, generally characterized by unusual solution compositions and, in particular, by saline and alkaline solutions. Nevertheless, some natural systems occur which are not easily classified with those discussed above or with other zeolitization environments described in the literature. These systems, although perhaps relatively uncommon, take on greater importance in some geological contexts and will therefore be considered in this chapter. We specifically refer to zeolitization systems linked to particular volcanic eruptive phenomena in which the emplacement temperature and the presence of water-bearing strata favored a rapid transformation of the glassy precursor. Relevant examples occur in central-southern Italy (Lenzi and Passaglia 1974, de'Gennaro et al. 1999a, de' Gennaro et al. 2000) and in the Canary Islands (Perez-Torrado et al. 1995). As a result of the difficulty in providing an accurate genetic classification of these zeolite deposits to known natural systems, several authors have ascribed such zeolite deposits to a *geoautoclave-type* system. This type of deposit refers to a partially insulated system which preserved temperatures and pressures above ambient for an extended period after eruption, while simultaneously experiencing minimal loss of solution thereby enhancing the opportunity for rock-water interactions without significant dilution and/or leaching.

Among known examples, the Neapolitan Yellow Tuff (NYT) is undoubtedly one of the most important and will be described in detail here. This formation represents the product of one of the largest and most powerful eruptions of Campi Flegrei (southern Italy). The NYT, dated at 12,000 years BP (Alessio et al. 1971, 1973), is a widespread pyroclastic deposit (trachyte to phonolite) on the periphery of Campi Flegrei, within the city of Napoli and on the Campanian Plain (Fig. 9). It is predominantly of

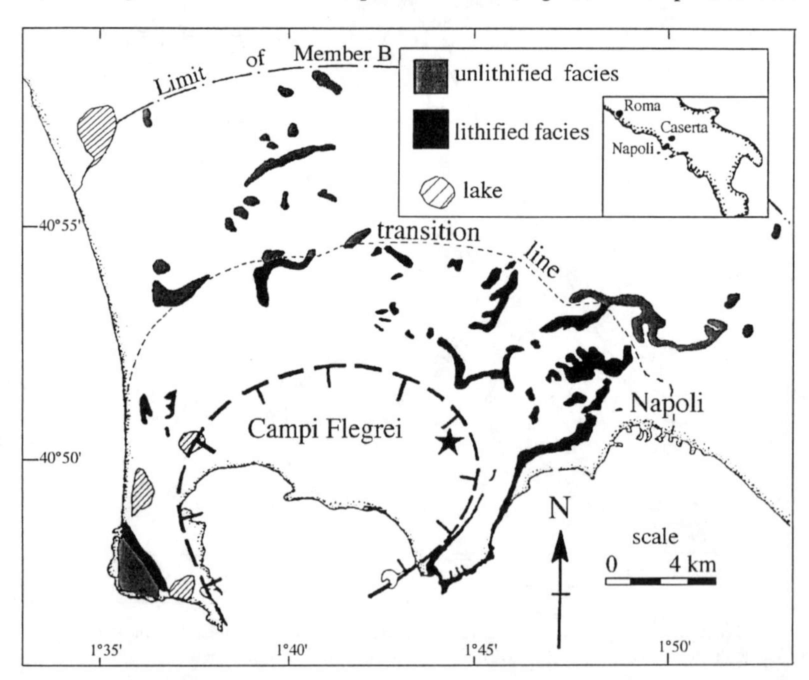

Figure 9. Distribution of outcrops of the Neapolitan Yellow Tuff (Scarpati et al. 1993); star represents the presumed vent location.

phreatomagmatic origin and consists of two different lithological members (A and B) distinguishable on the basis of their field characteristics and granulometric parameters. The lower Member A consists of 6 air fall units, interbedded with numerous ash layers and pumice lapilli fall units. Member B is characterized by a coarser grain size and can be found up to 14 km from the vent (Scarpati et al. 1993). The phreatomagmatic character of the NYT suggests a low emplacement temperature (≤100°C) for the deposit.

More than 50% by volume of the volcanoclastite was involved in diffuse zeolitization processes that produced phillipsite and, subordinately, chabazite and analcime. The mean content of these authigenic phases in the NYT generally exceeds 50 wt % and can locally reach 70-80 wt % (de'Gennaro et al. 1982, de'Gennaro et al. 1990). However, the zeolitization process active in the NYT is still debated because none of the hypotheses proposed to date seems to explain all observations.

The NYT occurs as both lithified and non-lithified facies. The former is yellow, whereas the latter (*pozzolana*) is gray and better preserves its primary depositional character. The lithified facies is consistently closer to the vent than the unlithified facies and also comprises the material accumulated within the caldera. In the transition area between the two facies a thick yellow tuff begins to show unlithified nuclei of different shapes, 2-5 m thick and 5-15 m wide.

de'Gennaro et al. (2000) proposed the following eruptive scenario for the NYT: the eruption probably occurred in a water-rich environment that gained access to the conduit with the formation of a phreatoplinian eruption column. This involved numerous explosions that gave rise to the laminated ash fall units and pyroclastic surges derived from partial collapse of the phreatoplinian eruption columns. A series of "dry" magmatic phases produced the pumice lapilli fall layers. Repeated eruption column collapses generated the numerous pyroclastic flows of Member B.

Post-depositional mineral-forming processes in the NYT have given rise to phillipsite, chabazite, analcime, amorphous hydrous iron oxides and, subordinately, smectite and an amorphous aluminosilicate phase (de'Gennaro et al. 1995). Zeolites of the NYT are particularly concentrated in Member B. de'Gennaro et al. (2000) proposed an alteration model for the NYT that accounts for the possible relationships between eruptive mechanisms and the authigenic mineral-forming processes involving the bulk of the formation. Figure 10 synthesizes the general features of the formation from a volcanological and mineralogical point of view. A progressive decrease in the thickness and lithification of the deposit with distance from the vent has been documented, whereas the zeolite content progressively increases from the bottom (unit A1) to the top of Member A. The gradual increase in zeolite content is also accompanied by progressive induration of the tuff. The upper part of Member B shows a progressive decrease from the bottom to the top in zeolite content and in the degree of lithification, giving rise to a thin pozzolanic cover. Conditions favorable for zeolitization (temperatures near 100°C and the presence of water) during emplacement of the volcanoclastite are confirmed by the occurrence of phreatomagmatic sedimentary structures (accretionary lapilli, soft-sediment deformation, vesiculated layers) indicating the emplacement of wet products with a water content ranging from 10 to more than 25 wt %.

The above considerations, as well as depositional features of the NYT, lead de'Gennaro et al. (2000) to hypothesize that the interaction of an alkali-trachytic to phonolitic volcanic glass with hot fluids permeating the deposit enhanced the zeolitization process. The leaching of glass produced a rise in pH to conditions above neutral and also lead to Na/K and Si/Al ratios in the solution favorable for phillipsite

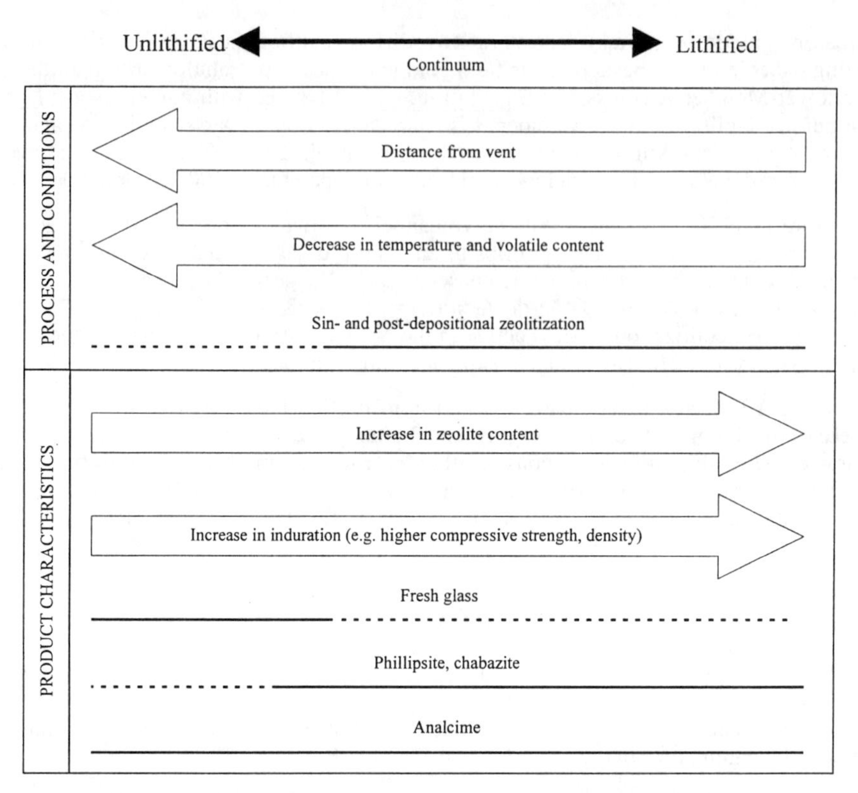

Figure 10. Zeolitization grade continuum, with associated geological and volcanological features, conditions, and products of the NYT.

crystallization. As the physico-chemical conditions of the system evolved, chabazite nucleation began, giving rise to chabazite overgrowths on phillipsite. The lack of authigenic feldspar, the final product of a series of metastable transformations, demonstrates that favorable conditions for authigenic mineralization ceased as a consequence of cooling of the deposit by normal thermal conduction.

On the basis of the results, de' Gennaro et al. (2000) concluded that zeolitization in the NYT took place soon after eruption in a well-insulated (closed) thermal system in the presence of hot aqueous solutions of hydromagmatic origin, whose ionic composition was controlled by equilibrium glass hydrolysis. The vertical and lateral variations in lithification grade reflect the effects of two factors: (1) fluctuating emplacement conditions and (2) thermal dispersion. Any change in physical conditions (water content, temperature, permeability) accounts for variations in zeolite content within Member B. Heat loss towards the atmosphere and substrate cooled the periphery of the deposit too quickly for zeolitization to occur, thereby reducing the degree of lithification towards the base and top of the sequence. In fact, in distal areas, where the NYT is less than 20-30 m thick, the deposit is not lithified, suggesting the possible existence of a thickness threshold value, below which thermal dispersion does not allow the full development of the zeolitization process.

The mineral assemblage formed by alteration of the phonolitic vitric component of the Roque Nublo ignimbrites (Gran Canaria, Spain) is very similar to that of the NYT and consists of chabazite, phillipsite, minor amounts of analcime, and subordinate smectite (Pérez-Torrado et al. 1995). An earlier hypothesis on the origin of these authigenic phases was suggested by Brey and Schmincke (1980) who concluded that the alteration occurred under diagenetic conditions and at temperatures <100°C. Their model requires circulation of meteoric waters through the deposit after emplacement, with consequent devitrification of the vitric component as follows: transformation to a palagonitic gel ⇒ increase in salinity and pH ⇒ precipitation of zeolites when the pH of the solution was about 9.5.

More recently, Pérez-Torrado et al. (1995) studied the same formation and demonstrated the crucial importance of eruption type and emplacement mechanisms of the deposit to the zeolitization process. They suggested the occurrence of phreatomagmatic eruptions, producing deposits with emplacement temperatures close to the condensation temperature of water vapor and a pyroclast transport mechanism involving poorly expanded, high-density flows. Loss of water vapor is limited in such deposits, providing a necessary condition for initiation of zeolitization. Furthermore, the lack of significant lateral or vertical gradation in alteration which could indicate the action of hydrothermal processes led Pérez-Torrado et al. (1995) to suggest a geo-autoclave mechanism to explain an alteration process closely linked to the geometry of the deposit and its mode of eruption. Other depositional products such as ash falls and surge deposits associated with the same formation contain different mineral assemblages (smectite and kaolinite), likely products of diagenesis in an open hydrologic system.

de'Gennaro et al. (1999a) studied 17 different pyroclastic deposits in the Campi Flegrei volcanic field (southern Italy) and also concluded that the eruptive and emplacement conditions were the primary controls on the zeolitization process. They hypothesized the existence of a secondary mineralization process consistent with a glass-fluid phase reaction in a nearly closed post-depositional system which took place immediately after emplacement, during the cooling of wet deposits. In this context, air-fall deposits characterized by limited water-magma interaction are unzeolitized, whereas hydromagmatic deposits can be distinguished on the basis of the efficiency of water-magma interaction. Highly expanded clouds allowing immediate separation of water vapor from the pyroclasts form a nearly unzeolitized deposit (tuff rings). Wet, poorly expanded and relatively cool deposits (tuff cones) are highly zeolitized.

In accordance with these observations, de'Gennaro et al. (1999a) termed zeolitization of these deposits as "zeolitization in phreatomagmatic pyroclastic deposits," summarized by the following criteria:

1. On the basis of laboratory experiments and paleomagnetic measurements, the temperature of emplacement ranges between 100° and 250°C (de'Gennaro et al. 1988, Incoronato 1990).
2. Temperature is a necessary but not sufficient controlling parameter of glass alteration;
3. The dominant factor in the zeolitization process is the presence of condensed water vapor trapped in the pyroclastic deposit.

SOME GENERAL REMARKS

It seems almost unnecessary to remark that in these types of genetic systems a prominent role is played by the pore solutions, whose composition is fundamental in directing mineral crystallization towards one phase or another. However, the evolution of the solution composition cannot be considered without taking into account the interactions occurring between the solution itself and the saline, alkaline lake sediments.

Obviously, if the sediments were constituted, *ab absurdo*, only by inert minerals, solution evolution would result in crystallization of salts on the basis of the solution composition and in a sequence opposite to the solubility products. In the context of the sediments and solution composition evolution in closed hydrologic systems, the reactivity of interacting glasses is of great importance. A necessary condition for zeolitization is the presence of alkaline or nearly neutral solutions interacting with the precursor sediments. If HCO_3^- prevails, such solutions in the beginning stages of alteration will leach alkali and alkaline-earth cations from glasses and/or minerals. Laboratory simulations (de'Gennaro et al. 2000) confirmed this assertion for trachytic glass reacting with acidic solutions (simulating eruption fluids) ranging between pH values of 1.7 and 5.1. Apart from starting solutions with very low pH (1.7), all systems quickly became neutral or alkaline. Laboratory tests were also carried out in connection with preparation of this chapter in order to evaluate the pH trends during interaction between trachytic or rhyolitic glass and HCO_3^--rich solutions from a geothermal field of Contursi (SA), southern Italy. These experiments also showed a rapid increase in pH, confirming that the interaction of such solutions with rhyolitic glass produces conditions favorable for zeolite crystallization.

It is clear that the role of the glassy precursor is to provide to the solution silica and alumina, whose activities are, however, determined by physico-chemical parameters which characterize the solution itself. The importance of the overall solution chemistry in a *closed hydrologic system* is illustrated by the fact that alteration of trachytic glasses, in some instances, gives rise to the crystallization of zeolites such as erionite and clinoptilolite whose high Si/Al ratios are distinct from those of the trachytic glass.

Zeolite formation in closed hydrologic systems is generally linked to the transformation of a volcanic precursor in lacustrine environments such as playa lakes or rift type systems. Nevertheless, we also include in this chapter on closed hydrologic systems, a discussion of zeolites forming under substantially different geological conditions. Phreatomagmatic eruptions can generate volcaniclastic deposits in which elevated temperatures and a liquid phase are preserved for sufficient time to give rise to zeolitization of the glassy precursor. The most important difference between lacustrine environments and phreatomagmatic deposits lies in the composition of the mineralizing solution. In the first environment, the solution is characterized by high salinity and pH which can progressively increase with time as a consequence of evaporative processes. In the second environment, the solution composition evolves as a function of the dissolution processes of the glassy volcanic precursor. Consequently, the resultant zeolites in saline, alkaline lakes can be very different and not necessarily bound by the composition of the precursor matter. In systems linked to phreatomagmatic eruptions zeolites will be more strictly related to the original glass and will be characterized by a Si/Al ratio comparable to or slightly lower than the corresponding glass.

FINAL REMARKS AND CONCLUSIONS

This chapter has examined recent occurrences of zeolites in closed hydrologic systems and emphasized several aspects of alteration in such systems that have not been considered in previous reviews. Within this broad scheme, basins may be further subdivided based on their tectonic context into closed basins in block-faulted regions in arid and semiarid environments and trough valleys associated with rifting. A feature common to both settings is the existence of saline, alkaline fluids that reacted with the prevailing volcanic glass precursor. The clear evidence is that some natural systems exist which cannot be strictly classified. Although, playa-lake and rift systems refer to well-defined geological systems, some systems occur on a smaller scale which can still be considered closed from the hydrologic point of view because they are partially or totally insulated from the external environment. This "insulation" allows the intrinsic features of the system to be maintained for a time sufficient to permit zeolitization. In others, it is possible to recognize the superimposition of two or more different mechanisms which illustrate the variation of geological conditions over time (Cochemé et al. 1996).

Other zeolite occurrences have been ascribed to a so-called geoautoclave system (Lenzi and Passaglia 1974, Garcia Hernandez et al. 1993, Perez-Torrado et al. 1995) but new data clarify this particular classification. This type of zeolitization can be considered neither as diagenesis, as it likely occurs at temperatures as high as 200°C, nor as zeolitization in an open hydrologic system because of the presence of nonzeolitized layers, above and below the zeolitized deposit, with the same chemical composition (Lenzi and Passaglia 1974). Complete zeolitization is therefore related to the high temperature and the high vapor pressure inside the rock unit immediately after its deposition. Hall (1998) demonstrated that the geoautoclave hypothesis is problematic because it requires a realistic geological mechanism for sustaining an excess pressure, namely an internal pressure greater that the lithostatic pressure. The only possible mechanism is a complete welding of the upper portion and the margins of the ash deposit. This phenomenon is quite rare in nature, and welding is generally most pronounced in the middle parts of deposits and not at the tops or bottoms. This illustrates the difficulty in creating a closed system capable of sustaining an excess internal pressure. Consequently, the zeolitized deposits so far attributed to a geoautoclave system must be re-considered, and they may be more correctly linked to the closed (or nearly closed) hydrologic system proposed by de'Gennaro et al. (2000) for the Neapolitan Yellow Tuff.

Finally, the possibility that zeolitized sequences may occur in as-yet unstudied volcaniclastic and epiclastic successions deposited in closed basins cannot be excluded. An example of such a system are the several endoreic (closed) basins of the carbonate Apennines in central-southern Italy. It is our understanding that thick pyroclastic deposits exist in this region that have been deeply altered to halloysite over the outcropping areas. However, a detailed study of the stratigraphy and alteration mineralogy of the entire volcaniclastic sequence has not been carried out. Paleoclimatic research on this area (Allocca et al. 1997) suggests the occurrence of a hot-arid period, 8000-4000 years BP, connected with the northward cyclic shifting of climatic zones. The increased evaporation rate and the decreased water supply to these basins may have provided conditions favorable for zeolitization. Discovery of zeolitized units in these basins would provide support for the paleoclimatic conclusions of Allocca et al. (1997). It appears worthwhile to investigate similar volcaniclastic and epiclastic deposits around the world for evidence of zeolitization. It would also be valuable to conduct further investigations into the diagenesis of basaltic tephra in saline, alkaline-lake deposits. In perhaps the only work on this subject, Sheppard (1996) showed that the mafic composition of the original glass in lacustrine deposits played a significant role in the diagenetic mineral assemblage and in the chemical composition of individual phases.

ACKNOWLEDGMENTS

The authors thank David L. Bish for his substantial improvement of the original draft, Richard A. Sheppard for his skillful review of the manuscript, Maurizio de'Gennaro for his encouragement and help during all stages of the work, and Paul H. Ribbe for editorial support. This work was supported by MURST: Programmi di Ricerca Scientifica di Rilevante Interesse Nazionale–COFIN '98.

REFERENCES

Alessio M, Bella F, Improta S, Belluomini G, Cortesi C, Turi B (1971) University of Rome Carbon-14 Dates IX Radiocarbon 13:395-411

Alessio M, Bella F, Improta S, Belluomini G, Cortesi C, Turi B (1973) University of Rome Carbon-14 Dates X Radiocarbon 15:165-178

Alietti A (1972) Polymorphism and crystal-chemistry of heulandites and clinoptilolite. Am Mineral 57:1448-1462

Allocca F, Amato V, de' Gennaro M, Ortolani F, Pagliuca S (1997) Evidenze geologiche di variazioni climatico-ambientale cicliche oloceniche nella Campania Occidentale. *In* D'Amico, Albore Livadie (eds) Le scienze della Terra e l'Archeometria. CUEN, Napoli, 61-65

Ataman G (1976) Occurrence of authigenic analcime in Turkey and probable relationship between genesis of zeolite series and plate tectonics. Yerbilimleri, Hacettepe University 1:9-23.

Bell TE (1981) A Jurassic closed basin in the Morrison Formation. Geol Soc Am Abstracts with Programs 13:406

Bell TE (1983) Deposition and diagenesis of the Brushy Basin and upper Westwater Canyon Members of the Morrison Formation in northwest New Mexico and its relationship to uranium mineralization. PhD dissertation, Univ California, Berkeley, 102 p

Brey G, Schmincke HU (1980) Origin and diagenesis of the Roque Nublo Breccia, Gran Canaria (Canary Island)—Petrology of Roque Nublo Volcanics, II. Bull Volcanol 43:15-33

Cochemé JJ, Lassauvagerie AC, Gonzalez-Sandoval J, Perez-Segura E, Munch P (1996) Characterization and potential economic interest of authigenic zeolites in continental sediments from NW Mexico. Mineral Deposita 31:482-491

Craig LC, Holmes CN, Cadigan RA, Freeman VL, Mullens TE (1955) Stratigraphy of the Morrison and related formations, Colorado Plateau region—A preliminary report. US Geol Surv Bull 1009-E: 125-168

Cronstedt AF (1756) Observation and description of a unknown kind of rock to be named zeolites. Kongl Vetenskaps Acad Handl Stockh 17:120-123 (in Swedish)

de' Gennaro M, Adabbo M, Langella A (1995) Hypothesis on the genesis of zeolites in some European volcaniclastic deposits. *In* Natural Zeolites '93. Ming DW, Mumpton FA (eds) Brockport, New York, 51-67

de' Gennaro M, Cappelletti P, Langella A, Perrotta A, Scarpati C (2000) Genesis of zeolites in the Neapolitan Yellow tuff: geological, volcanological and mineralogical evidence. Contrib Mineral Petrol 139:17-35

de' Gennaro M, Colella C, Franco E, Stanzione D (1988) Hydrothermal conversion of trachytic glass into zeolite. I. Reactions with deionized water: N Jahrb Mineral Monat 4:149-158

de' Gennaro, M. Colella C, Pansini M (1993) Hydrothermal conversion of trachytic glass into zeolite. 2. Reaction with high-salinity waters. Neues Jahrb Mineral Monat 3:97-110

de' Gennaro M, Franco E, Langella A, Mirra P, Morra V (1982) Le phillipsiti dei tufi gialli del napoletano. Period Mineral 51:287-310

de' Gennaro M, Incoronato A, Mastrolorenzo G, Adabbo M, Spina G (1999a) Depositional mechanisms and alteration processes in different types of pyroclastic deposits from Campi Flegrei volcanic field (Southern Italy). J Volcanol Geotherm Res 91:303-320

de' Gennaro M, Petrosino P, Conte MT, Munno R, Colella C (1990) Zeolite chemistry and distribution in a Neapolitan yellow tuff deposit. Eur J Mineral 2:779-786

Demant A, Cochemé J. J, Delpretti P, Piguet P (1988) Geology and petrology of the Tertiary volcanics of the northwestern Sierra Madre Occidental, Mexico. Bull Soc Geol France 8:737-748

Eugster HP (1970) Chemistry and origin of the brines of Lake Magadi, Kenya. Mineral Soc Am Spec Paper 3:215-235

Eyde TH (1982) Zeolite deposits in the Gila and San Simon Valleys of Arizona and New Mexico. New Mexico Bur Mines Mineral Resources Circ 182:65-71

Flood PG (1991) Prospecting for natural zeolites: exploration model based on an occurrence in 300 m.y. old rocks near Werris Creek, New South Wales, Australia. Trans Inst Min Metall (Sect B: Appl Earth Sci) 100:9-13

Flood PG, Taylor JC (1991) Mineralogy and geochemistry of Late Carboniferous zeolites, near Werris Creek, New South Wales, Australia. Neues Jahrb Mineral Monat 2:49-62

Gall Q, Hyde R (1989) Analcime in lake and lake-margin sediments of the Carboniferous Rocky Brook Formation, western Newfoundland, Canada. Sedimentology 36:875-887

Garcia Hernandez JE, Notario Del Pino JS, Gonzales Martin MM, Hernan Reguera F, Rodriguez Losada JA (1993) Zeolites in pyroclastic deposits in southeastern Tenerife (Canary Islands). Clays Clay Minerals 41:521-526

Gottardi G, Galli E (1985) Natural Zeolites. Springer-Verlag, Berlin, Heidelberg, 409 p

Gude AJ, 3rd, Sheppard RA (1988) A zeolitic tuff in a lacustrine facies of the Gila Conglomerate near Buckhorn, Grant County, New Mexico. US Geol Survey Bull 1763, 22 p

Hall A (1998) Zeolitization of volcaniclastic sediments: the role of temperature and pH. J Sedim Res 68:739-745

Hardie LA, Eugster HP (1970) The evolution of closed basin brines. Mineral Soc Am Spec Paper 3: 273-290

Hay RL (1964) Phillipsite of saline lakes and soils. Am Mineral 49:1366-1387

Hay RL (1966) Zeolites and zeolitic reactions in sedimentary rocks. Geol Soc Am Spec Paper 85, 130 p

Hay RL (1970) Silicate reactions in three lithofacies of a semi-arid basin, Olduvai Gorge, Tanzania. Mineral Soc Am Spec Paper 3:237-255

Hay RL (1976) Geology of the Olduvai Gorge. University of California Press, Berkeley, CA, 203 p

Incoronato A (1990) Determinazioni paleomagnetiche delle temperature di messa in posto di un deposito di Tufo Giallo Napoletano (Torregaveta, Campi Flegrei). (abstr) *In* Atti II Convegno Gruppo Nazionale Geofisica della Terra Solida, Roma, p 8

Kossowskaya AG (1973) Genetic association of sedimentary zeolites in the Soviet Union. Adv Chem Ser 121:200-208

Lenzi G, Passaglia E (1974) Fenomeni di zeolitizzazione nelle formazioni vulcaniche della regione Sabatina. Boll Soc Geol Ital 93:623-645

Mariner RH (1971) Experimental evaluation of authigenic mineral reactions in the Pliocene Moonstone Formation. PhD dissertation, Univ Wyoming, Laramie, WY, 133 p

Mariner RH, Surdam RC (1970) Alkalinity and formation of zeolite in saline alkaline lakes. Science 170:977-980

McPhie J (1983) Outflow ignimbrite sheets from late Carboniferous calderas Currabubula Formation, New South Wales, Australia. Geol Mag 120:487-503

Mezger K, Okrusch M (1985) Metamorphism of variegated sequence at Kallithea, Samos, Greece. Tschermaks mineral petrogr Mitt 34:67-82

Mumpton FA (1984) Zeolite exploration: the early days. *In* Proc Sixth International Zeolite Conf, Reno, Nevada. Olson D, Bisio A (eds) Butterworths, Guildford, UK, p 68-86

Munch P, Duplay J, Cochemé JJ (1996) Alteration of acid vitric tuffs interbedded in the Bacaurit Formation, Sonora State, Mexico. Contribution of transmission and analytical electron microscopy. Clays Clay Minerals 44:49-67

Obradovic J (1977) The review on the occurrences of zeolites in sedimentary rocks in Yugoslavia. Geol Anali Balk Pol 41:393-302

Obradovic J (1988) Occurrences and genesis of sedimentary zeolites in Serbia, Yugoslavia. *In* Kallò D, Sherry HS (eds) Occurrence, Properties and Utilization of Natural Zeolites, Akadémiai Kiadò, Budapest, p 59-69

Obradovic J, Dimitrijevic R (1987) Clinoptilolitized tuffs from Zlatokop, near Vranje, Serbia. Glas de l'Academie serbe des sciences et des arts: Classe des sciences naturelles et mathematiques 51:7-19

Papanikolaou D (1979) Unites tectoniques and phases de deformation dans l'ile de Samos, Mer Egee, Greece. Bull Soc Geol France 21:745-752

Perez-Torrado FJ, Martì J, Queralt I, Mangas J (1995) Alteration processes of the Roque Nublo ignimbrites (Gran Canaria, Canary Islands). J Volcanol Geotherm Res 65:191-204

Renaut RW (1993) Zeolitic diagenesis of late Quaternary fluviolacustrine sediments and associated calcrete formation in the Lake Bogoria Basin, Kenya, Rift Valley. Sedimentology 40:271-301

Scarpati C, Cole PD, Perrotta A (1993) The Neapolitan Yellow Tuff—A large volume multi-phase eruption from Campi Flegrei, southern Italy. Bull Volcanol 55:343-356

Sheppard RA (1989) Zeolitic alteration of lacustrine tuffs, western Snake River Plain, Idaho, USA. *In* Zeolites: Facts, Figures, Future. Jacobs PA, van Santen RA (eds) Elsevier Science Publishers BV, Amsterdam, Netherlands, p 501-510

Sheppard RA (1996) Diagenetic alteration of basaltic tephra in lacustrine deposits, southwestern Idaho and southeastern Oregon. US Geol. Surv. Open-File Report, 96-697, 25 p

Sheppard RA, Gude AJ III (1968) Distribution and genesis of authigenic silicate minerals in tuffs of Pleistocene Lake Tecopa, Inyo County, California. US Geol Survey Prof Paper 597, 38 p

Sheppard RA, Gude AJ III (1969a) Diagenesis of tuffs in the Barstow Formation, Mud Hills, San Bernardino County, California. US Geol Survey Prof Paper 635, 35 p

Sheppard RA, Gude AJ III (1969b) Chemical composition and physical properties of the related zeolites offretite and erionite. Am Mineral 54:875-886

Sheppard RA, Gude AJ III (1973) Zeolites and associated authigenic silicate minerals in tuffaceous rocks of the Big Sandy Formation, Mohave County, Arizona. US Geol Survey Prof Paper 830, 36 p

Sheppard RA, Gude AJ III (1974) Chert derived from magadiite in a lacustrine deposit near Rome, Malheur County, Oregon. US Geol Surv J Res 2:625-630

Sheppard RA, Gude AJ III (1982) Magadi-type chert: A distinctive diagenetic variety from lacustrine deposits. *In* Program and Abstracts, Workshop on Diagenesis in Sedimentary Rocks. US Geological Survey, Golden, Colorado, 38 p

Stamatakis MG (1986) Boron distribution in hot springs, volcanic emanations, marine evaporites and volcanic and sedimentary rocks of Cenozoic age in Greece. PhD dissertation, Univ Athens, Athens, Greece, 495 p

Stamatakis MG (1989a) A boron-bearing potassium feldspar in volcanic ash and tuffaceous rocks from Miocene lake deposits, Samos Island, Greece. Am Mineral 74:230-235

Stamatakis MG (1989b) Authigenic silicates and silica polymorphs in the Miocene saline-alkaline deposits of the Karlovassi Basin, Samos, Greece. Econ Geol 84:788-798

Surdam RC (1977) Zeolites in closed hydrologic system. Rev Mineral 4:65-91

Surdam RC, Eugster HP (1976) Mineral reactions in the sedimentary deposits of the Lake Magadi region, Kenya. Geol Soc Am Bull 87:1739-1752

Surdam RC, Mariner RH (1971) The genesis of phillipsite in Recent tuffs at Teels Marsh, Nevada. Geol Soc Am (abstr)

Surdam RC, Parker B (1972) Authigenic aluminosilicate minerals in the tuffaceous rocks of the Green River Formation, Wyoming. Geol Soc Am Bull 83:689-700

Surdam RC, Sheppard RA (1978) Zeolites in saline, alkaline-lake deposits. *In* Natural Zeolites, Occurrence, Properties, Use. Sand LB, Mumpton FA (eds) Pergamon Press, Oxford, UK, p 145-174

Surdam RC, Eugster HP, Mariner RH (1972) Magadi-type chert in Jurassic and Eocene to Pleistocene rocks, Wyoming: Geol Soc Am Bull 38:2261-2266

Theodoropoulos D (1979) Geological map of Greece 1:50.000 sheet of Neon Karlovassi. Inst Geol Mineral Research, Athens, Greece

Turner CE, Fishman NS (1991) Jurassic Lake T'oo'dichi': A large alkaline, saline lake, Morrison Formation, eastern Colorado Plateau. Geol Soc Am Bull 103:538-558

Whateley MKG, Querol X, Fernandez-Turiel JL, Tuncali E (1996) Zeolites in Tertiary coal from Çayirham mine, Beypazari, Turkey. Mineral. Deposita 31:529-538

Williamson BM (1987) Formation of authigenic silicate minerals in Miocene volcaniclastic rocks, Boron, California. University of California, Santa Barbara, MS Thesis, 89 p

8 Formation of Zeolites in Open Hydrologic Systems

Richard A. Sheppard

11647 West 37th Place
Wheat Ridge, Colorado 80033

Richard L. Hay

Department of Geosciences
University of Arizona
Tucson, Arizona 85721

INTRODUCTION

In contrast to those deposits formed by the alteration of volcanic ash in saline, alkaline lakes discussed in the chapter by Hay and Sheppard (this volume), large volumes of tuffaceous sediments around the world have been transformed to zeolites and other authigenic silicate minerals by the action of percolating water in open hydrologic systems. These include systems that have experienced significant chemical and hydrologic exchanges with the surrounding environment. Tephra sequences exposed to open-system alteration commonly show a vertical zonation of zeolites and other authigenic minerals that reflects the chemical changes in meteoric water moving through the system. Flow in an open hydrologic system can either be downward or have a downward component where meteoric water enters the system, resulting in a vertical or nearly vertical zonation of water composition and authigenic minerals. The original pyroclastic materials of tuffaceous sediments may have been laid down in the sea close to the volcanic sources, air-laid onto the land surface, or reworked into fluviatile and freshwater lacustrine environments. Open-system zeolite deposits are most common in nonmarine rocks, although many are known in sediments that were deposited in shallow marine environments. Zeolite deposits of the open-system type are commonly several hundred meters thick and can be traced laterally for several tens of kilometers. Many examples of zeolite deposits that formed from land-laid tephra have been recognized in the western United States, but most of the large marine deposits are in Japan and in southern and southeastern Europe. Because of their relatively large and commonly discontinuous areal extent, open-system deposits have not been studied in as much detail as some other types of zeolite deposits in sedimentary rocks; however, studies of several key areas have provided a sound basis for the current understanding of this type of zeolite body.

Zeolitic alteration can take place in tephra deposits when flowing or percolating ground water becomes chemically modified by hydrolysis or dissolution of vitric materials. The formation of clay minerals or palagonite results in the release of hydroxyl ions into the ground water, and the solution becomes increasingly alkaline and enriched in Na, K, and Si along its flow path. Zeolite minerals crystallize in addition to, or instead of, clay minerals where the cation to hydrogen ion ratio and other ionic activities are relatively high (Hay 1963; see chapter by Hay and Sheppard, this volume). Meteoric water entering the system moves either downward or with a downward component; hence, the zeolitic alteration zones are either horizontal or gently inclined. Zeolitic alteration zones in open-system deposits commonly cut across stratigraphic boundaries and show more-or-less vertical sequences of authigenic silicate minerals.

The reaction rate of volcanic glass is greatly increased in natural ground waters where the pH is raised through hydrolysis. In saline, alkaline lakes, where the pH is about 9.5,

1529-6466/00/0045-0008$05.00

tuffs can be wholly altered in 10^3 to 10^4 years (Taylor and Surdam 1981). Distinct zeolite zonation is often produced through alteration of vitric tuffs by high-pH waters; these waters can be progressively modified through rock-water interactions causing the ground water to become increasingly enriched in dissolved solids with depth. Such alteration commonly produces clinoptilolite, mordenite, chabazite, and phillipsite as the first-formed, shallow zeolites in open systems. Reaction of early formed zeolites with ground water that has been modified through water-rock interaction yields analcime and (or) potassium feldspar in deeper zones. Erionite is generally absent from zeolite deposits formed in open hydrologic systems. Although groundwater composition exerts considerable control on zeolite formation, the exact zeolite assemblage and the composition of individual zeolite species are controlled, at least in part, by the composition of the precursor tephra.

EXAMPLES OF ZEOLITE FORMATION IN SILICIC TEPHRA DEPOSITS

John Day Formation, Oregon

Numerous deposits of zeolitic tuff throughout the western United States and in other countries are likely to have formed in open hydrologic basins, but nonmarine accumulations of silicic tephra must reach thicknesses of several hundred meters for widespread zeolitic zoning to develop as described above. The John Day Formation of central Oregon is an excellent example of this type and illustrates well the sequence and conditions of authigenic mineral formation and the processes by which silicic glass is altered to clinoptilolite.

The John Day Formation (Late Eocene-Early Miocene) is a distinctive assemblage of tuffaceous sedimentary rocks, ignimbrites, and lava flows that is exposed over about 14,000 km^2 (Robinson et al. 1990). A western facies consists mainly of ignimbrites and lava flows, and an eastern and a southern facies are chiefly silicic tuffs and tuffaceous claystones. Zeolites have been recognized in the lower part of the formation over an area of about 10,000 km^2. Recent work by Bestland et al (1997) provides significant stratigraphic, temporal, and paleoenvironmental information about the eastern facies. They extended the base of the John Day Formation downward to include a lava flow and kaolinitic claystones that had been considered by Hay (1963) to be part of the underlying Clarno Formation. High-resolution $^{40}Ar/^{39}Ar$ dates show that the overlying 210 m of tuffaceous strata accumulated at average rates of 5-6 cm/1000 years and contain numerous clay-rich paleosols (Bestland et al. 1997). Pedogenic alteration is much less developed in the overlying strata because of more rapid sedimentation rates and decreased rainfall.

The John Day Formation is about 600-900 m thick where studied by Hay (1963) in the eastern facies. Where the top of the formation is eroded least, an upper zone, 300-450 m thick, contains either fresh (unaltered) glass or glass altered to montmorillonite and opal (Fig. 1). Clinoptilolite and montmorillonite, with lesser amounts of celadonite and opal, replace nearly all the glass in the lower 300-450 m of the formation. Potassium feldspar is locally common in the lower part of the zeolite zone, chiefly as a replacement of volcanic plagioclase. The upper surface of the zeolitic zone is nearly horizontal and cuts across stratification where the beds are folded (Fig. 1). The transition between the upper and lower zones is relatively sharp and is generally marked by a cavernous zone, 2-20 mm thick, in which glass shards have been dissolved. Zeolite replacement of shards took place by the crystallization of laths and plates of clinoptilolite, 0.01-0.1 mm long, in pseudomorphic cavities from which glass was dissolving or had already been dissolved.

Spheroids or concretions of zeolite-cemented tuff near the base of the upper, fresh-glass zone provide clear evidence that hydrolysis of glass to montmorillonite can help create the chemical environment in which glass will react to form zeolites. The concretions are 2-

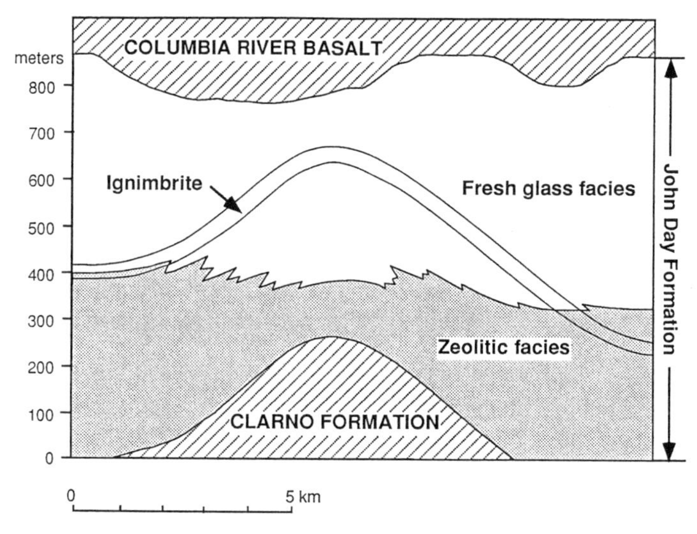

Figure 1. Diagrammatic cross section showing the pattern of zeolitic alteration in the folded John Day Formation north of Mitchell, Oregon. The line of section is north (left) to south (right). The fresh glass facies may contain authigenic montmorillonite and opal. Zeolitic alteration took place in the saturated zone, and the water table would have been at or near the top of the John Day Formation. Modified after Figure 4 of Hay (1963).

10 cm in diameter and lie in a matrix of fresh vitric ignimbrite ash. A pumice fragment altered to montmorillonite forms the nucleus of each concretion. These relationships suggest that the alteration of pumice to montmorillonite locally raised the pH and ionic activities to a level where the adjacent glass reacted to form clinoptilolite.

The replacement of rhyolitic glass by clinoptilolite in the John Day Formation involved gains of H_2O, Ca, and Mg and losses of Si, Na, K, and Fe. Open-system alteration is documented by chemical change in a rhyolitic ignimbrite near the middle of the formation. The altered vitric ignimbrite gained H_2O, Fe, Ti, Ca, and less commonly P. Some of the alkalis and Si were lost from most of the altered rhyolitic rocks, but these elements remained nearly constant in a few. The lower 210 m of the formation with clay-rich paleosols may have acted to some extent as a closed hydrologic system.

K-Ar dating of coexisting celadonite (24 Ma) and K-feldspar (22 Ma) suggests that silicate diagenesis may have been completed before the full thickness of the formation had been deposited. Zeolitic alteration probably had ceased before the time of deepest burial, about 15 Ma, when the Columbia River Basalt was erupted.

Yucca Mountain, Nevada

The diagenetic minerals and the mineral zonation are more complex in the Miocene tuffaceous rocks at Yucca Mountain than those in the John Day Formation. Yucca Mountain is located along the southwestern border of the Nevada Test Site (NTS), Nye County, Nevada, and is being considered as a potential site for an underground repository for high-level radioactive waste (Bish et al. 1982). Yucca Mountain is within the southwestern Nevada volcanic field, which is a faulted, dissected volcanic plateau composed mainly of silicic ash-flow tuffs and other volcaniclastic rocks and minor lava flows of Miocene age. The volcanic rocks are about 1.5-3 km thick and range in age from about 11.4 to 14.1 Ma. Several sources, including the Timber Mountain-Oasis Valley caldera complex (Byers et al. 1976) and the Calico Hills, lie just west, north, and northeast of Yucca Mountain and are the sources for the volcanic rocks. Yucca Mountain consists of many north- to northwest-trending, eastward tilted structural blocks that are repeated by westward-dipping, high-angle normal faults. The strata of Yucca Mountain dip mainly 5°-20° eastward.

Studies of the mineralogical and textural features of the Miocene volcanic rocks were carried out chiefly at the Los Alamos National Laboratory on core and cuttings from numerous drill holes. Diagenetic alteration is most conspicuous in nonwelded ash-flow tuff, bedded tuff, and locally in thin, nonwelded parts of densely welded ash-flow tuff. These nonwelded tuffs were originally vitric and were highly susceptible to alteration because of the instability of the glass in the ground water. The altered tuffs exhibit mappable zones of diagenetic minerals that can be correlated laterally across much of Yucca Mountain (Broxton et al. 1987).

Early mineralogical studies by Smyth and Caporuscio (1981) and Bish et al (1981) showed that the diagenetic mineral assemblages at Yucca Mountain are vertically zoned and progressively less hydrous with depth. Four diagenetic zones were recognized and assigned Roman numerals from I to IV (Fig. 2), analogous to those described by Iijima (1978) for thick, marine sequences of tuffaceous rocks in Japan. Broxton et al (1987) summarized the diagenetic mineralogy of these zones as follows. Zone I, the shallowest zone, is above the static water table and is characterized by widespread preservation of vitric material. Diagenetic minerals include smectite, opal, and minor heulandite or clinoptilolite. Zone II straddles the water table and is characterized by nearly complete replacement of glass by clinoptilolite, mordenite, opal-CT, and minor smectite, quartz, and potassium feldspar. Chabazite occurs locally with clinoptilolite in parts of Zone II to the south. Mordenite increases in abundance in the lower part of the zone and decreases in abundance from north to south. Zones III and IV are beneath the static water table. Zone III is characterized by analcime, quartz, potassium feldspar, and minor illite/smectite and calcite. Clinoptilolite, heulandite, and mordenite persist in the upper part of this zone and locally are replaced by analcime. Zone IV is characterized by albite, potassium feldspar, quartz, and minor smectite, chlorite, and calcite. Locally, the authigenic albite replaced analcime. The diagenetic zones transgress stratigraphic boundaries and rise in elevation and thin from south to north. In addition to the zeolites mentioned above, minor heulandite, erionite, phillipsite, and stellerite have been identified in fracture fillings chiefly in Zone I (Carlos et al. 1995).

Broxton et al (1987) showed that the chemical compositions of altered tuffs deviate significantly from the original compositions throughout Yucca Mountain. Na and K are generally depleted, and Ca and Mg are generally greatly enriched. Fe and Ti show smaller and variable compositional differences. Si is systematically depleted in the zeolitic tuffs, especially in the more Ca-rich zeolitic alteration in Zones I and II. The variable compositions of the zeolitic tuffs are evidence that crystallization of zeolites from glass in Zones I and II occurred in an open system. The chemical composition of tuffs in Zones III and IV, however, is similar to that of Zone II despite the replacement by analcime, potassium feldspar, and albite of the early-formed zeolites. This closure in chemical composition, despite the formation of more alkali-rich zeolites and feldspars, apparently was accomplished by the formation of calcite as a sink for calcium as the silicates became more alkali rich (Bish and Aronson 1993). As diagenesis progressed, the tuffs apparently became increasingly less permeable, and movement of pore waters was greatly restricted during the mineralogical transformations from Zone II to Zones III and IV.

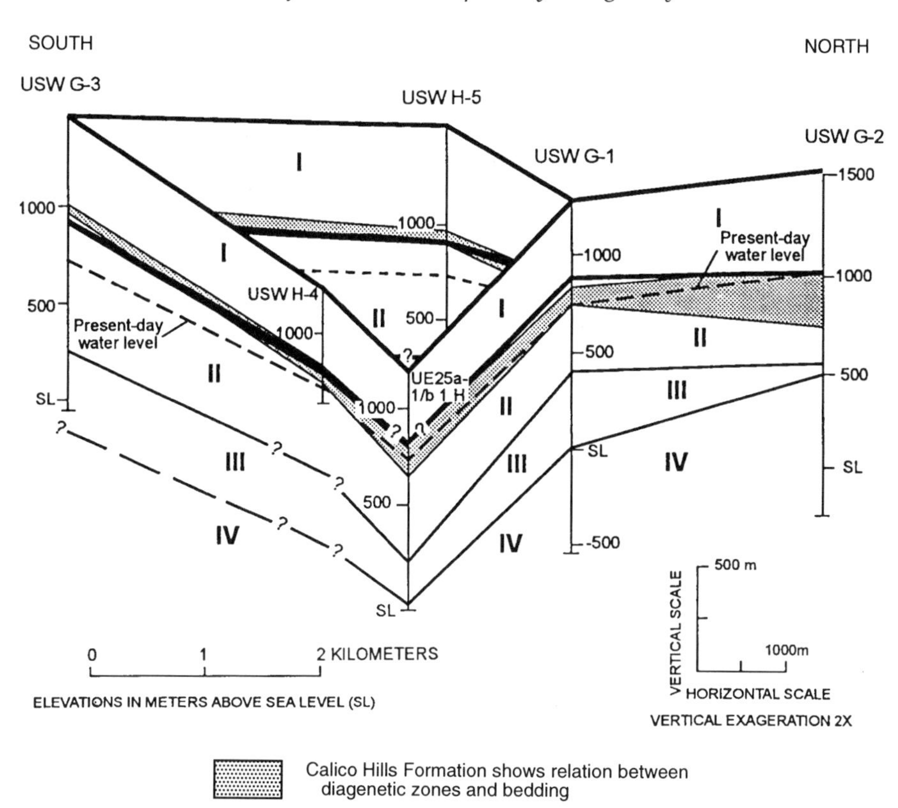

Figure 2. Fence diagram showing the distribution of diagenetic zones (Roman numerals), Yucca Mountain, Nevada. USW G-3, USW H-5, USW H-4, USW G-1, USW G-2, and UE25a-1/b1H are drill holes. The Calico Hills Formation is shown as an indication of how diagenetic zones can cross tuff units. Modified after Figure 3 of Broxton et al (1987).

Stratigraphic and structural considerations led Broxton et al (1987) to conclude that most of the zeolites at Yucca Mountain formed between 11.3 and 13.9 Ma. K/Ar dates by WoldeGabriel (1995) on clinoptilolite separated from the zeolitic tuffs showed a range of 2 to 13 Ma; and the dates increase with depth. This range was attributed to Ar loss from clinoptilolite during continued diagenesis, dehydration in the unsaturated zone, and contamination by authigenic feldspar and illite/smectite (I/S). Bish and Aronson (1993) concluded on the basis of the distribution of alteration minerals and K/Ar dates of I/S at the northern part of Yucca Mountain that hydrothermal alteration associated with the Timber Mountain volcanism about 10-11 Ma was responsible for mineralogical changes at depth, chiefly in Zone IV. Even some of the mordenite that formed at the expense of clinoptilolite in Zone II probably is due to a relatively high thermal regime near the Timber Mountain caldera (Sheppard et al. 1988).

Earlier workers recognized and described a vertical zonation of diagenetic minerals in the Tertiary silicic tuffs of the NTS, north and east of Yucca Mountain. Based on mineralogical data from drill holes, tunnels, and outcrops, Hoover (1968) described a downward succession of authigenic mineral assemblages characterized by glass, chabazite, clinoptilolite, mordenite, and then analcime and concluded that the alteration occurred in an open chemical system. Later, Moncure et al (1981) delineated three vertical diagenetic zones in the Tertiary tuffs beneath Pahute Mesa, north of Yucca Mountain. These authors demonstrated a downward succession of zones characterized by glass, clinoptilolite, and then analcime + albite. On the basis of the mineralogical descriptions provided in their paper, the lowest zone could be divided into an upper analcime zone and a lower albite zone, similar to the zonation at Yucca Mountain. Moncure et al (1981) found no change in the chemical composition of the altered tuffs with depth and concluded that the alteration occurred in a closed chemical system. At Yucca Mountain, alteration of tuff in Zones I and II occurred in an open system, but Zones III and IV probably formed in an essentially closed system characterized by increasing temperature at depth.

Southern Desatoya Mountains, Nevada

Widespread zeolitization has been recognized in Miocene (20-25 Ma) volcaniclastic rocks of the southern Desatoya Mountains of west-central Nevada. About 3000 m of predominately rhyolitic to andesitic ash-flow tuffs and minor lava, breccia, ash-fall tuff, and epiclastic rocks contain silicate mineral assemblages that formed diagenetically in an open system (Barrows 1980).

A vertical succession of diagenetic zones is characterized (from top to bottom) by fresh glass, montmorillonite, and then clinoptilolite (heulandite)-mordenite. Analcime was identified, but the paucity of occurrences did not permit its placement in a discrete zone. Volcanic minerals are generally fresh throughout the altered rocks except in the most severely altered zones in which plagioclase and pyroxene are partially to completely dissolved or replaced by diagenetic minerals. Barrows (1980) concluded that the diagenetic minerals formed chiefly through the breakdown of the unstable glassy components, and the original porosity and permeability of the volcaniclastic rocks were the major factors controlling the degree of alteration. She also determined that Si, Ca, Na, K, and H_2O were mobile during diagenesis. Andesitic rocks commonly gained Si and lost Na and K, whereas the more siliceous rocks commonly lost Si and Na and gained Ca. H_2O, of course, increased in all altered rocks.

White River sequence in Wyoming and adjacent states

The Oligocene White River sequence of the Great Plains and the northern Rocky Mountains consists mainly of volcaniclastic sediments that have been reworked and (or) altered to varying degrees by fluvial, eolian, pedogenic, and diagenetic processes. Lander (1990) and Lander and Hay (1993) studied the stratigraphy, diagenetic mineralogy, and chemistry of the White River sequence over an area of more than 500,000 km^2 in Montana, Wyoming, South Dakota, and Colorado. The sequence is commonly 100-200 m thick and consists chiefly of tuffaceous mudstone with about 2% discrete tuffs that were deposited over an interval of about 8 m.y. Vitroclastic material or its alteration minerals make up about 80-90% of the White River rocks. Most vitric material is rhyolitic, but some, especially in Montana and Wyoming, is dacitic to latitic. Most localities sampled by Lander and Hay (1993) contain some unaltered vitroclastic material.

The predominant diagenetic minerals in the White River rocks are smectite, clinoptilolite, opal-CT, and calcite. Clinoptilolite is the only zeolite recognized, and it occurs in mudstones, tuffs, and clastic dikes. The distribution of clinoptilolite is unusual in that it is patchy on both regional and local scales. Clinoptilolite-bearing mudstone increases in abundance from west to east, but the clinoptilolite-rich tuff is common only in Wyoming. The White River clinoptilolite has Si/Al ratios of 4.4-5.4 and major exchangeable cations of Na and K, although Ca-rich clinoptilolite has been recognized in calcretes and mudstones. Smectite is ubiquitous in the White River rocks and has detrital, pedogenic, and diagenetic origins.

The spatial pattern of zeolitic alteration in the White River sequence is complex. Most sampled localities have nonzeolitic intervals, and clinoptilolite was not detected at many localities. Clinoptilolite tends to be more abundant in the northern part of the Great Plains. In Wyoming, extensive deposits of clinoptilolite occur only at Teacup Butte and near Douglas. Tuffs in the Douglas area show a vertical zonation; tuffs in the upper part of the section are still vitric, but tuffs in the lower part are altered to smectite, clinoptilolite, and opal-CT (Lander 1990).

Although previous workers had suggested that clinoptilolite formed in soils or in lakes during White River sequence deposition, Lander and Hay (1993) proposed that the major factor controlling the genesis of clinoptilolite was the contact time between paleo-ground waters and the vitric material. They proposed that clinoptilolite in Wyoming formed as a result of long fluid contact times associated with major paleohydrologic discharge zones (Fig. 3), but that clinoptilolite in the northern Great Plains formed because low hydrologic head and low initial permeability resulted in long fluid residence times. The fluid residence time can be shortened by changes in the flow regime that occur in response to changes in topography. Lander and Hay (1993) showed that the late Tertiary denudation of the Great Plains and intermontane basins broke up relatively large-scale flow systems into several small-scale flow systems defined by stream drainages. The formation of clinoptilolite probably was retarded or halted by increases in hydrologic head and shorter flow paths resulting from uplift and erosion. As often occurs, the zeolite alteration changed the hydrologic role of vitroclastic rocks from aquifers to aquitards. Stable oxygen isotopic compositions of smectite, opal-CT, and calcite suggested that the diagenetic reactions took place at 27°-55°C.

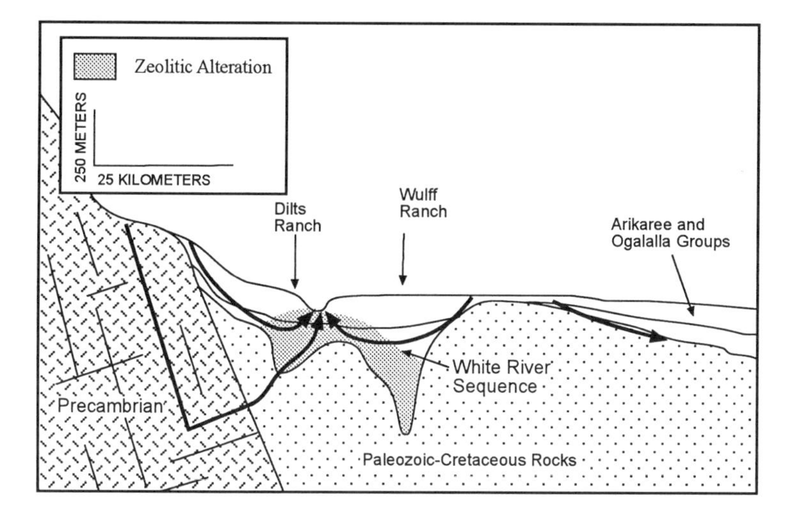

Figure 3. Schematic illustration of the hydrologic regime that may have led to the zeolitic alteration in White River rocks near Douglas, Wyoming. Line of section is southeast (left) to northwest (right). Modified after Figure 10 of Lander and Hay (1993).

ZEOLITE FORMATION IN MAFIC TEPHRA DEPOSITS

The formation of zeolites from mafic glasses in open hydrologic systems is somewhat more complex than it is from silicic glasses. Mafic glass (or sideromelane) reacts to form palagonite and zeolites in appropriate alkaline chemical environments. Palagonite can be viewed as a hydrous, iron-rich gel formed from mafic glass. It contains somewhat less Al, Si, Ca, Na, and K than the original glass. These components provide the raw materials for the crystallization of zeolites in pore spaces. Phillipsite and chabazite are probably the most common zeolites in low-silica tuffs, but natrolite, gonnardite, analcime and other low-silica zeolites are commonly recognized. In Hawaii, alkalic, low-silica tephra deposits only 10-20 m thick are zoned zeolitically to the same extent as silicic tephras that are 500-1000 m thick. Zonation at shallow depth reflects the highly reactive nature of alkalic, low-silica glass.

Koko Crater, Hawaii, illustrates the alteration of mafic tephra in an open-system environment. Koko Crater is one of a group of craters that formed on southern Oahu about 35,000 years ago. Its cone consists mainly of tuff and lapilli tuff of basanite composition (Basanite is a mafic igneous rock whose composition is close to that of basalt but which contains normative nepheline.) Hay and Iijima (1968) showed that an upper zone characterized by unaltered glass, opal, and montmorillonite overlies a lower zone of zeolitic palagonitic tuff. The upper zone is generally 12-20 m thick, and the transition between the zones is a few centimeters to a meter thick. Two zeolitic subzones were recognized within the zeolitic palagonite: an upper subzone that contains phillipsite and chabazite, and a lower subzone that contains phillipsite and analcime. A striking feature of the palagonitic alteration is the loss of about half of the Al from the mafic glass and incorporation of Al in interstitial zeolites.

The contact between zones roughly parallels the topography and cuts across stratification (Fig. 4), suggesting that the palagonitic alteration is a post-eruptive phenomenon related to percolating meteoric water in the vadose zone rather than to hydrothermal solutions. Palagonitic tuff abruptly grades downward into relatively unaltered tuff near sea level, presumably reflecting dilution of reactive alkaline meteoric water with sea water of lower pH. In this chemical environment, the mafic glass reacted more slowly than elsewhere. As it descended through the column of tephra, the meteoric water may have reached a pH as high as 9.0-9.5, which could account for the extensive alteration in less than 35,000 years and for the relatively sharp transition between fresh and palagonitic tuff. A high pH also could explain the mobility of Al, which is relatively soluble in the form of $Al(OH)_4^-$ under these conditions.

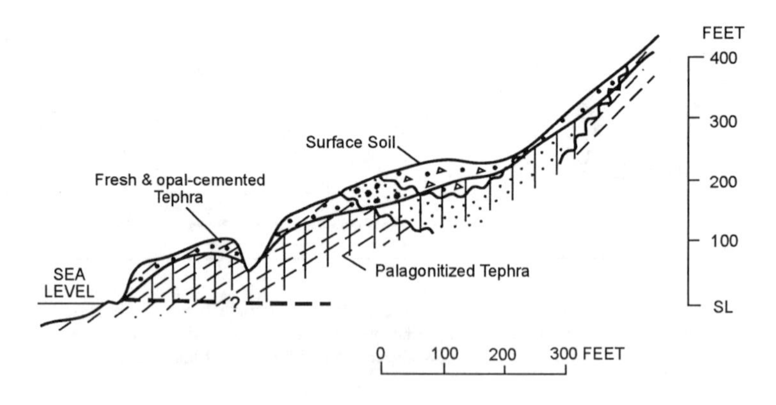

Figure 4. Cross section through the southeastern edge of the cone of Koko Crater, showing palagonitic alteration that transects stratification and ends abruptly near sea level (SL). Modified after Figure 3 of Hay and Iijima (1968).

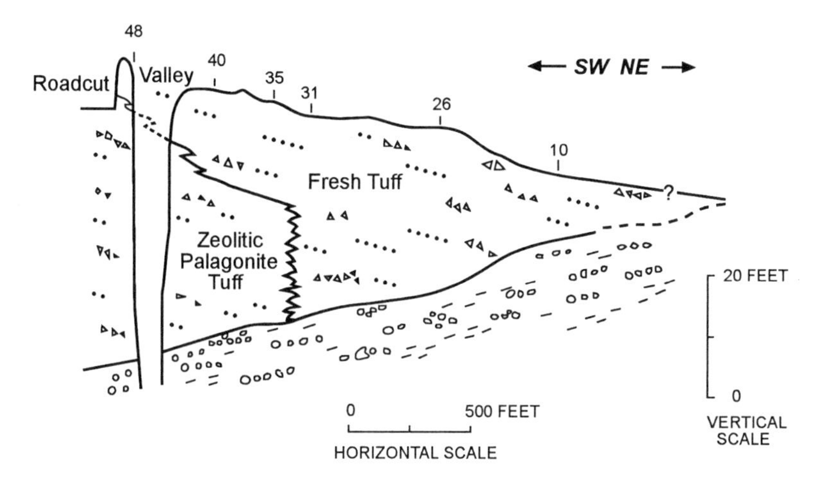

Figure 5. Salt Lake Tuff exposed along the northwest side of Moanalua Stream, at the northeastern-most extent of the Salt Lake Tuff. Numbers along the surface of the tuff represent thicknesses of measured sections. Boulder conglomerates, claystones, and reworked nephelinite ash layers underlie the Salt Lake Tuff. Modified after Figure 5 of Hay and Iijima (1968).

The alteration in the Salt Lake Tuff, also on Oahu, is related to the original thickness of tuff and to the position of the tuff relative to sea level in the Pleistocene (Hay and Iijima 1968). Where nephelinite tuff thins to the north and west, zeolitic palagonite intergrades laterally with fresh (unaltered) tuff. Fresh tuff commonly overlies palagonitic tuff in the transition zone (Fig. 5). Where the tuff is thinner than 9 m, it is unaltered except for surface weathering; where the tuff is thicker than 9 m, the upper 4-5 m are fresh, and the underlying material is palagonitic and cemented by zeolites. The vertical transition between fresh and palagonitic tuff is generally sharp. The chief zeolites in the Salt Lake Tuff are phillipsite, natrolite, gonnardite, and analcime. The Salt Lake Tuff was deposited when sea level was lower than at present, and the tuff was palagonitized and cemented by zeolites before the sea rose 6-8 m above the present level in late Pleistocene time.

ZEOLITE FORMATION IN ALKALIC TEPHRA DEPOSITS

The Neapolitan Yellow Tuff (tufo giallo napoletano), a widespread volcaniclastic unit in the Campania region of southern Italy, provides an example of the formation of zeolites in an open-system environment from tuff of potassic, trachytic composition (Sersale 1978). This tuff, which underlies the city of Naples, crops out over an area of about 200 km^2 and is about 20-120 m thick (Gottardi and Obradovic 1978). The zeolitic nature of the Neapolitan Yellow Tuff was first recognized by Norin (1955), who identified chabazite in the tuff. More recent studies have shown that the trachytic glass in the lower part of the formation is altered chiefly to phillipsite and chabazite with minor analcime (de' Gennaro et al. 1990; Passaglia et al. 1990). Although the mean zeolite content reported by de' Gennaro et al (1990) is 56%, parts of the tuff locally contain as much as 80% zeolite (Passaglia et al. 1990). Phillipsite is generally more abundant than chabazite. Scanning electron micrographs of the zeolitic tuff show that phillipsite crystallized before chabazite (de' Gennaro et al. 1990). Analcime crystallized after phillipsite, but its paragenetic relationship with chabazite is unknown. The upper part of the Neapolitan Yellow Tuff is essentially zeolite free and is known locally as "pozzolana." The transition from glassy tuff to zeolitic tuff is continuous, with the zeolite-rich tuff being characterized by increases in H_2O and Ca and decreases in Si, Al, and alkalis.

The Neapolitan Yellow Tuff was emplaced about 10,000-12,000 years ago. Isotopic evidence suggests that the zeolitic alteration lasted 4000-5000 years and is now thought to be inactive (Capaldi et al. 1971). Although Sersale (1978) and other investigators interpreted the zeolites in the tuff to have formed in an open hydrologic system, de' Gennaro et al (1995) suggested that the zeolites crystallized at elevated temperatures on the basis of the emplacement temperature (200°-300°C) of the tuff and on laboratory simulations of trachytic glass alteration.

COMPARISON OF ALTERATION IN AN OPEN HYDROLOGIC SYSTEMS WITH ALTERATION BY BURIAL DIAGENESIS AND GEOAUTOCLAVE

The mineralogical zoning in deposits of the open-system type is similar to the upper zones of burial diagenesis (see Iijima 1978), and the two types may be difficult to distinguish in thick sequences of silicic tephra. However, open-system type deposits generally exhibit a sharp rather than a gradational contact between fresh glass and zeolitic rock, the host rocks are commonly nonmarine rather than marine, and the alteration was short-lived, penecontemporaneous with deposition, and may not correspond to the time of deepest burial. Also, the individual zones of open-system type deposits generally are thinner (several hundred meters versus one to two kilometers thick) than those of the burial diagenesis type. Burial diagenetic zones (Zones I-IV) show a downward succession of authigenic zeolites and coexisting minerals that are characterized by decreasing hydration as a function of increasing temperature.

Aleksiev and Djourova (1975) coined the term "geoautoclave" to explain the genesis of clinoptilolite in ash-flow tuff from the Rhodope Mountains of southeastern Bulgaria, and Lenzi and Passaglia (1974) proposed a similar high-temperature genesis, without naming it, for the formation of chabazite in an ignimbrite of central Italy. Zeolite deposits formed by the geoautoclave process are difficult to distinguish from those formed by alteration in an open hydrologic system. Even though the alteration temperature of the geoautoclave type is thought by some investigators to be as high as 200°C (Gottardi 1989), the resulting authigenic mineralogy is similar to that formed during open-system alteration. Field relationships may provide evidence for distinguishing geoautoclave from open-system zeolitization. Djourova and Alexsiev (1995), studying clinoptilolite deposits in marine Oligocene ash-flow tuffs of the northeastern Rhodope Mountains, considered the zeolitic rocks to be an aqueous facies of welded ash-flow tuffs that were deposited on land. Due to rapid sedimentation in a marine environment and the thermal insulation of the vitric material, Djourova and Alexsiev (p. 26-28) inferred that much of the thermal energy remained in the sediments and that a large natural autoclave was created. The volcanic glass in this environment then reacted at the elevated temperature and pressure to form zeolites. Lenzi and Passaglia (1974) showed that a chabazite-rich ash-flow tuff in central Italy is underlain and overlain by air-fall tuff that lacks zeolites, although the two types of tuff are similar in chemical composition. The zeolitization of the ash-flow tuff was attributed to the relatively high temperature and vapor pressure within the tuff immediately after deposition. Factors such as the grain size of the vitric material need to be evaluated in determining the cause for the alteration. Fine-grained vitric material tends to alter more quickly and completely than coarse-grained material. Also, temperatures much above 100°C would tend to expel H_2O from the tuff, thereby producing a relatively dry environment that is not conducive to alteration of glass to zeolites.

More recently, Garcia Hernandez et al (1993) described phillipsite, chabazite, and analcime in phonolitic ash-flow tuff at Tenerife, Canary Islands, and attributed their formation to a geoautoclave mechanism. de' Gennaro et al (1995) re-examined the alteration of ash-flow tuffs from the Canary Islands, Germany, and Italy and concluded that the formation of chabazite, phillipsite, or chabazite + phillipsite was due to the relatively high emplacement temperature (100°C, up to as high as 600°C) and abundant fluid phase. Many of these zeolite deposits previously had been attributed to alteration in an open hydrologic system. de' Gennaro et al (1995) cited the lack of systematic vertical zoning of authigenic minerals in these deposits as evidence against alteration in an open hydrologic system.

SYNTHESIS OF OPEN-SYSTEM ZEOLITES

Wirsching (1976) simulated open-system alteration using rhyolitic glass as the starting material in low-temperature hydrothermal experiments. Experiments with 1 g of glass in 50 ml of 0.01 N NaOH at 150°C formed minor phillipsite and a trace of mordenite after 32 days. With increasing time, phillipsite and mordenite increased in abundance, and a trace of analcime was detected after 42 days. At the end of the 80-day experiment, analcime was the dominant phase. Analcime and alkali feldspar were the final products at higher temperatures.

Experiments by Höller and Wirsching (1980) in an open system reacting nephelinite in a solution of 0.01 N NaOH at 150°C showed that chabazite, phillipsite, and analcime crystallized. As the reaction continued, chabazite disappeared and analcime became the predominant zeolite after frequent changes of solution. Analcime was the only zeolite present at temperatures of 200°-250°C. These authors also found that the crystal shape of the zeolite was influenced by the solution composition, temperature, reaction time, and number of changes of solution.

On the basis of experimental studies starting with either basaltic or rhyolitic glass in an open system, Barth-Wirsching and Höller (1989) determined that the composition of the starting material, solution composition, and temperature are all important factors in determining which zeolite crystallizes. Material transport during the experiment is also an important factor and is a function of the chemical gradient, percolation velocity, and reaction time. Reactions with the starting material may change the pH of the solution and act as another factor that could affect the Si/Al ratio of the zeolite being formed.

ECONOMIC POTENTIAL OF OPEN-SYSTEM ZEOLITE DEPOSITS

Because many of the zeolite deposits of the open-system type are relatively thick, high grade, and exposed at or near the surface, they are being mined or are suitable for mining. Probable open-system clinoptilolite deposits are common in Tertiary silicic tephra of the United States; those that have been mined include Neogene nonwelded ash-flow tuff near Death Valley Junction, California (Sheppard 1985), Miocene Sucker Creek Formation of southeastern Oregon (Altaner and Teague 1993, Sheppard and Gude 1993), Oligocene tuff near Winston, New Mexico (Bowie et al. 1987), and Eocene Manning Formation of southeastern Texas (Bowie et al. 1987). Other probable open-system clinoptilolite-mordenite deposits in silicic tephra that have been mined elsewhere in the western hemisphere include Miocene ash-flow tuff near Oaxaca, Mexico (Mumpton 1973) and several Cretacaeous to Eocene tuff units near Havana and Tasajeras, Cuba (Rodriguez Fuentes et al. 1985). In southern Italy, numerous probable open-system phillipsite-chabazite deposits have been mined for more than 2000 years from Quaternary alkalic tuffs for use chiefly as dimension stone (de' Gennaro and Langella 1996). In eastern Europe, many clinoptilolite-mordenite deposits of probable open-system type have been mined or are mineable from Tertiary silicic tuffs, including clinoptilolite in Eocene deposits near

Metaxades, northeastern Greece (Tsirambides et al. 1993; Stamatakis et al. 1996); clinoptilolite and mordenite in marine tuffs of Oligocene age in the northeastern Rhodopes of southern Bulgaria (Djourova and Aleksiev 1995); clinoptilolite in Miocene tuffs of the Tokaj Hills of northeastern Hungary (Kalló et al. 1982); clinoptilolite and mordenite in Miocene tuffs of eastern Slovakia (Samajová 1988); clinoptilolite and mordenite deposits in tuffs of Cretaceous and Neogene age of Georgia (Skhirtladze et al. 1986); and clinoptilolite and mordenite in Miocene tuffs of the Carpathian Mountains of southwestern Ukraine (Samajová and Kuzvart 1994). Although some of the clinoptilolite and mordenite deposits of eastern Europe have been attributed to a genesis by geoautoclave or burial diagenesis, most or all of the above zeolite deposits in silicic tuffs may have formed at relatively low temperatures in an open hydrologic system.

Extensive clinoptilolite and mordenite deposits have been known in silicic tuffs of eastern China since 1972. The deposits occur in Jurassic and Cretaceous rocks, and at least 80 of them have been mined for use in cement and other agricultural and industrial applications (Quanchang 1994). Although definitive information is not available, many of these deposits may be of the open-system type. Zeolites in sedimentary rocks of Korea were first reported in 1976, but since then, many deposits of clinoptilolite and mordenite have been discovered and mined from Miocene silicic tuffs of southeastern Korea (Hwang 1994). The deposits are vertically zoned, and descriptions of the zoning and mineralogy suggest to us diagenetic alteration in an open system.

A recent discovery of zeolitic basaltic tephra in northeastern Jordan (Ibrahim and Hall 1996) is noteworthy for the occurrence of as much as 30% faujasite, a natural zeolite that is equivalent to the commercially important synthetic zeolite X, in an open-system deposit. The Quaternary basaltic tephra is vertically zoned, consisting of an upper zone (10-40 m thick) of unaltered glass, a middle zone (2-10 m thick) of smectite-bearing palagonite, and a lower zone (6-40 m thick) of zeolites. The lower zone contains about 20-60% zeolites and is characterized by phillipsite-chabazite in the upper part and phillipsite-faujasite in the lower part. At the Jabal Hannoun locality, about 6 m of zeolitic tephra average about 29% faujasite. Ibrahim and Inglethorpe (1996) showed that laboratory-scale mineral processing could upgrade the zeolite content to about 90% and the faujasite content to about 60%.

FUTURE RESEARCH

Recently, the zeolitic alteration of some tephra deposits, especially in southern Italy, that had been interpreted to have formed in an open hydrologic system, has been attributed by some investigators to the geoautoclave mechanism (de' Gennaro et al. 1995). The geoautoclave type of zeolite genesis needs additional study to determine the unresolved role and magnitude of elevated temperature in the crystallization of zeolites in ignimbrites. For example, Pérez-Torrado et al (1995) proposed a geoautoclave mechanism to account for zeolitization of the Roque Nublo ignimbrites on Gran Canaria, Canary Islands. These authors suggested that the ignimbrites were produced by phreatomagmatic eruptions and that the eruptive clouds (~100°C) were too poorly expanded to expel the water vapor during transport. Their evidence for the geoautoclave genesis and alteration at elevated temperatures is the presence of zeolites in the ignimbrites and apparent lack of zeolites in the associated ash-fall tuffs and tuffaceous epiclastic rocks. However, their own data (their Table 1 and Fig. 4) show the occurrence of zeolites, albeit minor, in these associated rocks. Future studies of the zeolitization of ignimbrites need to consider: (1) the vertical and lateral distribution of zeolites and coexisting minerals throughout the deposit; (2) the textural properties, especially grain size, of the vitric material because fine-grained ash is more susceptible to alteration than coarse-grained ash; and (3) oxygen-isotope analyses of the zeolites and coexisting authigenic minerals to help determine the temperature during alteration.

Another unsolved problem related to zeolite genesis in open hydrologic systems is the characterization and formation of analcime in mafic tephra. Much of the analcime in silicic tephra crystallized from early precursor zeolites such as clinoptilolite and mordenite (e.g. Broxton et al. 1987), but much of the analcime in basaltic tephra seems to have formed from glass or palagonite. Investigations of the diagenetic alteration of thick, mafic tephra deposits should pay particular attention to the physical and chemical properties of analcime, the paragenetic relationships with the coexisting silicate minerals, and the mechanism of genesis.

Because many investigations of open hydrologic systems are handicapped by the need to infer the composition of the fluid phase during zeolitic alteration, a search for an area undergoing present-day alteration would be instructive. Such an area might be the Phlegraean Fields or vicinity of Mount Vesuvius, southern Italy (de' Gennaro et al. 1995). Where zeolitization is in progress, the chemistry of the fluid, glass, and authigenic zeolites should be studied in addition to the temperature.

REFERENCES

Aleksiev B, Djourova EG (1975) On the origin of zeolitic rocks. C R Acad Bulg Sci 28:517-520

Altaner SP, Teague GA (1993) Mineralogy, chemistry, and genesis of tuffs of the Sucker Creek Formation (Miocene), eastern Oegon. *In* Zeo-Trip '93: An Excursion to Selected Zeolite and Clay Deposits in Sountheastern Oregon and Southwestern Idaho. FA Mumpton (ed) Int'l Comm on Natural Zeolites, Brockport, New York, p 14-29

Barrows KJ (1980) Zeolitization of Miocene volcaniclastic rocks, southern Desatoya Mountains, Nevada. Geol Soc Am Bull 91:199-210

Barth-Wirsching U, Höller H (1989) Experimental studies on zeolite formation conditions. Eur J Mineral 1:489-506

Bestland EA, Retallack G., Swisher CC III (1997) Stepwise climate change recorded in Eocene-Oligocene paleosol sequences from central Oregon. J Geol 105:153-172

Bish DL, Aronson JL (1993) Paleogeothermal and paleohydrologic conditions in silicic tuff from Yucca Mountain, Nevada. Clays & Clay Minerals 41:148-161

Bish DL, Caporuscio FA, Copp JF, Crowe BM, Purson JD, Smyth JR, Warren RG (1981) Preliminary stratigraphic and petrologic characterization of core samples from USW G-1, Yucca Mountain, Nevada. Los Alamos Nat'l Lab Rept LA-8840-MS, 66 p

Bish, DL, Vaniman, D.T, Byers, F.M, Jr, Broxton, DE (1982) Summary of the mineralogy-petrology of tuffs of Yucca Mountain and the secondary-phase thermal stability in tuffs. Los Alamos Nat'l Lab Rept LA-9321-MS, 47 p

Bowie MR, Barker JM, Peterson SL (1987) Comparison of selected zeolite deposits of Arizona, New Mexico, and Texas. Arizona Bur Geol Mineral Tech Spec Paper 4:90-105

Broxton DE, Bish DL, Warren RG (1987) Distribution and chemistry of diagenetic minerals at Yucca Mountain, Nye County, Nevada. Clays & Clay Minerals 35:89-110

Byers FM Jr, Carr WJ, Orkild PP, Quinlivin WD, Sargent KA (1976) Volcanic suites and related cauldrons of Timber Mountain-Oasis Valley caldera complex, southern Nevada. U S Geol Survey Prof Paper 919, 70 p

Capaldi G, Civetta L, Gasparini P (1971) Fractionation of the ^{238}U decay series in the zeolitization of volcanic ashes. Geochim Cosmochim Acta 35:1067-1072

Carlos B, Chipera S, Bish DL, Raymond R (1995) Distribution and chemistry of fracture-lining zeolites at Yucca Mountain, Nevada. *In* Natural Zeolites '93: Occurrence, Properties, Use. DW Ming, FA Mumpton (eds) Int'l Comm on Natural Zeolites, Brockport, New York, p 547-563

de' Gennaro M, Langella A (1996) Italian zeolitized rocks of technological interest. Mineral Deposita 31:452-472

de' Gennaro M, Adabbo M, Langella A (1995) Hypothesis on the genesis of zeolites in some European volcaniclastic deposits. *In* Natural Zeolites '93: Occurrence, Properties, Use. DW Ming, FA Mumpton (eds) Int'l Comm on Natural Zeolites, Brockport, New York, p 51-67

de' Gennaro, M, Petrosino, P, Conte, M.T, Munno, R, Colella, C (1990) Zeolite chemistry and distribution in a Neapolitan yellow tuff deposit. Eur J Mineral. 2:779-786

Djourova E, Aleksiev B (1995) Zeolitic rocks in the northeastern Rhodopes. *In* An Excursion to Selected Zeolite and Clay Deposits in the Eastern Rhodopes. B. Aleksiev (ed) Int'l Symp and Exhibition on Natural Zeolites, Sofia, Bulgaria, p 20-48

Garcia Hernandez JE, Notario del Pino JS, Gonzalez Martin MM, Hernan Reguera FH, Rodriguez Losada JA (1993) Zeolites in pyroclastic deposits in southeastern Tenerife (Canary Islands). Clays & Clay Minerals 41:521-526

Gottardi G (1989) The genesis of zeolites. Eur J Mineral 1:479-487

Gottardi G, Obradovic J (1978) Sedimentary zeolites in Europe. Fortschr Mineral 56:316-366

Hay RL (1963) Stratigraphy and zeolite diagenesis of the John Day Formation of Oregon. Univ Calif Publ Geol Sci 42:199-262

Hay RL, Iijima A (1968) Nature and origin of palagonite tuffs of the Honolulu Group on Oahu, Hawaii. Geol Soc Am Mem 116:331-376

Höller,H, Wirsching U (1980) Experiments on the hydrothermal formation of zeolites from nepheline and nephelinite. *In* Proc. 5th Int'l Conf on Zeolites. LVC Rees (ed) Heyden, London, p 164-170

Hoover DL (1968) Genesis of zeolites. *In* Nevada Test Site. E.B. Eckel (ed) Geol Soc Am Mem 110:275-284

Hwang J-Y (1994) Occurrence and utilization of natural zeolite in Korea. *In* Natural Zeolite and its Utilization. Japan Society for the Promotion of Science, Committee on Development of New Utilization of Minerals 111:129-141

Ibrahim K, Hall A (1996) The authigenic zeolites of the Aritayn Volcaniclastic Formation, northeast Jordan. Mineral Deposita 31:514-522

Ibrahim K, Inglethorpe SDJ (1996) Mineral processing characteristics of natural zeolites from the Aritayn Formation of northeast Jordan. Mineral Deposita 31:589-596

Iijima A (1978) Geologic occurrences of zeolites in marine environments. *In* Natural Zeolites: Occurrence, Properties, Use. LB Sand, FA Mumpton (eds) Pergamon Press, Elmsford, New York, p 175-198

Kalló D, Papp J, Valyon J (1982) Adsorption and catalytic properties of sedimentary clinoptilolite and mordenite from the Tokaj Hills, Hungary. Zeolites 2:13-16

Lander RH (1990) Geochemical and hydrologic impact of zeolitic alteration on tuffs in the White River Formation. Geol Soc Am Abstracts with Progr 22:A314

Lander RH, Hay RL (1993) Hydrologic control on zeolitic diagenesis of the White River sequence. Geol Soc Am Bull 105:361-376

Lenzi G, Passaglia E (1974) Fenomeni di zeolitizzazione nelle formazioni vulcaniche della regione sabatina. Boll Soc Geol Ital 93:623-645

Moncure GK, Surdam RC, McKague HL (1981) Zeolite diagenesis below Pahute Mesa, Nevada Test Site. Clays & Clay Minerals 29:385-396

Mumpton FA (1973) First reported occurrence of zeolites in sedimentary rocks of Mexico. Am Mineral 58:287-290

Norin E (1955) The mineral composition of the Neapolitan yellow tuff. Geol Rundschau 43:526-534

Passaglia E, Vezzalini G, Carnevali R (1990) Diagenetic chabazites and phillipsites in Italy: Crystal chemistry and genesis. Eur J Mineral. 2:827-839

Pérez-Torrado FJ, Martí J, Queralt I, Mangas J (1995) Alteration processes of the Roque Nublo ignimbrites (Gran Canaria, Canary Islands). J Volc Geotherm Res 65:191-204

Quanchang Z (1994) Utilization of natural zeolites in China. *In* Natural Zeolite and its Utilization, Japan Soc Promotion of Science, Committee on Development of New Utilization of Minerals 111:113-127

Robinson PT, Walker GW, McKee EH (1990) Eocene (?), Oligocene, and lower Miocene rocks of the Blue Mountains region. *In* Walker GW (ed) Geology of the Blue Mountains region of Oregon, Idaho, and Washington: Cenozoic geology of the Blue Mountains region. U S Geol Survey Prof Paper 1437:29-61

Rodriguez Fuentes G, Lariot Sanchez C, Romero JC, Roque Malherbe R (1985) Cuban natural zeolites: Morphological studies by electron microscopy. *In* Zeolites: Synthesis, Structure, Technology and Application. B Drzaj, S Hocevar, S Pejovnik (eds) Studies in Surface Science and Catalysis 24:375-384, Elsevier, New York

Samajová E (1988) Zeolites in tuffaceous rocks of the West Carpathians (Slovakia). *In* Occurrence, Properties and Utilization of Natural Zeolites. D Kalló, HS Sherry (eds) Akademiai Kiado, Budapest, p 49-57

Samajová E, Kuzvart M (1994) Zeolite deposits at the periphery of Tisza Basin (Carpathians, central Europe). *In* Natural Zeolite and its Utilization, Japan Soc Promotion of Science Committee on Development of New Utilization of Minerals 111:49-61

Sersale R (1978) Occurrences and uses of zeolites in Italy. *In* Natural Zeolites: Occurrence, Properties, Use. LB Sand, FA Mumpton (eds) Pergamon Press, Elmsford, New York, p 285-302

Sheppard RA (1985) Death Valley-Ash Meadows zeolite deposit, California and Nevada. *In* Clays and Zeolites—Los Angeles, California to Las Vegas, Nevada. 1985 Int'l Clay Conference Field Trip Guidebook, p 51-55

Sheppard RA, Gude AJ III (1993) Geology of the Sheaville clinoptilolite deposit, Sheaville, Oregon. *In* Zeo-Trip '93: An Excursion to Selected Zeolite and Clay Deposits in Southeastern Oregon and Southwestern Idaho. FA Mumpton (ed) Int'l Comm on Natural Zeolites, Brockport, New York, p 74-80

Sheppard RA, Gude AJ III, Fitzpatrick JJ (1988) Distribution, characterization, and genesis of mordenite in Miocene silicic tuffs at Yucca Mountain, Nye County, Nevada. U S Geol Survey Bull 1777, 22 p

Skhirtladze NI, Akhvlediani RA, Shubladze RL (1986) Silica-rich zeolite deposits of the Georgian SSR and their formation conditions. Izv Akad Nauk SSR, Ser Geol 8:130-133

Smyth JR, Caporuscio FA (1981) Review of the thermal stability and cation-exchange properties of the zeolite minerals clinoptilolite, mordenite, and analcime: Applications to radioactive waste isolation in silicic tuffs. Los Alamos Nat'l Lab Report LA-8841-MS, 30 p

Stamatakis MG, Hall A, Hein JR (1996) The zeolite deposits of Greece. Mineral Deposita 31:473-481

Taylor MW, Surdam RC (1981) Zeolite reactions in the tuffaceous sediments at Teels Marsh, Nevada. Clays & Clay Minerals 29:341-352

Tsirambides A, Filippidis A, Kassoli-Fournaraki A (1993) Zeolitic alteration of Eocene volcaniclastic sediments at Metaxades, Thrace, Greece. Appl Clay Sci 7:509-526

Wirsching U (1976) Experiments on hydrothermal alteration processes of rhyolitic glass in closed and "open" system. Neues Jahrb Mineral Mh, 203-213

WoldeGabriel G (1995) K/Ar dating of clinoptilolite, mordenite, and coexisting illite/smectite from Yucca Mountain, Nevada. *In* Natural Zeolites '93: Occurrence, Properties, Use. DW Ming, FA Mumpton (eds) Int'l Comm on Natural Zeolites, Brockport, New York, p 141-156

9 Zeolites in Burial Diagenesis and Low-grade Metamorphic Rocks

Minoru Utada

912-14 Futo, Itoh
Shizuoka 413-0231, Japan

INTRODUCTION

Many zeolite assemblages originated during burial and subsequent low-grade metamorphism of rocks, particularly those of volcaniclastic origin. Eskola (1939) first suggested this type of zeolitization and applied the term "zeolite facies" to assemblages formed under such low-temperature and low-pressure conditions. Although several definitions of the "zeolite facies" have been proposed over the years (e.g. Fyfe et al. 1958), the phrase as used here includes both low-grade metamorphic assemblages and zeolite assemblages formed successively during burial.

The regional occurrence of zeolites in sedimentary rocks was first recognized by Coombs (1954), who made detailed petrographic studies of Triassic tuffaceous greywackes and tuffs of andesitic to rhyolitic in composition in the Taringatura area of New Zealand. Coombs recognized a zonal arrangement of heulandite and analcime near the top of the sequence and heulandite and laumontite near the bottom. He also recognized a similar progressive zeolitization with depth in clastic sediments in other orogenic belts. The zeolite assemblages described by Coombs represent a rather advanced stage of alteration, as opposed to an earlier stage of zeolitization that was subsequently described in Neogene silicic volcaniclastic rocks from Japan. In the latter rocks, zones (from top to bottom) of fresh glass, clinoptilolite + mordenite, analcime + heulandite, laumontite, and albite were recognized (Utada 1965, 1970). Progressive zeolitization has also been reported as a result of the thermal effects of intrusive masses (see Seki et al. 1969) and around volcanic calderas (Utada and Ito 1989). Recently, Boles (1991) described the low-temperature formation of laumontite and stilbite in fractures and faults related to post-Jurassic folding and uplift of the Southland Syncline, New Zealand.

The physico-chemical conditions of zeolite crystallization in burial diagenesis and low-grade metamorphic environments can vary widely. Both borehole data (see Iijima 1995) and experimental results (Liou 1970; 1971a,b,c) have been useful in estimating the temperature and solution chemistry involved in these reactions, and formation conditions have recently been elucidated by means of fluid inclusion data, isotope studies, and radiometric dating. This chapter summarizes the present knowledge of zeolitization during burial diagenesis and low-grade metamorphism and focuses on the significance of this mode of occurrence in a variety of geologic settings.

TYPES OF ZEOLITIZATION

Burial zeolitization

Zeolitization during burial diagenesis and low-grade metamorphism typically progresses in sediments as the depth of burial increases. This progressive zeolitization can be divided according to appearance of the alteration zones and their relation to depth of burial, as schematically shown in Figure 1 and distributed as shown in Figure 2.

1529-6466/00/0045-0009$05.00

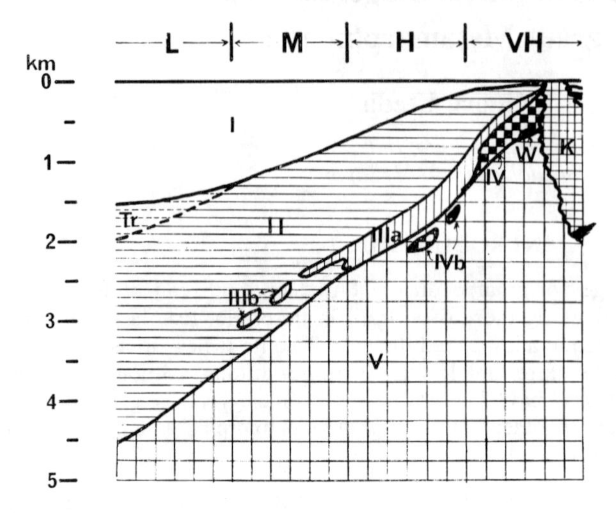

Figure 1. Relations between the types of zeolitic zoning, the successions of zones, the depth of burial in Neogene strata of Japan, and geothermal gradient (abbreviations as in the text). K: kaolin zone; Tr: transitional zone (after Utada 1971).

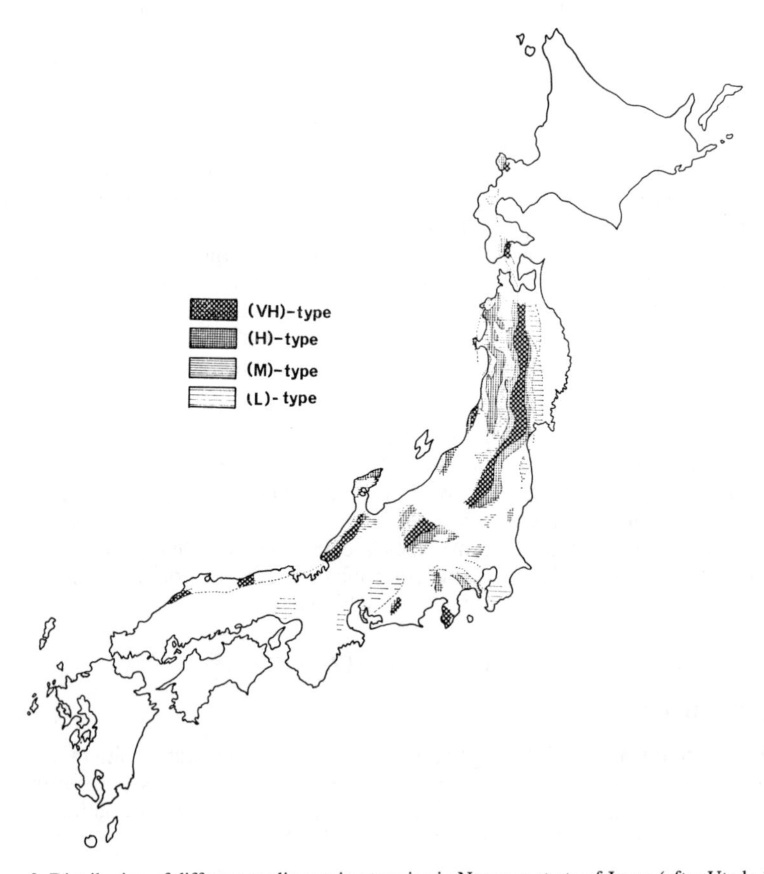

Figure 2. Distribution of different zeolite zoning terrains in Neogene strata of Japan (after Utada 1971).

The types of zeolitization may also be differentiated by their association with different geothermal gradients as shown in the volcanic-arc setting in Japan from the Miocene to the present (Utada 1970). The geothermal gradient of each type between the Neogene and Present can be classified as follows (°C/km): L-type (low), <30; M-type (moderate), 30-45; H-type (high), 45-60; and VH-type (very high), >60. Utada and Ito (1989) later showed that the VH-type includes the caldera-type of zeolitization. The four types are briefly summarized below:

- *L-type* Zones I, II and (V) appear successively from top to bottom, with a comparatively thick transitional zone between Zones I and II. Clinoptilolite is common, whereas mordenite is rare in the transitional zone.
- *M-type* Zones I, II, and V appear successively from top to bottom. Zone IIIb appears only locally. Zone II is very thick and clinoptilolite predominates over mordenite. Characteristically, the zone boundaries are nearly parallel to the stratigraphic boundaries. This type corresponds to the typical intermediate geothermal gradient type in this paper.
- *H-type* Zones I, II, III, and V appear successively from top to bottom and are nearly concordant to the stratification. The development of Zone IVb is local. This type corresponds to the typical high geothermal gradient zeolitization described above.
- *VH-type* Zones I, II, IIIa, IV, and V appear from top to bottom successively. They are usually oblique to the stratification.

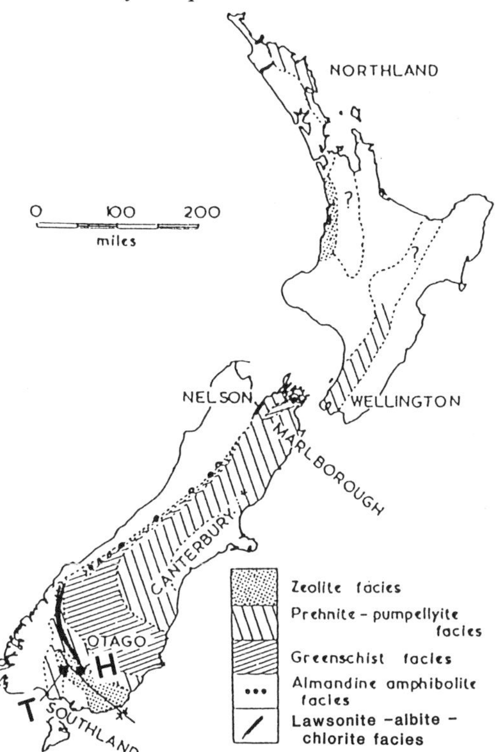

Figure 3. Generalized distribution of metamorphic facies in the metamorphic belt of New Zealand (modified from Coombs 1960). T: Taringatura Hills; H: Hokonui Hills.

Low geothermal-gradient zeolitization. As described above, Coombs (1954) documented advanced zeolitization in Triassic strata more than 9000 m thick in the Taringatura area, New Zealand (Fig. 3). In the upper parts of the sequence, where epiclastic volcanic sandstones and interbedded tuffs were more siliceous, volcanic glass in the ash beds has been replaced by heulandite or clinoptilolite and by analcime. Laumontite cement is present in some sandstones in the sequence, but in the lower half of the Taringatura sequence, where epiclastic detritus is largely of andesitic or basaltic andesite composition, laumontite is the dominant zeolite in vitric tuffs. It replaces earlier heulandite and also appears to be a product of albitization of plagioclase. Heulandite is present in some ash beds, especially those that are finer grained, but analcime has transformed to albite, its former presence being confirmed by pseudomorphs. Minor pumpellyite and prehnite were formed at the expense of laumontite, celadonite, and chlorite in the lower part of the section.

The distribution of authigenic minerals in this section is shown Figure 4. Coombs et al. (1959) and Coombs (1960) defined the zeolite facies to include at least all those assemblages produced under physical conditions in which the following mineral species are common: quartz + analcime, quartz + heulandite, and quartz + laumontite. They recognized a lower-grade heulandite and analcime + quartz "stage" and a higher-grade "stage" of laumontite, albite + quartz, and laumontite that formed sequentially after heulandite, but not in a simple zonal arrangement. They also proposed a higher-grade prehnite-pumpellyite metagreywacke facies characterized by the presence of quartz + prehnite or quartz + albite + pumpellyite + chlorite, in the absence of zeolites, jadeite, or lawsonite, that intervenes between the zeolite facies of the Murihiku sequence (Southland Syncline) and the greenschist facies of the Otago and Alpine Schist belt of New Zealand (Fig. 3).

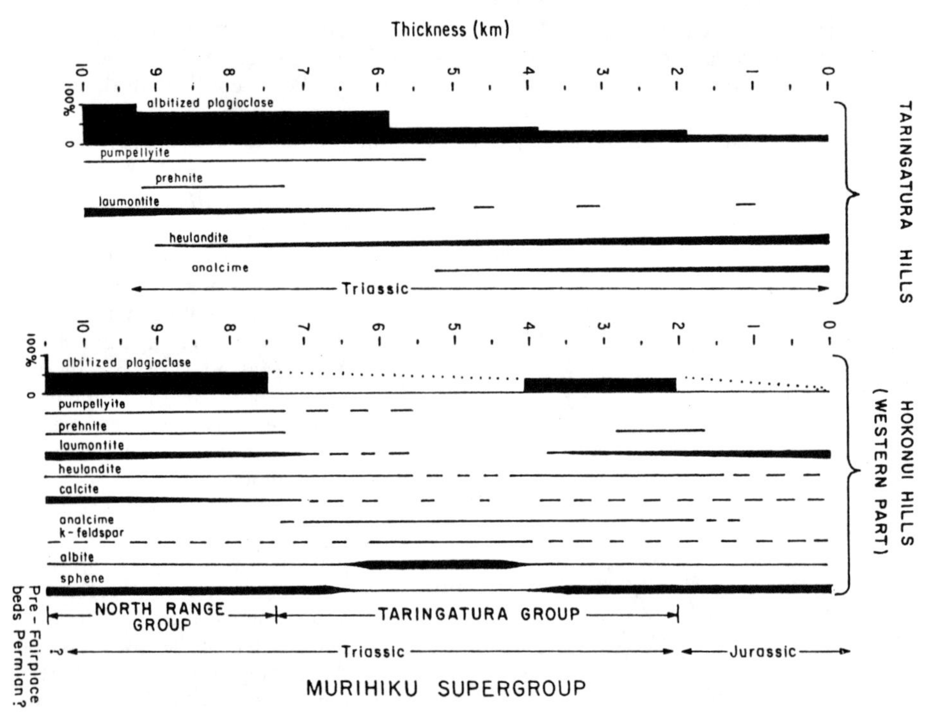

Figure 4. Mineral distributions in Taringatura Hills (Coombs 1954) and Hokonui Hills (Boles and Coombs 1977), New Zealand. Hokonui Hills "albitized plagioclase" dark area represents percentage of sandstones (0.2-0.5 mm average grain size) in which essentially all plagioclase grains are albitized. Hokonui Hills "albite" refers to authigenic albite in cavities and as cement.

Alteration shows a tendency to be more intense in coarser-grained beds. The distribution of new-formed minerals is likely the result of the effects of a thermal gradient estimated to be <25°C/km. Water-rock interaction will tend to be greater and fluid pressure lower in coarser grained and more permeable beds (Coombs 1993), thus promoting dehydration reactions at temperatures lower than those in finer grained, less permeable beds, in which $P(H_2O)$ approaches $P(load)$.

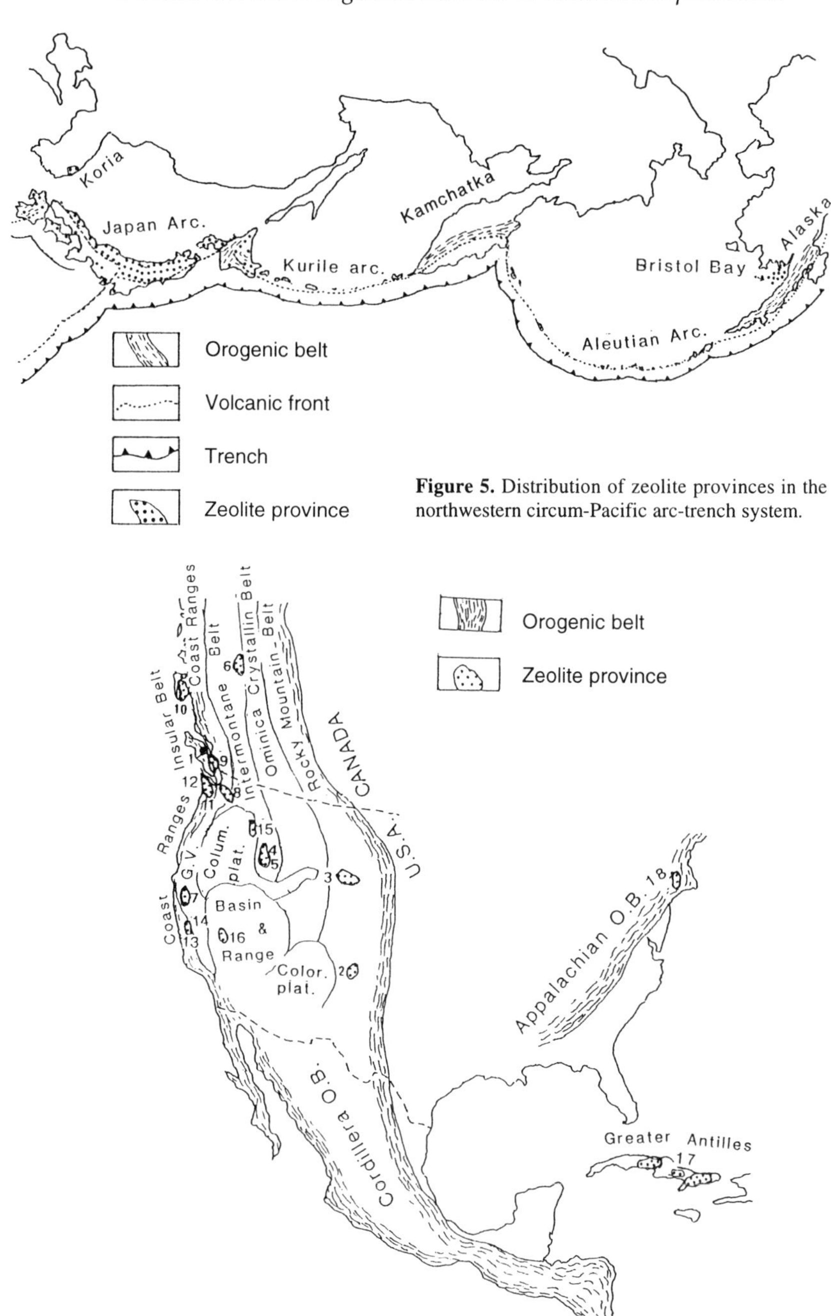

Figure 5. Distribution of zeolite provinces in the northwestern circum-Pacific arc-trench system.

Figure 6. Distribution of zeolite provinces in the northern Cordillera orogenic belt ("O.B." = orogenic belt; "Color. plat." = Colorado plateau; "Colum. Plat." = Columbia River plateau). Numbers on the figure are referred to in the text.

In another example of low geothermal-gradient zeolitization, Galloway (1974) studied diagenesis in three deep arc-related sedimentary basins of Alaska and Canada, specifically the Bristol, Queen Charlotte, and Grays Harbour-Chehalis Basins (Figs. 5 and 6). The sandstones are rich in feldspar and volcanic detritus, and present-day geothermal gradients range from 19-40°C/km. The sequence of diagenetic change in these basins is as follows:

- Stage 1: Formation of calcite pore-filling cement.
- Stage 2: Formation of clay coats and rims around detrital grains.
- Stage 3: Filling of remaining open pore spaces either by authigenic zeolites (usually laumontite) or chlorite or smectite.

Intermediate to high geothermal-gradient zeolitization. Progressive zeolitization has typically progressed in silicic volcaniclastic sediments in volcanic arcs in which the geothermal gradient is moderate to high. A typical example is found in the 2000 to 4000-m thick Neogene silicic volcaniclastic sediments in the "Green Tuff Region" of Japan (Fig. 7).

Utada (1965) recognized five diagenetic zones in the Neogene strata of the Shinjo Basin, northeast Honshu, (Fig. 7, Loc. 2) that yielded the assemblages of authigenic

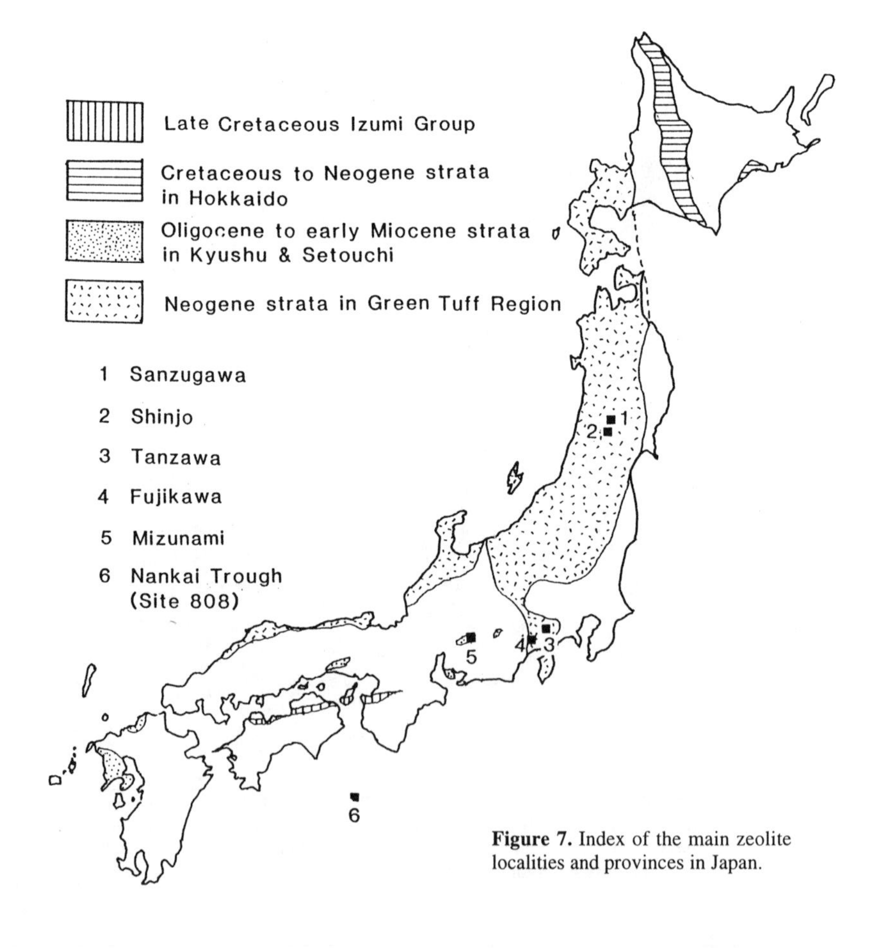

Figure 7. Index of the main zeolite localities and provinces in Japan.

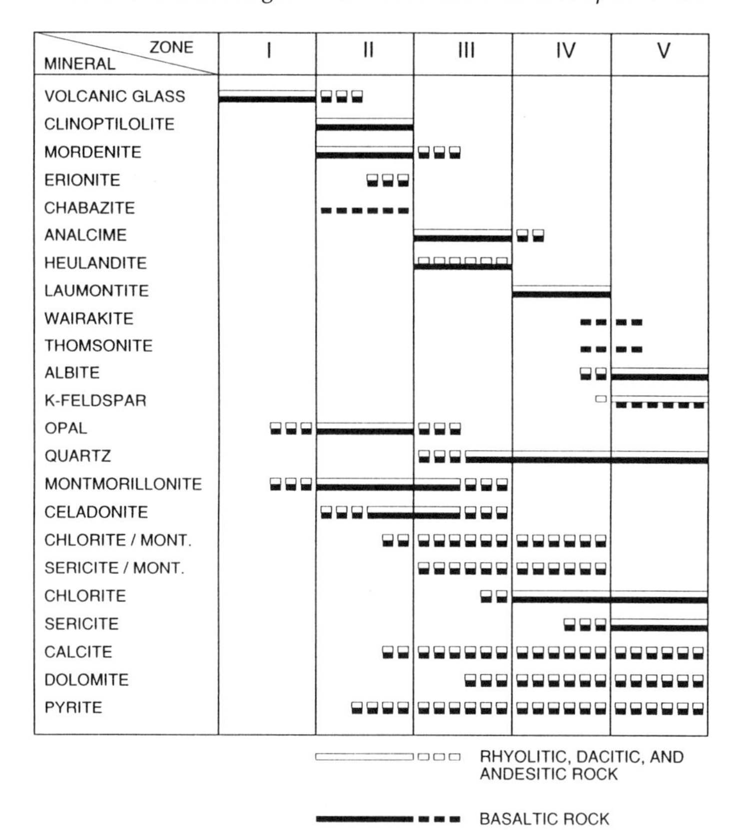

Figure 8. Assemblage of authigenic minerals in five diagenetic zones in the Neogene strata of Japan (modified from Utada 1965).

minerals shown in Figure 8. Zone I (fresh glass zone) in this illustration is characterized by the absence of zeolites and albite and the presence of unaltered volcanic glass. Zones II (clinoptilolite-mordenite zone), III (analcime-heulandite zone), and IV (laumontite zone) are respectively characterized by their particular authigenic zeolites. Zone IV appears over a rather narrow vertical area. Zone V (albite zone) is characterized by authigenic albite instead of zeolites. Utada (1970) also found that the diagenetic zones in andesitic volcaniclastic rocks are almost the same as those in silicic volcaniclastic sediments mentioned above. Smectite and chlorite predominate over zeolites, and stilbite is common in Zone II.

The vertical zoning of this type of zeolitization was recognized regionally in the Cretaceous to Neogene strata in Japan in both outcrops and in boreholes, which were emplaced, for oil and gas prospecting. Borehole studies by Iijima and collaborators have clarified the temperature-depth relation of each of these zones (see chapter by Iijima, this volume).

Based on regional studies, Utada (1970) added several zones to the intermediate to high geothermal gradient zeolitization sequence. A transitional zone was identified between Zone I and Zone II in some terrains. Based on the mode of occurrence of analcime, Zone III was subdivided into two subzones, a regional analcime-heulandite subzone (Zone IIIa) and a nodule-like or layer-like analcime subzone (Zone IIIb) whose mode of analcime occurrence is described below in the section on mode of occurrence and mineral reaction. Zone IV is also subdivided into two subzones, a regional laumontite subzone (Zone IVa) and a nodule-like or layer-like laumontite subzone (Zone IVb) whose mode of laumontite occurrence is described below.

Very high geothermal-gradient type zeolitization (caldera-type zeolitization). A volcanic caldera provides a special geological setting within volcanic arcs, having a very high geothermal gradient. Volcanic calderas are usually located along a volcanic front. The caldera-filling sediments consist mostly of volcanic glasses derived from pumice flows, pumice falls, ash falls, welded tuffs and their reworked sediments. These sediments also contain a large proportion of interstitial solutions. The combination of reactive original materials with a very high geothermal gradient provides conditions favorable for zeolitization.

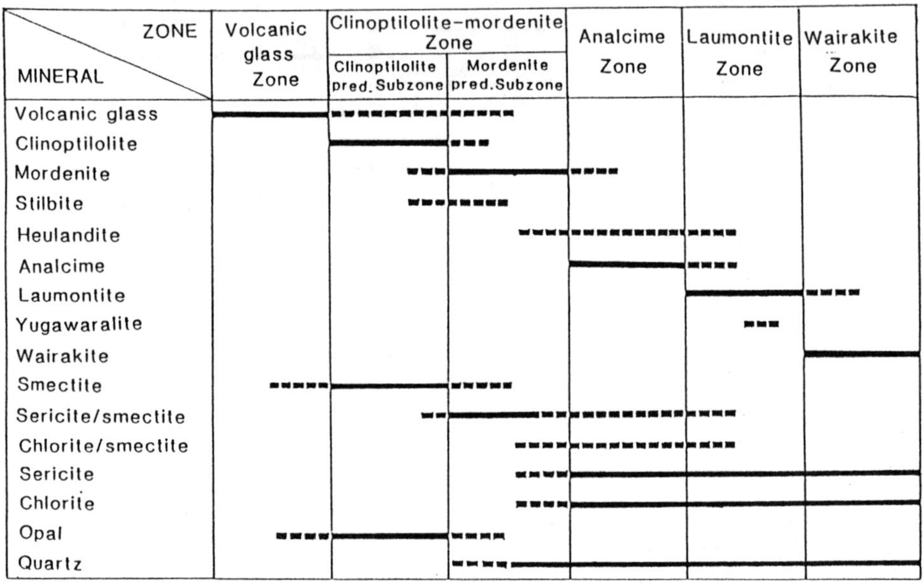

ZONE
MINERAL
Volcanic glass Zone
Clinoptilolite–mordenite Zone
Clinoptilolite pred. Subzone
Mordenite pred. Subzone
Analcime Zone
Laumontite Zone
Wairakite Zone
Volcanic glass
Clinoptilolite
Mordenite
Stilbite
Heulandite
Analcime
Laumontite
Yugawaralite
Wairakite
Smectite
Sericite/smectite
Chlorite/smectite
Sericite
Chlorite
Opal
Quartz

Figure 9. Mineral assemblages characteristic of caldera-type alteration in the Sanzugawa volcanotectonic depression of northeast Japan. Solid lines denote common occurrence, dashed lines denote rare occurrence (modified from Utada et al. 1999).

Several active geothermal areas in Japan are situated within volcanic calderas of late Miocene age to present, and caldera-type zeolitization has often been mistaken for localized hydrothermal alteration concentrated along vein systems. Recently, a typical example of caldera-type zeolitization was recognized in the Sanzugawa volcano-tectonic depression of late Miocene and Pliocene age in northeast Japan (Fig. 7, Loc. 1). Here, more than 80 holes were drilled during prospecting for geothermal energy by the Japanese government and a private company. Utada and Ito (1989) and Utada et al. (1999) divided the altered rocks into five mineralogical zones (Fig. 9), one of which was further subdivided into two subzones as follows:

- Volcanic glass zone
- Clinoptilolite-mordenite zone

    Clinoptilolite-predominant subzone

Mordenite-predominant subzone
- Analcime zone
- Laumontite zone
- Wairakite zone

Alteration in the Sanzugawa caldera consists of both calcic zeolite zones and sodic zeolite zones. The wairakite zone and the mordenite-predominant subzone are widespread. The distribution of each zone is similar to that of contact metamorphism (vide infra), although intrusive masses are not always found at the center. The average geothermal gradient in the Sanzugawa caldera at present is 175°C/km. Most volcanic calderas of late Miocene and younger age in Japan have had geothermal gradients >100°C/km (Okubo 1993).

Zeolitization due to the thermal effects of an entrusive mass

In the course of both progressive diagenesis and uplifting, many volcanic arcs have been intruded by plutonic masses, the thermal effects of which have promoted zeolitization within the surrounding sediments. Seki et al. (1969) described a typical example in Neogene basaltic lavas and sediments, both volcaniclastic and non-volcaniclastic, surrounding a quartz diorite complex in the Tanzawa Mountains, central Japan (Fig. 7, Loc. 3). They recognized five widespread alteration zones that are zonally distributed (Fig. 10) from the intrusive contact to the margin as follows:

- Zone V Amphibolite zone.
- Zone IV Actinolite green schist zone.
- Zone III Pumpellyite-prehnite-chlorite zone.
- Zone II Laumontite-mixed-layer clay mineral-chlorite zone.
- Zone I Stilbite (clinoptilolite)-vermiculite zone.

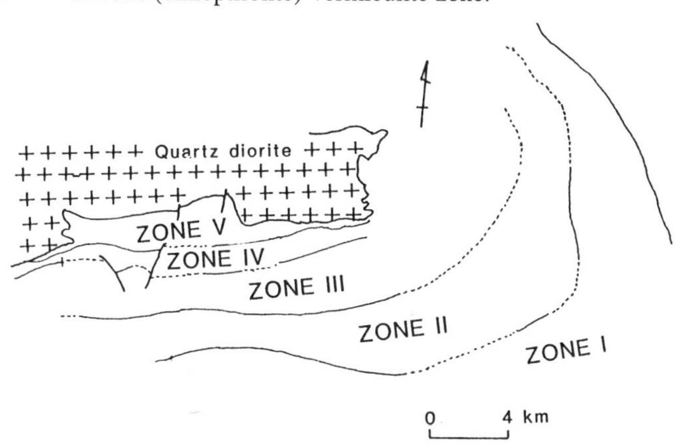

Figure 10. Distribution of five zones of contact metamorphism surrounding a quartz diorite complex in the southeastern Tanzawa Mountains, central Japan (simplified from Seki et al. 1969).

The mineral assemblage characteristic of each zone is shown in Figure 11. Vein minerals of hydrothermal origin are included in the zonation pattern that is characterized by calcium silicates. It is not clear whether the calcium silicates result from addition of calcium to interstitial solutions by albitization of calcic plagioclase or from other sources. The metamorphic rocks of Zones III, IV, and V have a distinct schistose structure subparallel to stratigraphy and were formed from originally mafic rocks of Miocene age.

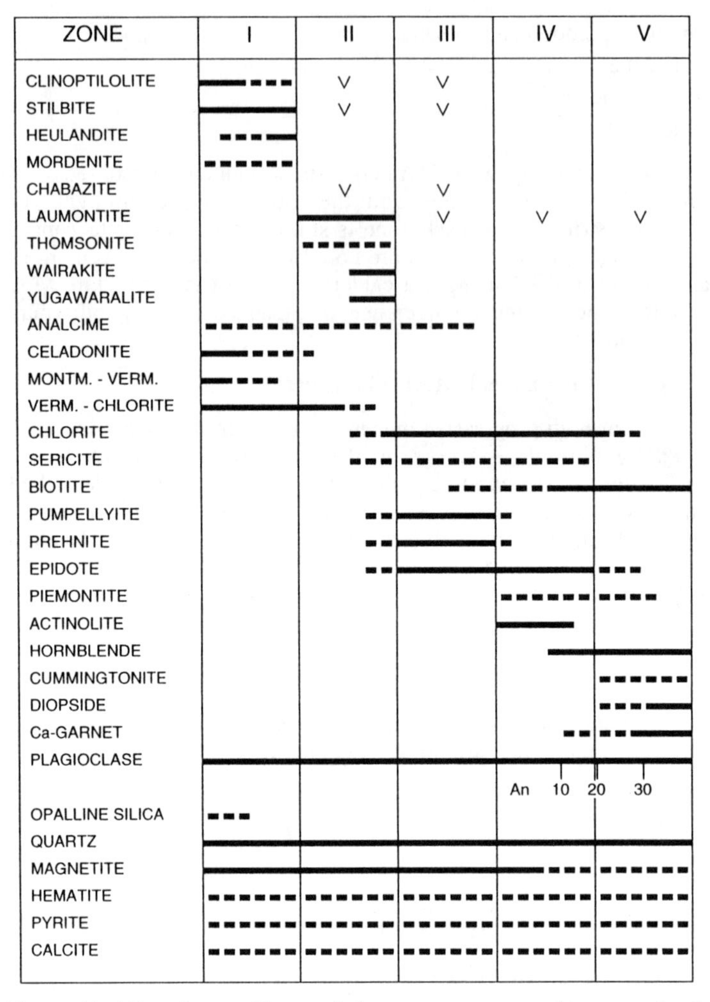

Figure 11. Mineral assemblages of the contact metamorphic zones in the southeastern Tanzawa Mountains, central Japan. Broken lines show that minerals occur rarely. "V" marks represent minerals found only as veins (Seki et al. 1969).

Laumontite in Zone II, however, is widespread in both Miocene volcaniclastic sediments and in Pleistocene conglomerates and sandstones.

Like the Tanzawa Mountains, the Neogene silicic volcaniclastic sediments in the Green Tuff Region were intruded by various rocks ranging from silicic to mafic in chemical composition and from small dikes to batholiths in scale. Here, metamorphic rocks with schistose structure are very rare. Although many varieties of mineral zoning have been noted (Utada 1973), the common type of zoning around small intrusive masses from contact to unaltered county rock is typically:

- Propylite zone (quartz-albite-chlorite-(sericite)-epidote)
- Laumontite zone
- Heulandite zone
- (Stilbite zone)

In the outer regions, these calcic zeolite zones interfinger with alkali zeolite zones of burial diagenetic origin and are commonly associated with hydrothermal alteration within and around the intrusive masses.

Zeolitization by contact metamorphism was also recognized in the Eocene arkosic sandstones and conglomerates that were intruded by "radial dikes" and stocks in the Spanish Peaks region (Fig. 6, Loc. 2) of southern Colorado (Utada and Vine 1984). Here, alteration zones are zonally distributed outwards from the intrusion as follows:

- Amphibole zone
- Prehnite-(pumpellyite) zone
- Authigenic plagioclase zone
- Laumontite zone
- Non-laumontite zeolite zone
- Weakly altered zone

The laumontite zone in the Spanish Peaks region is the most widespread, whereas the non-laumontite zeolite zone containing analcime, heulandite-clinoptilolite, stilbite, and mordenite occupies a very narrow area. Albitization of calcic plagioclase probably supplied calcium for the formation of laumontite under moderate to high temperatures.

Zeolitization during uplift

Boles (1991) recognized the occurrence of vein-filling zeolitization during uplifting in the Triassic and Jurassic strata of the Southland Syncline (Fig. 3, Loc. H) east of the Taringatura area, New Zealand. Superimposed upon the low geothermal-gradient zeolitization, zeolites crystallized in fractures and faults that were related to post-Jurassic folding and uplift. Laumontite and stilbite are common as fracture-filling minerals, and analcime, albite, chlorite, prehnite, pumpellyite, montmorillonite, halloysite, and carbonates occur locally. Replacement of heulandite by laumontite proceeded inwards from fracture surfaces. Calcic minerals were mainly found in andesitic rocks, whereas albite and/or analcime are predominant in rhyodacitic rocks. Cross-cutting relationships of veins show that stilbite has been formed between the formation of pre-date and post-date laumontite. Oxygen isotope data indicate that the crystallization temperatures of stilbites and laumontites were lower than the estimated temperature at the maximum burial.

Recently, another example of the zeolitization during an uplift stage was found in the Neogene Fujikawa Group of central Japan (Fig. 7, Loc. 4) by the writer. The Fujikawa Group, about 10,000 m thick, was deposited within a narrow zone and was uplifted, folded and faulted during only ten million years. Zeolites mainly occur as films on joint surfaces in clastic sediments. Locally, they also fill the breccia zones of faults. These zeolites probably formed near the surface during postburial diagenesis. Laumontite and stilbite are very common as joint-filling minerals and are zonally distributed as follows: laumontite ⇔ laumontite + stilbite ⇔ stilbite. Analcime, heulandite, calcite, and smectite are minor joint-filling minerals.

Miscellaneous zeolitization during burial diagenesis

Authigenic zeolites have been reported from bentonite deposits by many researchers (Kerr 1931, Crawford and Cowles 1932, Bramlette and Posnjak 1933, Kerr and Cameron 1936, Ames et al. 1958, Hayashi and Sudo 1957, Lougnan 1966, Pevear et al. 1980, Altaner and Grim 1990, Masuda et al. 1996). Judging from their modes of occurrence, most of the zeolites were probably formed by progressive zeolitization before or during bentonite formation. An exception to this is the author's discovery of zeolitization occurring after the formation of bentonite. In the Miocene Mizunami Group in central Japan (Fig. 7, Loc. 5), zeolites such as clinoptilolite-heulandite, mordenite, analcime, chabazite, and phillipsite, are independently distributed in strongly bentonitized sediments, without respect to any regional zeolite zones. Scanning electron microscope examination shows that the zeolites occur in micro-cavities of less than several tens of micrometers in diameter within the bentonite. They also fill minute cracks and veinlets in bentonites. Similar occurrences of zeolites can be found locally in Miocene marine argillaceous and limey sediments of Japan. Although the evidence is meager, the author considers that zeolites may have precipitated from alkaline waters, formed by closed-system reaction of the original interstitial waters with the surrounding bentonites, in local environments caused by "swelling-shrinkage" of bentonite.

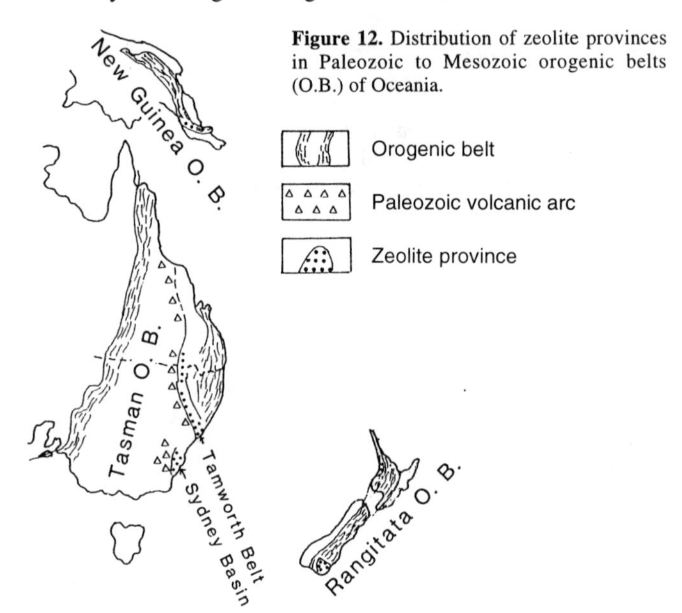

Figure 12. Distribution of zeolite provinces in Paleozoic to Mesozoic orogenic belts (O.B.) of Oceania.

DISTRIBUTION OF ZEOLITES IN DIFFERENT GEOLOGIC PROVINCES

Precambrian and Paleozoic orogenic belts

Although high-grade metamorphic rocks have formed in numerous orogenic belts of Precambrian and Paleozoic age, burial diagenetic and low-grade metamorphic formation of zeolites have seldom been reported in such rocks. Zeolites occur in late Paleozoic sediments in the Tasman orogenic belt of New South Wales, Australia (Fig. 12). Coombs (1958) first reported zeolitized tuffs from the Kuttung glacial beds. Subsequently, Crook (1961a) recognized three alteration zones—laumontite, prehnite-pumpellyite, and albite-epidote—in the Devonian and lower Carboniferous sediments of the Tamworth-Nundle district. Heulandite, analcime, and laumontite were also recognized in the Currabubula Formation of late Carboniferous age (Wilkinson and Whetten 1964) and in sediments of Upper Carboniferous and Permian age (Kisch 1968). Recently, Flood et al. (1995) added new zeolite localities to New South Wales and Queensland. There, clinoptilolite-Ca and mordenite formed in ignimbrites of late Carboniferous to Devonian age probably related to volcanic calderas. All of above-mentioned zeolite localities are thought to have been located in a Paleozoic volcanic arc that included volcanic calderas (Rutland 1976) as shown in Figure 12.

Late Paleozoic to Mesozoic orogenic belts

Rangitata orogenic belt *(Fig. 12).* As described above, the *zeolite facies* was established on the basis of Coombs' (1954) study of the Taringatura area of Southland, New Zealand. Low geothermal-gradient zeolitization is pervasive in Triassic and Jurassic strata in the Southland Syncline (Fig. 3). This area is believed to have been part of an arc-trench-ocean basin complex, fore-arc according to Carter et al. (1978), but back-arc according to Coombs et al. (1996).

New Guinea orogenic belt *(Fig. 12).* In the New Guinea orogenic belt, zeolites are abundant in the upper parts and prehnite is predominant in the lower parts of the Wahgi Valley sediments at depths >28,000 feet (Crook 1961b).

Insular belt of Cordillera. The Insular belt of the western Cordillera, Canada (Fig. 6, Loc. 1), is considered to be a volcanic arc of late Paleozoic to Mesozoic age which was located on the margin of the North American continent (Monger et al. 1972). Surdam (1973) recognized laumontite, prehnite, pumpellyite, epidote, analcime, and albite in pillow lavas and breccias of the Triassic Karmutsen Group. The metamorphosed rocks can be grouped into a laumontite-analcime zone and a pumpellyite-prehnite-epidote zone. A contact aureole around the intrusion of the Coast Range Batholith was recognized in the Karmutsen Group by Cho et al. (1987). Two major calcite-free assemblages are characteristic of the outer part of the aureoles: (1) laumontite-pumpellyite + epidote in the outmost parts, and (2) prehnite-pumpellyite + epidote closer to the inner parts of the aureole.

Alpine orogenic belt of Mesozoic to Cenozoic age

"Pietra Verde" in the Dinarides *(Fig. 13).* Middle Triassic strata are widely distributed in the outer Dinarides including Montenegro, Bosnia, Herzogovina, Croatia, Slovenia, and western Serbia. These strata consist mainly of volcaniclastic sediments that are colored green by diagenetic alteration. Obradovic (1979) recognized the following zonal distribution of authigenic minerals in the Dinarides, comparable to that of the Green Tuff region of Japan:

- Zone 1. Fresh glass and bentonites
- Zone 2. Clinoptilolite
- Zone 3. Clinoptilolite, mordenite, analcime, and leonhardite
- Zone 4. Analcime and leonhardite
- Zone 5. Albite

Figure 13. Distribution of zeolite provinces in the Alpine orogenic belt.

Volcanic activity resumed in the Dinarides during the Neogene, and many clinoptilolite and mordenite deposits were formed in volcaniclastic sediments of eastern Serbia. Obradovic and Kemanci (1975) studied core samples of boreholes penetrating the Neogene volcaniclastic rocks and found clinoptilolite, mordenite, and analcime within bentonitized rocks.

Northern zone of the western Alps. Burial diagenesis and low-grade metamorphism has affected the Taveyanne greywackes of Eocene to Oligocene age in the Helvetic nappes on the northern and western flanks of the Alps in France and Switzerland. Heulandite formed locally where the overburden of pre-alpine nappes was apparently least (Sawatski 1975). Laumontite developed in large quantities with albitized plagioclase over large areas, commonly with prehnite and/or pumpellyite (Martini and Vuagnat 1965, Martini 1968, Rahn et al. 1994). Higher-grade areas pass into the prehnite-pumpellyite and pumpellyite-actinolite facies.

Carpathian and Balkan Ranges. Clinoptilolite, mordenite, and rarely analcime were diagenetically formed from silicic volcaniclastic sediments in the Carpathian and Balkan ranges containing Cretaceous to Miocene volcanic arcs belonging to the Alpine orogenic belts. In the northern part of the East Slovakia basin, a deep drill hole penetrated a sequence of volcaniclastic-clastic sediments of middle Miocene age in which vertical zoning consists of clinoptilolite and cristobalite in the upper parts and analcime, quartz, and K-feldspar in the lower parts (Samajova and Kuzvart 1992). Aleksiev and Djourova (1977) recognized a similar vertical zonation in the Oligocene silicic volcaniclasts (partly ignimbrites) in the northeastern Rhodopes, Bulgaria. The zonation, from top to bottom, is as follows: fresh glass → K-clinoptilolite → Ca-clinoptilolite → analcime. Clinoptilolite-containing rocks are widely distributed in the entire northeastern Rhodopes and northern Greece, probably formed as a result of the intermediate to high geothermal gradient.

Crimea and Great Caucasus. Zeolites have been reported from the late Jurassic to Miocene strata of the Crimea and the Great Caucasus orogenic belts by many researchers (Rengarten 1950, Dzotsenidze and Shirtdadze 1953, Ermolova 1955, Buruzova 1964, Awakjan 1973, Mikhaylou and Silant'ev 1973, Kashkai and Babeav 1976). Clinoptilolite seems to be most pervasive and analcime is common. Heulandite, stilbite, and laumontite are rare.

Himalayan orogenic belt of Mesozoic to Cenozoic age

Alburz Range *(Fig. 14).* The Alburz Mountains, Iran, are the western extension of the Himalayan orogenic belt. The sequence of Eocene strata about 10,000 m thick is composed of submarine volcanic materials of silicic to intermediate composition. Iwao and Hushmand-Zadeh (1971) recognized three mineral zones of burial diagenesis and low-grade metamorphism, from top to bottom as follows: analcime → albite → prehnite-epidote-albite zones.

Great Sunda Archipelago. The Great Sunda Archipelago, Indonesia, is located at the southeastern extension of the Himalayan Orogenic belt and is in an arc-trench system at present. The Neogene formations contain a large amount of volcaniclastic sediments. Clinoptilolite and mordenite were formed from volcaniclastics probably by burial diagenesis (Ramlah 1992). Detailed studies, however, are still lacking.

Circum Western-Pacific orogenic belt of Cretaceous to Holocene age

Japanese Archipelago *(Fig. 5).* The Japanese Archipelago is located in a typical active plate margin and is one of the most productive zeolite provinces. Volcanic activity has become intense since the Cretaceous and has produced a large amount of volcaniclastic sediments. These have been deposited mainly in a back-arc basin and have been subjected to various types of diagenesis and hydrothermal alteration. The primary zeolite provinces in the Japanese archipelago (Fig. 5) are as follows:

- 1. The Izumi Group in Shikoku and Awaji Islands, and the Kii Peninsula of late Cretaceous age (Nakajima and Tanaka 1967, Nishimura et al. 1980).
- 2. The Cretaceous to Neogene strata in central Hokkaido (Iijima and Ohwa 1980).
- 3. The Oligocene to early Miocene strata in Northern Kyushu and the Paleo-Setouchi district (Utada 1970, Miki and Nakamuta 1985).
- 4. The Neogene formations in the "Green Tuff Region" (Utada 1970).

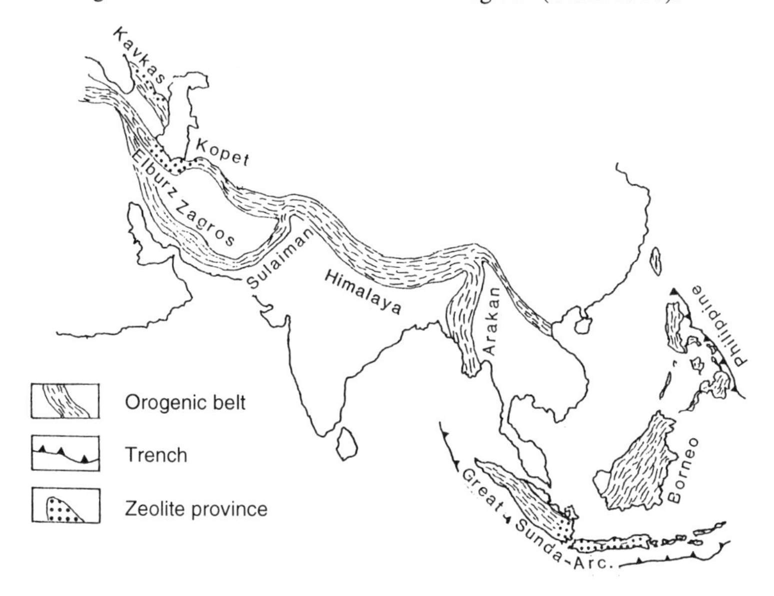

Figure 14. Distribution of zeolite provinces in the Himalayan orogenic belt.

As described above, zeolitization in the Green Tuff region of Neogene age includes all types of progressive zeolitization, contact metamorphism, and joint filling during uplifting. Among them, intermediate to high geothermal-gradient zeolitization is the most pervasive over the Green Tuff Region.

Zeolites have also been found in the Mesozoic and Cenozoic sediments under the seafloor surrounding the Japanese archipelago during drilling for petroleum and gas prospecting (see chapter by Iijima, this volume). Recently, Masuda et al. (1996) recognized intermediate to high geothermal-gradient zeolitization in the Nankai trough (Fig. 7, Loc. 6). They discussed possible mechanisms of zeolite formation within bentonitic cores.

Eastern Korean Peninsula. Before the opening of the Sea of Japan in middle Miocene time, the Green Tuff Region probably extended to the Korean Peninsula. The early to middle Miocene strata in the eastern part of the Korean Peninsula contain large volumes of volcaniclastic sediments corresponding to those of the Green Tuff Region. They were diagenetically altered to zeolites and bentonites (Kim and Moon 1978, Noh and Kim 1986). Hwang (1995) grouped the altered rocks into volcanic glass, clinoptilolite, and mordenite zones, corresponding to those of the Green Tuff Region.

Alaska Peninsula. The Bristol Basin is a back-arc trough which is located near the northeast margin of this system and which has been filled by detrital and volcaniclastic sediments of Eocene to Holocene age. Galloway (1974) recognized pore-filling laumontite formed in sandstones under a geothermal gradient of 19-40°C /km.

Cordillera orogen of Mesozoic to Quaternary age

Rocky Mountain belt *(Fig. 6).* Zeolites appear to be rare in pre-Mesozoic deposits of the Cordillera orogen. Laumontite was found in arkosic sandstones of the Sangre de Cristo Formation of Pennsylvanian and Permian age in the Spanish Peaks area, Colorado (Fig. 6, Loc. 2). This zeolitization may be related to the thermal effects of intrusive masses of Tertiary age (Hutchinson and Vine 1987). Contact metamorphism in this belt is more extensive in the Tertiary sandstones as described above in the section on zeolitization due to the thermal effects of an intrusive mass. Analcime and albite of probable diagenetic origin occur in thin tuffs of Jurassic age, which are distributed over wide areas of Idaho and Wyoming (Gulbrandsen and Cressman 1960) (Fig. 6, Loc. 3).

Intermontane Zone. The Intermontane Zone is thought to have been a fore-arc trough of Mesozoic age (Burchfiel and Davis 1972). The sediments consist of greywackes, conglomerates, sandstones, and shales attaining thickness of 10,000 to 20,000 m. Typical low geothermal gradient zeolitization is well documented in the following two examples. Triassic to Jurassic strata are distributed in the eastern area of the Cascade Range, Central Oregon. Dickinson (1962) recognized a zeolitization sequence in Jurassic greywackes and associated rocks corresponding to (1) crystallization of volcanic glass to heulandite (?); (2) local conversion of heulandite to laumontite; and (3) albitization of plagioclase and formation of pumpellyite, prehnite, and laumontite. Brown and Thayer (1963) also recognized laumontite-bearing rocks and the prehnite-pumpellyite facies in a thick sequence of volcaniclastics of late Triassic and early Jurassic age in the Aldrich Mountains (Fig. 6, Loc. 5).

The Sustut Group of late Cretaceous to early Tertiary age in British Columbia, Canada, consists of fluvial volcaniclastic sediments which are extensively zeolitized. Analcime, heulandite, laumontite, and albite are common, and prehnite veins are rare. Albite and analcime-quartz assemblages are interbedded with heulandite and laumontite-quartz assemblages, without vertical and horizontal zonations (Read and Eisbacher 1974) (Fig. 6, Loc. 6).

Great Valley. The Great Valley Sequence between the Coast Ranges and the Sierra Nevada Range (Fig. 6, Loc. 7) includes about 15,000 m of clastic sedimentary strata ranging from late Jurassic to late Cretaceous age. Laumontite appears in sandstones of lower Cretaceous age that were probably buried from 6000- to 10,000-m depth (Dickinson et al. 1969). Albitization of plagioclase and chloritization of biotite are also widespread in clastic rocks. Laumontite and heulandite formed as cement-fills in thick non-marine sediments of early Cretaceous age in the northern extension of this zone (Fig. 6, Loc. 8) (Miller and Ghent 1973, Ghent and Miller 1974). Heulandite and laumontite are also present in brackish sandstones of the Late Cretaceous Nanaimo Group at Vancouver Island and Gulf Island, British Columbia (Fig. 6, Loc. 9) (Stewart and Page 1974).

Pacific Coast region. Galloway (1974) studied diagenesis of sandstones in the Queen Charlotte Basin along the Pacific Coast of Canada. This basin was a fore-arc trough and was filled by Miocene to Pleistocene ferruginous clastic sediments about 4600 m thick. Zeolite pore-fillings were recognized in sandstones at burial depths from 1500 to 3000 m, under a geothermal gradient of 14-19°C/km. Similar zeolitization was reported in the Miocene to Quaternary sediments of the Chahalis basin (Fig. 6, Loc. 11) (Galloway 1974). Heulandite and laumontite occur mainly as cements or pore-fills in clastic sediments and as replacements of plagioclase, whereas analcime and albite occur selectively in volcaniclastic sediments. The present geothermal gradient of the Chahalis basin is only 11°C/km, and zeolite pore-fillings appear in the sequence deeper than 4000 m. Laumontite of similar occurrence has been reported from the Coast Ranges in the United States. It is extensively distributed in sandstones of Eocene to Miocene age at the Olympic Mountains, Washington (Fig. 6, Loc. 12) (Stewart 1974), and in Miocene sandstones at the Santa Cruz Mountains, California (Fig. 6, Loc. 13) (Madson and Murata 1970). This zeolitization is probably of the low geothermal-gradient type. Clinoptilolite, mordenite and heulandite are also found in Miocene tuffaceous sandstones in the area near the Olympic Mountains (Fig. 6, Loc. 14) (Murata and Whiteley 1973).

Basin and Range province. The Basin and Range and vicinity were a "volcanic arc" related to subduction in the Tertiary (Lipman et al. 1971). The Tertiary strata covering the Cordilleran orogen contain large amounts of volcaniclastic sediments that have been subjected to various modes of zeolitization. The Basin and Range Province was covered with non-marine volcanic materials such as lava flows, volcaniclastics, and sometimes ignimbrites of probable late Oligocene to Miocene age (Armstrong 1968). These volcanics lie on Cretaceous strata corresponding to the Great Valley Sequence, and the Columbia River Basalt covers the northern part. The total thickness of the Tertiary sequence exceeds 3000 m in Washington and 1600 to 5000 m in Oregon.

The Ohanapecosh Formation of probable early Miocene age in Washington state (Fig. 6, Loc. 15) contains laumontite and wairakite and varying amounts of albite, quartz, prehnite, epidote, carbonates, and clay minerals. Common mineral assemblages include laumontite-albite-quartz, wairakite-albite-quartz, and prehnite-albite-quartz (Fiske 1963). This assemblage may represent caldera-type alteration. In the Basin and Range Province, the Tertiary volcaniclastic sediments including ignimbrites cover the deformed and metamorphosed basement rocks. A large amount of volcaniclastic sediments was zeolitized by burial diagenesis, in both open and closed hydrologic systems. As a whole, clinoptilolite and mordenite are pervasive, whereas analcime is minor.

Yucca Mountain, in the southwestern Nevada volcanic field (Fig. 6, Loc. 16), is a potential site for a high-level radioactive waste repository in the United States. The geological setting consists of a complex of overlapping calderas, and vertical zonation in the distribution of zeolites and authigenic minerals appears to be related primarily to open-system diagenesis (Broxton et al. 1987, Bish and Aronson 1993) as described in Chapter on Open Hydrologic Systems.

Antilles Archipelago. The Antilles Archipelago consists largely of volcaniclastic and marine-clastic sediments of Cretaceous to early Miocene age. Otalora (1964) recognized the existence of analcime, laumontite, prehnite, pumpellyite, and other minerals in Cretaceous volcaniclastic sediments in Puerto Rico (Fig. 6, Loc. 17). Although he compared this zeolitization with the low geothermal gradient of the Taringatura area, it may belong to the intermediate to high geothermal-gradient type, judging from the distribution pattern of zeolites. In Cuba, clinoptilolite and mordenite occur in Cretaceous to Eocene strata (Aleksiev 1975). The fresh glass zone and the clinoptilolite-mordenite zone appear to be widespread.

Andean orogenic belt *(Fig. 15).* The terrain between the Pacific Ocean and the crystalline schist zone of the Andean orogenic belt consists largely of volcanic, volcaniclastic, and shallow-marine clastic sediments several tens of kilometers thick of Mesozoic to Cenozoic age. The sediments (Fig. 15, Loc. 1-8) have been subjected to

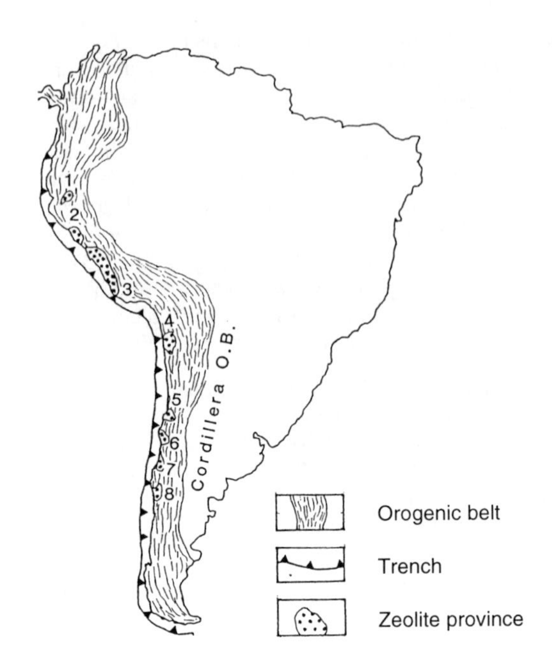

Figure 15. Distribution of zeolite provinces in the southern Cordillera orogenic belt.

burial diagenesis and low-grade metamorphism (Levy 1970, Aguirre and Atherton 1987, Levy et al. 1989). Levy et al. (1989) divided metamorphic rocks distributed from Ecuador to Chile into the zeolite facies, the prehnite-pumpellyite facies, the pumpellyite-actinolite facies, and the greenschist facies. The zeolite facies here includes stilbite, mordenite, laumontite, heulandite, chabazite, analcime, epistilbite, and wairakite. The alteration belongs mainly to the intermediate to high geothermal-gradient type of zeolitization, although it may result partly from hydrothermal activity.

Other areas

Although zeolites in burial diagenetic and low-grade metamorphic regimes typically formed in orogenic belts as described above, they also occur in other geologic settings. Rift valley systems are possible geological setting for progressive zeolitization, however, zeolites have been rarely reported from such geological settings, except for the formation of laumontite in lower Triassic arkosic sandstones of the Ruhuhu Basin, southern Tanzania (Wopfner et al. 1991). Also, laumontite occurs as a cement in Triassic arkose sandstones about 3000 m thick in Connecticut and Massachusetts (Fig. 6, Loc. 18) (Heald 1956). These are fluvial, lacustrine, or swamp deposits overlying crystalline schists.

Zeolites have also been reported from the sediments covering cratons and old orogens in the Russian platform by several researchers (Bushinski 1950, Kolbin and Pimburskaya 1955, Buryanova 1960). Thus, orogenic belts are not always the only zeolite province of burial diagenesis and low-grade metamorphism. Unknown zeolite provinces undoubtedly remain to be discovered, especially in the interior of continents. For instance, many zeolite deposits occur over a wide area of China (Zhang Quanchang 1995) and in Mongolia (Ivaanjav 1992), but their geological setting and genesis are uncertain.

MODE OF OCCURRENCE AND MINERAL REACTION

The mode of occurrence of authigenic minerals and the mineral reactions giving rise to these minerals differ for the various types of zeolite zoning, and an example of intermediate to high geothermal-gradient zeolitization in the Green Tuff Region is described below. Here, the original rocks were silicic volcaniclastics that were deposited in a shallow marine environment.

Fresh glass zone. Authigenic zeolites are absent in the earliest stage of burial diagenesis in the Green Tuff volcaniclastic sediments, and volcanic glass has subsequently dissolved slightly, without the formation of any authigenic minerals. Na, Ca, and Si, however, were removed from the glass, and H_2O was added. Smectite and opal-CT formed as authigenic minerals in this stage.

Clinoptilolite-mordenite zone. Clinoptilolite and/or mordenite formed after increasing depth of burial to several hundred meters. Clinoptilolite usually occurs as euhedral crystals replacing completely or partly dissolved volcanic glass and also as cements in interstitial pores. Mordenite occurs as fan-like aggregates replacing dissolved volcanic glass. These modes of occurrence may indicate a "micro-dissolution-precipitation" mechanism. Smectite, opal-CT, and locally celadonite are common co-existing minerals.

The distribution of authigenic minerals in the clinoptilolite-mordenite zone is not always homogeneous in the Green Tuff Region. Each authigenic mineral is commonly concentrated as spots, concretions, nodules, or layers. Therefore, micro-dissolution-precipitation may not always have taken place *in situ* but may have occurred through the following series of reactions: dissolution of volcanic glass → differential movement of chemical components → formation of authigenic minerals. The average temperature at the boundary between the fresh glass zone and the clinoptilolite-mordenite zone was estimated at 44°C from borehole temperatures of oil fields in Japan (Iijima 1995).

Analcime-heulandite zone. With increasing depth of burial, analcime and/or heulandite appear in place of clinoptilolite and mordenite. Newly formed minerals commonly occur as pseudomorphs of their precursors. In this zone, most of the associated smectite has transformed to mica through mixed-layer illite/smectite, and opal-CT has transformed to quartz.

The authigenic minerals in this zone are more heterogeneously distributed than in the clinoptilolite-mordenite zone and commonly form monomineralic nodules, which indicates differential movement of chemical components on a fairly large scale (Utada 1971). Transformation of clinoptilolite or mordenite of Zone II into analcime of Zone III, especially Zone IIIb, may necessitate some addition of sodium as shown in following equations, although the reactions may also be written without the addition of sodium:

$$\underset{\text{clinoptilolite}}{(Na,K)_2(Al_2Si_8O_{20})\cdot 7H_2O} + 2Na^+ \Rightarrow \underset{\text{analcime}}{Na_2(Al_2Si_4O_{12})\cdot 2H_2O} + \underset{\text{quartz}}{4SiO_2} + 5H_2O + 2(Na^+,K^+)$$

$$\underset{\text{mordenite}}{(Na_2,Ca)(Al_2Si_{10}O_{24})\cdot 8H_2O} + 2Na^+ \Rightarrow \underset{\text{analcime}}{Na_2(Al_2Si_4O_{12})\cdot 2H_2O} + \underset{\text{quartz}}{6SiO_2} + 6H_2O + 2(Na^+,Ca^{2+})$$

Additional sodium could have been provided by interstitial solutions, commonly of marine water origin (Utada and Minato 1971). Recently, Ogihara (1996) showed, based on chemical analyses, that analcime of Zone IIIa crystallizes after Na-K-clinoptilolite is modified into Na-clinoptilolite by ion exchange,. Conversely, the formation of heulandite necessitates the addition of calcium as shown in following equations:

$$(Na_2,K_2)(Al_2Si_8O_{20})\cdot 7H_2O + Ca^{2+} \Rightarrow Ca(Al_2Si_7O_{18})\cdot 6H_2O + SiO_2 + H_2O + 2(Na^+,K^+)$$

clinoptilolite heulandite quartz

$$(Na_2,Ca)(Al_2Si_{10}O_{24})\cdot 8H_2O + Ca^{2+} \Rightarrow Ca(Al_2Si_7O_{18})\cdot 6H_2O + 3SiO_2 + 2H_2O + (2Na^+,Ca^{2+})$$

mordenite heulandite quartz

Calcium may have been derived from interstitial solutions by the decomposition of carbonates or the albitization of calcic plagioclase. Alternatively, Ogihara and Iijima (1989) suggested that heulandite crystallizes after modification of Na-K-clinoptilolite into Ca-clinoptilolite by ion exchange. The temperature at the boundary between two zones was estimated to average 84°C (Iijima 1995).

Laumontite zone. More advanced diagenesis can be divided into Ca-silicate or Na-silicate trends. Although laumontite of burial diagenetic origin is uncommon in the Neogene formations of Japan, it occurs as comparatively coarse-grained, euhedral or subhedral crystals. The reaction of heulandite to laumontite can be written as:

$$Ca(Al_2Si_7O_{18})\cdot 6H_2O \Rightarrow Ca(Al_2Si_4O_{12})\cdot 4H_2O + 3SiO_2 + 2H_2O$$

heulandite laumontite quartz

Laumontite may also form from the Ca produced by albitization of Ca-plagioclase.

Albite zone. The albite zone is characterized by the presence of authigenic albite. Quartz, chlorite, and white mica are common coexisting minerals. The original texture has been largely changed into a mosaic-like texture. The common transformation of analcime to albite is not necessarily accompanied by distinct chemical changes, as shown in the following equation:

$$Na_2(Al_2Si_4O_{12})\cdot 2H_2O + 2SiO_2 \Rightarrow 2Na(AlSi_3O_8) + 2H_2O$$

analcime quartz albite

Iijima (1995) suggested that this transformation occurred at an average temperature of 123°C. This temperature is far lower than that suggested by experimental data for conditions $P(H_2O) = P(Total)$ (Thompson 1970, Bish and Aronson 1993).

GENETIC CONSIDERATIONS

Bore-hole temperatures

Although other factors affect mineral reactions under burial diagenesis and low-grade metamorphism (Hay 1966, Utada 1971, Boles and Coombs 1975), temperature is often the most important control. Borehole temperatures provide information about the temperature conditions of zeolitization, assuming that each zeolite is in equilibrium *in situ*, at present. Few data exist on borehole temperatures relating to low geothermal-gradient zeolitization other than those of Galloway (1974). In Japan, deep boreholes for petroleum and natural gas prospecting have provided many data. The relation between borehole temperature and zeolite zones was summarized by Iijima (1995) as shown in Figure 16. The temperatures of zeolite zone boundaries were averaged from eight boreholes in the terrain of intermediate to high geothermal gradient, which are described in the chapter by Iijima (this volume).

Homogenization temperature of fluid inclusions

In general, the crystallization temperatures of authigenic minerals can be estimated from homogenization temperatures of fluid inclusions within them. No fluid inclusion data, however, exist for zeolites formed by burial diagenesis and low-grade metamorphism, primarily because fluid inclusions within zeolite crystals are rare or absent. The scarcity of inclusions may stem from the fact that zeolites are permeable to water on the scale of their crystal structures. Aizawa (1990) estimated crystallization temperatures of 68-163°C for clinoptilolites from the homogenization temperature of fluid inclusions within authigenic calcite and quartz coexisting with clinoptilolite. These temperatures seem high, compared with 44-84°C borehole temperatures in similar geological settings shown by Iijima (1995).

Inference of temperature from isotopic data

Isotope data provide useful information concerning the origin and conditions of formation of authigenic minerals. In zeolites, however, several technical difficulties exist. One is that channel water within zeolites easily exchanges with surrounding water after crystallization. In addition, dehydration causes isotopic exchange between the zeolite framework and channel water (Karlsson and Clayton 1990, Karlsson 1995). Therefore, Murata and Whiteley (1973) measured the oxygen and carbon isotope composition of calcite associated with zeolites and concluded that the minerals were not in equilibrium with each other. Levy and O'Neil (1989) calculated isotopic temperatures of quartz coexisting with clinoptilolite-heulandite from Yucca Mountain as 40-100°C.

Stallard and Boles (1989) reported apparently reliable isotopic temperatures for zeolite crystallization in burial diagenesis and uplifting. Their data support the petrographic observation of co-crystallization of albite and laumontite in the groundmass of altered tuffs. The albite-quartz pair geothermometer temperatures are constrained in the range 145-170°C assuming a pore water $\delta^{18}O$ of +1.8 to +3.5‰, whereas laumontite-quartz temperatures are constrained in the range 139-162°C. Oxygen isotope values for fracture-filling laumontite from Hokonui Hill, New Zealand, are higher than those for laumontite in the groundmass. Values for stilbite veins are clearly higher than those for laumontite veins, suggesting that stilbite crystallized at lower temperatures than laumontite.

P-T conditions as measured by hydrothermal experimentation

As is common for rock-forming minerals, the stability fields of many zeolites have been investigated using hydrothermal techniques. However, the number of hydrothermal experiments is somewhat limited as a result of the relatively low-temperature conditions accompanying zeolite formation, giving rise to very slow reaction kinetics. Analcime-albite equilibria have been studied by Coombs et al. (1959) and by Campbell and Fyfe (1965). More recently, on the basis of several experiments, Thompson (1971) placed equilibrium for the reaction analcime + quartz $\Leftrightarrow$ albite + H_2O at 190±10°C at 2 kbar, at 170±10°C at 4 kbar, and at 150°C at 4.75±0.5 kbar ($P(H_2O) = P(Total)$). Liou (1971c) also showed a similar equilibrium diagram for the same reaction. These results are not completely consistent with the temperatures at the analcime-albite zone boundary found in boreholes (Fig. 16). The stability relations of laumontite, stilbite, heulandite, wairakite, prehnite, and lawsonite have been determined by Liou and other researchers (Liou 1970, 1971a,b; Thompson 1970, Liou et al. 1985, 1991; Cho et al. 1987). P-T grids for these and other phases have been compiled by Liou et al. (1985), Liou et al. (1991) and Frey et al. (1991). The stability fields of a variety of calcium zeolites and related minerals have been determined using hydrothermal experimentation. However, the temperatures for the heulandite-laumontite boundary obtained by Liou's experiments (Liou 1971d) are higher than those measured in bore holes. (Fig. 16); the disagreement between experimental measurements and borehole observation may be due to slow reaction kinetics at relatively

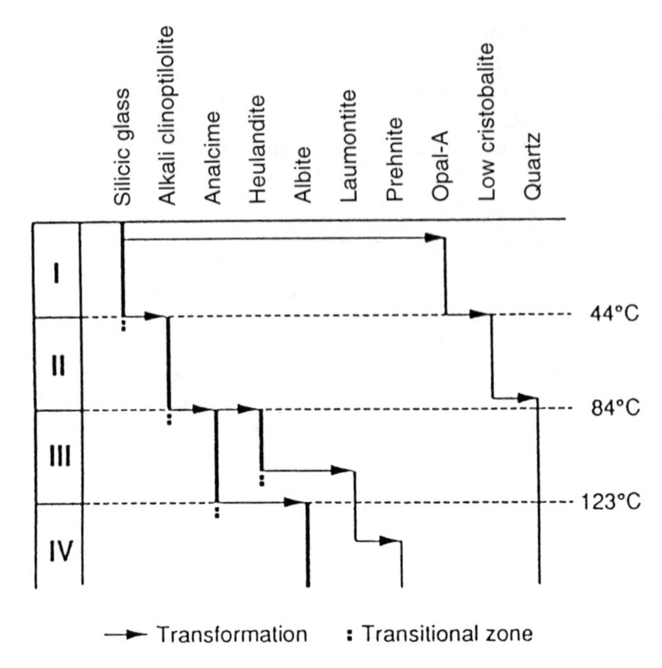

Figure 16. Present-day bore-hole temperatures at boundaries between zeolite zones in Japanese Islands (after Iijima 1995).

low temperatures. Formation of laumontite at very low temperatures (40-80°C) was reported in Sespe Hot Springs, California (McCulloh et al. 1981). This observation is in agreement with the low-pressure stability field experimentally determined by Cho et al. (1987). Unfortunately, the stability fields of clinoptilolite and mordenite, two of the most important early diagenetic zeolites, remain to be firmly established by experimental means. Recent studies by Wilkin and Barnes (1998, 1999) have begun to clarify clinoptilolite-analcime equilibria.

Chemistry of reacting solutions

The chemistry of reacting solutions is an important factor controlling the crystallization of zeolites. Judging from the sedimentary environment of the original sediments, the solutions relating to burial diagenesis and low-grade metamorphism are mainly of marine water origin, but they may be also fresh or brackish. The chemistry of these solutions, however, may have altered considerably during burial and uplift, as shown in data from Deep Sea Drilling Projects and others. At present, very few direct data are available on the chemistry of reacting solutions at the time of zeolite crystallization, except for active hydrothermal systems. The oxygen isotope composition of zeolite frameworks may become one of the most useful tools for constraining one aspect of the chemistry of reacting solutions. Activity diagrams, such as those of Browne and Ellis (1970), Bowers and Burns (1990), Chipera et al. (1995), Chipera and Bish (1997), and the chapter by Chipera and Apps (this volume), are also of great importance in this respect. The role of $P(CO_2)$in allowing or inhibiting crystallization of calcium zeolites was emphasized by writers such as Browne and Ellis (1970) and Thompson (1971).

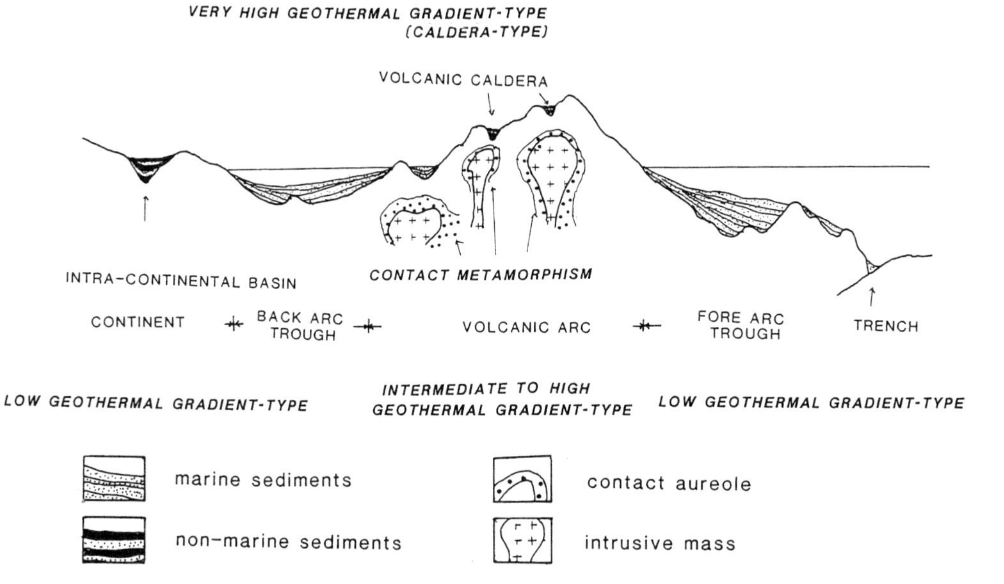

Figure 17. Relation between type of zeolitization and geological setting near a volcanic arc.

SUMMARY AND FUTURE PROBLEMS

Generally speaking, diagenetic and low-grade metamorphic zeolites are common in orogenic belts. Low geothermal-gradient zeolitization characterizes very thick sections, such as fore-arc troughs where the geothermal gradient is low. In volcanic arcs at active plate margins, intermediate to high geothermal-gradient zeolitization has progressed in sediments containing a large amount of vitric materials. Volcanic calderas located near the volcanic front form a special geological setting. Caldera-filling sediments consist mostly of volcanic glasses, and very high geothermal-gradient zeolitization occurs rapidly under the highest geothermal gradient. Contact metamorphism due to the thermal effects of intrusive masses is commonly superimposed on above-mentioned progressive zeolitization. The relation between types of zeolitization and geologic setting is schematically shown in Figure 17. The facies series of each type can be represented on a P-T diagram (Fig. 18). Despite considerable research over the past 40 years, our knowledge of the zeolite facies remains incomplete. Some remaining problems include; (1) The facies relations and reaction mechanisms for the early stages of the low geothermal gradient and the advanced stages of the intermediate to high geothermal gradient zeolitization remain obscure. (2) The effects of sourcc-rock variability on mineral reactions are incompletely understood. (3) Progressive zeolitization continuing to a lawsonite-bearing assemblage has been recognized only in the Bryneiya Group in southern New Zealand where there is a direct passage from laumontite to lawsonite assemblages (Landis 1974). Further research on the low-grade part of the Kanto Mountains in Japan, the Franciscan Group of California, the Andean orogenic belt of Chile, and the outer zone of western Alpine orogenic belt may reveal additional occurrences of the transition between zeolite- and lawsonite-bearing assemblages. (4) A clear need exists for further research on the genesis of zeolite deposits in the interior of continents, particularly in Russia, China and Mongolia. (5) There is a pressing lack of detailed information on the physico-chemical conditions of zeolitization, particularly direct measurement of temperature and water chemistry, using boreholes where zeolitization is progressing. Fluid inclusion, oxygen isotope, and K/Ar measurements may also be important in elucidating the physico-chemical conditions existing during zeolitization.

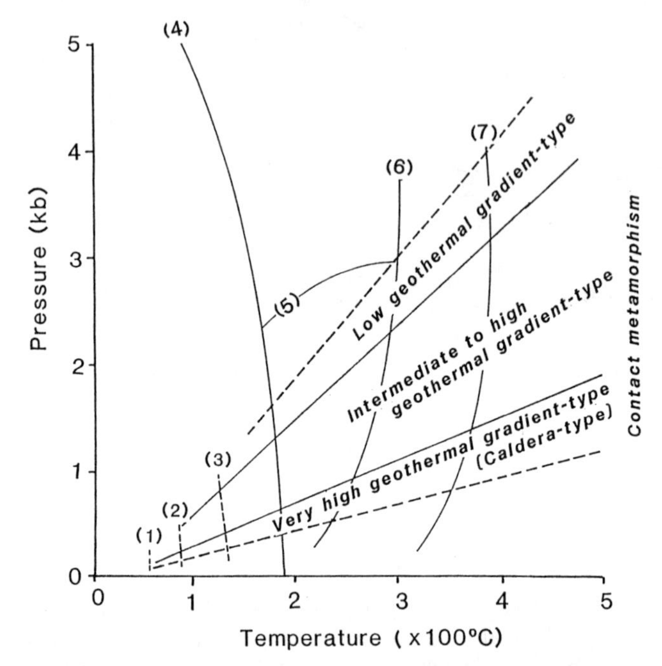

Figure 18. Estimated temperature-pressure conditions for each type of zeolitization (after Utada 1997):
(1) boundary between the fresh glass zone and the clinoptilolite zone (Iijima 1995)
(2) boundary between the clinoptilolite-mordenite zone and the analcime-heulandite zone
(3) boundary between the analcime-heulandite zone and the albite zone
(4) experimentally determined analcime-albite boundary (Thompson 1971)
(5) experimentally determined laumontite-lawsonite boundary (Liou 1971a)
(6) experimentally determined laumontite-wairakite boundary (Liou 1970)
(7) experimentally determined wairakite-anorthite boundary (Liou 1970)

ACKNOWLEDGMENTS

This manuscript benefited greatly from the comments of D. Coombs, F.A. Mumpton, and D.L. Bish, who are gratefully acknowledged for their assistance.

REFERENCES

Aguirre AL, Atherton MP (1987) Low-grade metamorphism and geotectonic setting of the Macuchi Formation, western Cordillera of Ecuador. J Metamorphic Geol 5:473-494

Aizawa J (1990) Paleotemperatures from fluid inclusion and coal rank of carbonaceous material of the Tertiary formations in Northwest Kyushu, Japan. J Assoc Mineral Petrol Econ Geol 85:145-154 (in Japanese)

Aleksiev B (1975) Zeolite rocks in Cuba. Compt rendus l'Acad Bulg Sci 28:1239-1240

Aleksiev B, Djourova EG (1977) Zeolitic rocks in the north-eastern Rhodopes. Ann Univ Sofia, Fac Geol Geogr 69:79-92

Altaner SP, Grim RE (1990) Mineralogy, chemistry, and diagenesis of tuffs in the Sucker Creek Formation (Miocene), eastern Oregon. Clays & Clay Minerals 38:561-572

Ames LL, Sand LB, Goldich SS (1958) A contribution on the Hector, California, bentonite deposit. Econ Geol 53:22-37

Armstrong RL (1968) Sevier orogenic belt in Nevada and Utah. Geol Soc Am Bull 79:429-458

Awakjan GS (1973) Die Zeolith-führenden Gestein in Nemberjanskij-Bezirk: Armenien und Aussichten für lhre Nutzbarmachung. Isw Akad Nauk Armjan SSR Nauki o Semle 26:6

Bish DL, Aronson JL (1993) Paleogeothermal and paleohydrologic conditions in silicic tuff from Yucca Mountain, Nevada. Clays & Clay Minerals 41:148-161

Boles JR (1977) Zeolite facies alteration of sandstones in the Southland syncline, New Zealand. Am J Sci 277:982-1012

Boles JR (1991) Diagenesis during folding and uplift of the Southland Syncline, New Zealand. New Zealand J Geol Geophys 34:253-259

Boles JR, Coombs DS (1975) Mineral reactions in zeolitic Triassic tuff, Hokunui Hills, New Zealand. Geol Soc Am Bull 86:163-173

Bowers TS, Burns RG (1990) Activity diagrams for clinoptilolite: Susceptibility of this zeolite to further diagenetic reactions. Am Mineral 75:601-619

Bramlette MN, Posnjak E (1933) Zeolitic alteration of pyroclastics. Am Mineral 18:167-171

Brown CE, Thayer TP (1963) Low-grade mineral facies in upper Triassic and lower Jurassic rocks of the Aldrich Mountains, Oregon. J Sed Petrol 33:411-425

Browne PRL, Ellis AJ (1970) The Ohaki-Broadlands hydrothermal area, New Zealand: Mineralogy and related geochemistry. Am J Sci 269:97-131

Broxton DE, Bish DL, Warren RG (1987) Distribution and chemistry of diagenetic minerals at Yucca Mountain, Nye County, Nevada. Clays & Clay Minerals 35:85-110

Burchfiel BC, Davis GA (1972) Structural framework and evolution of the southern part of the Cordilleran orogen, western United States. Am J Sci 272:97-118

Buruzova GI (1964) Contribution on the heulandite group of zeolites from the Paleogene of the southern part of USSR. Litol i polez iskop 4:66-80

Buryanova Ye Z (1960) Analcime- and zeolite-bearing sedimentary rocks of Tuva. Izuvo Akad Nauk SSSR Ser Geol 6:54-65

Bushinski GI (1950) Mordenite in marine sediments of Jurassic, Cretaceous and Paleogene ages. Dokl Akad Nauk SSSR 73:1271-1275

Campbell AS, Fyfe WS (1965) Analcime-albite equilibria. Am J Sci 263:807-816

Carter RM, Hicks MD, Norris RJ, Turnbull IM (1978) Sedimentation patterns in an ancient arc-trench-ocean basin complex, Carboniferous to Jurassic Rengitata orogen, New Zealand. *In* Sedimentation in Submarine Canyons, Fans, and Trenches. DJ Stanley, G Kelling (eds) Dowden, Hutchinson & Ross, Pennsylvania, p 340-361

Chipera SJ, Bish DL, Carlos BA (1995) Equilibrium modeling of the formation of zeolites in fractures at Yucca Mountain, Nevada. *In* Natural Zeolites '93: Occurrence, Properties, Use. Ming DW, Mumpton FA (eds) Int'l Comm Natural Zeolites, Brockport, New York, p 547-563

Chipera SJ, Bish DL (1997) Equilibrium modeling of clinoptilolite-analcime equilibria at Yucca Mountain, Nevada, USA. Clays & Clay Minerals 45:226-239

Cho M, Maruyama S, Liou JG (1987) An experimental investigation of heulandite-laumontite equilibrium at 1000 to 2000 bar P_{fluid}. Contrib Mineral Petrol 97:43-50

Coombs DS (1954) The nature and alteration of some Triassic sediments from Southland, New Zealand. Trans Royal Soc New Zealand 82:65-109

Coombs DS (1958) Zeolitized tuffs from the Kuttung Glacial Beds near Seaham, New South Wales. Austral J Sci 21:18-19

Coombs DS (1960) Lower grade mineral facies in New Zealand. *In* Rept 21st Int'l Geol Congr (Copenhagen) 13:339-351

Coombs DS (1993) Dehydration veins in diagenetic and very low-grade metamorphic rocks: Features of the crustal seismogenic zone and their significance to mineral facies. J Metamorphic Geol 11:389-399

Coombs DS, Ellis AJ, Fyfe WS, Taylor AM (1959) The zeolite facies, with comments on the interpretation of hydrothermal syntheses. Geochim Cosmochim Acta 17:53-107

Coombs DS, Cook NDJ, Kawachi Y, Johnstone D, Gibson IL (1996) Park Volcanics, Murihiku Terrane, New Zealand: Petrology, petrochemistry, and tectonic significance. New Zealand J Geol Geophys 39:469-492

Crawford AL, Cowles HD (1932) The Fuller's earth deposits near Aurora, Utah. Utah Acad Sci Proc 9: 55-60

Crook KAW (1961a) Vein minerals from the Tamworth and Parry Groups (Devonian and lower Carboniferous). Am Mineral 46:1017-1029

Crook KAW (1961b) Diagenesis in the Wahgi Valley sequence, New Guinea. Proc Royal Soc Victoria 74:77-81

Dickinson WR (1962) Petrography and diagenesis of Jurassic strata in central Oregon. Am J Sci 260: 481-500

Dickinson WR, Ojakanga RW, Stewart RJ (1969) Burial metamorphism of the late Mesozoic Great Valley Sequence, Cache Creek, California. Geol Soc Am Bull 80:519-526

Dzotsenidze GS, Shirtladze NE (1953) Horizon with analcime in the coal-bearing series in the Kutay-Gelet region. Sbarn Voprosy petrol mineral 1, Izd Akad Nauk SSSR 301-311

Ermolova EP (1955) Analcime and mordenite in the Oligocene and Miocene sediments of the western Transcaucassia. Trans Mineral Mus Akad Nauk SSSR 7:76-82

Eskola P (1939) Die Metamorphen Gestein. *In* Die Entstehung der Gestein. Barth TFW, Correns CW, Eskola P (eds) Springer, Berlin, p 263-407

Fiske RS (1963) Fifes Peak Formation. U S Geol Surv Prof Paper 93:27-37

Flood PG, Stephen G, Marx W (1995) Australian natural zeolite resources. *In* Proc Int'l Symp Mineral property and Utilization of Natural Zeolites Minato H (ed) Japan Sci Promot, Tokyo, p 158-169

Frey M, DeCapitani C, Liou JG (1991) A new petrogenetic grid for low-grade metabasites. J Metamorphic Geol 9:497-509

Fyfe WS, Turner FJ, Verhoogen J (1958) Metamorphic reactions and metamorphic facies. Geol Soc Am Mem 73, 253 p

Galloway W (1974) Deposition and diagenetic alteration of sandstone in northeast Pacific arc-related basins: implication for greywacke genesis. Geol Soc Am Bull 85:379-390

Ghent ED, Miller BE (1974) Zeolite and clay-carbonate assemblages in the Blairemore Group (Cretaceous) southern Alberta foothills, Canada. Contrib Mineral Petrol 44:313-329

Gulbrandsen RA, Cressman ER (1960) Analcime and albite in altered Jurassic tuff in Idaho and Wyoming. J Geol 68:458-464

Hay RL (1966) Zeolites and zeolitic reactions in sedimentary rocks. Geol Soc Am Spec Paper, 130 p

Hayashi H, Sudo T (1957) Zeolite-bearing bentonite. Mineral J 2:196-199

Heald M.T (1956) Cementation of Triassic arkoses in Connecticut and Massachusetts. Geol Soc Am Bull 67:1133-1154

Hutchinson RM, Vine DJ (1987) Alteration zones related to igneous activity. Spanish Peaks area, Las Animas and Huerfano counties, Colorado. Geol Soc Am, Centennial Field Guide, Rocky Mtn Sec, p 357-360

Hwang JY (1992) Occurrence and utilization of natural zeolite in Korea. Proc Int'l Symp: Mineral property and utilization of natural zeolites, p 128-142

Iijima A (1995) Zeolites in petroleum and natural gas reservoirs in Japan: A review. *In* Natural Zeolites '93: Occurrence, Properties, Use. DW Ming, FA Mumpton (eds) Int'l Comm Natural Zeolites, Brockport, New York, 99-114

Iijima A, Ohwa I (1980) Zeolitic burial diagenesis in Creta-Tertiary geosynclinal deposits of central Hokkaido, Japan. *In* Proc 5th Int'l Zeol Conf Heyden, London, p 139-148

Ivaanjav (1992) Natural zeolite deposits in Mongolia and their physico-chemical characters. Proc Int'l Symp "Mineral property and utilization of natural zeolites", p 143-149

Iwao S, Hushmand-Zadeh (1971) Stratigraphy and petrology of the low-grade regionally metamorphosed rocks of the Eocene formation in the Alborz Range, north of Tehran, Iran. J Japan Assoc Mineral Pet Econ Geol 65:265-285

Karlsson HR (1995) Application of oxygen and hydrogen isotopes to zeolites. *In* Natural Zeolites '93: Occurrence, Properties, Use. DW Ming, FA Mumpton (eds) Int'l Comm Natural Zeolites Brockport, New York, p 125-140

Karlsson HR, Clayton RN (1990) Oxygen and hydrogen isotope geochemistry of zeolites. Geochim Cosmochim Acta 54:1369-1386

Kashkai MA, Babeav IA (1976) Clinoptilolite from zeolitized tuffs of Azerbaidzhan. Mineral Mag 40: 501-511

Kerr PF (1931) Bentonite from Ventura, California. Econ Geol 26:153-168

Kerr PF, Cameron EN (1936) Fuller's earth of bentonitic origin from Techachapi, California. Am Mineral. 21:230-237

Kim JH, Moon HS (1978) Occurrence of zeolite in the Tertiary sediments. J Korean Inst Mining, Geol 11:59-68

Kisch HJ (1968) Zeolite facies and regional rank of bituminous coals. Geol Mag 103:415-422

Kolbin MF, Pimburskaja X (1955) Analcime from the sedimentary rocks of the B. Bogdo Mountain. Dokl Akad Nauk SSSR 100:155-158

Landis CA (1974) Stratigraphy, lithology, structure, and metamorphism of Permian, Triassic, and Tertiary rocks between the Mararoa River and Mount Snowdon, western Southland, New Zealand. J Royal Soc New Zealand 4:229-251

Levy B (1970) Burial metamorphic episodes in the Andean Geosyncline, central Chile. Geol Rund 59: 994-1013

Levy B, Aguirre L, Nyström JO, Padilla H, Vergara M (1989) Low-grade regional metamorphism in the Mesozoic volcanic sequences of the Central Andes. J Metamorphic Geol 7:487-495

Levy SS, O'Neil JR (1989) Moderate temperature zeolitic alteration in a cooling pyroclastic deposit. Chem Geol 76:321-326

Liou JG (1970) Synthesis and stability relations of wairakite, $CaAl_2Si_4O_{12}\cdot 2H_2O$. Contrib Mineral Petrol 27:259-282

Liou JG (1971a) P-T stabilities of laumontite, wairakite, and related minerals in the system $CaAl_2Si_2O_8$-SiO_2-H_2O. J Petrol 12:379-411

Liou JG (1971b) Synthesis and stability relations of prehnite $Ca_2Al_2Si_2O_{10}(OH)_2$. Am Mineral 56:507-531

Liou JG (1971c) Analcime equilibria. Lithos 4:389-402

Liou JG (1971d) Stilbite-laumontite equilibrium. Contrib Mineral Petrol 31:171-177

Liou JG, de Capitani C, Frey M (1991) Zeolite equilibria in the system $CaAl_2Si_2O_8$-$NaAlSi_3O_8$-SiO_2-H_2O. New Zealand J Geol Geogr 34:293-301

Liou JG, Maruyama S, Cho M (1985) Phase equilibria and mineral paragenesis of metabasites in low-grade metamorphism. Mineral Mag 49:321-333

Lipman PW, Prostka HJ, Christiansen RL (1971) Evolving subduction zones in the western United States, as interpreted from igneous rocks. Science 174:821-825

Lougnan FC (1966) Analcite in the Newcastle coal measure sediments of the Sydney Basin, Australia. Am Mineral 51:486-494

Madsen BM, Murata KJ (1970) Occurrence of laumontite in Tertiary sandstones of the Central Coast Ranges, California. U S Geol Surv Prof Paper 700-D, p D188-D195

Martini J (1968) Etude petrographique des Gres de Taveyanne entre Arve et Giffre (Haute Savoie, France). Schweiz mineral petrogr Mitt 48:539-654

Martini J, Vuagnat M (1965) Presence du facies a zeolites dans la formation des 'gres' de Taveyanne (Alpes francosuisse). Schweiz mineral petrogr Mitt 45:281-293

Masuda H, O'Neill JR, Jiang W-T, Peacor DR (1996) Relation between interlayer composition of authigenic smectite, mineral assemblages, I/S reaction rate and fluid composition in silicic ash of the Nankai Trough. Clays & Clay Minerals 44:443-449

McCulloh TH, Frizzell VA Jr, Stewart RJ, Barnes I (1981) Precipitation of laumontite with quartz, thenardite and gypsum at Sespe Hot Springs, Western Transverse Ranges, California. Clays & Clay Minerals 29:353-364

Mikhaylov AS, Silant'ev VN (1973) Heulandite like zeolite from the Georgian SSR spogilites. Zap Vses Mineral Obshch 102:707-709

Miki T, Nakamuta Y (1985) Zeolitic diagenesis of the Paleogene formations in the Munakata coal field, Fukuoka Prefecture. J Japan Assoc Mineral Pet Econ Geol 80:283-291 (in Japanese)

Millar BE, Ghent ED (1973) Laumontite and barianstrontian heulandite from the Blairmore Group (Cretaceous), Alberta. Can Mineral 12:188-192

Monger JWH, Souther JG, Gabrielse H (1972) Evolution of the Canadian Cordillera: A plate-tectonic model. Am J Sci 272:577-602

Murata KJ, Whiteley KR (1973) Zeolites in the Miocene Briones Sandstone and related formations of the Central Coast Ranges, California. J Res U S Geol Surv 1:255-265

Nakajima W, Tanaka K (1967) Zeolite-bearing tuffs from the Izumi Group in the central part of the Izumi Mountain Range, southwest Japan, with reference to mordenite-bearing tuffs and laumontite tuffs. J Geol Soc Japan 73:237-245 (in Japanese)

Nishimura T, Iijima A, Utada M (1980) Zeolitic burial diagenesis and basin analysis of the Izumi Group in Shikoku and Awaji Islands, southwest Japan. J Geol Soc Japan 86:341-351 (in Japanese)

Noh JH, Kim SJ (1986) Zeolites from Tertiary tuffaceous rocks in Yeongil area, Korea. *In* New Developments in Zeolite Science and Technology. Murakami Y, Iijima A, Vard JW (eds) Kodansha-Elsevier, Tokyo, p 59-66

Obradovic J (1979) Authigenic minerals in middle Triassic volcanoclastic rocks in Dinarides. Bull Mus Hist Nat'L, Ser A 34:13-36 (with English abstr)

Obradovic J, Kemenci R (1975) Diagenetic changes in Miocene pyroclastics from deep borings in the north of Backa (Yugoslavia). Ann Geol Penin Balkan 39:333-355

Ogihara S (1996) Diagenetic transformation of clinoptilolite to analcime in silicic tuffs of Hokkaido, Japan. Mineral Deposita 31:548-553

Ogihara S, Iijima A (1989) Clinoptilolite and heulandite transformation in burial diagenesis. *In* Zeolites, Facts, Figures, Future. Jacobs PA, van Santen RA (eds) Elsevier, Amsterdam, p 491-500

Okubo Y (1993) Temperature gradient map of the Japanese Islands. J Geotherm Res Soc Japan 15:1-21 (in Japanese)

Otalora G (1964) Zeolites and related minerals in Cretaceous rocks of east-central Puerto Rico. Am J Sci 262:726-734

Pevear DR, Williams VE, Mustoe GE (1980) Kaolinite, smectite, and K-rectorite in bentonites: relation to coal rank at Tulameen, British Columbia. Clays & Clay Minerals 28:241-254

Rahn M, Mullis J, Erdelbrock K, Frey M (1994) Very low-grade metamorphism of the Taveyanne greywacke, Glarus Alps, Switzerland. J Metamorphic Geol 12:625-641

Ramlah MS (1992) Outline of zeolite resources in Indonesia. *In* Proc Int'l Symp: Mineral Property and Utilization of Natural Zeolites. Minato H (ed) Japan Sci Promotion, Tokyo, p 150-157

Read PB, Eisbacher GH (1974) Regional zeolite alteration of the Sustut Group, north-central British Columbia. Can Mineral 12:527-541

Rengarten NV (1950) Laumontite and analcime of the Lower Jurassic sediments of the North Caucassus. Dokl Akad Nauk SSSR 70:485-489

Rutland RWR (1976) Orogenic evolution of Australia. Earth Sci Rev 12:161-196

Samajova E, Kuzvart M (1992) Zeolite deposits at the periphery of Tisza basin (Carpathians, Central Europe). Proc Int'l Symp: Mineral Property and Utilization of Natural Zeolites. Tokyo, p 48-61

Sawatzki G (1975) Étude géologique et minéralogique des flyschs à grauwackes volcaniques due synclinal de Thônes (Haute-Savoie, France). Arch Sci Genève 27:1-32

Seki Y, Oki Y, Matsuda T, Mikami K, Okumura K (1969) Metamorphism in the Tanzawa Mountains, Central Japan. J Japan Assoc Mineral Pet Econ Geol 61:1-75

Stallard ML, Boles JR (1989) Oxygen isotope measurement of albite-quartz-zeolite mineral assemblages, Hokunui Hills, Southland, New Zealand. Clays & Clay Minerals 37:409418

Stewart RJ (1974) Zeolite facies metamorphism of sandstone in the Western Olympic Peninsula, Washington. Geol Soc Am Bull 85:1139-1142

Stewart RJ, Page RJ (1974) Zeolite facies metamorphism of the late Cretaceous Nanaimo Group, Vancouver Island and Gulf Islands, British Columbia. Can J Earth Sci 11:280-284

Surdam RC (1973) Low-grade metamorphism of tuffaceous rocks in the Karmutsen Group, Vancouver Island, British Columbia. Geol Soc Am Bull 84:1911-1922

Thompson AB (1970) Laumontite equibria and the zeolite facies. Am J Sci 269:267-275

Thompson AB (1971) Analcime-albite equilibria at low temperatures. Am J Sci 271:79-92

Utada M (1965) Zonal distribution of authigenic zeolites in the Tertiary pyroclastic rocks in Mogami District, Yamagata Prefecture. Sci Paper Coll Gen Educ, Univ Tokyo 15:173216

Utada M (1970) Occurrence and distribution of authigenic zeolites in the Neogene pyroclastic rocks in Japan. Sci Paper Coll Gen Educ, Univ Tokyo 20:191-262

Utada M (1971) Zeolitic zoning of the Neogene pyroclastic rocks in Japan. Sci Paper Coll Gen Educ, Univ Tokyo 21:189-221

Utada M (1973) The type of alteration in the Neogene sediments relating to the intrusion of volcano-plutonic complexes in Japan. Sci Paper Coll Gen Educ, Univ Tokyo 23:167-216

Utada M (1977) Genetic condition of natural zeolites and related minerals. J Clay Sci Soc Japan 37:87-94

Utada M, Ito T (1989) Sedimentary facies of the MioPliocene volcanotectonic depressions along the volcanic front in northeast Honshu, Japan. *In* Sedimentary Facies in the Active Plate Margin. Taira A, Masuda F (eds) Terrapub, Tokyo, p 605-618

Utada M, Minato H (1971) Analcime nodule from Yusato, Shimane Prefecture the mode of occurrence and genetical relation to the migration of sodium. Sci Paper, Coll Gen Educ, Univ Tokyo 21:63-78

Utada M, Shimizu M, Ito T, Inoue A (1999) Alteration of caldera-forming rocks related to the Sanzugawa volcano-tectonic depression, northeast Honshu, Japan—with special reference to "caldera-type zeolitization." Resource Geol, Spec Issue 20:127-138

Utada M, Vine JD (1984) Zonal distribution of zeolites and authigenic plagioclase, Spanish Peaks Region, southern Colorado. *In* Proc 6th Int'l Zeolite Conf, Reno, Nevada, 1983. Butterworths, Surrey, UK, p 604-615

Wilkin RT, Barnes HL (1998) Solubility and stability of zeolites in aqueous solution: I. Analcime, Na-, and K-clinoptilolite. Am Mineral 83:746-761

Wilkin RT, Barnes HL (1999) Thermodynamics of hydration of Na- and K-clinoptilolite to 300 degrees C. Phys Chem Minerals 26:468-476

Wilkinson JFG, Whetten JT (1964) Some analcime bearing pyroclastic and sedimentary rocks from New South Wales. J Sed Petrol 34:543-553

Wopfner H, Markwort S, Semkiwa PM (1991) Early diagenetic laumontite in the lower Triassic Manda Beds of the Ruhuhu basin, Southern Tanzania. J Sed Petrol 61:65-72

Zhang Quanchang (1995) Zeolite deposits in China. *In* Proc Int'l Symp: Mineral Property and Utilization of Natural Zeolites. Minato H (ed) Japan Sci Promotion, Tokyo, p 143-148

10 Zeolites in Hydrothermally Altered Rocks

Minoru Utada

Futo 912-14 Itho
Shizuoka 413-0231, Japan

INTRODUCTION

Over the past 40 years, two universally accepted maxims of geology dealing with zeolite minerals have been shown to be incorrect: the first is that zeolites occur primarily as cavity fillings in basaltic igneous rocks, and the second is that these zeolites crystallized from so-called hydrothermal solutions, i.e. thermal waters from depth. We recognize today that enormous quantities of zeolites occur as low-temperature, low-pressure alteration products of pyroclastic material in sedimentary rocks, and that many of the zeolites filling vugs and cavities in basalts and traprocks are actually diagenetic in origin, having precipitated directly from ground water percolating through the rock mass (see Walker 1951, 1959, 1960a,b; Nashar and Davies 1960). This is not to say that zeolite deposits of hydrothermal origin do not exist—far from it. The extensive zones of zeolites surrounding hot-spring and geothermal activity in Yellowstone National Park, Wyoming; Wairakei, New Zealand; Reykjavik, Iceland; Pauzhetsk, Kamchatka, Russia; Lardello, Italy; and Takenoyu, Japan; and in the vicinity of magmatic ore deposits, such as the Kuroko deposits of northern Japan, suggest that waters arising and/or heated from below have indeed played an important role in their formation.

Hydrothermal alteration can be defined as the crystallization of mineral material or the alteration of preexisting mineral material that has taken place in the presence of hot solutions, the composition and temperature of which were wholly or partially the result of a deep-seated magmatic body. These reactions usually take place under comparatively high solution/solid ratios over a wide temperature range and under low-pressure conditions. Hydrothermal alteration is controlled by many factors, including solution chemistry, temperature, rock permeability and the composition and stability of the starting material. Although hydrothermal solutions are very diverse in composition, alkaline solutions are generally most favorable for formation of zeolites. The temperature range of hydrothermal alteration is typically equal to or greater than that characteristic of diagenesis to low-grade metamorphism. Generally, hydrothermal alteration rarely occurs over wide areas and is more typically localized around zones of circulating solutions. As a result, hydrothermal alteration is often superimposed upon diagenetic alteration zones.

This chapter describes the occurrence and genesis of zeolites in hydrothermally altered rocks, including hot spring deposits, ore-vein deposits, and deposits clearly relating to geothermal activity. It does not discuss the controversial genesis of zeolites in basalts and other basic igneous rocks.

TYPES OF HYDROTHERMAL ALTERATION

Hydrothermal alteration has been classified in a variety of ways by many researchers. For example, Meyer and Hemley (1967) classified hydrothermal alteration based on mineral assemblages described as propylitic, intermediate, argillic, advanced argillic, sericitic, and potassic. Zeolites, however, were not included in any of these categories. On the basis of studies of Cretaceous and Neogene formations in Japan, Utada (1980) classified hydrothermal alteration into three groups (alkaline, intermediate, acidic)

1529-6466/00/0045-0010$05.00

based on temperature, ion activity in solution, and the nature of the assemblage of neoformed silicate minerals (Table 1). The alkaline group of hydrothermal alteration is the only one of the three groups which includes zeolites. This type of alteration takes place at relatively high alkali and alkaline-earth activities and at high pH and is divided into a Na-silicate type and a Ca-silicate type, each of which is further subdivided into zones on the basis of temperature and the specific assemblages of zeolites present. In several deposits, a mixed Na-Ca-silicate type can also be recognized. To be more broadly applicable, a biotite zone has been added to the potassic type and the prehnite-pumpellyite and amphibole zones have been added to the Ca-Mg type, as the higher-temperature zones of the intermediate zone group. In addition, a chabazite-phillipsite zone has been added to the Ca-silicate type of the alkaline group as the lowest grade zone, and fibrous zeolite and anorthite zones have been added to this type as the intermediate grade and the highest-grade zones, respectively. These modifications are illustrated in Table 1. The following sections describe the typical occurrence, distribution, and geological setting of hydrothermal zeolites in the alkaline group.

Table 1. Types of hydrothermal alteration and mineral zones (modified from Utada 1980).

<table>
<tr><th colspan="2">TYPE</th><th colspan="9">ZONE</th></tr>
<tr><td rowspan="2">ACIDIC GROUP</td><td>SULFATE</td><td colspan="2">Alunite + opal</td><td colspan="7">Alunite + quartz</td></tr>
<tr><td>Al-SILICATE</td><td>Halloysite</td><td colspan="3">Kaolinite</td><td colspan="3">Dickite + nacrite</td><td colspan="2">Pyrophyllite</td></tr>
<tr><td rowspan="2">INTERMEDIATE GROUP</td><td rowspan="2">K-SILICATE</td><td colspan="2" rowspan="2">Smectite</td><td rowspan="2">Mixed-layer clay minerals</td><td colspan="2">Sericite</td><td colspan="3">K-feldspar</td><td>Biotite</td></tr>
<tr><td colspan="3">Propylite</td><td colspan="2">Prehnite-pumpellyite</td><td>Amphibole</td></tr>
<tr><td rowspan="2">ALKALINE GROUP</td><td>Ca-SILICATE</td><td>Chabazite + phillipsite</td><td>Stilbite + heulandite</td><td>Fibrous zeolite</td><td colspan="2">Laumontite</td><td colspan="3">Wairakite</td><td>Anorthite</td></tr>
<tr><td>Na-SILICATE</td><td>Clinoptilolite</td><td>Mordenite</td><td>Analcime</td><td colspan="6">Albite</td></tr>
</table>

Increasing Temperature ⇒

⇐ Increasing $\frac{\text{Alkali + Alkaline-Earth Ion Activity}}{\text{Hydrogen Ion Activity}}$

Calcium-silicate type

Ca-zeolites have formed in various geologic settings including volcanic arcs, oceanic ridges, hot spots, and tectonic zones. Within and around volcanic calderas which are usually situated near the volcanic front of arcs, hydrothermal alteration including Ca-zeolites often is superimposed upon caldera-type zeolitization (see the chapter on *Zeolites in Burial Diagenesis and Low-grade Metamorphic Rocks* by Utada, this volume). The Sanzugawa caldera of late Miocene to Pliocene age in northern Honshu, Japan, is an active geothermal area where a geothermal power plant is currently operating. Although caldera-filling rocks, composed mainly of rhyolitic volcaniclastics and tuffites, are widely subjected to caldera-type zeolitization (Utada et al. 1999), both intermediate and alkaline types of hydrothermal alteration are locally superimposed upon them near a fault zone in the northeast part of the caldera (Fig. 1). There, Ca-zeolite zones are distributed on surfaces associated with parallel faults with a NW-SE direction crossing the main fault (Minase Fault). The zonation of zeolites in these zones is, from the center to the margin of the hydrothermally altered zone (i.e. highest temperature to lowest), as follows; wairakite ⇒ laumontite + (yugawaralite) ⇒ stilbite + heulandite ⇒ chabazite. Laumontite and wairakite are common not only in this hydrothermal alteration

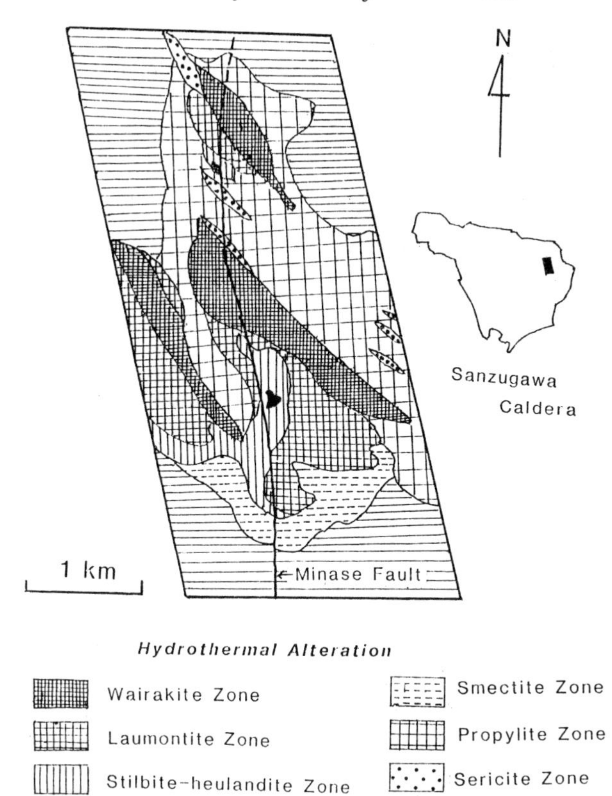

Figure 1. Distribution of alteration zones along the Minase Fault in the Sanzugawa Caldera in northern Honshu, Japan.

zone, but also in deep portions of surrounding areas which were subjected to caldera-type zeolitization. The modes of occurrence of these zeolites of both origins are very similar.However, the crystal size of the hydrothermal zeolites is usually larger than those of caldera-type zeolitization origin. In addition, the distribution of these two types of zeolites is significantly different. In particular, hydrothermal zeolites surround probable solution pathways such as faults, dikes, and porous layers. On the other hand, caldera-type zeolites are vertically distributed and laumontite and wairakite are found only in the lower parts of the caldera. In some cases, hydrothermal solutions appear to have flowed horizontally through porous layers within caldera-filling rocks. For example, in several horizons both laumontite and wairakite zones are superimposed upon the mordenite zones of caldera-type zeolitization in the Onuma geothermal area (Honda and Matsueda 1983) and the Sumikawa geothermal area (Inoue et al. 1999), northern Honshu, Japan. Ca-zeolites have also been recognized in basaltic igneous rocks in volcanic arcs, oceanic ridges and hot spots. Kristmannsdottir and Tomasson (1978) recognized a zonal distribution of mainly calcium zeolites from a pile of basaltic rocks in both low- and high-temperature geothermal areas of Iceland. The zeolites were vertically zoned (on the basis of neoformed minerals) from top to bottom in drill holes as follows: chabazite ⇒ mesolite/scolecite ⇒ stilbite ⇒ laumontite ⇒ analcime or wairakite. This type of hydrothermal zeolitization is common in basaltic amygdaloidal lavas, volcaniclastics and dolerite masses.

Table 2. Range of hydrothermal zeolites in two drill holes in the Yellowstone geothermal area, Wyoming. Y-1 (simplified from Honda and Muffler 1970); Y-2 (simplified from Keith and Melvin 1981).

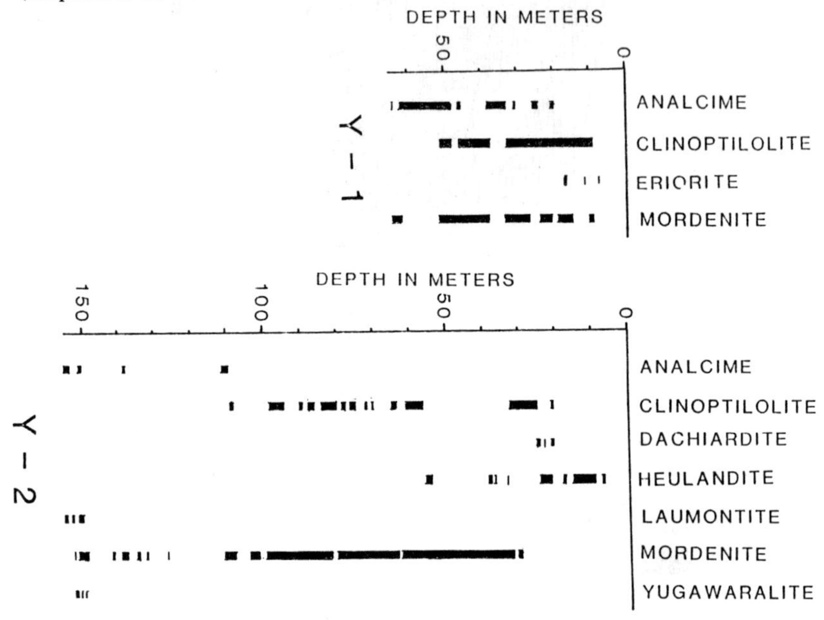

Sodium-calcium-silicate type

The Na-Ca-silicate type of hydrothermal zeolitization is characterized by the presence of both Ca- and Na-zeolites. An example is found in the active geothermal area at Yellowstone, Wyoming. The Yellowstone caldera is filled mainly by rhyolitic lavas, volcaniclastics, and a small amount of sedimentary rocks. Thus, the geological setting is very similar to that of the above-mentioned Sanzugawa caldera. After much study, Bargar and Keith (1995) summarized the systematic occurrences of analcime, clinoptilolite, dachiardite, erionite, heulandite, mordenite, wairakite, and yugawaralite in eleven drill holes. Erionite and dachiardite are particularly characteristic of this area. As an example, the distribution of zeolites in two drill holes (Y-1, Y-2) is shown in Table 2. According to Honda and Muffler (1970), sodium zeolites in drill hole Y-1 replace obsidian, glass shards, and plagioclase phenocrysts. They also fill interstitial spaces and secondary vugs. All of these zeolites, with the exception of erionite, appear over almost same depth range. In drill hole Y-2 (Keith and Melvin 1981), sodium zeolites appear zonally in nearly the same manner as in Y-1. Judging from these zeolites, clinoptilolite and mordenite may have been formed by the caldera-type zeolitization as in the Sanzugawa caldera. Another typical example of the Na-Ca-silicate type of zeolitization is found in a swarm of hydrothermal veins of the Takinoue geothermal area, northern Honshu, Japan, in which two geothermal power stations are operating. The mode of occurrence of zeolite veins is almost same to that of hydrothermal veins in metal and non-metal deposits. Most of

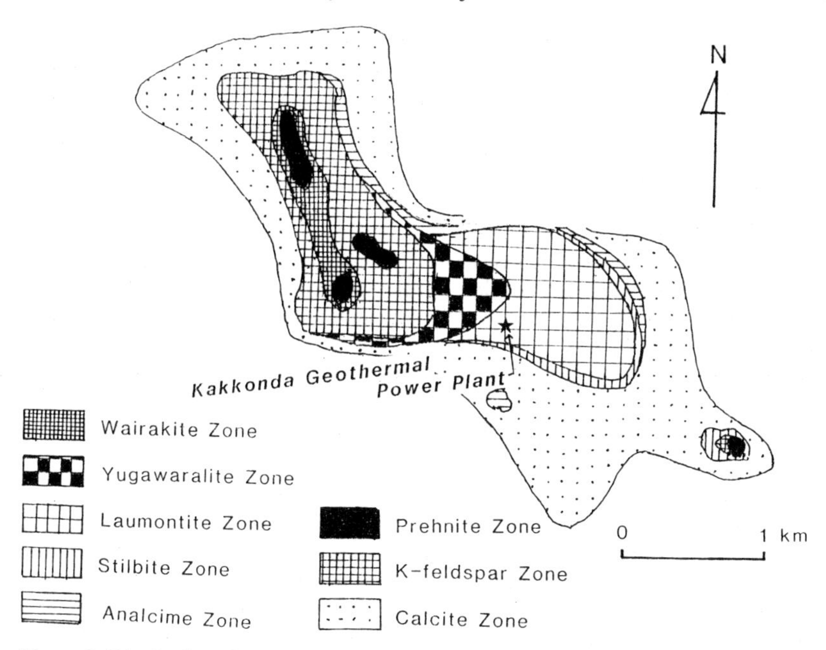

Figure 2. Distribution of zones of zeolites and related minerals in veins in the Takinoue (Kakkonda) geothermal area in northern Honshu, Japan (simplified from Koshiya et al. 1994).

Table 3. Range of authigenic minerals in hydrothermal alteration zones around Kuroko deposits in Japan (after Utada 1988).

	Hydrothermal Alteration Envelope						Diagenesis
	K-feldspar Zone	Sericite-chlorite Z.	Mixed layer clay Zone	Smectite Zone	Analcime Zone	Mordenite Zone	Clinoptilolite-mordenite Z.
Quartz	———	———	———	———	———	——— - - -	
Opal						- - - ———	———
K-feldspar	———						
Sericite	———	——— - - -	-				
Chlorite	———	———	- - -				
Sericite/smectite		- - -	———	- - -	- - -	- - -	
Chlorite/smectite		- - -	- - -	- - -	- - -		
Smectite			- - -	———	———	———	———
Analcime					———		
Mordenite						———	———
Ferrierite						- - -	
Clinoptilolite							———
Siderite			- - -	- - -	- - -	- - -	
Dolomite		- - -	- - -	- - -	- - -	- - -	
Calcite	- - -	- - -	- - -	- - -	- - -	- - -	- - -
Pyrite	- - -	- - -	- - -	- - -	- - -	- - -	
Gypsum	- - -	- - -					

the veins are mono-zeolitic, although in some veins, the zeolite is associated with quartz, clay minerals, or carbonate minerals. The width of a vein usually ranges from several millimeters to several centimeters. Koshiya et al. (1994) classified hydrothermal veins into eight mineral zones that contain both Ca- and Na-silicates, as shown in Figure 2. The hydrothermal veins were probably formed in several stages because they have crosscutting structures and occur in multiple orientations.

Sodium-silicate type

Na-rich alkali zeolites of hydrothermal origin usually occur in silicic rocks. A typical example is the hydrothermal alteration envelope surrounding the mid-Miocene Kuroko-type mineralization in Japan (Utada 1988). The alteration envelope is generally composed of six mineral zones that are defined by their specific hydrothermal mineral assemblage (Table 3, above). A succession of zones exists from the immediate vicinity of the ore body to the periphery, as shown in Figure 3. The K-feldspar zone has been found only around large ore bodies (Fig. 3a), whereas sericite is common around smaller ore bodies (Fig. 3b).

The alteration envelope can be divided into two parts, one formed in footwall rocks under the sea floor by intermediate hydrothermal alteration. The other part formed in hanging-wall rocks after ore deposition and is characterized by repeated clay zones and sodic zeolite zones that extend horizontally. The alteration envelope has a mushroom-like shape and extends laterally several kilometers from the center of alteration. Mordenite occurs, usually as fan-like aggregates of fibrous crystals, in the most peripheral zone, mainly replacing glass shards and filling interstitial spaces. Analcime occurs as subhedral aggregates or as pseudomorphs of mordenite aggregates, and ferrierite occasionally occurs in the mordenite zone.

Generally, hydrothermal veins of sodium zeolites are rarer than those of Ca-zeolites, but analcime veins are common in many types of host rocks ranging from sedimentary to volcaniclastic to igneous to metamorphic. Veins of mordenite, thought to be hydrothermal in origin, have chiefly been found in rhyolitic volcanic rocks, whereas veins of natrolite appear to be restricted to basaltic and alkali rocks. Ferrierite occasionally occurs as veins in altered rocks related to mercury mineralization (Yajima and Nakamura 1971), in addition to its occurrence in Kuroko-type deposits. Dachiardite(-Na) may also occur locally.

PROVINCES OF HYDROTHERMAL ZEOLITIZATION

Geothermal areas in volcanic arcs of orogenic belts

Most hydrothermal zeolites have been reported from active geothermal areas that are concentrated primarily in volcanic arcs of the Circum-Pacific Region and in the Alpine-Himalayan orogenic belt (Fig. 4).

Western Circum-Pacific archipelagos. Hydrothermal alteration has been well studied in 18 active geothermal areas of the Taupo Volcanic Zone, New Zealand (Fig. 4, Loc. 1), by several researchers (Steiner 1953, 1968, 1977; Browne and Ellis 1970; Henneberger and Browne 1988; Browne et al. 1989). In general, the hydrothermal alteration is of the intermediate type, according to the classification of Meyer and Hemley (1967), and zeolites are relatively rare. Even in the extensively zeolitized Wairakei geothermal area, mordenite, laumontite, and wairakite are sporadically distributed in only the upper portions of drill core taken from the body, whereas K-feldspar, albite, epidote and locally prehnite occur in the lower parts where temperatures are higher (Steiner 1977).

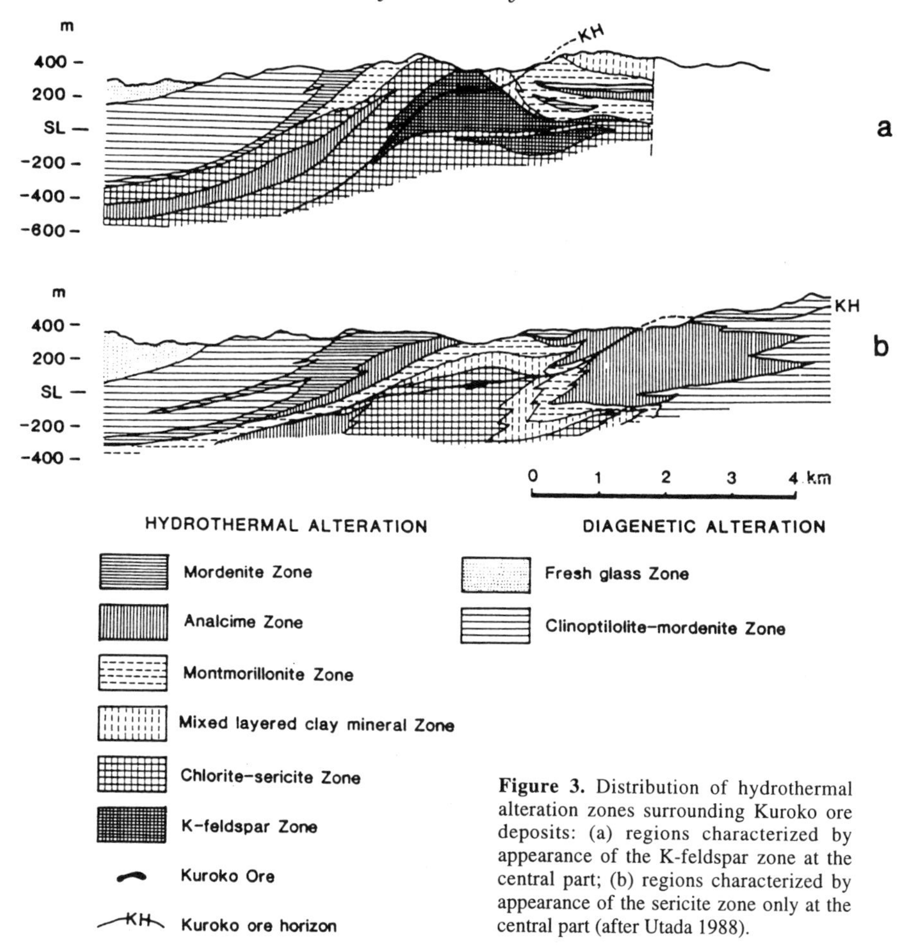

Figure 3. Distribution of hydrothermal alteration zones surrounding Kuroko ore deposits: (a) regions characterized by appearance of the K-feldspar zone at the central part; (b) regions characterized by appearance of the sericite zone only at the central part (after Utada 1988).

Reyes (1990) and Reyes et al. (1993) summarized the hydrothermal alteration in thirty active geothermal areas in the Philippines (Fig. 4, Loc. 2). The zeolitization is typically of the Ca-silicate type and is zoned from high to low temperatures as follows: epidote/clinozoisite ⇒ wairakite ⇒ laumontite ⇒ heulandite and stilbite zones.

Volcanic and hydrothermal activity has been extensive in the Japanese archipelago and vicinity since late Cretaceous time. However, zeolites of hydrothermal origin have rarely formed, with the exception of laumontite veins in Cretaceous to early Miocene rocks. In middle Miocene, zeolites of the typical sodium silicate-type were formed around the Kuroko deposits by submarine hydrothermal activity, as described above. This zeolitization is concentrated within a narrow zone (Kuroko Zone) of several ten kilometers in width extending from southwest Hokkaido to northeast Honshu and San-in district. The most intense zeolitization resulting from hydrothermal activity occurs in geothermal areas within and around volcanic calderas and volcanic zones of late Miocene to Quaternary age. The Green Tuff Region of northern Honshu is the most concentrated area of hydrothermal alteration in Japan (Fig. 4, Loc. 3). Especially along the volcanic front, many active geothermal areas are situated within volcanic calderas. Although mordenite, analcime, laumontite, and wairakite are commonly found in these areas, it is

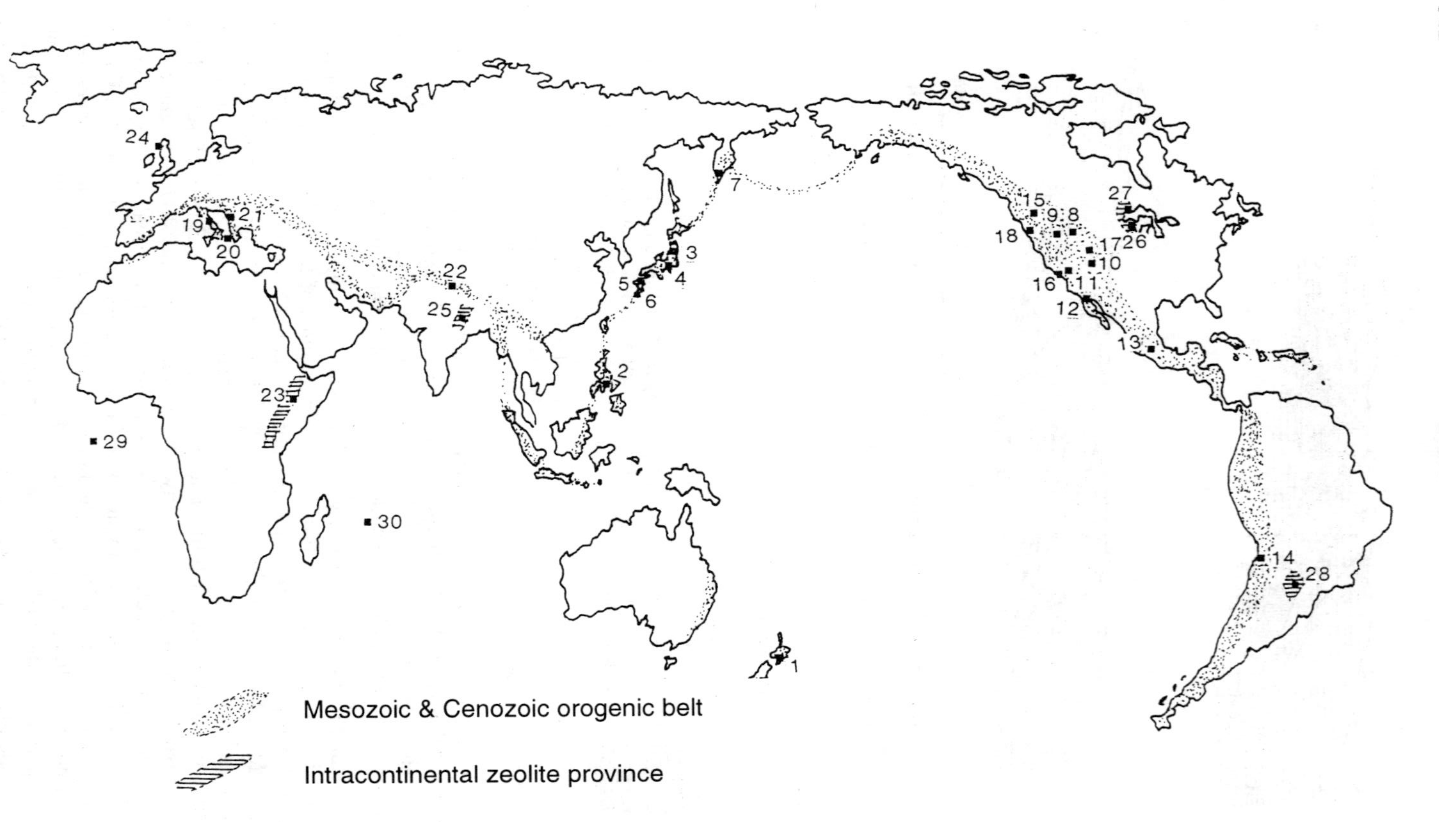

Figure 4. Distribution of hydrothermal zeolite provinces in the world.

often difficult to determine whether they formed by hydrothermal alteration or by caldera-type zeolitization . For instance, Seki et al. (1969) described the following zeolite distribution in drill holes from top to bottom in the Onikobe geothermal area: mordenite ⇒ laumontite ⇒ wairakite + yugawaralite. This zonal distribution is very similar to the caldera-type zeolitization in the Sanzugawa caldera (see the chapter on *Zeolites in Burial Diagenesis and Low-grade Metamorphic Rocks* by Utada, this volume). In addition to hydrothermal zeolite veins in the Takinoue (Kakkonda) geothermal area as described above, mordenite, laumontite, and wairakite are distributed in a large part of the Hachimantai caldera (Shimazu and Yajima 1973; Kubota 1979; Kimbara et al. 1982; Honda and Matsueda 1983; Kimbara 1983; Inoue et al. 1999). These zeolites formed by a combination of hydrothermal alteration and caldera-type zeolitization. However, these zeolites (heulandite, stilbite, thomsonite and yugawaralite) occur only locally and often in hydrothermal veins.

In central Honshu, the South Fossa Magna region (Fig. 4, Loc. 4) is the most extensive hydrothermally altered area. Many hot springs are distributed in the Hakone caldera of central Honshu. A variety of types of hot solutions are presently rising in several different systems in which various kinds of hydrothermal alteration are progressing. Hirano (1986) reported many hydrothermal zeolites including mordenite, thomsonite, dachiardite, stilbite, clinoptilolite, laumontite, and wairakite. In one drill hole (M.-116), these zeolites are zonally distributed from top to bottom (482 m) as follows: mordenite ⇒ mordenite + clinoptilolite (heulandite?) ⇒ laumontite ⇒ wairakite. These zeolites occur in fractures, as crack fillings, and as veinlets. Yugawaralite is found only in the original locality.

Both the northern portion (Tanzawa Mountains) and the southern portion (Izu peninsula) of the Hakone caldera are major provinces of hydrothermal zeolites. In the Tanzawa Mountains, a large number of zeolite veins (laumontite, chabazite, stilbite, wairakite, analcime, and garronite) formed in granitic masses and in their zeolitized contact aureoles (Seki 1969; Seki et al. 1971). On the other hand, 14 zeolites species were formed primarily in Pliocene andesites of the Izu peninsula. The zeolites occur mainly in veins filling conjugate faults and fractures and are distributed throughout the entire peninsula. These veins are distributed with a regional zonation from the center to the margin of the Izu peninsula as follows: prehnite ⇒ wairakite ⇒ laumontite ⇒ heulandite ⇒ stilbite ⇒ chabazite (Utada and Shimizu 1990).

The Hohi geothermal area in northern Kyushu is situated in a volcanic zone having a graben-horst structure (Beppu-Shimabara graben) (Fig. 4, Loc. 5). In this area, hydrothermal alteration around the Otake and the Hachobaru geothermal power plants has been studied in detail by Hayashi and others (Yamasaki et al. 1970; Hayashi 1973). They recognized Ca-silicate zeolitization (heulandite, laumontite, and wairakite) in altered Quaternary volcanic rocks, with the zeolites distributed in fractures nearly parallel to the main fault. Based on many drill-hole data, zeolites (clinoptilolite, mordenite stilbite, heulandite, laumontite, and wairakite) seem to be sporadically distributed over a large area of the Hohi volcanic zone (Takashima et al. 1985).

Hayashi and Taguchi (1987) reported an occurrence of Ca-silicate-type zeolitization (heulandite, laumontite, and wairakite) from a drill hole in the Yunotani geothermal area, situated in the Aso caldera of middle Kyushu (Fig.4, Loc. 6). In southern Kyushu, clinoptilolite and ferrierite were reported from the Kirishima geothermal area in the Aira caldera by Yoshida et al. (1983). Stilbite was found in the Ibusuki geothermal area in the Ata caldera (Kamitani et al. 1978). Kimbara and Okubo (1978) also reported clinoptilolite and mordenite from the upper part and wairakite from the lower part of an exploration hole drilled in this caldera.

Although many active geothermal areas are located in the Kamchatka-Kurile volcanic arc (Averyev et al. 1961; Noboko 1970), hydrothermal zeolitization is not well documented. However, nearly fifty authigenic minerals, including nine zeolites, have been identified in the Pauzhetka geothermal area, Kamchatka (Fig. 4, Loc. 7) (Noboko 1976). Zeolites appear mainly at the periphery of the geothermal area and are zonally distributed from top to bottom as follows: authigenic mineral-free ⇒ analcime ⇒ montmorillonite ⇒ mordenite or laumontite.

Active and fossil geothermal areas in the Cordillera region. Although geothermal activity has been common in the Cordillera region, zeolitization is not pervasive, with the exception of the Yellowstone National Park area in Wyoming (Fig. 4, Loc. 8) and the Beowawe geothermal area in Nevada (Fig. 4, Loc. 9). Hydrothermal zeolitization at Yellowstone is described above as a typical example of the Na-Ca-silicate type of zeolitization.

The Beowawe geothermal area, Nevada, is located along an extensive fault zone. The host rocks of this area consist of Quaternary alluvium, Miocene igneous rocks ranging from basalt to dacite, and Ordovician sedimentary rocks. Based on studies of three drill holes, Cole and Ravinsky (1984) reported the occurrence of five different zeolites which appear to be zonally distributed from shallow to deepest as follows: chabazite ⇒ thomsonite ⇒ heulandite ⇒ laumontite ⇒ wairakite. Epidote is sporadically distributed in deeper rocks. These minerals are characteristic of the Ca-silicate type of zeolitization.

In contrast to Beowawe and Yellowstone, only small amounts of wairakite have been reported from the Valles geothermal area in New Mexico (Fig. 4, Loc. 10) (Armstrong et al. 1995), and from the Geysers in California (Fig. 4, Loc. 11) (Steiner 1958). Similarly, wairakite was the only zeolite identified among a number of other authigenic minerals in the Cerro Prieto geothermal area of Baja California (Fig. 4, Loc. 12), studied in detail by Elders and others (1979, 1981). On the other hand, Cathelineau et al. (1985) described a zonal distribution of mordenite, laumontite and wairakite in drill holes of the Los Azufres geothermal area in Mexico (Fig. 4, Loc. 13).

The Cordilleran orogen in South America is also a region of concentrated geothermal activity, but reports of hydrothermal zeolites are very rare, except for an occurrence of laumontite from El Tatio, Chile (Fig. 4, Loc. 14) (Browne 1978).

Several examples of hydrothermal zeolitization from low-temperature geothermal areas are known in North America. For example, stilbite and chabazite occur as fracture linings and cementing granite cobbles in hot springs ranging from 20-94°C (Mariner and Presser 1995) in the Idaho batholith (Fig. 4, Loc. 15), a granitic complex of Cretaceous to Paleogene age. Laumontite formed at temperatures ranging from 43 to 89°C, associated with quartz, thenardite, and gypsum at Sespe Hot Springs, California (Fig. 4, Loc. 16) (McCulloh et al. 1981). Extensive laumontite fracture-fillings also occur in a quartz monzonite of Miocene age and in overlying sediments in central Colorado (Fig. 4, Loc. 17) (Sharp 1970).

Nine species of zeolites and related minerals were described from Eocene basaltic pillow lavas as replacements of plagioclase phenocrysts and as cavity, joint, and fracture fillings in the Siletz River Volcanics of the central Coast Range, Oregon (Fig. 4, Loc. 18), by Keith and Staples (1985). Isotopic data for quartz and calcite suggest that these zeolites were formed by submarine hydrothermal activity at temperatures of 60 to 70°C.

Alpine-Himalayan orogenic belt. Of the many active geothermal areas in the Alpine- Himalayan orogenic belt, only the hydrothermal alteration in the Larderello field, Italy (Fig. 4, Loc. 19), has been extensively investigated, but Cavarretta et al. (1980) described only rare wairakite in drill holes in the Upper Miocene to Pliocene sediments and flysch-type sequences of the Ligurian Nappes. Geothermal activity is also intense in the Aegean volcanic arc. K-mordenite, presumably of hydrothermal origin, occurs on the island of Samos (Fig. 4, Loc. 20) in veins and in the surrounding rhyolitic tuffs and flow-banded rhyolites (Pe-Piper and Tsolis- Katagas 1991). Hydrothermal zeolitization has not been reported from other active geothermal areas of the Alpine-Himalayan orogen in northern Hungary, northern Thailand, Sumatra, or Java. Submarine hydrothermal alteration of Tertiary age has been recognized in northeast Slovenia in andesitic pillow lavas, auto-brecciated lavas, and volcaniclastic deposits of the Mt. Smrekovic area (Fig. 4, Loc. 21). Here Kovic and Krosl-Kuscer (1986) reported the regional formation of hydrothermal albite, laumontite, and prehnite. Laumontite, analcime, stilbite, and yugawaralite were also found in veins and as fracture fillings.

The Tengchon area of southwestern China (Fig. 4, Loc. 22) may have been an area of geothermal activity in Cenozoic time. However, despite evidence of hydrothermal alteration in outcrop, only a small amount of chabazite and analcime have been found (Meixiang and Wei 1987).

Active and fossil geothermal areas in intracontinental areas

Africa. The Eastern African Rift Valley is a typical intracontinent active geothermal area. Gianelli and Teklemariam (1993) and Teklemariam et al. (1996) found heulandite and laumontite only in the upper parts of eight drill holes in the Aluto-Langano geothermal field of Ethiopia (Fig. 4, Loc. 23). Hydrothermal zeolitization has not been documented from other geothermal areas of the region.

Europe. Hydrothermal zeolites have been reported from a Tertiary granite on the Isle of Skye, northwest Scotland (Fig. 4, Loc. 24) (Ferry 1985).

India. The Tattapani hot spring area in central India (Fig. 4, Loc. 25) is an active intracontinental geothermal area in which stilbite and laumontite are common in non-volcanic Quaternary and Mesozoic sedimentary rocks (Shanker et al. 1987).

North America. Stoiber and Davidson (1959) reported the zeolite-prehnite-pumpellyite series in the alteration of the Portage Lake Series of the Keweenawan volcanic group on the south shore of Lake Superior, Michigan (Fig. 4, Loc. 26), and Jolly and Smith (1972) divided the rocks into laumontite, pumpellyite, and epidote zones. Laumontite and analcime are regionally distributed, suggesting that they are not of hydrothermal origin, whereas thomsonite, natrolite, and dachiardite locally occur in fracture fillings, veins, and amygdules. Schmidt (1990) identified 13 zeolite species from north-shore lavas mainly in veins and amygdules (Fig. 4, Loc. 27). In general, these zeolites are zonally distributed around veins or vein networks that may have acted as conduits for hydrothermal solutions. The mineral zonation is typically: central epidote-quartz vein $\Rightarrow$ laumontite $\Rightarrow$ thomsonite + heulandite $\Rightarrow$ stilbite $\Rightarrow$ stellerite $\Rightarrow$ calcite. The central veins include Fe-pumpellyite, prehnite, K-feldspar, and Cu-minerals such as native copper, chalcopyrite, cubanite, and covellite. The age of zeolitization is unknown.

South America. Similar zeolite assemblages have been recognized in Jurassic to Cretaceous plateau basalts and associated dikes and sills in the Parana Basin in southern Brazil (Fig. 4, Loc. 28) (Franco 1952). Murata et al. (1987) classified the zeolites into quartz-silicic zeolites, scolecite, and laumontite zones, from top to bottom regionally. These authors suggested that zeolitization may have been promoted by hydrothermal circulation at or near a continental spreading center, similar to that occurring today in Iceland.

Active and fossil geothermal areas in oceans. Extensive hydrothermal activity and mineralization have recently been reported from the Galapagos hydrothermal mounds (Hoffert et al. 1980; Honnorez et al. 1981) and from the Explorer Ridge (Grill et al. 1981) in the Pacific Ocean. However, zeolites of certain hydrothermal origin have not yet been recognized there. Zeolites in geothermal exploration drill holes were reported from basaltic igneous rocks in Ascension Island of the Mid-Atlantic Ridge (Nielson and Stiger 1996)(Fig. 4, Loc. 29) and Reunion Island in the Indian Ocean (Rancon 1985) (Fig. 4, Loc. 30).

CONDITIONS OF HYDROTHERMAL ZEOLITIZATION

Homogenization temperatures of fluid inclusions

The homogenization temperature of fluid inclusions is often used to determine the temperature of crystallization of both primary and secondary minerals. Unfortunately, the measurement of homogenization temperatures in zeolites has not been successful. The difficulty in obtaining homogenization temperatures from zeolites stems, in part, from the very fine-grained natures of zeolites and from the fact that zeolites interact with liquid water. It is unlikely that a liquid-vapor inclusion would remain for geologically significant times inside a zeolite crystal. Instead, the homogenization temperatures of inclusions in associated minerals, such as quartz and calcite, have been used based on the assumption that the zeolites and associated minerals were formed contemporaneously. Although many temperature estimates have been obtained for zeolite assemblages in geothermal areas, the temperature of crystallization for many zeolite assemblages has been difficult to establish quantitatively. Moreover, measured homogenization temperatures often do not coincide with present drill-hole temperatures, which suggests that the temperatures of hydrothermal solutions may have changed in a comparatively short time.

Drill-hole temperatures

Present-day drill-hole temperatures may also provide information on the temperature of formation of zeolite assemblages, but they do not always accurately reflect the actual temperature of crystallization (vide supra). In addition to the discordance between homogenization temperatures and present drill-hole temperatures described above, it is often observed that mineralogical zone boundaries are oblique to present-day isothermal lines obtained form drill hole temperatures. These facts may suggest that the physicochemical conditions of hydrothermal systems change over time scales shorter than those involved in diagenesis and metamorphism. Therefore, it is doubtful that the temperature of each zeolite zone obtained from active geothermal areas (Table 4) exactly reflects the formation temperatures of the zeolite constituents.

Chemistry of solutions from drill holes

The chemistry of solutions obtained from drill holes is potentially useful in estimating the pH, chemical environment, and temperatures prevailing at the time of crystallization of authigenic minerals. However, the distinct possibility exists that conditions have changed since the time of zeolite crystallization. Nevertheless, present-day conditions provide some useful information because the chemistry of hydrothermal solutions from zeolite-bearing drill holes is typically distinct from solutions in drill holes in which zeolites are absent. The temperature, pH, and chemical composition of major elements in representative hydrothermal solutions are listed in Table 4. As expected, most

Table 4. Temperature, pH, and concentration of major ions in waters obtained from active geothermal areas.

	T (°C)	pH/T	SiO_2 (mg/l)	Na^+ (mg/l)	K^+ (mg/l)	Ca^{2+} (mg/l)	Mg^{2+} (mg/l)	Total CO_2 (mg/l)	SO_4^{2-} (mg/l)	Cl^- (mg/l)	Ca^{2+}/ $Na^++K^++Ca^{2+}$
Calcium-silicate type											
Reykyanes (No.8), Iceland	270	6.27/270 °C	592	9,854	1,391	1,531	1.15	1,427	28.7	18,827	0.097
Reykyavic, Iceland	100	9.60/26 °C	130	48	1.2	1.5	0.24	51	27	19	0.030
Takenoyu (GSR-3), Japan		8.4	403	630.3	84	17.1	0.4	32.7	64.6	1,030	0.023
Beawawe (1-13), Nevada, USA	211	8.4	335	203	30	11	0.3	247	47	59	0.045
Sespe, California, USA	89	7.74	91	330	14	22	0.11	70	292	290	0.060
Banks, Idaho, USA	78	8.03	110	130	5.2	4.4	0.01	167	77	35	0.035
Tattapani (No.4), India	50	8.1	130	130	10	3	1	134	64	68	0.021
Kirkham, USA	44	9.3	66	67	1.5	1.8	<0.15	78	39	3.4	0.026
Sodium-calcium silicate type											
Yellowstone (Spouter Geyser), Wyoming	92	8.8	313	425	20	1	1	454	18	312	0.002
Broadlands (No. 9), New Zealand	270-276	8.45	805	930	203	2	-	140	4	1,142	0.002
Wairakei (No. 27), New Zealand	98	8.5	660	1,200	200	17.5	0.05	23	25	2,156	0.013

of hydrothermal solutions associated with zeolitization are alkaline, with the exception of discharge solutions from the Philippines (Reyes et al. 1993). Furthermore, as expected, the values of Ca/(Na+K+Ca) in hydrothermal solutions of the Ca-silicate type are generally larger than those of the Na-Ca-silicate and Na-silicate types. The wide range in Cl concentrations listed in Table 4 suggests various types of original solutions or varying degrees of dilution. However, most analyzed hydrothermal solutions are dilute, indicating that some or all of a given solution was meteoric in origin. As expected, the hydrothermal solution from the Reykjanes geothermal area of Iceland has nearly the same Cl concentration as seawater, suggesting a marine origin.

Isotopic data

The origin of hydrothermal solutions can be clarified by a variety of geochemical methods, and stable isotope geochemistry, particularly of hydrogen and oxygen, is particularly useful because the isotope ratios of both elements are significantly different between marine, fresh, saline, and magmatic waters. Typically, measured isotope ratios are compared with those of standard mean ocean water (SMOW) as follows; $\delta D = \{D/H(\text{measured})\}/\{D/H(\text{SMOW})\}$; $\delta^{18}O = \{^{18}O/^{16}O(\text{measured})\}/\{^{18}O/^{16}O(\text{SMOW})\}$. There has been considerable research on the use of stable isotopes to decipher the origins of hydrothermal solutions and minerals, and the interested reader is referred to the chapter in this volume on *Isotope Geochemistry of Zeolites* by Karlsson. Most $\delta D/\ \delta D^{18}O$ values of solutions from active geothermal areas plot close to the so-called meteoric water line (Taylor 1979), consistent with the dilute-solution characteristics mentioned above. On the other hand, $\delta D/\delta^{18}O$ values of solutions from active geothermal areas of New Zealand (Robinson 1974) and in ore minerals from the Kuroko deposits of Japan (Sakai and Matsubaya 1974; Ohmoto and Rye 1974) plot close to the value for SMOW, indicating a marine origin.

At present, there are no isotopic data for *proven* hydrothermal zeolites, but Sturchio et al. (1990) reported $\delta^{18}O$ values for quartz, calcite, and clay minerals in drill cores from the Yellowstone, geothermal area, Wyoming. Most of the $\delta^{18}O$ values for quartz and clay minerals were not in equilibrium with present conditions, but the calcite $\delta^{18}O$ values were close to isotopic equilibrium with the present hydrothermal solutions.

Hydrothermal experiments

Among the numerous laboratory hydrothermal experiments that have been carried out on the synthesis and stability of zeolite assemblages, those of Liou (1970 1971a, 1971b) are particularly useful in estimating the P-T conditions of hydrothermal zeolitization. From such experiments, he proposed the low-pressure formation temperatures of laumontite, wairakite, and anorthite to be <200°, 200-300°, and >330°C, respectively. Wirsching (1981) synthesized several Ca-zeolites in an open system at 100-250°C using rhyolitic glass, basaltic glass, nepheline, oligoclase, and three different solutions. She found that phillipsite and wairakite crystallized from runs using both the rhyolitic and basaltic glasses. However, heulandite and epistilbite formed only from rhyolitic glass, and scolecite and levyne formed only from basaltic glass. These data are roughly consistent with zeolite occurrences in active geothermal regions. In order to interpret hydrothermal zeolitization of the Na-silicate type around Kuroko deposits, Kusakabe (1982) carried out hydrothermal experiments using natural clinoptilolite and alkaline solutions at 100-400°C. Four mineral phases and two different alteration trends were recognized as follows: (1) clinoptilolite-mordenite-analcime-albite and (2) clinoptilolite-mordenite-albite. Analcime formed only under conditions of higher pH and high Na concentration above 150°C.

SUMMARY AND EXISTING PROBLEMS

Numerous studies have shown that zeolites have formed from alkaline hydrothermal solutions over a wide temperature range and under low-pressure conditions; however, much remains to be learned about the chemical environment and other conditions of their formation.

Hydrothermal zeolites appear to be concentrated in volcanic arcs of orogenic belts, especially in active geothermal areas in the western circum-Pacific archipelagos, the Cordilleran region, and the Alpine-Himalayan belt. Other important provinces of hydrothermal zeolites are oceanic ridges such as Iceland, hot spots such as Reunion Island, and intraplate valleys such as the Eastern African Rift Valley. However, little information is available on the occurrence of hydrothermal zeolites formed in ancient times, both in above regions and in others.

A particularly difficult problem remaining in this area of study is discrimination whether zeolites in active geothermal areas were formed by localized hydrothermal alteration or by some other more regional type of alteration such as caldera-type zeolitization. For instance, laumontite and wairakite in the Sanzugawa caldera are thought to have formed by both hydrothermal and caldera-type mechanisms, as described above. For discrimination of the origin of such zeolites, it is important to study in detail the mode of occurrence, the crystal size, and the relation between of zeolite distribution and stratigraphy and geologic structure.

Although some zeolite minerals in amygdaloidal basaltic rocks have been shown to have formed from groundwater or in burial diagenetic systems and others have been shown to have formed in hydrothermal systems, the genesis of the vast majority of these occurrences, including classic localities in Iceland, Ireland, Bohemia, Italy, Bulgaria, India, Nova Scotia, New Jersey, Colorado, Oregon, and Washington, is not well studied or accepted. Their genesis has long been assumed to be hydrothermal, although specific evidence for such an origin has often been lacking. Much more work is needed on such occurrences to establish the origin of these minerals.

Inasmuch as the physico-chemical conditions of the alteration zone of epithermal mineral deposits formed under about 300°C are similar to those found in areas of hydrothermal zeolitization, an effort should be made in all future investigations of such deposits to document the presence and paragenesis of zeolite minerals.

Few reliable data exist on homogenization temperatures of fluid inclusions, on stable isotope compositions, and K-Ar ages for zeolites. All of these data are very important because the conditions and timing of zeolitization may be different from those of the host rock, as they are for zeolites of other origins (see the chapter on *Zeolites in Burial Diagenesis and Low-grade Metamorphic Rocks* by Utada, this volume).

REFERENCES

Armstrong AK, Renault JR, Oscarson RL (1995) Comparison of hydrothermal alteration of Carboniferous carbonate and siliclastic rocks in the Valles caldera with outcrops from the Socorro caldera, New Mexico. J Vol Geotherm Res 67:207-220.

Averyev VV, Noboko SI, Pyp BI (1961) Contemporary hydrothermal metamorphism in regions of active volcanism. Doklady Acad Nauk SSSR 137:239-242

Bargar KH, Keith TEC (1995) Calcium zeolites in rhyolitic drill cores from Yellowstone National Park, Wyoming. *In* Natural Zeolites '93: Occurrence, Properties, Use. Ming DW, Mumpton FA (eds) Int'l Comm Natural Zeolites, Brockport, New York, p 69-86

Browne PRL (1978) Hydrothermal alteration active geothermal fields. Ann Rev Earth Planet Sci 6:229-250

Browne PRL, Courtney SF, Wood CP (1989) Formation rates of calc-silicate minerals deposited inside drill hole casing, Ngatamaeki geothermal field, New Zealand. Am Mineral 74:759-763

Browne PRL, Ellis AJ (1970) The Ohaki-Broadlands hydrothermal area, New Zealand: Mineralogy and related geochemistry. Am J Sci 269:97-131

Cathelineau M, Oliver R, Nieva D, Garfias A (1985) Mineralogy and distribution of hydrothermal mineral zones in Las Azufres (Mexico) geothermal field. Geothermics 14:49 57

Cavarretta G, Gianelli G, Puxeddu M (1980) Hydrothermal metamorphism in the Larderello geothermal field. Geothermics 9:297-314

Cole DR, Ravinsky LI (1984) Hydrothermal alteration zoning in the Beowave geothermal system, Eureka and Lander Counties, Nevada. Econ Geol 79:759- 767

Elders WA, Hoagland JR, McDowell SD, Cobo JM (1979) Hydrothermal mineral zones in the geothermal reservoir of Cerro Prieto. Geothermics 8:201-209

Elders WA, Hoagland JR, Williams AE (1981) Distribution of hydrothermal mineral zones in the Cerro Prieto geothermal field of Baja California, Mexico. Geothermics 10:245-253

Ferry JM (1985) Hydrothermal alteration of Tertiary igneous rocks from the Isle of Skye, northwest Scotland. Contrib Mineral Petrol 91:283-304

Franco RR (1952) Zeolitas dos Basaltos do Brasil Meridional (Genese e Paragenese) Univ Sao Paul, Fac Filos, 150, Mineral 10:1-53

Gianelli G, Teklemariam M (1993) Water-rock interaction processes in the Aluto-Langano geothermal field (Ethiopia). J Volc Geotherm Res 56:429-445

Grill EV, Chase RL, MacDonald RD, Murray JW (1981) A hydrothermal deposits from the Explorer Ridge in the northeast Pacific ocean. Earth Planet Sci Lett 52:142-150

Hayashi M (1973) Hydrothermal alteration in the Otake geothermal area, Kyushu. J Japan Geotherm Energy Assoc 10:9-46

Hayashi M, Taguchi S (1987) Yunotai geothermal field in Aso caldera, middle Kyushu. *In* Gold deposits and geothermal fields in Kyushu. Soc Mining Geol Japan, Guidebook 2:47-50

Henneberger RC, Browne PRL (1988) Hydrothermal alteration and evolution of the Ohakuri hydrothermal system, Taupo volcanic zone, New Zealand. J Vol Geotherm Res 34:211-231

Hirano T (1986) Hydrothermal alteration of volcanic rocks in the Hakone and northern Izu geothermal areas. Bull Hot Spring Res Inst Kanagawa Pref 17:73-166

Hoffert M, Person A, Courtois C, Karpoft AM, Trauth D (1980) Sedimentology, mineralogy and geochemistry of hydrothermal deposits from Holes 424A, 424B and 424C (Galapagos Spreading Center). Initial Reports DSDP 54:339-376

Honda S, Muffler LJP (1970) Hydrothermal alteration in core from research drill hole Y-1, Upper Geyser Basin, Yellowstone National Park, Wyoming. Am Mineral 55:1714-1737

Honda S, Matsueda H (1983) Wairakite at the Onuma geothermal field, Hachimantai district, northeastern Japan. Rept Undergr Res Lab Fac Mining, Akita Univ 48:1-10 (in Japanese)

Honnorez J, Von Herzen RP, Shipboard Scientific Party (1981) Hydrothermal mounds and young oceanic crust of the Galapagos Preliminary Deep Sea Drilling results, Leg 70. Geol Soc Am Bull 92:457-472

Inoue A, Utada M, Shimizu M (1999) Mineral-fluid interaction in the Sumikawa geothermal system, northeast Japan. Resource Geol Spec Issue 20:79-98

Jolly WT, Smith RE (1972) Degradation and metamorphic differentiation of the Keweenawan tholeiitic lavas of northern Michigan, USA. J Petrol 13:273-309

Kamitani M, Nakagawa S, Nishimura S, Sumi K (1978) Geological investigation of hydrothermal alteration haloes in Ibusuki geothermal field, Kagoshima Prefecture. Rept Geol Surv Japan 259:537-577 (in Japanese)

Keith EB, Melvin HB (1981) Hydrothermal alteration in research drill hole Y-2, Lower Geyser Basin, Yellowstone National Park, Wyoming. Am Mineral 66:473-490

Keith TEC, Staples LW (1985) Zeolites in Eocene basaltic pillow lavas of the Siletz River Volcanics, Central Coast Range, Oregon. Clay & Clay Minerals 33:135-144

Kimbara K (1983) Hydrothermal rock alteration and geothermal systems in the eastern Hachimantai geothermal area, Iwate Prefecture, northern Japan. J Japan Assoc Min Petr Econ Geol 78:479-490

Kimbara K, Ohkubo T (1978) Hydrothermal altered rocks found in an exploration bore-hole (No. SA-1), Satsunan Geothermal area, Japan. J Japan Assoc Min Petr Econ Geol 73:125-136 (in Japanese)

Kimbara K, Ohkubo T, Sumi K, Chiba Y (1982) Hydrothermal rock alteration of the Tamagawa welded tuff (Part 2). The upper stream areas of Kakkondagawa and Tamagawa Rivers, Iwate and Akita Prefectures-. J Japan Assoc Min Petr Econ Geol 77:86-93

Koshiya S, Okami K, Hayasaka Y, Uzawa M, Kikuchi Y, Doi N (1994) On the hydrothermal mineral veins developed in the Takinoue geothermal area, northeast Honshu, Japan. J Geotherm Res Soc Japan 16:1-24 (in Japanese)

Kovic P, Krosl-Kuscer N (1986) Hydrothermal zeolite occurrence from the Smrekovec Mt. Arca, Slovenia, Yugoslavia: New Development. *In* Zeolite Science. Murakami Y, Iijima A, Vard JW (eds), Kodansha-Elsevier, Tokyo, p 87-92

Kristmannsdottir H, Tomasson J (1978) Zeolite zones in geothermal area in Iceland. *In* Natural Zeolites, Occurrence, Properties, Use, p 277-284

Kubota Y (1979) Hydrothermal rock alteration in the northern Hachmantai geothermal field. J Japan Geotherm Energy Assoc 16:195-211 (in Japanese)

Kusakabe H (1982) An interpretation of zeolitic zoning around Kuroko ore deposits on the basis of hydrothermal experiments. Mining Geol 32:435-442

Liou JG (1970) Synthesis of stability relations of wairakite, $CaAl_2Si_4O_{12}{\cdot}2H_2O$. Contrib Mineral Petrol 27:259-282

Liou JG (1971a) P-T stabilities of laumontite, wairakite, lawsonite, and related minerals in the system $CaAl_2Si_2O_8$-SiO_2-H_2O. J Petrol 12:379-411

Liou JG (1971b) Synthesis and stability relations of prehnite, $Ca_2Al_2Si_3O_{10}(OH)_2$. Am Mineral 56:507-531

Mariner RH, Presser TS (1995) Conditions of stilbite and chabazite formation in hot springs of the Idaho batholith, central Idaho. *In* Natural Zeolites '93, p 87-97

McCulloh TH, Frizzell VA Jr, Stewart RJ, Barnes I (1981) Precipitation of laumontite with quartz, thenardite, and gypsum at Sespe hot springs, western Transverse Ranges, California. Clay & Clay Minerals 29:353-364

Meixiang Z, Wei T (1987) Surface hydrothermal minerals and their distribution in the Tengchong geothermal area, China. Geothermics 16:181-195

Meyer C, Hemley JJ (1967) Wall rock alteration. *In* Geochemistry of Hydrothermal Ore Deposits. Holt, Rinehart & Winston, p 166-235

Murata KJ, Milton LL, Roisenberg A (1987) Distribution of zeolites in lavas of southeastern Parana Basin, State of Rio Grande Do Sul, Brazil. J Geol 95:455 467

Nashar B, Davies M (1960) Secondary minerals of the Tertiary basalts, Barrington, New South Wales. Mineral Mag 32:480-491

Nielson DL, Stiger SG (1996) Drilling and evaluation of Ascension #1, a geothermal exploration well on Ascension Island, south Atlantic Ocean. Geothermics 25:543-560

Noboko SI (1970) Facies of hydrothermally altered rocks of Kamchatka-Kurile volcanic arc. Pacific Geol 2:23-27

Noboko SI (1976) The origin of hydrothermal solutions and related propylitization and argillization in the areas of tectonomagmatic activity. *In* Proc Int'l Conf on Water-Rock Interaction, Prague, 1974, p 184-195

Ohmoto H, Rye RO (1974) Hydrogen and oxygen isotope composition of fluid inclusions in the Kuroko deposits, Japan. Econ Geol 69:947-953

Pe-Piper G, Tsolis-Katagas P (1991) K-rich mordenite from late Miocene rhyolitic tuffs, Island of Samos, Greece. Clay & Clay Minerals 39:239-247

Rancon JPh (1985) Hydrothermal history of Piton des Neiges volcano (Reunion Island, Indian Ocean). J Vol Geotherm Res 26:297-315

Reyes AG (1990) Petrology of Philippine geothermal systems and the application of alteration mineralogy to their assessment. J Vol Geotherm Res 43:279-309

Reyes AG, Giggenbach WF, Saleras JM, Salonga ND, Vergara MC (1993) Petrology and geochemistry of Alto Peat, a vapor-cored hydrothermal system, Leyte Province, Philippines. Geothermics 22:479-519

Robinson BW (1974) The origin of mineralization at the Tui Mine, Te Aroha, New Zealand. Econ Geol 69:910-925

Sakai H, Matsubaya O (1974) Isotopic geochemistry of the thermal waters of Japan and its bearing on the Kuroko ore solutions. Econ Geol 69:974-991

Schmidt STh (1990) Alteration under conditions of burial metamorphism in the North Shore Volcanic Group, Minnesota—mineralogical and geochemical zonation. Heidelper Geowiss Abhand 41:1-309

Seki Y (1969) Garronite from the Tanzawa mountains, central Japan. J Japan Assoc Min Petr Econ Geol 61:241-249

Seki Y, Onuki H, Okumura K, Takashima I (1969) Zeolite distribution in the Katayama geothermal area, Onikobe, Japan. Japan J Geol Geogr 15:63-79

Seki Y, Oki Y, Onuki H, Odaka S (1971) Metamorphism and vein minerals of north Tanzawa mountains, central Japan. J Japan Assoc Min Petr Econ Geol 66:1-21

Shanker R, Thussu JL, Prasad JM (1987) Geothermal Studies at Tattapani hot spring area, Sarguja district, central India. Geothermics 16:61-76

Sharp WN (1970) Extensive zeolitization associated with hot springs in central Colorado. US Geol Surv Prof Paper 700-B, B14-B20

Shimazu M, Yajima J (1973) Epidote and wairakite in drill cores at the Hachmantai geothermal area, northeastern Japan. J Japan Asoc Min Petr Econ Geol 68:363-371

Steiner A (1953) Hydrothermal rock alteration at Wairakei, New Zealand. Econ Geol 48:1-13

Steiner A (1958) Occurrence of wairakite at The Geysers, California. Am Mineral 43:781

Steiner A (1968) Clay minerals in hydrothermally altered rocks at Wairakei, New Zealand. Clay & Clay Minerals16:193 213

Steiner A (1977) The Wairakei geothermal area, North Island, New Zealand: Its subsurface geology and hydrothermal rock alteration. New Zealand Geol Surv Bull 90:1- 133

Stoiber RE, Davidson ES (1959) Amygdule mineral zoning in the Portage Lake Lava Series, Michigan copper district. Econ Geol 54:1250-1277, 1444-1460

Sturchio NC, Keith TEC, Muehlenbachs K (1990) Oxygen and carbon isotope ratios of hydrothermal minerals from Yellowstone drill cores. J Vol Geotherm Res 40:23 37

Takashima I, Kimbara K, Sumi K (1985) Rock alteration and hydrothermal systems in the Hohi geothermal area, Kyushu, Japan. Rept Geol Surv Japan No. 254:185-241 (in Japanese)

Taylor HP, Jr (1979) Oxygen and hydrogen isotope relationships in hydrothermal mineral deposits. *In* Geochemistry of Hydrothermal Ore Deposits, 2nd edn. Barnes HL (ed) Wiley, New York, p 236-277

Teklemariam M, Battaglia S, Gianelli G, Ruggieri G (1996) Hydrothermal alteration in the Aluto-Langano geothermal field, Ethiopia. Geothermics 25:679- 702

Utada M (1980) Hydrothermal alterations related to igneous activity in Cretaceous and Neogene formations of Japan. Mining Geol, Spec Issue 8:67-83

Utada M (1988) Hydrothermal alteration envelope relating to Kuroko-type mineralization: A review. Mining Geol, Spec Issue 12:79-92

Utada M, Shimizu M (1990) Occurrence, distribution and genesis of zeolites in the Izu Peninsula, central Japan. Clay Sci 30:11-18 (in Japanese)

Utada M, Shimizu M, Ito T, Inoue A (1999) Alteration of caldera-forming rocks related to the Sanzugawa volcanotectonic depression, northeast Honshu, Japan with special reference to "caldera-type zeolitization." Resour Geol, Spec Issue 20:127-138

Walker GPL (1951) The amygdale minerals in the Tertiary lavas of Ireland. 1. The distribution of chabazite habits and zeolites in Garron plateau area, County Antrim. Mineral Mag 29:773-791

Walker GPL (1959) The amygdale minerals in the Tertiary lavas of Ireland. 2. The distribution of gmelinite. Mineral Mag 32:202-217

Walker GPL (1960a) The amygdale minerals in the Tertiary lavas of Ireland. 3. Regional distribution. Mineral Mag 32:504-527

Walker GPL (1960b) Zeolite zones and dike distribution in relation to the structure of the basalts of eastern Iceland. J Geol 68:515-528

Wirsching U (1981) Experiments on the hydrothermal formation of calcium zeolites. Clay & Clay Minerals 29:171-183

Yajima S, Nakamura T (1971) New occurrence of ferrierite. Mineral J 6:343-364

Yamasaki T, Matsumoto Y, Hayashi M (1970) The geology and hydrothermal alterations of Otake geothermal area, Kujo volcanic group, Kyushu, Japan. Geothermics Spec Issue 2:197-207

Yoshida T, Mukaiyama H, Nakagawa S (1983) Geothermal activities in the southwestern area of the Kirishima volcanoes, Kyushu, Japan. J Japan Geotherm Energy Assoc 20:111-128 (in Japanese)

11 Zeolites in Soil Environments

Douglas W. Ming

Mail Code SN2
NASA Johnson Space Center
Houston, Texas 77058

Janis L. Boettinger

Department of Plants, Soils, and Biometeorology
Utah State University
Logan, Utah 84322

INTRODUCTION

"The limited occurrence of these minerals has resulted in a scant knowledge about their properties in the soils." Zelazny and Calhoun (1977)

As pointed out by the above authors, the occurrence of natural zeolites in soils is rare and not well known, and only about 75 papers have been published describing the occurrence of zeolites in soils. These occurrences range from hot, humid soils in India to cold, arid soils in the Dry Valleys of Antarctica. Most of these reports describe residual zeolite phases that have persisted from the parent material during soil formation; however, there are reports of zeolites that have formed in soil environments. A variety of natural zeolites have been found in soils, including analcime, chabazite, clinoptilolite, gismondine, laumontite, mordenite, natrolite, phillipsite, and stilbite.

Over the past 25 years, natural zeolites have been examined for a variety of agricultural and environmental applications because of their unique cation-exchange, adsorption, and molecular sieving properties and their abundance in near-surface, sedimentary deposits. Natural zeolites have been used as soil conditioners, slow-release fertilizers, carriers for insecticides and herbicides, remediation agents in contaminated soils, and dietary supplements in animal nutrition (Ming and Allen, this volume; Pond 1995). These applications can result in direct or indirect incorporation of natural zeolites into soils. If these minerals are to be used effectively in these applications, it will be necessary to understand the long-term stability and effects of zeolites in soils or soil-like systems.

Several reviews have been published on the occurrences and properties of zeolites in soils (Zelazny and Calhoun 1977, Ming and Dixon 1987c, Ming and Dixon 1988, Ming and Mumpton 1989, Boettinger and Graham 1995). This chapter provides an overview and update on the occurrence of zeolites in soils and briefly describes methodology used for identifying and characterizing zeolites in soil environments.

OCCURRENCES IN SOIL

The zeolite group of minerals has been known for over one hundred years to exhibit the property of cation exchange, and when soils were also found to have ion-exchange properties, it was assumed that zeolites were major constituent (e.g. see Burgess and McGeorge 1926, Breazeale 1928). It was not until the introduction of modern X-ray diffraction (XRD) methods that this misconception was invalidated. However, in the past 20 years this group of minerals has indeed been reported in a variety of soils.

Ming and Mumpton (1989) first proposed a classification system for zeolites in soil

1529-6466/00/0045-0011$05.00

environments based on the origin of the zeolite, i.e. whether the zeolite was pedogenic (formed in the soil) or lithogenic (a residual mineral phase) from either volcanic or non-volcanic parent materials. Boettinger and Graham (1995) suggested slight modifications to the classification system, which we refine here. The types of occurrences in soil environments currently include (1) pedogenic zeolites in saline, alkaline soils of non-volcanic parent materials; (2) pedogenic zeolites in saline, alkaline soils of volcanic parent materials; (3) lithogenic zeolites inherited *in situ* from volcanic parent materials; (4) lithogenic zeolites inherited *in situ* from non-volcanic parent materials; (5) lithogenic zeolites from eolian or fluvial deposition; and (6) zeolites in other soil environments.

Most occurrences of zeolites in soils have had some influence from previous volcanic activity, e.g. tuffaceous parent materials, however, some zeolites have either been inherited into or formed in soils without previous influence by volcanic activity. Clinoptilolite is the most abundant zeolite found in soils, but, analcime, chabazite, gismondine, heulandite, laumontite, mordenite, natrolite, phillipsite, and stilbite have also been reported to occur in soils. In the following sections we briefly describe the several types of zeolite occurrences in soils.

Pedogenic zeolites in saline, alkaline soils of non-volcanic parent materials

Occurrences of pedogenic zeolites are almost always restricted to soils with saline, alkaline conditions (i.e. soils with pH > 7 and containing appreciable soluble salts). Unlike most sedimentary zeolites, volcanic parent materials are not necessary for zeolite formation in soils. Analcime appears to form most frequently in these soils, but chabazite, clinoptilolite, mordenite, natrolite, and phillipsite have also been reported to form in soils (Table 1).

Table 1. Pedogenic zeolites in saline, alkaline soils of non-volcanic parent materials.

Zeolite	*Soil locality*	*References*
Analcime	Burundi	Frankart & Herbillon (1970)
	California, USA	Schultz et al. (1964); Baldar & Whittig (1968); El-Nahal & Whittig (1973); Baldar (1968)
	India	Kapoor et al. (1980)
	Former U.S.S.R.	Travnikova et al. (1973)
	Kenya	Renaut (1993)
	Tanzania	Hay (1963); Hay (1976); Hay (1978)
Chabazite	Tanzania	Hay (1970); Hay (1978)
Clinoptilolite	Alberta, Canada	Goh et al. (1986); Spiers et al. (1984)
Mordenite	Chad	Maglione & Tardy (1971)
Natrolite	Kenya	Renaut (1993)
	Tanzania	Hay (1970); Hay (1978)
Phillipsite	Former U.S.S.R.	Travnikova et al. (1973)
	Tanzania	Hay (1964); Hay (1970); Hay (1978)

One of the first reported occurrences of zeolites in soils was discovered because of an unusual cation-exchange property of a soil in the San Joaquin Valley of California. The soil had an unusually high exchangeable sodium percentage (ESP) of up to 75%, but rice grown in the soil did not exhibit any sodium toxicity effects due the high ESP (Schultz et al. 1964). It became apparent that the sodium was easily removed during exchangeable-cation measurements in the laboratory but was not easily exchanged in the field. Schultz et al. (1964), Baldar and Whittig (1968), and El-Nahal and Whittig (1973) used XRD analyses to confirm that analcime was responsible for this unusual cation-exchange behavior.

The occurrence of analcime in the San Joaquin Valley was restricted to soils containing Na_2CO_3 with a pH above 9, and the zeolite was concentrated near the soil surface in the fine silt (2 to 5 μm) and coarse clay fractions (0.2 to 2 μm). Baldar and Whittig (1968) were able to effectively destroy the analcime in the soil with 0.5 M HCl treatments and then reprecipitate analcime by titrating the system above pH 9 with NaOH, followed by mild heating to 95°C for 14 days. The occurrence, distribution, and relative ease of synthesizing analcime led Baldar and Whittig (1968) to the conclusion that the zeolite was of pedogenic origin with no apparent influence from volcanic parent materials. Similar occurrences of analcime in soils have been reported in Burundi, India, the former U.S.S.R., Kenya, and Tanzania (see Table 1).

Renaut (1993) found up to 40% analcime and minor amounts of natrolite in exhumed paleosols on the former margin of saline, alkaline Lake Bogoria in the Kenya Rift Valley. The zeolitic sediments are up to 1 meter thick and occur in Late Quaternary fluviolacustrine siltstones, mudstones, and claystones. The sediment has many pedogenic features similar to those found in current soils, including prismatic soil structure and vertical to subvertical rootmarks. Fossilized root mats, calcareous rhizoliths, Fe-Mn concretions, and secondary concentrations of opaline silica further indicate that these sediments were pedogenically altered. The amount of analcime decreases with increasing paleosol depth, and decreases with increasing distance from the lake. The analcime occurred as subhedral and euhedral crystals (0.5 to 2.5 μm) filling former root channels. Renaut (1993) concluded that analcime probably formed from reaction of detrital silicates with Na_2CO_3-rich pore waters, which moved upward in response to evapotranspiration near the land surface. Analcime persists in these paleosols under the present hot and semi-arid climate and may be currently forming along the margins of the lake.

Phillipsite, chabazite, and natrolite have been reported to occur in relatively minor amounts in saline, alkaline soils, which have developed in non-volcanic parent materials (Hay, 1964, 1970, 1978). Their occurrence in these soils is likely due to the saline, alkaline nature of the soil environment and the low intensity of weathering and leaching.

Highly-siliceous zeolites (e.g. clinoptilolite, mordenite) have rarely been reported to *form* in soils. Spiers et al. (1984) reported the occurrence of clinoptilolite in a solodized solonetz (i.e. slightly leached sodic [high ESP] soil) that had developed on loam till in Alberta, Canada. These authors suggested that the weathering of smectites supplied Si into solution necessary for the formation of clinoptilolite. The chemistry of extracted soil solution was 'similar to that of pore waters associated with sedimentary clinoptilolite deposits. Thermodynamic modeling of the soil solution by SOLMNEQ predicted precipitation of clinoptilolite. Clinoptilolite occurred only in the zone of active leaching of Na and salts in the top 40 cm of the soil. No zeolites were detected in the deeper soil horizons. Clinoptilolite was concentrated in the coarse clay (1 to 2 μm) and silt (2 to 20 μm) fractions. Another highly siliceous zeolite, mordenite, has been reported to form in saline, alkaline sediments and soils at the edge of Lake Chad (Maglione and Tardy 1971).

Pedogenic zeolites in saline, alkaline soils of volcanic parent materials

As described by Hay and Sheppard (This volume), zeolites are widespread in geologic deposits as alteration products (i.e. authigenic minerals) of volcanic glass subjected to saline, alkaline conditions. For example, volcanic materials (e.g. glass) near the land surface in closed hydrologic basins (i.e. saline, alkaline lake deposits) will undergo dissolution, releasing Si, Al, and alkali and alkaline earth cations into solution. Precipitation occurs when the solution becomes saturated with respect to the zeolite. The type of zeolite that will form depends on a number of factors (Surdam and Sheppard, 1978; Chipera and Apps, this volume), including (1) cation ratios, (2) pH, (3) Si/Al atomic ratios, (4) activity of silica, (5) activity of water, and (6) salinity. Therefore, it is expected that zeolites will form in soils derived from volcanic sediments under saline, alkaline conditions (Table 2).

Table 2. Pedogenic zeolites in saline, alkaline soils of volcanic parent materials.

Zeolite	*Soil locality*	*References*
Analcime	Tanzania	Hay (1963); Hay (1976); Hay (1978)
Chabazite	Antarctica	Gibson et al. (1983)
	Tanzania	Hay (1970); Hay (1978)
Natrolite	Tanzania	Hay (1970); Hay (1978)
Phillipsite	California, U.S.A.	McFadden et al. (1987)
	Mississippi, U.S.A.	Raybon (1982)
	Tanzania	Hay (1964); Hay (1970); Hay (1978)

Analcime, phillipsite, natrolite, and chabazite are found in the saline, alkaline soils in the vicinity of Olduvai Gorge, Tanzania (Hay 1963 1970 1976 1978). These zeolites formed during the alteration of tuffaceous sediments near the land surface under saline-sodic conditions and they have persisted in a hot, semi-arid environment. Individual zeolite crystals are globular, euhedral, or anhedral, and they range between 0.005 to 0.1 mm in either length or diameter.

Phillipsite has formed in a basaltic rubble subsoil environment and desert pavement of the Cima volcanic field in southeastern California (McFadden et al. 1987). The zeolite formed under saline, alkaline conditions caused by eolian influx of highly alkaline sediments from playas, which interacted with water and the basaltic rubble to form euhedral, lath-shaped phillipsite crystals. The authors suggest that the source of Al and Si necessary for the formation of phillipsite was authigenic phyllosilicates, possibly a mixed-layer illite-smectite or illite-vermiculite that had formed in rubble samples.

Gibson et al. (1983) reported the occurrence of chabazite in a cold, sodic desert soil from Wright Valley in Antarctica. The area was known to have experienced past volcanic activity. These authors suggested that the chabazite must be authigenic because the 5 to 10 μm-sized crystals are euhedral, unabraded, and unfractured, which suggest *in situ* formation (Fig. 1). The zeolite occurs throughout the soil profile, including the permanently frozen zone. The presence of dissolution features and authigenic minerals throughout the profile suggests that chemical weathering may still be occurring, even in the permanently frozen zone.

Figure 1. Scanning electron microscope image of chabazite from the permanently frozen zone of a cold desert soil from the Wright Valley in Antarctica (Gibson et al. 1983).

Lithogenic zeolites inherited *in situ* from volcanic parent materials

Large deposits of authigenic zeolites formed as alteration products of volcanic glass in open hydrologic systems are very common around the world (Sheppard and Hay, this volume). Ground water percolating through columns of volcanic ash or glass becomes increasingly enriched in dissolved Si and Al as well as in various alkali and alkaline earth cations. As the ion activity products of zeolites are exceeded, zeolites (e.g. clinoptilolite) will precipitate from solution in areas where volcanic glass has dissolved. When these zeolite-rich sedimentary deposits are later exposed at the land surface, weathering and soil formation occur. Hence, the zeolites are incorporated into the soil as residual mineral phases (i.e. lithogenic). Because these types of zeolite-bearing deposits are very common and extensive, the majority of the reported occurrences of zeolites in soils result from lithogenic zeolites inherited *in situ* during soil formation. Clinoptilolite is the most abundant sedimentary zeolite; therefore, it is expected that clinoptilolite is the most abundant zeolite reported in soils (Table 3). However, there have also been reports of soils containing analcime, chabazite, gismondine, heulandite, mordenite, and stilbite inherited *in situ* from zeolite-rich, volcanic parent materials.

Clinoptilolite is widespread in calcareous soils formed on tuffaceous sediments of the Catahoula Formation in South Texas (Ming 1985, Ming and Dixon 1986). About 2 to 5 wt % clinoptilolite was found in the A and B horizons of a calcareous Mollisol (soil with thick A horizon and high-base saturation). Clinoptilolite concentrations progressively increased with soil depth to ~20 wt % in the C horizons. The zeolite was concentrated in the silt fractions and occurred as clusters of closely spaced laths that were 1 to 2 μm in thickness and 5 to 10 μm in length (Fig. 2). The smooth, angular, euhedral shape of individual crystals indicated their relative stability in the arid soil environment. Slightly pitted and fractured crystals in the A and B horizons suggested minimal weathering of the zeolite (Fig. 2). Clinoptilolite separated from the soil was dominantly Ca exchanged, which can be attributed to the calcareous nature of these soils.

A similar occurrence of clinoptilolite inherited *in situ* from volcanic-rich parent materials was also reported in West Texas (Jacob and Allen 1990). The zeolite occurred in sand, silt, and clay fractions of several soils along a topographic sequence in an arid to

Table 3. Lithogenic zeolites inherited *in situ* from volcanic parent materials.

Zeolite	*Soil locality*	*References*
Analcime	Italy	Portegies Zwart et al. (1975); Baroccio (1962)
	Japan	Morita et al. (1985)
	Washington, U.S.A.	Bockheim & Ballard (1975)
Chabazite	Italy	Portegies Zwart et al. (1975)
Clinoptilolite	Alabama, U.S.A.	Karathanasis (1982)
	Alaska, U.S.A.	Ping et al. (1988)
	Bulgaria	Atanassov et al. (1982); Atanassov & Do Vang Bang (1984)
	Hungary	Nemecz & Janossy (1988)
	Japan	Morita et al. (1985); Kaneko et al. (1971)
	Lebanon	Darwish et al. (1988)
	Mississippi, U.S.A.	Raybon (1982)
	Oregon, U.S.A.	Paeth et al. (1971)
	Romania	Asvadurov et al. (1978)
	Former U.S.S.R.	Gorbunov & Bobrovitsky (1973); Travnikova et al. (1973)
	Texas (southern), U.S.A.	Ming & Dixon (1986); Ming (1985)
	Texas (western), U.S.A.	Jacob & Allen (1990)
	Utah, U.S.A.	Southard & Kolesar (1978)
Gismondine	Arkansas, U.S.A.	Reynolds & Bailey (1990)
	Italy	Portegies Zwart et al. (1975)
	Japan	Morita et al. (1985)
Heulandite	India	Bhattacharyya et al. (1993); Bhattacharyya et al. (1999)
Mordenite	Hungary	Nemecz & Janossy (1988)
	Japan	Kaneko et al. (1971)
	New Zealand	Kirkman (1976)
Stilbite	California, U.S.A.	Reid et al. (1988)

semi-arid environment. The abundance and crystallinity of clinoptilolite in the fine clay fraction (<0.2 μm) varied with soil depth and landscape position. In the well-drained Aridisols (arid soils with subsoil development) of summit and backslope positions, fine clay-sized clinoptilolite was less abundant and less crystalline (i.e. broader XRD peaks) in the A horizon than in deeper horizons. There was no clinoptilolite in the fine-clay fraction of the upper 90 cm of the soil in the Entisol (weakly developed soil) in the concave footslope position, which suggested a higher degree of weathering in this wetter site. It

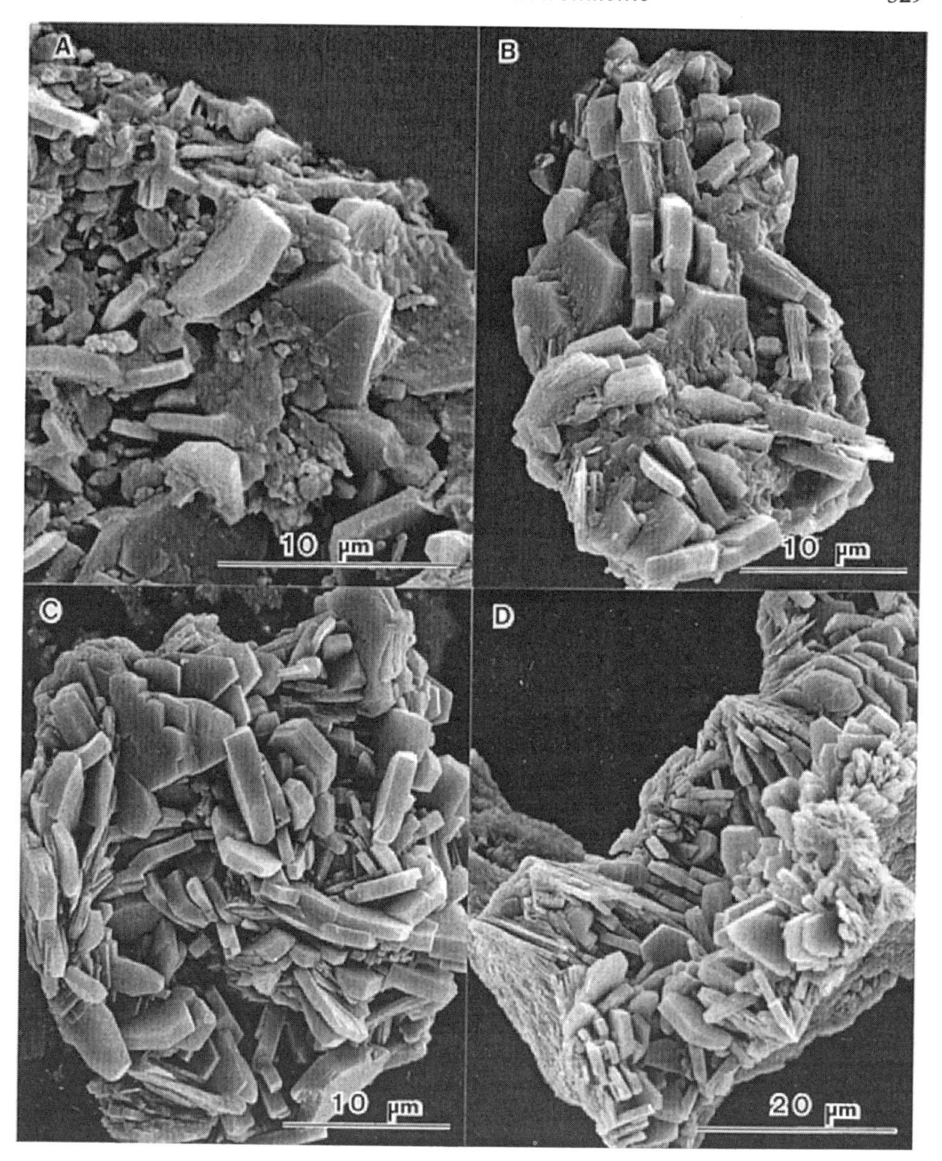

Figure 2. Scanning electron microscope images of clusters of clinoptilolite crystals from a Mollisol located in south Texas (Ming and Dixon 1986): (A) A horizon; (B) Bk1 horizon; (C) and (D) CBk1 horizons.

appears that the finer particle sizes of clinoptilolite are more susceptible to weathering in these upper soil horizons.

Other occurrences of soils containing lithogenic clinoptilolite inherited from volcanic parent materials, are listed in Table 3. The soils exhibit a wide range of weathering and pedogenesis, and they occur in a variety of climates. In general, lithogenic clinoptilolite in soils seems most persistent in larger particle sizes, in deeper horizons, in younger soils, on more recent geomorphic surfaces, in colder climates, and in drier environments.

In several occurrences of zeolites in soils, investigators did not distinguish between isostructural heulandite and clinoptilolite. In most of these cases, it is likely that clinoptilolite is the zeolite occurring in these soils, primarily because they are more likely to be inherited from common clinoptilolite-rich, sedimentary deposits (Ming and Dixon 1986). However, a very interesting occurrence of heulandite has been observed in Alfisols (soils with subsoil clay accumulation and high-base saturation), Vertisols (clay-rich soils capable of significant shrink-swell), and Inceptisols (in this case, intergrades to Vertisols) in India (Bhattacharyya et al. 1993 1999). These soils have developed upon the Deccan flood basalt of the Western Ghats in the state of Maharashtra. Heulandite is widespread in the vesicular or amygdular top of compound flows, although the zeolite is also common at the base of the highly amygdular part of compound flows (Phadke and Kshirsagar 1981, Sabale and Vishwakarma 1996). The zeolite has been inherited from the basalt during soil formation. Heulandite was found in almost all size fractions and composed up to 50 wt % of the weathered basalt C horizons of Vertisols and Inceptisols in the semi-arid zone (500-1000 mm annual rainfall) of the plateau (Bhattacharyya et al. 1993 1999). In high-altitude (~1100 meters) Alfisols of the humid region, partially-weathered heulandite grains were restricted to 5 to 10 wt % of the sand fraction (>50 μm). Bhattacharyya et al. (1999) attributed the persistence of the Alfisols in the tropical humid climate (>5000 mm annual rainfall) to the occurrence of the zeolite in these soils. The authors suggest that heulandite dissolution contributes base cations and Si into solution which facilitates the persistence of smectite in these acidic soils and prevents advanced weathering to kaolinitic or oxidic soils.

Two Ca-rich zeolites, chabazite and gismondine, appear to be stable in calcareous Mollisols weathered from alluvial materials interbedded with volcanic ash in South-Central Italy (Portegies Zwart et al. 1975). Analcime was also found to occur in these soils, however, the authors suggested that it was unstable in the soil environment because the soil solution was undersaturated with respect to analcime. Mordenite, which often occurs with clinoptilolite in sedimentary deposits, has been reported to occur in the clay fraction (<2 μm) of an acid (pH = 5.7) sandy loam soil in New Zealand (Kirkman 1976). Rarely are zeolites reported to persist in acid soils; however, an unusual occurrence of analcime has been observed in extremely acidic (pH around 2), "hydrothermal" soils on Mt. Baker in Washington (Bockheim and Ballard 1975). The zeolite was found to be concentrated in the silt and clay fractions (< 50 μm).

A rare occurrence of stilbite has been reported in the calcareous weathered rock C horizon of an arid Entisol near Red Rock Canyon in southern California (Reid et al. 1988). The occurrence of stilbite in the hydrothermally-altered basalt suggests that the zeolite was inherited *in situ* during soil formation.

Lithogenic zeolites inherited *in situ* from non-volcanic parent materials

Boettinger and Graham (1995) suggested the addition of this new category of zeolite occurrence in soils to the original classification system of Ming and Mumpton (1989). Previous reports of soils containing clinoptilolite/heulandite inherited from non-volcanic parent materials and recent reports of laumontite in California soils, also inherited from non-volcanic parent materials, warranted their separation into a distinct category of occurrence (Table 4).

As mentioned earlier, most reported occurrences of clinoptilolite in soils have been inherited from clinoptilolite-rich, volcanic parent materials. However it is often difficult to determine the origin of the zeolite, or there appears to be no volcanic influence on the soil parent material. For example, Brown et al. (1969) observed clinoptilolite and/or heulandite in soils derived from calcareous, fine-grained siliceous rock of the Upper Greensand Formation in England. There appeared to be no influence from volcanic materials. The

Table 4. Lithogenic zeolites inherited *in situ* from non-volcanic parent materials.

Zeolite	*Soil locality*	*References*
Clinoptilolite	England	Talibudeen & Weir (1972); Brown et al. (1969)
	Sicily	Bellanca et al. (1980)
Heulandite	Denmark	Nørnberg & Dalsgaard (1985); Nørnberg et al. (1985); Nørnberg (1990)
Laumontite	California, U.S.A.	Graham et al. (1988); Taylor et al. (1990); Boettinger (1988); Boettinger & Southard (1995); Liu (1988)*

*Not clear if parent material was volcanic or non-volcanic

authors suggested that the zeolite, which occurred mainly in the 1-20 μm size fraction, formed by diagenetic processes after biogenic opal formation and then was inherited by these soils. The high exchangeable K^+ in these soils has been attributed in part to the presence of clinoptilolite and/or heulandite (Talibudeen and Weir 1972). A similar occurrence of clinoptilolite has been observed in calcareous soils in Sicily (Bellanca et al. 1980).

Nørnberg et al. (1985) originally reported the occurrence of clinoptilolite in the A and C horizons of three Mollisols that had developed over Danian bryozoan chalk in northeastern Denmark. Later, Nørnberg (1990) concluded that the zeolite in these soils was heulandite and not clinoptilolite, based upon a Si/Al ratio of 2.88 for the soil zeolite. Heulandite in these soils was K-exchanged, which likely increased its thermal stability (to 600°C, similar to K-exchanged clinoptilolite). Nørnberg (1990) suggested that the heulandite formed under burial diagenetic conditions during chalk formation and was then inherited by the soil, where it became K-exchanged over time. The presence of heulandite in the A and C horizons of some soils, but not the B horizons, was not well understood. However, Nørnberg et al. (1985) suggested that the zeolite may have been added to the A horizon by sediments eroding from calcareous, chalk-derived soils at higher positions in the landscape. Heulandite occurred primarily in the silt (2 to 63 μm) and coarse clay (0.2 to 2 μm) fractions of these soils.

Laumontite has rarely been reported to occur in soils; however, the zeolite has been identified in several soils of southern California (Graham et al. 1988, Boettinger 1988, Taylor et al. 1990, Boettinger and Southard 1995). Taylor et al. (1990) identified laumontite throughout three Entisols in the San Gabriel Mountains of California. These three soils have formed on steep slopes in colluvium of anorthosite, granodiorite, and arkosic sandstone, respectively. Parent materials have been previously hydrothermally-altered by dilute, low-temperature solutions (McCulloh and Stewart 1982) and the laumontite that formed was inherited by the overlying soils. Soil pH ranged from 5.0 in the A horizon of the soil developed on arkosic sandstone to 7.2 in the Cr horizon of the anorthosite-derived soil. Laumontite was found in sand, silt, and clay fractions, however, the particle size of the zeolite depended on its grain size in the parent rock. Comparisons of laumontite grain morphology showed rounded, dissolution-pitted, and crusted weathering products on grains in the A horizons, whereas particles in the Cr horizons tended to be more angular, with fresher crystal faces (Fig. 3).

Boettinger and Southard (1995) identified laumontite in the silt and sand fractions of deep, calcareous B horizons in an Aridisol in the western Mojave Desert of California. The soil formed on weathered and coherent granitic saprolite, previously reported to have been subject to hydrothermal alteration (Dibblee 1963). Laumontite was absent from a soil

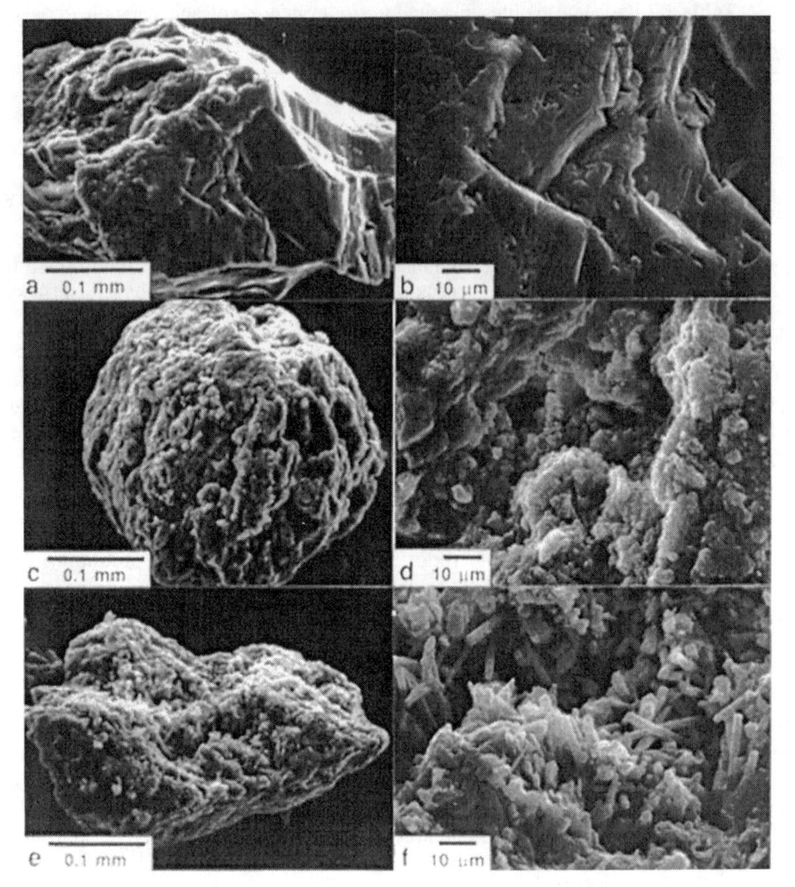

Figure 3. Scanning electron microscope images of sand-sized laumontite grains from an Entisol derived from anorthositic parent materials in the San Gabriel Mountains of southern California (Taylor et al. 1990). (a) angular grain from the Cr horizon; (b) up close surface image of grain in (a); (c) rounded grain from the A horizon, (d) up close surface image of grain in (c); (e) subangular grain from the A horizon; and (f) halloysite tubes on surface of grain in (e).

sampled 50 cm away (Boettinger 1988), suggesting that the zeolite was probably inherited from hydrothermally-altered plagioclase of the parent rock. Laumontite was also identified in two acidic soils (pH 4.8) derived from granitic rock that received additions of volcanic glass in Sequoia National Park in California's southern Sierra Nevada (Liu 1988). It is not clear whether laumontite was inherited as a hydrothermally altered phase in the granitic rock or as a diagenetic product of the volcanic ash.

Lithogenic zeolites from eolian or alluvial deposition

Zeolites have low specific gravities (1.9 to 2.2) and may therefore be removed from erosional surfaces rich in zeolites (e.g. sedimentary deposits) and carried considerable distances by wind or water before being deposited at a soil surface. In this type of soil occurrence, greater quantities of zeolites may be expected in the topsoil or A horizons and, in many cases, zeolites may be completely absent in the lower horizons (e.g. C horizons). Eolian and alluvial additions of chabazite, clinoptilolite, phillipsite, and stilbite have been

Table 5. Lithogenic zeolites from eolian or fluvial deposition.

Zeolite	*Soil locality*	*References*
Chabazite	Poland	Brogowski et al. (1980); Brogowski et al. (1983)
Clinoptilolite	Former U.S.S.R.	Gorbunov & Bobrovitsky (1973)
	Texas (southern), U.S.A.	Ming (1985)
	Utah, U.S.A.	Graham & Southard (1983)
Phillipsite	Poland	Brogowski et al. (1980); Brogowski et al. (1983)
Stilbite	Utah, U.S.A.	Jalalian & Southard (1986)

identified in soils (Table 5).

A Vertisol and a Mollisol derived from the Norwood Tuff Formation in northern Utah were reported to contain substantial quantities of heulandite or clinoptilolite (probably clinoptilolite) in the surface soil, progressively decreasing in quantity with increasing soil depth (Graham and Southard 1983). The zeolite was absent or nearly absent in the C horizons of these soils. Other unstable minerals, such as mica and amphiboles, were also found in higher quantities in the upper compared with the lower horizons. Because these unstable minerals were concentrated in the topsoil where weathering should be more intense, Graham and Southard (1983) suggested that the zeolite was introduced into these soils by eolian additions. Zeolites and the other weatherable minerals mostly occurred in the coarse clay and silt fractions. The authors suggest that the some of the eolian source materials may have originated from as far away as the basalt-rich Snake River Plain in Idaho (250 km to the northwest), or the source material for clinoptilolite may have been the nearby clinoptilolite-rich tuffs of the Salt Lake Formation.

In a similar occurrence, Jalalian and Southard (1986) identified stilbite in the medium-silt fraction (5-20 μm) A horizons of a Mollisol and an Alfisol in northern Utah. Stilbite, chlorite, amphiboles, and Ca-feldspar occurred together in the A horizon but were not found in the subsurface horizons. An abrupt change in mineralogy and texture suggested that the upper horizons resulted from recent additions of eolian material. The authors suggested that the zeolite-bearing eolian material had been transported from Lake Bonneville sediments and the Snake River Plain, mantling soils already developed in colluvium and residuum.

A recent study by Clausnitzer and Singer (1999) identified analcime, chabazite, offretite, and phillipsite, along with other mineral groups (e.g. feldspars, phyllosilicates, Fe and Ti oxides, pyroxenes, silica minerals, etc.), in respirable dust particles produced by farming activities from a soil located near Davis, California. The zeolites were identified in the dust using transmission electron microscopy (TEM) selected-area electron-diffraction (SAED) and energy dispersive X-ray spectroscopy (EDXS); however, no zeolites were detected in the source soil by XRD analysis. The original source of the zeolites in the soil is not known; however, it is likely that these particles may have been deposited at the soil surface as eolian additions from a distant source. This study suggests that zeolites may be widespread in soils, but in quantities below the detection limits of most standard soil mineralogical analytical procedures (e.g. XRD analysis).

Zeolites from zeolite-rich tuffs or soils may be eroded by water and transported hundreds of km before being deposited as alluvium. Clinoptilolite is commonly found in

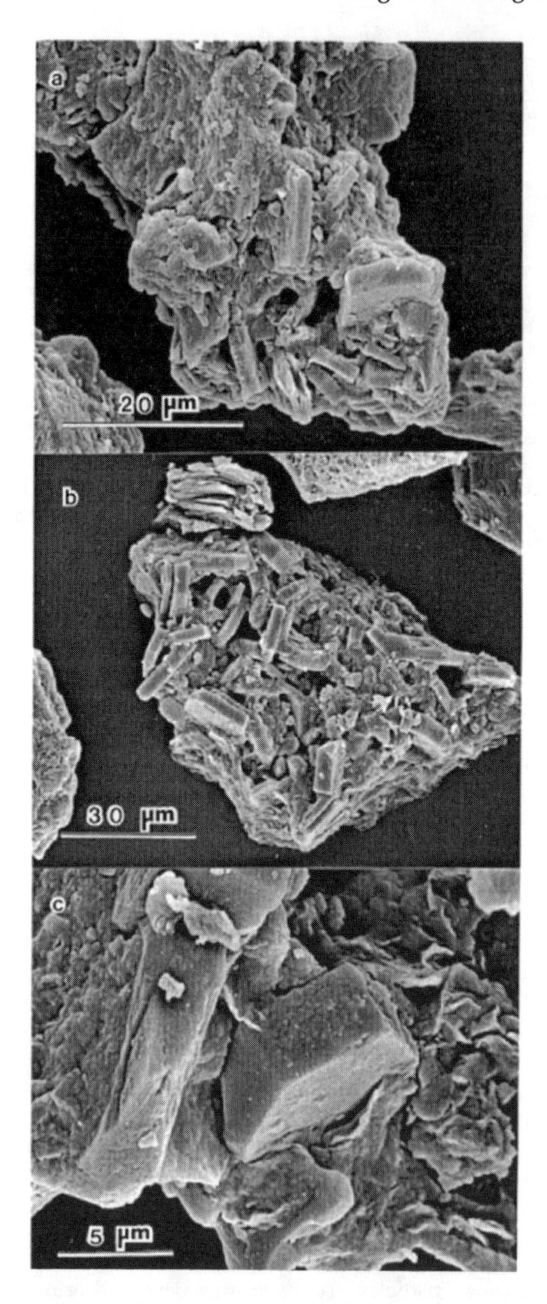

Figure 4. Scanning electron microscope image of a cluster of clinoptilolite crystals from the topsoil of an alluvial soil located along the Rio Grande River, south Texas (Ming 1985).

alluvial soils along the lower 200 to 300 km of the Rio Grande River that separates the United States and Mexico (Ming 1985). Ming (1985) identified clinoptilolite in an Entisol near Brownsville, Texas. The zeolite was found primarily as silt-sized particles and composed less than 5 wt % of the top soil. Morphological characteristics of clinoptilolite were similar to those found in the A and B horizons of soils that had formed on the clinoptilolite-rich Catahoula Formation in South Texas (Fig. 4). It is likely that the source of clinoptilolite in the alluvial soils along the Rio Grande River is the Catahoula and Jackson Formations, which intersect the river about 150 to 200 km upstream from Brownsville. It is also possible that the clinoptilolite may have been transported down the Rio Grande River from the clinoptilolite-rich tuffs in the Trans-Pecos Mountains in western Texas, nearly 1200 km from Brownsville. Interestingly, clinoptilolite was also identified in the muds on the bottom of the river, further indicating that the zeolite had been removed from an erosional surface and transported down river (Ming 1985).

Similar occurrences of clinoptilolite and heulandite are common in alluvial soils along the lower regions of the Rioni River in Russia (Gorbunov and Bobrovitsky 1973). Zeolites in these soils decreased with increasing soil depth and coexisted with kaolinite, "hydromica," chlorite, vermiculite, smectite, and randomly interstratified mica/smectite. The sources of these alluvial materials is thought to be a siliceous zeolitic tuff cut upstream by a tributary of the Rioni River. Brogowski et al. (1980, 1983) suggested the possibility of zeolites occurring in alluvial soils in Poland. Chabazite and phillipsite were tentatively identified on the basis of morphology by scanning electron microscopy (SEM); however, definitive XRD analyses were not reported. Brogowski et al. (1983) suggested that the zeolites were transported by the Vistula River from areas where these minerals are known to occur (e.g. Podkarpacie).

Zeolites in other soil environments

As with nearly any classification system, there are unusual cases or cases in which not enough information has been provided so that the occurrence can be placed into a specific category; hence, an "other" category was created in the classification system (Table 6). However, nearly all of the zeolite occurrences that we have examined fit into the previous five categories. An exception to this is an occurrence of clinoptilolite in a post-active acid sulfate Ultisol (soil with subsurface accumulation of clay and low-base saturation) in Maryland located on a bluff along the Potomac River (Fanning et al. 1983). Clinoptilolite was identified in both the oxidized and reduced zones. No volcanic ash is known to occur in the parent materials at this site and the origin of clinoptilolite in this soil is a mystery. The other exceptions are occurrences of laumontite in soils considered to be weathering products of plagioclase (Capdecomme 1952, Furbish 1965), where hydrothermal alteration of the parent materials was not mentioned.

Table 6. Zeolite occurrences in other soil environments.

Zeolite	*Soil locality*	*References*
Clinoptilolite	Maryland, U.S.A.	Fanning et al. (1983)
Laumontite	France	Capdecomme (1952)
	North Carolina, U.S.A.	Furbish (1965)

IDENTIFICATION OF ZEOLITES IN SOILS

Soils are very complicated mineralogical and chemical systems. A soil is defined as the unconsolidated mineral or organic matter on the surface of the earth that has been subject to and influenced by genetic and environmental factors of parent material, climate (including water and temperature effects), macro- and microorganisms, and topography, all acting over a period of time. A soil can be quite different from the material from which it was derived. Because of this complex system, it is sometimes difficult to detect and identify small quantities of minerals in soils, particularly zeolites. In the A and B horizons of most soils, the quantity of zeolites rarely exceeds 5 to 10 wt %, and special precautions and procedures may be required to concentrate or separate zeolites in soils for proper identification and characterization.

Separation and quantification techniques

Special precautions should be exercised when preparing a soil sample containing zeolites for subsequent mineralogical analyses. It is not uncommon for soil scientists to use chemical pretreatments (e.g. 1 M NaOAc buffered to pH = 5, 30% H_2O_2, and dithionite-citrate-bicarbonate) to remove cementing agents (e.g. carbonates, organic matter, and iron oxides, respectively) and to enhance dispersion of a sample before particle-size fractionation and subsequent mineralogical analyses. Zeolites react differently when subjected to various pH solutions. For example, zeolites with high Si/Al atomic ratios (e.g. clinoptilolite) can withstand acid treatments to a pH below 2 before structural degradation is detectable; however, zeolites with low Si/Al atomic ratios (e.g. analcime) tend to dissolve during acidic treatments around and below a pH of 5. Ming and Dixon (1987b) found that the above-mentioned chemical pretreatments caused no readily discernible modification to the structure of clinoptilolite found in soils of south Texas. In fact, pretreatments aided in the XRD identification of the zeolite as indicated by increases in peak intensities, probably due to concentrating clinoptilolite by removing calcite and allowing fractionation and separation of the zeolite from other minerals such as smectite. On the other hand, Baldar and Whittig (1968) found that analcime from soils of the San Joaquin Valley of California was destroyed by acid treatments and, hence, they avoided chemical pretreatments designed to enhance disaggregation and dispersion of the soil to eliminate any possible alteration of the analcime. Because pretreatments may have adverse effects on some zeolite minerals, it is possible that some zeolite occurrences in soils may have been missed if they were removed from the sample during chemical pretreatments.

Table 7. Exchangeable cations on zeolitic exchange sites and Si/Al atomic ratios of clinoptilolite separated from a calcareous Mollisol in South Texas (Ming and Dixon 1986).

Horizon	*Soil depth*	*Exchangeable cations* *Na*	*K*	*Mg*	*Ca*	*Si/Al atomic ratio*
	cm	$cmol_c\ kg^{-1}$				
A	0-25	2	7	5	127	5.1
Bk1	25-38	6	3	7	127	4.7
Bk2	38-66	13	2	7	124	4.7
BCk	66-89	29	3	7	123	4.8
CBk1	89-104	39	5	5	120	4.9
CBk2	104-137	40	5	5	121	5.1

In order to detect chemical and mineralogical differences of a zeolite that occurs throughout a soil profile, the separation of the mineral from the soil becomes an important aspect of the study. By separating the zeolite from the soil, important physical and chemical properties of the mineral can be studied without the interference of other minerals. Ming and Dixon (1987b) used the low specific gravity (≈ 2.16) and fine particle-size (2 to 50 μm) of clinoptilolite to separate the zeolite from soils of South Texas using gravity separation. Initially, representative soil samples were pretreated to remove carbonates, Fe-oxides, and organic matter to enhance particle separation of the clinoptilolite. Clinoptilolite was then separated from other silt-sized particles in the <2.28 sp. gr. heavy-liquid separate. Once clinoptilolite is separated from the soil, a variety of properties can be determined, including exchangeable cations and Si/Al atomic ratios (Table 7). In this example, clinoptilolite in a calcareous Mollisol in South Texas was predominantly Ca-exchanged, which can be attributed to the calcareous nature of the soil (Ming and Dixon 1986). The Si/Al atomic ratio of clinoptilolite did not vary significantly throughout the profile including the parent materials, suggesting that the zeolite was inherited in the soil from clinoptilolite-rich, volcanic parent materials.

A variety of methods may be used to quantify zeolites in soils. Quantification of a mineral in the soil aids in determining the distribution and stability of the mineral throughout a pedon. Most zeolites in soils are sufficiently crystalline to enable semi-quantitative estimations by XRD powder techniques. El-Nahal and Whittig (1973) compared the XRD peak heights of soil analcime from the San Joaquin Valley of California with those of a reference analcime, and they found the top 15 cm of the soil to contain approximately 22 wt % analcime. Pretreatments were avoided to prevent damage to the analcime crystals. Ming (1985) developed standard XRD curves for clinoptilolite in soils derived from the Catahoula Formation in southern Texas. Standard mixtures were prepared using a high-grade clinoptilolite from Castle Creek, Idaho, a Na-montmorillonite, and varying amounts of quartz, feldspar, and calcite to simulate the matrices of the sand, silt, and clay fractions of the soils. A regression equation (r^2 = 0.99) was established that estimated the percentage of clinoptilolite present from the XRD intensity ratios of the zeolite and an internal Al_2O_3 standard.

A cation exchange capacity (CEC) method based on the ion-sieving properties of clinoptilolite was developed for quantitative determination in soils of southern Texas (Ming and Dixon 1987a). Initially, both zeolitic and non-zeolitic exchange sites were exchanged with Na^+. The CEC of non-zeolitic exchange sites was determined by replacing Na^+ on these sites with tert-butylammonium ions, which are too large to fit into the channels of clinoptilolite (i.e. ion-sieving). Next, the CEC of clinoptilolite was determined by exchanging Na^+ off of the zeolitic exchange sites with NH_4^+. The amount of soil clinoptilolite was estimated by comparing the measured CEC of zeolitic exchange sites with the total estimated CEC for clinoptilolite (175 $cmol_c\ kg^{-1}$). Ming and Dixon (1987a) found a high correlation (r^2 = 0.96) between the abundance of clinoptilolite estimated using the CEC method and the abundance estimated by semi-quantitative XRD analysis (Table 8). The CEC procedure described by Ming and Dixon (1987a) was developed strictly to quantify clinoptilolite, and the procedure would have to be modified in order to quantify zeolites other than clinoptilolite. It should be noted that there are more than 60 zeolites that occur in nature and each has unique crystal structures, ion-sieving properties, cation selectivities, and cation-exchange capacities.

Table 8. Comparison of XRD and cation-exchange capacity (CEC) methods for determining clinoptilolite content in various soil separates for a calcareous Mollisol in South Texas (Ming and Dixon 1987a).

	Clinoptilolite content (%)							
	Sand (2.0-0.05 mm)		*Coarse silt (0.05-0.02 mm)*		*Fine & medium silt (0.02-0.002 mm)*		*Clay (<0.002 mm)*	
Horizon	*CEC*	*XRD*	*CEC*	*XRD*	*CEC*	*XRD*	*CEC*	*XRD*
A	1	0	2	3 (1)	20	29 (8)	1	2 (1)
Bk1	1	0	2	4 (2)	9	17 (6)	1	1 (1)
Bk2	2	3 (1)	4	7 (4)	6	12 (6)	1	2 (1)
BCk	6	10 (4)	8	14 (3)	21	38 (4)	1	2 (1)
CBk1	12	14 (6)	24	33 (8)	28	33 (5)	2	4 (1)
CBk2	12	16 (5)	24	30 (9)	45	47 (8)	2	3 (1)

Numbers in parenthesis represent a 95% confidence interval (n =4) for semi-quantitative XRD data only. CEC quantitative numbers are average of 2 determinations.

Identification and characterization methods

X-ray diffraction analysis. The most definitive method for identifying zeolites in soils is XRD analysis. For most reported occurrences of zeolites in soils, XRD patterns of zeolites consist of sharp, narrow diffraction peaks, suggesting highly ordered materials. Most of the primary XRD reflections for zeolites and those of other soil minerals do not coincide, making the identification of the zeolite relatively easy. However, several zeolites may be confused with other isostructural zeolites because their XRD patterns are nearly identical, e.g. isostructural phillipsite and merlinoite or isostructural clinoptilolite and heulandite.

To prevent a mistake in identifying the zeolite type in a soil, it may be necessary to use means other than XRD analysis to distinguish the two mineral phases. For example, it may be necessary to determine the Si/Al ratio and exchangeable-cation composition to distinguish isostructural zeolites, such as heulandite and clinoptilolite (see Bish and Boak, this volume). Jacob and Allen (1990) used Si/Al atomic ratios to identify clinoptilolite in Aridisols of the Texas Tans-Pecos volcanic field. Because the zeolite occurred in a calcareous soil, zeolitic exchange sites in the upper horizons had become Ca^{2+} exchanged, which greatly reduces the thermal stability of clinoptilolite and it behaves thermally like heulandite (e.g. Shepard and Starkey 1964). However, clinoptilolite in the lower horizons had both Na and Ca on zeolitic exchange sites and it was thermally stable after heating to 500°C. Jacob and Allen (1990) used a combination of thermal analysis and measurements of the Si/Al atomic ratio (between 5 and 6) and exchangeable cations to identify clinoptilolite. As mentioned earlier, Nørnberg (1990) showed by XRD analysis, Si/Al atomic ratios, exchangeable cations, and heat treatments that the zeolite occurring in Mollisols in Denmark was heulandite, and not clinoptilolite as previously reported based on XRD and heat treatments only (Nørnberg et al. 1985).

Cation-exchange characterization. Because of their unique cation-exchange properties, zeolites can cause abnormal cation-exchange behavior in a soil. These abnormal and unusually high cation-exchange capacities in soils have lead to several discoveries of zeolites in soils. As mentioned above, the abnormally high ESPs found in soils of the San Joaquin Valley in California were caused by the presence of analcime (Schultz et al. 1964, El-Nahal and Whittig 1973). Southard and Kolesar (1978) attributed an unusually high extractable K^+ and CEC in Utah soils to clinoptilolite. The clay (<2 μm) content of the A horizon was below 20%, however, the CEC was around 50 $cmol_c\ kg^{-1}$ (Table 9). An inverse trend was observed for the clay content and the CEC as soil depth increased, unlike the expected trend in most soils where the CEC increases with increasing clay content. Although the authors did not quantify the amount of clinoptilolite in the soil profile, the high cation-exchange capacities in the topsoil combined with the low clay content (15%) suggested that most of the CEC originates with the zeolite. A similar occurrence of clinoptilolite in a soil near Oxford, England, is responsible for the high K-supplying power of the soil (Brown et al. 1969, Talibudeen and Weir 1972).

Although important cation exchange information can be obtained from CEC measurements, standard methods to measure the CEC of a soil may not be appropriate for determining the CEC contribution from a zeolite. Each zeolite species has unique cation selectivity and ion sieving properties. Therefore, special methods may be necessary to characterize the cation-exchange properties of a zeolite in a soil. Ming et al. (1993) developed a method to determine the CEC and native cation composition for sedimentary and soil clinoptilolite. A K^+/Cs^+ exchange method (i.e. K-saturation/Cs-replacement) was used to determine CEC because clinoptilolite is more selective for K^+ than for Na^+, Ca^{2+}, or Mg^{2+}; Cs^+ was used as the replacement cation because it is selected over K^+.

Table 9. Selected chemical and physical properties of a Mollisol that contains clinoptilolite in northern Utah (Southard and Kolesar 1978). Note the inverse trend between clay content and CEC with depth.

(cm)	*%*	*-----------------cmol$_c$ kg^{-1}*	*-----------------*
0-30	15	52	42
30-65	19	47	32
65-90	21	44	24
90-115	20	40	23
115-145	30	35	8

For determination of native exchangeable cations, Cs^+ was used as the replacement cation because of clinoptilolite's high selectivity for Cs^+. The sum of exchangeable cations removed by this step nearly equaled the total CEC measured for various sedimentary and soil clinoptilolites. The high selectivity of clinoptilolite for Cs facilitated complete exchange of the cation-exchange sites in clinoptilolite.

Taylor et al. (1990) determined the CEC over different time frames for laumontite from a vein outcrop at the site of the anorthosite-derived Entisol in the San Gabriel Mountains in California. Only 17.3 cmol$_c$ kg^{-1} of native cations were extracted using Na^+ as the replacement cation (Table 10). Fewer native cations were replaced by other cations (i.e. Ba^{2+}, Sr^{2+}, Cs^+, NH_4^+). The theoretical or calculated CEC for laumontite based upon the mineral's Al/Si ratio is around 420 cmol$_c$ kg^{-1} (Ming and Mumpton 1989). The unexpectedly low measured CEC was attributed to apparently slow diffusion of cations from the one-dimensional channels of laumontite. Hence, laumontite did not contribute significantly to the CEC of these soils.

Table 10. Cumulative sums of native cations displaced by various replacement cations over time for a laumontite sample from the San Gabriel Mountains in California (Taylor et al. 1990).

Extracting solution	*3 days*	*Time of extraction* *10 days*	*17 days*
	------------------	cmol$_c$ kg^{-1}	------------------
NaCl*	7.3	11.5	17.3
$BaCl_2$	3.8	4.0	4.2
NH_4Cl	5.4	6.7	7.5
$SrCl_2$	2.1	2.4	2.6
CsCl	1.8	1.9	2.1

*Displaced cation values do not include native Na.

Other analytical techniques. Several techniques can complement XRD analysis in identifying zeolites in soils, including SEM, transmission electron microscopy (TEM), infrared analysis (IR), thermal analysis (TA), elemental analyses, nuclear magnetic resonance (NMR), and CEC measurements. Zeolites are very photogenic in sedimentary rocks (e.g. see Mumpton and Ormsby 1978, Ming and Mumpton 1989), however, little is known about the morphology of zeolites in soil environments. Because most of the zeolites occurring in soils have been inherited from the parent rock, which is usually zeolite-rich sedimentary material, the morphology of zeolites in soils would be expected to be similar to those found in sedimentary deposits. The morphologies of clinoptilolite occurring in the soils from south Texas are very similar to the usual lath-shaped morphology found in sedimentary deposits (see Fig. 2). As might be expected, the clusters of clinoptilolite inherited in the upper horizons appear to be "weathered" (e.g. etched crystals) and cemented with other soil materials. However, if the zeolite is forming in the soil environment, the morphology would likely be euhedral, with little evidence of dissolution pitting and etching, similar to the occurrence of chabazite in the soils of the Antarctica Dry Valleys (Gibson et al. 1983) (Fig. 1).

A variety of analyses can be performed on a transmission electron microscope, including morphological studies, electron diffraction, and high-resolution lattice imaging. Kirkman (1976) identified mordenite as long laths in a sandy loam soil on North Island, New Zealand, using electron diffraction. Clinoptilolite crystals separated from the A horizon of a calcareous Mollisol from southern Texas had a well-defined platy morphology, however, they were intimately associated with irregularly shaped particles with diffuse boundaries (Ming 1985). These irregular shaped particles were identified by electron diffraction to be smectite, suggesting a weathering transformation from clinoptilolite to smectite in the upper horizons of this soil (Fig. 5). No smectite was found associated with clinoptilolite crystals in the lower horizons (i.e. CB horizons).

Infrared spectroscopy is rarely used in the identification of zeolites, but IR can be used to study structural features of zeolites (Flanigen et al. 1971, Ward 1971). Infrared analysis is sensitive to the framework Si/Al compositions in which the asymmetric modes (900-1300 cm^{-1} for most zeolites) shift to lower frequencies with increasing Al contents (Flanigen et al. 1971). Ming (1985) examined the IR spectra of clinoptilolite separated from various soil horizons in southern Texas soils. Mid-infrared spectroscopy indicated slight or no variation in bond strengths for the (Si,Al)-O asymmetric and symmetric stretching modes of clinoptilolite separated from the various horizons, suggesting similar Si/Al atomic ratios for the zeolite throughout the soil profile. Ward and McKague (1994) developed non-destructive methods to differentiate between clinoptilolite and heulandite using static proton NMR, ^{27}Al magic angle spinning (MAS) NMR, and ^{29}Si MAS NMR techniques.

Thermal analysis (e.g. differential thermal analysis or DTA, differential scanning calorimetry or DSC, thermal gravimetric analysis or TGA) has been used occasionally to distinguish between zeolites with similar XRD properties (e.g. isostructural clinoptilolite and heulandite). However, variations in exchangeable cations will greatly influence the thermal behavior of zeolites. For example, Ca-exchanged clinoptilolite becomes thermally unstable around 250-350°C and behaves similarly to heulandite, which has Ca as its predominant exchange cation in nature. Conversely, K-exchanged clinoptilolite is thermally stable to near 700°C (Shepard and Starkey 1964). In order to insure proper identification of isostructural zeolites (i.e. in addition to XRD analysis), it is best to combine information on the Si/Al atomic ratio and exchangeable cation content with thermal analysis characterization.

CONCLUSIONS AND RECOMMENDATIONS FOR FUTURE STUDY

Reported occurrences of zeolites in soils are rare, but it is probable that many occurrences of zeolites in soils are overlooked because they are present in minor or trace amounts. We anticipate that zeolites are widespread in soils, particularly in areas near zeolite-rich tuffs. For example, Ming (1985) conducted a reconnaissance study through south Texas and found clinoptilolite in numerous soils derived from the Catahoula

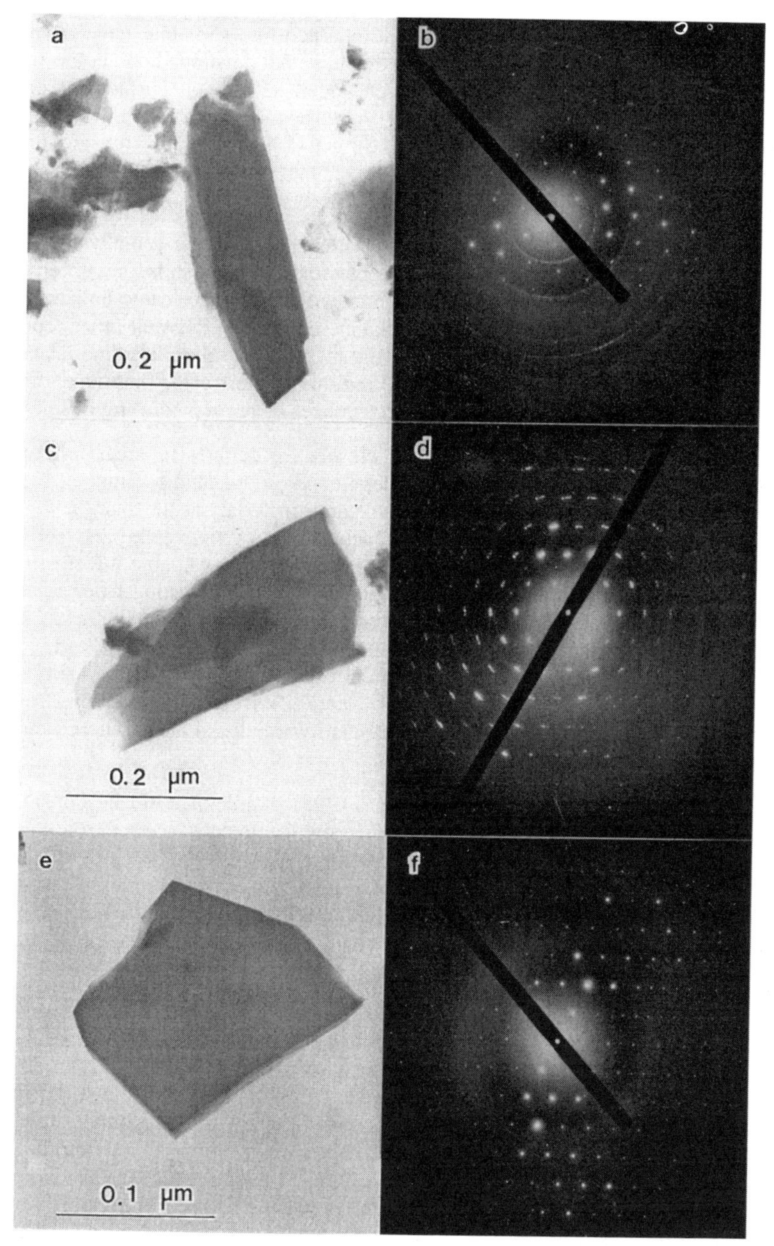

Figure 5. Transmission electron micrographs and electron diffraction patterns for clinoptilolite from a Mollisol located in south Texas (Ming 1985): (a) and (b) A horizon; (c) and (d) Bk2 horizon; and (e) and (f) CBk2 horizons.

Formation. These soils covered an area nearly 200 km long and 10 to 20 km wide. The zeolite content never exceeded 5 to 10% in the upper horizons (i.e. A and B horizons) and might have been missed if the objective of the reconnaissance study had not been to search for zeolites in these soils. Outcrops of zeolite-rich tuffs are common throughout the world, hence, it is quite likely that zeolites have been inherited into soils that have developed on these tuffaceous materials. Additionally, soils that have not developed on zeolite-rich parent materials might contain zeolites that have been added to the soil surface by eolian and fluvial deposition from nearby sources, e.g. outcrops of zeolite-rich tuffs or zeolite-containing soils. The study by Clausnitzer and Singer (1999) suggests that zeolites may be very common in soils as eolian additions; however, they might not be detected in soils by conventional identification methods such as XRD analysis.

Although we expect fewer occurrences in which zeolites have actually formed in the soil, it is likely that many occurrences of zeolite formation in soils are unreported. As described by Hay (1963 1964 1970 1978), zeolites form from volcanic and non-volcanic sediments at the land surface under saline, alkaline conditions. Several other reports have described the formation of zeolites in soils at the margins of saline, alkaline lakes. Saline, alkaline lakes and deposits occur in many arid regions in the world, therefore, there exists the potential for the discovery of additional occurrences of zeolites forming in soils.

We recommend that soil scientists look closely for zeolites in areas where zeolites might be expected to exist in soils, e.g. soils developed on volcanic sediments, soils near a zeolite-rich outcrop, or soils developing in volcanic materials at the margin of a saline, alkaline lake. Chemical and mineralogical characterization of the zeolite and soil, climatic conditions of soil formation, parent material , etc., will provide valuable information on the stability and persistence of zeolites in soils. Not only is this interesting from a pedological standpoint, but this information will also aid researchers as they evaluate the possibilities of adding zeolites to soils for agricultural and environmental applications. In addition, standard mineralogical analyses of soils often exclude analysis of the silt and sand fractions, in which most zeolites in soil occur. We also recommend expanding mineralogical investigations to larger size fractions, especially in areas where lithogenic or pedogenic zeolites may be expected to occur.

As suggested by Boettinger and Graham (1995), studies should be considered that examine zeolite distribution and weathering in soil sequences over chronological and climatic gradients. Studies should be conducted on the weathering of zeolites in a variety of environments, e.g. acid vs. neutral vs. alkaline soils. Additional studies of the chemical changes and stabilities of zeolites throughout a soil profile are warranted, i.e. comparison of zeolites in the parent material with those that have been inherited into the topsoil.

At present, we can only speculate on the stability of zeolites when added to a soil for an agricultural or environmental application. We expect that the mode of formation will shed some light on the expected stabilities of zeolites. Zeolites that readily form in saline, alkaline environments at the land surface, such as the more aluminous zeolites (e.g. analcime and phillipsite), will likely not persist very long if added to acidic soils. In fact, in very acidic environments (e.g. pH < 5), these zeolites may survive only a few days or weeks. On the other hand, more siliceous zeolites (e.g. clinoptilolite, mordenite) that formed under diagenesis and hydrothermal alterations at temperatures and pressures higher than in soil environments will likely be much more stable when introduced into soils. Lithogenic zeolites appear to weather more rapidly when in finer particle sizes and in near-surface horizons where weathering intensity is greatest in most soil environments. These zeolites may persist for years or up to, perhaps, millions of years, depending on the soil environment. No doubt, numerous factors such as zeolite type and particle size, soil pH, climatic conditions, etc., will have a major impact on the how long these zeolites may persist in soils. Because of the scant knowledge available on the occurrences of zeolites in soil environments, we are a long way from being able to adequately predict the behavior of zeolite additions into soils.

ACKNOWLEDGMENTS

We thank John Gruener for his helpful comments on the manuscript. We also thank Dr. Takashi Mikouchi for translation of articles from Japanese to English. This work was supported in part by NASA's Advanced Life Support program. We are very grateful to the guidance, wisdom, and encouragement provided by our colleague, Fred Mumpton.

REFERENCES

Asvadurov H, Popescu F, Constantinescu M (1978) Soluri si roci cu continut ridicat de potasiu schimbabil. An Inst Cerc Pedol Agrochim 43:115-123

Atanassov I, Dimitrov DN, Etropolski H (1982) General characteristics of soils developed upon zeolite rocks in the East Rodopa Mountain. Soil Sci Agrochem 17:64-70

Atanassov I, Do Vang Bang (1984) Minerals of fine dispersion and the evolution of the mineral portion of soils overlying zeolites. Pochvoznanie i Agrokhimiëĭa, 19:47-57

Baldar NA (1968) Occurrence and formation of soil zeolites. PhD dissertation, University of California, Davis, California, 150 p

Baldar NA, Whittig LD (1968) Occurrence and synthesis of soil zeolites. Soil Sci Soc Am Proc 32:235-238

Baroccio A (1962) Prove di concimazione su terreni vulcanici ad analcite. L'Agricoltura Italiana 7:207-217

Bellanca A, Di Caccamo A, Neri, R (1980) Mineralogia e geochimica di alcuni suoli della Sicilia Centro-Occidentale: Studio delle variazioni composizionali lungo profili pedologici in relazione ai litotipi d'orgine. Mineral Petrogr Acta 24:1-15

Bhattacharyya T, Pal DK, Deshpande SB (1993) Genesis and transformation of minerals in the formation of red (Alfisols) and black (Inceptisols and Vertisols) soils on Deccan basalt in the western Ghats, India. J Soil Sci 44:159-171

Bhattacharyya T, Pal DK, Srivastava P (1999) Role of zeolites in persistence of high altitude ferruginous Alfisols of the humid tropical Western Ghats, India. Geoderm 90:263-276

Bockheim JG, Ballard TM (1975) Hydrothermal soils of the crater of Mt. Baker, Washington. Soil Sci Soc Am Proc 39:997-1001

Boettinger JL (1988) Duripan genesis on granitic pediments of the Mojave Desert. MS thesis, Univ California, Davis, California, 195 p

Boettinger JL, Graham RC (1995) Zeolite occurrence in soil environments: An updated review. *In* Ming DW, Mumpton FA (eds) Natural zeolites '93: Occurrence, Properties, Use. Int'l Committee on Natural Zeolites, Brockport, New York, p 23-27

Boettinger JL, Southard RJ (1995) Phyllosilicate distribution and origin in Aridisols on the granitic pediment, Western Mojave Desert. Soil Sci Soc Am J 59:1189-1198

Breazeale JF (1928) Soil zeolites and plant growth. Agricultural Experiment Station Tech Bull No. 21:499-520, Univ Arizona, Tucson, Arizona

Brogowski Z, Dobrzanski B, Kocon JE (1980) Morphology of natural zeolites occurring in soil as determined by electron microscopy. Bull Acad Polonaise Sci, Ser Sci Terre 27:115-117

Brogowski Z, Dobrzanski B, Kocon JE, Zaniewska-Chlipalska (1983) The possibility of zeolites occurrence in the soils of Poland. Zeszyty Problemowe Postepow Nauk Rolniczych 220:489-494

Brown G, Catt JA, Weir AH (1969) Zeolites of the clinoptilolite-heulandite type in sediments of south-east England. Mineral Mag 37:480-488

Burgess PS, McGeorge WT (1926) Zeolite formation in soils. Science 64:652-653

Capdecomme L (1952) Laumontite du Pla des Aveillans (Pyrénées-Orientales). Bull Soc Hist Nat Toulouse 27:299-304

Clausnitzer H, Singer MJ (1999) Mineralogy of agricultural source soil and respirable dust in California. J Environ Qual 28:1619-1629

Darwish TM, Gradusov BP, Sfeyr S, Abdelnur L (1988) Mineralogy and chemical composition of finely dispersed fraction and properties of mountain soils of Lebanon. Pochvovedeniye 4:85-95

Dibblee TW Jr (1963) Geology of the Willowsprings and Rosamond quadrangles, California. U S Geol Surv Bull 1089-C, 255 p

El-Nahal MA, Whittig LD (1973) Cation exchange behavior of a zeolitic sodic soil. Soil Sci Soc Am Proc 37:956-958

Fanning DS, Rabenhorst MC, Wagner DP, Snow PA (1983) Soils-geomorphology field trip in Maryland (Guidebook). August, 1983, in conjunction with American Society of Agronomy Annual Meetings. Washington, DC, 119 p

Flanigen EM, Khatami H, Szymanski HA (1971) Infrared structural studies of zeolite frameworks. *In* Gould RF (ed) Molecular sieve zeolites-I. Adv Chem Ser 101:201-229

Frankart RP, Herbillon AJ (1970) Présence et genése d'analcime dans les sols sodiques de la Basse Rusizi (Burundi). Bull Groupe Franc Argiles 22:79-89
Furbish WJ (1965) Laumontite-loenhardite from Durham County, North Carolina. Southeastern Geol 6:189-200
Gibson EK, Wentworth SJ, McKay DS (1983) Chemical weathering and diagenesis of a cold desert soil from Wright Valley, Antarctica: An analog of martian weathering processes. Proc XIII Lunar Planet Sci Conf, Part 2, J Geophy Res 88:A912-A928
Goh TB, Pawluk S, Dudas MJ (1986) Adsorption and release of phosphate in chernozemic and solodized solonetzic soils. Can J Soil Sci 66:521-529
Gorbunov NI, Bobrovitsky AV (1973) Distribution, genesis, structure, and properties of zeolite. Pochvovedenive 5:93-101
Graham RC, Herbert BE, Ervin JO (1988) Mineralogy and incipient pedogenesis of Entisols in anorthosite terrane of the San Gabriel Mountains, California. Soil Sci Soc Am J 52:738-746
Graham RC, Southard AR (1983) Genesis of a Vertisol and an associated Mollisol in Northern Utah. Soil Sci Soc Am J 47:552-559
Hay RL (1963) Zeolitic weathering in Olduvai Gorge, Tanganyika. Geol Soc Am Bull 74:1281-1286
Hay RL (1964) Phillipsite of saline lakes and soils. Am Mineral 49:1366-1387
Hay RL (1970) Silicate reactions in three lithofacies of a semiarid basin, Olduvai Gorge, Tanzania. Mineral Soc Am Spec Paper 3:237-255
Hay RL (1976) Geology of the Olduvai Gorge. Univ California Press, Berkeley, California, 203 p
Hay RL (1978) Geologic occurrence of zeolites. *In* Sand LB, Mumpton FA (eds) Natural zeolites: Occurrence, Properties, Use. Pergamon Press, New York, New York, p 135-143
Jacob JS, Allen BL (1990) Persistence of a zeolite in tuffaceous soils of the Texas Trans-Pecos. Soil Sci Soc Am J 54:549-554
Jalalian A, Southard AR (1986) Genesis and classification of some Paleborolls and Cryoboralfs in Northern Utah. Soil Sci Soc Am J 50:668-672
Kaneko S, Shoji S, Masui J (1971) Zeolite in paddy soils. I. Occurrence. Nihon Dåojåo Hiryåogaku Zasshi 42:177-182
Kapoor BS. Singh HB, Goswami, SC (1980) Analcime in a sodic profile. J Indian Soil Sci 28, 513-515
Karathanasis AD (1982) Characteristics of naturally acid soil smectites. PhD dissertation, Auburn University, Auburn, Alabama, 215 p
Kirkman JH (1976) Clay mineralogy of Rotomahana sandy loam soils, North Island, New Zealand. N Z J Geol Geophy 19:35-41
Liu WC (1988) The sensitivity of selected soils from the Sierra Nevada to acidic deposition. PhD dissertation, Univ California, Riverside, 105 p
Maglione G, Tardy Y (1971) Néoformation pédogénétique d'une zéolite, la mordénite, associée aux carbonates de sodium dans une dépression interdunaire des bords du lac Tchad. CR Acad Sci Paris, Sér D 272:772-774
McCulloh TH, Stewart RJ (1982) Laumontite in the western Transverse Ranges, California: Mark of Neogene hydrothermal alteration coincident with transcurrent faults. Geol Soc Am Abstr with Programs, p 213-214
McFadden LD, Wells SG, Jercinovich MJ (1987) Influences of eolian and pedogenic processes on the origin and evolution of desert pavements. Geol 15:504-508
Ming DW (1985) Chemical and crystalline properties of clinoptilolite in South Texas soils. PhD dissertation, Texas A&M University, College Station, Texas, 257 p
Ming DW, Allen ER, Galindo C Jr, Henninger DL (1993) Methods for determining cation exchange capacities and compositions of native exchangeable cations for clinoptilolite. *In* Rodríguez Fuentes G, González JA (eds) Zeolites '91: Memoirs of the 3rd Int'l Conf Occurrence, Properties and Utilization of Natural Zeolites, April 9-12, 1991, p 31-35. Int'l Conf Center, Havana, Cuba
Ming DW, Dixon JB (1986) Clinoptilolite in South Texas soils. Soil Sci Soc Am J 50:1618-1622
Ming DW (1987a) Quantitative determination of clinoptilolite in soils by a cation-exchange capacity method. Clays & Clay Minerals 35:463-468
Ming DW (1987b) Technique for the separation of clinoptilolite from soils. Clays & Clay Minerals 35:469-472
Ming DW (1987c) Zeolites: Recent developments in soil mineralogy. Transactions, XIII Congress Int'l Society Soil Science, Vol V, Hamburg, Germany, p 371-382
Ming DW (1988) Occurrence and weathering of zeolites in soil environments. *In* Kalló D, Sherry HS (eds) Occurrence, properties, and utilization of natural zeolites. Akadémiai Kiadó, Budapest, Hungary, 699-715
Ming DW, Mumpton FA (1989) Zeolites in Soils. *In* Dixon JB, Weed SB (eds) Minerals in Soil Environments. Soil Science Society of America, Madison, Wisconsin, p 873-911

Morita Y, Ohsumi Y, Tanaka N (1985) On the nature, genesis and classification of eutrophic soils on the Coastal Hill (Mt. Komayama) in the Shonan District of Kanagawa Prefecture: (II) Distribution, clay mineral composition and properties of parent material of the eutrophic Brown Forest soil, and the assumption of its forming process. Bull For & For Prod Res Inst 333:67-91

Mumpton FA, Ormsby WC (1978) Morphology of zeolites in sedimentary rocks by scanning electron microscopy. *In* Sand LB, Mumpton FA (eds) Natural Zeolites: Occurrence, Properties, Use. Pergamon Press, New York, p 113-132

Nemecz E, Janossy AGS, Olaszi VJ (1988) Behavior of zeolitic minerals during soil formation. *In* Kalló D, Sherry HS (eds) Occurrence, properties, and utilization of natural zeolites. Akadémiai Kiadó, Budapest, Hungary, p 675-697

Nørnberg P (1990) A potassium-rich zeolite in soil development on Danian chalk. Mineral Mag 54:91-94

Nørnberg P, Dalsgaard K (1985) The origin of clay minerals in soils on Danian chalk. 5th Meeting European Clay Groups, Prague, p 553-561

Nørnberg P, Dalsgaard K, Skammelsen E (1985) Morphology and composition of three Mollisol profiles over chalk, Denmark. Geoderma 36:317-342

Paeth RC, Harward ME, Knox EG, Dyrness CT (1971) Factors affecting mass movement of four soils in the western Cascades of Oregon. Soil Sci Soc Am J 35:943-947

Phadke AV, Kshirsagar LK (1981) Zeolites and other cavity minerals in Deccan Trap Volcanics of Western Maharashra. Proc Symp Decades of Development in Petrology, Mineralogy, and Petrochemistry in India. Geological Survey of India, p 129-134

Ping CL, Shoji S, Ito T (1988) Properties and classification of three volcanic ash-derived pedons for Aleutian Islands and Alaska Peninsula, Alaska. Soil Sci Soc Am J 52:455-462

Pond WG (1995) Zeolites in animal nutrition and health: A review. *In* Ming DW, Mumpton FA (eds) Natural zeolites '93: Occurrence, Properties, Use. Int'l Committee on Natural Zeolites, Brockport, New York, p 449-457

Portegies Zwart R, Vink APA, Van Schuylenborgh J (1975) The weathering of zeolitic and non-zeolitic calcareous materials in a lacustrine plain in south-central Italy. Geoderma 14:277-295

Raybon SO (1982) Lithology and clay mineral variations in the middle phase of Paleocene Porters Creek Formation of Mississippi. MS thesis, Univ Mississippi, University, Mississippi, 101 p

Reid DA, Graham RC, Edinger SB, Bowen LH, Ervin JO (1988) Celadonite and its transformation to smectite in an Entisol at Red Rock Canyon, Kern County, California. Clays & Clay Minerals 36:425-431

Renaut RW (1993) Zeolitic diagenesis of late Quaternary fluviolacustrine sediments and associated calcrete formation in the Lake Bogoria Basin, Kenya Rift Valley. Sedimentology 40:271-301

Reynolds WR, Bailey MA (1990) Gismondine in a lamproite intrusive and soil system. Program and Abstracts, Clay Minerals Society, 27th Ann Mtg, Columbia, Missouri, Oct. 6-11, 1990, p 106

Sabale AB, Vishwakarma LL (1996) Zeolites and associated secondary minerals in Deccan volcanics: Study of their distribution, genesis and economic importance. National Symposium on Deccan Flood Basalts, India. Gondwana. Geol Mag 2:511-518

Schultz RK, Overstreet R, Barshad I (1964) Some unusual ionic exchange properties of sodium in certain salt-affected soils. Soil Sci 99:161-165

Shepard AO, Starkey HC (1964) Effect of cation exchange on the thermal behavior of heulandite and clinoptilolite. U S Geol Surv Prof Paper 475-D:89-92

Southard AR, Kolesar PT (1978) An exotic source of extractable potassium in some soils of Northern Utah. Soil Sci Soc Am J 42:528-530

Spiers GA, Pawluk S, Dudas MJ (1984) Authigenic mineral formation by solodization. Can J Soil Sci 64:515-532

Surdam RC, Sheppard RA (1978) Zeolites in saline, alkaline-lake deposits. *In* Sand LB, Mumpton FA (eds) Natural Zeolites: Occurrence, Properties, Use. 145-174. Pergamon Press, New York, p 145-174

Talibudeen O, Weir AH (1972) Potassium reserves in a 'Harwell' series soil. J Soil Sci 23:456-474

Taylor K, Graham RC, Ervin JO (1990) Laumontite in soils of the San Gabriel Mountains, California. Soil Sci Soc Am J 54:1483-1489

Travnikova LS, Gradusov BP, Chizhikova NP (1973) Zeolites in some soils. Pochvovedeniye 3:106-114

Ward JW (1971) Infrared spectroscopic studies of zeolites. *In* Gould RF (ed) Molecular Sieve Zeolites-I. Adv Chem Ser 101:380-404

Ward RL, McKague HL (1994) Clinoptilolite and heulandite structural differences as revealed by multinuclear magnetic resonance spectroscopy. J Phys Chem 98:1232-1237

Zelazny LW, Calhoun FG (1977) Palygorskite, (attapulgite), sepiolite, talc, pyrophyllite, and zeolites. *In* Dixon JB, Weed SB (eds) Minerals in Soil Environments. Soil Science Society of America, Madison, Wisconsin, p 435-470

12 Zeolites in Petroleum and Natural Gas Reservoirs

Azuma Iijima

JAPEX Research Center
Japan Petroleum Exploration Company
1-2-1 Hamada, Mihama-ku
Chiba 261-0025, Japan

INTRODUCTION

Zeolites have been reported in petroleum and natural gas reservoirs of a wide range of geological ages from many parts of the world. Such zeolite-bearing oil and gas reservoirs are not comparatively large in number and scale. However, where present in petroleum and natural gas reservoirs, zeolites can have profound effects on the quality of the reservoirs, affecting rock porosity and permeability. Zeolite-containing reservoir rocks vary widely lithologically, including not only porous, pelitic sedimentary rocks, but also volcanogenic rocks and even fractured intrusive and plutonic rocks.

Several different genetic types of zeolite occurrences have been distinguished and classified (e.g. Hay 1966, 1986, 1995; Iijima 1980, 1986), as Hay and Sheppard discussed and summarized in an earlier chapter (this volume). In addition, the organic origin theory for the formation of economic oil and gas resources seems to have been universally accepted, including the prominent kerogen decomposition theory summarized by Tissot and Welte (1978). Sedimentary basins filled by very thick sequences of sedimentary strata containing organic-rich source rocks and porous reservoir rocks are considered necessary for producing hydrocarbon resources. Such sedimentary basins develop primarily through rifting or in relation to arc formation at active plate boundaries, and they are often accompanied by heating and extensive volcanic activity. Most types of zeolite occurrence are consequently constrained to burial diagenesis, uplift-related late-stage diagenesis (Surdam and Boles 1979, Boles 1993), and hydrothermal alteration, as will be discussed in this chapter.

Zeolites in hydrocarbon reservoir rocks occur commonly as cements and grain replacements in the form of disseminated, microcrystalline aggregates and occasionally as macroscopic crystals in vugs and veins. It was in 1950s that Coombs and co-workers pioneered the petrologic study of rock-forming zeolites through their classic research on burial metamorphism of the Triassic geosynclinal deposits in South Syncline, New Zealand, and its interpretation based on hydrothermal synthesis (e.g. Coombs 1954, Coombs et al. 1959). The earliest-known report of zeolite occurrence in petroleum reservoirs, to the author's knowledge, is a laumontite cement in Miocene reservoir sandstone from a well in the San Joaquin Basin, California (Kaley and Hanson 1955). Zeolites, particularly laumontite, were subsequently found in arkoses (feldspathic sandstones principally derived from granitic rocks) and volcanic sandstones in arc-related, Northeast Pacific sedimentary basins (e.g. Murata and Whiteley 1973, Galloway 1974, 1979; Merino 1975). The zeolites and associated authigenic clay minerals forming in thick sedimentary columns were interpreted to be burial diagenetic minerals. Galloway (1974, 1979) reported that the porosity of reservoir sandstones tends to decrease gradually with an increase in burial depth. This was attributed not only to an increase in overburden but also to precipitation and replacement of the diagenetic minerals. At that time, the laumontite zone was regarded as an economic basement of exploration for oil and gas (McCulloh et al. 1973, McCulloh and Stewart 1979).

1529-6466/00/0045-0012$10.00

Boles and Coombs (1977) re-examined the zeolite assemblages in the sandstones of South Syncline, New Zealand, and confirmed that zeolites in thick piles of sediments were constrained not only by burial depth (simple increasing temperature and pressure) but also by P_{fluid}/P_{total} and the composition and movement of pore fluids. As will be discussed, this concept has strongly influenced the interpretation of laumontite occurrence in sandstone reservoirs. Surdam and his co-workers stressed that organic acids formed during decomposition of kerogen could play an important role in creating secondary porosity by dissolution of feldspar and laumontite (Surdam et al. 1984, 1989a,b). In fact, Coffman (1995) reported late-stage laumontite cementation and dissolution from some marginal fault-controlled oil fields of the Los Angeles basin. South Syncline of New Zealand has more recently yielded a new idea of zeolitization of geosynclinal deposits. Through the use of oxygen isotope analysis of laumontite and associated quartz in veins, Stallard and Boles (1989) and Boles (1991) showed that laumontite crystals filling fractures and the adjacent matrix of tuffs had been affected by meteoric water.

Many zeolite occurrences are in volcanogenic rocks composed mainly of unstable materials. Coombs (1954) recognized the vertical zonal distribution of zeolites in altered tuffs in the Triassic geosynclinal deposits of South Syncline of New Zealand, with shallower heulandite and analcime and deeper laumontite. Later, Utada (1965) and Iijima and Utada (1966) recognized the following vertical diagenetic zonation in silicic tuffs in the Neogene sections of Japan: unaltered glass shards ⇒ clinoptilolite and mordenite ⇒ analcime and heulandite ⇒ laumontite and albite, although burial-diagenetic laumontite is rare and localized. The surface distribution of alteration minerals was represented in regional zonal maps (Utada 1965, 1970). The vertical zeolite zonation which is formed by the present-day burial diagenesis of silicic tuffs in thick marine sections is dependent more on temperature than on burial depth (Iijima and Utada 1971, Iijima 1978, 1986, 1988a; Iijima and Ogihara 1995). Recently, it has been demonstrated that hydrothermal alteration can also play an important role in forming some volcanogenic and basement igneous petroleum and natural gas reservoirs in Japan, China, Vietnam, and Georgia (former USSR), as will be discussed later. Table 1 and Figure 1 summarize zeolite occurrences in hydrocarbon reservoirs and in sedimentary and volcanogenic rocks in sedimentary basins that are potentially hydrocarbon-productive.

DIAGENETIC ZEOLITES IN HYDROCARBON RESERVOIRS

Diagenetic zeolites occur not only in identified hydrocarbon reservoirs but also regionally in sedimentary and volcanogenic rocks in sedimentary basins that are potentially hydrocarbon-productive. Environments in which zeolites form can be classified into closed systems (primarily alkali-saline lake environments), open systems (percolating groundwater), and burial diagenesis, which constitutes the most common and regional occurrence of zeolites (Hay 1995, Hay and Sheppard, this volume). Hydrocarbon-producing resources most commonly develop in sedimentary basins filled by thick piles of sediments containing organic-rich deposits, and such basins are the most common host for zeolites. Volcanogenic reservoirs containing burial-diagenetic zeolites are largely confined to marine back-arc basins (Table 1).

Burial-diagenetic zeolites in volcanogenic reservoirs

Burial diagenesis of pyroclastic and volcaniclastic rocks is characterized by regional and burial-depth zonation of volcanic glass and the resulting alteration minerals, such as zeolites, feldspars, silica and clay minerals. Burial-depth zonation, with heulandite at the shallower depths overlapping laumontite-albitized plagioclase at greater depths, was first described by Coombs (1954) in the Southland Triassic of New Zealand, although the shallowest zone in

Table 1. Zeolite occurrences in hydrocarbon reservoirs and in sedimentary, volcanic, and igneous rocks in some depositional basins that are potentially oil-productive in the world.

Locality^{*1}	*Zeolites and other authigenic minerals*^{*1}	*Reservoir rocks*	*Age*	*Sediment. environ.*	*Basin type*^{*3}	*Controling factor of zeolite formation*	*Reference(s)*
U. S. A.							
1. Trap Spring oil field, Railroad Valley, Nevada	Zeolite	welded tuff	Neogene	T	R	open-system?	French and Freeman (1979)
2. San Emigdio area, San Joaquin Basin	Lm (28%), Hd, Ab, Ksp, Qz, C/S, Cc	volc. ss	Oligo.	D-s	F-l	temp. and volcanic material content	Bloch and Helmold (1955)
3. N. Tejon oil field, San Joaquin Basin	Lm (25%), Hd, Ab, Ksp, Qz, S, Cc	volc. ss	Oligo.– E Mio.	D-s	F-l	fluid mobility	Noh and Boles (1993)
4. Kettleman N. Dome, San Joaquin Basin	Lm^{+v}(8%), Ab, Ksp, Qz, Sp, S	dacitic ss	Mio.	M	F-l	burial diagenesis	Merino (1975)
5. Santa Fe Springs and Dominguez oil fields, Los Angeles Basin	Lm (15-23%), Ab, Qz, Cc	arkose	Tertiary	M	A-r	late hydrothermal alteration and fault-controlled	Coffman (1995)
6. Corpus Christi oil field, Texas Gulf Coast	Ac, K, Cc	volc. ss	Oligo.	C	F-l	burial diagenesis	Lynch (1966)
ARGENTINA							
7. Neuquén Basin	Ac, Hd	andesite tuff	E Jura.	D-s	R		Khatchikian (1983)
8. San Jorge Basin	Cp, Ac, Lm, C/S	volc. ss	Cret.	F	F-l R		Dunn et al. (1993)
JAPAN							
9. Yufutsu gas field region south-central Hokkaido	Lm^v, Pr^v, Qz^v, Cc^v	cg/granite, basic rocks	Eocene/ Creta.	F	F-a	hydrothermal alteration and fracturing	Agatsuma et al. (1996) J.N.O.C. (1998)
10. Sarukawa oil field, Akita Basin	Cp, Cr, S	acidic tuff	L Mio.	D-s	B-a	burial diagenesis	Uchida (1984)

11. Ayukawa oil and gas field, Akita Basin	Cp, Qz, S	vitric silicous siltstone	M Mio.	D-s	B-a	burial diagenesis	Okubo et al. (1996)
12. Ayukawa oil and gas field, Akita Basin	Ac^{+v}, Tn^{+v}, Ksp, Qz, Ds, S, S/C, K	dolerite	M Mio.	D-s	B-a	submarine hydro-thermal alteration	Okubo et al. (1996)
13. Yurihara oil and gas field, Akita Basin	Ac, Tn, Lm, Pn^{+v}, Pp, Ab, Ksp^{+v}, Qz^{+v}, Cc^{+v}, S, S/C, C	basalt	E-M Mio.	D-s	B-a R	submarine hydro-thermal alteration	Hoshi et al. (1992), Yagi (1992, 1994)
14. Yurihara oil and gas field, Akita Basin	Ac, Ksp, Qz	acidic tuff	M Mio.	D-s	B-a R	submarine hydro-thermal alteration	Yagi (1992, 1994)
15. Katakai gas field, Niigata Basin	Ab, Qz^{+v}, Cc^{+v}, Ak-Sd^{+v}, Ba, I, I/S, C, Ga, Py,	rhyolite	E-M Mio.	D-s	B-a R	submarine hydro-thermal alteration	Yamada and Uchida (1994)
16. Johban offshore gas field, Fukushima	Lm (2%), Ksp, Qz, K, Cc	feld. and rhyolitic ss	Oligo./ L Creta.	F, M	F-a	burial diagenesis an alkali pore fluids	Iijima (1995)
CHINA							
17. N Songliao Basin, NE China	Lm, Ab, Qz,	lithic arkose and feld. ss	E Creta.	F-L	C	burial diagenesis	Yang et al. (1991)
18. Zhou Kou Basin, Henan	Lm	sandstone	E Creta	F-L	R	burial diagenesis	Liu (1991
19. N Shaanxi slope, east Ordos Basin, Shaanxi	Lm, Ab, Qz, C, C/S, I, I/S, Cc	arkose	L Trias.	F-L	C	burial diagenesis	Liu et al. (1993)
20. Houshaoshan oil field, NE Junggar Basin	Ac, Ab, Ksp, I/S, Cc, Sd, Ak, Py, Halite	volcanic sandstone	L Perm.	F-L	F-l	closed system and burial diagenesis	Tang et al. (1997b)
21. Southern Junggar Basin, Xinjiang	Cp, Hd, Lm, Ab, Ksp, Qz, Cc, I	andesite-basaltic ss	E Trias.– L Perm.	F-L	F-l	burial diagenesis	Tang et al. (1997a)
22. Karamay oil field, NW Junggar Basin	Ac, Ab, Cc	andesite-rhyolitic cg	L Perm.	T	F-l	closed system and burial diagenesis	Chen (1992)
23. Karamay oil field, NW Junggar Basin	Hd, Md, Sb^{+v}, Cz Nl, Lm^{+v}, Cc^{+v}, C	basalt complex	Carbon.	M	F-l	hydrothermal alteration	*Ibid*, Liu (1986), Graham et al. (1990)

VIETNAM							
24. White Tiger offshore oil field, south Vietnam	Lm^{v}, Ab^{v}, Qz^{v}, C^{v}, Cc^{v}	granite	L Trias.– E Cret.		R	hydrothermal alteration and fracturing	Areshev et al. (1992)
GEORGIA, TRANSCAUCASIA							
25. Samgori oil field, Transcaucasia	Lm^{+v} (47%), Ac, Ab C, I, Cc	andesite-basalt tuff	M Eoc.	D-s	B-a	submarine hydrothermal alteration	Vernik (1990), Grynberg et al. (1993),
SLOVAKIA							
26. Trboviste area, E Slovakian Basin	Cp, Md, Ac, Cr, Qz, S, I	rhyolite tuff	L Mio.	M	B-a T	burial diagenesis	Reed et al. (1993a,b)
27. Pozdisovece area, E Slovakian Basin	Er, Of, Ksp	rhyolite tuff	L Mio.	E	B-a T	closed systen	Reed et al. (1993a,b)
28. Nizny Caj area, E Slovakian Basin	Md, S	andesite tuff	M Mio.	M	B-a T		Reed et al. (1993a,b)
29. Hrusov area, East Slovakian Basin	Lm, Er, Ac, Qz, K, I	volcanic ss	M Mio.	M	B-a T	hydrothermal alteration?	Reed et al. (1993a,b)

[*1] Ac: analcime, Cz: chabazite, Cp: clinoptilolite, Er: erionite, Hd: heulandite, Lm: laumontite (% in volume), Md: mordenite, Nl: natrolite, Of: offretite, Sb: stilbite, Ts: thomsonite, Ab: albite, Ksp: K-feldspar, Cr: cristobalite, Qz: quartz, Pn: prehnite, Pp: pumpellyite, Sp: Sphene, C: chlorite, C/S: chlorite-smectite mixed layer, K: kaolinite, I: illite, I/S: illite/smectite mixed layer, S: smectite, An-Sd: ankerite-siderite, Cc: calcite, Ba: barite, Ga: galena, ^{v}: vein occurrence only, $^{+v}$: occurrence as pervasive cement and in veins.

[*2] D-s: deep-sea, M: marine, C: sea coast and back-barrier, E: evaporitic, F: fluvial, F-L: fluvio-lacustrine, L: lacustrine, T: terrestrial.

[*3] A-r: arc-related basin, F-a: fore-arc basin, B-a: back-arc basin, B-a T: back-arc transitional basin, F-l: fore-land basin, F-l R: foreland rift basin, R: rift basin, C: cratonic basin.

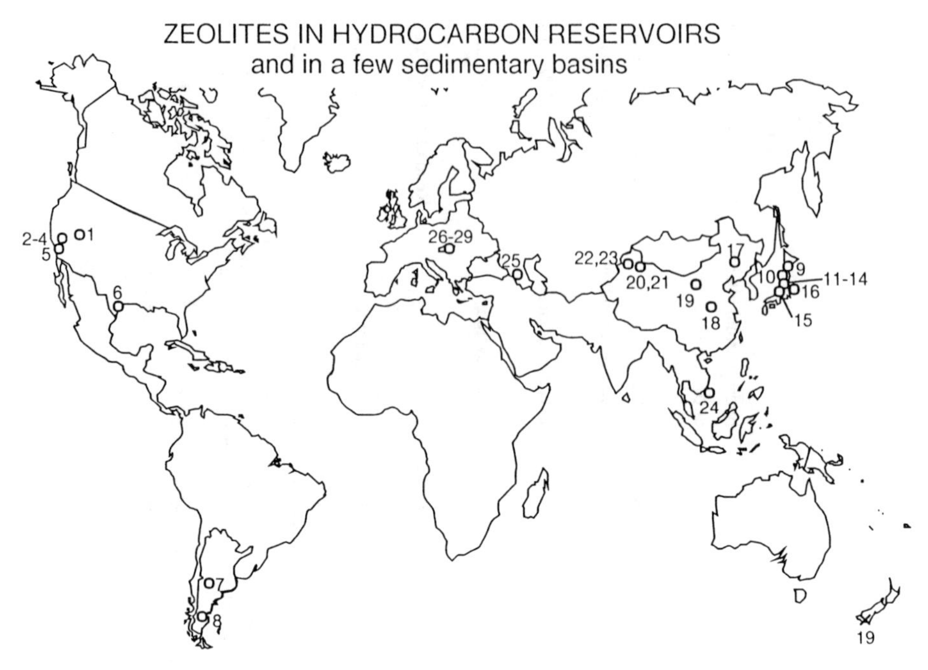

Figure 1. Distribution of zeolites in hydrocarbon reservoirs and in a few sedimentary basins in the world. Locality numbers refer to Table 1.

this deposit appears to be absent due to subsequent erosion (Iijima and Utada 1966). Zeolite zones formed by burial diagenesis of silicic tuffs are fully developed in Cenozoic and Cretaceous marine sequences in Japan. However, zeolite zones formed by burial diagenesis of mafic tephra are not well understood and few have been studied.

Zeolite zones formed by burial diagenesis of silicic tuffs in marine sequences

Several tens of exploration wells sponsored by the Ministry of International Trade and Industry (MITI) of Japan have been drilled in Japanese arc-related basins, including small oil and gas fields. Locations of the MITI-boreholes referred to in this section are shown in Figure 2. Based on the analysis of core and cutting samples recovered from the boreholes as well as surface zonal mapping, regional and vertically stacked zeolite zones formed by burial

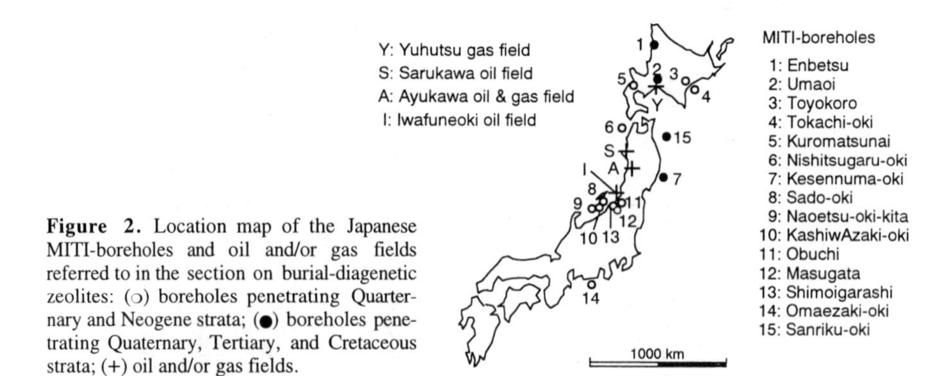

Figure 2. Location map of the Japanese MITI-boreholes and oil and/or gas fields referred to in the section on burial-diagenetic zeolites: (○) boreholes penetrating Quaternary and Neogene strata; (●) boreholes penetrating Quaternary, Tertiary, and Cretaceous strata; (+) oil and/or gas fields.

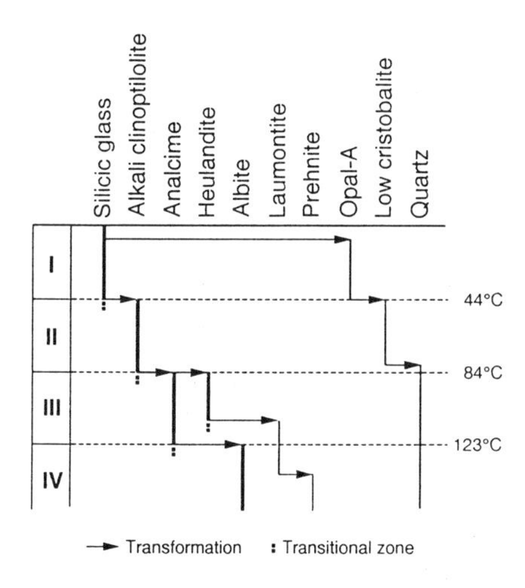

Figure 3. Zeolite and silica-mineral zoning formed by burial diagenesis of silicic tuffs in thick marine sequences. Zeolite zones are defined as the first occurrence of neoformed zeolites or albite. Neoformed zeolites or albite generally coexist with precursors in transitional zones 100-700 m thick. Temperatures at zeolite-zone boundaries were averaged from eight boreholes in which burial diagenesis is presently progressing (Iijima and Ogihara 1995).

diagenesis of silicic tuffs in thick marine Cenozoic and Cretaceous sequences have been documented in these basins (e.g. Iijima and Utada 1966, 1971; Utada 1970, Iijima 1978, 1988b, 1995). Four diagenetic zones extend from the surface downward: an unaltered silicic glass zone (zone I), an alkali-clinoptilolite zone (zone II), an analcime zone (zone III), and an albite zone (zone IV). Mordenite is frequently associated with clinoptilolite in zone II. Depending on the nature of the calcic zeolites present, zone III may be further subdivided into two subzones, an upper IIIa (heulandite) zone and a lower IIIb (laumontite) zone, although laumontite is very localized. Such zonation formed by two reaction series progressing as burial depth and temperature increase: (1) silicic glass in the tuffs hydrates to form alkali-clinoptilolite, which alters to analcime, which in turn alters to albite, and (2) alkali-clinoptilolite transforms to heulandite, which potentially alters to laumontite (Fig. 3).

Silicic glass fragments in zone I hydrate and alter to a mixture of smectite and opal at their margins. This reaction plays an important role in preserving the pseudomorphic shapes of subsequent zeolitized and albitized glass shards in the deeper zones. Very fine-grained glass shards and tiny pumice grains tend to change completely to aggregates of smectite and opal within zone I. Smectite transforms to mixed-layer clay minerals and alters finally to chlorite and/or illite with increased burial depth and temperature (Velde and Iijima 1988). Opal crystallizes to low-cristobalite at the top of zone II, which transforms to quartz in the lowermost part of zone II.

Transitional zones with a thickness of 100-700 m, in which the neo-formed zeolites or albite and their precursors coexist, are recognized at all zone boundaries. The top of each zeolite zone is defined as the first occurrence of neo-formed zeolite or albite. Dissolution structures, such as etch hillocks and etch pits, have been observed on the surfaces of glass fragments, alkali clinoptilolite (Ogihara and Iijima 1989, Ogihara 1996), and analcime (Watanabe et al. 1986) in the transitional zones at the boundaries between zones I and II, zones II and III, and zones III and IV, respectively. From these data, the above zeolitic reactions appear to occur by microdissolution-precipitation processes only in the transitional zones.

Clinoptilolite of zone II. Clinoptilolite of zone II occurs as euhedral thin tabular and prismatic crystals that fill voids of dissolved glass shards lined by a 1-3 μm thick film of smectite and low-cristobalite. The crystal size of clinoptilolite in five MITI boreholes (MITI-Naoetsu-oki-kita, MITI-Nishitsugaru-oki, MITI-Tokachi-oki, MITI-Kesennuma-oki, and MITI-Toyokoro) is nearly constant, 10-50 μm, irrespective of depth and geologic age of the host rocks, which varies from early Pleistocene to Late Cretaceous (Ogihara and Iijima 1989, 1990; Ogihara 1996). This indicates that the transformation of glass to clinoptilolite and the crystal growth of clinoptilolite occur only in the transitional zone in the uppermost part of zone II.

The chemical composition of clinoptilolite of zone II is variable, not only in different boreholes but also at different depths within one borehole. The Si/(Al+Fe) ratio of the clinoptilolite framework ranges from 3.9-5.1, and its variation at any depth is 0.4-0.9 in the five MITI-boreholes (Ogihara and Iijima 1989, 1990; Ogihara 1996). The Si/(Al+Fe) ratio of clinoptilolite changes randomly with burial depth throughout zone II, and its changing pattern is parallel to that of volcanogenic plagioclase composition. These observations appear to suggest that clinoptilolite composition is principally controlled by the composition of parent volcanic glass shards and has thereafter not undergone diagenetic modification during burial (Ogihara and Iijima 1989, 1990). The extra-framework cation ratio of clinoptilolite in zone II is also variable from borehole to borehole, but it tends to show one or two styles within the same borehole; i.e. Na-rich in the MITI-Kesennuma-oki, Na-K-rich in the Toyokoro, and Na-K rich in the upper part with K-rich in the lower part of the MITI-Naoetsu-oki-kita, MITI-Nishitsugaru-oki, and MITI-Tokachi-oki.

Transformation of clinoptilolite to analcime or heulandite. The transformation of clinoptilolite to analcime or heulandite has been investigated using core and cutting samples from the transitional zone between zones II and III in five MITI-boreholes (Ogihara and Iijima 1989, 1990; Ogihara 1996). Significant extra-framework cation exchange and dissolution of clinoptilolite have taken place in the transitional zone, prior to the precipitation of analcime and heulandite.

Na-K-clinoptilolite of zone II in the MITI-Toyokoro borehole begins to appear at ~1050 m, and analcime coexists with its precursor clinoptilolite between 2070 m and 2710 m (Ogihara 1996). From 2590 m to 2710 m, Na-clinoptilolite and Na-K-clinoptilolite coexist in the same voids of dissolved glass shards. From 2590 m to 2650 m, Na-K-clinoptilolite occurs in the central part of the voids, whereas Na-clinoptilolite is found as a thick rim surrounding Na-K-clinoptilolite (Fig. 4A). The boundary of the two types of clinoptilolite is clear, and a single crystal may actually be divided into two types (Fig. 4B). Analcime is confined to small crystals and is very rare in this depth range. At 2710 m, a great deal of analcime coexists with Na-clinoptilolite and minor Na-K-clinoptilolite; their compositional relation and formational sequence are shown in Figure 5. Analcime surrounded by Na-clinoptilolite shows obvious clear facets. When Na-clinoptilolite coexists with analcime in single voids of dissolved shards, the surfaces of Na-clinoptilolite show dissolution textures (Fig. 4C). At 2770 m where no clinoptilolite is detected, analcime fills not only the voids of glass shards but also intergranular space (Fig. 4D).

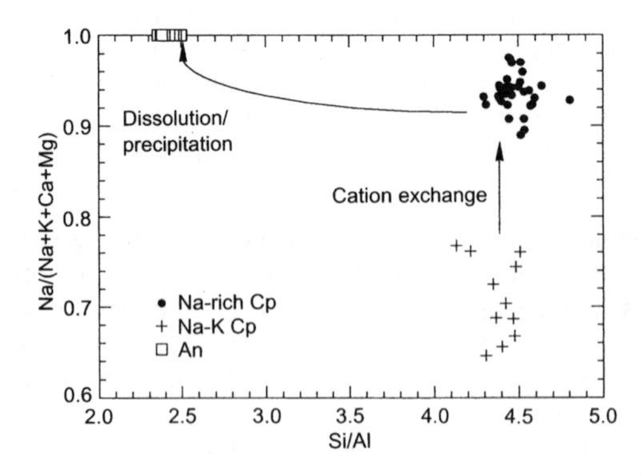

Figure 5. Relationships between Na/(Na+K+Ca+Mg) and Si/Al in Na-K clinoptilolite, Na-rich clinoptilolite, and analcime at 2710-m depth in the MITI-Toyokoro borehole, east Hokkaido, Japan. Arrows show the sequence of formation (Ogihara 1996).

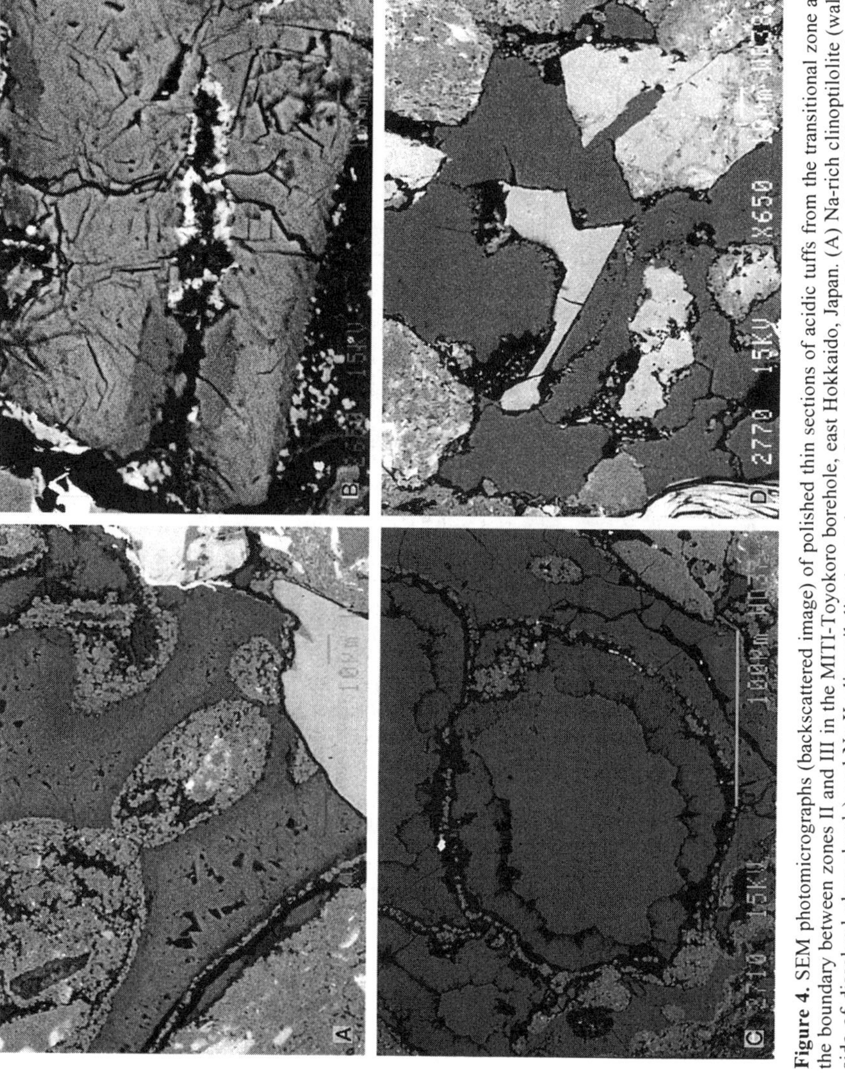

Figure 4. SEM photomicrographs (backscattered image) of polished thin sections of acidic tuffs from the transitional zone at the boundary between zones II and III in the MITI-Toyokoro borehole, east Hokkaido, Japan. (A) Na-rich clinoptilolite (wall side of dissolved glass shards) and Na-K clinoptilolite (central part of the glass shards: light gray) at 2650 m. (B) Single clinoptilolite crystals in a void of dissolved glass shard divided into Na-K type (light gray) and Na-rich type (gray) at 2650 m. (C) Faceted analcime (central part of voids of dissolved glass shards) surrounded by Na-rich clinoptilolite with dissolution texture at 2710 m. (D) Analcime (gray) filling voids of dissolved glass shards and cementing intergranular pores at 2770 m (Ogihara 1996).

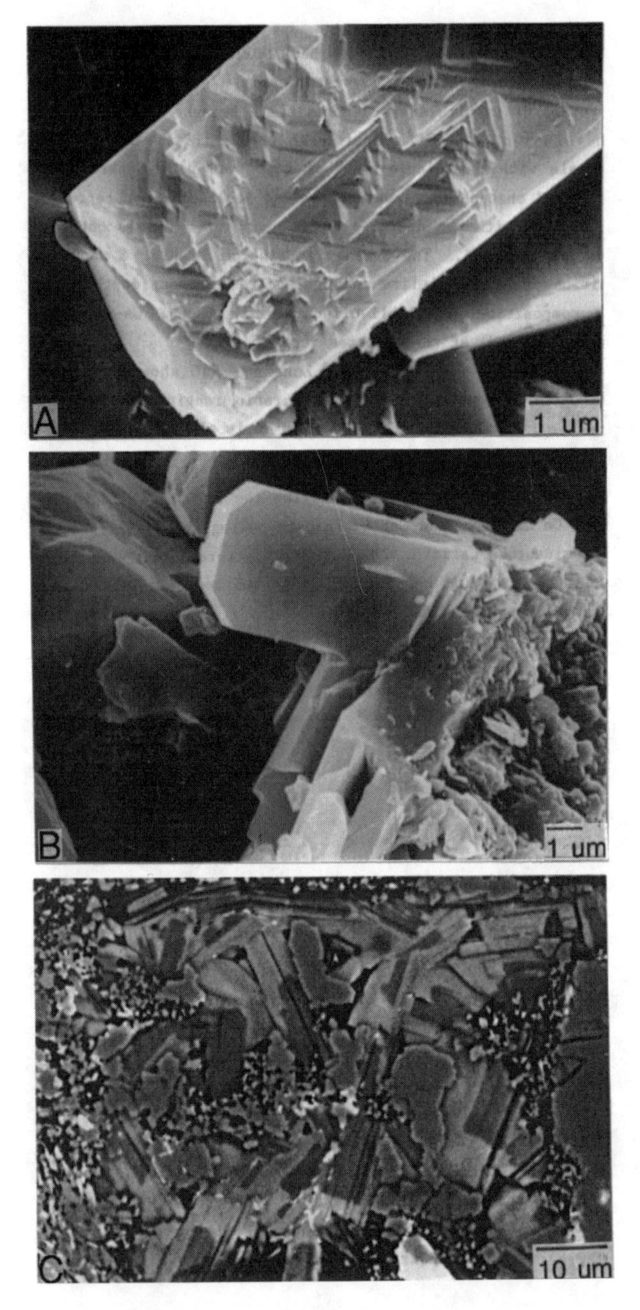

Figure 6. SEM photomicrographs of acidic tuffs from the transitional zone at the II-IIIa zone boundary in the MITI-Kesennuma-oki borehole, Japan. (A) "Etch hillock" texture showing dissolution of Na-clinoptilolite crystals, at 1650 m at the bottom of zone II. (B) A thin-tabular euhedral crystal of Ca-clinoptilolite overgrows the etched and dissolved Na-clinoptilolite crystals at 1650 m. (C) Backscattered image showing heulandite (light) overgrowing Ca-clinoptilolite (dark), both of which formed by a reaction of Na-clinoptilolite with pore water at elevated temperature. Fine-grained aggregates consist of authigenic quartz and chlorite at 1710 m at the top of zone III. Reprinted from *Zeolites: Facts, Figures, Future, Studies in Surface Science and Catalysis 49A*, S. Ogihara and A. Iijima (1989), with kind permission of Elsevier Science–NL.

Na-clinoptilolite of zone II in the MITI-Kesennuma-oki borehole occurs from 540 m to 1650 m (Ogihara and Iijima 1989). At 1650 m, Ca-clinoptilolite and Na-clinoptilolite coexist in the same voids of dissolved glass shards. The surfaces of some Na-clinoptilolite crystals in the marginal part of the voids show dissolution textures (Fig. 6A), and idiomorphic Ca-clinoptilolite crystals overgrow the dissolved Na-clinoptilolite (Fig. 6B). As a result, Ca-clinoptilolite tends to occupy the central part of the voids whereas Na-clinoptilolite occupies the marginal areas. From 1710 m to 1770 m, Ca-heulandite and Ca-clinoptilolite coexist as zoned crystals, and they fill the voids of dissolved glass shards (Fig. 6C). Most Ca-heulandite occurs as overgrowths on Ca-clinoptilolite cores although high-silica Ca-heulandite also comprises the cores. The zoned crystals of Ca-heulandite–Ca-clinoptilolite show a distinct thick-tabular habit with a length of more than 20-40 μm, which contrasts with thin tabular and smaller (10-20 μm) crystals of Na-clinoptilolite of zone II. The increase in crystal size results in a decrease in the number of crystals within a given dissolved glass shard void as compared with zone II, commonly inhibiting free crystal growth (Fig. 6C). Little open space remains among the zoned crystals, and, if available, it is frequently filled by tiny prismatic quartz crystals precipitated from the excess silica released by dissolved clinoptilolite. Zoned crystals of K-heulandite–K-clinoptilolite were found in the uppermost part of zone III in three other MITI-boreholes (Ogihara and Iijima 1990). The thick tabular zoned crystals are also significantly coarser grained than the thin tabular crystals of clinoptilolite in zone II (Fig. 7).

Figure 7. SEM photomicrographs showing the difference in crystal size of K-clinoptilolite of zone II and K-heulandite of zone IIIa in the MITI-Nishitsugaru-oki borehole, off Tsugaru, northeast Honshu, Japan. (A) Thin, tabular K-clinoptilolite filling a void of dissolved silicic glass shards at 1960 m. (B) Zoned crystals with K-clinoptilolite cores and K-heulandite overgrowths, at 2160 m (Ogihara and Iijima 1990).

The transformation of low-cristobalite to quartz precedes the change of clinoptilolite to analcime or heulandite (Iijima 1988b). For instance, low-cristobalite disappears at 1900 m, 170 m shallower than the appearance of analcime, in the MITI-Toyokoro borehole (Ogihara 1996). This fact suggests that the silica activity in pore water probably increased in the lower part of zone II with an increase in burial temperature and helped trigger the decomposition of clinoptilolite.

Analcime occurs as an alteration product of volcanic glass fragments in volcanic sandstones of the Oligocene Frio Formation of Gulf Coast Texas (Lynch 1996). Burial-diagenetic analcime is common in shore-zone and shelf sandstones in the growth-fault block 2, which contains highly saline formation water with a high Na^+/H^+ ratio. Lynch (1996) stated that lower-energy back-barrier sandstones include significant amounts of volcanic ash and clay that are replaced by syndepositional calcite and analcime. However, evidence to distinguish between burial-diagenetic and "syndepositional" analcimes was not described.

Temperature-time relationships of burial-diagenetic zeolite reactions. It is very important in petroleum geology to clarify the thermal history of sedimentary basins because paleo-geothermal gradients, maturity of kerogen in source rocks, and eroded overburden can be estimated from the thermal history. Temperature-time relationships of zeolitic reactions in burial diagenesis of silicic tuffs in marine sequences have been analyzed using the data from eleven MITI-boreholes (Iijima and Ogihara 1995). In eight boreholes, burial diagenesis is interpreted as being presently active. In other words, zeolite reactions are occurring in silicic volcaniclastic rocks at the present depth and temperature at the boundaries between zones I and II, II and III, and III and IV (Iijima and Utada 1971, Iijima 1988a,b; Ogihara and Iijima 1990, Iijima and Ogihara 1995). Quaternary to Neogene marine sequences in these boreholes have not undergone extensive tectonic deformation, local hydrothermal alteration, or igneous intrusion, so pore fluid is considered as inherited from trapped sea water (Iijima 1988a, b). Temperatures at the zone boundaries were estimated from the geothermal gradients calculated from corrected static bottom-hole temperatures. In contrast, diagenetic zoning in the MITI-Enbetsu and MITI-Kesennuma-oki boreholes was deformed by later tectonic movements and denudation of overburden (Iijima 1975, Ogihara and Iijima 1989). The zoning in the MITI-Kuromatsunai borehole was modified by recent geothermal activity as well (Iijima et al. 1984). The depth and temperature at the zone boundaries at the time of the deepest burial, at which the zeolitic reactions occurred, were estimated from the amount of denuded overburden and past geothermal gradient. Consequently, the estimated temperatures of these three boreholes are less accurate than those of the above eight boreholes in which diagenesis is presently occurring (Iijima et al. 1984, Iijima 1988b, Ogihara and Iijima 1989).

Three different times are considered in the temperature-time relations of zeolite reactions. The first time (T_{rock}) is simply the rock ages at the zone boundaries. The second time (T_{max}) is defined as the difference between T_{rock} and the time of maximum burial of the zone boundaries (t_m). In sequences in which burial diagenesis is presently active, t_m is zero; hence $T_{max} = T_{rock}$. Thus, T_{max} indicates the entire time between the deposition of tuffs at the zone boundaries and their zeolitization at maximum burial. The third time (T_{res}) is defined as the residence time of silicic glass, clinoptilolite, and analcime in zone I, II, and III, respectively. Table 2 shows depth, temperature, and T_{rock} of the volcaniclastic rocks at the boundaries between zones I and II, zones II and III, and zones III and IV at t_m in the eleven boreholes. T_{res} and maximum temperature of silicic tuffs in zone I, clinoptilolite in zone II, and analcime in zone III, are given in Table 3.

The temperature-time relationships of zeolitic reactions in burial diagenesis of silicic tuffs in marine sequences are illustrated in Figure 8. In the transformation of silicic glass (zone I) to alkali clinoptilolite (zone II), temperatures are not related to T_{rock}, T_{max}, and T_{res}, considering the small correlation coefficients, 0.247-0.265. The temperatures fall within a rather narrow range between 41° and 53°C as compared with the time span of 72.7 Myr (T_{rock}), and 42.7 Myr (T_{max} and T_{res}). This suggests that the transformation of silicic glass to alkali clinoptilolite in normal marine sequences depends not only on temperature but also on pore water chemistry, as inferred from the fact that silicic vitric tuffs in alkaline,

Table 2. Burial depth, temperature (T), and rock age (T_{rock}) at zonal boundaries of zeolitic zones formed by burial diagenesis of silicic tuffs in thick marine sequences in 11 MITI-boreholes in Japan.

MITI-borehole	*Boundary of zones I-II*			*Boundary of zones II-III*			*Boundary of zones III-IV*			
	Depth (m)	*T (°C)*	T_{rock} *(Ma)*	*Depth (m)*	*T (°C)*	T_{rock} *(Ma)*	*Depth (m)*	*T (°C)*	T_{rock} *(Ma)*	*Age of maximum burial (t_m, Ma)*
Masugata	1040	49	0.4	3480	88	3.1	4500	124	13.8	0
Ohbuchi	760	41	1.0	2900	84	5.7	4160	122	13.0	0
Shimoigarashi	1980	45	1.6	3500	91	3.7	4500	124	5.6	0
Nishitsugaru-oki	870	43	1.4	1670	81	3.0	-	-	-	0
Naoetsu-oki-kita	688	41	1.2	1988	81	4.3	-	-	-	0
Tokachi-oki	1415	41	7.8	-	-	-	-	-	-	0
Omaezaki-oki	1491	50	16.4	-	-	-	-	-	-	0
Sado-oki	1110	43	0.9	2140	80	9.4	-	-	-	0
Kuromatsunai	1350	53	4.4	2360	85	7.1	3450	122	13.2	0.5
Kesennuma-oki	1400	46	73.1	2310	76	77.3	-	-	-	30.0
Enbetsu	1100	46	7.0	2565	86	15.8	4860	120	84.0	2.0

Table 3. Residence time (T_{res}) and maximum temperature (T) of silicic glass of zone I, clinoptilolite of zone II, and analcime of zone III in marine sequences of 11 MITI-boreholes in Japan.

MITI-borehole	*Silicic glass of zone I*		*Clinoptilolite of zone II*		*Analcime of zone III*	
	T_{res} (myr)	*T (°C)*	*T_{res} (myr)*	*T (°C)*	*T_{res} (myr)*	*T (°C)*
Masugata	0.4	49	1.9	88	1.4	124
Ohbuchi	1.0	41	4.4	84	2.2	122
Shimoigarashi	1.7	45	1.1	91	0.7	124
Nishitsugaru-oki	1.4	43	1.2	81	-	-
Naoetsu-oki-kita	1.2	41	2.4	81	-	-
Tokachi-oki	7.8	41	-	-	-	-
Omaezaki-oki	16.4	50	-	-	-	-
Sado-oki	0.9	43	0.9	80	-	-
Kuromatsunai	3.9	53	3.0	85	3.2	122
Kesennuma-oki	43.1	46	6.7	76	-	-
Enbetsu	5.0	46	11.3	86	13.1	120

saline-lake deposits are altered to alkali clinoptilolite and other zeolites at near-surface conditions (Hay 1966, Surdam and Sheppard 1978). In addition, the wide compositional range of silicic glass may affect the temperatures separating zones.

In the transformation of alkali clinoptilolite (zone II) to analcime and heulandite (zone III), there is apparently an inverse relation between temperature and T_{rock} and T_{max}, although the correlation coefficients are not high (0.590 and 0.587). In addition, this apparent relation is strongly dependent on the point at 76°C and 73.1 Ma (T_{rock}) and 43.1 Myr (T_{max}) in the MITI-Kesennuma borehole, and considerable error is suspected in the temperature estimation in the borehole, as stated above. In fact, it is difficult to recognize any relation between temperature and T_{res}, where T_{res} is 6.7 Myr in the MITI-Kesennuma borehole, considering the small correlation coefficient (0.167). Four transformations occur at the boundary between zones II and III: (1) Na-clinoptilolite to analcime, (2) K-clinoptilolite to analcime, (3) Na-clinoptilolite to zoned Ca-clinoptilolite–Ca-heulandite, and (4) K-clinoptilolite to K-heulandite (Ogihara and Iijima 1989, 1990; Ogihara 1996). The wide temperature range from 76° to 91°C may reflect not only errors in temperature estimation but also the variety of chemically interconnected transformations at the boundary between zones II and III. The more reliable, rather limited temperature range between 80° and 91°C in the Neogene sequences, moreover, suggests that the dominant cations of clinoptilolites of zone II do not significantly affect the transformation temperature.

In the transformation of analcime to albite at the boundary between zones III and IV, temperatures vary inversely with the three measures of time, with high correlation coefficients, although the rate of temperature decrease is very small, i.e. 4°C for 78.4 Myr (T_{rock}), 76.4 Myr (T_{max}), and 12.4 Myr (T_{res}). This relationship is dependent on the point at 120°C and 84 Ma (T_{rock}), 86 Myr (T_{max}), and 13.1 Myr (T_{res}) in the MITI-Enbetsu borehole, although substantial error is suspected in temperature estimation in the borehole,

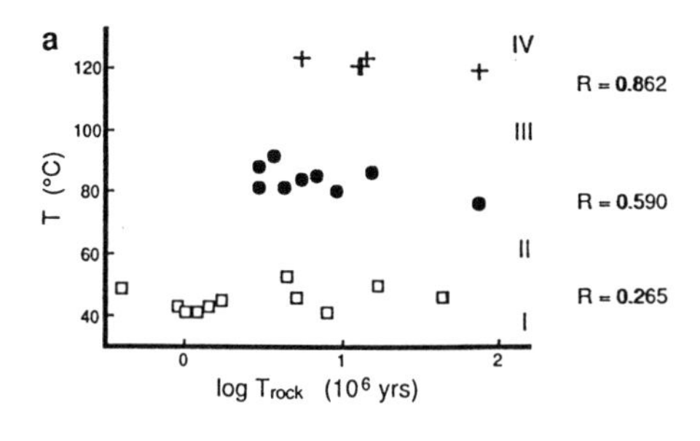

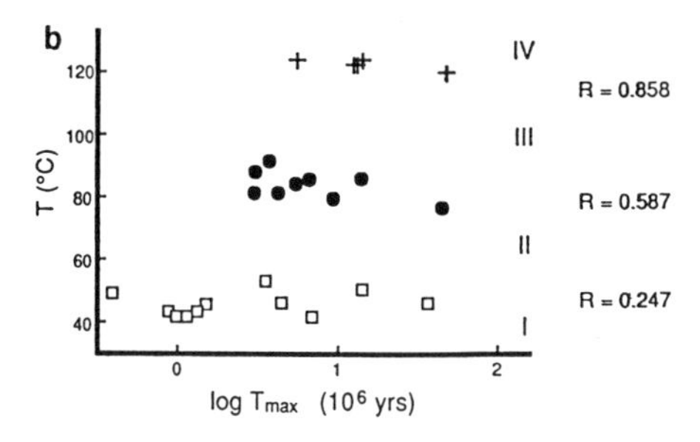

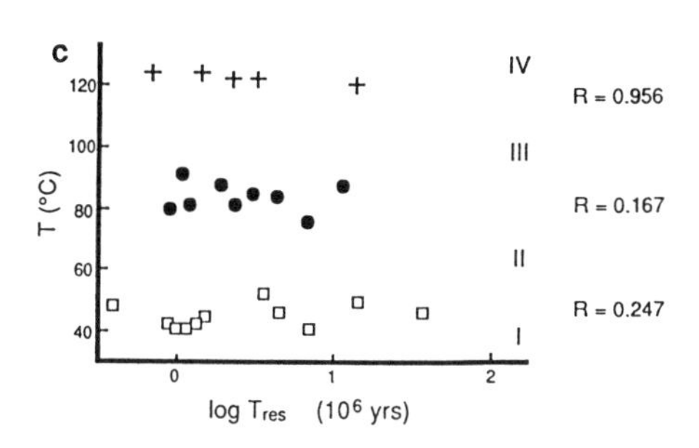

Figure 8. Temperature versus different types of time relationships of zeolitic reactions during burial diagenesis of silicic tuffs in marine sequences. (a) Relationship between temperature and rock ages (T_{rock}) of zeolite zones at the boundaries between zones I and II (□), between zones II and III (●), and between zones III and IV (+). (b) Relationship between temperature and rock ages at maximum burial (T_{max}) of zeolite zones at the boundaries between zones I and II (□), between zones II and III (●), and between zones III and IV (+). (c) Relationship between temperature and residence time (T_{res}) of silicic glass in zone I (□), of alkali clinoptilolite in zone II (●), and of analcime in zone (+) (Iijima 1995).

as stated above. More data are required to substantiate this relationship.

The temperature-T_{max} relationship is geologically more significant than the temperature-T_{rock} relationship, because zeolitic reactions in burial diagenesis do not proceed with decreasing temperatures during denudation (*i.e.* retrograde reactions do not appear to occur). The temperature-T_{res} relationship is most relevant, because analcime and heulandite of zone III and albite of zone IV are not directly transformed from silicic glass of zone I but from clinoptilolite of zone II and from analcime of zone III, respectively. However, the general trends of the three kinds of the temperature-time diagrams are somewhat similar to one another.

The zeolite zones in these boreholes are temperature dependent, i.e. the temperature ranges from 41° to 50°C, 44°C on an average at the boundary between zones I and II; it ranges from 80° to 91°C, 84°C on an average at the boundary between zones II and III; and it ranges from 122° to 124°C, 123°C on average at the boundary between zones III and IV, in the time span of 6-84 million years (Iijima and Ogihara 1995). Therefore, the zeolite zone boundaries formed by burial diagenesis of silicic tuffs in marine sequences may be useful as geothermometers.

Relation of burial-diagenetic zeolites to reservoir properties. The primary matrix porosity of volcaniclastic reservoirs is a function of the maximum burial depth. A schematic burial depth vs. matrix porosity diagram of coarse pyroclastic and volcaniclastic rocks in the Japanese Quaternary-Neogene sedimentary basins is shown in Figure 9. This relationship is crudely correlated to that of epiclastic coarse sandstones. In Figure 9, geothermal gradients of 2°, 3°, and 4°C per 100 m, zeolite zones, and oil maturity zones are shown. The porosity range in the mature zone (oil window) increases in size with an increase in geothermal gradient, due to a decrease in burial depth.

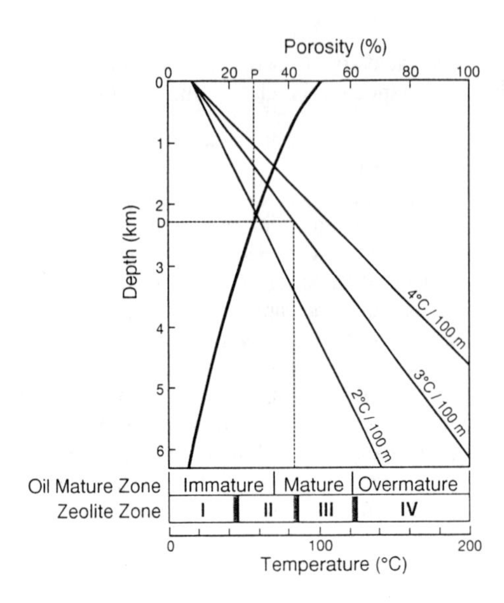

Figure 9. Burial depth vs. porosity diagram of coarser pyroclastic and volcaniclastic rocks in Tertiary marine sequences, showing geothermal gradients, oil mature zones, and zeolite zones I to IV formed by burial diagenesis of silicic volcaniclastic rocks. Depth (D) and porosity (P) for oil mature zone and zeolite zone can be obtained graphically: (1) draw a vertical line from the zone to an appropriate geothermal slope; (2) draw a horizontal line from the intersection between the above line and the slope toward the depth (D); and (3) draw a vertical line from the intersection between the horizontal line and the porosity-depth curve toward the porosity (P) (Iijima 1995).

Porous pyroclastic and volcaniclastic rocks of zone I retain as much as 40% porosity. For example, in the Iwahune-oki offshore oil and gas field of Niigata, volcanic sandstone reservoirs of the Lower Pliocene Nishiyama Formation have a porosity of 15-30%, and a permeability of 5-500 md (Uchida and Tada 1992).

Zeolitic rocks of zone II possess sufficient matrix porosity (20-40%) to serve as commercial reservoirs. The porosity consists of both the primary intergranular pores and the secondary intragranular pores of dissolved glass shards, the latter of which are composed principally of the intercrystalline space between clinoptilolite crystals (Fig. 10). The crystal size of clinoptilolite is constant throughout zone II, as stated above, so that the decrease in porosity with burial is due mainly to physical compaction. Dunn et al. (1993) argued that the occurrence of zeolite crystals and their morphology affect reservoirs of Cretaceous volcaniclastic sandstones in the San Jorge basin, Argentina. "Lath-shaped and tabular crystals of clinoptilolite typically line pores and form loose aggregates adjacent to pore throats, greatly reducing pore throats radii and increasing mean free path tortuosity. This morphology and occurrence reduce permeability, and provide mobile crystals, which are susceptible to "brush-piling" at pore throats and clogging flow. Clinoptilolite is commonly accompanied by smectite and smectite mixed-layer clays, the presence of which also reduces permeability." However, zeolite-bearing reservoirs are most common in zone II. For example, in the Sarukawa oil field of Akita, clinoptilolite-bearing tuff reservoirs of the Upper Miocene Funa -kawa Formation have a porosity of 15-31%, and a permeability of 80-360 md (J.A.N.G.M. and A.O.O.D. 1992). The pore-size distribution of the reservoirs measured by a mercury porosimeter shows a nearly normal distribution, with median pore radii of around 10,000 Å (Uchida 1984). In the MITI-Kashiwazaki-oki borehole, clinoptilolite-bearing volcanic sandstone reservoirs of the Upper Miocene Shiiya Formation retain a porosity of 25-33% (J.N.O.C. 1988a). In the Ayukawa oil and gas field of Akita, altered vitric siliceous rocks of the Middle Miocene Onnagawa Formation have sufficient porosity to act as reservoirs (Araki and Kato 1993). This porosity consists of intercrystalline pores of clinoptilolite filling voids of dissolved silicic glass shards in a matrix of clays and authigenic microcrystalline quartz of diatomite origin.

Porous tuffs of zone III also retain good porosity (15-35%). For instance, in the MITI-Kashiwazaki-oki borehole, a volcanic sandstone reservoir of the Middle Miocene lower Teradomari Formation within zone III has a porosity of 18-23% and a permeability of 4-32 md (J.N.O.C. 1988a). Intragranular pores of voids of dissolved glass shards are

Figure 10. SEM micrographs showing intergranular and secondary pores of dissolved glass shards in the reservoir acidic tuff of zone II in the Sarukawa oil field, Akita. (A) Microcrystalline aggregates of clinoptilolite filling voids of dissolved glass shards lined by thin films of smectite and cristobalite. Several intergranular pores are also seen. (B) Intercrystalline pores of tabular clinoptilolite, enlarging a part of A (Uchida 1984).

largely filled by analcime and heulandite in zone III, as stated above.

No albitized volcanogenic reservoirs have been recognized in zone IV, for they usually have very low porosities and permeabilities, due to large overburden and intense recrystallization. Albite-quartz rocks which are capable of acting as natural gas reservoirs in the Green Tuff formation of Japan are not burial-diagenetic but are of submarine hydrothermal origin, as will be discussed below.

Relationship between timing of burial-diagenetic zeolites and hydrocarbon accumulation. Kerogen in source rocks transforms with an increase in burial depth and temperature; and its transformation paths are different in three distinguished kerogen types (Tissot and Welte 1978), as shown in It is well known that the maturity of kerogen correlates crudely to reflectance of vitrinite (R_o), which is a function of temperature and time. In the Neogene sequence of Japan, the mature zone (oil window) exists between 0.5 and 1.3% R_o. Boundaries of the zeolite zones relative to the R_o-values are also drawn in Figure 11, with the lower half of zone II and all of zone III corresponding to the oil mature zone, i.e. the oil window.

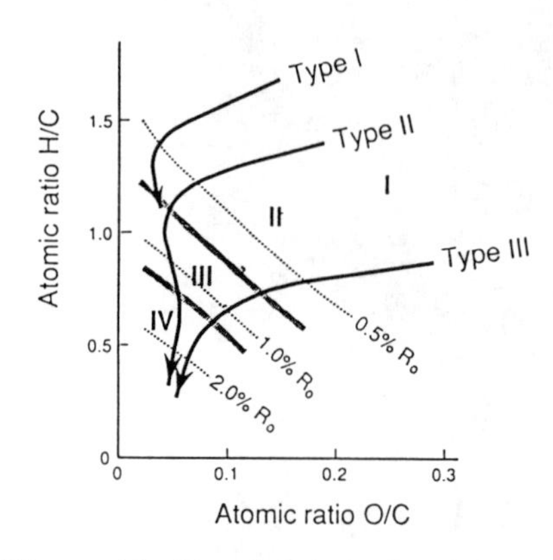

Figure 11. Van Krevelen's diagram of three types of kerogen with vitrinite reflectance (R_o) (simplified from Tissot and Welte 1978), showing zeolite zones I to IV formed by burial diagenesis of silicic volcaniclastic rocks in Quaternary–Cretaceous marine sequences (Iijima 1995).

Zeolitization occurs in volcaniclastic and volcanic rocks during burial diagenesis, whereas hydrocarbons are generated from organic-rich mudrocks. In Neogene sequences, the lower clinoptilolite zone (zone II) and the analcime zone (zone III) correspond to an oil window. The transformation of silicic glass to clinoptilolite occurs in a narrow interval at the top of zone II (Ogihara and Iijima 1989, 1990). Hence silicic glass fragments of volcaniclastic rocks are completely altered to clinoptilolite and other secondary minerals before nearby source rocks enter the oil window. Zeolites are formed during the process of burial of the host rocks with an increase in temperature, and reservoir structures are largely produced during uplift. Hydrocarbons generated in the source rocks migrate either upwards from the depth of the structures or laterally from nearby synclinal structures through porous permeable beds and accumulate in reservoir rocks. Therefore, the accumulation of hydrocarbons postdates zeolitization of reservoir rocks in burial diagenesis. The postdated accumulation of oil in clinoptilolite-bearing rock reservoirs is evidenced by a residual oil drop on the surfaces of clinoptilolite crystals in the tuff reservoir of the Sarukawa oil field, Akita (Fig. 12). Accumulation of hydrocarbons may predate the zeolitization of reservoir rocks in burial

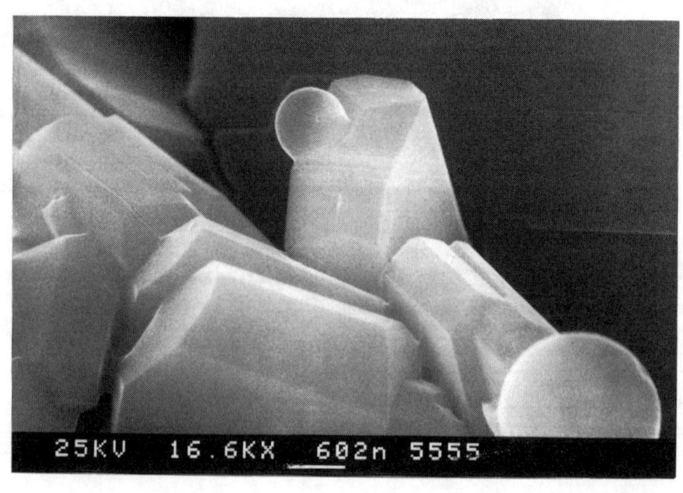

Figure 12. Residual oil drops on clinoptilolite crystals in the acidic tuff reservoir of zone II in the Sarukawa oil field, Akita, Japan (SEM micrograph courtesy of Takashi Uchida).

diagenesis in some stratigraphic traps and anticlinal traps formed by differential rates of subsidence, but no such case has been reported.

Zeolites formed by a combination of burial diagenesis and saline, alkaline pore-fluid diagenesis in marine-freshwater sequences

Regional zeolite zoning has recently been found in the Quaternary-Upper Cretaceous marine to freshwater sequences recovered in the MITI-Sohma-oki and MITI-Johban-oki boreholes in the Johban-oki offshore district, including the Iwaki-oki offshore gas field on the east side of northeastern Honshu (Iijima and Ogihara 1995) (Fig. 13). Zeolite zones in the two boreholes are shown on the temperature vs. depth diagram in Figure 14. The present geothermal gradients were obtained from corrected bottom-hole temperature measurements. Considering the limited extent of denudation, a few unconformities in the sequences do not significantly influence the zeolite zones, so that burial diagenesis is apparently presently taking place in the sequences.

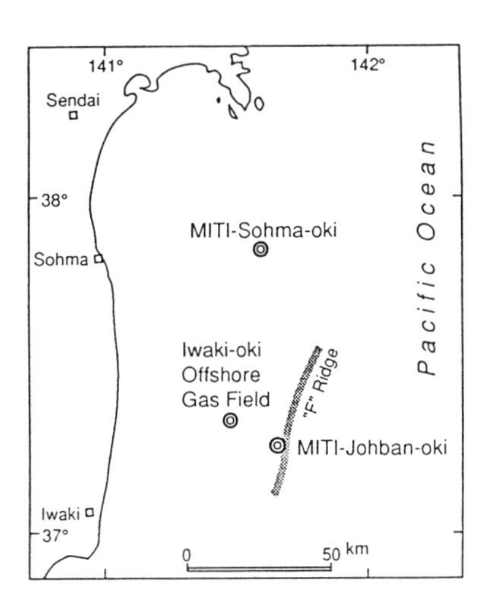

Figure 13. Locality map of the Johban-oki offshore district on the east side of northeast Honshu Island, Japan (Iijima 1995).

The Quaternary-Upper Cretaceous sequences are divided into a frèsh silicic glass zone lacking zeolites (zone 1), a clinoptilolite zone (zone II), and an analcime-heulandite zone (zone III). In zone II, silicic glass hydrates and alters to Na-K-clinoptilolite and scattered mordenite. In zone III, alkali clinoptilolite is transformed to analcime and heulandite. This zoning is the same as that produced by burial diagenesis of silicic tuffs in marine sequences. Nevertheless, the zoning was produced at much lower temperatures than in burial diagenesis (Fig. 14). Based on the present geothermal gradient, temperatures are estimated to be 21° and 34°C at the boundary between zones I and II, and 37° and 51°C at the boundary between zones II and III in the MITI-Johban-oki and MITI-Sohma-oki boreholes, respectively. In contrast, the mean temperatures in normal burial diagenesis are 44°C at the boundary between zones I and II and 84°C at the boundary between zones II and III (Iijima and Ogihara 1995). Also, silica mineral zones composed of a biogenic opal-A zone, an opal-CT zone, and a quartz zone show temperatures of 22-35°C at the boundary between opal-A and opal-CT zones, and 35-50°C at the boundary between opal-CT and quartz zones. These temperature values are lower than those found for normal burial diagenesis (Iijima 1988b). A past high geothermal gradient is unlikely, based on the fact that diagenetic transformation of organic matter is consistent with the present geothermal gradient; e.g. vitrinite reflectance R_0 = 0.5% corresponds to 67° and 73°C in the Upper Cretaceous of the MITI-Johban-oki and MITI-Sohma-oki boreholes, respectively. In thick marine sequences, 0.5% R_0 was measured in the middle part of the alkali clinoptilolite zone (zone II) under normal burial diagenesis, whereas this value was measured in the middle to lower part of the analcime-heulandite zone (zone III) in the Johban-oki offshore district.

Other peculiar diagenetic factors in the Johban-oki offshore district are the common presence of a zone of feldspar-dissolution and the occurrence of authigenic K-feldspar and laumontite in Paleogene-Upper Cretaceous lithic-feldspathic arenites in zone III (Fig. 14).

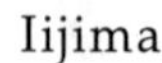

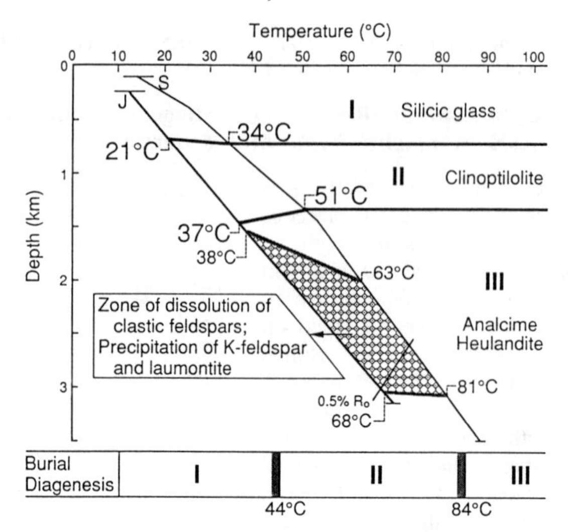

Figure 14. Temperature vs. depth diagram of the MITI-Sohma-oki (S) and MITI-Johban-oki (J) boreholes in the Johban-oki offshore district, northeast Honshu, Japan showing zeolite zones I to III, zone of dissolution of feldspars and carbonate cement, and zone of precipitation of K-feldspar and laumontite in lithic-feldspathic sandstones. Although the zones are mineralogically the same as those formed by burial diagenesis, the zeolite zones occur at much lower temperatures than those in burial diagenetic sequences. R_0: reflectance of vitrinite (modified from Iijima 1995).

In these arenites, clastic feldspar grains (oligoclase-andesine plagioclase predominates over orthoclase and microcline) and primary carbonate cements are extensively etched and dissolved to form secondary pores. The secondary pores are cemented to varying degrees with one to three kinds of authigenic minerals such as K-feldspar, laumontite, quartz, microcrystalline aggregates of kaolinite, and rare calcite. The sequence of precipitation of cement minerals is (1) K-feldspar, (2) quartz, (3) kaolinite and (4) calcite. The laumontite content is usually 1-2% in volume. Laumontite does not coexist with other authigenic minerals, but it may have formed after K-feldspar and before quartz. Carboxylic acids would be generated during maturation of organic matter (Surdam et al. 1984) contained in source shales and coal in the nearby syncline. The carboxylic acids along with pore water would be expelled by compaction from the source rocks into porous arenites, and they would migrate in the arenite beds up-dip and along strike to the adjacent anticlines. There, the carboxylic acids dissolved the early-stage carbonate cements and detrital feldspar grains, and produced secondary porosity. Surdam et al. (1989a) stated that organic acids, especially acetic acid caused by decomposition of kerogen, attack K-feldspar and carbonates at temperatures of 75-120°C. However, dissolution occurred at temperatures of 38-68°C at a depth of 1.6-3 km in the MITI-Johban-oki borehole, and at 63-81°C at a depth of 2-3 km in the MITI-Sohma-oki borehole. The absence of albitization of plagioclase is consistent with such lower-temperature alteration.

The above-stated low-temperature zeolitization of silicic glass was probably influenced by a saline, alkaline pore fluid, comparable to that in saline, alkaline-lake deposits (Hay 1966, Surdam and Sheppard 1978). The unusual alkali fluids are assumed to have been caused by the extensive dissolution of early-stage carbonate cements and detrital feldspar in the Upper Cretaceous and Paleogene lithic-feldspathic arenites. This dissolution resulted in concentration of Na^+, K^+, Ca^{+2}, H_4SiO_4, Al^{+3}, HCO_3^- and CO_3^{-2} in the pore waters of the arenites. Then, authigenic K-feldspar, laumontite, quartz, kaolinite, and calcite were precipitated successively from the concentrated pore fluids using the K^+, Ca^{+2}, H_4SiO_4, Al^{+3}, HCO_3^- and CO_3^{-2}. Pore fluids in the arenites likely changed from acid to alkaline, due to remaining Na^+, HCO_3^- and CO_3^{-2}. These alkaline fluids would migrate into the overlying porous marine sediments as compaction of the arenites proceeded, where they would mix with trapped seawater in the tuffaceous sediments. Such addition of Na^+, HCO_3^- and CO_3^{-2} to interstitial water is considered to have promoted the transformation of silicic glass to clinoptilolite and the transformation of clinoptilolite to analcime at much lower temperatures relative to normal burial diagenesis. Thus zeolite zonation in the Johban-oki offshore district maybe a result of a combination of burial diagenesis and saline, alkaline pore-fluid diagenesis.

Very recently, the same zeolite zoning has been found in the MITI-Sanriku-oki borehole in the Sanriku-oki offshore basin on the east side of northeastern Honshu (Fig. 2). Secondary pores formed by dissolution of carbonate cement and feldspar grains develop in gas reservoir feldspathic sandstones in the Middle Eocene to Upper Paleocene coal measures (J.N.O.C. 2000). Poikilotopic laumontite cement fills part of the secondary pores of the sandstones belonging to zone III (analcime-heulandite zone).

Relation of burial diagenesis-saline, alkaline pore-fluid diagenetic zeolites to reservoirs. Lower Miocene sandstones in the clinoptilolite zone (zone II) in the MITI-Johban-oki borehole comprise natural gas reservoirs having a porosity of 25-33% and a permeability of 49-1382 md (J.N.O.C. 1992). In the MITI-Sohma-oki borehole, Eocene-Upper Cretaceous, coarse-grained, lithic-feldspathic arenites in the analcime-heulandite zone (zone III) are effective gas reservoirs and register a porosity of 10-20% and a permeability of =1 md to 17 md, largely due to dissolution of carbonate cement and clastic feldspar grains (J.N.O.C. 1991).

Relationship between zeolitization by burial diagenesis-saline, alkaline pore-fluid diagenesis and hydrocarbon accumulation. Paleogene coal-bearing deposits in the Johban coal field extend to the Johban-oki offshore district. The Iwaki-oki offshore gas field produces methane and some condensates that are believed to be of coal-field gas origin. The MITI-Johban-oki borehole is located at the crest of a gentle and open, faulted, anticlinal structure on the "F" ridge in the Johban-oki district (Fig. 15). Natural gas produced from Lower Miocene sandstone reservoirs within zone II consists principally of thermogenic methane (J.N.O.C. 1992). Source rocks of the gas are not located in shallow, immature strata encountered in the borehole but in deep-seated Paleogene-Uppermost Cretaceous coal-bearing deposits and mudrocks in the synclinal structure between the MITI-Johban-oki borehole and the Iwaki-oki offshore gas field, about 17 km WNW of the borehole. The natural gas is believed to have migrated eastward from the synclinal structure and has accumulated in the anticline on the ridge since latest Pliocene, when the accumulation structure was completed and the source rocks were buried sufficiently to reach their maturity level (Fig. 15). At that time, the sequence of the borehole was shallower by only about 150 m, corresponding to 3°C lower than present temperatures. Hence, the zeolite zoning at that time was almost the same as the present, and the Lower Miocene sandstone reservoirs existed within the clinoptilolite zone (zone II).

Closed-system analcime (saline, alkaline-lake environments) in sandstone and conglomerate reservoirs

Upper Permian lacustrine oil shales are widespread in the Junggar basin, Xinjang, northwest China, and they comprise the principal oil source rocks of the Karamay oil field (Graham et al. 1990). The Upper Permian Lower Wuerhe and Pingdiquan Formations were deposited in fan-delta sequences within a lacustrine setting (Chen 1992, Tang et al.

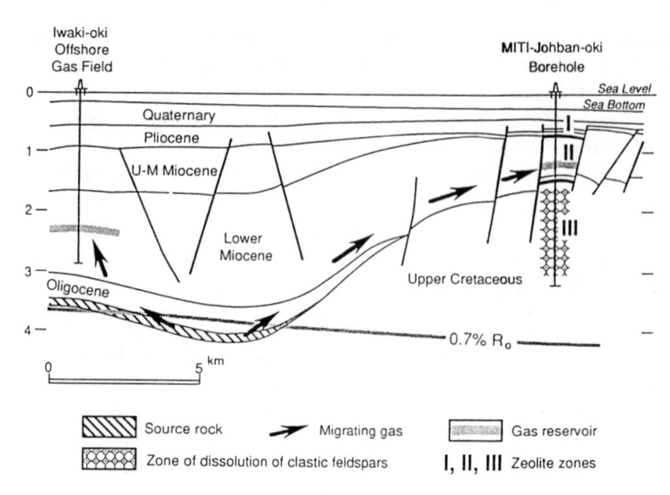

Figure 15. Cross-section through the MITI-Johban-oki borehole and the Iwaki-oki offshore gas field in the Johban-oki offshore district, northeast Honshu, Japan. Natural gas produced from Lower Miocene sandstone reservoirs has probably migrated from the Paleocene–uppermost Cretaceous coal-bearing deposits and mudrocks in the synclinal structure. R_0: reflectance of vitrinite (Iijima 1995).

1997a). The Lower Wuerhe volcanic conglomerates and the Pingdiquan sandstones constitute analcime-albite-cemented rock reservoirs.

In the No. 8 District of the Karamay oil field, northwestern Junggar basin (Chen 1992), the Lower Wuerhe volcanic conglomerates, 110-815 m thick, are encountered in the depth interval of 2600-3400 m and comprise the most representative oil-bearing layers in an area of 40 km^2. The conglomerates consist of poorly sorted, subangular, granule to cobble gravels composed of intermediate to silicic volcanic rock clasts. There are five reservoir conglomerate layers. Zeolites are rare in the uppermost layer, but they are evenly distributed in the other four layers. Abundant analcime and less heulandite fill intergranular pores of the conglomerates. Analcime occurs as euhedral crystals of less than 0.5 mm in size and is frequently associated with authigenic albite and calcite. Chen (1992) considered the zeolites to have had an exogenetic (diagenetic) origin. Volcanic glass contained in the volcanic conglomerates was probably hydrated and altered to clinoptilolite, which was then transformed to analcime (Chen 1992). Practically no primary pore space remains in analcime-free conglomerates because of physical compaction (Chen 1992). This strongly supports the interpretation that analcime precipitated within the intergranular pores of the conglomerates at a shallower depth. The analcime is best interpreted as having been formed by diagenesis in a closed system.

In the Huoshaoshan oil fields, northeastern Junggar basin (Tang et al. 1997a), the early diagenetic mineral assemblage in the Pingdiquan sandstones (volcanic and feldspathic litharenite) contains siderite, pyrite, analcime, albite, calcite, and quartz as well as trace amounts of halite. Late diagenetic minerals include K-feldspar, ankerite, and minor amounts of mixed-layer clay minerals. The cementation of siderite, analcime, albite, and calcite occluded the substantial porosity in the sandstones at an early diagenetic stage. The intercalated mudrocks also contains abundant analcime, albite, and microcrystalline dolomite of early diagenetic origin. Tang et al. (1997a) considered that early authigenic mineral formation (e.g. calcite versus analcime/albite) was controlled by alternating periodic fresh water and saline-alkaline water episodes in a lacustrine environment.

Authigenic albite is accompanied by analcime cement in the Upper Permian sandstones and conglomerates, and it is considered to have been formed by early diagenesis (Tang et al. 1997a). It is curious, however, that detrital feldspars and analcime were extensively dissolved, but the accompanying albite appears to remain unchanged. It is more reasonable to conclude that albite was probably formed by albitization of plagioclase grains and/or transformation of analcime during burial diagenesis, subsequent to the dissolution of analcime and feldspars. In District 7E of the Karamay oil field, analcime also occurs as intergranular cements in fluvial and non-volcanic conglomerates of the Triassic Kaxia Formation in the depth range of 1100-1300 m (Chen 1992).

Complicated occurrences of zeolites in tuffs and volcanic sandstones in the East Slovakian Tertiary basin reflect possibly multiple origins (Reed et al. 1993a,b). The common occurrence of erionite in rhyolitic tuffs and tuffaceous sandstones of the Pozdisovece area is very peculiar. Volcanic glass and plagioclase have transformed to erionite and K-feldspar. Evaporitic conditions in the northern basin during Karpatian and Badenian time (late Early to Middle Miocene) would have influenced sedimentary geochemistry and possibly resulted in silica-poor, alkaline erionite developing in the evaporitic areas in tuffs and sandstones (Reed et al. 1993b).

Relationship of analcime to oil reservoir properties. The Upper Permian analcime-cemented rock reservoirs have high secondary porosity due to dissolution of analcime and clastic feldspar grains (Chen 1992, Tang et al. 1997a). In the Lower Wuerhe volcanic conglomerates in the No. 8 District of the Karamay oil field, the analcime-rich zone substantially overlaps the areas of Class I (excellent) and Class II (fair) reservoirs, in which oil quality is relatively good and oil productivity is higher (Chen 1992). Porosity of the reservoir conglomerates is about 9%, more than 82% of which consists of secondary and corroded pores of analcime. Scanning electron microscope (SEM) observation reveals that the analcime crystals were extensively corroded to form a honeycomb structure composed of thin laminations of remaining analcime and corroded voids, despite preservation of their outline (Chen 1992). In the Pingdiquan sandstones, extensive dissolution of analcime cement and labile detrital feldspars occurred during burial diagenesis, resulting in significant secondary porosity enhancement in the sandstones and making them very good quality oil reservoirs (Tang et al. 1997a). The secondary porosity may have been induced by later groundwater dissolution of analcime (Chen 1992). Alternatively, however, the secondary porosity may have resulted from the generation of various organic acids due to maturation of the interbedded exceptionally organic-rich oil shales (Tang et al. 1997a).

Open-system zeolites in hydrocarbon reservoirs

The Trap Spring oil field in the Railroad Valley, Nye County, Nevada has its reservoir in the Tertiary Pritchards Station Tuff, an ash-flow tuff with a thickness of ~300 m (French and Freeman 1979). The tuff is densely to partially welded and zeolitized. This occurrence of zeolite suggests an open-system origin, like zeolites in thick tuffs of the John Day Formation of Oregon (Hay 1962, 1963). Unfortunately, zeolitic mineral species, detailed occurrence of zeolites, and genesis were not described.

Cooling joints and faults produce fractured reservoirs, and sealing is provided by the weathering clay layer at the top of the Pritchards Station Tuff. The source rock is either the lacustrine Sheep Pass Formation at the base of the Tertiary volcanic complex or the more deeply buried Mississippian black shale (French and Freeman 1979).

HYDROTHERMAL ZEOLITES IN VOLCANOGENIC RESERVOIRS

Occurrences of hydrothermal zeolites in volcanogenic reservoirs have been increasingly reported in the last decade, and at least five oil and/or gas fields have been described from hydrothermally altered, volcanogenic reservoirs (Table 1 and Fig. 1). They are the Ayukawa oil and gas field, the Yurihara oil and gas field, the Katakai gas field of northeast Honshu Island in Japan (Fig. 16), the Karamay oil field of Xinjang in northwest China, and the Samgori oil field in Georgia, Transcaucasia. In addition, hydrothermal laumontite commonly occurs as fracture-fills of fracture reservoirs in pre-sedimentary basement granites in the Yuhutu gas field, Hokkaido Island, northern Japan and in the White Tiger oil field, off Vietnam (Table 1 and Fig. 1).

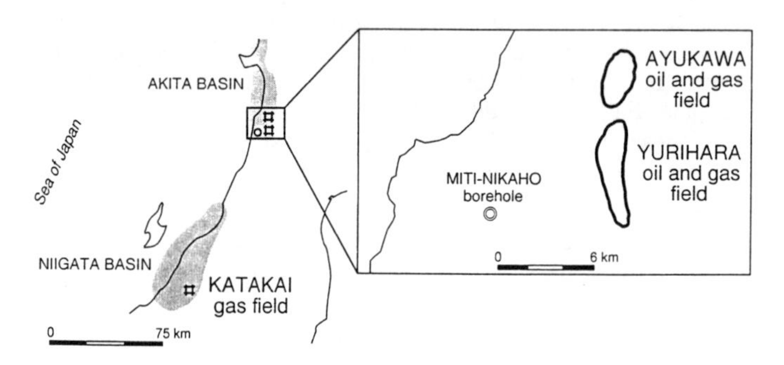

Figure 16. Locations of the Yurihara and the Ayukawa oil and gas fields, the MITI-Nikaho borehole of Akita, and the Katakai gas field of Niigata, northeast Honshu, Japan.

Zeolites and other alteration minerals, reservoir rock type, geological age, sedimentary environment, and sedimentary basin type of these fields are shown in Table 1. The parent rocks of the volcanogenic reservoirs range in chemical composition from mafic to felsic. All reservoir rocks except those of Karamay are Tertiary and have varying modes of emplacement such as subaqueous lava flows, lava domes, pyroclastic deposits, or shallow-seated dolerite sheets within back-arc or back-arc rift deep-sea basins. In the Karamay oil field, the reservoir basalt complex is Carboniferous and erupted in a fore-land basin as part of a volcanic arc (Liu 1986, Graham et al. 1990, Chen 1992). The basalt complex and overlying Permian volcanic conglomerate beneath a low-angle thrust fault were once regarded as economic basement for oil deposits, but they are now the major oil producing layers in the Karamay (Saito 1992).

Occurrences of hydrothermal zeolites in volcanogenic reservoirs

Zeolites occur in volcanogenic reservoirs in the form of (1) amygdules filling vesicles of lava rocks, (2) veins of various sizes filling fractures in lava, pyroclastics and dolerite, and (3) replacements of volcanic glass, plagioclase, and precursor zeolites. Zeolite species and zeolite assemblages in volcanogenic rocks are constrained essentially by the chemical compositions of host rocks and/or pore fluids, and temperature, although they are also more or less affected by P_{fluid}/P_{total}, permeability, and movement of pore fluids (Boles 1977, Iijima 1980, Gottardi and Galli 1985). Hydrothermal zeolite species are generally more abundant in mafic rocks than in felsic rocks. Also, the number of zeolite species tends to decrease at higher temperatures. Zeolite occurrences in the five fields are briefly summarized below.

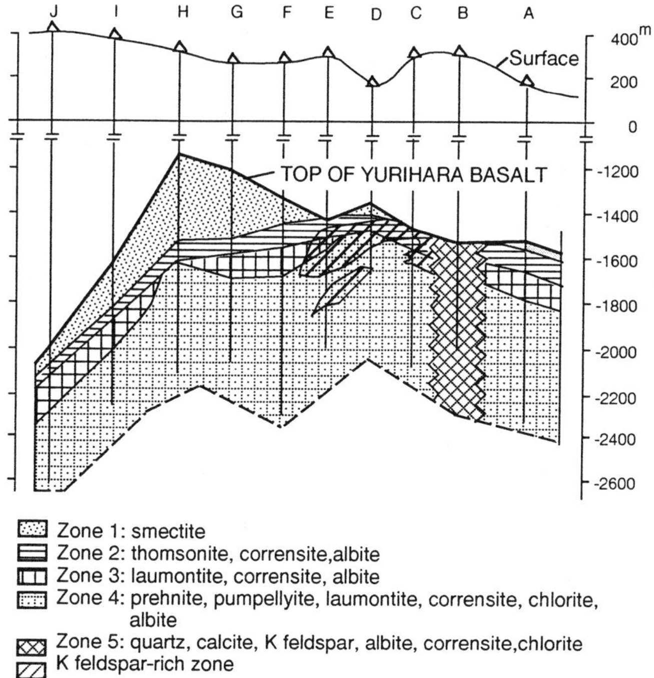

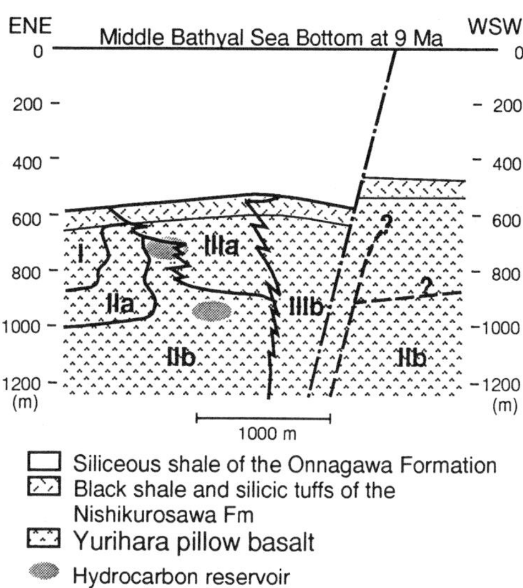

Figure 17. Two cross-sections of the Yurihara oil and gas field, Akita, Japan, showing alteration mineral zones in the Miocene submarine basalt complex. The alteration zones were formed by submarine hydrothermal activity below the deep-sea bottom of the Miocene back-arc basin at ~9 Ma, when diatom-rich siliceous shales of the Onnagawa Formation accumulated to about 500-m thickness. (Upper) North–south section covering the whole field. (Modified from Hoshi et al. 1992). (Lower) East northeast–west southwest section covering the northern half of the field. Zone I: zeolite zone (zones 2+3), subzones IIa and IIb: prehnite-pumpellyite-corrensite-albite zone (zone 4), subzones IIIa and IIIb: adularia-quartz-calcite zone (zone 5 and K-feldspar rich zone) (modified from Yagi 1992).

Yurihara oil and gas field, Akita, Japan. A vertically zonal arrangement of alteration minerals was reported in a deep-sea basalt complex of the Lower to Middle Miocene (Hoshi et al. 1992, Yagi 1992, 1994). Figure 17A shows a N-S section of the Yurihara field reconstructed from data from several wells. Hoshi et al. (1992) discriminated five alteration zones, four of which are layered in a vertical arrangement. Zone 1 is characterized by smectite, which occurs as vesicle- and fracture-fillings and as replacement of glassy groundmass, and develops in the uppermost section of the basalt, especially in the southern part of the field. Calcite is also associated with smectite in amygdules and veins. Zones 2 and 3 are respectively identified by the occurrence of thomsonite and laumontite in amygdules and veins. In both zones, analcime and calcite occur also in amygdules and veins. Smectite changes to mixed-layer chlorite/smectite, and calcic plagioclase is extensively albitized. Prehnite is also frequently found in veins and amygdules in zones 2 and 3. Zone 4 is widespread in the deeper section of the basalt and is characterized by prehnite and pumpellyite in amygdules and veins, with pumpellyite also replacing olivine. Analcime is present only in the upper part of zone 4, and laumontite and calcite are seen in amygdules and veins, although laumontite is extensively replaced by prehnite. Moreover, laumontite appears to have reacted with corrensite to form prehnite and minor pumpellyite, and mixed-layer chlorite/smectite re-crystallized to chlorite (Yagi 1994). Plagioclase is completely replaced by albite, prehnite, and pumpellyite (Yagi 1994). In zone 5, aggregates of quartz, calcite, K-feldspar, albite, corrensite, and chlorite completely replace the basalt. More precisely, zone 5 is a late-stage alteration product because it is distributed in a subvertical zone cutting the other zones. The subvertical zone, about 500 m wide, is probably along a late Middle Miocene growth-fault (Yagi 1992, 1994). K-feldspar and quartz veins are further impregnated into zone 2, zone 3, and the upper part of zone 4. The K-Ar age of the K-feldspar in veins is 8.5-9.6 Ma, which represents the age of the late K-metasomatism by hydrothermal solutions (Yagi 1994).

Temperature appears to have been the controlling factor in these alteration-mineral zones, especially the layered zoning (Hoshi et al. 1992), and it is estimated from the mineral assemblages to be 80-110°C in zone 1, 110-138°C in zone 2, 138-150°C in Zone 3, 150-220°C in zone 4, and 110-220°C in zone 5. Yagi (1994) estimated the formation temperature of chlorite in a chlorite-prehnite vein of zone 4 to be 190-215°C based on the substitution of Al in the tetrahedral sites of chlorite (Cathelineau and Nieva 1985). The homogenization temperatures of fluid inclusions in calcite of zone 5 were 165-245°C, and the salinity of the fluid is 3.0-3.3% NaCl (Yagi 1994).

Hoshi et al. (1992) and Yagi (1992) considered that the layered alteration-mineral zones may be attributed to submarine metamorphism followed by basalt eruption at about 15 Ma, comparable with mid-oceanic ridge metamorphism. They also concluded that the K-metasomatism of subvertical zone 5 was due to hydrothermal activity associated with a postulated later silicic volcanism event. The K-Ar age of K-feldspar in zone 5 indicates that K-metasomatism occurred at about 9 Ma at several hundreds of meters below the floor of the middle-bathyal Onnagawa sea (Yagi 1994). Nevertheless, the two sets of alteration patterns might have occurred sequentially more than several hundreds meters sub-bottom of the bathyal sea by a series of hydrothermal events. Zone I (zones 2+3), zone IIa (uppermost zone 4) and, zone IIb (zone 4) show subvertical zone boundaries adjacent to the upper part of Zone III (Zone 5), and zones 4 and 5 were formed at nearly the same temperature (Fig. 17B). In the early stage, hydrothermal fluids of seawater origin gradually percolated into and reacted with the basalt below thick unlithified muddy sediments and produced the layered alteration zones. In the later stage at about 9 Ma, the growth-fault opened the conduit for pressurized hydrothermal fluids to ascend and react quickly.

In contrast, only analcime replaces glass shards in the overlying silicic tuff reservoir in the northern part of the Yurihara field. Analcime is partly replaced by a microcrystalline aggregate of K-feldspar and quartz formed by later K-metasomatism alteration (Yagi 1994) (Fig. 18). Yagi (1992) considered that analcime in acidic tuff was formed by burial diagenesis. However, Iijima (1995) argued that analcime formed by hydrothermal alteration, on the basis of microcrystalline aggregates of euhedral analcime crystals not associated with quartz.

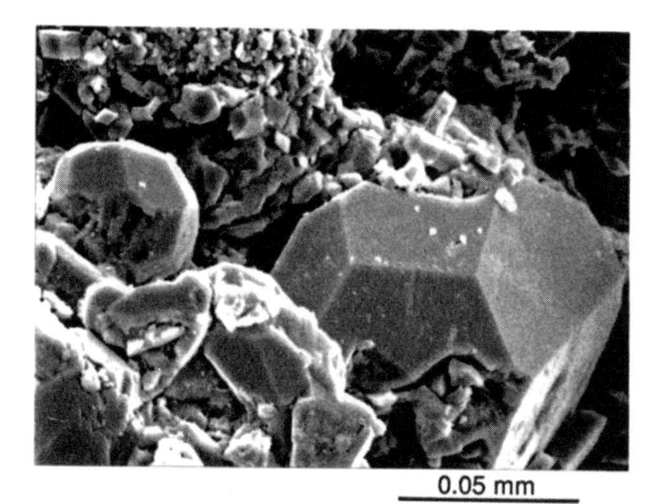

Figure 18. SEM photomicrograph showing secondary porosity in the reservoir silicic tuff of the Middle Miocene Nishikurosawa Formation in the Yurihara oil and gas field, Akita, northeast Honshu, Japan. Coarse analcime crystals were formed by hydrothermal alteration of silicic glass fragments and were dissolved by later hydrothermal fluids from which microcrystalline aggregates of K-feldspar (adularia) precipitated (courtesy of Mr. M. Yagi).

Ayukawa oil and gas field, Akita, Japan. Concentric alteration-mineral zones were found in the cross-section of two dolerite bodies intruding unlithified siliceous mudstone of the Middle Miocene Onnagawa Formation, i.e. from the outer margin to the central core; a saponite zone (zone I), a saponite-talc-corrensite zone (zone II), and a zeolite zone composed of a small amount of analcime and thomsonite (zone III), as shown in Figure 19 (Okubo et al. 1996). Saponite replaces olivine completely, clinopyroxene extensively, and plagioclase partly. Corrensite also replaces the mafic minerals in zone 3. Analcime and thomsonite occur as replacements of calcic plagioclase and microveins in central zone 3.

Secondary porosity formed by extensive dissolution of plagioclase and especially saponite, and it is concentrated in the upper part of the dolerite bodies. Parts of secondary pores were then filled by K-feldspar + quartz (rim cement), and dolomite (pore cement). A dolomite halo developed in tuffs and siliceous mudstones around the altered dolerite bodies. Dolomite occurs as microcrystalline cement in the matrix and veins; dolomite precipitation is interpreted to be synchronous with the dolomite pore cementation of the dolerite. A kaolinite halo inside the dolomite zone occurs in tuffs on the northeast side of the larger dolerite body.

All mineral zones in this oil and gas field are considered to have formed by a sequential series of moderate- to lower-temperature hydrothermal alteration during the cooling stage of the dolerite bodies within Middle Miocene soft diatomaceous muds below a middle-bathyal sea bottom. This conclusion is supported by concentrically zoned mineral distributions, superimposed on one another. Interstitial pore fluids that reacted with hot dolerite bodies with thicknesses of 140 and 160 m are undoubtedly of seawater origin (Okubo et al. 1996). The formation temperature of the central analcime-thomsonite zone is considered to be slightly lower than that of zone 2 (thomsonite-analcime) in the Yurihara

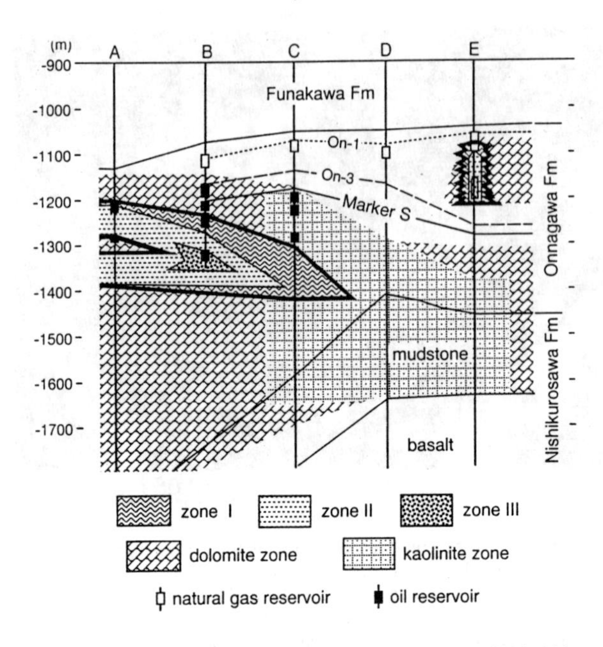

Figure 19. A section of the Ayukawa oil and gas field, Akita, northeast Honshu, Japan, showing hydrothermal-alteration mineral zones within two dolerite bodies (zones I, II, and III), and mudrocks of the Nishikurosawa Formation and siliceous shales of the Onnagawa Formation: dolomite zone and kaolinite zone surrounding the dolerite bodies are alteration halos. On-1, 3: marker beds, A–E: wells, zone I: saponite, zone II: saponite, talc, and corrensite, zone III: analcime and thomsonite (modified from Okubo et al. 1996).

field, because plagioclase is not albitized in the Ayukawa field.

Katakai gas field, Niigata, Japan. Although lacking authigenic zeolites, the geologic setting of hydrothermal alteration of the reservoir lava rocks of the "deep-sea Green Tuff facies" in the Katakai field is very similar to that in the Yurihara field. Yamada and Uchida (1994) showed that the "deep-sea Green Tuff facies" of the upper Lower to lower Middle Miocene in the Katakai field is made up of subaqueous rhyolite lavas and pyroclastic units plus minor basalt having a total thickness of over 1 km. Early-stage alteration mineral assemblages include illite-quartz-albite, illite-chlorite-quartz-albite, and chlorite-mixed-layer I/S-quartz-albite, which occur mainly as replacements of glassy groundmass, plagioclase, pyroxene, and hornblende. On the other hand, quartz, carbonates (calcite, ankerite and Mg siderite), chlorite, and barite precipitated extensively within fractures and vugs during later-stage silicification of the rhyolite lava rocks. Two different origins of hydrothermal fluids are suggested by homogenization temperatures of fluid inclusions in quartz and calcite: (1) a high-temperature (~160-240°C) origin involving high-salinity altered seawater, and (2) moderate-temperature (~70-150°C) alteration in low-salinity catagenetic water which was probably derived from dehydration of smectite in the overlying mudstone formation. Petroleum inclusions are rarely found in the moderate-temperature vein quartz. The K-Ar age of illite indicates that hydrothermal activity occurred around 12 Ma during sedimentation of deep-sea mudstones and turbidites of the lower Teradomari Formation at a sub-bottom depth of around 1 km. Yamada and Uchida (1994) concluded that the heat that promoted the hydrothermal alteration emanated from a shallow-seated silicic-magma reservoir.

Samgori oil field, Georgia, Transcaucasia. Three alteration-mineral assemblages can be distinguished in this Middle Eocene deep-sea pyroclastic formation having a thickness of about 600 m. These are: (1) an illite-chlorite-albite assemblage, (2) an analcime-calcite-chlorite assemblage, and (3) a laumontite-albite assemblage (Vernik 1990, Grynberg et al. 1993). Porous, silty to sandy tuffs and calcareous tuffites of good reservoir quality comprise the uppermost part of the pyroclastic formation and are altered to assemblages (1) and (2). Low permeability pelitic tuffs and tuffaceous marls are most widespread and are also altered to assemblages (1) and (2). Porous, laumontized tuffs are the main oil reservoirs and are present as isolated, well-stratified, 3-5-m thick lenses, stocks, and vein-like bodies within the matrix of impermeable tuffs and marls. Laumontite occurs as replacements of volcanic glass fragments and plagioclase, intergranular cements, and fracture-fillings. Plagioclase is also albitized. Laumontite reaches 50-80% of the volume of the tuffs.

The altered tuffs of all assemblages are andesite-basalt in composition (Vernik 1990, Grynberg et al. 1993). Therefore, the difference in the three alteration-mineral assemblages as well as the heterogeneous alteration pattern cannot be attributed to differences in chemical composition of parent rocks. However, the pyroclastic formation shows the chaotic sedimentary structure of a submarine-slide deposit, consisting of irregularly shaped and oriented blocks of sedimentary rocks with various geologic ages and laumontized tuff and the matrix tuff altered to analcime (Fig. 20). The laumontitic tuffs are of andesite-basalt composition, whereas the analcimic matrix tuffs may have been more silicic.

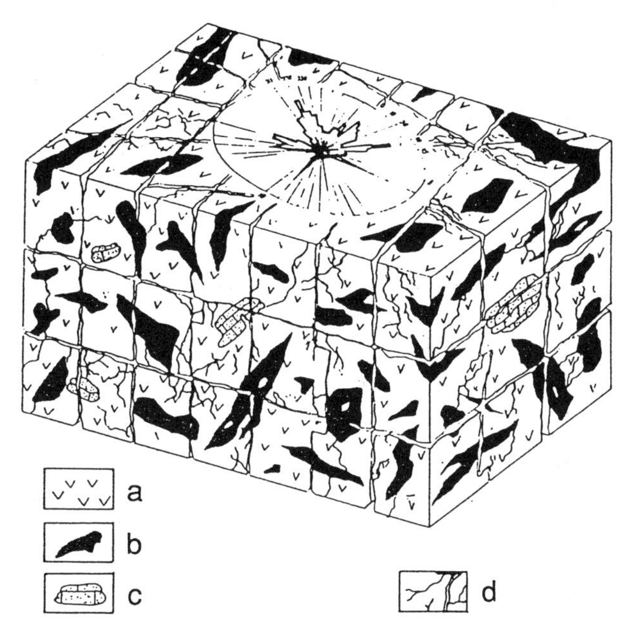

Figure 20. Fracture system of Middle Eocene volcanogenic-sedimentary reservoir of the Samgori oil field, Georgia. (a) volcanogenic-sedimentary complex, (b) laumontite tuff of andesite-basalt composition, (c) blocks of sedimentary rocks of various ages, (d) fractures. Rose diagram indicates fracture trends (modified from Grynberg et al. 1993).

Patton (1993) considered the alteration pattern to be a result of submarine hydrothermal alteration subsequent to the deposition of the pyroclastic formation, although he did not explain how the alteration pattern originated. Vernik (1990) stated that laumontization was a result of Ca-metasomatism. However, laumontization may have been formed by relatively later hydrothermal alteration, considering that laumontite fills fractures and that laumontized tuff occurs as stocks and vein-like bodies as if they cut the matrix with assemblages (1) and (2). The uppermost analcime zone may have formed at lower temperatures, like the analcime zone in the Yurihara basalt.

Karamay oil field, Xinjang, China. The Carboniferous basalt complex in the No. 1 District of the Karamay field has a thickness of over 900 m and covers an area of about 30 km^2. Zeolites occur largely as vesicle- and fracture-fillings in fractured vent facies and in surrounding pyroclastic facies composed of abundant basalt breccia (Chen 1992). The vesicle-filling zeolites were formed by reactions of volcanic glass with "magmatic water" (Chen 1992). Zeolite content attains over 30% by volume in the extensively zeolitized portion where zeolites precipitated in interstitial voids among breccia rubble.

Examination of drill cores permitted two fracturing stages to be distinguished. Earlier fractures are syn-sedimentary and are filled by zeolites of a "volcanic-hydrothermal" (higher-temperature hydrothermal) origin. Later fractures are concentrated in the pre-Permian weathering zone in the uppermost part of the basalt complex and are filled by zeolites of a "post-volcanic groundwater" (low-temperature hydrothermal) origin. There is a tendency for fractures with steeper dips to show larger width, and fractures with larger width have a lower zeolite-filling ratio.

Heulandite is most common in amygdules encrusted with chlorite. Other zeolites found in the basalt complex are chabazite, natrolite, mordenite, stilbite, and laumontite. Chen (1992) considered that the varied zeolite species were caused by a wide range in chemical composition of lavas. However, all of these zeolites are commonly recognized in basalts and their occurrence can be explained by a difference in formation temperature. Unfortunately, neither zeolite assemblages nor zeolite zonation were described. It is inferred from the description of Chen (1992) that the zeolites in the basalt complex of the Karamay field were formed by a series of hydrothermal episodes during and subsequent to the pre-Permian uplift.

Relation of hydrothermal zeolites to reservoir properties

Reservoir properties of hydrothermally altered volcanogenic reservoirs in the five oil and/or gas fields are summarized in Table 4. Reservoir properties are significantly affected by the nature of the hydrothermal alteration as well as by the original lithology of the reservoir rocks. Zeolitization may either improve or degrade the reservoir properties. Despite a small number of examples, hydrothermal zeolites seem to act negatively in lava and intrusive rock facies, whereas they may have positive effects in pyroclastic facies.

Lava and intrusive rock facies. Hydrocarbon reservoirs do not exist within zeolite zones in lava and intrusive rock facies, perhaps because zeolites tend to decrease the primary porosity by filling vesicles, vugs, or joint cracks. In the submarine rhyolite lavas of the Katakai field, natural gas traps occur in rigid but fragile rocks of the albite-quartz-ankerite assemblage around postulated conduits of the hydrothermal fluids (Hoshi et al. 1992, Yamada and Uchida 1994). The altered rhyolite reservoirs generally retain a porosity of 5-20% and a permeability of 1 md to 82 md in the depth interval between 3800 m and 5200 m (Sato 1984, Uchida 1984, Sato et al. 1992a). The porosity of albitized volcanic rock reservoirs originated from primary vugs, vesicles, and quenching fractures, and from secondary porosity formed by dissolution and tectonic fracturing (Sato et al. 1992b). Zeolites are absent in the Katakai field due to high-P(CO_2), moderate- to high-temperature (70-240°C) hydrothermal alteration (Yamada and Uchida 1994). In the submarine pillow lavas at Yurihara (Yagi 1994), hydrocarbon traps do not exist in the zeolite zones but are in the upper part of the prehnite-pumpellyite zone below a K-metasomatism halo near the

Table 4. Properties of hydrothermally altered volcanogenic and basement granitic reservoirs. Locality numbers refer to Table 1 and Figure 1.

Oil and/or gas field*	Host rocks	Principal alteration mineral	Porosity (%)	Permeability (md)	Primary porosity	Secondary porosity
9. Yufutsu	conglo./granite	laumontite			cooling joints	tectonic fractures and corroded voids
12. Ayukawa	dolerite	saponite, analcime	5 - 25	0.01 - 200	cooling joints	corroded voids
13. Yurihara	pillow basalt	prehnite, albite, laumontite, chlorite	20 - 25	40	vesicles, vugs and cooling cracks	volume decrease by mineral transformation
14. Yurihara	acidic tuff	analcime, adularia	20		intergranular pores	intercrystalline pores and corroded voids
15. Katakai	rhyolite lava	albite, ankerite	5 - 20	1 - 82	vugs and cracks	intercrystalline pores
23. Karamay	basalt breccia	heulandite, chlorite	av. 12.8	av. 20.3	cooling cracks	tectonic fractures and corroded voids
24. White Tiger	granite	laumontite	av. 2.77		cooling joints	tectonic fractures and hydrothermal corrosion
25. Samgori	andesite-basalt tuff	laumontite	8.4 - 20.7	av. 14.8	intergranular pores	tectonic fractures and intercrystalline pores

* 9: Agatsuma et al. (1996); J.N.O.C. (1998); 12: Okubo et al. (1996); 13: Hoshi et al. (1992); Yagi (1992, 1994); 14: Yagi (1992, 1994); 15: Hoshi et al. (1992); Yamada and Uchida (1994); 23: Chen (1992); 24: Areshev et al. (1992); 25: Vernik (1990); Grynberg et al. (1993).

conduit of the later, high-temperature (190-220°C) hydrothermal fluids (Fig. 17B). The improvement in reservoir properties was explained by Hoshi et al. (1992) as being a byproduct of hot fluids of seawater origin that penetrated and connected initially isolated vesicles and fractures. Secondary porosity was produced by mineral volume decreases during the transformations of laumontite and smectite to prehnite and chlorite, respectively, and higher-temperature alteration-minerals did not undergo subsequent diagenetic changes.

Traps exist in the upper marginal portion of the dolerite bodies of the Ayukawa field, where corroded plagioclase and saponite produce secondary porosity (Fig. 19). Okubo et al. (1996) concluded that heated acid pore-fluids of seawater origin reacted with plagioclase to form analcime and thomsonite in the central portion of the dolerite bodies. These fluids migrated upwards and corroded plagioclase and saponite in the upper portion.

Pyroclastic facies. Hydrocarbon reservoirs exist within zeolite zones in pyroclastic facies because the high primary porosity of glassy pyroclastic rocks is commonly preserved in zeolitized tuffs, as described by Utada (chapter on burial diagenesis, this volume). The reservoir silicic tuffs of the Yurihara field retain about 20% porosity, which consists mainly of intercrystalline pores among idiomorphic microcrystals of analcime and adularia (Yagi 1992, 1994). In addition, some corroded pores are found in analcime replaced by adularia (Fig. 18).

Analcime degrades reservoir quality in the pyroclastic formation of the Samgori oil field, Georgia (Vernik 1990), and hydrocarbon traps are present only in porous laumontized tuffs. The reservoir laumontized tuffs have variously shaped bodies which are enclosed and sealed by thick layers consisting of analcime-calcite-chlorite tuffs and illite-chlorite-albite tuffs (Grynberg et al. 1993) (Fig. 20). As a result, the laumontized tuff bodies were relatively isolated from the action of geostatic and geotectonic tension. Such sealing and isolation promoted the preservation of high productivity characteristics over a wide range of tensional conditions by preventing closure of fractures and microfractures within the reservoirs (Grynberg et al. 1993). Matrix porosity in these rocks is formed by intercrystalline micropores of over 10^{-7} m in diameter, and matrix permeability is less than 0.1 md. In high-porosity tuff reservoirs where intensive invasion of drilling mud occurs due to microfractures, effective porosity and fracture permeability increase to a range of 3.7-7% and 14.8-460 md, respectively (Grynberg et al. 1993). Regional tectonic fractures developed in the pyroclastic formation of the Adjara-Trialeti folded belt, and they are observed in the subsurface reservoir rocks as well. The subvertical open fractures can be grouped into four orientations, of which a NW orientation parallel to the fold axis is most common (Fig. 20). Varying fracture permeability and irregular distribution of laumontite matrix porosity preclude precise estimates of total reservoir pore volume and of total oil recovery by the statistical methods normally used for calculating reserves during field development (Grynberg et al. 1993).

In the Carboniferous basalt complex of the Karamay field, northwest China, an area of high zeolite content is superimposed upon not only the area of highest hydrocarbon concentration but also upon the area of relatively high production wells (Chen 1992). Zeolites are most highly concentrated in both the volcanic vent facies and the surrounding pyroclastic facies where the zeolite content is up to 50% by volume. In section, the high-zeolite zone is situated in the pre-Permian, fractured, weathering profile at the top of the basalt complex. Fractured and zeolitized basalts show the best oil-bearing properties, i.e. a maximum porosity of 29.6% (average of 12.8%) and an air permeability of 605 md. Porosity and permeability average 9.5% and 20 md in volcanic breccia, whereas they are 4.7% and 0.6 md in unfractured massive basalt. Zeolites contain abundant intercrystalline pores, pressure-solution cracks, and secondary corroded pores (Chen 1992). Chen (1992) stated that the secondary porosity of the zeolite-bearing rock reservoirs was formed by later groundwater corrosion of zeolites. Alternatively, organic acid-containing oil or formation water may have played an important role in the corrosion of some zeolites and generation of secondary porosity (Tang et al. 1997a).

Pre-sedimentary basement rock facies. The Yufutsu gas field was recently discovered in south-central Hokkaido, northern Japan (Agatsuma et al. 1996). The pay-zones exist in both Mid-Cretaceous pre-sedimentary granite and in the fluvial basal conglomerate of the Eocene coal measures. Reservoirs are related to tectonic fracture porosity, which is considered to have been reopened by younger inversion tectonics (J.N.O.C. 1998). Laumontite occurs commonly as fracture-filling veins whose orientations are the same as the reopened fractures (Agatsuma et al. 1996). Investigations on reservoir characteristics are now in progress.

The White Tiger oil field on the continental shelf of southern Vietnam has reservoirs in Early Cretaceous pre-sedimentary basement granite (Areshev et al. 1992). The granite has undergone severe alteration as a result of tectonic, hydrothermal, and surface weathering processes. Reservoirs having a total thickness of >1 km are related to cavernous fracture (fault) porosity in deep basement zones, and to "porous cavernous" fracturing at more shallow levels. Hydrothermal caverns and microcaverns that reach 55% of total porosity are formed by dissolution of feldspars, and laumontite occurs commonly as fracture-filling veins (Areshev et al. 1992).

Stratigraphic relationship between volcanogenic reservoirs and source rocks. Ages of volcanogenic reservoir rocks and source rocks in the above-stated five fields are given in Table 5. In all the fields except Samgori, the source rocks are younger than the volcanogenic reservoir rocks that are largely situated in the basal part of the basin fills. Hydrocarbons generated from deeply buried and mature source rocks in the neighboring synclinal or basinal structures migrated upwards and accumulated in traps in the anticlinal structures. The primary volcanogenic reservoirs in the Neogene oil and/gas fields of northeast Honshu, Japan occur in the over 1 km-thick "deep-sea Green Tuff facies" of Lower to Middle Miocene lava rocks in the basal part of the Tohoku deep-sea basin fills, with a total thickness of 6.3 km (Iijima 1995). The overlying thick deep-sea muddy sediments of the Middle to Upper Miocene form potential source rocks. The reservoir dolerite bodies of the Ayukawa field were intruded into the lower part of the muddy sediments (Okubo et al. 1996).

The primary reservoirs in the Karamay field of the western Junggar basin in northwest China are contained in a >900 m-thick Carboniferous basalt complex in the Keyiwu rift zone. Upper Permian lacustrine oil shales are the principal source rocks of the Karamay field (Graham et al. 1990). In contrast, the Middle Eocene, 600 m-thick reservoir tuffs of the Samgori oil field of Georgia, Trancaucasia, overlie source rocks of organic-rich illitic shales and limestones of the Lower Eocene and Paleocene, about 2 km-thick (Vernik 1990, Patton 1993).

Timing of hydrothermal activity relative to hydrocarbon accumulation. Ages of hydrothermal alteration and hydrocarbon accumulation in the five fields as well as in the Yufutsu gas field and the White Tiger oil field are also shown in Table 5. Hydrothermal alteration of the volcanogenic reservoirs preceded formation of the trap structures and subsequent hydrocarbon accumulation in all five fields. Submarine hydrothermal activity in the Neogene fields of northeast Japan had occurred by 8.5 Ma at the time of deposition of the source rock sequence. Formation of anticlinal traps and hydrocarbon accumulation in the traps occurred in Plio-Pleistocene time (Hoshi et al. 1992, Iijima 1995). For example, the Yurihara oil and gas field of Akita is situated in a faulted

Table 5. Ages of host rocks, hydrothermal alteration, hydrocarbon accumulation, and source rocks in volcanogenic and basement granitic reservoirs. Locality numbers refer to Table 1 and Figure 1.

*Oil and/or gas field**	*Age of host rocks (Ma)*	*Age of alteration (Ma)*	*Age of hydrocarbon accumulation (Ma)*	*Age of source rocks (Ma)*
9. Yufutsu	99.5 - 127	L Olig.-E Miocene	Late Mio.-Pliocene	M-L Eocene
12. Ayukawa	10 - 8	10 - 8	14 - 5	14 - 5
13-14. Yurihara	16.5 - 15	(15) - 8.5	14 - 5	14 - 5
15. Katakai	16.5 - 15.5	12	14 - 5	14 - 5
23. Karamay	Carboniferous	L Carbon.-E Perm.	Late Mesozoic	Early Permian
24. White Tiger	108	M Cretaceous	Late Mio.-Pliocene	Early Oligocene
25. Samgori	M. Eocene	Late Eocene	Late Oligocene	Paleocene-E Eoc.

*9: J.N.O.C. (1998), 12: Okubo et al. (1996), 13-14: Hoshi et al. (1992); Yagi (1992, 1994), 15: Hoshi et al. (1992); Yamada and Uchida (1994), 23: Chen (1992); Graham et al. (1990), 24: Areshev et al. (1992), 28: Vernik (1990); Patton (1993).

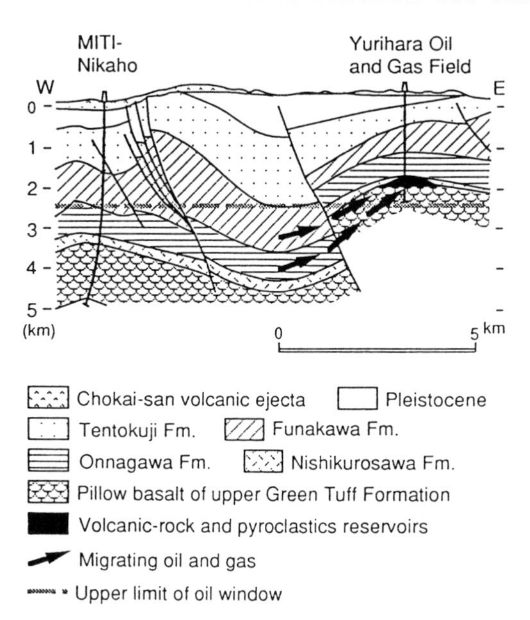

Figure 21. East-west cross-section through the Yurihara oil and gas field and the MITI-Nikaho borehole in Akita, northeast Honshu, Japan, showing the upper limit of the oil window and migration of oil and gas from a synclinal structure to basalt and tuff reservoirs of the field (modified from Araki and Kato 1993).

anticlinal structure (Fig. 21) which was produced tectonically in the Pliocene to early Pleistocene (Yagi 1992). According to Waseda and Omokawa (1990), mudrocks interbedded with the Yurihara Basalt and overlying siliceous shales of the Onnagawa Formation do not reach the maturity level necessary to be considered source rocks. Source rocks between 2500 m and 5000 m depth in the MITI-Nikaho borehole enter the oil window (J.N.O.C. 1988b). Consequently, oil and gas are believed to have been generated in the deeply buried source rocks of the Late to Middle Miocene Onnagawa and Funakawa Formations in the synclinal structure west of the Yurihara field, and oil and gas migrated eastward and accumulated in the reservoirs (Fig. 21). This implies that the migration and accumulation of oil and gas were much later events than the submarine hydrothermal alteration in the Late Miocene between 9 and 11 Ma (Yagi 1994).

Chen (1992) concluded that the vesicle-filling zeolites in the Carboniferous basalt complex of the No. 8 District of the Karamay oil field, northwest China, were related to "magmatic water", and that the early syn-sedimentary fractures were filled by "volcanic-hydrothermal" zeolites whereas the later pre-Permian weathering fractures and tectonic fractures were filled by "post-volcanic groundwater" zeolites. Both the amygdule zeolites and the early fracture-filling zeolites might have formed by the same hydrothermal alteration. Furthermore, it is possible that all zeolites in the Carboniferous basal complex are the products of a series of post-volcanic hydrothermal activities. On the other hand, anticlinal traps were formed by pre-Cretaceous compressional tectonic movement (Liu 1986). Oil migrated from the Upper Permian oil shales on the east of Karamay and accumulated in these traps during the late Mesozoic to Tertiary (Graham et al. 1990).

Submarine hydrothermal alteration in the Samgori field of Georgia occurred subsequent to deposition of the Middle Eocene pyroclastic formation (Patton 1993). Laumontization of the reservoir tuffs is considered to have occurred during a fracturing and uplift-stage in the Upper Eocene, as discussed above. Anticlinal traps began to develop during Late Eocene deposition and were completed prior to the Late Alpine (Miocene-Pliocene) deformation (Patton 1993).

Hydrothermal laumontite in the Yufutsu gas field of Hokkaido, Japan, precipitated in faults and fractures in Late Oligocene to early Miocene time when the horst structure containing the reservoirs formed below ~2.6 km-thick Upper Oligocene plus Eocene strata (J.N.O.C. 1998). Methane gas probably migrated westward from the deeply buried Eocene coal measures and accumulated in the reopened fracture reservoirs in Late Miocene to Pliocene (Agatsuma et al. 1996). It is likely that fracture-filling brittle laumontite might contribute to the reopening of reservoir fractures.

Hydrothermal alteration of Early Cretaceous granite in the White Tiger offshore oil field, southern Vietnam, occurred during Middle Cretaceous age, associated with tectonism and fracturing (Areshev et al. 1992). After prolonged stages of uplift, erosion, and weathering in latest Cretaceous to Eocene, thick strata of Oligocene and Neogene sediments covered the granite. Oil probably migrated from the surrounding deeply buried Early Oligocene source rocks and accumulated in the granite reservoirs in Late Miocene to Pliocene (Areshev et al. 1992).

Late hydrothermal alteration is likely to have a negative effect on accumulation of oil and gas. For instance, in the MITI-Kuromatsunai borehole where no oil and gas have been recognized, geothermal activity took place at about 0.5 Ma, when the structure was formed (Iijima et al. 1984). In the MITI-Yuri-oki-chubu borehole with a total depth of 5000 m (J.N.O.C. 1993), >1 km of mature source mudrocks of the Middle Miocene deep-sea sequence exist but significant hydrocarbons have not been found. Later hydrothermal activity is inferred from various occurrences, including (1) the common occurrence of carbonate and anhydrite-laumontite veins in the source mudrocks and underlying volcanic and volcaniclastic rocks of the Green Tuff Formation, (2) the unusual occurrence of mixed-layer I/S minerals with high smectite contents in the deeper part of the mudrock sequence, and (3) the unusual presence of diagenetic analcime in silicic tuffs in the depth interval between 2350 m and 4770 m corresponding to 81° and 155°C, respectively (J.N.O.C. 1993). The analcime has not been albitized because that interval has been flushed by hydrothermal solutions with low salinity.

LAUMONTITE IN SANDSTONE RESERVOIRS

Kaley and Hanson (1955) first reported the occurrence of laumontite in Miocene reservoir sandstone from a well in the San Joaquin basin, California. Since then, laumontite has been found not only in reservoir sandstones in many oil and/or gas fields, but also in sandstones from a number of potential hydrocarbon-producing sedimentary basins throughout the world.

It is well-known that laumontite in sandstones can form in a variety of ways, including burial diagenesis-low-grade metamorphism, late uplift-stage diagenesis, and hydrothermal alteration. In addition, laumontite in the Late Triassic arkose in the Ruhuhu basin, Tanzania, was formed in Late Triassic weathering profiles under near-surface conditions (Wopfner et al. 1991). Therefore, the formation temperatures of laumontite are interpreted to vary from near-surface to over 150°C.

Despite the differences in genesis, the occurrence of laumontite in sandstone is rather simple. Laumontite generally occurs as an intergranular cement and as a replacement of plagioclase. The zonal distribution with other zeolites, which develops typically in altered tuffs undergoing burial diagenesis or hydrothermal alteration, is commonly not recognized in laumontite-bearing sandstone, especially arkose, sequences. It is frequently difficult to judge the genesis of laumontite in sandstone using the literature on subsurface occurrences due to limited data. Therefore, the occurrence of laumontite in reservoir sandstones is treated here collectively.

Sedimentary environments of laumontite-bearing sandstones range from fluvial and lacustrine to deep-sea turbidite through shallow marine. Table 1 lists the occurrence of laumontite in reservoir sandstones and sandstones in potential oil-producing sedimentary basins.

Volcanic glass fragments and plagioclase are common precursors of laumontite in two types of sandstone (Boles 1993), namely volcanic sandstone and arkose. Volcanic sandstones consist largely of redeposited fragments of unstable volcanogenic materials, such as volcanic glass, mafic minerals, and plagioclase. Almost all volcanic sandstones are intercalated in thick sedimentary sequences of arc-related sedimentary basins associated with convergent plate margins, such as fore-arc basins, back-arc basins, and fore-land basins. Volcanic rock fragments were derived from either penecontemporaneous or older volcanic arcs within or on the sides of the basins. The arc-related basins containing laumontized volcanic sandstone and conglomerate reservoirs are widely distributed in younger orogenic belts in the Circum-Pacific region, e.g. the San Joaquin basin of California, the Ishikari coal basin of Hokkaido in northern Japan, and the San Jorge basin of Argentina. Other basins include the East Slovakian basin, a Tertiary back-arc transitional basin in the Alpine orogenic belt in Slovakia, and the Junggar basin, a Paleozoic-Mesozoic fore-land basin in Xinjang of northwest China.

Arkoses consist mainly of feldspar and quartz derived from granitic rocks and granitic gneisses. Laumontite-bearing arkosic sandstone reservoirs are found in either arc-related basins or cratonic basins within continents. Examples of the former are the Los Angeles basin of California, and examples of the latter are the North Songliao basin, the Ordos basin, and the Zhoukou basin of eastern China.

Formation of laumontite in reservoir sandstones

The formation of laumontite in sandstones has been well summarized by Boles and Coombs (1977) and by Surdam and Boles (1979). Laumontite is commonly formed by the transformation of not only various aluminosilicate minerals including heulandite, plagioclase, kaolinite, and smectite, but also of calcite (e.g. Coombs 1954, Zen 1961, Madsen and Murata 1970, Boles and Coombs 1977). Representative reactions forming laumontite are shown in Table 6. Laumontite is precipitating with thenardite, quartz, and gypsum from hot spring water of 43-89°C at orifices of Sespe Hot Springs and in fractures of Eocene arkose on Hot Spring Creek immediately downstream in the Transverse Ranges, southern California (McCulloh et al. 1981). Oxygen and hydrogen isotopes indicate that temperatures at the subsurface source of the spring water are 125-135°C at a depth of 3550-3900 m (McCulloh et al. 1981).

Laumontite in volcanic sandstone reservoirs. Coombs (1954) established a pattern of burial-depth zonation in Triassic andesitic tuffs and graywacke sandstones of Southland, New Zealand, with heulandite + volcanogenic Ca-plagioclase at shallower depths overlapping laumontite + albitized plagioclase at greater depths. This pattern of zonation was also recognized regionally in marine Tertiary reservoir sandstones in the southern part of the San Joaquin basin, California (Noh and Boles 1993, Bloch and Helmold 1995). Heulandite was considered to have altered from andesitic glass fragments, whereas laumontite transformed from heulandite and plagioclase. As stated above, silicic glass fragments in marine Tertiary rhyolitic to dacitic tuffs in Japan are altered to clinoptilolite, which is further altered to heulandite with an increase in burial depth and temperature. Clinoptilolite altered from glass fragments is also widespread in the shallow-burial zone of marine Miocene andesitic to rhyolitic sandstones in the central Coastal Ranges of California (Murata and Whiteley 1973). It remains unsolved whether heulandite can transform directly from andesitic glass or indirectly by way of clinoptilolite.

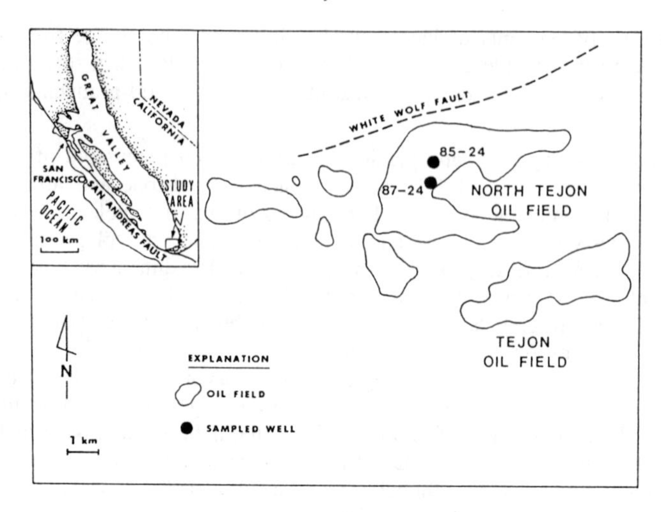

Figure 22. Index maps showing location of the North Tejon oil field and sample wells in the San Joaquin basin, California, USA (Noh and Boles 1993).

Considerable overlap exists in the above-stated pattern of burial-depth zonation in New Zealand and elsewhere (Surdam 1973, Boles and Coombs 1977, Surdam and Boles 1979). Noh and Boles (1993) investigated the overlap in marine Oligocene-Lower Miocene Vedder Sandstone in the North Tejon oil field in the San Joaquin basin, California (Fig. 22). They concluded that "early" laumontite was transformed from heulandite, whereas "later" laumontite was produced by albitization of plagioclase. The following is a summary of Noh and Boles (1993). The Vedder Sandstone is composed of volcanogenic feldspathic sandstones and contains less than 10% by volume rock fragments apart from altered glass fragments. Intermediate to mafic volcanic rocks dominate among rock fragments. Glass fragments make up to 10% by volume at depths shallower than 2743 m and are altered to heulandite and smectite. Plagioclase is more abundant than K-feldspar in almost all samples. Two types of plagioclase were identified, turbid albite to oligoclase (Ab > 80) of a granodiorite source, and clear, euhedral calcic plagioclase of a volcanic origin. The reservoir sandstones are now buried to almost their maximum depth and have experienced maximum burial temperatures. Figure 23 shows the paragenetic sequence and the depth occurrence of authigenic minerals in the reservoir sandstones in two wells from the North Tejon field. Heulandite occurs as both a replacement of glass shards and as an interstitial cement without obvious glass precursors. Heulandite aggregates replacing glass shards are rimmed by smectite films, whereas prismatic heulandite crystals stand on smectite coatings encrusting intergranular pores. Ca-rich heulandite is less than 10 μm in size and is rather silicic, with a Si/Al ratio of 3.79-4.05.

"Early" laumontite coexists with heulandite in a transitional zone between 2642 m and 2743 m. At the top of the transitional zone, less than 1% laumontite and 13% heulandite constitute an intergranular cement of the reservoir sandstones. The transformation of heulandite to laumontite takes place beginning precisely at a depth of 2703 m at about 90°C in one well, and continues to a depth of 2743 m. The incipient heulandite cement in the loosely cemented heulandite-bearing sandstones is an unstable remaining phase. The laumontite content is consistent with the expected volume of heulandite source material. "Early" laumontite occurs as large monocrystalline patches in which smectite is missing. Plagioclase is not albitized but is only partially dissolved in the transitional zone, and its

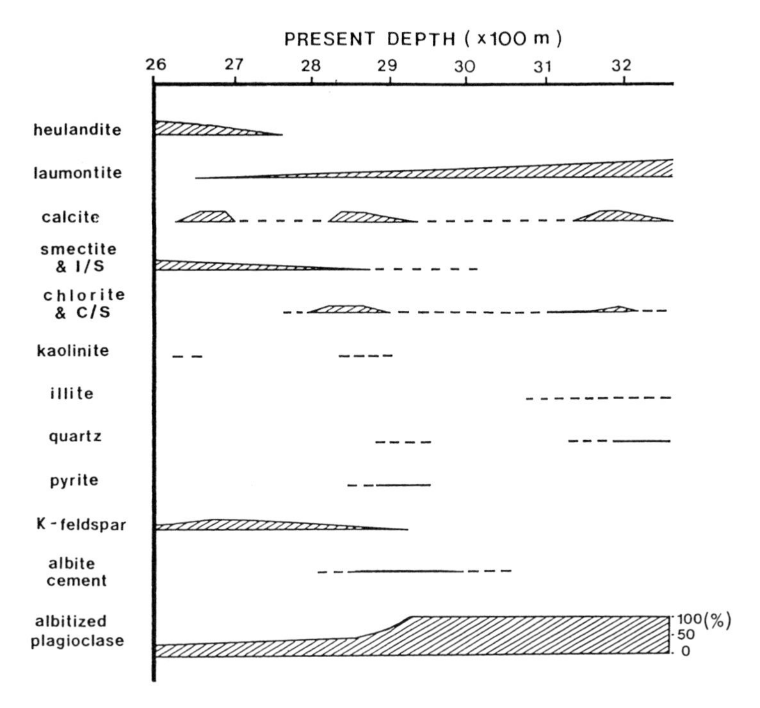

Figure 23. Schematic diagram showing the paragenetic sequence of diagenetic minerals and their relative abundance with present burial depth in the North Tejon oil field, California, USA (Noh and Boles 1993).

dissolution cavities are not filled by laumontite. This fact indicates that the transformation of plagioclase to albite and laumontite does not occur in the transitional zone.

Fluid mobility plays an important role in the formation of zeolites within the transitional zone, although on a very small scale. This is inferred from the following features observed in the transitional zone: (1) the heulandite content is lower in siltstone than in sandstone, and laumontite is absent in siltstone; and (2) zeolite cements are lacking in calcite-cemented sandstones, and K-feldspar is the principal cement with rare heulandite. Authigenic K-feldspar replaces plagioclase at a depth of 2703 m, and it occurs commonly as interstitial cement in fine-grained sandstone and siltstone to a depth of 2896 m. A simple dehydration decomposition of 100 volumes (V) of heulandite produces 62V laumontite and 24V quartz, when Al is preserved in equilibrium (Boles and Coombs 1977). However, 100V heulandite may produce 118V laumontite, if some Ca^{2+} and $Al(OH)_4^-$ in the pore fluid participate in the reaction. The latter case is probable at N. Tejon, considering the mode of occurrence. Ca^{2+} and $Al(OH)_4^-$ would originate from partial dissolution of plagioclase and smectite.

Heulandite is not present below 2743 m depth, but "late" laumontite appears as both a principal cement of reservoir sandstones and as a replacement of plagioclase, down to a depth of 3375 m. "Late" laumontite occurs as poikilotopic patches 0.2-10 mm in size, even larger than those of "early" laumontite. "Late" laumontite includes minute relics of replaced albite and chloritic clay. The laumontite content increases with depth below 3048 m, reaching a maximum of 25% at 3181 m, where laumontite replaces not only albitized plagioclase but also quartz. At a depth of 3408 m, laumontite fills fractures cutting clastic grains, indicating a late-stage origin.

Albitization of plagioclase begins to occur at a depth of 2835 m at a present temperature of about 94°C. Plagioclase is completely albitized at a depth of 3181 m at about 100°C. These temperature values are based on vitrinite reflectance (R_o) data and well-log measurements (Castano and Sparks 1974, Naeser et al. 1990). Albitized plagioclase is made up of numerous minute crystals of albite which tend to arrange in a regular manner with secondary micropores. This phenomenon supports Boles' (1982) conclusion that albitization of plagioclase involves a dissolution-precipitation mechanism. The micropores cause a dusty appearance in albitized plagioclase (Boles 1982).

There is a clear correlation between the presence of "late" laumontite and albitization of plagioclase in the Tejon area. This genetic relationship has commonly been observed around the world (Boles and Coombs 1977, Surdam and Boles 1979, Vavra 1989). The "late" laumontite-albite association suggests that laumontite acts as an important sink for Ca and Al released from albitization of plagioclase, as was first suggested by Coombs (1954). In this reaction (Reaction 2 in Table 6), 100V plagioclase (Ab_{70}) produces 70V albite and 62V laumontite. Silica derives from either the crystallization of chlorite/smectite to chlorite or I/S to illite (Boles 1982, Noh and Boles 1993) but may also result from quartz dissolution.

Tang et al. (1997b) reported the occurrence of clinoptilolite(?), heulandite, and laumontite in Permian-Triassic fluvial/lacustrine volcanic sandstones of andesitic to basaltic

Table 6. Chemical reactions of laumontite formation.

$$\underset{\text{heulandite}}{CaAl_2Si_7O_{18}{\cdot}6H_2O} \Leftrightarrow \underset{\text{laumontite}}{CaAl_2Si_4O_{12}{\cdot}4\,H_2O} + \underset{\text{quartz}}{3\,SiO_2} + 2\,H_2O \quad (1)$$

$$\underset{\text{plagioclase}}{NaAlSi_3O_8{\cdot}CaAl_2Si_2O_8} + \underset{\text{quartz}}{2\,SiO_2} + 4\,H_2O \Leftrightarrow \underset{\text{albite}}{NaAlSi_3O_8} + \underset{\text{+ laumontite}}{CaAl_2Si_4O_{12}{\cdot}4H_2O} \quad (2)$$

$$\underset{\text{plagioclase}}{NaAlSi_3O_8{\cdot}CaAl_2Si_2O_8} + \underset{\text{+ quartz}}{3\,SiO_2} + 2\,H_2O + Na^{+} \Leftrightarrow$$
$$\underset{\text{albite}}{2\,NaAlSi_3O_8} + \underset{\text{laumontite}}{0.5\,CaAl_2Si_4O_{12}{\cdot}4H_2O} + 0.5\,Ca^{+2} \quad (3)$$

$$\underset{\text{plagioclase}}{NaAlSi_3O_8{\cdot}CaAl_2Si_2O_8} + \underset{\text{+ quartz}}{SiO_2} + 6\,H_2O + 0.5\,Ca^{+2} \Leftrightarrow \underset{\text{laumontite}}{0.5\,CaAl_2Si_4O_{12}{\cdot}4H_2O} + Na^{+} \quad (4)$$

$$\underset{\text{calcite}}{CaCO_3} + 2\,Al^{+3} + \underset{\text{+ quartz}}{4\,SiO_2} + 7\,H_2O \Leftrightarrow \underset{\text{laumontite}}{CaAl_2Si_4O_{12}{\cdot}4H_2O} + CO_2 + 6\,H^{+} \quad (5)$$

$$\underset{\text{calcite}}{CaCO_3} + \underset{\text{kaolinite}}{Al_2Si_2O_5(OH)_4} + \underset{\text{+ quartz}}{2\,SiO_2} + 2\,H_2O \Leftrightarrow \underset{\text{laumontite}}{CaAl_2Si_4O_{12}{\cdot}4H_2O} + CO_2 \quad (6)$$

$$\underset{\text{calcite}}{15\,CaCO_3} + \underset{\text{montmorillonite}}{13.5\,Ca_{0.33}(Al_{3.34}Mg_{0.66})Si_8O_{20}(OH)_4} + 59\,H_2O \Leftrightarrow$$
$$\underset{\text{laumontite}}{19.5\,CaAl_2Si_4O_{12}{\cdot}4H_2O} + \underset{\text{chlorite}}{(Al_3Mg_9)(Al_3Si_5)O_{20}(OH)_{16}} + \underset{\text{+ quartz}}{25\,SiO_2} + 15\,CO_2 \quad (7)$$

1,2,3,5: Boles and Coombs (1977); 4,6: Zen (1961); 7: Madsen and Murata (1970).

composition in the southern Junggar basin, northwest China. Zeolites occur in these rocks as interstitial cements, associated with albitized plagioclase. Zeolite formation and albitization during burial are among the most pronounced diagenetic processes that have affected these sandstones. Heulandite is sporadically found in sandstones at shallower burial depth, whereas laumontite and heulandite coexist at deeper depths. The zeolite cement ranges in abundance from a trace to 10% and occurs largely in fine-grained sandstone beds of muddy lacustrine facies. Interbedded non-zeolitic coarse sandstones of fluvial and deltaic facies possess moderate to good reservoir quality with an average porosity of 18.2% in primary and secondary dissolution pores. In contrast, the zeolitic fine sandstones are well-cemented with poor reservoir quality.

Laumontite formation by burial diagenesis of arkoses. Laumontite is frequently found as a pervasive cement in arkoses in the deeper parts of sedimentary basins. In such cases, the overlying heulandite zone and the transitional zone are not seen. Laumontite-cemented arkosic sandstone reservoirs in the Upper Triassic Yanchang Formation on the North Shaanxi slope of the Ordos basin in the west of the North China platform are well documented by Liu et al. (1993), as summarized below. The Ordos basin forms a large-scale Mesozoic cratonic basin, and the North Shaanxi slope lies in the eastern part of the basin (Fig. 24), structurally forming a gently westward dipping monocline with a dip angle of about 0.5°. The Yanchang Formation is a fluvio-lacustrine system having a thickness <800 m and containing ten oil-bearing sandstone beds from Chang-1 to Chang-10. Large amounts of oil recovered comes mainly from Chang-2 to Chang-6. A common feature of these sandstone reservoirs is that they are fine arkoses consisting generally of 45-60% feldspar, 25-30% quartz, and <10% lithic fragments. The clasts are mostly moderately to well sorted and subangular to subrounded in shape, showing fairly good maturity in texture.

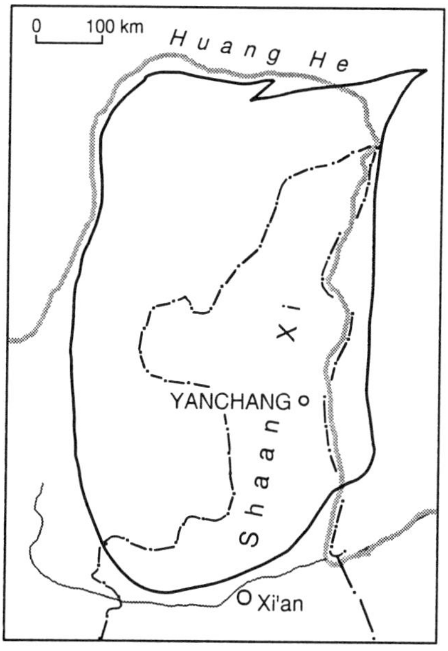

Laumontite is present in the sandstone reservoirs of Chang-6 and Chang-7 in the form of pore-filling cements, generally replaced by later carbonates. Laumontite also replaces plagioclase along cleavage and twinning planes so that plagioclase occurs in relics of pseudomorphs. The plagioclase associated with laumontite is almost exclusively albite. Liu et al. (1993) concluded that two generations of laumontite exist. In the earlier pore-filling stage, Ca and Si may have been derived primarily from hydration of volcanic matter including volcanic ash and volcanic plagioclase. In the later replacement stage, Ca was derived mainly from the albitization of plagioclase. However, it is unlikely that an earlier pore-filling stage was important in these rocks as volcanic material was rare. The lithic components of the arkose are chiefly fine-grained sedimentary rocks with only minor extrusive rocks. Therefore, it seems more likely that laumontite formation is related to the albitization of plagioclase in the later replacement stage.

Figure 24. Sketch map showing the geographic position of the Ordos basin in the western Chinese platform (simplified from Liu et al. 1993).

The temperatures of the Chang-6 and Chang-7 sandstone reservoirs in which laumontite is present are estimated to have been about 81-88°C when the Triassic Yanchang Formation reached its maximum depth of burial, based on vitrinite reflectance values (average R_o = 0.76-0.79). Proportions of illite in mixed-layer I/S in the reservoir sandstones are consistently in the range of 62-80%. Serpentine-like (7-Å material) and chlorite are present in roughly equal proportions, suggesting that the 7-Å material may be transforming into chlorite. Observations on clay mineral transformations indicate temperatures in the range of 70-100°C. The eroded overburden composed of the Jurassic and Cretaceous strata is ~1540 m by the R_o estimation (Fig. 25). A paleogeothermal gradient of about 2.7°C/100 m by the R_o estimation approximately equals the average present-day geothermal gradient of 2.72°C/100 m.

Regional development of laumontite in arkoses has also been observed in the deeper parts of arc-related sedimentary basins of the Circum-Pacific belt, e.g. in the Santa Barbara basin, the Tanner Banks area of the outer continental shelf, and Borderland of southern California, in Lower Cook Inlet and the offshore St. George basin of Alaska, and in the Otway basin of Australia. However, arkoses having pervasive laumontite cements formed by burial diagenesis do not constitute economic hydrocarbon reservoirs.

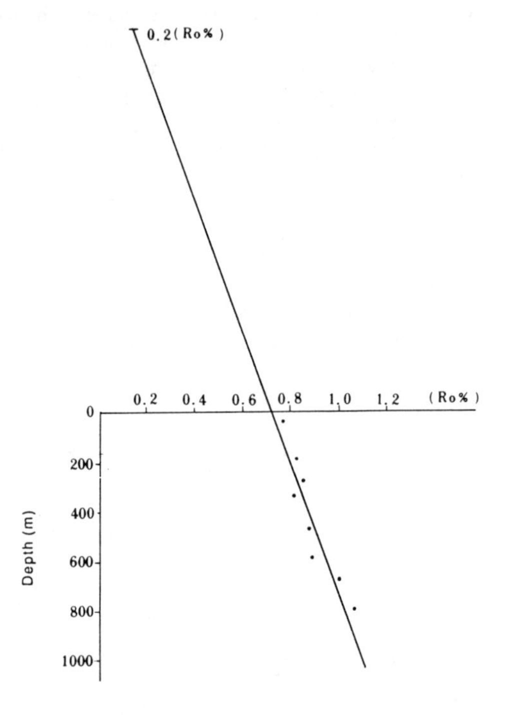

Figure 25. The paleogeothermal gradient (slope of the vitrinite reflectance-burial depth line) of the Upper Triassic on the Northern Shaanxi slope of the Ordos basin is about 2.7°C/100 m, and the eroded overburden is estimated to be ~1540 m by the R_o estimation method (Liu et al. 1993).

Late uplift-stage laumontite in arkosic reservoirs. In the classic zeolite localities of New Zealand, in which laumontite was first attributed to burial diagenesis–low-grade metamorphism (Coombs 1954), laumontite may have formed during uplift of the basin (Stallard and Boles 1989, Boles 1991). Laumontite is spatially associated with uplift-related faults and fractures, and isotopic data suggested that meteoric water may have been involved in its formation (Boles 1993). In addition, the common occurrence of laumontite filling fractures in sandstones indicates a late-stage origin (Boles 1993, Remy 1994). Late-stage laumontite formation and dissolution are important in arkosic sandstone reservoirs (Boles 1993). For example, laumontite cement is a recent "diagenetic" feature that developed after oil accumulation in the Los Angeles basin, California, as summarized below (Coffman 1995). The Los Angeles basin is a tectonically active Cenozoic depositional basin and one of the most highly explored basins in the United States, since oil production there began before the 1900s. The occurrence of laumontite in the Los Angeles basin appears to be structurally controlled because it is restricted, so far as is publicly known, to seven oil fields at the basin margin and/or aligned along faults (Fig. 26). Laumontite occurs as small, mm- to cm-

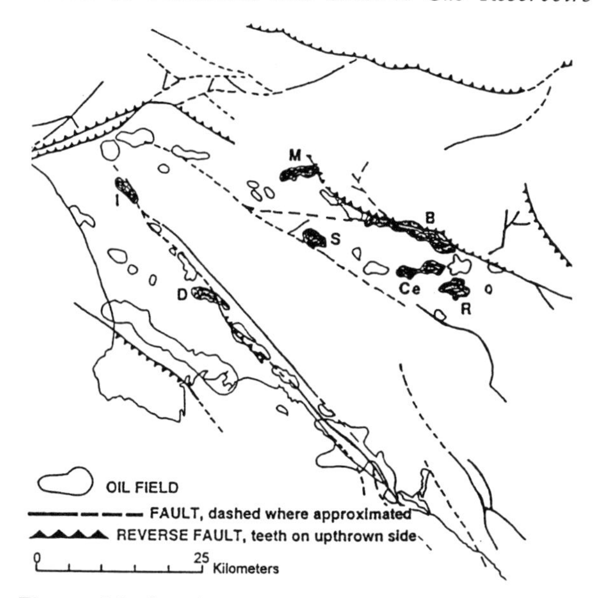

Figure 26. Location map of laumontite-bearing oil fields (shaded) and their relationship to major basin-bounding fault systems. Letters refer to the following oil fields: Brea Olinda (B), East Coyote (Ce), Dominguez (D), Inglewood (I), Montebello (M), Richfield (R), and Santa Fe Springs (S) (Coffman 1995).

sized, isolated patches in many oil fields. However, in both the Santa Fe Springs and Dominguez oil fields, laumontite appears as pervasive pore-filling and grain-replacing cements, and it is generally accompanied by extensive albitization of detrital plagioclase. The laumontite content is 15-23% by volume. Pore-cement laumontite occurs as a colorless, clear, continuous and uniform cement composed of interlocking, poikilotopic crystals of a few mm to 1-2 cm in size. Isolated hydrocarbon inclusions exist in some large laumontite crystals of 1 mm in size. In two wells in the Santa Fe Springs field, a laumontite-cemented zone and a calcite-cemented zone are separated by a non-cemented zone with a thickness of less than 1 cm. The contents of laumontite- and calcite-cements are nearly 20% by volume each, which is equal to the porosity of the non-cemented zone. Plagioclase is albitized in three zones. These facts suggest that rocks of each zone were exposed to similar conditions and underwent alteration to the same extent.

Detrital plagioclase in the Los Angeles basin has an average composition of about An_{20-25}. The laumontite content in laumontite-cemented reservoir sandstones of the Santa Fe Springs and Dominguez oil fields is very high, up to 17% higher than the amount of plagioclase theoretically required for direct conversion to laumontite. This indicates that additional Ca ions were required for laumontite crystallization in the laumontite-cemented sandstones. The existence of unaltered or only slightly altered plagioclase grains and the lower An content in the laumontite-cemented sandstones suggest that additions of Ca and Na were necessary for the observed laumontite content and the accompanied albitization of plagioclase, respectively. Assuming that Al was not conserved, the development of laumontite cement and replacement requires the addition of Al, Ca, and Na, as well as the migration of Al.

Volcanic rocks in association with altered carbonates, clays, and opaque minerals exist beneath the laumontite zone in the Dominguez field. The $^{87}Sr/^{86}Sr$ ratios of laumontite

(0.707335-0.708013), calcite cement (0.708577-0.708868), and deep-seated altered volcanic rocks (0.7028-0.7035) suggest that the Ca of the laumontite was possibly partially derived from the alteration of volcanic rocks and/or the albitization of detrital plagioclase in the deeper parts of the sedimentary basin. The primary pore fluid in the sediment was undoubtedly the contemporaneous sea water, which would be modified with increasing burial depth. We can only speculate on the fossil water composition from which laumontite crystallized, because water flooding as an oil recovery technique was prevalent in the Los Angeles basin by the 1960s. Ca-enriched fossil waters in albitized arkoses of the Stevens Sandstone were reported by Fisher and Boles (1990) from the Paloma oil field in the deep and central part of the San Joaquin basin, California, as shown in Figure 27, but they likewise can probably not be trusted as guides to Los Angeles basin water composition.

High heat flow and enhanced fluid mobility are observed in the margins of the Los Angeles basin where laumontite-cemented sandstones have been recognized. In the Santa Fe Springs field, the temperature is nearly 170°C at the present depth of about 4 km and it is 110-123°C at the top of the laumontite zone. Such high temperatures are probably a relatively recent phenomenon, in light of the fact that fission tracks (FT) in apatite are not yet totally annealed (Naeser et al. 1990). Annealing is almost certainly proceeding today and very likely had its onset at this site less than one million years. Apparently approximately six million years are required to anneal FT in apatite under these conditions (170°C, depth of 4 km). In contrast, FT in apatite are completely annealed at 140°C in the central synclinal part of the Los Angeles basin where the geothermal gradient is about 25°C/km.

Laumontite dissolution typically occurs under conditions of lowered pH or temperature, increased CO_2 partial pressure, or the presence of carboxylic acids (Crossey et al. 1984). The Los Angeles basin is filled by about 6 km of Neogene clastic strata with abundant interbedded source rocks, only 3 km of which have reached maturity with respect to oil generation. Episodic pulses of CO_2-rich and CO_2-poor fluids might have been released over a prolonged periods from the deep part of the basin, as byproducts of kerogen and oil generation. Heat convection and overburden pressure make the fluids migrate upward into the shallow-seated strata, and "diagenetic changes," including laumontite formation or dissolution, occurred where fluids interacted with the sediments (Coffman 1995). Hydrothermal alteration is a more appropriate descriptor than "diagenetic changes" for the laumontization in the Santa Fe Springs and the Dominguez fields, because laumontite crystallization is related to local higher-temperature fluids perhaps associated with faults and with stratigraphically older volcanic rocks.

Relation of laumontite to sandstone reservoir properties

Porosity and permeability are crucial properties of reservoir sandstones. Laumontite is the most frequently encountered and volumetrically abundant zeolite in sandstones, and it can locally cement 20% or more by volume of the rock, thereby creating barriers to flow in the reservoirs (Boles 1993). Laumontite-altered sandstones were once regarded as an economic basement for exploration for oil and gas, because the primary porosity of sandstones is essentially lost by cementation of burial-diagenetic laumontite (e.g. McCulloh et al. 1973, Galloway 1974, 1979; McCulloh and Stewart 1979). However, there are some occurrences where production has been reported locally beneath or near laumontite-bearing sandstones, including the Temblor Formation of the Kettleman North Dome (Merino 1975) and the Vedder Sandstone of the North Tejon oil field (Noh and Boles 1993) in the San Joaquin Valley, California, in the Yanchang Formation in the eastern Ordos basin, and in the Lower Cretaceous sandstones in North Songliao basin of China.

Secondary porosity enhancement. Surdam and coworkers (Crossey et al. 1984,

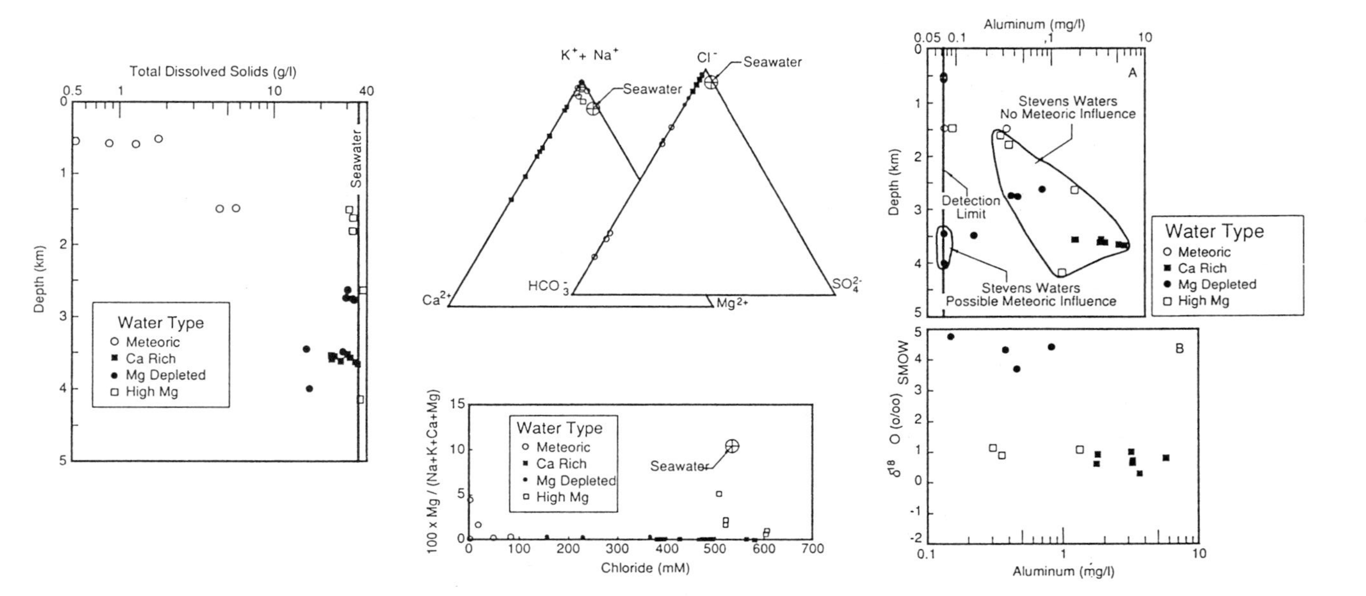

Figure 27. Diagrams showing the composition of wellhead water samples from 25 producing oil wells in the southern San Joaquin basin of California, USA. Ca-enriched waters from the deep Stevens sandstone of deep-marine turbidite origin are of particular interest. Points are identified by water type: (A) (left) total dissolved solids plotted against depth for sampled waters; (B) (center-upper) ternary diagrams showing relative molar proportions of major cations and anions; (center-lower) plot of relative molar abundance of magnesium, Mg/(Na+K+Ca+Mg), against chloride concentration (mM); (C) (right-upper) crossplot of measured dissolved Al concentrations against depth. Detection limit for Al is 0.075 mg l^{-1}; (right-lower) crossplot of Al and $\partial^{18}O$ for waters not showing evidence of meteoric influence. Reprinted and compiled from *Chemical Geology* 82:83-101, Fisher and Boles (1982), with kind permission of Elsevier Science–NL).

Surdam et al. 1984) concluded that organic acids released during decomposition of kerogen play an important role in creating secondary porosity by dissolution of feldspar and, locally, laumontite. The existence of carboxylic acid anions in oilfield water is generally attributed to thermal maturation of kerogen (e.g. Crossey et al. 1984). However, thermodynamic calculations indicate that the progressive degradation of kerogen is not controlled by thermal instability of organic compounds but is caused instead by the drastic decrease in oxygen fugacity with increasing burial depth, as Sato (1990) and Helgeson (1991) stressed. In some oil fields, the concentration of organic acids in oilfield water is low to nil. In other oil fields, the following have been reported: 10,000 ppm acetate; 4400 ppm propionate; 2500 ppm malonate; 700 ppm butanoate; 500 ppm oxalate; 400 ppm pentanoate; 100 ppm hexanoate; and 100 ppm heptanoate. In addition, small amounts of many other organic anions have been reported, without any clear temperature trend recognizable in the total content of organic anions (Helgeson et al. 1993).

The model for porosity enhancement by organic acids was discussed by Surdam et al. (1984). In this model, secondary porosity is enhanced by the timing of organic/inorganic reactions, the availability of fluid flux, the development of migration paths, and the primary porosity and permeability properties of potential reservoir sandstones. The most favorable conditions are a close association between organic-rich source rocks and reservoir sandstones as well as the partial preservation of primary porosity to provide a pathway for reactive fluids. Mass transfer is necessary for secondary porosity enhancement. Dissolved material must be removed beyond adjacent pores and pore throats in order to improve reservoir rock properties significantly by dissolution of clastic grains and/or authigenic cements. Dissolved aluminosilicates are ultimately precipitated as kaolinite. It is interesting to note that the only evidence of any aqueous Al complex remaining in the rock would be in the form of kaolinite (Surdam et al. 1984).

Prediction of reservoir quality. Bloch and Helmold (1995) discussed approaches to predicting reservoir quality in sandstones of the San Emigdio Mountains, a mature area in the southernmost San Joaquin basin, California. Laumontite cementation significantly affects reservoir quality, whereas calcite cementation is only locally important. The spatial distribution of heavily carbonate-cemented sandstones cannot be predicted in the San Emigdio area. In areas without laumontite cementation, reservoir quality can be crudely predicted from the depth-porosity relation, if the cement content is less than 10% by volume. In contrast, areas having laumontite cementation are expect to have low porosity, especially in the Zemorrian (late Early to Late Oligocene) sandstones which have been exposed to temperatures of over 104°C. These rocks have porosities of <6% if their laumontite content exceeds 10% by volume. Irrespective of its content (as much as 28% by volume), laumontite occurs only in sandstones exposed to such high temperatures and is almost solely restricted to the Zemorrian. The spatial distribution of laumontite likely matches the sedimentation area of glass shards and vitrophyric rock fragments of contemporaneous volcanic origin (cf. Noh and Boles 1993).

According to J.R. Boles (written comm.), however, laumontite in the North Tejon oil field occurs in all rock units up to and including the Upper Miocene at one place or another. Laumontite also occurs in all pre-Zemmorrian strata, down to and including gneisses of the pre-sedimentary basement where those are fractured. These occurrences are irrespective of volcanogenic detritus, but where such detritus is most abundant in the Oligo-Miocene Zemmorrian strata, laumontite (and heulandite) are both more common.

Relationship between timing of laumontite formation and hydrocarbon accumulation in sandstone reservoirs. Timing of zeolitization relative to oil emplacement is crucial in evaluating the effects of zeolites on hydrocarbon reservoirs (Boles

1993, Iijima 1995). Oil generation from kerogen is generally associated with a temperature range between about 75° and 120°C (Tissot and Welte 1978). Laumontite is stable under these conditions, and therefore its crystallization could possibly overlap hydrocarbon accumulation in the reservoirs. Isolated hydrocarbon inclusions in interstitial laumontite crystals in the Los Angeles basin demonstrate that laumontite cementation occurred during or after oil migration and accumulation in the arkosic reservoirs (Coffman 1995).

Most laumontite cements observed in reservoir sandstones formed late relative to hydrocarbon maturation and are related to hydrothermal events (Crossey et al. 1984). Laumontite cements in Cretaceous and Paleogene feldspathic sandstones in California may be related to thermal events associated with the migration of the Mendocino triple point. The triple point migrated northward beneath many of the West Coast basins during the Miocene and conceivably produced a thermal pulse that might have influenced hydrocarbon maturity in those basins and may also have produced late hydrothermal laumontite (Crossey et al. 1984). However, the very young to currently forming laumontite of the Los Angeles Basin oil fields does not fit this model (McCulloh et al. 1973).

In the Upper Triassic of the eastern Ordos basin, laumontite-bearing arkoses form hydrocarbon reservoirs primarily because the dissolution of laumontite has produced secondary porosity (Zhu 1985). Liu et al. (1993) considered that the formation of secondary porosity is mainly the result of extensive dissolution of carbonate and laumontite cements and detrital plagioclase caused by upward-migrating organic acid solutions produced by dehydrocarboxylation of organic matter at depth (Fig. 28). They suggested that the formation of laumontite must have been earlier than, or nearly concurrent with, the transformation of organic matter to hydrocarbons.

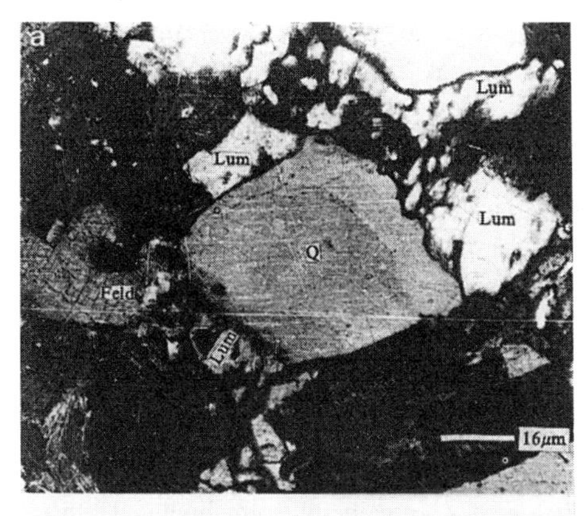

Figure 28. (a) Photomicrograph showing dissolution of laumontite cement in the arkose reservoir of the Upper Triassic at 643.48-m depth in well Yu 9, Ordos basin. Lum: laumontite, Q: quartz, Feld: feldspar. (b) SEM photomicrograph showing detrital plagioclase dissolution in the Upper Triassic arkose reservoir at 395.5-m depth in well He 121, Ordos basin. Feld: feldspar, P: pore. (Liu et al. 1993).

LOG EVALUATION AND ENHANCED OIL-RECOVERY OF ZEOLITE-BEARING RESERVOIR ROCKS

Log evaluation

Special evaluations are needed for zeolite-bearing rock reservoirs because zeolites possess quite different properties from typical sand grains and cementing minerals. For example, zeolite specific gravities are much lower than those of quartz, feldspar, and carbonates. Zeolites are characterized by loosely bound water and high cation-exchange capacities (CEC). These properties influence porosity estimations and log evaluation of the zeolite-bearing rock reservoirs. Therefore, zeolites in reservoirs impact a number of critical down-hole geophysical and electrical logging measurements. Dunn et al. (1993) investigated the influence of zeolites on petroleum evaluation of Cretaceous fluvial volcaniclastic sandstones in the San Jorge basin, Argentina. The sandstones are extensively altered to clays and zeolites, and the zeolite-bearing intervals are generally more productive. Reservoir sandstone cores contain laumontite, analcime, and clinoptilolite, whereas intervening unaltered intervals include abundant clays and minor carbonates. According to Dunn et al. (1993), log response and shaly sand analysis are summarized below.

Log response. (1) Porosity values given by all porosity logging tools are unusually high in the zeolite-bearing intervals. Density porosity data in the intervals are also unusually high as a result of the low densities of zeolites, 15-20% lower than quartz grains. (2) Sonic velocities of zeolite cements are usually low relative to values for sand grains and most common cements. (3) Hydrogen in zeolitic water is detected by neutron tools as porosity. Neutron tools frequently overestimate as much as 1% porosity for every 3% zeolite content. (4) The content of U, K, or Th in analcime and laumontite is too low to contribute to gamma-ray log response. Although clinoptilolite contains variable amounts of K, usually less than 3%, this K content does not significantly affect gamma-ray measurement calculations for clay volume.

Shaly sand analysis. Shaly sand analysis is an attempt to correct deviations from the ideal conductivity of a rock-water-oil system in a porous medium. Two principal difficulties in shaly sand analysis arise from the occurrence of zeolites: (1) effects on neutron tools used for porosity corrections and in-situ oil reserve calculation, and (2) effects of zeolites on the conductivity of the rock-water-oil system. Basically all common methods of shaly sand analysis calculate a log-derived volume (V_{clay}). At least one approach for evaluating V_{clay} uses the cation-exchange capacity (CEC). Zeolites have a high CEC, but their crystals rarely form continuous microporous conductive networks. Consequently, formation evaluation parameters derived from clay-rich intervals are not applicable to intercalated zeolite-bearing intervals. There are several methods to use log responses from adjacent shales in order to calculate the corrections for evaluating intercalated shale sands. An example of a depth record from a well drilled in the Samgori oilfield, Georgia, is shown in Figure 29 (Vernik 1990). Laumontite tuff, which is the oil reservoir, was penetrated between 2451 m and 2456 m and was characterized by distinct log responses. The porosity of the laumontite tuff was evaluated using the gamma-ray-neutron log and results were corrected for the calculated neutron porosity of matrix $ØN_{ma}$ = 0.18. The acoustic log data (interval P-wave velocity) also clearly indicated the location of the zone of intense laumontization by a dramatic decrease in velocity from about 5.0 km/sec to less than 4.0 km/sec. Note that the velocity reduction was detected using both conventional and low-frequency modifications of the acoustic log, so that the effect of well-bore enlargement, characteristic of the interval, appears to be minimal. The enlargement of the hole (caliper log) can be attributed to the anomalously low strength and cohesion of the laumontite rock. Temperature measurements in this well during the production stage clearly indicated that the oil produced came from the laumontite interval (Vernik 1990).

Zeolites and enhanced oil-recovery projects

Migration of fines, transformation of zeolites to other minerals, and crystallization of zeolites from high temperature and/or alkaline injection fluids are possible problems which make understanding zeolite stability an important future area of research (Boles 1993).

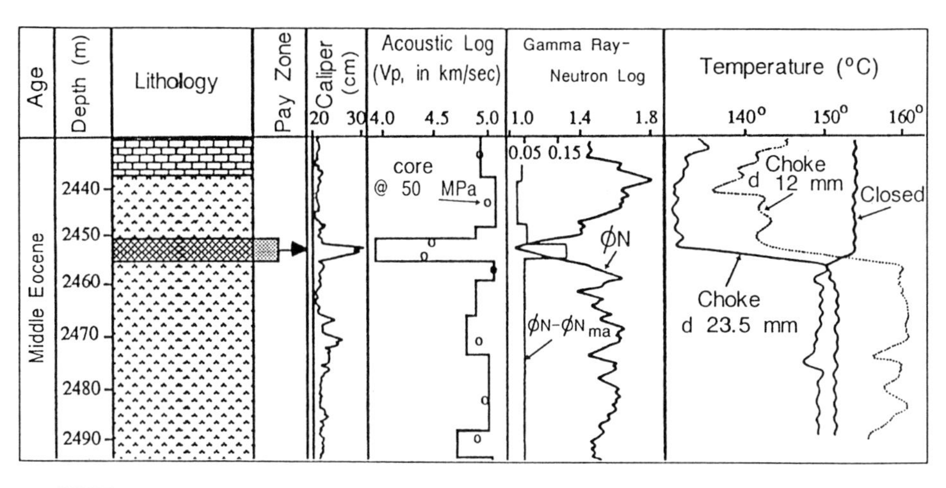

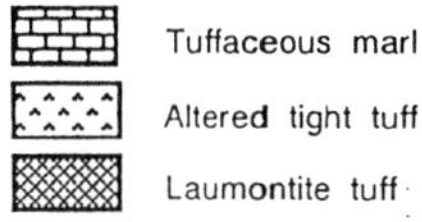

Figure 29. Part of the depth record, containing a laumontite tuff reservoir, from a borehole in the Samgori oil field, Georgia, showing caliper log, interval velocity from acoustic log, gamma-ray–neutron log, and temperature measurements during the production stage. Also indicated are P-wave velocity (V_p) values in core samples at a confining pressure (P_c) = 50 MPa (the overburden less hydrostatic pressure for a depth of 3 km), and neutron porosity from GNL. Calculated neutron porosity of matrix ($ØN_{ma}$) = 0.18 (Vernik 1990; reprinted by permission of the American Association of Petroleum Geologists).

Zeolite corrosion in reservoirs subjected to extensive water-injection can alter the rocks to hydrophilic or strongly hydrophilic. This tendency is undoubtedly effective in increasing the oil sweeping rate in the reservoirs (Chen 1992). Freshwater injection recovery has been continued for more than twenty years in No. 7E District of the Karamay oilfield, north-western China, in which analcime occurs as cements of Triassic sandstones and conglomerates (Chen 1992). Extensive corrosion of analcime was observed in core samples recovered from the prolonged water-injection portion of the closed #7172 well. The corroded analcime crystals have a thin lamellar structure, and whole analcimic aggregates display a pumice-like or a honeycomb-like structure. Analcime was not corroded in the shallower portion of the well where water injection was minimal. The porosity of the analcime-cemented reservoirs subjected to prolonged water-injection was improved 1-3%, which is significant in low-porosity intervals (Chen 1992).

Acid treatment recovery. Acid treatments were tested in analcime-cemented sandstone reservoirs of the Offshore Gulf Coast Tertiary, USA (Underdown et al. 1990). Analcime dissolves in concentrated hydrochloric acid and forms a silica gel, but acid treatments using acetic acid effectively prevent analcime dissolution. Hydrochloric acid treatments were tested in the analcime-cemented sandstone and conglomerate reservoirs in the Junggar Upper Permian of No. 8 District in the Karamay oilfield (Chen 1992). Analcime is commonly accompanied by calcite in the reservoirs, making the rocks suitable for acid treatment recovery. Hydrochloric acid treatments were also carried out in the zeolite-rich basalt reservoirs in the Junggar Carboniferous of No. 1 District in the Karamay field (Chen 1992). An increase in oil recovery after acid treatment was recognized in eleven out of the twelve water injection wells tested.

Sulfuric acid treatments were tested for the albite-analcime-rich sandstone and conglomerate reservoirs of the No. 8 District, and for the zeolite-calcite-rich basalt reservoirs of the No. 1 District in the Karamay field (Chen 1992). Concentrated sulfuric acid was poured into the dry reservoirs.

SUMMARY AND CONCLUSIONS

Table 7 provides a summary of zeolite-bearing oil and gas reservoirs with reference to their genetic type and zeolite formation temperature. Zeolite-bearing reservoir rocks are typically composed of arkosic and volcanic sandstones, volcanic and volcaniclastic rocks, or pre-sedimentary igneous basement rocks. Zeolites contained in the reservoirs are formed by either diagenetic or hydrothermal alterations.

Petroleum reservoirs are common in silicic volcaniclastic rocks and volcanic sandstones in thick marine sequences that have been subjected to burial diagenesis. Vertical zeolite zonation is formed through the following sequential transformations: silicic glass (zone I) alters to clinoptilolite (zone II), which alters to analcime (zone III) , which is transformed to albite (zone IV). The vertical zonation developed in Tertiary marine sequences is controlled primarily by increasing temperatures as burial depth increases, i.e. 41-53°C at the zone I-II boundary, 80-91°C at the zone II-III boundary, and 120-124°C at the zone III-IV boundary. Reservoirs in zones I and II typically are good quality with sufficient primary and intercrystalline pores, whereas reservoirs of zones III largely lose intergranular and intercrystalline pores through deep burial and recrystallization. No reservoirs have been found in zone IV. These reaction temperatures are significantly lowered by the interaction of silicic glass and clinoptilolite with a saline, alkaline solution generated by regional dissolution of detrital feldspars, as in the case of the Johban-oki gas field region. Precipitation of laumontite from such solutions can occur at much lower temperatures (38-81°C) than during normal burial diagenesis.

Laumontite-cemented arkoses formed by burial diagenesis are common in the deeper portions of arc-related sedimentary basins in the Circum-Pacific belt but they seldom constitute oil and gas reservoirs. However, laumontite-cemented arkosic reservoirs exist in some cratonic basins of China. Laumontite formation in these deposits is related to dissolution of feldspars and albitization of plagioclase over a temperature range of 80-140°C. Oil and gas reservoirs exist in laumontite-cemented arkoses where the reservoir quality has been improved by either tectonic fracturing or corrosion of laumontite cement and clastic feldspar grains due to organic-acid fluids, or by both.

Closed-system analcime occurs in the Lower Permian volcanic sandstone reservoirs in the Junggar basin of northwest China. The analcime has partly transformed to albite by subsequent burial diagenesis. Secondary corrosion of analcime and clastic feldspar grains by organic-acid fluids serves to improve the reservoir quality.

Hydrothermal zeolite-bearing oil and gas reservoirs have been recently discovered world-wide in arc-related basins. The altered reservoir rocks consist largely of volcanic and volcaniclastic rocks, but arkosic sandstones with laumontite cement and even pre-sedimentary igneous basement rocks with laumontite veins constitute oil and gas reservoirs. The host rocks range from Carboniferous to Middle Miocene in age, and their hydrothermal alteration occurred from Permo-Carboniferous to Recent. Except for Tertiary laumontite-cemented arkose reservoirs in the Los Angeles basin of California, where the laumontite precipitated in an existing oil-bearing environment in Recent time along basin-bounded faults, hydrothermal alterations typically occurred much earlier than the accumulation of oil and gas. Source rocks exist either in younger or older strata relative to

Table 7. A summary of zeolite-bearing hydrocarbon reservoirs with reference to their genetic type and formation temperature of zeolites.

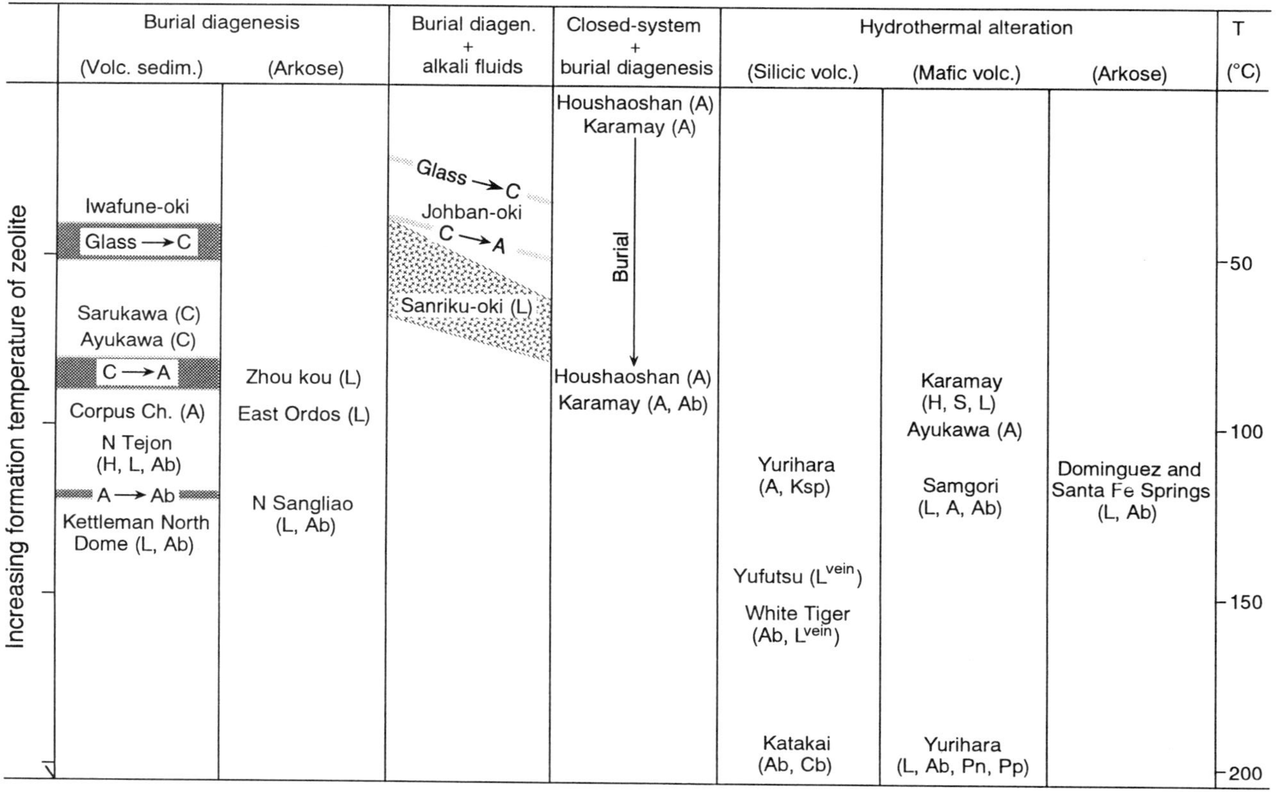

Burial diagenesis (Volc. sedim.)	Burial diagenesis (Arkose)	Burial diagen. + alkali fluids	Closed-system + burial diagenesis	Hydrothermal alteration (Silicic volc.)	Hydrothermal alteration (Mafic volc.)	Hydrothermal alteration (Arkose)	T (°C)
Iwafune-oki; Glass → C		Glass → C; Johban-oki C → A	Houshaoshan (A), Karamay (A); ↓ Burial				≤ 50
Sarukawa (C); Ayukawa (C); C → A; Corpus Ch. (A)	Zhou kou (L); East Ordos (L)	Sanriku-oki (L)	Houshaoshan (A), Karamay (A, Ab)		Karamay (H, S, L); Ayukawa (A)		50–100
N Tejon (H, L, Ab); A → Ab; Kettleman North Dome (L, Ab)	N Sangliao (L, Ab)			Yurihara (A, Ksp); Yufutsu (L^{vein})	Samgori (L, A, Ab)	Dominguez and Santa Fe Springs (L, Ab)	100–150
				White Tiger (Ab, L^{vein}); Katakai (Ab, Cb)	Yurihara (L, Ab, Pn, Pp)		150–200

Increasing formation temperature of zeolite

hydrothermally altered reservoir rocks that occupy a structurally or stratigraphically higher level, respectively. Hydrothermal alteration occurs in both continental and deep-sea environments. For example, in the Green Tuff region of northeast Japan, where volcanic and volcaniclastic reservoirs are superior, their hydrothermal alteration took place several hundreds meters below the Middle Miocene back-arc deep sea.

Hydrothermal zeolite precipitation generally has a negative effect on reservoirs, decreasing the porosity and permeability of the reservoir rocks. Subsequent tectonic fracturing commonly improves the reservoir quality, as in the case of Samgori, Karamay, Santa Fe Springs and Dominguez, Yufutsu, and White Tiger. In the Miocene Green Tuff reservoirs, higher-temperature alteration after zeolite formation produced good reservoirs. For example, the Yurihara basalt reservoirs exist in a prehnite zone, and the Katakai rhyolite reservoirs occur in an albite-quartz-carbonate zone. The decrease in solid volume during the transformation of laumontite to prehnite improves the reservoir quality at Yurihara, whereas the clay mineral-poor rigid brittle zone preserves primary porosity and improves reservoir quality by facilitating subsequent tectonic fracturing. Corrosion of analcime by later hydrothermal fluids improved the Yurihara analcimic tuff reservoir.

Special geophysical evaluation is needed for zeolitic rock reservoirs because of low specific gravity and high loosely bound water content of zeolites. These properties can greatly influence porosity estimations and log evaluation of the zeolitic rock reservoirs.

Zeolite corrosion in reservoirs subjected to extensive water-injection alters the rocks to hydrophilic or strongly hydrophilic, and this tendency is effective in increasing the oil sweeping rate in the reservoirs, as in the case of Karamay, where analcime cement was extensively corroded by prolonged water injection. Most zeolites are easily attacked by acids. Acid treatment recovery have been tested for some zeolitic rock reservoirs.

Zeolite-bearing oil and gas reservoirs are not comparatively large in number and scale. Clearly, altered volcanic rocks and volcanic sediments as well as pre-sedimentary basement rocks of arc-related basins have been not fully explored for oil and gas, and new discoveries are expected.

ACKNOWLEDGMENTS

I am grateful to the Japan National Oil Corporation for permitting the MITI-borehole data to be used in this chapter. I am indebted to Y. Yanagimoto, JAPEX Research Center and R. Matsumoto, University of Tokyo for collecting references, as well as T. Uchida and M. Yagi, JAPEX Research Center, for permitting the use of unpublished data. Also, I thank R.L. Hay, University of Illinois-Urbana, J.B. Boles, University of California at Santa Barbara, and D.L. Bish, Los Alamos National Laboratory, for critical reading of the manuscript and invaluable suggestions.

REFERENCES

Agatsuma T, Yokoi S, Inaba M (1996) Fracture analysis of Yufutsu gas field. J Japan Assoc Petrol Tech 61:145-50 [in Japanese, English abstr]

Araki N, Kato S (1993) A discovery of the Ayukawa oil and gas field, Akita Prefecture. J Japan Assoc Petrol Tech 58:119-127 [in Japanese, English abstr]

Areshev EG, Tran LD, Ngo TS, Shnip OA (1992) Reservoirs in fractured basement on the continental shelf of southern Vietnam. J Petrol Geol 15:451-464

Bloch S, Helmold KP (1995) Approaches to predicting reservoir quality in sandstones. Am Assoc Petrol Geol Bull 79:97-115

Boles JR (1977) Zeolites in low-grade metamorphic rocks. *In* Mineralogy and Geology of Natural Zeolites. FA Mumpton (ed) Rev Mineral 4:103-132

Boles JR (1982) Active albitization of plagioclase in Gulf Coast Tertiary. Am J Sci 282:165-180
Boles JR (1991) Diagenesis during uplift and folding of the Southland syncline, New Zealand. New Zealand J Geol Geophys 34:253-259
Boles JR (1993) Zeolite cements in hydrocarbon reservoirs. *In* Zeolite '93: Program and Abstr, 51-53
Boles,JR, Coombs DS (1977) Zeolite facies alteration of sandstones in the Southland Syncline, New Zealand. Am J Sci 277:982-1012
Castano JR, Sparks DM (1974) Interpretation of vitrinite reflectance measurements in sedimentary rocks and determination of burial history using vitrinite reflectance and authigenic minerals. Geol Soc Am Spec Paper 153:31-52
Cathelineau M, Nieva D (1985) A chlorite solid solution geothermometer, the Los Azufres (Mexico) geothermal system. Contrib Mineral Petrol 91:235-244
Chen G (1992) Zeolite minerals and their relation to oil and gas accumulation in the reservoir formations in Karamay Oil Field. Acta Petrolei Sinica 13:44-51 [in Chinese, English abstr]
Coffman RL (1995) Late-stage laumontite cementation and dissolution in the Los Angeles Basin, southern California. *In* Natural Zeolites '93: Occurrence, Properties, Uses. DW Ming, FA Mumpton (eds) Int'l Comm Natural Zeolites, Brockport, New York, p 39-49
Coombs DS (1954) The nature and alteration of some Triassic sediments from New Zealand. Royal Soc New Zealand Trans 82:65-109
Coombs DS, Ellis AJ, Fyfe WS, Taylor AM (1959) The zeolite facies, with comments on the interpretation of hydrothermal syntheses. Geochim Cosmochim Acta 17:53-107
Crossey LJ, Frost BR, Surdam RC (1984) Secondary porosity in laumontite-bearing sandstones. *In* Clastic Diagenesis. DA McDonald, RC Surdam (eds) Am Assoc Petrol Geol Memoir 37:225-237
Dunn TL, Destefano M, Decastelli OO (1993) Zeolites in petroleum evaluation of volcaniclastic sandstones, San Jorge Basin, Argentina. *In* Natural Zeolites '93: Occurrence, Properties, Uses. DW Ming, FA Mumpton (eds) Int'l Comm Natural Zeolites, Brockport, New York, p 83-84
Fisher JB, Boles JR (1990) Water-rock interaction in Tertiary sandstones, San Joaquin Basin, California: diagenetic controls on water composition. Chem Geol 82:83-101
French DE, Freeman KJ (1979) Volcanics yield another oil field. World Oil 188:58-63
Galloway WE (1974) Deposition and diagenetic alteration of sandstone in Northeast Pacific arc-related basins: implications for graywacke genesis. Geol Soc Am Bull 85:379-390
Galloway WE (1979) Diagenetic controls of reservoir quality in arc-related sandstones: implications for petroleum exploration. *In* Aspects of Diagenesis. PA Scholle, PR Schluger (eds) S E P M Spec Publ 26:17-43
Gottardi G, Galli E (1985) Natural Zeolites. Springer-Verlag, Berlin, Heidelberg, New York, Tokyo, 409 p
Graham SA, Brassell S, Carroll AR, Xiao X, Demaison G, McKnight CL, Liang Y, Chu J, Hendrix MS (1990) Characteristics of selected petroleum source rocks, Xianjang Uygur Autonomous Region, northwest China. Am Assoc Petrol Geol Bull 74:493-512
Grynberg ME, Papava D, Shengelia M, Takaishvili A, Nanadze S, Patton DK (1993) Petrophysical characteristics of the Middle Eocene laumontite tuff reservoir, Samgori field, Republic of Georgia. J Petrol Geol 16:313-322
Hay RL (1962) Origin and diagenetic alteration of the lower part of the John Day Formation near Mitchell, Oregon. *In* Petrologic Studies: A Volume in Honor of A.F. Buddington. AEJ Engel, HL James, BF Leonard (eds) Geol Soc Am, p 191-216
Hay RL (1963) Stratigraphy and zeolitic diagenesis of the John Day Formation of Oregon. Univ Calif Pub in Geol Sci 42:199-262
Hay RL (1966) Zeolites and zeolitic reactions in sedimentary rocks. Geol Soc Am Spec Papers 85:1-130
Hay RL (1986) Geologic occurrence of zeolites and some associated minerals. *In* New Developments in Zeolite Science and Technology. Y. Murakami, A. Iijima, J.W. Ward (eds) Kodansha-Elsevier, Tokyo, p 35-40
Hay RL (1995) New developments in the geology of natural zeolites. *In* Natural Zeolites '93: Occurrence, Properties, Uses. DW Ming, FA Mumpton (eds) Int'l Comm Natural Zeolites, Brockport, New York, p 3-13
Helgeson HC (1991) Organic/inorganic reactions in metamorphic processes. Can Mineral 29:707-737
Helgeson HC, Knox AM, Owens CE, Shock EL (1993) Petroleum, oil field waters, and authigenic mineral assemblages: are they in metastable equilibrium in hydrocarbon reservoirs? Geochim Cosmoschim Acta 57:3295-3339
Hoshi K, Saga H, Minowa H, Inaba M (1992) Alteration and reservoir properties of the Green Tuff rocks in Akita and Niigata oil fields. J Japan Assoc Petrol Tech 57:77-90 [in Japanese, English abstr]
Iijima A (1975) Effect of pore water to clinoptilolite-analcime-albite reaction series. J Fac Sci Univ Tokyo, Sec II 19:133-147
Iijima A (1978) Occurrence of natural zeolites in marine environments. *In* Natural Zeolites: Occurrence,

Properties, Uses. LB Sand, FA Mumpton (eds) Pergamon Press, Elmsford, New York, p 175-198
Iijima A (1980) Geology of zeolites and zeolitic rocks. Pure & Appl Chem 52:2115-2130
Iijima A (1986) Occurrence of natural zeolites. Clay Science 26:90-103. [in Japanese, English abstr]
Iijima A (1988a) Application of natural zeolites to petroleum exploration. *In* Occurrence, Properties and Utilization of Natural Zeolites. D Kallo, HS Sherry (eds) Akademiai Kiado, Budapest, p 29-37
Iijima A (1988b) Diagenetic transformation of minerals as exemplified by zeolites and silica—a Japanese view. *In* Diagenesis II, GV Chilingarian, KH Wolf (eds) Elsevier, Amsterdam, p 147-211
Iijima A (1995) Zeolites in petroleum and natural gas reservoirs in Japan: a review. *In* Natural Zeolites '93: Occurrence, Properties, Uses. DW Ming, FA Mumpton (eds) Int'l Comm Natural Zeolites, Brockport, New York, p 99-114
Iijima A, Ogihara S (1995) Temperature-time relationships of zeolitic reactions in burial diagenesis in marine sequences. *In* Natural Zeolites '93: Occurrence, Properties, Uses. DW Ming, FA Mumpton (eds) Int'l Comm Natural Zeolites, Brockport, New York, p 115-123
Iijima A, Utada M (1966) Zeolites in sedimentary rocks, with reference to the depositional environments and zonal distribution. Sedimentology 7:327-357
Iijima A, Utada M (1971) Present-day zeolitic burial diagenesis of the Neogene geosynclinal deposits in the Niigata Oil Field, Japan. *In* Molecular Sieve Zeolites. Advances in Chemistry Series 101:342-349
Iijima A, Aoyagi K, Kazama T (1984) Diagenetic zeolite zone modified by recent high heat flow in MITI-Kuromatsunai hole, southwest Hokkaido, Japan. Proc 6th Int'l Zeolite Conf, Reno, Nevada, 1983. HS Sherry (ed) Butterworths, Surrey, UK, p 595-603 [in Japanese]
J.A.N.G.M, A.O.O.D (1992) Oil and Natural Gas Resources in Japan, a Revised Edition. Japanese Association of Natural Gas Mining and Association of Offshore Oil Development, Tokyo, 520 p [in Japanese]
J.N.O.C (1988a) Survey report of the MITI-Kashiwazaki-oki borehole—Fundamental survey of domestic oil and gas in 1987. Japan National Oil Corporation, Tokyo, 80 p [in Japanese]
J.N.O.C (1988b) Survey report of the MITI-Nikaho borehole—Fundamental survey of domestic oil and gas in 1987. Japan National Oil Corporation, Tokyo, 132 p [in Japanese]
J.N.O.C (1991) Survey report of the MITI-Sohma-oki borehole—Fundamental survey of domestic oil and gas in 1990. Japan National Oil Corporation, Tokyo, 115 p [in Japanese]
J.N.O.C (1992) Survey report of the MITI-Johban-oki borehole—Fundamental survey of domestic oil and gas in 1991. Japan National Oil Corporation, Tokyo, 125 p [in Japanese]
J.N.O.C (1993) Survey report of the MITI-Yuri-oki-chubu borehole—Fundamental survey of domestic oil and gas in 1992. Japan National Oil Corporation, Tokyo, 110 p [in Japanese]
J.N.O.C (1998) Survey report of the MITI-Umaoi borehole—Fundamental survey of domestic oil and gas in 1997. Japan National Oil Corporation, Tokyo, 59 p [in Japanese]
J.N.O.C (2000) Survey report of the MITI-Sanriku-oki borehole—Fundamental survey of domestic oil and gas in 1998: Japan National Oil Corporation, Tokyo, 49 p [in Japanese].
Kaley ME, Hanson RF (1955) Laumontite and leonhardite cement in Miocene sandstone from a well in San Joaquin Valley, California. Am Mineral 40:923-925
Khatchikian A (1983) Log evaluation of oil-bearing igneous rocks. World Oil 197(7):79-98
Liu H (1986) Geodynamic scenario and structural styles of Mesozoic and Cenozoic basins in China. Am Assoc Petrol Geol Bull 70:377-395
Liu Y (1991) Diagenesis of the Lower Cretaceous in the Tanzhang-Shenqui Region. *In* The pore Structure of Reservoir in Clastic Rocks, its Origin and Control over the Oil and Gas Migration. Di S, Zhu Z (eds) Northwest University Press, Xi'an, p 138-162 [in Chinese, English abstr]
Liu Y, Zhou D, Li T (1993) A discussion on the boundary between diagenesis and metamorphism with reference to zeolite facies. Island Arc 2:262-272
Lynch FL (1996) Mineral/water interaction, fluid flow, and Frio Sandstone diagenesis: evidence from the rocks. Am Assoc Petrol Geol Bull 80:486-504
Madsen BM, Murata KJ (1970) Occurrence of laumontite in Tertiary sandstones of the Central Coast Ranges, California. U S Geol Surv Prof Paper 700-D:D188-D195
McCulloh TH, Stewart RJ (1979) Subsurface laumontite crystallization and porosity destruction in Neogene sedimentary basins (abstr). Geol Soc Am Abstr Programs 11:475
McCulloh TH, Cashman SH, Stewart RJ (1973) Diagenetic baselines for interpretive reconstructions of maximum burial depths and paleotemperatures in clastic sedimentary rocks. *In* A Symposium in Geochemistry: Low Temperature Metamorphism of Kerogen and Clay Minerals. Pacific Sect S E P M, Los Angeles, p 18-46
McCulloh TH, Frizzell VA Jr, Stewart .J, Barnes, I (1981) Prediction of laumontite with quartz, thenardite, and gypsum at Sespe Hot Springs, Western Transverse Ranges, California. Clays & Clay Minerals 29:353-364
Merino E (1975) Diagenesis in Tertiary sandstones from Kettleman North Dome, California. I. Diagenetic

mineralogy. J Sedim Petrol 45:320-336
Murata KJ, Whiteley KR (1973) Zeolites in the Miocene Briones Sandstone and their related formations of the Central Coast Ranges, California. J Res U S Geol Surv 1:255-265
Naeser ND, McCulloh TH, Crowley KD, Reaves CM (1990) Thermal history of the Los Angeles basin: Evidence from fission-track analysis (abstr) . Am Assoc Petrol Geol Bull 74:728
Noh JH, Boles JR (1993) Origin of zeolite cements in the Miocene sandstones, North Tejon oil fields, California. J Sedim Petrol 63:248-260
Ogihara S (1996) Diagenetic transformation of clinoptilolite to analcime in silicic tuffs of Hokkaido, Japan. Mineralium Deposita 31:548-553
Ogihara S, Iijima A (1989) Clinoptilolite to heulandite transformation in burial diagenesis. *In* Zeolites: Facts, Figures, Future. PA Jacobs, RA van Santen (eds) Studies in Surface Science and Catalysis. Elsevier, Amsterdam 49A:491-500
Ogihara S, Iijima A (1990) Exceptionally K-rich clinoptilolite-heulandite group zeolites from the offshore boreholes off northern Japan. Eur J Mineral 2:819-826
Okubo,S, Hoshi K, Kato K, Suzaki T (1996) Dolerite reservoir and its alteration in the Ayukawa oil and gas field, Akita Prefecture, Japan. J Japan Assoc Petrol Tech 61:61-70 [in Japanese, English abstr]
Patton DK (1993) Samgori field, Republic of Georgia: critical review of island-arc oil and gas. J Petrol Geol 16:153-168
Reed JK, Gipson M Jr, Neese DG (1993a) Hydrocarbon potential of sandstone reservoirs in the Neogene east Slovakian Basin, Part 1: a petrographic examination of lithology, porosity, and diagenesis. J Petrol Geol 16:89-108
Reed JK, Gipson M Jr, Vass D (1993b) Hydrocarbon potential of sandstone reservoirs in the Neogene east Slovakian Basin, Part 2:zeolites and clay minerals. J Petrol Geol 16:223-236
Remy RR (1994) Porosity reduction and major controls on diagenesis of Cretaceous—Paleocene volcaniclastic and arkosic sandstone, Middle Park Basin, Colorado. J Sedim Res A64:797-806
Saito T (1992) Oilfield development in China. TRC Special Publ #1, Japan National Oil Corporation, Tokyo, 134 p
Sato M (1990) Thermochemistry of the formation of fossil fuels. *In* Fluid-Mineral Interactions: A Tribute to H.P. Eugster. RJ Spencer, I-M Chou (eds) Geol Soc Am Spec Publ 2:271-283
Sato O (1984) Rock facies and pore spaces on volcanic-rock reservoirs—especially rhyolite reservoirs in the Minami Nagaoka gas field. J Japan Assoc Petrol Tech 49:11-19 [in Japanese, English abstr]
Sato S, Takasaki M, Kawamoto T, Hasegawa K, Watanabe T (1992a) Lithofacies and reservoir characteristics of rhyolite core samples from the Minami Nagaoka gas field, Japan. Abstr 29th Int'l Geol Congr, Kyoto, Japan 3:810
Sato S, Sekiguchi K, and Watanabe T (1992b) Reservoir quality of rhyolite in the Minami Nagaoka gas field, Japan. Abstr 29th Int'l Geol Congr, Kyoto, Japan 3:810
Stallard ML, Boles JR (1989) Oxygen isotope measurements of albite-quartz-zeolite mineral assemblages, Hokonui Hills, Southland, New Zealand. Clays & Clay Minerals 37:409-418
Surdam RC (1973) Low-grade metamorphism of tuffaceous rocks in the Karmutsen Group, Vancouver Island, British Columbia. Geol Soc Am Bull 84:1911-1922
Surdam RC, Boles JR (1979) Diagenesis of volcanic sandstones. *In* Aspects of Diagenesis. PA Scholle, PR Schluger (eds) S E P M Spec Publ 26:227-242
Surdam RC, Sheppard RA (1978) Zeolites in saline, alkaline-lake deposits. *In* Natural Zeolites: Occurrence, Properties, Uses. LB Sand, FA Mumpton (eds) Pergamon Press, Elmsford, New York, p 145-175
Surdam RC, Boese SW, Crossey LJ (1984) The chemistry of secondary porosity. *In* Clastic Diagenesis. DA McDonald, RC Surdam (eds) Am Assoc Petrol Geol Bull Memoir 37:127-149
Surdam RC, Crossey LJ, Hagen ES, Heasler HP (1989a) Organic-inorganic interactions and sandstone diagenesis. Am Assoc Petrol Geol Bull 73:1-23
Surdam RC, Dunn TL, Heasler HP, MacGowan DB (1989b) Porosity evolution in sandstone/shale system. *In* Burial Diagenesis. IE Hutchon (ed) Mineral Assoc Can Short Course Handbook 15:61-134
Tang Z, Parnell J, Longstaffe F.J (1997a) Diagenesis of analcime-bearing sandstones: the Upper Permian Pingdiquan Formation, Junggar Basin, northwest China. J Sedim Res 67:486-498
Tang Z, Parnell J, Longstaffe F.J (1997b) Diagenesis and reservoir potential of Permian- Triassic fluvial/lacustrine sandstones in the southern Junggar Basin, northwestern China. Am Assoc Petrol Geol Bull 81:1843-1865
Tissot BP, Welte DH (1978) Petroleum Formation and Occurrences. Springer-Verlag, Berlin, 638 p
Uchida T (1984) Properties of pore systems and their pore-size distributions in reservoir rocks. J Japan Assoc Petrol Tech 49:29-40 [in Japanese, English abstr]
UchidaT, Tada R (1992) Pore properties of sandstones—Part II. Applications of pore-size distribution to natural sandstones. J Japan Assoc Petrol Tech 57:213-222 [in Japanese, English abstr]
Underdown DR, Hickey JJ, Karia SK (1990) Acidization of analcime-cemented sandstones. Gulf of Mexico

Soc Petrol Eng Reprint 20624:97-100

Utada M (1965) Zonal distribution of authigenic zeolites in the Tertiary pyroclastic rocks in Mogami district, Yamagata Prefecture. Tokyo Univ Coll Gen Educ Sci Paper 15:173-216

Utada M (1970) Occurrence and distribution of authigenic zeolites in the Neogene pyroclastic rocks in Japan. Tokyo Univ Coll Gen Educ Sci Paper 20:191-262

Vavra CL (1989) Mineral reactions and controls on zeolite-facies alteration in sandstone of the central Transantarctic Mountains, Antarctica. J Sedim Petrol 59:688-703

Velde B, Iijima A (1988) Comparison of clay and zeolite mineral occurrences in Neogene age sediments from several deep wells. Clays & Clay Minerals 36:337-342

Vernik L (1990) A new type of reservoir rock in volcaniclastic sequences. Am Assoc Petrol Geol Bull 74:830-836

Waseda A, Omokawa M (1990) Generation, migration and accumulation of hydrocarbons in the Yurihara oil and gas field. Res Rept JAPEX Research. Center 6:1-16 [in Japanese, English abstr]

Watanabe Y, Utada M, Iijima A (1986) Geology and zeolitic alteration in the Itaya zeolite deposit, Yamagata Prefecture, northeast Japan. *In* New Developments in Zeolitic Science and Technology, Y Murakami, A Iijima, JW.Ward (eds) Kodansha-Elsevier, Tokyo 51-58

Wopfner H, Markwort S, Semkiwa P.M (1991) Early diagenetic laumontite in the lower Triassic Manda Beds of the Ruhuhu Basin, southern Tanzania. J Sedim Petrol 61:65-72

Yagi M (1992) Characteristics of hydrothermal alteration and its effects on oil reservoirs related to Miocene volcanism in the Yurihara oil and gas field, northern Honshu, Japan. Res Rept JAPEX Research Center 8:27-79

Yagi M (1994) Regional metamorphism and hydrothermal alteration related to Miocene submarine volcanism ("green tuff") in the Yurihara oil and gas field, Japan. The Island Arc 2:240-261

Yamada Y, Uchida T (1994) Characteristics of hydrothermal alteration and secondary porosities occurred in volcanic rocks, the Katakai Green Tuff region. JAPEX Research Center Research Reports 10:1-27 [in Japanese, English abstr]

Yang B, Lin Z, Gu S (1991) Diagenesis of laumontite-bearing sandstones in lower part of Lower Cretaceous of North Songliao Basin. Oil & Gas Geology 12:1-9 [in Chinese, English abstr]

Zen E-an (1961) The zeolite facies—an interpretation. Am J Sci 259:401-409

Zhu G (1985) Formation of low permeability sandbodies and secondary pore sandbodies in the Upper Triassic Yanchang Series of southwestern San-Gan-Ning basin. Acta Sedimentologica Sinica 3:1-17 [in Chinese, English abstr]

13

Thermal Behavior of Natural Zeolites

David L. Bish and **J. William Carey**

Los Alamos National Laboratory
Hydrology, Geochemistry, and Geology, MS D469
Los Alamos, New Mexico 87545

INTRODUCTION

The open framework structures of zeolites, containing variable amounts of extra-framework cations and water molecules, are very responsive to changes in temperature and/or water-vapor pressure. Several coupled changes occur in water content and structure whenever a zeolite is subjected to a change in environment. The structural changes include modifications in the unit-cell size and geometry, movement of extraframework cations, and even statistical breakage of the Al-Si framework. The water content in some zeolites is a smoothly varying function of temperature and water-vapor pressure, whereas in others there appear to be distinct transitions between different hydration levels. The response of zeolites to changes in temperature and/or water-vapor pressure is thus a very important aspect of their behavior and has a bearing on subjects ranging from their industrial applications to their identification. For example, numerous zeolites are used in gas adsorption, as selective catalysts, and as molecular sieves, and a detailed understanding of their *short-term* (e.g. overnight) thermal behavior and appropriate activation temperatures is crucial. Short-term thermal behavior has also been suggested as a means to distinguish between similar zeolites, such as clinoptilolite from heulandite (Mumpton 1960) and barrerite from stellerite (Passaglia 1980). In addition, de'Gennaro and Colella (1989) showed that the zeolite content in mixtures such as zeolitic tuffs can be determined through the details of the bulk-sample water-vapor desorption behavior. In contrast, other applications (e.g. the performance of clinoptilolite in a high-level radioactive waste repository) require that the *long-term* (10^2 to 10^5 years) behavior of these minerals under elevated-temperature conditions be understood. Because many thermal reactions of interest are at least partially kinetically limited, effects that may not be important during short-term heating can become increasingly important during long-term heating. In some instances, long-term heating experiments show results very different from those of shorter duration (e.g. Bish 1990a). Several publications provide information on the thermal behavior of individual zeolite species (e.g. Gottardi and Galli 1985; Colella 1999); rather than duplicate this information, we will provide an overview and analysis of zeolite thermal behavior using several individual zeolites as examples.

A thorough understanding of the linked structural changes and water sorption/desorption processes in zeolites can be realized through the development of a full structural and thermodynamic description of the water sorption process. The data for such descriptions exist in the form of studies of zeolite-H_2O equilibria and in detailed structural studies by X-ray and neutron diffraction. Zeolite-H_2O equilibria provide the water content of zeolites as function of temperature and water-vapor pressure. Detailed studies of the structural effects accompanying hydration/dehydration processes permit evaluation of the stresses induced by changes in environmental conditions that ultimately lead to structural modification or breakdown. Some of the important structural factors governing zeolite stability include the amount and type of extraframework cations (i.e. their ionic potential or Z/r), the framework Al/Si ratio, the connectivity of the structure

1529-6466/00/0045-0013$05.00

(e.g. "flexibility"), the presence or absence of H_2O, and the time, temperature, and rate of heating. Some "robust" zeolites can be heated to high temperatures and can be completely dehydrated while maintaining their original structure, whereas many zeolites undergo complex structural transformations upon heating and dehydration, often leading to the destruction of the original structure.

A full thermodynamic description provides the basis for evaluating the effect of water-vapor pressure on zeolite stability and for characterizing the amount of water and energy consumed or liberated by zeolites as a function of changes in pressure and temperature. Zeolite-H_2O equilibria are important in a variety of environments. These include diagenetic and low-grade metamorphic occurrences in which zeolite stability may be used to determine the conditions of metamorphism; pollution abatement in which zeolites are present naturally or placed in the environment to act as sorptive barriers to contaminant migration; and industrial settings in which zeolites are used as catalysts, molecular sieves, and cation exchangers.

The objective of this chapter is to summarize existing understanding of the thermal behavior of natural zeolites, emphasizing the intimate link between crystal structure and dehydration/rehydration behavior. The chapter will focus first on a classification of the thermal behavior of natural zeolite structures, describing the variety of different volumetric and structural changes occurring on heating or dehydration of natural zeolites. Although the natural zeolites are comprised of structurally similar Al-Si frameworks, subtle variations in their structures and extraframework cation compositions give rise to significant differences in their thermal behavior. Both thermally induced (i.e. thermal expansion) and dehydration-induced structural effects can be observed in zeolites, usually occurring together, and it is a challenge to separate these effects. A second focus will be on the development of a general description of the thermodynamic behavior of zeolites in order to create a common framework for analysis of zeolites and to facilitate the comparison of the thermodynamic properties of zeolites. Clinoptilolite will be used as an example and other zeolite data from the literature will be used as available. The thermodynamic treatment of zeolites differs in some important aspects from other mineral solid solutions. In essence, the mixing of H_2O occurs among vacancies. Most other volatiles are only weakly adsorbed in the presence of H_2O. In addition, the sites of mixing can vary due to rearrangements in the zeolite extraframework sites, and the zeolites can respond by volumetric changes. The approach taken here will combine a macroscopic description with microscopic data on atomic positions as available. However, the microscopic (crystal structure) environment is so complex that we can only use it as a guideline rather than employing specific criteria such as site assignments and occupancies.

The most significant feature of natural zeolites is the presence of an open aluminosilicate framework that creates, effectively, an internal surface area. This internal surface responds in almost all respects the way an external surface does: it contains diffusely charge-compensated species and consequently sorbs volatile species; and it readily exchanges charged species, sorbed to its surface, with the environment (e.g. gaseous and solution species). However, the crystalline nature of zeolites yields an internal surface area that is characterized by a regular structural environment (ignoring Al-Si disorder), so that the largely unpredictable heterogeneity of an external surface does not exist. Rather, an atomically regular internal surface occurs, greatly simplifying the analysis of sorbed species-surface interactions. Another important difference is that for most zeolites, the internal structure (i.e. channels and cages) is sufficiently confining and limited in volume that "multi-layer" absorption and other transitions from sorbed species to liquid-like conditions typically do not occur.

Previous thermodynamic descriptions of zeolites include the classic work of Barrer (1978). Barrer's work, like many others, emphasized measured values and differences observed among zeolites rather than the development of a full thermodynamic description. These data do provide the basis for an equilibrium description, but Barrer's purpose was not to provide the kind of formulation that lends itself to studies of natural equilibria.

VOLUMETRIC AND STRUCTURAL CHANGES ON DEHYDRATION

Alberti and Vezzalini (1984) divided the volumetric and structural effects of heating zeolites into three convenient categories: (1) reversible dehydration accompanied, in some cases, by rearrangement of the extraframework cations and residual water molecules, with little or no modification of the framework or unit-cell volume; (2) complete or nearly complete reversible dehydration accompanied by a large distortion of the framework and a significant decrease in unit-cell volume; and (3) reversible dehydration at low temperature, usually accompanied by large modifications in the framework, followed by irreversible changes due to breaking of T-O-T (tetrahedral cation-oxygen-tetrahedral cation, e.g. Al-O-Si) bonds prior to complete dehydration (see Table 1 for a summary of zeolite classifications). A single zeolite structural type may occur in several categories depending on the nature of the tetrahedral framework, the charge on the framework, Al-Si order/disorder in the framework, and the amount and type of extraframework cations. A large amount of new research on the thermal behavior of natural zeolites has been conducted in the past decade, much of it concentrating on the *in situ* (structural state at a particular temperature and pressure) behavior of zeolites. Those studies employing sufficiently intense radiation sources, such as a synchrotron, provide time-resolved information on temperature-induced structural modifications (e.g. Artioli 1997).

Table 1. Zeolites classified according to thermal stability.

Category-1	*Category-2*	*Category-3*
Chabazite	Ca/Na-clinoptilolite	Heulandite
Mordenite	Natrolite	Barrerite
Analcime	Mesolite	Stilbite
Erionite	Scolecite	Stellerite
K-clinoptilolite	Gismondine	Phillipsite
	Laumontite	Harmotome
	Yugawaralite	Thomsonite
		Edingtonite
		Gonnardite

Terminology

Many terms have been used to describe dehydration in zeolites including phase transition, phase transformation, structural breakdown, structural collapse, etc. Zeolite hydration/dehydration as described in this review is divided between continuous processes (a small change in temperature or water-vapor pressure leading to a generally reversible change in water content or framework) and discontinuous processes (a small change in temperature or water-vapor pressure leading to an irreversible loss of sorption capacity and an irreversible change in structure). Thermodynamically, the continuous processes are not phase transitions (accompanied by a symmetry change and a discontinuity in second-order thermodynamic variables) or phase transformations (accompanied by a discontinuity in first-order thermodynamic variables). These continuous processes are probably best referred to as just that, i.e. the continuous response of the zeolite to changes in environmental variables.

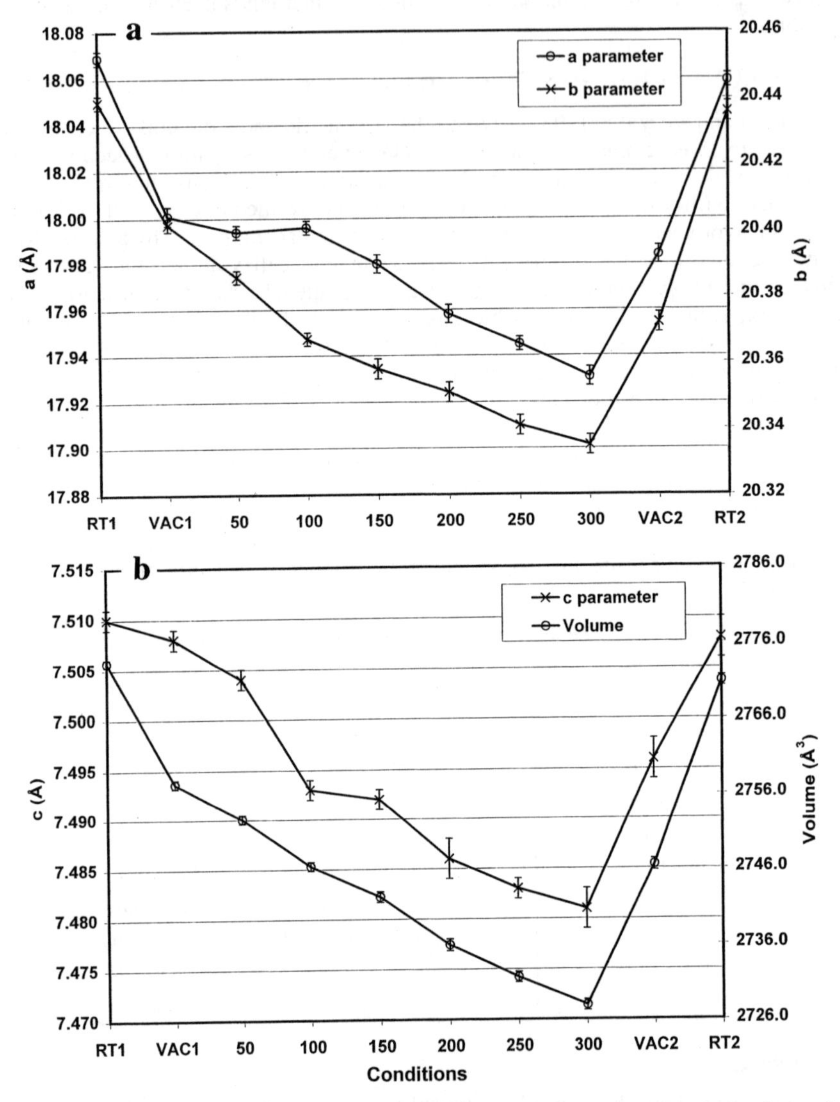

Figure 1. (a) *a* and *b* unit-cell parameters; and (b) *c* unit-cell parameter and unit-cell volume of mordenite from Custer City, Idaho, showing minor decrease in volume (1.6%) with increasing temperature (°C) and the reversibility of the reaction (Bish 1990b). RT1 represents the sample examined under room conditions before heating, VAC1 represents examination at room temperature under vacuum, data up to 300°C represent examination at-temperature under vacuum, VAC2 represents examination under vacuum at room temperature after heating, and RT2 represents examination under room conditions after heating. Equilibrium time between measurements was 30 min. Error bars for unit-cell parameters are plotted but are often smaller than the symbol.

The discontinuous processes are phase transformations involving first-order discontinuities in thermodynamic variables (i.e. discontinuities in the Gibbs free energy, volume, enthalpy, etc.). We can distinguish two types of discontinuities: a *structural collapse* in which T-O-T bonds are broken but the structure is still recognizable using diffraction data as "similar" to the original zeolite (collapsed structures also typically retain some sorption capacity); and *structural breakdown* resulting in the complete loss of the zeolite structure (usually with the loss of all sorption capacity).

Category-1 transformations

Category-1 transformations are exemplified by reversible dehydration accompanied, in some cases, by rearrangement of the extraframework cations and residual water molecules, with little or no modification of the framework or unit-cell volume. The thermal behavior of chabazite and mordenite are examples of this category, showing little decrease in volume below 300°C *in vacuo* (and even more subtle changes at room humidities). For example, the a, b, and c unit-cell parameters in a natural mordenite from Custer City, Idaho, decreased by only 0.5, 0.3, and 0.2%, respectively, on heating from room temperature to 300°C *in vacuo* (Figs. 1a and 1b; Bish 1990b). Much of the decrease in a, b, and volume occurred on evacuation at room temperature. Importantly, the unit-cell parameters returned to their pre-heating values within 30 min after exposure to room temperature and humidity (23°C, 15% relative humidity), illustrating that re-expansion and rehydration are not kinetically limited in the same way as they are for other zeolites such as analcime. Sedimentary chabazite from Christmas, Arizona, displayed similar thermal behavior, with the a parameter and unit-cell volume decreasing by 0.5% and 0.9%, respectively, on heating from room temperature to 300°C (Fig. 2a). Hydrothermal chabazite from the Gila River, New Mexico, showed respective decreases of 0.8 and 1.69% for the a parameter and unit-cell volume at 300°C (Fig. 2b; Bish 1989). Indeed, chabazite showed its greatest volume decrease on evacuation at room temperature, and

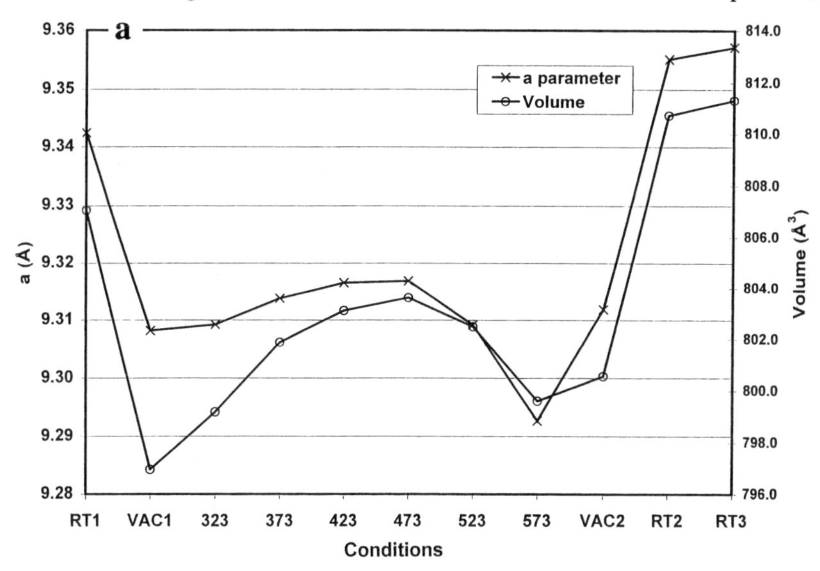

Figure 2. (a) Unit-cell parameters of chabazites from Christmas, Arizona; and (b) [next page] from Gila, New Mexico. Both show the bulk of volume decrease on evacuation, a reversible reaction, and the effects of thermal expansion. Conventions as in Figure 1; T in °C; error bars (not plotted) are comparable to size of symbols; RT3 represents examination under room conditions 17 and 25 d after heating the Gila and Christmas material, respectively (Bish 1989).

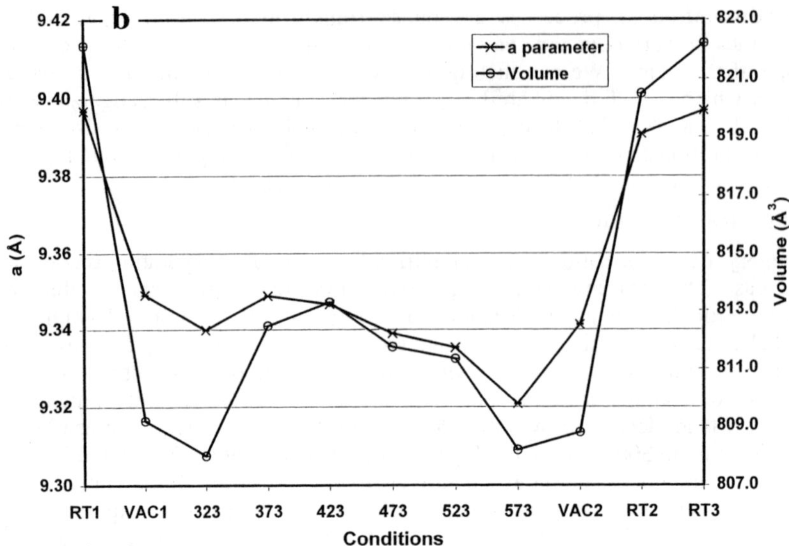

Figure 2, continued.

we see the combined effects of thermal expansion and dehydration in chabazite up to 300°C. The Gila River chabazite experienced a volume decrease on evacuation and on heating to 50°C but then underwent a volume increase to 150°C. It thereafter experienced a subtle volume decrease up to 300°C. Both chabazite samples quickly returned to their pre-heating unit-cell parameters after re-exposure to room conditions. Butikova et al. (1993) showed that dehydration of Ca-chabazite produced a more-distorted Al-Si framework due to migration of Ca^{2+} ions within structural cages. They pointed out that dehydration of Ca^{2+} ions results in charge localization and local charge imbalances.

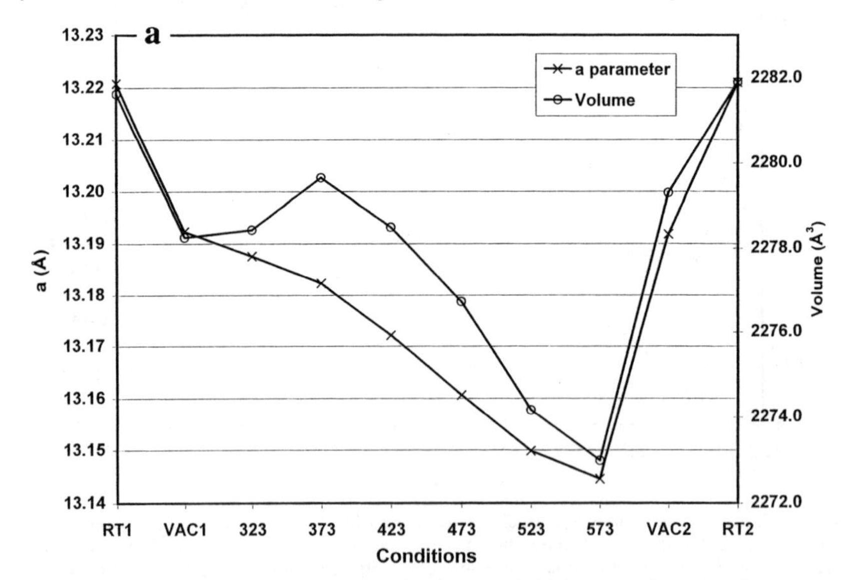

Figure 3. (a) *a* unit-cell parameter and cell volume; and (b) [next page] *c* unit-cell parameter and cell volume of erionite from Eastgate, Nevada, illustrating thermal expansion of *c*. Conventions as in Figure 1; T in °C; error bars (not plotted) are comparable in size to the symbols.

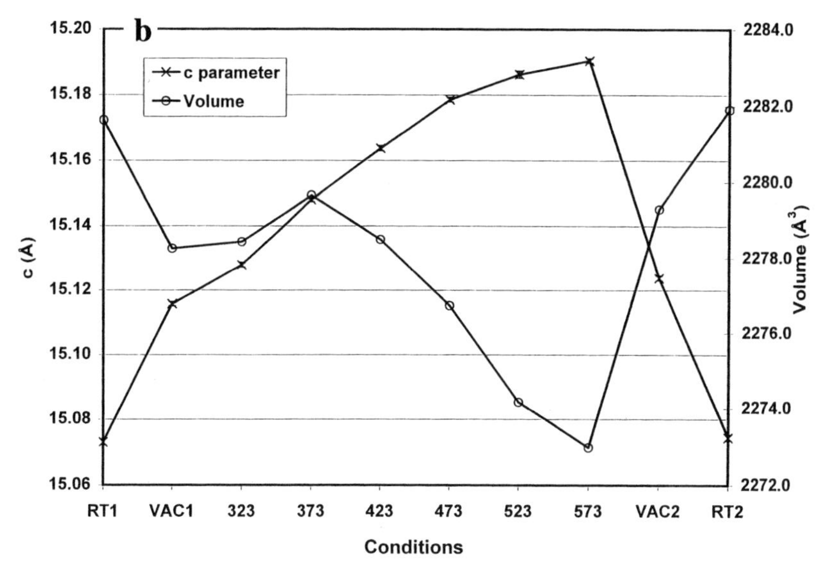

Figure 3, continued.

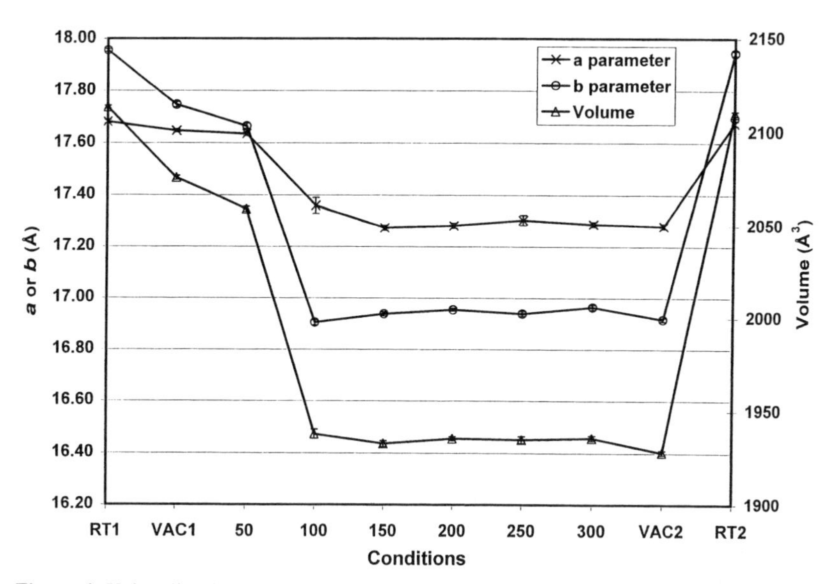

Figure 4. Unit-cell volume and *a* and *b* unit-cell parameters of Na-clinoptilolite from Castle Creek, Idaho, showing significant decrease in volume upon evacuation and increasing temperature (8.4%) and reversibility of the reaction after re-exposing to room conditions (Bish 1984). Conventions as in Figure 1, temperature in °C.

Erionite from Eastgate, Nevada, is an example of a zeolite that has a slightly different thermal behavior. The *a* parameter decreased by only 0.58% on heating to 300°C (Fig. 3a), but the *c* parameter actually *increased* by 0.78%, giving a volume decrease of 0.38% (Fig. 3b; Bish 1989). Like mordenite and chabazite, the erionite sample quickly returned to its pre-heating unit-cell parameters after re-exposure to room conditions. The type of extraframework cation in erionite has only a minor effect on the thermal behavior: the *a* and *c* unit-cell parameters and the volume of the Ca-exchanged sample changed by -0.71%, +0.78%, and -0.65%, those for the K-exchanged sample changed by -0.37% and +0.04%, and -0.69%, and those for the Na-exchanged sample changed by -0.70%, +0.93, and -0.47%, respectively. Erionite and chabazite are examples of the few natural zeolites in which thermal expansion effects are evident at or below 300°C and that are not masked by dehydration-induced decreases in unit-cell parameters. Observation of unit-cell expansion along *c* in erionite likely results from relative structural rigidity along *c**.

The extraframework cation can have a dramatic effect on a zeolite's thermal behavior, in some cases changing a Category-1 zeolite to one of Category-2. For example, K-exchanged clinoptilolite experiences only a ~1.6% volume decrease upon heating to 300°C, whereas Na-exchanged clinoptilolite undergoes an 8.4% volume decrease (Fig. 4; Bish 1984), placing it in Category-2. The long-term thermal stability of clinoptilolite is also greatly dependent on extraframework cation (Bish 1990a, see below).

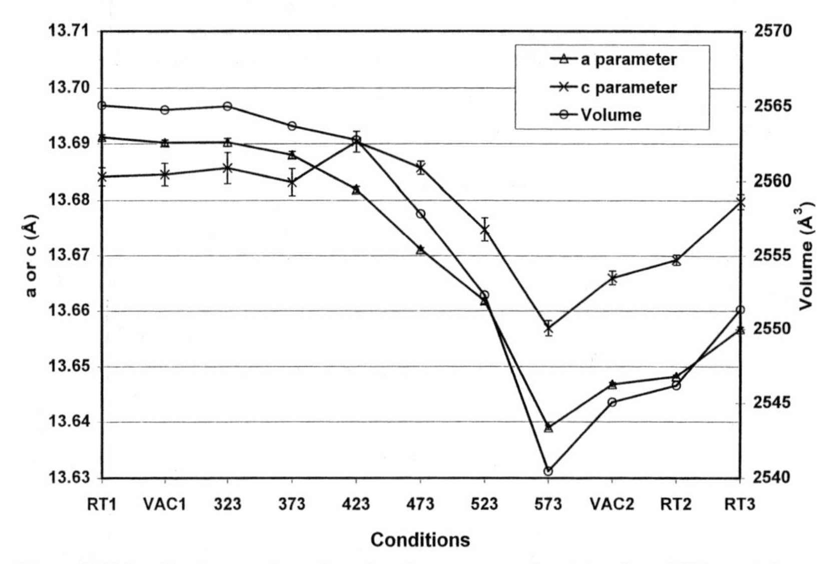

Figure 5. Unit-cell volume and *a* and *c* unit-cell parameters of analcime from Wikieup, Arizona, showing minor decrease in volume upon evacuation and increasing temperature (0.96%) and lack of reversibility of the reaction after re-exposing to room conditions. Conventions as in Figure 1; RT3 represents examination under room conditions 7 days after heating.

Different structure types can also give rise to major differences in the kinetics of the dehydration/hydration process. Analcime is an example of a zeolite that can be dehydrated with little or no consequent structural effects but for which rehydration and re-expansion are very sluggish (Chipera and Bish 1991). Upon heating to 300°C *in vacuo*, resulting in almost complete dehydration, analcime decreases in volume by only 0.6-1.0% (Fig. 5). Rehydration and associated volume expansion are very slow under room conditions and are estimated to take about 11 years in H_2O at room temperature (Chipera and Bish 1991). The inability of analcime to rehydrate readily is not due to thermally induced structural modifications but is instead due to the inability of water to diffuse through its small structural ports (8- and 6-tetrahedral rings). Rehydration is considerably more rapid at elevated temperatures in liquid water. Although rehydration of analcime is very sluggishly reversible, it is classified as a Category-1 zeolite as little modification of the framework or decrease in unit-cell volume occurs on heating.

Category-2 transformations

In Category-2 zeolites, more than for Category-1 zeolites, dehydration/rehydration and the amount of H_2O contained in structural cavities affect the zeolite's molar volume. The volume changes observed with Category-1 and Category-2 zeolites result from a combination of minor thermal expansion and variable dehydration-induced contraction. The three-dimensional connectivity of the structure determines the relative importance of expansion and contraction and may also give rise to anisotropic behavior upon heating (as seen to a minor extent with Category-1 zeolites). The effects of structure are exemplified by contrasting the thermal behavior of Na-clinoptilolite with that of mordenite shown above. Although mordenite typically undergoes a relatively isotropic volume decrease <2% upon heating and dehydration, clinoptilolite can undergo a large, very anisotropic decrease in unit-cell dimensions (e.g. Fig. 4, Na-clinoptilolite); the difference in behavior is related to differences in structural "flexibility" and connectivity.

The clinoptilolite-heulandite group of minerals exemplifies a number of the factors controlling zeolite thermal behavior. Bish and Boak (this volume) described the compositional variations observed in clinoptilolite-heulandite minerals, including variations in thermal behavior between the two minerals. Recently, the Zeolite Subcommittee of the Commission on New Minerals and Mineral Names of the International Mineralogical Association (IMA) (Coombs et al. 1997) defined heulandite "as the zeolite mineral series having the distinctive framework topology of heulandite and the ratio Si:Al < 4.0. Clinoptilolite is defined as the series with the same framework topology and Si:Al ≥ 4.0." Heulandite is also typically Ca rich. The structural details of the dehydration/volume-decrease process for Category-2 zeolites are typified by some forms of clinoptilolite, studied in detail by Armbruster and Gunter (1991) and Armbruster (1993). They examined two different heulandite-clinoptilolite samples at 100 K in the hydrated state and in several partially hydrated states, and the structures of the sample with 25, 7, and 5 H_2O molecules per 72 oxygens were determined. They identified a variety of extraframework cation sites in the hydrated sample, including (Na,K)1 within the ten-membered ring (M1), Ca2 within the eight-membered ring (M2), K3 at the edge of the ten-membered ring (M3), and Mg4 in M4 in the center of the ten-membered ring (Fig. 6a). Six H_2O molecules coordinate Mg4, and the remaining cation sites are coordinated by a combination of framework oxygens and H_2O molecules. Sites (Na,K)1, K3, and Mg4 are sufficiently close that they cannot be occupied simultaneously, as are the two Ca2 sites in the eight-membered ring. Armbruster (1993) identified only two H_2O sites that are not directly coordinating cations; thus, dehydration will directly affect most extraframework cations. Dehydration of Ca-K heulandite-clinoptilolite (or essentially any heulandite-clinoptilolite) leads to an anisotropic decrease in unit-cell dimensions, most pronounced along *b* (e.g. Fig. 4). Dehydrating the sample to 7 H_2O molecules per 72 oxygens results first in loss of the two H_2O molecules not coordinated to cations (the most weakly bonded) and then causes a significant decrease in the population of the other H_2O molecules. Accompanying dehydration is a migration of the (Na,K)1, Ca2, and Mg4 sites towards the K3 site near the cavity wall. The 5-H_2O sample showed further migration of (Na,K)1 and Ca2 to K3 and significant changes in T-O-T angles, producing large compression of the channel system (Fig. 6b). Although the amount of H_2O and the nature of the dehydration reaction appear to be controlled largely by the hydration energy of extraframework cations, the structural responses to dehydration are related both to loss of H_2O molecules and to movement of extraframework cations and their subsequent interactions with the framework oxygen atoms. Hambley and Taylor (1984) and

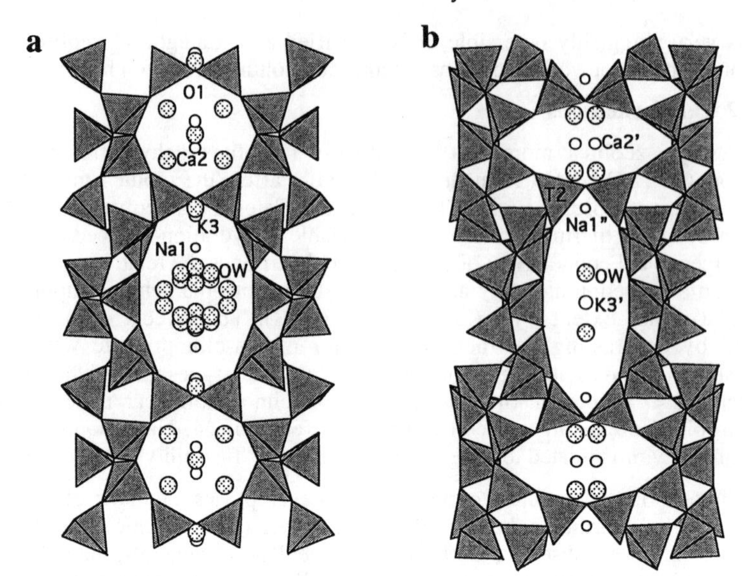

Figure 6. Projection of the heulandite-clinoptilolite structure along [001], showing larger ten-membered ring and the smaller eight-membered ring; (a) natural sample; (b) heat-collapsed phase. Small open circles labeled Na1, Ca2, and K3 represent the positions of extraframework cation sites (Na1", Ca2', and K3' in the heat-collapsed phase), large dotted circles represent the positions of statistically occupied water molecule sites, and O1 and T2 represent the O1 bridging oxygen and the T2 tetrahedron, respectively. Note that dehydration produces a loss of H_2O molecules, migration of extraframework cations, and collapse of both rings (modified from Armbruster and Gunter 1991).

Armbruster (1993) showed that thermal stability of clinoptilolite and heulandite is related both to extraframework cations and Al-Si substitutions. The T2 tetrahedron in all clinoptilolite-heulandites is enriched in Al compared with the other tetrahedra, and the underbonded O1 atom is bonded both to T2 and Ca2. Thus, the Ca-O1 interaction is dependent on the Al population of T2. Armbruster and Gunter (1991) suggested that in high-Al samples (heulandites), the Ca-O1 interaction is strong and that Ca remains localized around O1, even after dehydration. In low-Al samples (clinoptilolites), however, the Ca-O1 interaction is much weaker, and Ca may easily diffuse to other channel positions such as K3 during dehydration. Koyama and Takéuchi (1977) suggested that large cations (or cations with low Z/r) such as K^+ in K3 prevent the structure from collapsing, whereas smaller cations such as Ca^{2+} and Na^+ are too small to keep the channels expanded. Armbruster (1993) supported this conclusion from site occupancy data for cations occurring in the C-channel; K^+ holds the channel open whereas Na^+ allows it to collapse during dehydration. The difference in thermal behavior between Na- and Ca-exchanged forms is related to the much stronger interaction between Ca and O1, resulting in random breakage of the T-O-T oxygen bridge in heulandite (e.g. Alberti and Vezzalini 1983a). It also appears that the shift of O1 and breakage of the T-O-T oxygen bridge is related to the extraframework cation content of the zeolite, which is a function of the total Al-for-Si substitution.

Natrolite undergoes similar anisotropic, heating-induced changes, transforming to so-called metanatrolite at ~300°C (Belitsky et al. 1992, found the reaction to metanatrolite at ~ 160°C and attributed the relatively low-temperature change to the vacuum in their X-ray camera). Upon heating from room temperature to 350°C, the *a*, *b*, and *c* unit-cell parameters decrease abruptly by 12.5, 10.2, and 2.9%, respectively (Alberti and Vezzalini 1983b). Alberti and Vezzalini (1983b) showed that the large decreases in the *a* and *b* unit-cell parameters result from ~11° rotations of the tetrahedral chains, and that the smaller decrease in *c* is due to twisting of the chains. This transformation in natrolite is accompanied by movement of the extraframework Na atoms to positions in metanatrolite originally occupied by H_2O in natrolite. Belitsky et al. (1992) also documented the occurrence of several other thermal transformations, including both α- and β–metanatrolite and x-metanatrolite. In spite of the large volumetric changes accompanying dehydration, rehydration and re-expansion remain possible in natrolite to the temperature at which the structure breaks down, about 775°C (Van Reeuwijk 1974). Interestingly, the only major water-loss event in natrolite is at about 330°C, illustrating that H_2O molecules are held very tightly, comparable to those in analcime. The behavior of Na-natrolite contrasts markedly with that of K-exchanged natrolite, studied by Yamazaki and Otsuka (1986). They showed that K-natrolite contains ~60% of the water molecules in Na-natrolite and it dehydrates at ~150°C. However, the structure of K-exchanged natrolite is stable to >1000°C, in comparison with ~775°C for Na-natrolite. In contrast, Belitsky et al. (1992) suggested that the α-to-β metanatrolite temperature decreased with increasing ionic radius of univalent exchangeable cations. The thermal behavior of natrolite illustrates the combined factors that control thermal reactions: K-for-Na exchange results in a lower dehydration temperature due to the lower ionic potential of K^+, but the larger ionic radius of K^+ yields a natrolite framework with greater thermal stability but a lower α-to-β metanatrolite transition temperature.

Belitsky et al. (1992) conducted a comprehensive study of the behavior of natrolite and edingtonite under conditions of elevated temperature and pressure, using both penetrating (H_2O) and non-penetrating, high molecular-weight pressure media. As expected, they found that additional H_2O molecules entered extraframework sites under elevated H_2O pressures, and these extra H_2O molecules caused deformation of the framework and anisotropic expansion. Under H_2O pressure, natrolite underwent a phase transition to natrolite II at 7.5 kbar, with a 6% increase in unit-cell volume. Further increase in pressure caused the transition to natrolite III at 12.5 kbar, with a 3% decrease in unit-cell volume. No phase transitions were observed up to 50 kbar with non-penetrating liquids and the unit-cell volume continuously decreased. Although these results do not have direct bearing on zeolite thermal behavior, they provide important information on the behavior of zeolites under elevated $P(H_2O)$. The occurrence of additional H_2O molecules in the natrolite structure under elevated H_2O pressure causes the framework to be modified in order to increase the size of the extraframework regions. NMR spectra indicated increased translational diffusivity of H_2O molecules in natrolite II.

Scolecite and mesolite have the same framework topology as natrolite, and both minerals experience phase transitions on heating similar to those seen in natrolite. Ståhl and Hanson (1994) performed a real-time X-ray powder diffraction (XRD) study of dehydration in both minerals using synchrotron radiation. Scolecite dehydration begins at ~137°C as one of three H_2O molecules [O(2W)] is expelled. The site occupancy of O(2W) is reduced to one-half at 207°C, leading to migration of Ca, a rearranged Ca coordination polyhedron, and symmetry change; this phase is commonly referred to as metascolecite. Continued heating above 207°C does not cause further H_2O loss, and unit-cell parameters do not vary significantly. The structure is known to break down above ~350°C (Van Reeuwijk 1974). Similarly, mesolite dehydration becomes significant at ~177°C, with the loss of half of O(4W) (this H_2O molecule corresponds to O(2W) in scolecite). Significant structural changes begin at ~214°C with further and complete loss of O(4W), and the previously well-defined Na, Ca, and vacant extraframework cation sites become disordered at an order/disorder transition that causes a reduction in the *b* unit-cell dimension by a factor of three. Above 250°C, mesolite is known as metamesolite, and structural breakdown occurs above ~320°C. Although the dehydration process is similar in scolecite and mesolite, the phase transition to the "meta" phase in scolecite is rapid because it requires cation movement only within individual structural channels. The transition in mesolite takes place over an extended temperature range, as it requires extraframework-cation migration between channels as opposed to within one channel. Although both of these zeolites can be classified as Category-2 zeolites based on their behavior below the breakdown temperatures, their thermal reactions are not reversible above about 350°C and they could equally well be classified as Category-3 zeolites.

The Ca-zeolite, gismondine, was shown by Van Reeuwijk (1974) to dehydrate in five stages up to 350°C. Although these stages are reversible, gismondine is known to exhibit remarkable structural changes even at room temperature upon long-term vacuum dehydration. Vezzalini et al. (1993) found that one-hour dehydration under vacuum at room temperature caused no significant changes in the tetrahedral framework but resulted in a minor volume decrease (0.6%), a change in the arrangement of H_2O molecules, and a change in space group from $P2_1/c$ to $P2_1$. Their results after one-hour evacuation are clearly not representative of an end state because continued evacuation at room temperature (for 24 hours) caused a further loss of ~55% of the total water and a volume decrease of ~17%. These changes resulted in significant deformation of the tetrahedral framework and the structural channels and caused a change in symmetry to $I2_12_12_1$ (Vezzalini et al. 1993). In spite of these large structural changes, Van Reeuwijk (1974) showed that dehydration in gismondine is completely reversible, likely due to the relative "flexibility" of the gismondine framework and the lack of broken T-O-T bonds in the dehydrated material. The results of Vezzalini et al. (1993) emphasize that heating is not a prerequisite for dehydration and volume reduction in zeolites. Simply lowering the water-vapor pressure at room temperature can lead to dehydration in many zeolites.

Laumontite is an interesting zeolite that shows a dramatic loss in water content at room temperature upon reduction of relative humidity from near saturation. The lower water-content form is commonly referred to by the obsolete name "leonhardite." Indeed, partially Na- and K-exchanged "leonhardites" may not hydrate completely to laumontite due to the greater number of extraframework cations and the lower amount of H_2O in the structural channels (e.g. Baur et al. 1997; Stolz and Armbruster 1997). Below the temperature of structural collapse (~350°C), laumontite exhibits several reversible transitions. Ståhl et al. (1996) performed a dynamic study of the dehydration of laumontite up to 311°C, beginning with a sample submerged in water at 37°C and collecting a set of diffraction data every five minutes as a function of temperature. Their results were similar to those of Armbruster and Kohler (1992), but their study gave a clear picture of the dehydration path for laumontite, something that is not gained by studying samples at intervals of several hours to days or by examining quenched samples. They showed that the first response to heating was a loss of the H_2O molecules not directly bonded to Ca cations [W(1)], and further heating reduced the number of H_2O molecules from the 18/unit cell of laumontite to the 13/unit cell of "leonhardite." The structural response was a gradual reduction in volume, by ~3.3%, with minimal changes in the tetrahedral framework. Further heating resulted in continued loss of H_2O molecules, gradually decreasing the number of H_2O molecules coordinated to Ca^{2+} but causing only minor unit-cell changes. Above ~150°C, laumontite exhibits changes in unit cell similar to those outlined above for erionite, namely a contraction of *a* and *c* but an increase in *b*. When the H_2O coordination sphere around Ca^{2+} drops significantly below six, Ståhl et al. (1996) speculated that the Ca^{2+} ions interact with the framework oxygen atoms, causing what the authors termed a "structural collapse" that occurs at ~ 350°C. X-ray diffraction photographs given by Van Reeuwijk (1974) show a structural change at ~350°C, with what appears to be complete structural breakdown at ~850°C. The thermal reactions exhibited by laumontite below 350°C are reversible, and in this temperature range it may be considered a Category-2 zeolite. However, it cannot be dehydrated completely without the occurrence of irreversible structural changes that apparently begin at ~350°C, as soon as significant dehydration of W(8) begins. Indeed, the thermogravimetric analysis of laumontite shown by Gottardi and Galli (1985) reveals significant water losses above 400°C, suggesting that hydroxyl groups exist in the high-temperature structure.

Artioli et al. (2001) conducted a similar study of the Ca-zeolite yugawaralite using time-resolved synchrotron X-ray powder diffraction data. They found only minor changes in Ca^{2+} coordination polyhedra and minor distortions of the tetrahedral framework as water evolved from the structure up to ~422°C. The unit-cell volume gradually *increased* from ~117 to ~400°C. This is an unusual example of a significant range (almost 300°C) over which thermal expansion occurs while a zeolite dehydrates. Above this temperature, they found a marked decrease in water content, accompanied by a change in the Ca^{2+} coordination polyhedra and a large decrease in unit-cell volume. They concluded that the first-order phase transition at 422°C results from the rearrangement of the extraframework Ca^{2+} ions; as soon as the Ca^{2+} coordination falls below seven, the Ca^{2+} ions increase their coordination by forming new bonds with the tetrahedral framework O atoms. This phase transition is accompanied by a large decrease in unit-cell volume and a change in space group symmetry from *Pc* to *Pn*. This is the same mechanism operating in other Ca-rich zeolites such as laumontite, scolecite, and mesolite, and all of these zeolites may be considered Category-3 zeolites when heated above their transition temperatures. The studies of Ståhl et al. (1996) and Artioli et al. (2001) illustrate the wealth of information that can be obtained in real-time *in situ* studies of the dynamics of the dehydration process. Their results demonstrate that heating and/or changes in H_2O vapor pressure modify the amount and arrangement of H_2O molecules in the zeolite structure, which modify the extraframework-cation positions, which in turn can give rise to changes in the tetrahedral framework. Through such studies, one gains a great appreciation for the importance of water molecules in maintaining the open zeolite structures with which we are familiar.

Category-3 transformations

Category-3 zeolites are characterized by reversible dehydration at low temperature (below ~200°C), usually accompanied by large modifications in the framework, followed by irreversible changes due to breaking of T-O-T bonds prior to complete dehydration. Volume contraction (and dehydration) is typically only partially reversible or even irreversible (i.e. Category-3) if dehydration also involves modification of the tetrahedral framework. For example, heulandite, barrerite, stilbite, and stellerite exhibit structural transformations at temperatures <400°C, involving migration of extraframework cations, rotations of fundamental structural units, and, most importantly, breakage of some T-O-T linkages. The transformation in heulandite to heulandite B has been studied extensively and is a good example of the types of modifications occurring in Category-3 zeolites. As described above, heulandite has a larger Ca content and a greater substitution of Al-for-Si than does clinoptilolite. These differences result in migration of extraframework cations, structural relaxation, and subsequent strong Ca-O1 interactions in heulandite (after removal of H_2O molecules surrounding the Ca ions, e.g. those H_2O molecules shown in Fig. 6b adjacent to Ca). On heating, the fundamental polyhedral unit of heulandite composed of four- and five-membered rings typically rotates, with a consequent shift in tetrahedral oxygen positions and considerable change in the shape of the two major channels in the structure (e.g. Alberti and Vezzalini 1983a). Both the eight- and ten-membered ring channels become much more elliptical in shape and the free minimum diameters are reduced considerably (see Fig. 6b). Although Alberti (1973) found no evidence for random breakage of T-O-T bridges in heulandite from the Faroer Islands, Alberti and Vezzalini (1983a) documented the breakage of T-O-T bridges, both in the Faroer Islands and Nadap, Hungary, heulandites, due to a large shift in the position of O1. It is the breakage of T-O-T bridges that is the precursor to structural breakdown and yields a partially reversible or irreversible transformation with heating. As a result of the important structural changes taking place in heulandite at relatively low temperatures, the evolution of H_2O is significantly different than it is from clinoptilolite, with abrupt water losses occurring in two discrete steps below 300°C. The most distinct H_2O loss, beginning at ~205°C, is undoubtedly associated with the structural transition to heulandite B at ~230°C (Bish 1988).

Barrerite exhibits thermal effects slightly different than the related minerals stilbite and stellerite, and it undergoes several different thermal transformations involving large volume decreases, only partial reversibility, and, at the highest temperatures, random breakage of T-O-T bridges. Rotation of the fundamental structural units in barrerite gives rise to the large volume decrease and results in a very different channel geometry than exists in the unheated material (Alberti and Vezzalini 1978). Interestingly, Alberti and Vezzalini (1978) demonstrated the presence of hydroxyl groups in the barrerite structure at elevated temperatures using single-crystal X-ray diffraction methods. They speculated that all zeolites exhibiting "water" losses at temperatures 500°C contain hydroxide ions that probably form through reaction of intracrystalline H_2O molecules with the interrupted tetrahedral framework. Although the frameworks of stellerite and stilbite are identical to that in barrerite, they exhibit significantly different thermal behaviors due to their different H_2O arrangements arising from their distinct extraframework cation compositions. Alberti et al. (1978) showed that stellerite undergoes a rotation of its fundamental framework structural unit on dehydration, a phenomenon common to all platy zeolites examined by them (heulandite, barrerite, stilbite, and stellerite). This rotation of the fundamental structural unit lowers the symmetry from *Fmmm* to *Amma*. T-O-T linkages are randomly broken upon rotation, yielding OH^- groups at the broken oxygen bridges. These broken oxygen bridges are much less abundant in stellerite than in barrerite or stilbite. Passaglia (1980) showed that the differences in thermal behavior are due to the large influence of the extraframework cations in these three minerals. Na-exchanged forms of these zeolites (e.g. barrerite) are stable to significantly higher temperatures than Ca-exchanged forms (e.g. stellerite). K- and Rb-exchanged forms of these minerals remain nearly unaffected to temperatures as high as 900°C, whereas Mg and Ca forms exhibit a progressive contraction with destruction of the structure as low as 350°C. The driving forces for contraction and structural destruction thus include the interactions between dehydrated extraframework cations and framework oxygen atoms. The dynamic study of stilbite by Cruciani et al. (1997) confirmed that stilbite undergoes only minor framework distortion at low temperatures, but significant framework collapse, volume contraction, and T-O-T bond breakage occur at ~147°C. As with laumontite and yugawaralite, the collapse of the stilbite structure is related to the shift of extraframework Ca^{2+} ions in order to achieve more optimum coordination after the loss of H_2O molecules.

Fewer data exist on the structural effects of heating phillipsite and harmotome, but both dehydrate in multiple steps, apparently as a consequence of the existence of discrete water sites in their structures. Rykl and Pechar (1991) suggested that phillipsite transforms to wairakite + quartz at about 200°C but did not comment on the reversibility of the phillipsite-to-wairakite transformation (clearly if the reaction produces quartz, it will not be reversible under non-hydrothermal conditions). Phillipsite shows a mass loss of ~6% from 100-400°C during the transformation to wairakite, but it continues to lose water to temperatures well above 600°C, strongly supporting the presence of hydroxyl groups in the thermally altered structure. Garcia et al. (1992) showed that dehydration of phillipsite was accompanied by a migration of Al^{+3} ions toward the surfaces of crystallites at temperatures as low as 400°C, a phenomenon that may be important in many zeolites upon dehydration and has also been shown to occur during steam treatment of other zeolites (e.g. mordenite, O'Donovan et al. 1995). Indeed, heating larger specimens of zeolite may induce a phenomenon known as self-steaming without the use of a hydrothermal steam reactor (e.g. Carvalho et al. 1997). Garcia et al. suggested that, as tetrahedral Al migrates as a result of thermal breakdown of phillipsite, amorphous Al-rich zones and hydroxyl groups are created. Few detailed structural data as a function of temperature have been published for thomsonite, edingtonite, and gonnardite, although Van Reeuwijk (1974) documented changes in XRD patterns accompanying dehydration. Belitsky et al (1992) conducted a low-temperature NMR study of edingtonite documenting the ordering of H_2O molecules but no structural data were included. Mesolite appears to undergo Category-3 transformations at temperatures as low as 300°C, inasmuch as rehydration after heating to this temperature is not possible (Van Reeuwijk 1974). Thomsonite exhibits only minor volume contraction upon heating to <300°C, but it undergoes significant structural rearrangement above 320°C (Van Reeuwijk 1974). Gonnardite exhibits a significant dehydration event <100°C, accompanied by a marked decrease in the *a* and *b* unit-cell parameters, and the structure is destroyed at temperatures >300°C (Van Reeuwijk 1974). Dehydration of edingtonite causes reversible decreases in the *a* and *b* axes and expansion along the *c* axis before destruction at about 400°C. The driving forces for these Category-3 transformations appear to be dehydration-induced contraction and the shifts of extraframework cations and H_2O. As with Category-1 and -2 transformations, the structural transformations are also correlated with extraframework cation and the three-dimensional connectivity of the structure, as well as the Al/Si ratio. Thus the presence of H_2O or hydroxyl may be necessary for the occurrence of these transformations.

Long-term transformations

Although most of the dehydration processes in Category-1 and Category-2 zeolites are reversible over short times, reactions involving breaking of tetrahedral T-O-T linkages are partially or completely irreversible. Because of the sluggish nature of Al and Si diffusion and the strength of T-O bonds, a number of these transformations are kinetically limited. For example, although clinoptilolite is not affected by short-term heating to 500°C, long-term (>1 yr) heating at 200°C causes irreversible changes to some Na-clinoptilolites (Bish 1990a). Clinoptilolites characterized by low Si/Al ratios, high Na, and low K contents undergo a transformation to a phase resembling heulandite B during long-term heating at 200°C. Although Na-clinoptilolite appears to exhibit Category-3 behavior during long-term heating, similar to that observed for heulandite during short-term heating, the transformations do not appear to be identical. The Na-clinoptilolites that underwent a long-term transformation at 200°C were unaffected by overnight heating to 480°C, whereas samples that were destroyed by overnight heating to 480°C were unaffected by long-term heating at 200°C. These differences are undoubtedly associated with the greater abundance of Na^+ ions (twice as many required as divalent cations), the inability of the smaller Na^+ ions to maintain expansion of the channels, and the large degree of structural distortion and volume decrease occurring below 100°C for Na-clinoptilolite.

Other structural/dehydration studies

There was a great increase in the use of *in situ*, time-resolved studies of thermal processes in zeolites during the 1990's. The work of Ståhl and Hanson (1994), Ståhl et al. (1996), Artioli 1997, Cruciani et al. (1997), and Artioli et al. (2001), discussed above, are excellent examples illustrating the detailed information on the structural effects of heating that can be obtained using synchrotron facilities. Such studies provide the opportunity to monitor structural changes on an almost continuous basis as a function of changing environmental parameters such as temperature. However, most studies of natural zeolite structures are generally done under only partially saturated to dry conditions, and the partial pressure of H_2O is seldom controlled. This is ironic as many uses of natural zeolites, particularly those involving cation exchange, operate under water-saturated conditions. Clearly there is a great need for further information on the structures of zeolites, particularly under conditions where the extraframework sites are occupied by H_2O molecules.

Structural studies of zeolites under water-saturated conditions

Thermodynamic analyses of zeolite dehydration (described below) rely on accurate information for the maximum amount of water contained in the particular zeolite (maximum water loading, or x_{max}). However, the high-water-content region is experimentally difficult to analyze, particularly by gravimetric methods, due to the occurrence of water condensation (i.e. liquid H_2O is present in the powder sample between crystallites). As a consequence, x_{max} is estimated in most studies by extrapolation of the data to 100% RH (e.g. Carey and Bish 1996). To investigate such problems, Bish and Carey (2000) measured XRD data for Na- and Ca-clinoptilolite as a function of RH from 0 to ~100%, as well as for samples wet with liquid H_2O. Using Rietveld refinement methods, they showed that the clinoptilolite a, b, and c unit-cell parameters steadily decreased with decreasing humidity (the c parameter for the Na-exchanged sample increased slightly, primarily between 20 and 0% RH), whereas the β parameter increased, resulting in a steady decrease in volume with decreasing humidity to 0% RH for all samples. The Na-exchanged sample showed no significant change in volume going from wet conditions to 100% RH, whereas the Ca-exchanged sample exhibited a minor decrease in volume at 100% RH, compared with the value obtained from the wet sample. Carey and Bish (1996) were unable to determine whether this minor change was real because of the possibility of small changes in temperature during the experiment.

Likewise, refined water-site occupancies for Na-exchanged clinoptilolite showed no evidence for an increase in structural water when the sample was wet, whereas the data for Ca-exchanged clinoptilolite suggested a minor, but perhaps statistically insignificant, increase in water content when the sample was wet, compared with 100%-RH data. Variations in water occupancies at high relative humidities occurred primarily in the W(1) and W(5) sites in Ca-exchanged clinoptilolite, and the W(5) site was responsible for much of the change in water content in Na-exchanged clinoptilolite. The XRD/Rietveld-derived water contents showed the same trends with relative humidity as the TGA data, although the Rietveld-derived values were systematically less than those obtained by TGA, probably due to the difficulty in locating all water molecules with high displacement parameters. These results suggest that maximum H_2O contents assumed in gravimetric extrapolation are valid.

The question of whether clinoptilolite continues to hydrate under saturated conditions remains. However, there is precedent for temperature-induced changes under water-saturated conditions. Stahl et al. (1996) showed that the unit-cell volume of laumontite steadily decreased while the W(1) site almost completely dehydrated as temperature was increased in *water-saturated* laumontite powder. This observation notwithstanding, it appears in most cases that the most important environmental factor giving rise to structural changes in zeolites at relatively low temperatures (i.e. less than about 300°C) is the water-vapor pressure. For most zeolites, the most important variable controlling dehydration is the relative humidity at the temperature of interest. Unfortunately, the behavior of zeolites under high $P(H_2O)$ or under saturated conditions is presently very poorly documented and warrants further research.

ENERGETICS OF DEHYDRATION

The essential characteristics of the hydration behavior of zeolites have been known since the mineral group was first identified: zeolites produce abundant water upon heating under atmospheric conditions, with many, but not all, varieties capable of reversibly evolving and resorbing the water. Some zeolites, like analcime, do not exchange water at room temperature; whereas others, like clinoptilolite and faujasite, dehydrate continuously. Yet others show distinct steps in their dehydration behavior, e.g. laumontite and heulandite. These differing types of behavior reflect the differing energetics of H_2O in the many varieties of zeolite structure and composition.

Many earlier analyses of natural zeolites by thermogravimetric and calorimetric methods suggested that several specific "types" of H_2O exist. These have been labeled variously as "zeolitic," "loosely held," "structural," "crystal," "tightly bound," "external," etc. (e.g. Milligan and Weiser 1937; Van Reeuwijk 1974; Knowlton et al. 1981). These labels suggest both differing apparent energies of H_2O as well as differing apparent structural roles of H_2O. To clarify this terminology, we suggest that there are only three useful distinctions that can be made about H_2O in zeolites:

1. H_2O that varies in content as a continuous function of temperature and pressure
2. H_2O that changes discontinuously at a unique temperature for a given pressure
3. H_2O that is sorbed to external surfaces

The first type, continuously varying H_2O, is *characteristic* of zeolites (e.g. Gottardi and Galli 1985) and reflects the observation that most, if not all, zeolites lose or gain H_2O in response to small changes in temperature or pressure over an extended temperature range. This type of H_2O can be properly referred to as *zeolitic*. Zeolitic H_2O encompasses many of the terms cited above and could be modified to express relative energies by "loosely held zeolite H_2O" or "high-energy H_2O".

The second type of H_2O is similar in nature to that found in hydrates (e.g. gypsum, $CaSO_4 \cdot 2H_2O$) and reflects the observation that dehydration occurs over narrow temperature intervals in some zeolites (e.g. step-like dehydration in laumontite). This type can be referred to as *hydrate* H_2O. However, it is unclear whether any zeolites truly possess hydrate H_2O. In order to demonstrate the presence of hydrate H_2O as distinct from zeolitic H_2O, it is necessary to demonstrate that the H_2O resides in a well-defined site and for a given pressure the site is either *completely occupied* or *completely empty* depending on the temperature. Such relationships are difficult to establish by conventional thermal analysis and require equilibrium studies.

The third type of H_2O is externally sorbed to the crystal and may be referred to as *external*. This type is present in quantities much smaller than the H_2O present within the structure for any reasonable zeolite grain size. Most workers believe that external H_2O does not contribute significantly to thermal analysis data (e.g. Van Reeuwijk 1974)

Zeolite dehydration reactions are critically dependent on the partial pressure of H_2O and temperature. Small changes in water vapor pressure or in temperature can produce dehydration/hydration reactions. The consequences of this continuous response of zeolites to changes in the environment are not always fully appreciated. Most zeolites are not fully hydrated at room conditions, but continue to hydrate as relative humidity is increased to 100%. In fact, some zeolites will continue to hydrate at 100% RH as temperature falls (e.g. Carey and Bish 1996). Zeolites do not show rapid changes in water content that coincide with the boiling point of liquid water. In fact, we argue that no zeolites show true hydrate-like behavior with a phase transformation discontinuity in water content, unit-cell parameters, etc., occurring at a unique temperature for a given pressure. Rather all zeolites respond to changes in pressure or temperature with continuous changes in water content or unit-cell parameters, until the point of structural collapse or breakdown.

A significant amount of work (e.g. Barrer and Cram 1971, Bischke et al. 1988; Bish 1988; Armbruster 1993) has shown that H_2O molecules in at least some zeolites are held with a more or less continuous range of energies. Furthermore, this range of energies is a function of temperature and zeolite composition. The H_2O molecules that evolve at lower temperatures are not as closely associated with extraframework cations and high-energy frameworks sites as are the remaining H_2O molecules. In addition, many zeolites provide evidence for a continuous rearrangement of extraframework cations and H_2O molecules with concomitant structural relaxation as dehydration proceeds (e.g. clinoptilolite, Armbruster and Gunter 1991; Armbruster 1993). The net result is that H_2O evolution in these zeolites occurs in a pseudo-continuous manner, despite the fact that a structural snapshot of H_2O molecules may suggest specific site occupancies with discrete energetic steps.

However, the observation that some zeolites lose H_2O as a continuous function of temperature and pressure does not necessarily require that the H_2O is present in a range of energies. Many thermodynamic treatments of H_2O in zeolites such as Langmuir sorption theory (e.g. Barrer 1978) or adsorption potential theory (e.g. Dubinin and Astakhov 1972) predict continuous dehydration of H_2O present in the zeolite with a single energy (see below). A range of energetic values of H_2O can only be demonstrated though calorimetric or phase equilibria (including gravimetric equilibrium and TGA modeling described below) studies that yield the enthalpy of dehydration as a function of H_2O content.

The nature of zeolite dehydration (and the total amount of H_2O) is strongly dependent on the extraframework cations, in particular their hydration energy. Zeolites containing high hydration-energy cations, such as Li^+, contain significantly more H_2O than those containing low hydration-energy cations, such as K^+, and those with high hydration-energy cations also generally retain their H_2O to higher temperatures (e.g. Bish 1988).

The temperature of dehydration and structural collapse is also critically dependent on the partial pressure of H_2O. Simonot-Grange et al. (1968) showed that heulandite transforms to heulandite-B near 300°C at elevated $P(H_2O)$, but at low $P(H_2O)$, it undergoes this transformation close to 200°C. Similarly, Van Reeuwijk (1974) illustrated that the transformation of natrolite to α-metanatrolite occurs below 300°C at water vapor pressures <0.02 bar (i.e. ~65% relative humidity at 25°C), but the transformation temperature is elevated to >450°C at a water vapor pressure of 5 bar.

In general, unless very high heating rates are used, *dehydration* approaches an equilibrium process, with rates depending on crystallite size, packing density, the partial pressure of H_2O, and, to a lesser extent, the size of the structural cavities (e.g. Van Reeuwijk 1974). *Rehydration*, however, can be kinetically limited by the size of the structural apertures as well as by the other factors listed for dehydration. Chipera and Bish (1991) showed that analcime, having relatively small structural apertures, rehydrates very sluggishly in contrast with clinoptilolite, erionite, or mordenite, which have relatively large apertures and rehydrate quickly. Figure 7 shows a plot of the degree of rehydration of a dehydrated analcime as a function of rehydration time and temperature. Clearly, the rehydration of analcime is very sluggish and would take about 4000 days in liquid H_2O at 23°C to reach the original degree of hydration. Contrast this behavior with the results shown in Figures 1 and 4, which show that after dehydration, mordenite and clinoptilolite regain their original volume within 30 min after being returned to room conditions, without immersion in water. Of course rehydration may not be reversible if dehydration-induced structural changes have modified the zeolites ability to interact with H_2O molecules. Such structural changes can include breaking of T-O-T linkages and migration of extraframework cations so that they are strongly interacting with the O atoms of the framework.

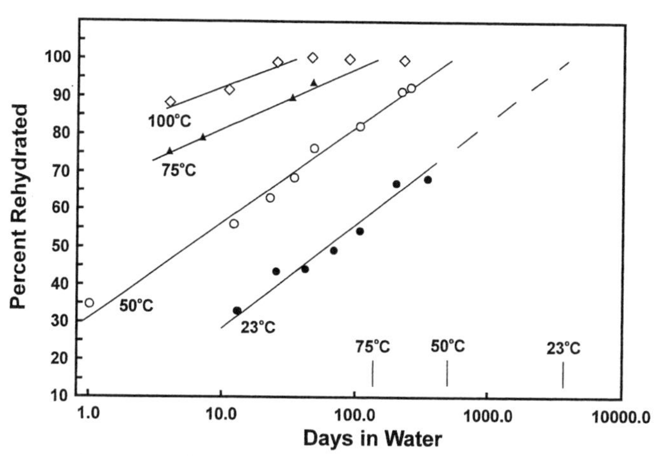

Figure 7. Hydration of dehydrated analcime from Wikieup, Arizona, immersed in liquid water (total pressure = 1 atm), as a function of temperature and log of rehydration time. Vertical bars at bottom of figure represent approximate number of days to reach full rehydration at 75, 50, and 23°C, based on extrapolation (Chipera and Bish 1991).

The purpose of the next section is to provide a quantitative framework for comparing the energetics of H_2O in zeolites. For example, can a quantitative explanation for step-like versus continuous dehydration reactions be developed? Is it possible to quantify the observation that water appears to be essential to the formation of zeolites and answer the question of the stability or metastability of zeolites with respect to feldspar-silica assemblages? At the present time, there are few zeolites for which it is possible to answer these questions. These questions can be addressed by analysis of zeolite dehydration/rehydration using a thermodynamic framework. The following section develops a thermodynamic description of zeolite-H_2O interactions, provides examples of the application of the method, and suggests that standard thermogravimetric data can be used to provide constraints on possible thermodynamic values for dehydration reactions in zeolites.

Thermodynamic relations

The thermodynamics of water sorption in zeolites depends on several characteristics of the zeolite host, including sorption capacity, framework configuration, framework response, cation type, and cation mobility. Because the zeolite framework is potentially highly responsive to the sorption of H_2O, including structural changes (see above) and cation mobility, the concept of mixing of H_2O on discrete sites is made more ambiguous. A useful method of avoiding these complications at the outset of a thermodynamic analysis is to emphasize the H_2O rather than the zeolite structure.

Using such an approach, the equilibria can be considered as the equilibrium between H_2O in the fluid/vapor phase (V) and H_2O in the zeolite (Zeo). According to the definition of equilibrium, the chemical potential of H_2O in the vapor phase ($\mu_{H_2O}^{V}$) must equal the chemical potential of the H_2O component in the zeolite ($\mu_{H_2O}^{Zeo}$):

$$\mu_{H_2O}^{V} = \mu_{H_2O}^{Zeo} \tag{1}$$

Expanding this equation to illustrate standard states and the equilibrium constant yields:

$$\Delta\mu_{Hy} = \Delta\mu_{Hy}^{o} + RT\ln\left(a_{H_2O}^{Zeo} / f_{H_2O}^{V}\right) = \Delta\mu_{Hy}^{o} + RT\ln K = 0 \tag{2}$$

in which the subscript *Hy* refers to the hydration reaction of Equation (1) (a transfer of H_2O from the atmosphere to the zeolite), R is the gas constant, T is temperature, $a_{H_2O}^{Zeo}$ is the activity of H_2O in the zeolite, $f_{H_2O}^{V}$ is the fugacity of H_2O in the vapor phase, and K is the equilibrium constant. Equation (2) suggests several important relationships. At equilibrium,

$$d\ln\left(f_{H_2O}^{V}\right) = d\ln\left(a_{H_2O}^{Zeo}\right) \tag{3}$$

so that any series of measurements of the amount of H_2O in a zeolite at a known fugacity (or partial pressure of H_2O under ideal-gas conditions) provides the basis for a thermodynamic description of the system. The difficulty, of course, is developing a thermodynamic formulation that is valid beyond the conditions measured.

The most elegant approach to developing a thermodynamic formulation is to integrate Equation (3) from zero vapor pressure to the maximum experimental pressure at several different temperatures. However, this approach requires equilibrium measurements at sufficiently low H_2O content to perform the integration properly. In general, zeolites are so hygroscopic that they are substantially hydrated at the lowest H_2O pressures near room temperature. For example, in the work of Carey and Bish (1996), clinoptilolite was 50% hydrated at less than 0.2 mbar (1% RH). In the absence of such detailed measurements, however, there are several other approaches to evaluating the equilibrium relationship in Equation (2). A general method, illustrated by Carey and Bish (1996), is to perform a least-squares fit to all of the available equilibrium data. For example, a series of measurements of the equilibrium H_2O content of zeolite as a function of temperature and pressure is available. These data can be used to obtain a general description of the thermodynamics of hydration.

The thermodynamic relations can be simplified by using inverse temperature and pressure as the independent variables and the ratio of the chemical potential to temperature as the equilibrium condition[1]:

---

[1] The relation G/T has been referred to as the Planck function by Guggenheim (1967, p 24). Equation (4) is readily derived from the usual Gibbs free energy relation (dG = -SdT + VdP) by substitution of the

$$d\left(\Delta\mu/T\right) = \Delta\overline{H}d(1/T)_P + \Delta\overline{V}/T\, d(P)_{1/T} \tag{4}$$

where ΔH is the enthalpy of reaction and ΔV is the volume of reaction. At equilibrium, $\Delta\mu/T$ is zero. Incorporating the equilibrium statement for the system (Eqn. 2) into Equation (4) and integrating to the temperature and pressure of interest yields:

$$\Delta\mu_{Hy}/T = \tag{5}$$

$$\Delta\mu^o_{Hy}/T_o + \Delta\overline{H}^o_{Hy}\left(1/T - 1/T_o\right) - \iint T^2 \Delta\overline{C}_{p,Hy} d(1/T)d(1/T) + \int \frac{\overline{V}^{Zeo}_{H_2O}}{T} dP + R\ln\left(a^{Zeo}_{H_2O}/f^V_{H_2O}\right) = 0$$

The standard state for H_2O in the vapor phase is the ideal gas at the temperature of interest and 1 bar. The standard state of H_2O in zeolite is discussed below.
The volume of hydration is given by $\overline{v}^{Zeo}_{H_2O}$, the partial molar volume of H_2O in zeolite. The constant-pressure heat capacity is $\Delta C_{p,Hy}$, the enthalpy of hydration is ΔH^o_{Hy}, and T_O is the reference-state temperature.

A set of equilibrium measurements can be used to extract, by a least-squares regression, the unknown values of Equation (5), namely

$$\Delta\overline{H}^o_{Hy},\ \Delta\overline{C}_{p,Hy},\ \overline{v}^{Zeo}_{H_2O},\ \Delta\mu^o_{Hy},$$

and the activity coefficient of H_2O in zeolite. In practice, it is useful to make as many direct measurements (e.g. heat capacity measurements) as possible to reduce the number of unknowns. In many cases it is possible to make assumptions concerning the values of some of these quantities.

The partial volume of hydration, $\overline{v}^{Zeo}_{H_2O}$, can be obtained (or constrained) directly from XRD data (see structural section above). This value reflects the volume of the zeolite as a function of H_2O content. Bish (1984) and Bish and Carey (2000) demonstrated some of the significant changes in volume that can occur on hydration or dehydration. For example, they measured a molar volume change of approximately 12 cm^3/mol-clinoptilolite during dehydration from 100% to 0.2% relative humidity *at room temperature*. Dehydration was accompanied by the loss of about 8 H_2O/mol-clinoptilolite, for an approximate partial molar volume of hydration of 1.5 cm^3/mol-H_2O. However, at moderate pressures, the energetic contribution of the volume of hydration is quite small (e.g. at 1000 bars, a 1.5 cm^3/mol-H_2O results in only a 0.15 kJ/mol energetic effect). The volumetric contribution to energy is generally far smaller than uncertainties in other quantities of the thermodynamic description. Although significant volume changes accompany the dehydration of most zeolites, for most thermodynamic calculations, it is adequate to assume that the volume of hydration is zero.

Heat capacity measurements of hydrous zeolites are difficult to make because the measurement generally consists of contributions from the enthalpy of dehydration. Reliable low-temperature data (below room temperature) exist, but very little data are available otherwise. In the absence of such data, simplifying assumptions can be made based on studies of other, more stable, hydrous phases. Carey (1993) reviewed a variety of data and presented results on the zeolite-like phase cordierite. In summary, much of the difference in heat capacity between sorbed H_2O and the gas phase can be related to weak vibrational modes (bonding) for sorbed H_2O compared with rotational and translational freedom in the vapor-phase molecule. With this model, the difference in heat

---

independent variable (1/T) for (T). The Planck function has the advantage that changes of state are expressed directly in terms of the measurable quantities of enthalpy and volume.

capacity between sorbed and free H_2O is constrained to lie between zero and 3*R (these relations were derived from considerations of the statistical mechanics of sorbed volatile species). Carey (1993) found that the difference for sorbed H_2O in cordierite was ~R.

These constraints allow calculation of the maximum effect of heat capacity on the thermodynamics. For example, a heat capacity difference of 3*R and a maximum temperature of 250°C results in an energetic stabilization of 1.7 kJ/mol-H_2O. For many purposes, it is adequate to set the heat capacity to zero with likely errors on the order of 1-2 kJ/mol of H_2O.

With the assumption that the heat capacity and volume of hydration are constant, known values, Equation (5) can be rearranged to give the unknown quantities (left-hand side) in terms of measurable or assumed quantities (right-hand side):

$$-\frac{\Delta\mu^o_{Hy}}{RT_o} - \frac{\Delta\overline{H}^o_{Hy}}{R}\left(\tfrac{1}{T} - \tfrac{1}{T_o}\right) - \ln\left(\gamma^{Zeo}_{H_2O}\right) = \ln\left(x^{Zeo}_{H_2O}\right) - \ln\left(f^V_{H_2O}\right) - \frac{\Delta\overline{C}_{p,Hy}}{R}\left(\ln\left(\tfrac{T}{T_o}\right) + \left(\tfrac{T_o}{T} - 1\right)\right) + \frac{\overline{V}^{Zeo}_{H_2O}}{R}(P-1) \tag{6}$$

The parameter, $x^{Zeo}_{H_2O}$, is the quantity of H_2O in the zeolite. The units for x are arbitrary (e.g. g H_2O/g anhydrous zeolite); although the units chosen obviously affect the thermodynamic values. The units can be converted from one to another as they differ only by some constant value.

The final step in this process is to transform the non-ideal activity, γ, into a form that has compositional dependence (i.e. it is unlikely that γ has a constant value). The form for γ is arbitrary, but polynomial expressions are frequently found useful. For example, γ can be represented as

$$RT\ln\left(\gamma^{Zeo}_{H_2O}\right) = W_1x + W_2x^2 + W_3Tx + W_4Tx^2 + \ldots, \tag{7}$$

where the polynomial is of arbitrary length and form and the *W* terms are polynomial coefficients. As seen below, the *W* terms without T correspond to excess enthalpies of mixing, whereas the *W* terms with T correspond to excess entropies of mixing. With this assumption, Equation 6 becomes:

$$-\left(\left(\frac{\Delta\mu^o_{Hy} - \Delta\overline{H}^o_{Hy}}{RT_o}\right) + \frac{\Delta\overline{H}^o_{Hy}}{RT} + \frac{W_1}{RT}x + \frac{W_2}{RT}x^2 + \frac{W_3}{R}x + \frac{W_4}{R}x^2 \ldots\right) = \ln\left(x^{Zeo}_{H_2O}\right) - \ln\left(f^V_{H_2O}\right) - \frac{\Delta\overline{C}_{p,Hy}}{R}\left(\ln\left(\tfrac{T}{T_o}\right) + \left(\tfrac{T_o}{T} - 1\right)\right) + \frac{\overline{V}^{Zeo}_{H_2O}}{R}(P-1) \tag{8}$$

or, in terms of regression coefficients,

$$A + \frac{B}{T} + \frac{Cx}{T} + \frac{Dx^2}{T} + Ex + Fx^2 \ldots = \ln\left(x^{Zeo}_{H_2O}\right) - \ln\left(f^V_{H_2O}\right) - \frac{\Delta\overline{C}_{p,Hy}}{R}\left(\ln\left(\tfrac{T}{T_o}\right) + \left(\tfrac{T_o}{T} - 1\right)\right) + \frac{\overline{V}^{Zeo}_{H_2O}}{R}(P-1) \tag{9}$$

permitting a general means of regressing the equilibrium data.

In several studies, it has been found useful to introduce the concept of a sorption capacity (θ), a function of the maximum sorption capacity, x_{max}. In general, x_{max} is a function of temperature and pressure. This parameter is useful because it represents an estimate of the number of sites available for mixing H_2O. In some cases, this may improve the quality of the fitting algorithm. The parameter x_{max} is generally introduced by defining:

$$\theta = \frac{x}{x_{max}} \tag{10}$$

and substituting θ for x in Equations (8) and (9). This solution model is a modified form of Langmuir sorption, with non-ideality in the solution given by:

$$RT\ln\left(\gamma_{H_2O}^{Zeo}\right) = -RT\ln(1-\theta) + W_1x + W_2x^2 + W_3Tx + W_4Tx^2 + ... \tag{11}$$

This transforms the relations above to

$$A + B/T + C\theta/T + D\theta^2/T + E\theta + F\theta^2 .. = \ln K - \frac{\Delta\overline{C}_{p,Hy}}{R}\left(\ln(T/T_o) + (T_o/T - 1)\right) + \frac{\overline{V}_{H_2O}^{Zeo}}{R}(P-1) \tag{12}$$

where

$$\ln K = \left(\frac{\theta}{(1-\theta)f_{H_2O}^{v}}\right) \tag{13}$$

The parameter x_{max} can be estimated experimentally by determining the sorption capacity (typically the H_2O content at 25°C and 100% RH). Alternatively, x_{max} can be treated as an unknown and determined by non-linear regression:

$$A + B/T + Cx/T + Dx^2/T + Ex + Fx^2 ... - \ln\left(x_{max} - x_{H_2O}^{Zeo}\right) = \ln\left(x_{H_2O}^{Zeo}\right) - \ln\left(f_{H_2O}^{V}\right) - \frac{\Delta\overline{C}_{p,Hy}}{R}\left(\ln(T/T_o) + (T_o/T - 1)\right) + \frac{\overline{V}_{H_2O}^{Zeo}}{R}(P-1) \tag{14}$$

However, Carey and Bish (1996) found it difficult to regress x_{max} because its value is only significant to the regression as $x_{H_2O}^{Zeo}$ approaches x_{max} (i.e. near saturation). These are difficult measurements because of the problem of condensation in experimental apparatus near saturation.

In these equations, the standard state for H_2O is the ideal gas at the temperature of interest and 1 bar. The standard state for sorbed H_2O as represented by Equation (7) corresponds to the hypothetical condition where $\gamma_2 = x = 1$ at the temperature of interest. In the case of Equation (11), the standard state corresponds to a Henry's-law extrapolation to complete saturation of clinoptilolite at the temperature of interest.

The use of θ represents a simplified treatment of mixing but is well established in the zeolite literature (e.g. Barrer 1978). The use of x avoids the assumptions of the Langmuir model and allows a more general treatment of the mixing properties in zeolites. Carey and Bish (1996) were able to obtain equally satisfactory results using Equations (9) and (12) in their analysis of clinoptilolite.

APPLICATION OF THE THERMODYNAMIC RELATIONS

The thermodynamic relations shown in Equations (9) and (12) are readily applied to a set of equilibrium data for a zeolite-H_2O system. These data are generally obtained by (thermo)gravimetric methods in which the mass of zeolite is determined as a function of temperature and H_2O pressure or the pressure of a closed system is measured following the introduction of a known quantity of H_2O vapor. Each experimental determination provides a value for x or θ at a given temperature and pressure. These values are substituted into the right-hand sides of Equations (9) or (12) and provide the basis for performing a regression for the coefficients A, B, C, D, ... These coefficients can be related to standard thermodynamic properties as follows:

$$\Delta\mu^o_{Hy} = -R(T_oA + B) \quad \text{Standard state chemical potential of hydration} \quad (15)$$

$$H^o_{Hy} = -RB \quad \text{Standard state enthalpy of hydration} \quad (16)$$

$$S^o_{Hy} = RA \quad \text{Standard state entropy of hydration} \quad (17)$$

$$W_1 = -RC \quad \text{Non-ideal enthalpy term} \quad (18)$$

$$W_2 = -RD \quad \text{Non-ideal enthalpy term} \quad (19)$$

$$W_3 = -RE \quad \text{Non-ideal entropy term} \quad (20)$$

$$W_4 = -RF \quad \text{Non-ideal entropy term} \quad (21)$$

The partial and integral molar values can then be obtained from the relations:

$$\overline{H} = \partial(\mu/T)/\partial(1/T) \quad \text{Partial molar enthalpy} \quad (22)$$

$$\tilde{H} = 1/\theta \int_0^\theta \overline{H} d\theta \quad \text{Integral molar enthalpy} \quad (23)$$

$$\overline{S} = -(\mu/T - \overline{H}/T) \quad \text{Partial molar entropy} \quad (24)$$

$$\tilde{S} = 1/\theta \int_0^\theta \overline{S} d\theta \quad \text{Integral molar enthalpy} \quad (25)$$

For example, using Equation (12) and assuming a temperature-independent value of $\Delta\overline{C}_{p,Hy}$, the partial and integral molar properties as a function of temperature, pressure, and H_2O content are given by:

$$\Delta\overline{H}_{Hy} = \Delta\overline{H}^o_{Hy} + W_1\theta + W_2\theta^2 + \Delta\overline{C}_{p,Hy}(T - T_o) \quad (26)$$

$$\Delta\tilde{H}_{Hy} = \Delta\overline{H}^o_{Hy} + W_1/2\,\theta + W_2/3\,\theta^2 + \Delta\overline{C}_{p,Hy}(T - T_o) . \quad (27)$$

$$\Delta\overline{S}_{Hy} = \Delta\overline{S}^o_{Hy} - W_3\theta - W_4\theta^2 + \Delta\overline{C}_{p,Hy}\ln(T/T_o) - R\ln\left(\frac{\theta}{(1-\theta)P}\right) \quad (28)$$

$$\Delta\tilde{S}_{Hy} = \quad (29)$$

$$\Delta\overline{S}^o_{Hy} - W_3/2\,\theta - W_4/3\,\theta^2 + \Delta\overline{C}_{p,Hy}\ln(T/T_o) - R/\theta(\theta\ln(\theta) + (1-\theta)\ln(1-\theta) - \theta\ln(P))$$

and the relation

$$G = H - TS. \quad (30)$$

Example calculation

As an example application of the method, consider a small data set for analcime obtained by gravimetric methods by Balgord and Roy (1973). They performed equilibrium studies of analcime-H_2O at 10-mbar H_2O pressure between 25 and 400°C (complete dehydration). They found their data were reversible at temperatures above 200°C (Table 2).

For this example, data were regressed using Equation (12) with the assumption that $\Delta\overline{C}_{p,Hy}$ = R (8.314 J/mol-K) and $\overline{V}^{Zeo}_{H_2O} = 0$. A more complete example would consider the

Table 2. Equilibrium data for analcime-H_2O obtained by Balgord and Roy (1973) at an applied H_2O-vapor pressure of 7.9 torr (0.010 mbar). θ is defined by Equation (10) with x_{max} taken as 1 H_2O per formula unit (pfu) of analcime and lnK is given by Equation (13).

T (K)	*θ*	*lnK*
499	0.515	4.6282
509	0.432	4.2934
522	0.382	4.0872
547	0.261	3.5271
565	0.170	2.9780
599	0.090	2.2560
625	0.059	1.8039
649	0.027	0.9939
676	0.017	0.4865

statistical significance of each of the regressed coefficients (A, B, C, D, ...), but for simplicity the results for three coefficients (A, B, and C) are presented (Note: for limited data sets, as in this example, the choice of the number of regression coefficients can have a significant impact on the results.). The raw data are shown in Figure 8 in a conventional θ vs. T plot. However, as indicated by the form of Equation (12), it is quite useful to replot the data as lnK vs. 1/T (Fig. 9) or θ. In Figure 9, the intercept at 1/T = 0 provides an estimate of A, the slope at $\theta = 0$ provides an estimate of B, and any curvature in the data requires at least one non-ideal term (i.e. at least one of C, D, E ...). The data do show curvature and the results of a regression for three terms are given in Table 2. A comparison of the raw data with the regression results, illustrated in Figures 8 and 9, shows a close fit to the data. However, the regression results should be used with caution because data at only a single pressure are available and there are no data for $\theta > 0.55$. These limitations are perhaps not as serious in the case of analcime, because of the relatively well-defined sites for H_2O and Na in the structure.

Table 3. Linear regression results for the analcime-H_2O data of Balgord and Roy (1973). The calorimetric data of Johnson et al. (1982) are given for comparison.

Regressed parameters	A (unitless)	B (K)	C (K)	r^2
	-14.8746	10244.5	-1144.41	0.996
Partial molar quantities	$\Delta\mu^o_{Hy}$ (J/mol)	$\Delta\overline{H}^o_{Hy}$ (J/mol)	$\Delta\overline{S}^o_{Hy}$ (J/mol-K)	W_1 (J/mol)
	-48304	-85178	-123.68	9515.2
Integral quantities	$\Delta\tilde{G}_{Hy}$ (kJ/mol)	$\Delta\tilde{H}_{Hy}$ (kJ/mol)	$\Delta\tilde{S}_{Hy}$ (J/mol-K)	
(This study)	-43.5	-80.4	-123.7	
(Johnson et al. 1982)	-44.9	-84.9	-133.8	

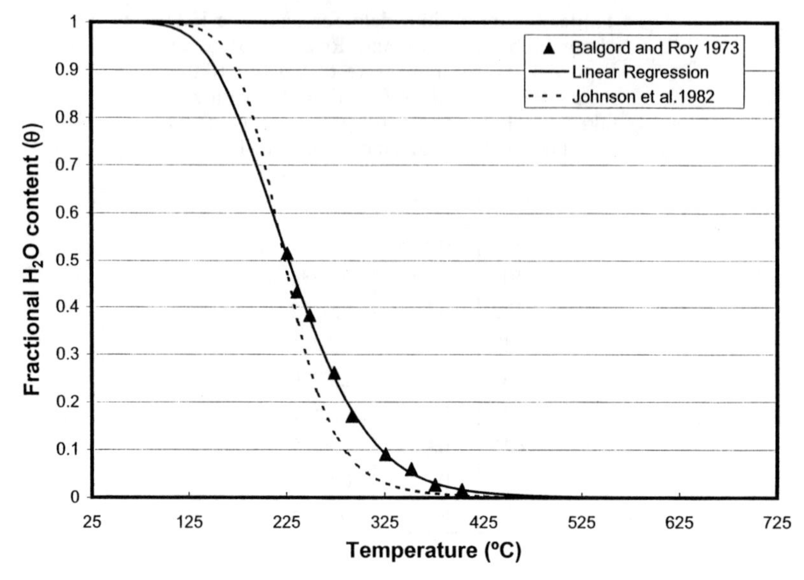

Figure 8. Gravimetric equilibrium data for analcime (Balgord and Roy 1973) plotted as fractional water content (θ) vs. temperature (°C). The plot shows the fit to the data obtained by regression of Equation (12) and the calculated equilibrium relation predicted by the calorimetric study of Johnson et al. (1982) at Balgord and Roy's experimental P of 0.01 bar.

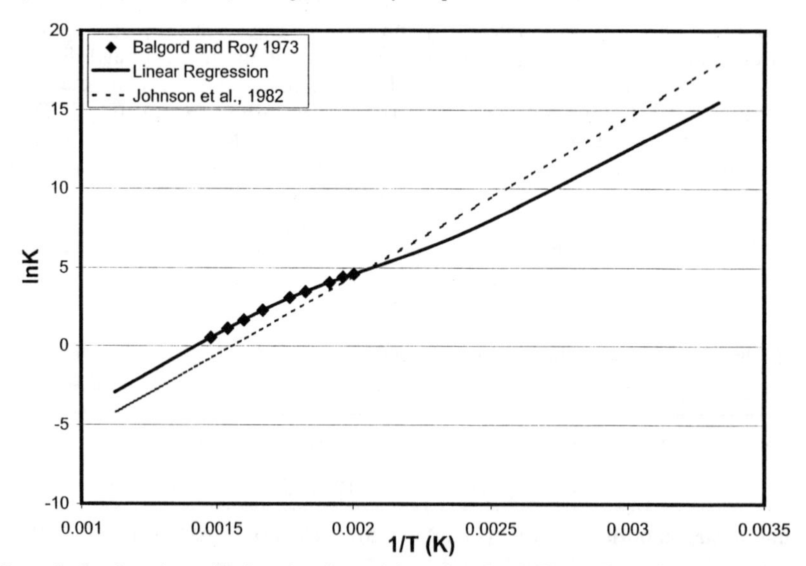

Figure 9. Gravimetric equilibrium data for analcime plotted as ln K (see Eqn. 13; P = 0.01 bar) vs. reciprocal temperature in Kelvin. The plot shows the fit to the data obtained by regression of Equation (12) and the predicted relation from the calorimetric study of Johnson et al. (1982).

Comparison of the thermodynamic values obtained by regression of the Balgord and Roy (1973) data with the HF solution and adiabatic calorimetric data of Johnson et al. (1982) in Table 3 shows close agreement. The thermogravimetric data indicate that the analcime-H_2O system has a small degree of non-ideality.

Limitations of the method

The thermodynamic formulation is limited to a reversible hydration process. In many zeolites, for example, heulandite, dehydration processes are irreversible. Once the thermal/dehydration condition is exceeded, the zeolite no longer hydrates or dehydrates according to the same physical processes. A different equation of state would follow such a transition. We know of no thermodynamic formulations for multiple states. In addition, it is quite possible that there are very few zeolites for which a true completely reversible hydration/dehydration is possible. For example, Yang et al. (2001) discussed the possible role of these effects on the differing enthalpies of hydration determined by different methods. Note that Carey and Bish (1996) developed thermodynamic relations describing complete hydration/dehydration but using only reversible data. Thus the Carey and Bish (1996) equations cannot correctly describe the entire dehydration process; at temperatures greater than those at which reversible dehydration occurs, their equations represent estimates of the hydration energetics. To adhere strictly to the limitations of the experiments, these results should only be used within the reversible dehydration range.

Summary of some thermodynamic data for hydration of zeolites

The thermodynamics of hydration in zeolites have been obtained by a variety of methods (Table 4). These include measurements by immersion of dehydrated zeolites in water (immersion calorimetry); measurements by dissolution of zeolites in HF acid; measurements by transposed-temperature drop calorimetry; and phase equilibria experiments (e.g. gravimetric studies).

The large majority of data are calorimetric measurements of the integral enthalpy of hydration. These values range from a minimum of -41.8 kJ/mol (cordierite, a dense zeolite-like phase) to a maximum of -84.9 kJ/mol (analcime). Barrer and Cram (1971) and Coughlan and Carroll (1975) conducted extensive immersion calorimetry studies of the factors affecting hydration energetics including exchangeable-cation type, structure type, and Al/Si ratio. Barrer and Cram (1971) demonstrated that on a *per gram of zeolite basis*, the enthalpy of hydration becomes less exothermic with increasing radius of the cation:

$$Li^+ > Na^+ > K^+ > Rb^+ > Cs^+ \quad \text{and} \quad Mg^{2+} > Ca^{2+} > Sr^{2+} > Ba^{2+}$$

with the divalent sequence more exothermic than the monovalent sequence. They also demonstrated that for samples having a range of Al/(Al+Si) ratios but having the same faujasite structure, an increase in framework polarity (i.e. an increase in Al/(Al+Si) ratio) results in a more exothermic enthalpy of hydration. Coughlan and Carroll (1975) were able to demonstrate a linear relationship between the enthalpy of hydration on a *per cation basis* and the inverse radius of the cation *for a given structure*.

Interestingly, the enthalpy of hydration data on a *per molecule of* H_2O basis (Table 4) do not correlate simply with cation type. Carey and Navrotsky (1992) observed, however, that a correlation could be developed between the enthalpy of hydration and the ratio of the framework charge to the number of H_2O molecules at saturation (expressed as #Al/#H_2O or the ratio of the number of Al to H_2O per formula unit) in Na-zeolites. This ratio normalizes for the effects of increased sorption capacity for zeolites with small-radius cations. This analysis is extended to K-, Ca-, and Mg-zeolites in Figure 10. K-zeolites show a very regular relation between #Al/#H_2O and the enthalpy of hydration, whereas the data for Ca and Mg are somewhat less regular. In general, higher #Al/#H_2O results in increased exposure of H_2O molecules to cations and therefore more exothermic enthalpies of hydration. At a given #Al/#H_2O, the energetics of hydration per H_2O decrease in the order $Mg^{2+} \approx Ca^{2+} > Na^+ > K^+$ .The summary of enthalpy of hydration data in Table 4 shows that there are significant differences between the results obtained by different methods. The techniques listed in Table 4 include:

Table 4.

Material	$\Delta\widetilde{G}_{Hy}$	$\Delta\widetilde{H}_{Hy}$	$\Delta\widetilde{S}_{Hy}$	$\Delta\overline{H}_{Hy}$	*Hydration and Dehydration Conditions*	*Method*	*Ref.*
	(kJ/mol-H_2O)	(kJ/mol-H_2O)	(J/mol-H_2O/K)	at $\theta = 0.25$			
H_2O	-8.57	-44.02	-118.9	n.a.			R et al. 79
Cordierite	-9.5	-41.8	-108.2	0 (ideal)	RT to 784°C	PE & TTDC	C 95
Analcime	-44.9	-84.9	-133.8	n.d.	RT to 600°C	HF	J et al. 82
Ca-Chabazite		-71.3		n.d.	RT, 78% RH to 360°C, vacuum	IC	BC 71
Ca-Chabazite		-78.6		-87.8	23°C, 50% RH to 700°C	TTDC	S et al. 99
K-Chabazite		-66.8		n.d.	RT, 78% RH to 360°C, vacuum	IC	BC 71
Li-Chabazite		-69.1		n.d.	RT, 78% RH to 360°C, vacuum	IC	BC 71
Na-Chabazite		-70		n.d.	RT, 78% RH to 360°C, vacuum	IC	BC 71
Clinoptilolite		-75.8		n.d.	25°C, 50% RH to 500°C, 18% loss of HC	TTDC	Y et al. 01
Ca-Clinoptilolite	-36.13	-76.92	-136.8	-90.9	25°C, 100% RH to 248°C, 0.2 mbar	PE	CB 96
Ca-Clinoptilolite		-74.3		-87.6	25°C, 100% RH to 245°C, 0.4 mbar	IC	CB 97
Ca-Clinoptilolite		-75.3		-92.8	RT, 78% RH to 300°C, vacuum	GAC	P et al. 01**
Ca-Clinoptilolite		-71.3		n.d.	25°C, 50% RH to 500°C, 60% loss of HC	TTDC	Y et al. 01
K-Clinoptilolite	-25.53	-67.78	-141.7	-75.1	25°C, 100% RH to 248°C, 0.2 mbar	PE	CB 96
K-Clinoptilolite	-24.9	-81.0	-188.3	-87.8	115°C, 0.5 bar to 300°C, 35 bar	PE	WB 99
K-Clinoptilolite		-66.4		-77.3	25°C, 100% RH to 245°C, 0.4 mbar	IC	CB 97
K-Clinoptilolite		-63.0		-74.1	RT, 78% RH to 300°C, vacuum	GAC	P et al. 01**
K-Clinoptilolite		-74.9		n.d.	25°C, 50% RH to 500°C, 7% loss of HC	TTDC	Y et al. 01
Na-Clinoptilolite	-29.68	-74.19	-149.3	-83.5	25°C, 100% RH to 218°C, 0.2 mbar	PE	CB 96
Na-Clinoptilolite	-26.9	-84.8	-194.3	-91.4	115°C, 0.5 bar to 300°C, 35 bar	PE	WB 99
Na-Clinoptilolite		-66.0			RT, 78% RH to 360°C, vacuum	IC	BC 71
Na-Clinoptilolite		-67.4		-77.9	25°C, 100% RH to 245°C, 0.4 mbar	IC	CB 97
Na-Clinoptilolite		-67.0		-78.9	RT, 78% RH to 300°C, vacuum	GAC	P et al. 01**
Na-Clinoptilolite		-74.2		n.d.	25°C, 50% RH to 500°C, 19% loss of HC	TTDC	Y et al. 01

Sodic-Clinoptilolite		-65.5		n.d.	RT 50% RH to 627°C in helium	HF	J et al. 91
NaK-Clinoptilolite		-72.3		n.d.	25°C, 50% RH to 500°C, 9% loss of HC	TTDC	Y et al. 01
Ca-Ferrierite		-62.9		n.d.	RT, 78% RH to 360°C, vacuum	IC	BC 71
Li-Ferrierite		-60.4		n.d.	RT, 78% RH to 360°C, vacuum	IC	BC 71
Na-Ferrierite		-61.3		n.d.	RT, 78% RH to 360°C, vacuum	IC	BC 71
Ca-Heulandite		-78.3		n.d.	23°C, 55% RH to 702°C, argon	TTDC	K et al. 01*
Ca-Laumontite		-79.9		n.d.	? to 702°C	TTDC	K et al. 96a
Ca-Leonhardite		-84.1		n.d.	RT, 100% RH to 702°C	TTDC	K et al. 96a
CaKNa-Leonhardite		-72.8		n.d.	? to 702°C	TTDC	K et al. 96b
Na-Mordenite		-70.3			RT, 78% RH to 360°C, vacuum	IC	BC 71
NaCa-Mordenite	-33.5	-73.8	-134.8	n.d.	RT 50% RH to 627°C in helium	HF	J et al. 92
Ca-Stilbite		-75.9		n.d.	23°C, 55% RH to 702°C, argon	TTDC	K et al. 01
Ca-A		-71.2		n.d.	RT, 78% RH to 360°C, vacuum	IC	BC 71
K-A		-73.3		n.d.	RT, 78% RH to 360°C, vacuum	IC	BC 71
Li-A		-69.6		n.d.	RT, 78% RH to 360°C, vacuum	IC	BC 71
Mg-A		-71.2		n.d.	RT, 78% RH to 360°C, vacuum	IC	BC 71
Mg-A		-70.8		n.d.	RT, 78% RH to 360°C, vacuum	IC	CC 76
Na-A		-69.9		n.d.	RT, 78% RH to 360°C, vacuum	IC	BC 71
Na-A		-69.9		n.d.	RT, 78% RH to 360°C, vacuum	IC	CC 76
Cs-L		-55.9		n.d.	RT, 78% RH to 360°C, vacuum	IC	CC 76
K-L		-60.9		n.d.	RT, 78% RH to 360°C, vacuum	IC	CC 76
Li-L		-65.4		n.d.	RT, 78% RH to 360°C, vacuum	IC	CC 76
Na-L		-61.6		n.d.	RT, 78% RH to 360°C, vacuum	IC	CC 76
Rb-L		-59.7		n.d.	RT, 78% RH to 360°C, vacuum	IC	CC 76
Ca-X		-72.2		n.d.	RT, 78% RH to 360°C, vacuum	IC	BC 71
K-X		-68.1		n.d.	RT, 78% RH to 360°C, vacuum	IC	BC 71
Li-X		-69		n.d.	RT, 78% RH to 360°C, vacuum	IC	BC 71

Table 4, concluded.

Mg-X		-68.3	n.d.	RT, 78% RH to 360°C, vacuum	IC	BC 71
Mg-X		-67.4	n.d.	RT, 78% RH to 360°C, vacuum	IC	CC 76
Na-X	n.c.	n.c.	-64.0	20°C, 79% RH to 280°C, vacuum	PE	D et al. 66
Na-X		-68.7	n.d.	RT, 78% RH to 360°C, vacuum	IC	BC 71
Na-X		-68.6	n.d.	RT, 78% RH to 360°C, vacuum	IC	CC 76
Ca-Y		-65.1	n.d.	RT, 78% RH to 360°C, vacuum	IC	BC 71
Ca-Y		-65.7	n.d.	23°C, 50% RH to 500-550°C	TTDC	YN 00
K-Y		-62.1	n.d.	RT, 78% RH to 360°C, vacuum	IC	BC 71
K-Y		-57.8	n.d.	23°C, 50% RH to 500-550°C	TTDC	YN 00
Li-Y		-61.3	n.d.	RT, 78% RH to 360°C, vacuum	IC	BC 71
Li-Y		-67.2	n.d.	23°C, 50% RH to 500-550°C	TTDC	YN 00
Mg-Y		-67.5	n.d.	RT, 78% RH to 360°C, vacuum	IC	BC 71
Mg-Y		-64.7	n.d.	RT, 78% RH to 360°C, vacuum	IC	CC 76
Na-Y		-61.6	n.d.	RT, 78% RH to 360°C, vacuum	IC	BC 71
Na-Y		-61.4	n.d.	RT, 78% RH to 360°C, vacuum	IC	CC 76
Na-Y		-65.2	n.d.	23 °C, 50% RH to 500-550°C	TTDC	YN 00

n.d. : not determined
n.a.: not applicable
n.c. : not calculated by the authors
HC: hydration capacity
RH: relative humidity
RT: room temperature

* Kiseleva et al. (2001) estimated the enthalpy of transition of heulandite to heulandite-B as 2.5 kJ/mol/TO_2 based on measurements of Drebushchak (1990), suggesting a revised enthalpy of hydration of -74.4 kJ/mol.

** Petrova et al. (2001) revised values based on linear fit of data (in contrast to the squared-term fit used in the paper).

Other studies of the thermodynamics of hydration in zeolites:

Simonot-Grange & Cointot (1969)—phase equilibria studies of heulandite
Tarasevich et al. (1988)—phase equilibria and calorimetric study of clinoptilolite
Valueva (1995)—immersion calorimetry data for natural heulandite and clinoptilolite
Yamanaka et al. (1989)—phase equilibria of clinoptilolite, mordenite, erionite, chabazite, and phillipsite

Key to Methods

GAC	Gas-adsorption calorimetry
HF	Hydrofluoric acid calorimetry
IC	Immersion calorimetry
PE	Phase equilibria
TTDC	Transposed-temperature drop calorimetry

Key to References

BC 71	Barrer and Cram (1971)
C 95	Carey (1995)
CB 96	Carey and Bish (1996)
CB 97	Carey and Bish (1997)
CC 76	Coughlan and Carroll (1976)
D et al. 66	Dubinin et al. (1966)
J et al. 82	Johnson et al. (1982)
J et al. 91	Johnson et al. (1991)
J et al. 92	Johnson et al. (1992)
K et al. 96a	Kiseleva et al. (1996a)
K et al. 96b	Kiseleva et al. (1996b)
K et al. 01	Kiseleva et al. (2001)
P et al. 01	Petrova et al. (2001)
R et al. 79	Robie et al. (1979)
S et al. 99	Shim et al. (1999)
WB 99	Wilkin and Barnes (1999)
Y et al. 01	Yang et al. (2001)
YN 00	Yang and Navrotsky (2000)

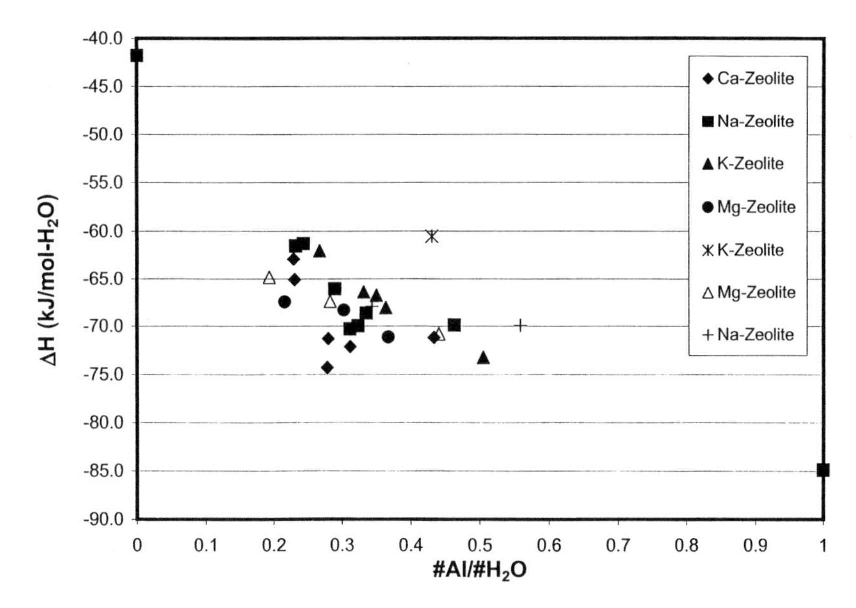

Figure 10. Molar enthalpy of hydration as a function of the ratio of the saturated water content to the Al content of the framework, as a function of cation type. Data are from Barrer and Cram (1971), Coughlan and Carroll (1976), Carey and Navrotsky (1986), Johnson et al., (1992), and Carey and Bish (1997). The data for zeolite-L do not fit the trend observed for the other zeolites shown in the figure (analcime, chabazite, clinoptilolite, cordierite, ferrierite, mordenite, zeolite-A, zeolite-X , and zeolite-Y).

1. *Immersion calorimetry.* This method provides measurements of integral enthalpies of hydration as a function of degree of hydration. In this method, a sample is dehydrated, sealed in a glass bulb and then exposed to water in the calorimeter by breaking the bulb (e.g. Barrer and Cram 1971). Some of the difficult parameters to control include the amount of H_2O adsorbed by the sample (requires knowledge of the amount of H_2O in the sample while immersed in water); accurate measurement of the initial state of dehydration; and the potential for irreversible changes in the zeolite structure during dehydration such that the sample does not readsorb its full capacity of H_2O during immersion. The advantages of the method include that it is a direct measurement of the enthalpy of hydration.
2. *Phase equilibria.* Enthalpy of hydration can be determined from phase equilibria measurements by the isoteric method in which data are plotted as lnP (or lnK) vs. 1/T at constant hydration state (e.g. Dubinin et al. 1966). The slope of these lines gives the enthalpy of hydration. Alternatively, the same data can be linearly regressed to fit a thermodynamic formulation to provide enthalpies as well as the Gibbs free energy and entropy of hydration (e.g. Carey and Bish 1996). The choice of isoteric vs. linear regression is governed in part by the amount of available data. The isoteric approach requires sufficient data to define equilibrium constants at constant composition. One of the difficulties associated with this method includes the indirect nature of the measurement that requires fitting the data to extract the enthalpy of hydration. The advantages of the method include the possibility of demonstrating equilibrium for all data (precluding inadvertent measurements on irreversibly modified structures); the measurement of the state of hydration during the experiment; and the determination of other thermodynamic properties (e.g. Gibbs free energy of hydration) of the

zeolite-H_2O system.

3. *Transposed-temperature drop calorimetry.* Enthalpy of hydration data are obtained by measurements of the heat evolved from a sample dropped at room temperature into a calorimeter maintained at 700-800°C (e.g. Kiseleva et al. 1996a). One of the difficulties associated with this measurement includes enthalpy effects due to irreversible structural changes at 700-800°C (the effect depends on whether collapse is exothermic or endothermic). Petrovic and Navrotsky's (1997) data suggest moderately exothermic effects of 1-3 kJ/TO_2, which would lead to calculated enthalpies of hydration that are not as negative. Some of the advantages of the method include that the initial state of the zeolite can be well determined (e.g. 25°C, in a constant RH environment); and that in most cases the measurement records the complete dehydration of the structure.
4. *Gas-adsorption calorimetry.* Enthalpies of hydration data are obtained by exposing a bed of zeolite to differing H_2O-vapor pressures within a calorimeter (e.g. Petrova et al. 2001). Some of the difficulties of this method include kinetic limitations on attainment of equilibrium (slow kinetics require a very sensitive and stable calorimeter). Some of the advantages of the method include the capacity to measure partial molar enthalpies of hydration directly.
5. *Hydrofluoric acid calorimetry.* Enthalpies of hydration are obtained by measuring the difference in the enthalpy of solution in HF acid at about 70°C of a hydrated and a dehydrated sample (e.g. Johnson et al. 1982). Difficulties of this method include the potential for irreversible changes in the dehydrated sample and the requirement that the enthalpy of hydration is found as the difference between large heat of solution effects. One of the advantages includes the measurement of an essentially completely hydrated sample.

All of the integral enthalpy values listed in Table 4 refer, nominally, to the average enthalpy of hydration per mole of H_2O during complete hydration of the zeolite structure. One of the notable differences in results includes the transposed-temperature drop calorimetry observation that there is no correlation between enthalpies of hydration and cation charge density (Z/r; e.g. Yang et al. 2001) in contrast to results by other methods (e.g. immersion calorimetry, Barrer and Cram 1971). Barrer and Cram (1971) and Yang et al. (2001) discussed some of the possible reasons for different results from different methods. These differences may be summarized as follows:

- Differing "fully" saturated hydration states. For example, Carey and Bish (1996, 1997) use a saturation state of 25°C, 100% RH; Barrer and Cram (1971) use 25°C, 80% RH; Kiseleva et al. (1996a) use 25°C, 50% RH. Other factors being equal, higher relative humidities used to define "full saturation" will result in lower integral enthalpies of hydration (per mole of H_2O) either because more H_2O is assumed to be involved in the enthalpic effect or because of measurements of H_2O with lower partial enthalpies. Note that Carey and Bish (1996) show that clinoptilolite is only at about 90% of total sorption capacity at 50% RH.
- Differing "fully" dehydrated states. Some studies probably achieve complete dehydration (e.g. Kiseleva et al. 2001 used 702°C) but at a cost of irreversible structural collapse. Other studies used less severe dehydration conditions, but at the potential cost of failing to achieve complete dehydration. Still other studies were limited to conditions where equilibrium could be demonstrated. Other factors being equal, failure to completely dehydrate samples will most likely yield values of the integral enthalpy of hydration that are less exothermic (not as negative).
- Different zeolite samples with different structures, different Al/(Al+Si) ratios, and different exchangeable-cation compositions. Note that in many cases, cation-

exchanged samples are not end-members because of incomplete exchange.

- Some dehydrated samples may not have an equilibrium distribution of H_2O (Barrer and Cram 1971). At low degrees of hydration, the high energy of the remaining sorption sites may limit the rate at which H_2O molecules migrate to an equilibrium configuration. This effect would lead to measurements of the integral enthalpy of hydration that are less exothermic.
- Some measurements are made on samples that have undergone irreversible changes in hydration capacity. The effect on the integral enthalpy of hydration depends on the measurement method and the enthalpy of the irreversible change.

Unfortunately, Barrer and Cram's (1971) recommended study of the enthalpy of hydration by a variety of methods using a single zeolite sample has not yet been realized. Such a study would help resolve the differences shown in Table 4 and would permit a more effective utilization of the strengths/advantages of the various techniques of investigating zeolite-H_2O energetics.

The most significant impact of these differences is that there is still not an accepted measure of the magnitude of the *total* energetic stabilization imparted by H_2O on the zeolite structure. This is one of the reasons that we are as yet unable to answer the question as to whether zeolites have true stability fields relative to other aluminosilicate assemblages. Essentially this question of stability depends on the trade-off between a less stable, open aluminosilicate framework (i.e. zeolite framework) with the stabilizing effect of H_2O vs. a denser aluminosilicate framework (i.e. feldspar framework) without the capacity to adsorb H_2O.

There are relatively few equilibrium studies that provide a full description of the thermodynamics of hydration. By necessity, such studies must have a phase equilibrium component. In general, the phase equilibrium studies have shown that more loosely bound zeolitic-H_2O (whether because of cation-charge density or openness of the aluminosilicate framework) is associated with more negative entropies of hydration. In general terms, the thermodynamic values can be related to more definite physical behavior as follows. The Gibbs free energy of hydration reflects the affinity of the zeolite for water. All zeolites have greater affinity for H_2O compared with the free energy of condensation (i.e. zeolites are desiccants). Among the zeolites, greater affinities for H_2O are associated with increased framework charge density and higher-charged exchangeable cations (i.e. divalent cations attract more H_2O than monovalent cations). The enthalpy of hydration reflects the amount of H_2O lost for a given change in temperature. Zeolites with high enthalpies of hydration dehydrate over narrow temperature intervals. For example, analcime retains its H_2O more effectively than other zeolites. High enthalpies of hydration are similarly associated with high framework charge density and cation type. The entropy of hydration is reflected in the rate of acceleration of dehydration as temperature increases. A high negative entropy of hydration indicates that an increase in temperature will increasingly destabilize the H_2O in zeolites. Based on available data, it appears that Na-exchanged forms have the greatest susceptibility to thermal changes. The practical consequences of these thermodynamic relations will be explored below in the section on "Modeling Thermal Behavior."

Hydration processes and cation exchange

Cation-exchange and hydration processes are intimately connected because cation exchange is accompanied by changes in H_2O content (e.g. Barrer and Cram 1971). These changes can be quite significant (e.g. almost 1 H_2O molecule for every 2 cations exchanged between Na- and K-clinoptilolite, Carey and Bish 1996). A second, smaller energetic consequence is the change in the average energy per H_2O molecule upon cation exchange (see Table 4). A simple calculation illustrates these effects. Pabalan (1994) measured cation exchange in clinoptilolite for Na-, K-, and Ca-exchanged end-members. Clinoptilolite is selective for K relative to Na, with a modest free energy of exchange (ΔG = 7980 J/mol-cation):

$$NaAlSi_{4.26}O_{10.51}(3.49\ H_2O) + K^+(aq) \Leftrightarrow$$
$$KAlSi_{4.26}O_{10.51}(3.04\ H_2O) + Na^+(aq) + 0.45\ H_2O \qquad (35)$$

Note that Na-clinoptilolite loses H_2O, so that the cation-exchange reaction is actually impeded by the decreased H_2O content of K-clinoptilolite. Using the data of Carey and Bish (1996), the energetic cost of losing 0.45 H_2O to K-clinoptilolite is about 13500 J/mol-cation. One can then calculate an effective distribution coefficient in the absence of changes in H_2O of 21500 J/mol-cation, changing the distribution coefficient from 25 to 5800! Clinoptilolite would be even more selective for K^+, if not for the exchange of H_2O.

MODELING THERMAL BEHAVIOR OF ZEOLITES

The thermodynamic relations as summarized in Equations (5) and (12) provide a framework for outlining the possible behavior of zeolite-H_2O systems as a function of temperature and pressure. It is convenient to present this analysis in the form of synthetic thermogravimetric curves. This permits comparison of thermodynamic data with the great quantity of published data on zeolites in the form of TGA curves (e.g. Gottardi and Galli 1985). In particular, the thermodynamic formulation of Equation (12) can be used to investigate the thermodynamic implications of the TGA data.

The synthetic TGA curves were calculated with the assumptions of equilibrium dehydration occurring at an H_2O-vapor pressure of 0.03 bar. The actual conditions in an experiment could be as high as 1 bar and at least as low as 0.00003 bar (roughly 0.1% RH) depending on the heating rate, sample size, and the flux and relative humidity of any purge gas. Experience in our laboratory suggests that many TGA experiments are conducted with an effective vapor pressure near 0.03 bars (i.e. 100% RH at room temperature; see below). The synthetic TGA curves were calculated using Equation (12) and with different combinations of thermodynamic data for the Gibbs free energy of hydration, the enthalpy of hydration, and non-ideal mixing parameters (A, B, C, D, etc.). These values were chosen to span the range of available measurements (Table 4). In these calculations, the "base-case" zeolite was assumed to have an integral Gibbs free energy of hydration of -30 kJ/mol, an integral enthalpy of hydration of -70 kJ/mol, and an integral entropy of hydration of -134 J/mol-K, dehydrating at equilibrium into a 0.03 bar [$P(H_2O)$] atmosphere. In the discussion, "significant dehydration" is defined as the approximate temperature at which the zeolite loses >5% H_2O.

The effect of H_2O pressure. Higher H_2O pressure results in an increase in the onset temperature of dehydration (Fig. 11). Pressures ranging from 0.0003 bar (~1% RH at 25°C) to 1 bar (100% RH at 100°C) were investigated. A pressure of 0.0003 bar yields a sharply falling TGA curve that lacks the typical initiation period in which temperature rises without significant dehydration. For a pressure of 0.03 bar (~100% RH at 25°C), dehydration is not significant until about 100°C; for a pressure of 0.5 bar, dehydration begins at 150°C; and for 1 bar or for a variable pressure that starts near 0 and increases gradually to 1, dehydration is not significant until about 170°C. Note that the shape of the TGA curves (the S-shape) is not sensitive to the assumed H_2O-vapor pressure. Calculations made with variable pressure that starts near 0.003 bar and increases gradually to 1 bar at 150°C are not appreciably different from the 1-bar case.

The effect of the Gibbs free energy of hydration. The effect of differing Gibbs free energy of hydration is shown in Figure 12. These calculations were made for a Gibbs free

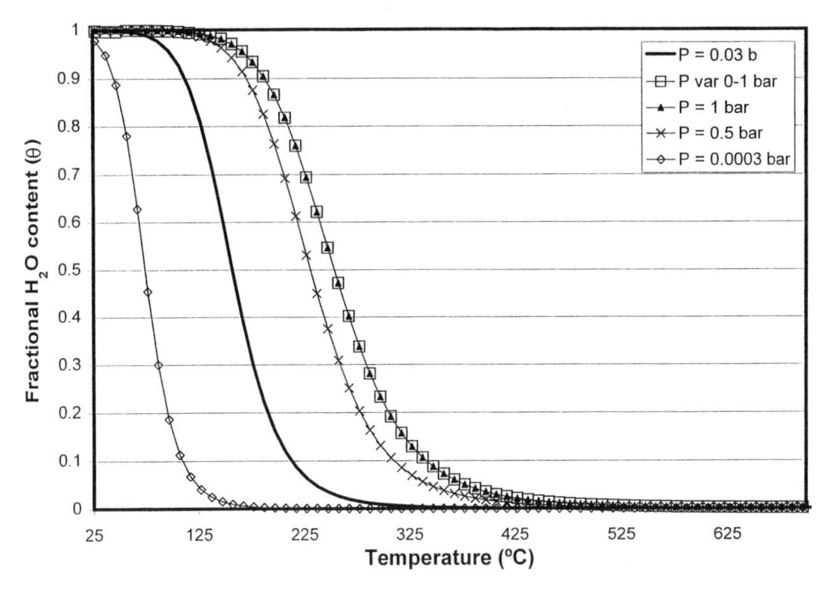

Figure 11. Synthetic thermogravimetric curves showing the effect of pressure on thermogravimetric data. The curves were calculated with Equation (12) and each assumes the same integral value of the Gibbs free energy and enthalpy of hydration (-30 and -70 kJ/mol, respectively; P = 0.03 bar).

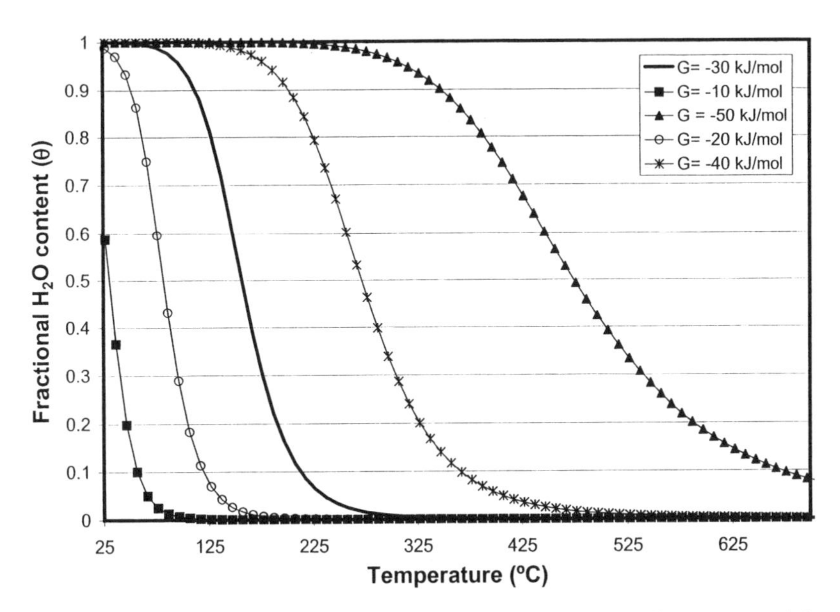

Figure 12. Synthetic thermogravimetric curves showing the effect of differing values of the Gibbs free energy of hydration on thermogravimetric data. The curves were calculated with Equation (12) and each assumes the same integral value of the enthalpy of hydration (-70 kJ/mol; P = 0.03 bar).

energy of hydration varying between -10 kJ/mol (near that of water or cordierite) to -50 kJ/mol (greater than analcime). For the calculations, the enthalpy of hydration was kept constant at the base-case value of -70 kJ/mol and the entropy of hydration was calculated from G = H – TS. The effect on the TGA curves is quantitatively the same as a change in H_2O pressure (see Eqn. 12). For base-case zeolites with Gibbs free energies of hydration of -10, -20, -30, -40, and -50 kJ/mol, dehydration becomes significant at <25, 40, 100, 185, and 315°C, respectively. The results in Figures 11 and 12 show that the effects of differing H_2O-vapor pressure and Gibbs free energy of hydration cannot be separated. As discussed below, TGA data are considerably more useful if a known H_2O-vapor pressure purge gas can be applied.

The effect of the enthalpy of hydration. The enthalpy of hydration controls the steepness of the dehydration curve (Fig. 13). The beginning of dehydration is similar for base-case zeolites with enthalpies of hydration of -40, -55, -70, and -90 kJ/mol with values covering a 110°C range (190, 130, 100, and 80°C, respectively). However, the slope of the dehydration curve is strongly affected, and the temperature interval for dehydration between 5 and 95% is 760, 230, 135, and 85°C for the four cases. The zeolite with the most negative enthalpy of hydration dehydrates over the narrowest temperature interval.

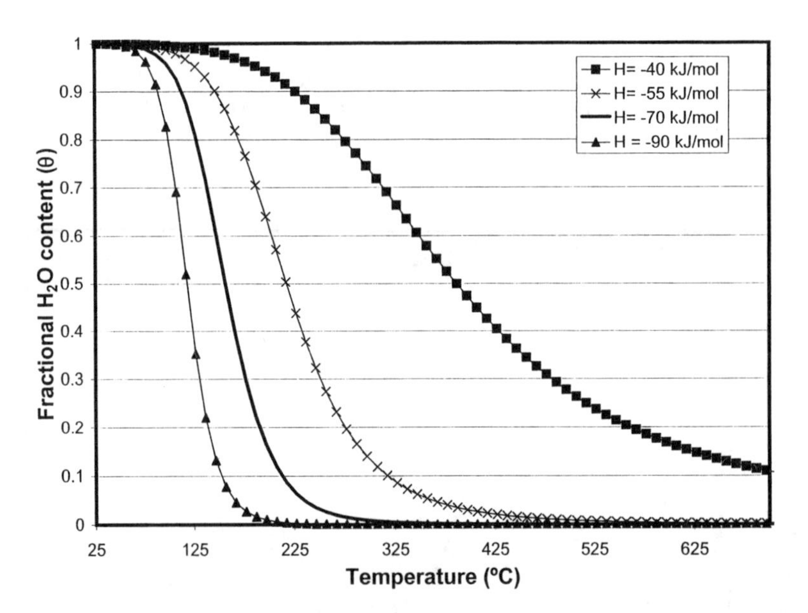

Figure 13. Synthetic thermogravimetric curves showing the effect of differing values of the enthalpy of hydration on thermogravimetric data. The curves were calculated with Equation (12) and each assumes the same integral value of the Gibbs free energy of hydration (-30 kJ/mol; P = 0.03 bar).

The effect of non-ideal enthalpy of mixing. Non-ideality in enthalpy of hydration is given by the W_1 and W_2 terms of Equation (11) and is illustrated in Figure 14. These calculations were conducted assuming identical integral values of the Gibbs free energy of hydration, enthalpy of hydration, and entropy of hydration as for the base-case. Figure 14 demonstrates the tremendous variability in dehydration behavior that is available even for a constant total (i.e. integral) enthalpy of dehydration. Knowledge of the partial molar values is essential to describing the behavior of zeolite-H_2O systems.

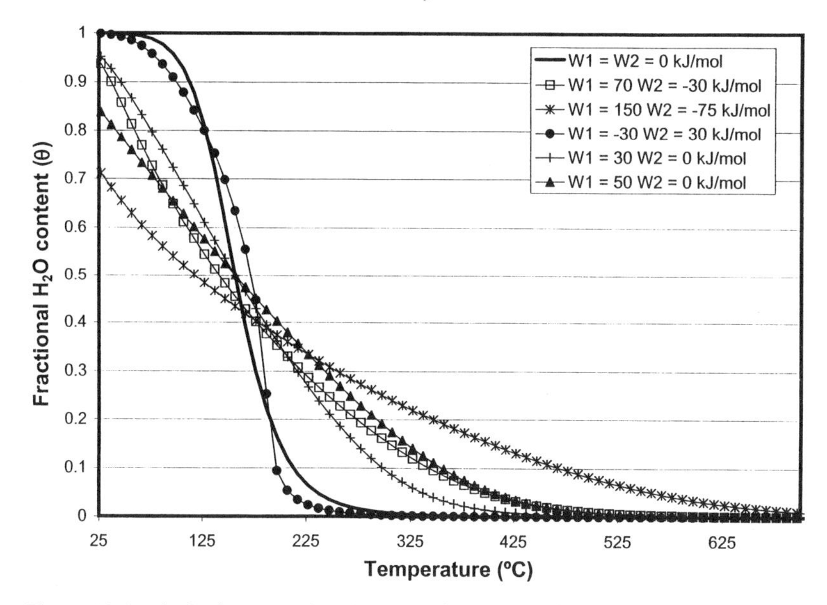

Figure 14. Synthetic thermogravimetric curves showing the effect of non-ideal enthalpy on thermogravimetric data. The curves were calculated with Equation (12) and each assumes the same integral value of the Gibbs free energy and enthalpy of hydration (-30 and -70 kJ/mol, respectively; P = 0.03 bar). An ideal mixing curve is also shown for reference.

Unfortunately, there are very few studies that provide estimates of these non-ideality terms. Three of the curves in Figure 14 represent a range of clinoptilolite-like behaviors as observed by Carey and Bish (1996, 1997) with paired values of W_1 and W_2 of (150, -75), (70, -30), and (50, 0) in kJ/mol. These curves are all convex to the temperature axis and are markedly different than the ideal-mixing case that shows an S-shaped pattern. Other paired values of W1 and W2 produce approximately linear curves (30, 0) or concave curves to the temperature axis (-30, 70). Interestingly, it was not possible to generate TGA curves that had a higher temperature of dehydration initiation without producing unstable behavior in the thermodynamic equations (discontinuity or jump in the degree of dehydration; in other words such relations predict a solvus in the system).

The effect of non-ideal entropy of mixing. Non-ideality in entropy is illustrated in Figure 15. The non-ideality terms are given by W_3 and W_4 of Equation (11). These calculations were conducted for identical integral values of the Gibbs free energy of hydration, enthalpy of hydration, and entropy of hydration as for the base-case. There are no readily available data for constraining the possible values for non-ideality in the entropy of mixing. The results produce curves that are similar in some respects to the non-ideal enthalpy of mixing results (Fig. 14). The effects include curves that are convex and concave to the temperature axis. Figure 15 also illustrates a calculation in which a discontinuity (or solvus) occurs in the zeolite-H_2O system (W_3 = -100, W_4 = 75 in J/mol-K) at T ≈ 160°C and P = 0.03 bar. Non-ideal entropies of mixing are capable of rendering zeolites highly retentive of H_2O at high temperatures.

Combined non-ideal enthalpy and entropy of mixing. A combination of non-ideal entropy and enthalpy yields several other possible forms of behavior (Fig. 16). These include flattening of the dehydration profiles or increased curvature and the initiation of double S-curves.

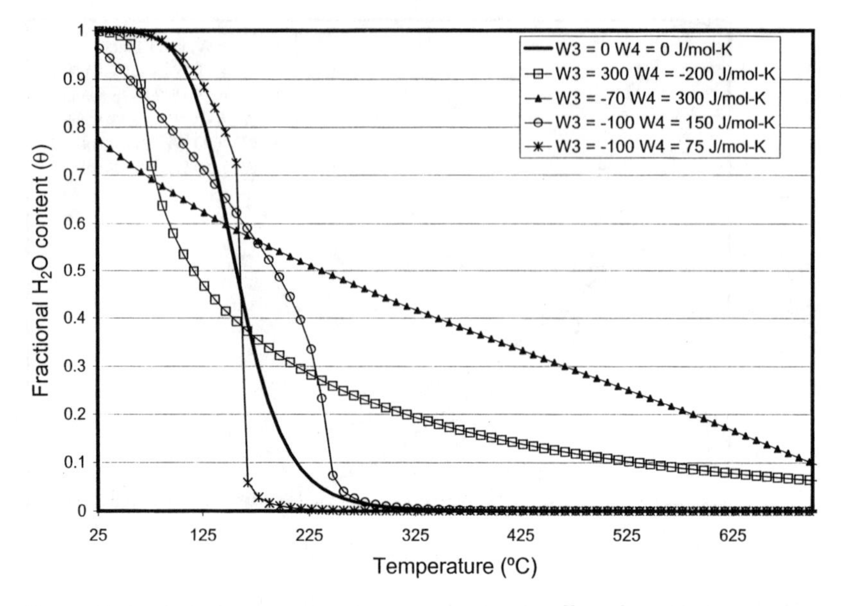

Figure 15. Synthetic thermogravimetric curves showing the effect of non-ideal entropy on thermogravimetric data. The curves were calculated with Equation (12) and each assumes the same integral value of the Gibbs free energy and enthalpy of hydration (-30 and -70 kJ/mol, respectively; P = 0.03 bar). An ideal mixing curve is also shown for reference.

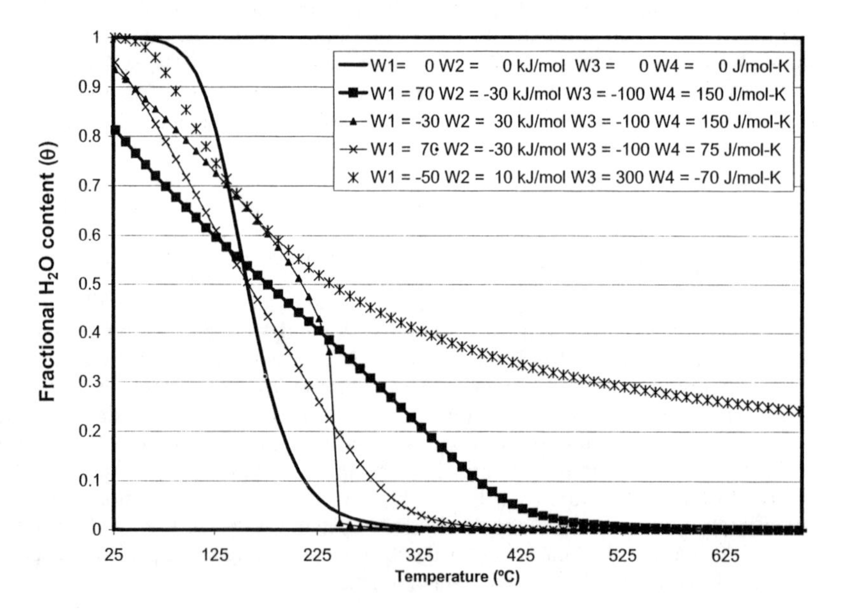

Figure 16. Synthetic thermogravimetric curves showing the effect of combined non-ideal enthalpy and non-ideal entropy on thermogravimetric data. The curves were calculated with Equation (12) and each assumes the same integral value of the Gibbs free energy and enthalpy of hydration (-30 and -70 kJ/mol, respectively; P = 0.03 bar). An ideal mixing curve is also shown for reference.

Steps in TGA data. One feature that is striking in examining all of these possibilities is that none possess the step-like characteristics of many zeolite dehydration curves. The above calculations appear to rule out the possibility that a non-ideal formulation on a single type of H_2O site can explain step-like behavior, at least with the use of polynomial-type non-ideal mixing terms as in Equation (12). It is possible, however, to model step-like behavior by summing two or more distinct hydration types. Figures 17 and 18 show two summed dehydration curves for zeolitic H_2O with differing Gibbs free energy, enthalpies, or both. These calculations were made for ideal-mixing of two equally abundant sites and represent combinations of zeolite energetics taken from the previously investigated ideal mixing cases (Figs. 12 and 13). A combination of Gibbs free energies of hydration of -20 and -40 kJ/mol (Fig. 17) produces a distinct and well-resolved step in the TGA curve. Less distinctly separated Gibbs free energy of hydration values would produce less distinctly resolved steps. A combination of enthalpies of hydration of -55 and -90 kJ/mol fails to produce a marked step in the curves, although the differential TGA curve does show the effect (Fig. 18). Finally a combination of H_2O sites differing in both the Gibbs free energy of hydration and enthalpy of hydration can provide very large step functions (Fig. 19).

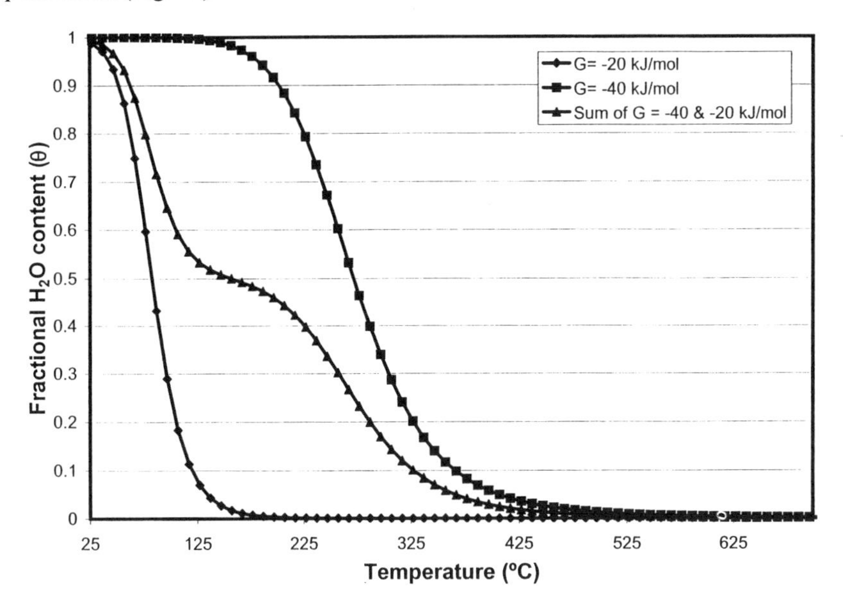

Figure 17. Synthetic thermogravimetric curves showing the combined effect of two sites having differing Gibbs free energy of hydration on thermogravimetric data. The curves were calculated with Equation (12), assuming an enthalpy of hydration of -70 kJ/mol (P = 0.03 bar).

Deriving thermodynamic properties from TGA curves

The theoretical calculations of Figures 11-19 suggest the possibility that TGA data may be used to investigate the thermodynamic properties of zeolites in at least a semi-quantitative way. The key assumption in using actual TGA measurements is that the measured dehydration occurs near equilibrium. As shown below, it is also useful to have a known H_2O pressure for the experiments, but an unknown pressure does not preclude useful analyses.

One approach to a thermodynamic analysis is to recast the θ vs. T plots of Figures

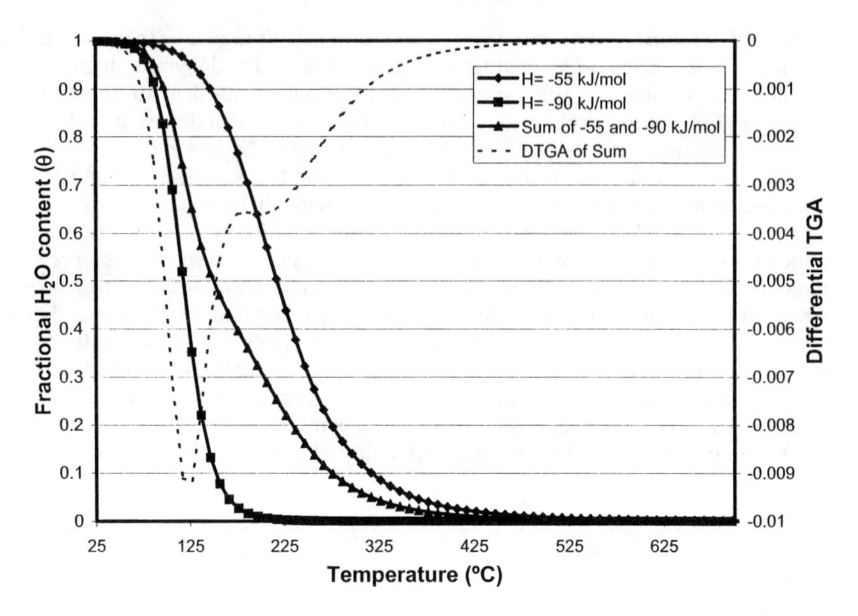

Figure 18. Synthetic thermogravimetric curves showing the combined effect of two sites having differing enthalpy of hydration on thermogravimetric data. The curves were calculated with Equation 12, assuming a Gibbs free energy of -30 kJ/mol (P = 0.03 bar). A first derivative thermogravimetric curve of the combined site data is also shown.

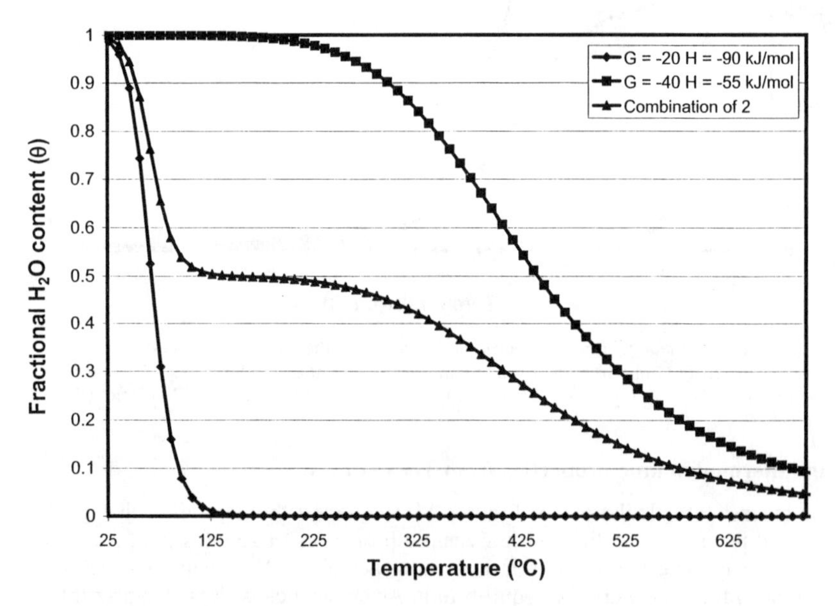

Figure 19. Synthetic thermogravimetric curves showing the combined effect of two sites having differing Gibbs free energy and enthalpy of hydration on thermogravimetric data. The curves were calculated with Equation (12) (P = 0.03 bar).

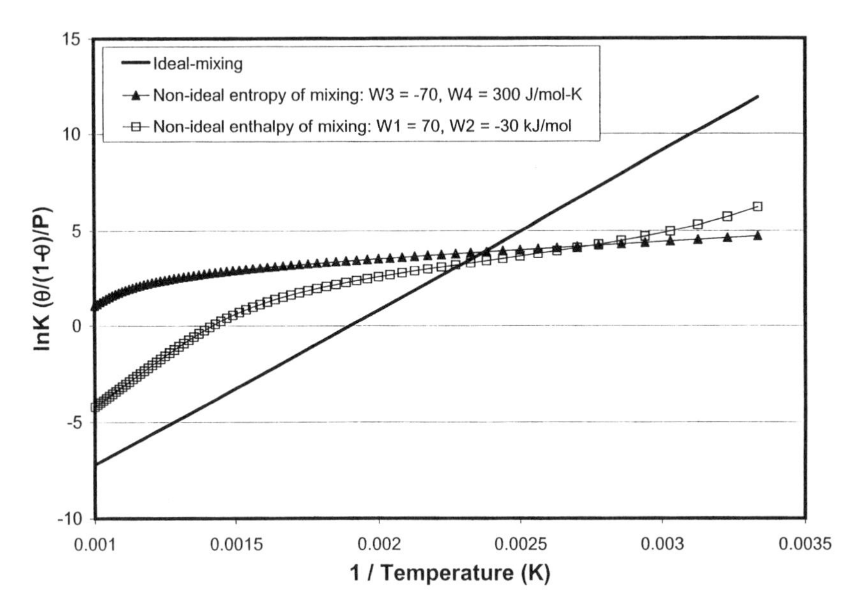

Figure 20. Synthetic thermogravimetric curves showing the effect of non-ideality, plotted as lnK (see Eqn. 13) vs. reciprocal temperature in Kelvin. The curves were calculated with Equation (12) and each assumes the same integral value of the Gibbs free energy and enthalpy of hydration (-30 and -70 kJ/mol, respectively; P = 0.03 bar). Linear trends indicate ideal mixing and non-linear curves demonstrate non-ideality.

11-19 to a lnK vs. 1/T(K) plot with lnK defined by Equation (13) (cf. Carey and Bish 1996). Figure 20 shows the results for three synthetic TGA curves having ideal-mixing, non-ideal enthalpy, and non-ideal entropy. Plotting the TGA data in this way provides a quick characterization of the zeolite-H_2O system: (1) linear behavior demonstrates ideal mixing; (2) curvature in the data requires non-ideal mixing terms; (3) the minimum number of non-ideal terms required is shown by the number of inflections in the curve; (4) at high temperature (low values of θ) the slope trends towards $\Delta H°$ (if $\Delta Cp = 0$); and (5) the $1/T = 0$ intercept is $\Delta S°$ (if $\Delta Cp = 0$).

In addition to these qualitative observations, the data in Figure 20 can be used quantitatively by regression to obtain the thermodynamic properties as described above for Equations (9) and (12). For experimental TGA data, this approach requires the assumption that the TGA data represent an equilibrium state of dehydration and that the effective H_2O pressure can be estimated. The assumption of equilibrium can only be rigorously tested in a reversed experiment. However, the following example shows that the assumption does hold well for clinoptilolite. In addition, the kinetics of dehydration are likely to be sufficiently rapid to preclude a significant lag except for rapid scan rates or large, packed samples. These factors can be accounted for by conducting experiments at a few scan rates and sample sizes to find where the system ceases to show evidence of kinetic factors. The pressure assumption is also not critical. For ideal systems, the equilibrium constant is independent of pressure. Thus, a variable pressure poses no difficulty in deriving thermodynamic properties from ideal systems.

As an example, Figure 21 shows TGA data for a Ca-clinoptilolite sample that is the same as that used by Carey and Bish (1996). The data were collected at a scan rate of 10°C/min from 25 to 1000°C (using a purge gas with about 0.003 bar H_2O-vapor

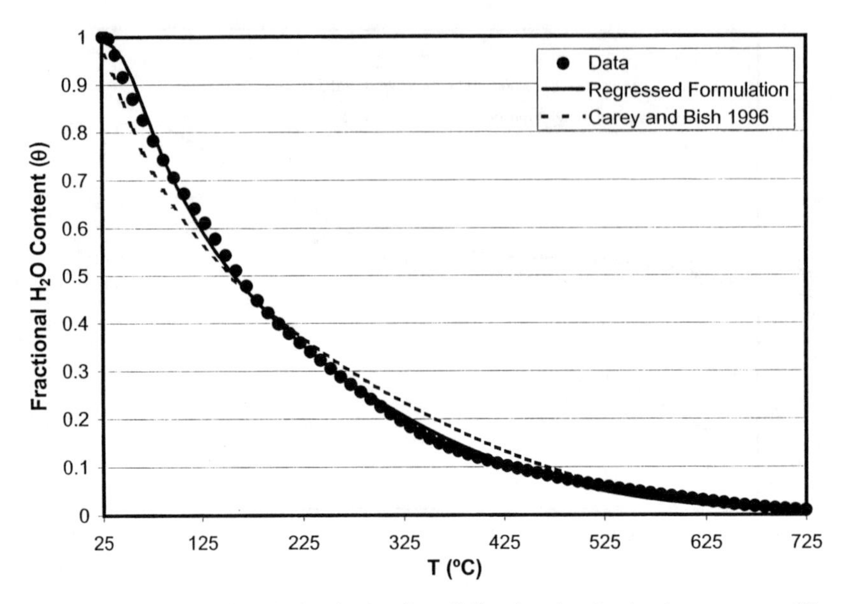

Figure 21. Thermogravimetric data for Ca-clinoptilolite plotted as fractional water content (θ) vs. temperature (°C). The plot shows the fit to the data obtained by regression of Equation (12) (P = 0.03 bar) and the predicted dehydration curve from the equilibrium study of Carey and Bish (1996; P = 0.03 bar).

pressure). The data were recast to lnK form by assuming that the initial H_2O content represented the maximum weight content and by assuming that the zeolite was completely dehydrated at 800°C. It was further assumed that the Ca-clinoptilolite dehydrated at equilibrium with an atmosphere of 0.03 bar. The lnK vs. 1/T plot (Fig. 22) shows that the data are not linear, suggesting non-ideal behavior. The data were regressed (ignoring the first three low-temperature values) using Equation (12) with the C (W_1) and D (W_2) parameters. The results of the regression are shown in Figures 21 and 22. The derived values in Figure 22 for Ca-clinoptilolite compare favorably with those obtained by Carey and Bish (1996) in an equilibrium gravimetric study. Note that the predicted TGA curves based on Carey and Bish (1996) confirm that Ca-clinoptilolite dehydrates at an effective H_2O pressure of 0.03 bar.

The success of this example should not be taken as proof that this method will work for all TGA data. The fit is not good at low temperatures where the TGA curve has higher values of θ than are predicted by the equilibrium studies. This may result from an initial kinetic lag in dehydration. At temperatures >125°C, the TGA data appear to be close in value to those predicted by Carey and Bish (1996). Note, however, that the TGA curve is more nearly linear than that calculated from Carey and Bish (1996; Fig. 22). A regression of the data ignoring values collected at <125°C failed to produce thermodynamic values that were close to Carey and Bish (1996), in part because this regression does not capture the curvature. Carey and Bish's (1996) results were used in Figures 21 and 22 to determine the effective "equilibrium" H_2O-vapor pressure of 0.03. This is nearly 100% RH at 25°C and was found to be a good approximation for many but not all TGA experiments examined with this method.

Figure 23 shows the results for a series of Ca-clinoptilolite samples derived by cation exchange from Castle Creek, Idaho, clinoptilolite. These data show the effect of differing

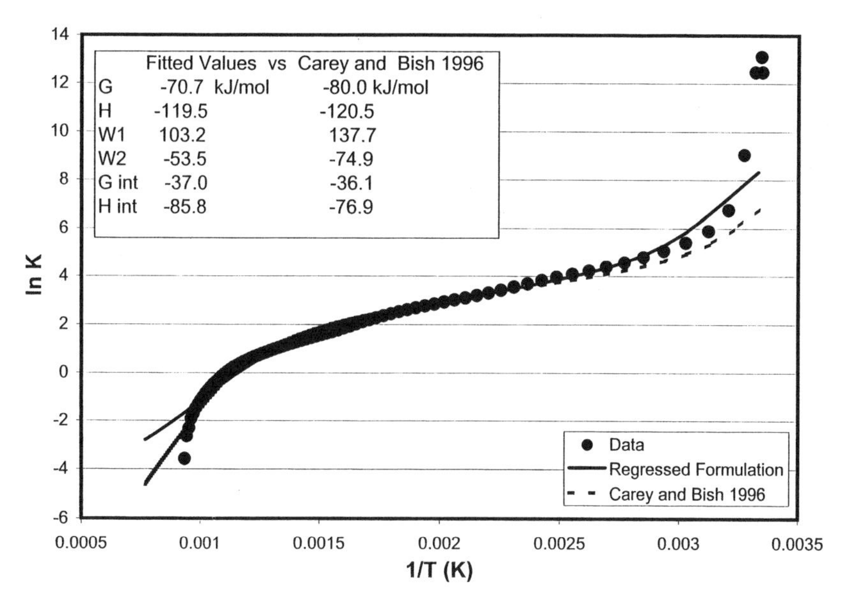

Figure 22. Thermogravimetric data for Ca-clinoptilolite plotted as lnK (see Eqn. 13; with P = 0.03 bar) vs. reciprocal temperature in Kelvin. The plot shows the fit to the data obtained by regression of Equation (12) and the predicted dehydration curve from the equilibrium study of Carey and Bish (1996; P = 0.03 bar). A comparison of the thermodynamic values obtained by regression and obtained by Carey and Bish is shown in an inset to the figure.

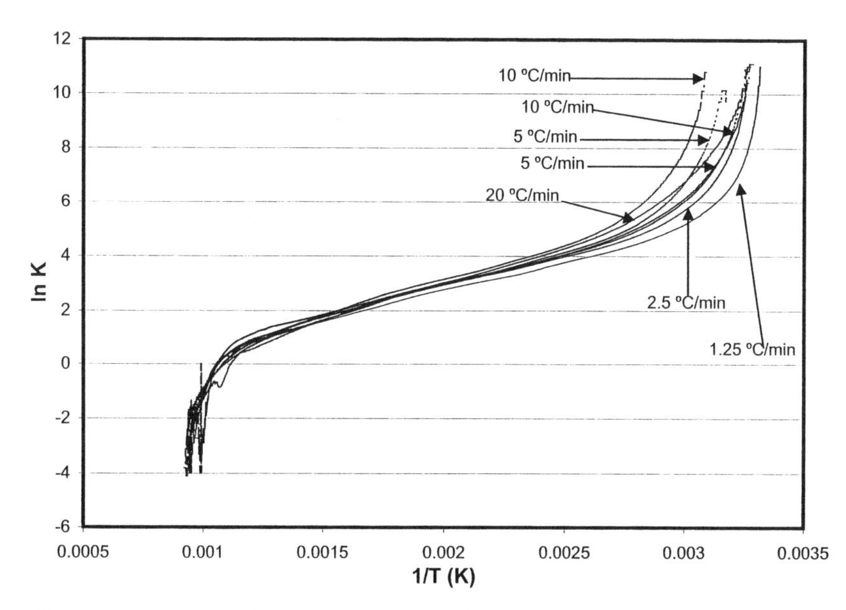

Figure 23. A series of TGA experiments on Ca-clinoptilolite at differing scan rates. The data are plotted as lnK (see Eqn. 13; with P = 0.03 bar) vs. reciprocal temperature in Kelvin. The scatter in the low-temperature data is reduced at high temperature. At the highest temperatures the data become noisy because of small H_2O contents.

scan rates on the lnK values (all collected under a dry N_2 purge gas). The data show that there is probably a low-temperature kinetic effect, but near 125°C the results are in substantial agreement. In particular, data at 2.5, 5, and 10°C/min show good consistency and may reflect a range of effective scan rates for making these calculations. The noise at high temperature (low values of lnK) shows the problems encountered as the zeolite nears complete dehydration. The integral Gibbs free energy of hydration and enthalpy of hydration derived at an H_2O-vapor pressure of 0.03 bar are -40.6 and -91.9 kJ/mol. Comparison with Table 4 suggests that these, like the results in Figure 22, are probably too negative.

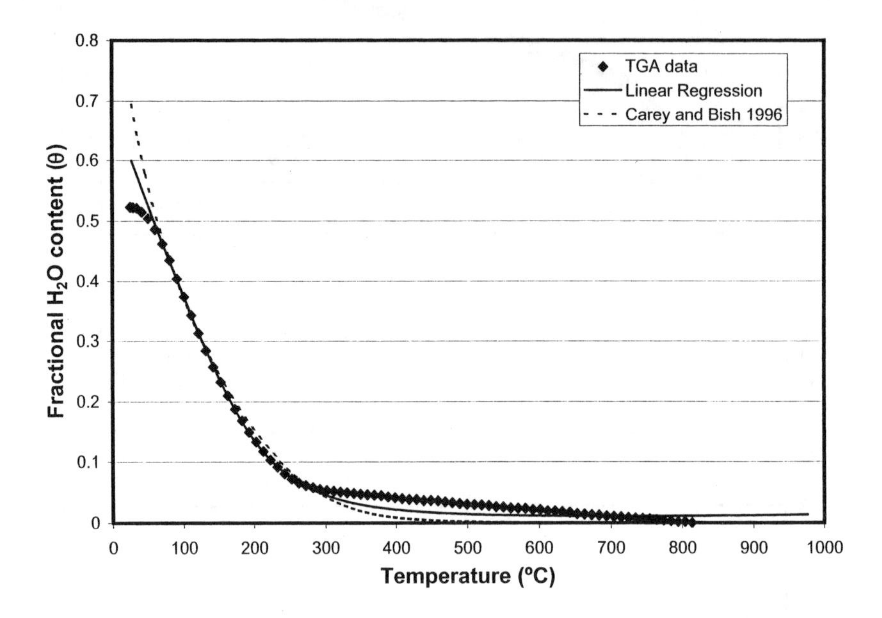

Figure 24. Thermogravimetric data for K-clinoptilolite plotted as fractional water content (θ) vs. temperature (°C). The plot shows the fit to the data obtained by regression of Equation (12) (P = 0.03 bar) and the predicted dehydration curve from the equilibrium study of Carey and Bish (1996; P = 0.03 bar).

The results must also be considered in terms of irreversible structural modifications. TGA data for K-clinoptilolite (the same material used by Carey and Bish 1996) are given in Figure 24 (collected at 10°C/min under an approximate H_2O-vapor pressure of 0.0006 bar). Here the curve predicted by Carey and Bish (1996) differs sharply from the TGA results at temperatures >550°C. In this case, the TGA data show greater H_2O retention than predicted by Carey and Bish (1996). This is most likely a result of structural modifications. In modeling these data, it is necessary to exclude the TGA data at temperatures greater than the irreversible structural modification. In addition, it was found that in order to have the low-temperature region agree with Carey and Bish (1996) an effective experimental pressure of 0.0025 bar was needed.

In summary, it appears that TGA data can be used at least in a semi-quantitative way to investigate the character of zeolite-H_2O systems and in favorable circumstances can be used to obtain quantitative measures. TGA data are readily recast into a lnK vs. 1/T form allowing the possibility of estimating thermodynamic values or at least constraining them. In addition, the application of a fixed H_2O-vapor pressure to TGA experiments creates the potential of extracting thermodynamic values at known conditions.

SUMMARY

The thermal behavior of zeolites is governed by complex interactions among the framework and the cations and water molecules that occur in the channel systems. It is conceptually useful to separate these effects into

(1) modifications of the framework including changes in the unit cell, structural collapse, and structural breakdown;

(2) interactions and movement of cations and H_2O within the channel system; and

(3) dehydration behavior. The first two effects are traditionally studied by X-ray or neutron diffraction methods and the third by thermal analysis.

Volume changes on heating differ among zeolites depending on the tetrahedral framework and the type of extraframework cation. Alberti and Vezzalini (1984) defined three categories of volumetric response to thermal dehydration:

Category-1 in which reversible dehydration occurs with little or no modification of the framework or unit-cell volume;

Category-2 in which complete or nearly complete reversible dehydration is accompanied by large distortion of the framework and significant decrease in unit-cell volume; and

Category-3 in which reversible dehydration at low temperature is usually accompanied by large modifications in the framework and is followed by irreversible changes due to breaking of T-O-T bonds prior to complete dehydration.

A single zeolite structural type may show differing categories of behavior particularly as a function of differing types of extraframework cations and the temperature to which it is heated.

Most zeolites undergo dehydration-induced reductions in volume, accommodated primarily within the channel systems, although minor thermal expansion can be seen with certain zeolites having robust, rigid frameworks (e.g. Category-1 zeolites such as erionite, chabazite, or yugawaralite) that do not necessarily collapse when H_2O molecules are removed. Changes in unit-cell parameters are often anisotropic due to anisotropy in the framework or channels. Thermally induced structural modifications are driven initially by loss of H_2O molecules and the concomitant migration of extraframework cations, both of which often give rise to significant contraction of the framework. Many structural changes begin as extraframework cations migrate in order to achieve optimal bonding when H_2O molecules leave the structure. Structural contraction is accommodated by rotations of the fundamental structural units. Given sufficiently severe compression and accompanying rotations, tetrahedral cation-oxygen linkages may be broken, particularly if strong interactions exist between extraframework cations and framework oxygen atoms. Breaking of tetrahedral cation-oxygen bonds not only gives rise to a marked irreversibility in the observed thermal transformations, it also explains the sluggish nature of some thermal reactions and the observed differences between the results of short- and long-term heating.

A variety of changes occur in the channel systems of zeolites during dehydration, including:

(1) extraframework cation-H_2O interactions, dominated by the hydration energy of the cation;

(2) H_2O-tetrahedral (framework) oxygen interactions;
(3) extra-framework cation-tetrahedral oxygen interactions;
(4) extraframework cation movement and consequent strain on T-O-T bridges; and
(5) H^+ attack on T-O-T bridges and consequent breakage.

These interactions result in modification of site occupancies as well as the movement of extraframework cations and the remaining H_2O to energetically more favorable sites. As a consequence, dehydration behavior is strongly dependent on the nature of the extraframework cation and the sizes of the structural channels. A variety of structural studies have shown that dehydration in many zeolites is accompanied by relocation of extraframework cations and the remaining H_2O

The amount of H_2O in most if not all zeolites varies continuously as a function of temperature and pressure until the point of irreversible structural change. Such H_2O is termed *zeolitic* and is characteristic of the dehydration behavior of zeolites. Some zeolites may have *hydrate* H_2O in which H_2O is lost at a unique temperature for a given pressure (perhaps laumontite). However, there are no studies that have definitively established true hydrate behavior in a zeolite.

The temperature and pressure of dehydration is governed by the energetics of the H_2O in the zeolite channels. In many cases, the H_2O molecules are held with a range of semi-discrete energies arising from their interactions with extraframework cations in a variety of diverse crystallographic sites. For example, H_2O molecules coordinating Ca^{2+} are held more strongly and to higher temperatures than those coordinating K^+ ions due to the higher hydration energy of Ca^{2+}. Likewise, an H_2O molecule coordinating a given cation on the M1 site in clinoptilolite will behave differently from an H_2O molecule coordinating the same cation on M2, M3, or M4. In addition, as dehydration proceeds the environment of the H_2O molecules changes as extraframework cations and H_2O molecules migrate so that the energetics of zeolite-water interactions change constantly during dehydration. Therefore, one must consider the instantaneous zeolite structure under any given condition in order to evaluate the thermal behavior; e.g. a structural change taking place at elevated temperatures or under reduced H_2O vapor pressure cannot be understood based on the room-temperature structure.

A thermodynamic model of zeolitic H_2O based on a modified form of Langmuir sorption is capable of modeling and predicting the dehydration behavior of most zeolites. The model is readily applied to equilibrium studies of zeolite-H_2O systems and allows the extraction of thermodynamic properties by simple linear regression. The model may also be applied to TGA data and can facilitate semi-quantitative comparisons of the dehydration behavior of many zeolites.

The thermodynamic model combined with experimental data has several implications. There is no "magic" temperature at which a zeolite will dehydrate (e.g. the boiling point of water). If the water-vapor pressure surrounding a zeolite is changed at constant temperature, the zeolite will change its water content accordingly. The many observations that a small change in pressure or temperature causes a small change in water content in a zeolite show that the zeolite water content varies continuously with temperature and pressure. This contrasts with the situation for pure water or for a true hydrate such as gypsum. If water and vapor are coexisting at equilibrium (i.e. conditions on the boiling curve) and the temperature and/or pressure are changed by a small amount and held there, then all of the water must either evaporate or condense (depending on the path taken from the boiling curve). The most important single variable in zeolite dehydration reactions is the relative humidity. This is because relative humidity is a combined function of temperature and water-vapor pressure. Constant relative humidity processes will tend to result in relatively small changes in H_2O content of zeolites. An example process is the heating of zeolites along the boiling curve (both temperature and H_2O pressure increase and zeolite dehydration is minimized). On the other hand, processes that occur with changing relative humidity will either strongly hydrate or dehydrate the zeolite. An example process is the heating of zeolites in a TGA experiment in which the H_2O pressure stays relatively constant and the zeolite dehydrates rapidly.

Although many studies investigate the thermal behavior of zeolites from either a structural or a thermal-analysis approach, a complete understanding requires an integrated view of all of these processes. Significant recent progress has been made in this regard through the development of integrated thermal analysis-diffraction studies (e.g. Bish and Carey 2000) and *in situ*, time-resolved diffraction studies of thermal processes in zeolites (e.g. Ståhl et al. 1996 and Artioli et al. 2001).

SUGGESTIONS FOR FURTHER WORK

The thermal behavior of natural zeolites has been extensively studied, but it is remarkable that little data exist for zeolites under H_2O pressures above ~0.03 bar (~100% RH at 25°C). Conditions between ~0.03 bar and saturation are experimentally difficult to achieve, but calorimetric and structural studies in this $P(H_2O)$ regime should prove very useful and would be more applicable to understanding the behavior of zeolites in saturated systems. Additional general calorimetric data on the energetics of interactions occurring in zeolite extraframework sites would, of course, also be valuable.

Much progress has been made in the development of thermodynamic models of dehydration and hydration, but new data are needed to determine whether water is essential to the formation of zeolites. In other words, we still do not know whether zeolites have a true stability field relative to feldspar-quartz assemblages. This question is difficult to answer because of the experimental difficulty of either measuring complete dehydration without structural collapse (e.g. gravimetric, phase-equilibria studies) or of separating the energetics of structural collapse from a complete dehydration reaction (e.g. transposed-temperature drop calorimetry). As suggested by Barrer and Cram (1971), a study of a single zeolite by the full variety of available methods may allow the separation of the energetics of structural collapse from H_2O stabilization.

There is a suggestion that some zeolites, e.g. laumontite, show true hydrate-like dehydration behavior such as that exhibited by gypsum. This hypothesis could be tested by careful equilibrium and structural measurements that determine whether H_2O is truly lost at a univariant reaction. In addition, it would be useful to explore the relationship between reversible step-like behavior and the structural collapse reactions in some zeolites that appear to coincide with a rapid phase of H_2O loss (e.g. heulandite).

ACKNOWLEDGMENTS

We are grateful to S. Chipera for continued assistance with heating experiments, to M. Gunter for original copies of Figure 6, and to D. Ming and R. Pabalan for constructive reviews.

REFERENCES

Alberti A (1973) The structure type of heulandite B (heat-collapsed phase). Tschermaks mineral petrogr Mitt 19:173-184

Alberti A, Vezzalini G (1978) Crystal structures of heat-collapsed phases of barrerite. *In* Natural Zeolites. Occurrence, Properties, Use. Sand LB, Mumpton FA (eds) Pergamon Press, Elmsford, New York, p 85-98

Alberti A, Vezzalini G (1983a) The thermal behavior of heulandites: A structural study of the dehydration of Nadap heulandite. Tscher mineral petrogr Mitt 31:259-270

Alberti A, Vezzalini G (1983b) How the structure of natrolite is modified through the heating-induced dehydration. Neues Jahrb Mineral Mh 135-144

Alberti A, Vezzalini G (1984) Topological changes in dehydrated zeolites: Breaking of T-O-T bridges. *In* Proc Sixth Int'l Zeolite Conf, Reno, Nevada, 1983. D Olson, A Bisio (eds) Butterworths, Guildford, UK, p 834-841

Alberti A, Rinaldi R, Vezzalini G (1978) Dynamics of dehydration in stilbite-type structures; stellerite phase B. Phys Chem Minerals 2:365-375

Armbruster T (1993) Dehydration mechanism of clinoptilolite and heulandite: Single-crystal X-ray study of Na-poor, Ca-, K-, Mg-rich clinoptilolite at 100 K: Am Mineral 78:260-264

Armbruster T, Gunter, ME (1991) Stepwise dehydration of heulandite-clinoptilolite from Succor Creek, Oregon, U.S.A.: A single-crystal X-ray study at 100 K. Am Mineral 76:1872-1883

Armbruster T, Kohler T (1992) Rehydration and dehydration of laumontite: A single-crystal X-ray study at 100 K. Neues Jahrb Mineral Mh 385-397

Artioli G (1997) *In situ* powder diffraction studies of temperature induced transformations in minerals. Nucl Instr Meth Phys Res B 133:45-49

Artioli G, Ståhl K, Cruciani G, Gualtieri A, Hanson JC (2001) *In situ* dehydration of yugawaralite. Am Mineral 86:185-192

Balgord WD, Roy R (1973) Crystal chemical relationships in the analcite family. II. Influence of temperature and P_{H_2O} on structure. Adv Chem Ser 121:189-199

Barrer RM (1978) Zeolites and Clay Minerals as Sorbents and Molecular Sieves. Academic Press, London, 497 p

Barrer RM, Cram PJ (1971) Heats of immersion of outgassed and ion-exchanged zeolites. *In* EM Flanigen, LB Sand (eds) Molecular Sieve Zeolites-II, p 105-131. American Chemical Society, Washington, DC

Baur WH, Joswig W, Fursenko BA, Belitsky IA (1997) Symmetry reduction of the aluminosilicate framework of LAU topology by ordering of exchangeable cations: The crystal structure of primary leonhardite with a primitive Bravais lattice. Eur J Mineral 9:1173-1182

Belitsky IA, Fursenko BA, Gabuda SP, Kholdeev OV, Seryotkin YuV (1992) Structural transformations in natrolite and edingtonite. Phys Chem Minerals 18:497-505

Bischke SD, Chemburkar R, Brown LF, Travis BJ, Bish DL (1988) The extraction of site-energy distributions from temperature-programmed desorption spectra. *In* Proc Am Inst Chem Eng Ann Meet, Washington, DC, Nov. 28-Dec. 2, 1988, p 6

Bish DL (1984) Effects of exchangeable cation composition on the thermal expansion/contraction of clinoptilolite: Clays Clay Minerals 32:444-452

Bish DL (1988) Effects of composition on the dehydration behavior of clinoptilolite and heulandite: in Occurrence, Properties and Utilization of Natural Zeolites. D Kallo, HS Sherry (eds) Akademiai Kiado, Budapest, p 565-576

Bish DL (1989) Determination of dehydration behavior of zeolites using Rietveld refinement and high-temperature X-ray diffraction data: Geol Soc Am Ann Meet Abstr with Program 21:A73

Bish DL (1990a) Long-term thermal stability of clinoptilolite: The development of a "B" phase: Eur J Mineral 2:771-777

Bish DL (1990b) Thermal stability of zeolitic tuff from Yucca Mountain, Nevada: High Level Radioactive Waste Management, Proc First Int'l High-Level Radioactive Waste Management Conf 1:596-602

Bish DL, Carey JW (2000) Coupled X-ray powder diffraction and thermogravimetric analysis of clinoptilolite dehydration behavior. *In* Natural Zeolites for the Third Millennium. Colella C, Mumpton FA (eds) De Freda Editore, Naples, p 249-257

Butikova IK, Shepelev YF, Smolin YI (1993) Structure of hydrated and dehydrated (250°C) forms of Ca-chabazite. Crystallogr Rep 38:461-463

Carey JW (1993) The heat capacity of hydrous cordierite above 295 K. Phys Chem Minerals 19:578-583

Carey JW (1995) A thermodynamic formulation of hydrous cordierite. Contrib Mineral Petrol 119:155-165

Carey JW, Bish DL (1996) Equilibrium in the clinoptilolite-H_2O system. Am Mineral 81:952-962

Carey JW, Bish DL (1997) Calorimetric measurement of the enthalpy of hydration of clinoptilolite. Clays Clay Minerals 45:826-833

Carey JW, Navrotsky A (1992) The molar enthalpy of dehydration of cordierite. Am Mineral 77:930-936

Carvalho AP, Brotas de Carvalho M, Pires J (1997) Degree of crystallinity of dealuminated offretites determined by X-ray diffraction and by a new method based on nitrogen adsorption. Zeolites 19: 382-386

Chipera SJ, Bish DL (1991) Rehydration behavior of a natural analcime, *In* Program & Abstracts, Clay Minerals Soc, 28th Ann Meet, Houston, Texas, p 29

Colella C (1999) Use of thermal analysis in zeolite science and applications. *In* Characterization Techniques of Glasses and Ceramics. J Ma Rincon, M Romero (eds) Springer-Verlag, Berlin-Heidelberg, p 112-137

Coombs DS, Alberti A, Armbruster T, Artioli G, Colella C, Galli E, Grice JD, Liebau F, Mandarino JA, Minato H, Nickel EH, Passaglia E, Peacor DR, Quartieri S, Rinaldi R, Ross M, Sheppard RA, Tillmanns E., Vezzalini G (1997) Recommended nomenclature for zeolite minerals: Report of the Subcommittee on Zeolites of the International Mineralogical Association, Commission on New Minerals and Mineral Names. Can Mineral 35:1571-1606

Coughlan B, Carroll WM (1976). Water in ion-exchanged L, A, X, and Y zeolites: a heat of immersion and thermogravimetric study. J Chem Soc–Faraday Trans I 72:2016-2030

Cruciani G, Artioli G, Gualtieri A, Ståhl K, Hanson JC (1997) Dehydration dynamics of stilbite using synchrotron X-ray powder diffraction. Am Mineral 82:729-739

dé Gennaro M, Colella C (1989) Use of thermal analysis for the evaluation of zeolite content in mixtures of hydrated phases. Thermochim Acta 154:345-353

Dubinin MM, Kadlec O, Zulal A (1966). Adsorption equilibria of water on NaX zeolite. Coll Czech Chem Comm 31:406-414

Dubinin MM, Astakhov VA (1971). Description of adsorption equilibria of vapors on zeolites over wide ranges of temperature and pressure. *In* EM Flanigen, LB Sand (eds) American Chemical Society, Washington, DC, Molecular Sieve Zeolites-II 102:69-85

Drebushchak UA (1990) Thermogravimetric investigation of the phase transition zeolite heulandite at dehydration. Thermochim Acta 159:377-381

Garcia J, Gonzales M, Caceres J, Notario J (1992) Structural modifications in phillipsite-rich tuff induced by thermal treatment: Zeolites 12:664-669

Gottardi G, Galli E (1985) Natural Zeolites. Springer-Verlag, Berlin, 409 p

Guggenheim EA (1967). Thermodynamics: An Advanced Treatment for Chemists and Physicists. North-Holland, Amsterdam, 390 p

Hambley TW, Taylor JC (1984) Neutron diffraction studies on natural heulandite and partially dehydrated heulandite: J Solid State Chem 54:1-9

Johnson GK, Flotow HE, O'Hare PAG, Wise WS (1982) Thermodynamic studies of zeolites: analcime and dehydrated analcime. Am Mineral 67:736-748

Johnson GK, Tasker IR, Jurgens R, O'Hare PAG (1991) Thermodynamic studies of zeolites: clinoptilolite. J Chem Thermodyn 23:475-484

Johnson GK, Tasker IR, Flotow HE, O'Hare PAG, Wise WS (1992) Thermodynamic studies of mordenite, dehydrated mordenite, and gibbsite. Am Mineral 77:85-93

Kiseleva I, Navrotsky A, Belitsky I, Fursenko B (1996a) Thermochemistry and phase equilibria in calcium zeolites. Am Mineral 81:658-667

Kiseleva I, Navrotsky A, Belitsky I, Fursenko B (1996b) Thermochemistry of natural potassium sodium calcium leonhardite and its cation-exchanged forms. Am Mineral 81:668-675

Kiseleva I, Navrotsky A, Belitsky I, Fursenko B (2001) Thermochemical study of calcium zeolites —heulandite and stilbite. Am Mineral 86:448-455

Knowlton GD, White TR., McKague HL (1981) Thermal study of types of water associated with clinoptilolite: Clays Clay Minerals 29:403-411

Koyama K, Takéuchi Y (1977) Clinoptilolite: The distribution of potassium atoms and its role in thermal stability: Z Kristallogr 145:216-239

Milligan WO, Weiser HB (1937) The mechanism of the dehydration of zeolites: J Phys Chem 41: 1029-1040

Mumpton FA (1960) Clinoptilolite redefined: Am Mineral 45:351-369

O'Donovan AW, O'Connor CT, Koch KR (1995) Effect of acid and steam treatment of Na-mordenite and H-mordenite on their structural, acidic and catalytic properties. Micropor Mater 5:185-202

Pabalan RT (1994) Thermodynamics of ion exchange between clinoptilolite and aqueous solutions of Na^+/K^+ and Na^+/Ca^{2+}. Geochim Cosmochim Acta 58:4573-4590

Passaglia E (1980) The heat behavior of cation exchanged zeolites with the stilbite framework: Tschermaks mineral petrogr Mitt 27:67-78

Petrova N, Mizota T, Fugiwara K (2001) Hydration heats of zeolites for evaluation of heat exchangers. J Therm Anal Calor 64:157-166

Petrovic I, Navrotsky A (1997) Thermochemistry of Na-faujasites with varying Si/Al ratios. Micropor Mater 9:1-12

Robie RA. Hemingway BS, Fisher JR (1979) Thermodynamic Properties of Minerals and Related Substances at 298.15 K and 1 Bar (10^5 Pascals) Pressure and at Higher Temperature. U S Geol Surv Bull 1452, 456 p

Rykl D, Pechar F (1991) Thermal decomposition of natural phillipsite: Zeolites 11:680-683

Shim S-H, Navrotsky A., Gaffney TR, MacDougall JE (1999) Chabazite: Energetics of hydration, enthalpy of formation, and effect of cations on stability. Am Mineral 84:1870-1882

Simonot-Grange MH, Cointot A (1969) Evolution des proprietes d'adsorption de l'eau par la heulandite en relation avec la structure cristalline. Bull Soc Chim France 421-427

Simonot-Grange M-H, Watelle-Marion G, Cointot A (1968) Caractères physico-chimiques de l'eau dans la heulandite. Étude diffractométrique des phases observées au cours de la dés hydratation et de la réhydratation: Bull Soc Chim France 7:2747-2754

Ståhl K, Artioli G, Hanson JC (1996) The dehydration process in the zeolite laumontite: a real-time synchrotron X-ray powder diffraction study. Phys Chem Minerals 23:328-336

Ståhl K, Hanson J (1994) Real-time X-ray synchrotron powder diffraction studies of the dehydration process in scolecite and mesolite. J Appl Crystallogr 27:543-550

Stolz J, Armbruster T (1997) X-ray single-crystal structure refinement of Na,K-rich laumontite, originally designated 'primary leonhardite.' Neues Jahrb Mineral Mh 131-144

Tarasevich YI, Polyakov VE, Badekha LI (1988) Structure and localization of hydrated alkali, alkaline earth and transition metal cations in clinoptilolite determined from ion-exchange, calorimetric and spectral measurements. *In* D Kallo, HS Sherry (eds) Occurrence, Properties and Utilization of Natural Zeolites. Akademiai Kiado, Budapest, p 421-430

Valueva G (1995) Dehydration behaviour of heulandite-group zeolites as a function of their chemical composition. Eur J Mineral 7:1411-1420

Van Reeuwijk LP (1974) The thermal dehydration of natural zeolites: Meded Landbouwhogeschool Wageningen 74:1-88

Vezzalini G, Quartieri S, Alberti A (1993) Structural modifications induced by dehydration in the zeolite gismondine. Zeolites 13:34-42

Wilkin RT, Barnes HL (1999). Thermodynamics of hydration of Na- and K-clinoptilolite to 300°C . Phys Chem Minerals 26:468-476

Yamanaka S, Malla PB, Komarneni S (1989) Water sorption and desorption isotherms of some naturally occurring zeolites. Zeolites 9:18-22

Yamazaki A, Otsuka R (1986) The thermal behavior of K-exchanged forms of natrolite. Thermochim Acta 109:237-242

Yang S, Navrotsky A (2000) Energetics of formation and hydration of ion-exchanged zeolite Y. Microporous Mesoporous Mater 37:175-186

Yang S, Navrotsky A, Wilkin, R (2001) Thermodynamics of ion-exchanged and natural clinoptilolite. Am Mineral 86:438-447

14 Cation-Exchange Properties of Natural Zeolites

Roberto T. Pabalan and **F. Paul Bertetti**

Center for Nuclear Waste Regulatory Analyses
Southwest Research Institute
6220 Culebra Road
San Antonio, Texas 78238

INTRODUCTION

Zeolite minerals are crystalline, hydrated aluminosilicates of alkali and alkaline earth cations characterized by an ability to hydrate/dehydrate reversibly and to exchange some of their constituent cations with aqueous solutions, both without a major change in structure. Because of their ion-exchange, adsorption, and molecular sieve properties, as well as their geographically widespread abundance, zeolite minerals have generated worldwide interest for use in a broad range of applications. Examples of these applications are discussed in other chapters of this book. Of particular interest in this chapter are the cation-exchange properties of zeolite minerals. Due to the favorable ion-exchange selectivity of natural zeolites for certain cations, such as Cs^+, Sr^{2+}, and NH_4^+, these minerals have been studied for potential use in the treatment of nuclear wastewaters (Howden and Pilot 1984; Baxter and Berghauser 1986; Robinson et al. 1995; Pansini 1996), municipal and industrial wastewaters (Kallo 1995; Pansini 1996), and acid mine drainage waters (Bremner and Schultze 1995; Zamzow and Schultze 1995). Natural zeolites have also been studied for potential use in the remediation of sites contaminated with fission products such as ^{90}Sr and $^{135,137}Cs$ (Leppert 1988; Valcke et al. 1997a; Valcke et al. 1997b) and in the remediation of soils contaminated with heavy metals (Ming and Allen, this volume). Additional interest resulted from the potential siting of a high-level nuclear waste repository at Yucca Mountain, Nevada, which is underlain by diagenetically altered, zeolite-rich (clinoptilolite, heulandite, and mordenite) rhyolitic tuffs (Broxton et al. 1986; Broxton et al. 1987) that could serve as barriers to radionuclide migration to the accessible environment.

Zeolites consist of three-dimensional frameworks of $(Si,Al)O_4$ tetrahedra where all oxygen ions of each tetrahedron are shared with adjacent tetrahedra. The presence of Al^{3+} in place of Si^{4+} in the structure gives rise to a deficiency of positive charge in the framework. The net negative charge is balanced by cations, principally Na^+, K^+, and Ca^{2+}, less frequently Li^+, Mg^{2+}, Sr^{2+}, and Ba^{2+}, which are situated in cavities within the structure. Zeolite structures are remarkably open, and void volumes of dehydrated zeolites of almost 50% are known.

The following idealized general formula for natural zeolites has been proposed (Gottardi 1978; Gottardi and Galli 1985):

$$(M_x^+, M_y^{2+})\ [Al_{(x+2y)}\ Si_{n-(x+2y)}O_{2n}] \cdot mH_2O\ ,$$

where M^+ represents monovalent cations with stoichiometry x, and M^{2+} represents divalent cations with stoichiometry y. Cations within the first set of parentheses are the exchangeable cations. Those within the brackets are the structural cations because, with oxygen, they make up the framework of the structure. The value of m gives the number of water molecules in the structure and provides an idea of the volume of the channels relative to the total volume. Normally, this number does not exceed half the number of framework oxygens, and $n/2 < m < n$ (Gottardi 1978). The (Si + Al):O ratio of a zeolite is

1529-6466/00/0045-0014$10.00

1:2, and the number of tetrahedral Al^{3+} is equal to the sum of positive charges (x + 2y) of the exchangeable cations in the idealized formula. Thus, the theoretical cation-exchange capacity (CEC) is primarily a function of the charge density of the anionic structure, i.e. the degree of substitution of Al^{3+} for Si^{4+} in its tetrahedral framework. The greater the Al^{3+} substitution, the more cations are needed to maintain electrical neutrality and, hence, the higher the CEC. For example, a study by Zamzow et al. (1990) on heavy metal sorption on natural zeolites showed a linear correlation between the Si/Al ratio of the zeolite and the ion-exchange capacity for Pb^{2+}, in the order phillipsite > chabazite > erionite > clinoptilolite > mordenite. The expected CECs for several zeolite minerals based on their isomorphic substitution are listed in Table 1.

Table 1. Cation-exchange capacity (CEC) of zeolite minerals based on the number of equivalents of exchangeable cations or the number of moles of Al^{3+} in the chemical formula. Values are given in milliequivalents per gram of solid (meq/g).

Zeolite	*Typical Unit-Cell Formula**	*CEC (meq/g)*
Analcime	$Na_{16}(Al_{16}Si_{32}O_{96})\cdot 16H_2O$	4.5
Chabazite	$Ca_2(Al_4Si_8O_{24})\cdot 12H_2O$	3.9
Clinoptilolite	$(Na,K)_6(Al_6Si_{30}O_{72})\cdot 20H_2O$	2.2
Erionite	$NaK_2MgCa_{1.5}(Al_8Si_{28}O_{72})\cdot 28H_2O$	2.8
Faujasite	$Na_{20}Ca_{12}Mg_8(Al_{60}Si_{132}O_{384})\cdot 235H_2O$	3.6
Ferrierite	$(Na,K)Mg_2Ca_{0.5}(Al_6Si_{30}O_{72})\cdot 20H_2O$	2.3
Heulandite	$(Na,K)Ca_4(Al_9Si_{27}O_{72})\cdot 24H_2O$	3.2
Laumontite	$Ca_4(Al_8Si_{16}O_{48})\cdot 16H_2O$	4.3
Mordenite	$Na_3KCa_2(Al_8Si_{40}O_{96})\cdot 28H_2O$	2.2
Natrolite	$Na_{16}(Al_{16}Si_{24}O_{80})\cdot 16H_2O$	5.3
Phillipsite	$K_2(Ca_{0.5},Na)_4(Al_6Si_{10}O_{32})\cdot 12H_2O$	4.5
Wairakite	$Ca_8(Al_{16}Si_{32}O_{96})\cdot 16H_2O$	4.6

*taken from Gottardi and Galli (1985)

The exchangeable cations of a zeolite are only loosely held in the anionic framework and, to a first approximation, can be removed or exchanged easily by washing the zeolite with a concentrated solution of another cation. In practice, the ion-exchange behavior of zeolites also depends on other factors, including: (1) the framework topology (channel configuration and dimensions); (2) ion size and shape (polarizability); (3) charge density of the anionic framework; (4) ionic charge; and (5) concentration of the external electrolyte solution (Barrer 1978). The diffusion character of a zeolite depends on the number of channels and their spatial configuration.[1] All other factors remaining equal, cations diffuse faster through zeolites with three-dimensional channel systems than those

[1] The kinetics of ion exchange is controlled by diffusion of ions within the crystal structure. This topic is outside the scope of this chapter, but extensive literature on the subject is available. A review of ion-exchange kinetics is given by Helfferich and Hwang (1991). Other useful references include Brooke and Rees (1968, 1969), Breck (1974), and Barrer (1980).

with one- or two-dimensional channel systems. The size of the ion, as well as the channel dimensions, determine whether or not a given cation will fit into a particular framework. For example, chabazite exhibits complete exclusion of the ions La^{3+}, $(CH_3)_4N^+$, and $(C_2H_5)_4N^+$, due to the large size of these cations (Breck 1974). Analcime, which has a dense structure and nonintersecting channels and cavities, exhibits very limited ion exchange of its Na^+ for other cations, particularly at room temperature. To some extent, high temperatures can offset the effect of large ionic radius. Thus, at elevated temperatures, Na^+ in analcime can be completely replaced by K^+, Ag^+, Tl^+, NH_4^+, and Rb^+ (Vaughn 1978).

Size considerations and the effect of the rigid nature of the zeolite framework (i.e. the nearly fixed pore volumes) can also explain the steric limitations and ion sieve properties exhibited by zeolites, such as those observed by Barrer et al. (1967). In their studies on ion exchange between Na-clinoptilolite and various alkyl-ammonium cations, Barrer et al. (1967) observed that ions small enough to enter the two main channels of clinoptilolite [e.g. NH_4^+, $CH_3NH_3^+$, $C_2H_5NH_3^+$, $(CH_3)_2NH_2^+$, and n-$C_3H_7NH_3^+$] exchange completely with Na^+. Those small enough to enter the 10-ring channel, but too large to penetrate the 8-ring channel, are only partially exchanged [e.g. $(CH_3)_3NH^+$, n-$C_4H_9NH_3^+$, and iso-$C_3H_7NH_3^+$], whereas the largest ions [e.g. $(CH_3)_4$ NH_3^+ and tert-$C_4H_9NH_3^+$] are totally excluded.

Many zeolites contain several crystallographically distinct sets of sites that can be occupied by exchangeable cations, and each site may exhibit different selectivities and ion-exchange behavior. The number of available exchange sites commonly exceeds the number of negative charges to be neutralized. Hence, the anionic charge of the framework may be neutralized when only some of the sites are occupied or when some of the sites are only partially occupied, and the occupancy factors may vary with the nature of the neutralizing cation (Barrer 1980, 1984). In addition, the entering ion does not necessarily take the position of the leaving ion (Sherry 1971; Cremers 1977). Thus, zeolites may exhibit a high degree of cationic disorder, both in terms of unoccupied sites and in terms of different distributions of cations of different kinds among the available sites.

This site heterogeneity in the zeolite is likely to manifest itself in compositional variations of the selectivities and activity coefficients of the zeolite components. As a consequence, it is difficult to predict multicomponent equilibrium exchange relations from binary data alone (Fletcher et al. 1984). Although it is theoretically possible to evaluate the contribution of the component site groups to the overall thermodynamics of exchange and to the overall equilibrium constant (Barrer 1978), Townsend (1984) questioned whether one should use measurements of ion-exchange equilibria to infer details of a particular heterogeneous site model for the exchanger, unless independent measurements that provide information on the structure and site heterogeneity are also applied to the material (e.g. X-ray diffraction, neutron diffraction, nuclear magnetic resonance, electron microscopy). Thermodynamic measurements are concerned with changes in macroscopic physical properties of the system under study, and it is difficult to infer from thermodynamic data alone the fundamental mechanisms that underlie the observed ion-exchange behavior.

Nevertheless, thermodynamic formulations, if properly conceived, provide firm and systematic bases for understanding ion-exchange behavior and its dependence on various parameters, and these formulations serve as tools for predicting exchange equilibria under conditions not previously studied. The basic thermodynamic formulations for ion exchange are based on principles developed long ago by researchers on inorganic exchange materials, especially clays (Vanselow 1932; Gapon 1933; Kielland 1935;

Gaines and Thomas 1953). These formulations are still widely used in current ion-exchange literature, irrespective of the nature of the exchanger under study. A review of the thermodynamics of ion exchange is given in the following section.

THERMODYNAMICS OF ION EXCHANGE

Ion-exchange isotherm

For a binary ion exchange involving cations A^{z_A+} and B^{z_B+}, the basic reaction may be written as

$$z_B A^{z_A+} + z_A BL_{z_B} \Leftrightarrow z_A B^{z_B+} + z_B AL_{z_A} \ , \tag{1}$$

where z_A+ and z_B+ are the valences of the respective cations, and L is defined as a portion of zeolite framework holding unit negative charge.[2] Anions are also present in the aqueous solution and maintain electroneutrality in that phase.[3]

The binary exchange equilibrium can be described conveniently by an ion-exchange isotherm, which is a plot of the equilibrium concentration of an exchanging ion in solution against the equilibrium concentration of that same ion in the zeolite at constant temperature and solution normality. The isotherm is usually plotted in terms of the equivalent cation fraction of the ion in solution against that in the solid (Dyer et al. 1981). The equivalent cationic fractions of A^{z_A+} and B^{z_B+} ($\bar{E}_A$ and $\bar{E}_B$, respectively) in the zeolite phase are defined as

$$\bar{E}_A = \frac{z_A \bar{n}_A}{z_A \bar{n}_A + z_B \bar{n}_B}; \quad \bar{E}_B = \frac{z_B \bar{n}_B}{z_A \bar{n}_A + z_B \bar{n}_B} \ , \tag{2}$$

where $\bar{n}_A$ and $\bar{n}_B$ are the number of moles of ions A^{z_A+} and B^{z_B+}, respectively, in the zeolite. Equivalent cationic fractions in the aqueous solution can be defined similarly as:

$$E_A = \frac{z_A n_A}{z_A n_A + z_B n_B}; \quad E_B = \frac{z_B n_B}{z_A n_A + z_B n_B} \ , \tag{3}$$

where n_A and n_B are the number of moles of A^{z_A+} and B^{z_B+}, respectively, in the aqueous phase.

The ion-exchange isotherm can then be plotted from the equilibrium values of $\bar{E}_A$ and E_A (or $\bar{E}_B$ and E_B). For the binary exchange reaction shown in Equation (1),

$$E_B = 1 - E_A; \quad \bar{E}_B = 1 - \bar{E}_A \ . \tag{4}$$

Thus, the isotherm fully defines the equilibrium at a specified temperature and solution normality.[4]

---

[2]Other studies represent the ion-exchange reaction as Equation (1b),

$$z_B A^{z_A+} + z_A z_B B_{(1/z_B)} L \Leftrightarrow z_A B^{z_B+} + z_A z_B A_{(1/z_A)} L \tag{1b}$$

which can be traced to the work by Gapon (1933), is an equally valid way of expressing the binary exchange reaction (Townsend 1986). However, Equations (1) and (1b) have different reaction stoichiometries, resulting in different ion-exchange equilibrium constants and thermodynamic functions. These differences should be kept in mind when interpreting published data on ion exchange. Detailed discussions of these alternate forms are provided by Townsend (1986) and Grant and Fletcher (1993).

[3]In systems where the cation exhibits a strong tendency to form aqueous complexes with some anions in solution, the selectivity of the zeolite for the cation is strongly dependent on the nature of the co-anion. See, for example, Loizidou and Townsend (1987a).

[4]Equations for calculating equivalent cationic fractions and the associated uncertainties from measured solution

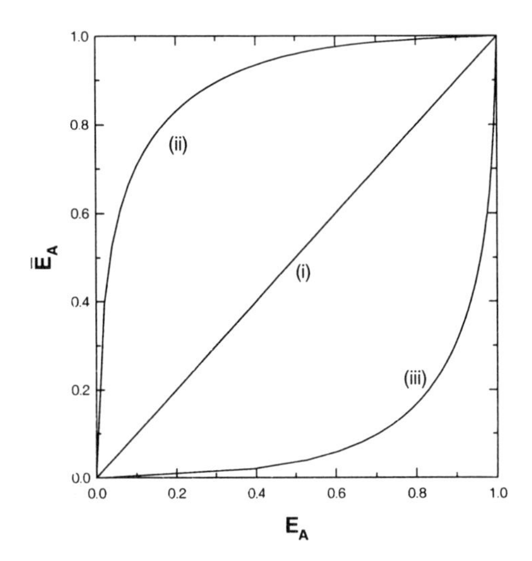

Figure 1. Examples of isotherms for the ion-exchange reaction shown in Equation (1).

Isotherm plots are convenient illustrations of the preference or selectivity of a zeolite for ion A^{z_A+} relative to ion B^{z_B+}. For example, a zeolite with equal preference for ions A^{z_A+} and B^{z_B+} will exhibit an isotherm shown by line (1) in Figure 1. Isotherm (2) illustrates the case where the zeolite prefers A^{z_A+} over B^{z_B+}, whereas isotherm (3) is for the case where B^{z_B+} is preferred over A^{z_A+}. In some cases, the zeolite exhibits a preference for A^{z_A+} over B^{z_B+} at one composition range and a reversed selectivity at another composition range. An example of selectivity reversal is shown in Figure 2 in which the zeolite prefers A^{z_A+} (Sr^{2+}) at low $\bar{E}_A$ and B^{z_B+} (Na^+) at high $\bar{E}_A$.

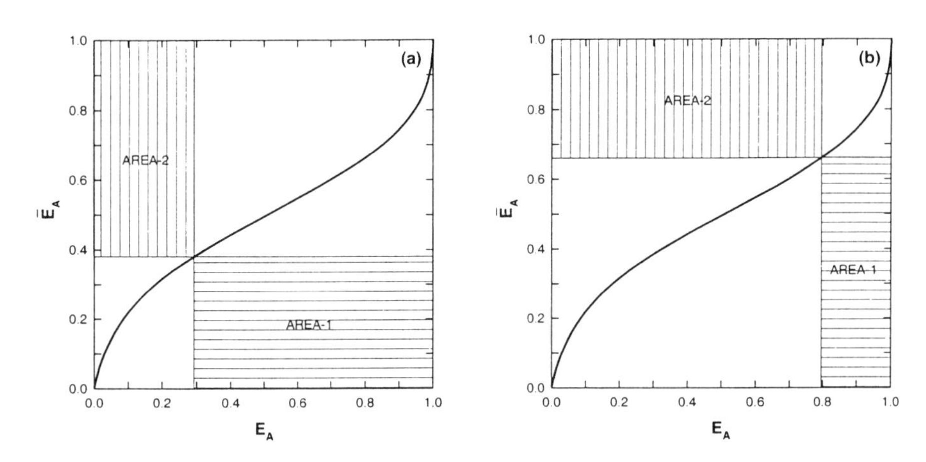

Figure 2. Ion-exchange isotherm showing selectivity reversal. The isotherm is based on data from Pabalan and Bertetti (1999) on ion exchange involving clinoptilolite and 0.5 N Sr^{2+}/Na^+ solutions (Cl^- anion). The zeolite prefers A^{z_A+} (Sr^{2+}) at low E_A (Fig. 2a) and B^{z_B+} (Na^+) at high E_A (Fig. 2b). Area-1 equals $[(1-E_A)\cdot\bar{E}_A]$, and Area-2 equals $[E_A\cdot(1-\bar{E}_A)]$.

---

concentrations of A^{z_A+} and B^{z_B+} and other experimental parameters are given in the appendix.

Selectivity coefficient

The preference of a zeolite for ion A^{z_A+} relative to ion B^{z_B+} can be expressed in terms of a selectivity coefficient, α, which is defined as

$$\alpha = \frac{\overline{E}_A n_B}{\overline{E}_B n_A} \ . \tag{5}$$

From Equation (3), $n_A = (z_A n_A + z_B n_B)E_A/z_A$ and $n_B = (z_A n_A + z_B n_B)E_B/z_B$. It follows that

$$\alpha = \frac{z_A}{z_B} \cdot \frac{\overline{E}_A E_B}{\overline{E}_B E_A} \tag{6}$$

or, from Figure 2,

$$\alpha = \frac{z_A}{z_B} \cdot \frac{\text{Area - 1}}{\text{Area - 2}} \ . \tag{7}$$

It is apparent from Figure 2 that α can vary with the level of exchange ($\overline{E}_A$) (Dyer et al. 1981). The conditions for selectivity at a specified $\overline{E}_A$ are:

(1) when $\alpha > (z_A/z_B)$, the zeolite is selective for A^{z_A+};

(2) when $\alpha = (z_A/z_B)$, the zeolite exhibits no preference; and

(3) when $\alpha < (z_A/z_B)$, the zeolite is selective for B^{z_B+}.

The examples illustrated in Figure 2 are based on experimental data on Sr^{2+}/Na^+ ion exchange ($A^{z_A+} = Sr^{2+}$; $B^{z_B+} = Na^+$; $z_A/z_B = 2$). The calculated value of α is 2.97 for the data point (solid circle) shown in Figure 2a; thus, the zeolite prefers Sr^{2+} over Na^+ at this value of $\overline{E}_{Sr}$. The value of $\overline{E}_{Sr}$ is higher and the calculated α is 0.99 for the data point in Figure 2b, i.e. the zeolite is selective for Na^+ relative to Sr^{2+}.

Concentration-valency effect

The total concentration of the aqueous solution does not have a large effect on the selectivity of the zeolite for a particular ion for exchange reactions in which the exchange ions have equal charges (e.g. Na^+ and K^+). However, when the exchange ions have different valences (e.g. Na^+ and Ca^{2+}), the ion-exchange behavior of the system depends strongly on the total concentration of the aqueous solution, and the selectivity of the zeolite for the ion of higher valence becomes progressively greater with increasing dilution. This 'concentration-valency' effect can arise universally from high dilution of the electrolyte solution independent of the exchanger phase (Barrer and Klinowski 1974a). This effect explains why Ca^{2+} can be removed from dilute aqueous solutions by synthetic ion exchangers, such as water softeners, whereas an exhausted exchanger in Ca-form can be regenerated into Na-form with concentrated NaCl solution. Figure 3 shows an example of a concentration-valency effect for ion exchange involving clinoptilolite and Sr^{2+}/Na^+ solutions at various solution concentrations. With increasing dilution, the isotherm becomes more rectangular and clinoptilolite exhibits increasing selectivity for Sr^{2+} relative to Na^+.

Equilibrium constant, Gibbs free energy, and Vanselow coefficient

The thermodynamic equilibrium constant, $K_{(A,B)}$, for the ion-exchange reaction represented by Equation (1) is given by:

$$K_{(A,B)} = \frac{(\bar{a}_A)^{z_B} (a_B)^{z_A}}{(a_A)^{z_B} (\bar{a}_B)^{z_A}} \ , \tag{8}$$

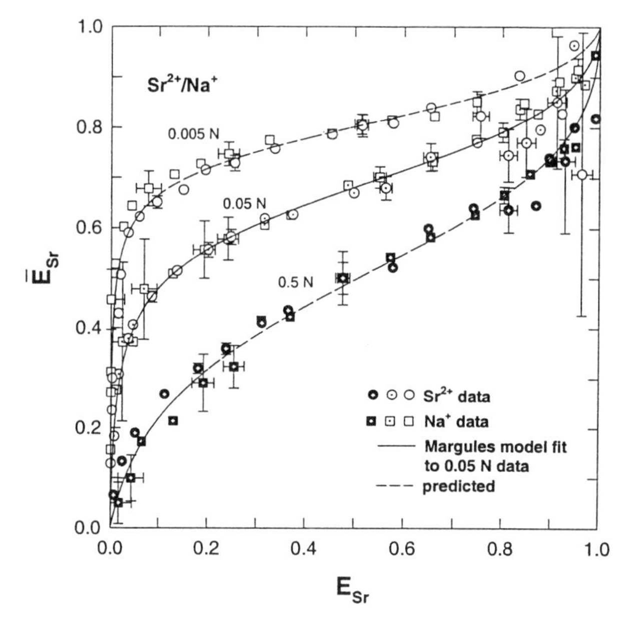

Figure 3. Isotherms for ion exchange at 298 K between clinoptilolite and Sr^{2+}/Na^+ solutions (0.005, 0.05, and 0.5 N; Cl^- anion). Circles and squares were calculated from Sr^{2+} and Na^+ analytical data, respectively. Some error bars, based on Sr^{2+} and Na^+ analytical uncertainties, are shown. The solid curve was fit to the 0.05 N isotherm data using a Margules solid-solution model. The dashed curves represent predicted values. Data from Pabalan and Bertetti (1999).

where a_i represents activities of the aqueous species and $\bar{a}_i$ represents activities of the zeolite components. The standard Gibbs free energy, $\Delta G^\circ_{(A,B)}$, of ion exchange can be calculated from

$$\Delta G^\circ_{(A,B)} = -RT \ln K_{(A,B)} , \tag{9a}$$

where R is the gas constant and T is temperature (Kelvin). Values of $K_{(A,B)}$ and $\Delta G^\circ_{(A,B)}$ provide an estimate of the selectivity of the zeolite for A^{z_A+} relative to B^{z_B+}. However, Colella (1996) pointed out that these thermodynamic parameters give unequivocal indications of selectivity only for ion-exchange pairs that exhibit isotherms like (2) and (3) in Figure 1. Selectivity reversals exhibited in some systems, such as shown in Figure 2, are not evident from values of $K_{(A,B)}$ and $\Delta G^\circ_{(A,B)}$ alone.

It should be noted that some studies report the standard free energy in terms of the equivalents involved in the exchange reaction, i.e.

$$\Delta G^\oplus_{(A,B)} = \frac{-RT \ln K_{(A,B)}}{z_A z_B} . \tag{9b}$$

The zeolite may be considered as a solid solution of two components AL_{z_A} and BL_{z_B} (Ekedahl et al. 1950; Freeman 1961; Barrer and Klinowski 1977), where L_{z_A} is the amount of anionic framework associated with an A^{z_A+} ion and carrying anionic charge z_A-, and L_{z_B} is the amount of framework associated with B^{z_B+} and carrying anionic charge z_B-. The number of moles of AL_{z_A} and BL_{z_B} are then respectively equal to the total number of moles $\bar{n}_A$ and $\bar{n}_B$ of ions A^{z_A+} and B^{z_B+} in the zeolite, and solid phase compositions can be described in terms of cationic mole fractions, $\bar{X}_A$ and $\bar{X}_B$, defined as:

$$\bar{X}_A = \frac{\bar{n}_A}{\bar{n}_A + \bar{n}_B}; \quad \bar{X}_B = \frac{\bar{n}_B}{\bar{n}_A + \bar{n}_B} \quad . \tag{10}$$

Equation (8) may be expanded to give:

$$K_{(A,B)} = \frac{(\bar{X}_A)^{z_B}(M_B)^{z_A}}{(\bar{X}_B)^{z_A}(M_A)^{z_B}} \cdot \frac{(f_A)^{z_B}}{(f_B)^{z_A}} \cdot \frac{(\gamma_B)^{z_A}}{(\gamma_A)^{z_B}} \tag{11}$$

or,

$$K_{(A,B)} = K_{v(A,B)} \cdot \frac{(f_A)^{z_B}}{(f_B)^{z_A}} \ , \tag{12}$$

where $K_{v(A,B)}$ is the Vanselow corrected selectivity coefficient (Vanselow 1932; Townsend 1986) defined by

$$K_{v(A,B)} = \frac{(\bar{X}_A)^{z_B}(M_B)^{z_A}}{(\bar{X}_B)^{z_A}(M_A)^{z_B}} \cdot \frac{(\gamma_B)^{z_A}}{(\gamma_A)^{z_B}} \tag{13}$$

and M_A and M_B are the molarities of A^{z_A+} and B^{z_B+} in the aqueous phase.[5] The quantities γ_A and γ_B are single-ion activity coefficients for the aqueous cations and account for nonideal behavior in the aqueous phase. The quantities f_A and f_B are rational (i.e. in terms of mole fractions) activity coefficients for the zeolite components and account for nonideality in the zeolite phase.

Evaluation of $K_{(A,B)}$, f_A, f_B, and $\Delta G^\circ_{(A,B)}$

It is necessary to define the standard states of the various components to allow the evaluation of $K_{(A,B)}$, f_A, and f_B from experimental data using the Gibbs-Duhem relation. The usual standard state for the aqueous electrolyte solution external to the zeolite phase is that of a hypothetical, one molar solution referenced to infinite dilution. It has been normal practice to follow Gaines and Thomas (1953) for the exchanger phase and make the standard state for each exchanging cation the appropriate homoionic form of the zeolite in equilibrium with an infinitely dilute solution of the same cation (Sposito 1981). With that standard state, the criterion for ideal behavior in the zeolite solid solution is that $\bar{a}_i = \bar{X}_i$ for all $\bar{X}_i$.

The evaluation of $K_{(A,B)}$, f_A, and f_B from experimental data results from an appropriate integration of the Gibbs-Duhem relation (Argersinger et al. 1950). Together with Equation (13), the Gibbs-Duhem equation gives the following expressions for calculating the equilibrium constant and zeolite phase activity coefficients:

$$\ln K_{(A,B)} = \int_0^1 \ln K_{v(A,B)} \, d(\bar{E}_A) \ , \tag{14}$$

$$\ln f_A^{z_B} = -\bar{E}_B \ln K^*_{v(A,B)} + \int_{\bar{E}_A}^1 \ln K_{v(A,B)} \, d(\bar{E}_A) \ , \tag{15}$$

and

---

[5] Equations for calculating $\ln K_{v(A,B)}$ and the associated uncertainty from measured concentrations of A^{z_A+} and B^{z_B+} in solution are given in the appendix.

$$\ln f_B^{z_A} = \overline{E}_A \ln K^*_{v(A,B)} - \int_0^{\overline{E}_A} \ln K_{v(A,B)} \, d(\overline{E}_A) , \quad (16)$$

where the superscript * refers to the value of $K_{v(A,B)}$ at a particular $\overline{E}_i$. Note that although $K_{v(A,B)}$ is defined in terms of $\overline{X}_i$, Equations (14) to (16) also require values of the charge fraction $\overline{E}_i$. The relationships between $\overline{X}_i$ and $\overline{E}_i$ are shown by

$$\overline{X}_A = \frac{\overline{E}_A / z_A}{\overline{E}_A / z_A + \overline{E}_B / z_B} ; \quad \overline{X}_B = \frac{\overline{E}_B / z_B}{\overline{E}_A / z_A + \overline{E}_B / z_B} . \quad (17)$$

Values of $K_{(A,B)}$, $\ln f_A$, and $\ln f_B$ can be determined by graphical integration of the plot of $\ln K_{v(A,B)}$ versus $\overline{E}_A$, or analytically by integrating a best-fit equation to the ion-exchange data.[6]

The above thermodynamic formulations are valid for conditions where electrolyte imbibition by the solid is negligible and where changes in zeolite water activity are insignificant. The salt concentration at which imbibition for zeolites is negligible depends on the intracrystalline channel dimensions. The concentration often cited is <0.5 M, although Dyer et al. (1981) give a value of <1.0 M. The water activity terms are not significant (Barrer and Klinowski 1974a) for most cases of ion-exchange equilibria. However, for ion exchange involving concentrated electrolyte solutions, terms can be included in the equations to account for the effects of sorbed or imbibed solvent and of imbibed salts (Gaines and Thomas 1953; Townsend 1986; Grant and Fletcher 1993).

Triangle rule

If ion-exchange experiments involving cations A^{z_A+}, B^{z_B+}, and C^{z_C+} were made pairwise, self-consistency of the derived thermodynamic constants and Gibbs energies can be assessed by applying the 'triangle rule' (Howery and Thomas 1965), which can be represented by

$$[K_{(A,B)}]^{z_C} \times [K_{(B,C)}]^{z_A} \times [K_{(C,A)}]^{z_B} = 1 \quad (18)$$

and

$$z_C \Delta G^\circ_{(A,B)} + z_A \Delta G^\circ_{(B,C)} + z_B \Delta G^\circ_{(C,A)} = 0 \quad (19)$$

where $K_{(i,j)}$ and $\Delta G^\circ_{(i,j)}$, respectively, are the equilibrium constant and Gibbs free energy for ion exchange involving cations i and j. Note that $K_{(i,j)} = [K_{(j,i)}]^{-1}$ and $\Delta G^\circ_{(i,j)} = -\Delta G^\circ_{(j,i)}$.

Although the triangle rule can be used to predict the values of K and ΔG° for the third member of a set of three binary ion-exchange reactions, it is not possible to predict ion-exchange selectivity as a function of the zeolite composition (Townsend et al. 1984). Thus, it is essential to obtain isotherm data at one constant solution normality for each ion-exchange system of interest. With the data at one solution concentration, it would be possible to predict the ion-exchange isotherms over a large range of solution compositions and concentrations.

Systems that exhibit incomplete exchange

In some zeolites and for certain entering ions, A^{z_A+}, the exchange reactions reach a

---

[6]Equations (14) to (16) involve integrations from end-member compositions of the zeolite. However, data points near the extreme ends of ion-exchange isotherms can have large uncertainties (Laudelot 1987; Pabalan 1994), which can introduce significant errors in the results of the integration. An alternative approach to deriving ion-exchange thermodynamic parameters based on integrating isotherm data from two non-endmember compositions was recently presented by Ioannidis et al. (2000).

always possible to produce the homoionic A-zeolite, and it is necessary to 'normalize' the isotherm (Barrer et al. 1973). This procedure involves dividing all values of $\overline{E}_A$ by the maximal value observed experimentally to give normalized values, i.e.

$$\overline{E}^N_A = \frac{\overline{E}_A}{\overline{E}_{A,max.}} \quad . \tag{20}$$

This procedure does not affect the solution activity correction but does affect the calculation of the equilibrium constant and zeolite activity coefficients. A normalized Vanselow selectivity quotient, $K^N_{v(A,B)}$, is expressed in terms of the normalized cationic mole fractions, and the essential step in obtaining the thermodynamic equilibrium constant then involves evaluating (Barrer et al. 1973) the integral

$$\int_0^1 \ln K^N_{v(A,B)} \, d(\overline{E}^N_A) \ . \tag{21}$$

Equations (14) to (16) are still applicable, but normalized parameters must be used throughout. For example, Equation (15) becomes

$$\ln f^{z_B}_A = -\overline{E}_B \ln K^{N*}_{v(A,B)} + \int_{\overline{E}^N_A}^1 \ln K^N_{v(A,B)} \, d(\overline{E}^N_A) \ . \tag{22}$$

This procedure is necessary to conform to the definition of the exchanger phase standard state given previously. In effect, normalization results in the B^{z_B+} ions that are not involved in the exchange reaction being regarded as part of the exchanger framework (Dyer et al. 1981). These ions can still affect the ion-exchange equilibrium, but these effects are accounted for in f_A and f_B.

Townsend et al. (1984) noted that it is possible to apply the triangle rule successfully even in cases of partial ion exchange. However, they cautioned that great care must be exercised when applying the triangle rule to such systems because the standard states for all three exchange reactions must be consistent. As an example they noted that if data on ion exchange involving mordenite and Na^+/Cd^{2+} and Na^+/Pb^{2+} are used to predict $\Delta G°_{(Cd,Pb)}$, a value of −11.44 kJ/mol is obtained, whereas if the NH_4^+/Cd^{2+} and NH_4^+/Pb^{2+} data are used, a predicted $\Delta G°_{(Cd,Pb)}$ equal to 4.02 kJ/mol is obtained. The large difference arises from the difference in the values of $\overline{E}_{Cd,max}$ and $\overline{E}_{Pb,max}$. Townsend et al. (1984) concluded that *a priori* prediction of $\Delta G°_{(A,B)}$ values when partial exchange occurs is inadvisable.

Activity coefficients of aqueous ions

It is apparent from the preceding equations that an evaluation of equilibrium constants and zeolite phase activity coefficients from experimental data involves activity correction for the aqueous phase. Ion-exchange studies by Fletcher and Townsend (1985) demonstrated the importance of correctly evaluating aqueous solution activity coefficients for accurate interpretation of exchange equilibria, particularly on systems with mixed background anions. In principle, γ's, which account for nonideal behavior in the aqueous solution, can be calculated from well-established electrolyte solution theories (Bronsted 1922a,b; Guggenheim 1935; Scatchard 1936, 1968; Glueckauf 1949; Pitzer 1973, 1991). The ion-interaction model developed by Pitzer and coworkers was used to calculate the activity coefficients of aqueous ions for the examples presented in this chapter.[7] This model has the advantage of having a large database of parameters at 298 K

[7]The ion-interaction model uses a molal (moles/kilogram H_2O) concentration scale, whereas the solution concentrations in this chapter are expressed in normality (equivalents/liter solution) or molarity (moles/liter solution).

and above, and it can be used for high ionic-strength aqueous solutions. The ion-interaction model has been used successfully in many applications to multicomponent systems over wide ranges of solution composition, concentration, and temperature.

For details on the ion-interaction model, the papers by Pitzer (1973, 1987, 1991) and the references cited therein should be consulted. The model gives expressions for single-ion activity coefficients, which are more convenient to use for complex mixed electrolytes than the use of mean activity coefficients and electrically neutral differences of activity coefficients (Pitzer 1991). However, it should be remembered that single-ion activity coefficients cannot be measured independently because of electroneutrality constraints. Ion-interaction parameters used in this study are listed in Table 2.

Table 2. Ion-interaction parameters used in this study to calculate aqueous activity coefficients using the Pitzer equations.*

	Single Electrolyte Parameters		
Electrolyte	$\beta^{(0)}$	$\beta^{(1)}$	C^{φ}
NaCl	0.0765	0.2664	0.00127
KCl	0.04835	0.2122	–0.00084
CsCl	0.03478	0.03974	–0.000496
LiCl	0.1494	0.3074	0.00359
NH_4Cl	0.0522	0.1918	–0.00301
$CaCl_2$	0.3159	1.614	–0.00034
$SrCl_2$	0.2858	1.6673	–0.00130
$NaNO_3$	0.0068	0.1783	–0.00072
KNO_3	–0.0816	0.0494	0.00660
$Ca(NO_3)_2$	0.2108	1.409	–0.02014

Mixture parameters

$\theta_{Na,K}$ = –0.012 $\psi_{Na,K,Cl}$ = –0.0018 ψ_{Na,K,NO_3} = –0.0012	$\theta_{Na,Ca}$ = 0.07 $\psi_{Na,Ca,Cl}$ = –0.007 ψ_{Na,Ca,NO_3} = –0.007	$\theta_{Na,Sr}$ = 0.051 $\psi_{Na,Sr,Cl}$ = –0.0021	$\theta_{K,Sr}$ = 0.0149 $\psi_{K,Sr,Cl}$ = –0.0201
$\theta_{Na,Cs}$ = –0.03886 $\psi_{Na,Cs,Cl}$ = –0.00135	$\theta_{K,Cs}$ = 0.0 $\psi_{K,Cs,Cl}$ = –0.0013	$\theta_{Na,Li}$ = 0.012 $\psi_{Li,Na,Cl}$ = –0.003	$\theta_{K,Ca}$ = 0.032 $\psi_{K,Ca,Cl}$ = –0.025
θ_{Na,NH_4} = 0.004 $\psi_{Na,NH_4,Cl}$ = 0.0005	θ_{K,NH_4} = –0.065 $\psi_{K,NH_4,Cl}$ = 0.036		

*Values were taken mostly from Pitzer (1991). ψ_{Na,Ca,NO_3} is based on the activity coefficient data of Smith et al. (1993). $\theta_{Na,Sr}$ and $\psi_{Na,Sr,Cl}$ are from Reardon and Armstrong (1987); $\theta_{K,Sr}$ and $\psi_{K,Sr,Cl}$ are from Kim and Frederick (1988); θ_{Na,NH_4} and $\psi_{Na,NH_4,Cl}$ are from Maeda et al. (1989); θ_{K,NH_4} and $\psi_{K,NH_4,Cl}$ are from Maeda et al. (1993).

concentrations in this chapter are expressed in normality (equivalents/liter solution) or molarity (moles/liter solution). For convenience, molal activity coefficients were not converted to a normal or molar basis. However, the errors contributed to the derived ion-exchange thermodynamic parameters or to the calculated isotherm values are small relative to errors introduced by analytical uncertainties (Pabalan 1994).

Activity coefficients of zeolite components

Various models have been proposed to represent the activity coefficients of exchangeable ions or solid solutions (Elprince and Babcock 1975; Elprince et al. 1980; Chu and Sposito 1981; Grant and Sparks 1989; Pabalan 1994; Morgan et al. 1995; Mehablia et al. 1996). Two models that have been used in ion-exchange studies are the Margules and Wilson equations. These equations have been widely applied to describe nonideal behavior in both solid and liquid solutions (Pitzer 1995).

Margules model. The Margules model has been used successfully in studies of ion-exchange equilibria involving zeolite minerals (Pabalan 1994; Shibue 1998). In this model, the molar excess Gibbs energy, $\overline{G}^E$, for a zeolite solid solution with two components, AL_{z_A} and BL_{z_B}, is represented by the equation

$$\frac{\overline{G}^E}{RT} = \overline{X}_A\overline{X}_B(\overline{X}_B W_A + \overline{X}_A W_B), \tag{23}$$

where W_A and W_B are empirical parameters that are functions only of temperature and pressure. From Equation (23) and the Gibbs-Duhem relation, the zeolite activity coefficients, f_A and f_B, can be expressed in terms of W_A and W_B as

$$\ln f_A = \overline{X}_B^2[W_A + 2\overline{X}_A(W_B - W_A)]; \quad \ln f_B = \overline{X}_A^2[W_B + 2\overline{X}_B(W_A - W_B)]\ . \tag{24}$$

Analogous expressions can be derived for solid solutions with three or more components (Grant and Sparks 1989; Mukhopadhyay et al. 1993).

The Vanselow selectivity coefficient, $K_{v(A,B)}$, can then be represented by

$$\ln K_{v(A,B)} = \ln K_{(A,B)} + z_A\overline{X}_A^2[W_B + 2\overline{X}_B(W_A - W_B)] - z_B\overline{X}_B^2[W_A + 2\overline{X}_A(W_B - W_A)]\ . \tag{25}$$

Values of $K_{(A,B)}$, W_A, and W_B can be derived by nonlinear regression of Equation (25) to isotherm data. If the zeolite phase behaves ideally, $f_A = f_B = 1$ and $\overline{G}^E = 0$ for all values of $\overline{X}_A$ and $\overline{X}_B$, and $K_{(A,B)} = K_{v(A,B)}$. Values of W_A and W_B are zero for ideal solid solutions.

Wilson model. The Wilson model also has been applied successfully to studies of ion-exchange equilibria (e.g. Elprince and Babcock 1975; Shallcross et al. 1988; de Lucas et al. 1992; Mehablia et al. 1996; Shibue 1999; Ioannidis et al. 2000). The molar excess Gibbs energy for a two-component zeolite solid solution is taken to be

$$\frac{\overline{G}^E}{RT} = -\overline{X}_A \ln(\overline{X}_A + \Lambda_{AB}\overline{X}_B) - \overline{X}_B \ln(\overline{X}_B + \Lambda_{BA}\overline{X}_A), \tag{26}$$

where Λ_{AB} and Λ_{BA} are empirical parameters. The measure of nonideality in this model is the departure of the parameters from 1.0. For a solid solution that behaves ideally, $\Lambda_{AB} = \Lambda_{BA} = 1.0$. The corresponding activity coefficients for the binary solid solution are

$$\ln f_A = -\ln(\overline{X}_A + \Lambda_{AB}\overline{X}_B) + \overline{X}_B\left(\frac{\Lambda_{AB}}{\overline{X}_A + \Lambda_{AB}\overline{X}_B} - \frac{\Lambda_{BA}}{\overline{X}_B + \Lambda_{BA}\overline{X}_A}\right) \tag{27a}$$

and

$$\ln f_B = -\ln(\overline{X}_B + \Lambda_{BA}\overline{X}_A) + \overline{X}_A\left(\frac{\Lambda_{BA}}{\overline{X}_B + \Lambda_{BA}\overline{X}_A} - \frac{\Lambda_{AB}}{\overline{X}_A + \Lambda_{AB}\overline{X}_B}\right) \tag{27b}$$

The Wilson equation for multicomponent systems can be generalized to

$$\frac{\overline{G}^E}{RT} = -\sum_{i=1}^{m} \overline{X}_i \ln\left(\sum_{j=1}^{m} \Lambda_{ij}\overline{X}_j\right) \tag{28}$$

and the activity coefficient, f_k, of zeolite component k is given by

$$\ln f_k = 1 - \ln\left(\sum_{j=1}^{m} \Lambda_{kj}\overline{X}_j\right) - \sum_{i=1}^{m}\left(\frac{\Lambda_{ik}\overline{X}_i}{\sum_{j=1}^{m}\Lambda_{ij}\overline{X}_j}\right), \tag{29}$$

where Λ_{ii}, Λ_{jj}, Λ_{kk}, etc. are equal to 1.0 and the other Λ_{ij} are just the binary parameters. The absence of parameters beyond the binary terms makes the Wilson model attractive for application to ternary or more complex mixtures. However, Pitzer (1995) pointed out that interactions of three different species in a mixture do occur, and a provision for their representation is needed when they are significant.

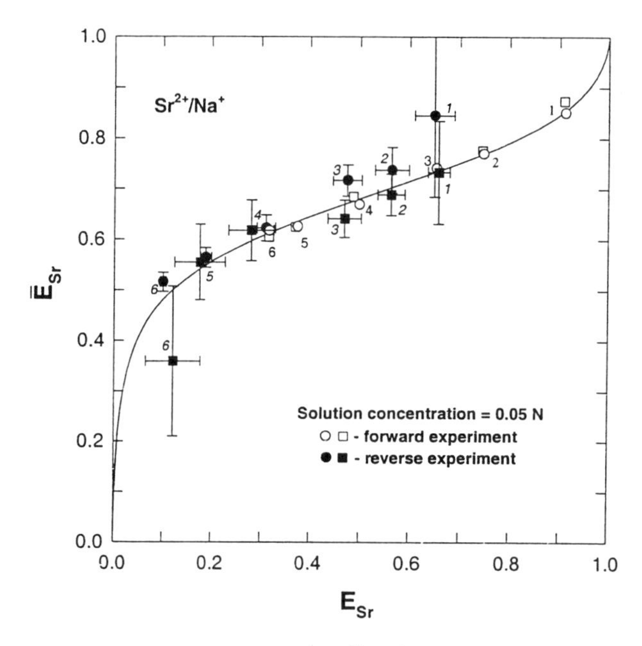

Figure 4. Isotherm data from forward and reverse Sr^{2+}/Na^+ experiments on clinoptilolite at 0.05 N solution concentration. Circles and squares were calculated from solution concentrations of Sr^{2+} and Na^+ data, respectively. The symbols are numbered to indicate matching forward and reverse data points. The curve is calculated using a Margules solid-solution model. Data from Pabalan and Bertetti (1999).

FACTORS TO CONSIDER IN EVALUATING ION-EXCHANGE DATA

Before ion-exchange experimental data can be subjected to thermodynamic treatment, it is important that the exchange reaction be shown to be reversible. An example of demonstrated reversibility is shown in Figure 4 for ion exchange involving clinoptilolite and Sr^{2+}/Na^+ solutions. However, most ion-exchange studies do not present data on reversibility.

The thermodynamic models used in evaluating exchanger phase activity coefficients essentially involve integrating Gibbs-Duhem type equations. The precision of parameters derived from such models relies on the acquisition of accurate experimental data, particularly at the extreme ends of the isotherm plot. Small errors in the analysis of low concentrations of an ion can have a dramatic effect on the shapes of plots of the quantity $\ln K_{v(A,B)}$ versus composition. This is likely a primary explanation for the many discrepant results in the literature for a given system (Townsend 1986).

Other factors also cause discrepant data being obtained for different studies on a given exchange.[8] It was common practice in past binary exchange measurements to analyze for the concentration of one ion only and to infer the concentration of the other ion by differences. However, for some zeolites hydronium exchange also takes place concomitantly with the other exchange reaction (Drummond et al. 1983). When this occurs, the exchange becomes a ternary exchange process and serious errors may result in the calculation of selectivity coefficients, especially at the extreme ends of isotherms.

A similar problem may occur for systems that exhibit incomplete exchange relative to a certain cation. During the course of an experiment the cation remaining in the solid phase may eventually exchange out, changing the nature of the exchange to a ternary reaction. For example, in Townsend and Loizidou's (1984) study on Na^+/NH_4^+ equilibria, clinoptilolite that had already been maximally exchanged for Na^+ released traces of K^+ when equilibrated with the mixed Na^+/NH_4^+ solutions. Thus, a third component was added to the exchange reaction. The K^+ concentration was low enough in this particular case that the reaction remained essentially binary (Townsend and Loizidou 1984).

Another potential source of error in ion-exchange experiments is dissolution of the zeolite framework. Zeolites and other aluminosilicates are not very resistant to strong acids and bases. Under acidic conditions, aluminum is preferentially dissolved, whereas silicon is preferentially dissolved in alkaline solutions. However, under neutral pH conditions, dissolution of aluminum and silicon is at a minimum and the effect of zeolite dissolution on ion-exchange equilibria usually can be neglected.

Therefore, in careful studies of ion-exchange equilibria, it is advisable to analyze for each exchanging ion in both the aqueous solution phase and the solid phase. This may be supplemented by analysis of other cations for systems that exhibit incomplete exchange or for systems in which more than two cations appear to participate in the exchange reactions. In cases where there is potential dissolution of the solid, aluminum and silicon concentrations in the aqueous phase may be analyzed to determine the extent of dissolution. The pH of the aqueous solution may also be controlled or monitored and other analytical techniques may be used. For example, scanning electron microscopy may be used to detect any visible evidence of mineral dissolution, and a more sophisticated technique such as ^{27}Al solid-state magic-angle spinning nuclear magnetic resonance spectroscopy may be used to characterize the aluminum sites in the zeolite structure.

Ion exchange studies on some natural zeolites, such as clinoptilolites, present special problems. Clinoptilolite minerals used in ion-exchange experiments are usually zeolitized volcanic tuff specimens. These commonly contain mineral impurities such as quartz, feldspar, smectite, and unaltered volcanic glass, and in some cases, other zeolites, halite, and calcite. Soluble salts and carbonate minerals, if not eliminated before conducting the experiments, can later dissolve and invalidate the assumption of binary exchange. In addition, thermodynamic treatment requires that the CEC be known. Previous studies have estimated the CEC from the Al^{3+} concentration in the zeolite determined by chemical analysis (e.g. Townsend and Loizidou 1984), from the concentration of

---

[8]Lehto and Harjula (1995) provided an excellent discussion of experimental factors that require careful consideration in order to obtain reliable and consistent ion-exchange results.

exchangeable cation(s) in the zeolite determined by chemical analysis (e.g. Barrer and Townsend 1976b), or from the observed maximum levels of exchange (e.g. Ames 1964a,b). Each of these methods has drawbacks caused by impurities in clinoptilolite samples. If the first method is used, chemical analysis of clinoptilolite will overestimate CEC in cases where feldspars or other aluminosilicate minerals are present and result in higher Al^{3+} content. If the second method is used, chemical analysis will overestimate the amount of exchangeable cations Na^+, Mg^{2+}, and Ca^{2+}, hence CEC, in cases where impurities such as halite or carbonate minerals are present. On the other hand, CECs determined by the third method are sensitive to the method of pretreatment used (Semmens and Seyfarth 1978).

These problems can be minimized by careful characterization (e.g. analysis of mineralogical composition) or pretreatment (e.g. dissolution of soluble minerals or physical separation/purification) of zeolite specimens. Unfortunately, little attention has been given in many ion-exchange studies to the methods used in the preparation of the materials, or, in some cases, to their mineralogical and chemical composition. Thus, comparisons of experimental results and related thermodynamic quantities derived by various investigators can be complicated and, in more than a few cases, not very useful.

ION-EXCHANGE STUDIES ON NATURAL ZEOLITES

Literature information on the ion-exchange properties of the more common natural zeolites is summarized in this section. Information on a few synthetic zeolites, notably synthetic mordenite, is included for comparison. Equilibrium constants and Gibbs free energies taken from the literature are listed in tables. However, as noted previously, caution should be used when comparing published thermodynamic values because of the different materials (e.g. degrees of impurity, preparation method) used in the experiments, different temperatures, and different methods for calculating activity coefficients of aqueous cations. In most cases, the quality of the experimental data and the derived thermodynamic parameters have not been evaluated for this review. The original references should be consulted for details on experimental and analytical procedures and zeolite characterization.

Chabazite $[Ca_2(Al_4Si_8O_{24})\cdot 12H_2O]$[9]

Chabazite was one of the first zeolites to be studied for its ion-exchange properties. Ames (1961) used a column apparatus to load natural zeolite samples with Cs^+ and competing cations to determine selectivity of the zeolites and to gain insight into the mechanisms responsible for a particular selectivity sequence. Natural chabazite crystals from Nova Scotia, Canada, hand-picked to greater than 95% purity, were exposed to solutions with a total concentration of 0.01 N at 298 K. Ames' results indicated a selectivity sequence of $Cs^+ > K^+ > Na^+ > Li^+$ for alkali metals and $Ba^{2+} > Sr^{2+} > Ca^{2+} > Mg^{2+}$ for alkaline earths.

In extensive follow-up studies on the ion exchange of alkali metal cations and alkaline-earth cations in zeolites, Ames (1964a,b) conducted experiments on AW-500, a commercially available Na-form prepared from natural chabazite from a deposit near Bowie, Arizona. The CEC of the chabazite was determined to be 2.2 meq/g. Ion-exchange experiments were conducted at 298 and 343 K with a total solution normality of 1.0 N. Isotherms are presented in Ames (1964a) for the K^+/Na^+ system. Equilibrium constants and Gibbs free energy were determined for K^+/Cs^+, K^+/Na^+, Na^+/Cs^+, Na^+/Sr^{2+} Na/Ca^{2+}, and Ca^{2+}/Sr^{2+} systems (Ames 1964a,b). The values are listed in Table 3.

---

[9]The idealized formula listed with the zeolite name is taken from Gottardi and Galli (1985).

Table 3. Literature values of equilibrium constants and Gibbs free energies for ion-exchange reactions involving chabazite

Ion-Exchange Reaction	*ln* $K_{(A,B)}$	$\Delta G^{\circ}_{(A,B)}$ *(kJ/mol)**	*Reference*	*Remarks*
$Li^+ + NaL \Leftrightarrow Na^+ + LiL$	−2.91	7.20	Barrer et al. (1969)	T = 298 K; Nova Scotia, Canada chabazite
$Na^+ + KL \Leftrightarrow K^+ + NaL$	−1.98	4.91	Ames (1964a)	T=298 K; AW-500; Bowie, Arizona chabazite
$K^+ + NaL \Leftrightarrow Na^+ + KL$	2.70	−6.69	Barrer et al. (1969)	T = 298 K; Nova Scotia, Canada chabazite
$K^+ + NaL \Leftrightarrow Na^+ + KL$	2.1	−5.2	Torracca et al. (1998)	T = 298 K; Avellino, Italy chabazite; normalized isotherm
$Rb^+ + NaL \Leftrightarrow Na^+ + RbL$	2.26	−5.61	Barrer et al. (1969)	T = 298 K; Nova Scotia, Canada chabazite
$Cs^+ + NaL \Leftrightarrow Na^+ + CsL$	3.93	−9.7	Perona (1993)	T = 296 K; Ionsiv
$Cs^+ + NaL \Leftrightarrow Na^+ + CsL$	3.39	−8.40	Ames (1964a)	T=298 K; AW-500; Bowie, Arizona chabazite
$Cs^+ + NaL \Leftrightarrow Na^+ + CsL$	4.22	−10.46	Barrer et al. (1969)	T = 298 K; ; Nova Scotia, Canada chabazite; normalized isotherm
$Cs^+ + NaL \Leftrightarrow Na^+ + CsL$	4.56; 4.27; 3.90	−11.3; −11.1; −10.8	Dyer and Zubair (1998)	T = 298 K; 313 K; 333K; Bowie, Arizona chabazite
$Cs^+ + NaL \Leftrightarrow Na^+ + CsL$	2.59	−7.39	Ames (1964a)	T=343 K; AW-500; Bowie, Arizona chabazite
$Cs^+ + KL \Leftrightarrow K^+ + CsL$	2.8; 2.7; 2.7	−6.9; −6.9; −7.4	Dyer and Zubair (1998)	T = 298 K; 313 K; 333K; Bowie, Arizona chabazite
$Cs^+ + KL \Leftrightarrow K^+ + CsL$	1.49	−3.69	Ames (1964a)	T=298 K; AW-500; Bowie, Arizona chabazite
$Cs^+ + KL \Leftrightarrow K^+ + CsL$	1.49	−4.25	Ames (1964a)	T=343 K; AW-500; Bowie, Arizona chabazite
$Cs^+ + RbL \Leftrightarrow Rb^+ + CsL$	1.8; 1.8; 1.7	−4.5; −4.8; −4.8	Dyer and Zubair (1998)	T = 298 K; 313 K; 333K; Bowie, Arizona chabazite
$Cs^+ + 0.5MgL_2 \Leftrightarrow Mg^{2+} + CsL$	6.78; 5.73; 5.67	−16.8; −14.9; −15.7	Dyer and Zubair (1998)	T = 298 K; 313 K; 333K; Bowie, Arizona chabazite
$Cs^+ + 0.5CaL_2 \Leftrightarrow 0.5Ca^{2+} + CsL$	5.17; 5.03; 4.88	−12.8; −13.1; −13.5	Dyer and Zubair (1998)	T = 298 K; 313 K; 333K; Bowie, Arizona chabazite

$Cs^+ + 0.5BaL \Leftrightarrow 0.5Ba^{2+} + CsL$	4.52; 4.30; 4.12	–11.2; –11.2; –11.4	Dyer and Zubair (1998)	T = 298 K; 313 K; 333K; Bowie, Arizona chabazite
$Cs^+ + 0.5SrL_2 \Leftrightarrow 0.5Sr^{2+} + CsL$	5.81; 5.80; 5.49	–14.4; –15.1; –15.2	Dyer and Zubair (1998)	T = 298 K; 313 K; 333K; Bowie, Arizona chabazite
$NH_4^+ + NaL \Leftrightarrow Na^+ + NH_4L$	2.5	–6.1	Torracca et al. (1998)	T = 298 K; Avellino, Italy chabazite; normalized isotherm
$NH_4^+ + NaL \Leftrightarrow Na^+ + NH_4L$	1.67	–4.14	Barrer et al. (1969)	T = 298 K; Nova Scotia, Canada chabazite
$Ag^+ + NaL \Leftrightarrow Na^+ + AgL$	2.43	–6.02	Barrer et al. (1969)	T = 298 K; Nova Scotia, Canada chabazite
$Tl^+ + NaL \Leftrightarrow Na^+ + TlL$	4.42	–10.96	Barrer et al. (1969)	T = 298 K; Nova Scotia, Canada chabazite
$Mg^{2+} + 2NaL \Leftrightarrow 2Na^+ + MgL_2$	–7.6	19	Perona (1993)	T = 296 K; Ionsiv
$Ca^{2+} + 2NaL \Leftrightarrow 2Na^+ + CaL_2$	–3.82	9.46	Perona (1993)	T = 296 K; Ionsiv
$Ca^{2+} + 2NaL \Leftrightarrow 2Na^+ + CaL_2$	–0.323	0.800	Ames (1964b)	T=298 K; AW-500; Bowie, Arizona chabazite
$0.5Ca^{2+} + NaL \Leftrightarrow Na^+ + 0.5CaL_2$	0.08; 0.12; 0.22	–0.2; –0.3; –0.6	Dyer and Zubair (1998)	T = 298 K; 313 K; 333K; Bowie, Arizona chabazite
$Ca^{2+} + 2NaL \Leftrightarrow 2Na^+ + CaL_2$	–0.3	0.75	Barrer et al. (1969)	T = 323 K; Nova Scotia, Canada chabazite
$0.5Ca^{2+} + KL \Leftrightarrow K^+ + 0.5CaL_2$	–1.3; –1.1; –0.94	3.3; 2.9; 2.6	Dyer and Zubair (1998)	T = 298 K; 313 K; 333K; Bowie, Arizona chabazite
$0.5Ca^{2+} + RbL \Leftrightarrow Rb^+ + 0.5CaL_2$	–1.7; –1.5; –1.4	4.3; 4.0; 3.9	Dyer and Zubair (1998)	T = 298 K; 313 K; 333K; Bowie, Arizona chabazite
$0.5Ca^{2+} + CsL \Leftrightarrow Cs^+ + 0.5CaL_2$	–2.1; –1.9; –1.7	5.2; 5.0; 4.6	Dyer and Zubair (1998)	T = 298 K; 313 K; 333K; Bowie, Arizona chabazite
$Ca^{2+} + MgL_2 \Leftrightarrow Mg^{2+} + CaL_2$	1.3; 1.2; 1.3	–3.2; –3.2; –3.5	Dyer and Zubair (1998)	T = 298 K; 313 K; 333K; Bowie, Arizona chabazite
$Ca^{2+} + SrL_2 \Leftrightarrow Sr^{2+} + CaL_2$	0.57; 0.61; 0.61	–1.4; –1.6; –1.7	Dyer and Zubair (1998)	T = 298 K; 313 K; 333K; Bowie, Arizona chabazite
$Ca^{2+} + BaL_2 \Leftrightarrow Ba^{2+} + CaL_2$	0.32; 0.35; 0.36	–0.8; –0.9; –1.0	Dyer and Zubair (1998)	T = 298 K; 313 K; 333K; Bowie, Arizona chabazite
$Sr^{2+} + 2NaL \Leftrightarrow 2Na^+ + SrL_2$	–1.84	4.56	Perona (1993)	T = 296 K; Ionsiv
$Sr^{2+} + 2NaL \Leftrightarrow 2Na^+ + SrL_2$	–1.24	3.07	Ames (1964b)	T=298 K; AW-500 (Bowie, Arizona) chabazite
$0.5Sr^{2+} + NaL \Leftrightarrow Na^+ + 0.5SrL_2$	0.04; 0.19; 0.22	–0.1; –0.5; –0.6	Dyer and Zubair (1998)	T = 298 K; 313 K; 333K;

				Bowie, Arizona chabazite
$Sr^{2+} + 2NaL \Leftrightarrow 2Na^{+} + SrL_2$	−0.13	0.33	Barrer et al. (1969)	T = 323 K; Nova Scotia, Canada chabazite
$0.5Sr^{2+} + KL \Leftrightarrow K^{+} + 0.5SrL_2$	−1.4; −1.2; −1.1	3.4; 3.1; 3.1	Dyer and Zubair (1998)	T = 298 K; 313 K; 333K; Bowie, Arizona chabazite
$0.5Sr^{2+} + RbL \Leftrightarrow Rb^{+} + 0.5SrL_2$	−1.9; −1.7; −1.5	4.8; 4.3; 4.1	Dyer and Zubair (1998)	T = 298 K; 313 K; 333K; Bowie, Arizona chabazite
$0.5Sr^{2+} + CsL \Leftrightarrow Cs^{+} + 0.5SrL_2$	−2.4; −1.9; −1.6	5.9; 4.9; 4.4	Dyer and Zubair (1998)	T = 298 K; 313 K; 333K; Bowie, Arizona chabazite
$Sr^{2+} + MgL_2 \Leftrightarrow Mg^{2+} + SrL_2$	−4.24; −4.07; −4.08	10.5; 10.6; 11.3	Dyer and Zubair (1998)	T = 298 K; 313 K; 333K; Bowie, Arizona chabazite
$Sr^{2+} + CaL_2 \Leftrightarrow Ca^{2+} + SrL_2$	−2.2; −2.1; −2.0	5.4; 5.4; 5.4	Dyer and Zubair (1998)	T = 298 K; 313 K; 333K; Bowie, Arizona chabazite
$Sr^{2+} + CaL \Leftrightarrow Ca^{+} + SrL_2$	−0.992	2.46	Ames (1964b)	T=298 K; AW-500; Bowie, Arizona chabazite
$Sr^{2+} + BaL_2 \Leftrightarrow Ba^{2+} + SrL_2$	0.08; 0.15; 0.18	−0.2; −0.4; −0.5	Dyer and Zubair (1998)	T = 298 K; 313 K; 333K; Bowie, Arizona chabazite
$Ba^{2+} + 2NaL \Leftrightarrow 2Na^{+} + BaL_2$	0.0	0.00	Barrer et al. (1969)	T = 298 K; Nova Scotia, Canada chabazite
$Pb^{2+} + 2NaL \Leftrightarrow 2Na^{+} + PbL_2$	1.4	−3.4	Torracca et al. (1998)	T = 298 K; Avellino, Italy chabazite
$Pb^{2+} + 2NaL \Leftrightarrow 2Na^{+} + PbL_2$	0.57	−1.42	Barrer et al. (1969)	T = 298 K; Nova Scotia, Canada chabazite
$Cu^{2+} + 2NaL \Leftrightarrow 2Na^{+} + CuL_2$	−2.5	6.1	Colella et al. (1998)	T = 298K; Riano, Italy chabazite; normalized isotherm
$Zn^{2+} + 2NaL \Leftrightarrow 2Na^{+} + ZnL_2$	−3.3	8.2	Colella et al. (1998)	T = 298K; Riano, Italy chabazite; normalized isotherm

*Values from Barrer et al. (1969), given in kilocalories per equivalent, were converted to kiloJoules per mole.

Barrer et al. (1969) studied the exchange of a variety of cations with Na^+ in a natural chabazite from Nova Scotia, Canada. Na-chabazite was prepared by repeated exchange of the natural material with NaCl solutions. The CEC for the Na-chabazite was calculated to be 3.49 meq/g. Experiments were conducted at 298 K, except for exchanges of Sr^{2+} and Ca^{2+} for Na^+, which were conducted at 323 K. Cs^+ exhibited incomplete exchange with Na^+, reaching only 0.84 of the CEC ($\bar{E}_{Cs,max.} = 0.84$). Isotherms for all experiments conducted are presented in the paper. A general selectivity sequence, based on the standard free energies of exchange, is given as $Tl^+ > K^+ > Ag^+ > Rb^+ > NH_4^+ > Pb^{2+} > Na^+ = Ba^{2+} > Sr^{2+} > Ca^{2+} > Li^+$. Heats of exchange were also determined by calorimetry. Gibbs free energy values determined by Barrer et al. (1969) and equilibrium constants derived from those values are listed in Table 3.

Dyer and Zubair (1998) produced ion-exchange isotherms for several homoionic forms of a purified natural chabazite from a deposit near Bowie, Arizona. K^+, Rb^+, Cs^+, Mg^{2+}, Ca^{2+}, Sr^{2+}, and Ba^{2+} were exchanged with radiolabeled Cs^+, Sr^{2+}, and Ca^{2+} in solutions of 0.01 N concentration at temperatures of 298, 313, and 333 K. The authors reported that reversible ion exchange was exhibited in all cases[10] and temperature had little effect on the nature of the isotherms. Some slight differences in thermodynamic values were attributed to differences in the Si/Al ratio of the chabazites used for the study. The derived thermodynamic values are very similar to those from previous studies. Thermodynamic data from Dyer and Zubair (1998) are listed in Table 3.

The potential to remove Pb^{2+} from wastewaters through ion exchange in natural chabazite, obtained from San Mango sul Calore, Italy, was investigated by Colella and Pansini (1988). The mean value of CEC for the chabazite used in their study was 1.70 meq/g. The authors concluded that chabazite was selective for Pb^{2+}, but high concentrations of Na^+ interfered with exchange.

Colella et al. (1998) investigated the exchange of Cu^{2+} and Zn^{2+} with Na^+ using purified chabazite obtained from chabazite-rich tuff near Riano, Italy. The purified chabazite was converted to a Na-form through exchange with NaCl solution. The measured CEC of the chabazite was 3.37 meq/g. Isotherms were generated from experiments conducted at 298 K in solutions of Cu^{2+}, Zn^{2+}, and Na^+ at a total concentration of 0.1 N (NO_3^- co-anion). Results for Cu^{2+}/Na^+ and Zn^{2+}/Na^+ exchange indicated that the ion-exchange reactions were reversible. However, exchange was not complete for Zn^{2+}, with an $\bar{E}_{Zn,max.}$ equal to 0.83. Thermodynamic values determined by Colella et al. (1998) are given in Table 3.

Torracca et al. (1998) studied the ion exchange of Na^+ with Pb^{2+}, NH_4^+, K^+, Ca^{2+}, and Mg^{2+}. Torracca et al. (1998) used chabazite prepared from a chabazite-rich tuff outcrop near Avellino, Italy. The chabazite was purified and converted to the Na-form through contact with NaCl solutions. The CEC for the chabazite was determined to be 2.75 meq/g. Isotherms were produced through experiments conducted at 0.1 N solution concentration and 298 K. Exchange kinetics were slow for Ca^{2+} and Mg^{2+}, an observation that agrees with results from previous studies (Barrer et al. 1969). Although the ion-

---

[10]However, the reported reversibility of the ion-exchange reactions is not consistent with some of the tabulated Gibbs free energies of ion exchange. The Gibbs free energies for three reverse reactions: (1) $2CsL \rightarrow SrL_2$ and $SrL_2 \rightarrow 2CsL$, (2) $2CsL \rightarrow CaL_2$ and $CaL_2 \rightarrow 2CsL$, and (3) $CaL_2 \rightarrow SrL_2$ and $SrL_2 \rightarrow CaL_2$, do not sum to zero and, instead, sum to several kiloJoules per mole. Dyer suggested (personal communication with A. Dyer, August 9, 2000) that multication exchange may have occurred, possibly because (1) none of the samples is truly homoionic such that Na^+ impurity may have played a role, and (2) the hydronium ion usually plays a role in zeolite ion exchange.

exchange reactions were reversible, incomplete exchange was observed for Na^+/K^+ and Na^+/NH_4^+. A selectivity sequence of $NH_4^+ > K^+ > Pb^{2+} > Na^+$ was determined. Thermodynamic quantities for Na^+/NH_4^+, Na^+/K^+, and Na^+/Pb^{2+} systems as determined by Torracca et al. (1998) are provided in Table 3.

Chabazite has been used at Oak Ridge National Laboratory to clean up various waste streams containing radioactive Sr^{2+} and Cs^+ (Robinson et al. 1991; Perona 1993; Robinson et al. 1995; DePaoli and Perona 1996). Perona (1993) modeled isotherm data to derive thermodynamic parameters for Na^+/Sr^{2+}, Na^+/Cs^+, Na^+/Ca^{2+}, and Na^+/Mg^{2+} ion exchange. The data were based on experiments using a Na-exchanged natural chabazite from a deposit near Bowie, Arizona (commercially available from Ionsiv), conducted at 296 K and solution concentrations of 0.01 N or below. Perona (1993) extended the binary system data to model a five-component system. Values of the ion-exchange equilibrium constants derived by Perona (1993) and the associated Gibbs free energies are listed in Table 3.

Clinoptilolite [$(Na,K)_6(Al_6Si_{30}O_{72})\cdot 20H_2O$]

Numerous ion-exchange studies have been performed on clinoptilolite, primarily because of its ability to extract ^{137}Cs from radioactive waste solutions and the NH_4^+ ion from municipal wastewater streams. The earliest comprehensive investigations into the selectivity of clinoptilolite for inorganic ions were conducted by Ames (1960, 1961) using column experiments and clinoptilolite from Hector, California. These studies demonstrated the high selectivity of clinoptilolite for Cs^+ and established the selectivity series $Cs^+ > Rb^+ > K^+ > Na^+ > Li^+$ for the alkali elements, and $Ba^{2+} > Sr^{2+} > Ca^{2+} > Mg^{2+}$ for the alkaline-earth elements, on the basis of the power of the different cations to compete with Cs^+. Other studies during that period evaluated the use of clinoptilolite in extracting ^{137}Cs and ^{90}Sr from radioactive wastes (Mercer 1960; Nelson et al. 1960; Mathers and Watson 1962; Tomlinson 1962). Because clinoptilolite also exhibits relatively high selectivity for NH_4^+, some studies evaluated its use in the treatment of municipal wastewater (Mercer 1966; Ames 1967; Mercer et al. 1970).

Semmens and Seyfarth (1978) published isotherm data on ion exchange between Na-clinoptilolite, prepared using zeolite material from Buckhorn New Mexico, and the heavy metal ions Ba^{2+}, Cd^{2+}, Cu^{2+}, Pb^{2+}, and Zn^{2+}. They reported good reversibilities for the exchange couples Na^+/Ba^{2+}, Na^+/Cd^{2+}, and Na^+/Cu^{2+}, but not for Na^+/Pb^{2+} and Na^+/Zn^{2+}. They showed that heavy metals are concentrated well by clinoptilolite at low solution fractions of the heavy metals ($E_A < 0.1$), and they established the selectivity sequence $Pb^{2+} \approx Ba^{2+} >> Cu^{2+}, Zn^{2+}, Cd^{2+} > Na^+$. The reported isotherms were not complete, i.e. isotherm points were derived to heavy-metal solution fractions typically <0.6. Therefore, it is not possible to tell from the data whether full exchange between Na-clinoptilolite and the heavy metals could be achieved. The isotherm curves for barium and cadmium exchange level off at $\bar{E}_A \approx 0.6$-0.8, suggesting partial exchange for the Na^+/Ba^{2+} and Na^+/Cd^{2+} couples.

Also, the results of Semmens and Seyfarth (1978) indicated that the ion-exchange capacity of clinoptilolite depends significantly on the method used to pretreat the samples. For example, the ion-exchange capacity, determined from the amount of NH_4^+ eluted from the NH_4-form of the clinoptilolite by exchange with NaCl solution, tends to increase with repeated capacity determinations on the same zeolite sample. This result has important implications on the manner in which ion-exchange experiments on clinoptilolite are conducted if reproducible results are to be achieved and if valid extrapolation of experimental data to other clinoptilolite samples is desired.

Blanchard et al. (1984) conducted batch experiments to study the removal of NH_4^+ and heavy metals from waters by Na-clinoptilolite. The results show that clinoptilolite has a good selectivity for the NH_4^+ ion. Based on the isotherm data, they determined the selectivity series $Pb^{2+} > NH_4^+, Ba^{2+} > Cu^{2+}, Zn^{2+} > Cd^{2+} > Co^{2+}$.

Zamzow et al. (1990) conducted column experiments and measured the ion-exchange loading of heavy metals and other cations on clinoptilolite material from Owyhee County, Idaho, and from Ash Meadows, Nevada. The clinoptilolite samples were primarily in the Na- or Ca-form, but potassium and magnesium were also present in the zeolite. The loading tests were done by passing one liter of a 0.1 M solution of the cation of interest through a 30-cm long by 1-cm diameter glass column containing the zeolite powder. The measured heavy metal loading values, which ranged from 0 meq/g for mercury and 1.6 meq/g for lead, were used to determine the following selectivity series: $Pb^{2+} > Cd^{2+} > Cs^+ > Cu^{2+} > Co^{2+} > Cr^{3+} > Zn^{2+} > Ni^{2+} > Hg^{2+}$.

In addition to column experiments, Ames (1964a,b) also conducted batch experiments on clinoptilolite and alkali and alkaline-earth cations and presented ion-exchange isotherms for the couples Cs^+/K^+, Na^+/K^+, Cs^+/Na^+, Ca^{2+}/Na^+, Sr^{2+}/Na^+, and Sr^{2+}/Ca^{2+}. The ion-exchange equilibrium constants and Gibbs free energies derived by Ames (1964a,b) from the isotherm data are listed in Table 4. Additional thermodynamic values reported by Ames (1968) for Cs^+/K^+ and Cs^+/Na^+ exchange are also listed in the table. Other early studies on clinoptilolite ion exchange were reported by Frysinger (1962) for Na^+/Cs^+, and by Howery and Thomas (1965) for the binary mixtures of Na^+/Cs^+, Na^+/NH_4^+, and NH_4^+/Cs^+. The results of Frysinger (1962) demonstrated that clinoptilolite is selective for Cs^+ relative to Na^+, whereas Howery and Thomas (1965) noted a selectivity sequence for clinoptilolite of $Cs^+ > NH_4^+ >> Na^+$. The equilibrium constants and Gibbs free energies reported by Frysinger (1962) and Howery and Thomas (1965) are included in Table 4.

Chelishchev et al. (1973) presented isotherm data for ion exchange involving clinoptilolite and mixtures of Na^+/Cs^+, Na^+/Rb^+, Na^+/K^+, Na^+/Li^+, and Na^+/Sr^{2+}. Their results show that the exchange reactions are all reversible, and the selectivity series is $Cs^+ > Rb^+ > K^+ > Na^+ > Sr^{2+} > Li^+$. Thermodynamic data derived from their isotherm data are listed in Table 4.

Townsend and co-workers conducted a number of ion-exchange studies using clinoptilolite samples from Hector, California. Barrer and Townsend (1976b) studied the exchange equilibria between the NH_4^+-form of clinoptilolite and copper ammine and zinc ammine metal complexes. The exchange isotherms indicated maximum exchange limits of 88% and 73% for the copper and zinc, respectively. Reversibility tests showed good reversibility for the copper ammine + clinoptilolite system, but not for zinc ammine + clinoptilolite. The method used to test for reversibility in the latter case involved drying the zeolite sample at 353 K before measuring the reverse isotherm points, and this apparently allowed the zinc ions to enter exchange sites not accessible to the complexed species at ambient temperatures (Barrer and Townsend 1976b). Thus the reverse isotherm points were at higher values of $\bar{E}_{Zn}$ than for the forward isotherm points. The results of Barrer and Townsend (1976b) also showed that ammination improves the selectivity of clinoptilolite for copper and zinc, and that the zeolite shows increasing selectivity for the complexed transition metal ion with decreasing concentration. Values of equilibrium constants and Gibbs free energy of exchange were not calculated by the authors because the activity coefficients of the complexed ions in solution were not known and partly because the systems exhibited incomplete exchange (Barrer and Townsend 1976b).

Townsend and Loizidou (1984) published an isotherm for Na^+/NH_4^+ exchange in

Table 4. Literature values of equilibrium constants and Gibbs free energies for ion-exchange reactions involving clinoptilolite.

Ion-Exchange Reaction	$\ln K_{(A,B)}$	$\Delta G^{\circ}_{(A,B)}$ (kJ/mol)	Reference	Remarks
$Li^+ + NaL \Leftrightarrow Na^+ + LiL$	−2.34	5.74	Chelishchev et al. (1973)	T = 295 K
$Na^+ + CsL \Leftrightarrow Cs^+ + NaL$	−3.15	8.46	Frysinger (1962)	T = 323 K
$Na^+ + CsL \Leftrightarrow Cs^+ + NaL$	−2.58	7.47	Frysinger (1962)	T = 348 K
$Na^+ + KL \Leftrightarrow K^+ + NaL$	−2.54	6.29	Ames (1964a)	T = 298 K; Hector, California clinoptilolite; CEC = 1.7 meq/g
$K^+ + NaL \Leftrightarrow Na^+ + KL$	3.22	−7.98	Pabalan (1994)	T = 298 K; Death Valley Junction, California clinoptilolite; CEC = 2.04 meq/g; aqueous activity coefficients calculated using the Pitzer equations
$K^+ + NaL \Leftrightarrow Na^+ + KL$	3.68	−9.03	Chelishchev et al. (1973)	T = 295 K
$2K^+ + CaL_2 \Leftrightarrow Ca^{2+} + 2KL$	2.22	−5.50	Vucinic (1998a)	T = 298 K; Serbia clinoptilolite; incomplete exchange of Ca^{2+} by NH_4^+; isotherm data normalized
$Rb^+ + NaL \Leftrightarrow Na^+ + RbL$	3.91	−9.60	Chelishchev et al. (1973)	T = 295 K
$Cs^+ + NaL \Leftrightarrow Na^+ + CsL$	4.14	−10.2	Chelishchev et al. (1973)	T = 295 K
$Cs^+ + NaL \Leftrightarrow Na^+ + CsL$	3.28	−8.07	Ames (1968)	T = 296 K; Pierre clinoptilolite; CEC = 1.32 meq/g
$Cs^+ + NaL \Leftrightarrow Na^+ + CsL$	3.86	−9.56	Ames (1964a)	T = 298 K; John Day, Oregon clinoptilolite; CEC = 2.0 meq/g
$Cs^+ + NaL \Leftrightarrow Na^+ + CsL$	3.95	−9.79	Ames (1964a)	T = 298 K; Hector, California clinoptilolite; CEC = 1.7 meq/g
$Cs^+ + NaL \Leftrightarrow Na^+ + CsL$	4.02	−10.1	Howery and Thomas (1965)	T = 303K; CEC = 2.04±0.02 meq/g
$Cs^+ + NaL \Leftrightarrow Na^+ + CsL$	3.21	−9.16	Ames (1964a)	T = 343 K; John Day, Oregon clinoptilolite; CEC = 2.0 meq/g
$Cs^+ + NaL \Leftrightarrow Na^+ + CsL$	2.94	−8.38	Ames (1964a)	T = 343 K; Hector, California clinoptilolite; CEC = 1.7 meq/g
$Cs^+ + KL \Leftrightarrow K^+ + CsL$	1.55	−3.81	Ames (1968)	T = 296 K; Hector, California clinoptilolite; CEC = 1.7 meq/g

$Cs^+ + KL \Leftrightarrow K^+ + CsL$	1.30	–3.22	Ames (1964a)	T = 298 K; Hector, California clinoptilolite; CEC = 1.7 meq/g
$Cs^+ + NH_4L \Leftrightarrow NH_4^+ + CsL$	1.69	–4.27	Howery and Thomas (1965)	T = 303K; CEC = 2.04±0.01 meq/g
$NH_4^+ + NaL \Leftrightarrow Na^+ + NH_4L$	1.63	–4.04	Townsend and Loizidou (1984)	T = 298 K; Hector, California clinoptilolite; CEC = 2.19; isotherm data normalized to $\overline{E}_{NH_4,max}$ of 0.765
$NH_4^+ + NaL \Leftrightarrow Na^+ + NH_4L$	1.65	–4.09	Jama and Yucel (1990)	T = 298 K; Western Anatolia clinoptilolite; isotherm data normalized to $\overline{E}_{NH_4,max}$ of 0.68
$NH_4^+ + NaL \Leftrightarrow Na^+ + NH_4L$	2.14	–5.40	Howery and Thomas (1965)	T = 303K; CEC = 2.04±0.01 meq/g
$NH_4^+ + NaL \Leftrightarrow Na^+ + NH_4L$	2.07	–5.73	Barrer et al. (1967)	T = 333 K; Hector, California clinoptilolite; CEC = 1.83
$NH_4^+ + KL \Leftrightarrow K^+ + NH_4L$	–0.163	0.403	Jama and Yucel (1990)	T = 298 K; Western Anatolia clinoptilolite; isotherm data normalized to $\overline{E}_{NH_4,max}$ of 0.50
$2NH_4^+ + CaL_2 \Leftrightarrow Ca^{2+} + 2NH_4L$	4.98	–12.3	Jama and Yucel (1990)	T = 298 K; Western Anatolia clinoptilolite; isotherm data normalized to $\overline{E}_{NH_4,max}$ of 0.64
$2NH_4^+ + CaL_2 \Leftrightarrow Ca^{2+} + 2NH_4L$	4.868	–12.07	Vucinic (1998a)	T = 298 K; Serbia clinoptilolite; complete exchange reported, but used a lower CEC (1.5, instead of 2.1 meq/g) to calculate isotherms
$Ca^{2+} + 2NaL \Leftrightarrow 2Na^+ + CaL_2$	–0.161	0.400	Ames (1964b)	T = 298 K; Hector, California clinoptilolite; CEC = 1.7 meq/g
$Ca^{2+} + 2NaL \Leftrightarrow 2Na^+ + CaL_2$	–1.65	4.09	Pabalan (1994)	T = 298 K; Death Valley Junction, California clinoptilolite; CEC = 2.04 meq/g; aqueous activity coefficients calculated using the Pitzer equations
$Ca^{2+} + 2KL \Leftrightarrow 2K^+ + CaL_2$	–8.50	21.1	Pabalan and Bertetti (1999)	T = 298 K; Death Valley Junction, California clinoptilolite; CEC = 2.04 meq/g; aqueous activity coefficients

				calculated using the Pitzer equations
$Sr^{2+} + 2NaL \Leftrightarrow 2Na^{+} + SrL_2$	1.19	–2.92	Chelishchev et al. (1973)	T = 295 K
$Sr^{2+} + 2NaL \Leftrightarrow 2Na^{+} + SrL_2$	–1.14	2.83	Pabalan and Bertetti (1999)	T = 298 K; Death Valley Junction, California clinoptilolite; CEC = 2.04 meq/g; aqueous activity coefficients calculated using the Pitzer equations
$Sr^{2+} + 2NaL \Leftrightarrow 2Na^{+} + SrL_2$	0.255	–0.632	Ames (1964b)	T = 298 K; Hector, California clinoptilolite; CEC = 1.7 meq/g
$Sr^{2+} + 2KL \Leftrightarrow 2K^{+} + SrL_2$	–6.52	16.2	Pabalan and Bertetti (1999)	T = 298 K; Death Valley Junction, California clinoptilolite; CEC = 2.04 meq/g; aqueous activity coefficients calculated using the Pitzer equations
$Sr^{2+} + CaL_2 \Leftrightarrow Ca^{2+} + SrL_2$	0.113	–0.280	Ames (1964b)	T = 298 K; Hector, California clinoptilolite; CEC = 1.7 meq/g
$Cd^{2+} + 2NaL \Leftrightarrow 2Na^{+} + CdL_2$	–2.23	5.53	Loizidou and Townsend (1987a)	T = 298 K; Hector, California clinoptilolite; CEC = 2.19; isotherm data normalized to $\overline{E}_{Cd,max.}$ of 0.656; Cl^{-} co-anion
$Cd^{2+} + 2NaL \Leftrightarrow 2Na^{+} + CdL_2$	–1.83	4.54	Loizidou and Townsend (1987a)	T = 298 K; Hector, California clinoptilolite; CEC = 2.19; isotherm data normalized to $\overline{E}_{Cd,max.}$ of 0.656; NO_3^{-} co-anion
$Cd^{2+} + 2NaL \Leftrightarrow 2Na^{+} + CdL_2$	0.388	–0.962	Torres (1999)	T = 298 K; CEC = 1.60 (measured); 1.92 (calculated); isotherm data not normalized
$Pb^{2+} + 2NaL \Leftrightarrow 2Na^{+} + PbL_2$	3.085 (at 0.1 N) 3.215 (at 0.5 N)	–7.64 (at 0.1 N) –7.96 (at 0.5 N)	Loizidou and Townsend (1987b)	T = 298 K; Hector, California clinoptilolite; CEC = 2.19; isotherm data normalized to $\overline{E}_{Pb,max.}$ of 0.795; NO_3^{-} co-anion

clinoptilolite and compared their derived thermodynamic constants with those determined by other workers. Their results showed that clinoptilolite exhibits a high preference for NH_4^+ over Na^+. However, the exchange is incomplete, with $\bar{E}_{NH4,max.}$ equal to 0.765, indicating that the theoretical exchange capacity estimated from the chemical analysis of the zeolite cannot be attained with NH_4^+. Townsend and Loizidou (1984) noted that the partial exchange of NH_4^+ for Na^+ observed in their experiment is in contrast to results of Howery and Thomas (1965), who observed full exchange of the Na^+ in their zeolite by NH_4^+. However, Howery and Thomas (1965) did not publish their isotherm data. On the other hand, Barrer et al. (1967) observed full exchange of Na^+ by NH_4^+ in their experiments at a temperature of 333 K. The Gibbs free energy reported by Townsend and Loizidou (1984), listed in Table 4, is different from that of Howery and Thomas (1965). Townsend and Loizidou (1984) suggested that this difference could be due to the variability of the mineral samples.

Jama and Yucel (1990) published isotherm data on ion exchange involving clinoptilolite from Western Anatolia and mixtures of Na^+/NH_4^+, K^+/NH_4^+, and Ca^{2+}/NH_4^+. The results showed that clinoptilolite exhibits very high preference for NH_4^+ over Na^+ and Ca^{2+}, but not over K^+. Consistent with the results of Townsend and Loizidou (1984), Jama and Yucel (1990) observed that full replacement of the cations by NH_4^+ was not achieved, with $\bar{E}_{NH4,max.}$ equal to 0.68, 0.50, and 0.64 for Na-, K-, and Ca-clinoptilolite, respectively. Jama and Yucel (1990) also noted that the exchange reactions were not strictly binary, i.e. when the Na-, K-, or Ca-forms of clinoptilolite were reacted with the mixed cation/ammonium ion solutions, other cations (e.g. K^+, Ca^{2+}, and Mg^{2+}) were also detected in solution. However, these other cations amounted to less than 5% of the total cation equivalents in solution. Equilibrium constants and Gibbs free energies reported by Jama and Yucel (1990) are listed in Table 4.

Data on ion exchange between clinoptilolite material from Serbia and aqueous mixtures of K^+/Ca^{2+} and NH_4^+/Ca^{2+} were presented by Vucinic (1998a). In contrast to the results of Jama and Yucel (1990) and Townsend and Loizidou (1984), Vucinic (1998a) reported that complete exchange was achieved between Ca^{2+} and NH_4^+. However, Vucinic (1998a) used a CEC value of 1.50 meq/g to calculate the isotherm points, instead of 2.10, which is the CEC measured by the author from exchange with NH_4^+ and from the Al^{3+} content of the clinoptilolite. The author also reported that only partial exchange was achieved between Ca^{2+} and K^+, with $\bar{E}_{K,max.}$ of about 0.68, which is inconsistent with the complete exchange observed by Pabalan and Bertetti (1999). Equilibrium constants and Gibbs free energies reported by Vucinic (1998a) for a temperature of 298 K are listed in Table 4.

Isotherms for the exchange of Cd^{2+} into the Na- and NH_4-forms of clinoptilolite were published by Loizidou and Townsend (1987a) based on experiments using chloride or nitrate as the co-anion. Their results indicate that the ion-exchange reaction is reversible for the Cd^{2+}/Na^+ couple, but not for the Cd^{2+}/NH_4^+ couple. Only partial exchange was observed for the two couples, with $\bar{E}_{Cd,max.}$ equal to 0.656 for the Cd^{2+}/Na^+ exchange and 0.810 for the Cd^{2+}/NH_4^+ reaction. Also, clinoptilolite was observed to be more selective for cadmium when nitrate is the co-anion rather than chloride. The observed effect of co-anion type on selectivity was attributed to the tendency of cadmium to form aqueous complexes with the chloride ion. The observed incomplete exchange for the Cd^{2+}/Na^+ couple is consistent with more recent data from Torres (1999). Thermodynamic data for the Cd^{2+}/Na^+ exchange reactions reported by Loizidou and Townsend (1987a) and Torres (1999) are listed in Table 4.

Loizidou and Townsend (1987b) also studied ion exchange between clinoptilolite (Na- and NH_4-forms) and Pb^{2+} in solutions with nitrate as the co-anion. These experiments showed that the Pb^{2+}/Na^+ exchange is reversible, whereas the Pb^{2+}/NH_4^+ exchange is irreversible. The irreversible behavior was argued to be due primarily to the ternary rather than binary nature of the exchange. For example, the NH_4-form of clinoptilolite used in the experiments still had Na^+ in its structure, which apparently exchanged out when reacted with the Pb^{2+}/NH_4^+ solutions. Thermodynamic data for the Pb^{2+}/Na^+ exchange reaction reported by Loizidou and Townsend (1987b) are included in Table 4.

Pabalan (1994) and Pabalan and Bertetti (1999) conducted ion-exchange experiments involving homoionic Na- and K-clinoptilolite, prepared using zeolite material from Death Valley Junction, California, and aqueous mixtures of K^+/Na^+, Ca^{2+}/Na^+, Sr^{2+}/Na^+, Sr^{2+}/K^+, and Ca^{2+}/K^+. The isotherm data were used to derive equilibrium constants and Gibbs free energies for the exchange reactions. The values, listed in Table 4, are significantly different from those derived by other investigators. The difference is mainly due to the different activity coefficient model used for the aqueous ions (Pabalan 1994). Pabalan and Bertetti (1999) checked the self-consistency of the Gibbs free energies for binary ion exchange using the triangle rule (Eqn. 19) on exchanges involving: (1) K^+/Na^+, Ca^{2+}/Na^+, and Ca^{2+}/K^+; and (2) K^+/Na^+, Sr^{2+}/Na^+, and Sr^{2+}/K^+. The value of $\Delta G^\circ_{(Ca,K)}$ calculated from $\Delta G^\circ_{(Ca,Na)}$ minus $2\Delta G^\circ_{(K,Na)}$ equals 20.1±0.2 kJ/mol, which compares relatively well with the value 21.1±0.5 kJ/mol derived from the K^+/Ca^{2+} experiment. The value of $\Delta G^\circ_{(Sr,K)}$ calculated from $\Delta G^\circ_{(Sr,Na)}$ minus $2\Delta G^\circ_{(K,Na)}$ equals 18.8±0.2 kJ/mol, significantly higher than the value 16.2±0.3 J/mol derived from the K^+/Sr^{2+} experiment. The authors argued that because experiments on K^+/Ca^{2+} and K^+/Sr^{2+} ion exchange, conducted at 0.05 N only, are more limited compared with the Na^+/K^+, Na^+/Ca^{2+}, and Na^+/Sr^{2+} experiments that were conducted at 0.005, 0.05, and 0.5 N, the derived $\Delta G^\circ_{(Ca,K)}$ and $\Delta G^\circ_{(Sr,K)}$ are not as well-constrained as $\Delta G^\circ_{(K,Na)}$, $\Delta G^\circ_{(Ca,Na)}$, and $\Delta G^\circ_{(Sr,Na)}$. The authors suggested that K^+/Ca^{2+} and K^+/Sr^{2+} experiments at other solution concentrations are needed to better constrain the regression of Equation (25) and the subsequent calculation of $\Delta G^\circ_{(Sr,K)}$ and $\Delta G^\circ_{(Ca,K)}$. The isotherm data of Pabalan (1994) and Pabalan and Bertetti (1999) were also used to derive parameters for the Margules solid-solution model, predict ion-exchange equilibria as a function of total solution concentration, and calculate aqueous compositions based on zeolite analysis. These calculations and the isotherm data from the two studies are presented in a later section.

Several other references have useful information on the ion-exchange properties of clinoptilolite. These references include White (1988), Dyer and Jozefowicz (1992), Chmielewska-Horvathova and Lesny (1992), Malliou et al. (1994), Pode et al. (1995), Tsukanova et al. (1995), Tarasevich et al. (1996), de Barros et al. (1997), Vucinic (1998b), Cooney et al. (1999), Loizidou et al. (1992), Ali et al. (1999), and Faghihian et al. (1999).

Erionite [$NaK_2MgCa_{1.5}(Al_8Si_{28}O_{72})\cdot 28H_2O$]

The ion-exchange properties of erionite were studied by Sherry (1979). Natural erionite from Jersey Valley, Nevada, was converted to the Na-form by reacting with 1.0 N NaCl solutions at 298 K and 363 K. It was determined that two K^+ ions per unit cell, probably located in the cancrinite cages of the zeolite structure, could not be exchanged even after exhaustively exchanging with NaCl solutions, most likely due to steric hindrance. Isotherms for ion exchange with Li^+, K^+, Rb^+, Cs^+, Ca^{2+}, Sr^{2+}, and Ba^{2+} were determined at a total solution concentration of 0.1 N and at temperatures of 278 K and 298 K. Complete replacement of Na^+ by K^+, Rb^+, and Cs^+ was observed, but only incomplete exchange of Na^+ by Ca^{2+}, Sr^{2+}, and Ba^{2+} was attained. Sherry (1979) concluded that it is extremely difficult to replace all the exchangeable cations (those in large erionite cages) in natural erionite with divalent cations. The isotherm shapes of the

K^+/Na^+, Rb^+/Na^+, and Cs^+/Na^+ exchange reveal the strong preference of erionite for K^+, Rb^+, and Cs^+ ions over Na^+. Isotherms for Ca^{2+}/Na^+, Sr^{2+}/Na^+, and Ba^{2+}/Na^+ show that erionite prefers Ca^{2+}, Sr^{2+}, and Ba^{2+} ions over Na^+ at low loading levels (i.e. at low values of $\bar{E}_A$), but this preference for the divalent cations drastically decreases with increasing loading, even reversing at high loadings. The selectivity series exhibited by erionite at low loadings is $Rb^+ > Cs^+ \geq K^+ > Ba^{2+} > Sr^{2+} > Ca^{2+} > Na^+ > Li^+$. Equilibrium constants and Gibbs free energies were not calculated by Sherry (1979) from the isotherm data.

Chelishchev and Volodin (1977) conducted isotherm experiments on K^+/Na^+, Cs^+/Na^+, Li^+/Na^+, and Rb^+/Cs^+ exchange using natural erionite from Georgia (of the former U.S.S.R.) at 293 K and total solution concentration of 0.1 N. The isotherm shapes are similar to those observed by Sherry (1979). However, in contrast to the results of Sherry (1979), Chelishchev and Volodin's data indicate that erionite has higher selectivity for Cs^+ than for Rb^+. The selectivity series for erionite determined by Chelishchev and Volodin (1977) is $Cs^+ > Rb^+ > K^+ > Na^+ > Li^+$. Equilibrium constants and Gibbs free energies reported by Chelishchev and Volodin (1977) are listed in Table 5.

Ames (1964a,b) also studied the ion-exchange properties of natural erionite using material from Pine Valley, Nevada, that is 90% or higher in purity and has a CEC of 2.2 meq/g. Isotherms for Na^+/K^+, Cs^+/Na^+, Ca^{2+}/Na^+, Sr^{2+}/Na^+, and Sr^{2+}/Ca^{2+} exchange were determined at 298 K and 1.0 N solution concentration. In contrast to the results of Sherry (1979), the results of Ames (1964b) indicate that full exchange between Na^+ and the divalent cations Sr^{2+} and Ca^{2+} was achieved. The equilibrium constants and Gibbs free energies reported by Ames (1964a) are given in Table 5.

Ferrierite [$(Na,K)Mg_2Ca_{0.5}(Al_6Si_{30}O_{72})\cdot 20H_2O$]

Ahmad and Dyer (1984) conducted experiments to determine the ease of replacement of Na^+, K^+, and Mg^{2+} ions in ferrierite by counterions (Na^+, K^+, NH_4^+, Ca^{2+}, and Mg^{2+}) and mixtures of counterions (Na^+/K^+, Na^+/NH_4^+, K^+/NH_4^+, Ca^{2+}/Mg^{2+}). The experiments used natural ferrierite from Lovelock, Nevada, purified by several sedimentations in a water column and sieving through a 150-mesh sieve. The CEC of the natural ferrierite, determined from the amount of K^+, Na^+, Ca^{2+}, and Mg^{2+} leached by a 1 M ammonium acetate solution, was 0.81 meq/g, which is less than the value of 1.88 meq/g[11] calculated by Ahmad and Dyer (1984) from the unit-cell composition of the ferrierite. The authors noted that 0.24 Mg^{2+} ions per unit cell were not available for exchange, possibly due to the occurrence of some of the Mg^{2+} in small cavities of the ferrierite structure. The results indicate that ferrierite is selective for NH_4^+ and K^+ and, to lesser degrees, for Na^+, Ca^{2+}, and Mg^{2+}, but the selectivity for specific cations is diminished by the presence of other cations. Ahmad and Dyer (1984) also prepared a K-form of ferrierite from the 'as received' natural material, and they reacted the K-ferrierite with aqueous solutions of NH_4^+, Na^+, Rb^+, Ag^+, Cs^+, Tl^+, Mg^{2+}, Ca^{2+}, Sr^{2+}, and Ba^{2+} (with Cl^- or NO_3^- as the co-anion). The amount of K^+ released from K-ferrierite by successive treatments of the counterions decreased in the order $Rb^+ > NH_4^+ \approx Cs^+ > Tl^+ > Ag^+ > Ba^{2+} > Mg^{2+} \approx Na^+ > Sr^{2+} > Ca^{2+}$.

Ahmad and Dyer (1988) conducted a more detailed study on the ion-exchange properties of ferrierite from Lovelock, Nevada. Isotherms for ion exchange between K-ferrierite and Na^+, Rb^+, Cs^+, Ag^+, Tl^+, NH_4^+, Mg^{2+}, Ca^{2+}, Sr^{2+}, and Ba^{2+} were determined at a total solution concentration of 0.1 N and at temperatures in the range 297 to 353 K.

---

[11]The CEC value of 1.08 meq/g given in Ahmad and Dyer (1984) is incorrect (pers. comm., A. Dyer, August 9, 2000).

Table 5. Literature values of equilibrium constants and Gibbs free energies for ion-exchange reactions involving erionite.

Ion-Exchange Reaction	$\ln K_{(A,B)}$	$\Delta G^{\circ}_{(A,B)}$ (kJ/mol)	Reference	Remarks
$Li^{+} + NaL \Leftrightarrow Na^{+} + LiL$	−2.34	5.70	Chelishchev and Volodin (1977)	T = 293 K; Georgia (of the former U.S.S.R.) erionite
$Rb^{+} + NaL \Leftrightarrow Na^{+} + RbL$	3.04	−7.41	Chelishchev and Volodin (1977)	T = 293 K; Georgia (of the former U.S.S.R.) erionite
$K^{+} + NaL \Leftrightarrow Na^{+} + KL$	2.04	−4.97	Chelishchev and Volodin (1977)	T = 293 K; Georgia (of the former U.S.S.R.) erionite
$K^{+} + NaL \Leftrightarrow Na^{+} + KL$	2.40	−5.94	Ames (1964a)	T = 298 K; Pine Valley, Nevada erionite; CEC = 2.2 meq/g
$Cs^{+} + NaL \Leftrightarrow Na^{+} + CsL$	4.52	−11.0	Chelishchev and Volodin (1977)	T = 293 K; Georgia (of the former U.S.S.R.) erionite
$Cs^{+} + NaL \Leftrightarrow Na^{+} + CsL$	3.49	−8.65	Ames (1964a)	T = 298 K; Pine Valley, Nevada erionite; CEC = 2.2 meq/g
$Cs^{+} + NaL \Leftrightarrow Na^{+} + CsL$	2.86	−8.16	Ames (1964a)	T = 343 K; Pine Valley, Nevada erionite; CEC = 2.2 meq/g
$Ca^{2+} + 2NaL \Leftrightarrow 2Na^{+} + CaL_2$	0.0488	−0.12	Ames (1964b)	T = 298 K; Pine Valley, Nevada erionite; CEC = 2.2 meq/g
$Sr^{2+} + 2NaL \Leftrightarrow 2Na^{+} + SrL_2$	−0.231	0.57	Ames (1964b)	T = 298 K; Pine Valley, Nevada erionite; CEC = 2.2 meq/g
$Sr^{2+} + CaL_2 \Leftrightarrow Ca^{2+} + SrL_2$	−0.277	0.69	Ames (1964b)	T = 298 K; Pine Valley, Nevada erionite; CEC = 2.2 meq/g

The results indicate that the ion-exchange reactions were reversible within the composition and temperature ranges examined except for Ag^+/K^+ exchange, which had unusual features at high values of $\bar{E}_{Ag}$,Rb^+, Cs^+, and Tl^+ were preferred by ferrierite over K^+. In all cases except Ag^+, an increase in temperature increased the extent of replacement of K^+ by the incoming cation, although the temperature effect for Tl^+ is very small. The isotherm data were extrapolated to obtain the maximum equivalent fraction of the ingoing cation in the zeolite. Equilibrium constants and Gibbs free energies were derived from the normalized isotherm data. The Gibbs free energies at 297 K reported by Ahmad and Dyer (1988) and the corresponding equilibrium constants are listed in Table 6. From the Gibbs free energies, the authors determined the following affinity sequence for ferrierite: $Tl^+ \approx Cs^+ > Rb^+ > K^+ > NH_4^+ > Ag^+ > Na^+ > Ba^{2+} > Sr^{2+} > Ca^{2+} > Mg^{2+}$.

Townsend and Loizidou (1984) also used ferrierite from Lovelock, Nevada, in their ion-exchange experiments involving Na^+/NH_4^+ mixtures. As in the studies by Ahmad and Dyer (1984, 1988), the natural material was purified by a sedimentation procedure to remove the fines. A homoionic Na-ferrierite was prepared by exchange with $NaNO_3$ solution at 298 K and also at 343 K. The isotherm for NH_4^+/Na^+ exchange was determined at 298 K and a total solution concentration of 0.1 N. Although the ferrierite had a CEC of 1.80 meq/g based on the Al^{3+} content of the zeolite, a maximum level of exchange, $\bar{E}_{NH4,max.}$, of only 0.499 was achieved. The NH_4^+/Na^+ isotherm shape is similar to the NH_4^+/Na^+ results for mordenite. However, in contrast to the mordenite results, the solution analyses confirmed that the ion exchange with ferrierite was ternary in nature; K^+, along with Na^+, was released from the zeolite upon reaction with the NH_4^+/Na^+ solutions. Townsend and Loizidou (1984) noted that at low $\bar{E}_{NH4,max.}$ NH_4^+ exchanged principally with Na^+. It was only as $\bar{E}_{NH4,max.}$ increased that significant quantities of K^+ were also involved in the exchange reaction. No thermodynamic parameters for this system were derived because of the ternary nature of the exchange reactions.

Loizidou and Townsend (1987a) reported isotherms for ion exchange between Na- and NH_4-ferrierite, prepared from natural ferrierite from Lovelock, Nevada, and mixtures of Pb^{2+}/Na^+ and Pb^{2+}/NH_4^+ (0.1 or 0.5 N; NO_3^- co-anion). Maximum levels of exchange, $\bar{E}_{Pb,max.}$, determined for systems with Na^+ and NH_4^+ as the counter-cations were 0.506 and 0.486, respectively. The Pb^{2+}/NH_4^+ solutions analyzed after exchange contained only $Pb^{2+}$and NH_4^+ ions plus a trace of Ca^{2+}, indicating that the reaction is essentially binary. On the other hand, the Pb^{2+}/Na^+ exchange was irreversible due to the ternary nature of the exchange reaction. K^+ ions in the zeolite, which had not been removed by exhaustive exchanges with $NaNO_3$ during the preparation of the 'homoionic' Na-ferrierite, were released into solution when the zeolite was reacted with Pb^{2+}/Na^+ mixtures. Thermodynamic values for Pb^{2+}/NH_4^+ exchange reported by Loizidou and Townsend (1987b) are listed in Table 6.

In a separate study, Loizidou and Townsend (1987b) also measured isotherms for Cd^{2+}/Na^+ and Cd^{2+}/NH_4^+ exchange at 298 K (0.1 N; Cl^- or NO_3^- co-anion). The results show that both exchange reactions are reversible. Only traces of Ca^{2+} and no K^+ were found in solution after Cd^{2+}/Na^+ exchange, and negligible quantities of Ca^{2+} and K^+ were found in solution after Cd^{2+}/NH_4^+ exchange. Maximum levels of exchange, $\bar{E}_{Cd,max.}$, determined for the Cd^{2+}/Na^+ and Cd^{2+}/NH_4^+ systems are 0.339 and 0.280, respectively, independent of the co-anion. Similar to the other zeolites, ferrierite shows stronger selectivity for Cd^{2+} when NO_3^- is the co-anion rather than Cl^- due to the complexation of Cd^{2+} with Cl^-. In addition, the Na-ferrierite shows a higher preference for Cd^{2+} than the NH_4-ferrierite. Thermodynamic parameters reported by Loizidou and Townsend (1987b), listed in Table 6, indicate that Cd^{2+} is not preferred over Na^+or NH_4^+ by ferrierite.

Table 6. Literature values of equilibrium constants and Gibbs free energies for ion-exchange reactions involving ferrierite.

Ion-Exchange Reaction	$\ln K_{(A,B)}$	$\Delta G^{\circ}_{(A,B)}$ (kJ/mol)*	Reference	Remarks
$Na^{+} + KL \Leftrightarrow K^{+} + NaL$	−0.089	0.22	Ahmad and Dyer (1988)	T = 297 K; Lovelock, Nevada ferrierite; normalized isotherm
$Rb^{+} + KL \Leftrightarrow K^{+} + RbL$	1.38	−3.42	Ahmad and Dyer (1988)	T = 297 K; Lovelock, Nevada ferrierite; normalized isotherm
$Cs^{+} + KL \Leftrightarrow K^{+} + CsL$	1.98	−4.90	Ahmad and Dyer (1988)	T = 297 K; Lovelock, Nevada ferrierite; normalized isotherm
$NH_4^{+} + KL \Leftrightarrow K^{+} + NH_4L$	0.948	−2.34	Ahmad and Dyer (1988)	T = 297 K; Lovelock, Nevada ferrierite; normalized isotherm
$Ag^{+} + KL \Leftrightarrow K^{+} + AgL$	0.33	−0.82	Ahmad and Dyer (1988)	T = 297 K; Lovelock, Nevada ferrierite; normalized isotherm
$Tl^{+} + KL \Leftrightarrow K^{+} + TlL$	1.99	−4.91	Ahmad and Dyer (1988)	T = 297 K; Lovelock, Nevada ferrierite; normalized isotherm
$Mg^{2+} + 2KL \Leftrightarrow 2K^{+} + MgL_2$	−3.82	9.44	Ahmad and Dyer (1988)	T = 297 K; Lovelock, Nevada ferrierite; normalized isotherm
$Ca^{2+} + 2KL \Leftrightarrow 2K^{+} + CaL_2$	−3.63	8.96	Ahmad and Dyer (1988)	T = 297 K; Lovelock, Nevada ferrierite; normalized isotherm
$Sr^{2+} + 2KL \Leftrightarrow 2K^{+} + SrL_2$	−2.79	6.88	Ahmad and Dyer (1988)	T = 297 K; Lovelock, Nevada ferrierite; normalized isotherm
$Ba^{2+} + 2KL \Leftrightarrow 2K^{+} + BaL_2$	−2.02	4.98	Ahmad and Dyer (1988)	T = 297 K; Lovelock, Nevada ferrierite; normalized isotherm
$Pb^{2+} + 2NH_4L \Leftrightarrow 2NH_4^{+} + PbL_2$	−1.15	2.84	Loizidou and Townsend (1987a)	T = 298K; Lovelock, Nevada ferrierite; isotherm data normalized to $\overline{E}_{Pb,max.}$ of 0.486
$Cd^{2+} + 2NaL \Leftrightarrow 2Na^{+} + CdL_2$	−1.36 (Cl^{-} co-anion) −1.65 (NO_3^{-} co-anion)	3.36 (Cl^{-} co-anion) 4.10 (NO_3^{-} co-anion)	Loizidou and Townsend (1987b)	T = 298K; Lovelock, Nevada ferrierite; isotherm data normalized to $\overline{E}_{Cd,max.}$ of 0.339
$Cd^{2+} + 2NH_4L \Leftrightarrow 2NH_4^{+} + CdL_2$	−2.95 (Cl^{-} co-anion) −2.40 (NO_3^{-} co-anion)	7.32 (Cl^{-} co-anion) 5.96 (NO_3^{-} co-anion)	Loizidou and Townsend (1987b)	T = 298K; Lovelock, Nevada ferrierite; isotherm data normalized to $\overline{E}_{Cd,max.}$ of 0.280

*Values from Ahmad and Dyer (1988) and Loizidou and Townsend (1987a,b), given in kiloJoules per equivalent, were converted to kiloJoules per mole.

Heulandite [$(Na,K)Ca_4(Al_9Si_{27}O_{72})\cdot 24H_2O$]

Ames (1968) conducted K^+/Na^+, Cs^+/K^+, and Cs^+/Na^+ ion-exchange experiments using samples of heulandite (>95% pure) from Bay of Fundy, Nova Scotia. The material, with grain size in the range 0.25 to 0.50 mm, had a CEC of 0.43 meq/g, much lower than the CECs measured by Ames (1964a,b; 1968) for Hector and Pierre clinoptilolite (1.70 and 1.32 meq/g, respectively). The lower CEC of the 0.25 to 0.50 mm heulandite was explained as probably due to crystal 'stacking faults,' which presumably reduced the accessible intracrystalline exchange sites in the heulandite sample. The CEC of the heulandite increased to 1.02 meq/g when ground to <200 mesh (Ames 1968). The isotherm points for the K^+/Na^+ couple lie along a trend coincident with the K^+/Na^+ isotherm for clinoptilolite reported in the same study. The Cs^+/Na^+ isotherm indicates that heulandite prefers Cs^+ over Na^+, similar to clinoptilolite, although heulandite is less selective for Cs^+ compared with clinoptilolite. On the other hand, the data for Cs^+/K^+ exchange show that heulandite is selective for K^+ relative to Cs^+, in contrast to the preference of clinoptilolite for Cs^+ relative to K^+. The sigmoidal-shaped isotherms presented by Ames (1968) for both Cs^+/K^+ and Cs^+/Na^+ exchange suggest that selectivity reversal occurs at an intermediate heulandite composition. Thermodynamic parameters reported by Ames (1968) are listed in Table 7.

Table 7. Equilibrium constants and Gibbs free energies for ion exchange at 296 K involving a heulandite specimen from Bay of Fundy, Nova Scotia (CEC = 0.43 meq/g). Data from Ames (1968).

Ion-Exchange Reaction	$\ln K_{(A,B)}$	$\Delta G^\circ_{(A,B)}$ (kJ/mol)
$K^+ + NaL \Leftrightarrow Na^+ + KL$	2.43	–5.98
$Cs^+ + NaL \Leftrightarrow Na^+ + CsL$	1.36	–3.35
$Cs^+ + KL \Leftrightarrow K^+ + CsL$	–0.75	1.85

Ion-exchange studies using Siberian heulandite and Bulgarian clinoptilolite were conducted by Filizova (1974). The author determined the selectivity series K > Rb > Na > Li > Sr > Ba > Ca for heulandite, and the sequence Rb > K > Na > Ba > Sr > Ca > Li for clinoptilolite.

Al'tshuler and Shkurenko (1990, 1997) presented experimental data on ion exchange involving heulandite (CEC = 2.20 meq/g) and Li^+, Na^+, K^+, Rb^+, Cs^+, NH_4^+, Ca^{2+}, and Pb^{2+} from which they determined the selectivity series $Cs^+ > Rb^+ > Pb^{2+},K^+ > NH_4^+ > Ca^{2+} > Na^+ > Li^+$, which is somewhat inconsistent with the selectivity series determined by Filizova (1974). Al'tshuler and Shkurenko (1992) also conducted microcalorimetric measurements at 303 K of the heat effects of heulandite ion exchange. Their results show that the enthalpies of ion exchange decrease in the order $Na^+ > NH_4^+ > K^+ > Rb^+ > Cs^+$. In addition, Al'tshuler et al. (1996) reacted heulandite with binary mixtures of NaCl, $NiCl_2$, $CuCl_2$, $ZnCl_2$, and $MnCl_2$, and determined the selectivity series $Mn^{2+} > Na^+ > Zn^{2+} > Cu^{2+} > Ni^{2+}$.

Laumontite [$Ca_4(Al_8Si_{16}O_{48})\cdot 16H_2O$]

A natural laumontite from Bernisdale, Isle of Skye, Scotland, was examined by Dyer et al. (1991) as a candidate material for treating aqueous nuclear wastes. The material was ground and sieved to a 150–240 mesh fraction and was converted to a maximum Ca-exchanged form by repeated contact with 1 M $CaCl_2$ solution. The CEC of the zeolite, measured at different pHs, is 2.38, 3.73, 3.94, and 6.18 meq/g at a pH of 5, 8, 9.5, and 11.4, respectively. Isotherms for NH_4^+/Ca^{2+}, K^+/Ca^{2+}, Cs^+/Ca^{2+}, and Sr^{2+}/Ca^{2+} exchange were determined at room temperature and a total solution concentration of 0.01 N. The results show that laumontite prefers Ca^{2+} over K^+, NH_4^+, Na^+, and Cs^+, but the zeolite is selective for Sr^{2+} relative to Ca^{2+}. The isotherms for Na^+/Ca^{2+}, Cs^+/Ca^{2+}, NH_4^+/Ca^{2+}, K^+/Ca^{2+} exchange are very similar.

Mordenite [$Na_3KCa_2(Al_8Si_{40}O_{96})\cdot 28H_2O$]

Both natural and synthetic forms of mordenite have been studied extensively, and the use of synthetic mordenites in industry is common. However, there are distinct differences in the ion-exchange behavior of natural and synthetic mordenites. Because of these differences, several studies of synthetic mordenite are also discussed in this section.

Ames (1961) used a column apparatus to load natural zeolite samples with Cs^+ and competing cations to determine the selectivity of the zeolites and to gain insight into the mechanisms responsible for a particular selectivity sequence. Natural mordenite crystals from Nova Scotia, Canada, hand-picked to greater than 95% purity, were exposed to solutions of alkali and alkaline-earth cations at a total concentration of 0.01 N and a temperature of 298 K. Ames' results indicated a selectivity sequence of $Cs^+ > K^+ > Na^+ > Li^+$ for alkali metals and $Ba^{2+} > Sr^{2+} > Ca^{2+} > Mg^{2+}$ for alkaline earths.

In more extensive studies of the ion exchange of alkali metal cations and alkaline-earth cations in zeolites, Ames (1964a,b) conducted experiments on two types of mordenite: (1) AW-300, commercially available in the Na-form but prepared from natural mordenite, and (2) Zeolon, a synthetic mordenite also commercially available in the Na-form. The CECs of the commercial mordenites were determined to be 1.6 and 1.9 meq/g for the AW-300 and Zeolon, respectively. Ion-exchange experiments were conducted at 298 and 343 K with a solution total normality of 1.0 N. Isotherms are presented in Ames (1964a) for Na^+/Cs^+, K^+/Cs^+, K^+/Na^+ for AW-300 and K^+/Na^+ for Zeolon. The isotherms for Na^+/Cs^+ and K^+/Cs^+ exchange on AW-300 presented by Ames (1964a) suggest that incomplete exchange occurred. In contrast, the synthetic Zeolon mordenite exhibited complete exchange. The difference in the exchange behavior of AW-300 and Zeolon was explained by Ames (1964a) as due to the presence of 'stacking faults,' present in AW-300 but not in Zeolon. Thermodynamic data generated by Ames (1964a,b) for AW-300 and Zeolon are listed in Table 8.

Lu et al. (1981) produced ion-exchange isotherms for NH_4^+/Ag^+ and NH_4^+/K^+ systems using natural and synthetic mordenite. Experiments were conducted at 0.22 N total solution concentration and 298 K using a near homoionic NH_4-form of mordenite. Calculated CECs for the natural and synthetic mordenite were 2.17 and 2.19 meq/g, respectively. Differences in calculated ion-exchange equilibrium constants and Gibbs free energies between the natural and synthetic mordenite were attributed to 'stacking faults' within the natural mordenite. Thermodynamic values derived from the Lu et al. (1981) experiments are listed in Table 8.

Townsend and Loizidou (1984) investigated Na^+/NH_4^+ exchange equilibria for natural mordenite from Lovelock, Nevada. Ion-exchange experiments were conducted at 298 K and a solution concentration of 0.1 N (NO_3^- co-anion). A near homoionic Na-form of mordenite, with a calculated CEC of 2.11 meq/g based on the Al^{3+} content, was exchanged with NH_4^+. The results indicated that the exchange was reversible but incomplete, with only a 50.1% maximal level of exchange achieved. This result was in marked contrast to results for synthetic mordenite, which showed exchange to 100% of the calculated CEC (Barrer and Klinowski 1974b). The authors suggested that the differences in exchange properties between natural and synthetic mordenite may be due to 'stacking faults' within the natural mordenite crystal structure as originally proposed by Ames (1964a) and by Lu et al. (1981). Thermodynamic values derived by Townsend and Loizidou (1984) are listed in Table 8.

In follow-up studies of ion exchange in natural zeolites, Loizidou and Townsend (1987a,b) investigated the exchange of Pb^{2+} and Cd^{2+} on Na- and NH_4-forms of natural mordenite from Lovelock, Nevada. In one study (Loizidou and Townsend 1987b), near homoionic Na- and NH_4-forms of mordenite were exchanged with Pb^{2+}-bearing solutions at two different concentrations, 0.1 and 0.5 N (NO_3^- co-anion). Incomplete exchange was observed for both Na- and NH_4-forms of the zeolite. The Na-form exhibited a higher selectivity for Pb^{2+} than the NH_4-form. As with Na^+/NH_4^+ exchange studied previously (Townsend and Loizidou 1984), Pb^{2+}/Na^+ and Pb^{2+}/NH_4^+ exchanges were reversible, but values of $\bar{E}_{Pb,max.}$ for the Na^+ and NH_4^+ counter-ions were found to be 0.490 and 0.517, respectively, values that are probably equal within experimental uncertainty. Likewise, values of $\bar{E}_{Pb,max.}$ were independent of the total normality of the solutions. At similar solution concentrations, the Na-mordenite was more selective for Pb^{2+} than the NH_4^+-form.

In contrast to the Pb^{2+} study, Loizidou and Townsend (1987a) observed that Cd^{2+} was preferred by neither the Na- nor the NH_4-form of natural mordenite. In their study, Cd^{2+}-bearing solutions at a total concentration of 0.1 N and a temperature of 298 K were exchanged with Na- and NH_4-forms of natural mordenite. The effect of using Cl^- or NO_3^- as the co-anion was also investigated. Although the exchanges were reversible, incomplete exchange was again observed, with the values for $\bar{E}_{Cd,max.}$ reaching only 0.334 and 0.327 for the Na- and NH_4-forms, respectively. Interestingly, although the choice of co-anion affected the resulting thermodynamic values calculated for the exchange of Cd^{2+} (mordenite was less selective for Cd^{2+} when Cl^- was the co-anion), it did not affect the value of $\bar{E}_{Cd,max.}$. Differences in selectivity due to co-anion type were attributed to differences in aqueous complexation of Cd^{2+}. Thermodynamic values from Loizidou and Townsend (1987a,b) are listed in Table 8.

Liang and Hsu (1993) studied the sorption of Cs^+ and Sr^{2+} on natural mordenite and interpreted the results using a Freundlich isotherm model. Temperature (298 and 363 K) and pH (2B12) effects were investigated. The results indicated that temperature did not have a significant effect for Cs^+ exchange, but an increase in temperature did enhance the sorption of Sr^{2+}. Sorption of both Cs^+ and Sr^{2+} was reduced at low pH, probably due to ion-exchange competition from H^+. Cs^+ was preferentially exchanged over Sr^{2+}

Barrer and Klinowski (1974b) measured isotherms for a synthetic mordenite (Zeolon), which was prepared in Na- and NH_4-forms for their experiments. Experiments were conducted for the Na^+/Cs^+, NH_4^+/K^+, NH_4^+/Na^+, NH_4^+/Li^+, NH_4^+/Ca^{2+}, NH_4^+/Sr^{2+}, and NH_4^+/Ba^{2+} systems at 298 K using a total solution concentration of 0.05 N. All reactions were reversible except those involving Ca^{2+} and Sr^{2+}. A thermodynamic affinity sequence of $Cs^+ > K^+ > NH_4^+ > Na^+ > Ba^{2+} > Li^+$ was established. The selectivity sequence for alkali metals was similar to other zeolites, provided that normalization was used to compensate for incomplete exchange. Comparison of Na^+/Cs^+ exchange between natural and synthetic mordenite showed that the selectivity for Cs^+ relative to Na^+ was greater in the synthetic sample.

Barrer and Townsend (1976a) conducted an ion-exchange study of Zeolon (NH_4-form) and several transition metals. Experiments were conducted at 298 K and a solution total normality of 0.08 N. None of the metals exchanged completely, with maximum

Table 8. Literature values of equilibrium constants and Gibbs free energies for ion-exchange reactions involving mordenite.

Ion-Exchange Reaction	$\ln K_{(A,B)}$	$\Delta G^{\circ}_{(A,B)}$ (kJ/mol)*	Reference	Remarks
$H^+ + NaL \Leftrightarrow Na^+ + HL$	1.67	–3.86	Wolf et al. (1978)	T = 278 K; synthetic
	1.67	–4.07		T = 293 K; synthetic
$H^+ + KL \Leftrightarrow H^+ + KL$	–1.19	2.76	Wolf et al. (1978)	T = 274 K; synthetic
	–1.36	3.31		T = 293 K; synthetic
$Li^+ + HL \Leftrightarrow H^+ + LiL$	–5.3	13.0	Golden & Jenkins (1981)	T = 298 K; H-Zeolon
$Li^+ + NaL \Leftrightarrow Na^+ + LiL$	–3.0	7.5	Golden & Jenkins (1981)	T = 298 K; Na-Zeolon
$Na^+ + KL \Leftrightarrow K^+ + NaL$	–1.94	4.82	Ames (1964a)	T = 298 K; Zeolon
$Na^+ + KL \Leftrightarrow K^+ + NaL$	–3.0	7.4	Ames (1964a)	T = 298 K; AW-300
$K^+ + NaL \Leftrightarrow Na^+ + KL$	2.45	–5.58	Wolf et al. (1978)	T = 278 K; synthetic
	2.22	–5.42		T = 293 K; synthetic
$K^+ + NH_4L \Leftrightarrow NH_4^+ + KL$	0.456	–1.129	Lu et al. (1981)	T = 298 K; synthetic
$K^+ + NH_4L \Leftrightarrow NH_4^+ + KL$	0.539	–1.335	Lu et al. (1981)	T = 298 K; natural
$Cs^+ + NaL \Leftrightarrow Na^+ + CsL$	3.37	–8.36	Ames (1964a)	T = 298 K; Zeolon
$Cs^+ + NaL \Leftrightarrow Na^+ + CsL$	0.596	–1.475	Ames (1964a)	T = 298 K; AW-300 equilibrium may not have been attained
$Cs^+ + NaL \Leftrightarrow Na^+ + CsL$	3.30	–8.18	Barrer & Klinowski (1974b)	T = 298 K; Zeolon
$Cs^+ + NaL \Leftrightarrow Na^+ + CsL$	2.91	–8.29	Ames (1964a)	T = 343 K; Zeolon
$Cs^+ + NaL \Leftrightarrow Na^+ + CsL$	1.02	–2.92	Ames (1964a)	T = 343 K; AW-300
$Cs^+ + KL \Leftrightarrow K^+ + CsL$	–0.569	1.41	Ames (1964a)	T = 298 K; AW-300; equilibrium may not have been attained
$Cs^+ + KL \Leftrightarrow K^+ + CsL$	1.52	–3.76	Ames (1964a)	T = 298 K; Zeolon
$Cs^+ + KL \Leftrightarrow K^+ + CsL$	1.41	–4.02	Ames (1964a)	T = 343 K; Zeolon
$NH_4^+ + LiL \Leftrightarrow Li^+ + NH_4^+L$	4.44	–11.0	Barrer & Klinowski (1974b)	T = 298 K; Zeolon
$NH_4^+ + NaL \Leftrightarrow Na^+ + NH_4L$	1.52	–3.76	Barrer & Klinowski (1974b)	T = 298 K; Zeolon
$NH_4^+ + NaL \Leftrightarrow Na^+ + NH_4L$	1.86	–4.590	Townsend & Loizidou (1984)	T = 298 K; NO_3^- co-anion; Lovelock, Nevada mordenite; normalized isotherm

$NH_4^+ + KL \Leftrightarrow K^+ + NH_4^+L$	–0.467	1.16	Barrer & Klinowski (1974b)	T = 298 K; Zeolon
$Ag^+ + NH_4L \Leftrightarrow NH_4^+ + AgL$	0.785	–1.946	Lu et al. (1981)	T = 298 K; synthetic
$Ag^+ + NH_4L \Leftrightarrow NH_4^+ + AgL$	0.724	–1.793	Lu et al. (1981)	T = 298 K; natural
$Ca^{2+} + 2NaL \Leftrightarrow 2Na^+ + CaL_2$	–0.370	0.916	Ames (1964b)	T = 298 K; Zeolon
$Sr^{2+} + 2NaL \Leftrightarrow 2Na^+ + SrL_2$	–1.45	3.59	Ames (1964b)	T = 298 K; Zeolon
$Sr^{2+} + CaL \Leftrightarrow Ca^{2+} + SrL_2$	–1.15	2.85	Ames (1964b)	T = 298 K; Zeolon
$Ba^{2+} + 2NH_4L \Leftrightarrow 2NH_4^+ + BaL_2$	–4.14	10.2	Barrer & Klinowski (1974b)	T = 298 K; Zeolon
$Cd^{2+} + 2NH_4L \Leftrightarrow 2NH_4^+ + CdL_2$	–2.3 (Cl^-) –0.99 (NO_3^-)	5.78 (Cl^-) 2.46 (NO_3^-)	Loizidou & Townsend (1987a)	T = 298 K; Lovelock, Nevada mordenite; normalized isotherm
$Cd^{2+} + 2NaL \Leftrightarrow 2Na^+ + CdL_2$	–0.896 (Cl^-) –0.870 (NO_3^-)	2.22 (Cl^-) 2.16 (NO_3^-)	Loizidou & Townsend (1987a)	T = 298 K; Lovelock, Nevada mordenite; normalized isotherm
$Co^{2+} + 2NaL \Leftrightarrow 2Na^+ + CoL_2$	–0.46	1.1	Golden & Jenkins (1981)	T = 298 K; Na-Zeolon
$Co^{2+} + 2NH_4L \Leftrightarrow 2NH_4^+ + CoL_2$	–3.57	8.84	Barrer & Townsend (1976b)	T = 298 K; NH_4-Zeolon, normalized isotherm
$Cu^{2+} + 2NH_4L \Leftrightarrow 2NH_4^+ + CuL_2$	–3.24	8.04	Barrer & Townsend (1976b)	T = 298 K; NH_4-Zeolon, normalized isotherm
$Mn^{2+} + 2NH_4L \Leftrightarrow 2NH_4^+ + MnL_2$	–3.11	7.72	Barrer & Townsend (1976b)	T = 298 K; NH_4-Zeolon, normalized isotherm
$Ni^{2+} + 2NH_4L \Leftrightarrow 2NH_4^+ + NiL_2$	–3.64	9.02	Barrer & Townsend (1976b)	T = 298 K; NH_4-Zeolon, normalized isotherm
$Pb^{2+} + 2NaL \Leftrightarrow 2Na^+ + PbL_2$	3.750 (at 0.1N) 2.841 (at 0.5N)	–9.29 (at 0.1N) –7.04 (at 0.5N)	Loizidou & Townsend (1987b)	T = 298 K; NO_3^- co-anion; Lovelock, Nevada mordenite; normalized isotherm
$Pb^{2+} + 2NH_4L \Leftrightarrow 2NH_4^+ + PbL_2$	–2.66	6.59	Loizidou & Townsend (1987b)	T = 298 K; NO_3^- co-anion; Lovelock, Nevada mordenite; normalized isotherm
$Zn^{2+} + 2NH_4L \Leftrightarrow 2NH_4^+ + ZnL_2$	–3.60	8.92	Barrer & Townsend (1976b)	T = 298 K; NH_4-Zeolon, normalized isotherm

*Values from Loizidou and Townsend (1987a,b), Barrer and Klinowski (1974b), and Barrer and Townsend (1976b), given in kiloJoules per equivalent, were converted to kiloJoules per mole.

levels of exchange less than 50%. The isotherm results were normalized for thermodynamic analysis using values of $\bar{E}_{Co,max.}$, $\bar{E}_{Mn,max.}$, $\bar{E}_{Zn,max.}$, $\bar{E}_{Ni,max.}$, and $\bar{E}_{Cu,max.}$ equal to 0.467, 0.470, 0.470, 0.416, and 0.483, respectively. Varying the pH (from 4 to 7) and the co-anion (acetate, formate, and chloride) did not appear to affect exchange significantly. A thermodynamic affinity sequence was established as $Mn^{2+} > Cu^{2+} > Co^{2+}$ $Zn^{2+} > Ni^{2+}$ for NH_4-mordenite. Barrer and Townsend (1976a) made full use of the triangle rule to estimate ion-exchange affinities of synthetic mordenite and to compare exchange between Na- and NH_4-mordenite. Thermodynamic values derived by Barrer and Townsend (1976a) from the experimental data are given in Table 8.

Suzuki et al. (1978) determined cation-exchange isotherms for several alkali metal and alkaline-earth cations, H^+, and NH_4^+. The experiments used a Na-form of synthetic mordenite and were conducted at 298 K using solutions with a total normality of 0.1 N. The authors determined that synthetic mordenite preferred monovalent cations relative to divalent cations. A general selectivity sequence of $Cs^+ > NH_4^+$ K^+ $H^+ > Ba^+ > Sr^{2+} \approx Ca^{2+} > Rb^+ > Mg^{2+}$ was postulated. The shape of the isotherms presented suggests that ion exchange was incomplete for several cations.

Isotherms of K^+/Na^+, H^+/Na^+, and H^+/K^+ exchange on a Na-form of synthetic mordenite are presented in Wolf et al. (1978). Experiments were conducted at 274, 278, and 293 K using a total solution concentration of 0.6 N, and they illustrate the temperature effects on ion exchange in synthetic mordenite. The CEC for the mordenite was determined to be 2.13 meq/g. Thermodynamic values determined by Wolf et al. (1978) are provided in Table 8.

Golden and Jenkins (1981) generated isotherms for synthetic mordenite (Zeolon) for Na^+/Li^+, Na^+/Co^{2+}, and H^+/Li^+ exchange. The experiments were conducted at 298 K with a total solution concentration of 0.1 N. Reversibility was demonstrated in all three systems, although the exchange of Co^{2+} was incomplete. The results of thermodynamic calculations generally agreed with results from other investigations (e.g. Barrer and Klinowski 1974b), and the triangle rule was successfully applied for Na^+/Li^+ and Na^+/Co^{2+}. The thermodynamic data of Golden and Jenkins (1981) are listed in Table 8.

Kuznetsova et al. (1998) recently described the ion-exchange properties of synthetic mordenite based on a strong electrolyte model. Reasonable agreement with the equilibrium constants derived by Barrer and Klinowski (1974b) was achieved.

Other papers regarding ion exchange on mordenite that may be of interest include Gradev and Gulubova (1982), an investigation of the uptake of ^{134}Cs, ^{137}Cs, ^{89}Sr, and ^{90}Sr on natural clinoptilolite and mordenite, and Grebenshchikova et al. (1973a,b), who investigated the uptake of Pu^{4+} and Th^{4+} on a Na-form of mordenite at low pH.

Phillipsite [$K_2(Ca_{0.5},Na)_4(Al_6Si_{10}O_{32})\cdot 12H_2O$]

Ion exchange of Cs^+ and Sr^{2+} for Na^+ in three types of phillipsite—sedimentary, hydrothermal, and synthetic—was studied by Adabbo et al. (1999) at 298 K and 0.1 N solution concentration. The sedimentary phillipsite material was prepared from a phillipsite-rich rock from Marano, Italy, whereas the hydrothermal phillipsite was obtained from cavities and vugs of a basalt sample from Vesuvius, Italy. The synthetic phillipsite was prepared hydrothermally from oxide mixtures. The measured CECs are 3.55, 4.71, and 3.60 meq/g for the sedimentary, hydrothermal, and synthetic phillipsite, respectively. For comparison, the CECs calculated from the aluminum content of the zeolites are 3.66, 4.70, and 3.52 meq/g, for the sedimentary, hydrothermal, and synthetic phillipsite, respectively. The results indicate that the Cs^+/Na^+ and Sr^{2+}/Na^+ exchange reactions are reversible for the three types of phillipsite and the total CEC of the three zeolites is available for Cs^+ and Sr^{2+} exchange. The sedimentary and synthetic phillipsites, characterized by a higher Si/Al ratio than the hydrothermal zeolite, display good selectivity for Cs^+ and moderate selectivity for Sr^{2+}, whereas the more aluminous hydrothermal phillipsite has lower selectivity for Cs^+ and higher selectivity for Sr^{2+} compared with the sedimentary and synthetic phillipsites. The Cs^+/Na^+ isotherm for hydrothermal phillipsite exhibits a selectivity reversal, not observed for the sedimentary and synthetic phillipsites, at an $\bar{E}_{Cs,max.}$ of about 0.73. The Sr^{2+}/Na^+ isotherms for the sedimentary, synthetic, and hydrothermal zeolite exhibit selectivity reversal at $\bar{E}_{Cs,max.}$ equal to 0.15, 0.18, and 0.69, respectively. The equilibrium constants and Gibbs free energies reported by Adabbo et al. (1999) are listed in Table 9.

Isotherm data on ion exchange between phillipsite and Pb^{2+}/Na^+ and Pb^{2+}/K^+ solutions (NO_3^- co-anion) at 0.1 N total solution concentration were reported by Pansini et al. (1996). The Na- and K-phillipsite were obtained by purifying phillipsite-rich tuff from Chiaiano, Naples, Italy, and exchanging with 0.5 M NaCl or KCl solutions. The measured CEC of the phillipsite material is 3.30 meq/g. The results show that Pb^{2+} exchange for Na^+ and K^+ is reversible. Phillipsite exhibits a strong affinity for Pb^{2+} relative to Na^+, with the Pb^{2+}/Na^+ isotherm lying above the diagonal over the whole composition range. The Pb^{2+}/K^+ isotherm shows a selectivity reversal at an $\bar{E}_{Pb}$ of about 0.37. Below this value, phillipsite exhibits a moderate selectivity for Pb^{2+}. Thermodynamic parameters reported by Pansini et al. (1996) are listed in Table 9.

Colella et al. (1998) used Na-phillipsite obtained by purifying phillipsite-rich tuff from Marano, Italy, and exhaustively exchanging with NaCl solutions. The measured CEC of the zeolite is 3.30 meq/g. Isotherms for Cu^{2+}/Na^+ and Zn^{2+}/Na^+ exchange were determined at 298 K and 0.1 N solution concentration (NO_3^- co-anion). The results indicate that the Zn^{2+}/Na^+ exchange is reversible, but the Cu^{2+}/Na^+ exchange is irreversible—the isotherm shows an obvious hysteresis loop. Incomplete exchange was observed for both Cu^{2+}/Na^+ and Zn^{2+}/Na^+ reactions, with $\bar{E}_{Cu,max.}$ and $\bar{E}_{Zn,max.}$ equal to 0.76 and 0.73, respectively. The isotherms show that phillipsite has poor selectivity for both Cu^{2+} and Zn^{2+}. The thermodynamic parameters for Zn^{2+}/Na^+ exchange reported by Colella et al. (1998) are listed in Table 9.

Chelishchev et al. (1984) reported isotherm data on K^+/Na^+, Rb^+/Na^+, Cs^+/Na^+, and Li^+/Na^+ exchange based on experiments at 295 K and 1 N solution concentration using phillipsite material from Georgia (of the former U.S.S.R.). The isotherms are very similar to those of erionite reported by Chelishchev and Volodin (1977). The results show that phillipsite is selective for Cs^+, Rb^+, and K^+, but not for Li^+. The selectivity series for phillipsite determined by Chelishchev et al. (1984) is $Cs^+ > Rb^+ > K^+ > Na^+ > Li^+$. The equilibrium constants reported by the authors and the Gibbs free energies calculated from those values are listed in Table 9.

Thermodynamic parameters derived by Ames (1964a,b) for Na^+/K^+, Cs^+/Na^+, Cs^+/K^+, Ca^{2+}/Na^+, Sr^{2+}/Ca^{2+}, and Sr^{2+}/Na^+ exchange on phillipsite are also included in Table 9. The isotherm experiments by Ames (1964a,b) used a phillipsite material (90% or higher in purity; CEC equal to 2.3 meq/g) from Pine Valley, Nevada, and were conducted at 298 K and 0.1 N solution concentration.

APPLICATION OF THERMODYNAMIC MODELS

As noted in a previous section, thermodynamic models provide systematic bases for understanding ion-exchange behavior and serve as tools for predicting exchange equilibria under conditions not previously studied. Several examples are presented in this section to illustrate the use of an ion-exchange model, based on the Margules equation for

Table 9. Literature values of equilibrium constants and Gibbs free energies for ion-exchange reactions involving phillipsite.

Ion-Exchange Reaction	$\ln K_{(A,B)}$	$\Delta G^{\circ}_{(A,B)}$ (kJ/mol)*	Reference	Remarks
$Li^{+} + NaL \Leftrightarrow Na^{+} + LiL$	−2.60	6.38	Chelishchev et al. (1984)	T = 295 K; phillipsite from Georgia (of the former U.S.S.R.)
$Na^{+} + KL \Leftrightarrow K^{+} + NaL$	−1.66	4.1	Ames (1964a)	T = 298 K; Pine Valley, Nevada phillipsite; CEC = 2.3 meq/g
$K^{+} + NaL \Leftrightarrow Na^{+} + KL$	2.90	−7.11	Chelishchev et al. (1984)	T = 295 K; phillipsite from Georgia (of the former U.S.S.R.)
$Rb^{+} + NaL \Leftrightarrow Na^{+} + RbL$	3.54	−8.68	Chelishchev et al. (1984)	T = 295 K; phillipsite from Georgia (of the former U.S.S.R.)
$Cs^{+} + NaL \Leftrightarrow Na^{+} + CsL$	4.01	−9.84	Chelishchev et al. (1984)	T = 295 K; phillipsite from Georgia (of the former U.S.S.R.)
$Cs^{+} + NaL \Leftrightarrow Na^{+} + CsL$	2.82	−7.0	Adabbo et al. (1999)	T = 298 K; synthetic phillipsite; CEC = 3.60 meq/g
$Cs^{+} + NaL \Leftrightarrow Na^{+} + CsL$	0.83	−2.1	Adabbo et al. (1999)	T = 298 K; hydrothermal phillipsite from Vesuvius, Italy; CEC = 4.71 meq/g
$Cs^{+} + NaL \Leftrightarrow Na^{+} + CsL$	3.27	−8.1	Adabbo et al. (1999)	T = 298 K; sedimentary phillipsite from Marano, Italy; CEC = 3.55 meq/g
$Cs^{+} + NaL \Leftrightarrow Na^{+} + CsL$	3.27	−8.1	Ames (1964a)	T = 298 K; Pine Valley, Nevada phillipsite; CEC = 2.3 meq/g
$Cs^{+} + NaL \Leftrightarrow Na^{+} + CsL$	2.59	−7.4	Ames (1964a)	T = 343 K; Pine Valley, Nevada phillipsite; CEC = 2.3 meq/g
$Cs^{+} + KL \Leftrightarrow K^{+} + CsL$	1.67	−4.1	Ames (1964a)	T = 298 K; Pine Valley, Nevada phillipsite; CEC = 2.3 meq/g
$Cs^{+} + KL \Leftrightarrow K^{+} + CsL$	1.61	−4.6	Ames (1964a)	T = 343 K; Pine Valley, Nevada phillipsite; CEC = 2.3 meq/g
$Ca^{2+} + 2NaL \Leftrightarrow 2Na^{+} + CaL_2$	−1.81	4.5	Ames (1964b)	T = 298 K; Pine Valley, Nevada phillipsite; CEC = 2.3 meq/g
$Sr^{2+} + 2NaL \Leftrightarrow 2Na^{+} + SrL_2$	−2.66	6.6	Adabbo et al. (1999)	T = 298 K; synthetic phillipsite; CEC = 3.60 meq/g

$Sr^{2+} + 2NaL \Leftrightarrow 2Na^+ + SrL_2$	–0.97	2.4	Adabbo et al. (1999)	T = 298 K; hydrothermal phillipsite from Vesuvius, Italy; CEC = 4.71 meq/g
$Sr^{2+} + 2NaL \Leftrightarrow 2Na^+ + SrL_2$	–1.97	4.8	Adabbo et al. (1999)	T = 298 K; sedimentary phillipsite from Marano, Italy; CEC = 3.55 meq/g
$Sr^{2+} + 2NaL \Leftrightarrow 2Na^+ + SrL_2$	–2.23	5.5	Ames (1964b)	T = 298 K; Pine Valley, Nevada phillipsite; CEC = 2.3 meq/g
$Sr^{2+} + CaL_2 \Leftrightarrow Ca^{2+} + SrL_2$	–0.53	1.3	Ames (1964b)	T = 298 K; Pine Valley, Nevada phillipsite; CEC = 2.3 meq/g
$Pb^{2+} + 2NaL \Leftrightarrow 2Na^+ + ZnL_2$	3.13	–7.8	Pansini et al. (1996)	T = 298 K; phillipsite from Chiaiano, Naples, Italy; CEC = 3.3 meq/g
$Pb^{2+} + 2KL \Leftrightarrow 2Na^+ + PbL_2$	–1.7	4.2	Pansini et al. (1996)	T = 298 K; phillipsite from Chiaiano, Naples, Italy; CEC = 3.3 meq/g
$Cd^{2+} + 2NaL \Leftrightarrow 2Na^+ + CdL_2$	–3.7	9.2	Colella et al. (1998)	T = 298 K
$Zn^{2+} + 2NaL \Leftrightarrow 2Na^+ + ZnL_2$	–5.29	13.1	Colella et al. (1998)	T = 298 K; phillipsite from Marano, Italy; CEC = 3.30 meq/g

*Values from Adabbo et al. (1999), Pansini et al. (1996), and Colella et al. (1998), given in kiloJoules per equivalent, were converted to kiloJoules per mole. The $\Delta G^{\circ}_{(A,B)}$ values listed for Chelishchev et al. (1984) were calculated from the equilibrium constants reported in the reference.

zeolite solid solutions and the Pitzer equations for aqueous-phase activity coefficients, to represent and predict ion-exchange equilibria. The examples focus on clinoptilolite, for which selected sets of experimental data taken from the literature were used to derive parameters for the Margules equation and to calculate ion-exchange isotherms as functions of solution composition and concentration. Where ion-exchange data are available, calculated values are compared with measured values. The ion-exchange model also was used to predict cation concentrations in the aqueous solution based on the composition of the zeolite in equilibrium with the solution.

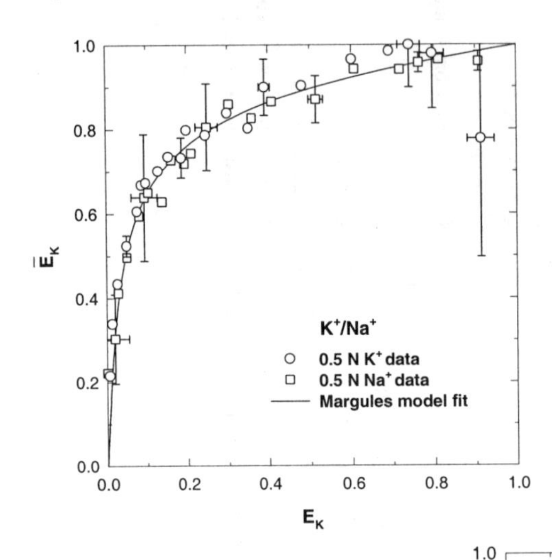

Figure 5. Isotherm data for ion exchange at 298 K between clinoptilolite and K^+/Na^+ solutions (0.5 N; Cl^- anion). Circles and squares were calculated from solution concentrations of K^+ and Na^+, respectively. Some error bars, based on K^+ and Na^+ analytical uncertainties, are shown. The curve was fit to the isotherm data using a Margules solid-solution model. Data are from Pabalan (1994).

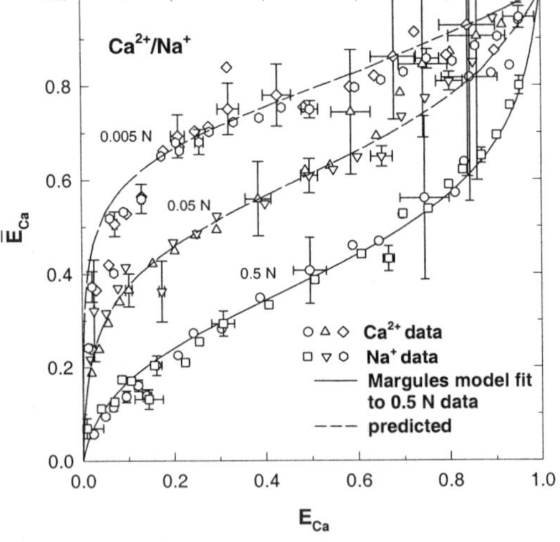

Figure 6. Isotherm data for ion exchange at 298 K between clinoptilolite and Ca^{2+}/Na^+ solutions (0.005, 0.05, and 0.5 N; Cl^- anion). The isotherm points were calculated from solution concentrations of Ca^{2+} or Na^+. Some error bars, based on Ca^{2+} and Na^+ analytical uncertainties, are shown. The solid curve was fit to the 0.50 N isotherm data using a Margules solid-solution model. The dashed curves represent predicted values. Data are from Pabalan (1994).

Margules model parameters

Values of $\ln K_{v(A,B)}$ versus $\overline{X}_A$ were calculated from isotherm data on clinoptilolite for the exchange couples K^+/Na^+, Ca^{2+}/Na^+, Sr^{2+}/Na^+, Sr^{2+}/K^+, and Ca^{2+}/K^+ taken from the studies of Pabalan (1994) and Pabalan and Bertetti (1999). The isotherm data are shown in Figures 3 to 8, and, as examples, $\ln K_{v(A,B)}$ versus $\overline{X}_A$ for two systems, Sr^{2+}/Na^+ and Sr^{2+}/K^+, are plotted in Figures 9 and 10. Using the calculated values of $\ln K_{v(A,B)}$ versus $\overline{X}_A$, Equation (25) was used to derive the ion-exchange equilibrium constant,

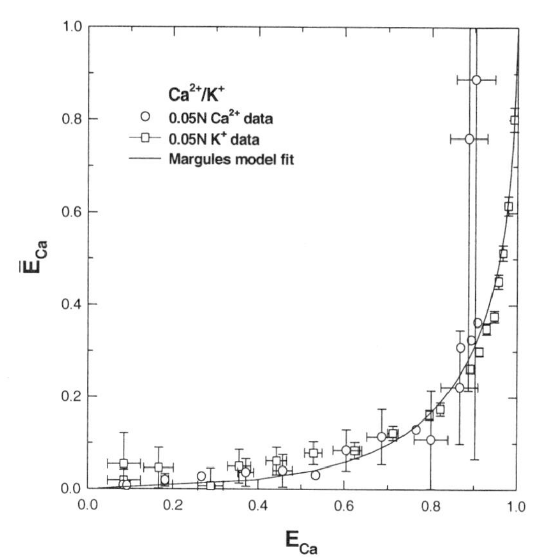

Figure 7. Isotherm data for ion exchange at 298 K between clinoptilolite and Ca^{2+}/K^+ solutions (0.05 N; Cl^- anion). Circles and squares were calculated from Ca^{2+} and K^+ analytical data, respectively. Some error bars, based on Ca^{2+} and K^+ analytical uncertainties, are shown. The curve was fit to the isotherm data using a Margules solid-solution model. Data are from Pabalan and Bertetti (1999).

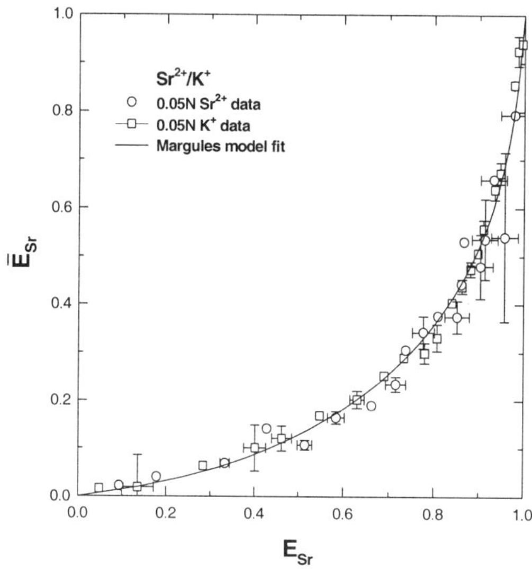

Figure 8. Isotherm data for ion exchange at 298 K between clinoptilolite and Sr^{2+}/K^+ solutions (0.05 N; Cl^- anion). Circles and squares were calculated from Sr^{2+} and K^+ analytical data, respectively. Some error bars, based on Sr^{2+} and K^+ analytical uncertainties, are shown. The curve was fit to the isotherm data using a Margules solid-solution model. Data from Pabalan and Bertetti (1999).

$K_{(A,B)}$, and the Margules parameters, W_A and W_B. In the regression, ln $K_{v(A,B)}$ was weighted inversely proportional to the square of its estimated overall uncertainty (Pabalan 1994). Where data at different total solution concentrations are available, data at only one solution concentration were used in the regression, and data at other concentrations were used to check model predictions. The derived equilibrium constants and Margules parameters are listed in Table 10. The solid curves in Figures 3 to 8, as well as in Figures 9 and 10, represent the Margules model fits to the isotherm data.

Experimental data for NH_4^+/Na^+ exchange at 0.1 N total solution concentration from Townsend and Loizidou (1984) and Jama and Yucel (1990) were used to calculated ln $K_{v(NH_4,Na)}$ versus X_{NH_4} and to derive the equilibrium constant and Gibbs free energy for the exchange reaction (Table 10). As discussed in a previous section, both sets of authors

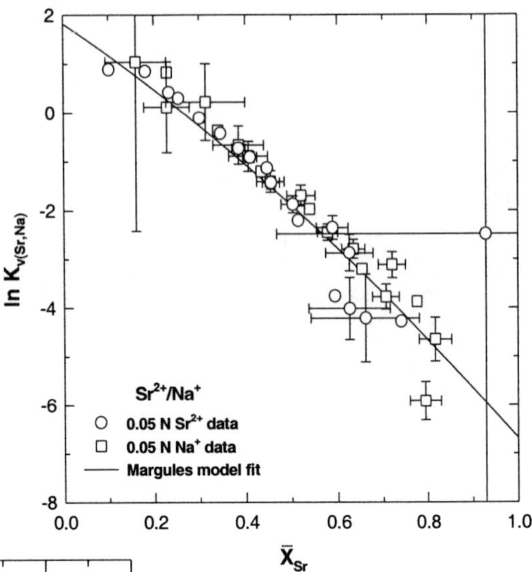

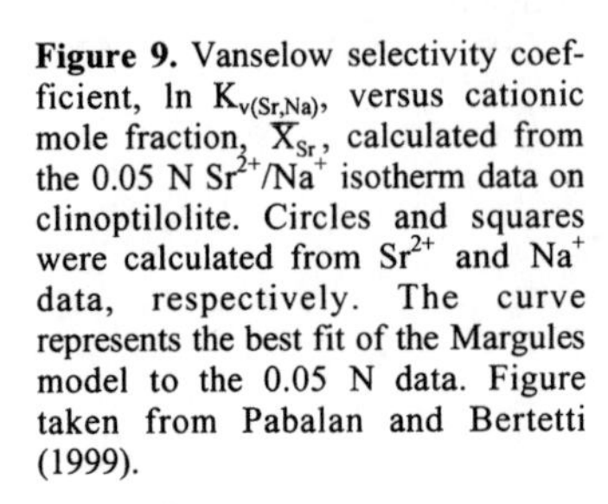
Figure 9. Vanselow selectivity coefficient, ln $K_{v(Sr,Na)}$, versus cationic mole fraction, $\bar{X}_{Sr}$, calculated from the 0.05 N Sr^{2+}/Na^+ isotherm data on clinoptilolite. Circles and squares were calculated from Sr^{2+} and Na^+ data, respectively. The curve represents the best fit of the Margules model to the 0.05 N data. Figure taken from Pabalan and Bertetti (1999).

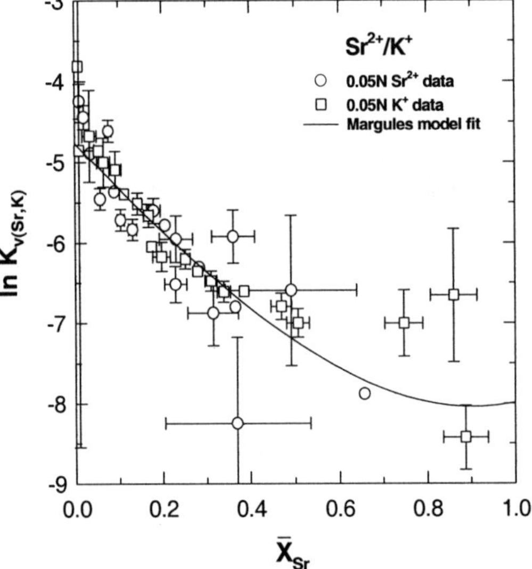

Figure 10. Vanselow selectivity coefficient, ln $K_{v(Sr,K)}$, versus cationic mole fraction, $\bar{X}_{Sr}$, calculated from the 0.05 N Sr^{2+}/K^+ isotherm data on clinoptilolite. Circles and squares were calculated from Sr^{2+} and K^+ data, respectively. The curve represents the best fit of the Margules model to the 0.05 N data. Figure taken from Pabalan and Bertetti (1999).

observed incomplete exchange of NH_4^+ for Na^+, and the isotherm data were therefore normalized to the reported $\bar{E}_{NH4,max.}$ before the thermodynamic parameters were derived. The original and normalized NH_4^+/Na^+ isotherm points from the two references are plotted in Figure 11a. The solid curve in Figure 11a represents the best-fit of the Margules model to the two sets of data. Experimental data from White (1988), also at 0.1 N, are compared in Figure 11b with the isotherm calculated from the model. There is very good agreement between experimental and calculated values.

Data on NH_4^+/K^+ ion exchange at 0.1 N total solution concentration from Jama and Yucel (1990) are plotted in Figure 12. These data were normalized to the reported $\bar{E}_{NH4,max.}$ of 0.50 and were used to derive the equilibrium constant and Gibbs free energy for the exchange reaction (Table 10). The isotherms calculated from the model compare well with experimental data, as shown in Figure 12.

Table 10. Values of Margules parameters, equilibrium constants, and Gibbs free energies of ion exchange derived from isotherm data*

Cations (A,B)	W_A	W_B	$\ln K_{(A,B)}$	$\Delta G^{\circ}_{(A,B)}$ (kJ/mol)	Reference
K^+,Na^+	−0.989	−1.41	3.22±0.03	−7.98±0.08	Pabalan (1994)
Cs^+,Na^+	−1.62	−1.45	4.30±0.24	−10.7±0.6	this study
Li^+,Na^+	−0.52	−0.88	−2.12±0.07	5.26±0.18	this study
NH_4^+,Na^+	−0.51	−0.58	1.64±0.05	−4.07±0.13	this study
Ca^{2+},Na^+	−2.67	−1.22	−1.65±0.08	4.09±0.20	Pabalan (1994)
Sr^{2+},Na^+	−2.68	−3.27	−1.14±0.07	2.83±0.18	Pabalan and Bertetti (1999)
Cs^+,K^+	−1.83	−0.706	1.04±0.12	−2.58±0.30	this study
NH_4^+,K^+	−0.324	−0.496	−0.645±0.173	1.60±0.43	this study
Ca^{2+},K^+	−2.02	−0.591	−8.50±0.18	21.1±0.5	Pabalan and Bertetti (1999)
Sr^{2+},K^+	−1.75	−0.731	−6.52±0.10	16.2±0.3	Pabalan and Bertetti (1999)

*W_A, W_B and $\ln K_{(A,B)}$ were derived from weighted regression of Equation (25) to the $\ln K_{v(A,B)}$ versus $\overline{X}_A$ data. The 1σ errors in $\ln K_{(A,B)}$ and $\Delta G^{\circ}_{(A,B)}$ are also listed.

Isotherm data from Ames (1968), Howery and Thomas (1965), and Chelishchev et al. (1973), plotted in Figure 13, were used to derive the thermodynamic parameters listed in Table 10 for Cs^+/Na^+ exchange on clinoptilolite. As discussed earlier, homovalent cation exchange is not significantly affected by variations in total solution concentration. One reason for this is the minor variation in solution activity coefficient ratios for homovalent exchange reactions over a large concentration range (e.g. 0.0005 N to 1.0 N). Therefore, isotherm data for homovalent exchanges, like the Cs^+/Na^+ system, collected at different total solution concentrations can be expected to be very similar. However, as seen in Figure 13, data from several investigators exhibit a greater than expected variability, possibly due to differences in the clinoptilolite samples and in the experimental protocols.

One way of dealing with variability in data between different studies is to use all the available data to generate a single model regression to represent the system isotherm. The solid curve in Figure 13 represents the best-fit of the Margules model to all isotherm data from the three Cs^+/Na^+ studies. The Margules fit not only does an adequate job of representing the 'average' Cs^+/Na^+ exchange isotherm, but it also captures the variability of data between experiments. Curves representing model calculations that account for 2σ uncertainties in $\ln K_{(Cs,Na)}$ alone (dashed lines) or in all model parameters ($\ln K_{(Cs,Na)}$, W_{Cs}, and W_{Na}) (dotted lines) are also plotted in the figure. Note that the isotherms accounting for the ±2σ variation in parameter uncertainty capture the variability of data

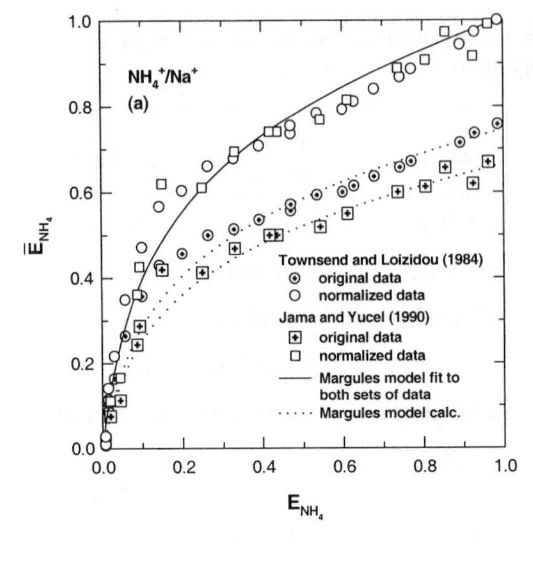

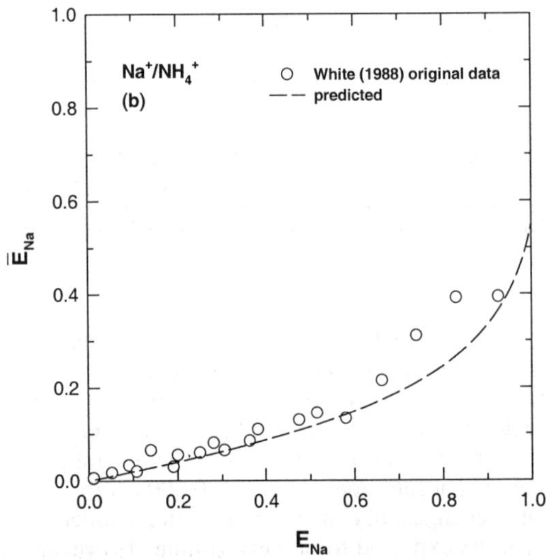

Figure 11. Isotherm data for ion exchange between clinoptilolite and 0.1 N NH_4^+/Na^+ solutions. The data from Townsend and Loizidou (1984) and Jama and Yucel (1990), plotted in Figure 11a, were normalized and regressed to derive the equilibrium constant and Margules parameters. A comparison of the calculated isotherm with data from White (1988) is shown in Figure 11b.

in the three experiments. Thus, although it is difficult to determine which of the three sets of experiments is 'correct,' the 2σ uncertainties in the Margules model parameters adequately represent the uncertainties that arise due to differences among experiments. Therefore, bounding isotherms can be generated where uncertainty due to inter-laboratory variability is high.

Experimental results from Chelishchev et al. (1973) on the binary system Li^+/Na^+ are plotted in Figure 14. The Margules model (solid curve) fits the isotherm data well. Data on K^+/Cs^+ exchange from Chelishchev et al. (1973) and Ames (1968) are plotted in Figure 15. The two data sets do not agree well, even considering the difference in total solution concentration. The Margules model could fit either data set equally well but would result in different thermodynamic parameters. For the purposes of this study, the model was fit to both sets of isotherm data. The derived equilibrium constants and Gibbs

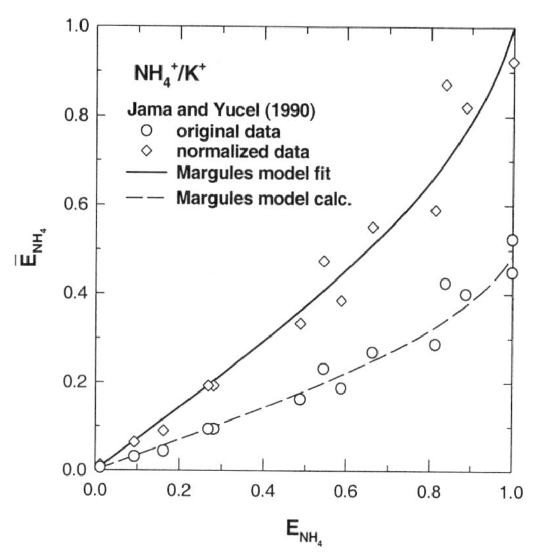

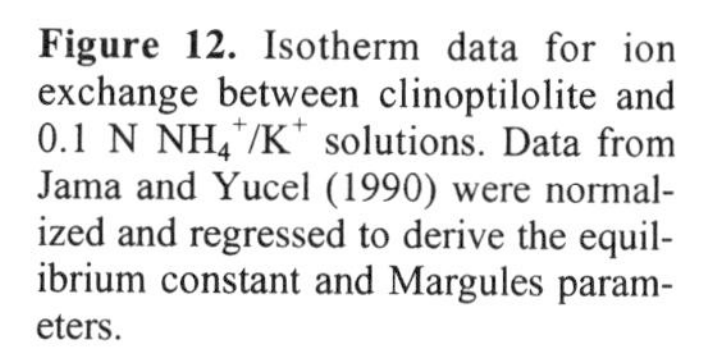

Figure 12. Isotherm data for ion exchange between clinoptilolite and 0.1 N NH_4^+/K^+ solutions. Data from Jama and Yucel (1990) were normalized and regressed to derive the equilibrium constant and Margules parameters.

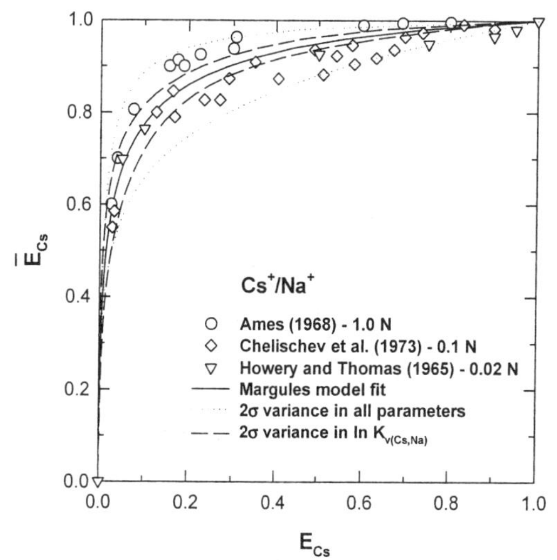

Figure 13. Isotherm data for ion exchange between clinoptilolite and Cs^+/Na^+ solutions. Data from Ames (1968), Chelishchev et al. (1973), and Howery and Thomas (1965) were regressed to derive the equilibrium constant and Margules parameters. The solid curve is the best fit to all the isotherm data. The dashed curves represent model calculations in which ln $K_{(Cs,Na)}$ was varied between its $\pm 2\sigma$ variance. The dotted curves represent model calculations in which all model parameters [ln $K_{(Cs,Na)}$, W_{Cs}, and W_{Na}] were varied between their $\pm 2\sigma$ variance. The $\pm 2\sigma$ isotherm envelope is able to represent data variation among different studies.

free energies for Li^+/Na^+ and K^+/Cs^+ exchange are listed in Table 10.

Excess Gibbs energies calculated from the Margules parameters and Equation (23) for several binary clinoptilolite solid solutions are plotted in Figures 16 and 17. These figures illustrate the relative deviation from ideal behavior of the zeolite solid solutions. All the systems studied exhibit some degree of nonideal behavior. Where the degree of nonideality is not large, the solid phase activity coefficients are close to 1.0 and the activities of the zeolite components can be approximated by their mole fraction (i.e. $\bar{a}_i \approx \bar{X}_i$). For example, Figure 16 indicates that the nonideality for the NH_4^+/Na^+ system is relatively small, with a maximum of –0.34 kJ/mol. Figure 18 compares NH_4^+/Na^+ ion exchange data (shown previously in Fig. 11) with isotherms calculated either using the Margules model, i.e. explicitly accounting for nonideal behavior, or assuming an ideal solid solution. The comparison shows that, for this particular case, the data can be

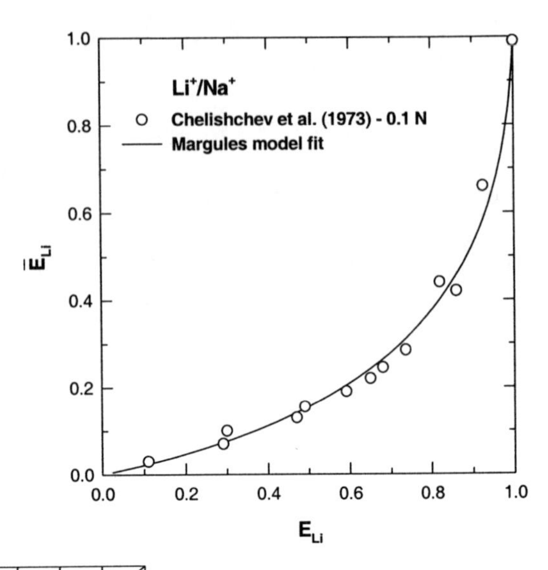

Figure 14. Isotherm data for ion exchange between clinoptilolite and Li^+/Na^+ solutions. Data from Chelishchev et al. (1973) were regressed to derive the equilibrium constant and Margules parameters. The solid curve is the best fit to the isotherm data.

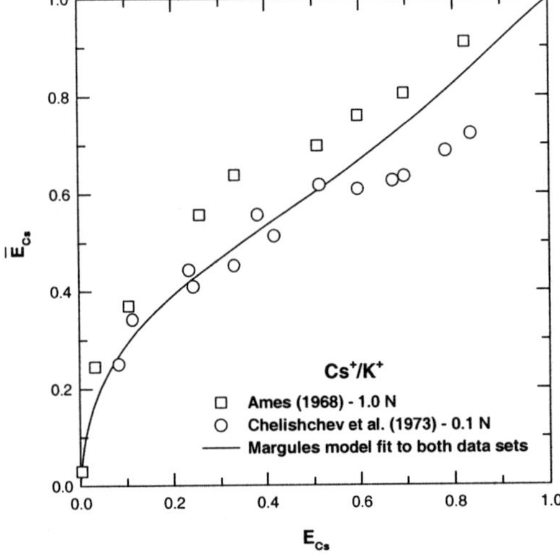

Figure 15. Isotherm data for ion exchange between clinoptilolite and K^+/Cs^+ solutions. Data from Ames (1968) and Chelishchev et al. (1973) were regressed to derive the equilibrium constant and Margules parameters. The solid curve is the best fit to both sets of isotherm data.

adequately represented by an ideal solid solution model. In contrast, Figure 16 indicates that the Sr^{2+}/Na^+ system is highly nonideal. The Sr^{2+}/Na^+ isotherm calculated without explicitly accounting for nonideality effects, shown in Figure 19, agrees poorly with experimental data.

Isotherms as functions of solution composition and concentration

Thermodynamic models facilitate the calculation of ion-exchange equilibria, particularly for exchange involving heterovalent cations for which the isotherm shapes depend strongly on the total solution concentration. From the known ion-exchange equilibrium constant, $K_{(A,B)}$, and Margules solid solution parameters, W_A and W_B, values of $K_{v(A,B)}$ can be calculated for a given zeolite composition, $\overline{X}_A$, using Equation (25). If the total solution normality is known and aqueous activity coefficients can be calculated,

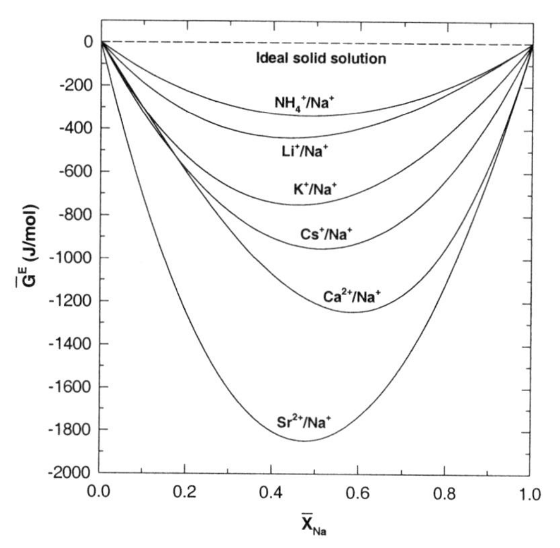

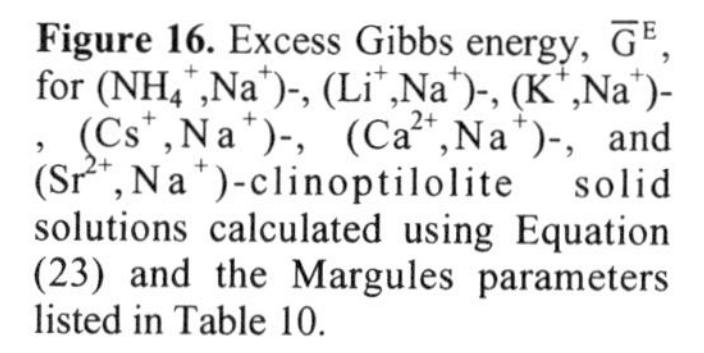

Figure 16. Excess Gibbs energy, $\overline{G}^E$, for (NH_4^+,Na^+)-, (Li^+,Na^+)-, (K^+,Na^+)-, (Cs^+,Na^+)-, (Ca^{2+},Na^+)-, and (Sr^{2+},Na^+)-clinoptilolite solid solutions calculated using Equation (23) and the Margules parameters listed in Table 10.

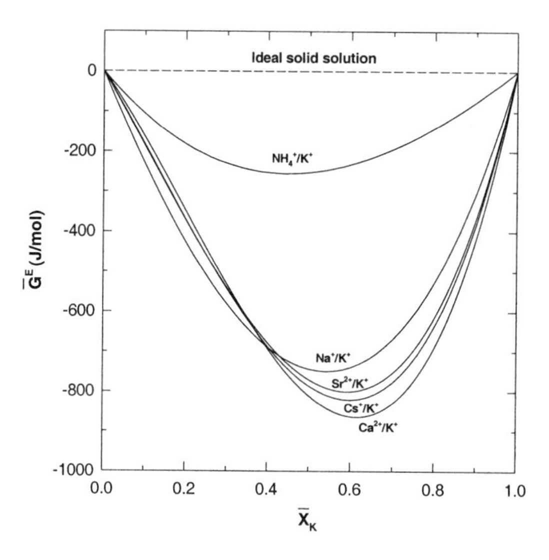

Figure 17. Excess Gibbs energy, $\overline{G}^E$, for (Cs^+,K^+)-, (Na^+,K^+)-, (NH_4^+,K^+)-, (Ca^{2+},K^+)- and (Sr^{2+},K^+)-clinoptilolite solid solutions calculated using Equation (23) and the Margules parameters listed in Table 10.

then the composition of the aqueous solution in equilibrium with the zeolite can be solved from Equation (13). Isotherm values for the exchange couples Sr^{2+}/Na^+ and Ca^{2+}/Na^+ predicted using the Margules model (dashed curves) are compared in Figures 3 and 6 with experimental data. The figures show very good agreement between measured and calculated values.

Figures 20, 21, 22, and 23 show isotherms for ion exchange involving Ca^{2+}/Na^+, Sr^{2+}/Na^+, Sr^{2+}/K^+, and Ca^{2+}/K^+ mixtures, respectively, and clinoptilolite as functions of total solution concentration. The isotherms were calculated using Margules parameters derived from the isotherm data of Pabalan (1994) and Pabalan and Bertetti (1999). Isotherm data from other sources are also plotted for comparison. Data from Ames (1964b) on Sr^{2+}/Na^+ exchange at 1.0 N agree well with the calculated isotherm (Fig. 21), although his data on Ca^{2+}/Na^+ do not agree as well with the calculated values (Fig. 20).

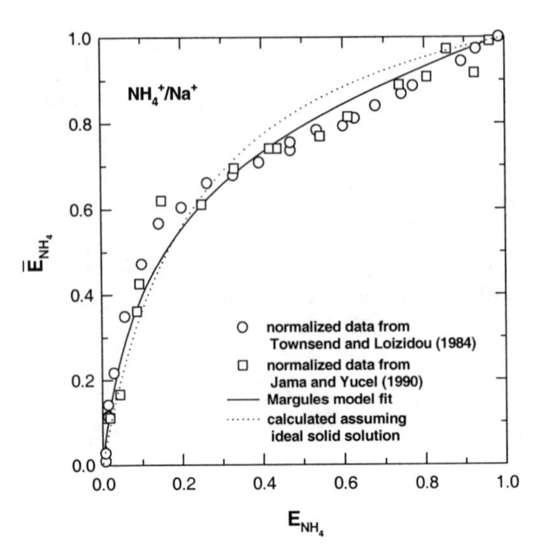

Figure 18. The symbols are normalized NH_4^+/Na^+ isotherm data from Figure 11(a). The solid curve is the isotherm calculated using the Margules model. The dotted curve is the isotherm calculated assuming that (NH_4^+,Na^+)-clinoptilolite behaves as an ideal solid solution.

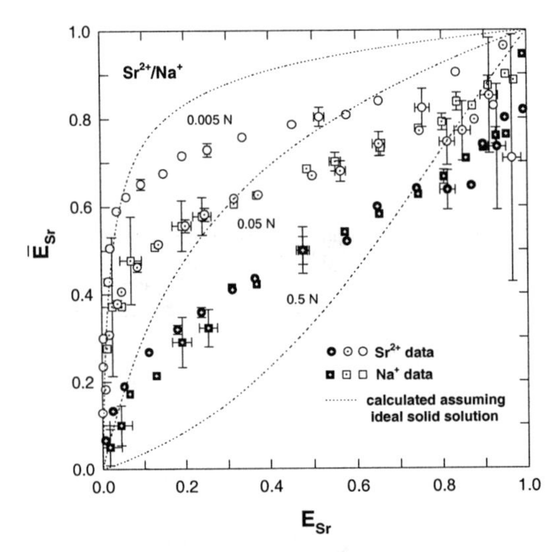

Figure 19. Sr^{2+}/Na^+ isotherm data (shown by symbols; taken from Fig. 3) compared with isotherms (shown by dotted lines) calculated assuming that (Sr^{2+},Na^+)-clinoptilolite behaves as an ideal solid solution.

Values of $\overline{E}_{Sr}$ from Chelishchev et al.'s (1973) Sr^{2+}/Na^+ isotherm points are lower than the calculated values (Fig. 21), whereas $\overline{E}_{Ca}$ values from White's (1988) Ca^{2+}/K^+ data, which show quite a bit of scatter, are generally higher than the calculated values (Fig. 23).

Isotherms change little with a change in solution concentration for homovalent exchange reactions. For example, in Figure 24, the K^+/Na^+ isotherm calculated for a solution concentration of 1.0 N is essentially the same as the isotherm at 0.005 N. Comparison of calculated values with experimental data from Ames (1964a) at 1.0 N and from White (1988) at 0.1 N shows good agreement.

Aqueous composition calculated from zeolite analysis

An interesting application of ion-exchange models is the calculation of groundwater

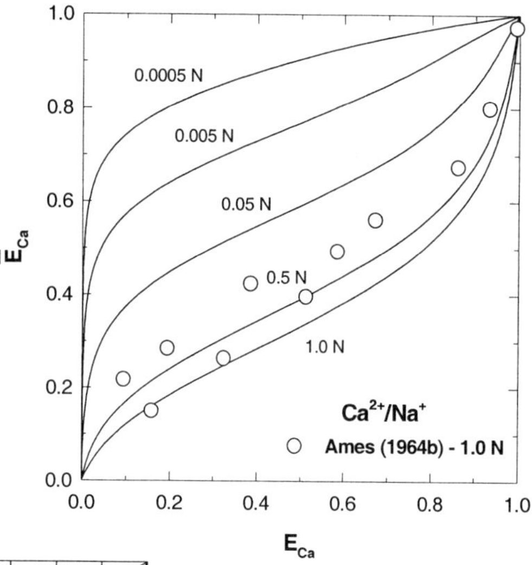

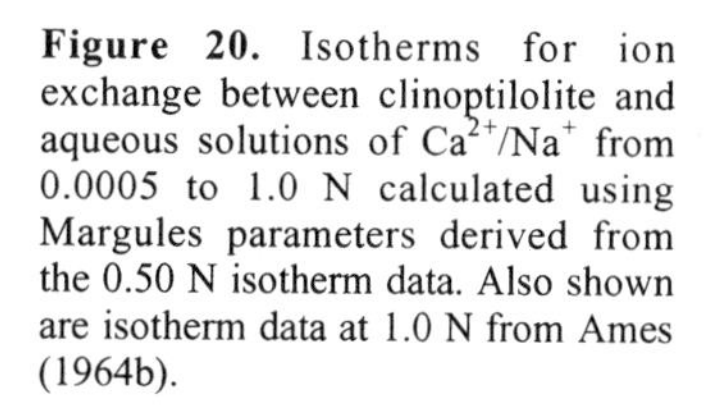
Figure 20. Isotherms for ion exchange between clinoptilolite and aqueous solutions of Ca^{2+}/Na^{+} from 0.0005 to 1.0 N calculated using Margules parameters derived from the 0.50 N isotherm data. Also shown are isotherm data at 1.0 N from Ames (1964b).

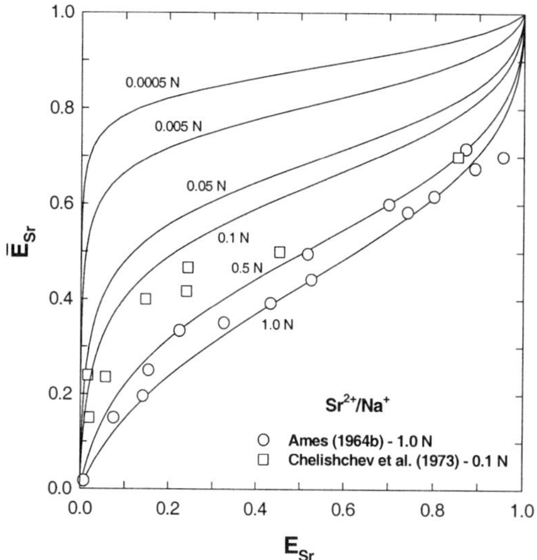

Figure 21. Isotherms for ion exchange between clinoptilolite and aqueous solutions of Sr^{2+}/Na^{+} from 0.0005 to 1.0 N calculated using Margules parameters derived from the 0.05 N isotherm data. Also shown are isotherm data at 1.0 N from Ames (1964b) and at 0.1 N from Chelishchev et al. (1973).

composition based on analytical data on the coexisting zeolite phase (Pabalan and Bertetti 1999). For example, it is difficult to obtain samples of groundwater from hydrologically unsaturated rock units, such as those surrounding the potential nuclear waste geologic repository at Yucca Mountain, Nevada. Although efforts have been made to extract aqueous solutions by high-pressure triaxial compression of Yucca Mountain rock samples (Yang et al. 1988, 1996; Peters et al. 1992), there are large variabilities in the chemical composition of solutions derived using this technique and some rock samples do not yield sufficient water for chemical analysis. Aqueous samples have also been extracted from unsaturated soils and sands using ultracentrifugation techniques (Edmunds et al. 1992; Puchelt and Bergfeldt 1992). It is uncertain whether compositions of water extracted from rock pores by ultracentrifugation or by high-pressure squeezing methods accurately represent the compositions of *in situ* water. Water extracted by these methods may have compositions different from those of *in situ* water due to several possible processes

(Peters et al. 1992): (1) dilution of pore solutions by water desorbed from hydrated minerals like zeolites and clays; (2) dissolution reactions due to increased mineral solubility and/or higher carbon dioxide concentration at higher pressures; (3) membrane filtration by clays and zeolites; and (4) ion exchange with zeolites and clays.

Thermodynamic models for ion-exchange equilibria may help reduce uncertainties associated with sampling and analysis of groundwater in unsaturated stratigraphic horizons by constraining the water cation chemistry. Data from analysis of zeolite composition, such as those derived by electron microprobe techniques, combined with estimates of the total normality of the aqueous phase, can be used to calculate cation concentrations in the groundwater. Ideally, the zeolite sample should be obtained by dry coring methods to minimize altering its composition. The ability of thermodynamic models to estimate the cationic composition of the aqueous phase based on the chemical composition of the zeolite is evaluated in this section.

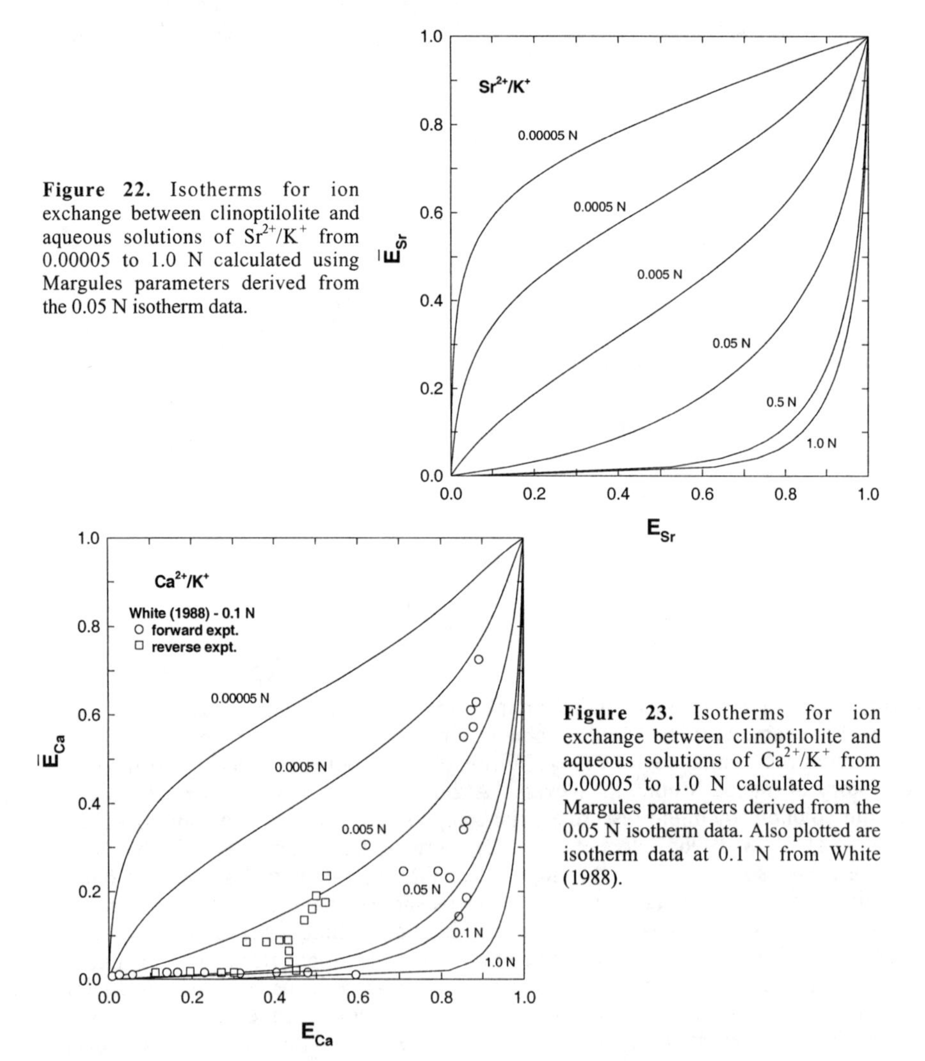

Figure 22. Isotherms for ion exchange between clinoptilolite and aqueous solutions of Sr^{2+}/K^+ from 0.00005 to 1.0 N calculated using Margules parameters derived from the 0.05 N isotherm data.

Figure 23. Isotherms for ion exchange between clinoptilolite and aqueous solutions of Ca^{2+}/K^+ from 0.00005 to 1.0 N calculated using Margules parameters derived from the 0.05 N isotherm data. Also plotted are isotherm data at 0.1 N from White (1988).

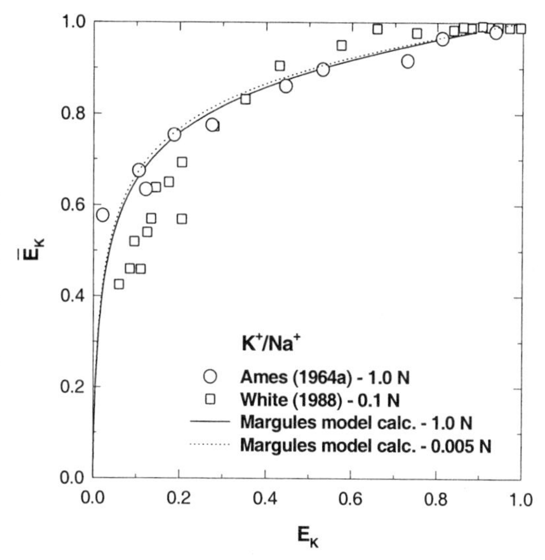

Figure 24. Isotherms for ion exchange between clinoptilolite and K^+/Na^+ calculated for 0.005 N and 1.0 N solution concentration using Margules parameters derived from the 0.5 N isotherm data. Also plotted are isotherm data at 1.0 N from Ames (1964a) and at 0.1 N from White (1988).

Equations for calculating solution composition from zeolite data. The ratio of the aqueous activities of two cations, A^{z_A+} and B^{z_B+}, participating in an ion-exchange reaction such as that represented by Equation (1) can be written as

$$\frac{(a_B)^{z_A}}{(a_A)^{z_B}} = \frac{(M_B)^{z_A}(\gamma_B)^{z_A}}{(M_A)^{z_B}(\gamma_A)^{z_B}} = K_{v(A,B)} \frac{(\bar{X}_B)^{z_A}}{(\bar{X}_A)^{z_B}} \tag{30}$$

Substituting $K_{v(A,B)}$ from Equation (25) and writing the zeolite composition in terms of $\bar{X}_A$ only, the aqueous activity ratio can be represented in terms of $K_{(A,B)}$, W_A, and W_B as

$$\frac{(a_B)^{z_A}}{(a_A)^{z_B}} = \frac{(1-\bar{X}_A)^{z_A}}{(\bar{X}_A)^{z_B}} \exp\Big\{ \ln K_{(A,B)} + z_A\bar{X}_A^2[W_B + 2(1-\bar{X}_A)(W_A - W_B)]$$

$$- z_B(1-\bar{X}_A)^2[W_A + 2\bar{X}_A(W_B - W_A)]\Big\}. \tag{31}$$

Because the total normality of a binary aqueous solution, TN, can be calculated from

$$TN = z_A M_A + z_B M_B , \tag{32}$$

the ratio of the molarities of A^{z_A+} and B^{z_B+} can be written in terms of the molarity of either A^{z_A+} only or B^{z_B+} only. Thus,

$$\frac{(M_B)^{z_A}}{(M_A)^{z_B}} = \frac{[(TN - z_A M_A)/z_B]^{z_A}}{(M_A)^{z_B}} = \frac{(a_B)^{z_A}(\gamma_A)^{z_B}}{(a_A)^{z_B}(\gamma_B)^{z_A}} \tag{33}$$

For binary ion-exchange reactions involving monovalent cations ($z_A = z_B = 1$), one can derive from Equation (33) expressions for M_A and M_B that are fairly simple:

$$M_A = \frac{TN}{(1+k)} \; ; \; M_B = TN - M_A \tag{34}$$

where

$$k = \frac{(a_B)^{z_A}(\gamma_A)^{z_B}}{(a_A)^{z_B}(\gamma_B)^{z_A}} \tag{35}$$

For exchange reactions where $z_A = 2$ and $z_B = 1$, the solution compositions can be solved using the quadratic formula

$$M_A = \frac{-c \pm \sqrt{c^2 - 4bd}}{2b} \; ; \; M_B = TN - 2M_A \quad (36)$$

where

$$b = 1; \; c = \frac{-4TN - k}{4} \; ; \; d = \left(\frac{TN}{2}\right)^2 \quad (37)$$

Using Equations (31) to (37), together with the equilibrium constants and Margules parameters derived from the ion-exchange experiments, values of M_A and M_B can be calculated if the value of TN is known or can be estimated by independent means. A value of TN is required as input into Equations (34), (36), and (37), and also in calculating the aqueous activity coefficient ratios, $(\gamma_A)^{z_B} / (\gamma_B)^{z_A}$. For a specific value of TN, a range of values corresponding to the activity coefficient ratios for solutions of almost pure A^{z_A+} to almost pure B^{z_B+} can be calculated. The aqueous activity coefficient ratios do not change very much for a particular TN value. For example, for $NaCl/CaCl_2$ solutions at TN = 0.5 N, $(\gamma_{Na^+})^2 / (\gamma_{Ca^{2+}})$ changes by only ~6%, from 1.953 to 2.056 over a range of $M_{Na^+} / M_{Ca^{2+}}$ ratio from 0.498/0.001 to 0.002/0.249. The variation in activity coefficient ratios is smaller at lower total normalities or for solutions of homovalent cations (e.g. Na^+/K^+). For instance, for $NaCl/CaCl_2$ solutions at TN = 0.05 N, $(\gamma_{Na^+})^2 / (\gamma_{Ca^{2+}})$ changes only ~0.3%, from 1.496 to 1.501, over a range of $M_{Na^+} / M_{Ca^{2+}}$ from 0.0498/0.0001 to 0.0002/0.0249. Likewise, for NaCl/KCl solutions at TN = 0.5 N, $\gamma_{Na^+} / \gamma_{K^+}$ changes by less than 3%, from 1.066 to 1.039, over a range of M_{Na^+} / M_{K^+} from 0.499/0.001 to 0.001/0.499. Therefore, although aqueous activity coefficient ratios are required to calculate solution compositions from zeolite data, for a given solution normality an activity coefficient ratio for a solution of median composition (e.g. equal normalities of A^{z_z+} and B^{z_B+}) can be used without resulting in a large error in predicted solution composition.

Groundwaters are typically dilute. For example, ionic strengths of Yucca Mountain saturated-zone groundwaters calculated by Turner et al. (1999), based on the comprehensive water chemistry data of Perfect et al. (1995), average about 0.019 ±0.099 M. Ionic strengths of porewaters from the hydrologically unsaturated zone of Yucca Mountain calculated from the data of Yang et al. (1996) average about 0.008±0.004 M. For these and other dilute groundwaters, it may be sufficient to use an activity coefficient ratio equal to one and obtain reasonable predictions of solution compositions.

Measured versus predicted solution compositions in binary systems. Pabalan and Bertetti (1999) equilibrated clinoptilolite powders with Na^+/K^+ and Na^+/Ca^{2+} solutions and recovered the powders using standard filtration methods. The zeolites were contacted only with the solutions with which they equilibrated to avoid altering the clinoptilolite compositions during the filtration step. This method is particularly important for the Na^+/Ca^{2+} system where, due to the concentration-valency effect, the zeolite composition may change significantly if the solid is contacted with deionized water (Stumm and Morgan 1996). Table 11 presents the chemical compositions of several clinoptilolite powders before and after equilibration with aqueous NaCl-KCl, $NaCl\text{-}CaCl_2$, $NaNO_3\text{-}KNO_3$, and $NaNO_3\text{-}Ca(NO_3)_2$ solutions at a total normality of 0.05 or 0.5 N. The zeolite compositions were determined by inductively coupled plasma emission spectrometry, subsequent to lithium metaborate fusion and dissolution in an HCl/HNO_3 matrix.

Table 11. Chemical composition of clinoptilolites before and after equilibration with ($Na^+ + K^+$) and ($Na^+ + Ca^{2+}$) solutions (from Pabalan and Bertetti, 1999).

Aqueous Mixtures	wt % SiO_2	wt % TiO_2	wt % Al_2O_3	wt % Fe_2O_3	wt % MgO	wt % Na_2O	wt % K_2O	wt % CaO	Total wt %
Na-form (unreacted)	68.28	0.04	11.28	0.40	0.16	6.36	0.66	0.02	87.20
0.5 N Na/K/Cl	67.81	0.06	11.17	0.36	0.17	1.91	7.52	0.00	89.00
	66.95	0.06	10.98	0.36	0.17	1.23	8.55	0.00	88.30
	65.82	0.06	10.75	0.35	0.16	0.55	9.29	0.01	86.99
	67.80	0.06	10.91	0.34	0.17	0.81	9.34	0.00	89.44
0.05 N Na/K/Cl	67.48	0.06	10.63	0.35	0.17	2.35	6.20	0.01	87.25
	67.58	0.04	10.55	0.25	0.25	0.55	8.43	0.00	87.65
	67.48	0.06	10.66	0.35	0.17	0.56	9.29	0.01	88.58
	67.34	0.06	10.58	0.35	0.17	0.32	9.64	0.00	88.46
0.05 N Na/K/NO_3	68.00	0.06	10.78	0.35	0.17	4.83	2.66	0.00	86.85
	68.00	0.06	10.77	0.38	0.18	2.24	6.71	0.00	88.34
	67.30	0.06	10.79	0.35	0.17	1.52	7.59	0.00	87.78
	66.76	0.06	10.49	0.35	0.17	0.78	8.70	0.01	87.32
	65.65	0.06	10.46	0.34	0.16	0.46	8.85	0.01	85.99
	67.00	0.06	10.72	0.40	0.18	0.17	9.60	0.00	88.13
0.5 N Na/Ca/Cl	68.46	0.06	11.25	0.36	0.18	5.00	0.61	1.32	87.24
	67.02	0.06	10.95	0.4	0.19	4.43	0.62	1.79	85.46
	66.13	0.06	10.57	0.36	0.18	3.79	0.61	2.14	83.84
	67.17	0.06	10.73	0.36	0.19	3.03	0.63	3.10	85.27
	67.32	0.06	10.74	0.36	0.18	2.47	0.63	3.60	85.36
0.05 N Na/Ca/Cl	67.62	0.06	11.10	0.36	0.18	3.57	0.63	2.56	86.08
	67.54	0.05	11.00	0.36	0.20	2.55	0.58	3.52	85.80
	68.03	0.06	11.08	0.36	0.19	2.12	0.61	3.91	86.36
	68.62	0.06	11.33	0.36	0.19	1.72	0.62	4.36	87.26
	67.41	0.06	11.05	0.35	0.20	1.39	0.55	4.51	85.52
0.05 N Na/Ca/NO_3	69.09	0.06	11.30	0.36	0.19	5.10	0.62	1.21	87.93
	66.08	0.06	10.89	0.35	0.18	3.28	0.60	2.66	84.10
	65.74	0.06	10.86	0.38	0.18	2.55	0.62	3.46	83.85
	66.17	0.06	10.98	0.37	0.19	1.90	0.60	4.13	84.40
	65.44	0.06	10.78	0.35	0.19	1.02	0.55	4.92	83.31

The measured Na_2O and K_2O weight percents given in Table 11 for the zeolites equilibrated with Na^+/K^+ solutions are listed in Table 12. These values were used to calculate $\overline{X}_K$ and, from Equation (31), a_{Na}/a_K. The measured Na_2O and CaO weight percents given in Table 11 for the zeolites equilibrated with Na^+/Ca^{2+} solutions are listed in Table 13 and were used to calculate $\overline{X}_{Ca}$ and $(a_{Na})^2/a_{Ca}$. The $\overline{X}_K$ values given in Table 12 were normalized to the molar amounts of Na^+ and K^+ only, whereas the $\overline{X}_{Ca}$ values in Table 13 were normalized to the molar amounts of Na^+ and Ca^{2+} only. These calculations assumed that the small amounts of Ca^{2+} and Mg^{2+} present in the zeolite did not participate in the exchange process for the Na^+/K^+ system, and that K^+ and Mg^{2+} did not participate in the exchange for the Na^+/Ca^{2+} mixture. This assumption is justified by the zeolite compositions given in Table 11 which show that, within analytical uncertainty, the MgO and CaO contents of the zeolites used in the Na^+/K^+ ion-exchange experiments remained equal to the MgO and CaO contents of the unreacted Na-clinoptilolite. Likewise, the K_2O and MgO contents of the zeolites from the Na^+/Ca^{2+} experiments remained equal to those of the Na-clinoptilolite.

The composition of solutions in equilibrium with the zeolites were predicted from the values of a_{Na}/a_K and $(a_{Na})^2/a_{Ca}$ using Equations (34) to (37) and ln $K_{(A,B)}$, W_A, and W_B given in Table 10. Aqueous activity coefficients required for the calculations were derived from the Pitzer equations and the parameters listed in Table 2. Values of $\gamma_{Na^+}/\gamma_{K^+} = 1.05$ and 1.007 and $(\gamma_{Na^+})^2/(\gamma_{Ca^{2+}}) = 1.98$ and 1.50 were used for 0.5 and 0.05 N NaCl/KCl solutions, respectively. Activity coefficient ratios of 1.019 and 1.49 were used for 0.05 N mixtures of $NaNO_3/KNO_3$ and $NaNO_3/Ca(NO_3)_2$, respectively. Table 12 compares the predicted aqueous concentrations of Na^+ and K^+ with values measured using ion-selective electrodes, and Table 13 compares predicted and measured Na^+ and Ca^{2+} solution concentrations. Measured and predicted concentrations are also compared in Figure 25.

The uncertainties in predicted Na^+, K^+, and Ca^{2+} solution concentrations were propagated assuming a ±10% error in the measured weight percents of Na_2O, K_2O, and CaO, but neglecting regression errors for ln $K_{(A,B)}$, W_A, and W_B, and errors due to the use of a single value of activity coefficient ratio at a given solution normality. The results listed in Tables 12 and 13 and plotted in Figure 25 show that the predicted solution compositions agree very well with measured values, mostly within analytical uncertainty. This agreement demonstrates that it is possible to predict with confidence the cationic composition of an aqueous solution based on chemical analysis of zeolites with which it equilibrated, at least for simple two-cation systems for which the solution normality is known.

The solution concentration may be uncertain for natural aqueous systems. Therefore, it may not be possible to calculate accurate aqueous activity coefficients. In some cases, especially when solutions are dilute, it may be expeditious to use activity coefficient ratios equal to 1.0. For comparison, Tables 12 and 13 list K^+, Ca^{2+}, and Na^+ concentrations calculated using activity coefficient ratios equal to 1.0. For aqueous solutions of K^+ and Na^+, both monovalent cations, the solution compositions predicted using activity coefficient ratios equal to 1.0 do not differ much from the previously calculated values. On the other hand, for Ca^{2+} and Na^+ ion exchange, which involves heterovalent cations, the two sets of predicted compositions have significant differences, ranging from 4 to 26% at 0.05 N and from 7 to 38% at 0.5 N. However, considering the estimated error propagated from the uncertainties in zeolite composition, the values calculated using activity coefficient ratios equal to 1.0 are reasonable, especially in light of the fact that the differences between predicted and actual solution compositions become smaller at lower solution concentrations.

Table 12. Measured K^+ and Na^+ solution concentrations versus values predicted from measured clinoptilolite Na and K content. Also tabulated are cationic mole fractions of K^+ in the zeolite, $\overline{X}_K$, and predicted solution activity ratios of Na^+ and K^+. Values in parentheses were derived using aqueous activity coefficient ratios equal to 1.0. From Pabalan and Bertetti (1999).

Aqueous Mixtures	wt % Na_2O	wt % K_2O	$\overline{X}_K$	a_{Na}/a_K	Calc. K^+ conc. (M)	Meas. K^+ conc. (M)	Calc. Na^+ conc. (M)	Meas. Na^+ conc. (M)
Na/K/Cl 0.5 N	1.91	7.52	0.721	5.906	0.076±0.019 (0.072)	0.061±0.002	0.424±0.019 (0.428)	0.444±0.009
	1.23	8.55	0.821	2.463	0.150±0.030 (0.144)	0.130±0.004	0.350±0.030 (0.356)	0.366±0.007
	0.55	9.29	0.917	0.7338	0.295±0.032 (0.288)	0.258±0.008	0.205±0.032 (0.212)	0.238±0.005
	0.81	9.34	0.884	1.207	0.233±0.034 (0.227)	0.209±0.006	0.267±0.034 (0.273)	0.298±0.006
Na/K/Cl 0.05 N	2.35	6.20	0.634	11.30	0.0041±0.0011 (0.0041)	0.0033±0.0001	0.0459±0.0011 (0.0459)	0.0465±0.0009
	0.55	8.43	0.910	0.8303	0.0274±0.0033 (0.0273)	0.0194±0.0006	0.0226±0.0033 (0.0227)	0.0306±0.0006
	0.56	9.29	0.916	0.7507	0.0287±0.0032 (0.0286)	0.0239±0.0007	0.0213±0.0032 (0.0214)	0.0271±0.0005
	0.32	9.64	0.952	0.3646	0.0367±0.0024 (0.0366)	0.0327±0.0010	0.0133±0.0024 (0.0134)	0.0174±0.0003
Na/K/NO_3 0.05 N	4.83	2.66	0.266	125.1	0.0004±0.0001	0.0003±0.00001	0.0496±0.0001	0.0476±0.0010
	2.24	6.71	0.663	9.181	0.0050±0.0013	0.0034±0.0001	0.0451±0.0013	0.0457±0.0009
	1.52	7.59	0.767	4.063	0.0100±0.0023	0.0069±0.0002	0.0401±0.0023	0.0404±0.0008
	0.78	8.70	0.880	1.262	0.0223±0.0033	0.0154±0.0005	0.0278±0.0033	0.0314±0.0006
	0.46	8.85	0.927	0.6237	0.0310±0.0031	0.0248±0.0007	0.0191 ±0.0031	0.0243±0.0005
	0.17	9.60	0.974	0.1799	0.0425±0.0015	0.0368±0.0011	0.0076 ±0.0015	0.0121±0.0002

Table 13. Measured Ca^{2+} and Na^+ solution concentrations versus values predicted from measured clinoptilolite Na and Ca content. Also tabulated are cationic mole fractions of Ca^{2+} in the zeolite, $\overline{X}_{Ca}$, and predicted solution activity ratios of Na^+ and Ca^{2+}. Values in parentheses were derived using aqueous activity coefficient ratios equal to 1.0. From Pabalan and Bertetti (1999).

Aqueous Mixtures	wt % Na_2O	wt % CaO	$\overline{X}_{Ca}$	$(a_{Na})^2/(a_{Ca})$	Calc. Ca^{2+} conc. (M)	Meas. Ca^{2+} conc. (M)	Calc. Na^+ conc. (M)	Meas. Na^+ conc. (M)
Na/Ca/Cl 0.5 N	5.0	1.32	0.127	5.868	0.052±0.012 (0.032)	0.051±0.002	0.395±0.025 (0.435)	0.409±0.008
	4.43	1.79	0.183	2.314	0.089±0.019 (0.061)	0.085±0.003	0.323±0.037 (0.377)	0.345±0.007
	3.79	2.14	0.238	1.005	0.124±0.022 (0.095)	0.118±0.005	0.252±0.043 (0.309)	0.260±0.005
	3.03	3.1	0.361	0.1885	0.184±0.018 (0.162)	0.180±0.007	0.133±0.036 (0.175)	0.155±0.003
	2.47	3.6	0.446	0.06527	0.208±0.013 (0.194)	0.199±0.008	0.083±0.026 (0.112)	0.100±0.002
Na/Ca/Cl 0.05 N	3.57	2.56	0.284	0.5259	0.0047±0.0016 (0.0035)	0.0039±0.0002	0.0406±0.0031 (0.0430)	0.0406±0.0008
	2.55	3.52	0.433	0.07686	0.0124±0.0026 (0.0107)	0.0105±0.0004	0.0252±0.0052 (0.0286)	0.0285±0.0006
	2.12	3.91	0.505	0.03232	0.0158±0.0023 (0.0143)	0.0136±0.0005	0.0184±0.0046 (0.0215)	0.0214±0.0004
	1.72	4.36	0.583	0.01291	0.0187±0.0018 (0.0175)	0.0173±0.0007	0.0126±0.0035 (0.0150)	0.0159±0.0003
	1.39	4.51	0.642	0.00658	0.0203±0.0013 (0.0194)	0.0194±0.0008	0.0094±0.0026 (0.0113)	0.0124±0.0002
$Na/Ca/NO_3$ 0.05 N	5.1	1.21	0.116	7.251	0.0005±0.0002	0.0005±0.00002	0.0490±0.0003	0.0496±0.0010
	3.28	2.66	0.309	0.3714	0.0059±0.0019	0.0051±0.0002	0.0382±0.0037	0.0392±0.0008
	2.55	3.46	0.429	0.0809	0.0122±0.0026	0.0101±0.0004	0.0257±0.0051	0.0303±0.0006
	1.9	4.13	0.546	0.01999	0.0174±0.0020	0.0159±0.0006	0.0153±0.0040	0.0177±0.0004
	1.02	4.92	0.727	0.00243	0.0220±0.0008	0.0217±0.0009	0.0060±0.0015	0.0069±0.0001

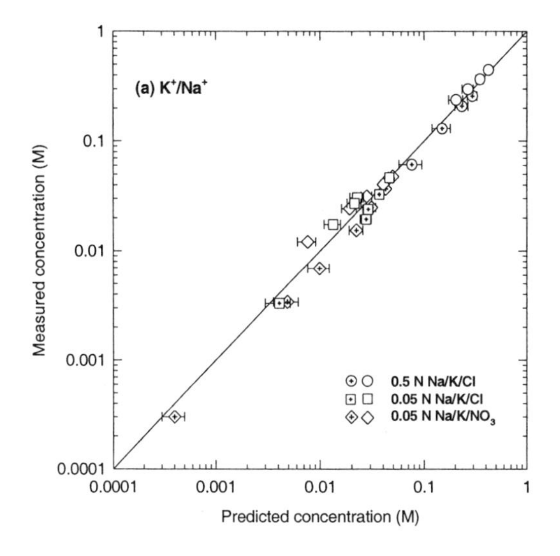

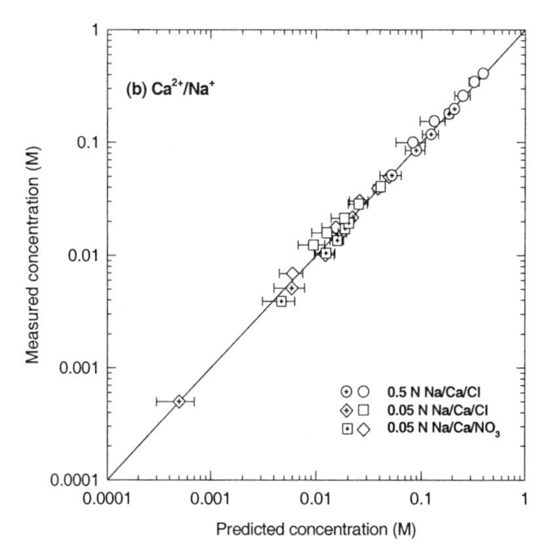

Figure 25. Comparison of measured Na^+, K^+, and Ca^{2+} concentrations in the aqueous phase with values calculated based on the composition of the coexisting zeolite phase. Open symbols are values for Na^+ concentrations. Cross-hair symbols in (a,b) are values for K^+ or Ca^{2+} concentrations. From Pabalan and Bertetti (1999).

ANION EXCHANGE ON SURFACTANT-MODIFIED ZEOLITES

The net negative structural charge of zeolites that results in the favorable ion-exchange selectivity for many cations also causes natural zeolites to have little or no affinity for anions, such as the oxyanions of toxic metals like chromate (CrO_4^{2-}), selenate (SeO_4^{2-}), and pertechnetate (TcO_4^-). Likewise, sorption affinity for actinides such as uranium and neptunium, which sorb primarily through a surface complexation mechanism, is also limited (Pabalan et al. 1998; Bertetti et al. 1998).

Cationic surfactants have been used to modify zeolite surfaces in attempts to enhance the sorptive capabilities of natural zeolites. Cationic surfactants in the aqueous phase associate in micelles due to a hydrophobic effect. In the presence of zeolites, the surfactant sorbs to the negatively charged zeolite surface and forms a bilayer of surfactant molecules similar to a lipid layer (Li and Bowman 1997; Sullivan et al. 1998). Surfactant sorption causes the surface charge to change from negative to positive and the organic carbon content of the zeolite to increase to about 5 wt %. The positive surface charge, which provides sites for sorption of anions, results when positively charged surfactant head groups are presented to the surrounding solution, where they are balanced by counterions that can be replaced by other anions in solution (Li and Bowman 1997; Sullivan et al. 1998). The relatively large surfactant molecules do not enter the zeolite channels or access internal cation-exchange sites. Therefore, some of the original CEC of the zeolite is retained for sorbing cations. In addition, the organic-rich surface layer provides a partitioning medium for sorption of nonpolar organics, such as chlorinated solvents and fuel components. Therefore, the surfactant-modified zeolite (SMZ) can simultaneously sorb inorganic cations, inorganic anions, and nonpolar organics (Haggerty and Bowman 1994; Bowman et al. 1995). The magnitude and stability of sorption of the surfactant on the zeolite and of anion exchange are influenced by the counterion (e.g. Cl^-, Br^-, or HSO_4^-) of the surfactant (Li and Bowman 1997).

Recent studies have shown that treatment of clinoptilolite with cationic surfactants, e.g. hexadecyltrimethylammonium-bromide (HDTMA), yields an SMZ that has a strong affinity for selenate and chromate, as well as nonpolar organics, such as benzene, 1,1,1-trichloroethane, and perchloroethylene, but that also has cation-exchange selectivity for heavy-metal cations, such as Pb^{2+} (Haggerty and Bowman 1994; Bowman et al. 1995). Additional studies investigating actinide sorption on SMZs indicate that surfactant modification also enhances the ability of clinoptilolite to sorb U^{6+}, particularly at pHs greater than 6 where U^{6+} sorption on unmodified zeolite is typically low due to formation of anionic U^{6+} aqueous carbonate complexes (Prikryl and Pabalan 1999). The enhanced sorption of U^{6+} is interpreted to be due to anion exchange with counterions on the external portion of a surfactant bilayer or admicelles (Prikryl and Pabalan 1999).

Although data on anion exchange on SMZs are limited compared with the data available on cation exchange on natural zeolites, published studies suggest that surfactant modification offers the potential to further enhance the industrial and environmental applications of natural zeolites.

RECOMMENDATIONS FOR FUTURE WORK

Although numerous studies have been conducted on ion-exchange equilibria involving natural zeolites, a systematic and comprehensive evaluation of ion-exchange data from different sources is not available. As shown by the tabulated values in this chapter, there are significant differences in the ion-exchange equilibrium constants and Gibbs free energies reported by different investigators for the same types of natural zeolites at the same temperatures. These differences could be due to the variability in composition and purity of mineral samples used in the experiments, but they also could be a result of the different approaches used in interpreting the ion-exchange data. An attempt was made in this chapter to evaluate literature data on clinoptilolite ion exchange and to provide a set of self-consistent thermodynamic parameters based on a specific approach for calculating solid solution and aqueous phase activity coefficients. The evaluation, however, is not comprehensive, and it would be useful to derive thermodynamic parameters for other binary systems and other zeolite minerals for which experimental data are available. Moreover, although multiple studies have been published for some binary cation mixtures (e.g. ion exchange between clinoptilolite and Cs^+/Na^+, NH_4^+/Na^+, Ca^{2+}/Na^+, and K^+/Na^+ solutions), data on other binary systems (e.g. mixtures with Mg^{2+} and Ba^{2+}) are not available. For some of these poorly understood binary systems, ion-exchange equilibrium constants and Gibbs free energies may be derived using the triangle rule. However, experimental data at one solution normality would still be required to constrain the shape of the isotherm. For binary systems with insufficient data to apply the triangle rule, it may be useful to evaluate the utility of correlation methods for predicting ion-exchange thermodynamic parameters.

Most published ion-exchange data on natural zeolites involve the exchange of two cations only. Little attention has been paid to the problem of understanding multicomponent ion-exchange equilibria, despite the fact that ion-exchange processes in natural systems generally involve more than two ions. Multicomponent ion-exchange studies are challenging. A large amount of experimental data is needed to completely characterize a ternary system and numerical analysis of the data is difficult. For a system with more than three ions, Grant and Fletcher (1993) concluded that the amount of data required to characterize the system is almost prohibitive. Nevertheless, experimental data on ternary and more complex mixtures are needed if progress is to be made in developing and evaluating thermodynamic models that can be used to predict ion-exchange equilibria in natural systems. Thermodynamic modeling of multicomponent ion-exchange equilibria is computationally more complex, but the basic principles are the same as those for binary exchange reactions. In particular, it would be useful to test the ability of the Wilson model, which does not require parameters beyond the binary terms, to predict ion-exchange equilibria in ternary and more complex mixtures.

Finally, several studies report ion-exchange data and thermodynamic parameters at temperatures other than 298 K. Some of the reported values are listed in the tables in this chapter. The temperature dependence of ion-exchange equilibria was not addressed in this chapter and a critical review of literature information on the enthalpies of ion-exchange reactions would be useful.

ACKNOWLEDGMENTS

The reviews by Doug Ming, Steven Grant, David Turner and John Russell are gratefully acknowledged. This work was funded by the Nuclear Regulatory Commission (NRC) under Contract No. NRC-02-97-009. This paper is an independent product of the Center for Nuclear Waste Regulatory Analyses and does not necessarily reflect the views or regulatory position of the NRC.

REFERENCES

Adabbo M, Caputo D, de Gennaro B, Pansini M, Colella C (1999) Ion exchange selectivity of phillipsite for Cs and Sr as a function of framework composition. Microporous Mesoporous Mater 28:315-324

Ahmad ZB, Dyer A (1984) Ion exchange in ferrierite, a natural zeolite. *In* D Naden, M Streat (eds) Ion Exchange Technology, p 519-532. Ellis Horwood Limited, Chichester, United Kingdom

Ahmad ZB, Dyer A (1988) Ion exchange in near-homoionic ferrierites. *In* D Kallo, HS Sherry (eds) Occurrence, Properties, and Utilization of Natural Zeolites, p 431-448. Akademiai Kiada, Budapest

Ali A, El-Kamash A, El-Sourougy M, Aly H (1999) Prediction of ion-exchange equilibria in binary aqueous systems. Radiochim Acta 85:65-69

Al'tshuler GN, Shkurenko GY (1990) Cation-exchange equilibrium in natural heulandite. Izv Akad Nauk SSSR, Ser Khim 7:1474-1477

Al'tshuler GN, Shkurenko GY (1992) Thermodynamics of cation-exchange on natural heulandite. Bull Russian Acad Sci, Chem Sci 41:592-593

Al'tshuler GN, Shkurenko GY (1997) Cation exchange in natural heulandite. Zh Fiz Khim 71:334-336

Ames LL Jr (1960) The cation sieve properties of clinoptilolite. Am Mineral 45:689-700

Ames LL Jr (1961) Cation sieve properties of the open zeolites chabazite, mordenite, erionite and clinoptilolite. Am Mineral 46:1120-1131

Ames LL Jr (1964a) Some zeolite equilibria with alkali metal cations. Am Mineral 49:127-145

Ames LL Jr (1964b) Some zeolite equilibria with alkaline earth metal cations. Am Mineral 49:1099-1110

Ames LL Jr (1967) Zeolitic removal of ammonium ions from agricultural and other wastewater. 13th Pacific Northwest Industrial Waste Conf, p 135-152. Washington State Univ, Pullman, Washington

Ames LL Jr (1968) Cation exchange properties of heulandite-clinoptilolite series members. *In* DW. Pearce, MR Compton (eds) Pacific Northwest Laboratory Annual Report for 1996 to the U.S.A.E.C. Division of Biology and Medicine, Volume II. Physical Sciences, p 54-59. BNWL-481-3. Pacific Northwest Laboratory, Richland, Washington

Argersinger WJ, Davidson AW, Bonner OD (1950) Thermodynamics and ion exchange phenomena. Trans Kansas Acad Sci 53:404-410

Barrer RM (1978) Cation-exchange equilibria in zeolites and feldspathoids. *In* LB Sand, FA Mumpton (eds) Natural Zeolites: Occurrence, Properties, Use, p 385-395. Pergamon Press, New York

Barrer RM (1980) Zeolite exchangers B some equilibrium and kinetic aspects. *In* LVC Rees (ed) Proc 5th Int'l Conf Zeolites, p 273-290. Heyden, London

Barrer RM (1984) Cation partitioning among sub-lattices in zeolites. Zeolites 4:361-368

Barrer RM, Klinowski J (1974a) Ion-exchange selectivity and electrolyte concentration. J Chem Soc Faraday Trans I 70:2080-2091

Barrer RM, Klinowski J (1974b) Ion exchange in mordenite. J Chem Soc Faraday Trans I 70:2362-2367

Barrer RM, Klinowski J (1977) Theory of isomorphous replacement in aluminosilicates. Philos Trans Royal Soc London A285:637-676

Barrer RM, Townsend RP (1976a) Transition metal ion exchange in zeolites. Part 1. Thermodynamics of exchange of hydrated Mn^{2+}, Co^{2+}, Ni^{2+}, Cu^{2+}, and Zn^{2+} ions in ammonium mordenite. J Chem Soc Faraday Trans I 72:661-673

Barrer RM, Townsend RP (1976b) Transition metal ion exchange in zeolites. Part 2. Ammines of Co^{2+}, Cu^{2+}, and Zn^{2+} in clinoptilolite, mordenite, and phillipsite. J Chem Soc Faraday Trans I 72:2650-2660

Barrer RM, Davies JA, Rees LVC (1969) Thermodynamics and thermochemistry of cation exchange in chabazite. J Inorg. Nucl Chem 31:219-232

Barrer RM, Klinowski J, Sherry HS (1973) Zeolite exchangers: Thermodynamic treatment when not all ions are exchangeable. J Chem Soc Faraday Trans II 69:1669-1676

Barrer RM, Papadopoulos R, Rees LVC (1967) Exchange of sodium in clinoptilolite by organic cations. J Inorg Nucl. Chem 29:2047-2063

Baxter SG, Berghauser DC (1986) The selection and performance of the natural zeolite clinoptilolite in British Nuclear Fuel's site ion exchange effluent plant, SIXEP. Waste Management '86: Proc Symp Waste Management, p 347-356, Univ of Arizona Board of Regents, Tucson, Arizona

Bertetti FP, Pabalan RT, Almendarez MG (1998) Studies of neptunium(V) sorption on quartz, clinoptilolite, montmorillonite, and γ-alumina. *In* E Jenne (ed) Adsorption of Metals by Geomedia, p 131-148. Academic Press, San Diego, California

Blanchard G, Maunaye M, Martin G (1984) Removal of heavy metals from waters by means of natural zeolites. Water Resources 18:1501-1507

Bowman RS, Haggerty GM, Huddleston RG, Neel D, Flynn M (1995) Sorption of nonpolar organics, inorganic cations, and inorganic anions by surfactant-modified zeolites. *In* DA Sabatini, RC Knox, JH Harwell (eds) Surfactant-enhanced Remediation of Subsurface Contamination, Am Chem Soc Symp Series 594:54-64

Breck DW (1974) Zeolite Molecular Sieves. John Wiley and Sons, New York

Bremner PR, Schultze LE (1995) Ability of clinoptilolite-rich tuffs to remove metal cations commonly found in acidic drainage. *In* DW Ming, FA Mumpton (eds) Natural Zeolites '93: Occurrence, Properties, Use, p 397-403. Int'l Comm Natural Zeolites, Brockport, New York

Bronsted JN (1922a) Calculation of the osmotic and activity functions in solutions of uni-univalent salts. J Am Chem Soc 44:938-948

Bronsted JN (1922b) Studies on solubility. IV. The principle of the specific interaction of ions. J Am Chem Soc 44:877-898

Brooke NM, Rees, LVC (1968) Kinetics of ion exchange. Part 1. Trans Faraday Soc 64:3383-3392

Brooke NM, Rees LVC (1969) Kinetics of ion exchange. Part 2. Trans Faraday Soc 65:2728-2739

Broxton DE, Bish DL, Warren RG (1987) Distribution and chemistry of diagenetic minerals at Yucca Mountain, Nye County, Nevada. Clays & Clay Minerals 35:89-110

Broxton DE, Warren RG, Hagan RC, Luedemann G (1986) Chemistry of Diagnetically-Altered Tuffs at a Potential Nuclear Waste Repository, Yucca Mountain, Nye County, Nevada. LA-10802-MS, Los Alamos National Laboratory, Los Alamos, New Mexico

Chelishchev NF, Volodin VF (1977) Ion exchange of alkali metals on natural erionite. Dokl Akad Nauk SSSR 237:122-125

Chelishchev NF, Berenshtein BG, Berenshtein TA, Gribanova NK, Martynova NS (1973) Ion-exchange properties of clinoptilolites. Dokl Akad Nauk SSSR 210:1110-1112

Chelishchev NE, Berenshtein BG, Novikov GV (1984) Ion-exchange equilibria of alkali metals on natural phillipsite. Dokl Akad Nauk SSSR 274:138-141
Chmielewska-Horvathova E, Lesny J (1992) Adsorption of cobalt on some natural zeolites occurring in CSFR. J Radioanal Nucl Chem B Letters 166:41-53
Chu S-Y, Sposito G (1981) The thermodynamics of ternary cation exchange systems and the subregular model. Soil Sci Soc Am J 45:1084-1089
Colella C (1996) Ion exchange equilibria in zeolite minerals. Mineral Dep 31:554-562
Colella C, Pansini M (1988) Lead removal from wastewaters using chabazite tuff. *In* WH Flank, TE Whyte (eds) Perspectives in Molecular Sieve Science, Am Chem Soc Symp Series 368, p 500-510
Colella C, de'Gennaro M, Langella A, Pansini M (1998) Evaluation of natural phillipsite and chabazite as cation exchangers for copper and zinc. Sep Sci Technol 33:467-481
Cooney EL, Booker NA, Shallcross DC, Stevens GW (1999) Ammonia removal from wastewaters using natural Australian zeolite. I. Characterization of the zeolite. Sep Sci Technol 34:2307-2327
Cremers A (1977) Ion exchange in zeolites. Molecular Sieves IIB 4th Int'l Conf, Am Chem Soc Symp Series 40:179-193.
de Barros MA, Machado NR, Alves FV, Sousa-Aguiar EF (1997) Ion exchange mechanism of Cr^{+3} on naturally occurring clinoptilolite. Brazil J Chem Eng 14:233-241
de Lucas A, Zarca J, Canizares, P (1992) Ion-exchange equilibrium of Ca^{2+} ions, Mg^{2+} ions, K^+ ions, Na^+ ions, and H^+ ions on Amberlite IR-120—Experimental determination and theoretical prediction of the ternary and quaternary equilibrium data. Sep Sci Technol 27:823-841
DePaoli SM, Perona JJ (1996) Model for Sr-Cs-Ca-Mg-Na ion-exchange uptake kinetics on chabazite. Am Inst Chem Eng J 42:3434–3441
Drummond D, DeJonge A, Rees LVC (1983) Ion-exchange kinetics in zeolite A. J Phys Chem 87:1967-1971
Dyer A, Zubair M (1998) Ion-exchange in chabazite. Microporous and Mesoporous Materials 22:135-150
Dyer A, Enamy H, Townsend RP (1981) The plotting and interpretation of ion-exchange isotherms in zeolite systems. Sep Sci Technol 16:173-183
Dyer A, Gawad ASA, Mikhail M, Enamy H, Afshang M (1991) The natural zeolite, laumontite, as a potential material for the treatment of aqueous nuclear wastes. J Radioanal Nucl Chem—Letters 154:265-276
Dyer A, Jozefowicz LC (1992) The removal of thorium from aqueous solutions using zeolites. J Radioanal Nucl Chem B Articles 159:47-62
Edmunds WM, Faye S, Gaye CB (1992) Solute profiles in unsaturated Quaternary sands from Senegal: Environmental information and water-rock interaction. *In* YF Kharaka, AS Maest (eds) Water-Rock Interaction, p 719-722. A.A. Balkema, Rotterdam, Netherlands
Ekedahl E, Hogfeldt E, Sillen LG (1950) Activities of the components in ion exchangers. Acta Chem Scand 4:556-558
Elprince AM, Babcock KL (1975) Prediction of ion-exchange equilibria in aqueous systems with more than two counter-ions. Soil Sci 120:332-338
Elprince AM, Vanselow AP, Sposito G (1980) Heterovalent, ternary cation exchange equilibria: NH_4^+-Ba^{2+}-La^{3+} exchange on montmorillonite. Soil Sci Soc Am J 44:964-969
Faghihian H, Marageh M, Kazemian H (1999) The use of clinoptilolite and its sodium form for removal of radioactive cesium, and strontium from nuclear wastewater and Pb^{2+}, Ni^{2+}, Cd^{2+}, Ba^{2+} from municipal wastewater. Appl Radiat Isot 50:655-660
Filizova L (1974) Ion-exchange properties of heulandite and clinoptilolite. Izv Geol Inst, Bulg Akad Nauk, Ser Rudni Nerudni Polezni Izkopaemi 23:311-325
Fletcher P, Franklin KR, Townsend RP (1984) Thermodynamics of binary and ternary ion exchange in zeolites: The exchange of sodium, ammonium, and potassium ions in mordenite. Phil Trans R Soc London, A 312:141-178
Fletcher P, Townsend RP (1985) Ion exchange in zeolites. The exchange of cadmium and calcium in sodium X using different anionic backgrounds. J Chem Soc Faraday Trans I 81:1731-1744
Freeman DH (1961) Thermodynamics of binary ion-exchange systems. J Chem Phys 35:189-191
Frysinger GR (1962) Caesium-sodium exchange on clinoptilolite. Nature 194:351-353
Gaines GL Jr, Thomas HC (1953) Adsorption studies on clay minerals. II. A formulation of the thermodynamics of exchange adsorption. J Chem Phys 21:714-718
Gapon YN (1933) Theory of exchange adsorption in soils. J Gen Chem USSR 3:144-163
Glueckauf E (1949) Activity coefficient in concentrated solutions containing several electrolytes. Nature 163:414-415
Golden TC, Jenkins RG (1981) Ion exchange in mordenite. Verification of the triangle rule. J Chem Eng. Data 26:366-367

Gottardi G (1978) Mineralogy and crystal chemistry of zeolites. *In* LB Sand, FA Mumpton (eds) Natural Zeolites: Occurrence, Properties, Use, p 31-44. Pergamon Press, New York

Gottardi G, Galli E (1985) Natural Zeolites. Springer-Verlag, Berlin

Gradev G, Gulubova I (1982) Cesium and strontium ion exchange on clinoptilolite and mordenite. I. Ion-exchangers. Equilibrium characteristics of the ion-exchange process. Yad Energ 16:64-72

Grant SA, Fletcher P (1993) Chemical thermodynamics of cation exchange reactions: Theoretical and practical considerations. *In* JA Marinsky, Y Marcus (eds) Ion Exchange and Solvent Extraction. A Series of Advances. Volume II, p 1-108. Marcel Dekker, New York

Grant SA, Sparks DL (1989) Method for evaluating exchangeable-ion excess Gibbs energy models in systems with many species. J Phys Chem 93:6265-6267

Grebenshchikova VI, Chernyavskaya NB, Andreeva NR (1973a) Sorption of tetravalent elements by mordenite. I. Sorption of thorium(IV). Radiokhimiya 15:308-311

Grebenshchikova VI, Chernyavskaya NB, Andreeva NR (1973b) Sorption of tetravalent elements by mordenite. II. Sorption of plutonium(IV). Radiokhimiya 15:761-766

Guggenheim EA (1935) The specific thermodynamic properties of aqueous solutions of strong electrolytes. Philos Mag 19:588-643

Haggerty GM, Bowman RS (1994) Sorption of inorganic anions by organo-zeolites. Environ Sci Technol 28:452-458

Helfferich FG, Hwang Y-L (1991) Ion exchange kinetics. *In* K Dorfner (ed) Ion Exchangers, p 1277-1309. Walter de Gruyter, Berlin

Howden M, Pilot J (1984) The choice of ion exchanger for British Nuclear Fuels Ltd.'s site ion exchange effluent plant. *In* D Naden, M Streat (eds) Ion Exchange Technology, p 66-73. Ellis Horwood Limited, Chichester, United Kingdom

Howery DG, Thomas HC (1965) Ion exchange on the mineral clinoptilolite. J Phys Chem 69:531-537

Ioannidis SA, Anderko A, Sanders S.J (2000) Internally consistent representation of binary ion-exchange equilibria. Chem Eng Sci 55:2687-2698

Jama MA, Yucel H (1990) Equilibrium studies of sodium-ammonium, potassium-ammonium, and calcium-ammonium exchanges on clinoptilolite zeolite. Sep Sci Technol 24:1393-1416

Kallo D (1995) Wastewater purification in Hungary using natural zeolites. *In* DW Ming, FA Mumpton (eds) Natural Zeolites '93: Occurrence, Properties, Use, p 341-350. Int'l Comm Natural Zeolites, Brockport, New York

Kielland J (1935) Thermodynamics of base-exchange of some different kinds of clays. J Soc Chem Ind, Lond 54:232-234

Kim H-T, Frederick J (1988) Evaluation of Pitzer ion interaction parameters of aqueous mixed electrolyte solutions at 25 °C. 2. Ternary mixing parameters. J Chem Eng. Data 33:278-283

Kuznetsova EM, Sinev AV, Krasovskii AL (1998) Description of constants of ion exchange equilibrium for singly-charged ions on mordenite. Vestnik Moskovskogo Univ Ser 2 Khimiya 39:159-162

Laudelot H (1987) Cation exchange equilibria in clays. *In* ACD. Newman (ed) Chemistry of Clays and Clay Minerals, p 225-236. John Wiley, New York

Lehto J, Harjula R (1995) Experimentation in ion exchange studies—The problem of getting reliable and comparable results. Reactive & Functional Polymers 27:121-146

Leppert DE (1988) An Oregon cure for Bikini Island? First results from the Zeolite Immobilization Experiment. Oregon Geol 50:140-141

Li Z, Bowman RS (1997) Counterion effects on the sorption of cationic surfactant and chromate on natural clinoptilolite. Environ Sci Technol 31:2407-2412

Liang T, Hsu C (1993) Sorption of cesium and strontium on natural mordenite. Radiochim Acta 61:106-108

Loizidou M, Townsend RP (1987a) Exchange of cadmium into the sodium and ammonium forms of the natural zeolites clinoptilolite, mordenite, and ferrierite. J Chem Soc Dalton Trans 1911-1916

Loizidou M, Townsend RP (1987b) Ion-exchange properties of natural clinoptilolite, ferrierite and mordenite: Part 2. Lead-sodium and lead-ammonium equilibria. Zeolites 7:153-159

Loizidou M, Haralambous KJ, Loukatos A, Dimitrakopoulou D (1992) Natural zeolites and their ion exchange behavior towards chromium. J Environ Sci Health A27:1759-1769

Lu G, Xu G, Zhao F (1981) Ion exchange in natural mordenite. Ranliao Huaxue Xuebao 9:311-319

Maeda M, Furuhashi H, Ikami J (1993) Evaluation of dissociation constants of ammonium ions in aqueous ammonium chloride and potassium chloride solutions and of pertinent higher-order parameters according to the Pitzer approach. J Chem Soc Faraday Trans I 89:3371-3374

Maeda M, Hisada O, Ito K, Kinjo Y (1989) Application of Pitzer's equations to dissociation constants of ammonium ion in lithium chloride-sodium chloride mixtures. J Chem Soc Faraday Trans I 85:2555-2562

Malliou E, Loizidou M, Spyrellis N (1994) Uptake of lead and cadmium by clinoptilolite. Sci Total Environ 149:139-44
Mathers WG, Watson LC (1962) A waste disposal experiment using mineral exchange on clinoptilolite. Atomic Energy of Canada, Ltd
Mehablia MA, Shallcross DC, Stevens GW (1996) Ternary and quaternary ion exchange equilibria. Solv Extract Ion Exch 14:309-322
Mercer BW (1960) The removal of cesium and strontium from condensate wastes with clinoptilolite. U S Atomic Energy Commission
Mercer BW (1966) Adsorption of Trace Ions From Intermediate Level Radioactive Wastes by Ion Exchange. Pacific Northwest Laboratory, Richland, Washington
Mercer BW, Ames LL Jr, Touhill CJ, Vanslyke WJ, Dean RB (1970) Ammonia removal from secondary effluents by selective ion exchange. J Water Pollution Control Fed 42:R95-R107
Morgan JD, Napper DH, Warr GG (1995) Thermodynamics of ion exchange selectivity at interfaces. J Phys Chem 99:9458-9465
Mukhopadhyay B, Basu S, Holdaway MJ (1993) A discussion of Margules-type formulations for multicomponent solutions with a generalized approach. Geochim Cosmochim Acta 57:277-283
Nelson JL, Mercer BW, Haney WA (1960) Solid fixation of high-level radioactive waste by sorption on clinoptilolite. U S Atomic Energy Commission, Washington, DC
Pabalan RT (1994) Thermodynamics of ion-exchange between clinoptilolite and aqueous solutions of Na^+/K^+ and Na^+/Ca^{2+}. Geochim Cosmochim Acta 58:4573-4590
Pabalan RT, Bertetti FP (1999) Experimental and modeling study of ion exchange between aqueous solutions and the zeolite mineral clinoptilolite. J Soln Chem 28:367-393
Pabalan RT, Turner DR, Bertetti FP, Prikryl JD (1998) Uranium(VI) sorption onto selected mineral surfaces. *In* E Jenne (ed) Adsorption of Metals by Geomedia, p 99-130. Academic Press, San Diego, California
Pansini M (1996) Natural zeolites as cation exchangers for environmental protection. Mineral Dep 31:563-575
Pansini M, Colella C, Caputo D, de'Gennaro M, Langella A (1996) Evaluation of phillipsite as cation exchanger in lead removal from water. Microporous Materials 5:357-364
Perfect DL, Faunt CC, Steinkampf WC, Turner AK (1995) Hydrochemical data base for the Death Valley region, California and Nevada. U S Geol Surv Open-File Report 94-305, U S Geological Survey, Denver, Colorado
Perona JJ (1993) Model for Sr-Cs-Ca-Mg-Na ion-exchange equilibria on chabazite. Am Inst Chem Eng J 39:1716-1720
Peters CA, Yang IC, Higgins JD, Burger PA (1992) A preliminary study of the chemistry of pore water extracted from tuff by one-dimensional compression. *In* YF Kharaka, AS Maest (eds) Water-Rock Interaction, p 741-744. A.A. Balkema, Rotterdam, Netherlands
Pitzer KS (1973) Thermodynamics of electrolytes, 1. Theoretical basis and general equations. J Phys Chem 77:268-277
Pitzer KS (1987) A thermodynamic model for aqueous solutions of liquid-like density. Rev Mineral 17: 97-142
Pitzer KS (1991) Ion interaction approach: Theory and data correlation. *In* KS Pitzer (ed) Activity Coefficients in Electrolyte Solutions, p 75-153. CRC Press, Boca Raton, Florida
Pitzer KS (1995) Thermodynamics. McGraw-Hill, New York
Pode R, Burtica G, Iovi A, Pode V, Mihalache T (1995) Ion exchange of Zn^{2+} and Cu^{2+} on clinoptilolite-type zeolite. Re. Chim (Bucharest) 46:530-533
Prikryl JD, Pabalan RT (1999) Sorption of uranium(VI) and neptunium(V) by surfactant-modified natural zeolites. *In* D Wronkiewicz, J Lee (eds) Scientific Basis for Nuclear Waste Management XXII, MRS Symp. Proc. 556, p 1035-1042. Materials Research Society, Warrendale, Pennsylvania
Puchelt H, Bergfeldt B (1992) Major and trace element concentrations in waters centrifuged from unsaturated soils. *In* YF Kharaka, AS Maest (eds) Water-Rock Interaction, p 751-752. A.A. Balkema, Rotterdam, Netherlands
Reardon EJ, Armstrong DK (1987) Celestite ($SrSO_4$(s)) solubility in water, seawater, and NaCl solutions. Geochim Cosmochim Acta 51:63-72
Robinson SM, Arnold WD, Byers CH (1991) Multicomponent ion exchange equilibria in chabazite zeolite. *In* DW Tedder, FG Pohland (eds) Emerging Technologies in Hazardous Waste Treatment II, Am Chem Soc Symp Series 468:133-152
Robinson SM, Kent TE, Arnold WD (1995) Treatment of contaminated wastewater at Oak Ridge National Laboratory by zeolites and other ion exchangers. *In* DW Ming, FA Mumpton (eds) Natural Zeolites '93: Occurrence, Properties, Use, p 579-586. Int'l Comm Natural Zeolites, Brockport, New York
Scatchard G (1936) Concentrated solutions of strong electrolytes. Chem Rev 19:309-327

Scatchard G (1968) The excess free energy and related properties of solutions containing electrolytes. J Am Chem Soc 90:3124-3217

Semmens MJ, Seyfarth M (1978) The selectivity of clinoptilolite for certain heavy metals. *In* LB Sand, FA Mumpton (eds) Natural Zeolites: Occurrence, Properties, Use, p 517-526. Pergammon Press, New York

Shallcross CD, Hermann CC, McCoy JB (1988) An improved model for the prediction of multicomponent ion exchange equilibria. Chem Eng Sci 43:279-288

Sherry HS (1971) Cation exchange on zeolites. *In* EM Flanigen, LB Sand (eds) Molecular Sieve Zeolites-I, Advances in Chemistry Series 101:350-379

Sherry HS (1979) Ion-exchange properties of the natural zeolite erionite. Clays & Clay Minerals 27:231-237

Shibue Y (1998) Cation-exchange properties of phillipsite (a zeolite mineral): The differences between Si-rich and Si-poor phillipsites. Sep Sci Technol 33:333-355

Shibue Y (1999) Calculations of fluid-ternary solid solution equilibria: an application of the Wilson equation to fluid-(Fe,Mn,Mg)TiO_3 equilibria at 600°C and 1 kbar. Am Mineral 84:1375-1384

Smith SN, Sarada S, Palepu R (1993) Activity coefficients of $NaNO_3$ in (Mg, Ca, Sr, and Ba)$(NO_3)_2$ + H_2O systems at 298 K by EMF methods. Can J Chem 71:384-389

Sposito G (1981) The Thermodynamics of Soil Solutions. Oxford University Press, Oxford

Stumm W, Morgan JJ (1996) Aquatic Chemistry—Chemical Equilibria and Rates in Natural Waters, 3rd edition. John Wiley & Sons, New York

Sullivan EJ, Hunter DB, Bowman RS (1998) Fourier transform Raman spectroscopy of sorbed HDTMA and the mechanism of chromate sorption to surfactant-modified clinoptilolite. Environ Sci Technol 32:1948-1955

Suzuki N, Saitoh K, Hamada S (1978) Ion exchange properties of a synthetic mordenite on alkali and alkaline earth metal ions. Radiochem Radioanal Lett 32:121-126

Tarasevich YI, Kardasheva M., Polyakov VE (1996) Ion-exchange equilibriums on clinoptilolite. Khim Tekhnol Vody 18:347-352

Tomlinson RE (1962) The Hartford program for management of high-level waste. U S Atomic Energy Commission

Torracca E, Galli P, Pansini M, Colella C (1998) Cation exchange reactions of a sedimentary chabazite. Microporous aMesoporous Materials 20:119-127

Torres J.C (1999) Ion exchange between Cd^{2+} solution and clinoptilolite material. 12th Int'l Zeolite Conf, p 2371-2377. Materials Research Society, Warrendale, Pennsylvania

Townsend RP (1984) Thermodynamics of ion exchange in clays. Phil Trans R Soc Lond A 311:301-314

Townsend RP (1986) Ion exchange in zeolites: Some recent developments in theory and practice. Pure Appl. Chem 58:1359-1366

Townsend RP, Fletcher, P, Loizidou M (1984) Studies on the prediction of multicomponent, ion-exchange equilibria in natural and synthetic zeolites. *In* D Olson, A Bisio (eds) Proc 6th Int'l Zeolite Conf, p 110-121. Butterworths, United Kingdom

Townsend RP, Loizidou M (1984) Ion-exchange properties of natural clinoptilolite, ferrierite, and mordenite. I. Sodium-ammonium equilibria. Zeolites 4:191-195

Tsukanova VM, Sharova NG, Pavlovskaya YA (1995) Study of the ion exchange sorption of aqua and hydroxo complexes of lead(II) on clinoptilolite in aqueous solutions at a different pH. Vestn S–Peterburg Univ Ser 4: Fiz, Khim 3:92-95

Turner DR, Pabalan RT, Prikryl JD, Bertetti FP (1999) Radionuclide sorption at Yucca Mountain, Nevada—A demonstration of an alternative approach for performance assessment. *In* D J Wronkiewicz, JH Lee (eds) Scientific Basis for Nuclear Waste Management XXII, Mater Res Soc Symp Proc 556:583-590

Valcke E, Engels B, Cremers A (1997a) The use of zeolites as amendments in radiocaesium- and radiostrontium-contaminated soils: A soil-chemical approach. 1. Cs-K exchange in clinoptilolite and mordenite. Zeolites 18:205-211

Valcke E, Engels B, Cremers A (1997b) The use of zeolites as amendments in radiocaesium- and radiostrontium-contaminated soils: A soil-chemical approach. 2. Sr-Ca exchange in clinoptilolite, mordenite, and zeolite A. Zeolites 18:212-217

Vanselow AP (1932) Equilibria of the base-exchange reactions of bentonites, permutite, soil colloids, and zeolites. Soil Sci 33:95-113

Vaughn DEW (1978) Properties of natural zeolites. *In* LB Sand, FA Mumpton (eds) Natural Zeolites: Occurrence, Properties, Use, p 353-371. Pergamon Press, New York

Vucinic D (1998a) The ion-exchange reactions on clinoptilolite. *In* S Ribnikar, S Anic (eds) Phys Chem '98 4th Int'l Conf. Fundam Appl Aspects Phys Chem, p 621-623. Society of Physical Chemists of Serbia, Belgrade, Yugoslavia

Vucinic DR (1998b) Thermodynamics of the ion-exchange reactions on calcium-clinoptilolite. *In* S Atak, G Onal, M Celik (eds) Innovations Mineral Coal Process, Proc 7th Int'l Mineral Process Symp, p 809-814. Balkema, Rotterdam, Netherlands

White KJ (1988) Ion-exchanges in Clinoptilolite. Univ of Salford, Salford, United Kingdom

Wolf F, Georgi K, Pilchowski K (1978) Ion exchange of monovalent cations on synthetic mordenite. I. Equilibrium and thermodynamics of the potassium ion/sodium ion, hydrogen ion/sodium ion and hydrogen ion/potassium ion exchanges. Z Phys Chem (Leipzig) 259:717-726

Yang IC, Rattray GW, Pei Y (1996) Interpretation of Chemical and Isotopic Data from Boreholes in the Unsaturated Zone at Yucca Mountain, Nevada. U S Geol Surv Water-Resources Investigations Report 96-4058, U S Geological Survey, Denver, Colorado

Yang IC, Turner AK, Sayre TM, Montazer P (1988) Triaxial-Compression Extraction of Pore Water from Unsaturated Tuff, Yucca Mountain, Nevada. U S Geol Surv Water-Resources Investigations Report 88-4189, U S Geological Survey, Denver, Colorado

Zamzow MJ, Eichbaum BR, Sandgren KR, Shanks DE (1990) Removal of heavy metals and other cations from wastewater using zeolites. Sep Sci Technol 25:1555-1569

Zamzow MJ, Schultze LE (1995) Treatment of acid mine drainage using natural zeolites. *In* DW Ming, FA Mumpton (eds) Natural Zeolites '93: Occurrence, Properties, Use, p 405-413. Int'l Comm Natural Zeolites, Brockport, New York

APPENDIX

Equations for calculating $\overline{E}_A$, E_A, $K_{v(A,B)}$, and associated uncertainties from experimental data

Values of $\overline{E}_A$ and E_A can be calculated from the measured zeolite mass (W, grams), solution volume (V, liters), cation-exchange capacity (CEC, equivalents per gram), and initial (i) and final (f) molar concentrations (M, moles/liter solution) of A^{z_A+} and B^{z_B+} using the equation

$$\overline{E}_A = \frac{z_A(M_{A,i} - M_{A,f})V}{W \cdot CEC}; \quad E_A = \frac{z_A M_{A,f}}{TN} \tag{A-1}$$

where TN is the total cation normality (equivalents per liter) of the aqueous phase. If measured concentrations of the competing cation B^{z_B+} are available, values of $\overline{E}_A$ and E_A can be independently calculated from the equation

$$\overline{E}_A = \frac{z_B(M_{B,f} - M_{B,i})V}{W \cdot CEC}; \quad E_A = 1 - \frac{z_B M_{B,f}}{TN}. \tag{A-2}$$

In a similar manner, one set of ln $K_{v(A,B)}$ values can be calculated from the A^{z_A+} analytical data and another set from the B^{z_B+} data using the respective equations

$$K_{v(A,B)} = \frac{(\overline{X}_A)^{z_B}\,[(TN - z_A M_A)/z_B]^{z_A}}{(1 - \overline{X}_A)^{z_A}(M_A)^{z_B}} \cdot \frac{(\gamma_B)^{z_A}}{(\gamma_A)^{z_B}} \tag{A-3}$$

and

$$K_{v(A,B)} = \frac{(1 - \overline{X}_B)^{z_B}\,(M_B)^{z_A}}{(\overline{X}_B)^{z_A}[(TN - z_B M_B)/z_A]^{z_B}} \cdot \frac{(\gamma_B)^{z_A}}{(\gamma_A)^{z_B}}. \tag{A-4}$$

From Pabalan (1994), the uncertainties in $\overline{E}_A$ and E_A calculated from the A^{z_A+} data (Eqn. A-1) are given by

$$\left(\frac{U_{\overline{E}_A}}{\overline{E}_A}\right)^2 = \left(\frac{U_{M_{A,i}}}{M_{A,i} - M_{A,f}}\right)^2 + \left(\frac{U_{M_{A,f}}}{M_{A,i} - M_{A,f}}\right)^2 + \left(\frac{U_V}{V}\right)^2 + \left(\frac{U_W}{W}\right)^2 + \left(\frac{U_{CEC}}{CEC}\right)^2 \tag{A-5}$$

and

$$\left(\frac{U_{E_A}}{E_A}\right)^2 = \left(\frac{U_{M_{A,f}}}{M_{A,f}}\right)^2 + \left(\frac{U_{TN}}{TN}\right)^2 . \tag{A-6}$$

The equation for the uncertainties in $\overline{E}_A$ calculated from the B^{z_B+} data (Eqn. A-2) is the same as Equation (A-5), except the molarities are those for B^{z_B+}. The uncertainty equation for E_A calculated from B^{z_B+} data is slightly different from Equation (A-6), and is given by

$$\left(\frac{U_{E_A}}{E_A}\right)^2 = \left(\frac{z_B M_{B,f}}{TN - z_B M_{B,f}}\right)^2 \cdot \left[\left(\frac{U_{M_{B,f}}}{M_{B,f}}\right)^2 + \left(\frac{U_{TN}}{TN}\right)^2\right] . \tag{A-7}$$

Note that Equation (A-5) has terms with concentration differences in the denominator. Where the difference in initial and final concentration is small, the uncertainty in the calculated parameter can be large. Therefore, errors in $\overline{E}_A$ calculated from the A^{z_A+} analysis become large as $\overline{E}_A$ approaches one. On the other hand, errors in $\overline{E}_A$ calculated from the B^{z_B+} data become large as $\overline{E}_A$ approaches zero. These trends explain why it is important to analyze the solution concentrations of both cations participating in the exchange reaction when constructing ion-exchange isotherms. In this manner, the ion-exchange isotherm is well-constrained throughout the entire composition range

Uncertainties in the selectivity coefficient, $K_{v(A,B)}$, derived using A^{z_A+} analytical data (Eqn. A-3) can be calculated from

$$\left(\frac{U_{K_v}}{K_v}\right)^2 = \left(\frac{z_A U_{\overline{X}_A}}{1 - \overline{X}_A} + \frac{z_B U_{\overline{X}_A}}{\overline{X}_A}\right)^2 + \left(\frac{z_A^2 U_{M_A}}{TN - z_A M_A} + \frac{z_B U_{M_A}}{M_A}\right)^2 + \left(\frac{z_A U_{\gamma_B}}{\gamma_B}\right)^2 + \left(\frac{z_B U_{\gamma_A}}{\gamma_A}\right)^2 . \tag{A-8}$$

The corresponding uncertainty equation for $K_{v(A,B)}$ calculated from the B^{z_B+} data (Eqn. A-4) can be derived by interchanging coefficients A and B in Equation (A-8). Also,

$$U_{\ln K_{v(A,B)}} = \frac{U_{K_{v(A,B)}}}{K_{v(A,B)}} \tag{A-9}$$

and

$$\frac{U_{\overline{X}_A}}{\overline{X}_A} = \frac{(z_A - z_B) U_{\overline{E}_A}}{z_A + (z_B - z_A)\overline{E}_A} + \frac{U_{\overline{E}_A}}{\overline{E}_A} . \tag{A-10}$$

15

Applications of Natural Zeolites in Water and Wastewater Treatment

Dénes Kalló

Chemical Research Center
Institute for Chemistry
Hungarian Academy of Sciences
Budapest, Hungary

INTRODUCTION

The world is faced with increasing demands for high-quality drinking water and for removal of contaminants from municipal, agricultural, and industrial wastewaters. Treatment is required to obtain drinking water from most natural resources as well as from wastewaters with varying amounts of impurities. These impurities may occur in a variety of forms including large particles such as microorganisms or suspended solids or as dissolved or colloidal inorganic and organic substances. This chapter provides an overview on the use of natural zeolites in removal of impurities from water or wastewater (Murphy et al. 1978, Tarasevich 1994, Kalló 1995).

Most technologies using natural zeolites for water purification are based on the unique cation-exchange behavior of zeolites through which dissolved cations can be removed from water by exchanging with cations on a zeolites exchange sites (see Pabalan and Bertetti, this volume). The most common cation in waters affecting human and animal health is NH_4^+. It can be removed by exchanging with biologically acceptable cations such as Na^+, K^+, Mg^{2+}, Ca^{2+} or H^+ residing on the exchange sites of the zeolite. Fortunately, many natural zeolites (e.g. clinoptilolite, mordenite, phillipsite, chabazite) are selective for NH_4^+ (*vide infra*), meaning that they will exchange NH_4^+ even in the presence of larger amounts of competing cations. Clinoptilolite and mordenite are also selective for transition metals (e.g. Cu^{2+}, Ag^+, Zn^{2+}, Cd^{2+}, Hg^{2+}, Pb^{2+}, Cr^{3+}, Mo^{2+}, Mn^{2+}, Co^{2+}, Ni^{2+}), which are often present in industrial wastes and can be very toxic even in concentrations as low as several mg/L. As emphasized in discussions of radioactive waste treatments, both clinoptilolite and mordenite have very high selectivities for Cs^+ and Sr^{2+} and can therefore be used to remove minute amounts of radioactive ^{137}Cs and ^{90}Sr from nuclear process wastewaters.

Organics are also often present in wastewaters in either dissolved or colloidal form as hydrocarbons, oxygen-containing compounds, halogenated derivatives, amines, humic acids, proteins, and lipids. Unfortunately, most organic molecules or particles are too large to penetrate into channels and cages to access the extraframework exchange sites of natural zeolites; thus their adsorption must take place on the surfaces of the zeolite-containing materials (e.g. external crystal surfaces of zeolites), which can have a surface area as large as several 10 m^2/g. The volumetrically most important natural zeolites on the surface of the Earth are of sedimentary origin, and zeolites in such deposits occur in clusters of crystals often having intercrystalline pore sizes of 10 to 1000 nm in diameter (i.e. rock pores). Colloids, enzymes or microorganisms as large as bacteria can be trapped within these intraparticle pores. As a results of the large surface area, which is accessible for adhering bacteria, natural zeolites can become effective biofilters when compared with particles having smaller total surface areas such as quartz sand beds (Tarasevich 1994, Baykal and Guven 1997).

1529-6466/00/0045-0015$05.00

The quality of water prior to treatment can vary considerably. Simple purification technologies, such as the use of chemical additives, may not meet demands for quality water. This is particularly so in cases when too much chemical is applied that may result in hazardous water (e.g. excess organic coagulants may increase the biological oxygen demand (BOD) in the effluent or surplus Al added for phosphate removal may become a harmful contaminant to humans). Fortunately, the use of natural zeolites has the potential to eliminate this problem by removing contaminants from the water via ion-exchange or adsorption processes. Furthermore, abundant deposits of natural zeolites in near-surface sedimentary deposits make their possible use in the treatment of waters and wastewaters very attractive.

PRODUCTION OF DRINKING WATER

Removal of NH_4^+ and other ions

Natural waters often contain a variety of impurities, including NH_4^+, heavy metals, As, H_2S, and humic acids. NH_4^+ is a common contaminant in natural waters, but its concentration in solution can be reduced below the recommended level for drinking water of <1 mg NH_4^+/L by ion-exchange processes. Most natural zeolites in their Na-exchanged form, i.e. Na-clinoptilolite, Na-mordenite, Na-phillipsite, Na-chabazite, and Na-erionite, are selective for NH_4^+ as shown experimentally by Ames (1960, 1961) and theoretically by Neveu et al. (1985).

Hlavay (1986) used a natural rhyolitic tuff containing 58 wt % clinoptilolite (Tokaj Hills, Hungary) to remove NH_4^+ from natural water. A 0.6- to 1.0-mm size fraction was packed into three columns, and two of the columns were used in series for water treatment while the third column was being regenerated. The exchange sites became exhausted after 470 bed volumes (BV) of the water, containing 1.6 mg NH_4^+/L, had passed through the columns. The dynamic ion-exchange capacity of the bed was 0.33 mg NH_4^+/g when the breakthrough concentration was 0.2 mg NH_4^+/L. The average NH_4^+ concentration in the effluent was 0.14 mg/L prior to reaching the breakthrough concentration. During the course of NH_4^+ removal, the chemical oxygen demand (COD), reflecting the total amount of oxidizable impurities, the humic acid concentration, and the Ca^{2+} and Mg^{2+} concentrations, did not significantly change.

Several other studies have been published on the use of clinoptilolite-rich rock for removing NH_4^+. Xu (1990) prepared distilled water (electrical conductivity <1 μS/m) by first removing NH_4^+ using a Chinese clinoptilolite (0.4-0.8 mm) and then passing the water through another commercial cation-exchange column. Linevich et al. (1990) produced drinking water from ground water by removing NH_4^+ (≤5.6 mg/L) via ion exchange with clinoptilolite-rich material (Georgian occurrence) and removing H_2S (≤1.5 mg/L) by oxidation with chlorine and sodium hypochlorite, produced electrolytically from a NaCl regeneration solution.

On some occasions, F^- exceeds 1 mg/L in drinking water supplies and must be removed before consumption. In an early use of a modified zeolite, Kravchenko et al. (1990) reacted a crushed clinoptilolite-rich rock with aluminum sulfate as a specific adsorbent for F^-. The clinoptilolite was modified by passing a 0.5% aluminum sulfate solution downwards (linear velocity of 10 m/h for 1.0-1.5 h) through a 2-m thick filter layer consisting of 1- to 3-mm clinoptilolite-rich grains from Trans-Carpathia, Ukraine. The filtration bed was then washed with raw artesian water until Al^{3+} in the effluent decreased to 0.5 mg/L. The fluoride content of artesian water containing 2-9 mg F^-/L was reduced to below 1 mg F^-/L after passing through the modified clinoptilolite-containing column. The F^- increased to 1.0-1.2 mg/L after passing 80 BV of the artesian water through the column, at which point filtration was stopped and regeneration was performed by treating the column again with the aluminum sulfate solution as described above. According to the authors, F^- is removed either by ion exchange, formation of aluminofluoride complexes bound to cations on exchange sites of clinoptilolite, or by molecular adsorption of fluoride salts; ion exchange seems an unlikely mechanism for this process.

Ion-exchange and filtration beds

The filtration efficiency of a sand bed can be increased by admixing porous zeolitic rock having a substantially larger surface area than the sand. For example, the surface area of a zeolite-rich rock is ~10 m^2/g, whereas a sand has a surface area of ~0.01 m^2/g. As might be expected, the increased surface area provides additional area for the adsorption of suspended solids, microorganisms, and other materials in solution. Ammonium ions removed from water may serve as a nutrient source for microorganisms (e.g. nitrifying bacteria) that adhere to crystal surfaces within macro pores of zeolitic rock. Nitrifying bacteria convert the NH_4^+ to NO_3^-, which can be efficiently removed from water (e.g. see Galindo et al. 2000). Thus, water purification can be achieved by a combination of ion-exchange, filtration, and/or microbiological processes using beds of zeolite-rich materials in addition to other unreactive materials.

A variety of microorganisms, such as *Escherichia coli*, poliovirus, coxackie virus, and bacteriophages, have been effectively removed from drinking water using an $Al_2(SO_4)_3$ coagulant and clinoptilolite-rich material from the Trans-Carpathia region of Ukraine (Grigorieva et al. 1988). The numbers of microorganisms decreased by 50% when clinoptilolite-rich material or coagulant alone were used to treat the water, whereas the two additives together removed 90% of all microorganisms during the same time period. Apparently, the efficient removal of microorganisms was caused by the strong bonding of microorganism-clinoptilolite-coagulant complexes (vide infra).

Natural phillipsite-rich tuff from Tenerife, Canary Islands, Spain was used in a percolation reactor at a constant flow rate of aqueous solution for removing indicator bacteria such as total *coliforms*, fecal *coliforms,* fecal *strepotcocci*, and dissolved organic matter. The amounts of oxidizable impurities were reduced as indicated by composition, BOD, COD, and total bacteria in the effluent (Garcia et al. 1992a). Fixation of bacteria and organic matter occurred not only on zeolite crystal surfaces, but also on the volcanic glass present in the zeolitic rock. The presence of acidic surface groups (i.e. functional groups) on zeolitic and glass surfaces account for the interaction (enhancement or inhibition) with polymeric materials secreted by the cells. Filtration beds consisting of such porous materials offer suitable colonization surfaces for bacteria and a potentially constant supply of nutrients to support the microbial activity from adsorbed and ion-exchanged inorganic and organic ions.

An apparatus for removal of NH_4^+, Fe^{2+}, Mn^{2+}, and As (in the form of AsO_4^{3-}) from drinking water has been disclosed in a Hungarian Patent and built in Hungary (Hosszú et al. 1983). The apparatus consists of two columns; the first containing two layers of clinoptilolite-rich materials—one layer of Na-clinoptilolite for NH_4^+ ion exchange and another layer of Mn-clinoptilolite for Fe and Mn removal (see below); and a second column filled with $Fe(OH)_3$-coated TiO_2 for the adsorption of arsenic. A mechanical filter is placed either between the two columns or after the second column to remove suspended solids that may not be retained by the different materials in the filtration and ion-exchange beds.

Hódi et al. (1994) evaluated a complex method of ion exchange and adsorption (similar to the method described above) for the removal of NH_4^+, arsenic, and humic impurities from natural water for the production of potable water. Efficient removal of NH_4^+ was achieved by ion exchange with Na-exchanged clinoptilolite-rich material from Tokaj Hills, Hungary. Humic substances were thereafter removed by adsorption on activated charcoal and, finally, arsenic impurities were removed by $Fe(OH)_3$ previously precipitated on an Al_2O_3 substrate. Influent concentrations for NH_4^+, Fe^{2+}, Mn^{2+}, AsO_4^{3-}, and COD of 0.86, 0.16, <0.02, 0.14, and 11.1 mg/L, respectively, were reduced to 0.1, <0.02, <0.02, 0.022, and 0.6 mg/L, respectively, in the treated effluent solution.

The mechanical strength of most zeolite-containing tuffs should be satisfactory for use as beds in columns for treating water. Unacceptably friable materials may be treated or altered to enhance their mechanical strength; for example, Aiello et al. (1984) altered the soft Neapolitan yellow tuff from southern Italy that contains phillipsite by subjecting the tuff to hydrothermal treatment with NaOH or KOH at 120-160°C. Pellets were prepared by compressing the treated material at 50 kg/cm^2 for >2 days. The ion-exchange capacity of the zeolite was essentially unaffected and the mechanical strength of the material was acceptable for use in ion exchange and filtration beds for water purification.

Pilot plants and full-scale plants

Several pilot plants and full-scale treatment plants have been constructed and are in operation throughout the world for treatment of natural waters with natural zeolites. Most of these plants use a variety of processes to treat the water, such as ion exchange and filtration as described above. For example, the cation-exchange and filtering properties of clinoptilolite-rich tuff from Georgia were utilized in a pilot plant to remove suspended particles and trace elements from conduit drinking water in Tbilisi, Georgia (Senyavin et al. 1986a). Cs^+, Sr^{2+}, and Cu^{2+} were removed from natural water of various sources and from synthetically produced model solutions containing around 1 mg/L of these cations. Solutions were passed through clinoptilolite-rich columns, 70 cm in length, at a linear velocity of 6 or 3 m/h. Breakthrough concentrations were registered after 25 h of solution passing through columns containing clinoptilolite rock particles of 0.05-0.2 cm and after 14 h in columns containing clinoptilolite rock particles of 0.3-0.5 cm.

A water reuse demonstration plant in Denver, Colorado, produced high-quality drinking water from secondary municipal wastewater effluents (Ray et al. 1985). The complex treatment involved lime clarification, 1- or 2-stage recarbonation (saturation with CO_2), pressure filtration, ion-exchange for NH_4^+ removal through a clinoptilolite column, first-stage adsorption on activated carbon, ozonation, second-stage adsorption on activated carbon, reverse osmosis, air stripping, and ClO_2 disinfection. Remarkably, water produced in this complex treatment system after two years of operation was lower in total organic carbon (TOC) content and contained fewer trace organics than potable water produced directly from mountain runoff (Rogers et al. 1987). A similar complex purification process for producing drinking water in Ukraine was reported by Kravchenko et al. (1994). The technology consisted of ozonation and multilayer filtration through clinoptilolite-rich material from the Trans-Carpathia area of Ukraine and activated carbon columns supplemented with Fe and Mn removal. Unfortunately, data on the performance of this system were not published.

The growth, composition, and filtering efficiency of algae layers covering the grains of a filter bed were determined in field experiments for the biological treatment of water from the Logan River in Utah, USA (McNair et al. 1987). In this process, termed Slow Sand Filtration (SSF), clinoptilolite (origin unpublished, cation-exchange capacity, CEC, = 1.7-1.9 meq/g) was added to a sand bed to increase surface area and to provide a nitrogen reservoir (via NH_4^+ exchange) for the algae covering the filter grains. Removal of organic and inorganic suspended solids was superior in clinoptilolite-amended SSF, even at filtration rates 2-4 times faster than conventional SSF rates. The growth of the algae coincided with an increase in ability of the amended filter to remove *Giardia lamblia* cyst-size particles. The zeolite-amended SSF system produced drinking water for longer periods of time at higher filtration rates than the conventional SSF that did not contain clinoptilolite. In addition to clinoptilolite, other zeolite-rich materials, including phillipsite and mordenite, have been suggested for use in these types of systems (Hulbert and Currier 1986).

Senyavin et al. (1986b) used clinoptilolite-rich tuff from different deposits in the former Soviet Union (Dzeg, Tedzam, Tshuguev, and Yagodnin deposits) instead of quartz sand as a filtration bed for drinking water purification. A decrease in the number of coliform bacteria, total bacteria, and phyto- and zooplankton was attained after three years of using clinoptilolite-rich materials in the finished water compared with other methods. The mechanical, physical, and chemical properties of clinoptilolite remained essentially unchanged during the three-year test period without regeneration.

Regeneration of NH_4^+-zeolite

Most of the technologies described above use the ion-exchange properties of the zeolite to remove NH_4^+. Thus it is fundamentally important to be able to remove NH_4^+ from exchange sites so the zeolite may be reused. An NH_4-exchanged zeolite can be exchanged or regenerated using 1 N NaCl or KCl solutions, repopulating the exchange sites with either Na^+ or K^+. The regeneration efficiency is increased when the pH of the regenerating solution is raised above 11, which can be attained by addition of lime (Oláh et al. 1988). High regeneration efficiency can be obtained under these conditions, even at high NH_4^+ concentrations in the regenerating solution.

Kalló (1990) achieved a regeneration efficiency of 80% by recycling 40 BV of 20 g KCl/L regenerating solution at pH = 7 through clinoptilolite-rich material. The exhausted regenerating solution contained 500 mg NH_4^+/L. The efficiency was increased to 88% and the NH_4^+ content of the regenerating solution was as high as 6000 mg/L at pH = 11-12. Nearly all dissolved NH_4^+ in solution was degassed from the regeneration solution at pH > 11 as NH_3 by air stripping. Hlavay (1986) achieved similar regeneration of NH_4^+-exchanged clinoptilolite-rich tuff by passing a solution of 20 g NaCl/L adjusted to pH > 11.5 by the addition of lime. Ammonia was stripped from the regenerating solution with air and the column was backwashed with 20 BV of de-ammoniated water in order to remove $CaCO_3$ and $Fe(OH)_3$ precipitates from the zeolite bed. No reduction in cation-exchange capacity was observed during 10 exchange-regeneration cycles.

Semmens et al. (1977) evaluated a biological method for regeneration of NH_4-exchanged clinoptilolite. Nitrifying bacteria converted NH_4^+ to NO_3^- on the surface of porous zeolitic tuff in oxygen-enriched air. The nitrification process results in the formation of acidic or H-exchanged clinoptilolite. Because nitrifying bacteria activity is reduced in acidic media, a neutral pH was maintained by a continuous addition of soda. Complete regeneration was achieved within two hours if the concentration of dissolved oxygen was >6 mg/L. The CEC of the clinoptilolite-rich tuff did not change after 40 exhaustion-regeneration cycles.

Ammonia can also be evolved from exchange sites by heating clinoptilolite between 350-450°C in a stream of air (Szymansky et al. 1960). After NH_3 devolatilization, H-exchanged sites are formed, and the clinoptilolite-rich material can again be used for the ion exchange of NH_4^+, which is highly selective over H_3O^+. Because of the high-energy costs and the technical difficulties associated with the heating process, this method is not generally used in water treatment.

Albertin et al. (1994) used a recycled 5% NaCl brine to remove NH_4^+ from spent zeolite columns. This method overcame difficulties encountered with the discharge of saline solutions used for regeneration of exhausted clinoptilolite-rich and phillipsite-rich columns producing drinking water in a pilot plant near Venice, Italy. The brine is stored in a tank, and NH_4^+ is removed from solution by break-point chlorination where chlorine is introduced until excess chlorine appears in the regenerating solution. At this point most of the NH_4^+ has been oxidized by the break-point chlorination process and is evolved from solution as N_2 gas.

MUNICIPAL AND AGRICULTURAL WASTEWATER TREATMENT

Removal of NH_4^+

Ammonium ion concentrations are at least one order of magnitude higher in municipal wastewaters than in natural waters and inorganic and organic impurities also occur in greater amounts. Furthermore, suspended solids (SS) are normally present in municipal wastewaters. Conventional biological wastewater treatment, which reduces the concentration of oxidizable impurities and produces secondary effluent, can be improved by aeration that results in nitrification of NH_4^+ by nitrifying bacteria. The nitrate is subsequently removed by denitrification under anaerobic conditions. Both steps are time consuming, require large treatment installations, and consume energy. Removal of NH_4^+ can be accomplished by ion exchange as described above. Permutite-type inorganic cation-exchange materials (i.e. synthetic aluminosilicates that are crystallographically amorphous) and the various organic ion-exchange resins have poor selectivities for NH_4^+; therefore, they have unacceptably low NH_4^+ exchange capacities and low regeneration efficiencies. Consequently, high operation costs and problems with the disposal of large volumes of brine are encountered. Clinoptilolite's high selectivity for NH_4^+ makes it promising for use in the removal of NH_4^+ from municipal wastewater. Several plants have been designed and built to treat municipal wastes using clinoptilolite-rich materials. A 27,000 m^3/d capacity plant at Lake Tahoe, California, used several hundred tons of clinoptilolite-rich tuff from the Hector deposit in California (Butterfield and Borgerding 1981), and even larger plants with 45,000 and 245,000 m^3/d capacities have been built in Virginia, USA (Gunn 1979). Several studies that illustrate the effective use of zeolite-rich materials for the removal of NH_4^+ and other impurities from municipal wastewaters are reported below.

Typically only a fraction of a zeolite's CEC is utilized before the breakthrough concentration of NH_4^+ is reached in a dynamic exchange system (i.e. flow-through, column exchange system). For example, Czárán et al. (1988) reported that only 0.2-0.3 mmol NH_4^+/g exchanged onto clinoptilolite-rich material under dynamic exchange conditions, instead of the theoretically maximum NH_4^+ exchange capacity of 1.1-1.4 mmol NH_4^+/g. The exchange capacity under dynamic exchange conditions depends on several factors, such as the particle size of clinoptilolite grains, NH_4^+ concentration and pH in the wastewater influent, flow rate of solution through columns, and contact time with the zeolite-rich material (e.g. length of exchange columns or filter beds). Smaller particle-size materials usually result in higher effective exchange capacities for NH_4^+; however, flow resistance through the bed often increases due to reduced permeability. Kalló (1990) suggested that the optimum particle size for clinoptilolite-rich material for NH_4^+ removal in these types of exchange systems is 0.5-2 mm. As might be expected, the effective NH_4^+ exchange capacity increases for zeolite-rich material as the NH_4^+ influent concentration increases (Kalló 1990). Horvathova (1986) suggested that an influent solution of pH 7 is optimum for NH_4^+ exchange under dynamic conditions. The effective NH_4^+ exchange capacity decreases as the flow rate increases through the column, and

Kalló (1990) recommended a flow rate of ~10 BV/h for optimum exchange of NH_4^+ in clinoptilolite-rich materials.

Although synthetic zeolites have higher total cation-exchange capacities, natural zeolites exhibit greater selectivity for NH_4^+. Metropoulos et al. (1993) compared NH_4^+ removal using Na-exchanged clinoptilolite-rich tuff from North Greece, mordenite-rich material from Nevada, USA, and natural ferrierite from Nevada, with Na-, K-, and Ca-forms of synthetic zeolite A (Carlo Erba, Italy). The Na-exchanged natural zeolites showed higher selectivities for NH_4^+ than synthetic zeolite A. Although the clinoptilolite-rich tuff had a lower CEC, it exhibited the best performance in NH_4^+ removal because of its high selectivity for NH_4^+ and its high ion-exchange rate.

Metabolic ammonia is a major pollutant in aquaculture systems, and natural zeolites along with other ion-exchange materials have been used to remove NH_4^+ from such polluted waters. For example, toxic NH_4^+ was removed from fish water tanks at the Seattle Aquarium in Seattle, Washington, by recycling the contaminated water through filter beds of clinoptilolite-rich material (Mumaw et al. 1981). The NH_4^+ concentrations in aquarium tank wastewater, which contained between 0.3-0.5 mg/L metabolic NH_4^+, was reduced to less than 0.1 mg NH_4^+/L by passing the water through a filter bed containing 850 kg of clinoptilolite-rich material at a flow rate of 190 L/min. The clinoptilolite-rich material retained 1 g NH_4^+/kg after about 10 days operation, at which point the clinoptilolite-rich filter bed was regenerated by back washing with saltwater.

Ciambelli et al. (1985a) determined the physicochemical characteristics and the NH_4^+ ion-exchange properties of Neapolitan yellow tuff from Italy, which contains mostly phillipsite and some chabazite, and compared it with clinoptilolite-rich tuff from near Hector, California. The Na^+/NH_4^+ cation-exchange isotherms indicated that phillipsite is more selective for NH_4^+ than clinoptilolite and the rate of cation exchange is higher for phillipsite than for clinoptilolite (Aiello and Nastro 1984). When Na-forms of the two zeolites were ion exchanged with solutions containing different competing cations, e.g. to simulate aquaculture systems, NH_4^+ uptake was found to be comparable for the two zeolites. However, phillipsite exhibited higher selectivity for NH_4^+ at equilibrium than clinoptilolite. Nevertheless, the practical application of phillipsite-rich tuff is rather limited because its mechanical strength is lower than that of most clinoptilolite-containing rock (see Liberti et al. 1986a).

Throughout the years, many methods have been studied for treating wastewaters associated with animal production, e.g. wastewater from cattle, swine, and poultry sewage pits or lagoons. Natural zeolites have been effectively used to treat these wastewaters. Passaglia and Azzolini (1994) reported the use of zeolite-rich tuff from Italy, containing different amounts of chabazite and phillipsite, for treating wastewater from swine sewage. Sewage wastewater was percolated through fixed beds of zeolite-rich materials in their original form (i.e. native exchange cations) and in their Na-exchanged forms after regeneration with NaCl. Effective NH_4^+ exchange capacities of zeolite-rich materials in their original form ranged from 0.4-0.9 meq NH_4^+/g. Sodium and Ca^{2+} were the primary cations removed by NH_4^+ from zeolite exchange sites. Potassium ions were also removed from the sewage wastewater, although more NH_4^+ was exchanged by the zeolites, likely due to its greater concentration in the wastewater suppressing the exchange of K^+. Effective NH_4^+ exchange capacities of Na-forms of the zeolites after regeneration with the Na solution increased to 1.2-1.7 meq NH_4^+/g.

Wastewater (150 m^3/d) from a pig farm with 10,000 pigs was treated in a cascade multi-step system using mechanical, chemical, and biological processes (Zubály et al. 1991). Large suspended particles were first removed by grating the sewage. Particles removed by the grate were composted and used as organic fertilizer. Suspended particles, colloids and dissolved organic and inorganic species not removed by the grate were then passed through a channel (six 20 m sections separated by barrages). Approximately 20 m^3 of clinoptilolite-rich tuff (3-10 mm) from the Tokaj Hills deposit in Hungary were placed in each section. The zeolite filters removed 100% of oils and fats, 98% of suspended solids, and 95% of dissolved organic and inorganic impurities from the wastewater. Zeolite filters successively removed these impurities for two years at which point they became exhausted. The exhausted zeolite material was used as a fertilizer. Effluent from the treated wastewater was used to water trees in an orchard operation. These studies suggest that zeolites may be used to treat these types of wastewaters; however, additional studies are needed to determine whether zeolites can efficiently and economically remove pollutants such as NH_4^+ from wastewaters of animal production in large feedlots.

Zeolite bed regeneration

As pointed out above, zeolite beds must be regenerated so that the zeolite can be economically reused once the beds have been exhausted after the treatment of municipal or agricultural wastewaters. For most applications, cations, e.g. Na^+, that easily exchange with NH_4^+ are used in the regeneration process. Several processes that have been successfully used to remove NH_4^+ from spent zeolite beds are described below.

A clinoptilolite-rich bed was effectively regenerated with solutions of 3% NaCl or $CaCl_2$ after the bed had been used to remove NH_4^+ from biologically treated wastewater (Linne and Semmens 1985). Some BV of treated water in which NH_4^+ had been removed was fed back through the outlet of the clinoptilolite-rich bed in order to remove precipitated impurities, which concentrated upstream in the bed, i.e. from the inlet to the outlet of the bed. This backwashing did not, however, affect the efficiency of NH_4^+ removal and aided in preventing plugging of the bed by the salt impurities.

A clinoptilolite-rich tuff of high NH_4^+ exchange capacity (2.18 meq NH_4^+/g) from near Death Valley Junction, California, was used in packed columns for removal of NH_4^+ from pond waters (Williford and Reynolds 1992). Although columns were regularly regenerated with salt solution the cation-exchange capacity decreased with use. The decrease was attributed to the presence of organic material, especially algae in the natural pond water, which partly coated the external surfaces of zeolite grains, thereby decreasing the accessibility of NH_4^+ to zeolite exchange sites. Frequent back flushing of the zeolite column was required in order to avoid fouling of the system by algae and other organics.

Phillipsite-rich tuff from Italy was used to remove >95% NH_4^+ from municipal wastewater (Ciambelli et al. 1985b). The phillipsite-rich tuff was exhausted after 350 BV of municipal wastewater had passed through the columns. Exhausted beds were regenerated with NaCl solution that resulted in complete Na^+ exchange and the zeolite's effective NH_4^+ exchange capacity increased after regeneration. The phillipsite exhibited no loss of NH_4^+ exchange capacity after 35 cycles of operation.

Homonnay et al. (1993) disclosed in a patent a process for the regeneration of NH_4-exchanged clinoptilolite-rich material with Na-hypochlorite solution containing 0.5-1 mg/L iodine or bromine as a catalyst, which accelerated the oxidation of NH_4^+. Nontoxic nitrogen compounds form during the course of this oxidative regeneration; however, the ion-exchange capacity of clinoptilolite is unaffected and the regenerating solution can be recycled. Unfortunately, the authors did not suggest a practical application for the regeneration process.

Biological regeneration of NH_4-clinoptilolite has also been suggested as a means of regenerating zeolite beds used to treat municipal wastewaters, similar to the regeneration described earlier for the manufacture of drinking water. A clinoptilolite-rich tuff filtration system was used to remove NH_4^+ from fish rearing ponds at the Eagle Creek National Fish Hatchery in Oregon, USA (Horsch and Holway 1984). Daily agitation of the filter bed with air prevented the conduits from plugging and the clinoptilolite-rich materials from fouling with organic and particulate matter. Nitrification (i.e. oxidation of NH_4^+ to NO_2^- to NO_3^- by nitrifying bacteria) began in the filtration system 15 days after start up of the treatment process, i.e. after the start of NH_4^+ removal from the rearing pond water. The NH_4^+ removal efficiency of clinoptilolite used as a biological filter was 89%. At no time did NH_4^+ or NO_2^- reach toxic levels. Oxidative bacterial regeneration of the clinoptilolite filter was accomplished within one week after it was taken off line from removing NH_4^+ from the pond water, making this process an alternative to brine regeneration of clinoptilolite.

Instead of regenerating NH_4^+-spent zeolite beds, the spent zeolite may be used as a fertilizer in agricultural applications (Mori and Tsuneyoshi 1986). An NH_4^+-spent mordenite (presumably from Itado, Japan) with an N content of 1 mg/g was applied to rice paddy fields at an application rate of 5 kg/m^2. The cation-exchange capacity of the soil was increased threefold by the amendment and the authors suggested that the zeolite may be used as a slow-release fertilizer (see Ming and Allen, this volume).

Preston and Alleman (1994) also reported the use of a clinoptilolite-rich material for NH_4^+ removal via ion exchange and as a biofilter medium for the oxidation of NH_4^+ by nitrifying bacteria. Nitrifying bacteria produce protons during the oxidation of NH_4^+, hence this process can increase solution acidity over time. The slightly alkaline NH_4^+ wastewater will usually neutralize the solution; however, neutralization of the acidity by the addition of lime may be required to maintain a slightly alkaline solution. The metabolism of nitrifying bacteria decreases with increasing acidity, eventually ending in bacteria death. Preston and Alleman (1994) found that this system remained viable through numerous cycles, and a single cycle may reach a maximum operation period of 60 h. They suggested that the negatively charged nitrifying bacteria are electrostatically bound or immobilized on zeolite crystal surfaces. The immobilization of nitrifying bacteria on clinoptilolite may enhance the nitrification process in wastewater treatment.

Simultaneous nitrification and denitrification have been suggested for removal of nitrogen from municipal wastewater (Halling-Soerensen and Hjuler 1992). A clinoptilolite-rich material was used as a support medium for microorganisms. The porous structure of the rock provided alternating aerobic and anoxic conditions, where an aerobic pattern developed as zeolitic grains contacted water high in dissolved oxygen that enhanced nitrification. Oxygen is consumed and H^+ is produced during the nitrification process, until anoxic conditions develop and denitrification begins. Simplified reactions for nitrification and denitrification can be expressed as follows:

$$Z^- NH_4^+ + 2\,O_2 \rightarrow Z^-H^+ + HNO_3 + H_2O \qquad \text{(nitrification)}$$

$$2\,HNO_3 + 5\,H_2 \rightarrow N_2 + 6\,H_2O \qquad \text{(denitrification)}$$

where Z^- is the zeolite substrate. Reducing agents other than H_2 (e.g. ethanol) may be used in the denitrification process. The biological reactors were operated for 6 months during which the influent wastewater contained between 30 and 5000 mg NH_4^+/L. These reactors removed a maximum of 13.5 kg N/m^{3}d with an efficiency of 95.8%. Nitrogen removal in this process is faster than with technologies using suspended cultures, i.e. without any support for the microorganisms.

Clinoptilolite has been used in combination with aerated mechanical sand filters for removal of peak concentrations of NH_4^+ from domestic wastewater (Baykal and Guven 1997). High-efficiency NH_4^+ removal was achieved when sufficient time was allowed for the development of nitrifying bacteria colonies. The effective NH_4^+ exchange capacity of clinoptilolite decreased 10% after 10 cycles of regular operation and regeneration.

The simple removal of N and P nutrients by sorption from secondary water of biologically treated municipal wastewater has been attempted using natural zeolites. Up to 62% NH_4^+ and 15% P were removed by adding 2-50 g/L powdered zeolitic rock (presumably clinoptilolite-rich rock from Australia) to the secondary wastewater (Komarowski et al. 1994).

Zeolites as coagulating agents

Natural zeolites may be used along with other compounds as coagulating agents to remove suspended solids from wastewater. This approach is especially attractive for the removal of microorganisms, e.g. in removal of bacteria from drinking water using clinoptilolite-rich tuff and $Al_2(SO_4)_3$ as described above (Grigorieva et al. 1988). Adherence of microorganisms to zeolite particles may also increase biological activity, such as in the ZeoFlocc process described in detail below.

Kvopkova et al. (1988) removed petroleum contaminants from the wastewater of a textile plant by a flocculation process. The textile plant wastewater effluent contained 12.6 mg/L petroleum products with a COD of 537 mg/L and was treated by three different coagulating agents: (a) 1.4 mmol/L $FeSO_4$ and 3.2 mmol/L CaO; (b) same as (a) plus 0.2 mg/L polyacrylamide; and (c) same as (a) plus 100 mg/L clinoptilolite-rich tuff from Eastern Slovakia and 2.5 mg/L diethylenetriamine-stearic acid condensate (C_{10}-C_{30} carboxylated or quaternary amines). Residual levels of the petroleum products and COD values were lowest in the effluent receiving the clinoptilolite treatment after sedimentation (see Table 1).

Table 1. Removal of petroleum products and chemical oxygen demand (COD) from wastewater of a textile plant treated with various flocculation agents (Kvopkova et al. 1988).

Flocculant dosage	*Petroleum products*	*COD*
	mg/L	*mg/L*
None added	12.6	537
(a) 1.4 mmol $FeSO_4$/L and 3.2 mmol CaO/L	3.2	192
(b) Same as in (a) and 0.2 mg polyacrylamide/L	2.7	172
(c) Same as in (a) and 100 mg clinoptilolite/L with 2.5 mg diethylenetriamine-stearic acid condensate/L	2.1	153

Complex treatment technologies

Municipal wastewaters often contain considerable amounts of phosphorous impurities, mainly as PO_4^{3-} in the range of 5-30 mg P/L, which can usually be removed in the form of insoluble compounds such as phosphates of multivalent metals. The solubility of di- and trivalent metal orthophosphates is relatively low, e.g. a few mg P/L depending on pH, temperature, and ion concentrations. The widely accepted method for removing phosphate is to precipitate phosphate by adding lime (e.g. $CaCO_3$) to the wastewater. Although this method is simple, it often results in a treated wastewater stream with some PO_4 because the solubility of $Ca_3(PO_4)_2$, formed from the lime addition, is not negligible. Calcium-exchanged zeolites may offer an effective alternative for the removal of phosphate from municipal wastewater. For example, Ca-clinoptilolite-rich tuff from Japan added to wastewater resulted in the precipitation of phosphates. Apparently, Ca^{2+} on the zeolite exchange sites reacted with solution phosphate to precipitate complex calcium phosphate salts associated with the zeolite (Kurita Water 1985).

Several more complex treatment processes have been designed that use natural zeolites for the removal of NH_4^+, phosphates, and other contaminants from wastewaters. Two of these processes (RIM-NUT and ZeoFlocc processes) are briefly described below.

RIM-NUT process. A promising technology was developed by Liberti et al. (1984) for removing NH_4^+ and phosphates from domestic secondary effluent (i.e. biologically oxidized, settled, and chlorinated effluent). Compounds in the secondary effluent serve as nutrients for microorganisms, which are responsible for creating eutrophic conditions in the wastewater (i.e. depletion of dissolved O_2 by microorganisms). A process termed RIM-NUT (from the Italian phrase for "removal of nutrients") was developed as a method for reducing the eutrophic potential of municipal and industrial wastewater by combining ion exchange from natural zeolites with precipitation processes to selectively remove NH_4^+ and/or phosphate ions from the wastewater. A pilot plant of 10 m^3/h capacity was built in the West Bari Sanitation Station in Italy and a flow diagram of the process is shown in Figure 1 (Liberti et al. 1986a).

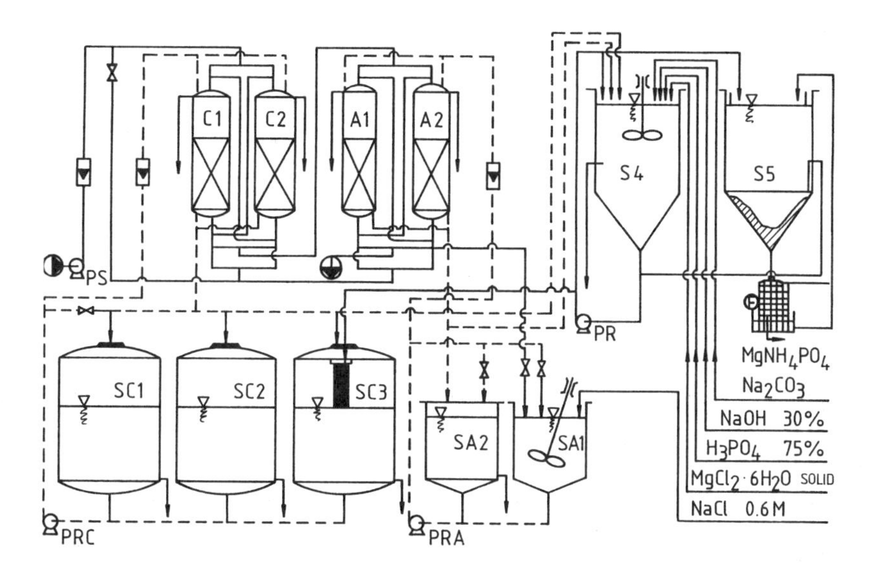

Figure 1. Flow sheet of the RIM-NUT process plant (C1 & C2 = cationic-exchange columns 1 & 2; A1 & A2 = anionic-exchange columns 1 & 2; SC1, SC2, & SC3 = cationic regeneration reservoirs 1, 2, & 3; SA1 & SA2 = anionic regeneration reservoirs 1 & 2; S4 & S5 = settler thickener tanks 4 & 5; F = filtration apparatus; solid line (—) = wastewater treatment streams; dashed lines (---) = 0.6 N NaCl regeneration streams; PS, PRC, PRA, & PR = pumps of different capacities (Liberti et al. 1986a).

The process is based on a combination of NH_4^+ and PO_4^{3-} ion-exchange reactions, where a zeolite (in this case clinoptilolite-rich tuff from the Anaconda Copper Company, Denver, Colorado) was used to remove NH_4^+ according to the following reaction:

$$Na^+\text{-}Z^- + NH_4^+ \Leftrightarrow NH_4^+\text{-}Z^- + Na^+$$

where Z^- is the negatively charged zeolite framework. A strongly basic anion-exchange resin (e.g. Amberlite IRA 458 from Rohm and Haas, Philadelphia, Pennsylvania, or Kastel A 501 D from Ausimont-Montedison, Milano, Italy) was used to exchange PO_4^{3-} from the wastewater as follows:

$$2\ R^+\text{-}Cl^- + HPO_4^{2-} \Leftrightarrow R_2^{2+}\text{-}HPO_4^{2-} + 2\ Cl^-$$

where R^+ is the anion-exchange resin.

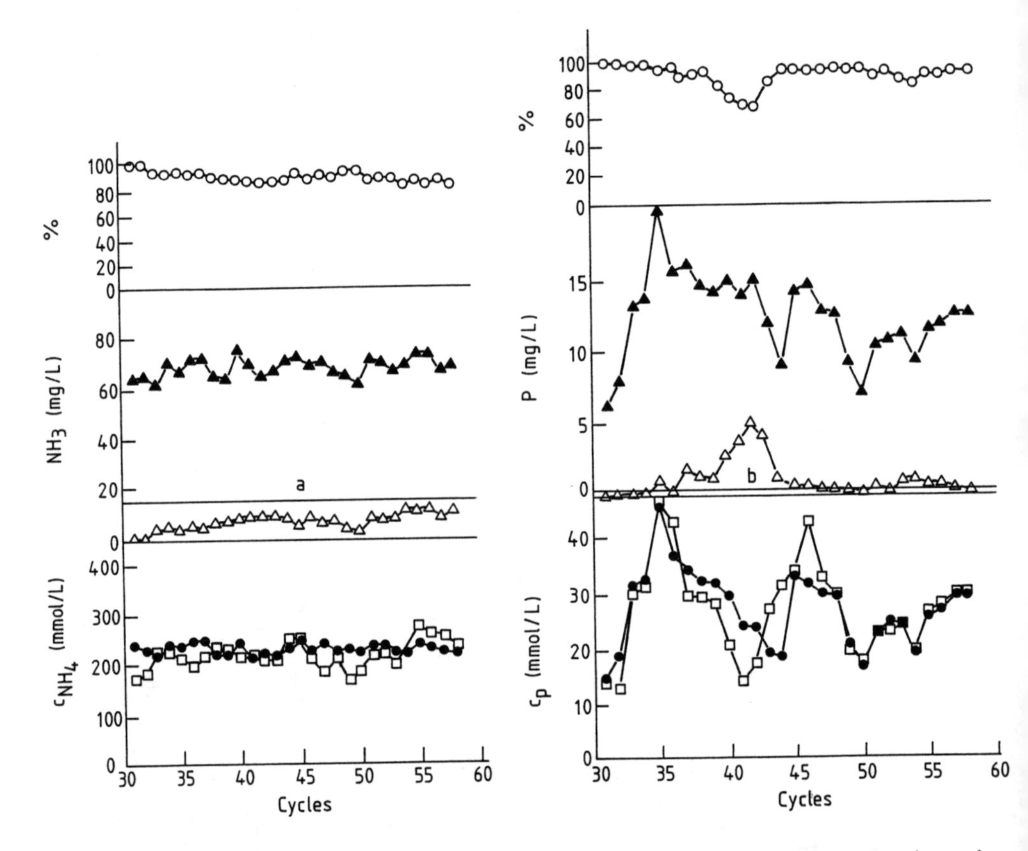

Figure 2 (left). Removal of NH_4^+ during the RIM-NUT process for successive exhaustion-regeneration cycles [Key for symbols: o = % removal of nutrient NH_4^+; ▲ = NH_4^+ concentration in secondary effluent; Δ = NH_4^+ concentration after RIM-NUT treatment; a = Italian regulatory limit for discharge of NH_4^+ into sea; ● = ion exchange capacity of clinoptilolite in service; □ = release of NH_4^+ in regeneration, C_{NH4} = exchange capacity with respect to NH_4 (mole NH_4^+/m^3 of exchanger) (Liberti et al. 1986b).]

Figure 3 (right). Removal of HPO_4^{2-} during the RIM-NUT process for successive exhaustion-regeneration cycles. [Key to symbols same as in Fig. 2 except for P instead of NH_4^+; line labeled b is the Italian regulatory limit for phosphorus discharge into lakes) (Liberti et al. 1986b).]

Two cation (C1 and C2 in Fig. 1) and two anion (A1 and A2 in Fig. 1) exchangers of 0.45 m^3 each were installed to ensure continuous operation by alternating between ion exchange and regeneration. Removal efficiencies for NH_4^+ and hydrophosphate are shown in Figures 2 and 3, respectively (Liberti et al. 1986b). The average phosphate removal from the wastewater was ≥95%. A major decrease of efficiency was observed during cycles 37-42 (Fig. 3) when the inlet concentrations of other anions were high, i.e. 105 mg NO_3^-/L and 55 mg SO_4^{2-}/L were present in the influent in addition to 15 mg PO_4^{3-}/L, and these anions likely retarded or slowed the exchange of HPO_4^{2-} on the anion-exchange resin.

Ion exchangers were regenerated with 0.6 N NaCl after passing 70 BV of wastewater for 3 h through the exchange columns. The regeneration of clinoptilolite was carried out four times with 6 BV of NaCl and the anion-exchange resin was regenerated with 2 BV of NaCl. The initial portions of the exhausted solutions, which contained NH_4^+ and HPO_4^{2-} in the highest concentrations, were accumulated in a settler tank (S4 in Fig. 1). Magnesium and phosphate salts (e.g. $MgCl_2$ and NaOH + H_3PO_4) were added to regeneration solutions (which contained NH_4^+ and HPO_4^{2-}) in order to establish a stoichiometric ratio of $Mg:NH_4:PO_4$ = 1:1:1 at pH = 9. A $MgNH_4PO_4$ precipitate formed according to the reaction:

$$Mg^{2+} + NH_4^+ + HPO_4^{2-} \rightarrow MgNH_4PO_4 + H^+$$

The end product ($MgNH_4PO_4$) resulted in the production of a slow-release premium quality fertilizer with 7% organic content, which originated from the suspended solids (SS) and bio-resistant organics in the wastewater. The product contained negligible traces of heavy metals and had no pathogenic hazards. Heavy metals, if present, were likely removed by clinoptilolite at the high pH of wastewater during the treatment. The negatively charged pathogenic bacteria, which adsorbed on positively charged quaternary ammonium groups of the strongly basic anion-exchange resin, were killed by osmotic shock during the elution of ion exchangers with the concentrated NaCl regeneration solution and by the alkalinity of the precipitating solution (Liberti et al. 1987). After six months of operation (treatment of 120,000 BV of municipal secondary effluent), both ion exchangers were fouled to some extent by strongly bound contaminants or ions. Libertí et al. (1987) reported that both clinoptilolite and the ion-exchange resin lost ~3% of their effective ion-exchange capacities. The economic feasibility of this process is still being evaluated and is primarily influenced by the price and amount of the additional chemicals required to produce the high-grade, premium fertilizer (Liberti et al. 1987). A second plant using the RIM-NUT process was built and evaluated in South Lyon, Michigan, USA (Liberti et al. 1988).

A less sophisticated method was suggested for simultaneous removal of NH_4^+ and PO_4^{3-} from wastewater effluents using phillipsite-rich tuff (Garcia et al. 1992b). Volcanic tuff with abundant phillipsite (Tenerife, Canary Islands, Spain) was studied as a water purifying bed for removing inorganic contaminants, pathogenic bacteria, and dissolved organic matter. The input and output concentrations were determined at constant flow rate. The percolation bed retarded about 70% of the total NH_4^+ and 14% of the total PO_4^{3-} after 10 days of operation. Phosphate removal was attributed to the precipitation of Ca and Mg phosphates. Calcium and Mg^{2+} were introduced into solution after being replaced at zeolite exchange sites by NH_4^+. The removal of soluble organic matter resulted in 20-50% reduction of COD and in complete removal of bacteria in the processed wastewater.

ZeoFlocc process. The ZeoFlocc process is based on the ability of microorganisms to adhere to zeolite-rich particles, thereby increasing the biological activity and efficiency of municipal wastewater treatment. ZeoFlocc is a patented process that derives its roots from zeolite (Zeo) and the flocculation (Flocc) of bacteria with the zeolite (Kiss et al. 1990). The ZeoFlocc technology is being used in Hungary, Germany, Austria, Switzerland, and Australia and the estimated wastewater treated by this technique is now over 400,000 m^3/d.

The increase of biological activity in the ZeoFlocc process can be attributed to the formation of bacteria-flocs around zeolite particles that result in a greater number of smaller bacteria flocs. The sizes of bacteria flocs without zeolite additions are between 0.4 to 2 mm, whereas the bacteria flocs associated with zeolite particles are smaller than 0.3 mm. As might be expected, the transport of oxygen and nutrients requires shorter time in the smaller flocs than in the larger ones. It is also likely that the nitrification process is accelerated because NH_4^+ is available as a nutrient source from zeolite exchange sites. As an example, the addition of 50 g of clinoptilolite-rich powder from Tokaj Hills, Hungary, with a particle size of 40-160 μm (note the small particle size) to 1 m^3 of wastewater increased the rate of the biological wastewater process by at least 25% (Oláh et al. 1991). Kallo (1995) conducted a comparison test of the ZeoFlocc process using clinoptilolite-rich materials, synthetic zeolites, perlite, and charcoal. The rates of oxygen consumption in the biological treatment of sludge wastewaters were 4, 4, 7, 32, and 77 mg O_2/Lh when the same relative amounts of synthetic zeolites A and X, perlite, charcoal, and clinoptilolite-rich tuff from Tokaj Hills, Hungary, respectively, were added to the wastewater.

Simultaneous removal of PO_4^{3-} during the ZeoFlocc process may be achieved by adding trivalent metal cations, particularly Fe^{3+}, to municipal wastewaters. For example, by adding 1.16-1.95 atoms of Fe as a concentrated solution of $FeClSO_4$, which is a secondary by-product of iron manufacturing called Ongroflok, and a suspended clinoptilolite powder from Tokaj Hills, Hungary, to the equivalent of 1 atom of P in the wastewater, the PO_4^{3-} content of the wastewater effluent was decreased from 15-20 mg/L to 1-1.5 mg/L (Oláh et al. 1989). Without the addition of the suspended zeolite, 1.7-2.5 atoms of Fe for each P atom in wastewater were required for similar phosphate removal.

The results of the ZeoFlocc process with clinoptilolite and Fe^{3+} additives for biological treatment plants of different capacities are summarized in Table 2 (Kalló 1995). The clinoptilolite and Fe^{3+} additions significantly decreased the effluent COD, BOD, NH_4^+, PO_4^{3-}, and suspended solids and increased NO_3^- for the various capacities compared with control tests without the additives. The significant increase in the NO_3^- content with zeolite addition suggests faster rates of nitrification and therefore an increase in biological activity compared with treatments not receiving zeolite additives. The nitrification rate was slowed by increasing the biological load from 0.04 to 0.14 kg BOD/kg d (e.g. see BOD values in Table 2 for treatment plants of 100 and 43 m^3/d capacity) and more NH_4^+ remained in the effluent (e.g. 10 and 1 mg NH_4^+/L for 43 and 100 m^3/d treatment plants, respectively). The remaining NH_4^+ can be removed from the effluent by ion exchange with clinoptilolite-rich materials as discussed above (Kalló 1995).

The changes in the removal of phosphates from wastewater treated with clinoptilolite and Fe^{3+} for a 2000 m^3/d treatment plant are shown over a 60-day period in Figure 4 (Oláh et al. 1991). Clinoptilolite and Fe^{3+} additions maintained the effluent P concentration below the limit allowed from biologically treated water during most of the 60-day ZeoFlocc process. The high phosphate removal efficiency in the ZeoFlocc process is due to retention of Fe (or Al) in zeolitic rock as oxides and/or hydroxides that react with the phosphate ions by specific adsorption processes. Phosphate concentrations in the effluent begin to increase two days after clinoptilolite and Fe^{3+} additions are stopped (see

Table 2. Mean composition of wastewater in biological treatment plants of various capacities for a 2-week period after establishment of steady-state conditions (Kallo 1995).

		Additives							*Susp.*
Capacity	*Sewage*[†]	*Clinoptilolite*	Fe^{3+}	*COD*[‡]	*BOD*[*]	NH_4^+	NO_3^-	PO_4^{3-}	*solid*
m^3/d		---mg/L---							
43	Inlet			161	89	32.5	1.6	4.0	88
	Control	-	-	69	16	28.6	16	3.0	27
	Test	65	17.5	44	9	10.0	30	0.4	15
100	Inlet			306	155	32.1	0.7	5.4	144
	Control	-	-	54	13	20.3	24	3.8	17
	Test	43	23.2	36	8	1.0	57	0.5	9
400	Inlet			83	32	22.3	2.3	2.9	22
	Control	-	-	37	8	0.8	53	2.2	10
	Test	35	16.0	28	5	0.6	85	1.0	5
1850	Inlet			411	194	41.1	0.6	5.5	69
	Control	-	-	68	19	6.4	28.2	3.6	35
	Test	44	20.9	63	18	2.8	42.1	0.9	18

[†] Inlet = raw sewage inlet to aerator; Control = sewage treated without additive after primary settling tank; Test = sewage treated with additive after primary settling tank
[‡] COD = Chemical Oxygen Demand [*] BOD = Biological Oxygen Demand

last 6 days in Fig. 4). When only Fe^{3+} is introduced in the biological treatment, PO_4^{3-} concentrations in the effluent begin to increase about 12 hours after the Fe^{3+} addition is stopped.

Recently, highly dispersed FeOOH-coated clinoptilolite has been found to be a more effective agent to add to the ZeoFlocc process for phosphate removal from wastewater (Illés et al. 1997). The FeOOH coating on clinoptilolite is achieved by dry grinding (in air) powders of $FeSO_4 \cdot 7H_2O$ (by-product of iron manufacturing), lime or limestone, and a clinoptilolite-rich material (Tokaj Hills, Hungary). The grinding results in the solid-state transformation of $FeSO_4 \cdot 7H_2O$ into a highly dispersed, FeOOH-coated clinoptilolite support. Kalló and Papp (1999) have shown that this material is effective in removing phosphate from wastewaters, and the biological activity for the FeOOH-coated clinoptilolite system is similar to the clinoptilolite and Fe^{3+} system.

Flocs produced with zeolite seeds are of higher bulk density than flocs without zeolite seeds; therefore, the zeolite/bacteria flocs settle faster, thereby reducing the concentration of suspended solids in the effluent leaving the secondary settling tank

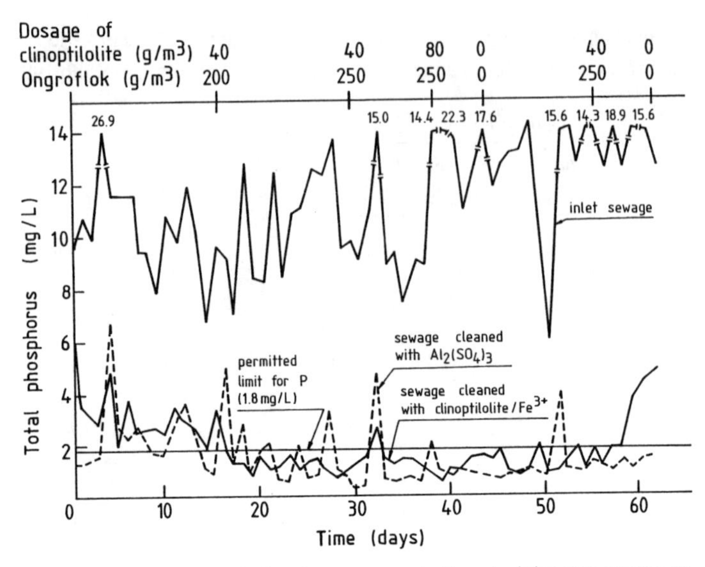

Figure 4. Comparison of phosphorus concentrations in inlet raw sewage to phosphorus concentrations in biologically treated effluents with either additions of clinoptilolite + Fe^{3+} or with aluminum sulfate (Kalló 1995).

(e.g. see the corresponding values for suspended solids in Table 2). It has been reported (Fujiwara et al. 1987) that the separation of sludge solids from a supernatant is facilitated by adding 0.3 wt % clinoptilolite to the sludge. The settling rate of the sludge was nearly twice as fast with the clinoptilolite flocculent as that observed with a commercial flocculent. Sludge removed during the ZeoFlocc process has a higher sedimentation rate than that removed in conventional biological treatment (Oláh et al. 1989, Mucsy 1992). The volume of zeolite-containing sludge, which is easier to dewater, is nearly 50% less than the sludge without zeolite additives under similar conditions. Composting of zeolite-containing sludge proceeds similarly to unamended sludge, but water leached through zeolite-containing sludge contains only 2/3 of the NH_4^+ and oxidizable compounds compared with water leached from unamended sludge (Papp 1992).

Kalló (2000) added 0, 10, 20, and 30 vol % composted sludge amended with clinoptilolite-rich tuff to a substrate consisting of 50% sand and 50% peat (Table 3). No elemental toxicities were observed in lettuce and cabbage grown in the sludge amended soils and plant productivity increased with increasing clinoptilolite/sludge content in the soil. Yields of onion, spinach, tomato, and bean were similarly increased by the addition of ZeoFlocc-treated sludge (Kalló 1995). Heavy-metal accumulation in plants grown in soils amended with ZeoFlocc-treated sludge did not exceed levels permitted for human consumption. The Zn plant tissue concentrations of several plant types grown in soils amended with ZeoFlocc-treated sludge are listed in Table 4. It is likely that Zn^{2+} and other heavy metals are exchanged onto zeolite exchange sites, where these heavy metals are very slowly released into solution at soil pHs of greater that 4 (Dyer 1988).

Clinoptilolite-rich tuff has been added directly to sewage sludge in attempts to reduce the release of heavy metals when the treated sludge is applied as a beneficial source of organic matter and plant nutrients (Weber et al. 1983). Varying amounts of anaerobically digested municipal wastewater sludge and clinoptilolite granules (< 0.8 mm) from the Washakie Basin deposit in Wyoming, USA, were added to soil. The uptake of Cu, Cd, Zn, Mo, and Ni by sorghum-sudangrass grown in the sludge and zeolite-

Table 3. Plant productivity of cabbage lettuce and common white cabbage grown in soil amended with varying amounts of sludge treated with clinoptilolite-rich materials as a by-product of the ZeoFlocc process (Kalló 2000).

Added sludge to soil	*Cabbage lettuce*	*Common white cabbage*
volume %	*g/plant*	*g/plant*
0	220	3760
10	237	4110
20	244	4600
30	257	4730

Table 4. Zinc content of vegetables grown in soil amended with varying amounts of sludge treated with clinoptilolite-rich materials during the ZeoFlocc process (Kalló 2000).

Added sludge	*Red pepper*	*Cabbage lettuce*	*Spinach*	*Radish*	*Onion*
volume %	-----------	-----------	*mg Zn/kg*	-----------	-----------
0	35	65	81	28	29
10	46	125	104	33	46
20	44	134	105	44	53
30	55	155	145	42	56

amended soils did not vary significantly from plants grown in sludge amended soils without zeolite amendments. The results presented here indicate that the addition of clinoptilolite-rich tuff along with the sludge is less effective for retention of heavy metal retardation when compared plants grown in soils where clinoptilolite has been added to the sludge during the ZeoFlocc process as described above.

REMOVAL OF HARMFUL METAL CATIONS FROM WATER

Natural zeolites can be used as ion exchangers for removal of *radioactive* cations such as Cs^+ and Sr^{2+} and *heavy-metal* cations such as Cu^{2+}, Cd^{2+}, Zn^{2+}, Ni^{2+} and Pb^{2+}. Radioactive Cs^+ and Sr^{2+} may be present in recycling waters of atomic power stations or as environmental contaminants after accidents at atomic power stations. Heavy- or transition-metal cations appear in wastewaters of many sources, e.g. in the electroplating industry, photographic material processing, tannery processing, coke manufacturing industry, or even in well water as iron and manganese. The zeolite-rich material chosen for use in removing these potentially harmful cations from wastewater must rely on their cation-exchange selectivities for the appropriate cation as well as their chemical, physical, and thermal stabilities.

Cesium and strontium

Ames (1959) concluded that clinoptilolite-rich tuff from Hector, California, was the most promising of 15 different zeolites tested for Cs^+ removal from solutions containing competing cations such as Al^{3+}, Fe^{3+}, Ba^{2+}, Sr^{2+}, Ca^{2+}, Mg^{2+}, Rb^+, K^+, NH_4^+, Na^+, and Li^+ in large concentrations. The concentrations of competing cations ranged from 0.5 to 6.0 N, whereas that of Cs^+ was as low as 1.75×10^{-8} N $^{137}Cs^+$. Cs^+ breakthrough was determined at linear flow rates of 0.9-10 m/h at 25 and 60°C. The high selectivity of clinoptilolite for Cs^+ was shown by the effective removal of Cs^+ in spite of the high concentration differences between Cs^+ and competing cations. Ames (1960) recorded the earliest application of clinoptilolite-rich material in removal of radioactive cations. He used clinoptilolite-rich tuff from the Hector deposit in California, to selectively remove ^{137}Cs and ^{90}Sr from the low-level radioactive wastewater of nuclear power plants. The advantage of natural zeolites over organic ion-exchange resins lies in their resistance to degradation in the presence of ionizing radiation and their low solubility. Zeolites can also be used for long-term storage of long-lived radioisotopes by drying a zeolite exchanged with radionuclide cations at 200°C and sealing the zeolite in a stainless steel container. Another radioisotope storage method is conversion of a radioisotope-loaded zeolite to glass (e.g. heating in air at 1100°C) having an extremely low radioisotope leaching rate (Sherman 1978).

Mercer and Ames (1978) presented a detailed description of the use of natural zeolites for the removal and fixation of radionuclides, including (1) removal of ^{137}Cs from high-level radioactive wastes; (2) decontamination of low- and intermediate-level wastes; and (3) fixation of radioactive wastes for long-term storage. They found that other natural zeolites, including chabazite, erionite, and mordenite, in addition to clinoptilolite, have high selectivities for Cs^+. Examples that illustrate these and other uses of zeolites for the removal of $^{137}Cs^+$ and $^{90}Sr^{2+}$ are given in Table 5.

Table 5. Applications of zeolites for the treatment of Cs^+ and Sr^{2+} radioactive wastes (Mercer and Ames 1978, Dyer 1984).

Type of effluent	*Location*	*Zeolite used*	*Isotopes removed*	*Plant details*
High-level radioactive waste	Hanford Nuclear Lab., Washington, USA	Linde AW-500 (chabazite)	^{137}Cs	8.4 m^3 zeolite bed processed millions of gallons of waste
Purification of product from above	Hanford Nuclear Lab., Washington, USA	Large-port mordenite	^{137}Cs, ^{90}Sr	Full-scale plant
Process condensate wastewater	Hanford Nuclear Lab., Washington, USA	Large-port mordenite	^{137}Cs, ^{90}Sr	
Low-level wastewater from fuel storage pond	Idaho National Engineering Laboratory, Idaho, USA	clinoptilolite (cp)	^{137}Cs, ^{90}Sr	4 _ 0.15 m^3 cp columns 1900 m^3 water treated/m^3 zeolite
Evaporator overheads and miscellaneous wastewater	Savannah River Plant, Aiken, South Carolina, USA	Linde AW-500 (Chabazite)	^{137}Cs	Treated 3077-10192 m^3 of overheads and 2019-1538 m^3 of miscellaneous wastewater/m^3 zeolite
Wastewater from fuel storage	British Nuclear Fuels, Sellafield, UK	clinoptilolite	^{137}Cs, ^{90}Sr	4000 m^3 water/d processed

Early success in using clinoptilolite-rich materials for Cs^+ and Sr^{2+} removal in the nuclear industry was not reflected by worldwide applications of the zeolite for this purpose. One of the reasons contributing to the low usage worldwide was the general unavailability of clinoptilolite-rich materials and the lack of availability of a high-grade, clinoptilolite-rich rock at that time. Since then, clinoptilolite deposits have been located throughout the world and some of these deposits contain high-grade clinoptilolite (e.g. the Itaya mine in Japan and several deposits in the western USA). Because of the increased availability of high-grade clinoptilolite, its current use has increased in the treatment of radioactive wastes (Dyer 1984). Clinoptilolite is used in the United Kingdom to treat pond waters at installations of the British Nuclear Fuels and Central Electricity Generating Board, and its use is becoming standard practice in some areas.

As noted above, clinoptilolite-rich material is attractive for treating radioactive wastes because of its high selectivity for Cs^+ and Sr^{2+}. The ion-exchange distribution coefficients for Cs^+ and Sr^{2+} in clinoptilolite when the counter ion is Ca^{2+} and the normality of the solution is 0.0003 are listed as follows (Nikashina and Zaborskaya 1977):

$$K_{Cs} = 5.5 \times 10^4 \text{ (meq } Cs^+/g_{clinoptilolite})/(\text{meq } Cs^+/\text{ml } H_2O) \text{ and}$$

$$K_{Sr} = 3 \times 10^3 \text{ (meq } Sr^{2+}/g_{clinoptilolite})/(\text{meq } Sr^{2+}/\text{ml } H_2O).$$

The high selectivity of clinoptilolite for Cs^+ and Sr^{2+} was instrumental in fostering its use for removal of radioactive isotopes from the discharge water of nuclear power plants in the former Soviet Union (Tarasevich 1981). This selectivity property was exploited in developing technology to decontaminate effluents produced from cleaning clothes in washing machines after the Chernobyl accident in 1986. Water with 10^{-4} to 10^{-5} Ci/L activity was filtered through a bed of clinoptilolite-rich granules from the Trans-Carpathian area of Ukraine. The removal of ^{137}Cs was 80-100% in static batch processes and 40-80% in dynamic flow-through column systems. The adsorption capacity for clinoptilolite-rich material was 4×10^{-6} Ci/kg under dynamic flow-through conditions (Zaitsev et al. 1995). In another application, nearly 500,000 tons of zeolitic-rich rock from the Trans-Carpathia area was used in an attempt decontaminate rivers polluted with these radioactive wastes after the Chernobyl accident.

Natural zeolites have also been used to reduce the migration of radioactive cations in contaminated soils. For example, the uptake of Sr^{2+} by *sorghum-sudangrass* from a Sr-contaminated soil amended with 32 t/ha of clinoptilolite-rich tuff from Washakie Basin, Fort Le Clede, Wyoming, USA, decreased from 38 to 24 mg Sr^{2+}/kg, and no reductions were detected in plant uptake for other cations such as Mg^{2+}, Na^+, Fe^{3+}, and Mn^{2+} (Weber et al. 1983). Other potential applications to reduce or eliminate radioactive Cs^+ and Sr^{2+} contamination by adding natural zeolites to soils are discussed by Ming and Allen (this volume).

Heavy metals

In most cases, it is cheaper to mine natural zeolites in near-surface deposits than to produce an equivalent quantity of synthetic zeolites because the price of natural zeolites is usually about 80-90% lower than that of the cheapest synthetic zeolites. As a result of the low cost of natural zeolites and the fact that their native exchangeable cations are relatively safe to humans, plants, and animals (e.g. Na^+, Ca^{2+}, Mg^{2+}, K^+), natural zeolites are especially attractive alternatives for removing undesirable heavy-metal ions from effluent wastewaters mainly of industrial origin (Kesraoui-Ouki et al. 1994). The exchange of multivalent metal ions can be achieved over a pH range between 3-6. However, the pH must be sufficiently low to ensure the solubility of metals (i.e. cationic form in solution) but high enough to minimize H^+ exchange instead of metal cation exchange onto zeolite exchange sites and to minimize destruction of zeolite structure by hydrolysis (Dyer 1988). Zeolites are highly selective for various heavy metals and have been considered suitable for the removal of heavy metals from wastewaters (Blanchard et al. 1984, Rustamov et al. 1991, Takasaka et al. 1991). Natural zeolites have been investigated for removal of Mn^{2+}, Fe^{2+}, Hg^{2+}, Cr^{3+}, Ag^{+}, Cu^{2+}, Cd^{2+}, Zn^{2+}, Pb^{2+}, Co^{2+}, and Ni^{2+} from various wastewaters.

Semmens and Martin (1988) investigated the selectivity of clinoptilolite (Anaconda Copper Company, Denver, Colorado) for Ba^{2+}, Pb^{2+}, Cu^{2+}, Cd^{2+}, and Zn^{2+}. They suggested that the CEC of clinoptilolite depends significantly on how the zeolite is pretreated and they recommended conditioning the zeolite with 0.5 N NaCl solutions for at least two exhaustion-regeneration cycles before measuring equilibrium effective CECs, which should approach theoretically expected CECs. Semmens and Seyfarth (1978) earlier observed a similar increase in CEC from 1.69 meq/g to 2.23 meq/g when clinoptilolite-rich tuff from near Buckhorn, New Mexico, USA, was conditioned by two sequential treatments with 0.5 N NH_4Cl and 1 N NaCl. Zamzow and Eichbaum (1990) also observed an increase in the effective CEC of conditioned clinoptilolite-rich material used to remove heavy metals such as Pb^{2+}, Cd^{2+}, Cu^{2+}, Co^{2+}, Zn^{2+}, Ni^{2+}, and Hg^{2+}. They reported that the concentrations of metal ions in the effluent of the regeneration solution were 30 times higher than in the contaminated wastewater.

Mn^{2+} has been removed from drinking water using simple ion exchange with natural zeolites (White et al. 1995). Clinoptilolite-rich tuff from Northern Romania was used in its native form and in modified forms by independent exchange with Li^{+}, Na^{+}, NH_4^{+}, K^{+}, Mg^{2+}, Ca^{2+}, Fe^{2+}, and Fe^{3+}. Clinoptilolite-rich materials exchanged with Li^{+} or Na^{+} were the most effective forms for Mn^{2+} removal and the Ca^{2+}-form performed the poorest in removing Mn^{2+}. Nearly 67% more Mn^{2+} was exchanged onto K-clinoptilolite than onto the natural exchange form. Ion-exchange distribution coefficients of Mn^{2+} (defined as Mn content of the zeolite in mg/kg divided by Mn content of solution in mg/L) for Li^{+}, Na^{+}, K^{+}, and Ca^{2+} were 900, 700, 280, and 100, respectively, when the initial Mn^{2+} concentration was 10 mg/L. These values decreased to 400, 510, 250, and 50 mg/L, respectively, if 10 mg/L Fe^{2+} was also present. Kinetics of ion exchange indicated that the rate-limiting step was film diffusion around the grains of zeolitic rock.

Misaelidis and Godelitsas (1995) investigated the removal of Hg^{2+} from aqueous solutions of low concentration (10 to 500 mg Hg^{2+}/L) with Na-exchanged clinoptilolite-rich tuff from Greece and Na-exchanged heulandite crystals. They found that Na-exchange of clinoptilolite and heulandite enhanced the uptake of Hg^{2+} by nearly 50%. For the highest solution concentration (500 mg Hg^{2+}/L), 43 mg Hg^{2+}/g was bound to Na-clinoptilolite, whereas 16 mg Hg^{2+}/g was bound to Na-heulandite. Mercury was thought to be bound by ion exchange, adsorption, and surface precipitation. However, Homonnai et al. (1996) reported that the NH_4-exchanged form of clinoptilolite-rich tuff from the Trans-Carpathian area in Ukraine had the capacity to adsorb only 4-7 mg Hg^{2+}/g, which likely reflects the higher selectivity of NH_4^{+} over Hg^{2+} on the exchange sites of clinoptilolite.

The high selectivity of clinoptilolite and mordenite for NH_4^{+} also suppresses the uptake of Zn^{2+} in solutions containing both NH_4^{+} and Zn^{2+}. For example, Kang (1989) found that more NH_4^{+} is bound to clinoptilolite-rich and mordenite-rich tuff than Zn^{2+} at concentrations $1\text{-}7 \times 10^{-3}$ N. The exchange of NH_4^{+} ranged from 0.44 to 0.50 meq/g and Zn^{2+} exchange ranged from 0.07 to 0.17 meq/g for clinoptilolite- and mordenite-rich materials from Korea that had been equilibrated with a 1×10^{-3} N NH_4^{+} and 7×10^{-3} N Zn^{2+} solution, respectively. The adsorption of Zn^{2+} is decreased by the presence of NH_4^{+}, but the exchange of NH_4^{+} is not affected by the presence of Zn^{2+}.

These findings are in accordance with the drastic reduction of Mn^{2+}, Cu^{2+}, and Ag^{+} exchange capacities of natural zeolites in the presence of NH_4^+ in wastewater (Otterstedt et al. 1989). They compared the uptake of these metals by clinoptilolite-rich tuff, mordenite-rich tuff, and synthetic zeolite A when the NH_4^+ concentration was ~50 mg/L. Synthetic zeolite A is not as selective for NH_4^+ as clinoptilolite or mordenite, and they found that synthetic zeolite A was ≥10 times more effective than clinoptilolite and ≥20 times more effective than mordenite in removal of Cu^{2+}. Synthetic zeolite A was ≥40 times more effective than clinoptilolite for removal of Ag^{+} and ≥15 times more effective than clinoptilolite for removal of Mn^{2+}. These differences in exchange of heavy metals are due to lower ion-exchange distribution coefficients between NH_4^+ and these metal ions for synthetic zeolite A than for clinoptilolite and mordenite.

Some authors, however, have suggested that Cu^{2+} removal from solutions with NH_4-exchanged clinoptilolite approaches 100%, presumably at lower concentrations than those usually encountered in wastewater (Kalita and Chelishchev 1995). Blanchard et al. (1984) achieved the simultaneous removal of NH_4^+ and heavy metals from drinking water by clinoptilolite-rich material (origin not reported, CEC of zeolitic rock was 2.2 meq/g). The concentration of heavy metals in the feed solution was 0.2 mg/L in addition to 2.6 mg NH_4^+/L. Heavy-metal concentrations in the effluent water after passing through the clinoptilolite rich material were significantly reduced. Breakthrough concentrations of several µg/L were obtained after passing 175, 220, 225, and 470 BV of wastewater through the clinoptilolite-rich bed for Cd^{2+}, Zn^{2+}, Cu^{2+}, and Pb^{2+}, respectively. A breakthrough NH_4^+ concentration of 0.15 mg/L was obtained after 480 BV had passed through the bed at a flow rate of 10 BV/h. Regeneration of spent clinoptilolite-rich beds was performed with 20 g NaCl/L at pH 4.0-4.5 in order to avoid precipitation of metal hydroxides.

It is well known that ion-exchange selectivities depend on the ion concentration of the metals in the wastewater to be treated. Cation-exchange selectivities for most transition or heavy-metal cations become higher compared with alkali or alkaline-earth cations by lowering the concentrations of transition and heavy-metal cations in solution (see Pabalan and Bertetti, this volume). Therefore, the concentration range of the metals to be removed in wastewater treatment will affect the sequence of ion-exchange selectivities and must be taken into consideration for effective treatment. Examples of cation-exchange selectivity sequences of clinoptilolite-rich materials for various heavy metals are listed in Table 6 (see references for details on effects of solution pH and ionic concentrations).

Table 6. Cation-exchange selectivity sequences of clinoptilolite-rich samples for heavy metals.

Selectivity sequence	*Reference*
Pb»Cd>Cu»Zn	Fujimori & Moriya (1973)
Pb>Cd>Zn>Cu	Semmens & Seyfarth (1978)
Pb>NH_4>Cu,Cd>Zn,Co>Ni>Hg	Blanchard et al. (1984)
Cs>Pb>Fe>Cu>Zn,Cd,Co>Ni>Mn>Cr	Horvathova & Kachanak (1987)
Pb>Cd>Cu>Co>Cr^{3+}>Zn>Ni>Hg	Zamzov & Eichbaum (1990)
Pb>Cu>Zn>Co>Ni	Kalita & Chelishchev (1995)

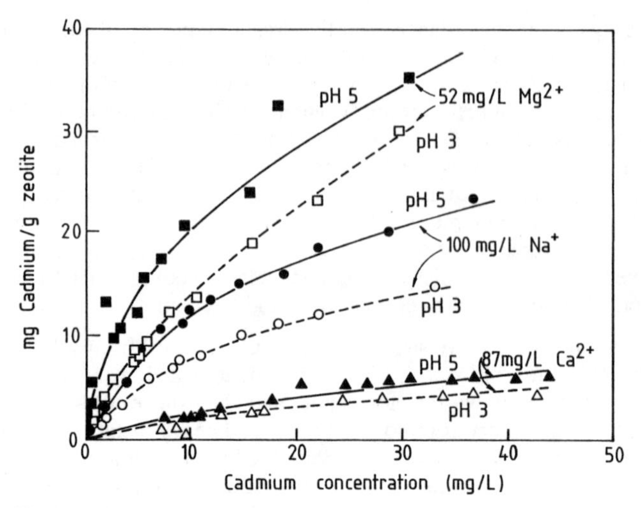

Figure 5. Cadmium ion exchange isotherms at 20°C for a clinoptilolite-rich tuff from the Anaconda Copper Company, Denver, Colorado, USA, in the presence of Mg^{2+}, Na^{+}, and Ca^{2+} competing cations (illustrated in mg/L on figure) at pH 3.0 and 5.0 [*x*-axis is Cd^{2+} in solution, *y*-axis is Cd^{2+} exchanged onto zeolite exchange sites (Semmens and Martin 1980)].

As mentioned earlier, the influence of pH is very important for efficient removal of heavy metals by ion exchange. A decrease in pH will decrease the effective CEC for heavy metals because H_3O^+ will compete for zeolite exchange sites (see Fig. 5; Semmens and Martin 1980). These authors also found that under similar solution concentrations and pH values selectivity coefficients for clinoptilolite-rich material (Anaconda Copper Company, Denver, Colorado) decreased in the order $Pb^{2+} > Ag^+ > Cd^{2+}$ in the presence of the same counter ions Mg^{2+}, Na^+, and Ca^{2+}. The exchange of Cd^{2+} and Pb^{2+} onto clinoptilolite exchange sites was least inhibited by Mg^{2+}, followed by Na^+ and Ca^{2+}. The exchange of Ag^+ was also least inhibited by Mg^{2+}, but Na^+ inhibited the exchange more than Ca^{2+}.

Heavy metals on saturated or exhausted zeolites can be effectively regenerated with a solution containing 20-25 g NaCl/L, which has been adjusted to a pH of 4-4.5 with HCl (Blanchard et al. 1984). The heavy-metal ions can be more effectively removed from high concentration regenerating solution than from the low heavy metal concentrations wastewater. The heavy metals are precipitated as hydroxides by increasing the pH to 6 or 7, depending on the type and concentration of heavy-metal ions in the regenerating solution.

An alternative to ion-exchange processes for the removal of heavy metals from wastewaters is the adsorption of heavy metals on zeolite surfaces coated with a suitable adsorbent. For example, soluble Fe^{2+} and Mn^{2+} are often present in well waters at concentrations of 0.1-2 mg/L, which is too high for human consumption. The removal of Fe^{2+} and Mn^{2+} can be accomplished simultaneously by adsorption of these ions on MnO_2 that has been precipitated onto the surfaces of zeolite particles embedded in the porous rock. The MnO_2 adsorbent must have high surface area and high mechanical strength. The zeolite-MnO_2 product is called "manganese zeolite" or "zeomangan" and various zeolites have been used in this process, including clinoptilolite-rich tuff from Fuatsui, Akita Prefecture, Japan (Torii 1978) and Tokaj Hills, Hungary (Polyák et al. 1995) and chabazite-rich tuff from Italy (Aiello et al. 1979). In addition to adsorption of Fe^{2+} and

Mn^{2+} on the MnO_2 coating, these bivalent cations may also be exchanged onto the exchange sites in the zeolite.

Polyák et al. (1995) have presented a method to prepare zeomangan. The zeolite is first exchanged into its K^+ form by ion exchange with 20 g KCl/L. Then up to 12-15% of the K^+ is exchanged by Mn^{2+} with a solution containing 169 g $MnSO_4$/L. The K,Mn-zeolite is percolated with a solution of 10 g $KMnO_4$/L. The resulting product is an MnO_2-coated zeolite where the reaction for the precipitation of MnO_2 is as follows:

$$3\ Mn^{2+} + 2\ MnO_4^- + 2\ H_2O \rightarrow 5\ MnO_2 + 4\ H^+$$

The oxygen content of the precipitated MnO_2 is slightly lower than that expected for stoichiometric MnO_2. The actual composition corresponds to the formula $MnO_{1.89\text{-}1.93}$. Consequently, the coating contains Mn^{2+} and some Mn^{3+}. The surface charge of the MnO_2 is dependent on the solution pH. The zero point of charge (pH_{zpc}) is pH 1.56; hence, surfaces become negatively charged above the pH_{zpc} and provide sites for the exchange of Mn^{2+} (and Cd^{2+}, Zn^{2+}, Cu^{2+}, and Pb^{2+} if they are present in the wastewater, e.g. Polyák et al. 1995).

The MnO_2 will also oxidize Fe^{2+} to Fe^{3+} according to the reaction:

$$MnO_2 + 2\ Fe^{2+} + 4\ H_2O \rightarrow 2\ Fe(OH)_3 + Mn^{2+} + 2\ H^+$$

Once oxidation occurs, Fe^{3+} precipitates as ferric hydroxide on the MnO_2-coated zeolite, thereby removing Fe from the water stream. Well water is treated by flowing the water stream through the MnO_2-coated zeolite. The capacity of the MnO_2-coated zeolite to extract Fe^{2+} and Mn^{2+} depends on their concentrations in the well water and the flow rate of water through the bed. Capacities to adsorb Fe^{2+} and Mn^{2+}on MnO_2-coated chabazite are summarized in Table 7 for different concentrations of Fe^{2+} and Mn^{2+} and flow rates through the bed.

The MnO_2-coated zeolite is usually regenerated by passing 10 BV of 0.05-0.1 N HCl, followed by washing the bed material with 10 BV of 10 g $KMnO_4$/L. When the

Table 7. Fe^{2+} and Mn^{2+} adsorption capacities on MnO_2-coated chabazite used to treat well waters until a breakthrough concentration of 0.1 mg/L was achieved (calculated from the data of Aiello et al. 1979).

Cation composition in influent		*Flow rate through*	**Adsorbed cations*	
Fe^{2+}	*Mn^{2+}*	*column*	*Fe^{2+}*	*Mn^{2+}*
------------*mg/L*-------------		*BV/h*[†]	----------*mmol/g*-------	
1	-	48	0.11	-
2.5	-	48	0.19	-
5	-	48	0.19	-
2	0.5	48	0.21	0.04
1.25	1.25	32	0.15	0.13
1.25	1.25	48	0.14	0.10
1.25	1.25	64	0.06	0.07

*Adsorbed on MnO_2-coated chabazite [†]BV/h = bed volumes/h

exhausted adsorbent is leached with the HCl solution at pH = 1.5-2, the adsorbed cations are removed in higher concentrations than in the original wastewater and can thus be more effectively precipitated by increasing the pH of the regenerating solution. The $KMnO_4$ solution reoxidizes the reduced MnO_2 on the zeolite support. Exhaustion-regeneration cycles can be repeated 5-8 times without loss of the adsorption capacity.

Many industries, such as photographic processing, electroplating, metal finishing, ore mining and mineral processing, coal mining and processing, and oil refining, have potential problems associated with heavy-metal contamination in wastewaters or in runoff waters. Effluents generally contain heavy metals in low concentrations in the presence of high salt concentrations; hence, the use of ion-exchange technologies must be evaluated on a case-by-case basis. It is also important to note that ion-exchange equilibria determined in batch or static systems differ from ion-exchange capacities measured in flow-through or dynamic systems (Semmens and Martin 1980). Ion-exchange equilibria and selectivities provide some information on the rates of ion exchange under dynamic conditions, but the rate of exchange is dependent on inter- and intra-particle diffusion. The former can be influenced by particle size of the zeolite and flow rate through the exchange column. Previous studies suggest that the optimum zeolite particle size is 0.3-0.8 mm and the flow rate through a column is 8-16 BV/h (Semmens and Martin 1980, Guangsheng et al. 1988, Kalló 1995).

In addition to the reduction of potential toxic metals, another objective of heavy-metal removal from wastewaters should be metal recovery and reuse. These metals are usually valuable to industry, but, in most cases, they cannot be economically recovered by conventional concentrating techniques. Technologies should be evaluated that recover heavy metals from regenerating solutions of ion-exchange processes to enhance their potential economic recovery.

INDUSTRIAL WASTEWATER TREATMENT

As mentioned earlier, because of their unique ion-exchange properties, adsorption capabilities, mechanical, chemical, and thermal resistance, and low price, natural zeolites may be applied for the treatment of wastewater and sewage produced by industry. As might be expected, ion-exchange selectivities of natural zeolites play an important role in removal of NH_4^+, and at low concentrations (below 10^{-3} N) they may be used in removal of heavy-metal cations from industrial wastewaters. Adsorption of organic and inorganic molecules is influenced by the molecular sieving properties of zeolites and additional adsorption may take place in pores in the zeolitic rock. The mechanical strength of most zeolitic rocks is satisfactory for most technical purposes such as filling in fixed bed ion exchangers or adsorbents, e.g. in trickling filter beds (Mercer and Ames 1978). Zeolite granules can often be manufactured from zeolite-rich rocks or tuffs by simple crushing and sieving; hence, expensive pelletization is not needed as in the case with synthetic zeolites, which are usually produced as fine powders with micrometer grain size. Natural zeolites are stable in the pH range of 4-9 and at solution temperatures up to the boiling point of water (or even above). They are not soluble in any organic solvent. They can often be dehydrated-hydrated many times without significant hysteresis. Some natural zeolites are stable at least up to 500°C in air, hydrogen, carbon monoxide, and other gases. Apart from airborne dust hazards, natural zeolites do not pose any environmental problems, and on the contrary are able to retard harmful substances in the environment as described earlier. The cost of mining natural zeolites is less than one fifth the cost of manufacturing the least expensive synthetic zeolite products. Hence, there are numerous properties that make natural zeolites attractive for treating industrial wastewaters. However, the optimum conditions for the use of natural zeolites in a given wastewater treatment application must be determined experimentally. Several practical and potential uses of natural zeolites in the treatment of industrial wastewaters are listed below.

Clinoptilolite-rich tuff from South Korea has been used to remove NH_4^+ from coke-oven wastewater (Ha 1987). The clinoptilolite was transformed into its Na-form in order to extend the effective capacity for NH_4^+ exchange. The NH_4^+ breakthrough concentration was reached after 55-60 BV of wastewater containing 20-30 meq NH_4^+/L passed through the column. The exhausted bed was regenerated with 2 N NaCl. The NH_4^+ exchange capacity of the clinoptilolite column decreased by less than 10% after four regenerations with fresh regenerating solution. Approximately 95% of the effective NH_4^+ CEC could be restored if the spent clinoptilolite was regenerated with 10 BV of NaCl. The first 20% of regenerating solution eluted from the spent clinoptilolite column contained high concentration of NH_4^+ (0.5-0.8 M) and was replaced with fresh regenerating solution. Ammonia may be removed from the spent regenerating solutions by stripping with air as described above.

Korobchanskii et al. (1987) exploited the ion-exchange properties of a clinoptilolite-rich tuff from the Trans-Carpathian region of Ukraine to remove NH_4^+ from wastewater from a coke oven after initially treating the wastewater with a series of processes including settling, sand filtration, and/or coagulation with $FeCl_3$. A breakthrough concentration of 5 mg NH_4^+/L was reached after passing 7 BV of the influent wastewater with a concentration of 700 mg NH_4^+/L at a rate of 2 BV/h through the clinoptilolite-containing column. Approximately 50-60% of the cation-exchange capacity of the 1- to 2- mm clinoptilolite grains was utilized under these dynamic conditions. Regeneration of spent clinoptilolite was achieved by passing a 10% solution of H_2SO_4 through the spent clinoptilolite column. The effective NH_4^+ exchange capacity remained unchanged at 8.2 mg NH_4^+/g after 3 sorption-desorption cycles. Ammonium sulfate was the by-product of the regeneration process.

Clinoptilolite-rich tuff from Georgia was used for the removal of mercury from chlorine production wastewater (Mamedov et al. 1985). Mercury contents of the wastewater to be treated varied between 6.4 and 52.0 mg/L. The lowest ratio of the Hg^{2+} in the treated effluent to the wastewater influent was 0.04 and the effective Hg^{2+} exchange capacity of the clinoptilolite-rich tuff was 11 mg Hg^{2+}/g (around 1/10 of CEC). The flow rate through the clinoptilolite column (particle size of about 1.6 mm) was 1.5 BV/h.

A Na-clinoptilolite from Georgia was used to remove Ag^+ by ion exchange from liquid wastes from photographic material processing (Rustamov and Makhmudov 1988). Liquid waste containing 0.14 meq Ag^+/L was passed through a Na-exchanged clinoptilolite column with a linear flow rate 27 m/h. Spent clinoptilolite columns were eluted with 1 N $NaNO_3$ solution at 1.8 m/h. Silver concentrations in regenerating solutions were 14 times higher than in the initial liquid waste.

Guangsheng et al. (1988) showed that Na-exchanged clinoptilolite-rich material from Nenjiang, China, could efficiently remove Cu^{2+} from electroplating effluents. The pH of the wastewater, which contained 20-30 mg Cu^{2+}/L, was adjusted to 4-5 for attaining the optimum conditions for Cu^{2+} exchange onto zeolite exchange sites. This process removed all of the Cu^{2+} from the electroplating wastewater. The effective Cu^{2+} exchange capacities decreased from 1.14 to 0.87 meq/g when the linear velocity was increased from 3 to 9 m/h. Approximately 90-97% of the exchanged Cu^{2+} was recovered during regeneration with 2 BV of a saturated NaCl solution. No reductions in the CEC were observed after 29 successive exhaustion-regeneration cycles.

Magnesium chloride and H_3PO_4 were used to flocculate particles in the effluents from a goatskin tannery in order to reduce sludge production and to produce sludge that is much easier to dewater (Bilotta and Vallero 1984). These flocculents were more efficient in removing COD, suspended solids, Cr, SO_4^{2-}, and NH_4^+ compared with conventional biological treatment methods. As might be expected, a clinoptilolite-rich tuff may be used to reduce the NH_4^+ concentration in the effluent after treatment with the flocculents. For example, an effluent with an NH_4^+ concentration of 90-100 mg/L was decreased to 25-30 mg NH_4^+/L after passing through a column containing clinoptilolite-rich materials. Most of the discussion in this chapter has been directed toward the use of clinoptilolite-rich materials. However, Colella et al. (1984) found in laboratory tests that phillipsite-rich tuff from near Naples, Italy, was more efficient than clinoptilolite-rich tuff for NH_4^+ removal from tannery wastewaters The phillipsite-rich tuff had a higher CEC of 2.28 meq/g compared with a CEC of 1.2 meq/g for the clinoptilolite-rich tuff. In addition to removing NH_4^+ from the tannery wastewater, the COD was reduced by 28% with phillipsite-rich tuff and by 24% with clinoptilolite-rich tuff compared with wastewater not treated with the zeolites.

A conventional nitrification-denitrification process was compared with an ion-exchange process using clinoptilolite-rich tuff from Slovakia for removal of NH_4^+ from shoe manufacturing wastewater (Chmielewska-Horvathova et al. 1992). The pilot plant consisted of 2 alternating pressure filters containing 70 L of clinoptilolite-rich material (0.3- to 1.0-cm grain size) and treated 20 m^3/d wastewater. Regeneration of spent clinoptilolite beds was achieved by passing NaCl solution through the columns and then, NH_3 was stripped from the spent regenerating solution by air. As an example, the operation costs for the ion-exchange/regeneration process are 22% higher than the biological nitrification-denitrification process for a plant of 6000 m^3/d capacity, but investment costs are 27% lower for the clinoptilolite ion-exchange process compared to the nitrification-denitrification process, not taking into account the high energy consumption of long time aeration required for the biological degradation.

Some organic impurities may be removed by adsorption on zeolites. For example, dichloropropanol ($ClCH_2$-(OH)CH-CH_2Cl) has been removed from aqueous solution by adsorption on mordenite-rich tuff from Eastern Crimea area of Ukraine (Kakhramanova et al. 1983). The adsorption takes place on the external surfaces of mordenite crystals because the size of hydrated dichloropropanol prevents penetration into the zeolite's channels to access extraframework sites. Adsorption of dichloropropanol was the highest (4.2 $mg/L_{mordenite}$) when mordenite-rich tuff was pre-treated with $[(C_2H_5)_2NH_2]^+Cl^-$, which exchanges onto the exchange sites in mordenite, and the wastewater stream was passed through the zeolite bed under dynamic flow conditions. Even traces of dichloropropanol can be removed from wastewater by passing the stream through a zeolite-rich bed.

Various toxic chlorinated organic compounds (e.g. trichloroethylene, chloroform, dichloroethane, perchloroethylene, epichlorohydrin, etc.) in wastewaters (1.0-1.3 g/L) have been removed by adsorption on the crystal surfaces of clinoptilolite-rich and mordenite-rich materials (Rustamov et al. 1992). Adsorption capacities of these zeolites for chlorinated organic compounds were increased by initially exchanging either amines or Cu^{2+} onto zeolite exchange sites. For example, clinoptilolite-rich material exchanged with $CH_3NH_2.HCl$ and Cu^{2+} increased the sorption of epichlorohydrin from 2.08 to 3.14 and 4.40 wt %, respectively. Mordenite-rich material exchanged with the sample ions increased the adsorption of epichlorohydrin from 1.56 to 2.70 and 3.28 wt %, respectively. For comparison, the sorption capacity of charcoal was 3.00 wt % under the same conditions. The charcoal lost around 75% of its adsorption capacity for epichlorohydrin after 10 regenerations, whereas clinoptilolite-rich material exchanged with $CH_3NH_2.HCl$ lost less than 20% of its adsorption capacity for epichlorohydrin.

Korean workers used an unspecified local zeolite (presumably clinoptilolite) to remove color from wastewater of a dyeing process (Doh and Park 1980) Their studies confirmed the earlier work of Iso et al. (1976) on the use of clinoptilolite-rich material to remove color from wastewaters produced in dyeing processes. Iso et al. (1976) reported that the amount of COD impurities adsorbed on clinoptilolite-rich materials at equilibrium was nearly equal to that adsorbed on granular active carbon, both of which were around 10 mg COD/g.

Although zeolites have a permanent positive framework charge, balanced by exchangeable cations, it is possible in some cases to use zeolites for removal of anionic contaminants by modifying the zeolite surfaces. Anionic impurities were removed from the water circulating system of a paper manufacturing company using clinoptilolite-rich tuff from Tokaj Hills, Hungary (Baumann and Heinzel 1996). Clinoptilolite was initially treated at 200-500°C with polyaluminum chloride, which is hydrolyzed on zeolitic surfaces to form a cationic polymer coating. Epichlorohydrin, epichlorohydrin derivatives, and/or dicyandiamid ($HN=C(NH_2)-NH-C{\equiv}N$) were effectively adsorbed on the treated zeolite substrate.

Wastewater produced in yeast production plants was more effectively treated in anaerobic reactors containing zeolitic tuff (41% clinoptilolite and 40% mordenite contents from Cuba) as a support medium than in reactors containing ceramic Raschig rings, PVC pellets, or limestone as support media (Sanchez and Roque-Malherbe 1987). An initial BOD of 20,000 mg/L was decreased by 65% after passing 10 BV of wastewater through the zeolite gravel bed, whereas wastewater passed through ceramic rings, PVC pellets, and limestone gravel decreased the BOD by 52, 40, and 34%, respectively. This supports earlier claims (see above) that the zeolite support media can increase the biological activity of bacteria, where the zeolite concentrates nutrient sources (i.e. NH_4^+) for bacteria and provides an excellent support medium for their growth.

CONCLUSIONS AND RECOMMENDATIONS FOR FUTURE WORK

The high selectivity of natural zeolites for NH_4^+ has prompted their use to treat a variety of natural waters and wastewaters from municipal, industrial, and agricultural sources. Natural zeolites also offer the possibility of removing minute amount of heavy-metal ions and radioactive Cs^+ and Sr^{2+} from wastewaters, even in the presence of competing alkali or alkaline-earth cations. In addition to intracrystalline cation exchange, the crystal surfaces of natural zeolites may act as adsorption sites or filters for organic molecules and microorganisms, which may contribute significantly in the treatment of wastewaters. In some cases, the technology of using natural zeolites has been elevated to the point where they are used to treat large-scale municipal and industrial wastewaters. However, numerous questions remain unanswered and additional areas of research must be pursued before natural zeolites may be used to their full potential to treat water and wastewater.

A promising area for the use of natural zeolites is in the adsorption of herbicides, insecticides, and fungicides so that the zeolite retards the migration of these agents into groundwater and runoff. Only recently have studies begun to examine this potential area (see Ming and Allen, this volume), and considerable work remains on the modification of zeolite surfaces to make them effective in adsorbing anionic or organic species. There are indications that natural zeolites, mainly clinoptilolite and mordenite, may adsorb traces of dioxin from air incineration plant emissions as well as from wastewaters (private communication from Metallurgische Gesellschaft/LURGI/ GmbH, Germany). Obviously, mechanisms and properties of dioxin adsorption on natural zeolites must be addressed. Another area that requires further studies is what to do with the heavy metals or other wastewater impurities removed by natural zeolites. New technologies need to be developed to economically recover and reuse these potentially valuable commodities.

Several of the recent, more complex water-treatment processes that use natural zeolites have potential for widespread use in wastewater treatment. For example, removal of phosphate from wastewater by adsorption on FeOOH-coated zeolites and the modification of natural zeolites as a substrate to produce more effectively bacteria flocs for biological treatment of wastewater are promising technologies. Studies are required to determine whether these technologies can be efficiently and economically used to treat wastewaters compared with more traditional wastewater-treatment processes.

There may be new technologies that use natural zeolites for the treatment of municipal, agricultural, and industrial wastewaters. For example, natural zeolites might be used in the treatment of recycled water in swimming pools to remove NH_4^+ and chloroamine formed after chlorination disinfection. Natural zeolites may be used in applications for the removal of NH_4^+ via ion exchange from wastewaters of fish-farming ponds or in transportation of live fish and subsequent release of Ca^{2+} and Mg^{2+} into solution where they will precipitate as carbonates in HCO_3^--rich water. New technologies that utilize natural zeolites may be developed to treat liquid manures produced during animal production. These are only a few examples and, no doubt, many other potential applications will become evident as we deal with the never-ending problem of treating contaminated waters and wastewaters. These potential applications and others require sound scientific research, including the thorough chemical, physical, and mineralogical characterization of the zeolite-rich materials used in these applications. Natural zeolites provide numerous possibilities to treat wastewaters with environmentally friendly materials and these unique minerals are far from being fully exploited.

REFERENCES

Aiello R, Colella C, Nastro A (1979) Natural chabazite for iron and manganese removal from water. *In* Townsend RP (ed) The Properties and Applications of Zeolites. The Chemical Society, Burlington House, London, p 258-268

Aiello R, Colella C, Nastro A, Sersale R (1984) Self-bonded phillipsite pellets from trachytic products. *In* Olson D, Bisio A (eds) Proc 6th Int'l Zeolite Conf. Butterworth, Guildford, UK, p 957-965

Aiello R, Nastro A (1984) Evaluation of phillipsite tuff for removal of ammonia from aquacultural wastewaters. *In* Pond WG, Mumpton FA (eds) Zeo-Agriculture: Use of Natural Zeolites in Agriculture and Aquaculture. Westview, Boulder, Colorado, p 239-244

Albertin P, Babato F, Bottin F, Ragazzo P, Navazio G (1994) Evaluation of natural zeolites for treatment of drinking water. Mater Eng (Modena, Italy) 5:283-287

Ames LL (1959) Zeolitic extraction of cesium from aqueous solutions. U S Atomic Energy Comm HW-62607, 25 p

Ames LL (1960) The cation sieve properties of clinoptilolite. Am Mineral 45:689-700

Ames LL (1961) Cation sieve properties of the open zeolites, chabazite, mordenite, erionite and clinoptilolite. Am Mineral 46:1120-1131

Baumann R, Heinzel G (1996) Removal of anionic impurities from circulation water of paper machines using cationized zeolite. European Patent #741,113, 7 p

Baykal BB, Guven DA (1997) Performance of clinoptilolite alone and in combination with sand filters for removal of ammonia peaks from domestic wastewater. Water Sci Technol 35:47-54

Bilotta G, Vallero P (1984) Removal of ammonium ion by chemical treatment and zeolites. Cuoio Pelli Mater Concianti 60:499-501

Blanchard G, Maunaye M, Martin G (1984) Removal of heavy metals from water by means of natural zeolites. Water Res 18:1501-1507

Butterfield OR, Borgerding J (1981) Tahoe Truckee Sanitation Agency: The first three years. TTSA Int'l Report, 26 p

Chmielewska-Horvathova E, Konecny J, Bosan Z (1992) Ammonia removal from tannery wastewaters by selective ion exchange on Slovak clinoptilolite. Acta Hydrochim Hydrobiol 20:269'l

Fujimori K, Moriya Y (1973) Removal and treatment of heavy metals in industrial wastewater. I. Neutralizing method and solidification by zeolite. Asahi Garasu Kogyo Gijutsu Shoreikai Kenkyu Hokuku 23:243-246

Fujiwara Y, Nomura M, Sato S, Motoya M, Igawa K (1987) Wastewater treatment by activated-sludge process employing zeolite as a settling for sludge. Japan Kokai Tokkyo Koho, JP #62,294,496, 5 p

Garcia JE, Gonzalez MM, Notario JS (1992a) Removal of bacterial indicators of pollutions and organic matter by phillipsite-rich tuff columns. Appl Clay Sci 7:323-333

Garcia JE, Gonzales MM, Notario JS, Arbelo CD (1992b) Treatment of wastewater effluents with phillipsite-rich tuffs. Environ Pollut 76:219-223

Galindo Jr. C, Ming DW, Morgan A, Pickering K (2000). Use of Ca-saturated clinoptilolite for ammonium from NASA's advanced life support wastewater system. *In* Colella C, Mumpton FA (eds) Natural Zeolites for the Third Millennium. De Frede Editore, Naples, Italy, p 363-371

Grigorieva LV, Salata OV, Kolesnikov VG, Malakhova LA (1988) Effectiveness of the sorptive and coagulational removal of enteric bacteria and viruses from water. Khim Tekhnol Vody 10:458-461

Guangsheng Z, Xingzheng L, Guangju L, Quanchang Z (1988) Removal of copper from electroplating effluents using clinoptilolite. *In* Kalló D, Sherry HS (eds) Occurrence, Properties and Utilization of Natural Zeolites. Akadémiai Kiadó, Budapest, Hungary, p 529-539

Gunn GA (1979) AWT plants makes wastewater potable. Water Wastes Engin 16:36-44

Ha, KS (1987) Removal and recovery of ammonium ion from wastewater by adsorption on natural zeolite. Korean J Chem Engin 4:149-153

Halling-Soerensen B, Hjuler H (1992) Simultaneous nitrification and denitrification with an upflow fixed bed reactor applying clinoptilolite as media. Water Treat 7:77-88

Hlavay J (1986) Selective removal of ammonium from waters by natural clinoptilolite. Hidrológiai Közlöny (in Hungarian) 66:348-355

Hódi M, Polyák K, Hlavay J (1994) Complex removal of pollutants from drinking water by ion exchange and adsorption. Hidrológiai Közlöny (in Hungarian) 74:104-134

Homonnai VI, Golub NP, Szekeres KY (1996) Adsorption of mercury(II) ions on natural clinoptilolite. Ekotekhnol Resursosberezhenie 1:64-66

Homonnay A, Juhász J, Somlyódy L, Szilágyi F (1993) Process for regeneration of ammonium-saturated ion exchanger. Hungarian Patent Appl (Hung. Teljes), #63,118, July 28, 1993, 13 pp

Horsch CM, Holway JE (1984) Use of clinoptilolite in salmon rearing. *In* Pond WG, Mumpton FA (eds) Zeo-Agriculture: Use of natural zeolites in agriculture and aquaculture. Westview Press, Boulder, Colorado, p 229-237

Horvathova E (1986) Ion-selective exchange on clinoptilolite-tuffite from Slovakia. Acta Hydrochim Hydrobiol 14:495-502

Horvathova E, Kachanak S (1987) Removal of heavy metals from water by natural clinoptilolite and semiquantitative determination of selectivity order of metal cations. Vodni Hospod B37:8-12

Hosszú S, Csete J, Inczédy J, Vígh G, Földi Polyák K, Olaszi V, Hajdu R, Horváth E, Hlavay J (1983) Apparatus for removal of arsenic, iron, manganese from drinking water. Hungarian Patent #188,886, 9 p

Hulbert MH, Currier JW (1986) Sand filter media and an improved method of purifying water. European Patent #175,956, 53 p

Illés G, Kalló D, Karácsonyi J, Kótai L, Papp J, Pálinkás G, Udvardy G (1997) Composition and process for the removal of the phosphate ion content of waters. Hungarian Patent Appl #P9701918; PCT Patent Appl #9800096, 13 p

Iso F, Shibata T, Okonogi T (1976) Studies on the treatment of dyeing wastewater. 2. Treatment of dyeing wastewater with waste sludge from aluminum production factories. Poll Control 11:8-13

Kakhramanova KhT, Zul'fugarov ZG, Mirzai DI, Annagiev, MKh (1983) Study of dichlorohydrin adsorption by natural and modified mordenite. Azerb Khim Zh 6:18-21

Kalita AP, Chelishchev NF (1995) Use of zeolite containing rocks for water purification. Razved Okhr Nedr 7:19-20

Kalló D (1990) Exploitation of ammonia with ion exchange. Unpublished research report. Hungarian Academy of Sciences, Budapest, Hungary

Kalló D (1995) Wastewater purification in Hungary using natural zeolites. *In* Ming DW, Mumpton FA (eds) Natural Zeolites '93: Occurrence, Properties, Use. Int'l Comm Natural Zeolites, Brockport, New York, p 341-350

Kalló D, Papp J (1999) Wastewater treatment with natural clinoptilolite: A new additive. *In* Kiricsi I, Pál-Borbély G, Nagy JB, Karge HG (eds) Stud Surf Sci Catal 125:699-706

Kalló D. (2000) Utilization of zeolites in environmental protection. *In* Memmi I, Hunziker JC, Panichi C (eds) A Geochemical and Mineralogical Approach to Environmental Protection. Univ Siena, Italy, p 33-58

Kang SJ (1989) Characterization of ammonium and zinc(2+) ion adsorption by Korean natural zeolites. Han'guk Nonghwa Hackhoechi 32:386-392

Kesraoui-Ouki S, Cheeseman CR, Perry R (1994) Natural zeolite utilization in pollution control: A review to applications to metal effluents. J Tech Biotechnol 59:121-126

Kiss J, Hosszú Á, Deák B, Kalló D, Papp J, Mészáros-Kis Á, Mucsy G, Oláh J, Urbányi G, Gál T, Apró I, Czepek G, Töröcsik F, Lovas A (1990) Process and equipment for removal of suspended material, biogenetic nutrients and dissolved metal compounds from sewage contaminated with organic and/or inorganic substances. European Patent #177,543, 10 p

Komarowski S, Yu Q, Jones P, McDougall A (1994) Removal of nutrients from secondary treated wastewater effluent using natural zeolite. *In* Pilkington NH, Norman H, Bayly RC (eds) Proc Aust Conf Biol Nutr Removal Wastewater, 2nd. Aust Water Wastewater Assoc, Artarmon, Australia, p 415-423

Korobchanskii VI, Grebennikova SS, Dobrovol'skaya LE, Koval IV, Novodvorskii AV (1987) Additional removal of ammonia from wastewater by ion exchange. Koks Khim 7:35-39

Kravchenko VA, Korostyshevskii AS, Kravchenko ND, Kozlovskaya VI, Baranov AI, Ovdei MN (1994) Self-contained devices for purification of drinking water. Vodosnabzh Sanit Tekh 5:31-32

Kravchenko VA, Kravchenko ND, Rudenko GG, Tarasevich Yu I (1990) Defluoration of natural waters with the aid of clinoptilolite. Khim Tekhnol Vody 12:647-649

Kurita Water Industries Ltd. (1985) Method of phosphate removal. Jpn Kokai Tokkyo Koho, JP #60 78,692, 5 p; #60 82,188, 6 p

Kvopkova O, Zilincik M, Drnec M (1988) Flocculants for wastewater treatment. Czech patent CS #256,080, 3 p

Liberti L, Boari G, Passino R (1984) Method for removing and recovering nutrients from wastewater: European Patent #114,038, 18 p; US Patent #4,477,355

Liberti L, Laricchinta A, Lopez A, Passino R (1987) The RIM-NUT process at West Bari for removal of nutrients from wastewater: Second demonstration. Res Conserv 15:95-111

Liberti L, Limoni N, Longobardi C, Lopez A, Passino R (1988) Field demonstration of the RIM-NUT process for nutrients recovery from municipal wastewater. Nucl Chem Waste Manage 8:83-86

Liberti L, Limoni N, Lopez A, Passino R (1986a) The RIM-NUT process at West Bari for removal of nutrients from wastewater: First demonstration. Res Conserv 12:125-136

Liberti L, Limoni N, Lopez A, Passino R (1986b) The 10 m^3 h^{-1} RIM-NUT demonstration plant at West Bari for removing and recovering N and P from wastewater. Water Res 20:735-739

Linevich SN, Sinev IO, Yablon'ko NA, Tret'yachenko VV, Tret'yachenko DV (1990) Water treatment technology for Borshchev city water supply. Vodosnabzh Sanit Tekh 8:18-20

Linne SR, Semmens MJ (1985) Studies on the ammonium removal and filtration performance and regeneration of clinoptilolite. *In* Proc Indus Waste Conf, 39th, p 757-770

Mamedov IA, Ibragimov ChSh, Muradova NM (1985) Regression model for removal of mercury from chlorine production wastewaters with natural clinoptilolite of Aidag area of Azerbaijan. Azerb Khim Zh 2:135-138

McNair DR, Sims RC, Sorensen DL, Hulbert MH (1987) Schmutzdecke characterization of clinoptilolite amended slow sand filtration. J Am Water Works Assoc 79:74-81

Mercer BW Jr, Ames LL (1978) Zeolite ion exchange in radioactive and municipal wastewater treatment. *In* Sand LB, Mumpton FA (eds) Natural Zeolites: Occurrence, Properties, Use, Pergamon, Oxford, UK, p 451-462

Metropoulos K, Maliou E, Loizidou M, Spyrellis N (1993) Comparative studies between synthetic and natural zeolites for ammonium uptake. J Environ Sci Health, Part A, A28:1507-15018

Misaelides P, Godelitsas A (1995) Removal of heavy metals from aqueous solutions using pretreated natural zeolitic materials: the case of mercury. Toxicol Environ Chem 51:21-29

Mori K, Tsuneyoshi K (1986) Soil treatment method. Japan Kokai Tokyo Koho, JP #61,264,087, 4 p

Mucsy G (1992) Grosstechnischer Einsatz von Zeolith auf Kläranlagen in Ungarn. Abwassertech 43:48-54

Mumaw L, Bruin W, Nightingale J (1981) Evaluation of a recirculating freshwater salmon rearing facility using clinoptilolite for ammonia removal. J World Maricul Soc 12:40-47

Murphy CB, Hrycyk O, Gleason WT (1978) Natural zeolite: Novel uses and regeneration in wastewater treatment. *In* Sand LB, Mumpton FA (eds) Natural zeolites: Occurrence, Properties, Use. Pergamon, Oxford, UK, p 471-478

Neveu A, Gaspard M, Blanchard G, Martin G (1985) Intracrystalline diffusion of ions in clinoptilolite as applied to sodium and ammonium ions. Water Res 19:611-618

Nikashina VA, Zaborskaya E Yu (1977) Equilibria and kinetical characteristics of clinoptilolite during the selective extraction of ions from aqueous solutions. *In* Tsitsishvili GV, Andronikashvili TG, Krupennikova AYu (eds) Proc Symp Clinoptilolite, Mecniereba, Tbilisi, Georgia, p 109-112

Oláh J, Papp J, Kalló D (1991) Upgrading the efficiency of biological sewage treatment by using zeolites. Hidrológiai Közlöny (in Hungarian) 71:70-76

Oláh J, Papp J, Mészáros-Kis Á, Mucsi G, Kalló D (1988) Removal of suspended solids, phosphate and ammonium-ions from communal sewage using clinoptilolite derivatives. *In* Kalló D, Sherry HS (eds) Occurrence, properties and utilization of natural zeolites, Akad Kiadó, Budapest, Hungary, p 511-520

Oláh J, Papp J, Mészáros-Kis Á, Mucsy G, Kalló D (1989) Simultaneous separation of suspended solids, ammonium and phosphate ions from wastewater by modified clinoptilolite. *In* Karge HG, Weitkamp J (eds) Stud Surf Sci Catal 46:711-719

Otterstedt JE, Schoeman B, Sterte J (1989) Effective zeolites remove ammonia from sewage. Kem Tidskr 101(13):49-56

Papp J (1992) Einsatzmöglichkeiten von Zeolith in der Abwassertechnik. Abwassertech 43:44-47

Passaglia E, Azzolini S (1994) Italian zeolites in wastewater purification: influence of zeolite exchangeable cations on NH_4 removal from swine sewage. Mater Eng (Modena, Italy) 5:343-355

Polyák FK, Hlavay J, Maixner J (1995) Surface properties of MnO_2 adsorbent prepared from clinoptilolite-rich tuff from Tokaj, Hungary. *In* Ming DW, Mumpton FA (eds) Natural Zeolites '93: Occurrence, Properties, Use. Int'l Comm Natural Zeolites, Brockport, New York, p 365-395

Preston KT, Alleman JE (1994) Co-immobilization of nitrifying bacteria and clinoptilolite for enhanced control of nitrification. *In* Proc Indus Waste Conf, 48th, p 407-412

Ray JM, Rogers SE, Lauer WC (1985) Denver's potable water reuse demonstration project: Instrument and control system. *In* Drake RAR (ed) Proc 4th IAWPRC Workshop. Pergamon, New York, p 489-496

Rogers SE, Peterson DL, Lauer WC (1987) Organic contaminants removal for potable reuse. J Water Pollut Control Fed 59:722-732

Rustamov SM, Makhmudov FT (1988) Concentration of silver and nickel ions from wastes on sodium-clinoptilolite. Zh Prikl Khim (Leningrad) 61:34-37.

Rustamov SM, Makhmudov FT, Bashirova ZZ, Zeinalova II (1991) Concentration of nonferrous metal ions from industrial liquid waste on clinoptilolite. Khim Tekhnol Vody 13: 851-853

Rustamov SM, Yagubov AI, Bashirova ZZ (1992) Adsorptive removal of organochloro compounds from wastewaters by modified zeolites. Zh Prikl Khim (St. Petersburg) 65:2716-2721

Sanchez E, Roque-Malherbe R (1987) Zeolite as support material in anaerobic wastewater treatment. Biotechnol Lett 9:671-672

Semmens MJ, Martin WP (1980) Studies on heavy metal removal from saline waters by clinoptilolite. AIChE Symposium Series 76:367-376. American Institute of Chemical Engineers, Washington, DC

Semmens MJ, Martin WP (1988) The influence of pretreatment on the capacity and selectivity of clinoptilolite for metal removal. Water Res 22:537-542

Semmens MJ, Seyfarth M (1978) The selectivity of clinoptilolite for certain heavy metals. *In* Sand LB, Mumpton FA (eds) Natural zeolites: Occurrence, properties, use, Pergamon, Oxford, UK, p 517-526

Semmens MJ, Wang JT, Booth AC (1977) Nitrogen removal by ion exchange: Biological regeneration of clinoptilolite. J Water Poll Cont Fed 49:2431-2444

Senyavin MM, Nikashina VA, Tyurina VA, Antonova OYa, Khristianova LA (1986a) Ion exchange and filtering properties of natural clinoptilolite in pilot plant. Khim Tekhnol Vody 8:49-51

Senyavin MM, Nikashina VA, Tyurina VA, Elenin SN, Ishchenko IG (1986b) Industrial tests of natural clinoptilolite. Khim Tekhnol Vody 8(6):52-56

Sherman JD (1978) Ion exchange separations with molecular sieve zeolites: *In* Adsorption and Ion Exchange Separations. Am Inst Chem Engin Symp Series 74(179), American Institute of Chemical Engineers, Washington, DC, p 98-116

Szymansky HA, Stamires DN, Lynch GR (1960) Infrared spectra of water sorbed on synthetic zeolites. J Optical Soc Am 50:1323-1328

Takasaka A, Inaba H, Matsuda Y (1991) Removal of cations from solution using Itaya zeolite. Nippon Kagaku Kaishi (5):618-622

Tarasevich YuI (1981) Natural sorbents in water treatment processes (in Russian). Nauk Domka, Kiev, Ukraine

Tarasevich YuI (1994) Natural, modified, and semisynthetic sorbents in water-treatment processes. Khim Tekhnol Vody 16:626-640

Torii K (1978) Utilization of natural zeolites in Japan. *In* Sand LB, Mumpton FA (eds) Natural Zeolites. Occurrence, Properties, Use. Pergamon, Oxford, UK, p 441-450

Weber MA, Barbarick KA, Westfall DG (1983) Application of clinoptilolite to soil amended with municipal sludge. *In* Pond WG, Mumpton FA (eds) Zeo-Agriculture: Use of natural zeolites in agriculture and aquaculture, Westview Press, Boulder, Colorado, USA, p 263-271

White DA, Franklin G, Bratt G, Byrne M (1995) Removal of manganese from drinking water using natural and modified clinoptilolite. Process Saf Environ Prot 73:239-242

Williford CW Jr, Reynolds WR (1992) Clinoptilolite removal of ammonia from simulated and natural catfish pond waters. Appl Clay Sci 6:277-291

Xu G (1990) Use of natural zeolite for ammonia removal in distilled water production. Shuichuli Jishu 16:456-459

Zaitsev VN, Kadenko IN, Vasilik LS, Oleinik VD (1995) The natural zeolite clinoptilolite as an adsorbent for removal a radionuclides and heavy metal salts. Izv Vyssh Uchebn Zaved, Khim Khim Tekhnol 38:40-45

Zamzov MJ, Eichbaum BR (1990) Removal of heavy metals and other cations from wastewater using zeolites. Separation Sci Tech 25:1555-1569

Zubály Z, Zubály Zné, Zubály E, Zubály Z Jr (1991) Installation for cleaning and removal of municipal, industrial or agricultural wastewaters especially of liquid manure in cascade system, Hungarian Patent #203707, 3 p

16 Use of Zeolitic Tuff in the Building Industry

Carmine Colella
Dipartimento di Ingegneria dei Materiali e della Produzione
Università Federico II
Piazzale V. Tecchio 80, 80125 Napoli, Italy

Maurizio de' Gennaro
Dipartimento di Scienze della Terra
Università Federico II
Via Mezzocannone 8, 80134 Napoli, Italy

Rosario Aiello
Dipartimento di Ingegneria Chimica e Materiali
Università della Calabria
87030 Rende (CS), Italy

INTRODUCTION

Zeolite-rich volcanic tuffs are widely distributed in almost every country of the world, where they are present in low-, medium-, or high-grade million-ton deposits. Even though the formation of the zeolite minerals may have followed different genetic paths, the zeolitic rocks have in common a matrix of finely crystalline zeolite that cements the other non-zeolitic particles and is responsible for the overall mechanical properties of the material. Zeolitic tuffs have been employed since pre-historic times in construction, mostly as dimension stone. This use is still the most common of natural zeolites in the building industry, although other applications have recently come to the forefront, such as lightweight aggregate or as additives for manufacturing blended cements.

Given the fact that much of this material is excavated and used locally and that the market demand is strongly affected by the trends of the building industry, estimates of the worldwide zeolitic tuff production for construction purposes are difficult to make. In Italy, where the use of zeolitic tuff as dimension stone is commonplace, about 75 quarries were reported in operation in 1992 with total production of about 3×10^6 tons per year (Aiello 1995). More recently, this production has decreased, due to the crisis of the building industry, and in 1998 it amounted to about 1.5×10^6 tons per year. In Japan, the production of tuff as dimension stone is currently about 4×10^5 tons per year (N. Kuchitsu, National Research Institute of Cultural Properties, Tokyo, Japan, pers. comm., 1996). Considering that zeolitic tuff is used as dimension stone in many other countries as well, e.g. Bulgaria, Cuba, Germany, Greece, Hungary, Mexico, Romania, and Turkey, the current worldwide zeolitic tuff consumption as dimension stone is about 3×10^6 tons per year.

Information is even more scarce on the use of natural zeolites as additions to cement, because the recourse to zeolitic tuff as a pozzolanic material is rather recent and is widespread only in those locations in which other pozzolanic materials, such as natural pozzolan, fly ash, silica fume, etc., are not readily available. The use of zeolitic tuffs as cement additives is common in Bulgaria, China, Cuba, Germany, Jordan, Russia, Turkey, United States, and Yugoslavia. It is of interest to note that in China, in 1989, about 7×10^7 tons of blended cement, about one third of the total cement production of the country, was manufactured with 10 to 30 wt % natural zeolite (Kasai et al. 1992). Mumpton (1996) also reported that in Serbia (Yugoslavia) about 10^5 tons per year of clinoptilolite-rich tuff are

1529-6466/00/0045-0016$05.00

mined for the production of pozzolanic cements. In 1991, the estimated consumption of local zeolitic tuff in Cuba for the preparation of pozzolanic cement and of *cemento romano* (Roman cement), which is a mixture of lime and zeolitic tuff, was about 3.7×10^5 tons (G. Rodriguez-Fuentes, Institute of Materials and Reagents, Faculty of Physics, University of Havana, Cuba, pers. comm., 1997). Furthermore, about 1×10^4 tons per year of clinoptilolite-rich tuff are currently used in two cement plants in Kuznetsk and Jakutsk (eastern Russia) for manufacturing blended cements (G.I. Ovcharenko, Altay Technical University, Barnaul, Russia, pers. comm., 1996). Therefore, a prudent estimate of the worldwide zeolitic tuff consumption in the cement industry is at least one order of magnitude greater than that estimated for dimension stone production (*vide supra*).

In all the uses related to constructions, either as cemented material or as powder, the zeolite in the tuff plays a fundamental role. It is the primary cementing agent in the rock; it is the adsorbent that controls the humidity and thermal levels inside excavations used as dwellings, deposits, and cellars; and it is the acidic reagent that in blended cements reacts with the lime released by Portland cement during hydration. Nevertheless, in construction applications, more than in other applications, the zeolitic tuff must be regarded in terms of its mineralogical, chemical, and macroscopic properties rather than simply as a poorly concentrated natural zeolite. The action of the tuff is by no means dependent only on the presence of zeolite, but it is the result of the synergy between the various other tuff constituents and depends also on the way they are assembled together (i.e. the tuff macroporosity, texture, and degree of induration).

HISTORICAL DEVELOPMENT OF ZEOLITIC TUFF USE IN CONSTRUCTION

Historical use of zeolitic tuff in Italy

The early times. The use of zeolitic tuff as a construction material has ancient origins in Italy. Tuffaceous rocks were frequently considered in ancient times as sites for human settlements. Ancient populations apparently appreciated the topographic and morphologic features of the volcanic formations, including those that were lithified (tuffs) and features such as steep walls, ridges, and plates which were often located in a topographically dominant position. Moreover they were attracted by the useful properties of the rock. For example, the rocks' reasonable resistance to erosion but with associated softness made them ideal for excavations for obtaining caves for houses, refuges (for instance the catacombs, excavated and inhabited by the early Christians in the Naples and Rome areas to escape the persecutions of the Romans), stables, cellars, and, locally, as underground roads. A fabulous account of this practice was transmitted by Homer's Odyssey, in which he reported a mysterious population, *Cimmerii*, who lived in excavated underground houses and communicated with each other by tunnels. According to the Roman historian Livy (*Ab Urbe condita* [From the foundation of Rome], III, 5) and the Greek geographer Strabo (*Geografica* [Geographical sketches] V, 241-245), the *Cimmerii's* town was located near the old Greek town of Cuma in the vicinity of Naples, where the geology is dominated by zeolitic tuff. The *Cimmerii's* legend, according to a modern interpretation (Annecchino 1960) is evidence of the very old practice of the inhabitants of the Phlegraean Fields near Naples of excavating caves in the tuff for living quarters, for producing short underground connecting roads, or for obtaining building material. The softness of the tuff favored its ancient use for architectural purposes or for carving memorial sculptures, including bas-reliefs and columns carved into phillipsite-rich Neapolitan yellow tuff in a few *hypogeums* (underground burial chambers) in Naples during the Greek period (4th to 3rd centuries B.C.) (Cardone 1990) and the memorial sculptures of *Matres* ["Mothers"] (Fig. 1) carved in chabazite-rich Campanian ignimbrite in the Caserta area (north of Naples) in 8th century B.C. (Di Girolamo 1968; de' Gennaro et al. 1984).

Living with the tuff. In south-central Italy, the role played by zeolitic tuff in the choice of sites of many cities and villages is undeniable (Scherillo 1977). Indeed, early Rome was founded in the 8th century B.C. on two hills, Campidoglio and Palatino, both of which are composed of zeolite-rich tuffaceous material. At about the same time (8th-6th centuries B.C.), the first Greek colonies in Italy, e.g. *Pithekoussai* (now Ischia), *Kyme* (now Cuma), *Dicaearchia* (now Pozzuoli), *Palaepolis* (the oldest settlement, from which Naples was born), were all founded on zeolitic tuffaceous hills or plates. Also, the Etruscans inhabited the zeolitic tuff hills in the volcanic area situated between the low Tuscany and the high Latium. There they founded many villages (e.g. Veio, Cerveteri, Pitigliano, and Chiusi) and excavated burial places for their dead (e.g. the famous necropolis of Tarquinia). But it was the ancient Romans who became specialists in the handling of volcanic zeolitic tuffs. They constructed many civil engineering works utilizing tuff. These applications included excavations in tuffaceous hills to construct roads. Two hill cuts are still preserved in the traces of the old *Via Domitiana* (Fig. 2a) and *Via Campana*, in the Phlegraean Fields. They constructed subterranean aqueducts and excavated underground passages that were locally notable for their impressive size (e.g. the *crypta romana* [Roman cave], which reaches a height of 23 m), pleasantness (e.g. the so-called "Antro della Sibilla Cumana" [Cuman Sybil's cave]), or for their audacity (e.g. the so-called "Grotta di Cocceio" [Cocceius' cave] in the Pozzuoli area, which is more than 1-km long and named after the Roman architect L. Cocceius Aucto, who also constructed other *cryptae* in the Naples area in thick zeolitic tuff formations; Fig. 2b) (Greco 1994). Even in ancient time, they exploited the favorable physical and chemical features of the tuff, employing the zeolitic tuff extensively as dimension stone and as aggregate in concrete.

Figure 1. Sculpture made of Campanian ignimbrite (8th century B.C.). From the collection of *Matres* [Mothers] in the Museo Provinciale Campano of Capua (Caserta, Italy). There are tens of these sculptures in the Museum and some of these are made of chabazite-rich ignimbrite.

The Roman times. The historical use of zeolitic tuff as dimension stone in southern Italy dates back 2700 years. The Greek colonizers used tuff as building blocks for constructing various masonry structures, houses and burial chambers. Unfortunately, few of these works remain in Naples, mainly because the Roman city was built over the ruins of the Greek city. The Romans used the tuff blocks for all kinds of construction and to meet both architectural and structural requirements (Cardone 1990).

The successful use of the tuff as dimension stone was enhanced by the introduction of a new type of cement, perhaps not invented by, but certainly improved by the Romans (namely, hydraulic mortar). This type of mortar was composed of lime mixed with a volcanic unlithified material called *pozzolan*, which was very common in the Neapolitan and Roman areas, and as we know now was the precursor of the zeolitized tuff, at least in the case of the Neapolitan yellow tuff (Scherillo 1955). A high-strength concrete (*opus*

Figure 2. Uses of zeolitic tuff in the Roman times in Italy: (a) view of the old *Via Domitiana* (1st century A.D.) through the excavation of Mount Grillo (height = 20 m) (Pozzuoli, Naples); visible is the so-called "Arco Felice", a support structure made of zeolitic tuff blocks and clay bricks, which served also as a bridge between the two sides of the hill (courtesy of E. Casale, Naples, Italy); (b) entrance of the so-called *Crypta neapolitana* [Neapolitan cave] connecting the old cities of Naples and Pozzuoli; it is a 700-m long tunnel excavated into the tuff hill at the end of the 1st century B.C.

caementicium), which was useful for the construction of large hydraulic projects (e.g. aqueducts, dams, harbors, bridges), was prepared by mixing lime, water, pozzolan, and/or powdered clay brick and pieces of zeolitic tuff (*caementa*). Many important facilities were constructed with this cement in Italy and in other provinces of the Roman Empire during the 1st and 2nd centuries A.D., such as the ports of Ostia, Pozzuoli, Civitavecchia, and Anzio; four bridges over the river Tiber; the aqueduct over the River Gard (Nimes, France); and the aqueduct of Segovia, Spain. According to the Roman architect Vitruvius, author of the fundamental work "De Architectura" [On Architecture], durable concretes can be prepared using crushed and graded zeolitic tuff as aggregate. This particularly favorable behavior was related probably to the presence of zeolites in the tuff and to their capability to act as a pozzolanic material. Examples of this practice are Hadrian's Villa in Tivoli near Rome, the Pantheon in Rome, and some amphitheaters in Rome and in other Italian provinces.

From the Middle Ages to the present time. The building techniques developed by the Romans were utilized almost unchanged during the first millennium and beyond. For example, the bridges of the *Acquedotto Carolino* [Charles III Bourbon Aqueduct] in Valle di Maddaloni near the Caserta Royal Palace were built with *opus mixtum* (tuff blocks, either zeolitized or unzeolitized, intercalated with clay brick lines) by the architect Luigi Vanvitelli in the second half of the 1700s. The use of zeolitic tuff in the Middle Ages became so extensive that entire villages were constructed using this stone, e.g. all the houses in the villages of Pitigliano, Sovana, and Bagnoregio and in the city of Orvieto in central Italy are made of chabazite-rich tuffs (Langella and Adabbo 1994).

During Roman times, the practice of using tuff blocks as "bare" stone was uncommon. The tuff walls were usually covered by plaster or decorated with marble and other valuable stones. Beginning in the Middle Ages, frequent use was made of tuff as a facing stone in recognition of its architectural value; thus considerable attention was paid to the cutting of the stones and to their dimensions. Examples of the use of bare blocks are numerous; monumental buildings constructed throughout Naples over the centuries and still preserved include castles (e.g. Castel dell'Ovo, 12th century, Figure 3a; Castel Nuovo, 13th century, Fig. 3b; Castel S. Elmo, 16th century) and churches (e.g. S. Domenico Maggiore, 13th century, Fig. 3c; Santa Chiara, 14th century, Fig. 3d). Today zeolitic tuffs are widely utilized in Italy for masonry and the external walls of houses, even though new building technologies (structures in concrete) and new building materials (brick made with cement and light-weight aggregates) are progressively reducing this use.

Uses of zeolitic tuff elsewhere

The use of tuff as dimension stone in other countries has not been as widespread as in Italy. This may depend partly on our lack of information about the zeolitic nature of individual building materials, but it is probably due primarily to the availability of other construction materials that better fulfilled the local economic requirements. The following is a survey of historic and/or present-day use of zeolitic tuff in construction in several countries.

Figure 3 (left). Historical monuments in Naples made with Neapolitan yellow tuff: (a) Castel dell'Ovo (12th century); (b) Castel Nuovo (Maschio Angioino) (13th century); (c) Basilica of S. Domenico Maggiore (13th century); (d) Basilica of Santa Chiara (14th century).

Figure 4. The Natural Science Museum in Kurdzali (northeastern Rhodopes, Bulgaria) that was made of clinoptilolite-rich tuff blocks.

Bulgaria. The clinoptilolite-, phillipsite- or analcime-rich tuffs of Srednogorian and clinoptilolite-rich tuffs from the northeastern Rhodopes (e.g. Scalna Glava village) are the zeolitized materials most widely used in Bulgaria. The oldest quarries, active from the 18th century, are near the Greek border. Djourova and Milakovska-Vergilova (1996) and M. Stamatakis (National University of Athens, pers. comm., 1997) reported that Bulgarian tuffstone cutters spread the use of these materials as dimension stone and also as carved ornamental stone. A massive, marine, clinoptilolite-rich tuff near Kurdzali (eastern Rhodopes) is nearly 100 m thick and has been mined as a pozzolan for many years. The same tuff also provided blocks for local construction (Mumpton 1996) (Fig. 4).

China. Although no information is available on the use of the local zeolitic tuffs as dimension stones, a considerable volume of these materials is known to be used as a pozzolan in the manufacture of blended cements in China (see the INTRODUCTION).

Cuba. Buildings made with natural zeolite rocks, about 150-200 years old, are reported to be in the Province of Las Villas, 350 km east of Havana (R. Roque-Malherbe, Instituto de Tecnologia Quimica, Valencia, Spain, pers. comm., 1996). The production of blended Portland cements, containing 10 to 20% clinoptilolite-rich tuff from deposits in the central part of the country, began during the 1980s (Urrutia Rodriguez and Gener Rizo 1991). Large amounts of clinoptilolite-rich tuff are used for manufacturing a low-cost hydraulic binder that is a Roman cement (a mixture of zeolitic rock and CaO) (Velazquez and Rabilero 1991), and for preparing lightweight aggregates for high-strength mortars and concretes (Gayoso Blanco et al. 1991).

Georgia. I. Akhvlediani (Department of Mineralogy and Petrology, Georgian Technical University, Tbilisi, Republic of Georgia, pers. comm., 1997) reports that the Hotel Iveria was built in Tbilisi during the 1970s from beautiful green analcime-rich tuffstones.

Germany. After Italy, Germany has the oldest and best-established tradition in the use of zeolitic tuff in the building industry. Rhenish tuff, locally called *trass*, has been used as a building stone in Germany and neighboring countries since ancient times. Two main types of *trass* have been recognized. One type, rich in chabazite and phillipsite and locally containing analcime, is quarried in the localities of Kruft, Kretz, and Plaidt in the volcanic region of the Eifel; the other type, rich in chabazite and locally called *leuzittuff* (tuff rich in leucite), crops out in the same region near Rieden (Sersale and Aiello 1964; Sersale et al. 1965; de' Gennaro et al. 1995). Whereas the *leuzittuff* has been used and is still used as a dimension stone, the chabazite-phillipsite tuff has been utilized since Roman times also for producing hydraulic mortars. Sersale and Aiello (1964) reported that this type of mortar was used for the construction of the Trier-Cologne aqueduct. The manufacturing of hydraulic mortars has a remarkable history beginning in the 17th century; however, only since 1929 has the *trass* been regularly utilized as an addition for blended cement (so-called *trasszemente*) (M. Tax, Heidelberger Zement AG, Heidelberg, Germany, pers. comm., 1997). Apart from traditional local use in houses (Fig. 5) which is still active, *trass* has been used for the construction of buildings since the Middle Ages. More than 200 monumental buildings were built in Germany using bare blocks of tuff. The greatest concentration is in Cologne, where 34 monuments and churches made of Rhenish *trass* have been recognized; among them are the Gross-St. Martin, Gürzenich, and St. Gereon (B. Fitzner, Geologisches Institut der Rheinisch-Westfälischen Technischen Hochschule, Aachen, Germany, pers. comm., 1996). Of remarkable architectural interest also are the S. Martin Minster in Bonn (Fig. 6) and the S. Quirinus Minster in Neuss (Fitzner and Lehners 1990).

Figure 5 (above). House made of zeolitic tuff blocks in Rieden, Germany (18th century).

Figure 6 (right). The S. Martin Minster in Bonn, Germany that was constructed with zeolitic tuff (12th–13th century).

Greece. Information on historical use of zeolitic tuff in Greece is scarce, but ancient populations likely had some use of tuff. For example, it is reported that some old houses on the southern end of the island of Milos were built of the local greenish-yellow zeolitic tuff (M. Stamatakis, National University of Athens, Greece, pers. comm., 1998). More recently, the use of the zeolitic tuff as dimension stone is rather common in northeastern Greece (Thrace) near the villages of Petrota and Metaxades. Here, extensive deposits of clinoptilolite- or mordenite-rich material are known to occur (Stamatakis et al. 1996). Zeolitic tuff has been used for constructing houses of Metaxades (Figs. 7 and 8) and five other villages in the area since about 1750. Other reported uses are in the construction of stables, irrigation channels, and dikes against the floods of the river Evros (A. Filippidis, Department of Mineralogy, Petrology, and Economic Geology, Aristotle University, Thessaloniki, Greece, pers. comm., 1996).

Hungary. The use of the local rhyodacitic tuff in the building industry in Hungary is widespread and dates back to the Middle Ages. Clinoptilolite- and mordenite-rich tuffs are quarried in the oenologically famous Tokaj Hills, in the northeastern part of the country. Here, large wine-cellars are carved into zeolitic tuff bodies several hundred meters thick.

Figure 7. A modern house built using clinoptilolite-rich tuff blocks in Metaxades, northeastern Greece.

Figure 8. A modern church built using clinoptilolite-rich tuff blocks in Metaxades, northeastern Greece.

The porosity of the cellar walls ensures some aeration, and humidity and temperature are stable. In this area, almost all the houses are built with zeolitic tuff cut in brick form (D. Kalló, Central Research Institute for Chemistry, Hungarian Academy of Sciences, Budapest, Hungary, pers. comm., 1995).

Japan. The use of zeolite tuff in construction in Japan is not as common as in Europe. Only the *Oya-ishi* ["stone of Oya"] tuff (Utsunomiya City, Tochigi Prefecture), which contains clinoptilolite and mordenite, has been employed in the building industry. The outcrop area is located 100 km north of Tokyo. Exploitation of this deposit began in the 5th century A.D. with the construction of stone chambers of tombs. A Buddhist temple was erected in this area in the 8th century using the *Oya-ishi* tuff, and statues of religious relevance and of very high artistic value were carved into the tuff in a cave owned by the Buddhist temple. The tuff was used in the 11th and 14th centuries for constructing a citadel and a Buddhist pagoda, respectively. Large-scale mining of the *Oya-ishi* tuff began in the 17th century, and the material was used in the 19th century for the construction of the Teikoku Hotel in Tokyo. A 50 m-high statue of Buddha was carved into the tuff some ten years ago in Utsunomiya City (Fig. 9). This statue has no particular religious background and was carved as a peace monument. Other much smaller statues, having very high artistic value, were carved about one thousand years ago in a cave owned by a Buddhist temple (Oya-Ji) in Utsunomiya City

National Research Institute of Cultural Properties, Tokyo, Japan, pers. comm., 1996).

Jordan. The zeolitized quaternary volcanic deposits of the Aritayn region in the northeastern area of Jordan are rich in phillipsite, chabazite, faujasite, and clay minerals. The tuffs are currently being used in Tell Rimah as a cement additive (Ibrahim and Hall 1996).

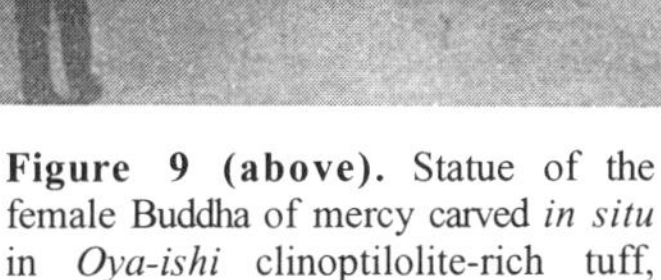

Figure 9 (above). Statue of the female Buddha of mercy carved *in situ* in *Oya-ishi* clinoptilolite-rich tuff, Oya, Japan (Mumpton 1988).

Figure 10 (right). Examples of the use of zeolitic tuff in Mexico: (a) bas-relief carved into zeolitic tuff ca. 2000 B.C., Mitla, Oaxaca (Mumpton 1988); (b) cathedral in Oaxaca that was constructed of zeolitic tuff quarried in the vicinity (Mumpton 1996).

Mexico. The use of zeolitic tuff in architectural decorations in Mexico has a very long history. Bas-reliefs, carved into zeolitic tuff by Zapotec Indians, ca. 2000 B.C., are still preserved at Mitla, Oaxaca, Mexico (Mumpton 1988) (Fig. 10a). Moreover many of the buildings of the Zapotec centers at Mitla and Monte Alban in southern Mexico and the old Spanish colonial city of Oaxaca are constructed of blocks of massive, green clinoptilolite- and mordenite-rich tuff, locally called *cantera verde* [green stone] (Grissom 1994; P. Bosch Giral, Departemento de Quimica, Universidad Autonoma Metropolitana-Iztapalapa, Mexico, pers. comm., 1997). This tuff is still being quarried today in the state of Oaxaca (e.g. at Etla, Tecoatlan, and Tejupan) for local use as dimension stone in walls, buildings (Fig. 10b), and other structures (Mumpton 1996).

Romania. The volcanic tuffs from Transylvania have been used by man since prehistoric times. Some Neolithic tools and others tools of the Roman times, made of zeolitic tuff, have been found in the surroundings of Gherla (northeast of Cluj) and in other villages in the area. Also, the city walls of the ancient city of Gelu (10th century), near the village of Dãbîca (Cluj area), are built of tuff (Fig. 11a). Many other masonry constructions and buildings constructed of tuff are widespread around Cluj, e.g. the Corneni Castle built in 1862 (Fig. 11b), a datcha (countryside house) and a farm in the locality of Pîglisa (Figs. 11c and 11d), and the war memorial of Luna de Jos (Bãrbat and Marton 1989; Bãrbat et al. 1991).

Russia. No information is at present available on the use of zeolitic tuff as a dimension stone in Russia. However, activity has been reported in the field of lightweight aggregates or in the manufacture of blended cements at the cement plants of Kuznetsk and Jakutsk (G.I. Ovcharenko, Altay Technical University, Barnaul, Russia, pers. comm., 1996).

Slovenia. Several thousand tons of clinoptilolite-rich tuff are quarried in northern Slovenia each month for use as a pozzolan near Celje and Zaloska Gorica and as a source of lightweight aggregate (Mumpton 1983).

Spain. Important deposits of phillipsite- and chabazite-rich tuffs have been recognized in the Canary Islands (Adabbo et al. 1991). The material is utilized for manufacturing pozzolanic cements, e.g. by the company Cementos Especiales de Las Palmas de Gran Canaria.

Turkey. Numerous Miocene tuffs of Anatolia have been altered by secondary mineralization to clinoptilolite and, to a less extent, erionite, chabazite, and phillipsite (Temel and Gündogdu 1996). The ignimbrites of Cappadocia (central Anatolia) are largely used locally as construction material, but the unaltered or slightly altered parts of these pyroclastic rocks are preferred over zeolitic tuffs because of their lower hardness (M.N. Gündogdu, Department of Geology, Hacettepe University, Beytepe-Ankara, Turkey, pers. comm., 1996). A caravanserai (inn) in the area of Konya (southwestern Cappadocia) is shown in Figure 12a. The inn was made with a local ignimbrite, either the unzeolitized gray *facies*, or the zeolitized reddish *facies* (P. Di Girolamo, Dipartimento di Scienze della Terra, Università Federico II, Napoli, Italy, pers. comm., 1996). Most zeolitic tuffs of Neogene age in western Anatolia are locally used as lightweight dimension stones in small villages, for instance Karain (Fig. 12b) and Tuzköy, which are situated around tuff outcrops (Gündogdu et al. 1996).

United States. Despite the large number of zeolitic deposits, especially in the western states (see, e.g. Sheppard 1971), zeolitized tuffs have been utilized only sporadically for building materials during the past century. Early settlers in southeast Oregon and northern Nevada found that erionite-rich tuffs could be cut, sawed, and nailed, so the material was used as a local building stone (Sheppard 1996), for example, for constructing ranch houses (Fig. 13) and out buildings (Mumpton 1973). In the early 1980s, the only known zeolite deposit that was being quarried for dimension stone was located a few miles southwest of Rome, Oregon, where a few hundreds tons per year of erionite-rich tuff was cut into facing stone for local use (Mumpton 1983). At present the quarry is abandoned (R.A. Sheppard, U.S. Geological Survey, Denver, Colorado, pers. comm., 1996). The use of zeolitic tuff as an addition for blended cements has also been reported. For example, zeolitic tuff from Tehachapi, California, is a principal constituent of the pozzolanic cement manufactured by the Monolith Portland Cement Company, Monolith, California (Mumpton 1983).

Yugoslavia. Extensive deposits of zeolitized tuffs have been recognized in Serbia. At Zlatokop (near Vranje), clinoptilolite-rich tuff was formed by alteration of dacitic glassy volcaniclastites in a lacustrine environment (Obradovic and Dimitrijevic 1987; Obradovic and Jovanovic 1987). These deposits are not mined at present, but the exceptionally large

Figure 11. Examples of the use of zeolitic tuff in Romania (Bârbat and Marton 1989):
(a) city walls of the ancient city of Gelu (10th century), near the village of Dăbîca (Cluj area);
(b) Corneni Castle (1862), Cluj;
(c) a datcha (country-side home);
(d) a farm near Pîglisa.

Figure 12. Examples of the use of zeolitic tuff in Turkey: (a) a caravanserai (inn) in the area of Konya (SW Cappadocia) made with local ignimbrite blocks (courtesy of P. Di Girolamo, Dipartimento di Scienze della Terra, Università Federico II, Napoli, Italy, 1996); (b) village library in Karain that contains an occasional block made from erionite–rich tuff (Mumpton 1981).

Figure 13. Abandoned ranch house in Jersey Valley, Nevada, that was constructed of blocks quarried of erionite–rich tuff (Mumpton 1996).

deposit of clinoptilolite-rich tuff near Vranska Banja is extensively mined for use in the production of pozzolanic Portland cements (Mumpton 1996).

ZEOLITIC TUFF AS DIMENSION STONE

Lithification of zeolitic tuff

Zeolitized tuffs typically form by the diagenetic alteration of volcanic glass in loosely consolidated pyroclasts after deposition. Lithification results from the interaction between glass and interstitial fluids (e.g. meteoric water, hydromagmatic water, and seawater). The products of alteration processes include crystalline zeolites, feldspars, and opal-CT or non-crystalline aluminosilicate gels and hydrated iron oxides. The lithified rock commonly preserves the structural and textural features of the precursor pyroclastic material. The aggregated product usually is characterized by softness, high porosity, and low bulk density.

The lithification process is not well understood. Lithification can arise from two different processes: (1) zeolitization under diagenetic conditions, e.g. the John Day Formation of Oregon (Hay 1963), and (2) alteration of pyroclastic flows (possibly due to hot fluids), e.g. *tufo giallo napoletano* [Neapolitan yellow tuff] of the Phlegraean Fields in Italy (de' Gennaro et al. 1990, 1999; de' Gennaro and Langella 1996). In the first process, lithification is due to the interaction of volcanic glass with meteoric waters percolating through the formation. This is demonstrated by the fact that only the lower-middle levels of the succession, where the crystallization of clinoptilolite has occurred, are lithic; whereas the topmost material in the deposit remains substantially unaltered

Figure 14. Type 1 tuff samples (see the lithological section): (a) Neapolitan yellow tuff, Marano (Naples, Italy); (b) Rhenish *trass* (Eifel region, Germany); (c) Campanian ignimbrite, Comiziano (Naples, Italy). Scale in centimeters.

and, therefore, unlithified. In the second process, only the middle part of the formation, which apparently remained hot for a longer time, has been transformed into phillipsite-rich tuff (de' Gennaro et al. 1990, 1999; de' Gennaro and Langella 1996; Langella et al., this volume).

Tuff lithology

The alteration processes described above give rise to many kinds of tuffs having very different structural and textural features. Therefore it is impossible to contain in a schematic lithological classification all of the zeolitic tuffs that have been found in nature. An attempt is made here to identify and characterize the four main tuff types employed for the production of dimension stone.

Type 1. Type 1 includes the tuffs connected with recent volcanism in south-central and insular Italy, Eifel in Germany, Canary Islands (Spain), and the Central Massif in France (Fig. 14). The genetic mechanism in all these areas is connected to an alteration of pyroclastic flows (see the previous section). The resulting tuff resembles a conglomerate in which the pumiceous and lithic clasts are embedded in a yellow cindery matrix. The matrix and the pumice grains are commonly deeply zeolitized. The rock usually exhibits low unit weight, γ ($\gamma < 13$ kN/m^3), elevated porosity, P (P ~50%), and low compressive strength, σ ($\sigma \leq 6$ MPa). The tuffs associated with the Oligocene-Miocene volcanism of northern Sardinia display somewhat different properties (P ≤ 25%; $\gamma > 18$ kN/m^3; $\sigma > 25$ MPa) because in this case the mechanism responsible for the lithification involved hot, highly mineralized fluids (M. de' Gennaro, Department of Earth Sciences, University of Naples Federico II, Italy, pers. comm., 1998).

Figure 14, cont'd. Type 1 tuff samples: (d) Red tuff with black scoriae, Palombara (Viterbo, Italy); (e) Yellow tuff, Tenerife (Canary Islands, Spain); (f) Campanian ignimbrite, San Mango (Avellino, Italy). Scale in centimeters.

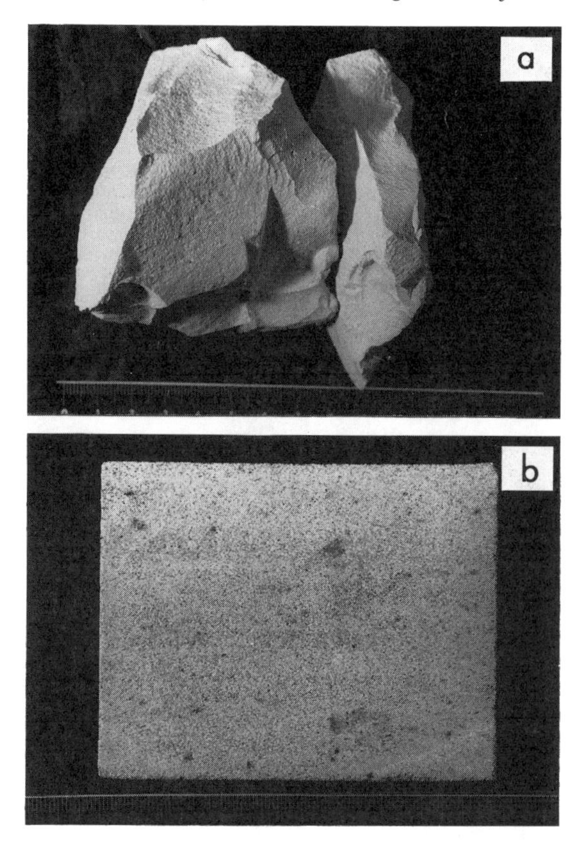

Figure 15. Type 2 tuff samples (see the lithological section) from European deposits: (a) Zlatokop (Serbia, Yugoslavia); (b) Metaxades (Thrace, Greece). Scale is in centimeters.

Type 2. Type-2 tuffs display a fine-grained matrix, they are usually very compact, and they generally contain millimeter-size lithic and pumiceous clasts. They are found in many areas of eastern Europe, e.g. Serbia (Yugoslavia), Thrace (northeastern Greece), Caucasus (Russia), and western and central Turkey (Fig. 15), and in a few localities in the United States, e.g. Nevada (Fig. 16a), and Australia (Gold Coast). Despite their apparent compactness, they have highly variable physical characteristics. Their porosity is between 20 and 45%, and the porosity of the tuff from Corrumbin, Gold Coast (Australia) is as low as 10%. The γ values are in the range 6-23 kN/m^3, and σ may reach values of 50 MPa or over.

Type 3. Type-3 tuffs are characterized by a fine-grained matrix, but they contain clasts having dimensions of several millimeters (typically 2-4 mm). Their genesis is connected to zeolitization under diagenetic conditions (see previous section). Porosity values are close to each other (P = 26-31%), γ ranges between 16 and 23 kN/m^3, and σ is less than 5 MPa. This type includes zeolitized tuffs that are widespread in Transylvania (Romania), the United States, e.g. in Oregon and Arizona (Figs. 16b and 16c), Oaxaca (Mexico), and Patagonia (Argentina).

Type 4. The characteristic features of Type-4 tuffs, defined as "pumiceous tuffs", is

Figure 16. Tuff samples from United States deposits: (a) Beatty, Nevada (type 2); (b) Christmas, Arizona (type 3); (c) Rome, Oregon (type 3); (d) Fish Creek Mountains, Nevada (type 4). Scale is in centimeters.

the centimeter size of the clasts and their abundance in the rock. Porosity is very variable (P = 16-40%), γ may exceed 15 kN/m^3, and σ values usually do not exceed 10 MPa. These tuffs are found in the United States, e.g. Nevada (Fig. 16d), and in Japan (Tochigi Prefecture).

Tuff mining techniques

Tuff mining in ancient times was undoubtedly initiated directly from the outcrop. Large, rough blocks were cut from the ridge, taking advantage of the presence of natural fractures such as columnar jointing. These blocks were successively split and hand shaped using suitable tools (e.g. hammer and chisel, axe). Mining of the Neapolitan yellow tuff in southern Italy was performed in galleries (i.e. underground tunnels) until open-pit mining became popular at the beginning of the 20th century (Dell'Erba 1923).

Mining in galleries. This technique has been utilized in the Naples area since ancient times. The *Cimmerii's* legend and the many *cryptae* found in the Phlegraean Fields (see the historical section) are probably evidence of this very old practice. Tuff mining in Naples was performed directly under the sites where buildings were constructed until the 18th century, so present-day Naples lies over a network of enormous excavations ("the underground city"). The excavations were commonly utilized as underground utilities (rain water reservoirs, cellars, etc.). Some of these caves were utilized in ancient times also as underground aqueducts and subterranean roads. Mining in galleries was utilized also when quarrying moved from the center to the periphery of the city because it allowed the preservation of surface soil for other human activities, e.g. agriculture. The entry of an old gallery is shown in Figure 17.

Open-pit mining. Open-pit mining is currently being used in Italy (Figs. 18a and 18b), Germany (Fig. 18c), Canary Islands (Fig. 18d), and Japan, where production of dimension stone is ongoing. Open-pit mining achieved a marked development with the introduction of circular saws, starting in the 1950s. The saws make three orthogonal cuts in two successive operations. The first cut, called *lining*, is performed perpendicular to the excavation front; the other two are performed jointly with two saws, which are perpendicular and parallel to the quarry plane, respectively. The standard dimensions of the tuff stones produced in Italy are: 10 × 25 × 40 cm, but selected dimensions can be easily obtained on request.

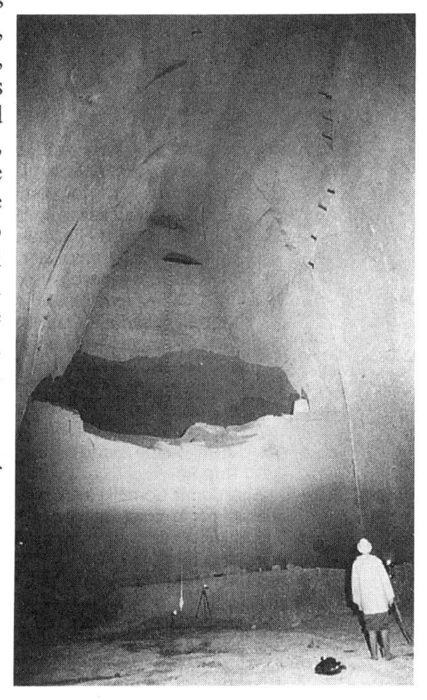

Figure 17. Gallery mining in Capodimonte, Naples, Italy. Entry of an old quarry of Neapolitan yellow tuff.

Table 1 lists the main exploited tuff deposits in the world, their relevant lithologies, and the types of zeolites present.

MECHANICAL AND PHYSICAL PROPERTIES

The principal mechanical and physical characteristics of zeolitized tuffs are listed in Tables 2 and 3. In particular, Table 2 refers to materials that are currently used as dimension stone, and Table 3 summarizes the properties of some zeolitic tuffs that are used sparingly or not at all. Note that

Figure 18 (below). Open-pit mining: (a) Yellow tuff, Riano (near Rome); (b) Red tuff with black scoriae, Palombara (near Viterbo, Italy); (c) Rhenish *trass* (Riedel, Germany); (d) Yellow tuff, Gran Canaria (Canary Islands, Spain).

some tuffs only one value is reported, which is not necessarily representative of the whole formation. The data in Table 2 illustrate that the properties of the materials utilized as dimension stones range widely, especially porosity (P) and compressive strength (σ). Compared with other building materials, zeolitized tuffs exhibit higher porosities and weaker mechanical properties, accounting for their inferior durability. It is noteworthy that, unlike other building materials, zeolitized tuffs show no strict relationship between P and σ, possibly due to the heterogeneity of the materials. Table 3 confirms these observations. Surprisingly, the materials coming from Russia and other states of the former U.S.S.R., in spite of their distribution over a wide territory, display somewhat similar physical properties.

Thermal behavior

Zeolitic tuffs are excellent materials for controlling their local environment, especially temperature and humidity. Zeolites adsorb and desorb water molecules reversibly as a function of tem-perature and/or humidity (Tchernev 1978; this volume). Thus, during natural heating/cooling cycles (for example, diurnal cycles), zeolitized materials can store both sensible heat and heat of adsorption. Buildings made with zeolitic tuffs are, in fact, cool during the day and warm at night because the zeolite removes heat from the environment by desorbing water molecules during the warmest hours of the day and returns this heat during the coolest hours of the night by re-adsorbing water.

Table 1. Main volcanogenic sedimentary zeolite deposits commercially mined for dimension stone*.

Country Rock Type	*Location*	*Area of deposit* (km²)	*Lithology*†	*Zeolites*‡
Bulgaria				
Volcaniclastic and epiclastic rocks	East Rhodopes	80	Type 2	Cp
Germany				
Leuzittuff	East Eifel	1	Type 1	Ch
Rhenish *trass*	East Eifel	7	Type 1	Ch>Ph
Greece				
Ignimbrite	Thrace	2	Type 2	Cp>>Mo
Hungary				
Tuff	Tokaj Area	100	Type 4	Cp>Mo
Tuff	Telkibanya	50	Type 4	Cp
Italy				
Orvieto-Bagnoregio ignimbrite	Vulsini District	1300	Type 1	Ch>>Ph
Red tuff with black scoriae	Vico Volcano	1250	Type 1	Ch>>Ph
Riano yellow tuff	Sabatini Mts.	1400	Type 1	Ch
Lionato tuff§	Albani Hills	1500	Type 1	Ph>>Ch
Logudoro ignimbrite	North Sardinia	36	Type 1	Cp>>Mo
Neapolitan yellow tuff	Phlegraean Fields	13	Type 1	Ph>>Ch>>An
Campanian ignimbrite	Campania Region	500	Type 1	Ch>>Ph
Japan				
Oya-ishi tuff	Utsonomya	15	Type 4	Cp>Mo
Mexico				
Cantera verde	Oaxaca	5	Type 3	Cp>>Mo
Romania				
Transylvanian tuff	Cluj-Napoca	–	Type 3	Cp

* Data of western European deposits taken from de' Gennaro et al. (1995) and Stamatakis et al. (1996). Source of data from Bulgaria is Djourova and Milakovska-Vergilova (1996); from Hungary is D. Kalló (Central Research Institute for Chemistry, Hungarian Academy of Sciences, Budapest, Hungary, pers. comm., 2000); from Japan is N. Kuchitsu (National Research Institute of Cultural Properties, Tokyo, Japan, pers. comm., 1996), from Mexico is M. A. Hernandez (Departemento de Investigacion en Zeolitas, Instituto de Ciencias, Univ. Autonoma de Puebla, Mexico, pers. comm., 1998); from Romania is Bãrbat and Marton (1989) and Bãrbat et al. (1991).

† See the section "Tuff lithology."

‡ In order of abundance. Legend: An = analcime; Ch = chabazite; Cp = clinoptilolite; Mo = mordenite; Ph = phillipsite.

§ *Lionato* = reddish-fawn, as the lion skin.

– No data available as regards the area of the deposits. Bedelean and Stoici (1984) reported that about 50 outcrops of zeolitized rocks had been investigated in Romania, most of them in Transylvania, in the vicinity of the city of Cluj-Napoca. This area, partially exploited for dimension stone production, extends for a few hundreds square kilometers.

The action of thermal regulation by the tuff is therefore coupled with that of the environmental humidity.

This favorable thermal behavior is perhaps the main reason for the historic use of zeolitic tuff since ancient times either for dwellings, refuges, cellars, etc. or as building

Table 2. Physical and mechanical properties of commercial zeolitized tuffs compared with some common building materials (see also Table 1).

Material	*Lithology*[*] *type*	γ (kN/m^3)	P (%)	σ (MPa)
Neapolitan yellow tuff[†]	1	10.3–14.1	42.7–61.0	0.68–11.9
Campanian ignimbrite[‡]	1	13	47.9–60.4	0.9–8.0
Lionato tuff[†§]	1	12.5–15.7	37–51	5.7–13.1
Riano tuff[†§]	1	12.5–14.2	44–52	6–12
Orvieto-Bagnoregio ignimbrite[†§]	1	10.7–13.4	45–57	2.5–6.8
Red tuff with black scoriae[†§]	1	12.8	49	3.9
Logudoro tuff[¬]	1	18.4–23.8	7.9–25.2	25.8–177
Leuzittuff, Ettringen[#]	1	16.6	30.7	19.95
Leuzittuff, Weiberner[#]	1	13.2	45.6	8.94
Trass (Roman tuff)[#]	1	13.5	43.4	5.32
Oya–ishi tuff[&]	4	16	n.d.	8.0
Clay brick	–	15.0–23.7[$]	7.7–41.8[$]	10–45[‰]
Basalt[@]	–	26.9–30.4	1–3	196–392
Limestone[@]	–	23.5–26.5	5–15	49–147
Sandstone[@]	–	17.6–26.5	4–20	39–127

Legend: P = porosity; γ = unit weight; σ = uniaxial compressive strength.
[*] See the section "Tuff lithology".
[†] Evangelista and Pellegrino (1990).
[‡] F. Rippa (Dipartimento di Ingegneria Geotecnica, Univ. Federico II, Napoli, Italy, pers. comm., 1996).
[§] Bianchetti et al. (1994).
[¬] G. Oggiano (Istituto di Scienze Geologico-Mineralogiche, Univ. di Sassari, Italy, pers. comm., 1996).
[#] Fitzner (1994).
[&] N. Kuchitsu (National Research Institute of Cultural Properties, Tokyo, Japan, pers. comm., 1996).
[$] Clews (1969).
[‰] Colombo (1990).
[@] Ippolito et al. (1980).

stone. The temperature-conditioning action of natural zeolites has recently been tested in Hungary in an experimental building, about 10 × 10 × 3 m in size, made with common clay bricks and without doors and windows. In summer, internal temperatures ranged between a maximum of 30°-33°C and a minimum of 16°-20°C. When the roof was covered with a 10-cm thick layer of 2-5-mm-size grains of clinoptilolite-bearing rock from Tokaj Hills, the range of the internal temperature was only between 22°-24°C during the day and 20°-22°C in the night (D. Kalló, Institute of Chemistry, Chemical Research Center, Hungarian Academy of Science, Hungary, pers. comm., 1995). Thermal regulation by natural zeolites can be partially hindered by the layer of plaster that covers most masonry walls; however, this effect is moderate as plasters have generally elevated porosities (~40%) which allow good permeability to water vapor.

The above-mentioned thermal behavior may be related to the specific thermal properties of the zeolitic tuff, compared with other building materials (see Table 4). Note that the thermal conductivity of zeolitized tuff makes it a better insulating material than concrete, clay brick, and other building materials (probably due to the high porosity and the presence of zeolitic water). As far as heat capacity is concerned, zeolitic tuff has a thermal storage capacity that is greater than that of the other non-adsorbent building materials as a result of water adsorption phenomena and the associated heat of adsorption (see Fig. 19).

Table 3. Physical and mechanical properties of some zeolitized tuffs occasionally utilized (Greece, Romania) or not utilized (Russia, Australia) as building stone.

Location	*Zeolites*	*Lithology** *Type*	γ (kN/m^3)	*P* (%)	σ (MPa)
Greece: Metaxades, Thrace†	Cp	2	15.80	36.24	25.9
Romania: Transylvania‡					
Cluj	Cp	3	16-17	27-31	3-4
Salaj	Cp	3	18-18.5	26-27	2.5-3
Bistrita-Nasaud	Cp	3	16.8-23	26	2.1-3.9
States of the former U.S.S.R.§					
Sokirniza, Carpathian, Ukraine	Cp	2	–	33.4	50.7
Tedzami, Caucasus, Georgia	Cp, Heu	2	–	33.0	–
Dzegvi, Caucasus, Georgia	Cp	2	6.4–7.5	34.3	78.8
I–Dag, Caucasus, Azerbaijan	Cp	2	9.5–10.3	30.9	47.9
Pegas, Kuzbuss, Russia	Cp, Heu	2	8.7–11.0	25.0	40.2
Kholinsky, Baikalian, Russia	Cp, Mo	2	8.0–9.3	19.0	40.2
Shivirtuy, Baikalian, Russia	Cp	2	6.6–7.8	44.0	–
Khonguruu, Yakutia, Russia	Cp, Heu	2	7.5–8.7	31.1	–
Chuguevsky, Far East, Russia	Cp, Mo	2	9.3–12.1	25.0	24.5
Lutogsky, Sakhalin, Russia	Cp	2	9.0–11.5	26.3	26.3
Jagodninsky, Kamchatka, Russia	Cp, Mo	2	8.0–9.5	25.8	–
Australia: Drummond, Queensland#	Cp	2	22.70	11.22	–

Legend: P = porosity; γ = unit weight; σ = uniaxial compressive strength; Cp = clinoptilolite; Heu = heulandite; Mo = mordenite.

* See the section "Tuff lithology".

† Original data obtained on a tuff block kindly provided by U. Lutat, Silver & Baryte Ores Mining Co. S.A., Athens, Greece.

‡ Bărbat and Marton (1989).

§ G. I. Ovcharenko (Chair of Building Materials, Altay Technical University, Russia, pers. comm., 1996). Note that the reported deposits cover the entire territory of the former U.S.S.R. from west to east.

Original data obtained on a tuff block kindly provided by B. McDougall, Currumbin Sand & Gravel Pty Ltd, Currumbin, Australia. This material is not suitable for dimension stones because it tends to split along preferential jointing, giving rise to irregular, roughly rhombohedral forms. It has therefore been impossible to measure its compressive strength.

Decay phenomena

Zeolitic tuff used as facing stone experiences remarkable deterioration by weathering due to atmospheric agents (Figs. 20 and 21). The decay takes the forms of exfoliation, flaking, and pulverization, as well as efflorescences, patinae, and dark-color spots. The type of deterioration and its extent depend on the stone properties, namely, mineralogical composition, textural features, and porosity, as well as on environmental conditions (climate or microclimate, atmospheric pollution, winds, and exposure). Unlike other building materials such as marble, there is a lack of published research on the weathering of zeolitic building materials. Most of the available data concern tuffs of lithological type 1

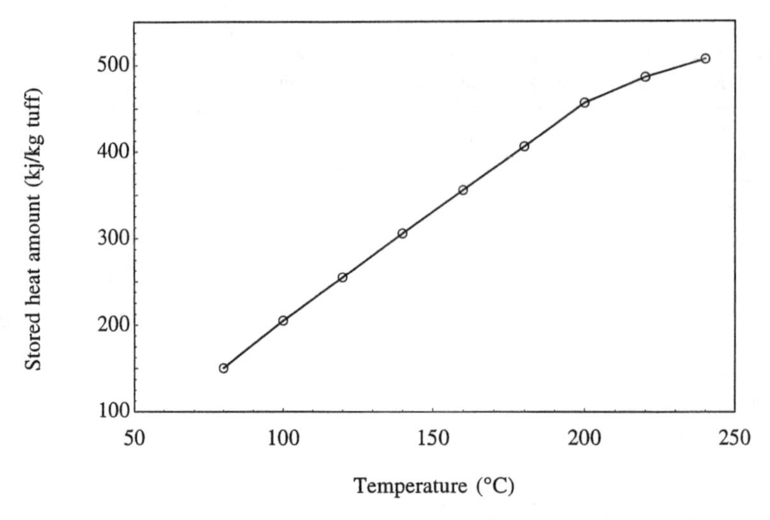

Figure 19. Adsorption heat measured on a tuff from Mercogliano, Avellino, Italy, as a function of dehydration temperature (from data reported in Table 3 in Aiello et al. 1984). The tuff contains about 80% chabazite.

(see above), especially those utilized in Italy and Germany. These data indicate that, irrespective of the environmental conditions, a determinant role in any type of decay, either of chemical or physical origin, is played by water.

Some results of field and simulation studies of tuffaceous rock decay, mainly limited to the Neapolitan yellow tuff and the Campanian ignimbrite, are reported below.

Chemical decay. Chemical decay is typical in urban environments, especially if the atmosphere is polluted by SO_2. Chemical decay commonly manifests itself as calcite and/or

Table 4. Thermal properties of some building materials at room temperature.

Material	*Thermal conductivity* (W/m·K)	*Specific heat* (kJ/kg·K)
Clay brick*	0.69	0.84
Concrete*	0.93	0.65
Granite*	1.70–3.98	0.84
Limestone*	0.69	0.92
Marble*	2.08–2.94	0.88
Zeolitic tuffs:		
Campanian ignimbrite†	0.32	1.09
Neapolitan yellow tuff†	0.32	–
La Rioja tuff†	0.38	–

* Data from Perry et al. (1984).

† Campanian ignimbrite contained 65 wt % chabazite, Neapolitan yellow tuff: 59 wt % phillipsite, 15 wt % chabazite, La Rioja tuff (Argentina) (Iñiguez Rodriguez 1992): roughly 80 wt % clinoptilolite. Data on thermal conductivity from Basile et al. (1992); data on specific heat from Scarmozzino et al. (1980).

Figure 20. Physical decay due to salt crystallization in tuff macropores in a marine environment: (a) Bastion of Castel dell'Ovo, Naples, Italy, that was built on a small tuff outcrop in the sea (the islet of Megaride); (b) a close-up of the west wall of Castel dell'Ovo that shows the strong decay of the Neapolitan yellow tuff blocks.

gypsum patinae and encrustations. de' Gennaro et al. (1993) interpreted the formation mechanisms of the patinae as follows: the alteration process begins with the dissolution of the hardened mortar between blocks through the action of atmospheric CO_2:

$$\underset{\text{lime mortar}}{CaCO_{3(s)}} + CO_{2(aq)} + H_2O_{(l)} \rightarrow Ca^{2+}{}_{(aq)} + 2\ HCO_3^-{}_{(aq)} \qquad (1)$$

The next step may be the carbonation of leached Ca^{2+} on the surface of the tuff block (formation of calcite patina, Eqn. 3), which is favored by a sharp increase in pH caused by the interaction (ion exchange) between solution and zeolite (Z^-) (Eqn. 2):

$$\underset{\text{zeolite}}{Z^-(Me^+)_{(s)}} + H_2O_{(l)} \leftrightarrow Z^-(H^+)_{(s)} + OH^-{}_{(aq)} + \underset{\text{leached cation}}{(Me^+)_{(aq)}} \qquad (2)$$

$$Ca^{2+}{}_{(aq)} + HCO_3^-{}_{(aq)} + OH^-{}_{(aq)} \leftrightarrow \underset{\text{calcite patina}}{CaCO_{3(s)}} + H_2O_{(l)} \qquad (3)$$

The final step may be the transformation of the calcite patina into gypsum patina (Eqn. 6) through an initial dissolution step (Eqn. 5), which is favored by a pH decrease caused by the dissolution of atmospheric SO_2 in water (Eqn. 4):

$$SO_{2(aq)} + H_2O_{(l)} \leftrightarrow HSO_3^-{}_{(aq)} + H^+{}_{(aq)} \quad (4)$$

$$\underset{\text{calcite patina}}{CaCO_{3(s)}} + H^+{}_{(aq)} \leftrightarrow Ca^{2+}{}_{(aq)} + HCO_3^-{}_{(aq)} \quad (5)$$

$$2\, Ca^{2+}{}_{(aq)} + 2\, HSO_3^-{}_{(aq)} + O_{2(aq)} + 4\, H_2O_{(l)} \leftrightarrow \underset{\text{gypsum patina}}{2[CaSO_4 \cdot 2H_2O]_{(s)}} + 2\, H^+{}_{(aq)} \quad (6)$$

Figure 21. Physical and chemical decay due to meteoric water and rising moisture: (a) the Beverello tower of Castel Nuovo, Naples; (b) a close-up of (a) that shows the strong decay of the Neapolitan yellow tuff blocks.

An increase or (especially) a decrease in solution pH during the above dissolution-recrystallization process may be responsible for the dissolution and partial corrosion of zeolites and other phases that act as binders in the tuff (see the section on the lithification process in the tuff), thus causing the disaggregation of the block. Scanning electron micrographs in Figure 22 clearly illustrate eroded crystal edges of phillipsite and chabazite in a decayed tuff from Castel Nuovo (Naples, Italy).

Figure 22. Scanning electron micrographs of zeolite crystals from a Neapolitan yellow tuff sample from Castel Nuovo that have eroded edges: (a) chabazite and (b) phillipsite.

Physical decay. Alternate wetting/drying cycles (particularly if the waters contain soluble salts, e.g. marine water) appear to be more important in causing decay than do thermal fluctuations and ice formation. The formation of ice is particularly disruptive in highly porous materials, such as zeolitic tuffs, in areas characterized by frequent thermal fluctuations around 0°C. On the contrary, fluctuations above 0°C and up to ~80-90°C (the maximum temperatures reachable under the direct action of the sun), which normally cause stresses, do not present a problem for zeolitic tuffs. In fact, the thermal expansion of the material is balanced by shrinkage due to reversible water loss from the zeolite (Fig. 23).

The main cause of the disaggregation of zeolitic tuffs in Italy therefore appears to be related to salt crystallization, due to the evaporation of water, which may generate internal stresses resulting in the formation of microcracks (Fig. 24). The effects of salt crystallization are particularly devastating if the tuff masonry is located near the sea (e.g. Castel dell'Ovo in Naples, see Fig. 20). Here, the disintegrating action of the crystallizing salt is further promoted by strong marine winds, which continuously remove superficial disintegrated layers and expose new material to the action of the sea water.

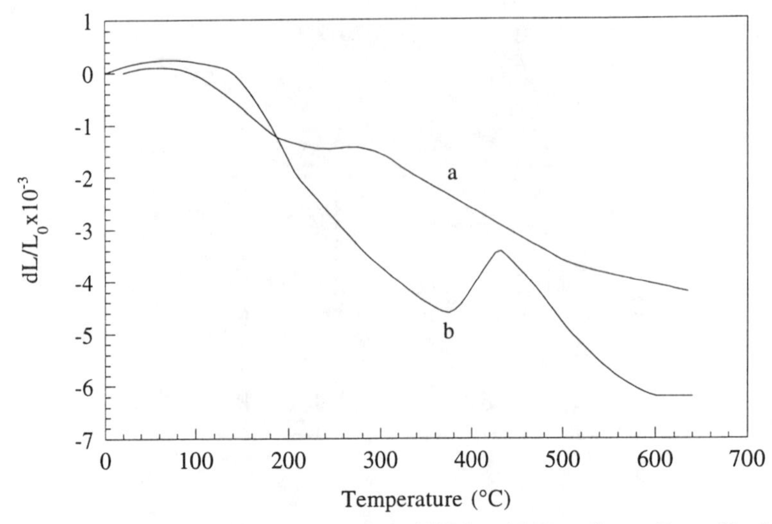

Figure 23. Thermodilatometric curves of (a) phillipsite–rich Neapolitan yellow tuff and (b) chabazite–rich Campanian ignimbrite (Marino et al. 1993). Note in the temperature range 20°-100°C that the dimensions of the test pieces do not change substantially (dL is the thermal shrinkage of the test piece relative to its original length L_0).

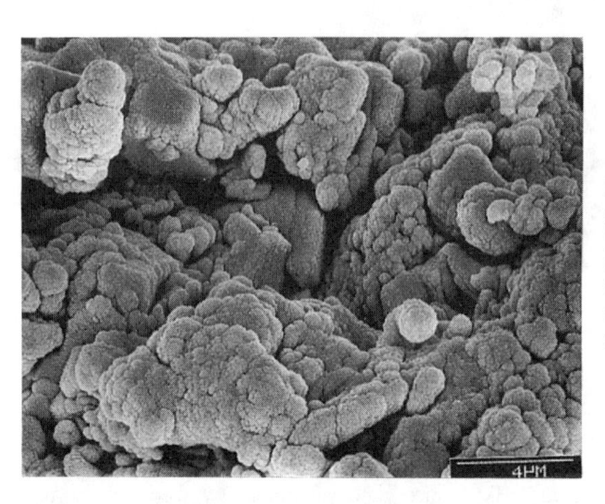

Figure 24. Scanning electron micrograph of a fragment of Neapolitan yellow tuff from the surface of a wall of Castel dell'Ovo that shows an extended recrystallization of soluble salts on the tuff constituents and the possible formation of microcracks.

Preservation and restoration treatments. Because of the above-described decay phenomena, buildings made of zeolitic tuffs often require conservation or restoration treatments (Aiello 1995). As far as conservation is concerned, polymeric protective agents were proposed by Aurisicchio et al. (1982). Experiments with thermosetting acrylates on zeolitic tuff were not completely successful, chiefly because of the low water-vapor permeability, the large difference in the mechanical and physical properties of treated and untreated parts of the masonry, and changes in color of the blocks with the time. The more recent use of fluorinated polymers reported by Granato et al. (1994) appears more promising, mainly because of the presence of fluorine, which ensures high stability of the treated material towards heat, light, and atmospheric pollutants.

Decayed parts of old tuff blocks may be restored using special concretes prepared by mixing crushed zeolitic tuff with cement containing appropriate amounts of the pulverized tuff. Sersale and Frigione (1986) obtained good results, both from a mechanical and an aesthetic point of view, using special concretes for the Neapolitan yellow tuff.

ZEOLITIC TUFF AS A CONSTITUENT OF LIGHTWEIGHT BUILDING MATERIALS

Tuffs as lightweight aggregates

The term "aggregate" indicates a class of incoherent natural materials having different grain sizes that may be used in the manufacture of concrete or as raw materials for dimension blocks through compaction, either in the presence of binders or not. Two kinds of lithified materials can be obtained: ordinary tuff and lightweight products. The latter (*ceteris paribus*) have unit weights γ (see the section "Tuff lithology") only one third that of the former. The unit cost is usually low. Other important features, such as low bulk density, good sound-proofing, and heat insulation, depend on the high porosity of the material.

On the basis of the source of the raw material (if natural or industrial) and specific weight, lightweight aggregates can be grouped as follows (Anonymous 1977; McCarl 1983):

(1) Natural lightweight aggregates: scoria, volcanic ash, pumice, and diatomite. These volcanic or sedimentary materials display a high natural porosity and, therefore, a low bulk density. They are also inexpensive.

(2) Expanded lightweight aggregates: expanded clay, expanded schist, and expanded slate. These materials are obtained by heating the crushed rock at temperatures of about 1100°C. Expansion as much as two to three times is mostly due to the rapid evolution of water vapor or carbon dioxide evolving from organic matter naturally present or added to the rock.

(3) Man-made lightweight aggregates: coke powder, industrial scoria, fly ash, and bottom ash (i.e. ash deposited at bottom of furnace), and expanded slag. These materials are secondary products of the electric and metallurgic industries. They are usually fine grained and characterized by a low specific weight.

(4) Ultra-lightweight aggregates: expanded perlite, expanded vermiculite, and expanded zeolite. These materials are obtained by heating the crushed mineral or rock at temperatures between 400° and 1400°C. Expansions of as much as 15 to 30 times are due to rapid water vapor evolution.

Similar to perlite and other hydrated volcanic glasses, zeolitic tuff fragments can be expanded into low-density pellets for direct use as lightweight aggregate in mortars and concretes, e.g. for producing pre-compacted building blocks. The temperature required for expanding zeolitic tuffs is significantly higher than that needed for perlite or other expandable materials (about 1200°C vs. about 800°C), but the expanded zeolite products are considerably stronger and more resistant to abrasion (Mumpton 1978).

Stojanovic (1972) reported densities as low as 0.8 g/cm^3 and porosities as high as 65% for expanded clinoptilolite pellets from several Serbian deposits after firing at 1200°-1400°C. Similar materials were prepared by Ishimaru and Ozata (1975) from Japanese zeolites at 1250°C. Bush (1974) found that high-grade clinoptilolite from the Barstow Formation, California, expanded four to six times when heated for 5 min at 1150°-1250°C, and Torii (1978) reported that about 1500 tons per year of mordenite tuff was used in Japan to manufacture lightweight bricks (bulk density: 0.7-1.0 g/cm^3) of high physical strength and high acid and alkali resistance.

Gaioso Blanco et al. (1991) used expanded Cuban clinoptilolite-rich tuffs (bulk density averaging 0.9-1.0 g/cm^3) to prepare special cement mortars to be used in the construction of steel-reinforced concrete boats. The hardened mortars displayed compressive strengths as high as 55 MPa and a pH of 11.8, which preserved steel from corrosion for at least two years.

A method for preparing bricks from a mixture of natural zeolite and aplite that takes advantage of the gas-foaming property of the zeolite was proposed by Koizumi and Goto (1993). Expanded clinoptilolite-tuff powders obtained by thermal treatment at 600°C for 4 hr and aplite powders were mixed with water. The paste, which set for one day at 80°C to allow the gas-foaming property of the zeolite to produce a porous product, was treated at 1140°C for one hour to increase its strength. The final products were reported to have porosities as high as 22%.

Tuffs for manufacturing foamed or cellular materials

The possible use of zeolitized materials for manufacturing lightweight building materials arises from the thermal behavior of zeolites. They can store large amounts of thermal energy upon dehydration at temperatures of 400°C or greater. When mixed with water and cement, dehydrated zeolite powders generate large amounts of heated air through water adsorption, with resultant foaming and volume expansion. A lightweight material with elevated porosity is obtained upon hardening.

Kasai et al. (1973) prepared foaming agents by calcining clinoptilolite at about 550°C. The calcined zeolite was mixed with equal parts of water and dolomite plaster, molded, and hardened for 2 hr in an autoclave to produce a product having a bulk density of 0.75 g/cm^3 and a compressive strength of about 4.7 MPa. Escalona Ledea (1991) prepared cellular concretes by hydrothermally treating mixtures of clinoptilolite-heulandite-rich tuffs, lime, and gypsum in an autoclave (pressure: 10-15 bar) for 7-5 hr. The resulting materials displayed bulk densities of 0.5-0.6 g/cm^3 and compressive strengths of 5-7 MPa. More recently Fu et al. (1995) reported a zeolite-based lightweight concrete containing 50-80 wt % natural zeolite (type not specified in the patent) from western Canada (heat-treated in the range of 400°-600°C), 20-50 wt % Portland cement, and 0-5 wt % lime. After 3-6 hr autoclave curing at 170°-180°C, the density and strength of the product were in the range of 0.4-1 g/cm^3 and 2-10 MPa, respectively. The density and compressive strengths of untreated zeolite were 1-1.3 g/cm^3 and 10-30 MPa, respectively. Recently, a material named *Sibeerfoam* was developed in Siberia by the expansion of clinoptilolite-rich rocks in the range 1120°-1220°C in the presence of a suitable gas-producing material, such as 0.3-1 wt % silicon carbide. *Sibeerfoam*, which is prepared in the form of blocks or granules, is characterized by uniform closed porosity (0.5-5 mm), low density (0.2-0.9 g/cm^3), and high compressive strength (≤18 MPa) (Kazantseva et al. 1997).

ADDITION OF ZEOLITIC TUFF TO BLENDED CEMENTS

Pozzolan and pozzolanic materials

Pozzolan (from *pulvis puteolana*, which is a powder from Puteoli, today's Pozzuoli, near Naples) is an incoherent volcaniclastic rock formed by rapid cooling of a liquid magma, ejected during a hydromagmatic eruption. Most pozzolans are essentially glassy, but some may include crystalline fragments, such as K-feldspars (as in the Neapolitan pozzolans) or leucite (as in the Roman pozzolans). They are very porous materials as a result of the rapid evolution of volatiles in the course of rapid cooling. At present the name pozzolan suits any kind of finely grained, mainly glassy volcanic ash.

Pozzolans react readily with lime as a result of the large surface area and non-crystalline nature of their grains and their siliceous composition. They form hydrated calcium silicates, which, upon hardening, display high mechanical resistance either in aerial or in sub-aqueous environments. These same properties account for the ease with which such materials transform into zeolite (see the section on the lithification process of zeolitic tuffs). The reactivity towards lime was appreciated by the Romans who used natural pozzolans for manufacturing hydraulic mortars (see the historical section). Today pozzolan is a fundamental constituent of pozzolanic cements.

The absence or shortage of natural pozzolans has led to the search for materials that display *pozzolanic activity*, namely the capability to react easily with lime. Many non-crystalline siliceous materials behave as pozzolans; the most common are *fly ash* (a secondary product of coal combustion in electric power plants), *silica fume* (a by-product of the silicon industry), and *fired clay* or *ground clay brick*, which were used 3000 years ago by the Phoenicians in mixtures with lime for manufacturing nearly impermeable plasters that were useful for covering the walls of reservoirs (Sersale 1991).

Zeolitized tuff displays excellent pozzolanic activity. This behavior has been exploited, unconsciously, since at least the beginning of this century. Blended Portland cements containing 25% of a natural pozzolan from Tehachapi, California, later recognized by Mumpton (1983) as a clinoptilolite-rich tuff, were used in 1912 for the construction of the 386-km-long Los Angeles aqueduct (Drury 1954). Zeolitized tuffs have also been used, beginning in the first decades of the 20th century, for producing commercial blended cements, e.g. in the United States (Monolith Portland Cement Co.) (Mumpton 1983), in the Canary Islands (Cementos Especiales S.A.) (Sersale 1959), and in Germany (M. Tax, Heidelberger Zement AG, Heidelberg, Germany, pers. comm., 1997). At the end of the 1950s, many tuffs were recognized to contain zeolites (e.g. see Sersale 1958). Research was then initiated on the pozzolanic properties of zeolites. Sersale (1960) compared the zeolitized tuffs from Italy (Neapolitan yellow tuff and *lionato* tuff), Germany (Eifel), and Gran Canaria (Spain) with a typical Phlegraean, non-zeolitic pozzolan in a standardized test for measuring pozzolanic activity. According to this test, proposed by Fratini (1950) and then accepted as European Standard EN 196/5 (ECS 1987), mixtures of powdered Portland cement clinker (65%) and pozzolanic material (35%) are mixed with distilled water (solid-to-liquid ratio = 1/50) at 40°C for 8 days. At the end of the experiment, Ca^{2+} and OH^- concentrations in the contact solution are measured and reported in a plot of the lime solubility vs. alkalinity. Points representing Portland cement clinker should fit the curve but frequently lie above it because supersaturated $Ca(OH)_2$ solutions often form. On the contrary, the points representing the evaluated materials (Fig. 25) are all below the equilibrium $Ca(OH)_2$ solubility curve, which means that some of the lime resulting from the hydrolysis of the clinker was fixed by the pozzolanic materials. All of the tuffs tested displayed better pozzolanic behavior than the pozzolan itself, demonstrating that, at least for the zeolites tested (phillipsite and/or chabazite in various assemblages), the zeolitization of the original incoherent volcaniclastic material does not reduce its capacity to "fix" CaO. On the contrary, it increases the reactivity of the aluminosilicate framework.

Use of zeolitized tuff in pozzolanic cement

The manufacture of pozzolanic cements is based on both economic and technical considerations. The partial replacement of Portland cement clinker with less expensive additives, namely pozzolan and related materials, is a consistent money saver. The manufacture of blended cements actually allows a 40% fuel savings and increases cement production 1.5 to 2 times without altering the quality and performance of the end products (Sersale 1992). These benefits result because pozzolanic materials, by reacting with the lime produced by hydration of the calcium silicates, form binding compounds.

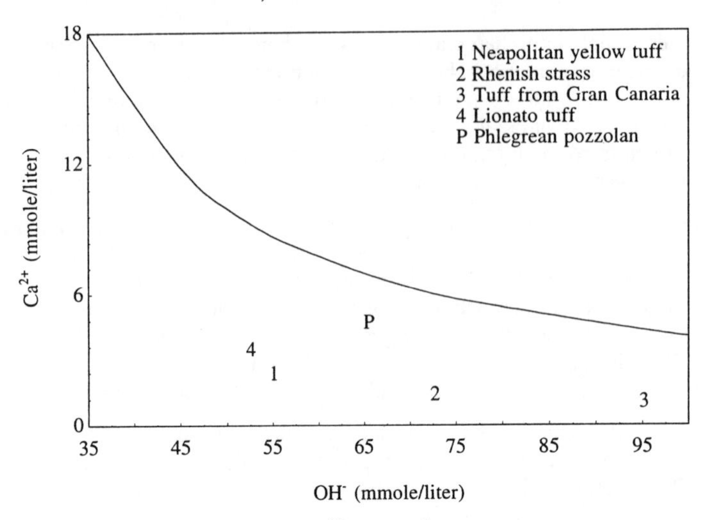

Figure 25. Concentrations of Ca^{2+} and OH^- in the contact solution of a mixture of powdered Portland cement and various pozzolanic materials. Points under the equilibrium $Ca(OH)_2$ solubility curve indicate pozzolanic behavior (Sersale 1960).

The presence of such additives in the blend also involves substantial technical advantages, including:

(1) improved resistance of the hardened mortars or concretes to chemical attack, especially by sulfate, due to a decrease in the amount of CaO that forms as a result of the hydrolysis of the hydrated calcium silicates (see Fig. 25);

(2) a reduction in the amount and rate of heat evolution during hydration (beneficial in casting large sections of concrete), due to the minimization of calcium aluminate and calcium alumino-ferrite contents in the blend;

(3) a reduction of the alkali level of the blend, thereby minimizing the risk of alkali-silica reactions. These reactions occur during concrete setting between hydroxyl ions in the pore solution and the reactive components of the aggregate that leads to the formation of an alkali silicate gel. It also sets up expansion forces that lead to the deterioration of the concrete (Sersale 1995).

Among the mineral additives that can be utilized for the production of pozzolanic cements, zeolitic tuffs are of great industrial interest because zeolite-bearing cements have properties comparable to those made with natural pozzolan. As shown in Figure 25, some zeolitic tuffs behave like natural pozzolan.

Influence of blending compositions on strength of mortars. Sersale (1985) reported the compressive strengths of hardened mortars made by mixing Portland cement clinker with various Italian zeolitic tuffs (mostly those reported in Table 1) and a typical Phlegraean pozzolan as a function of the amount of the pozzolanic material in the blend. Zeolitic tuffs contained chabazite or phillipsite (and locally analcime) or a mixture of them. The total zeolite content ranged between 50 and 80 wt %. The strength of mortar specimens was measured according to ISO Recommendation R 679-1968 (ISO 1968). The compressive strengths of mortars prepared with zeolitic tuff were greater than those made with the same amounts of pozzolan, particularly at later ages (Fig. 26). Similar results have been obtained by other authors with different zeolites. For example, Kasai et al. (1992) reported the performances of experimental blended cements containing clinoptilolite-rich or

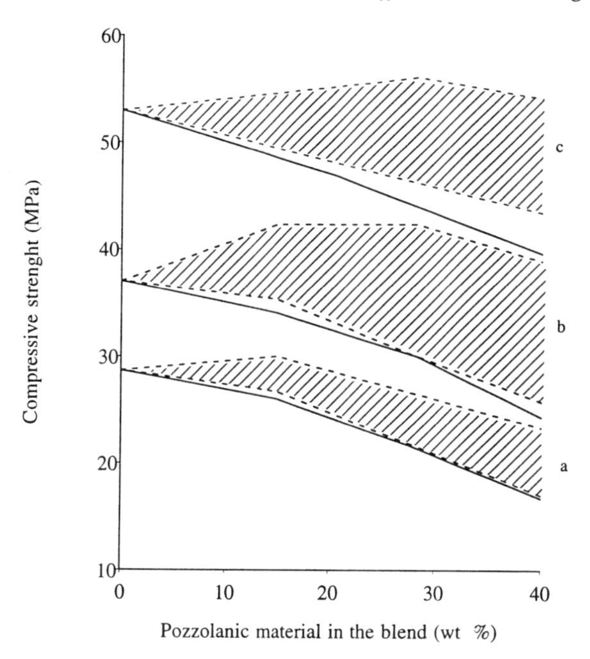

Figure 26. The compressive strength of mortars as a function of clinker replacement and curing time: (a) 3 days; (b) 7 days; (c) 28 days. Solid line refers to cement blends prepared using a Phlegraean pozzolan; bands represent the range of values for cement blends prepared using tuffs from various eruption districts (Sersale 1995).

mordenite-rich tuffs and compared them with reference blended cements manufactured using fly ash or silica fume. Compressive strength values were in the range 32-49 MPa and depended on the water/cement ratio. Clinoptilolite gave better values than mordenite and required less water to obtain the same workability. In general, the strength of the clinoptilolite-containing mortars was lower than that of standard mortars after brief aging but reached comparable strength values at 28 days. A pozzolanic cement in Cuba is produced from local zeolitized tuffs and named PP-350, developing a compressive strength of about 40 MPa after 28 days of curing (Urrutia Rodriguez and Gener Rizo 1991).

The compressive strength of hardened cement mortars containing zeolitized tuffs is evidently due to the great reactivity of the zeolitic matrix of the tuffs, compared with that of non-zeolitic pozzolan and other pozzolanic materials. Due to their microporous nature, zeolites in the cementing matrix of the tuffs are apparently capable of combining readily with lime, resulting in rapid and thorough crystallization of abundant, low-base, non-crystalline hydrated calcium silicates that have good binding properties (Sersale 1995), probably by means of a diffusion-controlled topochemical reaction (Drzaj et al. 1978).

Influence of blending components on alkali-silica reactions. The use of ground zeolitic tuff as a blending admixture to cement mortars and concretes also appears to be particularly attractive in minimizing the risk of expansion and cracking resulting from alkali-silica reactions. Figure 27 shows the effect of using pozzolan or zeolitic tuff on expansion abatement in mortar bars prepared using Pyrex glass as the aggregate, according to the test method C227-87 (ASTM 1988a). The pozzolanic materials are those mentioned in the previous section (Sersale 1985). With the exception of one sample containing analcime, the same level of clinker replacement and age of zeolite-bearing mortars minimized expansion better than did Portland cement-pozzolan mortars (Fig. 27). This behavior confirms the great reactivity of zeolites and their ability to incorporate free alkalis, thereby minimizing the reaction between hydroxyl ions in the pore water of the concrete and certain forms of silica, which may be present in significant quantities in aggregates (Sersale 1995).

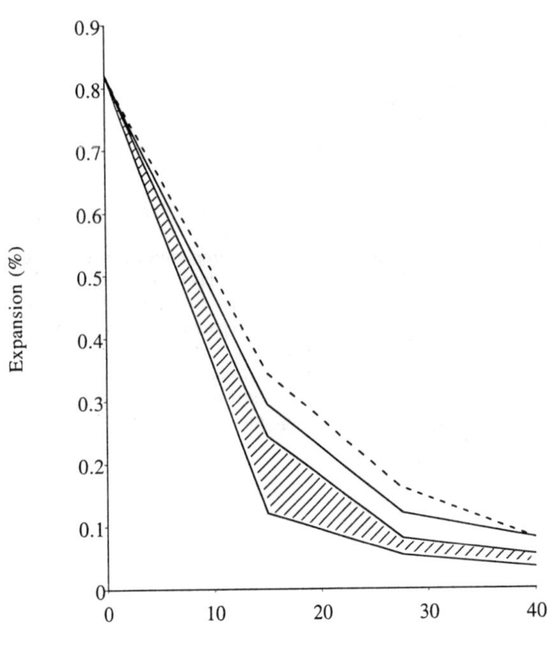

Figure 27. Expansion abatement of mortars aged 180 days as a function of clinker replacement. Solid and dashed lines refer to cement blends prepared using a Phlegraean pozzolan and a zeolitic tuff containing analcime, respectively; bands represent the range of values for cement blends prepared using tuffs from various eruption districts (Sersale 1995).

Influence of blending composition on heat evolution during setting. Sanchez de Rojas et al. (1993) measured by calorimetry the heat evolved during the hydration of several blended cements, and they compared the heat evolved for different pozzolanic materials such as natural pozzolan, zeolitized tuff, diatomaceous earth, opaline rock, fly ash, and rice husk ash. They compared the heat evolved for different Portland clinker/pozzolanic material ratios with that evolved by a typical Portland cement. The results demonstrated that zeolitized tuff and pozzolan, both of trachytic composition (possibly from the Canary Islands), behave very similarly. Both gave a hydration heat after 5 days of about 300 J/g, compared with about 350 J/g for a Portland cement for a 70/30 Portland cement/pozzolan mixture. This reduction is the result of (1) the reduction of evolved heat due to a reduced content of calcium aluminate and alumino-ferrite in the blend; and (2) the heat production from the reaction between pozzolanic material and the lime derived from the hydration of the clinker constituents.

Influence of blending composition on workability. The correct application of a mortar requires control of its rheological properties, including so-called workability, which is a function of the mortar composition and the water content. Any addition to the cement decreases its workability. Moreover, the higher the water content of the mortar, the greater its fluidity but also the lower the strength of the final hardened mortar. Workability is measured, according to ASTM Specifications C-230-83 (ASTM 1988b), in terms of an increase in flow and is expressed as a percentage of the original base diameter of a mortar placed on a motor-driven flow table. Figure 28 shows flow values as a function of the amount of cement replacement for mortars prepared with Phlegraean pozzolan and Neapolitan yellow tuff (Sersale 1995). The slightly greater decrease in workability of mortars prepared with zeolitic tuff, compared with those made with the pozzolan, has little significance because workability can be easily increased by the proper addition of a water-reducing admixture (Ramachandran 1984).

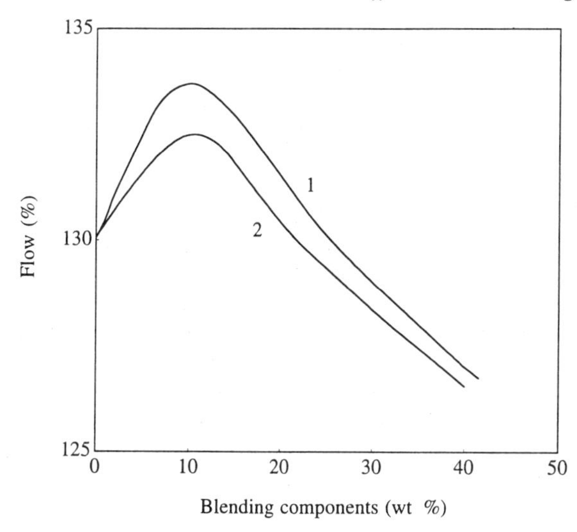

Figure 28. Flow values as a function of clinker replacement. Mortars prepared with blends of
(1) cement–Phlegraean pozzolan (non-zeolitic), and
(2) cement-Neapolitan yellow tuff (zeolitic)
(Sersale 1995).

Use of zeolitized tuff in high-alumina cements

The addition of zeolitic tuffs (containing mainly clinoptilolite, gismondine, levyne, and offretite) to high-alumina cement was reported by Ding et al. (1995) and Fu et al. (1996) to prevent strength reduction of mortars, due to the conversion of hexagonal hydrated calcium aluminate to hydrogarnet. This phenomenon, which normally occurs at high temperatures and relative humidity, has caused high alumina cements to be banned in virtually every country in the world for use in structural concrete. The prevention of the conversion of the hexagonal hydrated calcium aluminate to hydrogarnet in the presence of natural zeolites is explained by the preferential formation of strätlingite, a hexagonal hydrated calcium aluminosilicate.

CONCLUSIONS AND RECOMMENDATIONS FOR FUTURE WORK

The present survey, although incomplete due to the difficulty in locating reliable data in the international scientific literature, has emphasized that the use of zeolitic tuff in the building industry is not only the most ancient among the various applications for sedimentary zeolites but is definitely also the most widespread. Present worldwide consumption is estimated to be some ten million tons per year either as dimension stone (e.g. in Bulgaria, Cuba, Germany, Greece, Hungary, Italy, Japan, Mexico, Romania, and Turkey) or as additives in blended cements (e.g. in Bulgaria, China, Cuba, Germany, Jordan, Russia, Turkey, United States, and Yugoslavia). More limited, but of potential interest, is the use of zeolitic tuffs as constituents of lightweight building materials.

Future directions of research on the use of zeolitic tuffs as dimension stones should focus on their durability. Zeolitic tuff stones are readily obtained by cutting of rock; they are inexpensive, and they have the ability, as pointed out above, to control humidity and temperature inside a house. One weak characteristic of zeolitic tuff stones is that they are only moderately resistant to weathering from atmospheric agents (e.g. sulfuric acid weathering). Specific investigations are also needed on the preservation and restoration treatments of zeolitic materials utilized in building archaeological and historical sites. Research should focus on the basic and technological aspects for the use of zeolitic tuff in manufacturing blended cements. Additional research is needed to explain why some zeolitic materials have a greater reactivity than glassy pozzolan. Studies are needed to determine the mechanisms for reactions involving zeolites, e.g. lime-zeolite interaction during hydration and hardening of pozzolanic cements and hydraulic lime mortars, and the influence of zeolites in reducing the risk of the alkali-aggregate reactions. And finally, studies are still needed to find the best zeolite type and the best zeolite/cement ratio to obtain the maximum benefits in the use of zeolitic tuffs in the cement industry.

ACKNOWLEDGMENTS

The authors are indebted to many colleagues all over the world who have supplied with great willingness abundant information and data. The following persons are gratefully acknowledged: I. Akhvlediani, P. Bosch Giral, P. Di Girolamo, A. Filippidis, B. Fizner, M.N. Gündogdu, M.A. Hernandez, D. Kalló, N. Kuchitsu, U. Lutat, B. McDougall, G. Oggiano, G. I. Ovcharenko, F. Rippa, G. Rodriguez-Fuentes, R. Roque-Mahlerbe, R.A. Sheppard, M. Stamatakis, and M. Tax. The authors are particularly grateful to F.A. Mumpton for critical reading of the text and for suggestions. D. Fiorentino, who produced the photographs, is also gratefully acknowledged.

REFERENCES

Adabbo M, de' Gennaro M, Granato A, Munno R, Petrosino P (1991) Zeolitized tuffs from Gran Canaria, Canary Islands. *In* C Colella (ed) Atti 1° Convegno Naz.le Scienza e Tecnologia delle Zeoliti, p 61-71, De Frede, Napoli, Italy

Aiello R (1995) Zeolitic tuffs as building materials in Italy: A review. *In* DW Ming, FA Mumpton, Natural Zeolites '93, Occurrence, Properties, Use, p 589-602, Int'l Comm Natural Zeolites, Brockport, New York

Aiello R, Nastro A, Colella C (1984) Solar-energy storage through water adsorption/desorption cycles in zeolitic tuffs. Thermochim Acta 79:271-278

Annecchino R (1960) Storia di Pozzuoli e della zona flegrea. Comune di Pozzuoli (Napoli), p 40-42

Anonymous (1977) An introduction to lightweight aggregates. Industrial Minerals 114:29-33

ASTM (1988a) Standard specification for flow table for use in tests of hydraulic concrete, ASTM C230-83. *In* 1988 Annual Book of ASTM Standards, p 126-130, American Society for Testing Materials, Philadelphia, Pennsylvania

ASTM (1988b) Standard test method for potential alkali reactivity of cement-aggregate combinations (mortar-bar method), ASTM C227-87. *In* 1988 Annual Book of ASTM Standards, p 121-125, American Society for Testing Materials, Philadelphia, Pennsylvania

Aurisicchio S, Masi P, Nicolais L, Evangelista A, Pellegrino A (1982) Polymer impregnation of tuff. Polymer Composites 3:25-130

Bărbat A, Marton A (1989) Tufurile Vulcanice Zeolitice. Editura Dacia, Cluj Napoca, Romania, 236 p

Bărbat A, Bengeanu M, Marton A (1991) Modern trends in turning to best account the volcanic tuffs of Transylvania. Results obtained. *In* The Volcanic Tuff from the Transylvanian Basin, p 383-390, Cluj-Napoca, Romania

Basile A, Cacciola G, Colella C, Mercadante L, Pansini M (1992) Thermal conductivity of natural zeolite-PTFE composites. Heat Recovery Systems & CHP 12:497-503

Bedelean I, Stoici SD (1984) Zeolites. Ed. Tehnica, Bucarest, Romania, 227 p (in Romanian)

Bianchetti PL, Lombardi G, Matini S, Meucci C (1994) The volcanic rocks of the monuments of the forum and Palatine (Rome): Characterization, alteration, and results of chemical treatments. *In* AE Charola, RJ Koestler, G Lombardi (eds) Proc Int'l Meeting on Lavas and Volcanic Tuffs (Easter Island, Chile), p 83-105, ICCROM (Int'l Centre Study Preserv Restor Cultural Prop), Sintesi Grafica, Roma, Italy

Bush AL (1974) National self-sufficiency in lightweight aggregate resources. Minutes—25th Ann Meet Perlite Inst, Colorado Springs, Colorado, April 18-23, p 974

Cardone V (1990) Il tufo nudo nell'architettura napoletana. CUEN (Cooperativa Universitaria Editrice Napoletana), Napoli, Italy, p 63-72

Clews FH (1969) Heavy Clay Technology. Academic Press, London, 481 p

Colombo G (1990) Manuale dell'Ingegnere. Hoepli, Milano, Italy, C292-C294

de' Gennaro M, Langella A (1996) Italian zeolitized rocks of technological interest. Mineral Deposita 31:452-472

de' Gennaro M, Colella C, Aiello R, Franco E (1984) Italian zeolites 2. Mineralogical and technical features of Campanian tuff. Industrial Minerals 204:97-109 de' Gennaro M, Petrosino P, Conte M, Munno R, Colella C (1990) Zeolite chemistry and distribution in a Neapolitan yellow tuff deposit. Eur J Mineral 2:779-786

de' Gennaro M, Fuscaldo MD, Colella C (1993) Weathering mechanisms of monumental tuff-stone masonries in downtown Naples. Sci Technol Cultural Heritage 2:53-62

de' Gennaro M, Adabbo M, Langella A (1995) Hypothesis on the genesis of zeolites in some European volcaniclastic deposits. *In* DW Ming, FA Mumpton (eds) Natural Zeolites '93. Occurrence, Properties, Use, p 51-67, Int'l Comm Natural Zeolites, Brockport, New York

de' Gennaro M, Incoronato A, Mastrolorenzo G, Adabbo M, Spina G (1999) Depositional mechanisms and alteration processes in different types of pyroclastic deposits from Campi Flegrei volcanic field (southern Italy). J Volcan Geoth Res 91:303-320

Dell'Erba L (1923) Il tufo giallo napoletano. Studio scientifico-tecnico esteso alle cave e frane. Pironti, Napoli, 289 p

Di Girolamo P (1968) Petrografia dei tufi campani: il processo di pipernizzazione (tufo → tufo pipernoide → piperno). Rend Acc Sci Fis Mat (Naples) 35:329-394

Ding J, Fu Y, Beaudoin JJ (1995) Strätlingite formation in high-alumina cement-zeolite systems. Adv Cem Res 7:71-178

Djourova EG, Milakovska-Vergilova ZI (1996) Redeposited zeolitic rocks from the NE Rhodopes, Bulgaria. Mineral Deposita 31:523-528

Drury FW (1954) Pozzolans in California. Calif Mineral Information Service 7:6

Drzaj B, Hocevar S, Slokan M, Zajc A (1978) Kinetics and mechanism of reaction in the zeolitic tuff-CaO-H_2O systems at elevated temperature. Cement Conc Res 8:711-720

ECS (1987) Methods for testing cement. Pozzolanicity test for pozzolanic cements. European Standard EN 196/5, European Committee for Standardization, Bruxelles, Belgium, 11 p

Escalona Ledea I (1991) Hormigon ligero de silicato a partir de tobas zeolitizadas. *In* Rodriguez-Fuentes, JA Gonzales (eds) Proc Zeolites '91: 3rd Int'l Conf Occurrence, Properties and Utilization of Natural Zeolites, Part 2, p 214-215, International Conference Center Publications, La Havana, Cuba

Evangelista A, Pellegrino A (1990) Caratteristiche geotecniche di alcune rocce tenere italiane. *In* G. Barla (ed) Atti 3° Ciclo di Conferenze di Meccanica e Ingegneria delle Rocce, p 2/1-3/31, SGE Editoriali, Padova, Italy

Fitzner B (1994) Volcanic tuffs: The description and quantitative recording of their weathered state. *In* AE Charola RJ Koestler, G Lombardi (eds) Proc Int'l Meeting on "Lavas and Volcanic Tuffs" (Easter Island, Chile), p 33-51, ICCROM (Int'l Centre Study of Preserv and Restor CulturalProp), Sintesi Grafica, Roma, Italy

Fitzner B, Lehners L (1990) Rhenish tuff. A widespread, weathering-susceptible natural stone. *In* DG Price (ed) Proc. 6th Int'l Congr, Int'l Assoc Engin Geol, p 3181-3188, Balkema, Rotterdam

Fratini N (1950) Ricerche sulla calce d'idrolisi nelle paste di cemento. Nota II. Proposta di un saggio per la valutazione chimica dei cementi pozzolanici. Ann Chim 40:461-469

Fu Y, Ding J, Beaudoin JJ (1995) Zeolite-based lightweight concrete products and binding materials. US Patent 5,494,513, Oct. 1995, 21 p

Fu Y, Ding J, Beaudoin JJ (1996) Zeolite-based additives for high alumina cement products. Advn Cem Bas Mater 3:37-42

Gaioso Blanco RA, De Jongh Caula E, Gil Izquierdo, C (1991) Uso de aridos ligeros naturales zeolitizados en morteros y hormigones de alta resistencia. *In* G Rodriguez-Fuentes, JA Gonzales (eds) Proc Zeolites '91: 3rd Int'l Conf. Occurrence, Properties and Utilization of Natural Zeolites, Part 2, p 203-207, International ConferenceCenter Publications, La Havana, Cuba

Granato A, Apicella A, Montanino M (1994) Utilization of fluorinated polymers in surface protection of zeolite-bearing tuff. Mater Engin. 5:329-342

Greco G (1994) The Greeks on the shores of ancient Campania. Int'l Assoc Sedimentologists, De Frede, Napoli, Italy, 47 p

Grissom CA (1994) The deterioration and treatment of volcanic stone: Review of the literature. *In* AE Charola, RJ Koestler, G Lombardi (eds) Proc Int'l Meeting on Lavas and Volcanic Tuffs (Easter Island, Chile), p 3-29, ICCROM (Int'l Centre Study of Preserv and Restor CulturalProp), Sintesi Grafica, Roma, Italy

Gündogdu MN, Yalçin H, Temel A, Clauer N (1996) Geological, mineralogical and geochemical characteristics of zeolite deposits associated with borates an the Bigadiç, Emet and Kirka neogene lacustrine basins, western Turkey. Mineral Deposita 31:492-513

Hay RL (1963) Stratigraphy and zeolitic diagenesis of the John Day Formation of Oregon. Univ Calif Publ Geol Sci 42:99-262

Ibrahim K, Hall A (1996) The authigenic zeolites of the Aritayn volcaniclastic formation, north-east Jordan. Mineral Deposita 31:514-522

Iñiguez Rodriguez AM (1992) Natural zeolite resources in Argentina: First studies for their application. *In* Proc Int'l Symp Mineral Prop, Utilization of Natural Zeolites, p 221-231, University of Kyoto, Japan
Ippolito F, Nicotera P, Lucini P, Civita M, de Riso R (1980) Geologia tecnica per ingegneri e geologi, p 82-113, ISEDI, Milano, Italy
Ishimaru H, Ozata K (1975) Ceramic foam. Japan. Kokai 75,028,510, 5 p
ISO (1968) Methode d'essais méchanique des ciments. Resistance à la compression et à la flexion du mortier plastique (Methode Rilem-Cembureau). Norme Europeenne R 679, EN 196/1, Int'l Standard Organization, Geneva, Switzerland, 24 p
Kasai J, Urano T, Kuji T, Takeda W, Machinaga O (1973) Inorganic foaming agents by calcination of natural silica-high zeolites. Japan. Kokai 73,066,123, 5 p
Kasai Y, Tobinai K, Asakura E, Feng N (1992) Comparative study of natural zeolites and other inorganic admixtures in terms of characterization and properties of mortars. *In* VM Mahlotra (ed) Proc 4th Int'l Conf Fly Ash, Silica Fume, Slag, and Natural Pozzolans in Concrete 1:615-634, American Concrete Institute, Detroit, Michigan
Kazantseva LK, Belitsky IA, Fursenko BA (1997) Zeolite-containing rocks as raw materials for Sibeerfoam production. *In* G Kirov, L Filizova, O Petrov (eds) Natural Zeolites: Sofia '95, p 33-42, Pensoft, Sofia, Bulgaria
Koizumi M, Goto Y (1993) Preparation of porous building materials from a mixture of Ohyaishi zeolite and aplite. *In* Prog Abstracts Zeolite '93: 4th Int'l Conf Occurrence, Properties, and Utilization of Natural Zeolites, p 137-138, Int'l Comm Natural Zeolites, Brockport, New York
Langella A, Adabbo M (1994) Field guide to the tuff deposits of northern Latium. 10th Int'l Zeolite Conf, Supplement to Boll AIZ No. 3, De Frede, Napoli, Italy, 45 p
Marino O, Pansini M, Vitale A, Dal Vecchio S, Colantuono A, Mascolo G (1993) Caratteristiche di assobimento capillare di acqua e comportamento termodilatometrico di tufi zeolitici. *In* Programma e riassunti dei lavori, 2nd National Meeting "Scienza e Tecnologia delle Zeoliti", p 107-110, University of Modena, Modena, Italy
McCarl H (1983) Aggregates-lightweight aggregates. In SJ Lefond (ed) Industrial Minerals and Rocks, p 81-95, Soc Mining Eng Am Inst Mining Metall Petrol Engin, New York
Mumpton FA (1973) World-wide deposits and utilization of natural zeolites. Indus Minerals 73:2-11
Mumpton FA (1978) Natural zeolites: A new industrial mineral commodity. *In* LB Sand, FA. Mumpton (eds) Natural Zeolites: Occurrence, Properties, Use, p 3-27, Pergamon Press, Elmsford, New York
Mumpton FA (1981) Zeolites and mesothelioma. *In* R Sersale, C Colella, R. Aiello (eds) 5th Int'l Conference on Zeolites, Recent Progress Reports and Discussion, p 259-285, Giannini, Napoli, Italy
Mumpton FA (1983) Commercial utilization of natural zeolites. *In* SJ Lefond (ed) Industrial Minerals and Rocks 2:1418-1426, Soc. Mining Eng Am Inst Mining Metall Petrol Engin, New York
Mumpton FA (1988) Development of uses for natural zeolites: A critical commentary. *In* D Kallò, HS Sherry (eds) Occurrence, Properties and Utilization of Natural Zeolites, p 333-366, Akadémiai Kiadó, Budapest
Mumpton FA (1996) The natural zeolite story. *In* C Colella (ed) Atti 3° Congresso Nazionale AIMAT (Assoc Italiana d'Ingegneria dei Materiali) 1:31-64, De Frede, Napoli, Italy
Obradovic J, Dimitrijevic R (1987) Clinoptilolitized tuffs from Zlatokop, near Vranje, Serbia. GLAS de l'Académie Serbe des Sciences et des Arts. Classe des Sci Naturelles et Math 51:7-19
Obradovic J, Jovanovic O (1987) Some characteristics of the sedimentation in Neogene Valjevo-Mionica Lake Basin (Serbia). GLAS de l'Académie Serbe des Sciences et des Arts. Classe des Sci Naturelles et Math 51:53-63
Perry RH, Green D, Maloney JO (eds) (1984) Perry's Chemical Engineers' Handbook. McGraw-Hill Int'l Editions, New York, p 3-146–3-260
Ramachandran VS (1984) Cement Admixtures Handbook. Properties, Science and Technology. p 116, Noyes Publ, Park Ridge, New Jersey
Sanchez de Rojas M, Luxan MP, Frias M, Garcia N (1993) The influence of different additions on portland cement hydration heat. Cement Conc Res 23:46-54
Scarmozzino R, Aiello R, Santucci A (1980) Chabazite tuff for thermal storage. Solar Energy 24:415-416
Scherillo A (1955) Petrografia chimica dei tufi flegrei. II: Tufo giallo, mappamonte, pozzolana. Rend Accad Sci Fis Mat (Naples) [4] 22:345-370
Scherillo A (1977) Vulcanismo e bradisismo nei Campi Flegrei. *In* Atti del Conv. Int'l "I Campi Flegrei nell'archeologia e nella storia", p 81-116, Accad Naz dei Lincei, Roma, Italy
Sersale R (1958) Genesi e costituzione del tufo giallo napoletano. Rend Accad Sci Fis Mat (Naples) [4] 25:81-207
Sersale R (1959) Analogie costituzionali fra il tufo giallo napoletano e il tufo giallo della Gran Canaria. Rend Accad Sci Fis Mat (Naples) [4] 26:441-452

Sersale R (1960) Analogie genetiche e costituzionali tra tufi vulcanici a comportamento "pozzolanico". Silicates Indus 11:449-459
Sersale R (1985) Natural zeolites as constituents of blended cements. *In* B Drzaj, S Hocevar, S Pejovnik (eds) Zeolites, Synthesis, Structure, Technology and Application, p 523-530, Elsevier, Amsterdam
Sersale R (1991) La storia della calce dall'antichità ai nostri giorni. L'Industria Italiana del Cemento 61: 56-62
Sersale R (1992) Advances in portland and blended cements. *In* Proc. 9th Int'l Conf Chem Cement, I, p 261-302, National Council Cement and Building Materials, New Delhi, India
Sersale R (1995) Zeolite tuff as a pozzolanic addition in the manufacture of blended cements. *In* DW Ming, FA. Mumpton (eds) Natural Zeolites '93: Occurrence, Properties, Use, p 603-612, Int'l Comm Natural Zeolites, Brockport, New York
Sersale R, Aiello R (1964) Costituzione e reattività del "trass" renano. L'Industria Italiana del Cemento 34:747-760
Sersale R, Frigione G (1986) Sul recupero di manufatti architettonici in tufo giallo napoletano. *In* G Biscontin (ed) Proc Nat'l Conf "Manutenzione e conservazione del costruito fra tradizione e innovazione," p 505-509, Libreria Progetto Editore, Padova, Italy
Sersale R, Aiello, R Vero, E (1965) Costituzione e reattività del trass "leucitico" (renano). L'Industria Italiana del Cemento 35:513-522
Sheppard RA (1971) Zeolites in Sedimentary Deposits of the United States—A Review. *In* RF Gould (ed) Molecular Sieve Zeolites - I, Adv Chem Ser 101:279-310
Sheppard RA (1996) Occurrences of erionite in sedimentary rocks of the western United States. U S Geol Surv Open-File Report 96-018, 24 p
Stamatakis MG, Hall A, Hein JR (1996) The zeolite deposits of Greece. Mineral Deposita 31:473-481
Stojanovic D (1972) Zeolite-containing volcanic tuffs and sedimentary rocks in Serbia. *In* Proc Serb Geol Soc for 1968, 1969, and 1970, p 9-20, Belgrade
Tchernev DI (1978) Solar energy application of natural zeolites. *In* Sand LB, Mumpton FA (eds) Natural Zeolites: Occurrence, Properties, Use, p 479-485, Pergamon Press, Elmsford, New York
Temel A, Gündogdu MN (1996) Zeolite occurrences and the erionite-mesothelioma relationship in Cappadocia, central Anatolia, Turkey. Mineral Deposita 31:539-547
Torii K (1978) Utilization of natural zeolites in Japan. *In* Sand LB, Mumpton FA (eds) Natural Zeolites: Occurrence, Properties, Use, p 441-450, Pergamon Press, Elmsford, New York
Urrutia Rodriguez F, Gener Rizo M (1991) Empleo de cementos mezclados con tobas zeoliticas cubanas en el hormigon prefabricado. *In* G Rodriguez-Fuentes, JA Gonzales (eds) Proc. Zeolites '91: 3rd Int'l Conf Occurrence, Properties and Utilization of Natural Zeolites, Part 1, p 198-202, International Conference Center Publications, La Havana, Cuba
Velazquez E, Rabilero A (1991) Planta para el procesamiento de zeolitas y calizas. Tecnologia de producción. *In* G Rodriguez-Fuentes, JA Gonzales (eds) Proc Zeolites '91: 3rd Int'l Conf Occurrence, Properties and Utilization of Natural Zeolites, Part 2, p 104-106, International Conference Center Publications, La Havana, Cuba

17 Natural Zeolites in Solar Energy Heating, Cooling, and Energy Storage

Dimiter I. Tchernev

The Zeopower Company
75 Middlesex Avenue
Natick, Massachusetts 01760

INTRODUCTION

This chapter describes the use of zeolites in solar energy storage and in solar energy heating and cooling applications. This chapter concentrates on natural zeolites, but considerable work has also been done with synthetic zeolites, especially zeolite 13X. The chapter begins with a review of energy storage applications of natural zeolites, both for short-term (day-to-night) and long-term (seasonal) storage. It then discusses the use of zeolites in heating and cooling cycles. Open cycles are considered first, where the zeolite is open to the ambient air and is used mainly for air dehumidification followed by evaporative cooling. This is followed by a discussion of closed-cycle systems, where the zeolite and the working refrigerant gas are sealed in an airtight container, which excludes all other impurity gases. The pressure in the closed-cycle systems is not ambient but is determined by the condensation and evaporation pressures of the refrigerant.

The chapter concludes with a brief description of the needs and directions for future research in this field. For this reason it describes the thermodynamic analysis of open and closed cycles, the approximations involved in the analysis, and the appropriate application of the Clausius-Clapeyron equation to determine the isosteric heat of adsorption for various gases on natural and modified or synthetic zeolites.

HISTORY

Solar energy is intermittent by nature. The sun shines during the day whereas the need for heat in homes is largest at night, when the ambient temperature is lowest. To overcome the time lag between supply and demand, people have used day-to-night averaging of solar energy since very early times. Throughout human history, homes in hot climates were built with thick walls of large thermal mass that warmed slowly during the day (keeping the home cool) and released the stored heat during the night. Zeolitic tuffs have been used for centuries as building blocks for the construction of homes in Italy, Turkey, and other places, and anecdotal evidence suggests that this practice resulted in dry houses with large thermal mass (see Colella et. al., this volume).

STORAGE

Modern solar homes use solar collectors to provide hot air or water during the day and gravel or water tank storage to produce day-to-night averaging. The size of such storage tanks is large (10 to 30 m^3 for 10^6 kJ capacity) and efforts to reduce it have generated considerable research during the last 30 years. The focus for liquid storage has been on phase-change materials, utilizing the heat of fusion of water or hydrated salts. The first building that relied on phase-change material storage was a solar-heated house constructed in 1947 in Dover, Massachusetts (Telkes 1974). The heat store contained 21 tons of Glauber's salt ($Na_2SO_4 \cdot 10H_2O$), a salt hydrate chosen for its low cost and high heat of fusion. This salt, unfortunately, is difficult to work with because it melts

1529-6466/00/0045-0017$05.00

incongruently.

Although ice storage produced during off-peak hours is being used for cold storage to reduce utility costs of modern air conditioning systems (Dorgan and Elleston 1994), the problems of most phase-change materials, such as phase separation, reliable and reproducible nucleation centers, and the prevention of the super cooling of the liquid have not yet been completely resolved. The problem of super cooling of salt hydrates can be remedied through the addition of a nucleating agent, which is a small amount of material that never melts, although at an increased cost of the phase change material. Salt hydrates have high heat of fusion due to their high water content. The major disadvantage of salt hydrates is that many melt incongruently or semi-congruently, i.e. they melt to a saturated aqueous phase and a solid phase, which is the anhydrous salt (incongruent melting) or a lower hydrate of the same salt (semi-congruent melting) (Lane 1985).

During melting, the heavier solid salt settles out (phase segregation). Upon freezing, this salt at the bottom of the container does not recombine with the saturated solution to form the original hydrate, and the latent heat of the material is reduced. The process is progressive and irreversible, unless special provisions are made to counteract this phase segregation. Many methods to thicken the phase change material or prevent phase separation have not achieved commercial success. However, polymeric additives are available that form a gel matrix to keep the solid phase that separates from incongruent or semi-congruent hydrates in contact with the saturated solution.

Alternatively, thickening agents using clay or thixotropic gels permit the commercial application of Glauber's salt at reduced storage capacity (Calmac Manufacturing Corporation 1982). Stabilization techniques that appear to be proven so far include: mechanical agitation, microencapsulation, and gelation including use of cross-linked polymers. Each of the stabilization methods has inherent penalties. Thickening or gelling additives decrease the latent heat of fusion and add to the cost. Mechanical agitating equipment adds expense, reduces reliability, and requires power to operate. Micro-encapsulated salts may expand upon melting and rupture the container seals through crystal growth. Finally, most salt hydrates are highly corrosive, and special attention must be paid to their containerization. The presence of the container gives raise to analytical difficulties due to the nonlinearity of the governing equations and the time-dependence of the liquid-solid interface when heat transfer is accompanied by a phase change of the material (Hamdan and Elwerr 1996). Rock storage avoids most of the problems of phase-change materials.

During testing in 1968 of an Australian solar home with a gravel storage bed, Close and Dunkle (1977) observed an increase of 25% in heating storage capacity when humid air was circulated through the gravel bed. They concluded correctly that the effect was due to water adsorption by the gravel pieces. They proposed the use of beds of adsorbent materials for latent heat (heat of adsorption) storage and predicted an order of magnitude reduction of the storage bed's size if it used silica gel or LiBr-soaked gravel.

Expanding on this idea, Shigeishi et al. (1979) proposed the use of the latent heat of adsorption of synthetic zeolites for solar energy storage. They compared activated alumina and silica gel with synthetic zeolites 4A, 5A, and 13X and determined that after drying at 150°C the energy storage density of zeolites 4A and 13X is 1020 and 1370 kJ/kg, respectively; whereas, that of activated alumina is 523 kJ/kg, silica gel is 991 kJ/kg, and water at 80°C is only 250 kJ/kg. The use of zeolites results in significant (4 to 5 times) reduction of storage weight. Although silica gel has a large storage density, it must be cooled to <30°C before it begins to adsorb large quantities of water vapor, thus limiting its practical use. See Table 1.

Table 1. Energy storage density of natural and synthetic zeolites and other materials.

Zeolite	*Water vapor adsorption (kJ/kg)*	*Waterhydration (kJ/kg)*	*Energy volume density (MJ/m^3)*
Natural zeolites			
Chabazite tuff (Italy)	193-400	—	268
Clinoptilolite tuff (Turkey)	200-500	123-229	350
Chabazite tuff (Baikal)		100-400	
Synthetic zeolites			
4A	1020		
13X	1370	284	
Others			
Silica gel	991		
Water (80°C)	250	$LiBr/H_2O$	Water
		620	251

For a solar house in the Netherlands, Wijsman et al. (1979) determined that if flat-plate solar air collectors with 80°C maximum temperature are used, the stored energy density per unit weight of an adsorbent bed is 7 to 11 times larger than that of a conventional rock bed. They also recognized the fact that the adsorbent bed provides hot, but very dry air, which, although useful for agricultural drying, is not comfortable in a living space. Therefore, they reversed the usual airflow to blow the exhausted ventilation air from the house through the adsorbent bed and used it to pre-heat the incoming fresh ventilation air through an air-to-air heat exchanger.

Scarmozzino et al. (1980) first proposed the use of natural zeolites for thermal storage. A calculated energy density of 193 kJ/kg for the zeolite (chabazite-rich tuff containing 65% chabazite from southern Italy) activated at 80°C (similar to that of water at 80°C) suggested that this value could be increased if concentrating collectors with 120°C output temperature were used. The energy volume density of 268 MJ/m^3 is only slightly larger than that of water, i.e. 251 MJ/m^3 at 80°C.

Gopal et al. (1982) measured the dynamic properties of the zeolite-water storage systems for synthetic zeolites 4A, 5A, and 13X. They determined calorimetrically the rate of heat generation during adsorption for zeolites activated at 250°C in vacuum and showed that it ranged from 70 kJ/min kg initially to 10 kJ/min kg near saturation. In the mid-adsorption range, a 1000-kg zeolite system could generate 50,000 kJ/min, which is far in excess of normal heating requirements, however they warned of the difficulty of establishing vacuum conditions for a working system.

Ülkü (1986) measured the energy storage and heat pump applications for a natural clinoptilolite from the Anatolia region of Turkey. She determined that the energy density varies with regeneration temperature from 200 kJ/kg at 100°C to 500 kJ/kg at 200°C and that the mean values for the thermal conductivity, specific heat, and density of the material are 0.606 W/m°C, 1.108 kJ/kg°C, and 1525 kg/m^3, respectively. The volumetric energy density for a 80°C activation temperature is about 350 MJ/m^3, which is about 30% higher than that of the chabazite-rich tuff from Italy.

Aiello et al. (1988) studied the behavior of an Italian chabazite-rich tuff in open solar energy storage systems. Open storage systems can be used not only with solar energy but also when waste heat or combustion gases are available for the activation of the zeolite.

The warm and dry air coming from the activated zeolite bed can then be utilized for industrial drying processes or to preheat combustion air for furnaces. The reported energy density after activation at 200°C was 400 kJ/kg and the dynamics of the heat release process in the zeolite bed was considerably faster than in closed systems. The temperature of the outgoing air increases with time to a peak value that is determined more by the relative humidity of the incoming air than by the activation temperature. For incoming air at 25°C and 70% relative humidity, the outgoing air temperature peaks in time at above 60°C. The zeolite activation temperature determines only how broad the peak is whereas the air temperature at the peak is determined by the water vapor available in the incoming air.

The input energy during the activation process of a zeolite bed is divided in two parts; one part is used to raise the temperature of the dry zeolite bed (specific heat part) and the other part provides the heat of desorption of the water vapor. The thermal efficiency of the energy storage system, also called storage efficiency, is the ratio of the energy that can be withdrawn from storage divided by the energy amount put into storage. Generally, all energy storage systems suffer standby losses, although the latent heat (heat of adsorption) losses are considerably smaller than the sensible (specific) heat losses.

If, after activation, the process is reversed immediately and the heat losses from the bed are small, almost all of the input energy will be recovered and thermal efficiencies over 90% can be achieved. This is the case of industrial waste-heat recovery. If, however, the zeolite is allowed to cool to ambient temperature, as during day-to-night storage and especially during annual storage, before the process is reversed the sensible (specific) heat portion of the stored energy will be lost. Only the heat of adsorption will be recovered and then only if sufficient water vapor is provided. These sensible heat losses reduce the thermal (storage) efficiency of the system to <60%.

In many climate zones the dew point of the ambient air at night is low and insufficient water vapor is available to satisfy the adsorption needs for energy regeneration. Boiling liquid water is not a solution to this problem because its heat of vaporization is very high. This problem is even worse in some summer-to-winter zeolite storage systems that have been proposed (Close and Dunkle 1977; Shigeishi et al. 1979). Once the zeolite is cooled to room temperature, such systems have no further standby thermal energy losses over any length of time. During winter however, when the ambient air is extremely dry, it is difficult to provide water vapor required for adsorption unless an inexpensive waste heat source is available for the conversion of liquid water into vapor.

Selvidge and Miaoulis (1990) investigated reversible hydration reactions for use in long-term energy storage to avoid the water-vapor supply problem. They evaluated various materials including activated alumina, silica gel, synthetic zeolites 3A, 4A, 5A, and 13X, natural zeolite ZLD-9000 (chabazite), and LiBr, CaO, $CaCO_3$ and $CaSO_4$ salts. After drying the materials between 100°C and 350°C they determined that zeolite 13X has the highest heat of hydration for the zeolites, about 284 kJ/kg zeolite. LiBr salt at 620 kJ/kg, however, surpassed all other materials. Fischer et al. (1992) studied thermochemical energy storage with zeolite 13X using adsorption of water vapor from air with a dew point temperature of 25°C. They determined an energy storage density of 615 kJ/kg at a power output of 10 kW and 60% thermal efficiency.

Valueva et al. (1988) determined the heats of rehydration of natural chabazite from the Baikal region of Russia (after heating to various temperatures) to be between 100 and 400 kJ/kg depending on the activation temperature. Yörükogullari and Orhun (1997) also investigated the hydration energy storage capacity of natural clinoptilolite from Balikesír,

Turkey, and provided a detailed thermodynamic analysis of the hydration process. They determined that cation exchange of the natural zeolite with Ca^{2+} increased the heat of hydration from 123 kJ/kg to 229 kJ/kg, whereas Na and K cations resulted in heats of immersion of 185 and 99 kJ/kg, respectively.

Considerably more work is required to identify natural zeolites for energy storage. Their water adsorption isosteres must be determined as well as the effect of ion exchange on their adsorption capacity and heat of adsorption, as was done for Turkish zeolites above. The selection of adsorption or hydration method for solar energy storage depends more on the local climate conditions and the time duration of storage rather than the zeolites available. For short storage times adsorption is advantageous, whereas, for long storage times and in very dry climates, the hydration method is definitely superior.

COOLING

Sorption cooling is a well-known technology, having been demonstrated by Faraday almost 150 years ago. Sorption cooling can be divided into two major systems: liquid/gas or *absorption* systems and solid/gas or *adsorption* systems. Sorption cooling systems may also be open cycles, in which ambient air is dehumidified by the adsorbent, cooled back to room temperature, and then further cooled by water evaporation and closed cycles. The adsorbent and refrigerant gas are sealed in air-tight containers and the cooling or heating effects are brought out of the container via a heat exchanger and/or a heat transfer fluid. Zeolites have been used in both open and closed cycles.

Open-cycle cooling

Throughout human history, civilizations residing in hot and dry climates have used the evaporation of water to reduce the ambient air temperature. In the ancient Persian and Roman empires porous ceramic vessels containing water were stored in low-level rooms of the house. Here the water seeping through the pores evaporated on the outer surface of the vessel and cooled it and the surrounding air. Cool water (or wine) was available from the vessels. The ambient ventilation air was first drawn through this room to be cooled, and, thereafter, through the rest of the house finally being exhausted by natural draft through a chimney on the roof. Even today, people residing in the desert southwest of the United States use evaporative (swamp) coolers in their homes. The hot and dry desert air is drawn through the cooler as water trickles down a porous wetted surface. The heat of vaporization of water cools the incoming desert air and increases its humidity.

Such a cooling process will not operate in areas with high humidity because very little additional water will evaporate in the humid air (as the spraying of water in the air at the 1996 Olympic Games in Atlanta proved beyond any doubt). Therefore, during the past 70 years considerable effort has been devoted to develop systems that will dehumidify the ambient air.

Dehumidifying systems bring the air in contact with a liquid or solid desiccant. The liquids are usually water solutions of various salts such as LiBr, LiCl, $CaCl_2$, or even table salt, NaCl. Such concentrated solutions absorb additional water vapor from the air, which passes through a spray tower, thereby reducing its humidity. The diluted solutions are concentrated again by heating them and evaporating some of the water before they are reused. Recently, some more exotic liquids, such as triethylene and propylene glycols, have been proposed, but their high cost and the potential of vapor migration into the house has limited prospective applications. Löf (1955) proposed the use of solar energy to concentrate triethylene glycol solutions that are used to dry ambient air. The dry, desert-like air is then cooled with an evaporative cooler. Dunkle (1965) expanded the concept further and applied it under the climate conditions in western Australia.

Mullick and Gupta (1975) reported a system employing $CaCl_2$ and water in which the solution is concentrated by solar energy in a collector. A 10.5-kW LiBr system constructed and tested at Colorado State University was described by Patnaik et al. (1990). The heat and mass transfer relationships for an open-flow system with LiCl solution in a solar collector were developed in the former U.S.S.R. by Kakabaev and Khandurdyev (1969) and used by Collier (1979) for cooling-system performance simulations in five cities in the United States.

Solid adsorbent systems require that the desiccant be regenerated by heating periodically before it is reused. The oldest method for providing continuous operation is to divide the adsorbent in two separate containers, called packed beds or columns, and to desorb each bed alternately while the other one adsorbs moisture from the air. Such systems, utilizing zeolite pellets, have been available on the market for many years ranging in size from less than 1 kg (providing dry air for ozone generators to the hot tub industry) to hundreds of tons (for drying natural gas before it is pumped through a pipe line). Attempts have been made to use packed beds of desiccants for solar cooling, but the size and cost of the containers have been too high for practical use, and the amount of electric power needed to blow the air through the bed has been very large.

Another method for continuous operation of solid desiccants was developed after World War II in Sweden by Carl Munters who produced a variety of corrugated material sheets in which the cross-fluted structure greatly increased the contact surface area between the humid air and the desiccant material. The sheets are wound into desiccant wheels, which create a compact cross-flow configuration of open channels in which the air comes in contact with the enlarged adsorbing surface. The operating principle of a rotary dehumidifier is shown in Figure 1. The desiccant wheel slowly rotates first through a drying sector in which moisture is taken from the humid air, and thereafter through a reactivation sector, in which heated air discharges moisture from the wheel, thus resulting in a continuous dehumidification process. The corrugated sheets are usually made from metal, paper, or special plastics and impregnated with desiccant materials, such as LiBr and other salts, silica gel, or zeolites. Rotary dehumidifiers have been commercially available in the United States since World War II from companies such as Munters Cargocaire, Engelhard/ICC Technologies, and Airflow Co., with moisture removal capacities ranging from 1 kg water/hr to >1000 kg/hr. They are used extensively to control the humidity of warehouses, hotels, supermarkets, electronic assembly plants, food and pharmaceutical manufacturing facilities, etc. (Banks 1990). To provide evaporative cooling, however, they must reduce the humidity of the air to very low levels (dew point lower than 32°F or 0°C) and this is most easily achieved by the use of zeolites.

The process of air drying in a rotary dehumidifier is almost adiabatic; therefore, the heat of adsorption that is released during drying significantly increases the temperature of the passing air and the adsorption surfaces. If the hot dry air is cooled at this stage by evaporating water, it will become as warm and humid as it was before entering the rotarywheel. For this reason, a second wheel with large thermal mass, usually made of metal, is added as a rotary regenerative heat exchanger after the desiccant wheel. This heat exchanger reduces the temperature of the hot dry air before it is cooled further by water evaporation and preheats the air used for regenerating the desiccant wheel.

The first open-cycle solar cooling system utilizing rotary solid desiccant, the Solar-MECTM (Munters Environmental Control), was constructed in the late 1970's by the Institute of Gas Technology of Chicago, Illinois (Macriss et al. 1981). It consisted of a desiccant wheel made from corrugated paper and impregnated with hygroscopic salt followed by the regenerative heat-exchanger wheel made from aluminum honeycomb. The need to use zeolites became obvious immediately and they developed successfully,

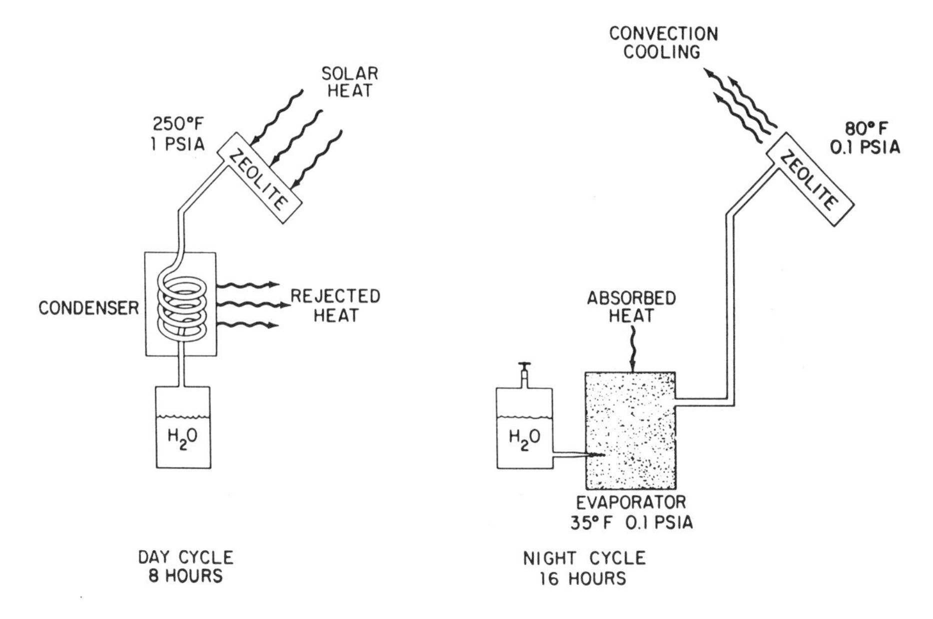

Figure 1. Operation of a rotary desiccant wheel dehumidifier. The desiccant dehumidifies incoming moist air, and the preheated exhaust air reactivates the desiccant wheel.

with a paper manufacturer, a zeolite containing paper with more than 50% synthetic zeolite added to the paper pulp before extrusion. The Zeopower Company suggested that they use natural zeolites and provided them with chabazite-rich powder from a deposit near Bowie, Arizona. Their paper manufacturer successfully molded the chabazite-rich powder into paper. Samples of the paper were provided to The Zeopower Company for testing. These samples exhibited 60% of the adsorption properties of the bulk zeolite, despite the dilution resulting from the presence of paper. The use of corrugated paper for construction of the rotary adsorber resulted in mechanical instability (wobbling) during rotation and poor seals between the different sectors and the periphery of the wheel. Considerable progress has been made during the past 20 years in construction materials, and desiccant wheels presently are made mostly from specialized plastic or extruded zeolite ceramic honeycomb. Engelhard/ICC presently markets titanium silicate molecular sieve rotary-wheel cooling systems that can be regenerated at temperatures as low as 60°C from various waste heat sources (Grzanka 1996).

Van den Bulck et el. (1985) described the design theory for rotary heat and mass exchangers, and Van den Bulck (1988) reported the performance characteristics of open-cycle solid-desiccant systems. Extruded matrix wheels or impregnated honeycomb rotors made from hydrophobic zeolites are being used presently for the removal of volatile organic compounds (VOC) and solvents from the exhaust stream of painting shops and plastic manufacturing facilities to decrease pollution. Drohan (1992) compared the performance of different rotors made from activated carbon and hydrophobic zeolites for the removal of low concentration VOC from high volume airflow. He determined that activated carbon rotors function well at low relative humidity (below 55%), and high solvent concentration (above 100 ppm). Hydrophobic zeolites resist water adsorption until the relative humidity exceeds 90% and function well even at very low solvent concentration (from several hundred ppb up to 200 ppm).

Closed-cycle cooling

The first closed-cycle, continuous-cooling sorption system, using water as the sorbent and ammonia as the refrigerant, was constructed by Ferdinand Carrè in 1859. Carl Munters and Baltzar von Platen developed the liquid/gas ammonia absorption refrigerator in the early 1920s. This gas-fired unit was mass-produced by Electrolux and is still widely used in many third-world countries and in motor homes in the United States. Although some domestic refrigerators of this type are still in operation in this country, no new units can be installed in homes because of the danger of poisoning from ammonia leaks.

Another closed-cycle absorption system utilizing water vapor for refrigerant and LiBr solution as the sorbent is widely used for large size commercial and industrial air-conditioners energized by natural gas. This system requires water-cooling towers to reject heat to the ambient air at relatively low temperatures and, therefore, is used mostly in office buildings or shopping centers in which regular maintenance and trained personnel are available to assure proper operation of the cooling tower.

Solid/gas adsorption systems have been proposed and developed with silica gel, activated alumina, or carbon as the adsorbents and water, different alcohols, ammonia, or hydrocarbons as the refrigerant gas (Muradov and Shadeyev 1971; Shadeyev and Umarov 1972; Delgado et al. 1982; Vasiliev et al. 1992). Because silica gel, activated alumina, and carbon are surface adsorbents, their isotherms, as well as those of the liquid/gas systems, are linear, i.e. they obey the Arrhenius equation (Tchernev 1979). Zeolites on the other hand have extremely non-linear (Type I) isotherms, and the interaction between the adsorbed gas and the internal fields in the cavities of the solid are very complex (Dubinin and Astakhov 1971). Meunier (1978) reviewed the experimental attempts to use sorption cycles for refrigeration with solar energy as the only power source, and he concluded that zeolites with water vapor or alcohol as the working gas are superior to other liquid or solid sorbent-gas combinations.

Tchernev proposed two approaches in 1973 to explore the use of zeolites for the cooling of buildings with solar energy, based on the non-linearity of zeolites isotherms. The temperature of the zeolite is almost uniform in the first approach and the zeolite is alternately heated and cooled, resulting in cyclic desorption and adsorption. For the second approach, a temperature gradient is established across a solid zeolite body resulting in a continuous desorption from the hot end and adsorption at the cold end, with continuous diffusion simultaneously taking place along the concentration gradient in the zeolite (Tchernev 1974).

In the temperature-gradient experiments zeolite pellets of 6 mm and 50 mm diameter were hot-pressed to densities as high as 95% of theoretical crystallographic density (Tchernev 1976). Pressure differences of 1/10 atmosphere (75 mm Hg) were obtained within minutes with water vapor as the refrigerant and temperature gradients of less than 100°C. The technology of producing large areas of solid zeolite bodies for solar collectors and sealing them along the edges, however, appeared too difficult at that time. Additionally, leaks around bad seals short-circuit the operation of the system; hence the main research effort was concentrated on the uniform temperature approach.

Because of their non-linear isotherms, zeolites can convert small changes in temperature into large variations in pressure (Tchernev 1978). Pressure ratios approaching 100 to 1 can be achieved with temperature differentials as small as 150°C. The high pressure ratio permits closed-cycle zeolite adsorption systems to operate with an air-cooled condenser, eliminating the need for a cooling tower and reducing the initial cost and operation complexity of the zeolite cycle.

Considering a number of potential polar gases as part of the solid-gas refrigeration system, and expanding on the work done by Meunier (1978), Tchernev (1980) determined that of all possible candidates in the temperature range of interest (0 to 10°C evaporator temperatures) water vapor delivers the highest cycle efficiency. Most zeolites adsorb the same amount of refrigerant (about 20% by weight) (Tchernev 1977), and water has the highest heat of vaporization of all candidate refrigerants (2330 kJ/kg vs. 1160 kJ/kg for alcohols and ammonia and 230 kJ/kg for halocarbons). In addition, water vapor is a refrigerant that is chemically stable in the presence of zeolites at 200°C, whereas some of the other refrigerants undergo catalyzed chemical reactions, such as methanol on synthetic zeolite ZSM-5 (Olson et al. 1981).

Operating principle of the intermittent zeolite cycle

Figure 2 illustrates the principles of operation of a zeolite solar-cooling system. The zeolite is sealed in a blackened, airtight container, which is irradiated by the sun. During the daytime part of the cycle, the zeolite is heated to a maximum temperature of 120°C (248°F). Water vapor desorption first begins at about 40°C zeolite temperature. This desorption increases the water vapor pressure in the container from 5 to 55 mm Hg, corresponding to a water dew point increase from 0 to 40°C. During this part of the cycle most of the thermal energy entering the container is consumed by the heat capacity of the

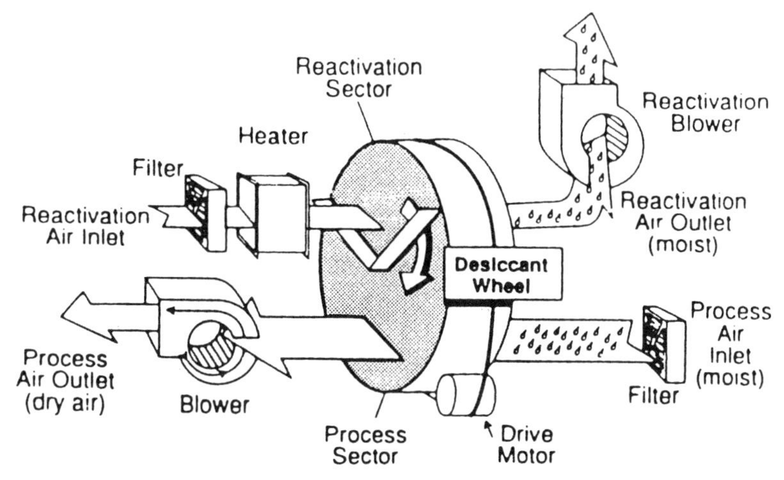

Figure 2. The basic solar zeolite cycle. During the 8-hr-day portion of the cycle, the heated zeolite desorbs water vapor that is condensed into liquid and stored. During the 16-hr night, the cool zeolite adsorbs water vapor from the evaporator, producing cooling.

zeolite, augmented by the adsorbed water vapor. The main desorption takes place along a constant pressure isobar determined by the condenser temperature; for example, 55 mm Hg (1 psia) at 40°C. The vapor begins to liquefy as the heat of condensation is rejected to the outside air, and liquid water flows into a storage tank.

The heat input of the system is determined by integrating the total energy entering the zeolite container during the day. Some of the thermal energy is used again by the heat capacity of the zeolite, but most of it goes into the desorption of water, with heats of desorption ranging from 3000 to 5000 kJ/kg. The potential cooling output generated equals the heat of vaporization of water vapor (at 40°C it is about 2400 kJ/kg) that was liquefied. Dividing the cooling output by the heating input provides the efficiency of the zeolite cycle, typically between 35% and 50%, which is reduced further by collector losses.

During the nighttime part of the cycle, the zeolite is cooled by convection and radiation to ambient temperature and can adsorb water vapor at low pressure. Liquid water from the storage tank is introduced into the evaporator where it absorbs heat from the space to be cooled and is converted into vapor. The zeolite adsorbs the water vapor produced by the evaporator, maintaining the pressure at 5 mm Hg (0.1 psia) while the water boils at less than 1°C, and heat of adsorption is rejected to the atmosphere. At the end of the nighttime cycle, the zeolite has loaded all the water it can adsorb at 5 mm Hg pressure and is ready for the beginning of a new day cycle.

Tchernev (1978) first described a solar refrigerator that operated on the zeolite/water cycle (Fig. 3). It utilized natural chabazite from a deposit near Bowie, Arizona, and, with a zeolite loading of 50 kg/m^2, was capable of producing 9 kg of ice per day for each square meter of collector area with an overall efficiency of 15%. In comparison, Trombe and Foex (1958) produced 5 kg/m^2 ice per day with an ammonia/water refrigerator utilizing parabolic concentrating solar collectors. A number of other groups have since tested solar refrigerators utilizing zeolites and obtained similar results. Meunier et al. (1979) determined theoretically an overall solar efficiency of 17% for the synthetic zeolite 13X-water cycle. Guilleminot et al. (1979, 1980) and Guilleminot and Meunier

Figure 3. The solar zeolite adsorption refrigerator. The 0.75 m^2 zeolite collector produces 6.75 kg of ice per day.

(1981) described the theoretical and experimental evaluation of solar refrigerators, again utilizing the synthetic zeolite 13X/water cycle. Using synthetic zeolite 13X and water, Dupont et al. (1982) produced between 7 and 8 kg ice/m^2 per day with a solar refrigerator in Montpellier, France, and Pointe a Pitre, Caribbean, with an overall efficiency of 10 to 14%. The ice production was increased to 12.5 kg/day when an east-west reflector was used to increase the solar flux incident on the solar collector by 50%.

Fejes et al. (1988) investigated a solar refrigerator utilizing Hungarian Ca^{2+}-exchanged clinoptilolite. They projected a water desorption capacity of 1 kg/m^2 or about 7 kg of ice production per day for each square meter of solar collector. Brossard et al. (1991) described a solar refrigerator, which used a Cuban natural zeolite and methanol as the refrigerant.

Eichengruen and Winter (1993) compared the thermal behavior and periodic performance of the zeolite-water adsorption refrigerator with the results of numerical simulation of its dynamic behavior. They determined that both simulation and experimental studies of the dynamics indicate a strong dependence on the thickness of the zeolite layer. In India, Phadke (1984) investigated the use of natural clinoptilolite from western Maharashtra Province for solar refrigeration. Poredos (1994) described a simple synthetic zeolite-water adsorption refrigerator and showed that the measured experimental parameters compare favorably with the predictions of a parametric model for the adsorption cycle.

Schwarz (1994) described a refrigerated trolley for serving food on German passenger trains. The synthetic zeolite 5A/water system is regenerated at the station where the water in the trolley's evaporator is frozen. The trolley is then loaded with food and the unit is rolled on the train where cooling continues for a few hours. Phelouzat et al. (1994) described a slightly different approach to food refrigeration and ice making. At the beginning of the cycle the system is open to the atmosphere; after the food has been inserted, the container is closed and the air is removed. As the pressure is reduced, the water contained in the food or in a reservoir begins to evaporate and the water vapor is adsorbed by a synthetic zeolite. This process chills the food or the water in the reservoir, resulting in refrigerated meals or chilled water. After the cooling cycle is completed, the container is opened, the food is removed, and the zeolite is regenerated before being reused during the next cycle. Using the same principle, Takasaka and Matsuda (1986) reported the vacuum freeze-drying of food using natural clinoptilolite from Itaya, Japan. The food is packed in alternating layers with zeolite, the air removed, and as water evaporates the food is frozen and dried at the same time. Circulating air at low pressure can shorten the drying time substantially because it promotes heat and mass transfer.

Figure 4. Walk-in modular solar refrigerator being tested in Kenitra, Morocco. The unit produces 100 kg of ice on a sunny day.

A modular walk-in zeolite refrigerator that used chabazite-rich tuff from a deposit near Bowie, Arizona (Fig. 4), was tested in Morocco (Tchernev 1995). The four-module unit has a refrigerated volume of 15.4 m^3 and internal dimensions of $2 \times 4.8 \times 1.6$ m. The collector area of each module is 2.9 m^2 and it produces 25 kg ice/day per collector, for a system total of 100 kg ice/day at an ambient temperature of 40°C. Each module is separately evacuated and sealed, thereby improving the reliability of the system. The modular nature of this system allows the construction of large on-site refrigeration units for perishable products in areas without available electric supply. Tchernev (1984) described a solar milk cooler using natural chabazite from a deposit near Bowie, Arizona. The cooler uses a zeolite collector of 1.5 m^2 area, and, on a sunny day, it can cool 60 liters of milk (two, thirty-liter milk cans) from 35 to 15°C for a total cooling load of 1200 kcal/day. The cooler contains a water basin having a volume of 165 liters, which is cooled during the nighttime to about 7°C. After the containers at 35°C are added during the daytime, the temperature of the combined system approaches 15°C. The cooler will treat the milk from a few cows every day, preventing spoilage in areas where electricity is unavailable.

Water is the refrigerant and storage medium in solar refrigeration systems because part of the water is frozen into ice every night. The ice slowly melts during the daytime while maintaining constant temperature in the refrigerated space. For example, the refrigerated-volume temperature remains constant at +3°C if the ice melting temperature is reduced to -2°C (by salt addition) and the door remains closed. For lower temperatures, as in deep freezers, alcohols are better refrigerants because the vapor pressure of water drops rapidly with temperature and (below the knee of the Type I isotherm at 1 mm Hg) the adsorption capacity of the zeolite is reduced considerably.

Tchernev (1995) reported freezer temperatures of -20°C with the Bowie, Arizona, chabazite and methanol as the refrigerant. Brossard et al. (1991) described the use of Cuban zeolites with methanol; however, the selection of the proper zeolite is of utmost importance in as much as alcohols are catalytically decomposed by most zeolites at 150°C to 200°C (e.g. the first step of methanol conversion to gasoline, Olson et al. 1981). For this reason hydrophobic zeolites with high Si/Al ratios or activated carbon may be preferred for freezer applications utilizing alcohols. Many researchers have performed thermodynamic analyses of the closed-cycle solar adsorption refrigerator (Guilleminot 1978; Meunier and Mischler 1979; Chang and Roux 1985; Hajji et al. 1991); this topic is discussed in more detail below.

ZEOLITE HEATING/COOLING SYSTEMS

Intermittent solar heating/cooling systems

A complete zeolite solar system for heating and cooling of buildings is shown in Figure 5. Tchernev (1978) described this system and Meunier et al. (1979) described a similar system. The main building block of such a system is the solar zeolite collector (Tchernev 1981a). In the integrated zeolite collector (Tchernev 1981b), the condenser and evaporator are combined into a single heat exchanger, which is connected to an external water loop. During the daytime, the water vapor desorbed from the solar-heated zeolite is condensed in this heat exchanger, and the heat of condensation is rejected to the external water loop, thus providing hot water for various heating needs. At nighttime, water in the condenser/evaporator unit evaporates, and the vapor is adsorbed by the zeolite. The external loop provides the necessary heat of vaporization, producing chilled water for use in air conditioning (Tchernev 1982). Each integrated solar zeolite collector has an area of 1.5 m^2 and on a sunny day will provide about 1.5 kwh/m^2 of heating and cooling output (Tchernev 1995). Pairs of such collectors were tested in Natick, Massachusetts; Golden, Colorado; and Tucson, Arizona; test data are presented in detail elsewhere (Tchernev 1983) and the results are summarized in Table 2. Eight such collectors were used to cool an insulated room used to house electronic equipment in the desert. The unit was field-tested in Jebel Ali, United Arab Emirates, and successfully maintained an indoor temperature of <30°C, even under ambient outdoor temperatures >50°C (Tchernev 1995).

The integrated zeolite collector, however, was developed mainly for the residential

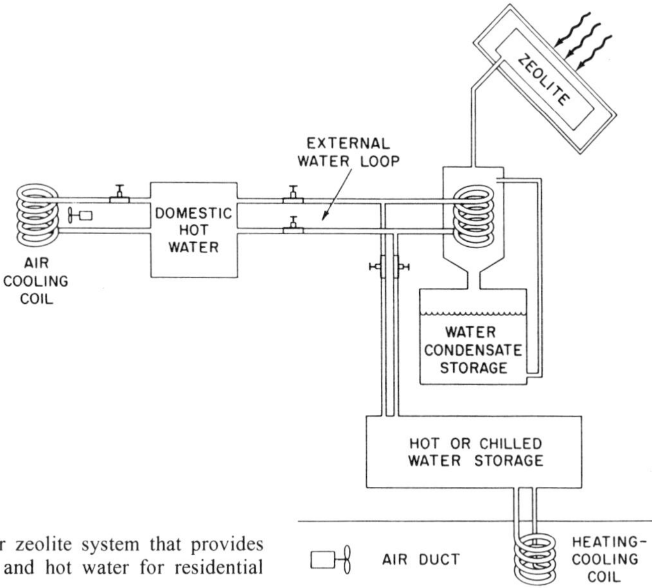

Figure 5. A solar zeolite system that provides heating, cooling, and hot water for residential buildings.

Table 2.. Zeolite solar-cooling performance (Tchernev 1983).

	Solar energy input = 20,000 to 26,000 kJ/m²day	
	Collection efficiency 50 to 60% Solar energy collected and delivered to the zeolite system: 10,000 to 13,000 kJ/m²day	
	Cooling produced	
Ice Box	**Milk Cooler**	**Integrated Collector**
7-9 kg/m²day at 0°C 3,000 kJ/m²day	60 liter milk at 7°C 750 kJ/m²day	Cooling at 7°C 5,500 kJ/m²day
	Overall system efficiency	
14 to 15%	4 to 5%	20 to 25%
	Zeolite system efficiency	
25 to 30%	7 to 8%	40 to 50%

home market. For a typical single-family home in the United States, 60 m^2 of collector area will provide heating, cooling, and hot water needs in almost any climate area. Such a home, shown in Figure 6, was built and tested in Denver, Colorado (Tchernev 1995). Six arrays of six collectors each having a total surface area of 54 m^2 were installed at ground level with 240 m^3 of insulated storage tanks for hot/chilled water that were buried in the ground between the arrays. The system operated continuously for a year and provided more than 90% of all energy needs for the house.

Heat pumps with energy regeneration

Integrating the zeolite in a solar collector results in a simple intermittent system in

Figure 6. A solar heated and cooled house in Denver, Colorado. The adsorption/desorption bed uses a chabazite-rich tuff from a deposit near Bowie, Arizona.

which the desorption/adsorption process is controlled by the daily solar cycle. The efficiency of such a system however is controlled by the collection efficiency, which is typically between 50 and 80%. The inherent losses of the adsorption process, in which the heat of adsorption is rejected to the ambient, result in overall efficiencies of about 30% on cooling and 60% on heating. Considerably higher efficiencies can be obtained with a regenerative adsorption heat pump, however, at the expense of increasing the complexity of the system. Because the zeolite cycle is inherently intermittent, and in order to provide continuous operation, the zeolite is divided into at least two separate sealed containers that are indirectly heated by a high-temperature heat-transfer oil. In this configuration, one of the containers is always in the desorption part of the cycle and the other is always in the adsorption part. This configuration permits a significant exchange of thermal energy from the zeolite being cooled (adsorber) to the zeolite being heated (desorber), thereby reducing the energy consumption by almost half.

The concept of energy regeneration is schematically represented in the energy-balance diagram of Figure 7. The heat transfer oil in the loop leaving the reversible gear pump is preheated by the adsorbing zeolite, which is being cooled from the previous desorption mode. The adsorber slowly cools while continuously adsorbing refrigerant vapor from the evaporator at low pressure. The generated heat of adsorption plus the specific heat of the zeolite and container preheat the heat-transfer oil to a temperature that decreases in time from 200° to about 50°C.

The preheated oil leaving the adsorber is warmed by the heater (a gas boiler or concentrating solar collector) to about 200°C before entering the zeolite container in the desorption mode. Here, the zeolite is heated from about 50°C to about 200°C while the refrigerant is desorbed at a sufficiently high pressure for it to liquefy at the condenser temperature. The heat-transfer oil leaves the desorbing zeolite considerably cooler at the beginning of the desorption cycle, and its temperature increases to about 200°C at the end of the cycle. It is further cooled to about 40°C in an oil heat exchanger before it closes the loop of the reversible gear pump. After this portion of the cycle is completed, the gear pump direction is reversed. The desorbing zeolite becomes the adsorber, the direction of oil flow through the heater is reversed, and the adsorbing zeolite begins desorbing the refrigerant. The cycle is repeated again and again during the operation of the system.

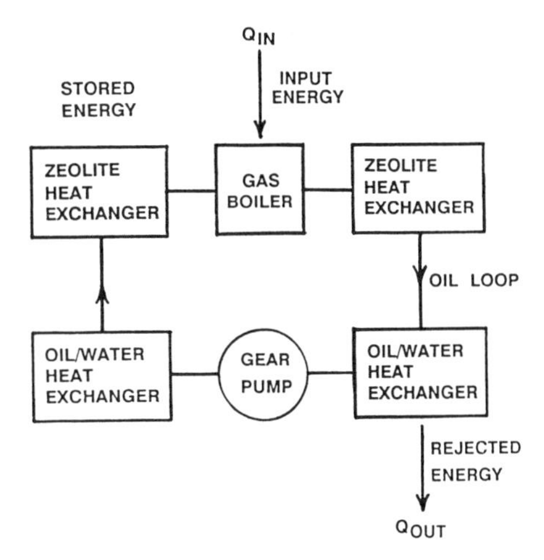

Figure 7. The energy balance diagram and regeneration principle. The necessary input energy provided by the source, Q_{in}, equals the energy rejected to the outside, Q_{out}, plus any energy losses, Q_{loss}.

A simple energy balance indicates that the heater has to provide the energy rejected by the heat exchanger plus any additional losses from the system. The energy stored in the zeolite, however, is recycled from adsorber to desorber, back and forth, and this energy regeneration increases the efficiency of the cycle from 35% to about 65% (Tchernev and Clinch 1989b). If the zeolite/hot oil heat exchanger is constructed such that the heat transfer from the oil to the zeolite is faster than the heat flow along the path of the oil, a temperature front will be created in the zeolite. This temperature front will slowly propagate along the length of the heat exchanger in the direction of motion of the oil. Such a heat exchanger with a propagating temperature front can regenerate >80% of the thermal energy from the adsorber to the desorber, and the efficiency of the cycle becomes closer to the theoretical efficiency. This type of heat exchanger has demonstrated energy regeneration efficiencies of 75 to 80% (Tchernev and Emerson 1990).

The field of regenerative adsorption heat pumps has gained considerable interest among researchers around the globe. Meunier (1985, 1986b), Guilleminot et al. (1987), and their coworkers in France reported a theoretical analysis of adsorption heat pumps, and they performed experimental studies of cascading cycles using the zeolite/water and activated carbon/methanol pairs for heating and cooling in which they demonstrated a 50% increase of cooling efficiency. In cascading cycles, first presented and classified by Alefeld (1983), the cooling sorption system consists of at least two separate stages: a high temperature stage (usually zeolite/water) and a low temperature stage, which utilizes the waste heat of the high temperature stage thus increasing the cycle efficiency.

Alefeld et al. (1981), Maier-Laxhuber et al. (1983), Rothmeyer et al. (1983), and Peters et al. (1986) in Germany investigated theoretically and experimentally adsorption heat pumps and heat transformers with the zeolite/water combination and constructed a 1 to 5 kW heat pump prototype utilizing 17 kg synthetic zeolite NaX. The term "heat transformer" was introduced by Nesselmann (1933) because the increase of the waste heat source temperature generated by the zeolite is similar to the voltage increase in an electrical transformer. In a heat transformer application, the zeolite is cooled to the waste-heat source temperature with the water-vapor pressure determined by the ambient temperature. When water vapor at the waste heat source temperature is added to the zeolite, the generated heat of adsorption raises the zeolite temperature to a higher level, where the waste heat is recovered and utilized.

Sakoda and Suzuki (1984), Suzuki (1992), and coworkers in Japan investigated both silica gel/water and synthetic zeolite NaX/water pairs for solar adsorption heat pumps and automotive cooling applications. They estimated that a typical air conditioning system of 3.5 kW capacity will require 570 kg of silica gel in a packed bed with a regeneration temperature of 100°C. Restuccia et al. (1988), Cacciola and Restuccia (1994), Cacciola et al. (1996), and their coworkers in Italy investigated the use of both synthetic zeolites (3A, 4A, 5A, and 13X) and natural clinoptilolite for heat pumps and heat transformers and determined that zeolite 13X, with a heat output of 0.262 kWh/kg, is most suitable for heat pump applications and a 5 kW system requires 20 kg of zeolite.

Gonzales et al. (1991) studied the use of Cuban natural zeolites, mostly clinoptilolite and mordenite, for air conditioning and dehumidification applications with energy regeneration, utilizing direct and indirect evaporation. They constructed and successfully demonstrated the performance of a natural zeolite air conditioning system. Zhang and Guo (1988) studied the use of natural clinoptilolite and mordenite and synthetic zeolites A, X, and Y in heat pump and solar applications in China, and in Turkey, Ülkü (1986, 1992, 1994), Ülkü and Mobedi (1989), and coworkers have focused their work on local natural zeolites (mostly clinoptilolite) and water vapor for use in energy storage and heat pumps. They constructed and tested a system containing 25 kg of clinoptilolite from the Bigadic region of Turkey. A cooling efficiency of only 33.5% was achieved and they suggested that the low-grade clinoptilolite tuff would require beneficiation (i.e. increase the clinoptilolite content) to increase its adsorption capacity.

A regenerative zeolite adsorption heat pump with a propagating temperature front, based on the zeolite/water cycle, was designed, constructed, and performance tested by Tchernev and Clinch (1989a). The heat pump was operated by a hot fluid (150 to 200°C), which can be supplied by concentrating solar collectors, waste heat, or any type of fossil fuel. This heat pump has two operating capacities and cycles between the two to match its output to the actual load. The cooling output at high speed is 10.5 kw with an efficiency of 55% and 2.5 kw at low speed with 110% efficiency. The heating output is 30 kw at an efficiency of 150% and 5.8 kw at 190% efficiency. The seasonal cooling efficiency in Dallas, Texas, is 100%, whereas in Chicago, Illinois, it is 130%. The seasonal heating efficiencies are 184% in Dallas and 143% in Chicago.

To keep the initial cost of this heat pump comparable with competing electrical units, small, regenerative, zeolite heat exchangers with fast dynamic properties were developed (Tchernev 1993). Both fast heat transfer (high zeolite thermal conductivity) and mass transfer (fast vapor diffusion in the zeolite) were achieved by controlling the physical properties (high density and porosity) during the preparation of solid zeolite bodies. Sintered zeolite tiles made of chabazite from a deposit near Bowie, Arizona, had a thermal conductivity of 0.17 W/m°C with a diffusion coefficient of 3×10^{-6} cm^2/sec at 40°C. The high thermal conductivity resulted in an increase in zeolite cooling output from 53 w/kg of zeolite in 1986 to 116 w/kg in 1991, thus reducing the projected final cost of the manufactured heat pump to about $115/kw (in 1991 dollars). This cost is competitive with that of electrically driven vapor-compression heat pumps; however, further reduction in weight and size are possible if the cycle time of the system is made shorter. A large number of researchers are presently working on the improvement of the cycle dynamics and shorter cycle times can be achieved soon (Ülkü 1994; Cacciola et al. 1996; Restuccia et al. 1999).

With the increased concern over the ozone-depletion effect that chlorinated fluorocarbons (CFCs) produce in the atmosphere and the phase-out of their production, considerable interest has been focused on the zeolite/water system as a replacement for the conventional vapor-compression cooling systems. The automobile industry over the past five years, especially in Europe and Japan, has conducted feasibility studies to evaluate the use of the waste heat of the internal combustion engine as the heat source for the zeolite air-conditioning cycle. The typical automobile air-conditioner has a capacity of about 3 kW and weighs about 25 kg. A zeolite system that replaces it must therefore use no more than 10 kg zeolite because evaporator, condenser, containers, pumps, and fans contribute the remainder of the system's weight. Schwarz (1991) reviewed different sorption methods and the application of synthetic zeolite NaA-based processes to automobile air conditioning. Schwarz et al. (1991) described a 3 kw zeolite-water system for cooling the passenger compartment in motor vehicles. Suzuki (1992, 1993) projected a cooling output of 2800 w/kg for an automobile adsorption air-conditioning unit with 2 kg of zeolite in a zeolite-water system of 3 kw capacity, utilizing the waste heat of the engine with a 60 sec cycling time for adsorption and desorption. The estimates of Suzuki do not seem to be realistic inasmuch as cycle times presently are about 5 to 6 min and power densities are about 300 w/kg (Höppler 1992, 1993). Therefore, a 3.5-kW adsorption system will require 10 to 15 kg of zeolite, have a total weight of 45 to 50 kg, and a volume of 75 to 100 liters according to Höppler (1993). However, Gentner et al. (1993, 1994) found the performance of a zeolite automobile air conditioning unit inferior to a refrigerant R134a-based conventional unit on a unit mass and unit volume basis. Furthermore, Davias (1995) concluded that very high efficiency heat exchangers would be needed for the recovery of the exhaust waste heat of automobiles because the available power is at low temperatures, especially during city driving. Tchernev (1999) confirmed these results and concluded that at idle and city driving conditions there is insufficient exhaust waste heat available to operate a zeolite/water adsorption air conditioning system for automotive vehicles, even if the system is very efficient and has a high Coefficient of Performance (COP). The waste heat available from the radiator cooling circuit is sufficient even at idle, but only at temperatures lower than the exhaust heat. A considerable portion of the present work in the field is focused on reducing the cycle time (especially during adsorption) and further progress in this direction is expected to achieve a cycle time of 1 to 2 min in the near future.

ENERGY CONSIDERATIONS

The equilibrium condition of a gas adsorbed on a microporous adsorbent can be described by three parameters; (1) the adsorbent temperature; (2) the gas pressure (or its dew point temperature); and (3) the mass fraction of the adsorbed gas (kg gas/kg dry adsorbent), also referred to as adsorbed weight percentage or filling factor. These properties are usually presented as (1) isotherms, i.e. weight percent as a function of gas pressure at constant adsorbent temperature, (2) isobars, i.e. weight percent as a function of adsorbent temperature at constant gas pressure, and (3) isosteres, i.e. gas pressure (or dew point) as a function of adsorbent temperature at constant mass fraction or weight percent.

Isotherms for the adsorption of polar gases on zeolites are usually extremely nonlinear, (i.e. Type I isotherms). The adsorbed weight percent rises very rapidly with pressure and saturates at a value almost independent of pressure but strongly dependent on temperature. The solubilities of gases in liquids and the adsorption of gases on surface adsorbents, such as silica gel or activated alumina, depend exponentially on H/RT, where H is the energy of solution or adsorption, R is the gas constant, and T is the absolute temperature (i.e. they obey the Arrhenius equation). Adsorption of polar gases on zeolites, on the other hand, has been shown by Dubinin and Astakhov (1971) to depend exponentially on at least the second, and as high as the fifth, power of H/RT. This extreme nonlinearity of thermal activation makes zeolites uniquely well suited for solar energy storage and heat pump applications because they can adsorb large quantities of refrigerant gas at low pressures.

Isobar diagrams are most often used to represent experimental data because under laboratory conditions it is convenient to keep the pressure or dew point temperature of the gas constant while changing the zeolite temperature and recording its weight. For energy considerations and thermodynamic analysis, however, the adsorption isosteres are most convenient to use because under equilibrium conditions they provide significant information about the heat of adsorption of the refrigerant gas on the zeolite. The equilibrium isosteres are usually presented in two equivalent forms; lines of constant adsorbed wt % in a *ln*P vs. -1/T diagram, or in a dew point temperature vs. zeolite temperature diagram. The latter form is usually preferred because both abscissa and ordinate axes are calibrated in temperature. For example, the isosteres for water

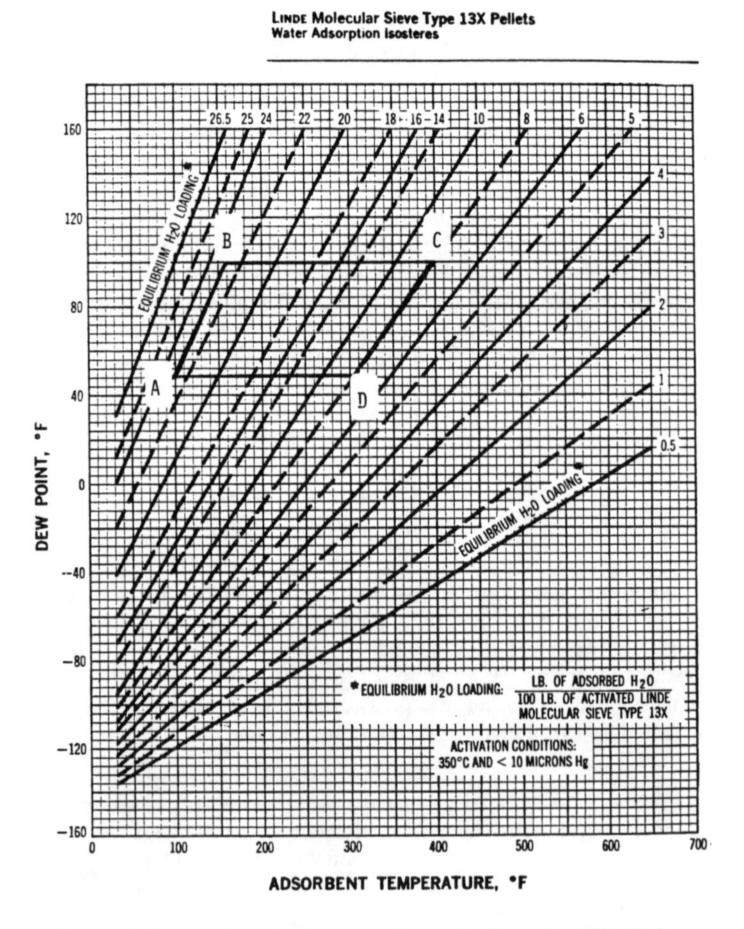

Figure 8. Water adsorption isosteres for synthetic zeolite 13X (Union Carbide, Linde Molecular Sieves Division, Data Sheets). See text for discussion on Points A, B, C, and D.

adsorption on synthetic zeolite 13X are shown in Figure 8. As the zeolite is heated from point A to points B and C, water vapor is desorbed at 40°C (100°F) dew point temperature and the adsorbed mass fraction changes from 23 to 8 wt %. When the zeolite is cooled from point C to points D and A, water vapor is adsorbed at 7°C (45°F), and the adsorbed mass fraction returns to 23 wt %.

The isosteres in both representations are straight lines because at equilibrium they must satisfy the Clausius-Clapeyron equation. Their slope is always larger than 1 because the adsorbed gas phase is in a lower energy state than the gas vapor, in order for the adsorption process to take place. The isosteric heat (enthalpy) of adsorption (H) is given by the Clausius-Clapeyron equation:

$$d\,ln P/dT = H(T)/RT^2, \qquad (1)$$

where P is the gas pressure, R is the gas constant, and T is the zeolite temperature. Because the same equation applies to the liquid/gas phase at the dew point with the latent heat of vaporization (L) replacing H, it is easy to show (Breck 1974; Meunier and Mischler 1979; Ülkü 1986) that the slope of the isostere is equal to the ratio H/L, i.e. the isosteric heat of adsorption (H) at any temperature and pressure is the latent heat of vaporization (L) of the gas at the same temperature and pressure multiplied by the slope of the isostere. This fact permits rapid thermodynamic analysis of any adsorption process as long as the adsorption isosteres of the material are known, which unfortunately is not the case for most natural and many synthetic zeolites.

In Figure 8 the slope of the 8% isostere is 1.85 but the 23% isostere has a slope of 1.15; therefore, the isosteric heat of adsorption will be 1.15 or 1.85 times the heat of vaporization of water found in the steam tables, which at these temperatures is about 2400 kJ/kg of water. For example, the isosteric heat of adsorption at point C of Figure 8 will be 1.85 × 2400 = 4440 kJ/kg of water, whereas at point B the heat of adsorption will be 1.15 × 2400 = 2760 kJ/kg of water. To calculate the energy storage density discussed above, the mass average of the heat of adsorption between points B and C must be determined and then multiplied by the weight percent desorbed between these two points. Unfortunately, this average is difficult to calculate because the heat of adsorption is a nonlinear function of the amount adsorbed and declines with increasing water adsorption. Ülkü (1986) showed that for clinoptilolite from Turkey, the isosteric heat of adsorption could be fit well to a fourth-order polynomial of the weight percent adsorbed.

The effect of this nonlinearity is also visible in Figure 8, in which the spacing between isosteres varies widely. For example, for the 120°F (50°C) dew point line, the spacing between the 20 and 24% isosteres is 85°F (47°C), whereas the spacing between the 10 and 18% isosteres is 95°F (53°C). Therefore, for about the same temperature change twice as much desorption (8% between 150 and 200°C vs. 4% between 71 and 118°C) can be obtained depending on the adsorbent temperature. The spacing between isosteres is an important design parameter for both solar energy storage and cooling applications. For storage, closely spaced isosteres with high heat of adsorption are required in order to obtain large energy densities without the use of very high temperatures and the resulting losses. For solar cooling, closely spaced isosteres with low heat of adsorption are required in order to increase the efficiency of the cooling cycle. For this reason, more than 20 years ago Tchernev (1974) defined a Figure Of Merit parameter as the weight % desorbed per °C or as kg water/kg zeolite °C. When plotted as a function of zeolite temperature at constant condensation pressure, this parameter should be larger than 0.001 (or 1% desorbed for every 10°C) over a reasonably large temperature range in order for a zeolite to be useful for a practical application. For synthetic zeolite 13X in Figure 8 in the range between 18% and 10%, the parameter is 8%/53°C or 0.0015 kg/kg°C (the slope of the isostere is 1.5), and the heat of adsorption is about 4000 kJ/kg water, resulting in an energy density of 6 kJ/kg zeolite °C or 20 times the specific heat of water. The large value of the Figure Of Merit parameter explains why this particular zeolite exhibits the largest energy storage density for heat sources with temperatures between 150 and 200°C (Restuccia et al. 1988). For solar cooling applications, however, this zeolite is less well suited because the Figure Of Merit parameter is only 0.0085 kg/kg°C in the temperature range of interest between 70 and 120°C. Natural chabazite from a deposit near Bowie, Arizona, on the other hand, has a parameter of 0.0015 at temperatures <150°C and isosteres with a low slope of 1.15, making it preferable for solar cooling applications (Tchernev 2000). In Figure 8, if the zeolite is heated from point A to point B along the isostere, the heat energy supplied is used almost entirely to heat the zeolite with its adsorbed water to the temperature at point B because almost no desorption takes place. The specific heat of the dry zeolite is about 1.2 kJ/kg°C, whereas that of the adsorbed water is closer to that of ice, (2.1 kJ/kg°C multiplied by the weight % of the adsorbed water) because the adsorbed molecules are not as free to move as they are in the liquid state. Desorption of water begins at point B at a zeolite temperature T_z (also called initiation temperature) determined by the condensation pressure (or dew point) of the water vapor and the slope of the isostere. Zeolites having low isosteric heats of adsorption (low slope of the isostere) are more efficient for cooling applications because desorption starts at lower temperatures and less energy is required to heat the zeolite from point A to B before useful desorption takes place. Chang and Roux (1985) and the ensuing discussion (Meunier 1986b, Chang and Roux 1986) emphasized the importance of the slope of the isosteres in determining the initiation temperature of the desorption process of a solar zeolite refrigerator. They specifically addressed the differences between synthetic zeolite 13X and natural chabazite from Bowie, Arizona, and the effect of the lower initiation temperature for the chabazite on improving the cycle efficiency.

In Figure 8, from point B to C the input energy is divided into three parts: the specific heat of the dry zeolite, the specific heat of the decreasing weight % of adsorbed water, and the heat of adsorption of the desorbed water vapor. The first part is almost independent of temperature. The specific heat of the adsorbed water can be represented with reasonable accuracy by the average adsorbed wt % between points B and C, multiplied by the specific heat of ice. The heat of adsorption is the most difficult to estimate without lengthy numerical computer simulations because of the constantly changing slopes of the isosteres. For most polar gases adsorbed on zeolites, however, the Figure Of Merit decreases with increasing zeolite temperature, whereas the isosteric heat of adsorption increases. The product of these two quantities remains almost constant and has the appearance of an effective specific heat, which can be used in an energy-balance calculation with reasonable degree of accuracy (Tchernev 1992). If at point C the process is reversed immediately and the losses to the ambient are small, almost all of the input energy will be recovered and, when combined with energy regeneration, the heat source has to provide only sufficient energy to cover the losses from the system. Thermodynamic analysis of the zeolite adsorption cycle was based in early work on the theory of Dubinin and Astakhov (1971) in combination with the Clausius-Clapeyron equation (Guilleminot 1978; Meunier et al. 1979; Simonot-Grange et al. 1980). Meunier and Mischler (1979) determined the desorption initiation temperature and calculated the COP of the zeolite/water cycle, and Grenier et al. (1982) compared it with other possible solar cycles. Alefeld et al. (1981) and Maier-Laxhuber et al. (1983) analyzed the zeolite/water cycle not only for cooling but also for heat storage, heat pump, and heat transformer applications (where the zeolite cycle is used to raise the waste heat source temperature). Sakoda and Suzuki (1984) applied this analysis to the silica gel/water system. Meunier (1985) performed second-law and entropy analysis of the zeolite/water cycle and Meunier (1986a) proposed a double-effect zeolite/water and active carbon/methanol cascading cycle. A similar double-effect cycle, combining the zeolite/water and LiBr/water cycles, was proposed by Alefeld et al. (1981); they projected that this will double the combined cooling COP from 0.6 to about 1.2.

Guilleminot et al. (1987) included the case of non-uniform temperature in the heat and mass transfer analysis of the adsorption cycle. Hajji et al. (1991) performed a dynamic analysis of the solar adsorption refrigerator. Their numerical simulation calculates the Figure Of Merit (adsorptivity function in their nomenclature) of 0.001 kg H_2O/kg zeolite °C, in good agreement with the 0.0015 value for natural chabazite.

The importance of the dynamics of heat and mass transfer of the zeolite cycle has become more evident as research emphasis has shifted from cycle COP to cycle time and the size and cost of the system. The cycle time for solar energy applications was set by nature at 24 hr and ample time does exist in which to complete the heat and mass transfer during desorption and adsorption. With the advent of heat regeneration techniques, especially the zeolite heat exchanger with propagating temperature front (Tchernev and Emerson 1988a), the efficiency of the adsorption cycle was greatly increased and the importance of the adsorption/desorption dynamics was recognized. Tchernev (1993) discussed the dynamics of regenerative zeolite heat exchangers and determined that desorption is heat-flow limited; whereas adsorption is usually limited by thermally activated diffusion. Further research on propagating temperature fronts (Tchernev and Emerson 1988b; Tchernev and Clinch 1989a; Shelton et al. 1989; Jones 1992; Ludwig 1995; Pons et al. 1996) in zeolite- and activated carbon-containing regenerative heat exchangers has indicated that fast heat and mass transfer in and out of the adsorbent is crucial for economic viability. Early research in the field of regenerative zeolite heat exchangers was conducted with zeolite beds packed with beads or pellets, where the heat transfer was the limiting step due to the poor thermal contact between the pellets themselves and between the pellets and the walls of the heat exchanger. From studies on the effect of the geometry and shape of the solid adsorbent on the heat transfer between the zeolite and the heat transfer fluid, Cacciola et al. (1992) concluded that planar geometry and good contact with the metal plate increase the heat transfer rate and allow more powerful and compact adsorption systems. Such a planar geometry has been used exclusively for the design and construction of high efficiency serpentine-like regenerative zeolite heat exchangers for heat pumps with propagating temperature fronts as shown in Figure 9 (Tchernev and Emerson 1988a; Tchernev and Clinch 1989b). Ülkü (1994) constructed various heat pumps, compared the theoretical predictions with experimental

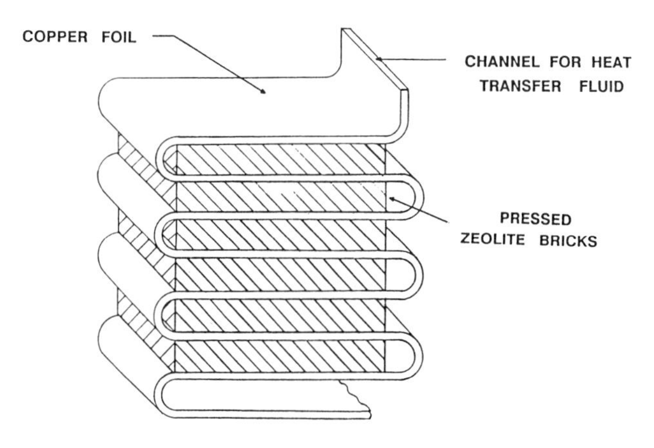

Figure 9. (a) Diagram of a serpentine-zeolite/oil heat exchanger.

Figure 9. (b) Actual heat exchanger containing 5 kg of zeolite.

results, and found good agreement between theory and experiments for adsorption cycles with clinoptilolite-, synthetic zeolite 13X- and silica gel-water as well as activated carbon-alcohol systems. Sami and Tulej (1992) described the performance of a synthetic zeolite/water system where the cylindrical zeolite pellets are embedded in a finned tube-type heat exchanger and nickel foam is used to enhance the thermal conductivity of the zeolite pellets. They determined that the cooling capacity ranged between 50 and 70 kJ/kg zeolite and increased with evaporator temperature (or water vapor pressure). This increase should be expected from a diffusion-limited process because the adsorption rate is proportional to the gas pressure. Strauss et al. (1992) studied experimentally the heat and mass transfer for synthetic zeolite granules and solid layers. The kinetic adsorption curves fit reasonably well to the time$^{1/2}$ law, which allowed them to conclude that the mass transport process is controlled mainly by the isothermal adsorption, i.e. diffusion, and that significant improvements can only be achieved by thinner zeolite layers. The same conclusions were reached by Eichengruen and Winter (1994) who determined that the adsorption process is slower than the desorption process for a zeolite thickness of 10 mm and that the dynamics of the process show a strong dependence on the thickness of the zeolite layer, in agreement with Tchernev (1993). To reduce the adsorbent volume and to increase the cooling power, Poyelle et al. (1996) constructed the adsorber from a new composite material consisting of a mixture of synthetic zeolite NaX and Expanded Natural Graphite (ENG). The composite cylinder has a thermal conductivity 100 times larger than the conventional packed bed with improved wall heat-transfer coefficient. The mass transfer resistance, however, is also increased 100 to 1000 times, and, despite the provision of 2-mm holes every 6 mm to improve mass transfer, the desorption time constant is shorter than the adsorption time constant, indicating that mass diffusion is limited. With the experimental evidence presented above pointing to mass diffusion as the limiting factor for shorter cycle times and smaller adsorption equipment, research has been focused recently on experimental measurements of the mass-transfer adsorption and desorption kinetics of water in zeolites.

Barrer and Fender (1961) reported the intracrystalline diffusion coefficients of water in natural chabazite, gmelinite, and heulandite. More recent measurements by nuclear magnetic resonance (NMR) or neutron scattering techniques yield the equilibrium self-diffusivity on a microscopic scale, which is of little value in systems containing concentration gradients. New and accurate experimental macroscopic techniques, described next, are used today to determine the intracrystalline diffusivity defined by Fick's first law under finite concentrations and gradients. For example, Grenier et al. (1995) reported mass diffusion values in single-crystal zeolites and pellets by pressure or volume steps while monitoring the surface temperature with an infrared detector of high sensitivity and short (10^{-3} s) time constant. Bourdin et al. (1996) described a frequency response method for the study of mass-transfer kinetics in zeolites in which the temperature of the sample was measured during volume and pressure modulation. The method distinguishes between the different physical processes that limit the mass-transfer rate; i.e. heat transfer inside and outside the zeolite, intracrystalline diffusion rate, and presence of a surface barrier. These methods confirmed the value of the diffusion coefficient for water determined by Barrer and Fender (1961) and Tchernev (1993) as about 3×10^{-10} m^2/s. Further improvements of these experimental techniques will contribute significantly to the understanding of the different physical phenomena that limit the reduction of the cycle time of practical adsorption systems and, hopefully, lead to the reduction of their size and cost. With the recognition of the importance of short diffusion lengths for fast adsorption processes on zeolites, research is correctly being conducted on the direct deposition of thin films of zeolites on different substrates. Mintova and Valtchev (1996) reported the *in situ* deposition of synthetic zeolite A on various vegetable fibers. Cacciola et al. (1996) synthesized zeolites on metal plates made of aluminum, nickel, titanium, copper, and stainless steel with the goal to reduce the heat-transfer resistance between the zeolite and the heat exchanger wall of adsorption systems. Their experimental results indicate that by significantly reducing the cycle time, adsorption systems with higher specific power can be designed. Bratton et al. (1996) described an adsorbent material of use in refrigeration and air conditioning units. It consisted of a porous metal substrate, coated with zeolite crystals (zeolite type not specified), which was capable of adsorbing and desorbing a working fluid with improved heat and mass transfer properties. More progress should be made in this field in the future because many researchers are concentrating their efforts on better understanding of the dynamic properties of adsorption and developing new methods to reduce its cycle time.

SUGGESTED FUTURE RESEARCH

A major research need in this field is the characterization of the adsorption and desorption properties of natural zeolites. Usually, deposits of natural zeolites are tested by X-ray diffraction methods for mineralogical identification and purity; however, a clinoptilolite from Turkey does not necessarily have the same adsorption properties as one from Sheaville, Oregon. As Tchernev (1980) reported, the adsorption properties of natural zeolites with the same crystal structure vary more between deposits of different location than between zeolites with different crystal structures and, therefore, characterization of each individual zeolitic tuff is necessary. Because water adsorption is the phenomenon most often used in solar energy applications, it is important that natural zeolite deposits should be characterized at least for water adsorption. Many other gases exist, however, whose adsorption properties on natural zeolites are of interest. Tchernev (1992, 1997) studied the adsorption of H_2S, N_2, CO, and CO_2 on chabazite from a deposit near Bowie, Arizona, and on synthetic zeolites 5A, NaY, and 13X. Some studies of methanol adsorption on natural and synthetic zeolites have been published by Simonot-Grange et al. (1980), Delgado et al. (1982), Meunier (1986a), Brossard et al. (1991), and Tchernev (1995). However, almost no adsorption data are available on the adsorption of other organic or inorganic gases on natural zeolites.

The question of long-term chemical stability of the adsorbed gas under repeated thermal cycling has not been answered; because zeolites are good catalysts, many adsorbed gases, such as CFCs and methanol, decompose during desorption and sometimes destroy the zeolite in the process. Considerably more experimental adsorption data are needed if natural zeolites are to be considered for use in practical applications. The availability of adsorption isosteres of different zeolite/gas pairs will be very useful for the determination of the heats of adsorption and the energy storage capabilities of different natural zeolites.

More work also needs to be done on the modification of the adsorption properties of natural zeolites by ion exchange as was reported by Yörükogullari and Orhun (1997). Another area of interest for future research is the modification of natural zeolites from hydrophilic to hydrophobic by chemical treatment or ion exchange. Such modification, if successful, will permit the use of natural zeolites in air-pollution applications and for the removal of VOCs from industrial exhaust, a field where they are not used at present.

Once the equilibrium adsorption properties have been established, the dynamics of adsorption and desorption must be determined. A large need exists for measurements of thermal conductivity, porosity, and diffusion constants for different gases on natural zeolites. The excellent work that Ülkü (1986) has done with a Turkish clinoptilolite should be extended to other deposits. Such information should be easier to obtain using the new experimental techniques of pressure and volume steps or frequency response. These data are necessary in order to establish if any barriers to gas diffusion exist and if the natural zeolite could be used in fast heat exchangers for heat pump applications.

Research is also needed on methods to produce solid zeolite bodies of different shapes and dimensions with or without the use of binders. Such methods are needed both for the pelletization of natural zeolites for commercial use and for the construction of zeolite heat exchangers for solar or waste heat application. Of special importance are methods that will permit the deposition of thin layers of natural zeolite on metal surfaces. Although scientists have been reasonably successful in synthesizing zeolite NaA directly on metal foils in autoclaves, this method cannot be used for natural zeolites because they have been already crystallized by nature. Therefore, work is needed on methods to attach natural zeolite particles to metal surfaces with good adhesion and good thermal contact for fast heat transfer. Such work, if successful, will permit rapid temperature cycling without physical damage due to the mismatch of thermal expansion coefficients between the zeolite and the metal. Furthermore, if binders are used, they should not block the pore openings of the zeolite particles and slow the adsorption rate. If such research efforts are successful, a whole new field of use will be opened for the temperature control of catalytic reactions and for very fast heat exchangers. Such heat exchangers will be small in size, weight, and price and will permit the use of natural zeolites in high cooling power density systems. For example, the volume and weight of waste-heat adsorption cooling systems is a major factor in the acceptability of the use of natural zeolites for air conditioning by the automobile industry.

REFERENCES

Aiello R, Nastro A, Giordano G, Colella C (1988) Solar energy storage by natural zeolites II. Activation and heat recovery in open systems. *In* D Kallo, HS Sherry (eds) Occurrence, Properties and Utilization of Natural Zeolites, p 763-772. Akademiai Kiado, Budapest

Alefeld G (1983) Double-effect, triple-effect and quadruple effect absorption machines. *In* Proc 16th Int'l Congr Refrigeration II:951-956, Paris, France

Alefeld G, Bauer HC, Maier-Laxhuber P, Rothmeyer M (1981) A zeolite heat pump, heat transformer and heat accumulator. *In* HS Stephens and B Jarvis (eds) Proc Int'l Conf Energy Storage, Brighton, UK. p 61-72. BHRA Fluid Engineering, Cranfield, UK

Banks NJ (1990) Desiccant dehumidifiers in ice arena. Ashrae Transactions 96:1269-1271

Barrer RM, Fender BEF (1961) The diffusion and sorption of water in zeolites. J Phys Chem Solids 21: 1-24

Bourdin V, Grenier Ph, Meunier F, Sun LM (1996) Thermal frequency response method for the study of mass-transfer kinetics in adsorbents. Am Inst Chem Engin J 42:700-712

Bratton GJ, Buck KD, De Villiers NT (1996) Adsorbent material for use in refrigeration and air conditioning unit. Int'l Pat. Appl. No. WO 96 24,435, Aug. 15, 1996

Breck DW (1974) Zeolite Molecular Sieves. Wiley, New York, 771 p

Brossard LF, Barreda E, Vazquez L, Abreu A. (1991) Sistema de enfriamiento de agua potable, zeolita-methanol-energia solar. *In* Program & Abstracts, Zeolite '91: 3rd Int'l Conf Natural Zeolites, p 161. International Conference Center, Havana, Cuba

Cacciola G, Cammarata G, Fichera A, Restuccia G (1992) Advances on innovative heat exchangers in adsorption heat pumps. Sci Tech Froid 1:239-245

Cacciola G, Restuccia G (1994) Progress on adsorption heat pumps. Heat Recovery Systems & CHP 14:409-420

Cacciola G, Restuccia G, Muller JCM, Jansen JC, van Bekkum H (1996) Zeolite synthesized on metal for more efficient adsorption machines. *In* Proc Int'l Absorption Heat Pump Conf Montreal, p 609-616. Natural Resources, Quebec, Canada

Calmac Manufacturing Corporation (1982) Bulk storage of PCM: Report to Argonne National Laboratory, NTIS PB82-805862. National Technical Information Service, Springfield, Virginia

Chang SC, Roux JA (1985) Thermodynamic analysis of a solar zeolite refrigeration system. J Solar Energy Engin 107:189-195

Chang SC, Roux JA (1986) Discussion of thermodynamic analysis of a solar zeolite refrigeration system. J Solar Energy Engin 108:257

Close DJ, Dunkle RV (1977) Use of adsorbent beds for energy storage in drying of heating systems. Solar Energy 19:233-238

Collier RK (1979) The analysis and simulation of open cycle absorption refrigeration system. Solar Energy 23:357-366

Davias M (1995) Limits of a single-effect adsorption system for automobile air conditioning. Rev Gen Term 34:154A-165A

Delgado R, Choisier A, Grenier P, Ismail I, Meunier F, Pons M (1982) Etude du cycle intermittent charbon actif-methanol en vue de la realisation d'une machine a fabriquer de la glace fonctionnant a l'energie solaire. Sci Tech Froid, Proc Int'l Inst Refrigeration Meeting, Israel, 1982, Edit. I.I.R., p 181-187

Dorgan CE, Elleson JS (1994) ASHRAE's new design guide for cool thermal storage. ASHRAE J 36:29-34

Drohan D (1992) Different routes to VOC control. Pollution Engin 24:30-33

Dubinin MM, Astakhov VA (1971) Description of adsorption equilibria of vapors on zeolites over wide ranges of temperature and pressure. *In* RF Gould (ed) Molecular Sieve Zeolites-II, Adv Chem Ser 102:69-85

Dunkle RV (1965) A method of solar air conditioning. Mech Chem Engin Trans, I E Austr 1:73-78

Dupont M, Guilleminot JJ, Meunier F (1982) Etude de glacieres solaires utilisant le cycle intermittent jour-nuit zeolite 13X-eau en climat tempere et en climat tropical. Sci Tech Froid, Proc Int'l Inst Refrigeration Meeting, Israel, 1982, Edit. I.I.R., p 189-196

Eichengruen S, Winter ERF (1993) Zeolite/water sorption aggregate for cold generation. DKV-Tagungsber 2:227-243

Eichengruen S, Winter ERF (1994) Zeolite/water adsorption refrigerating units. Ki Luft-Kältetechn 30: 114-118

Fejes P, Hannus I, Kiricsi I (1988) Investigation of Hungarian natural zeolites for use in solar cooling. *In* D Kallo, HS Sherry (eds) Occurrence, Properties and Utilization of Natural Zeolites, p 773-780. Akademiai Kiado, Budapest

Fisher S, Hauer A, Holst S, Schoelkopf W (1992) Thermochemical energy storage with low temperature heat for space heating. Sol. World Congr, Proc Bienn Congr Int'l Solar Energy Soc 1991, p 1769-1773

Gentner H, Winter ERF, Höppler R (1993) Vapor compression engine and adsorption cooling aggregate for vehicle air conditioning. DKV-Tagungsber 20:245-266

Gentner H, Winter ERF, Höppler R (1994) Zeolite/water or R 134a for automobile air conditioning? Ki Luft-Kältetech 30:288-293

Gonzales R, Denis A, Isla J, Leal M (1991) Empleo de zeolitas naturales en aconditionamiento de aire por enfriamiento evaporativo indirecto. *In* Program & Abstracts, Zeolite '91: Third Int'l Conf on Natural Zeolites, p 160. International Conference Center, Havana, Cuba

Gopal R, Hollebone BR, Langford CH, Shigeishi RA (1982) The rates of solar energy storage and retrieval in a zeolite-water system. Solar Energy 28:421-424

Grenier P, Meunier F, Pons M, Brandon B, Merigoux J (1982) Les differentes possibilites d'application du couple zeolithe 13X-H_2O pour le froid solaire en fonction du type de captation de l'energie solaire. *In* Sci Tech Froid, Intl Inst Refrigeration meeting, Israel, 1982, Edit. I.I.R., p 197-204

Grenier P, Bourdin V, Sun LM, Meunier F (1995) Single-step thermal method to measure intracrystalline mass diffusion in adsorbents. Am Inst Chem Engin J 41:2047-2057

Grzanka LE (1996) Big savings in aisle one—controlling humidity in supermarkets is possible using waste heat from refrigeration. Engineering Systems 13

Guilleminot JJ (1978) Application of an Intermittent Cycle to Refrigeration. Thesis, Univ Dijon, France

Guilleminot JJ, Meunier F, Mischler B (1979) Utilisation d'un cycle intermittent zeolithe 13X-H_2O pour la refrigeration solaire. *In* Proc XV Int'l Congr Refrigeration, Venezia, E1-88:1-4, Venice, Italy

Guilleminot JJ, Meunier F, Mischler B (1980) Etudes des cycles intermittants a adsorption solide pour la refrigeration solaire. Rev Phys Ap Fr 15:441-452

Guilleminot JJ, Meunier F (1981) Ethude experimentale d'une glaciere solaire utilisant le cycle zeolithe 13X-eau. Rev Gen Term Fr 239:825-835

Guilleminot JJ, Meunier F, Pakleza J (1987) Heat and mass transfer in a non-isothermal fixed bed solid adsorbent reactor: A uniform pressure-non-uniform temperature case. Int'l J Heat Mass Transfer 30:1595-1606

Hajji A, Worek WM, Lavan Z (1991) Dynamic analysis of a closed-cycle solar adsorption refrigerator using two adsorbent-adsorbate pairs. J Solar Energy Engin. 113:73-79

Hamdan MA, Elwerr FA (1996) Thermal energy storage using a phase change material. Solar Energy 56:183-189

Höppler R (1992) Eine Sorptionskälteanlage für die PKW-Klimatisierung basierend auf den Naturstoffen Zeolith und Wasser. DKV-Statusber. Dtsch Kälte-Klimatesch Ver 12:59-62

Höppler R (1993) Sorptionanlagen für die PKW-Beheizung/Klimati-sierung basierend auf den Naturstoffen Zeolith und Wasser. DKV-Statusber. Dtsch Kälte-Klimatesch Ver.14:31-34

Jones JA (1992) Sorption refrigeration research at JPL/NASA. Sci Tech Froid 1:143-152

Kakabaev A, Khandurdyev A (1969) Absorption solar refrigeration unit with open regeneration of solution. Geliotechnika 5:69-72

Lane GA (1985) PCM science and technology: The essential connection. ASHRAE Trans 91, Part 2, p 1897-1909

Löf GOG (1955) Cooling with solar energy. *In* Proc Congr Solar Energy, Tucson, Arizona, p 171-189

Ludwig J (1995) Adsorptionswärmepumpe mit 2 Adsorbern. Brennstoff-Wärme-Kraft 47-3:94-96

Macriss RA, Wurm J, Zawacki TS, Kinast JA (1981) Solar-MECtm development program. *In* Proc Ann DOE Active Heating and Cooling Contractors' Review Meeting, Washington, DC, p 3-13 to 3-16

Maier-Laxhuber P, Rothmeyer M, Alefeld G (1983) Zeolite heat pump and heat storage. *In* HS Stephens, GW Warren (eds) Int'l Conf Energy Storage, Stratford-Upon-Avon, UK, p 205-210. BHRA Fluid Engineering, Cranfield, UK

Meunier F (1978) Utilisation des cycles à sorption pour la production de froid par l'energie solaire. Proc Colloque Cahiers de l'AFEDES 5:57-67

Meunier F, Mischler B (1979) Solar cooling through cycles using microporous solid adsorbents. *In* KW Böer, BH Glenn (eds) Proc Int'l Solar Energy Soc SUN II, Atlanta, Georgia, p 676-680. Pergamon Press, New York

Meunier F, Mischler B, Guilleminot JJ, Simonot MH (1979) On the use of a zeolite 13X-H_2O intermittent cycle for the application to solar climatization of buildings. *In* KW Böer, BH Glenn (eds) Proc Int'l Solar Energy Soc SUN II, Atlanta, Georgia, p 739-743. Pergamon Press, New York

Meunier F (1985) Second law analysis of a solid adsorption heat pump operating on reversible cascade cycles: Application to the zeolite-water pair. Heat Recovery Systems 5:133-141

Meunier F (1986a) Theoretical performance of solid adsorbent cascading cycles using the zeolite-water and active carbon-methanol pairs: Four case studies. Heat Recovery Systems 6:491-498

Meunier F (1986b) Discussion of thermodynamic analysis of a solar zeolite system. J Solar Energy Engin 108:257

Mintova S, Valtchev V (1996) Deposition of zeolite A on vegetable fibers. Zeolites 16:31-34

Mullick SC, Gupta MC (1975) Solar air conditioning using adsorbents. *In* Second Workshop on the Use of Solar Energy For the Cooling of Buildings, Los Angeles, California, August 4-6, 1975

Muradov Dzh, Shadeyev O (1971) Testing of a solar adsorption refrigerator. Geliotechnica 3

Nesselmann K (1933) Zur Theorie der Wärmetransformation. Wiss Veröffentl Siemens Konzern 12:89

Olson DH, Kokotailo GT, Lawton SL, Meier WM (1981) Crystal structure and structure-related properties of ZSM-5. J Phys Chem 85:2238-2243

Patnaik S, Lenz TG, Löf GOG (1990) Performance studies for an experimental solar open-cycle liquid desiccant air dehumidification system. Solar Energy 44:123-135

Peters H, Brückner P, Najork H (1986) Möglichkeit zur Anwendung des Systems Wasser/Zeolith 5A in periodischen Adsorptionswärmepumpen und solargetriebenen Kälteanlagen. Luft- und Kältetechnik 3:154-159

Phadke AV (1984) Natural occurrence of clinoptilolite from western Maharashtra: Its utilization in refrigeration using solar energy. Proc Indian Natl Sci Acad 50, Part A, p 479-482

Phelouzat JL, Rey R, Noguera R (1994) Study of a rapid chilling unit for prepared meals based on vacuum and absorption of water by zeolite. Sci Tech Froid 3:749-755

Pons M, Laurent D, Meunier F (1996) Experimental temperature fronts for adsorptive heat pump applications. Appl Thermal Engin 16:395-404

Poredos A (1994) Research on adsorption refrigeration processes. Sci Tech Froid 3:279-237

Poyelle F, Guilleminot JJ, Meunier F, Soidé I (1996) Experimental tests of a gas fired adsorptive air conditioning system. *In* Proc Int'l Absorption Heat Pump Conf Montreal, p 221-229. Natural Resources, Quebec, Canada

Restuccia G, Cacciola G, Quagliata R (1988) Identification of zeolites for heat transformer, chemical heat pump and cooling systems. Int'l J Energy Res 12:101-111

Restuccia G, Freni A, Cacciola G (1999) Adsorption beds of zeolite on aluminum sheets. *In* Proc Int'l Heat Pump Conf, Munich, Germany, p 343-347. ZAE Bayern

Rothmeyer M, Maier-Laxhuber P, Alefeld G (1983): Design and performance of zeolite-water heat pumps. *In* Proc 16th Congr Refrigeration 5:703-707. Paris, France

Sakoda A, Suzuki M (1984) Fundamental study on solar powered adsorption cooling system. J Chem Engin Japan 17:52-57

Sami SM, Tulej P (1992) Analysis of zeolite adsorption cycle using air cooled heat exchangers. Sci Tech Froid 1:135-142

Scarmozzino R, Aiello R, Santucci A (1980) Chabazitic tuff for thermal storage. Solar Energy 24:415-416

Schwarz J (1991) Sorptionstechnik, Alternative zu den Alternativen. Ki Luft-Kältetech 27:127-132

Schwarz J, Winter ERF, Maier-Laxhuber P, Soltes J (1991) Adsorptionssysteme mit dem Stoffpar Wasser/Zeolith. DKV Tagungsber 18:203-216

Schwarz J (1994) Refrigeration of food in passenger trains with water/zeolite adsorption systems. Ki Luft-Kältetech 30:536-540

Selvidge M, Miaoulis IN (1990) Evaluation of reversible hydration reactions for use in thermal energy storage. Solar Energy 44:173-178

Shadeyev O, Umarov GYa (1972) On the temperature distribution in a solar refrigerator generator and the thermal conductivity of the adsorbent. Geliotechnica 8:34-38

Shelton SV, Wepfer WJ, Miles DJ (1989) Square wave analysis of the solid-vapor adsorption heat pump. Heat Recovery Systems & CHP 3:233-247

Shigeishi RA, Langford CH, Holleborne BR (1979) Solar energy storage using chemical potential changes associated with drying of zeolites. Solar Energy 23:489-495

Simonot-Grange MH, Guilleminot JJ, Setier JC, Meunier F (1980) The liquid-gas-zeolite stationary state and the sorption cycle-compared sorption of methanol, ethanol and water by zeolite 13X. *In* LVC Rees (ed) Proc 5th Int'l Conf Zeolites, p 832-840. Heyden, London

Strauss R, Schallenberg K, Knoche KF (1992) Measurements of the kinetics of water vapor adsorption into solid zeolite layers. Sci Tech Froid 1:246-250

Suzuki M (1992) Application of adsorption cooling systems to automobiles. Sci Tech Froid 1:153-158

Suzuki M. (1993) Application of adsorption cooling systems to automobiles. Heat Recovery Syst. & CHP 13:335-340

Takasaka A, Matsuda Y (1986) Vacuum freeze drying of food using natural zeolites. *In* Y Murakami, A Iijima, JW Ward (eds) Proc 7th Int'l Zeolite Conf Tokyo, p 1041-1046. Elsevier, New York

Tchernev DI (1974) Solar energy cooling with zeolites. *In* S Sargent (ed) Proc NSF/RANN Workshop on Solar Collectors for Heating and Cooling of Buildings, New York, p 262-266. Univ Maryland, College Park, Maryland

Tchernev DI (1976) Solar energy application of natural zeolites. *In* F de Winter, JW de Winter (eds) Proc of ERDA Second Workshop on the Use of Solar Energy for the Cooling of Buildings, p 307-319. ALTAS Corp, Santa Clara, California

Tchernev DI (1977) Final Report, NSF/RANN Grant AER 74-09038. Lincoln Laboratory, Massachusetts Institute of Technology, Lexington, Massachusetts, Appendix, p 27

Tchernev DI (1978) Solar energy application of natural zeolites. *In* LB Sand, FA Mumpton (eds) Natural Zeolites: Occurrence, Properties, Use, p 474-485. Pergamon Press, Elmsford, New York

Tchernev DI (1979) Solar refrigeration utilizing zeolites. *In* Proc 14th Intersociety Energy Conversion Engineering Conf, p 2070-2073. Am Chem Soc, Washington, DC

Tchernev DI (1980) The use of zeolites for solar cooling. *In* LVC Rees (ed) Proc 5th Int'l Conf Zeolites, Naples, Italy, p 788-794. Heyden, London

Tchernev DI (1981a) The development of low cost integrated zeolite collector. Final Report for Work Performed under DOE Contract AC03-78CS32117. The Zeopower Co, Natick, Massachusetts

Tchernev DI (1981b) Integrated solar zeolite collector for heating and cooling. *In* Proc 1981 Ann Meet ISES, p 520-524. Am Section of Int'l Solar Energy Soc, Newark, Delaware

Tchernev DI (1982) Solar air conditioning and refrigeration systems utilizing zeolites. Sci Tech Froid, IIR Meeting, Israel, 1982, p 205-211

Tchernev DI (1983) Use of natural zeolites in solar refrigeration. *In* WG Pond, FA Mumpton (eds) Zeo-Agriculture, Use of Natural Zeolites in Agriculture and Aquaculture, p 273-280. Westview Press, Boulder, Colorado

Tchernev DI (1984) Use of natural zeolites in solar refrigeration. Abstr of Sel Solar Energy Technology 6: 21-24. The United Nations, University of Tokyo

Tchernev DI, Emerson D (1988a) High-efficiency regenerative zeolite heat pump. ASHRAE Transactions 14:2024-2032

Tchernev DI, Emerson D (1988b) Closed cycle zeolite regenerative heat pump. *In* F Moser (ed) Proc 2nd Int'l Workshop on Research Activities on Advanced Heat Pumps, Graz, Austria, p 79-88. Inst Chem Engin Techn Univ Graz, Austria

Tchernev DI, Clinch MJ (1989a) Closed cycle zeolite regenerative heat pump. *In* AH Fanney, KO Lund (eds) Solar Engineering, p 347-351. Am Soc Mech. Eng, San Diego, California

Tchernev DI, Clinch MJ (1989b) A closed-cycle regenerative zeolite gas heat pump. *In* T Cramer (ed) Proc 1989 Int'l Gas Res Conf, p 684-692. Government Institutes, Inc, Rockville, Maryland

Tchernev DI, Emerson D (1990) Closed cycle zeolite regenerative heat pump. *In* S Deng (ed) Heat Transfer Enhancement and Energy Conservation, p 747-756. Hemisphere Publ Corp, New York

Tchernev DI (1992) Final Report, NSF/SBIR Grant ISI-9160143. The Zeopower Company, Natick, Massachusetts, p 18-21

Tchernev DI (1993) Dynamics of regenerative zeolite heat exchangers. *In* R von Ballmoos (ed) Proc 9th Int'l Zeolite Conf, p 675-682. Butterworth-Heinmann, Stoneham, Massachusetts

Tchernev DI (1995) Zeolites in solar energy and refrigeration applications: A review of Zeopower Company work for the past 20 years. *In* DW Ming, FA Mumpton (eds) Natural Zeolites '93: Occurrence, Properties, Use, p 613-622, Int'l Comm. Natural Zeolites, Brockport, New York

Tchernev DI (1997) Waste heat/solar zeolite power systems. *In* G Kirov, L Filizova, O Petrov (eds) Proc Sofia Zeolite Meeting '95, p 11-18. Pensoft, Sofia-Moscow

Tchernev DI (1999) A waste heat driven automotive air conditioning system. *In* C Schweigler, S Summerer, H-M Hellman, F Ziegler (eds) Proc Int'l Sorption Heat Pump Conf Munich, Germany, p 65-70. ZAE Bayern, Munich, Germany

Tchernev DI (2000) Evaluation of the desorption properties of natural zeolites suitable for thermal energy conversion. *In* C Colella, FA Mumpton (eds) Natural Zeolites '97, De Frede Editore, Napoli, Italy (in press)

Telkes M (1974) Storage of solar heating/cooling. Symp. *In* Solar Energy Applications, ASHRAE Annual Meeting, Montreal, Canada, p 34-39

Trombe F, Foex M (1958) Production de glace a l'aide de l'energie solaire. Colloque Int'l du CNRS, Montiouis, France, p 469-473

Ülkü S (1986) Natural zeolites in energy storage and heat pumps. *In* Y Murakami, A Iijima, JW Ward (eds) Proc 7th Int'l Zeolite Conf, Tokyo, p 1047-1054. Elsevier, New York

Ülkü S (1992) Adsorption heat pumps (refrigerators). Sci Tech Froid 1:102-108

Ülkü S (1994) Adsorbents in refrigeration. Sci.Tech.Froid 3:155-163

Ülkü S, Mobedi M (1989) Zeolites in heat recovery. *In* PA Jacobs, RA van Santen (eds) Zeolites: Facts, Figures, Future, p 511-518. Elsevier, Amsterdam

Valueva GP, Belitsky IA, Seryotkin YV, Pavlychenko VS (1988) Natural chabazite: Heats of rehydration and X-ray study in relation to H_2O contents at room temperature. *In* D Kallo, HS Sherry (eds), Occurrence, Properties and Utilization of Natural Zeolites, p 282-289. Akademiai Kiado, Budapest

Van den Bulck E, Mitchell JW, Klein SA (1985) Design theory for rotary heat and mass exchangers. Int'l J Heat Mass Transfer 28:1575-1595

Van den Bulck E (1988) Performance characteristics of open-cycle solid desiccant heat transformers. *In* F Moser (ed) Proc 2nd Int'l Workshop Res Activities on Adv Heat-Pumps, Graz, Austria, p 69-77. Inst Chem Engin Techn Univ Graz, Austria

Vasiliev LL, Gulko NV, Khaustov VM (1992) Solid adsorption refrigerators with active carbon-acetone and carbon-ethanol pairs. SciTech Froid 1:109-117

Wijsman AJTM, Oosterhaven R, den Ouden C (1979) Development of a thermal storage system based on the heat of adsorption of water in hygroscopic materials. *In* Proc Int'l Solar Energy Soc SUN II, Atlanta, Georgia, p 619-622. Pergamon Press, New York

Yörükogullari E, Orhun Ö (1997) Increasing the effective energy storage capacity of natural zeolite of Balikesir-Turkey. *In* G Kirov, L Filizova, O Petrov (eds) Proc Sofia Zeolite Meeting '95, p 161-164. Pensoft, Sofia

Zhang Q, Guo B (1988) Water sorption capacity of clinoptilolite and mordenite. Dizki Kezue 2:147-154

18 Use of Natural Zeolites in Agronomy, Horticulture, and Environmental Soil Remediation

Douglas W. Ming

Mail Code SX, NASA Johnson Space Center
Houston, Texas 77058

Earl R. Allen

Zeoponic Gardens, P.O. Box 83
Mason City, Illinois 62664

> *"In the next century, it is possible that a huge industry will develop based on natural zeolites. Soil conditioning by zeolites might lead to greater agricultural production. Control of toxic materials in waste water by zeolites might rescue some stressed aquasystems. There just might be gold for geochemists in them thar zeolite beds."* J. V. Smith (1988)

INTRODUCTION

The use of natural zeolites to improve plant productivity or as a remediation agent in environmental protection has the potential of becoming a "huge" industry as pointed out in the above quote by Smith (1988). This potential industry is based on the unique chemical and physical properties of natural zeolites (e.g. high cation-exchange capacities, cation selectivity, molecular sieving) and their widespread occurrence in sedimentary deposits derived from volcanic materials. A variety of potential applications have been examined for natural zeolites, including use as soil conditioners, slow-release fertilizers, soilless substrates, carriers for insecticides and pesticides, and remediation agents in contaminated soils. However, although numerous applications have been suggested or examined, today there are only a few commercial markets for natural zeolites in the horticultural, agronomic, and environmental protection industries. Allen and Ming (1995) suggested that there may be several reasons that the commercial use of natural zeolites has been slow to develop, including (1) the lack of studies that focus on deriving the economic benefits of zeolite applications; (2) the need to develop products and formulations that meet a specific agronomic, horticultural, or environmental use, instead of a one-size-fits-all approach; (3) the need to fully characterize the zeolite or zeolite-containing material before it is utilized; and (4) the lack of sound scientific research to support the proposed uses of natural zeolites.

We agree with J.V. Smith that the potential applications for natural zeolites are huge, provided the proper research is conducted to fully use their unique properties. In this chapter, we provide an overview on research that has been conducted that is related to agronomic, horticultural, and environmental applications of natural zeolites. It is also the goal of this chapter to point out areas of needed research in the hope that it will encourage horticulturists, agronomists, environmentalists, mineralogists and geochemists to find the *"gold"* in "*them thar zeolite beds*."

BACKGROUND

It is not the purpose of this chapter to describe in detail the unique properties of natural zeolites that makes them attractive for applications (see other chapters in this volume, e.g. Armbruster and Gunter, Passaglia and Sheppard, Pabalan and Bertetti);

1529-6466/00/0045-0018$05.00

however, it is useful to briefly describe the important properties. Probably one of the most important properties for their potential use is the ability to freely exchange cations within the zeolite structure with cations in solution. Zeolites are one of the most effective natural cation exchangers, and they have cation-exchange capacities (CECs) that commonly range from 200 to 300 $cmol_c\ kg^{-1}$, two to three times higher than the CECs of smectites found in soils. Because of the arrangement of Al and Si in the three-dimensional framework of SiO_4 and AlO_4 tetrahedra and the channels and cages that are created in this framework, each zeolite has unique selectivities for various cations. For example, clinoptilolite has a selectivity sequence of

$$Cs > Rb > K > NH_4 > Ba > Sr > Na > Ca > Fe > Al > Mg > Li$$ (Ames 1960).

Other unique properties are low densities of crystals (1.9 to 2.2 $Mg\ m^{-3}$) and low bulk densities (e.g. 0.8 to 1.5 $Mg\ m^{-3}$) of zeolitically altered volcanic sediments.

The use of natural zeolites to enhance plant growth was first reported in Japan during the 1960s (Minato 1968). Clinoptilolite-rich tuff was added with N fertilizers to rice fields to improve the availability of N in paddy soils. Since those initial studies, a variety of horticultural, agricultural, and environmental applications have been described (see reviews of Barbarick and Pirela 1984; Allen and Ming 1995; Ming and Allen 2000). In the following sections, we provide examples of applications that use the unique properties of natural zeolites.

SOIL CONDITIONING FOR PLANT GROWTH

Natural zeolites have been added to soils in attempts to improve the soil's physical and chemical properties for plant growth. The addition of a zeolite to a soil will increase the soil's overall CEC and, therefore, result in an increase in the soil's nutrient-holding capacity. The addition of natural zeolites to soils generally also increases the pH of the soil. Most natural zeolites from sedimentary deposits have crystals that are cemented by glass and other phases, yielding mined material that can be crushed and sized into a desired particle-size range, e.g. sand-sized particles (0.05 to 2 mm). Therefore, zeolites can be added to a soil in a desirable particle size to improve the soil's physical properties (e.g. water infiltration, water-holding capacity).

Chemical properties

The addition of natural zeolites to soils increases both the soil's CEC and pH in most cases. For example, clinoptilolite-rich tuff was amended to coarse, acidic Podzolic soils (sandy and sandy loam, highly leached top horizons) in the Ukraine at rates of 0 to 37.5 tonne ha^{-1} (Mazur et al. 1984). The addition of 35 tonne ha^{-1} clinoptilolite-rich tuff increased the CEC of the topsoil from 6.1 $cmol_c\ kg^{-1}$ to 11.2 $cmol_c\ kg^{-1}$ and increased the soil pH from 5.2 up to 7, depending on the amount of zeolite added (Table 1). Yields of potato, barley, and wheat increased from 2.8 to 79.0%, depending on the crop, fertilizer treatment, and clinoptilolite addition as compared with fertilized and unfertilized controls in these Ukrainian soils.

Suwardi et al. (1994) characterized the mineralogy and chemistry of 22 natural zeolite deposits from Japan and Indonesia for their potential use as soil conditioners. These deposits consisted of either clinoptilolite only, mordenite only, or mixtures of clinoptilolite and mordenite. Four of the zeolite-rich materials were selected and added individually to three different soils – a red-yellow podzolic soil (well-drained soil formed under warm-temperate to tropical, humid climates, under deciduous or coniferous forest vegetation), a sandy soil, and a volcanic-ash soil. An application rate of 50 $Mg\ ha^{-1}$ of 1-mm sieved clinoptilolite-rich tuff (approximately 82 wt % clinoptilolite) increased the CEC of the red-yellow podzolic soil from 10.6 $cmol_c\ kg^{-1}$ to 14.6 $cmol_c\ kg^{-1}$, and the

Table 1. Cation-exchange capacity (CEC) and pH of soils amended with zeolite-rich materials.

Zeolite*	Zeolite locality	Zeolite CEC[†]	Soil type[‡]	Zeolite amendment rate[#]	CEC of soil + zeolite	pH of soil + zeolite	Reference
		$cmol_c\ kg^{-1}$			$cmol_c\ kg^{-1}$		
Cp	Ukraine	-	Podzolic	0 to 37.5 tonne ha^{-1}	6.1 to 11.2	5.2 to 7.0	Mazur et al. (1984)
Cp	Oregon, USA	160	Sand	0 wt %	0.08	5.4	Huang & Petrovic (1994)
				10 wt %	15.59	6.6	
Cp	Fukushima, Japan	164	Sandy	0 Mg ha^{-1}	6.45	NR	Suwardi et al. (1994)
				50 Mg ha^{-1}	11.2		
Cp	Wakayama, Japan	164	Volcanic-ash	0 Mg ha^{-1}	40.4	NR	Suwardi et al. (1994)
				50 Mg ha^{-1}	41.0		
Cp	Oregon, USA	166.3	Glacial till	0 wt %	11.89	NR	Katz et al. (1996)
				25 wt %	38.8		
				50 wt %	77.5		
Cp	Oregon, USA	166.3	Marine clay	0 wt %	17.12	NR	Katz et al. (1996)
				25 wt %	44.8		
				50 wt %	73.0		
Cp/Mo	San Andres, Cuba	141	Gley Solonetz	0 to 15 tonne ha^{-1}	8.9 to 10.6	4.0 to 4.5	Soca et al. (1991)
Mo	Wakayama, Japan	159	Sandy	0 Mg ha^{-1}	6.45	NR	Suwardi et al. (1994)
				50 Mg ha^{-1}	10.7		
Mo	Wakayama, Japan	159	Volcanic-ash	0 Mg ha^{-1}	40.4	NR	Suwardi et al. (1994)
				50 Mg ha^{-1}	42.2		

* Cp = clinoptilolite-rich material; Mo = mordenite-rich materials

[†] Cation-exchange capacity of zeolite-rich material

[‡] Podzolic = soils with thin, dark surface horizons formed under deciduous forest in humid temperate climates; Gley Solonetz = alkaline soils that have developed under poor drainage

[#] units for amendment rates varied from researcher to researcher

NR = Not reported

addition of the same zeolite to the sandy soil nearly doubled the soil's CEC from 6.45 to 11.2 $cmol_c\ kg^{-1}$ (Table 1). However, the addition of the clinoptilolite-rich tuff to the volcanic soil resulted in minimal change to the soil's CEC, increasing it from 40.4 $cmol_c\ kg^{-1}$ to 41.0 $cmol_c\ kg^{-1}$; the slight increase is likely due to the already high CEC of the volcanic soil. Tomato plants grown in zeolite-amended soils produced ~30 wt % more aboveground biomass compared with tomatoes grown in untreated soils. Addi-tional selected studies on the effects to soil pH and CEC by adding natural zeolites are listed in Table 1.

Physical properties

Natural zeolites have several unique physical properties that make them attractive as soil additives to improve the physical properties of the soil. For example, bulk densities of zeolitically altered volcanic tuffs and pyroclastics range from about 0.8 to 2.5 $Mg\ m^{-3}$, which reflects the porous nature and cementation of the vitreous parent material. The cemented zeolitic tuff is lightweight and can be easily crushed and sieved into a specified size range, which can be added to soils to improve physical properties of soils, such as water infiltration, water availability or water holding capacity, and aeration. Most of these types of zeolite applications have been directed toward improving the physical properties of turfgrass greens on golf courses.

Petrovic (1990) found in laboratory studies that the optimum particle size of clinoptilolite added to golf course sand was between 0.1 to 1 mm in order to maximize benefits for water infiltration, water availability, and aeration. He then illustrated these properties in a field study where creeping bentgrass was grown in sandy media (quartz sand) amended with clinoptilolite, sphagnum peat, or sawdust at 5, 10, and 20 vol %. All of the amendments, except for the 10 and 20 vol % sawdust amendments, resulted in superior establishment of the creeping bentgrass compared with the unamended quartz sand. Peat-amended sand retained significantly more water than the clinoptilolite- or sawdust-amended treatments; however, all of the treatments retained significantly more water than the sand-only treatment.

Huang and Petrovic (1995) reported that the water available to plants increased when clinoptilolite particles decreased in size and the proportion of clinoptilolite increased in a sand medium, which was amended with six different particle sizes at rates of 5 and 10 wt % clinoptilolite-rich tuff from Oregon, U.S.A. (Fig. 1). Plant-available water for sand amended with 5 and 10 wt % clinoptilolite with a particle size of >1 mm was near 6 $g\ kg^{-2}$; whereas plant-available water for sand amended with 5 and 10 wt % of <0.047 mm clinoptilolite was approximately 10 and 17 $g\ kg^{-2}$, respectively. The saturated hydraulic conductivity, which ranged from 3 to 48 $cm\ hr^{-1}$, decreased with increasing clinoptilolite addition and decreasing particle size.

Huang and Petrovic (1996) used lysimeters in a greenhouse study to evaluate water consumption of creeping bentgrass in sand amended with clinoptilolite. Shoot-growth rates were 26 to 60% greater in sand amended with 10 wt % clinoptilolite-rich tuff from Oregon, U.S.A., compared with unamended sand. Evapotranspiration rates were 6% higher in creeping bentgrass grown on sand amended with 10 wt % clinoptilolite. They concluded that sand-based putting green turf could benefit from a 10 wt % clinoptilolite amendment by increasing shoot-growth rate without a substantial increase in evapotranspiration rate.

Although few studies have addressed the use of zeolite-rich materials to enhance the physical properties of a sandy soil (e.g. golf green), it appears that zeolite additions (depending on particle size) will increase the soil's water holding capacity while slightly decreasing the saturated hydraulic conductivity. Further studies are required to evaluate the effect of zeolites on soil physical properties.

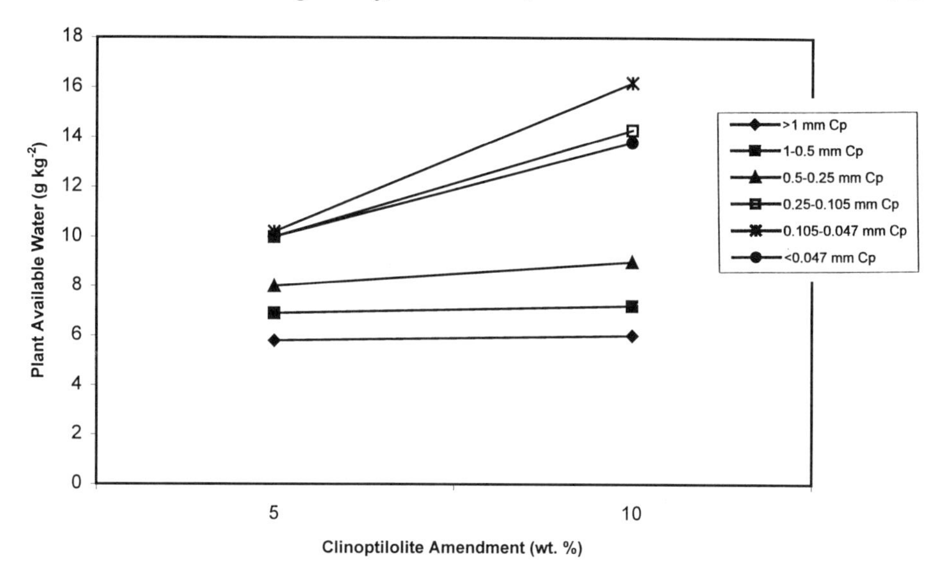

Figure 1. Plant-available water in sand amended with various particle sizes and application rates of clinoptilolite-rich tuff from a deposit in Oregon, U.S.A. (Huang and Petrovic 1995).

FERTILIZER-USE EFFICIENCY AND PREVENTION OF NUTRIENT LEACHING

Because of their unique cation-exchange and selectivity properties, the use of natural zeolites has been suggested to increase fertilizer efficiency and to reduce the leaching of nutrients. Plant-essential elements may be exchanged onto zeolite exchange sites, where the nutrients can be subsequently slowly released for plant uptake. Not only can this exchange process act as a slow-release fertilizer, it will delay or reduce leaching out of the plant root zone, thereby reducing migration of these nutrients into ground water or runoff and decreasing the potential for environmental pollution. Volatilization of gaseous N (e.g. as NH_3 or N_2) can also be reduced if NH_4^+ is exchanged onto zeolite exchange sites so that it is unavailable for conversion to these gaseous phases via microbial processes.

Fertilizer-use efficiency

The fertilizer industry is continuously searching for new technologies and methods that help increase the efficiency of fertilizers added to soils. Many of the fertilizers used in field agriculture, e.g. anhydrous ammonia, have poor fertilizer-use (or nutrient-use) efficiency; and rarely does the fertilizer-use efficiency of conventional fertilizers exceed 50% for most crops. Zeolite additions may aid in increasing fertilizer-use efficiency. In this application, zeolites are added along with soluble fertilizers, such as anhydrous ammonia, urea, potash, and ammonium sulfate. Exchange sites within zeolite structures act as sinks for dissolved cationic nutrients and, as mentioned earlier, zeolite additions can enhance the soil's nutrient-holding capacity by increasing its CEC. Over time, these nutrients are exchanged off of zeolite sites where they become available for plant uptake. The capability to release nutrients over time generally improves the efficiency of added fertilizers. Clinoptilolite has been the principal zeolite examined for these application, but phillipsite, chabazite, and erionite have also been evaluated.

The use of natural zeolites to improve fertilizer-use efficiency is not new, in fact, natural zeolites were first added to fields in Japan in the 1960s (Minato 1968). Researchers were able to improve the plant-available nitrogen for growing rice by 63% when about 80 tonnes ha^{-1} of clinoptilolite-rich tuff and a NH_4-fertilizer were added to coarse-textured soil. Shortly after the use of natural zeolites in soils was introduced in Japan, plant growth experiments were begun in the former Soviet Union in the 1970s (Tsitsishvili et al. 1984). For example, the addition of clinoptilolite-rich tuff at a rate of 400 kg ha^{-1} (particle size of 2-5 mm) from the Dzegvi deposit in Georgia, along with NH_4NO_3, superphosphate, and K fertilizers, increased fertilizer-use efficiency by 50% in the production of geraniums when compared with common fertilizer practices for the production of geraniums without the addition of clinoptilolite (Marshania and Equania 1984).

Several other experiments have been conducted on the use of zeolites to increase fertilizer-use efficiency since the early experiments in Japan and the former Soviet Union (see Table 2). Bouzo et al. (1994) evaluated field yields of sugarcane grown in several different soils amended with different application rates of a clinoptilolite-rich tuff (approx. 85-95% clinoptilolite, CEC ≈ 150 cmol$_c$ kg^{-1}) from a deposit near Tasajeras, Cuba. The results were mixed, depending on the soil type (Entisol, Inceptisol, or Oxisol), application rate of the zeolite (0 to 30 tonne ha^{-1}), and the method for applying the zeolite (i.e. broadcast or banded application). The largest increases in plant production occurred in the Oxisol (highly weathered soil with low CECs), where the sugarcane yield doubled from 38.0 tonne ha^{-1} in the unamended soil to 75.4 tonne ha^{-1} in the soil banded with an application of 3 tonne ha^{-1} clinoptilolite-rich tuff. An addition of 6 tonne ha^{-1} clinoptilolite-rich tuff to the Oxisol (banded application near the seed row) nearly tripled the sugarcane yield to 110.9 tonne ha^{-1}. Both of these soils received N-P-K fertilizers. The significant increases in plant production are likely due to the ability of the zeolite to increase the CEC or nutrient-holding capacity of the highly weathered soil. Additions of clinoptilolite-rich materials to the Entisol and Inoeptisol increased sugarcane production to a lesser extent and, in one case (30 tonne ha^{-1} clinoptilolite addition to the Inceptisol), yields were reduced when compared with the unamended soil. Although the production of sugarcane varied, it was found that the amount of nitrogen fertilizer could be reduced by 50% in the Entisol and Inceptisol without yield reductions when compared with the unamended soils with the higher N treatments.

Reduced nutrient leaching in coarse-textured soils

Fertilizer-use efficiency and reduced leaching are a function of each other; i.e. reduced leaching of fertilizers by zeolite additions will likely increase fertilizer-use efficiency and vice versa. Natural zeolite additions may also reduce leaching of fertilizers into environmentally sensitive groundwater, rivers, and lakes.

Golf greens consist almost entirely of inert sand, so researchers have been examining a variety of amendments (e.g. sphagnum peat, clay minerals, perlite) that might increase the nutrient-holding capacity of the sand and reduce nutrient leaching and volatilization. Recently, the use of natural zeolites as an amendment to golf greens has become attractive for reducing nutrient leaching. Huang and Petrovic (1994) evaluated the effect of adding 10 wt % clinoptilolite-rich tuff (CEC = 160 cmol$_c$ kg^{-1}, Oregon, U.S.A) on NO_3^- and NH_4^+ leaching in simulated sand-based putting greens in lysimeter tests. Four rates of N fertilizer (0, 98, 196, and 293 kg N ha^{-1} as $(NH_4)_2SO_4$) were applied to an acid-washed quartz sand. The leaching of NO_3^- and NH_4^+ was 86 and 99% lower, respectively, when compared with the leaching in the unamended sand, and N use efficiency was improved by 16 to 22% with the addition of the clinoptilolite-rich tuff to the sand, depending on the N treatment. This increase in fertilizer-use efficiency and reduced

Table 2. Selected studies in which zeolites were added with fertilizers to soils to increase fertilizer efficiency and crop yields.

Zeolite/Locality*	Soil type[†]	Fertilizer added (kg/ha)	Crop	Zeolite amendment rate[‡]	Yield[‡]	Reference
Cp	Mollisol	$(NH_4)_2SO_4$	sorghum	0 tonne ha^{-1}	3.1 g pot^{-1}	Pirela et al. (1984)
Washakie Basin, Wyoming, USA				0.5	3.0	
				2	3.2	
				8	3.0	
Cp	Mollisol	urea	cucumber	0 tonne ha^{-1}	8.0 tonne ha^{-1}	Pirela et al. (1984)
Washakie Basin, Wyoming, USA				4	8.7	
				8	9.0	
Cp/Mo	silty clay loam	$NH_4H_2SO_4$	rice	0 kg acre^{-1}	82.44 kg acre^{-1}	Ahn et al. (1984)
Korea				50 kg acre^{-1}	87.5	
Cp	NR	100 N/	wheat, seed yield	0 tonne ha^{-1}	2.95 tonne ha^{-1}	Iskenderov & Mamedova (1988)
Azerbaijan		100 P/		5	3.32	
		90 K		10	3.52	
				20	3.42	
Cp	high clay	urea	brachiaria decumbens	0 g per 360 kg soil	63.8 tonne ha^{-1}	Crespo (1989)
Tasajeras, Cuba				60 g	88.4	
				120 g	103.6	
				180 g	143.8	
Cp	Oxisol	120 N/	sugarcane	0 tonne ha^{-1}	38.0 tonne ha^{-1}	Bouzo et al. (1994)
Tasajeras, Cuba		50 P_2O_5/		3	75.4	
		180 K_2O		6	110.9	
Cp	Entisol, sandy loam		wheat, seed yield	0 (No fertilizer)	10.42 g pot^{-1}	Tsadilas et al. (1997b)
Greece		$(NH_4)_2SO_4$		0 (w/ fertilizer)	13.71	
				6 tonne ha^{-1}	15.50	
				15	9.60	
				30	14.80	
				45	18.20	
				60	17.70	

* Cp = Clinoptilolite-rich materials; Mo = Mordenite-rich materials

[†] Mollisol = soil with thick A horizon and high base saturation; Oxisol = highly weathered soil; Entisol = weakly developed soil

[‡] units for amendment rates and yields varied from researcher to researcher

NR = Not reported

leaching would be expected because the clinoptilolite addition increased the CEC nearly 200 times, from 0.08 $cmol_c\ kg^{-1}$ for the unamended sand to 15.59 $cmol_c\ kg^{-1}$ for the sand amended with 10 wt % clinoptilolite-rich tuff.

Composting of biosolid wastes from municipal sewage treatment has become a popular method for their disposal, instead of land filling, ocean dumping, and incineration. Bugbee and Elliott (1998) have suggested that biosolid compost can be combined with various amendments and used as a potting soil for greenhouses and nurseries They evaluated the leaching of N and P from biosolid compost amended with peat, sand, bark, coconut processing waste, sawdust, calcined clay, vermiculite, or clinoptilolite-rich tuff. The clinoptilolite-containing compost, which consisted of a mixture of 50 vol % compost, 30 vol % sphagnum peat, and 20 vol % clinoptilolite-rich tuff, reduced the total N leached through the pot to 0.4 mg compared with 1.4 g leached through the pot of the control mix, which consisted of the biosolid compost, sphagnum peat, and sand. The zeolite-containing medium, however, produced the lowest total biomass for the production of Black-Eyed Susan when compared with mixtures of the other amendments and compost. The reason for the reduced growth of Black-Eyed Susan was not known.

The above examples illustrate the potential of using zeolites to protect against the leaching of fertilizers into environmentally sensitive waters, although related effects on plant productivity can vary. Additional studies that have examined the use of natural zeolites to reduce nutrient leaching through coarse-textured soils are listed in Table 3.

SLOW-RELEASE FERTILIZATION

Zeolites have the potential to act as a slow-release fertilizer. Slow-release is a term that is interchangeable with delayed-release, controlled-release, controlled-availability, slow-acting, and metered-release. Slow-release is generally used to describe the slow dissolution rate of a soluble fertilizer. However, zeolites are relatively insoluble in comparison with traditional slow-release fertilizers (i.e. the aluminosilicate framework is relatively not affected when contacted with water); slow-release fertilization from zeolites involves the slow exchange of plant nutrients from exchange sites. Two primary mechanisms can be employed to achieve the slow-release of nutrients using zeolites: (1) slow-release fertilization by ion exchange and (2) slow-release fertilization by a combination of mineral dissolution and ion exchange. Slow-release fertilization by ion exchange involves the exchange (or nutrient loading) of a zeolite with a nutrient cation. The nutrient-exchanged zeolite is then added to a soil or soilless substrate, where the zeolite's exchangeable cations are "slowly" replaced by another cation in solution. Cations released from the zeolite exchange sites are available for plant uptake. Slow-release fertilization by a combination of mineral dissolution and ion exchange involves the addition of a mineral fertilizer (e.g. phosphate rock) along with the nutrient-loaded zeolite. As the mineral fertilizer begins to dissolve in water, cations released during dissolution (e.g. Ca^{2+} from phosphate rock) are available for either exchange at zeolite exchange sites or uptake by the plant.

Clinoptilolite has been the primary zeolite studied for use in slow-release fertilization because of its widespread abundance in nature and its selectivity for certain cations, i.e. K^+ and NH_4^+. Chabazite, phillipsite, and mordenite have also been examined for use as slow-release fertilizers. Below, we provide several examples of the two mechanisms described above for the slow release of plant-essential nutrients using zeolites.

Table 3. Selected studies in which zeolites were added with fertilizers to soils or horticultural potting mixtures to reduce nutrient leaching.

Zeolite added*	Zeolite locality	Zeolite CEC[†]	Soil type[‡]	Fertilizer added	Zeolite amendment rate	Amount of nutrient leached	Reference
		$cmol_c$ kg^{-1}					
Cp	Kuykendall, Texas, USA	114	Entisol, loamy sand	0.2 g NH_4^+-N kg^{-1}	0 g kg^{-1}	168.4 mg NH_4^+-N	MacKown & Tucker (1985)
					12.5	133.3 mg NH_4^+-N	
					25	83.6 mg NH_4^+-N	
					50	29.4 mg NH_4^+-N	
Er	Rome, Oregon, USA	157	Entisol, loamy sand	0.2 g NH_4^+-N kg^{-1}	0 g kg^{-1}	168.4 mg NH_4^+-N	MacKown & Tucker (1985)
					12.5	88.2 mg NH_4^+-N	
					25	63.1 mg NH_4^+-N	
					50	11.6 mg NH_4^+-N	
Cp	Teague Mineral, Oregon, USA	160	golf course sand (acid-washed)			NO_3-N/NH_4-N (mg pot^{-1})	Huang & Petrovic (1994)
				98 kg N ha^{-1}	0	20.52/0.25	
				98	10 wt %	18.72/0.22	
				196	0	31.64/1.90	
				196	10 wt %	16.58/0.1	
				293	0	141.84/11.0	
				293	10 wt %	18.27/0.15	
Cp	Tasajeras, Cuba	150	Entisol			Residual N in Soil	Bouzo et al. (1994)
				0 kg urea-N ha^{-1}	0 kg ha^{-1}	53.90 mg N kg^{-1}	
				50	0	58.58 mg N kg^{-1}	
				25	3	123.2 mg N kg^{-1}	
				25	6	157.85 mg N kg^{-1}	
Cp	NR	NR	municipal biosolid compost and peat	6 g Osmocote pot^{-1}	0	1.4 g total N pot^{-1}	Bugbee & Elliott (1998)
				(9 N-6 P-12 K)	20 vol. %	0.4 g total N pot^{-1}	

* Cp = clinoptilolite-rich material, Er = erionite-rich material [†] Cation-exchange capacity of zeolite-rich material
[‡] Entisol = weakly developed soil NR = Not reported

Fertilization by ion exchange

Most of the early studies (i.e. in Japan and the former Soviet Union) added zeolites to soils without nutrient loading (i.e. zeolites containing only native exchangeable cations) to improve the soil's physical and chemical properties. Since that time, several studies have been conducted exchanging a zeolite with a plant nutrient cation to be used as a slow-release fertilizer. As might be expected, these studies have primarily focused on loading clinoptilolite with NH_4^+-N and K^+ because of the zeolite's pronounced selectivity for K^+ and NH_4^+-N over Ca^{2+}, Mg^{2+}, and Na^+. It is therefore "difficult" to remove NH_4^+ and K^+ from zeolite exchange sites by these less selective cations, hence, NH_4^+ and K^+ are "slowly released" over time.

Hershey et al. (1980) chose a clinoptilolite-rich material naturally high in exchangeable K (160 $cmol_c\ kg^{-1}$ of K^+ out of a total CEC of 213 $cmol_c\ kg^{-1}$) as a potential slow-release K fertilizer for the growth of chrysanthemums. A soilless potting medium consisting of 1.5 liters of 30 vol % silica sand, 35 vol % sphagnum peat, and 35 vol % redwood sawdust was fertilized with an equivalence of 3 g of K from either K-clinoptilolite or KNO_3 fertilizer and 5 g dolomite, 2 g $CaCO_3$, 2 g superphosphate fertilizer (0-20-0), and 0.3 g $Ca(NO_3)_2$. Substrates were leached with a Hoagland's nutrient solution without K (see Hoagland and Arnon (1950) for Hoagland's nutrient solution recipe). Over 90% of the K from the KNO_3 source had been leached out of the potting soil after 3 liters of leachate were collected; however, the clinoptilolite exhibited a slow-release of K^+ and only about 10% of the K on zeolite exchange sites had been leached. The authors reported that a single application of 50 g clinoptilolite per 1.5 liters of potting medium produced 3-month yields of chrysanthemums equal to those obtained in the potting medium without zeolite watered daily with a nutrient solution containing 234 mg K/liter, which is 6.7 times more K than the potting medium fertilized with K-clinoptilolite.

Clinoptilolite-rich tuff from near Bowie, Arizona, U.S.A., was exchanged with either K^+ or PO_4^{3-} at two concentrations and added to a 60 vol % sphagnum peat and 20 vol % perlite potting mix to formulate a soilless medium containing 20 vol % nutrient-treated clinoptilolite-rich tuff (Williams and Nelson 1997). It was assumed that K^+ would exchange onto the zeolite and the PO_4^{3-} would adsorb along broken edge sites of the zeolite structure. The K requirements for the growth of chrysanthemum in the soilless medium were met by the K-treated zeolite, which produced similar dry mass and tissue K concentrations as the control plants that received complete fertilization. The low-rate K-treated clinoptilolite addition (2838 µg/ml) also resulted in 23% less leaching of K than the control substrates. Plants that relied on P from P-treated clinoptilolite had lower dry mass and tissue P levels (< 0.23 wt % P) compared with the controls. Less P (10-18%) was leached from the P-treated clinoptilolite media compared with the control, which leached 37% of the applied P. Carlino et al. (1998) reported similar results for chrysanthemum growth and nutrient leaching in potting medium amended with N-treated clinoptilolite-rich tuff in addition to K- and P-treated clinoptilolite-rich tuff.

The above examples illustrate the possibility of achieving slow-release fertilization using nutrient-charged zeolites for horticultural potting mixtures; however, few studies have examined the potential of using nutrient-loaded zeolites in field agriculture. Recently, Perrin et al. (1998b) determined the growth rates and total nitrogen balance (i.e. N fertilizer-use efficiency) for corn grown in a sandy over loamy soil, which was an Aridisol that contained 81% sand, amended with various particle sizes (< 0.25 mm, 0.25-2 mm, and 2-4 mm) and application rates (112, 224, 336 kg N ha^{-1}) of NH_4-exchanged clinoptilolite-rich tuff (NH_4-CEC = 165 $cmol_c\ kg^{-1}$) from Cache County, Utah, U.S.A.. Nitrogen-use efficiency ranged from 95.2 to 72.0% in NH_4-clinoptilolite amended soils

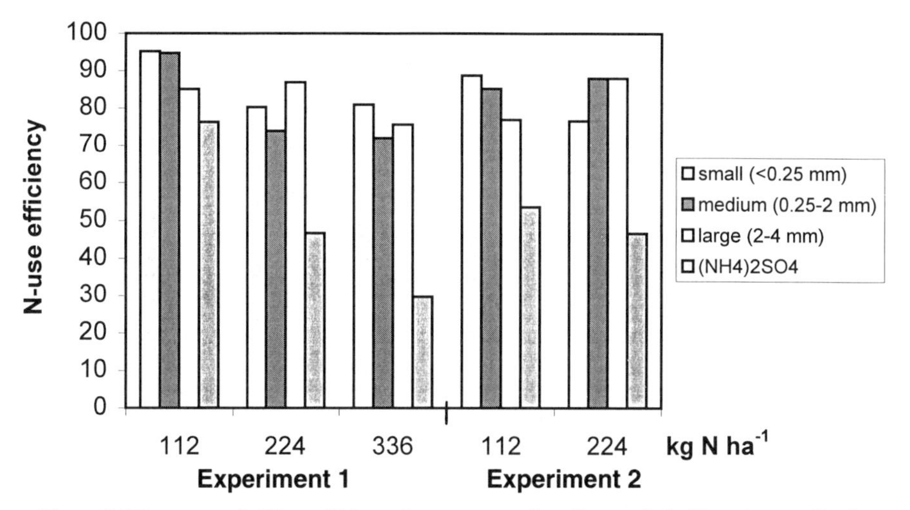

Figure 2. Nitrogen-use fertilizer efficiency for corn grown in soils amended with various application rates and size fractions of an NH_4^+-exchanged clinoptilolite from Cache County, Utah, U.S.A., compared with N-use efficiency of corn grown in the same soil fertilized with $(NH_4)_2SO_4$ as a nitrogen source (Perrin et al. 1998b). Experiments 1 and 2 were duplicates, except that the N fertilizer was banded in Experiment 2.

after 42 days of plant growth, whereas, the N-use efficiency in soils fertilized with $(NH_4)_2SO_4$ ranged only from 29.7 to 76.3 (Fig. 2). No significant differences in corn growth were observed among the N fertilizers. This study illustrates the potential of using NH_4-exchanged zeolites as a slow-release N fertilizer to improve N-use efficiency, while sustaining normal corn production. Perrin et al. (1998a) continued the studies above and found that after 40 days of simulated leaching experiments, 93 to 97% of total N applied as $(NH_4)_2SO_4$ leached through the control quartz sand compared with 24.9 to 72.1% total N leached through the same quartz sand amended with NH_4^+-exchanged clinoptilolite (as described above). The particle size of the added clinoptilolite-rich materials had a significant impact on the amount of N leached through quartz sand. Large NH_4^+-clinoptilolite particles (2-4 mm) had significantly less leaching of total N (24.9-32.5% leaching) than the medium- (0.25-2 mm) and small-sized (<0.25 mm) NH_4^+-clinoptilolite amended sands, which had leaching rates of 53.4 to 68.4% total N and 44.4 to 72.1% total N, respectively. These leaching rates reflect the expected lower exchange rates in the larger particle sized clinoptilolite-rich materials because more time is required for intra-particle diffusion.

Zeolites may have the ability to "protect" NH_4^+ on zeolite exchange sites from microbiological conversion of NH_4^+ to NO_3^- because nitrifying bacteria are too large to fit into the channels and cages within the zeolite structure where NH_4^+ resides on exchange sites. MacKown (1978) reported that nitrification rates decreased in a loamy sand by 11% and in a silty clay loam by 4% when using an equivalent rate of 30 tonne ha^{-1} of NH_4-exchanged clinoptilolite-rich tuff from a deposit near Tilden, Texas, U.S.A.. Reduced nitrification rates were attributed to the retention by clinoptilolite of NH_4^+ on the exchange sites, thereby protecting NH_4^+ from conversion by the nitrifying bacteria.

Other plant-essential cations may be exchanged onto zeolite exchange sites, e.g. Ca^{2+}, Mg^{2+}, Zn^{2+}, Cu^{2+}, Fe^{2+}, and Mn^{2+}. However, most natural zeolites are not particularly selective for these divalent cations and, as might be expected, they are usually more difficult to exchange onto zeolite exchange sites. Although zeolites exchanged with these secondary and micronutrient cations may have the potential to act as slow-release fertilizers, little research has been directed toward this application. Mitov et al. (1995) suggested that clinoptilolite and synthetic zeolites may act as slow-release fertilizers for Cu^{2+}, Zn^{2+}, Co^{2+}, and Mn^{2+}, and Uren and Qing (1997) reported that Mn^{2+}-exchanged clinoptilolite and synthetic zeolites acted as slow-release fertilizer when compared with soluble $MnSO_4$ fertilizers. Additional selected studies that used nutrient-exchanged zeolites as slow-release fertilizers are listed in Table 4.

Fertilization by mineral dissolution and ion exchange

Slow-release fertilization has also been achieved by a combination of zeolite ion exchange and mineral dissolution (e.g. of apatite-rich phosphate rock). In addition to the slow release of K^+, NH_4^+, and other cations exchanged onto zeolite exchange sites, the dissolution of apatite-rich phosphate rock, for example, is enhanced by the exchange of dissolved Ca^{2+} onto zeolite exchange sites. This idea of increasing the dissolution rate of a phosphate phase by creating exchange sites or a "sink" for Ca^{2+} is not new. Moller and Mogensen (1953) used an ion-exchange resin (Na-exchange synthetic aluminosilicate resin called Nylite) to estimate the available P in soils nearly 50 years ago.

Lai and Eberl (1986) were the first to examine the possibility of achieving slow-release fertilization by zeolite ion exchange and apatite dissolution. They mixed a phosphate rock from Florida (mainly carbonate apatite) with untreated and treated (NH_4^+, H^+, or Na^+) clinoptilolite-rich tuff from the Mud Hills deposit near Barstow, California (CEC = 180 $cmol_c\ kg^{-1}$), at a ratio of 5 parts clinoptilolite to 1 part apatite (by weight). The addition of NH_4^+-, H^+-, or Na^+-exchanged clinoptilolite-rich tuff significantly increased solution P concentrations when compared with phosphate rock without zeolite additions (Fig. 3). Phosphorus release was also affected by solution pH; P released from the H-exchanged clinoptilolite-rich tuff and phosphate rock (highly acidic solution) was much higher than with the other treatments.

Based on the original studies by Lai and Eberl (1986), other studies have illustrated slow-release fertilization by zeolite ion exchange and mineral dissolution (Table 5). Allen et al. (1993) expanded the idea of Lai and Eberl (1986) by examining the solubility and cation-exchange relationships in mixtures of phosphate rock and NH_4- and K-exchanged clinoptilolite. Nutrient release in these systems can be represented by the following simplified chemical reactions in which the chemical formula for fluorapatite ($Ca_5(PO_4)_3F$) is used as the phosphate rock and clinoptilolite is represented by Cp:

$$0.5\ Ca_5(PO_4)_3F + 1.5\ H_2O = 2.5\ Ca^{2+} + 1.5\ HPO_4^{2-} + 0.5\ F^- + 1.5\ OH^- \quad (1)$$

$$0.5\ Ca^{2+} + (NH_4^+\text{-Cp}) = NH_4^+ + (0.5\ Ca^{2+}\text{-Cp}) \quad (2)$$

$$0.5\ Ca^{2+} + (K^+\text{-Cp}) = K^+ + (0.5\ Ca^{2+}\text{-Cp}) \quad (3)$$

The above reactions proceed until an equilibrium state is approached, at which point the reactions stop unless another sink, such as nutrient uptake by plant roots, removes any of the reaction products in Equations (1-3), again driving the dissolution and ion-exchange reactions.

Batch equilibration experiments were used to investigate the effect of phosphate source, equivalent fraction of exchangeable K^+ and NH_4^+ on clinoptilolite exchange sites, and the clinoptilolite-to-phosphate rock ratio on solution NH_4-N, P, K, and Ca concentrations (Allen et al. 1993). Clinoptilolite-rich material from a marker tuff in the San Miguel lignite deposit in South Texas, U.S.A., was concentrated and sized to 2 to 50 μm, exchanged into either its K^+- or NH_4^+-form, and then combined with either a high-reactivity phosphate rock (francolite, CO_3/PO_4 ratio = 0.232) from North Carolina,

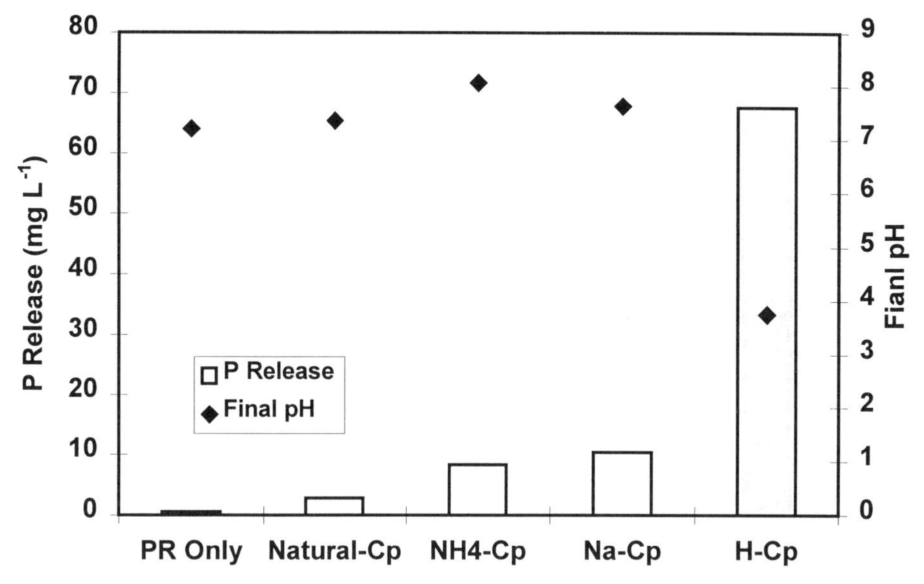

Figure 3. Phosphorus release into solution and final solution pH of a mixture of clinoptilolite-rich tuff from the Mud Hills, California, U.S.A., deposit and a phosphate rock from Florida, mixed at a rate of 5:1 clinoptilolite:phosphate rock (Lai and Eberl 1986) (Cp = clinoptilolite-rich material; PR = phosphate rock).

U.S.A., or a low-reactivity phosphate rock (francolite, CO_3/PO_4 = 0.093) from Tennessee, U.S.A. Mixtures containing the high-reactivity phosphate rock had greater solution concentrations of NH_4^+ and K^+ than mixtures with the low-reactivity phosphate rock. As expected, solution P concentrations increased with increasing clinoptilolite-to-phosphate rock ratios (Table 6), indicating that more exchange sites (i.e. sinks) are available for Ca^{2+} from solution, thereby driving the dissolution of the phosphate rock (see Eqs. 1-3). Solution concentrations of N, P, K, and Ca and the ratios of these nutrients in solution varied predictably with the type of phosphate rock, the clinoptilolite/phosphate rock ratio, and the equivalent fraction of K^+ and NH_4^+ on clinoptilolite exchange sites, suggesting that appropriate mixtures of clinoptilolite and phosphate rock can be formulated to meet specific plant needs. Rates of NH_4, K, and P release in these same systems were best described by a power function equation (Allen et al. 1995a). Initial release rates and cumulative release increased at higher clinoptilolite to phosphate rock ratios or if the more reactive phosphate rock was used (Allen et al. 1996). Nutrient-release rates were also affected by changes in the relative charge fractions of NH_4^+ and K^+ (Fig. 4), due to the ion-selectivity characteristics of clinoptilolite (Allen et al. 1996), i.e. clinoptilolite is more selective for K^+ than for NH_4^+.

Clinoptilolite-rich materials have been the focus of most researchers, probably because of its widespread abundance in near-surface deposits. Phillipsite-rich rocks, however, have also been examined as possible exchange "sinks" for enhancing the dissolution of phosphate rocks. In a study similar to the study of Lai and Eberl (1986), Mnkeni et al. (1994) evaluated the effectiveness of a phillipsite-rich rock (CEC = 148 $cmol_c$ kg^{-1}) from the Mbeya region of Tanzania to enhance the dissolution of a reactive phosphate rock of sedimentary origin from Minjingu, Tanzania, and a unreactive phosphate rock of igneous origin from Panda, Tanzania. The phosphate rocks were combined with the phillipsite-rich tuff in ratios (weight) of 1:1, 1:10, and 1:100

Table 4. Selected studies in which nutrient-loaded (exchanged) zeolites were added to soils to act as slow-release fertilizers.

Zeolite added* (nutrient exchanged)	Zeolite locality	Soil type	Zeolite amendment rate	Zeolite or fertilizer added	Crop productivity	Reference
NH_4-Ph/NH_4-Ch	Tufino, Italy (30% Ph and 19% Ch)	medium-textured soil	200 g m^{-2} (note: Tufino and S. Mango Sul Calore tuffs were statistically grouped together to report yields)		Radish	Langella et al. (1995)
				no zeolite	28.5 g $plant^{-1}$	
				untreated zeolites	29.56 g $plant^{-1}$	
				NH_4-zeolites	29.67 g $plant^{-1}$	
	S. Mango Sul Calore, Italy (9% Ph and 61% Ch)				Sugar beet	
				no zeolite	567.0 g $plant^{-1}$	
				untreated zeolites	645.75 g $plant^{-1}$	
				NH_4-zeolites	680.75 g $plant^{-1}$	
K-Ph/K-Ch	Tufino, Italy (30% Ph and 19% Ch)	medium-textured soil	200 g m^{-2}		Radish	Langella et al. (1995)
				no zeolite	28.5 g $plant^{-1}$	
				untreated zeolites	29.56 g $plant^{-1}$	
				K-zeolites	29.39 g $plant^{-1}$	
	S. Mango Sul Calore, Italy (9% Ph and 61% Ch)				Sugar beet	
				no zeolite	567.0 g $plant^{-1}$	
				untreated zeolites	645.75 g $plant^{-1}$	
				K-zeolites	674.63 g $plant^{-1}$	
NH_4-Cp	Mud Hills, California, USA			$(NH_4)_2SO_4$	Radish (root)	Lewis et al. (1984)
		medium-textured soil	0 g kg^{-1}	150 mg N kg^{-1}	8.5 g $plant^{-1}$	
			5.0	0	13.5 g $plant^{-1}$	
		coarse-textured soil	0 g kg^{-1}	200 mg N kg^{-1}	7.6 g $plant^{-1}$	
			6.7	0	11.6 g $plant^{-1}$	
NH_4-Cp	Belia Bair, Eastern Rhodopes, Bulgaria				Barley (6 cuttings)	Manolov & Stoilov (1997)
		none	100% Cp substrates	0.1 wt % N Cp	100.3 g pot^{-1}	
				0.25 wt % N Cp	81.02	
				0.45 wt % N Cp	69.86	
				0.80 wt % N Cp	61.33	
			Control (alluvial soil)	soil only	48.25	

* Ph = phillipsite-rich material; Ch = chabazite-rich material; Cp = clinoptilolite-rich material

Table 5. Selected studies on slow-release fertilization via zeolite ion-exchange and mineral dissolution of phosphate rock (PR).

Zeolite added* (nutrient exchanged)	Zeolite locality	Phosphate rock locality	Zeolite:PR	Crop grown	Reference
NH_4^+-Cp, H^+-Cp	Mud Hills,	Florida, USA	Various ratios	none	Lai & Eberl (1986)
Na^+-Cp & Native Cp	California, USA				
Native Cp	Ash Meadows Deposit, California. USA	Cargill, Ontario, Canada	1:1 (wt.)	corn	Chesworth et al. (1987)
NH_4^+-Cp	Washakie Basin, Wyoming, USA	North Carolina, USA	Various ratios	sorghum-sudangrass	Barbarick et al. (1990)
NH_4^+-Cp	Washakie Basin, Wyoming, USA & Craven Creek, South Dakota, USA	North Carolina and Idaho, USA	Various ratios	sorghum-sudangrass	Eberl et al. (1995)
NH_4^+-Cp and K^+-Cp (Ratio of 3:1 NH_4^+:K^+)	San Miguel Lignite Mine, Texas, USA & Fort LeClede Deposit, Wyoming, USA	North Carolina & Tennessee, USA	5:1 (vol.)	wheat	Allen et al. (1995b)
Natural-Cp	NR	North Carolina, USA	≅7.2:1	none	He et al. (1999)
NH_4^+-Ph/Ch	Chiaiano Quarry, Naples, Italy	Maroc, Italy	Various ratios	none	Franchini-Angela et al. (1994)
Natural Ph	Rukwa rift valley, Tanzania	Minjingu & Panda, Tanzania	Various ratios	maize	Mnkeni et al. (1994)
NH_4^+-Ph	Ariatain, Jordan	Rusiafa, Jordan	Various ratios	none	Dwairi (1998)

* Cp = clinoptilolite-rich material; Native Cp = clinoptilolite-rich material in its native exchange form; Ph = phillipsite-rich material; Ch = Chabazite-rich material; PR = phosphate rock

NR= Not reported

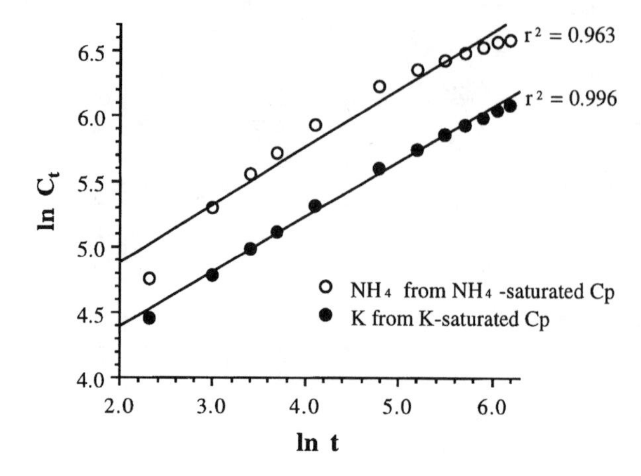

Figure 4. Power-function model of K and NH_4 release from San Miguel clinoptilolite (SM Cp) leached with 0.01 M $CaCl_2$ (t = time in minutes; C_t = cumulative K or NH_4 release at time t in micromoles) (Allen et al. 1996).

Table 6. Fraction of P dissolved, solution concentrations of P, Ca, and K, and solution pH values after 96 hours of equilibration for various ratios of San Miguel clinoptilolite (Cp) to North Carolina phosphate rock (PR) (K-exchanged clinoptilolite quantities were kept constant for each treatment; water:clinoptilolite ratio = 40:1; system closed to the atmosphere) (Allen et al. 1993).

Cp:PR ratio	P dissolved	*Solution concentration*			Solution pH
		P	Ca	K	
	g kg⁻¹		*mM*		
1:1	0.01	0.015	0.014	1.86	8.2
2:1	0.02	0.030	0.009	1.38	8.6
3:1	0.05	0.044	0.008	1.14	8.7
5:1	0.13	0.069	0.006	1.01	8.9
10:1	0.35	0.094	0.004	0.82	9.0
20:1	0.87	0.116	0.002	0.70	9.0
LSD (0.05)	---	0.012	0.006	0.04	0.2

zeolite:phosphate rock and added to a sandy clay loam soil (Inceptisol). Triple superphosphate (40 mg P kg^{-1}) was used as a P source to compare with the phosphate-rock treatments for the growth of maize in greenhouse pot experiments. The zeolite addition had little effect on the plant uptake of P from the unreactive Panda phosphate rock; however, the plant uptake of P nearly doubled in the soil containing the more reactive Minjingu phosphate rock and the phillipsite-rich tuff at a ratio of 1:100.

Most of the research on zeolite ion exchange and mineral dissolution has been directed toward zeolite-apatite reactions. However, other soluble minerals or phases may be precipitated along with zeolites (i.e. occluded salts) or phases may be precipitated by reactions with cations removed from zeolite exchange sites and other elements removed from waste streams. The RIM-NUT process (a term derived from *rim*ozione dei *nut*rienti = removal of nutrients) was developed as a method for reducing the eutrophic potential (i.e. potential for depletion of O_2 in water by microorganisms) of municipal and industrial wastewater by combining ion-exchange from zeolites and precipitation processes to selectively remove ammonium and/or phosphate ions (Liberti et al. 1979, 1981; Ciambelli et al. 1988; Liberti et al. 1995, 1999). In a more recent version of the process (Liberti et al. 1995), wastewater is passed through columns of clinoptilolite-rich tuff and a gel-type, strongly basic anion resin (e.g. Amberlite IRA 458) to selectively remove NH_4^+ and $H_2PO_4^-$, respectively. Exhausted columns (both anion resin and clinoptilolite) are regenerated with 0.6 M NaCl. Effluents from the regeneration, which contain NH_4^+ and $H_2PO_4^-$, are then added to a Mg salt under controlled pH conditions to produce a high-grade $MgNH_4PO_4{\cdot}6H_2O$ fertilizer. Phillipsite-rich tuff from Italy has also been successfully demonstrated to remove NH_4^+ from wastewaters using in the RIM-NUT process (Ciambelli et al. 1988). The first full-scale installation of the RIM-NUT process was recently designed and funded to treat 11,000 m^3/d of mixed municipal and industrial wastewater in Manfredonia, Italy (Liberti et al. 1999). Although the plant has been primarily designed to remove NH_4^+, phosphate from a nearby chemical factory will be used for the production of $MgNH_4PO_4{\cdot}6H_2O$ fertilizer.

Slow-release fertilization may also be obtained by zeolite ion exchange and mineral dissolution of nutrient salts precipitated on the zeolite. The salt phase either precipitates on zeolite crystal surfaces or precipitates as an occluded salt in the channels or cavities of the zeolite crystal. Plant-essential elements are then released into solution as the precipitated salts begin to dissolve and, as described above, additional plant-essential cations (e.g. K^+, NH_4^+) are "slowly" released into solution via ion exchange from zeolite exchange sites. Salts that precipitate on the surfaces of zeolite crystals are quickly dissolved with water and those occluded in the zeolite's extraframework sites are more slowly dissolved due to kinetic effects. Barrer and Meier (1958) first reported the occlusion of NaCl in zeolite X and $AgNO_3$ in zeolite A, where the occluded $AgNO_3$ could be easily extracted (i.e. dissolved) in hot water. The amount of salt that a zeolite can occlude varies depending on the concentration of the salt solution, the type of cation and anion, the cation and anion density, and the zeolite framework structure (Breck 1974) as well as the size of the salt anion.

Phosphates and nitrates are the primary salt phases occluded in zeolites for potential applications in agriculture. Garcia Hernández et al. (1993) treated a phillipsite-rich tuff from Tenerife, Canary Islands, with different concentrations of KH_2PO_4 in a reactor to precipitate KH_2PO_4. Although no plant growth studies were conducted, nutrient release rates suggested that the KH_2PO_4 occluded salts within the K-exchanged phillipsite tuff can act as a slow-release fertilizer for K and P. Notario del Pino et al. (1994) conducted a follow-on greenhouse study in which alfalfa was grown in a coarse-textured mixture of a soil and basaltic ash (1:4 v:v) amended with the Tenerife phillipsite-rich tuff treated with KH_2PO_4. Plant tissue P was consistently higher in alfalfa grown in phillipsite-rich tuff treated with KH_2PO_4 when compared with plants grown in substrates with KH_2PO_4 fertilizer only. There were, however, no significant differences in dry-matter production for alfalfa grown in either the substrate amended with KH_2PO_4-treated phillipsite-rich tuff or the substrate fertilized only with KH_2PO_4 fertilizer.

Park and Komarneni (1997) used molten salt treatments of KNO_3 and NH_4NO_3 to occlude these salts in erionite (Shoshone, California, CEC = 177 $cmol_c\ kg^{-1}$), clinoptilolite (Castle Creek, Idaho, CEC = 181 $cmol_c\ kg^{-1}$), chabazite (Christmas, Arizona, CEC = 216 $cmol_c\ kg^{-1}$), and phillipsite (Pine Valley, Nevada, CEC = 267 $cmol_c\ kg^{-1}$). Zeolite and salt mixtures were heated just above the melting points of the two salts (350°C for KNO_3 and 185°C for NH_4NO_3) for 4 hours. The occlusion of salts in these zeolites was confirmed by infrared analyses and the amount of occluded nitrate content of KNO_3-treated zeolites was greatest in erionite > clinoptilolite > chabazite > phillipsite, and for NH_4NO_3-treated zeolites, the occluded nitrate content was greatest in phillipsite >

erionite > chabazite > clinoptilolite. Because the zeolites were exposed to a higher temperature for the molten KNO_3 treatment, chabazite and phillipsite did undergo partial decomposition, although, clinoptilolite and erionite were unaffected by the treatment. In a follow-on study, Park and Komarneni (1998) examined the total uptake of N (i.e. occluded and exchanged NO_3 and exchanged NH_4^+) for the same suite of zeolites that had been treated with molten NH_4NO_3. As might be expected the total N content for the molten NH_4NO_3-treated zeolites was considerably higher than the total N content of NH_4^+-exchanged zeolites. A simulated soil solution was passed through the N-loaded zeolites to determine release rates of NH_4^+ and NO_3^-. Both NH_4-exchanged and molten NH_4NO_3-treated zeolites exhibited similar NH_4^+ release rates, where NH_4^+ was slowly and steadily released over the 30-day leaching experiment. A greater percentage of total NO_3^- was released compared with the percentage of NH_4^+, which suggested that the occluded NO_3^- is more readily available for release than exchangeable NH_4^+ (although some NH_4^+ will be released during the dissolution of the occluded NH_4NO_3). Based upon these studies, it is clear that the zeolite's nutrient content (in this case N) can be significantly increased by the addition of occluded salts and that the occluded salts behave similarly to the slow steady release of exchangeable NH_4^+. Obviously more research is needed to determine the capabilities of occluded salt zeolites to act as slow-release fertilizers for plant growth.

Although no studies on nutrient release were conducted, Pode et al. (1998) initially exchanged clinoptilolite with Mn^{2+}, Fe^{2+}, or Zn^{2+} and then treated the metal-exchanged zeolite with molten NH_4NO_3. They determined that the NH_4NO_3 had precipitated in the zeolite after treatment, probably as occluded salts although no direct data were presented. Pode et al. (1998) suggested that the product may be used as a slow-release fertilizer for micronutrients (and no doubt N).

ZEOPONIC PLANT-GROWTH SUBSTRATES

Zeoponic definition and background

The development of zeoponic substrates has emerged as one of the leading research topics in the area of using natural zeolites in the agricultural and horticultural industries. Parham (1984) first used the term *zeoponics* to describe an artificial soil consisting of zeolites, peat, and vermiculite that Bulgarian researchers had developed. Since that time, the definition of zeoponic plant growth systems has evolved to be defined as the cultivation of plants in artificial soils having zeolites as a major component (Allen and Ming 1995).

Petrov et al. (1982) developed zeoponic substrates composed of specific size ranges of clinoptilolite-rich material, vermiculite, and/or peat supplemented with N and P fertilizers (NH_4NO_3, $(NH_4)_2SO_4$, and superphosphate). Strawberries and peppers were successfully grown in the zeoponic substrates. Cuban scientists developed zeoponics for a variety of vegetable and ornamental crops (Rivero and Rodríguez Fuentes 1988; Perez et al. 1991; Rivero et al. 1991a,b). They have also successfully advanced zeoponics from basic research to the commercial level as evidenced by the development of commercial tomato and cucumber operations utilizing zeoponic substrates (pers. comm., Rodríguez Fuentes, 1991).

Zeoponic research in the United States has focused on developing synthetic soils with desirable physical properties that supply a balanced diet of plant nutrients for many growth cycles without fertilizer additions. Allen et al. (1995b) monitored wheat dry-matter production and nutrient uptake in zeoponic mixtures composed of apatite, NH_4- and K-exchanged clinoptilolite, and acid-washed quartz sand. Clinoptilolite-rich materials from the Fort LeClede deposit in Wyoming and from the San Miguel lignite mine in Texas and two phosphate rock materials (apatite) from North Carolina and Tennessee were factorially combined to form four zeoponic mixtures. Sand was added in varying proportions as an inert component to give a series of synthetic soils containing from 1 to 100 vol % of the zeoponic mixtures. Dry-matter yields were consistent through five cuttings if at least 25 vol % of the synthetic soil was composed of the zeoponic mixture, which included the high reactivity North Carolina apatite. If 10 vol % or less of the synthetic soil consisted of the zeoponic mixture, K deficiency limited wheat dry-matter production. Calcium deficiency limited wheat dry-matter production if the low reactivity Tennessee apatite was used.

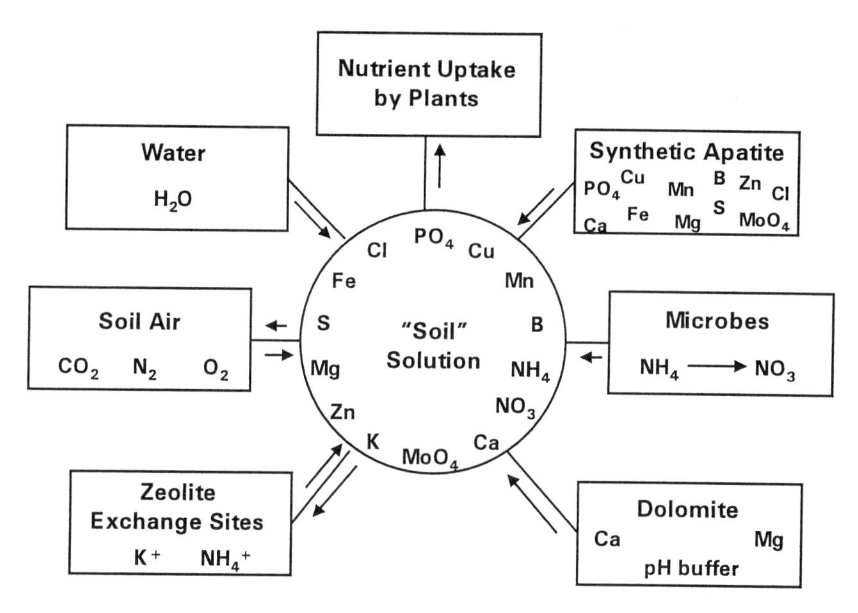

Figure 5. Dynamic equilibria for National Aeronautics and Space Administration's (NASA) zeoponic plant growth system. Plant growth nutrients are slowly released from synthetic, nutrient-substituted apatite by dissolution reactions and from the zeolite (clinoptilolite) by ion-exchange reactions. The reactions in soil solution (i.e. nutrient release) are driven towards the root-soil interface by the uptake of nutrients by the plant (Ming and Allen 1999).

Ming et al. (1995) improved the clinoptilolite and phosphate rock systems by substituting synthetic hydroxyapatites for phosphate rock. These synthetic hydroxyapatites had Mg, S, and the plant-essential micronutrients incorporated into their structures in addition to Ca and P (Golden and Ming 1999). A range of elements can be incorporated into the structure and it is possible to modify the composition of the synthetic hydroxyapatite in order to meet a range of requirements for plant growth. Mixtures of K- and NH_4-exchanged clinoptilolite and nutrient-substituted synthetic hydroxyapatite have the capability to provide all of the essential plant-growth nutrients. Ca, P, Mg, S, and the micronutrients are made available to the plant by the dissolution of the synthetic, nutrient-substituted hydroxyapatite, and K and NH_4-N are made available by ion exchange with Ca at the zeolite exchange sites (Fig. 5). Wheat was grown in a zeoponic substrate, which consisted of K- and NH_4-exchanged clinoptilolite from the Fort LeClede deposit in Wyoming and a synthetic, nutrient-substituted hydroxyapatite, and watered with de-ionized water (Ming et al. 1995). Plant tissue tests indicated that clinoptilolite-synthetic hydroxyapatite substrates supplied sufficient levels of essential plant elements (i.e. N, P, K, Ca, Mg, S, Fe, Mn, Zn, Cu, Mo, B, and Cl) for the intensive vegetative growth of wheat. A natural phosphate rock (apatite) from North Carolina was also evaluated in this study and was found to supply sufficient levels of the essential plant-growth elements, except for Mg. Dry matter production for clinoptilolite-synthetic apatite, clinoptilolite-natural apatite, clinoptilolite watered with nutrient solution, and quartz sand watered with nutrient solution were 10.65, 12.28, 5.06, and 5.46 g pot^{-1}, respectively.

The first seed yields reported for wheat grown in clinoptilolite-synthetic hydroxyapatite substrates diluted with a potting mix were nearly 70% lower than seed produced from a control potting mix watered with nutrient solution (Gruener et al. 2000). Phosphorus content in tissue of plants grown in zeoponic substrates after 30 days was considerably higher (1.89-2.41 wt %) than normally expected for wheat at this stage of growth (0.3-0.6 wt %). They contributed the low seed yield to several factors, including (1) above-normal contents of P in the plant tissue; (2) a NH_4-N source; and (3) a wheat variety (var., 'Super Dwarf') that has a low-yield. In a subsequent test (Henderson et al. 2000), plant-tissue P concentrations in wheat (var., 'USU Apogee') were lowered into the expected nutrient range by the addition of dolomite ($CaMg(CO_3)_2$), ferrihydrite ($Fe_2O_3 \cdot 9/5H_2O$), and nitrifying bacteria to the zeoponic substrate containing synthetic hydroxyapatite. Seed yields were approximately 30% greater in the dolomite-amended zeoponic substrate inoculated with nitrifying bacteria compared with a peat-vermiculite-perlite control substrate watered with _-strength Hoagland's nutrient solution. It is thought that the dolomite addition reduced the solubility of the synthetic hydroxyapatite due to a common-ion effect (i.e. Ca^{2+}) and the nitrifying bacteria aided in the conversion of NH_4^+ to NO_3^- that appears to enhance seed production. The addition of dolomite reduced the P content from 1.4 wt % in wheat grown in substrates without dolomite additions to 1.1 wt % in wheat grown in substrates containing dolomite. The addition of the nitrifying bacteria further reduced the P plant tissue content to approximately 0.8 wt %. Grain yields increased from approximately 4 g pot^{-1} without the nitrifying bacteria to an average of approximately 8 g pot^{-1} with the bacteria addition.

Steinberg et al. (2000) conducted a side-by-side comparison of wheat (var., USU-Apogee) growth and yield in a hydroponic system versus a zeoponic system in a controlled-environment chamber. The hydroponic culture was a nutrient-film technique with a modified $^1/_2$-strength Hoagland's nutrient solution, and the zeoponic substrate consisted of K- and NH_4-saturated clinoptilolite, synthetic hydroxyapatite, dolomite, and nitrifying bacteria described by Henderson et al. (2000). The zeoponic substrate was diluted with 70 vol % porous ceramic soil and watered with de-ionized water in a microporous tube irrigation system that maintained the soil matric potential at -0.5 kPa (Steinberg and Henninger 1997). Temperature, light, and relative humidity were kept constant throughout the experiment at 23°C, 1700 μmol m^{-2} s^{-1}, and 70%, respectively, with a 24 hour photoperiod. The seed yield and harvest index at 64 days for plants grown in the zeoponic substrate were 1.3±0.2 kg m^{-2} and 37% versus 1.8±0.3 kg m^{-2} and 51% for plants grown in the hydroponic system. The amount of wheat seed produced in the zeoponic substrate was equivalent to approximately 200 bushels $acre^{-1}$. Optimal concentrations of N and P were found in plant tissue of wheat grown in the zeoponic substrate; however, the Ca content (0.43 wt %) was lower than expected (1.0 wt %) for plants at this stage of growth.

Space applications

Plants are being considered as an important component of regenerative life-support systems (i.e. air, water, solid waste regeneration, and food production) for long-duration space missions (e.g. space stations, planetary outposts). Both the Russian Space Agency (RSA) and the U. S. National Aeronautics and Space Administration (NASA) have been conducting investigations on growing plants in space using zeoponic substrates (Ming et al. 1995; Morrow et al. 1995; Ivanova et al. 1997).

A nutrient-charged zeolite substrate was developed by Bulgarian scientists for plant-growth experiments in the SVET Space Greenhouse aboard the RSA's Mir Space Station (Ivanova and Petrova 1993; Ivanova et al. 1994, 1997). The substrate is termed 'Balkanine' and consists of a clinoptilolite-rich tuff from the "Beli Plast" deposit in Bulgaria. The clinoptilolite-rich tuff was treated with solutions of plant-nutrient salts (e.g. $(NH_4)_2SO_4$, NH_4-superphosphate). The dry-weight biomass of radish and Chinese cabbage plants grown in the 'Balkanine' on Mir were 2 to 5 and 5 to 10 times smaller, respectively, than ground controls (Ivanova et al. 1993). During a visit to the Mir Space Station by a NASA astronaut, 'Super Dwarf' wheat was grown from seed in the Balkanine; however, the first wheat experiments failed to set seed (pers. comm., F. Salisbury, Professor, Plant, Soils, and Biometerology Department, Utah State University, Logan, Utah, 1997). The last wheat experiment conducted on the Mir Space Station produced seed, which has been returned to Earth and the space-grown wheat has produced two generations of wheat in plant-growth chambers at Utah State University (pers. comm., G. Bingham, Professor, Space Dynamics Laboratory, Logan, Utah, 2000).

Zeoponic substrates developed by NASA for space flights consisted of K- and NH_4-exchanged clinoptilolite and either synthetic or natural apatite. Dwarf wheat (variety 'Super Dwarf') and rapid cycling "Wisconsin Fast Plants" (*Brassica rapa*) were grown for eight days in zeoponic substrates on the Space Shuttle in February 1995 (Morrow et al. 1995). The growth and development of both plant species in orbit appeared normal and were similar to those of plants grown in ground controls, although the wheat grown in space was shorter in height. Due to the short time in space (8 days), it was difficult to fully interpret the effectiveness of zeoponic substrates for plant growth in microgravity environments. Plans are currently underway to use zeoponic substrates in plant-growth units that will be onboard the International Space Station, which is scheduled to be operational by 2002-2003.

ZEOLITES AS CARRIERS FOR HERBICIDES, INSECTICIDES AND OTHER ORGANIC COMPOUNDS

Zeolites have the potential to be slow-release carriers for herbicides, insecticides, and other organic compounds, although most of these compounds are too large to access the channels and cages of natural zeolites. Organic molecules too large to fit into zeolite channels, however, can be adsorbed onto charged sites on the crystal's surface (e.g. see Cadena and Cazares 1995). Surface (or edge) adsorption is dependent on several factors, but pH is the most important of those factors (i.e. pH-dependent charged sites).

Several studies were conducted in the 1970s and 1980s to examine the possible use of natural zeolites as carriers for herbicides, insecticides, and fungicides. Yoshinaga et al. (1973) suggested that clinoptilolite could be used as a substrate for benzyl phosphorothioate to control stem blasting in rice, and Hayashizaki and Tsuneji (1973) suggested that clinoptilolite might be an effective adsorbent for benthiocarb, which could be used as a herbicide to control weeds in rice paddy fields. Zhang (1982) reported that zeolites were being used in the 1970s in China to improve the efficiency of pesticides and herbicides.

Because of increasing concerns of protecting the environment from chemicals migrating into groundwater or runoff, studies on utilizing natural zeolites as carriers for insecticides, herbicides, and fungicides have increased with regularity over the past few years (Table 7). García Hernández et al. (1995) examined the adsorption and desorption of oxamyl (a systematic, broad spectrum carbamate used as a pesticide) by a phillipsite-rich tuff from Terenife (Carnary Islands). Adsorption of oxamyl followed a first-order rate law and the isotherm showed two well-defined stages that suggested two mechanisms for oxayml adsorption on zeolite surfaces. Release kinetics of oxamyl from phillipsite-rich tuff also followed a first-order rate law with two stages of release. The second stage of release is slow (shallow slope), which suggests that oxamyl-phillipsite may act as a slow-release herbicide and reduce leaching of the carbamate into groundwater.

In recent years, the golf industry has been forced to consider new innovative methods of restricting or retarding the migration of fertilizers and chemicals into the groundwater. A major concern is groundwater contamination of the fungicide, metalaxyl (N-(2,6-dimethylphenyl)-N-(methoxyacetyl) alanine methyl ester), which is used to control *Pythium app.* diseases in turfgrass. Petrovic et al. (1998) addressed this problem by amending a fine sandy loam soil (Alfisol, pH = 6.5) or sand alone with peat, brewery wastes, sewage sludge, or clinoptilolite-rich material (CH zeolite, Teague Mineral Products, Adrian, Oregon, U.S.A.). Penncross creeping bentgrass was sodded and established on all of the treatments. Metalaxyl was applied at a rate of 0.64 g per 4.5 m^2, which was the area of a lysimeter constructed for the leaching experiments. The percentage of cumulative metalaxyl leached through the substrate decreased as the content of the zeolite increased; however, only small percentages (ranging from 0.22 to 1.22%) of the total metalaxyl applied to all of the amended sands and the unamended soil were accounted for in the leachates.

Bio-insecticides (i.e. the use of microbes against insect pests) are an environmentally attractive alternative to using organic chemicals for insecticides. One problem of applying bio-insecticides under field conditions is that the microbes are not stable and degrade rapidly in UV radiation. Kvachantiradze et al. (1999) suggested the use of clinoptilolite-rich tuff from the Tedzami deposit in Eastern Georgia as an additive to photostabilize *Bacillus thuringiensis*, i.e. to protect the bacteria from UV radiation. A *Bacillus thuringiensis*-clinoptilolite complex (Bt-clinoptilolite complex) was made by shaking solutions of the bacteria and zeolite for 24 hours in order to ensure adsorption of the delta-endotoxin of the bacteria by clinoptilolite. Flour moth larva mortality rate was 100% when treated with the Bt-clinoptilolite complex after 12 hours of irradiation, whereas the mortality rate of larva subjected to unprotected *B. thuringiensis* was 90% (Fig. 6). The larva mortality rate was 83.3% in the Bt-clinoptilolite complex compared with 63.3% in unprotected *B. thuringiensis* at the end of 36 hours of sunlight. These results are encouraging for using zeolites as a photostablizer for bio-insecticides.

ENVIRONMENTAL SOIL REMEDIATION

Soils throughout the world have been contaminated with heavy metals (e.g. Cu, Cd, Pb, Zn) and radionuclides (e.g. ^{137}Cs, ^{90}Sr). Because zeolites have high CECs, unique ion selectivities and physical characteristics, and chemical stability, they may be effectively added to contaminated soils to reduce the leaching or uptake by plants of these environmentally unfavorable elements. These cations usually exchange onto zeolite exchange sites; therefore, it is important to choose a zeolite that is highly selective for these cations and that will be stable in the soil environment.

Remediation of soils contaminated with radioactive nuclides

Based upon their unique selectivity for Cs^+ and Sr^{2+}, zeolite-rich rocks have been

Table 7. Selected studies that utilize natural zeolites as carriers for herbicides, insecticides, fungicides, and other organic compounds.

Zeolite type*	Zeolite locality	Suggested application	Adsorbent/chemical	Reference
Ph	Tenerife, Canary Islands	Slow-release pesticide	Oxamyl	García Hernández et al. (1995)
Heu-Cp & Mo Synthetic zeolite Na-Y	La Pita & San Ignacio, Cuba	Substrate for the synthesis of herbicide	2,4-dichlorophenoxyacetic acid	Lami et al. (1999)
Cp	NR	Control stem blasting in rice	Benzyl phosphorothioate	Yoshinaga et al. (1973)
Cp	NR	Adsorption substrate for herbicide	Benthiocarb	Hayashizaki & Tsuneji (1973)
Cp	Tedzami deposit, Georgia	Photostabilizer for bio-insecticide	*Bacillis thuringiensis*	Kvachantiradze et al. (1999)
Cp	Teague Mineral, Oregon, USA	Adsorption of a fungicide to prevent migration into groundwater	Metalaxyl	Petrovic et al. (1998)
Cp	Teague Mineral, Oregon, USA	Inactivation of *Salmonella typhimurium* in soils amended with poultry manure	None	Ricke et al. (1995)
Cp	Khakordzula deposit, Georgia	Substrate for pesticide	Organic phosphorous	Sikharulidze & Putsikina (1993)
Cp	Nizny Hrabovec, Slovakia	Substrate for a slow-release insecticide	Synthetic pyrethroide ((R,S)-α-cyano-3-phenoxybenzyl-IR, S-*cis*, *trans*-3-(2,2-dichlorovinyl)-2,2-dimethyl cyclopropanocarboxylate)	Sopková & Janoková (1998)

* Ph = phillipsite-rich material; Cp = clinoptilolite-rich material; Heu-Cp = heulandite/clinoptilolite-rich material, Mo = mordenite-rich material

NR = Not reported

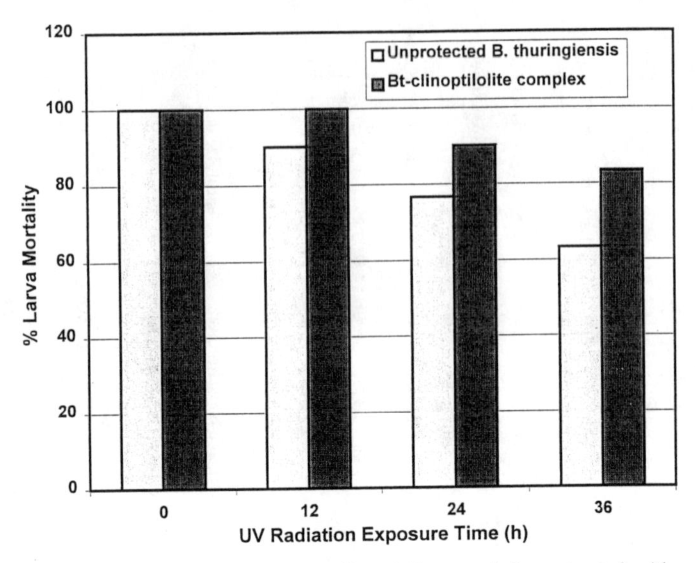

Figure 6. Comparison of the mortality of flour moth larva treated with a *Bacillus thuringiensis* bio-insecticide unprotected from UV radiation or a clinoptilolite-rich tuff from the Tedzami deposit in Eastern Georgia treated with *Bacillus thuringiensis* (Bt-clinoptilolite complex) (Kvachantiradze et al. 1999). The Bt-clinoptilolite complex provides some protection from UV radiaion.

studied as remediation agents for trapping radioactive Cs and Sr from contaminated soils due to nuclear fallout (e.g. soils around the Chernobyl nuclear plant in the Ukraine), contact with water from reactor cooling reservoirs, or radioactive waste spills. The primary purpose of using natural zeolites is to immobilize radionuclides in the soil and to reduce or prevent the uptake of radionuclides by plants, which may enter into the food chain and eventual be consumed by humans.

Studies were conducted over 30 years ago that used a clinoptilolite-rich tuff (predominately the Na form) from Hector, California, to determine whether addition of zeolites to contaminated soils would reduce the uptake of $^{90}Sr^{2+}$ and $^{137}Cs^{+}$ in plants growing in these soils (Nishita et al. 1968; Nishita and Haug 1972). Clinoptilolite additions reduced ^{90}Sr contents in bean plants by 48 to 70, 54 to 77, and 44 to 77% in leaves, stems, and fruits, respectively, compared with plants grown in control soils (Nishita and Haug 1972). Surprisingly, the zeolite addition had no effect on the $^{137}Cs^{+}$ uptake by the bean plants. Similar results were reported for the uptake of $^{90}Sr^{2+}$ and $^{137}Cs^{+}$ by barley and clover grown in soils amended with either untreated or Ca-exchanged clinoptilolite-rich tuff (Nishita and Haug 1972). Because Sr^{2+} appeared to be exchanged more readily onto zeolite exchange sites than Cs^{+}, which contradicts the expected selectivity of Cs^{+} over Sr^{2+} in clinoptilolite, the authors suggested that Cs^{+} may have been adsorbed by soil colloids and thus clinoptilolite was ineffective in exchanging Cs^{+}.

Several other studies have also shown beneficial effects of the "trapping" of Cs^{+} by zeolite-rich materials added to contaminated soils. The fallout of radionuclides from the Chernobyl nuclear plant accident in 1986 was quite extensive. Zeolites were considered for numerous treatments related to the accident (summarized by Chelishchev 1995). One of those treatments was to add clinoptilolite-rich tuff (60 wt % clinoptilolite) to soils at rates of 10-50 tonne acre^{-1} to decrease the radionuclide content in plant tissue by a factor of 2-3 for ^{137}Cs and 50-70% for ^{90}Sr.

In response to Chernobyl fallout throughout the western countries of the former Soviet Union, and Eastern and Northern European countries, including Great Britain, Campbell and Davies (1997) conducted greenhouse experiments to evaluate the effectiveness of clinoptilolite-rich material (CEC = 185 $cmol_c\ kg^{-1}$) on the uptake of Cs^+ from two soils (a lowland loam and a upland peat). Twelve treatments were prepared by spiking the soils with 0, 10, 20, and 40 mg $Cs^+\ kg^{-1}$ soil and either with or without 10 wt % clinoptilolite-rich material. The Cs^+ content of ryegrass grown in clinoptilolite-amended loam soil with the addition of 40 mg $Cs^+\ kg^{-1}$ soil was 5.08 mg kg^{-1} as compared with 150 mg $Cs^+\ kg^{-1}$ in ryegrass grown in the loam soil without clinoptilolite. The effect was more pronounced in the peat soil in which ryegrass grown in clinoptilolite-amended soil spiked with 40 mg $Cs^+\ kg^{-1}$ soil was 22.6 mg kg^{-1} compared with 1864 mg kg^{-1} for ryegrass grown in peat soil without clinoptilolite. Jones et al. (1999) evaluated several amendments (ammonium ferric hexacyanoferrate, bentonite, clinoptilolite, potassium chloride) on the uptake of $^{137}Cs^+$ in native plants grown in contaminated upland soils in the British Isles (Cwm Eigiau and Corney Fell). Although the source and properties of the clinoptilolite were not reported in this study, clinoptilolite-rich material broadcast on Cs-contaminated soils at rates of 500 and 1500 g m^{-2} showed the greatest and most consistent trend in reducing $^{137}Cs^+$ uptake in native plants.

Several researchers have expressed concern about the use of zeolite additions to immobilize radionuclides or reduce their uptake in plants. Prister et al. (1993), for example, stated that numerous experiments conducted in Ukraine and Belarus immediately after the Chernobyl accident demonstrated that the effectiveness of soil treatments such as humates and zeolites is low and they do not appreciably reduce radionuclide movement. Adriano et al. (1997) also expressed concern that zeolites may not be appropriate additives for remediation of Cs-contaminated soils, primarily because insufficient data exist to warrant their use. They conducted a case study at the Savannah River Site near Aiken, South Carolina, U.S.A., to evaluate Cs^+ uptake by cabbage and collard greens grown in lake-bottom sediments of a reactor cooling effluent reservoir. Several treatment methods were examined, including high K fertilization amendment, clinoptilolite-rich amendment (applied at a rate of 22.4 tonne ha^{-1}), clean soil cover over sediment, geotextile fabric plus a clean soil cover, and a biobarrier plus clean soil cover. The uptake of $^{137}Cs^+$ of plants grown in clinoptilolite-amended sediment was not significantly different from the $^{137}Cs^+$ uptake of plants grown in sediments alone. However, the clinoptilolite application rate used in this study was low. The incorporation of the geotextile fabric plus clean soil cover and the biobarrier plus clean soil cover were effective in reducing plant uptake of Cs^+ from the sediment.

A summary of selected studies on the possible use of zeolites to remediate soils contaminated with radioactive cations is provided in Table 8.

Remediation of soils contaminated with heavy metals

Physical and chemical methods that have been either tested or suggested for the decontamination or treatment of soils containing excessive levels of heavy metals include dilution, covering with clean soil, soil flushing, soil washing, electrokinetics, vaporization, solidification, vitrification, and amendments such as pH adjustment, organic matter, Fe, Mn, zeolites, carbonates, sulfides, and phosphates (Brown 1997). Colella (1996) pointed out that data on selectivities for heavy metals in natural zeolites are scarce, sometimes conflicting, and often poorly comparable to each other. He stated that most natural zeolites are fairly selective for monovalent heavy metal cations of low

Table 8. Selected studies on the immobilization of radioactive cations (Cs and Sr) in soils amended with zeolites.

Zeolite type*	Zeolite locality	Radioactive cations	Soil type	Zeolite amendment rate	Plants grown	Reference
Cp	Hector, California, USA	$^{90}Sr^{2+}$ $^{137}Cs^{+}$	loam & sandy loam soils	0, 0.5, 1, 2, 4, 6 wt %	beans	Nishita et al. (1968)
Cp	Hector, California, USA	$^{90}Sr^{2+}$ $^{137}Cs^{+}$	sandy loam, loam, and muck soils	17.5 or 35 g layers below 1,400 g sand and 3,500 g soil	beans & barley	Nishita & Haug (1972)
Cp (60 %)	NR	$^{90}Sr^{2+}$ $^{137}Cs^{+}$	contaminated soils from Chernobyl accident	10-50 tonne ha^{-1}	vegetables	Chelishchev (1995)
Cp (CEC = 185 $cmol_c$ kg^{-1})	NR	Cs^{+}	loam and peat soils artificially contaminated with Cs	0 and 10 wt %	ryegrass	Campbell & Davies (1997)
Cp-rich material	NR	$^{137}Cs^{+}$	Cs-contaminated soils	22.4 tonne ha^{-1}	cabbage & collards	Adriano et al. (1997)
Cp	NR	$^{137}Cs^{+}$	peat and bog soils	0, 500, 1500 g m^{-2}	native plants	Jones et al. (1999)

* Cp = clinoptilolite-rich material

NR = Not reported

charge density (e.g. Ag^+), and they tend to prefer divalent cations with lower hydration energy (e.g. Pb^{2+} over Cd^{2+}). Pabalan and Bertetti (this volume) describe the ion exchange behaviors of natural zeolites for heavy metal cations elsewhere in this book.

The use of natural zeolites to immobilize or reduce leaching and plant uptake of heavy metals in soils has been met with varying degrees of success. Scientists from the former Soviet Union were some of the first to examine the use of natural zeolites to remediate soils contaminated with heavy metals (e.g. Gudushauri et al. 1980; Tsitsishvili et al. 1980). Tsitsishvili et al. (1980) reported that half as much Pb^{2+} was taken up by maize grown in soils purposely contaminated with $Pb(NO_3)_2$ and amended with clinoptilolite-rich tuff from the Khekordzula deposit in Georgia as compared with maize grown in soils without the zeolite amendment.

Studies on the potential "trapping" of heavy metals by naturals zeolites have become common over the past 10 years (Table 9). Chlopecka and Adriano (1996) found that agricultural lime (mixture of dolomite and calcite), clinoptilolite-rich material, "Fe-rich" (Fe oxide, perhaps poorly crystalline ferrihydrite, an industrial trademark product of E. I. Du Pont de Nemours), and phosphate rock (apatite; North Carolina) decreased exchangeable Zn in an acidic, silt loam soil (Alfisol; pH = 5.4) and, in several instances, reduced the uptake of Zn in maize and barley. Gworek et al. (1998) traced the fate of Cd in heavy loam and light loam soils of the Upper Silesia Industrial Region in Poland that were amended with clinoptilolite-rich tuff (90% clinoptilolite). Oats, ryegrass, lettuce, and beets were grown in these clinoptilolite-amended soils (2 wt % clinoptilolite-rich tuff) spiked with 20 mg Cd per kg of soil. Depending on the plant species and soil, Cd uptake decreased between 19.4 and 81.4% in plants grown on the clinoptilolite-amended soils compared with plants grown on soils receiving no clinoptilolite treatments.

Tsadilas et al. (1997a) compared the effect of clinoptilolite and lime (calcium oxide) additions on soil pH and Cd adsorption in four acidic soils (Alfisols) of Greece. Clinoptilolite-rich tuff (approx. 70-85% clinoptilolite) from near Drama in Northern Greece slightly raised the pH of these soils (e.g. pH of 4.69 to 4.92 for 0 and 10 wt % clinoptilolite additions, respectively, in one soil) as well as slightly increasing the CECs (e.g. 12.0 to 15.0 $cmol_c\ kg^{-1}$ for 0 and 10 wt % clinoptilolite additions, respectively, in one soil). In Cd adsorption studies, soil clinoptilolite mixtures were shaken with solutions containing a range of Cd concentrations, from 0 to 100 mg Cd L^{-1}. Again, as might be expected, Cd adsorption of the soil increased with increasing amounts of amended clinoptilolite-rich tuff (Fig. 7). Liming also increased the capacity of the soils to adsorb Cd. Both lime and clinoptilolite applications reduced Cd release or desorption from the soils.

Phillips (1998) compared the ability of red mud (an alkaline byproduct of the aluminum industry remaining after digestion of bauxite with NaOH during extraction), clinoptilolite (Zeolite Australia Limited), and a calcium phosphate ($Ca(H_2PO_4)_2$) to adsorb Cu, Pb, and Zn from silica sand at application rates of 0, 5, 10, and 20 wt %. Red mud amended sand had the highest adsorption of Cu, Pb, and Zn; however, this could have been in part due to the high pH of the mixture (over pH = 9 in some experiments), which may have caused hydrolysis and precipitation of Cu, Pb, and Zn hydroxides. The clinoptilolite-amended sand adsorbed or exchanged more Cu, Pb, and Zn than the $Ca(H_2PO_4)_2$.

Iskandar and Adriano (1997) indicated that the most promising and cost-effective methods for remediating soils contaminated with metals are soil washing with dilute acid solutions or chelating agents, natural attenuation, and enhanced bioremediation. The authors indicated that little data are available on the use of ion-exchange materials such

Table 9. Selected studies on the immobilization of heavy metals in soils amended with natural zeolites.

Zeolite type*	Zeolite locality	Heavy metal(s)	Soil type	Zeolite amendment rate	Plant types grown	Reference
Heu	Pegasskoye desposit, Kemerovo Oblast	Cd, Pb, & Zn	Industrially polluted Chernozem[†] near zinc plant	0.5, 1, 3.5, 5 wt %	beets & tomatoes	Baydina (1996)
Unnamed zeolite	Ai-Dag deposit, Azerbaijan	Cd, Pb, Zn	Podzolic[‡] soil	10 wt %	barley, strawberry, & cherry	Mineyev et al. (1989)
Cp	Colorado, USA	Zn	acidic, silt loam soil contaminated or spiked with various levels of Zn	1.5 wt %	maize, barley, & radish	Chlopecka & Adriano (1996)
Cp (90% Cp)	NR	Cd	upper horizons of anthropogenic soils in Poland	2 wt %	lettuce, beet, ryegrass, & oats	Gworek et al. (1998)
Cp	Australia	Cu, Pb, Zn	silica sand	0, 5, 10, 20 wt %	none	Phillips (1998)
Cp	Drama, Northern Greece	Cd	four soils ranging from pH = 4.2-6.8	0, 1, 5, 10 wt %	none	Tsadilas et al. (1997a)
Fau and Ph	Jordan	Pb, Cd, Ni	sandy soil artificially contaminated with Pb, Cd, & Ni	0-50 wt %	none	Shanableh & Kharabsheh (1996)

* Heu = heulandite-rich material; Cp = clinoptilolite-rich material; Fau = faujasite-rich material; Ph = phillipsite-rich material

[†] Chernozem soils have thick, nearly black, organic-matter rich surface horizons and occur in cool climates under prairie vegetation

[‡] Podzolic soils are highly leached and occur in temperate, humid climates under coniferous or mixed coniferous and deciduous forests

NR = Not reported

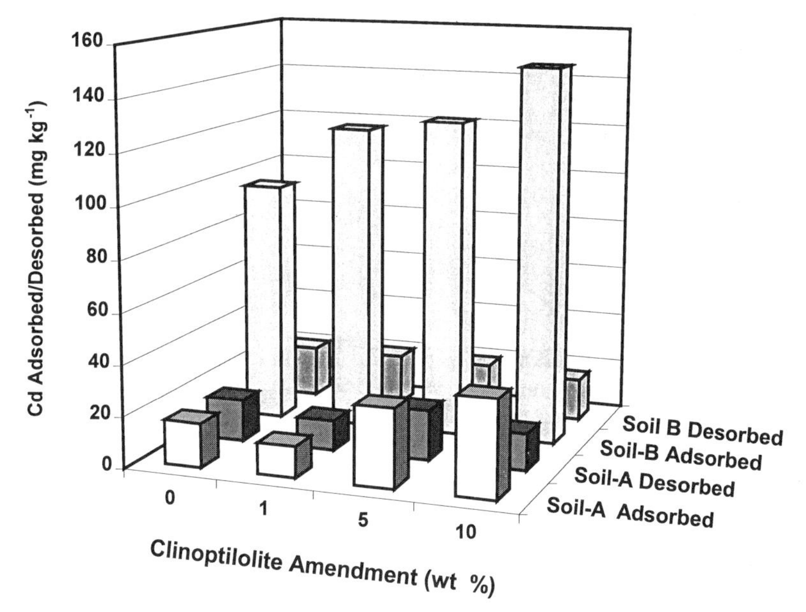

Figure 7. Adsorption and desorption of Cd by two acidic soils (A & B) artificially spiked with Cd and amended with clinoptilolite-rich tuff from a deposit near Drama in Northern Greece, at rates of 0, 1, 5, and 10 wt % (Tsadilas et al. 1997a).

as zeolites. They further suggested that the ion-exchange treatment may not be a cost-effective or efficient method when dealing with large volumes of highly contaminated soils because of the large volume of ion-exchange material needed for remediation.

Although the focus of this chapter is on the agricultural, horticultural, and environmental applications of natural zeolites, synthetic zeolites have also been examined as potential soil remediation agents for heavy metals (e.g. see Gworek 1992a,b,c,d; Querol et al. 1997; Lin et al. 1998; García-Sánchez et al. 1999).

COMMERCIAL APPLICATIONS FOR NATURAL ZEOLITES

The commercial uses of natural zeolites as slow-release fertilizers, zeoponic substrates, and agents for soil conditioning and remediation are becoming increasingly attractive, primarily due to advances in understanding the behavior and effects of zeolites in these systems. However, the current world-wide commercial market for natural zeolites in agricultural and environmental applications is very small compared with other commercial applications for natural zeolites, e.g. in the building industry (see Collela et al., this volume). Zeolite markets in plant growth and environmental protection have not flourished for a variety of reasons, including: the use of natural zeolites is often not cost effective; natural zeolites may provide no improvements over other established practices; there is a lack of research demonstrating marketable increases in production or environmental protection; transportation costs are high; and many zeolite materials and properties are poorly characterized. There have been few (if any) studies that examine the cost effectiveness of using natural zeolites. Another major problem is the lack of funding for conducting studies that apply zeolites to real situations or specific applications. Most of the companies that sell natural zeolites are small and hence, do not have the capital for research and development. Another problem is the difficulty in evaluating a zeolite's performance in a particular application because there are often many factors that control its performance (e.g. complex chemical and physical properties of the soil to which the zeolite is added, complex solution chemistry, etc.).

There are a few companies worldwide that market natural zeolites or zeolite-based products to the golf and horticultural industries. Recently, these companies have focused on selling natural zeolites as soil conditioners to the turfgrass industry. Turfgrass is considered a "high cash" crop and the industry is driven by environmental demands to reduce fertilizer runoff (i.e. non-point source pollution) and increase fertilizer-use efficiency. Natural zeolites are attractive for this application, mainly because their addition to sandy soils can reduce migration of fertilizers and chemicals, such as fungicides, into the groundwater and runoff. Slow release of N, K and other essential plant-growth elements increases fertilizer-use efficiency, which may save money over time. Nutrient-enriched clinoptilolite-rich material is currently being sold as a slow-release fertilizer and soil conditioner for the turfgrass and horticultural industries. The product, called ZeoPro™, is produced and marketed by a company in Colorado, U.S.A. ZeoPro™ consists of a K- and NH_4-exchanged clinoptilolite-rich tuff from the St. Cloud mine, Winston, New Mexico, U.S.A., and a synthetic hydroxyapatite, which supplies Ca and P (micronutrients may also be supplied by the synthetic hydroxyapatite). The nutrient content in one ZeoPro™ product is 0.1 wt % N, 0.05 wt % P, and 0.6 wt % K, which is low compared with other conventional fertilizers. Allen and Andrews (1997) reported that sand amended with 10 vol % ZeoPro™ gave greater turfgrass establishment rates compared with sand control plots and sand plots amended with untreated clinoptilolite. Plots were fertilized with 1 lb/1000 ft^2 of N as 20-10-20 (N-P-K) plus trace elements and then sown with Penncross creeping bentgrass, which was fertilized biweekly during the growth period with 0.5 lb/1000 ft^2 of N as 20-10-20 plus trace elements. Grass clippings showed an 81% increase in dry weight at the end of the fifth week for plots amended with ZeoPro™ compared with the control plots. In another experiment, Miller (2000) found that bermudagrass grown in golf green sand amended with 8.5 vol % ZeoPro™ had the highest turfgrass quality compared with nine other amendments (three clinoptilolite-rich materials, one diatomaceous earth, one calcined diatomaceous earth, three porous ceramic, and one peat), a native soil, and the unamended sand.

Many different zeolite-rich materials are sold by various trade names (e.g. Ecolite™, EcoSand™). Most of these materials are not treated to enhance their nutrient content for plant growth but are added to golf green sands and horticultural potting media as soil conditioners. Little research has been published to support their effectiveness as a soil conditioner, likely due to several of the obstacles for the commercialization of natural zeolite products mentioned earlier in this section. Another area in which natural zeolites have been used is the horticulture industry, where they have been used in zeoponic applications for growing flowers and vegetables.

CONCLUSIONS AND RECOMMENDATIONS FOR FUTURE WORK

Several commercial applications of zeolites in enhancing plant growth show promise, including using zeolites: (1) as substrates for plants in greenhouses; (2) as substrates for potting mixes for house plants; (3) as fertilizers and soil conditioners on golf greens; (4) in field agriculture applications as slow-release fertilizers; (5) as slow-release fertilizers, insecticides, and herbicides in areas where environmental issues may be a concern (e.g. sandy soils near surface water or aquifers); and (6) as remediation agents for environmental problems (e.g. heavy metal and radionuclide contamination). Progress has been made in the last five years on the utilization of natural zeolites in plant growth; particularly in the areas of zeoponic plant growth systems and slow-release fertilization for golf greens. As pointed out by Allen and Ming (1995), slow-release fertilization, zeoponic substrates, and soil conditioning techniques have great commercial potential because of demands for increased fertilizer-use efficiency and environmental protection.

Future work must address a number of concerns and unanswered questions connected with the use of natural zeolites in plant growth and environmental protection, such as cost effectiveness, whether zeolites improve over established practices, and the issue of high transportation costs. In addition, the field urgently needs sound basic and applied research to counteract our lack of understanding of many zeolite materials and properties. Liberti et al. (1999) pointed out that several of these problems must be addressed before the RIM-NUT process is economical; one of these concerns is the economic commercial production of zeolite products. The cost of mined zeolites must be similar to the costs of other mined soil amendments and fertilizers. Others have pointed out that the effectiveness of zeolite amendments to soils is highly variable because of the variability in soil properties from site to site. Adriano et al. (1997) and others suggested that zeolites are not appropriate additives for remediation of contaminated soils because insufficient data exist to warrant their use. No doubt, considerable research and development is needed to warrant their use in plant growth applications and environmental protection.

Natural zeolite products and formulations must be developed that meet the particular needs of specific agronomic and horticultural crops and environmental applications. The "one-size-fits-all" approach does not necessarily work for all crops and environmental problems. Properties (CEC, exchangeable cations, particle size, zeolite purity, impurities, location) of the natural zeolite-containing material (e.g. tuff) *must* be fully characterized prior to conducting research or using the zeolitic material for a specific application.

With sound scientific research and product development, the possible applications for natural zeolites are very promising. As J.V. Smith so eloquently stated,

"*There just might be gold for geochemists in them thar zeolite beds*"

assuming that the "gold" is the benefits that will be reaped from their utilization, and not for the cost of the zeolitic material.

ACKNOWLEDGMENTS

We thank John Gruener and Prof. Dave Bish for their helpful comments on the manuscript. We also thank Drs. I-Ching Lin and Chi-Y. Shih for translation of articles from Chinese to English. This work was supported in part by NASA's Advanced Life Support program. We are very grateful to the guidance, wisdom, encouragement, and articles provided by our colleague, Dr. Fred Mumpton, the 'Dean' of natural zeolites.

REFERENCES

Adriano DC, Albright J, Whicker FW, Iskandar IK, Sherony C (1997) Remediation of metal- and radionuclide-contaminated soils. *In* Iskandar IK, Adriano DC (eds) Remediation of Soils Contaminated with Metals. Science Reviews, Northwood, Middlesex, UK, p 27-45

Ahn SB, Cho SJ, Kang JS (1984) Effect of zeolite as a amendment for sandy paddy. Han'guk T'oyang Piryo Hakhoechi 17:381-388

Allen E, Andrews R (1997) Space age soil mix uses centuries-old zeolites. Golf Course Manage 65:61-66

Allen ER, Hossner LR, Ming DW, Henninger DL (1993) Solubility and cation exchange relationships in mixtures of phosphate rock and saturated clinoptilolite. Soil Sci Soc Am J 57:1368-1374

Allen ER, Hossner LR, Ming DW, Henninger DL (1996) Comparison of kinetic models describing nutrient release form mixtures of clinoptilolite and phosphate rock. Soil Sci Soc Am J 60:1467-1472

Allen ER, Ming DW (1995) Recent progress in the use of natural zeolites in agronomy and horticulture. *In* Ming DW, Mumpton FA (eds) Natural zeolites '93: Occurrence, Properties, Use. Int'l Comm Natural

Zeolites, Brockport, New York, p 477-490
Allen ER, Ming DW, Hossner LR, Henninger DL (1995a) Modeling transport kinetics in clinoptilolite-phosphate rock systems. Soil Sci Soc Am J 59:248-255
Allen ER, Ming DW, Hossner LR, Henninger DL, Galindo C Jr (1995b) Growth and nutrient uptake of wheat in clinoptilolite and phosphate rock substrates. Agron J 87:1052-1059
Ames LL (1960) The cation sieve properties of clinoptilolite. Am Mineral 45:689-700
Andronikashvili T, Tsitsishvili G, Kardava M, Gamisonia M (1999) The effect of organic-zeolite fertilizers on microbial landscape of soil. Izv Akad Nauk Gruz, Ser Khim 25:243-250
Barbarick KA, Lai TM, Eberl DD (1990) Exchange fertilizer (phosphate rock plus ammonium-zeolite) effects on sorghum-sudangrass. Soil Sci Soc Am J 54:911-916
Barbarick KA, Pirela HJ (1984) Agronomic and horticultural uses of zeolites: A review. *In* Pond WG, Mumpton FA (eds) Zeo-Agriculture: Use of Natural Zeolites in Agriculture and Aquaculture. Westview Press, Boulder, Colorado, p 93-103
Barrer RM, Meier WM (1958) Salt occlusion complexes of zeolites. J Chem Soc 1958:299-304
Baydina NL (1996) Inactivation of heavy metals by humus and zeolites in industrially contaminated soil. Eurasian Soil Sci 28:96-105
Bouzo L, Lopez M, Villegas R, Garcia E, Acosta JA (1994) Use of natural zeolites to increase yields in sugarcane crop minimizing environmental pollution. Trans XV Congress Int'l Soc Soil Sci 5a:695-701
Breck DW (1974) Zeolite Molecular Sieves. John Wiley & Son, New York, 771 p
Brown KW (1997) Decontamination of polluted soils. *In* Iskandar IK, Adriano DC (eds) Remediation of Soils Contaminated with Metals. Science Reviews, Northwood, Middlesex, UK, p 47-66
Bugbee GJ, Elliott GC (1998) Leaching of nitrogen and phosphorus from potting media containing biosolids compost as affected by organic and clay amendments. Bull Environ Contam Toxicol 60:716-723
Cadena F, Cazares E (1995) Sorption of benzene, toluene, and o-xylene from aqueous solution on surfaces of zeolitic tuffs modified with organic cations. *In* Ming DW, Mumpton FA (eds) Natural Zeolites '93: Occurrence, Properties, Use. Int'l Comm Natural Zeolites, Brockport, New York, p 309-324
Campbell LS, Davies BE (1997) Experimental investigation of plant uptake of caesium from soils amended with clinoptilolite and calcium carbonate. Plant and Soil 189:65-74
Carlino JL, Williams KA, Allen ER (1998) Evaluation of zeolite-based soilless root media for potted chrysanthemum production. Hort Technology 8:373-378
Chelishchev NF (1995) Use of natural zeolites at Chernobyl. *In* Ming DW, Mumpton FA (eds) Natural Zeolites '93: Occurrence, Properties, Use. Int'l Comm Natural Zeolites, Brockport, New York, p 525-532
Chen J, Gabelman WH (2000) Morphological and physiological characteristics of tomato roots associated with potassium-acquisition efficiency. Scientia Horticulturae 83:213-225
Chesworth W, van Straaten P, Smith P, Sadura S (1987) Solubility of apatite in clay and zeolite-bearing systems: Applications to agriculture. Appl Clay Sci 2:291-297
Chlopecka A, Adriano DC (1996) Mimicked *in situ* stabilization of metals in a cropped soil: Bioavailability and chemical form of zinc. Environ Sci Technol 30:3294-3303
Ciambelli P, Corbo P, Liberti L, Lopez A (1988) Ammonium recovery from urban sewage by natural zeolites. *In* Kalló D, Sherry HS (eds) Occurrence, Properties, and Utilization of Natural Zeolites. Akadémiai Kiadó, Budapest, Hungaria, p 501-509
Colella C (1996) Ion-exchange equilibria in zeolite minerals. Mineral Deposita 31:554-562
Crespo G (1989) Effect of zeolite on the efficiency of the N applied to Brachiaria decumbens in a red ferrallitic soil. Cuban J Agric Sci 23:207-212
Dwairi IM (1998) Renewable, controlled and environmentally safe phosphorus release in soils from mixtures of NH4+-phillipsite tuff and phosphate rock. Environ Geol 34:293-296
Eberl DD, Barbarick KA, Lai TM (1995) Influence of NH_4-exchanged clinoptilolite on nutrient concentrations in sorghum-sudangrass. *In* Ming DW, Mumpton FA (eds) Natural Zeolites '93: Occurrence, Properties, Use. Int'l Comm Natural Zeolites, Brockport, New York, p 491-504
Franchini-Angela M, Boero V, Rinaudo C (1994) Tuffo giallo Napolitano as possible phosphorus source for plant nutrition. *In* Proc 2nd Conveg Nazionale del'Assoc Italiana Zeoliti, Modena. E Passaglia (ed) Mater Engineer (Modena, Italy) 5:321-328
Garcia Hernández JE, Notario del Pino JS, Arteaga Padrón IJ, González Martín MM (1993) Phosphate and potassium fixation on a phillipsite-rich tuff as a slow-release fertilizer. Treatments with KH_2PO_4. Agrochimica 37:1-11
García Hernández JE, Notario del Pino JS, Gonzalez Martin MM, Díaz Díaz R, Febles González EJ (1995) Natural phillipsite as a matrix for a slow-release formulation of oxamyl. Environ Pollut 88:355-359
García-Sánchez A, Alastuey A, Querol Z (1999) Heavy metal adsorption by different minerals: Application to the remediation of polluted soils. Sci Total Environ 242:179-188

Golden DC, Ming DW (1999) Nutrient-substituted hydroxyapatites: Synthesis and characterization. Soil Sci Soc Am J 63:657-664

Gruener JE, Ming DW, Henderson KE, Carrier C (2000) Nutrient uptake of wheat grown in diluted clinoptilolite-natural/synthetic apatite substrates. *In* Colella C, Mumpton FA (eds) Natural Zeolites for the Third Millennium. DeFreda Editore, Napoli, Italy, p 427-439

Gudushauri TsN, Brouichek PI, Maisuradze GV, Gvakharia VG (1980) Studies of adsorption ability of clinoptilolite and soil with respect to lead (II). *In* Krupennikova AYu (ed) Proc Symp Utilization of Natural Zeolites in Agriculture, Sukhumi, 1978. Metzniereba Publ House, Tbilisi, p 152-158

Gworek B (1992a) Effect of zeolites on cadmium intake by plants. Archiwum Ochrony Strodowiska, p 149-156

Gworek B (1992b) Effect of zeolites on the nickel uptake by plants. Pol J Soil Sci 25:127-133

Gworek B (1992c) Inactivation of cadmium in contaminated soils using synthetic zeolites. Environ Pollut 75:269-271

Gworek B (1992d) Lead inactivation in soils by zeolites. Plant Soil 143:71-74

Gworek B, Borowiak M, Kwapisz J (1998) Effect of zeolite-bearing rocks upon inactivation of cadmium in soils. Roczniki Gleboznawcze, p 71-78

Hayashizaki T, Tsuneji N (1973) Acaricidal composition containing lime-nitrogen. Japon Kokai 73,031,888

He ZL, Baligar VC, Martens DC, Ritchey KD, Elrashidi M (1999) Effect of byproduct, nitrogen fertilizer, and zeolite on phosphate rock dissolution and extractable phosphorus in acid soil. Plant Soil 208: 199-207

Henderson KE, Ming DW, Carrier C, Gruener JE, Galindo Jr C, Golden DC (2000) Effects of adding nitrifying bacteria, dolomite, and ferrihydrite to zeoponic plant growth substrates. *In* Colella C, Mumpton FA (eds) Natural Zeolites for the Third Millennium, DeFreda Editore, Napoli, Italy, p 441-447

Hershey DR, Paul JL, Carlson RM (1980) Evaluation of potassium-enriched clinoptilolite as a potassium source for potting media. Hort Science 15:87-89

Hoagland DR, Arnon DI (1950) The water-culture method for growing plants without soil. Calif Agric Exp Station, Univ California, Berkeley, California, Circular 347, 32 p

Huang ZT, Petrovic AM (1994) Clinoptilolite zeolites influence on nitrate leaching and nitrogen use efficiency in simulated sand based golf greens. J Environ Qual 23:1190-1194

Huang ZT, Petrovic AM (1995) Physical properties of sand affected by clinoptilolite zeolite particle size and quanity. J Turfgrass Manage 1:1-15

Huang ZT, Petrovic AM (1996) Clinoptilolite zeolite effect on evapotranspiration rate and shoot growth rate of creeping bentgrass on sand base greens. J Turfgrass Manage 1:1-9

Iskandar IK, Adriano DC (1997) Remediation of soils contaminated with metals—a review of current practices in the U.S.A. *In* Iskandar IK, Adriano DC (eds) Remediation of Soils Contaminated with Metals. Science Reviews, Northwood, Middlesex, UK, p 1-26

Iskenderov Ish, Mamedova SN (1988) The utilization of natural zeolite in Axerbaijan SSR for increasing yield of wheat. *In* Kalló D, Sherry HS (eds) Occurrence, Properties, and Utilization of Natural Zeolites. Akadémiai Kiadó, Budapest, Hungary, p 717-720

Ivanova M, Petrova K (1993) Effect of increased potassium nitrate on the expression of heterosis in the hybrid B73 x Mo17 grown in zeolite. Rastenievud Nauki 30:39-41

Ivanova T, Sapunova S, Dandolov I, Ivanov Y, Meleshko G, Mashinsky A, Berkovich Y (1994) 'Svet' space greenhouse onboard experiment data received from 'Mir' station and future prospects. Adv Space Res 14:343-346

Ivanova T, Stoyanov I, Stoilov G, Kostov P, Sapunova S (1997) Zeolite gardens in space. *In* Kirov G, Filizova L, Petrov O (eds) Natural Zeolites—Sofia '95. Pensoft Publishers, Sofia, Bulgaria, p 3-10

Ivanova TN, Bercovich YuA, Mashinskiy AL, Meleshko GI (1993) The first "space" vegetables have been grown in the "Svet" greenhouse using controlled environmental conditions. Acta Astronautica 29:639-644

Jones DR, Paul L, Mitchell NG (1999) Effects of ameliorative measure on the radiocesium transfer to upland vegetation in the UK. J Environ Radioact 44:55-69

Katz LE, Humphrey DN, Jankauskas PT, DeMascio FA (1996) Engineering soils for low-level radioactive waste disposal facilities: Effects of additives on the adsorptive behavior and hydraulic conductivity of natural soils. Hazardous Waste & Hazardous Mater 13:283-306

Kvachantiradze M, Tvalchrelidze E, Kotetishvili M, Tsitsishvili T (1999) Application of clinoptilolite as an additive for the photostabilization of the *Bacillus thuringiensis* formulation. Stud Surf Sci Catal 125:731-735

Lai TM, Eberl DD (1986) Controlled and renewable release of phosphorus in soils from mixtures of phosphate rock and ammonium-exchanged clinoptilolite. Zeolites 6:129-132

Lami L, Casal B, Cuadra L, Merino J, Alvarez A, Ruiz-Hitsky E (1999) Synthesis of 2,4-D ester herbicides. Green Chem 1:199-204

Langella A, de'Gennaro M, Colella C, Buondonno A (1995) Effects of phillipsite- or chabazite-rich tuff addition to soil on the growth and yield of Beta vulgaris and Raphaanus sativus. *In* Proc 3rd National Conf. Science & Technology of Zeolites, Cetraro, Italy. R. Aiello (ed) Italian Zeolite Assoc, Naples, Italy, p 277-285

Lewis MD, Moore IFD, Goldsberry KL (1984) Ammonium-exchanged clinoptilolite and granulated clinoptilolite with urea as nitrogen fertilizers. *In* Pond WG, Mumpton FA (eds) Zeo-Agriculture: Use of Natural Zeolites in Agriculture and Aquaculture. Westview Press, Boulder, Colorado, p 105-111

Liberti L, Boari G, Passino R (1979) Phosphates and ammonium removal from secondary effluents by selective ion exchange with production of a slow-release fertilizer. Water Res 13:63-73

Liberti L, Boari G, Petruzzelli D, Passino R (1981) Nutrient removal and recovery from wastewater by ion exchange. Water Res 15:337-342

Liberti L, Boghetich G, Lopez A, Petruzzelli D (1999) Application of microporous materials for the recovery of nutrients from wastewaters. *In* Misaelides P, Macásek F, Pinnavaia TJ, Colella C (eds) Natural microporous materials in environmental technology. NATO Adv Sci Ser E-362:253-270. Kluwer Academic Publishers, Dordrecht, The Netherlands, p 253-270

Liberti L, Lopez A, Amicarelli V, Boghetich G (1995) Pollution-abatement technologies by natural zeolites: The Rim-Nut process. *In* Ming DW, Mumpton FA (eds) Natural zeolites '93: Occurrence, Properties, Use. Int'l Comm Natural Zeolites, Brockport, New York, p 351-362

Lin C-F, Lo S-S, Lin H-Y, Lee Y (1998) Stabilization of cadmium contaminated soils using synthesized zeolite. J Hazard Materials 60:217-226

MacKown CT (1978) Role of Mineral Zeolites as Soil Amendments. PhD dissertation, University of Arizona, Tucson, Arizona

MacKown CT, Tucker TC (1985) Ammonium nitrogen movement in a coarse-textured soil amended with zeolite. Soil Sci Soc Am J 49:235-238

Manolov I, Stoilov B (1997) Changes in nutritional properties of zeolite substrates during their exploitation. *In* Kirov G, Filizova L, Petrov O (eds) Natural Zeolites—Sofia '95. Pensoft Publishers, Sofia, Bulgaria, p 83-92

Marshania II, Equania (1984) The effect of using natural zeolites and nitrogen-based fertilizers in geranium. *In* SlovZeo '84: Conference on the Study and Use of Natural Zeolites. Vysoke Tatry, Czechoslovakia, Part 2. Czech Sci Tech Soc, Kosice, Slovakia, p 163-174

Mazur GA, Medvid GK, Grigora TI (1984) Use of natural zeolites for increasing the fertility of light-textured soils. Pochvovedenie 10:70-77

Miller GL (2000) Physiological response of bermudagrass grown in soil amendments during drought stress. HortSci 35:213-216

Minato H (1968) Characteristics and uses of natural zeolites. Koatsugasu 5:536-547

Mineyev VG, Kochetavkin AV, Van Bo N (1989) Use of natural zeolites to prevent heavy-metal pollution of soils and plants. Agrokhimiya 8:89-95

Ming DW, Allen ER (1999) Zeoponic substrates for space applications: Advances in the use of natural zeolites for plant growth. *In* Misaelides P, Macásek F, Pinnavaia TJ, Colella C (eds) Natural microporous materials in environmental technology. NATO Adv Sci Ser E-362:157-176. Kluwer Academic Publishers, Dordrecht, The Netherlands

Ming DW, Allen ER (2000) Recent advances in the United States in the use of natural zeolites in plant growth. *In* Colella C, Mumpton FA (eds) Natural Zeolites for the Third Millennium, DeFreda Editore, Napoli, Italy, 417-426

Ming DW, Barta DJ, Golden DC, Galindo Jr C, Henninger DL (1995) Zeoponic plant-growth substrates for space applications. *In* Ming DW, Mumpton FA (eds) Natural Zeolites '93: Occurrence, Properties, Use. Int'l Comm Natural Zeolites, Brockport, New York, p 505-513

Mitov M, Manev S, Lazarov D (1995) Possibilities for use of zeolites as carriers of some microelements. God Sofii Univ "Kliment Okhridski," Kim Fak 88:31-36

Mnkeni PNS, Semokam JMR, Kaitaba EG (1994) Effects of Mapagoro phillipsite on availability of phosphorus in phosphate rock. Trop Agri (Trinidad) 71:249-253

Moller J, Mogensen T (1953) Use of an ion exchanger for determining available phosphorus in soils. Soil Sci 76:297-306

Morrow RC, Duffie NA, Tibbitts TW, Bula RJ, Barta DJ, Ming DW, Wheeler RM, and Porterfield DM (1995) Plant response in the ASTROCULTURE™ flight experiment unit. SAE Technical Paper Series #951624, SAE International, San Diego, California

Nishita H, Haug RM (1972) Influence of clinoptilolite on Sr-90 and Cs-137 uptake by plants. Soil Sci 114:149-157

Nishita H, Haug RM, Hamilton M (1968) Influence of minerals on Sr-90 and Cs-137 uptake by bean plants. Soil Sci 105:237-243

Nissen LR, Lepp NW, Edwards R (2000) Synthetic zeolites as amendments for sewage sludge-based compost. Chemosphere 41:265-269

Notario del Pino JS, Arteaga Padron IJ, Gonzalez Martin MM, Garcia Hernandez JE (1994) Response of alfalfa to a phillipsite-based slow-release fertilizer. Commun Soil Sci Plant Anal 25:2231-2245

Parham WE (1984) Future perspectives for natural zeolites in agriculture and aquaculture. *In* Pond WG, Mumpton FA (eds) Zeo-Agriculture: Use of Natural Zeolites in Agriculture and Aquaculture. Westview Press, Boulder, Colorado, p 283-285

Park M, Komarneni S (1997) Occlusion of KNO_3 and NH_4NO_3 in natural zeolites. Zeolites 18:171-175

Park M, Komarneni S (1998) Ammonium nitrate occlusion vs. nitrate ion exchange in natural zeolites. Soil Sci Soc Am J 62:1455-1459

Perez JE, Rivero L, Arozarena N (1991) Estudio del contenido de nutrients en sustratos Nerea. *In* Rodríguez Fuentes G, González JA (eds) Zeolites '91: Memoirs 3rd Int'l Conf Occurrence, Properties and Utilization of Natural Zeolites. International Conference Center, Havana, Cuba, p 3-7

Perrin TS, Boettinger JL, Drost DT, Norton JM (1998a) Decreasing nitrogen leaching from sandy soil with ammonium-loaded clinoptilolite. J Environ Qual 27:656-663

Perrin TS, Drost DT, Boettinger JL, Norton JM (1998b) Ammonium-loaded clinoptilolite: A slow-release nitrogen fertilizer for sweet corn. J Plant Nutrition 21:515-530

Petrov GS, Petkov IA, Etropolski HI, Dimitrov DN, Popov NN, Uzunov AI (1982) Substrate for cultivation of agricultural crops and rooting of green cuttings in greenhouses and in open air. U.S. Patent 4,337,078, 4 p

Petrovic AM (1990) The potential of natural zeolite as a soil amendment. Golf Course Manage 58:92-93

Petrovic MA, Barrett WC, Larsson-Kovach L-M, Reid CM, Lisk DJ (1998) Downward migration of metalaxyl fungicide in creeping bentgrass sand lysimeters as affected by organic waste, peat, and zeolite amendments. Chemosphere 37:249-256

Phillips IR (1998) Use of soil amendments to reduce nitrogen, phosphorous and heavy metal availability. J Soil Contam 7:191-212

Pirela HJ, Westfall DG, Barbarick KA (1984) Use of clinoptilolite in combination with nitrogen fertilization to increase plant growth. *In* Pond WG, Mumpton FA (eds) Zeo-Agriculture: Use of Natural Zeolites in Agriculture and Aquaculture. Westview Press, Boulder, Colorado, p 113-122

Pode R, Pode V, Iovi A, Herman S (1998) Possibilities for natural zeolites usage in fertilizers technology. Sci Technol Environ Prot 5:67-73

Prister BS, Perepelyatnikov GP, Perepelyatnikova LV (1993) Countermeasures used in the Ukraine to produce forage and animal food products with radionuclide levels below intervention limits after the Chernobyl accident. Sci Total Environ 137:183-198

Querol X, Alastuey A, López-Soler A, Plana F (1997) A fast method for recycling fly ash: Microwave-assisted zeolite synthesis. Environ Sci Technol 31:2527-2533

Ricke SC, Pillai SD, Widner KW, Ha SD (1995) Survival of *Salmonella typhimurium* in soil and liquid microcosms amended with clinoptilolite compounds. Biores Tech 53:1-6

Rivero L, Pérez JE, Rodríguez G, Morales V, Soca M (1991a) Growth & nutrient uptake of wheat in a zeoponic system. *In* Rodríguez Fuentes G, González JA (eds) Zeolites '91: Memoirs of the 3rd Int'l Conf Occurrence, Properties and Utilization of Natural Zeolites. International Conference Center, Havana, Cuba, p 20-22

Rivero L, Rodríguez Fuentes G (1988) Cuban experience with the use of natural zeolites substrates in soilless culture. Proc Int'l Congr Soilless Culture, Wageningen, Netherlands, p 405-416

Rivero L, Rodríguez G, Morales V (1991b) Cultivos intensivos de hortalizas con neuvos sistemas de zeoponicos. *In* Rodríguez Fuentes G, González JA (eds) Zeolites '91: Memoirs of the 3rd Int'l Conf Occurrence, Properties and Utilization of Natural Zeolites. International Conference Center, Havana, Cuba, p 23-26

Shanableh A Kharabsheh A (1996) Stabilization of Cd, Ni, and Pb in soil using natural zeolite. J Hazard Mater 45:207-217

Sikharulidze NS, Putsikina EB (1993) Study of chemical resistance of phosphorus organic pesticides on cation modified clinoptilolites. *In* Abstracts Symposium—Natural Zeolites '93, Tbilisi, Georgia. Metsniereba Publishing House, Tbilisi, Georgia, p 42

Smith JV (1988) Book review: Occurrence, Properties and Utilization of Natural Zeolites. Geochem Cosmochim Acta 52:3026

Soca M, Ruz R, Peña CE (1991) Determinacion de la dosis de zeolita natural en el pasto Sorghum vulgare y suelo solonetizado gleysoso. *In* Rodríguez Fuentes G, González JA (eds) Zeolites '91: Memoirs of the 3rd Int'l Conf Occurrence, Properties and Utilization of Natural Zeolites. International Conference Center, Havana, Cuba, p 44-46

Sophia Knox A, Adriano DC (1998) Metal stabilization in contaminated soil from Upper Silesia, Poland using natural zeolite and apatite. Institute for International Cooperative Environmental Research, Florida State University, Tallahassee, Florida, p 1180-1186

Sopková A, Janoková E (1998) An insecticide stabilized by natural zeolite. J Therm Anal Calorim 53: 477-485

Steinberg SL, Henninger DL (1997) Response of the water status of soybean to changes in soil water potentials controlled by the water pressure in microporous tubes. Plant Cell Environ. 20:1506-1516

Steinberg SL, Ming DW, Henderson KE, Carrier C, Gruener JE, Barta DJ, Henninger DL (2000) Wheat response to differences in water and nutritional status between zeoponic and hydroponic growth systems. Agron J 92:353-360

Suwardi, Goto I, Ninaki M (1994) The quality of natural zeolites from Japan and Indonesia and their application effects for soil amendment. J Agri Sci, Tokyo Nogyo Daigaku 39:133-148

Tsadilas CD, Dimoyiannis D, Samaras V (1997a) Effect of zeolite application and soil pH on cadmium sorption in soils. Commun Soil Sci Plant Anal 28:1591-1602

Tsadilas CD, Voulgarakis N, Theophilou N (1997b) Zeolite influence on nitrogen uptake by wheat. *In* Zeolite '97: 5th Int'l Conf Occurrence, Properties, and Utilization of Natural Zeolites: Progr Abstr. Ischia, Naples, Italy, p 301-303

Tsitsishvili GV, Andronikashvili TG, Kvantaliani AS, Maisuradze GV, Mdivani AL, Machavariani MSh, Gvakharia VG (1980) Perspectives of utilization of tuffs containing clinoptilolite to prevent lead accumulation in agricultural crops. *In* Krupennikova AYu (ed) Proc Symp Utilization of Natural Zeolites in Agriculture, Sukhumi, 1978. Metzniereba Publ House, Tbilisi, Georgia, p 159-163

Tsitsishvili GV, Andronikashvili TG, Kvashali NPh, Bagishvili RM, Zurabashvili ZA (1984) Agricultural applications of natural zeolites in the Soviet Union. *In* Pond WG, Mumpton FA (eds) Zeo-Agriculture: Use of Natural Zeolites in Agriculture and Aquaculture. Westview Press, Boulder, Colorado, p 211-218

Uren NC, Qing ZQ (1997) Manganese-enriched zeolites as potential manganese source in soil. *In* Zeolite '97: 5th Int'l Conf Occurrence, Properties, and Utilization of Natural Zeolites: Progr Abstr. Ischia, Naples, Italy, p 311-313

Williams KA, Nelson PV (1997) Using precharged zeolite as a source of potassium and phosphate in a soilless container medium during potted chrysanthemum production. J Am Soc Hort Sci 122703-708

Yoshinaga E et al. (1973) Organophosphate-containing agricultural and horticultural granule formulation. U.S. Patent 3,708,573

Zhang Q (1982) Agricultural applications of natural zeolites in the People's Republic of China. *In* Zeo-Agriculture '82: A Conference on the Use of Natural Zeolites in Agriculture and Aquaculture: Progr Abstr. Rochester, New York, p 46